Foundations and Interpretation of Quantum Mechanics

In the Light of a Critical-Historical Analysis of the Problems and of a Synthesis of the Results

With a Foreword by Giorgio Parisi

Foundations and Interpretation of Quantum Mechanics

In the Light of a Critical-Historical Analysis of the Problems and of a Synthesis of the Results

With a Foreword by Giorgio Parisi

Gennaro Auletta

University " La Sapienza " (Rome I), Italy

Published by

World Scientific Publishing Co. Pte. Ltd.
5 Toh Tuck Link, Singapore 596224
USA office: 27 Warren Street, Suite 401-402, Hackensack, NJ 07601
UK office: 57 Shelton Street, Covent Garden, London WC2H 9HE

British Library Cataloguing-in-Publication Data
A catalogue record for this book is available from the British Library.

Revised edition 2001
Reprinted 2009

FOUNDATIONS AND INTERPRETATION OF QUANTUM MECHANICS

ISBN-13 978-981-02-4039-4
ISBN-10 981-02-4039-2
ISBN-13 978-981-02-4614-3 (pbk)
ISBN-10 981-02-4614-5 (pbk)

Printed in Singapore by World Scientific Printers

Alla vita che cresce
e si impone intorno a me

History is the most fundamental science,
for there is no human knowledge
which cannot loose its scientific character
when men forget the conditions under which it originated,
the question to which it answered,
and the functions it was created to serve.
A great part of the mysticism
and superstition of educated men
consists of knowledge which has broken loose
from its historical moorings
Benjamin FARRINGTON
quoted in [SCHRÖDINGER 1952, : 479]

The first process, therefore, in the effectual study of the sciences,
must be one of simplification and reduction
of the results of previous investigations
to a form in which the mind can grasp them
J. C. MAXWELL
quoted in [*JAUCH* 1968, : 67]

SCHEME OF THE BOOK

FOREWORD

Quantum mechanics is one of the greatest intellectual achievements of this century. As an effect of its discovery, the very concept of physical reality was changed, the words 'observation', 'measurement', 'prediction', 'state of the system' acquired a new and deeper meaning.

Probability was not unknown in physics: it was introduced by Boltzmann in order to control the behaviour of a system with a very large number of particles. It was the missing concept in order to understand the thermodynamics of macroscopic bodies, but the structure of the physical laws remained still deterministic. The introduction of probability was needed as a consequence of our lack of knowledge of the initial conditions of the system and of our inability to solve an enormous number of coupled non linear differential equations. If we were both infinitely able experimentalists and infinitely able mathematicians, probability would be useless in classical physics: it is only a tool which allows imperfect beings, with a bounded brain like us, to control the behaviour of many particle systems.

In quantum mechanics, the tune is different: if we have 10^6 radioactive atoms *no* intrinsic unknown variables decide which of them will decay firstly. What we observe experimentally seems to be an irreducible random process. The original explanation of this phenomenon in quantum mechanics was rather unexpected. All atoms have the same probability of having decayed: only when we observe the system we select which atoms have decayed in the past. In spite of the fact that this solution seems to be in contrast with common sense, it is the only possible one in the framework of the conventional interpretation of quantum mechanics. The old problem of the relations among the observer and the observed object, discussed for centuries by philosophers, had a unexpected evolution and now it must be seen from a new, completely different perspective.

It is remarkable that all these conceptual changes were not due to reflections aimed to understanding better the *philosophical* problems, but came out naturally as a byproduct of the effort to explain the properties of the radiation emitted and absorbed in various conditions, particularly the spectrum of the hydrogen atom. Heisenberg, de Broglie, Pauli, Dirac and many others invented a formalism that was able to explain and predict the experimental data and this formalism led, beyond the very intention of the men who constructed it, to this conceptual revolution.

Once established, quantum mechanics became a wonderful and extremely powerful tool. The properties of the different materials, the whole chemistry, became for the first time objects that could be predicted from the theory and not only phenomenological rules deduced from experiments. The technological discovery that shaped the second half of this century, the transistor (i.e. the basis of all the modern electronics and computers) could not be invented without a deep command of quantum mechanics.

In spite of this remarkable success, quantum mechanics remains mysterious. It is not only the problem of explaining its meaning without using advanced mathematics that forbids a simple exposition of its properties to the layman. The rules are weird: the fundamental objects are complex amplitudes and probabilities are the modulus square. A scientist like Feynman, who contributed to a new formulation of quantum mechanics and made some of the crucial steps to extend quantum mechanics into the relativistic domain, wrote once *nobody understands quantum mechanics*. Also Poliakov, one of the greatest living theoretical physicists said in a lecture that *eventually someone has to explain why the probability is the modulus square of a complex amplitude.*

These statements may seem rather strange: each year quantum mechanics is taught in thousands of university courses, computations based on quantum mechanics are routinely compared to experiments and found to be correct with incredible precision. However, the meaning of Feynman's and Poliakov's statements becomes more clear if we ask the professors of quantum mechanics about the meaning, the interpretation of quantum mechanics. If we don't ask which are the rules for predicting the output of an experiment, but which is the *nature* of the external world, which is the *meaning* (if any) of an objective physical reality, what happens before, during and after measurements, the opinions become quite diverse and some dissatisfaction is often present in our answers. Somebody may even tell you: *I do not believe that the world is done in this way, but the computations turns out to be accurate.*

Of course we could take the position that the only thing we can know is a set of rules (i.e. quantum mechanics) to deduce the results of the experiments and any further question is only a metaphysical irrelevant question which does not belong to the field of physics. However this answer would be considered unsatisfactory by most physicists. In trying to give an answer to these questions, many different schools arose: in standard books one usually finds the *classic* Copenhagen interpretation and sometimes the puzzling many worlds interpretation, which brings the positions of the Copenhagen school to a logic but paradoxical conclusion.

Many different interpretations have been proposed, starting from de Broglie's pilot wave and arriving to the most recent stochastic quantum mechanics or to the consistent history approach. These different and quite diverse approaches are normally ignored in the manuals of quantum mechanics, but they are very interesting and viable alternatives to the usual Copenhagen interpretation.

The progresses in recent years have not only concentrated on the problems of interpretation that could be (wrongly) dismissed by some people as metaphysical, because they are beyond experimental tests. In the last twenty years, the whole complex of problems connected to quantum mechanics and the meaning of measurements started to be studied from a new perspective. Real, not only *Gedanken* experiments began to be done on some of the most elusive properties of quantum mechanics, i.e. the existence of correlations among spatially separated systems that could not be explained using the traditional concept of probability. The precise quantum mechanical meaning of measurements started to be analyzed in a more refined way (e.g. quantum non demolition measurements were introduced) and various concepts from statistical mechanics and other fields of physics began to be used.

This is not only an academic or philosophical problem. The possibility of constructing a quantum computer, that would improve the speed of present day computers by an incredible factor, is deeply rooted in these achievements. It is now clear that a quantum computer can solve problems, which on conventional computers take a time exploding as exponent of some parameter (e.g. the factorization into primes of a number of length N), in a time which is only a polynomial in N. The technical problems to be overcome for constructing a quantum computers are not easy to solve, but this result has an high conceptual status, telling us how deeply quantum mechanics differs from classical mechanics. Another possible practical application of these developments is quantum cryptography, in which a message is transmitted in such a way that it cannot be read without interfering with it. Another quantum-information puzzling phenomenon, the teleportation, has been recently proved experimentally to exist and it is a very active area of experimental research.

All these aspects of quantum mechanics form a gigantic field in which a very large amount of work was done, but unfortunately this rich and deep activity remains confined to specialized reviews and unknown to most physicists. Of course many books present these new progresses, but they deal with only a part of the area. In this unique book, Gennaro Auletta reviews and summarizes the main achievements of the whole field, not only a part of it. One of the aims of the author was in fact completeness, in the sense of bringing to the attention of the reader all the different points of view that have been presented till now, discussing their relative advantages and disadvantages. The immense bibliography quoted in the book (which can be found at the end) is a proof of the attention which the author has devoted also to many interesting, but forgotten papers. In this way we obtain a very rich, deep and precise picture of the status of the art which looks like a mosaic. Such a comprehensive presentation was lacking till now and this book fills very well this lacuna. At the end of the book the author presents his original synthesis of the problem of the foundation of quantum mechanics and a deep personal outline of the interpretation of quantum mechanics.

The book is written in a condensed, but clear style, where not only mathematical objects, but also the basic interpretative concepts are well defined, avoiding (as far as possible) statements which are unclear or open to many interpretations. This quest for precision and clarity, which is important everywhere, is crucial in this domain, especially in all problems connected to the foundation and the interpretation of quantum mechanics, in which many non-mathematical considerations must be done. One could arrive to wrong conclusions if one does not proceed in a careful way. In order to avoid to fall in such a trap the author has followed some strict methodological principles which are stated at the beginning of the book (e.g. *Do not infer ontological conclusions from formal premisses without specific and extraformal motivations*). The philosophical skills of the author helped him in this difficult task and also for this reason this book is quite different from any other published on the subject.

Giorgio Parisi

Contents

INTRODUCTION

Historical Premise

The great theoretical and experimental development of Quantum Mechanics (= QM) in the last twenty years has largely changed the face of this discipline, and above all has permitted for the first time the testing of a number of postulates and assumptions which from the beginning were made but never proved.

The early days Characteristic of the early days of QM was a rather rhapsodical development of the theory: the fathers of QM, on the one hand, offered a lot of genial solutions to particular problems, which were partly disconnected from each other; and, on the other, they elaborated great generalizations which disturbed the community of physicists for their abstractness and for the gap and sometimes the open conflict with traditional, classical physics or the experience of the ordinary macroscopic world. Hence one spoke and speaks [*FINKELSTEIN* 1996, 183] of a gap between quantum physics practice and its philosophical formulation. This is a historical fact and not a judgement upon physicists like Bohr, Heisenberg, Born, von Neumann, Dirac, and many others.

Further developments Already in the thirties, but particularly after the second world war, the community of physicists, mathematicians and philosophers strived to give a better foundation to the theory and interpret it. Often the results were not resolutive, but without this enormous work we would never have reached the understanding we have today.

Theoretically the turning point was between the end of the fifties and the beginning of the sixties: the theorem of Gleason [1] and later Bell's theorem were able — due to conceptual clarity and the quantification of problems — to determine some questions until the theoretical possibility to test for the first time basic assumptions and hypothesis of QM. In the second half of the seventies the development of refined conceptual and technical instruments, particularly in the domain of measurement problem, permitted the translation of that possibility into factuality.

So then, in the eighties a 'big bang' of the theory began: the experience has confirmed an incredible number of statements of the theory, and QM has revealed to us a wealth of new discoveries and possibilities; but the development was also followed by some fragmentation of the theory. A lot of areas began to develop partially independent from each other, and each approach generally reconstructs a theory in which some parts of QM are lost. This situation is a little uncomfortable because many areas overlap with each other, a lot of identical questions are given different names and 'destructive interference' is often produced.

[1] As pointed out by Busch and co-workers [*BUSCH et al.* 1995a, 1].

Aims and Character of the Book

Aims Perhaps the moment has come to attempt a more systematic and global work at the *theoretical* level. This is not an easy job, and surely the book does not want to be the final answer to this problem. Moreover, it is difficult to foresee what point of unification can be attained, and the problem is posed whether or not QM is able to undergo such a systematic formulation like classical mechanics. The aim of the book is rather a preparation of the task. An effort is necessary in order to find some points of convergence between different areas and approaches, without sacrificing the richness of theoretical and experimental results to the unity of a point of view. The ultimate justification for this approach is to be found in an operational point of view which will be developed later. In line with an operational point of view, the logical structure of the work is not an axiomatic one. In fact we do not stipulate at the beginning postulates or principles, from which we may derive some conclusions, but we formulate principles and postulates during the examination, with the aim of answering specific problems arising from the context of QM itself. Such a procedure is typical of a logic[2] and an epistemology[3] intended as *open* systems and not as closed ones. Shimony names such an 'open' methodology 'dialectical'.

And here a word of warning is necessary: The book has no pretension of globalism. It cannot reproduce all the results of QM nor all formalisms and results which are to be found in handbooks. The literature quoted is surely very partial too: Already in 1978 there were one million of articles and books available on the problem of Einstein/Podolsky/Rosen Paradox alone [CANTRELL/SCULLY 1978]. Surely more than that was published after that date. Hence the task of completeness is ruled out.

Structure In order to accomplish this partial task we need some combination of historical and theoretical approaches. The *structure* of the book partly reflects these different perspectives. The different parts of the book correspond to *thematic* areas, while the succession of chapters in each part reflects to a certain extent the *chronological* order in which the problems arise. But in such a work the historical analysis as such must already be a critical one; therefore the historical organization is dominated by the *theoretical* aim. In the critical connection between theory and history of QM one should also see a pedagogical purpose: the errors and *faux pas* of a science are an integral part of its teaching — we shall come back to this point in the final conclusions of the book. Without such a critical history a science breaks down into technique.

We can understand this point better if we see what the problems constituting the preliminary task to which this book is devoted are. The two points which we need as preliminary ones to the accomplishment of the task of a global understanding of the theory are a better development of QM's *foundations* and a first global *interpretation*. The book aims at both goals.

Some more on history Before we discuss these two aspects, some words on the historical aspect are useful. The *historical* perspective is here needed to understand how some problems were posed and partly resolved so as to condition the later developments. In fact, as we shall see in the following, almost all areas and problems of QM have originated from discussions in the twenties and thirties.

Obviously the history of QM also has an interest in itself. For the first period of the theory — from Planck to the thirties — a lot of work was already done. More is to be accomplished for

[2]For an exposition of this subject see [CELLUCCI 1993] and also [CELLUCCI 1990] [CELLUCCI 1992].
[3]See [SHIMONY 1981, 5].

the later developments of QM. Due to the tumultuous development of the latter years, it is still impossible to write an accomplished history of QM: one needs more historical distance, while we are still in the middle of the tempest. But we also need instruments *now*.

Some Preliminary Epistemological Considerations

Interpretation When one speaks of QM's *interpretation*, one refers to a *philosophical* interpretation. One may think here of the title of the famous Jammer's book *The Philosophy of Quantum Mechanics* or of some of Born's locutions[4]. But one thing is the philosophical interpretation, another is the physical one: a theory cannot only present or represent phenomena[5]. Here we are concerned with both types of interpretation. As a preliminary, to clarify the question — we shall return to the problem by discussing it from the 'inside' of the theory itself [see sections 6.2 and 6.3] —, let us say that a physical theory consists in the first instance of a *Formalism* (= F) and of a set of *Correspondence Rules* (= CR) or bridges between the theory and the experience such that it is possible to assign an empirical meaning to the terms of F.[6] As such F is a mathematical calculus devoid of empirical meaning, and its terms, charged with evocatives like 'particle', 'wave' and so on, have no meaning apart from that resulting from the place they occupy in F itself. Hence CRs are necessary for each physical theory and are an intrinsic part of it. But in F, a lot of terms occur for which we cannot find a satisfactory correspondence in the empirical reality. Such terms are named *theoretical terms*. It is a vexing question if it is possible to eliminate completely all theoretical terms so that the physical theory contains only empirically meaningful terms — which was the original program of neopositivism[7]. Sometimes one tries to do it by using a *model* of the theory and specific logical rules[8]. Generally this transformation happens at the expense of the fertility of the theory: the constraint of a logical coherence results in a loss of explanatory and predictive power. Hence the fertility of a model for QM is an open question, and we shall return to it later [see part III]. However, it is clear that no isolated propositions of a physical theory can be tested or interpreted and that only the theory itself can give the context for such an examination[9].

Hence we can say that the doubts that QM raised from the beginning, and to a certain extent still raises, are due to the absence of a global physical interpretation of it. As we shall see, the Copenhagen interpretation was partly unsound, too much epistemologically characterized and incomplete to be satisfactory. And one may wonder with Mittelstaedt [*MITTELSTAEDT* 1998, 1] that a physical theory after 70 years is almost still in this situation.

Foundations But what about the *philosophical* interpretation? It enters to the extent that such an analysis — and generally a physical theory, but often implicitly — requires a lot of metatheoretical assumptions and principles, so that without a philosophical work we could not arrive at a satisfactory *physical interpretation*, i.e. some conceptual framework built upon CRs[10]. And here we are faced with the other problem, namely the *foundations* of the theory: upon which

[4]For example in [BORN 1961, 456].

[5]On this point see also [*VAN FRAASSEN* 1991, 9–10].

[6]Examination of this problem can be found in [PRUGOVEČKI 1967] [*VON WEIZSÄCKER* 1971a, 231] [*JAMMER* 1974, 9–16] [BUNGE 1965] [DIEKS 1989, 1401] [CUSHING 1991].

[7]See examination in [*P. SMITH* 1981] [BERGSTRÖM 1984] [DILWORTH 1984].

[8]On this point see [CRAIG 1953].

[9]See [QUINE 1951] [*KUHN* 1957] [*KUHN* 1962].

[10]A point stressed by Ferrero and Santos [FERRERO/SANTOS 1997, 766–69].

principles, assumptions, postulates is the theory founded? And are those which are factually acknowledged the best ones or perhaps the only ones? In this sense we see in this book an essay of *applied philosophy*[11].

The first three parts of the book discuss these three fundamental aspects which will be further developed in the other parts and chapters of the book: the formalism, the interpretation (the first and more basic interpretations from which all others generally stem) and the foundations (a first examination of the most important problems). In the next section we give a more detailed presentation of the book's contents.

Presentation of the Content

After a short *introductory chapter* [*chapter 1*] which reports some basic concepts of classical mechanics, the *first part* begins with the first steps of the theory of the black-body problem [*chapter 2*]. Its central aim is to develop the formalism F, which represents the basis of all subsequent examinations. Chapter *3* is dedicated to the basic formalism of the theory.

The *second part* is devoted to the first interpretation of the theory, the Copenhagen interpretation. We shall see the first attempts to solve the interpretative difficulties and how the Copenhagen interpretation was born as an attempt to escape all unsound consequences of the first interpretations of Schrödinger, Heisenberg, and Einstein.

The *third part* is devoted to the first attempts at a foundation. Different problems are examined: the possibility of an axiomatization of QM, the problem of the Hilbert spaces and of the representation of arbitrary observables by operators, and the features of quantum probability. Finally *chapter 12* discusses the theory of the geometric phase.

The other parts of the book are devoted to specific areas and problems of the theory. The two problems which in the last years were mainly discussed are the measurement and the non-locality problems; therefore they also cover a large part of the book [part IV and IX, respectively]. From the fourth part to the sixth part the examination is centred on the relationship between the quantum superposition and the apparent absence of the latter by macroscopical apparatus or generally by macroscopic bodies.

The *fourth part* handles, as said, what is probably the most important thematic area of the theory: the *measurement* problem. After a first examination of some basic concepts in an introductory chapter, this part can be divided in two major blocks: in the first one [*chapters 14–17*], we examine different general proposals which are given for the problem; in the second one, [*chapters 18–20*], we synthesize specific contributions which are given for the measurement theory.

The *fifth part* is centered on the relationship between microworld and macroworld. Two problems are of interest here: is the passage from a superposition (given an observable) to an eigenstate due only to a measurement? In more general terms, what is the relationship between microworld and macroworld? Is there perhaps a sharp boundary line? The traditional starting point of the problem is the Schrödinger cat [see the Introduction to the V part], which by Schrödinger himself was understood as a paradox showing the unsoundness of the theory. And with a 'generation' of a Schrödinger cat we shall end the part [*chapter 24*]. As we shall see actual models

[11] As wished, for example, by Auroux [*AUROUX* 1990].

of decoherence are investigated and an estimate of the decoherence time is also given [*chapter 21*].

The *sixth part* is a discussion of a specific aspect of the relationship between microphysics and macrophysics: the possibility of coherent histories. After a short analysis of some proposals in this regard [*chapter 25*], the 'delayed choice' and its very important consequences for the interpretation are analysed theoretically and experimentally [*chapter 26*]. This part is practically an ideal development of the problematic of time operator treated in chapter 10: in fact *chapter 27* is devoted to the problem of reversibility/irreversibility. The irreversibility which quantum systems seem to show by a measurement-like interaction will be examined here.

The following three parts are devoted to the problem of the ontological interpretation of microrealities and its consequences.

The *seventh part* is centered on the wave/particle dualism, the problem and the mystery of QM. Different proposals are analysed: that both are always real — de Broglie's proposal [see *chapter 28*, where also recent proposals and experiments are reported] — and that none can be before a measurement — three-valued logic and later Heisenberg interpretation [*chapter 29*]. Finally a new interpretation is proposed, based on recent experiments, particularly in quantum optics and using techniques such as the Positive Operator Valued Measure [*chapter 30*].

While the seventh part is more centered on the problematic of the wave (in the binome Wave/Particle), the *eighth part*, and particularly the last chapter, treats the problem of Localization and hence of the particle, but on the other hand also of the position operator, and in this sense it is also the complement to the sixth part. It is specifically dedicated to the Hidden-Variable theories, an attempt to interpret QM in a deterministic way. The starting point is represented by the Einstein/Podoslky/Rosen thought-experiment, aiming at showing the incompleteness of the theory [*chapter 31*]. Then Bohm's proposals in order to complete the theory and some confutation at logical level are analysed [*chapter 32*]. A more concrete refutation, the Bell theorem, will be the subject of the next part, because it goes behind the problem of Hidden Variables, opening an interesting new field: that of the non-locality of QM. The last chapter of this part concerns the development of stochastic ideas treated in chapter 18.

The *ninth part* is devoted to the other main problem of the theory: the *non-locality* problem, which originally was developed from the Hidden-variables problematic, but in the last years has gone much further than the original discussion. As we shall see the Bell inequalities are a turning point in the development of QM in the seventies and eighties. Other non-local effects, like the tunnelling or Aharonov/Bohm effect, are analysed in the last chapter of this part.

The *tenth part* is devoted to the most recent QM domain — and surely one of the most promising ones: Quantum information. As we shall see, by using the means of information theory it is possible not only to open new theoretical areas and to find new technological applications, but also to reformulate, in a rich and more synthetic form, almost all the fundamental results of QM.

The *eleventh part* consists of the concluding discussion of the book. *Chapter 45* is dedicated to the problem of foundations. *Chapter 46* is dedicated to an attempt of a physical interpretation of QM.

Methodological Principles Followed in this Book

Throughout the Book, the following methodological principles have been used:

M1 Do not seek solutions to physical problems which are not in the physics' frame.

M2 Introduce new hypothesis and explanations only if they produce or stimulate new conceptual or experimental results.

M3 Do not reject acknowledged theories if a great conceptual gain determined by the proposed new theories is not evident[12].

M4 Do not formulate *ad hoc* hypothesis.

M5 Do not infer ontological conclusions from formal premisses without specific and extraformal motivations.

M6 Do not solve paradoxes with other paradoxes.

M7 Respect Ockham's principle of economy.

M8 Do not try to solve everything at all costs but accept that there can be open questions.

M9 Face the problems systematically (in all experimental and conceptual aspects and by examining the relevant literature -- to the best of our knowledge -- of it).

Technical Instructions

Finally some technical instructions about the reading of the book can be useful.

The *starred* chapters, sections or subsections can be postponed during a first reading. This signifies in no way that their content is less important, but only that they require more mathematical and conceptual instruments than the rest or that they presuppose some formalism which is treated extensively in the following. Anyway, references to these places are always given.

Definitions, corollaries, postulates, principles, and theorems which are *between horizontal lines* become constituent parts of the theory itself (sometimes the justification of their relevance is not given by their presentation but in an another place, due to the necessity of discussing them in a more general context). This does not signify that definitions, and so on, which are not between horizontal lines, are necessarily rejected by the author: it can be that we use some theoretical element which is not a constituent part of the theory (for example it has only a mathematical

[12] As a specific aspect one can also say that common sense judgments should not be discounted without clear and positive reasons [SHIMONY 1981, I, 6].

meaning or is part of classical mechanics), or which is later integrated in a more general definition, and so on. Anyway it is always explicitly said which is the case.

Squared equations are the main mathematical results of the theory.

Citations with names in *italic* mean 'book' (in plain text mean articles). As we have already said, the bibliography does not intend to be complete. However, it covers the period until the beginning of 1999.

The progressive numeration of experiments (for example One-Hole Exp. 1, One-Hole Exp. 2, and so on), refers to ideal experiments that may have or have not been performed, or also differently performed. Each time it will be specified if they are only proposed or performed experiments. Theoretical studies and models are generally not considered as experiments.

In the footnotes further expositions, developments and commentaries are reported.

Proofs and examples are reported in smaller characters. We could not prove extensively all results: we made a selection following the importance of the subject and of the proof itself and following its pedagogical value. We always, however, provide references to the articles or books in which the proofs can be found.

Acknowledgments

Prof. Giorgio Parisi was the first to propose to the author the present enterprise in 1994 and to have encouraged him in the work. The many very fruitful discussions which the author had with him in the last three years were fundamental. The positive contributions of Prof. Parisi are both on a general level and about specifical problems. The contributions of the latter type are too many to be enumerated here. On the general level, in particular the methodology and the structure of the book owe very much to Prof. Parisi.

The author is also privileged to be acquainted with Dr. Mauro Fortunato, a specialist in Quantum Optics. It suffices to say that during the stay of Dr. Fortunato in Germany, the author wrote to him practically every day in order to discuss a lot of problems and questions, the results of which are in this book (in footnote are remembered specifical contributions only).

The author is also very grateful to Prof. Marcello Cini, who patiently read the book and contributed positively concerning many very specialized matters; to Prof. Giovanni Ciccotti, whose interests are known to be also in the philosophical domain, so that we could meet in a 'middle way': his critic and caustic *forma mentis* was necessary to the author in order to work in a more rigorous way; to Dr. Valeria Mosini for some specifical contributions.

The author would like also to thank the following persons for their help in correcting the manuscript's English: two students of his at the Gregorian University in Rome, Davide Lees, a seminarian at the *Redemptoris Mater* Diocesan Seminary in Rome, and Ross S. J. Crichton, a seminarian at the Pontifical Scots College in Rome; and Mr Dominic Walters, English language teacher at the University of L'Aquila (Italy). Last but not least, the author thanks Miss Annamaria Ingelfinger of the Herder Buchhandlung in Rome for her practical help.

Naturally the author alone is responsible for errors and omissions. He would however be appreciative were his readers to write to him in order to indicate any such omissions and errors. The e-mail address is:

md0509@mclink.it

Gennaro AULETTA

Chapter 1

SHORT REVIEW OF CLASSICAL CANONICAL FORMALISM

In that which follows we give a short review of some classical concepts.

Spaces For each physical system of n degrees of freedom we distinguish a coordinate configuration space $\Re^n$ $\{q_1, q_2, \ldots, q_n\}$ and a momentum configuration space $\Re^n$ $\{p_1, p_2, \ldots, p_n\}$, where the q_j's ($j = 1, \ldots, n$) are the generalized coordinates and the p_j's ($j = 1, \ldots, n$) the generalized momenta, the variables pertinent to a description of a classical system. For a system $\mathcal{S}$ with n degrees of freedom the phase space Γ is the set $\Re^{2n}$ $\{q_1, q_2, \ldots, q_n, p_1, p_2, \ldots, p_n\}$.

Definition The variable p [1] (together with the energy E and the angular momentum J) is named the *dynamic variable* of a system $\mathcal{S}$; q (together with the time t and the angular position ϕ) is named *kinematic variable* of $\mathcal{S}$. The reason for these denominations is that the dynamic variables are pertinent to the description of the evolution of $\mathcal{S}$, while the kinematic ones express the localization of $\mathcal{S}$.

A well defined classical system $\mathcal{S}$ is described by the Hamiltonian H, the energy function ($H = p^2/2m + V$, where V is the potential energy and m is the mass), with the relative canonical equations (= Eqs.). The Hamiltonian is a function of q and p, named *conjugate variables*, and components q_k and p_k with their time derivatives $\dot{q_k}, \dot{p_k}$ are correlated by canonical equations:

$$\dot{q_k} = \frac{\partial H}{\partial p_k}, \qquad \dot{p_k} = -\frac{\partial H}{\partial q_k}, \tag{1.1}$$

which in Poisson brackets can be written:

$$\dot{q_k} = \{H, q_k\}, \qquad \dot{p_k} = \{H, p_k\}, \tag{1.2}$$

where the Poisson brackets for two arbitrary functions $f(q,p), g(q,p)$ are defined by

$$\{f, g\} = \sum_k \left(\frac{\partial f}{\partial p_k} \frac{\partial g}{\partial q_k} - \frac{\partial f}{\partial q_k} \frac{\partial g}{\partial p_k} \right). \tag{1.3}$$

---

[1] For that which follows see the exposition in [*Lev LANDAU/LIFSTITS* 1976a, 193–201] .

Poisson brackets have the following properties:

$$\{f,g\} = -\{g,f\}, \tag{1.4a}$$

$$\{f,C\} = 0, \tag{1.4b}$$

$$\{Cf + C'g, h\} = C\{f,h\} + C'\{g,h\}, \tag{1.4c}$$

$$\{f,\{g,h\}\} + \{g,\{h,f\}\} + \{h,\{f,g\}\} = 0, \tag{1.4d}$$

$$\frac{\partial}{\partial t}\{f,g\} = \left\{\frac{\partial f}{\partial t}, g\right\} + \left\{f, \frac{\partial g}{\partial t}\right\}, \tag{1.4e}$$

where C, C' are some constants.

Action In classical mechanics[2] the path of a particle is determined by the *Principle of least action*:

Principle 1.1 (Least action) *A classical system moving from an arbitrary space position a occupied at time t_1 to another arbitrary position b, occupied at time t_2, follows the path by which the action has the least possible value:*

$$\delta \mathrm{S} = \delta \int_{t_1}^{t_2} L(q,\dot{q},t)dt = 0, \tag{1.5}$$

where $L(q,\dot{q},t)$ is the *Lagrangian function* which can be also be expressed by:

$$L = \frac{d\mathrm{S}}{dt} = \sum_j p_j\dot{q}_j - H(q,p,t) = T_E - V(q), \tag{1.6}$$

where $T_E = \sum_j \frac{p_j\dot{q}_j}{2}$ is the Kinetic energy and $V(q)$ is the potential energy, and where S in Eq. (1.5) is the *action*:

$$\mathrm{S} = \int_{t_1}^{t_2} L(q,\dot{q},t)dt. \tag{1.7}$$

The principle can be expressed through the *Lagrange equation*:

$$\frac{d}{dt}\left(\frac{\partial L}{\partial \dot{q}}\right) - \frac{\partial L}{\partial q} = 0, \tag{1.8}$$

which is the equivalent of Newton's second law.

Hamiltonian and Lagrangian The relationship between action and Hamiltonian[3] is given by:

$$d\mathrm{S} = \sum_j p_j dq_j - Hdt, \tag{1.9}$$

or in terms of the Hamilton/Jacobi Eq. by:

$$\frac{\partial \mathrm{S}}{\partial t} + H\left(q_1,\ldots,q_n;\frac{\partial \mathrm{S}}{\partial q_1},\ldots,\frac{\partial \mathrm{S}}{\partial q_n};t\right) = 0; \tag{1.10}$$

and, between Hamiltonian and Lagrangian, by:

$$H(p_j,q_j) = \sum_j p_j\dot{q}_j - L(q_j,\dot{q}_j). \tag{1.11}$$

[2]For that which follows see [*Lev LANDAU/LIFSTITS* 1976a, 27–31] .
[3]See [*Lev LANDAU/LIFSTITS* 1976a, 203, 214].

Liouville equation The evolution of a system[4] is subject to the Liouville Eq. (continuity Eq.), which is:

$$\frac{\partial \rho}{\partial t} = \sum_k \left(\frac{\partial H}{\partial q_k} \frac{\partial \rho}{\partial p_k} - \frac{\partial H}{\partial p_k} \frac{\partial \rho}{\partial q_k} \right) = \{H, \rho\}, \tag{1.12}$$

where $\rho(p, q, t) d^{3n} p d^{3n} q$ is the density of representative points that at time t are contained in the infinitesimal volume element $d^{3n} p d^{3n} q$ in Γ. The Eq. (1.12) says that the density of representative points in Γ is a constant, which can also be expressed by:

$$\frac{d\rho}{dt} = 0. \tag{1.13}$$

[4]For this point see [*K. HUANG* 1963, 63–65].

Part I

BASIC FORMALISM AND EXTENSIONS OF QUANTUM MECHANICS

Introduction to Part I

- This part begins with an historical examination of the first steps of the theory [*chapter 2*].
- *Chapter 3* starts with the interpretative difficulties regarding the new discoveries (waves or particles?) and with the classical two-slit experiment, and in a short presentation of the historical development from Heisenberg's article of 1925 until the formalisms of groups and propagators, synthesizes the mathematical basis of the theory (which we term Basic-QM) — which will be used in the following. Obviously, these results are basic only relative to later developments and not in themselves (for their time they were of foundational type).
- *Chapter 4* is dedicated to the relativistic developments of the theory (in a very short synthesis) and particularly to the localization problem — from which important contributions of QM stochastic theories were later developed.
- *Chapter 5* is devoted to quantum optics, which, in the last twenty years, has acquired an enormous weight in the testing of the theory.

Chapter 2

WHERE THE PROBLEMS BEGIN

Contents Section 2.1 briefly reconstructs the contents of the quantum postulate and how it was formulated. Section 2.2 analyses its physical consequences by comparing QM and classical mechanics. Section 2.3 finally returns to the quantum postulate for an initial examination of it.

2.1 The Quantum Postulate

The Quantum Postulate (= QP), formulated for the first time in 1913 by Bohr [BOHR 1913], is a generalization — at the level of atomic model — of Planck's solution of the classical problem of a black body's emission. It presents two aspects which can be synthesized as follows[1]:

- QP1 postulates (in agreement with Planck's theory) that a radiation quantum has a frequency equal to its energy divided by Planck's constant h.
- QP2 postulates that an atomic system can exist in particular or discrete states, each of which corresponds to a definite energy of the system, and it emits energy quanta when it jumps from one level to another

2.1.1 Exposition of QP1

The Black-Body problem The Problem faced by Planck was the emission of a black-body. Let us consider a hollow body with internal surface at constant and uniform temperature T: the continuous motion of electrical charges — consequence of the temperature itself — produces an unceasing emission of electromagnetic waves from the different elements $d\mathbf{S}$ of the internal surface $\mathbf{S}$. The waves are partly reflected and partly absorbed from the different surface elements $d\mathbf{S}$. One might expect that this mutual energy exchange between all $d\mathbf{S}$ reaches an equilibrium point after which it ends. Experimentally this is exactly what happens. But at the level of mathematical formalism the problem is more complex. If the *spectrum* of black body is

$$f(\nu) = \frac{\partial U(\nu, T)}{\partial \nu}, \tag{2.1}$$

where $U(\nu, T)$ is the energy density at temperature T and ν is a determinate frequency, the Rayleigh/Jeans formula is:

$$f(\nu)d\nu = \frac{8\pi}{c^3} k_{\mathrm{B}} T \nu^2 d\nu, \tag{2.2}$$

[1]On this point see [*SCHIFF* 1955, 4] [WIGNER 1983a, 260–61].

where k_{B} is Boltzmann's constant and c is the speed of light. This was a consequence of the classical energy equipartition law. The problem arises because the total (classical) intensity I_c of emitted radiation of a black body — calculated over all frequencies — is infinite, which is contrary to experimental evidence and — in the limits of a determinate cavity certainly — impossible:

$$I_c = \int_0^\infty d\nu \frac{dI(\nu)}{d\nu} = \frac{2\pi}{c^2} k_{\mathrm{B}} T \int_0^\infty d\nu \nu^2 = \infty, \tag{2.3}$$

where $[dI(\nu)/d\nu]d\nu = [cf(\nu)/4]d\nu$. This situation is named *ultraviolet catastrophe* — it is clearer from figure 2.1. The problem which derives from this is named *Black-body Problem.*

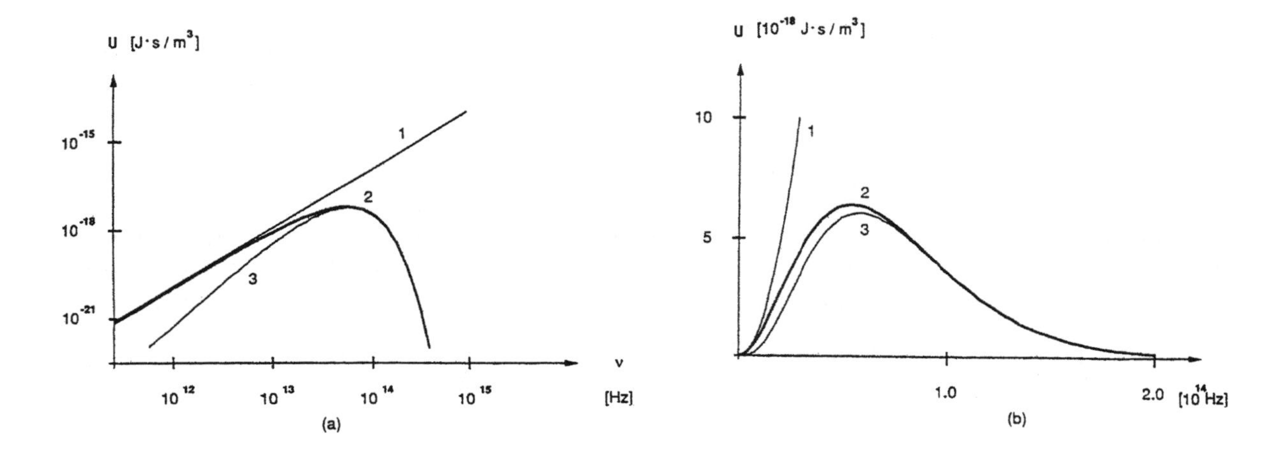

Figure 2.1: Black-body radiation intensity corresponding to the formula of Rayleigh/Jeans (1), Planck (2), and Wien (3) — $E_\lambda = \lambda^{-5} f(\lambda T)$, where wavelength λ instead of frequency ν is considered —, in (a) logarithmic scale and (b) linear scale — from [*BIALYNICKI-B. et al.* 1992, 7].

Planck's solution Planck proposed to consider the black-body cavity as a collection of linear harmonic resonators[2]. If H_{B}^N represents the Boltzmann entropy of the total system, we have:

$$H_{\mathrm{B}}^N = k_{\mathrm{B}} \ln w_E. \tag{2.4}$$

Supposing that the physical system is a collection of N resonators of total energy E_N, the equation links H_{B}^N to the quantity w_E , which comes to represent here the number of different ways in which E_N may be distributed among the resonators. In the continuum frame of classical physics E_N is an infinitely divisible quantity, which would have the consequence of an infinite number of permutations w_E and of an — impossible — infinite entropy. But Planck proposed [PLANCK 1900a] [PLANCK 1900b] to treat E_N as finitely divisible, i.e. as consisting of a finite number P_ϵ of discrete energy elements ϵ, each of them having a definite fixed value for each frequency ν:

$$E_N = P_\epsilon \epsilon. \tag{2.5}$$

[2]For historical expositions of that which follows see [*KUHN* 1978, 97–110] [*JAMMER* 1966, 7–16].

Hence he assumed that the energy density of radiation of frequency ν is proportional to the average energy $\overline{E}$ of a resonator having frequency ν and temperature T, given by:

$$\overline{E} = \frac{\epsilon}{e^{\epsilon/k_B T} - 1}. \tag{2.6}$$

We are now able to calculate w_E, which becomes:

$$w_E = \frac{(N + P_\epsilon - 1)!}{(N-1)!P_\epsilon!}. \tag{2.7}$$

Hence Planck could express the entropy in terms of the average energy $\overline{E}$:

$$H_B^N = k_B N \left[\left(1 + \frac{\overline{E}}{\epsilon}\right) \log\left(1 + \frac{\overline{E}}{\epsilon}\right) - \frac{\overline{E}}{\epsilon} \log \frac{\overline{E}}{\epsilon} \right]. \tag{2.8}$$

QP1 We said already that Planck assumed that, for each value of ν, the elements ϵ have fixed values. The explicit formula is:

$$\epsilon = h\nu, \tag{2.9}$$

where h is the Planck's constant — necessary for agreement with experimental results:

$$h = 6.55 \times 10^{-27} \text{erg/s}. \tag{2.10}$$

Consequently we can state the following result:

Postulate 2.1 (QP1) *The energy of the Black-body system can assume only discrete values for each* ν*:*

$$\boxed{E = h\nu, 2h\nu, 3h\nu, \ldots nh\nu, \ \ldots} \tag{2.11}$$

where $n = 1, 2, \ldots$.

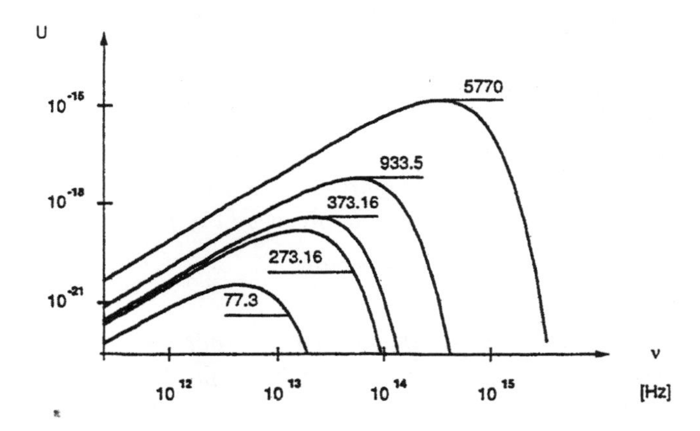

Figure 2.2: Planck's radiation curves in logarithmic scale for the temperature of liquid nitrogen, melting ice, boiling water, melting aluminium, and the solar surface — from [*BIALYNICKI-B. et al.* 1992, 8].

Planck finally calculated the quantized energy density:

$$U(\nu, T) = 3D \left(\frac{8\pi\nu^2}{c^3}\right) \frac{h\nu}{e^{h\nu/k_B T} - 1}, \tag{2.12}$$

which is the product of the average energy [Eqs. (2.6) and (2.9)] and the modes of permutation of the energy [see figure 2.2].

2.1.2 Exposition of QP2

Planck and Sommerfeld Planck and the scientific circles of his time considered the quantization of energy only a mathematical device for solving calculation problems[3]. However, it was Planck himself who again gave the hint, which led to the new theory's turning point. According to the classical theory of Gibbs the probability of finding the representative point for the state of a system in the element $dqdp$ of the phase space is given by $e^{-E/k_B T}dqdp/\int e^{-E/k_B T}dqdp$. Now Planck [PLANCK 1912] initially proposed[4] to consider as a multiple of $h\nu$ not only the average energy of an oscillator, but energy itself; he then proposed the following expression which became the bridge to QP2:

$$\int\int_E^{E+\epsilon} dqdp = h. \tag{2.13}$$

From this Sommerfeld formulated [SOMMERFELD 1912] the fundamental hypothesis that in every elementary process the atom gains or loses a definite amount of action S, given by [see Eq. (1.7)]:

$$\mathrm{S} = \int_0^\tau L dt = \frac{h}{2\pi}, \tag{2.14}$$

where τ is the time of duration of the process.

QP2 Sommerfeld's hypothesis was assumed and developed by Bohr. At that time Rutherford's planetary atom-model was affected by the great problem of instability: a negatively charged particle (the electron) revolving around a positively charged nucleus should continuously radiate energy and rapidly fall, in a spiral trajectory, in the nucleus[5]. Bohr searched for a solution to the stability problem in Planck's hypothesis. Hence, following Sommerfeld, he postulated [BOHR 1913, 874] that[6]:

Postulate 2.2 (QP2) *For an atomic system there exists a discrete set of permissible (stationary) stable orbits characterized by energy values $E_1, E_2, \ldots$, in which the angular momentum of an electron (correspondent to the quantization of energy) is quantized: $\mathbf{J} = n\hbar$, and which are governed by ordinary laws of mechanics. As long as the electron remains in one of these, no energy is radiated. The energy of stationary states can be obtained from the preceding quantization rule and by the mechanical equilibrium condition (the electromagnetic force is equal to the centripetal force), and, for the hydrogen atom, it is given by the formula:*

$$E_n = \frac{-2\pi^2 me^4}{h^2 n^2}. \tag{2.15}$$

The energy is emitted (or absorbed) during the transition from one stationary state to another in a discontinuous way — an electron is said to jump from one level to another —, so that the amount of energy emitted (or absorbed) is quantised in accordance with Planck's Law (2.11) *and Sommerfeld's hypothesis* (2.14)*: for arbitrary j, k, we have $\Delta E = E_k - E_j$.*

It is worth mentioning that Bohr's solution does not really solve the problem of atomic instability: he assumes that some orbits are stationary but he cannot show why because the electron's trajectories are still classical[7].

[3] As pointed out by Murdoch [*MURDOCH* 1987, 3].

[4] See also [*JAMMER* 1966, 42–45].

[5] See [*MURDOCH* 1987, 16–17].

[6] For formulations and comments see also [*BIALYNICKI-B. et al.* 1992, 14] [*LUDWIG* 1954, 33] [*FINKELSTEIN* 1996, 144, 171].

[7] In other words we do not already have the Schrödinger equation.

2.2 Physical consequences of the Quantum Postulate

By assuming QP, physics underwent an enormous break with the classical way of treating the physical world. To understand the problem we must begin by seeing with what a description of a system is, according to classical physics.

2.2.1 Classical Physical Description

Classical determinism As mentioned[8] in the Introduction [p. 3], a physical *Theory* consists in an interpretation (built upon CRs) and in a formalism F. F is the instrument by means of which we are able to give a definition of a physical system $\mathcal{S}$. The *definition* of $\mathcal{S}$ consists of its dynamical laws (i.e. the laws which govern the evolution of $\mathcal{S}$ in time, which are related to quantities such as the energy, and more generally to dynamic variables) and in its kinematic laws (i.e. the laws which localize $\mathcal{S}$ in space and time). On the other hand, the CRs are the bridge for observing the physical quantities, or variables, which F handles: the *observation* consists in the measurement of some of the physical variables of $\mathcal{S}$. So that there are two aspects of the *description* of a system $\mathcal{S}$: definition and observation. More precisely, we classically intend with 'definition' a *complete definition* and with 'observation' a *complete measurement* of one of the *states* of $\mathcal{S}$, i.e. a definition and a measurement of physical variables of $\mathcal{S}$ such that allows us (in the framework of classical physics) to determine univocally all future evolution and all past situations of $\mathcal{S}$. For this, we also speak of *complete description* .

Hence classically it is presupposed that each state is *defined* by the sum of all its physical (contingent) *properties* (= values of the variables of the system in that, not necessarily relativistic, space-time frame) and it is presupposed that each property is absolutely determined (either certain or impossible) [see postulate 6.3: p. 105]. But, to avoid a process *in infinitum*, only those properties which allow a prediction about all other physical properties of the same are pertinent to a description of a system. But, on the other hand, since a system is defined by its variables and properties, and, again, the variables are determined by the definition, we are already forced to assume an *operational point of view*[9]: What a physical system is, is ultimately a matter of decision, i.e. of practical purposes, and what the variables which the theory defines in reality are, is the result of operative actions. Hence we prefer to renounce to the term *variable* for the more operational term *observable*, by leaving aside for now an in depth analysis of the problem [see p. 32]. In general terms we say that an observable consist in a determinate relationship between the set of preparatory instruments and the results of a measurement (i.e. its values) by a measurement device [see chapter 13]. It is also worth mentioning here that absolutely sharp values of observables are an idealization — we have a sharp value only if the observable is dispersion-free [see definition 3.5: p. 54], but, as we shall see, this is not generally the case [see chapters 18 and 33]. Hence a *property* is a result of measurement of a system that one obtains whenever one measures the observables of the system in a certain state, and a *classical state* is a collection of all the results of all possible measurements or at least of a canonically accepted set of results of measurements[10].

Resuming we have the following Definition of classical complete physical description[11]:

[8] For what follows see also [*DRIESCHNER* 1979, 94–95, 106, 112–14].

[9] We shall discuss this subject extensively in chapter 18.

[10] On the last point see [*SCHROECK* 1996, 13–14].

[11] For formulations, comments and examples see [BIRKHOFF/VON NEUMANN 1936, 2] [STRAUSS 1937, 46–47] [*JAUCH* 1968, 91] [*VARADARAJAN* 1968, 3].

Definition 2.1 (C. Complete Description) *A description of a physical system $\mathcal{S}$ is complete if and only if (= iff), having determined by observation a state of $\mathcal{S}$, we are able, by definition of $\mathcal{S}$, i.e. by a rule of the evolution of $\mathcal{S}$, to individuate all positions (past and future) occupied by $\mathcal{S}$ in its curve in the phase space Γ.*

In other words a classical description of an isolated physical system $\mathcal{S}$ is complete iff, on the basis of the definition of $\mathcal{S}$ and of the observation of at least one of $\mathcal{S}$' states in a certain moment, it is possible to determine univocally all future evolution and all past situations of $\mathcal{S}$. Hence a classical physical system has only one possible future and only one possible past. Physical systems of this type are said to be *predictable*[12].

A complete definition of a well defined classical system $\mathcal{S}$ may be represented by the Hamiltonian H — the energy function. The canonical Eqs. [see Eqs. (1.1)] have unique solutions for all t, so that, if we can measure exactly the values of the initial position $\mathbf{q}_i$ and of the initial momentum $\mathbf{p}_i$, and we can determine H (by knowing the forces acting on the system) for a determinate state in a moment t_0 (initial conditions), we know all $\mathbf{q}(t), \mathbf{p}(t)$ for all states for all t, and hence obtain a complete description of the system. Its evolution is then guaranteed by the Liouville Eq. (1.12).

Some related considerations If the definition of $\mathcal{S}$ is necessarily a dynamic one (because it governs the evolution of $\mathcal{S}$ in time), we must carefully distinguish the pair of concepts kinematic/dynamic, that concern the variables themselves — objects of the definition and of the observation of $\mathcal{S}$ — from the pair of concepts definition/observation, which are the two sides of the description of $\mathcal{S}$. But it is right to say that the discontinuity of $\mathcal{S}$ postulated by QP1 and QP2 is a dynamic one because E is a dynamical variable.

Moreover, Bohr (particularly in the twenties) identifies the concept of *causality* with that of *dynamics*[13]. But by *causality* we mean the relation between two — not necessarily immediately — following states of a physical system when it is possible to describe completely the latter. Hence causality includes both definition and observation and (*a fortiori*) dynamic and kinematic variables[14].

2.2.2 Classical Measurement Problem and Continuity

As already explained, classical physics assumes that variables have always determined values [see postulate 6.3: p. 105] and that one can approximate the 'objective' value arbitrarily by a measurement, so as to obtain a complete description [see also postulate 6.4: p. 105]:[15]

Postulate 2.3 (Increasing Measurement Precision) *It is in principle possible to increase a measurement precision beyond any given quantity of uncertainty, so that at the end the uncertainty $\longrightarrow 0$.*

But, even if we assume the preceding postulate and assume that classical systems are completely describable, a problem arises in measurements, the classical measurement problem:

[12] See [*BOHM/HILEY* 1993, 24–26].

[13] For example see [BOHR 1928, 153].

[14] See also [BOHR 1929, 204] [BOHR 1949, 211]. Nagel [NAGEL 1945, 441–42] pointed out that 'causality' is not eliminated in QM, but only means something different from what it means in classical mechanics, due to the CCRs [see Eq. (3.14)]. This is right. Nevertheless a general problem regarding the concept of causality as such can be put forward [see subsection 26.4.4].

[15] See, for example, [*TORALDO DI F.* 1976, 44–48].

Proposition 2.1 (Classical Measurement Problem) *It is impossible to measure with infinite precision two conjugate variables simultaneously.*

We can see the problem by returning to the problem of both conjugate variables [BOHR 1935b , 146–47].[16] In order to measure $\mathbf{q}$ on a physical system $\mathcal{S}$ we need an instrument rigidly attached to the measurement apparatus $\mathcal{A}$ defining the spatial frame; when we wish to measure $\mathbf{p}$ we need, on the contrary, an instrument not rigidly attached to $\mathcal{A}$ such that, knowing the momentum of the instrument before and after the measurement, we are able — applying the law of conservation of momentum —, to determine $\mathbf{p}$ of $\mathcal{S}$ exactly. Because the elastic coefficient of the apparatus can be arbitrarily made large or small, we see how for classical physics it is possible to reduce the uncertainty arbitrarily. On the other hand, if we knew only one of the conjugate variables, we would have only an incomplete measurement of a state of $\mathcal{S}$ and hence an incomplete description of $\mathcal{S}$.

Now 'classical' physics knows how to solve the problem by means of another postulate, the Postulate of continuity of physical systems[17]:

Postulate 2.4 (Continuity) *Each classical physical system has a continuous dynamical evolution in time.*

The meaning of postulate 2.4 is the following: if we know, **for example**, the value of $\mathbf{q}$ at a certain instant t_1 (when $\mathcal{S}$ is in a state 1) and $\mathbf{p}$ at another successive instant t_2 (when $\mathcal{S}$ is in a state 2), on the basis of Eq. (1.1) and of the postulated continuity of S, it is possible to determine with sufficient precision $\mathbf{q}$ at t_2 and $\mathbf{p}$ at t_1, obtaining at the end complete information about the states 1 or 2 and, in consequence, the searched complete description of $\mathcal{S}$.

2.2.3 Quantum Mechanics' Specific Problem

The problem faced by a physics based on QP is clear if we consider that now the condition posed by Postulate 2.4 — necessary to solve Classical Measurement problem [proposition 2.1] — comes to lack [BOHR 1935b , 147]. As a consequence, we can say that in QM the finiteness of Planck's constant makes a continuous reduction of the error impossible — hence we also do not accept postulate 2.3.

2.3 Examination of the Quantum Postulate

The character of QP1 and only partly of QP2 is the recognition and the generalization of a fact: Bohr thought that microphysical systems actually present *quantum jumps*, an aspect which totally distinguishes them from macrophysical classical systems [see chapter 8].

But Bohr never pointed out any physical mechanism able to account for that radical difference or able to lead one system back to the other. Here, for the first time, we are faced with the problem of a *physical interpretation* of the quantum theory.

This is why QP appears an unconditionated and absolute postulate that is in clear conflict with classical mechanics, which, at this time, is still presupposed as a basic point of reference. A lot of criticism addressed to Bohr's[18] is also based on these grounds.

[16]For further expositions see [BOHR 1935b , 697–98] [BOHR 1949, 219–20]; for comments see also [*MURDOCH* 1987, 84, 157, 168].

[17]On this point see [*FOLSE* 1985, 67, 89, 93].

[18]See for example [BOHM/BUB 1966a, 457] [DEWITT 1971, 177–78].

On the other hand the enormous physical consequences — partly examined and partly still to be seen — of the whole QP — consequences which Bohr himself acknowledged, as we shall later see [in chapter 8] — exclude the possibility of a purely Formal Interpretation (where the theory is conceived as a technical device), interpretation that Planck still believed in for QP1.

In conclusion it is worth mentioning that the great philosophical and scientific revolutions normally begin with the unsatisfaction with a partial element of old theories[19].

[19] One might think to the Copernican revolution, where the hint was given by mathematical and aesthetical unsatisfaction with the Ptolemaic system [*KUHN* 1957].

Chapter 3

BASIC QUANTUM MECHANICS

Contents In this chapter we shall very briefly give a picture of the formal basis of the theory. The specific problems of the interpretation and of the foundations can be understood only in relation to these formal aspects. Readers who have already knowledge on the matter can bypass this chapter.

In section 3.1 we give a short account of the discussion which brings, from QP, to the first independent formalism. In section 3.2 we briefly discuss the first attempts of an independent formalism. In section 3.3 we analyse Heisenberg's matrix mechanics, the first element of the new formalism while in section 3.4 we analyse Schrödinger's wave mechanics. In section 3.5 we report the density matrix formalism. Section 3.6 is devoted to unitary transformations, while in section 3.7 the formalism of distribution and of characteristic functions is presented. In the section 3.8 we briefly report the different particle statistics. Section 3.9 is devoted to the difficult spin formalism. In section 3.10 we treat the concepts of symmetries and of groups in the frame of QM. Finally, section 3.11 is devoted to different types of propagators.

3.1 Antecedents and First Experimental Results

Einstein In 1905, when the significance of QP1 [postulate 2.1: p. 19] began to appear, Einstein [EINSTEIN 1905] proposed the hypothesis of light-quanta: the nature of light was not wavelike — as was generally believed at that time — but corpuscular[1]. Einstein used this hypothesis to explain the photoelectric effect: individual light quanta — photons — transmit kinetic energy to individual electrons ejecting them from the atoms of the metal. Einstein's formula is the following:

$$\frac{1}{2}m_e v^2 + E_i = h\nu, \tag{3.1}$$

where the first term of the Left Hand-Side (= LHS) is the kinetic energy of the electron ejected from metal, E_i is the binding (initial) energy of the electron in the metal — we take the absolute value of of the potential V, since the sum of potential and kinetic energy for a bounded particle is negative —, and ν is the frequency of incoming electromagnetic radiation. Bohr [BOHR 1949, 202] considered Einstein's 1905 article very important because it introduced a concept of quantum which goes beyond a Formal Interpretation [p. 24] of Planck's discovery.

[1] Einstein successively reproposed the hypothesis in many papers [EINSTEIN 1906][EINSTEIN 1909] [EINSTEIN 1917]. See also [COMBOURIEU/RAUCH 1992, 1409–11].

This was the first attempt at a purely Corpuscular Interpretation of all microphysics — at that time the corpuscular nature of particles which constitute the matter was universally taken for granted[2].

Particles or Waves? But Einstein's hypothesis was not unanimously accepted. In 1910 Lorentz [LORENTZ 1910] showed that interference and diffraction effects of the light cannot be explained on the assumption that light-quanta are mutually independent, indivisible, point-like entities. Lorentz's paper influenced Bohr[3].

But there were also other papers which returned to the Corpuscular Interpretation. In 1923 — shortly before QM was born — Duane [DUANE 1923] proposed a corpuscular explanation of diffraction, but only of Fraunhofer Diffraction — with a plate located at an infinite distance — and not of Fresnel diffraction — where a plate is at a finite distance.

Surely there was also experimental evidence for the Corpuscular Interpretation. In 1921 Compton [COMPTON 1923] investigated the scattering of x-rays' after their collision with electrons: the wavelength of scattered x-rays λ_s was longer than the wavelength of incident x-rays λ_i — this is the *Compton Effect*. The change of wavelength can be explained with the assumption that photons of incident energy $E_i = h\nu_i$ and incident momentum $\mathbf{p}_i = h/\lambda_i$, collide with electrons of the target and are successively deflected with reduced energy $E_s = h\nu_s$ and reduced momentum $\mathbf{p}_s = h/\lambda_s$, and the assumption that there is a transfer of energy $E_d = E_i - E_s$ and of momentum $\mathbf{p}_d = \mathbf{p}_i - \mathbf{p}_s$ to the electrons so that the Law of Conservation of Energy and Law of Conservation of Momentum are respected for the total system (photons + electrons). In theory the situation is the same as that of elastic collisions between point-like particles in classical mechanics[4].

Particles and Waves But the difficulties increased when in 1924–1925 de Broglie — symmetrically to Einstein's thesis of a corpuscular nature of radiation — proposed that the matter, and particularly the electrons, should be considered as wavelike[5]. Hence the following equation:

$$\boxed{\mathbf{p} = \frac{h}{\lambda}} \tag{3.2}$$

which relates wavelength and momentum and which before had only be applied to waves, began to be applied also to matter. De Broglie's hypothesis received experimental confirmation in the Davisson/Germer electron diffraction experiment of 1927 [DAVISSON/GERMER 1927].[6]

The situation was now very difficult because there was evidence at the same time of a corpuscular nature and of a wavelike nature of microphysical entities (constituents of both radiation or matter). This is the problem of Wave/Particle Dualism (= WP-Dualism), the heart and the only mystery of QM [*FEYNMAN et al.* 1965, 1–1]. To see it, we will briefly go through Young's famous two-slit-experiment, (= Two-Slit-Exp.) in QM context, as discussed by Feynman[7].

---

[2]See [*FINKELSTEIN* 1996, 167–68]. Later on [in sections 6.4 and23.1], we shall discuss other attempts.

[3]On this point see [*MURDOCH* 1987, 14, 20–21].

[4]A recent experiment developed by Grangier , Roger and Aspect [GRANGIER *et al.* 1986a] confirms clearly the particle-like nature of the light [see also section 28.6].

[5]For historical analysis see [*JAMMER* 1966, 244–52] [COMBOURIEU/RAUCH 1992, 1405, 1411–13] .

[6]The wave-like nature of atoms [TAKUMA *et al.* 1995], of electrons [TONOMURA *et al.* 1989] [TONOMURA 1995] and neutrons [S. WERNER 1995] has been also recently experimentally confirmed.

[7]For a complete mathematical analysis of the experiment see [WHEELER 1978b, 14–29] .

Two-Slit-Exp. 1 In the first version of the Two-Slit-Exp. (= Two-Slit-Exp. 1) we fire electrons with an electric gun toward a wall with two slits [*FEYNMAN et al.* 1965, : 1–4/1–6]: some will pass through and reach another wall (the screen) with a detector, which serves as backstop. If we leave both slits open, the electrons show a wavelike nature with an interference pattern [see figure 3.1]; but if we close in turn slit 1 and slit 2 the electrons show a corpuscular nature, i.e. without interference patterns.

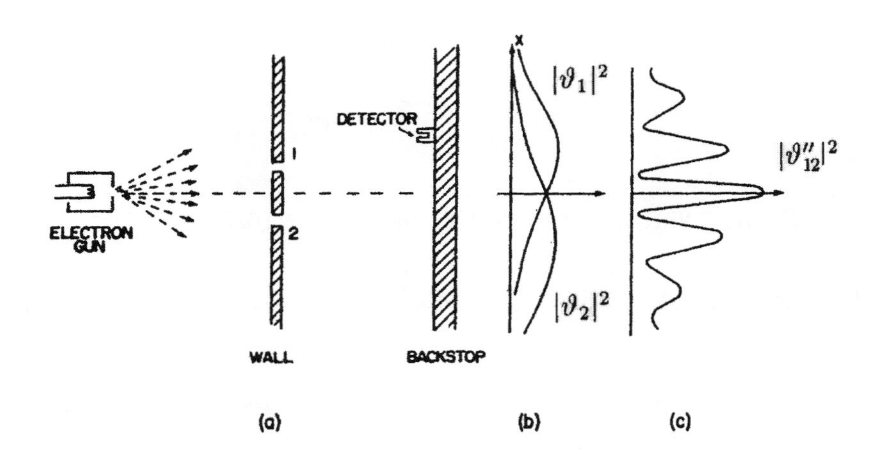

Figure 3.1: Interference experiment with electrons (Two-Slit-Exp. 1) — from [*FEYNMAN et al.* 1965, 1–4].

If we symbolise here the amplitude of the wave by ϑ — leaving aside for the moment what this amplitude may be [see postulate 3.5: p. 38] — and if there is no interference, we have that the square modulus of this amplitude is simply the sum of the square moduli of the two components:

$$|\vartheta'_{12}|^2 = |\vartheta_1|^2 + |\vartheta_2|^2, \tag{3.3}$$

where ϑ'_{12} is the amplitude of the total wave, which passes through slits 1, 2, and $\vartheta_j, j = 1,2$ is the amplitude of the partial wave passing through slit j.

But if there is interference we have a different result:

$$|\vartheta''_{12}|^2 = |\vartheta_1 + \vartheta_2|^2 = |\vartheta_1|^2 + |\vartheta_2|^2 + \mathrm{Re}(\vartheta_2^*\vartheta_1) + \mathrm{Re}(\vartheta_1^*\vartheta_2) \tag{3.4}$$

where the last two terms are named *crossed terms*.

In other words: in the case (3.3) the distribution of marks on wall II would be in the shape of a 'rose' as one would get on a gun shooting target. In the case (3.4) we would have interference fringes like those produced by acoustic or optical waves.

Two-Slit-Exp. 2 We change now the Two-Slit-Exp. by adding a very strong light source behind the wall I between the two slits — Two-Slit-Exp. 2 [*FEYNMAN et al.* 1965, 1–7] [see figure 3.2].

Since electric charges scatter light, we can see, when an electron passes wall I, through which slit it passes. Now, even if we leave both slits open we have Eq. (3.3). It seems that we must

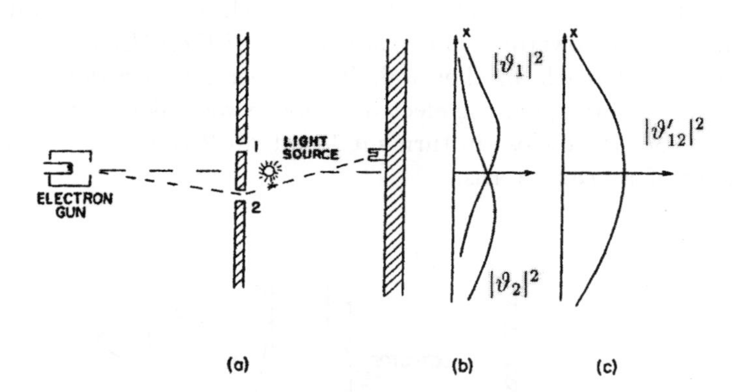

Figure 3.2: Two-Slit-Exp.s 2-3 – from [*FEYNMAN et al.* 1965, 1–7].

conclude that acquiring information on the electrons' path changes the distribution of these on the screen[8].

Two-Slit-Exp. 3 We may try to avoid the effect by reducing the intensity of the light, i.e. by making the light dimmer (Two-Slit-Exp. 3) [*FEYNMAN et al.* 1965, 1–8]. But as a result the flashes emitted by the light source when an electron is intercepted do not get weaker. What happens now is that sometimes an electron arrives to the detector on wall II without being intercepted by the light source, i.e. without flash. This is so because the light shows a corpuscular quantum nature too, and what changes from Two-Slit-Exp. 2 to Two-Slit-Exp. 3 is not the energy of the individual photons — which is quantised and hence cannot diminish gradually [see QP1: p. 19] — but only the rate at which they are emitted. At the end, if we distinguish the distribution of the electrons that are seen from the distribution of electrons not seen, and calculate the two respective square moduli of amplitudes, we find that in the first case we have a situation of type (3.3) and in the second case we have a situation of type (3.4).

Experimental contextuality In conclusion we can assume a Principle of experimental contextuality:

Principle 3.1 (Experimental Contextuality) *Both electrons and photons behave sometimes particle-like and sometimes wave-like depending on the experimental frame in which we detect the microentities.*

We shall analyse principle 3.1 later [see particularly chapter 26]. From Two-Slit-Exp.s 1-3 also seems to follow an indetermination hypothesis:

Proposition 3.1 (Indetermination) *The microentities have in themselves no determinate nature — wave-like or corpuscular — before an experiment is done,*

[8] What happens is that the light scattering off the electrons changes the centre-of-mass wave function, removing the correlation between both paths [KWIAT *et al.* 1992, 7729] [see also section 8.4].

which is a stronger assumption which we do not necessarily accept. We shall return to the whole problem in part VII. However, as we shall see in the next section, related ideas have been already proposed in the mid 1920's.

3.2 First Attempts at an Independent Formalism

Limiting assumption and Correspondence At the beginning of quantum theory, in absence of an independent formalism, there was only a qualitative treatment[9]. Already in 1906 Planck proposed the following assumption [PLANCK 1906]:

Postulate 3.1 (Limiting Assumption) *Classical mechanics is a limit case of microphysics when* $h \longrightarrow 0$.

From this assumption Bohr could formulate a Correspondence Principle between QM and classical physics [BOHR 1920]:[10]

Principle 3.2 (Correspondence) *The elements of quantum theory have analogous elements in classical physics.*

The reason for principle 3.2 is to be found in the fact that classical mechanics appeared to be a complete theory, while the new mechanics was nothing more than an ensemble of a small number of assumptions and experiments (like that of Compton or that of Davisson/Germer). So, for filling the gaps, recourse to classical concepts was necessary.

Bohr/Slater/Kramers Note, nevertheless, that such a correspondence between a discontinuous frame and a continuous one can only be statistical. Hence in 1924, Bohr, Slater and Kramers [BOHR *et al.* 1924] felt the necessity to give a more systematical account of the problem in statistical terms[11]. According to the authors, the quantum jumps have as a consequence the renouncement to causal connection, because causality presupposes continuity [see p. 22 postulate 2.4: p. 23]. Hence in place of causal connections we only have statistical connections, and the conservations laws (of momentum and of energy) have only statistical meaning but no individual commitment.

On the other hand the radiation field was interpreted as a ghost-field, not something real but at the same time not a pure probability: a probability unfurling in space (and time) as a real wave. We will not discuss this point here: we shall return later to this and similar interpretations [see chapter 29 and particularly subsection 29.2.1]. Here we only aim to show why the interpretation was rejected.

Bothe and Geiger [BOTHE/GEIGER 1924] [BOTHE/GEIGER 1925a] [BOTHE/GEIGER 1925b showed experimentally that, in the Compton-effect experiments, the chances that coincidences due to the Law of momentum conservation are accidental are of the order 10^{-5}, which is a clear refutation of the hypothesis of a statistical conservation law. On the other hand the idea of interpreting QM as a statistical theory has further developments [see section 6.5].

[9] As pointed out by Heisenberg himself [HEISENBERG 1929, 109]; historical analysis in [*JAMMER* 1966, 110, 208].

[10] See also [*LUDWIG* 1954, 31].

[11] Historical reconstruction and analysis in [*JAMMER* 1974, 188–191]. See also [COMBOURIEU/RAUCH 1992, 1413–14]. An in depth discussion, at the theoretical level, about the interaction of light with atoms is in [CARMICHAEL 1997].

3.3 Matrix Mechanics and Observables

3.3.1 A Non-Commutative Multiplication Rule

The serious problem in the application of quantum conditions to the classical model of the atom was that it gave false energy values as result, so that in 1925 Bohr's young assistant, Werner Heisenberg, felt the necessity of a mathematical formalism independent from the classical one and able to face the problems arisen from microphysics. This work [HEISENBERG 1925, 384–87] [DIRAC 1925, 643–44] was the real birth of QM.[12] In particular, Heisenberg began to think that **q** and **p** are not to be understood as position and momentum in the sense of classical mechanics: it was not possible to assign to the electron a point position in space as function of time. But we can always assign a radiation to the electron; and by means of the correspondence principle [principle 3.2: p. 29] Heisenberg thought that between the observed frequencies and amplitudes of electrons there must be the same relationship as between classical frequencies and amplitudes. If we were to apply classical theory to microphysics, for the η–th harmonic component we would have following relationship:

$$\nu(n,\eta) = \eta \cdot \nu(n) = \eta \frac{1}{h}\frac{dE}{dn}; \tag{3.5}$$

while in a quantized theory [see QP2: p. 20] we have:

$$\nu(n, n-\eta) = \frac{1}{h}[E(n) - E(n-\eta)], \tag{3.6}$$

subject to following requirement:

$$\nu(n,\eta) + \nu(n,\zeta) = \nu(n, \eta + \zeta). \tag{3.7}$$

The problem now is to find correct transition probability amplitudes for QM. Now, consider a time dependent quantity $\xi_n(t)$, which in classical theory can be Fourier-expanded in the form:

$$\xi_n(t) = \sum_\eta \xi(n,\eta) = \sum_\eta \vartheta_\eta(n) e^{2\pi \imath \nu(n,\eta)t}, \tag{3.8}$$

where the η–th component $\xi(n,\eta)$ has amplitude $\vartheta_\eta(n)$. In the Quantum theory we take as representative of $\xi_n(t)$ the set of $\{\vartheta(n, n-\eta)e^{2\pi\imath\nu(n,n-\eta)t}\}$. Now we represent classically the quantity $\xi_n^2(t)$ in the following manner:

$$\xi_n^2(t) = \left[\sum_\zeta \vartheta_\zeta(n) e^{2\pi\imath\nu(n,\zeta)t}\right] \sum_\eta \vartheta_{\eta-\zeta}(n) e^{2\pi\imath\nu(n,\eta-\zeta)t}. \tag{3.9}$$

We can assume that in QM we correspondingly have, for the η–th term:

$$\vartheta(n, n-\zeta)\vartheta(n-\zeta, n-\eta)e^{2\pi\imath\nu(n,n-\zeta)t}e^{2\pi\imath\nu(n-\zeta,n-\eta)t}, \tag{3.10}$$

which by Eq. (3.7) is equal to:

$$\left[\sum_\zeta \vartheta(n, n-\zeta)\vartheta(n-\zeta, n-\eta)\right] e^{2\pi\imath\nu(n,n-\eta)t}. \tag{3.11}$$

[12] Historical reconstruction in [*JAMMER* 1966, 210–15].

Hence the required quantum correspondent to $\xi_n^2(t)$ is

$$\sum_{\zeta} \vartheta(n, n-\zeta)\vartheta(n-\zeta, n-\eta), \tag{3.12}$$

by means of which we can now represent the quantum transition amplitudes. However the multiplicative rule expressed in Eq. (3.12) is non commutative — i.e. $ab - ba \neq 0$ for some arbitrary elements a, b. In fact generally for two different amplitudes ϑ, ϑ', we would have:

$$\sum_{\eta=-\infty}^{+\infty} \vartheta(n, n-\eta)\vartheta'(n-\eta, n-\zeta) \neq \sum_{\eta=-\infty}^{+\infty} \vartheta'(n, n-\eta)\vartheta(n-\eta, n-\zeta). \tag{3.13}$$

On the basis of Heisenberg's article, Dirac could work at the problem of orbital motion in a Hydrogen atom [DIRAC 1926a].

3.3.2 Observables as Operators

Commutation relations Heisenberg's work was further developed also by Born and Jordan[13]. They understood Heisenberg's sets governed by the non-commutative multiplication rule — and hence **q** and **p** — as matrices [BORN/JORDAN 1925, 126–41]. For example 2 × 2 for the plane (vectors 2 × 1). The identity matrix (the matrix with all ones on the diagonal and all zeros elsewhere, i.e. the **1**-matrix) is written $\hat{I}$. But mathematically a matrix is the same as an operator (**for example** by means of a matrix one performs a rotation of a vector, i.e. an operation)[14].

This was the birth of Matrix Mechanics , which was the beginning of the new mechanics: Quantum Mechanics. Following up this result they arrived [BORN/JORDAN 1925, 137] to the basic *canonical commutation relations* (= CCRs) of QM — further developed by Dirac [DIRAC 1925, 648] [DIRAC 1926a, 561] —, which are the QM equivalent of the Poisson brackets of classical mechanics: they too respect conditions (1.4a)–(1.4d). If we label the components of the variables with index j and k, we have[15]:

$$\boxed{[\hat{q}_j,\ \hat{p}_k] := \hat{q}_j\hat{p}_k - \hat{p}_j\hat{q}_k = \imath\hbar\delta_{jk} = \imath\hbar\hat{I};\ \ [\hat{q}_j, \hat{q}_k] = 0;\ \ [\hat{p}_j, \hat{p}_k] = 0} \tag{3.14}$$

where the hat symbolizes an operator, $\hbar = h/2\pi$ and δ is the Kroneker symbol[16]. A pair of observables which satisfies the relations (3.14) is said [*BUSCH et al.* 1995a, 57] to be a *Heisenberg pair.*

So that we have the classical relationship between observables if we substitute the CCRs (3.14) with Poisson brackets and make the product symmetric (it is analogous to the Ehrenfest theorem). Hence we speak of the two observables **p** and **q** as *non-commuting observables.* This expresses in a more precise way the fact that in QM we cannot measure both at the same time [section 2.2].[17]

---

[13]Historical analysis of that which follows in [*JAMMER* 1966, 215–27].

[14]See [*VON NEUMANN* 1932, 130–31] [*VON NEUMANN* 1955, 247–49].

[15]Formulations in [*DIRAC* 1930c, 87] [*SCHIFF* 1955, 175].

[16]The operatorial formalism will be discussed in the following pages.

The Dirac function, the continuous counterpart of the Kroneker symbol, wass introduced in [DIRAC 1926c, 625–26].

[17]Note that, since operators are not necessarily defined on the whole Hilbert space of the system they pertain but maybe on a dense subspace [*J. FUCHS/SCHWEIGERT* 1997, 9], the commutators are not necessarily defined.

Measurement and definition Hence we substitute to the Classical Measurement Problem [proposition 2.1: p. 23] a QM Measurement Problem:

Theorem 3.1 (QM Measurement Problem) *It is impossible to measure exactly, at the same time, two non-commuting observables.*

In this sense one also speaks [BIRKHOFF/VON NEUMANN 1936, 1] of the non-commutativity of observation. But Eq. (3.14) also has a consequence for the definition of a system. In classical physics it is sufficient to assign values by mean of observation (measurement) to $\mathbf{p}, \mathbf{q}$, in order to obtain, with an Hamiltonian, a complete description of the system [see subsection 2.2.1]. But in QM, theorem 3.1 has the following corollary [DIRAC 1926c, 623] as a consequence:

Corollary 3.1 (QM Definition Problem) *It is impossible to univocally define the state of a quantum system by assigning initial values to its observables, because the CCRs rule out the possibility of determining the simultaneous values of non-commuting observables.*

Note, additionally, that in QM we have $\mathcal{M}(\mathbf{p}) \sqcap \mathcal{M}(\mathbf{q}) \neq \mathcal{M}(\mathbf{q}) \sqcap \mathcal{M}(\mathbf{p})$, where $\mathcal{M}$ symbolizes a measurement and $\sqcap$ the succession of two measurements (this can be extended to any number of non-commuting observables). The QM Definition Problem has no classical counterpart at all.

In conclusion, while in classical mechanics it is possible to define a state of a system as a catalogue of the (determined) values of the different variables pertaining to it, this is not possible in QM, which causes serious problems in the theoretical determination of what a QM state is and in the practical observation of it [see sections 28.4 and 28.5].

Linear Operators The mathematical formalism was developed further by Born, Heisenberg and Jordan [BORN *et al.* 1926] , and by Born and Wiener in collaboration.As we have said Born and Wiener interpreted QM physical variables as *linear* operators, i.e. linear transformations of given functions [BORN/WIENER 1926, 214–216]:

Definition 3.1 (Linear operator) *A* linear operator $\hat{O}$ *is an operator such that, given two arbitrary functions* f, g *, and a number* c*, the two following conditions are satisfied:*

$$\hat{O}(f+g) = \hat{O}f + \hat{O}g, \tag{3.15a}$$

$$\hat{O}(cf) = c\hat{O}f. \tag{3.15b}$$

Since the definition is a purely mathematical one, it is not part of the theory itself. Now, since a physical observable cannot have imaginary values, not all linear operator are good for QM. We must add the requirement that the operators be *self-adjoint* or Hermitian [BORN/WIENER 1926, 221–22]. We remember that the *adjoint* (transposed conjugate) $\hat{O}^\dagger$ operator of $\hat{O}$ is the unique operator which satisfies $(f, \hat{O}g) = (\hat{O}^\dagger f, g)$ for all f, g, where $(,)$ is the scalar product. $\hat{O}$ is *self-adjoint* iff $\hat{O} = \hat{O}^\dagger$.[18] Note that an operator is symmetric if it is equal to its transposed ($\hat{O} = \hat{O}^{\mathrm{T}}$), and that a symmetric operator is Hermitian only if it is real because then $\hat{O}^{\mathrm{T}} = \hat{O}^\dagger$.

So QM proposes the postulate that follows[19]:

[18] For a mathematical exposition see [*KOLMOGOROV/FOMIN* 1980, 230]

[19] See formulation in [*Lev LANDAU/LIFSTITS* 1976b, 26–27].

Postulate 3.2 (Operators Postulate) *Each QM observable can be represented by a linear self-adjoint operator.*

A later examination will show that this postulate cannot be assumed in its whole generality [see section 7.4, subsection 7.5.1, and sections 10.2–10.4 for a discussion of this point]. But it is important to remember that an observable and an operator are not the same thing: an observable is defined, in the end, by operational methods [see subsection 2.2.1], while an operator is a mathematical entity[20]. But for practical purposes, in what follows we shall generally speak of an observable in terms of the corresponding operator.

This was a very important change, because in classical physics the variables are normally represented by functions.

3.4 Wave Mechanics

3.4.1 Waves

Schrödinger equation We spoke already of de Broglie's assumption [Eq. (3.2)]. But until 1926 nobody thought that, if radiation and matter are both wavelike, there a wave equation for particle like electrons must also exist. This was the work of Schrödinger, the beginner of Wave Mechanics, the second chapter of the formalization of QM.[21] Searching for an expression of the quantum theory equivalent to classical physics he formulated (for a free particle of mass m) the — later named — *Schrödinger equation* [SCHRÖDINGER 1926a, 103, 106, 118–19]:

$$\boxed{\imath\hbar\frac{\partial\psi}{\partial t}=-\frac{\hbar^2}{2m}\frac{\partial^2\psi}{\partial x^2}} \tag{3.16}$$

where ψ is the wave function. Eq. (3.16) is the equivalent of the classical one:

$$\frac{\partial^2\psi}{\partial t^2}=v^2\frac{\partial^2\psi}{\partial x^2}, \tag{3.17}$$

where v is the wave velocity.

The time dependent Schrödinger equation for interacting particles (with potential V) is the following:

$$\boxed{\imath\hbar\frac{\partial\psi(\mathbf{r},t)}{\partial t}=\left[-\frac{\hbar^2}{2m}\nabla^2+V(\mathbf{r},t)\right]\psi(\mathbf{r},t)} \tag{3.18}$$

We can separate the time-part from the space-part, so as to obtain:

$$\psi(\mathbf{r},t)=f(\mathbf{r})e^{\imath Et/\hbar}, \tag{3.19}$$

where the space part is:

$$\left[-\frac{\hbar^2}{2m}\nabla^2+V(\mathbf{r})\right]f(\mathbf{r})=Ef(\mathbf{r}). \tag{3.20}$$

[20] As pointed out by Kraus [*KRAUS* 1983, 8].

[21] On this point see the following handbooks: [*SCHIFF* 1955, 22–24, 30–31] [*Lev LANDAU/LIFSTITS* 1976b, 23–24].

We notice that, as in the classical theory, the ψ function can be analysed in an amplitude component ϑ and a phase component φ:

$$\psi(\mathbf{r}) = \vartheta(\mathbf{r}) e^{\frac{\imath}{\hbar}\varphi(\mathbf{r})} \tag{3.21}$$

Generally, each wave function ψ can be expanded as a linear combination of the eigenfunctions of a given observable, in the following general form:

$$\psi(\mathbf{r}) = \sum_j c_j b_j(\mathbf{r}), \tag{3.22}$$

where the c_j's are some complex coefficients and the b_j's vectors (the basis of ψ).[22]

Hamiltonian and momentum We have seen that to each observable we associate an operator [postulate 3.2: p. 33]. Now on the basis of the Schrödinger Eq. (3.16) and of the classical Eq.:

$$E = \frac{\mathbf{p}^2}{2m}, \tag{3.23}$$

the Hamiltonian can be interpreted in the following manner [BORN/WIENER 1926, 222]:

$$\boxed{\hat{H}\psi := \imath\hbar\frac{\partial}{\partial t}\psi} \tag{3.24a}$$

which makes the Schrödinger Eq. an equivalent of the classical Hamilton/Jacobi Eq. [see Eq. (1.10)].

The momentum — for the coordinate representation [see subsec-Different Representations] — can be interpreted as follows:

$$\boxed{\hat{\mathbf{p}}\psi := -\imath\hbar\nabla\psi} \tag{3.24b}$$

Eigenvalues Schrödinger began to think of the wave as a superposition of eigenfrequencies [SCHRÖDINGER 1926d, 138]. The introduction of operators by Born and Wiener led Schrödinger to interpret the energy-levels (encountered by the harmonic oscillator problem) as eigenvalues of the energy operator that acts on the function ψ. He began with the general eigenvalue problem [SCHRÖDINGER 1926b, 169]. Generally we say that for an operator $\hat{O}$, the number o_k is the k–th eigenvalue of $\hat{O}$ if there exists a non-zero vector ψ, called *eigenvector*, such that:

$$\hat{O}\psi_k = o_k\psi_k. \tag{3.25}$$

In practice an operator acts on its eigenvectors simply as a multiplier (i.e. as the most elementary transformation) — while generally an operator can be understood as a transformation of any vector in a given space into another vector of the same space — **for example** by rotation. Only the eigenvalues correspond to the values of classical variables, which, as said above, are always represented by functions. This means that, if the operator acts on a state which is not an eigenstate, then one cannot assign values to that same operator. This is a mathematical fact, but it is related to the most important feature of QM: superposition [see subsection 3.4.6].

In the case of the energy operator [Eq. (3.24a)] we can write the eigenvalue Eq. for the energy in the following form:

$$\boxed{\hat{H}\psi_k = E_k\psi_k} \tag{3.26}$$

[22] As we will explain in the next subsection we prefer to write (3.22) as follows: $|\psi\rangle = \sum_j c_j|b_j\rangle$.

which is the time-independent Schrödinger Eq. In other words, a system which is in an eigenstate of the Hamiltonian, if not perturbed, will remain in that eigenstate.

Working on the problem of the harmonic oscillator Schrödinger found for it a solution of the eigenvalue problem [SCHRÖDINGER 1926d, 137–38].[23] Choosing for the potential of the harmonic oscillator $V = 1/2Kx^2$, and writing Eq. (3.18) in the following manner:

$$-\frac{\hbar^2}{2m}\frac{d^2 f}{dx^2} + \frac{1}{2}Kx^2 f = Ef, \tag{3.27}$$

we can rewrite it in the form:

$$\frac{d^2 f}{d\xi^2} + (g - \xi^2)f = 0, \tag{3.28}$$

where:

$$\xi = \eta x; \quad \eta^4 = \frac{mK}{\hbar^2}; \quad g = \frac{2E}{\hbar\omega}. \tag{3.29}$$

We then obtain the Hermite orthogonal functions as eigenfunctions:

$$f(\xi) = H_n(\xi)e^{-\frac{1}{2}\xi^2}, \tag{3.30}$$

where H_n is the n–th Hermite polynomial; and as eigenvalues (the energy levels) we have:

$$E_n = \hbar\omega\left(n + \frac{1}{2}\right), \qquad n = 0, 1, 2, \ldots, \tag{3.31}$$

where $\omega = 2\pi\nu$ is the angular frequency[24].

3.4.2 Dirac's algebra

Dirac developed a very simple algebra, successively named *Dirac Algebra*, to treat QM problems [DIRAC 1926d].[25] Dirac introduces the general concept of *state vector* $|\psi\rangle$, named ket vector (or simply *ket*), and of its dual, the bra vector (or simply *bra*): $\langle\psi|$, so that to each ket corresponds a state of a ψ–function, and to each bra its complex conjugate. Hence we have a perfect isomorphism with the Schrödinger equation, and we can rewrite Eq. (3.25) in the following way:

$$\hat{O}|j\rangle = o_j|j\rangle. \tag{3.32}$$

Its complex conjugate can be expressed as follows:

$$\langle j|\hat{O} = o_j\langle j|. \tag{3.33}$$

While the product $|b_j\rangle\langle b_k|$ — or, more simply: $|j\rangle\langle k|$ — is a linear operator that can act on kets, the product $\langle b_k|b_j\rangle$ — or, more simply: $\langle k|j\rangle$ — is a number: it is the scalar or inner product. The relation between the (one-dimensional) wave function $\psi(x)$ and the ket $|x\rangle$ is $\psi(x) = \langle x|\psi\rangle$, where we have: $\langle b_k|b_j\rangle = \int b_k^* b_j dx$. We shall symbolize an initial state by $|i\rangle$ and a final state by $|f\rangle$.

Dirac Algebra is an attempt to find a formulation of QM which is neutral with respect to Matrix Mechanics and Wave Mechanics. But the vector state is not a classical state, i.e. a state in which both observables of a canonical pair have determinate values. In fact it was Dirac himself that posed the QM Definition Problem [see corollary 3.1: p. 32]. As we shall see [in sections 13.4 and 18.1], the state vector is to be understood as a preparation procedure[26].

[23]For a historical reconstruction see [*JAMMER* 1966, 264–66]; exposition in [*SCHIFF* 1955, 67–72].

[24]Note that Matrix Mechanics and Wave Mechanics are respectively centred on observables and on states, the two fundamental and irreducible aspects of the theory [see subsection 9.4.1].

[25]See also the exposition in [*DIRAC* 1930c, 15–45].

[26]On the latter point see also [*HAAG* 1992, :4] [*PERES* 1993, 222].

3.4.3 Hilbert Spaces

Hilbert space As a consequence of Wave Mechanics, Hilbert spaces were introduced in QM. We will start here by giving a general definition of a Hilbert space[27]: A Hilbert space $\mathcal{H}$ is an Euclidian linear strictly positive inner product (or scalar product) space (generally over the field of complex numbers F) which is complete with respect to the metric generated by the inner (scalar) product, which is separable and which can have a finite or an infinite number of dimensions. Its elements are vectors. Given the importance of the square modulus of coefficients, normally the Hilbert space of QM is l^2 or L^2 (depending on whether we assume Matrix Mechanics or Wave Mechanics), where (in l^2) we have:

$$\sum_j |c_j|^2 \ < \ \infty, \tag{3.34}$$

where the c_j's are the components of an arbitrary vector in the Hilbert space. So we assume $\mathcal{H} = L^2(\Re^3)$ and, in what follows, by $\mathcal{H}$ we always mean $L^2(\Re^3)$ unless otherwise specified.

From the self-adjointness of QM operators the *completeness condition* is also derived:

$$\sum_j \langle b_j | b_j \rangle = 1. \tag{3.35}$$

The QM Hilbert spaces are *orthonormal* [SCHRÖDINGER 1926b, 177]:

- a space is *orthogonal* if a basis of non-zero vectors $\{|b_k\rangle\}$ can be found such that $\langle b_j | b_k \rangle = 0$ for $j \neq k$;
- *normal* if each element $|\psi\rangle$ has norm 1: $\parallel \psi \parallel = 1$.

Theoretically, all vectors with the same direction but with different lengths (eigenvalues) are considered as representing the same state (apart from a phase factor) — later we shall see in more detail what a quantum state can be [section 3.5]. Hence it is always possible to choose vectors which are normalized.

If $\{|b_j\rangle\} \subset \mathcal{H}$ is a basis (a complete orthonormal set) of $\mathcal{H}$, then any $|\psi\rangle \in \mathcal{H}$ can be expressed as the Fourier series $|\psi\rangle = \sum \langle b_j|\psi\rangle|b_j\rangle$, with $\parallel \psi \parallel^2 = \sum_j |\langle b_j|\psi\rangle|^2$. Now in the representation diagonal to a determinate operator, to each dimension of the Hilbert space corresponds an eigenvector.

Matrix and operator We have seen that we associate an operator and a matrix to each observable [SCHRÖDINGER 1926e, 146–48]. Being $\{|b_j\rangle\}, j = 1, \ldots, m, n, \ldots$ an orthonormal complete set of eigenvectors in a Hilbert space, the relationship between the operator and matrix elements is given by[28]:

$$O_{nm} = \int \langle b_n | \hat{O} b_m \rangle dq, \tag{3.36}$$

where $\hat{O}$ is an arbitrary operator acting on a vector $|\psi\rangle = \sum c_j |b_j\rangle$ and O_{nm} is an element of the matrix associated with the operator.

[27] For the mathematical aspects see [*KOLMOGOROV/FOMIN* 1980, 52, 144, 154] [*PRUGOVEČKI* 1971, 30–33, 101–102]; an easy exposition in [*BUB* 1974, 8–15].

[28] See [*Lev LANDAU/LIFSTITS* 1976b, 50] .

Maximality If a QM operator $\hat{O}$ can vary in the space of quadratically integrable functions from $-\infty$ to $+\infty$, it is self-adjoint; but if it can vary only from 0 to $+\infty$, it is only Hermitian and not self-adjoint, unless the vector $|\psi(\xi)\rangle$ on which it operates satisfies the boundary condition $|\psi(0)\rangle = 0$. But, if so, it has no eigenvectors at all. In von Neumann's terminology [*VON NEUMANN* 1932, 130] [*VON NEUMANN* 1955, 247] is only *maximal* and not *hypermaximal*[29]. Clearly, if energy has a lower bound, it can be maximal but not hypermaximal [on this point see also subsection 7.5.1 and section 10.2].

Compound systems In the case of a composed system (for example a multiparticle one), the Hilbert space of the composed system is a tensor product: let the subsystems be $\mathcal{S}_1, \mathcal{S}_2$ and the respective Hilbert spaces $\mathcal{H}_1, \mathcal{H}_2$. Then the global Hilbert space of system $\mathcal{S}_{12} = \mathcal{S}_1 + \mathcal{S}_2$ is given by: $\mathcal{H}_{12} = \mathcal{H}_1 \otimes \mathcal{H}_2$.

3.4.4 Equivalence Between Matrix Mechanics and Wave Mechanics

Schrödinger already proved in 1926 the first element of the equivalence between Matrix Mechanics and Wave Mechanics: he saw the operator

$$\frac{\partial}{\partial q_j}\hat{q}_j - \hat{q}_j\frac{\partial}{\partial q_j}, \tag{3.37}$$

which follows from CCRs between $\hat{\mathbf{p}}$ and $\hat{\mathbf{q}}$ as an identity operator, i.e. an operator which applied to an eigenfunction of $\hat{q}_j$, reproduces the same function [SCHRÖDINGER 1926e, 146].

But the greatest results are due to Jordan, Klein and Wigner [P. JORDAN/KLEIN 1927] [P. JORDAN/WIGNER 1928], who proved such equivalence[30]. Von Neumann treated the mathematical aspects [VON NEUMANN 1929], showing that Heisenberg's matrix mechanics — focused on series and sums (l^2) — and Schrödinger wave mechanics — focused on continuous functions and derivations (L^2) — are operator calculi on isomorphic realizations of the same Hilbert space, and this thanks to the famous functional analysis theorem of Riesz/Fischer[31].

3.4.5 Values and Observables

Measurement results and probability We owe to Hilbert and von Neumann the central Quantization Algorithm[32]:

Postulate 3.3 (Quantization Algorithm) *The possible measurement results on the observable O are the eigenvalues of the associated operator $\hat{O}$.*

The reason is clear: we expect that a measurement of an observable gives a determined value which mathematically, as we have seen, is an eigenvalue of the operator associated with the measured observable. However, note that in general we never have perfectly determined

[29] See also [J. COOPER 1950] [*JAMMER* 1974, 237].

[30] For the history see [*JAMMER* 1966, 273, 329–30].

[31] For mathematical aspects see [*KOLMOGOROV/FOMIN* 1980, 152] [*PRUGOVEČKI* 1971, 101–109].

[32] See expositions in [*DIRAC* 1930c, 36–37][*Lev LANDAU/LIFSTITS* 1976b, 25–27] [*REDHEAD* 1987b, 6–8].

values [see chapter 18], so that in general we must understand values only in an unsharp and approximate sense.

The Statistical Algorithm is also fundamental:

Postulate 3.4 (Statistical Algorithm) *Given that a QM system is completely defined by a vector* $|\psi\rangle$ [Eq. (3.22)], *the probability of having a determinate result — an eigenvalue* o_j *of an arbitrary observable* $\hat{O}$ *— is given by:*

$$\boxed{\wp(o_j, \psi) = |c_j|^2} \tag{3.38}$$

where the coefficient c_j *is given by* [see Eq. (3.22)]:

$$c_j = \langle b_j | \sum_k c_k |b_k\rangle = \langle b_j|\psi\rangle, \tag{3.39}$$

and the eigenvector $|b_j\rangle$ *of* $\hat{O}$ *corresponds to the value* o_j.

This postulate provides the possibility of performing a measurement statistic and, by means of a generalization, allows us to interpret the amplitudes of Eqs. (3.3) and (3.4) as probability amplitudes. Such step thanks to Einstein, who established that the square modulus of the wave amplitude is the probability of finding the particle in a given position.

One can synthesize this point with the following Probabilistic assumption[33]:

Postulate 3.5 (Probabilistic Assumption) *The square modulus of the wave function* $\psi(\mathbf{r}, t)$ *is the position probability density* $\wp(\mathbf{r}, t)$ *of the associated particle:*

$$\wp(\mathbf{r}, t) = |\psi(\mathbf{r}, t)|^2. \tag{3.40}$$

Eq. (3.40) is therefore immediately subjected to following normalization condition:

$$\int |\psi(\mathbf{r}, t)|^2 d^3r = 1, \tag{3.41}$$

i.e. the probability of finding the particle somewhere in the whole space must be 1.

Spectrum As we have already seen, the eigenvalues of each quantity that can be measured are *real* [see postulate 3.2: p. 33] and its eigenvectors form a *complete set* [see Eq. (3.35)].[34] The *Spectrum*[35] Υ of an observable $\hat{O}$ is the ensemble of its eigenvalues. It can be discrete or continuous. If the microentity is free, the spectrum of the energy eigenvalues is continuous; if bounded, the spectrum is discrete. If the spectrum is continuous, there are also no precise values[36] but only finite intervals. If the spectrum is bounded, an operator is bounded. Formally

[33] Exposition in [*SCHIFF* 1955, : 25].

[34] It is only in this specific sense that one speaks of 'observable' [*DIRAC* 1930c, 36–37].

[35] For examination of this problem see [*SCHIFF* 1955, 34–37] [*HOLEVO* 1982, 45, 49, 52].

an operator is bounded if

$$\forall|\psi\rangle \in \mathcal{H}, \exists c, \parallel \hat{O}|\psi\rangle \parallel \leq c \parallel \psi \parallel, \tag{3.42}$$

where c is some constant and $\forall$ means 'for all' and $\exists$ 'there is almost one'.

In an infinite-dimension Hilbert space not all operators admit eigenvectors. To handle these cases we need spectral measure and the spectral theorem [see subsection 3.5.4]. But before this we must introduce projectors.

3.4.6 Superposition Principle

Assumption 3.5 came from studies regarding collision problems. This is closely connected with the Superposition Principle [3.3: p. 39] because

- one cannot clearly separate distinct paths of the particles due to interference of type (3.4);
- the fact that the wave fucntion $\psi(\mathbf{r})$ is complex, as we have seen above, push one to consider it only as a probability amplitude and not as a probability as such [see again Eq. (3.4)].

The principle A quantum system $\mathcal{S}$ is normally not in an eigenstate relative to an observable $\hat{O}$. In this case we speak of *superposition*. Here we recall the Two-Slit-Exp. 1 [p. 26]. We saw in Eq. (3.4) that in some experimental arrangements the total distribution (total state) is not simply the sum of the partial distributions, i.e. of the 'classical' states: in that case we have not localization in (through) slit 1 or localization in (through) slit 2, but there can be combinations of both (or of more, if there are more alternatives) possibilities (*crossed terms*). In a more general form we say that, when such a combination of different states holds, we have *superposition*. Hence Dirac generalized that experiment in the Superposition Principle [*DIRAC* 1930c, 11–18]:[37]

Principle 3.3 (Superposition) *If a QM system $\mathcal{S}$ can be in a state $|b_1\rangle$ and also in a state $|b_2\rangle$, then it can be in each linear combination of both: $|\psi\rangle = c_2|b_1\rangle + c_2|b_2\rangle$ (where c_1, c_2 are complex numbers).*

Suppose that $\mathcal{S}$ is in $|b_1\rangle$ and that a measurement of observable $\hat{O}$ gives a particular result o_1; if it is in the state $|b_2\rangle$ another measurement of $\hat{O}$ gives another result o_2. Both $|b_1\rangle$ and $|b_2\rangle$ are eigenstates of $\hat{O}$. Then, every linear combination $c_2|b_1\rangle + c_2|b_2\rangle$ describes a state in which a measurement can give either o_1 or o_2. But, given the Statistical algorithm [postulate 3.4: p. 38], the result of a measurement can never be an intermediate combination such as $(o_1 + o_2)/2$.

It is clear that, relative to an observable, if a state is a superposition, then, relative to the conjugate observable, it may be an eigenstate — we will return later to possible intermediate situations [see section 18.1]. This is evident from the CCRs [Eqs. (3.14)].

Though the Superposition principle is an immediate consequence of the fact that the Schrödinger Eq. is homogeneous and linear[38], its enormous importance for the whole theory justifies a specific assumption.

---

[36]I.e. no projectors [see subsection 3.5.3].

[37]For different formulations and discussions see [*JAUCH* 1968, 106] [*Lev LANDAU/LIFSTITS* 1976b, 22] [*D'ESPAGNAT* 1976, 9] [*BELTRAMETTI/CASSINELLI* 1981, 8] .

[38]See [*D'ESPAGNAT* 1995, 39].

In general we must distinguish between *coherent* superposition (case in which we can see interference), and an *incoherent* one (where the interference terms average out to zero)[39]. In what follows, when we speak of 'superposition' without further specifications, we intend a coherent one.

Experiments The experimental verification of the principle of linear superposition of probability amplitudes of coherent QM states was proposed by Wigner [WIGNER 1963] for the z–spin components of a particle moving in the x direction, specified in a theoretical model with neutron interferometry [see section 3.9 and subsection 5.3.1] by Eder and Zeilinger [EDER/ZEILINGER 1976] and finally realized by Badurek, Zeilinger himself, and other co-workers [SUMMHAMMER *et al.* 1982] [BADUREK *et al.* 1983b].[40]

3.5 States, Density Matrices and Projectors

3.5.1 Density Operator

Definition The concept of density matrix or *density operator* (also called 'statistical operator') was introduced for the first time by Lev Landau [Lev LANDAU 1927] and mathematically treated by von Neumann [*VON NEUMANN* 1932] [*VON NEUMANN* 1955].[41] It is very useful to treat the states in terms of density matrices instead of wave vectors if either 1. the initial state is not fully specified; or 2. the final measurement is performed on a subsystem. Generally, as we shall see, it is very useful for all operations of tracing out.

A density operator $\hat{\rho}$ is a non-negative (Hermitian) trace-one class operator, where the trace Tr is the sum of the diagonal elements of a matrix[42]. A trace operator is any operator $\hat{O}$ which has a finite trace norm:

$$|\mathrm{Tr}(\hat{O})| = \mathrm{Tr}[(\hat{O}\hat{O}^\dagger)^{\frac{1}{2}}] = \sum_{n=1}^{\infty} o_n < \infty, \tag{3.43}$$

and an arbitrary operator $\hat{O}$ pertains to the trace-one class if $\mathrm{Tr}(\hat{O}) = 1$. One adds the requisite that $\hat{\rho}$ is a positive semi-definite, that is $\langle\psi|\hat{\rho}|\psi\rangle \geq 0$ for all $|\psi\rangle \in \mathcal{H}$ of the system $\mathcal{S}$. We define the density operator $\hat{\rho} \in \mathcal{H}^{\mathrm{Tr}=1;\geq 0}$ for a pure state by the following formula [see Eq. (3.22)]:

$$\boxed{\hat{\rho} := |\psi\rangle\langle\psi| = \sum_j |c_j|^2 |b_j\rangle\langle b_j| + \sum_{j\neq k} c_j c_k^* |b_j\rangle\langle b_k|} \tag{3.44}$$

where $|\psi\rangle = \sum_j c_j |b_j\rangle$.

Pure states and mixtures This formula can be useful to distinguish between pure states and mixtures [*VON NEUMANN* 1932, 162–66] [*VON NEUMANN* 1955, 306–313].[43] In fact, we certainly have a *mixed state* (specifically: a completely mixed state) if the second sum (diagonal or interference terms) is zero: the probability of the outcomes are given by the squared moduli [see the Statistical algorithm: p. 38].

[39] On this point see [*VARADARAJAN* 1968, 117, 160] [GREENBERGER 1984].

[40] A more global exposition can be found in [SUMMHAMMER *et al.* 1983].

[41] One in depth exposition can be found in [FANO 1957].

[42] Definitions in [*DAVIES* 1976, 3] [*KRAUS* 1983, 8] [*BELTRAMETTI/CASSINELLI* 1981, 291–92] [*PITOWSKY* 1989b, 53].

[43] For a handbook exposition see [*SCHIFF* 1955, 379–80].

We symbolize with $\hat{\rho}$ the pure states and with $\tilde{\hat{\rho}}$ the mixtures. In the case of pure states we have:

$$\mathrm{Tr}(\hat{\rho}) = 1, \tag{3.45a}$$

and

$$\mathrm{Tr}(\hat{\rho}^2) = \sum_k w_k^2 = 1, \tag{3.45b}$$

where the w_k's are the eigenvalues (weights) of $\hat{\rho}$. Hence:

$$\hat{\rho}^2 = \hat{\rho}. \tag{3.45c}$$

While Eq. (3.45a) is still valid for a mixture, in the latter case, instead of Eq. (3.45b), we have:

$$\mathrm{Tr}(\tilde{\hat{\rho}}^2) \leq 1, \tag{3.46}$$

so that we obtain $\tilde{\hat{\rho}}^2 \neq \tilde{\hat{\rho}}$. In what follows we distinguish in an informal and practical way between pure states and mixtures — below we shall give more formal definitions.

A state of a system $\mathcal{S}$ is pure if, for a given subdivision, each subsystem of $\mathcal{S}$ has the same features as $\mathcal{S}$; it is a mixture if each subensemble is statistically distinct from the others[44].

A pure state is represented by a vector, or better a ray, (because, as has already been said, for a possible phase factor c we take vector $|\psi\rangle$ to represent the same state) in Hilbert space. As we shall see, while the distinction between superposition/eigenstate is relative to some observable, the distinction pure state/mixture is unrelated to observables [see subsection 9.4.1].

Convexity The fact that the density operator can express both pure states and mixtures is due to the *convexity* of $\mathcal{H}^{\mathrm{Tr}=1;\geq 0}$:[45]

Theorem 3.2 (Convexity Condition) *The space of the density operator is convex, i.e. if some* $\hat{\rho}_1, \hat{\rho}_2 \in \mathcal{H}^{\mathrm{Tr}=1;\geq 0}$ *and* $0 < w < 1$ *for some function* w *(which signifies a probability weight), then we have:*

$$w\hat{\rho}_1 + (1-w)\hat{\rho}_2 \in \mathcal{H}^{\mathrm{Tr}=1;\geq 0}. \tag{3.47}$$

On the basis of the convexity condition we have the following general formula for writing a state of a system:

$$\boxed{\hat{\rho} = w\hat{\rho}_1 + (1-w)\hat{\rho}_2} \tag{3.48}$$

We say that $\hat{\rho}$ is pure iff for some w_1, w_2 we have:

$$\hat{\rho}_1 = w_1\hat{\rho}, \quad \hat{\rho}_2 = w_2\hat{\rho}. \tag{3.49}$$

---

[44] See [*D'ESPAGNAT* 1976, 62–64].

[45] Formulations and formalism in [J. PARK 1968, 213–214] [*DAVIES* 1976, 2] [*HELSTROM* 1976, 13–14, 92–93] [*BELTRAMETTI/CASSINELLI* 1981, 6] [*KRAUS* 1983, 18–20] [*BUSCH et al.* 1995a, 35].

Degrees of mixtureness It is possible to speak of different degrees of mixtureness by using the following parameter:

$$\Xi = 1 - \mathrm{Tr}(\hat{\rho}^2). \tag{3.50}$$

As we know, in the case of a pure state we have $\mathrm{Tr}(\hat{\rho}^2) = \mathrm{Tr}(\hat{\rho}) = 1$ and hence $\Xi = 0$. In the case of a mixture we have:

$$0 < \Xi < 1. \tag{3.51}$$

We must distinguish two cases: total mixture and partial mixture. If we write $\mathrm{Tr}(\hat{\rho}) = \sum_{j,k} \rho_{jk} |j\rangle\langle k|$, we have a total mixture if

$$\rho_{jk} = |c_j|^2 \delta_{jk}, \tag{3.52}$$

and hence

$$\hat{\rho} = \sum_j |c_j|^2 |j\rangle\langle j|. \tag{3.53}$$

In the case of a proper mixture, Ξ approaches 1 as the number of mixed elements increases.

For example: Let there be a mixture such that $\hat{\rho}_1 = \frac{1}{2}|0\rangle\langle 0| + \frac{1}{2}|1\rangle\langle 1|$. It follows that $\hat{\rho}_1^2 = \frac{1}{4}|0\rangle\langle 0| + \frac{1}{4}|1\rangle\langle 1|$. In conclusion $\mathrm{Tr}(\hat{\rho}_1^2) = \frac{1}{4} + \frac{1}{4} = \frac{1}{2}$.

Now let there be another mixture in the following form: $\hat{\rho}_2 = \frac{1}{3}|0\rangle\langle 0| + \frac{1}{3}|1\rangle\langle 1| + \frac{1}{3}|2\rangle\langle 2|$. We have: $\mathrm{Tr}(\hat{\rho}_2^2) = \frac{1}{9} + \frac{1}{9} + \frac{1}{9} = \frac{1}{3}$. And we see that $\mathrm{Tr}(\hat{\rho}_2^2) < \mathrm{Tr}(\hat{\rho}_1^2)$.[46]

As we shall see the parameter Ξ is strictly related to von Neumann entropy [see section 42.1].

3.5.2 Entangled States

We see now an important property of pure states (and of some mixtures) of compound systems. Let us start by formulating the following definition:

Definition 3.2 (Isolated System) *We say that a system $\mathcal{S}$ is* isolated *if*

- *it is dynamically independent from every other system $\mathcal{S}'$, i.e. there is no interaction Hamiltonian ($\hat{H}_{\mathcal{SS}'} = 0$),*
- *and if it is probabilistically independent* (= separable) *from every other system.*

We say that a compound system is factorisable *in relation to some subsystems if the these subsystems are isolated from one other.*

Note that in QM dynamic independence is only a necessary condition of an isolated system, while in classical mechanics there is an equivalence between the two. In other words, the independence of the second condition from the first is a typical feature of QM [see also subsection 3.7.2].

We now introduce the concept of *entanglement*[47] — formulated for the first time by Schrödinger [SCHRÖDINGER 1936]:[48]

[46] See [*SCHROECK* 1996, 62–63].

[47] We discuss here only the formal aspects of the problem, postponing [to parts VIII and IX] the analysis of the nature of the entanglement and of its ontological significance, if any.

[48] The article was written as an answer to Einstein/Podolsky/Rosen [see chapter 31 and section 34.1]. A distinction between classical and QM entanglement is to be found in [J. PARK 1968, 224]. For a general discussion see subsection 42.8.3.

Definition 3.3 (Entanglement) *An entangled state is a state of a compound system whose subsystems are not probabilistically independent.*

Entanglement is a natural feature of the Schrödinger Eq. for multiparticle systems. However, note that entanglement is not another name for superposition in a multiparticle system, as has often been proposed[49]: entanglement is rather like superposition plus no-factorization of states.

Therefore we can prove following theorem:

Theorem 3.3 (Entanglement and Factorization) *An entangled state can never be factorised and a factorised one can never be written as an entangled one.*

We will prove first part of the theorem for the case of a two-dimensional Hilbert space. Generalization can then follow by induction.

Proof [50]

An entangled state in a two-dimensional Hilbert space can always be written in the form:

$$|\Psi\rangle = c|j\rangle_1 \otimes |k\rangle_2 + c'|j'\rangle_1 \otimes |k'\rangle_2, \tag{3.54}$$

where $|j\rangle_1$ and $|j'\rangle_1$ are orthogonal kets in the space of particle 1's states, and the same for $|k\rangle_2$ and $|k'\rangle_2$ for particle 2. The scalars c and c' are both non-zero. In what follows we generally omit the sign $\otimes$. Now we prove that Eq. (3.54) cannot be rewritten as the factorised expression:

$$|\Psi\rangle = |m\rangle_1|n\rangle_2, \tag{3.55}$$

where $|m\rangle_1$ and $|n\rangle_2$ are, respectively, in the spaces of the states of particles 1 and 2.

Let us assume that it is possible. If so, we can write $|m\rangle_1$ as a linear superposition of $|j\rangle_1$ and $|j'\rangle_1$, and $|n\rangle_2$ as a linear superposition of $|k\rangle_2$ and $|k'\rangle_2$:

$$|m\rangle_1 = \eta|j\rangle_1 + \eta'|j'\rangle_1, \tag{3.56a}$$

$$|n\rangle_2 = \zeta|k\rangle_2 + \zeta'|k'\rangle_2. \tag{3.56b}$$

Then:

$$|m\rangle_1|n\rangle_2 = \eta\zeta|j\rangle_1|k\rangle_2 + \eta\zeta'|j\rangle_1|k'\rangle_2 + \eta'\zeta|j'\rangle_1|k\rangle_2 + \eta'\zeta'|j'\rangle_1|k'\rangle_2. \tag{3.57}$$

Comparing Eqs. (3.54) and (3.57) we see that:

$$\eta\zeta' = 0; \quad \eta'\zeta = 0, \tag{3.58}$$

which implies either $c = \eta\zeta = 0$ or $c' = \eta'\zeta' = 0$, which is contrary to our assumption. Q.E.D.

In conclusion, in QM there can be systems which are non-factorisable, though there is no interaction Hamiltonian between them. In this case they are only probabilistically and not causally interdependent. Hence entanglement can persist also when the systems become dynamically independent (which is a form of saying that there is correlation without interaction)[51].

---

[49] See for example [GREENBERGER *et al.* 1993, 22].

[50] See [GREENBERGER *et al.* 1990, 1139].

[51] See also [J. PARK 1968, 222–23]. A generalization which includes the case of entanglement of non-orthogonal, but distinct, component states is to be found in [SANDERS 1992]

3.5.3 Projectors

Definition Von Neumann [*VON NEUMANN* 1932, 131–32] [*VON NEUMANN* 1955, 247–49] introduced the concept of the projection operator $\hat{P}$ [see figure 3.3 for examples of projectors]:[52]

Definition 3.4 (Projectors) *Projectors are one-dimensional (take only values* $\{0,1\}$*), self-adjoint, idempotent* ($\hat{P}^2 = \hat{P}$) *operators.*

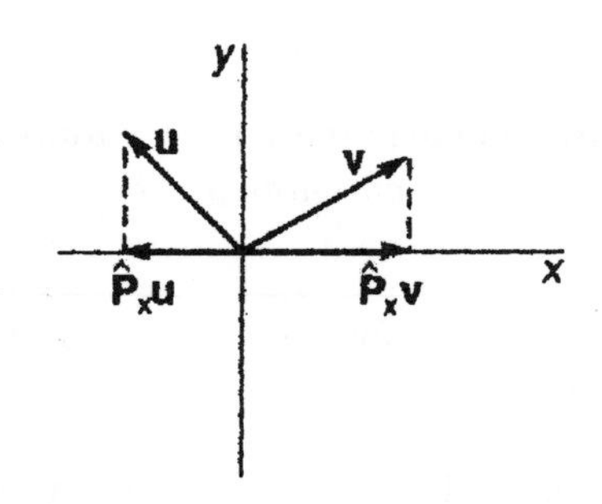

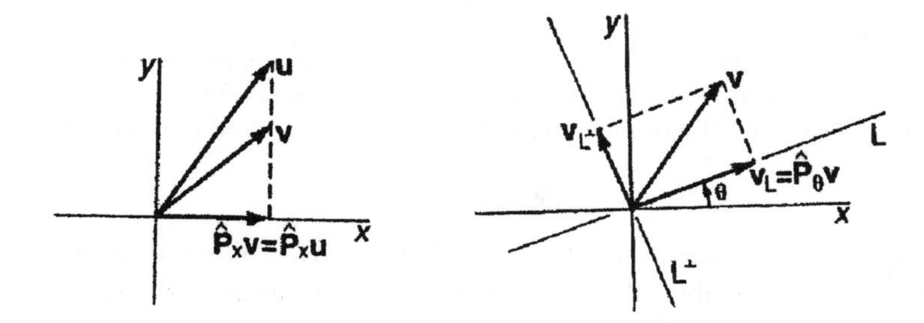

Figure 3.3: Examples of projectors. (c) shows the action of a projector $\hat{P}_\theta = \begin{pmatrix} \cos^2\theta & \cos\theta \cdot \sin\theta \\ \cos\theta \cdot \sin\theta & \sin^2\theta \end{pmatrix}$ on a vector $\mathbf{v} = \mathbf{v}_L + \mathbf{v}_{L^\perp}$ — from [*HUGHES* 1989, 16, 17, 20].

This definition is certainly part of Basic-QM (the formalism of projectors is fundamental). But the concept of projection presents some problems of interpretation [see subsection 18.1.2]: hence, so as to stress this point, we do not explicitly integrate it into the theory.

The relationship between projectors and density operators is given by the definition:

$$\boxed{\hat{\rho} := \sum_k w_k \hat{P}_{b_k} = \sum_k |b_k\rangle w_k \langle b_k|} \tag{3.59}$$

If $\hat{\rho}$ is a pure state all w_k's, except one, are zero and we have: $\hat{\rho} = \hat{P}$; otherwise we have a mixture. So we can state following theorem, to be accredited to Von Neumann

[52] For definition and formalism see [*PRUGOVEČKI* 1971, 200–202][*HOLEVO* 1982, 47–48].

Theorem 3.4 (I von Neumann) *Formally, to each pure state (superposition or eigenstate) corresponds a projection operator defined as:*

$$\hat{P}_{b_k} = |b_k\rangle\langle b_k|. \tag{3.60}$$

As we shall see the converse is true only if there are no Superselection Rules [see subsection 10.1.4].[53] Hence projectors are extreme elements of the convex set $\mathcal{H}^{\mathrm{Tr}=1;\geq 0}$ [see Eqs. (3.48) and (3.49)]: they are characterized by *orthogonality.*

Their domain is the whole Hilbert space $\mathcal{H}$ of a system. Their range is only the one-dimensional subspace spanned by $|b_k\rangle$. So they also are characterized by *completeness*:

$$\sum_k \hat{P}_{b_k} = \hat{I}, \tag{3.61}$$

where $\hat{I}$ is the identity operator, so that $\sum_k \hat{P}_{b_k}$ acting on an arbitrary vector $|\psi\rangle$ just reproduces $|\psi\rangle$ itself.

Action of projectors The significance of the operator is clearer if we allow it to act on an arbitrary state $|\psi\rangle$ which it projects onto one of its components:

$$\hat{P}_{b_j}|\psi\rangle = |b_j\rangle\langle b_j|\psi\rangle = c_j|b_j\rangle, \tag{3.62}$$

i.e. acting on $|\psi\rangle$ it projects out the component $c_j|b_j\rangle$ in the expansion according to Eq. (3.22).

3.5.4 Spectral Theorem and Consequences

In a finite-dimensional Hilbert space every (discrete) observable admits the following *Spectral representation*[54]:

$$\hat{O} = \sum_{j=1}^{k} o_j \hat{P}_{o_j}, \tag{3.63}$$

where the $\hat{P}_{o_j}$'s are the projection operators corresponding to the k orthogonal subspaces $\mathcal{H}_k$ determined by k distinct eigenvalues.

In an infinite-dimensional Hilbert space we have an equation analogous to Eq. (3.63) for compact operators only. For a more general case we can use a continuous Eq. analogous to Eq. (3.63). This is the content of the *Spectral theorem.* First of all we need to write a projection for the case in which we have continuous values:

$$\hat{P}(do) = \left[\sum_k \delta(o - o_k)\hat{P}_{o_k}\right] do \tag{3.64}$$

by means of which we formulate [*PRUGOVEČKI* 1971, 231–36, 242, 250] [*HOLEVO* 1982, 52–64] the theorem[55]:

---

[53] See [BELTRAMETTI *et al.* 1990, 91].

[54] See [*BUB* 1974, 17] [JAUCH 1976, 136] .

[55] Further formulations and comments in [*BELTRAMETTI/CASSINELLI* 1981, 14, 293–96] [BUSCH *et al.* 1992, 950] [*PERES* 1993, 99–106] [*SCHROECK* 1996, 32–33].

Theorem 3.5 (Spectral Theorem) *For each observable $\hat{O}$ there is always a spectral representation given by*

$$\hat{O} = \int o\hat{P}(do). \tag{3.65}$$

It is the Spectral theorem which gives us the possibility of representing each QM state with a density operator because from it follows [*DAVIES* 1976, 4]:

Corollary 3.2 (Convex Decomposition) *Every density operator admits a convex decomposition* [see Eq. (3.47)].

This decomposition is normally not unique[56]. But in the case of orthogonal states it is unique iff there is no degeneracy (i.e. if all weights of the mixture have different values) [BELTRAMETTI 1985, 86]:[57]

Corollary 3.3 (Non-Uniqueness of the Convex Decomposition) *The decomposition of a density matrix is generally not unique. This is always the case unless orthogonal states are considered and there is no degeneracy.*

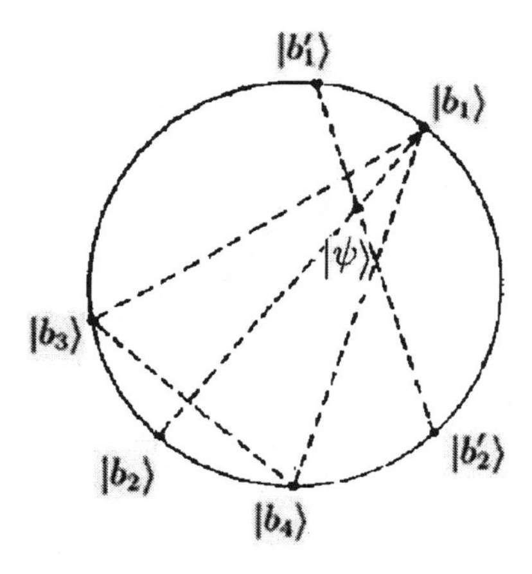

Figure 3.4: Different decompositions of a non-pure state $|\psi\rangle$ — from [*BELTRAMETTI/CASSINELLI* 1981, 33] .

We show it by means of an **example.** Let us imagine an arbitrary non-pure state represented by vector $|\psi\rangle$ [see figure 3.4]. We can decompose it for example either by the two orthogonal pure

[56] Which forbids the ignorance interpretation of QM [*BELTRAMETTI/CASSINELLI* 1981, 8–11] [see also subsection 6.6.1].

[57] See also [*BELTRAMETTI/CASSINELLI* 1981, 33] .

states $|b_1\rangle, |b_2\rangle$, or by the two non-orthogonal pure states $|b'_1\rangle, |b'_2\rangle$. By iterated convex combinations we can decompose $|\psi\rangle$ into more than two states: for instance $|b_1\rangle, |b_3\rangle, |b_4\rangle$. This can be represented by means of a plane which passes through the centre of a Poincaré sphere: with a Poincaré sphere we can represent pure states as points on its surface and mixed states as points inside. A convex combination of pure states is represented by the points of a segment they determine. The centre of the sphere can be decomposed into infinitely many pairs of orthogonal states (there are infinitely many diameters through it). It is represented by the matrix $\hat{I}/2$, and has a doubly degenerate eigenvalue. Now $|\psi\rangle$ can be decomposed into a family of pure states if it lies inside the polygon (i.e. the convex hull) determined by that family.

In the case of non-unique decomposition the situation is the following: given a large number n of identical copies of $\mathcal{S}$, prepare nw_1 of them in state $|b_1\rangle$ and nw_2 of them in state $|b_2\rangle$, mix them up and choose one exemplar at random. Non-unique decomposition means that this preparation is indistinguishable from infinitely many others of a similar kind involving the mixture of other pairs of pure states (possibly with other weights).[58]

3.6 Unitary Transformations

3.6.1 Beginnings

The beginnings of a QM Transformation Theory are to be accredited to Born, Heisenberg and Jordan [BORN *et al.* 1926, 163–66].[59] The first problem was finding a canonical transformation which from observables $\hat{\mathbf{p}}, \hat{\mathbf{q}}$ allows one to pass to new observables $\hat{\mathbf{p}}', \hat{\mathbf{q}}'$ by respecting:

$$\hat{\mathbf{p}}\hat{\mathbf{q}} - \hat{\mathbf{q}}\hat{\mathbf{p}} = \hat{\mathbf{p}}'\hat{\mathbf{q}}' - \hat{\mathbf{q}}'\hat{\mathbf{p}}' = \imath\hbar. \tag{3.66}$$

A general transformation which satisfies this condition is given by

$$\hat{\mathbf{P}} = \hat{U}\hat{\mathbf{p}}\hat{U}^{-1}, \tag{3.67}$$

$$\hat{\mathbf{Q}} = \hat{U}\hat{\mathbf{q}}\hat{U}^{-1}, \tag{3.68}$$

where $\hat{U}^{-1}$ is the inverse of $\hat{U}$. Such transformation is unitary, i.e. one for which: $\hat{U}^{-1} = \hat{U}^\dagger$ or $\hat{U}\hat{U}^\dagger = \hat{I}$.

The transformation theory was further developed by Jordan [P. JORDAN 1926, 199]. London [LONDON 1926, 199] understood such a transformation in terms of a rotation of axes in Hilbert space. Further developments were carried by Dirac [DIRAC 1926a] [DIRAC 1926b] [*DIRAC* 1930c] and again by Jordan [P. JORDAN 1927].

In conclusion, we have following general result[60]:

Theorem 3.6 (Stone) *Every unitary operator can be represented in the form:*

$$\hat{U} = e^{\imath\hat{O}}, \tag{3.69}$$

where $\hat{O}$ is a Hermitian operator.

We do not integrate this theorem in QM because, though valid, it is a mathematical theorem, i.e. one not specifically characterizing QM.

[58]For other developments regarding pure states and mixtures see also [HADJISAVVAS 1981], and for some implication of the non-unique decomposition property see also subsection 42.4.2.

[59]For handbook expositions see [*DIRAC* 1930c, 103–107] [*Lev LANDAU/LIFSTITS* 1976b, 56–58]. Historical analysis in [*JAMMER* 1966, 310–22].

[60]Formulations in [*HOLEVO* 1982, 62, 105] [*SCHROECK* 1996, 36–37, 48].

3.6.2 Time Evolution of the Wave function and states

We now introduce a particular realization of the Stone theorem: the unitary time-evolution operator:

$$\boxed{\hat{U}_t = e^{-\frac{\imath}{\hbar}\hat{H}t}} \tag{3.70}$$

where $\hat{H}$ is the Hamiltonian of an arbitrary system $\mathcal{S}$. This operator's function is to transform an arbitrary state vector of $\mathcal{S}$ at an initial moment t_0 into a vector at an arbitrary moment t (time translation):

$$|\psi_t\rangle = \hat{U}_t|\psi_{t_0}\rangle, \tag{3.71}$$

so that the time-evolution of the wave function has a unitary and deterministic character. For other possible realizations of Stone Theorem see below [subsection 3.10.2].

3.6.3 Schrödinger, Heisenberg and Dirac Pictures

Now we study various possible pictures of the evolution of a QM system[61].

Schrödinger and Heisenberg Pictures We have just seen the time evolution of the wave function (or of states). The transformation in time of an observable $\hat{O}$ is something different[62]. What is important here is to preserve the mean value $\langle\psi|\hat{O}|\psi\rangle$ [see subsection 3.7.1]. So that we have two possible cases, a *passive* transformation, given by Eq. (3.71):

$$\boxed{\hat{\rho} \mapsto \hat{\rho}_t = \hat{U}_t\hat{\rho}\hat{U}_t^\dagger = e^{-\frac{\imath}{\hbar}\hat{H}t}\hat{\rho}_0 e^{\frac{\imath}{\hbar}\hat{H}t}} \tag{3.72}$$

which corresponds to the *Schrödinger picture*, where states evolve while observables are kept fixed; and an *active* transformation:

$$\boxed{\hat{O} \mapsto \hat{O}_t = \hat{U}_t^\dagger\hat{O}\hat{U}_t} \tag{3.73}$$

which corresponds to the *Heisenberg picture*, where observables evolve while states are kept fixed. The concepts of active and passive transformation are another way of expressing the fact that, while states are vectors, observables are operations on vectors in a Hilbert space [subsection 3.4.3]. Passive and active unitary transformations are related to each other, so that we are also able to pass from the Schrödinger picture to the Heisenberg picture:

$$\hat{U}_t\hat{\rho}(\hat{O})\hat{U}_t^\dagger = \hat{\rho}(\hat{U}_t^\dagger\hat{O}\hat{U}_t). \tag{3.74}$$

The concepts of active and passive transformation can be interpreted in the following way: the density matrix (in the Schrödinger picture) can be treated as an ordinary operator (in the Heisenberg picture) evolving backwards in time[63].

We can express both pictures in a very simple way using Dirac algebra. We can write the Schrödinger equation in the Schrödinger picture [Eq. (3.71)] as follows:

$$\imath\hbar\frac{d}{dt}|\psi_S(t)\rangle = \hat{H}|\psi_S(t)\rangle, \tag{3.75}$$

[61] For handbook expositions of the following see [*SCHIFF* 1955, 169–73] [*DIRAC* 1930c, 108–15]. See also [*DAVIES* 1976, 25] [*BELTRAMETTI/CASSINELLI* 1981, 53, 57] [*KRAUS* 1983, 42] [*PERES* 1993, 221–22].

[62] We again have duality between states and observables which, historically, depends on the formulation of Wave and Matrix Mechanics respectively, but which has a deeper justification [see subsection 9.4.1].

[63] See [COLLETT 1988, 2235].

where the subscript S means 'Schrödinger picture'. We can pass to a Heisenberg picture by means of the following substitutions:

$$|\psi_H(t)\rangle := |\psi_S(0)\rangle = e^{\imath\hat{H}t/\hbar}|\psi_S(t)\rangle, \quad (3.76a)$$
$$\hat{O}_H := e^{\imath\hat{H}t/\hbar}\hat{O}_S e^{-\imath\hat{H}t/\hbar}, \quad (3.76b)$$

where the subscript H here means 'Heisenberg picture'. As result we have the following equation of motion in the Heisenberg picture (evolution of an observable $\hat{O}$):

$$\frac{d\hat{O}_H}{dt} = \frac{\partial\hat{O}_H}{\partial t} + \frac{1}{\imath\hbar}[\hat{O}_H, \hat{H}]. \quad (3.77)$$

Dirac Picture We can also speak of a third picture, the Dirac or *interaction picture* , in which we have:

$$\boxed{\hat{H} = \hat{H}_0 + \hat{H}_i,} \quad (3.78)$$

i.e. we divide the Hamiltonian in two parts: one ($\hat{H}_0$) which does not explicitly depend on time (for example the kinetic energy), the other, the interaction Hamiltonian ($\hat{H}_i$), which may depend on time (the potential energy). We have following Eqs. (where subscript 'I' stands for 'interaction'):

$$|\psi_I(t)\rangle := e^{\imath\hat{H}_0 t/\hbar}|\psi_S(t)\rangle, \quad (3.79a)$$

$$\hat{O}_I(t) = e^{\imath H_0 \frac{t}{\hbar}}\hat{O}_S e^{-\imath\hat{H}_0 t/\hbar}, \quad (3.79b)$$

so that interaction and Heisenberg pictures are the same when $\hat{H}_i = 0$.

The interaction picture is very useful for all forms of interaction or of scattering because it distinguishes among the interaction time, the time before the interaction is switched on and the time after is switched off. For these last two states we have an initial vector $|\psi^i\rangle$ and a final vector $|\psi^f\rangle$ which are constant.

Conservation Law for States and Observables The evolution law for the state is given by[64]:

$$\boxed{\imath\hbar\frac{d}{dt}\hat{\rho} = [\hat{H}, \hat{\rho}]} \quad (3.80)$$

which is the law corresponding of the classical Liouville equation [see Eq. (1.12)]. If the Hamiltonian and the state commutes (if Eq. (3.80) is equal to 0), then the state is conserved.

Eq. (3.80) gives the conservation law of an arbitrary observable $\hat{O}$ by substituting $\hat{\rho}$ with:

$$\boxed{\imath\hbar\frac{d}{dt}\hat{O} = [\hat{H}, \hat{O}]} \quad (3.81)$$

3.6.4 Different Representations

Transforms We spoke of different forms of Eq. (3.25) depending on the particular operator which is used. Hence, among others, we can especially distinguish a momentum representation, a coordinate representation, and an energy representation.

---

[64]For a handbook exposition of the following see [*SCHIFF* 1955, 380].

One can go from the coordinate representation to the momentum representation and *vice versa* by means of the following Fourier transforms[65]:

$$\hat{U}_F : \psi(\mathbf{r}, t) \mapsto \tilde{\psi}(\mathbf{k}, t) = (2\pi\hbar)^{-\frac{3}{2}} \int_{\Re^3} d\mathbf{r} e^{-\frac{i}{\hbar}\mathbf{k}\cdot\mathbf{r}} \psi(\mathbf{r}, t) \,, \tag{3.82a}$$

$$\hat{U}_F^{-1} : \tilde{\psi}(\mathbf{k}, t) \mapsto \psi(\mathbf{r}, t) = (2\pi\hbar)^{-\frac{3}{2}} \int_{\Re^3} d\mathbf{k} e^{-\frac{i}{\hbar}\mathbf{r}\cdot\mathbf{k}} \tilde{\psi}(\mathbf{k}, t) \,, \tag{3.82b}$$

where $\mathbf{k}$ is the propagation vector and we have the following relationship between the j-th components of momentum and position:

$$\hat{p}_j = \hat{U}_F^{-1} \hat{q}_j \hat{U}_F, \tag{3.83}$$

where $\psi(\mathbf{r}) \in L^2(\Re^3), \mathbf{r} \in \Re^3$ and $\tilde{\psi}(\mathbf{k}) \in L^2(\Re^3), \mathbf{k} \in \Re^3$. Obviously (in one dimension) $\tilde{\psi}(k) = \langle k|\psi\rangle$. Two functions like $\psi(\mathbf{r}), \tilde{\psi}(\mathbf{k})$ are called a *Fourier pair.*

There are also transforms from the momentum representation to the energy representation and *vice versa* [*HOLEVO* 1982, 132], given by the following Eq.:

$$\int_{-\infty}^{+\infty} |\tilde{\psi}(\mathbf{k})|^2 dk^3 = \sqrt{\frac{m}{2}} \left[\int_0^{+\infty} |\tilde{\psi}(\sqrt{2mE})|^2 \frac{dE}{\sqrt{E}} + \int_0^{+\infty} |\tilde{\psi}(-\sqrt{2mE})|^2 \frac{dE}{\sqrt{E}} \right] = \int_0^{+\infty} |\tilde{\tilde{\psi}}(E)|^2 dE, \tag{3.84}$$

where

$$\tilde{\tilde{\psi}}(E) = \left(\frac{m}{2E}\right)^{\frac{1}{4}} \begin{bmatrix} |\tilde{\psi}(\sqrt{2mE})\rangle \\ |\tilde{\psi}(-\sqrt{2mE})\rangle \end{bmatrix} \tag{3.85}$$

describes the transformation from the momentum representation to the energy representation for a free particle.

Representations In what follows we give explicit expressions only for coordinate and momentum representations[66]. In the *coordinate representation* we have:

$$-\imath\hbar\nabla\psi(\mathbf{r}) = \hat{\mathbf{p}}\psi(\mathbf{r}), \tag{3.86}$$

where the momentum operator is that of Eq. (3.24b). For the position operator in coordinate representation we have an eigenvalue equation:

$$\hat{\mathbf{q}}\psi(\mathbf{r}) = q\psi(\mathbf{r}). \tag{3.87}$$

In the *momentum representation* we can have a position operator which behaves as follows:

$$\imath\hbar\frac{\partial\tilde{\psi}(\mathbf{k})}{\partial\mathbf{p}} = \hat{\mathbf{q}}\tilde{\psi}(\mathbf{k}). \tag{3.88}$$

Analogous to the position in coordinate representation, the momentum in momentum representation is a simple multiplier:

$$\hat{\mathbf{p}}\tilde{\psi}(\mathbf{k}) = k\tilde{\psi}(\mathbf{k}). \tag{3.89}$$

The possible representations are as many as the observables introduced or considered [see for example p. 380].

[65]See [*PRUGOVEČKI* 1984b, 11] [*BUSCH et al.* 1995a, 59] .

[66]Exposition in [*SCHIFF* 1955, 46, 53–54] [*Lev LANDAU/LIFSTITS* 1976b, 64–67] .

3.7 Distribution and Characteristic Functions

3.7.1 Expectation Value

Classical expectation In classical probability theory[67] the *average expectation value* or *mean value* [for the energy see Eq. (2.6)] of a random variable ξ [see definition 11.4: p. 195] is:

$$\overline{\xi} \equiv \langle \xi \rangle = \int_x \xi d\mathcal{F}_{<x}, \tag{3.90}$$

where the $\mathcal{F}_{<x}$ is called the *distribution function* and is defined as the probability that a random variable ξ takes a value $< x$: $\mathcal{F}_{<x} =: \wp\{\xi < x\}$, so that it consists in an integration of probabilities.

Quantum expectation The QM correspondent[68] — but note that QM as such has no classical probability structure due to the CCRs [Eq. (3.14)] and to the superposition principle [3.3: p. 39] — average expectation value or mean value of an observable $\hat{O}$ in a state $\hat{\rho}$ is given by:

$$\boxed{\left\langle \hat{O} \right\rangle := \mathrm{Tr}(\hat{O}\hat{\rho}) = \mathrm{Tr}(\hat{\rho}\hat{O})} \tag{3.91}$$

We can also express this as the sum of all eigenvalues of the operator each multiplied by the corresponding probability $|c_j|^2$:

$$\left\langle \hat{O} \right\rangle = \sum_j |c_j|^2 o_j. \tag{3.92}$$

As we know this formula does not give the result of any single measurement [see postulate 3.4: p. 38].

If $\hat{\rho} = |\psi\rangle\langle\psi|$ is a pure state of an individual system, then the expectation value is equivalent to the inner product, that is to say:

$$\left\langle \hat{O} \right\rangle = \langle \psi | \hat{O} | \psi \rangle, \tag{3.93}$$

where the form of the RHS of the (3.93) is due to the fact that the operator is self-adjoint [see p. 32]. For statistical ensembles we need an average of the expectation values for states $|b_j\rangle$ [FURRY 1936, 394]:

$$\left\langle \hat{O} \right\rangle = \sum_j w_j \langle b_j | \hat{O} | b_j \rangle. \tag{3.94}$$

The mean value of each observable is given by [VON NEUMANN 1927c] [WIGNER 1932, 749]:

$$\left\langle \hat{O} \right\rangle = \hat{O} e^{-\frac{\hat{H}}{k_B T}}. \tag{3.95}$$

Use of projectors We can use projectors[69] and say that the probability that the result of measurement of $\hat{O}$ lies in the set Υ is simply:

$$\mathrm{Tr}\left[\hat{P}_{\hat{O}}(\Upsilon) \cdot \hat{\rho}\right]. \tag{3.96}$$

[67]Formalism in [*GNEDENKO* 1969, 125, 132, 165] [*BUB* 1974, 34] [JAUCH 1976, 128]. See also [*GUDDER* 1988b, 5].

[68]Expositions in [*DE BROGLIE* 1957, 20] [*SCHIFF* 1955, 379] [JAUCH 1976, 136]. See also [*KRAUS* 1983, 10] [*SCHROECK* 1996, 59].

[69]For that which follows see [LUDWIG 1961, 151] [*BELTRAMETTI/CASSINELLI* 1981, 12, 23] [*PERES* 1993, 71, 187–88] [*SCHROECK* 1996, 144–146, 168].

For a system in state $|\psi\rangle$, we can express the probability of having o_j as the result of a measurement by means of $\hat{P}$ — if the eigenvalue is not degenerate —:

$$[\wp(o_j)]^{\psi} = |\langle b_i|\psi\rangle|^2 = \mathrm{Tr}(\hat{P}_{b_j} \cdot \hat{P}_{\psi}). \tag{3.97}$$

Hence we say that projectors define *sharp properties* of a system. Joint probability distributions of compatible observables are given by:

$$\wp(\hat{O}_1, \hat{O}_2) = \mathrm{Tr}[\hat{\rho}\hat{P}_{O_1}(\Upsilon_1)\hat{P}_{O_2}(\Upsilon_2)], \tag{3.98}$$

and similarly subsequential measurement results (o_1 for observable $\hat{O}_1$ and o_2 for $\hat{O}_2$) are expressed by the *Wigner formula*:

$$\wp(o_1, o_2) = \mathrm{Tr}[\hat{P}_{o_2}\hat{P}_{o_1}\hat{\rho}\hat{P}_{o_1}]. \tag{3.99}$$

More generally we can express transition probabilities from a generic state $\hat{\rho}$ to a state $\hat{P}_{\psi}$ as:

$$\mathrm{Tr}(\hat{\rho}\hat{P}_{\psi}), \tag{3.100}$$

which is a generalization of the Malus Law $\wp = \cos^2(\mathbf{n}'_f - \mathbf{n}_i)$, i.e. the probability for a transition from an initial polarisation direction to a final one. On the basis of this one arrives to state conditional probability (*von Neumann formula*), i.e. the probability of obtaining a state $\hat{\rho}_t$ when the initial state is $\hat{\rho}_0$

$$\hat{\rho}_t = \frac{(\hat{P}\hat{\rho}_0\hat{P})}{\mathrm{Tr}(\hat{\rho}_0\hat{P})}, \tag{3.101}$$

which can be generalized further when also the event represented by $\hat{P}$ has already occurred[70]:

$$\mathrm{Tr}(\hat{\rho}_t\hat{P}_t) = \frac{\mathrm{Tr}(\hat{\rho}_0\hat{P}\hat{P}_t\hat{P})}{\mathrm{Tr}(\hat{\rho}_0\hat{P})}, \tag{3.102}$$

if $\mathrm{Tr}(\hat{\rho}_0\hat{P}_t) \neq 0$.

Partial trace In the case of compound systems[71], it is always possible to isolate a subsystem by a *partial trace*.

For example, supposing that in a system $\mathcal{S} = \mathcal{S}_1 + \mathcal{S}_2$ (where the Hilbert space can be written as $\mathcal{H} = \mathcal{H}_1 \oplus \mathcal{H}_2$) we wish to exclude $\mathcal{S}_2$. First we write the global density matrix of the system by $\hat{\rho}_{\mathcal{H}_1\otimes\mathcal{H}_2}$; then we choose an observable pertaining to $\mathcal{H}_1$ only, say $\hat{O}_{\mathcal{H}_1}$ and perform the trace on $\mathcal{H}_2$, i. e. the sum of all expectation values of $\hat{O}_{\mathcal{H}_1}$ for all possible states of subsystem $\mathcal{S}_2$, which gives the reduced density matrix for $\mathcal{S}_{12}$:

$$\mathrm{Tr}_{\mathcal{H}_1}(\hat{O}_{\mathcal{H}_1} \cdot \mathrm{Tr}_{\mathcal{H}_2}\hat{\rho}_{\mathcal{H}_1\otimes\mathcal{H}_2}) = \mathrm{Tr}[(\hat{O}_{\mathcal{H}_1} \otimes \hat{I}_{\mathcal{H}_2})\hat{\rho}_{\mathcal{H}_1\otimes\mathcal{H}_2}]. \tag{3.103}$$

For arbitrary $\hat{\rho}_{\mathcal{H}_1}, \hat{\rho}_{\mathcal{H}_2}$ we can write more easily:

$$\hat{\rho}^{1+2}_{\mathrm{Tr}_2} = \mathrm{Tr}_2(\hat{\rho}^1 \otimes \hat{\rho}^2) = \mathrm{Tr}\hat{\rho}_2 \cdot \hat{\rho}_1. \tag{3.104}$$

In general if the compound system is a pure state, the partial trace gives rise to a mixture — an *improper mixture* (because the compound system remains in a pure state).

[70] Formulation in [*GUDDER* 1988b, 54].

[71] Expositions in [*KRAUS* 1983, 67–68] [*SCHROECK* 1996, 65–68, 108–112]. See also [*HOLEVO* 1982, 78] .

3.7.2 (Un)correlations

Though we have distinguished superposition from entanglement, we shall use the term *correlation* to indicate any type of no-factorization: of probability distributions for some observable in the case of superposition, and of states in the case of entanglement. By using earlier formulas for expectation values, we can treat the problem of correlation in the case of entanglement on a formal plane[72].

Two subsystems $\mathcal{S}_1, \mathcal{S}_2$ are factorised or *uncorrelated* or non-entangled [see also definition 17.1: p. 264] if the expectation value for the joint measurement of observables pertaining to each of them always factorises:

$$\begin{aligned}\mathrm{Tr}(\hat{\rho} \cdot \hat{O}_1 \otimes \hat{O}_2) &= \mathrm{Tr}(\hat{\rho} \cdot \hat{O}_1 \otimes \hat{I})\mathrm{Tr}(\hat{\rho} \cdot \hat{I} \otimes \hat{O}_2) \\ &= \mathrm{Tr}(\hat{\rho}_1 \hat{O}_1)\mathrm{Tr}(\hat{\rho}_2 \hat{O}_2). \end{aligned} \tag{3.105}$$

Now imagine that we have a random number generator which produces numbers $r = 1, \ldots, n$ with probability $\wp_r$ and that the two preparing devices for the two subsystems $\mathcal{S}_1, \mathcal{S}_2$ are then set according to the result r. The expectation of the measurement of $\hat{O}_1, \hat{O}_2$ will be:

$$\sum_{r=1}^{n} \wp_r \mathrm{Tr}(\hat{\rho}_1^r \hat{O}_1)\mathrm{Tr}(\hat{\rho}_2^r \hat{O}_2) = \mathrm{Tr}(\hat{\rho} \cdot \hat{O}_1 \otimes \hat{O}_2). \tag{3.106}$$

$\hat{\rho}$ is a convex combination of product states. Now expectation values for this state no longer factorise. We call a density operator *Classically Correlated* (= Cl-Correlated) if it can be approximated (in trace norm) by density operators of the form (3.106). Quantum states which are not Cl-Correlated are called *EPR-Correlated* [see chapter 31]. Any probability measure on a product space is Cl-Correlated — in the context of C*–algebraic QM theory [HAAG/KASTLER 1964] [see subsection 9.3.1] if one of two subsystems of a composite systems is classical (commutative algebra of the observables), then all states of the composite system are Cl-Correlated.

3.7.3 Characteristic Function and Variance

Characteristic and distribution functions An important mathematical tool is the *characteristic function*[73]. Classically the characteristic function of a random variable ξ is defined as the expectation of the random variable $e^{\imath\eta\xi}$, where here η stands for a real parameter. If $\mathcal{F}_{<x}$ is the distribution function of ξ [Eq. (3.90)], the characteristic function is:

$$\chi_\xi(\eta) = \int e^{\imath\eta\xi} d\mathcal{F}_{<x}(\xi). \tag{3.107}$$

A distribution function is uniquely determined by its characteristic function:

$$\mathcal{F}_{<x}(\xi) = \frac{1}{2\pi} \lim_{y \longrightarrow -\infty} \lim_{c \longrightarrow \infty} \int_{-c}^{+c} \frac{e^{-\imath\eta y} - e^{-\imath\eta\xi}}{\imath\eta} \chi_\xi(\eta) d\eta, \tag{3.108}$$

where the limit in y is evaluated with respect to any set of points y that are points of continuity for the function $\mathcal{F}_{<x}(\xi)$.

---

[72] See [R. WERNER 1989, 4277].

[73] Mathematics in [*GNEDENKO* 1969, 219, 227] [*GUDDER* 1988b, 30].

Variance As such, $\chi_\xi(\eta)$ is the mean value of $e^{\imath\eta\xi}$ [Eq. (3.90)]. But the n–th derivative of the characteristic function, calculated at $\eta = 0$ gives the n–th *moment* of the variable[74]:

$$\chi_\xi^{(n)}(0) = \imath^n \langle \xi^n \rangle , \tag{3.109}$$

so that the first derivative is the mean or expectation value of ξ [eq (3.91)], the second moment the variance of ξ. The QM *variance* or *dispersion* of observable $\hat{O}$ is symbolised by $(\Delta\hat{O})^2$ and it is defined as the expectation of the square of the deviation of $\hat{O}$ from the expectation value of $\hat{O}$:

$$\boxed{(\Delta\hat{O})^2 = \left\langle \left(\hat{O} - \langle\hat{O}\rangle\right)^2 \right\rangle = \langle \hat{O}^2 \rangle - \langle \hat{O} \rangle^2} \tag{3.110}$$

We define:

Definition 3.5 (Dispersion-free) *A* Dispersion-free *observable or state is one that has a variance of zero; hence the expectation values of the corresponding operator coincide with its eigenvalues*

We shall later see that dispersion-free observables and states are not possible in the frame of QM [see chapter 32].

Characteristic function and density matrix Given an observable $\hat{O}$, the density operator $\hat{\rho}$ is uniquely determined by its characteristic function[75], the latter being the Fourier transform of the density operator itself:

$$\boxed{\chi^{\hat{\rho}}(\eta) = \hat{U}_F\hat{\rho} : \hat{\rho}(\hat{O}) \mapsto \mathrm{Tr}\left[\hat{\rho}(\hat{O})e^{\eta\hat{O}^\dagger - \eta^*\hat{O}}\right]} \tag{3.111}$$

where $\hat{O}^\dagger + \hat{O}$ is proportional to the approximate position and $\hat{O}^\dagger - \hat{O}$ is proportional to the approximate momentum; these are also named *quadratures* [see Eq. (5.3)], both of zero order, i.e. what one obtains by substituting c-numbers (complex numbers) to position and momentum [see Eqs. (5.43) and (5.3)].

The first moment of $\chi^{\hat{\rho}}(\eta)$ gives the mean value of the state, and in terms of the second moment we can express the correlation function of the state.

3.8 Particles Statistics and Pauli Exclusion Principle

There are two types of particle statistics: Bose/Einstein statistics, for bosons (photons for example), i.e. particles with integer spin [see following section], described by symmetric wave-functions (they are symmetric under interchange of coordinates), and Fermi/Dirac statistics, for fermions (electrons for example), i.e. particles with half-integer spin, described by antisymmetric wave functions (antisymmetric under interchange of coordinates)[76].

Let us now see the different distributions. In the case of *bosons* the individuality of the particles is of no importance (if we interchange the place of two bosons, we always obtain the

[74]Mathematics in [*GNEDENKO* 1969, 165–70, 185–86] [*HOLEVO* 1982, 69, 235–39]. See also [*SCHIFF* 1955, 60] [*BUB* 1974, 34–35] [JAUCH 1976, 128] [*BUSCH et al.* 1995a, 134–35] .

[75]Formalism in [*DAVIES* 1976, 39–44] [*HOLEVO* 1982, 231–39] [*WALLS/MILBURN* 1994, 62].

[76]See [*DIRAC* 1930c, 209] [*SCHIFF* 1955, 368–69] [*K. HUANG* 1963, 179–85, 472] [*JAUCH* 1968, 277–78] [*DRIESCHNER* 1979, 62–63]..

same statistics): only the number of particles is important. If we have k particles which have to be distributed in N 'boxes', we obtain $N+k-1$ possible 'places' and $N-1$ separations between particles, from which we have that the total possibility of distribution for choosing k elements between the $k+N-1$ 'elements' is:

$$\binom{N+k-1}{k}. \tag{3.112}$$

In the case of *fermions* we can have only one particle in each 'box' (energy-levels for electrons in atoms): this is the *Pauli Exclusion Principle.* Then the possible ways to distribute k particles in N boxes is given by:

$$\binom{N}{k}. \tag{3.113}$$

Different proposals for deducing one statistics from the other have been made[77].

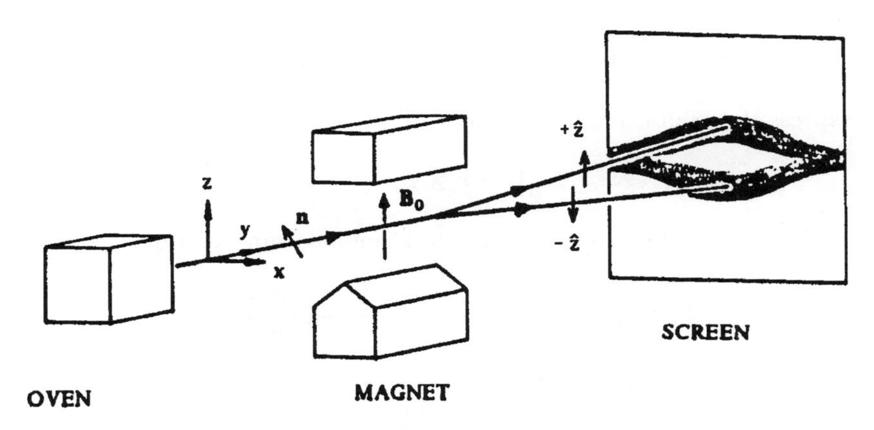

Figure 3.5: Scheme of Stern/Gerlach Experiment (= SGM Exp.). A typical explanation states that due to the magnetic field an initial polarisation $\mathbf{n}$ is changed into 'spin-up' ($+z$) or 'spin-down' ($-z$), relative to the main field direction $\mathbf{B}_0$ (of an inhomogeneous magnetic field) if the particle is found in the upper or lower part respectively of the deflected beam — from [BUSCH/SCHROECK 1989, 811].

3.9 Spin Theory

Spin and angular momentum The start of spin theory is in [PAULI 1927]. We owe the discovery of the intrinsic magnetic momentum of microentities to Stern-Gerlach-Magnet Exp.s [GERLACH/O. STERN 1922a] [GERLACH/O. STERN 1922b] [GERLACH/O. STERN 1922c] — in the case of massless particles (photons) one speaks of *helicity*[78]. A beam of identically prepared silver atoms is sent through a magnetic field oriented so that the gradient of the field is constant and perpendicular to the beam axis. The emerging silver atoms are captured by a screen whose orientation is perpendicular to the initial beam axis [see figure 3.5]. The result is

[77]See [CUFARO-P. *et al.* 1984c]. A recent work (also useful for review) is [COSTANTINI/GARIBALDI 1989].
[78]See [*SCHROECK* 1996, 158–59].

another aspect of quantization: the silver atoms assume only two orientations, spin up ↑ and down ↓.

Hence in QM we have[79] a total angular momentum which is the sum of the orbital angular momentum **L** and of the spin momentum $\hat{\sigma}$:

$$\hat{\mathbf{J}} = \hat{\mathbf{L}} + \hat{\sigma}, \tag{3.114}$$

where the spin operator $\hat{\sigma}$ has the same commutation relationship as $\hat{\mathbf{J}}$ and $\hat{\mathbf{L}}$, which in tensorial form is:

$$\boxed{[\hat{J}_i, \hat{J}_k] = \imath\hbar e_{ikl}\hat{J}_l} \tag{3.115}$$

or, introducing raising and lowering operators $\hat{L}_{\pm} = \hat{L}_x \pm \imath\hat{L}_y$:[80]

$$[\hat{L}_z, \hat{L}_{\pm}] = \pm\hat{L}_{\pm}, \; [\hat{L}_+, \hat{L}_-] = 2\hat{L}_z. \tag{3.116}$$

The relationship between orbital momentum and spin is:

$$\hat{\mathbf{L}} = \frac{1}{2}\hbar\hat{\sigma}, \quad \hat{\sigma} = (\hat{\sigma}_x, \hat{\sigma}_y, \hat{\sigma}_z). \tag{3.117}$$

Classically we express the orbital momentum by:

$$\mathbf{L} = \boldsymbol{r} \times \boldsymbol{p}, \tag{3.118}$$

so that in QM we have three equations for the orbital momentum [see Eq. (3.24b)]:

$$\hat{L}_x = \hat{q}_y\hat{p}_z - \hat{q}_z\hat{p}_y = -\imath\hbar\left(y\frac{\partial}{\partial z} - z\frac{\partial}{\partial y}\right), \tag{3.119a}$$

$$\hat{L}_y = \hat{q}_z\hat{p}_x - \hat{q}_x\hat{p}_z = -\imath\hbar\left(z\frac{\partial}{\partial x} - x\frac{\partial}{\partial z}\right), \tag{3.119b}$$

$$\hat{L}_z = \hat{q}_x\hat{p}_y - \hat{q}_y\hat{p}_x = -\imath\hbar\left(x\frac{\partial}{\partial y} - y\frac{\partial}{\partial x}\right). \tag{3.119c}$$

Therefore, in general, the description of the state of a particle must not only give the probabilities of the different positions in space (by mean of three continuous variables x, y, z), but also the probability of the different spin orientation, and therefore must include a discrete spin variable [see figure 3.6].

If $|\psi(\mathbf{r}, \sigma)\rangle$ is such a vector, it represents an ensemble of coordinate functions which correspond to different values of $\hat{\sigma}$: the functions are the components of spin. Therefore,

$$\int |\psi(\mathbf{r}, \sigma)|^2 d\mathrm{V} \tag{3.120}$$

represents the probability that the particle has a particular value of spin in a volume V. The probability of being in a volume element dV but with an arbitrary spin value is:

$$d\mathrm{V}\sum_{\sigma} |\psi(\mathbf{r}\sigma)|^2. \tag{3.121}$$

[79] Exposition in [*SCHIFF* 1955, 76–82, 199] [*Lev LANDAU/LIFSTITS* 1976b, 110–122, 240–48] .
[80] For a more formal exposition see [*J. FUCHS/SCHWEIGERT* 1997, 20] .

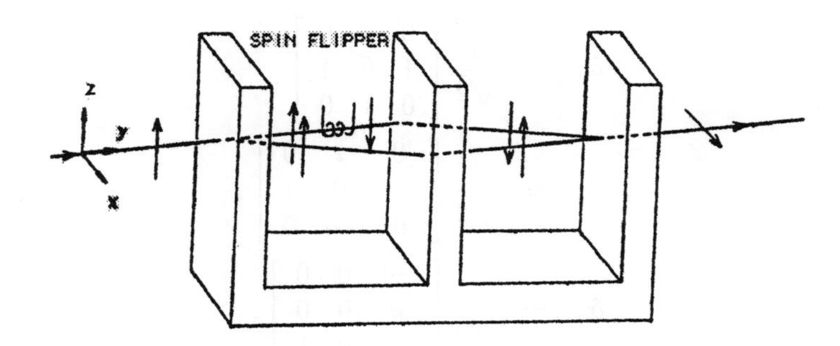

Figure 3.6: Scheme of spin superposition in single crystal neutron interferometry [see section 5.3.1]. The arrows indicate incoming polarisation in the z direction and outgoing polarisation typically in the $x-y$ plane (here in the x direction) — from [BUSCH/SCHROECK 1989, 812].

While the orbital angular momentum operator acts on the coordinates, the spin operator $\hat{\sigma}$ acts only on the spin variable [see Eq. (3.114)]:

$$\hat{\sigma}\psi(\sigma_1) = \sum_{\sigma_2} \sigma_{12}\psi(\sigma_2), \tag{3.122}$$

where the σ_{12}'s are the elements of the associated matrix [see Eq. (3.36)]; but, in the place of integration, we have now a sum over discrete elements:

$$\sigma_{12} = \sum_{\sigma_1} \langle \psi_{\sigma_2} | \hat{\sigma} | \psi_{\sigma_1} \rangle, \tag{3.123}$$

where $\psi_{\sigma_1}(\sigma)$ and $\psi_{\sigma_2}(\sigma)$ are the eigenfunctions of the operator $\hat{\sigma}_z$, corresponding to eigenvalues σ_1, σ_2. Here each function corresponds to the state in which the particle has a determinate value of $\hat{\sigma}_z$, i.e., of all the components of the wave function, only one is non-zero. The matrix is of rank $m_{\hat{\sigma}} = (2s+1)$, i.e. for each experiment there are always $(2s+1)$ possible outcomes for a particle of spin number s.

Spin matrices In what follows[81] we show the calculations for $s = 0, \frac{1}{2}, 1$. The wave function for zero spin ($s = 0$) has only one component ($m_{\hat{\sigma}} = 1$), hence: $\psi(0)$ and the spin operator give zero: $\hat{\sigma}_z\psi = 0$. The wave function with $s = \frac{1}{2}$ has two components: $m_{\hat{\sigma}} = \pm\frac{1}{2}$, therefore one has: $\psi(1/2)$ and $\psi(-1/2)$:

$$\psi = \begin{pmatrix} \psi_1 \\ \psi_2 \end{pmatrix} \equiv \begin{pmatrix} \psi(1/2) \\ \psi(-1/2) \end{pmatrix}. \tag{3.124}$$

This situation is represented by the Pauli spin matrices [PAULI 1927, 614]:

$$\boxed{\hat{\sigma}_x = \begin{bmatrix} 0 & 1 \\ 1 & 0 \end{bmatrix}; \ \hat{\sigma}_y = \begin{bmatrix} 0 & -\imath \\ \imath & 0 \end{bmatrix}; \ \hat{\sigma}_z = \begin{bmatrix} 1 & 0 \\ 0 & -1 \end{bmatrix}} \tag{3.125}$$

[81]Exposition in [*SCHIFF* 1955, 206–210] [*Lev LANDAU/LIFSTITS* 1976b, 249–53] .

For $s = 1$ we have three possible values $m_{\hat{\sigma}} = -1, 0, +1$. The matrices are the following:

$$\hat{\sigma}_x = \frac{1}{\sqrt{2}} \begin{bmatrix} 0 & 1 & 0 \\ 1 & 0 & 1 \\ 0 & 1 & 0 \end{bmatrix}, \tag{3.126a}$$

$$\hat{\sigma}_y = \frac{1}{\sqrt{2}} \begin{bmatrix} 0 & \imath & 0 \\ -\imath & 0 & \imath \\ 0 & -\imath & 0 \end{bmatrix}, \tag{3.126b}$$

$$\hat{\sigma}_z = \frac{1}{\sqrt{2}} \begin{bmatrix} -1 & 0 & 0 \\ 0 & 0 & 0 \\ 0 & 0 & 1 \end{bmatrix}. \tag{3.126c}$$

Operators and spin The Hilbert space[82] for spin-1/2 systems is C^2. Any operator (not only an observable) $\hat{O}$ acting on C^2 can be written as:

$$\hat{O}(\mathbf{n}) = \frac{1}{2}(\mathbf{1} + \mathbf{n} \cdot \hat{\sigma}), \qquad \mathbf{n} \cdot \hat{\sigma} = n_x \hat{\sigma}_x + n_y \hat{\sigma}_y + n_z \hat{\sigma}_z. \tag{3.127}$$

We can write the density operators in the same form:

$$\hat{\rho} = \frac{1}{2}(\mathbf{1} + \mathbf{n} \cdot \hat{\sigma}), \qquad x \in \Re^3, \| \mathbf{n} \| \leq 1 \tag{3.128}$$

One dimensional projections are of the form

$$\hat{P}_x = \frac{1}{2}(\mathbf{1} + \mathbf{n} \cdot \hat{\sigma})\; \mathbf{n} \in \Re^3 \quad \| \mathbf{r} \| = 1 \tag{3.129}$$

The probability of obtaining $\hat{P}_{\mathbf{n}}$ when this is in state $\hat{\rho}_{\mathbf{r}}$ is given by [see Eq. (3.100) and comments]:

$$\mathrm{Tr}(\hat{P}_{\mathbf{r}} \cdot \hat{P}_{\mathbf{n}}) = \frac{1}{2}(\mathbf{1} + \mathbf{r} \cdot \mathbf{n}) = \frac{1}{2}(1 + \cos\theta_{r,n}) = \cos^2 \frac{\theta_{r,n}}{2}, \tag{3.130}$$

which, for example, gives the transition probability from one polariser orientation state to another — a *polariser* is an apparatus which only allows the transmission of particles with a determinate orientation and absorbs all the others[83].

Polarisation By using the Poincaré sphere [see figure 3.7] and letting $\mathbf{e}_i$ $(i = 1, 2, 3)$ denote some Cartesian coordinate system, we obtain [BUSCH/SCHROECK 1989, 820]:

$$|\mathbf{n}\rangle = |\theta, \phi\rangle = e^{-\imath\frac{\phi}{2}} \cos\frac{\theta}{2}|\mathbf{e}_3\rangle + e^{\imath\frac{\phi}{2}} \sin\frac{\theta}{2}|-\mathbf{e}_3\rangle, \tag{3.131}$$

where $|\mathbf{e}_3\rangle$ and $|-\mathbf{e}_3\rangle$ are north and south poles of the sphere or the third cartesian axis. The two poles represent circular polarisation, the equatorial points the linear polarisation while vector $|\mathbf{n}\rangle$ represent an arbitrary elliptic polarisation. Mixtures are represented by points in the interior of the sphere.

Light is normally in a superposition of linear polarisations [see figure 3.8 for examples].

[82] Formalism in [BUSCH/SCHROECK 1989, 818–19] [*SCHROECK* 1996, 161–68]: .

[83] In [BUSCH/SCHROECK 1989, 819] a coherent superposition of pure spin states is shown — see also [SUMMHAMMER *et al.* 1983] . Let $\mathbf{e}_i$ $(i = 1, 2, 3)$ denote some Cartesian coordinate system. Adding up eigenvectors of $\hat{P}_{\mathbf{e}_3}$ yields an eigenvector of $\hat{P}_{\mathbf{e}_1}$: $|\psi_{\mathbf{e}_3}\rangle + |\psi_{-\mathbf{e}_3}\rangle = \sqrt{2}|\psi_{\mathbf{e}_1}\rangle$.

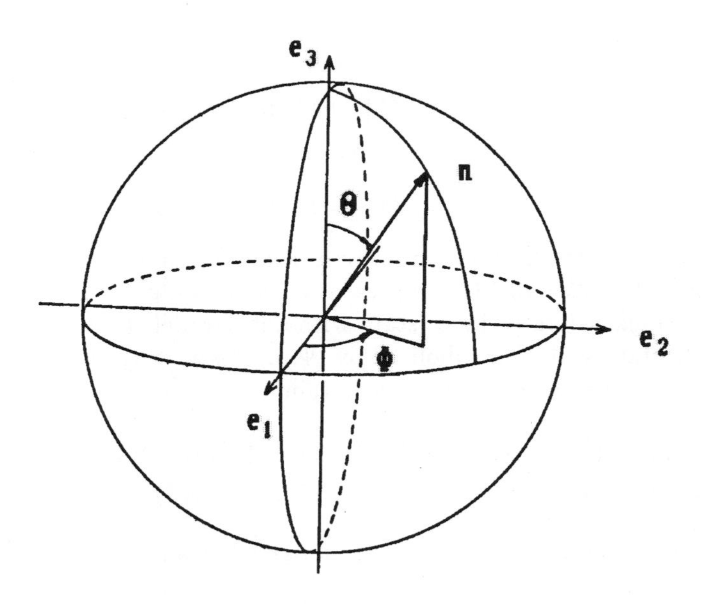

Figure 3.7: The set of states for a spin-$\frac{1}{2}$ system and for helicity on the Poincaré sphere ($\mathbf{e}_j$, $\mathbf{j} = \mathbf{1}, \mathbf{2}, \mathbf{3}$ are unitary vectors). North and south poles represent circular polarisation; the equatorial points linear polarisation. Arbitrary antipodal points $\mathbf{n}, -\mathbf{n}$ on the surface represent the orthogonal spin-up and spin-down for projectors $\sigma \cdot \mathbf{n}$ of spin onto this direction or states of general elliptic polarisation — from [BUSCH/SCHROECK 1989, 820].

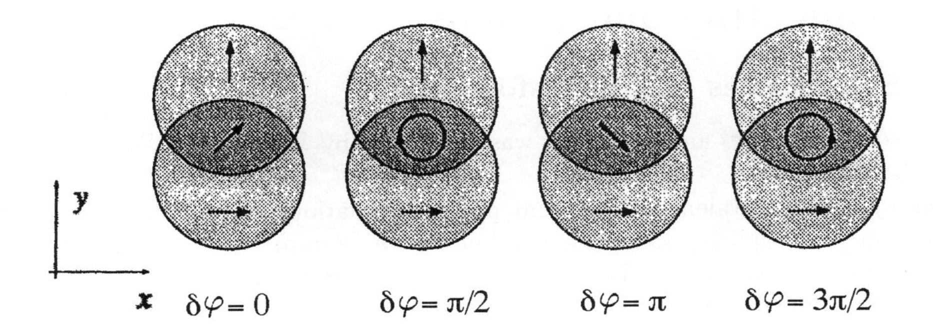

Figure 3.8: Overlapping light beams with opposite polarisations. For simplicity, the beams have been drawn with sharp boundaries and they are supposed to have equal intensities, uniformly distributed within these boundaries. Depending on the phase difference $\delta\varphi$, one may have, in the overlapping part of the beams, linearly, circularly, or, in general, elliptically polarised photons — from [*PERES* 1993, 8].

Singlet and triplet For a two-particle system with spin–$\frac{1}{2}$ we have four possible combinations of spin[84]:

$$(\uparrow\uparrow), \tag{3.132a}$$

$$(\uparrow\downarrow) + (\downarrow\uparrow), \tag{3.132b}$$

$$(\downarrow\downarrow), \tag{3.132c}$$

$$(\uparrow\downarrow) - (\downarrow\uparrow), \tag{3.132d}$$

where between the brackets the first arrow represents the first particle, the second arrow the second particle. Eqs. (3.132a)–(3.132c) represent a *triplet state* — i.e. a state with parallel spin —, Eq. (3.132d) a *singlet state* — i.e. a state with antiparallel spins and total sum zero. In the following the single state will be symbolised by $|\Psi_0\rangle$. If we study the possible combinations for all three directions x, y, z in singlet state, we obtain:

$$\begin{array}{c}(\uparrow,\uparrow,\uparrow;\downarrow,\downarrow,\downarrow) \vee (\downarrow,\uparrow,\uparrow;\uparrow,\downarrow,\downarrow) \quad \vee \quad (\uparrow,\downarrow,\uparrow;\downarrow,\uparrow,\downarrow) \vee (\uparrow,\uparrow,\downarrow;\downarrow,\downarrow,\uparrow) \quad \vee \\ (\downarrow,\downarrow,\uparrow;\uparrow,\uparrow,\downarrow) \vee (\downarrow,\uparrow,\downarrow;\uparrow,\downarrow,\uparrow) \quad \vee \quad (\uparrow,\downarrow,\downarrow;\downarrow,\uparrow,\uparrow) \vee (\downarrow,\downarrow,\downarrow;\uparrow,\uparrow,\uparrow),\end{array}$$

where left of the semicolon are the three values of particle 1 and right the three values of particle 2.

If we desire **for example** to calculate all possibilities on the assumption that we have value $\uparrow$ for the y component of particle 1 and value $\uparrow$ for the z component of particle 2, we obtain:

$$(\uparrow,\uparrow,\downarrow;\downarrow,\downarrow,\uparrow) \vee (\downarrow,\uparrow,\downarrow;\uparrow,\downarrow,\uparrow), \tag{3.133}$$

which together are only $\frac{1}{4}$ of all possible cases. Now suppose that θ_{jk} is the angle between an arbitrary direction j of the first particle and an arbitrary direction k of the second, which for the sake of simplicity we suppose to be $= \frac{\pi}{2}$. The probability that the first particle has value $\uparrow$ for y is $\frac{1}{2}$. The conditional probability that we have $\uparrow$ for the z component of the second particle is given by $\frac{1}{2}\sin^2(\frac{1}{2}\theta_{yz})$ [see Eq. (3.130)], which is exactly $\frac{1}{4}$, as expected.

3.10 Symmetries and Groups

3.10.1 Symmetries and Transformations

The theory of symmetries and of groups was developed by Wigner and Weyl [*WIGNER* 1959] [*WEYL* 1931].[85]

Symmetry is an equivalence of different physical situations. The hallmark of symmetry is the impossibility of acquiring some physical knowledge. Symmetries are based on two notions: *Transformations* and *Invariance*. The Wigner theorem [*WIGNER* 1959, 233], which is generally mathematical, is fundamental:

Theorem 3.7 (Wigner) *All vectors $|j\rangle, |k\rangle$ for which there is a mapping $|j\rangle \mapsto |j'\rangle, |k\rangle \mapsto |k'\rangle$ which preserves the inner product: $|\langle j'|k'\rangle|^2 = |\langle j|vk\rangle|^2$, are subject to transformations which are either linear and unitary or antilinear and antiunitary.*

[84] Formalism in [*SCHIFF* 1955, 375–77].

[85] Formalism and discussion in [*PERES* 1993, 215–20, 223, 239–40, 257–59] [*HOLEVO* 1982, 102] [*DAVIES* 1976, 21–22] [*HAAG* 1992, 27].

Transformations which are part of continuous groups (translations and rotations) can only be unitary. Continuous transformations, as we shall see, have generators. Under discrete transformations, space reflection is unitary but time reversal is antiunitary. The reason for this last fact — as we shall see [see section 10.2] — is that in Basic-QM the time observable is not an operator.

3.10.2 Groups

A *Group*[86] is a mathematical structure which has a set of elements $\{a, b, c, \ldots\}$, which includes the identity element $\mathbf{1}$, a unique complement a' for each element a and a binary operation $\circ$ such that following equations hold:

$$(a \circ b) \circ c = a \circ (b \circ c), \tag{3.134a}$$

$$a \circ \mathbf{1} = a = \mathbf{1} \circ a, \tag{3.134b}$$

$$a \circ a' = \mathbf{1} = a' \circ a. \tag{3.134c}$$

If the group is commutative, i.e. $a \circ b = b \circ a$, it is called an *Abelian group.*

3.10.3 Galilei Group

Generals A Galilean group[87] $\mathcal{G}(b, \mathbf{a}, \mathbf{v}, \mathbf{R})$ includes translations in space ($\mathbf{a}$) and in time (b), three dimensional rotations ($\mathbf{R}$) and boosts, i.e. momentum translations ($\mathbf{v}$):

$$\mathbf{r} \mapsto \mathbf{r}' = \mathbf{R}\mathbf{r} + \mathbf{v}t + \mathbf{a}, \tag{3.135a}$$

$$\mathbf{k} \mapsto \mathbf{k}' = \mathbf{R}\mathbf{k} + m\mathbf{v}, \tag{3.135b}$$

$$t \mapsto t' = t + b. \tag{3.135c}$$

Position and momentum The *covariance* conditions (imprimitivity relations) for position observables are[88]:

$$\hat{U}_{\mathbf{a}}\hat{P}_{\mathbf{r}}\hat{U}_{\mathbf{a}}^{-1} = \hat{P}_{\mathbf{r}+\mathbf{a}}, \tag{3.136a}$$

$$\hat{U}_{\mathbf{v}}\hat{P}_{\mathbf{r}}\hat{U}_{\mathbf{v}}^{-1} = \hat{P}_{\mathbf{r}}, \tag{3.136b}$$

$$\hat{U}_{\mathbf{R}}\hat{P}_{\mathbf{r}}\hat{U}_{\mathbf{R}}^{-1} = \hat{P}_{\mathbf{R}\mathbf{r}}; \tag{3.136c}$$

and for momentum observables they are:

$$\hat{U}_{\mathbf{a}}\hat{P}_{\mathbf{k}}\hat{U}_{\mathbf{a}}^{-1} = \hat{P}_{\mathbf{k}}, \tag{3.137a}$$

$$\hat{U}_{\mathbf{v}}\hat{P}_{\mathbf{k}}\hat{U}_{\mathbf{v}}^{-1} = \hat{P}_{\mathbf{k}+m\mathbf{v}}, \tag{3.137b}$$

$$\hat{U}_{\mathbf{R}}\hat{P}_{\mathbf{k}}\hat{U}_{\mathbf{R}}^{-1} = \hat{P}_{\mathbf{R}\mathbf{k}}. \tag{3.137c}$$

where $\hat{P}_{\mathbf{r}}, \hat{P}_{\mathbf{k}}$ are projectors for position $\mathbf{r}$ and wave number $\mathbf{k}$ respectively. There are many possible ways of satisfying these conditions; but, on the basis of Stone Theorem [Eq. (3.69)], we can write:

$$\boxed{\hat{U}_{\mathbf{a}} := e^{-\imath\mathbf{a}\cdot\hat{\mathbf{p}}}} \tag{3.138}$$

[86] See [*JAUCH* 1968, 137–38] [*HUGHES* 1989, 39].

[87] Exposition in [*PERES* 1993, 225–26, 232–33, 246–247] [*SCHIFF* 1955, 188–97] [*PRUGOVEČKI* 1984b, 10].

[88] See [*BUSCH et al.* 1995a, 55–59].

and:

$$\boxed{\hat{U}_{\mathbf{v}} := e^{\imath \mathbf{v}\cdot\hat{\mathbf{q}}}} \tag{3.139}$$

Time translation, as we know, is represented by [Eq. (3.70)]:

$$\hat{U}_t := e^{\imath \hat{H} t}. \tag{3.140}$$

Space- and time-displacement groups are Lie groups. A *Lie group*[89] is a continuously connected group in which the parameters of the product of two elements are continuous, differentiable functions of the parameters of the elements.

Rotation For the rotation group $\hat{U}_{\mathbf{R}}$ we have the following: a rotation $\mathbf{R}$ about an axis $\mathbf{n}$, through an angle ϕ, in a cartesian frame is given by[90]:

$$\mathbf{R} = \mathbf{1} - \phi \sum_k n_k \hat{\sigma}_k, \tag{3.141}$$

where the $\hat{\sigma}_k, k = x, y, z$ are the spin matrices for three dimensions — in two dimensional case the matrices are Pauli spin matrices (3.125) — which here represent the angular momentum operator $\hat{\mathbf{J}}$, which obeys the following CCR:

$$[\hat{\mathbf{J}}_x, \hat{\mathbf{J}}_y] = -\hat{\mathbf{J}}_z,\ [\hat{\mathbf{J}}_y, \hat{\mathbf{J}}_z] = -\hat{\mathbf{J}}_x,\ [\hat{\mathbf{J}}_z, \hat{\mathbf{J}}_x] = -\hat{\mathbf{J}}_y, \tag{3.142}$$

which constitutes the Lie algebra of the rotation group.

The corresponding unitary transformation is [see again theorem 3.6: p. 47]:

$$\boxed{\hat{U}_{\mathbf{R}} = e^{\frac{1}{2}\phi\cdot\hat{\mathbf{J}}}} \tag{3.143}$$

Momentum, Hamiltonian and angular momentum are called *generators of the group symmetry*. Here we understand much better the distinction between dynamic and kinematic variables [p. 9]: the generators of the symmetry transformations are precisely the *dynamic variables*, while position, time and angle come into the formalism only as coordinates of the reference frame used to define these symmetries.

3.11 Scattering matrix and Propagators

A later (of the 1950's), but very important development is that of propagators, i.e. functions and operators which express transitions in space and time (or space-time) from some 'location' to another. They are very useful for some problems that we shall later discuss.

3.11.1 Green Function

We can write the evolution of a system in space and time in the following form[91]:

$$\psi(\mathbf{r}', t) = \imath \int \mathrm{G}(\mathbf{r}', t'; \mathbf{r}, t)\psi(\mathbf{r}, t) d^3 r, \tag{3.144}$$

[89] On this point and that which follows see [*J. FUCHS/SCHWEIGERT* 1997, 51–52, 144–58].

[90] See [*HOLEVO* 1982, 148–54] [*JAUCH* 1968, 247–67] [*BELTRAMETTI/CASSINELLI* 1981, 30–44].

[91] Expositions regarding the subject of this subsection in [*SCHIFF* 1955, 300–307] [*PRUGOVEČKI* 1971, 520–42].

where G is the *Green function.* Eq. (3.144) makes no distinction between a forward propagator and a backward one. For forward propagation we define the *retarded Green function* or propagator as follows:

$$\mathrm{G}^{+}(\mathbf{r}',t';\mathbf{r},t) = \mathrm{G}(\mathbf{r}',t';\mathbf{r},t) \text{ for } t'>t, \tag{3.145a}$$
$$= 0 \text{ for } t'<t; \tag{3.145b}$$

and for backward propagation we introduce an *advanced Green function*:

$$\mathrm{G}^{-}(\mathbf{r}',t';\mathbf{r},t) = -\mathrm{G}(\mathbf{r}',t';\mathbf{r},t) \text{ for } t'<t, \tag{3.146a}$$
$$= 0 \text{ for } t'>t; \tag{3.146b}$$

The Green functions are subject to the following condition:

$$\left(\imath\frac{\partial}{\partial t'}-\frac{1}{\hbar}\right)\mathrm{G}^{+}(\mathbf{r}',t';\mathbf{r},t)=\delta^{3}(\mathbf{r}'-\mathbf{r})\delta(t'-t). \tag{3.147}$$

An explicit expression for the Green function can be found in the case in which $V=0$. We denote it by G_0 and name it *interaction-free Green function*:

$$\mathrm{G}_0(\mathbf{r}',t';\mathbf{r},t)=-\imath\left[\frac{m}{2\pi\imath\hbar(t'-t)}\right]^{\frac{3}{2}}e^{\frac{\imath m|\mathbf{r}'-\mathbf{r}|^2}{2\hbar(t'-t)}}. \tag{3.148}$$

The Green functions can be expanded by means of Eq. (3.148). Relationships between Green functions and Hamiltonians are given by the following equations:

$$\mathrm{G}^{\pm} = \left(\imath\frac{\partial}{\partial t}-\frac{1}{\hbar}\hat{H}\right)^{-1}, \tag{3.149a}$$
$$\mathrm{G}_0^{\pm} = \left(\imath\frac{\partial}{\partial t}-\frac{1}{\hbar}\hat{H}_0\right)^{-1}, \tag{3.149b}$$

where H_0 is the interaction-free Hamiltonian.

3.11.2 Path Integrals

Unlike classical mechanics, in QM there a principle of least action [see Eq. (1.5)] does not exist: In QM each trajectory from an arbitrary point a to an arbitrary point b contributes to the total amplitude. Hence the probability $\wp(b,a)$ is the absolute square $\wp(b,a)=|\mathrm{K}(b,a)|^2$ of an amplitude $\mathrm{K}(b,a)$ — it is another form of the superposition principle [see Eq. (3.4) and principle 3.3]. This amplitude is the sum of the contribution $\vartheta[\mathbf{q}(t)]$ of each path [FEYNMAN 1948]:[92]

$$\mathrm{K}(b,a)=\sum\vartheta[\mathbf{q}(t)]. \tag{3.150}$$

The contribution of each path has a phase proportional to the action S:

$$\vartheta[\mathbf{q}(t)]\propto e^{\frac{\imath}{\hbar}\mathrm{S}[\mathbf{q}(t)]}. \tag{3.151}$$

[92]Expositions in [FEYNMAN/HIBBS 1965, 28–37] [*GUDDER* 1988b, 63–65]. See also [RONCADELLI 1996].

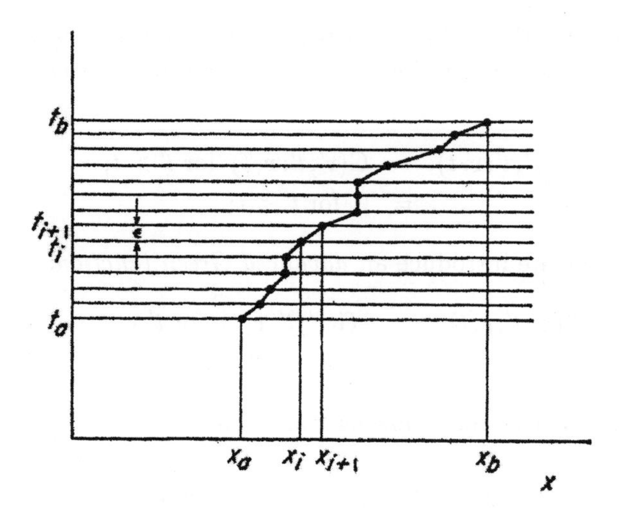

Figure 3.9: The sum over paths is defined as a limit, in which at first the path is specified only by its coordinate x at a large number of specified times separated by very small intervals ϵ. The path sum is then an integral over all these specific coordinates. Then, so as to achieve the correct measure, the limit is taken as $\epsilon \longrightarrow 0$ — from [FEYNMAN/HIBBS 1965, 33].

We now formulate an integral which is understood as the limit of the sum over small time intervals ϵ:

$$\mathrm{K}(b,a) \simeq \int\int\cdots\int \varphi[\mathbf{q}(t)]dq_1 dq_2\cdots dq_{n-1}. \tag{3.152}$$

The normalization factor is N^{-1} where

$$N = \left(\frac{2\pi\imath\hbar\epsilon}{m}\right)^{\frac{1}{2}}. \tag{3.153}$$

With this factor the limit exists. Hence we can write:

$$\mathrm{K}(b,a) = \lim_{\epsilon\longrightarrow 0}\frac{1}{N}\int\int\cdots\int e^{\frac{\imath}{\hbar}\mathrm{S}(b,a)}\frac{dq_1}{N}\frac{dq_2}{N}\cdots\frac{dq_{n-1}}{N}, \tag{3.154}$$

where $\mathrm{S}(a,b) = \int_{t_a}^{t_b} L(\dot{\mathbf{q}},\mathbf{q},t)dt$ is the line taken on the trajectory passing through the points q_i [see figure 3.9].

For practical purposes we might possibly need other formulations. A more general form, which is called *path integral*, is the following:

$$\mathrm{K}(b,a) = \int_a^b e^{\frac{\imath}{\hbar}\mathrm{S}(b,a)}Dq(t). \tag{3.155}$$

Suppose now that we have two events in succession such that $\mathrm{S}(b,a) = \mathrm{S}(b,c) + \mathrm{S}(c,a)$. We can write:

$$\mathrm{K}(b,a) = \int e^{\frac{\imath}{\hbar}\mathrm{S}(b,c)+\frac{\imath}{\hbar}\mathrm{S}(c,a)}Dq(t), \tag{3.156}$$

which can be reformulated as:

$$\mathrm{K}(b,a) = \int_{q_c}\int_c^b e^{\frac{\imath}{\hbar}\mathrm{S}(b,c)}\mathrm{K}(c,a)Dq(t)dq_c, \tag{3.157}$$

in order to finally arrive to:

$$\mathrm{K}(b,a) = \int_{q_c} \mathrm{K}(b,c)\mathrm{K}(c,a)dq_c. \tag{3.158}$$

In conclusion the path-integral method is a formal instrument for calculating probability amplitudes for evolutions between different experimental results — the events 'a' and 'b' in Eqs. (3.156)–(3.158) — because there are not 'paths' in the classical sense of the word. Recently an application of the path-integral method to the expansion of the wave function in eigenstates of observables which are not defined at a given instant but over a period of time — which is the case, for example, by tunnelling [see section 41.3] —, for which problem no traditional solution exists, has been developed by Sokolovski [SOKOLOVSKI 1998].

3.11.3 Scattering Matrix

The scattering matrix (= S matrix) $\hat{S}_{fi}$ was first introduced in [WHEELER 1937] and worked out in [HEISENBERG 1942].[93] Suppose that in the presence of a scattering centre (whatever this may be) the evolution $\hat{U}_t$ is generated by the full Hamiltonian $\hat{H}$, whereas, if the scattering centre were not present, the unperturbed evolution $\hat{U}_t^0$ would be generated by the free Hamiltonian $\hat{H}_0$. One can then assume that 'far' form the scattering centre the evolution is almost free. Therefore, far away, incoming and outgoing state vectors $|\psi^i\rangle, |\psi^f\rangle$ evolve freely and we assume that for each observable $\hat{O}$ and for each interpolating state $|\psi\rangle$ we have:

$$\lim_{t \longrightarrow -\infty} \left| \mathrm{Tr}\left[\hat{U}_t^0 \hat{\rho}^0(\psi^i)(\hat{U}_t^0)^\dagger \hat{O}\right] - \mathrm{Tr}\left[\hat{U}_t \hat{\rho}(\psi)\hat{U}_t^\dagger \hat{O}\right]\right| = 0, \tag{3.159a}$$

$$\lim_{t \longrightarrow +\infty} \left| \mathrm{Tr}\left[\hat{U}_t^0 \hat{\rho}^0(\psi^f)(\hat{U}_t^0)^\dagger \hat{O}\right] - \mathrm{Tr}\left[\hat{U}_t \hat{\rho}(\psi)\hat{U}_t^\dagger \hat{O}\right]\right| = 0. \tag{3.159b}$$

If we define the Møller Matrices as follows:

$$\hat{\mathcal{U}}_\pm := \lim_{t \longrightarrow \pm\infty} \hat{U}_{-t}\hat{U}_t^0 \tag{3.160}$$

we can define in the following form the conditional probability that the detected state vector $|\psi^f\rangle$ will come from the prepared $|\psi^i\rangle$:

$$\begin{aligned} \wp(\psi^f|\psi^i) &= |\langle \hat{\mathcal{U}}_+\psi^f|\hat{\mathcal{U}}_-\psi^i\rangle|^2 \\ &= |\langle \psi^f|\hat{\mathcal{U}}_+^*\hat{\mathcal{U}}_-\psi^i\rangle|^2 \\ &= |\langle \psi^f|\hat{S}^{fi}\psi^i\rangle|^2. \end{aligned} \tag{3.161}$$

Another form of writing this is the following[94]:

$$\hat{S}^{fi} = \lim_{t_2 \longrightarrow \infty} \lim_{t_1 \longrightarrow -\infty} \langle \psi^f|\hat{U}(t_2,t_1)|\psi^i\rangle, \tag{3.162}$$

where $\hat{U}(t_2,t_1)$ is the time evolution operator from t_1 to t_2. It is also possible to rewrite (in one-dimensional form) the S–matrix in terms of the Green function[95]:

$$\hat{S}^{fi} = \imath \lim_{t' \longrightarrow \infty} \lim_{t \longrightarrow -\infty} \int d^3r' d^3r \mathrm{G}(x',t';x,t)(\psi^f(x'))^*\psi^i(x). \tag{3.163}$$

[93]Exposition in [*EMCH* 1972, 4–6]

[94]See [*SCHWEBER* 1961, 318].

[95]See [*BJORKEN/DRELL* 1964, 87] .

Chapter 4

RELATIVISTIC QUANTUM MECHANICS

Contents In the following we do not wish to give a full exposition of relativistic QM — a too great subject to be treated in a single chapter. In general our work is not concerned with relativistic problems. We only wish to give some formalisms which will be useful in the following chapters for discussing specific problems, and to briefly indicate some problems which will lead to the successive analysis — particularly the localization problem.

The chapter contains a brief introduction to the Poincaré group [section 4.1], followed by an exposition of the basic Klein/Gordon and Dirac Eqs. [section 4.2]. Then the second quantization — for bosonic and fermionic fields — [section 4.3] and the relativistic scattering [section 4.4] are briefly treated. Finally, we give a short introduction to the localization problem in relativistic QM [section 4.5].

4.1 Poincaré Group

The problem of all mechanics is that of *relativity*, i.e. of transformation between different reference frames and the individuation of invariant or covariant elements. Basic-QM is founded, as we have seen [section 3.10], on Galilei relativity or Galilei group. The introduction, at the beginning of the century, of Einstein's special relativity theory, based on the invariance, in every reference frame, of the speed of light and on the Lorentz transformations, made it necessary to extend QM so as to give it a relativistic formulation.

The Poincaré Group[1] $\mathcal{Q}$ includes Lorentz coordinate and time transformations and the Lorentz boost, and, instead of the non-relativistic mass, we have a quantity which does not commute with all the others.

The Lorentz coordinate and time transformations from a frame $(x, y, z; t)$ to another frame $(x', y', z'; t')$, where the second frame has a constant speed $\mathbf{v}$ relative to the first, in a direction parallel to the x–axis, are:

$$x' = \frac{x - \mathbf{v}t}{\sqrt{1 - \frac{\mathbf{v}^2}{c^2}}}, \quad y' = y, \quad z' = z, \quad t' = \frac{t - \frac{\mathbf{v}}{c^2}x}{\sqrt{1 - \frac{\mathbf{v}^2}{c^2}}}, \tag{4.1a}$$

[1]On what follows see [*PERES* 1993, 249–52, 254] [*PRUGOVEČKI* 1984b, 72–73, 83–84] [*SCHWEBER* 1961, 36–53] [*HAAG* 1992, 10–11, 26–27, 38].

of which the inverse transformations are:

$$x = \frac{x' + \mathbf{v}t'}{\sqrt{1 - \frac{\mathbf{v}^2}{c^2}}}, \quad y = y', \quad z = z', \quad t = \frac{t' + \frac{\mathbf{v}}{c^2}x'}{\sqrt{1 - \frac{\mathbf{v}^2}{c^2}}}. \tag{4.1b}$$

For Lorentz transformations of the velocity, imagine a third system which moves with speed $\mathbf{v}^{(1)}$ relative to the frame $(x, y, z; t)$ and $\mathbf{v}^{(2)}$ relative to the frame $(x', y', z'; t')$ — which has always speed $\mathbf{v}$ relative to $(x, y, z; t)$. Then relative to the frame $(x, y, z; t)$ we have:

$$v_x^{(1)} = \frac{v_x^{(2)} + \mathbf{v}}{1 + v_x^{(2)}\frac{\mathbf{v}}{c^2}}, \tag{4.2a}$$

$$v_y^{(1)} = \frac{v_y^{(2)}\sqrt{1 - \frac{\mathbf{v}^2}{c^2}}}{1 + v_x^{(2)}\frac{\mathbf{v}}{c^2}}, \tag{4.2b}$$

$$v_z^{(1)} = \frac{v_z^{(2)}\sqrt{1 - \frac{\mathbf{v}^2}{c^2}}}{1 + v_x^{(2)}\frac{\mathbf{v}}{c^2}}; \tag{4.2c}$$

and relative to the frame $(x', y', z'; t')$:

$$v_x^{(2)} = \frac{v_x^{(1)} - \mathbf{v}}{1 - v_x^{(1)}\frac{\mathbf{v}}{c^2}}, \tag{4.3a}$$

$$v_y^{(2)} = \frac{v_y^{(1)}\sqrt{1 - \frac{\mathbf{v}^2}{c^2}}}{1 - v_x^{(1)}\frac{\mathbf{v}}{c^2}}, \tag{4.3b}$$

$$v_z^{(2)} = \frac{v_z^{(1)}\sqrt{1 - \frac{\mathbf{v}^2}{c^2}}}{1 - v_x^{(1)}\frac{\mathbf{v}}{c^2}}. \tag{4.3c}$$

As has already been said above, relativistically the mass is not a constant but a function of velocity:

$$m = \frac{m_0}{\sqrt{1 - \frac{\mathbf{v}^2}{c^2}}}, \tag{4.4}$$

where m_0 is the rest mass.

The different parameters of $\mathcal{Q}$ correspond to all possible 'placements' and 'displacements' — four for space-time, three for boosts, one for mass, one for rotation. The general element $(a; \Lambda)$ of $\mathcal{Q}$ consists in a spacetime translation by the 4–vector a^μ [Eqs. (4.1a) or (4.1b)] and a proper Lorentz transformation Λ. Each proper Lorentz transformation Λ can be uniquely decomposed into the product of a rotation $\mathbf{R}$ and a boost Λ_ν. Hence the transformation for position is:

$$\mathbf{r} \mapsto \mathbf{r}' = a + \Lambda\mathbf{r}, \ \mathbf{r} \in \mathcal{Z}(1,3), \ \Lambda = \Lambda_\nu \mathbf{R}, \tag{4.5a}$$

where $\mathcal{Z}(1,3)$ is Minkowsky space; and the transformation for momentum is:

$$\mathbf{k} \mapsto \mathbf{k}' = \Lambda\mathbf{k} = \Lambda_\nu(k^\circ, \mathbf{R}\mathbf{k}). \tag{4.5b}$$

$\mathcal{Q}$ is a continuous group and so only contains unitary transformations. Therefore, a Lorentz mapping which preserves the inner product, in the momentum representation has the form [see theorem 3.6: p. 47]:

$$\hat{U}_L(a, \Lambda) : \tilde{\psi}(\mathbf{k}) \mapsto \tilde{\psi}'(\mathbf{k}) = e^{\frac{i}{\hbar}a\cdot\mathbf{k}}\tilde{\psi}(\Lambda^{-1}\mathbf{k}). \tag{4.6}$$

But in the case of interactions, i.e. when we have an interaction Hamiltonian, one must modify $\mathbf{q}$ (generator of spatial translations) or $\mathbf{k}$ (generator of boosts) or both in such a way that $[\hat{p}_j, k_k] = \imath\delta_{jk}\hat{H}/c^2$ is valid. But in this way we obtain a privileged position for a determinate particle because we must assign it a precise position in space. Hence the existence of observables which obey Poincaré algebra is not a guarantee of Lorentz-invariance.

4.2 Klein/Gordon and Dirac Equations

4.2.1 Klein/Gordon Equation

The first attempt at constructing a Relativistic QM was the Klein/Gordon Eq. [O. KLEIN 1926].[2] As we have said, we need, first of all, a relativistic equivalent of the Schrödinger Eq. [Eq. (3.16); see also Eq. (3.23)]:

$$\imath\hbar\frac{\partial}{\partial t}|\psi\rangle = \left(\sqrt{-\hbar^2c^2\nabla^2 + m^2c^4}\right)|\psi\rangle := \hat{H}|\psi\rangle. \tag{4.7}$$

It is possible to square Eq. (4.7)

$$\hat{H}^2 = \hat{\mathbf{p}}^2 c + m^2c^4, \tag{4.8}$$

so as to obtain:

$$-\hbar^2\frac{\partial^2}{\partial t^2}|\psi\rangle = (-\hbar^2\nabla^2c^2 + m^2c^4)|\psi\rangle, \tag{4.9}$$

which is the Klein/Gordon Eq.:

$$\boxed{\left[\Box + \left(\frac{mc}{\hbar}\right)^2\right]|\psi\rangle = 0} \tag{4.10}$$

where $\Box$ is the Dalembertian:

$$\Box = \frac{\partial^2}{\partial x^2} + \frac{\partial^2}{\partial y^2} + \frac{\partial^2}{\partial z^2} - \frac{1}{c^2}\frac{\partial^2}{\partial t^2}. \tag{4.11}$$

4.2.2 Dirac Equation

But in this way there are also negative-energy solutions of Eq. (4.8): $\hat{H} = -\sqrt{\hat{\mathbf{p}}^2c^2 + mc^4}$, which seem nonsensical in the light of relativistic equivalence between energy and mass: $E = mc^2$. Another difficulty is the impossibility of defining probability densities with definite positive expression (also negative probabilities come in anyway). So it is better to find some first order equation[3].

Dirac tried to solve this last problem by finding an Eq. with a Hamiltonian linear in the space derivative in the same way that Eq. (3.16) is in the time derivative, with the following result [DIRAC 1928a] [DIRAC 1928b]:

$$\boxed{\imath\hbar\frac{\partial}{\partial t}|\psi\rangle = \frac{\hbar c}{\imath}\left(\alpha_1\frac{\partial}{\partial x_1}|\psi\rangle + \alpha_2\frac{\partial}{\partial x_2}|\psi\rangle + \alpha_3\frac{\partial}{\partial x_3}|\psi\rangle\right) + \beta mc^2|\psi\rangle \equiv \hat{H}|\psi\rangle} \tag{4.12}$$

[2]Exposition in [*BJORKEN/DRELL* 1964, 4–6] [*SCHWEBER* 1961, 54–60].

[3]Exposition in [*BJORKEN/DRELL* 1964, 6–8, 17–18] [*SCHWEBER* 1961, 65–85].

where $\alpha_k, (k = x, y, z)$ and β are Dirac matrices:

$$\alpha_k = \begin{bmatrix} 0 & \hat{\sigma}_k \\ \hat{\sigma}_k & 0 \end{bmatrix}; \beta = \begin{bmatrix} 1 & 0 \\ 0 & -1 \end{bmatrix} \tag{4.13}$$

and where the $\hat{\sigma}_k$'s in α_k are 2×2 Pauli spin matrices [Eq. (3.125)] and the 1's in β are 2×2 unit matrices.

Rewriting it by expressing the symmetry between ct and x^j, we obtain:

$$\imath\hbar \left(\gamma^0 \frac{\partial}{\partial x^0} + \gamma^1 \frac{\partial}{\partial x^1} + \gamma^2 \frac{\partial}{\partial x^2} + \gamma^3 \frac{\partial}{\partial x^3} \right) |\psi\rangle - mc|\psi\rangle = 0, \tag{4.14}$$

where $\gamma_0 = \beta$ and $\gamma_j = \beta\alpha_j, j = 1, 2, 3$.

4.2.3 Hole Theory

But the Dirac Eq. does not solve the problem of negative energy solutions. The problem also exists, to a certain extent, for non-relativistic QM [see section 10.2]. Therefore, both the Klein/Gordon Eq. and Dirac Eq. have been later used by physicists: the first is used for spinless particles (bosons: π and K mesons) and the Dirac Eq. for spin particles[4].

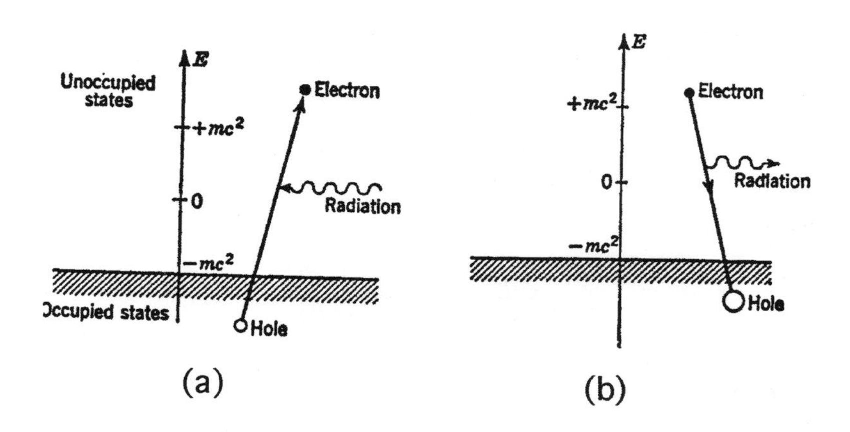

Figure 4.1: (a) Pair production: a negative-energy electron is excited to a positive-energy state by radiation.
(b) Pair annihilation: a positive-energy electron falls into a negative-energy hole emitting radiation — from [*BJORKEN/DRELL* 1964, 65].

Dirac tried to solve the problem of negative-energy solutions by proposing, at the interpretational level, the *Hole theory* [DIRAC 1930b].[5] According to the Pauli exclusion principle [see section 3.8], he filled up the negative-energy levels with electrons and defined the *vacuum state* as the state with all negative-energy levels filled and all positive-energy levels empty. In this state it is possible for a negative-energy electron to absorb radiation and be excited into a positive-energy

[4]On this point see [*BJORKEN/DRELL* 1964, 184].
[5]Discussion of the problem in [*BJORKEN/DRELL* 1964, 64–65, 69, 91–92, 97] .

state. If this occurs, we have an electron of charge $-|e|$ and energy $+E$ and, in addition, a hole in the negative-energy sea; and the hole registers the absence of an electron with charge $-|e|$ and energy $-E$. For an observer relative to a vacuum this situation would be interpreted as the presence of a particle of charge $+|e|$ and energy $+E$: the *positron*. In this case we therefore speak of *pair production* while, if a positive energy electron falls into a negative-energy hole, we speak of *pair annihilation*. Hence the destruction of a positron can be understood as the creation of a negative- energy electron and *vice versa* [see figure 4.1]. In conclusion, we can identify the amplitude for producing a positron in position $\mathbf{r}$ and propagating forward in space-time until $\mathbf{r}'$ and the amplitude for a negative-energy electron to propagate back from $\mathbf{r}'$ and to be destroyed in $\mathbf{r}$.

In order to treat the problem formally we need creation and annihilation operators, which can account for these processes of particle creation and destruction. We shall treat this problem in what follows.

4.3 Field Formalism

The procedure that we use here is called *second quantization*. While in the first quantization (Basic-QM) classical particle coordinates are replaced by quantum mechanical operators acting on wave functions, in second quantization we take the equivalent of Schrödinger Eq. — i.e. the Klein/Gordon or the Dirac Eq. —, we interpret it as a field equation and then we impose quantum conditions on the field amplitudes (i.e. they are treated as operators).

4.3.1 Klein/Gordon Field

We start by treating the case of boson fields, known as *Klein/Gordon fields*[6]. For a Klein/Gordon field $\hat{\varphi}(x)$ we define a Lorentz-invariant functional $\Lambda(\hat{\varphi}, \partial\hat{\varphi}/\partial x^\mu)$, called *Lagrangian density*:

$$\Lambda\left(\hat{\varphi}, \frac{\partial\hat{\varphi}}{\partial x^\mu}\right) := \frac{1}{2}\left(\frac{\partial\hat{\varphi}}{\partial x^\mu}\frac{\partial\hat{\varphi}}{\partial x_\mu} - m^2\hat{\varphi}^2\right), \tag{4.15}$$

thus obtaining a field Lagrangian function as the volume integral of the density:

$$L_{\hat{\varphi}} := \int_{-\infty}^{+\infty} d^3x \Lambda\left(\hat{\varphi}, \frac{\partial\hat{\varphi}}{\partial x^\mu}\right). \tag{4.16}$$

By means of the density we can write the equivalent of Eq. (4.10) — the quantum relativistic Lagrange Eq. —:

$$\boxed{\frac{\partial}{\partial x^\mu}\frac{\partial\Lambda}{\partial(\partial\hat{\varphi}/\partial x^\mu)} - \frac{\partial\Lambda}{\partial\hat{\varphi}} = 0} \tag{4.17}$$

We symbolize the average of $\partial\hat{\varphi}(\mathbf{x},t)/\partial t$ over the k–th cell in tridimensional space with $\dot{\hat{\varphi}}_k(t)$ and rewrite the Lagrangian as:

$$L_{\hat{\varphi}} = \int d^3x\Lambda = \sum_k \Delta V_k \overline{\mathcal{L}}(\dot{\hat{\varphi}}_k(t), \hat{\varphi}_k(t), \hat{\varphi}_{k\pm s}, \ldots), \tag{4.18}$$

where ΔV_k are cells of tridimensional space.

[6]Exposition in [*BJORKEN/DRELL* 1965, 13–16, 28]. See also [*PARISI* 1988, 284–85].

Now we can define the momentum $\hat{\pi}$, conjugate to $\hat{\varphi}(\mathbf{x},t)$, which, in continuous language, is given by:

$$\hat{\pi}(\mathbf{x},t) = \frac{\partial \Lambda(\hat{\varphi}, \dot{\hat{\varphi}})}{\partial \dot{\hat{\varphi}}(\mathbf{x},t)}, \tag{4.19}$$

while canonical momenta are given by:

$$\hat{p}_k(t) = \frac{\partial L_{\hat{\varphi}}}{\partial \dot{\hat{\varphi}}(t)} = \Delta \mathrm{V}_k \frac{\partial \overline{\mathcal{L}}_k}{\partial \dot{\hat{\varphi}}_k(t)} \equiv \Delta \mathrm{V}_k \hat{\pi}_k(t). \tag{4.20}$$

We define a corresponding Hamiltonian:

$$\hat{H} = \sum_k p_k \dot{\hat{\varphi}}_k - L_{\hat{\varphi}} = \sum_k \Delta \mathrm{V}_k (\hat{\pi}_k \dot{\hat{\varphi}}_k - \overline{\Lambda}_k). \tag{4.21}$$

The CCRs, in quantised form, are the following:

$$[\hat{\varphi}_i(\mathbf{x},t), \hat{\varphi}_j(\mathbf{x}',t)] = 0, \tag{4.22a}$$

$$[\hat{\pi}_i(\mathbf{x},t), \hat{\pi}_j(\mathbf{x}',t)] = 0, \tag{4.22b}$$

$$[\hat{\pi}_i(\mathbf{x},t), \hat{\varphi}_j(\mathbf{x}',t)] = \imath \delta_{ij} \delta^3(\mathbf{x} - \mathbf{x}'). \tag{4.22c}$$

In the case of Klein/Gordon fields we can establish the following commutation relationships (Bose CCRs, with commutators) between the creation operator $\hat{a}^\dagger$ and the annihilation operator $\hat{a}$ [see formula (3.112)]:

$$\boxed{[\hat{a}_i^\dagger, \hat{a}_j^\dagger] = [\hat{a}_i, \hat{a}_j] = \mathbf{0}, \ [\hat{a}_i, \hat{a}_j^\dagger] = \imath \delta_{ij}} \tag{4.23}$$

in terms of which we can define the total energy and momentum for the free Klein/Gordon field:

$$\hat{H} = \frac{1}{2} \int d^3k \omega_k [\hat{a}_k^\dagger \hat{a}_k + \hat{a}_k \hat{a}_k^\dagger], \tag{4.24a}$$

$$\hat{\mathbf{p}} = \frac{1}{2} \int d^3k \mathbf{k} [\hat{a}_k^\dagger \hat{a}_k + \hat{a}_k \hat{a}_k^\dagger], \tag{4.24b}$$

where in both cases the integral is made on each cell k.

By representing a free field by annihilation and creation operators satisfying Bose CCRs, we obtain formal equivalence to the QM harmonic oscillator, which can be formulated either in terms of position/momentum or in terms of amplitude/phase[7]. In fact the product $\hat{a}^\dagger \hat{a}$ specifies the number of particles (photons) present in an eigenstate of the field and for this reason it is called the *number operator* and is formalized as $\hat{N}$. Hence, in analogy with Eq. (3.31), the Hamiltonian of the Klein/Gordon field can be written as:

$$\hat{H} = \hbar \omega \left(\hat{a}^\dagger \hat{a} + \frac{1}{2} \right) \equiv \hbar \omega \left(\hat{N} + \frac{1}{2} \right). \tag{4.25}$$

[7]See [*SCHIFF* 1955, 182–83] [*BUSCH et al.* 1995a, 82–83].

4.3.2 Dirac Field

When treating fermions, the case in which we have a *Dirac Field*, we need another form of second quantization: instead of commutators we use here anticommutators, because fermions — which obey to Fermi/Dirac statistics [see section 3.8] — cannot occupy the same energy level[8]. Hence we have the following Fermion CCRs:

$$(\hat{a}^{\dagger})^2 = \mathbf{0}, \tag{4.26a}$$

$$\hat{a}_j^{\dagger}\hat{a}_k^{\dagger} + \hat{a}_k^{\dagger}\hat{a}_j^{\dagger} = \hat{a}_j\hat{a}_k + \hat{a}_k\hat{a}_j = \mathbf{0}, \tag{4.26b}$$

so that $\hat{a}_j^{\dagger}\hat{a}_k^{\dagger} = -\hat{a}_k^{\dagger}\hat{a}_j^{\dagger}$. We can write $[\hat{a}_j, \hat{a}_k]_+ := \hat{a}_j\hat{a}_k + \hat{a}_k\hat{a}_j$ and call it the anticommutation brackets. More synthetically we have that:

$$\boxed{[\hat{N}_j, \hat{N}_k] = 0} \tag{4.27}$$

where:

$$\hat{N}_k = \hat{a}_k^{\dagger}\hat{a}_k. \tag{4.28}$$

By imposing the second quantization, the field amplitudes again become operators satisfying:

$$[\hat{\varphi}(\mathbf{x}, t), \hat{\varphi}(\mathbf{x}', t)]_+ = 0, \tag{4.29a}$$

$$[\hat{\varphi}^*(\mathbf{x}, t), \hat{\varphi}^*(\mathbf{x}', t)]_+ = 0, \tag{4.29b}$$

$$[\hat{\varphi}(\mathbf{x}, t), \hat{\varphi}^*(\mathbf{x}', t)]_+ = \delta^3(\mathbf{x} - \mathbf{x}'). \tag{4.29c}$$

This formalism is the same as that of a many-body Schrödinger Eq.

4.4 Relativistic Scattering

We see now some application to scattering problems of the above formalism[9] — for the sake of simplicity we discuss the one-dimensional case.

First of all we need a relativistic S–matrix subject to a condition analogous to Eq. (3.147):

$$\sum_{\lambda=1}^{4}\left[\gamma_\mu\left(\imath\frac{\partial}{\partial x'_\mu} - eA^\mu(x')\right) - m\right]_{\alpha\lambda} \hat{S}_F(x' - x) = \delta_{\alpha\beta}\delta^4(x' - x), \tag{4.30}$$

and whose inverse Fourier transform in momentum space is given by:

$$\hat{S}_F(x' - x) = \int \frac{d^4p}{(2\pi)^4} e^{-\imath p\cdot(x'-x)} \hat{S}_F(p), \tag{4.31}$$

which has the following simple algebraic form:

$$\hat{S}_F(p) = (\gamma \cdot p - m_0 + \imath\eta)^{-1}. \tag{4.32}$$

Feynman propagators for electromagnetic radiation are symbolized by $\hat{D}_F(x - y)$ with their Fourier transform $\hat{D}_F(k)$ which has the following algebraic form:

$$\hat{D}_F^{\mu\nu}(k) = \frac{g^{\mu\nu}}{k^2 - \imath\eta}. \tag{4.33}$$

[8] Exposition in [*BJORKEN/DRELL* 1965, 44–48, 52] [*SCHWEBER* 1961, 218–31] [*PERES* 1993, 138–45].
[9] Formalism in [*BJORKEN/DRELL* 1964, 89–120] . See also [MILLS 1993, 70].

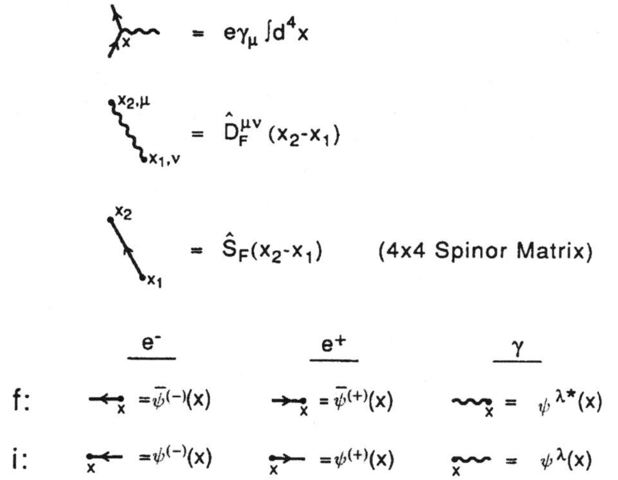

$= e\gamma_\mu \int d^4x$

$= \hat{D}_F^{\mu\nu}(x_2 - x_1)$

$= \hat{S}_F(x_2 - x_1)$ (4x4 Spinor Matrix)

	e^-	e^+	γ
f:	$=\overline{\psi}^{(-)}(x)$	$=\overline{\psi}^{(+)}(x)$	$=\psi^{\lambda *}(x)$
i:	$=\psi^{(-)}(x)$	$=\psi^{(+)}(x)$	$=\psi^{\lambda}(x)$

Figure 4.2: The Feynman graph prescription (a) in coordinate space and (b) in momentum space. Particles present in the initial state i are represented by lines arriving to a dot, and particles present in the final state f are represented by lines leaving a dot. For the antiparticles it is the contrary — from [MILLS 1993, 68, 70].

Second order electron scattering is given by the formula[10]:

$$-\imath e^2 \int d^4x_1 d^4x_2 \overline{\psi}_f^{(+)}(x_2) \not{A}(r_2) \hat{S}_F(x_2 - x_1) \not{A}(x_1) \psi_i^{(+)}(x_1), \tag{4.34}$$

where $\not{A} = \gamma^\mu A_\mu = \gamma^0 A^0 - \boldsymbol{\gamma} \cdot \mathbf{A}$ ($\mathbf{A}$ is the electromagnetic vector potential), $\psi^{(+)}$ denotes a positive energy electron and the symbol $\overline{\psi}$ is named the adjoint spinor and is $= \psi^\dagger \gamma_0$, and $\psi^\dagger \psi := \hat{\rho} = \sum_{\sigma=1}^4 \psi_\sigma^* \psi_\sigma$ [see figure 4.2 for a general understanding of Feynman's graphs].

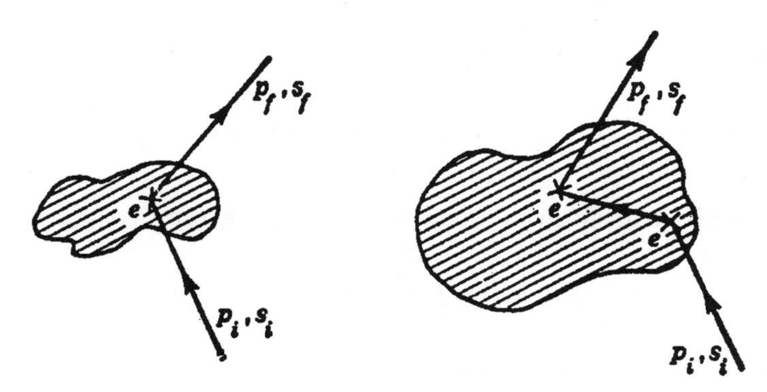

Figure 4.3: Coulomb scattering of electrons — from [*BJORKEN/DRELL* 1964, 108].

We now give some examples of scattering. The Coulomb scattering of an electron [see figure 4.3] is given by:

$$\hat{S}_{fi} = -\imath e \int d^4x \overline{\psi}_f^{(+)}(x) \not{A}(x) \psi_i^{(+)}(x). \tag{4.35}$$

[10] See [*BJORKEN/DRELL* 1964, 9, 24].

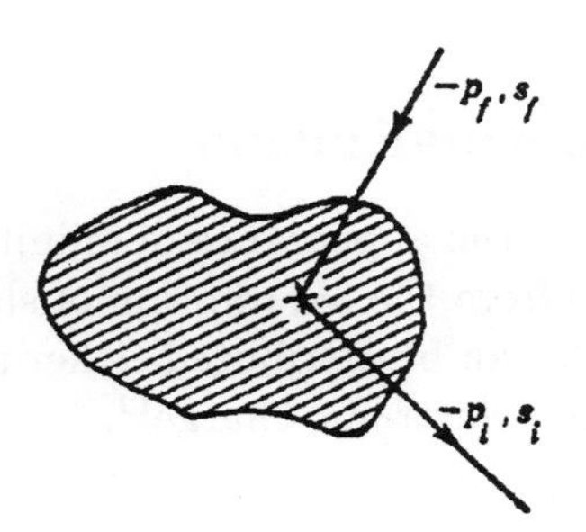

Figure 4.4: Coulomb scattering of positrons — from [*BJORKEN/DRELL* 1964, 107].

The scattering of positrons in a Coulomb field [see figure 4.4] is given by the expression:

$$\hat{S}_{fi} = \imath e \int d^4x \overline{\psi}_f(x) \not{A}(x) \psi_i^{(-)}(x), \tag{4.36}$$

where the incoming state is in the future and is interpreted as a negative energy electron, symbolized by $\psi_i^{(-)}$.

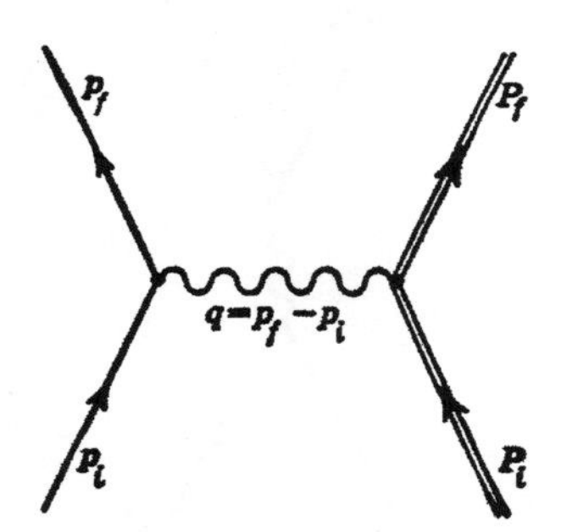

Figure 4.5: Electron–proton scattering — from [*BJORKEN/DRELL* 1964, 111].

A second-order interaction between an electron and a proton with the exchange of photon is given by:

$$\hat{S}_{fi}^{(2)} = -\imath e^2 \int d^4x d^4y \overline{\psi}_f(x) \not{A}(x) \hat{S}_F(x-y) \not{A}(y) \psi_i(y), \tag{4.37}$$

where the electromagnetic potential is generated by the current of the proton. A fourth-order interaction with a two-photon exchange between an electron and a positron ('between' x and w, and between y and z) can be written [*BJORKEN/DRELL* 1964, 148] [see figures 4.5 and 4.6]:

$$\begin{aligned} \hat{S}_{fi}^{(4)} &= -(-\imath e)^4 \int d^4w d^4x d^4y d^4z [\overline{\psi}_f^{(+)}(x) \gamma_\mu \imath \hat{S}_F(x-y) \gamma_\nu \psi_i^{(+)}(y)] \imath \times \\ & \hat{D}_F(x-w) \imath \hat{D}_F(y-z) [\overline{\psi}_{i'}^{(-)} \gamma^\nu \imath \hat{S}_F(z-w) \gamma^\mu \psi_{f'}^{(-)}(w)]. \end{aligned} \tag{4.38}$$

4.5 Localization

4.5.1 The Problem of Renormalization

The great problem of Relativistic-QM is the appearance of infinities (non convergent integrals) in the calculations. The problem derives from the assumption of sharp localization of points of the field[11]. All infinities in Relativistic-QM can be attributed either to the infinity in electromagnetic mass or to the infinity appearing in the electronic charge[12].

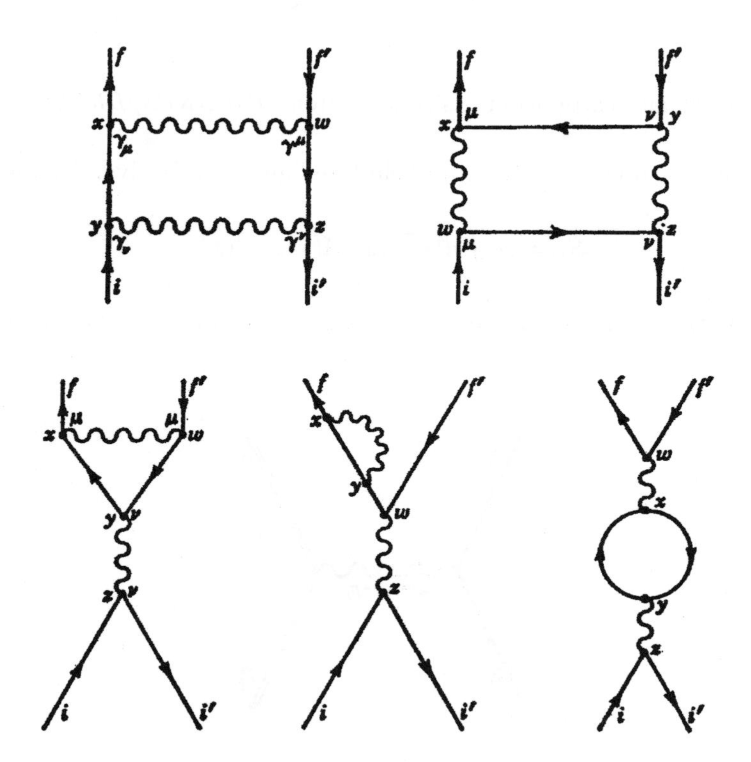

Figure 4.6: Some fourth-order graphs for electron–positron scattering — from [*BJORKEN/DRELL* 1964, 148].

Renormalization faces this problem by applying Path-integral methods to Relativistic-QM: it consists in 'integrating out' the variables which describe the field's degrees of freedom (the ideal oscillators of which the field is composed) leaving only the particle variables. One introduces an arbitrary function and a new parameter (the *cutoff*) with the aim of eliminating the infinities. As cutoffs approach infinity, the theory becomes identical to the standard theory.

We cannot here discuss the technical details of the problem of renormalization[13]. Nevertheless, we can face the problem in general terms with the following considerations[14]. Take a divergent

[11]Discussion of the problem in [SCHWEBER 1993, 145] [*BJORKEN/DRELL* 1964, 154].

[12]On this point see [L. BROWN 1993b, 15].

[13]For a more recent approach see [WILSON/KOGUT 1974].

[14]See [MILLS 1993, 62–63]. The article is in general a very good and easy exposition of the subject matter. For a historical reconstruction see also [DRESDEN 1993]

integral

$$f(\xi) = \int_1^\infty \frac{d\eta}{\xi + \eta}. \tag{4.39}$$

In fact the integrand goes like $1/\eta$ for large η; it goes to zero, but not fast enough to keep the integral from diverging logarithmically. On the face of it, the function $f(\xi)$ is undefined for every ξ. Here ξ and η are playing the roles of energy variables, and the integral represents one of the sums over virtual states we have to deal with. By comparing $f(\xi)$ with f at $\xi = 0$ and by exporting ξ we obtain the following convergent integral:

$$f(\xi) - f(0) = -\xi \int \frac{d\eta}{\eta(\xi + \eta)} = f_c(\xi), \tag{4.40}$$

so that we can write:

$$f(\xi) = C_\infty + f_c(\xi), \tag{4.41}$$

where C_∞ is an infinite constant. With this method we can also treat the infinities in Relativistic QM.

Dirac was not satisfied with the procedure of renormalization, at a physical level, because a gap results between relativistic QM and non-relativistic QM, and, at a mathematical level, because it is not completely sound to discard infinitely large quantities [DIRAC 1978, 3–5].

4.5.2 The Problem of Sharp Localization

We have seen that the problem of divergent integrals comes from sharp localization. Renormalization is a possible formal answer. But the problem of sharp localization remains: is it sound to assume it? The answer is no[15]. Take two operators, $\hat{O}_1(\mathbf{r})$ and $\hat{O}_2(\mathbf{r}')$. If they act on points $\mathbf{r}, \mathbf{r}'$ separated by a space-like distance, they commute. But this is in no way true for related annihilation and creation operators, which means that states are not sharply localized. In fact it is impossible to associate an operator to the field position [see Eq. (4.22a)]. If we try, on the other hand, to construct an operator $\hat{\mathbf{q}} = \imath\nabla_{\mathbf{p}}$ (equivalent to the non-relativistic position operator), it is non-symmetric and hence non-Hermitian [see p. 32]:

$$\begin{aligned} \langle\psi_1|\hat{\mathbf{q}}\psi_2\rangle &= \imath\int_{\Re^3} \frac{d^3k}{k_0}\overline{\psi}_1(\mathbf{k})\nabla_{\mathbf{p}}\psi_2(\mathbf{k}) \\ &= \int_{\Re^3} \frac{d^3k}{k_0}\left[\left(-\imath\nabla_{\mathbf{p}} + \frac{\imath\mathbf{k}}{\mathbf{k}^2 + m^2c^2}\right)\overline{\psi}_1(\mathbf{k})\right]\psi_2(\mathbf{k}) \\ &\neq \langle\hat{\mathbf{q}}\psi_1|\psi_2\rangle \end{aligned} \tag{4.42}$$

for two different $\tilde{\psi}_1(\mathbf{k}), \tilde{\psi}_2(\mathbf{k})$.[16]

To develop a relativistic QM without sharp localization and without Renormalization is Prugovečki's aim [PRUGOVEČKI 1984a] [*PRUGOVEČKI* 1984b]. We shall discuss his proposal in chapter 33.

---

[15] Discussion of the problem in [*SCHWEBER* 1961, 60–62] [*PRUGOVEČKI* 1984b, 78–82].

[16] The impossibility of a photon position operator is treated in [KRAUS 1977b] [*BUSCH et al.* 1995a, 93–94] — see also section 10.3 and particularly the Hegerfeldt Theorem [p. 187].

Chapter 5

QUANTUM OPTICS

Introduction and contents Quantum Optics (= Q-Optics) is the domain of QM which has been most developed and which has been the basis of most experiments which have changed QM in the last twenty years. It consists in the study of all interaction phenomena between radiation and matter and of the electromagnetic field, by the quantization of this field or of the matter field. In the study of the electromagnetic field, there is a very important feature: the Maxwell Eqs. of classical free-space field have exactly the same form as the Heisenberg Eqs. for Bose-type operators of the field.

As in the preceding chapter, we do not presume to give a full account of this domain — handbooks are quoted which do. We wish only to write down some formalism which will be useful in the following. After a short introduction to the treatment of the states in Q-Optics and a specifical analysis of optical interference — intensity and visibility —, and of the corpuscular aspect of Q-Optics, i.e. the photon count [section 5.1], section 5.2 is devoted to the very important — also for foundational problems — theme of Q-Optics in phase space. Finally a short introduction to some basic apparatus and methods used in Q-Optics is furnished [section 5.3]: it is intended as a mean to better follow and understand many of the experiments reported in the following chapters.

5.1 General Formalism

5.1.1 States

The following considerations are not always restricted to Q-Optics, but Q-Optics has developed the best analysis of the problems.

It is convenient to distinguish three states among others:

- *Fock* (*number*) states, where the number of photons is known with certainty and the phase is completely unknown [FOCK 1932][1]: this is a consequence of the Uncertainty Principle [see Eq. (7.17)]. We remember here that each wave can be decomposed into a phase and an amplitude [see Eq. (3.21)]. The Fock states are eigenstates of the number operator $\hat{N}_k$ [Eq. (4.28)]:

$$\hat{a}_k^\dagger \hat{a}_k |n_k\rangle = n_k |n_k\rangle. \tag{5.1}$$

[1]Exposition in [*EMCH* 1972, 6–9] [*WALLS/MILBURN* 1994, 10–11].

We call *occupation number* the number of quanta in a generic QM state. Note that when we have a Fock state, the field can be decomposed into a sum of creations and annihilations of single particles, so that the field formalism and the particle formalism turn out to be the same[2].

- *Coherent States*, where the phase is known better — and not the number of photons — [GLAUBER 1963a] [GLAUBER 1963c] [YUEN 1976]; this is always a consequence of the Uncertainty Principle. Coherent states are eigenstates of the annihilation operator $\hat{a}$. They are the states most similar to classical points in phase space because they produce least entropy [see also below subsection 5.1.5].[3]

  An initially coherent state of an oscillator remains coherent at all later times, and in the case of a system of oscillators, the wave packets retain their minimum uncertainty character at all times, and their centres move along trajectories which are precisely those of classical theory [GLAUBER 1966].

- *Squeezed States*: a general class of minimum uncertainty states[4].

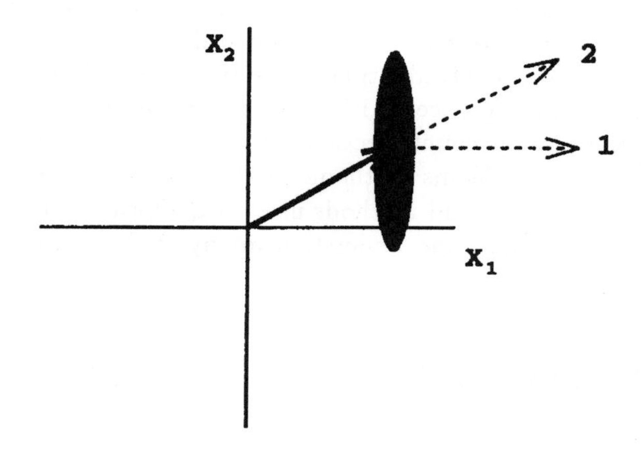

Figure 5.1: A phase convention for squeezed states. Direction 1 is the direction of squeezing, direction 2 is the direction of coherent excitation — from [*WALLS/MILBURN* 1994, 43].

The coherent state is a particular case of a squeezed state when the variances of the quadratures [YUEN 1976] — the adimensional position and momentum operators, components of the field mode — are equal.In other words, we have a coherent state iff we have a minimum-uncertainty state [see chapter 7, and for a better formulation of the minimum see Eq. (19.26)] such that:

$$\Delta\hat{X}_1 = \Delta\hat{X}_2 = 1. \tag{5.2}$$

[2]See for example [*PARISI* 1988, 290, 292].
[3]See [*WALLS/MILBURN* 1994, 12–13] [*HOLEVO* 1982, 117, 143–48]. See also [ZUREK *et al.* 1993].
[4]Exposition in [*WALLS/MILBURN* 1994, 15–18].

In analogy to the quantum harmonic oscillator, we write the position quadrature and the momentum quadrature in the following form[5] [see also Eq. (3.111) and comments]:

$$\hat{X}_1 = \frac{1}{\sqrt{2}}(\hat{a}^\dagger + \hat{a}), \quad \hat{X}_2 = \frac{\imath}{\sqrt{2}}(\hat{a}^\dagger - \hat{a}), \tag{5.3}$$

which obey quadrature CCRs [see figures 5.1 and 5.2]:

$$[\hat{X}_1, \hat{X}_2] = \imath\hbar\hat{I}. \tag{5.4}$$

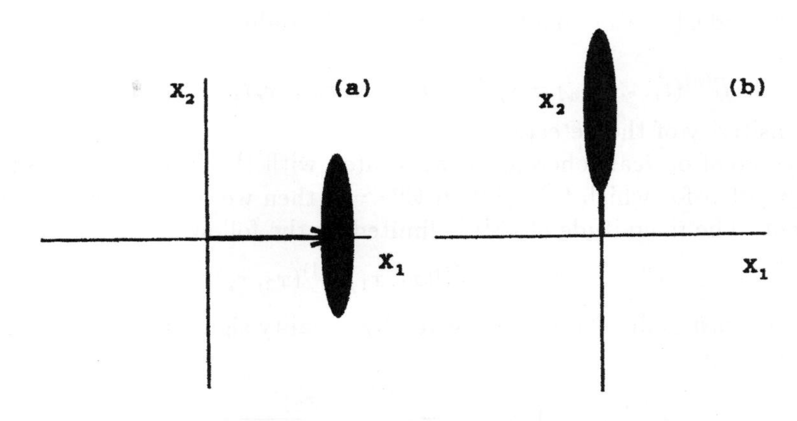

Figure 5.2: Phase-space of amplitude and phase squeezed states. (a) The quadrature carrying the coherent excitation is squeezed. (b) The quadrature out of phase with the coherent excitation is squeezed — from [*WALLS/MILBURN* 1994, 44].

Now by means of quadratures it is possible to decompose the annihilation and creation operators into a real and an imaginary part:

$$\hat{a} = \frac{1}{\sqrt{2}}(\hat{X}_1 + \imath\hat{X}_2), \quad \hat{a}^\dagger = \frac{1}{\sqrt{2}}(\hat{X}_1 - \imath\hat{X}_2). \tag{5.5}$$

Remembering here the Hamiltonian for boson fields [Eq. (4.25)] we can write:

$$\hat{H} = \hbar\omega\left(\hat{a}^\dagger\hat{a} + \frac{1}{2}\right) = \frac{1}{2}\hbar(\frac{\hat{p}^2}{m} + \omega^2\hat{q}^2). \tag{5.6}$$

We can now define a *two-photon operator* $\hat{b}$ such that

$$\hat{b} = c_1\hat{a} + c_2\hat{a}^\dagger, \tag{5.7}$$

$$|c_1|^2 - |c_2|^2 = 1. \tag{5.8}$$

Then $\hat{b}$ obeys the CCR:

$$[\hat{b}, \hat{b}^\dagger] = \hat{I}, \tag{5.9}$$

and we can write Eq. (5.7) as

$$\hat{b} = \hat{U}\hat{a}\hat{U}^\dagger. \tag{5.10}$$

The eigenstates of $\hat{b}$ are called *two-photon coherent states* and are closely related to squeezed states.

[5] See [*BUSCH et al.* 1995a, 83].

5.1.2 Intensity and Visibility

One can restate the problem of interference [see p.s 27– 28] in terms of Q-Optics [GLAUBER 1963a].[6] If the first order (which is sufficient for classical interference experiments) correlation function between the electromagnetic field in space-time points $x_i(\mathbf{r}, t)$ and $x_j(\mathbf{r}, t)$ is [see Eq. (3.91)]:

$$\mathrm{C}^{(1)}(x_i, x_j) = \mathrm{Tr}[\hat{\rho} E^{(-)}(x_i) E^{(+)}(x_j)], \tag{5.11}$$

and the nth–order correlation function is given by:

$$\mathrm{C}^{(n)}(x_1, \ldots, x_n, x_{n+1} \ldots, x_{2n}) = \mathrm{Tr}[\hat{\rho} E^{(-)}(x_1) \ldots E^{(-)}(x_n) E^{(+)}(x_{n+1}) \ldots E^{(+)}(x_{2n})], \tag{5.12}$$

then, if we consider an ideal n–atom photon detector, we introduce the n–fold delayed coincidence rate as follows:

$$R^{(n)}(t_1, \ldots, t_n) = s_D^n \mathrm{C}^{(n)}(r_1 t_1, \ldots r_n t_n, r_n t_n \ldots r_1 t_1), \tag{5.13}$$

where s_D is the sensitivity of the detector.

The highest degree of optical coherence is associated with the fringes of maximum visibility $\mathcal{V}$, i.e. with the largest n for which $\mathrm{C}^{(n)} \neq 0$. If $\mathrm{C}^{(n)} = 0$ then we have no fringes and hence the fields are incoherent. The magnitude of $\mathrm{C}^{(1)}$ is limited by the following inequality:

$$|\mathrm{C}^{(1)}(x_1, x_2)| \leq [\mathrm{C}^{(1)}(x_1, x_1) \mathrm{C}^{(1)}(x_2, x_2)]^{\frac{1}{2}}. \tag{5.14}$$

The best possible contrast (full coherence) is given by equality sign. If we introduce the normalized correlation function:

$$\mathrm{C}_N^{(1)}(x_1, x_2) = \frac{\mathrm{C}^{(1)}(x_1, x_2)}{[\mathrm{C}^{(1)}(x_1, x_1) \mathrm{C}^{(1)}(x_2, x_2)]^{\frac{1}{2}}}, \tag{5.15}$$

then we have full coherence iff:

$$|\mathrm{C}_N^{(1)}(x_1, x_2)| = 1. \tag{5.16}$$

By using the detector's transition probability for absorbing a photon at position $\mathbf{r}$ and at time t

$$\wp_{fi} = |\langle f | E^{(+)}(\mathbf{r}, t) | i \rangle|^2, \tag{5.17}$$

we can express the average field intensity I as:

$$\begin{aligned} I(\mathbf{r}, t) = \sum_f \wp_{fi} &= \sum_f \langle i | E^{(-)}(\mathbf{r}, t) | f \rangle \langle f | E^{(+)}(\mathbf{r}, t) | i \rangle \\ &= \langle i | E^{(-)}(\mathbf{r}, t) E^{(+)}(\mathbf{r}, t) | i \rangle. \end{aligned} \tag{5.18}$$

We can now write $\mathcal{V}$ in the following form[7]:

$$\boxed{\mathcal{V} := \frac{I_{\mathrm{Max}} - I_{\mathrm{min}}}{I_{\mathrm{Max}} + I_{\mathrm{min}}} = |\mathrm{C}_N^{(1)}| \frac{2(I_1 I_2)^{\frac{1}{2}}}{I_1 + I_2}} \tag{5.19}$$

so that maximum of visibility is equal to maximum optical coherence [see figure 5.3].

The even-ordered correlation functions (for example the second-order ones) contain no phase information and represent a measure of the fluctuations in the photon number. The odd-ordered correlation functions will contain information about the phase fluctuations of the electromagnetic field.

[6] An exposition of this point can be found in [*WALLS/MILBURN* 1994, 29–35, 44]. See also [*SCHROECK* 1996, 207–208, 211–12].

[7] For a stochastic formulation see [*SCHROECK* 1996, 179].

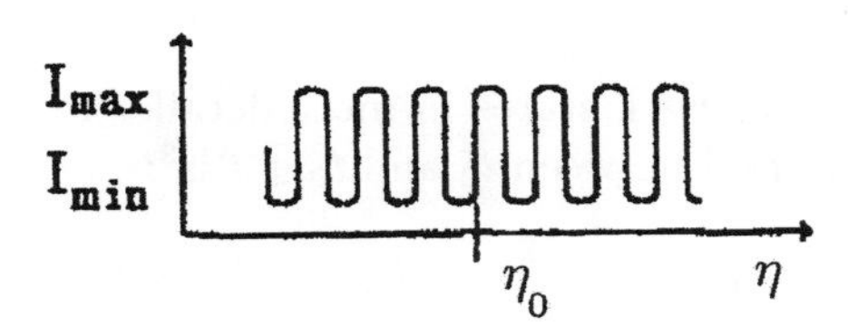

Figure 5.3: Interference fringes. η is a (multi-)parameter representing phase changes, time delays or position shifts of a basic wave — from [*SCHROECK* 1996, 208].

5.1.3 Interaction Between Electromagnetic and Electron Wave fields

The interaction between an electromagnetic field [see subsection 4.3.1 and an electron wave field, the Dirac field [see subsection4.3.2], is given by the following Hamiltonian[8]:

$$\hat{H} = \frac{1}{2m}(\hat{\mathbf{p}} - e\mathbf{A})^2 + eV(\mathbf{r}) + \hat{H}_{\text{free}}, \tag{5.20}$$

where $\hat{H}_{\text{free}}$ is the Hamiltonian of a free field and $V(\mathbf{r})$ is the Coulomb potential. When the electron wave field is quantized, the Hamiltonian becomes:

$$\hat{H} = \hat{H}_e + \hat{H}_{\gamma e} + \hat{H}_{\text{free}}, \tag{5.21}$$

where

$$\hat{H}_e = -\frac{\hbar^2}{2m}\nabla^2 + eV(\mathbf{x}) \tag{5.22}$$

refers to the free motion of the electron and $\hat{H}_{\gamma e}$ describes the interaction of the electron with the light field.

The interaction between a two-level atom with a single mode field is given by the Hamiltonian:

$$\hat{H} = \hat{H}_0 + \hat{H}_i, \tag{5.23}$$

where:

$$\hat{H}_0 = \hbar\omega\hat{b}^\dagger\hat{b} + \hbar\omega\hat{\sigma}_z \tag{5.24a}$$

$$\hat{H}_i = \hbar\Omega(\hat{b}\hat{\sigma}^+ + \hat{b}^\dagger\hat{\sigma}^-), \tag{5.24b}$$

where Ω is the vacuum Rabi frequency, and $\hat{\sigma}^+ = |e\rangle\langle g|, \hat{\sigma}^- = |g\rangle\langle e|$ are raising and lowering operators, respectively [corresponding to those defined at p. 56], which acts on ground $|g\rangle$ and excited $|e\rangle$ states as follows:

$$\hat{\sigma}^+|e\rangle = 0, \quad \hat{\sigma}^+|g\rangle = |e\rangle, \tag{5.25a}$$

$$\hat{\sigma}^-|e\rangle = |g\rangle, \quad \hat{\sigma}^-|g\rangle = 0. \tag{5.25b}$$

This is the simplest form of atom–field interaction and is called the *Jaynes/Cummings Model* [CUMMINGS 1965].

[8] Exposition in [*WALLS/MILBURN* 1994, 199–204].

5.1.4 Photon Count

We now return to the analysis of Fock states in more details. In classical theory the probability of n–photon counts in the interval between t_0 and $t_0 + t$ is[9]:

$$\wp_n(t_0, t) = \frac{1}{n!}[s_D t \overline{I}(t_0, t)]^n e^{-s_D t \overline{I}(t_0,t)}, \tag{5.26}$$

where s_D is always the sensitivity of the counter and $\overline{I}$ is the mean intensity during the counting interval, given by:

$$\overline{I}(t_0, t) = \frac{1}{t}\int_0^{t_0+t} I(t')dt'. \tag{5.27}$$

But since $\overline{I}(t_0, t)$ may vary from one interval to the next, $\wp_n(t)$ is a time average of $\wp_n(t_0, t)$ over a large number of different starting times:

$$\wp_n(t) = \langle \wp_n(t_0, t) \rangle = \left\langle \frac{[s_D \overline{I}(t_0, t)t]^n}{n!} e^{-s_D \overline{I}(t_0,t)t} \right\rangle. \tag{5.28}$$

The QM formula is similar in form to Eq. (5.28):

$$\wp_n(t) = \langle \wp_n(t_0, t) \rangle = \left\langle : \frac{[s_D \overline{I}(t_0, t)t]^n}{n!} e^{-s_D \overline{I}(t_0,t)t} : \right\rangle, \tag{5.29}$$

where :: denotes normal ordering of operators and where:

$$\begin{aligned} \overline{I}(t) &= \frac{1}{t}\int_0^t I(t')dt' \\ &= \frac{1}{t}\int_0^t E^{(-)}(\mathbf{r}, t')E^{(+)}(\mathbf{r}, t')dt'. \end{aligned} \tag{5.30}$$

5.1.5 Coherent States

Coherent states $|\alpha\rangle$ can be formalized as follows

$$|\alpha\rangle = e^{-\frac{|\alpha|^2}{2}} \sum_{n=0}^{\infty} \frac{\alpha^n}{\sqrt{n!}} |n\rangle, \tag{5.31}$$

where $\alpha = \langle \hat{a} \rangle$. If $|0\rangle$ represent the vacuum, we have that[10]:

$$|\alpha\rangle = e^{\alpha \hat{a}^\dagger - \alpha^* \hat{a}} |0\rangle. \tag{5.32}$$

We can express every state $\hat{\rho}$ in terms of coherent states $|\alpha\rangle$:

$$\hat{\rho} = \int \wp(\alpha)|\alpha\rangle\langle\alpha| d^2\alpha. \tag{5.33}$$

Now if a system is in the following coherent state[11]:

$$\hat{\rho} = |\alpha_0\rangle\langle\alpha_0|, \tag{5.34}$$

[9] Exposition in [*WALLS/MILBURN* 1994, 47–51].
[10] See [*GARDINER* 1991, 108–109].
[11] Exposition in [*WALLS/MILBURN* 1994, 58–59].

then

$$\wp(\alpha) = \delta^{(2)}(\alpha - \alpha_0). \tag{5.35}$$

If we have two independent sources the weight function is:

$$\wp(\alpha) = \int \delta^2(\alpha - \alpha_1 - \alpha_2)\wp_1(\alpha_1)\wp_2(\alpha_2)d^2\alpha_1 d^2\alpha_2. \tag{5.36}$$

We can now express the second order correlation function in terms of $\wp(\alpha)$:

$$\mathrm{C}_N^{(2)}(0) = 1 + \frac{\int \wp(\alpha)[(|\alpha|^2) - \langle|\alpha|^2\rangle]^2}{[\int \wp(\alpha)|\alpha|^2 d^2\alpha]^2}. \tag{5.37}$$

The scalar product of two coherent states $|\alpha\rangle, |\beta\rangle$ is given by:

$$\langle\alpha|\beta\rangle = e^{\alpha^* - \frac{1}{2}|\alpha|^2 - \frac{1}{2}|\beta|^2}. \tag{5.38}$$

No two coherent states are actually orthogonal to each other. However, if α and β are significantly different from each other ($|\alpha - \beta| \gg 1$), the two states are almost orthogonal.

5.2 Quantum Optics in Phase Space

Here we consider some methods which have widespread application in Q-Optics. As we have seen, unlike classical mechanics [see chapter 1], Basic-QM is not represented in a phase space. The greatest problem stems from the CCRs, which seem to forbid a simultaneous representation of momentum and position. But, as we shall see, it is possible to construct some representations in phase space and also a corresponding group. The price of this is that we are forced to introduce negative probabilities. In the following we introduce several quasi-probability distributions, i.e. distributions in which negative probabilities figure. Though historically the Wigner function was the first one to be formulated, for the sake of simplicity we will start with the Q-function.

5.2.1 Q-function

The Q-function — studied for the first time by Husimi [HUSIMI 1937] [12] — is particularly interesting for coherent states:

$$\boxed{\mathrm{Q}(\alpha, \alpha^*) = \frac{1}{\pi}\langle\alpha|\hat{\rho}|\alpha\rangle} \tag{5.39}$$

It represents the expectation value of the density matrix on a coherent state $|\alpha\rangle$. This expression gives, for a coherent state $|\alpha_0\rangle$, that:

$$\mathrm{Q}(\alpha, \alpha^*) = \frac{1}{\pi}e^{-|\alpha - \alpha_0|^2}. \tag{5.40}$$

Though the Q-function behaves as a probability function, not all positive normalizable Q-functions correspond to positive definite normalizable density operators, so that, as has already been said, it is a quasi-probability.

[12]Exposition in [*GARDINER* 1991, 117–19]. For the whole subject of this section see also [CAHILL/GLAUBER 1969].

5.2.2 P-function

The P-function[13], introduced by Glauber [GLAUBER 1963b], is interesting for ensembles of coherent states:

$$\boxed{\hat{\rho} = \int d^2\alpha \mathrm{P}(\alpha, \alpha^*)|\alpha\rangle\langle\alpha|} \tag{5.41}$$

The relationship between P-function and Q-function is given by:

$$\mathrm{Q}(\alpha, \alpha^*) = \frac{1}{\pi}\int d^2\beta e^{-|\alpha-\beta|^2}\mathrm{P}(\beta, \beta^*). \tag{5.42}$$

5.2.3 Characteristic Function

We can also write simply the characteristic function for the electromagnetic field [see Eq. (3.111) and Eq. (5.3)]:[14]

$$\chi(\eta) = \mathrm{Tr}[\hat{\rho}e^{\eta\hat{a}^\dagger - \eta^*\hat{a}}]. \tag{5.43}$$

The characteristic function is the inverse of the Weyl transform $\hat{U}_{WT}$:

$$\chi^{\hat{U}_{WT}} = (2\pi)^{-1}\int \hat{U}_{WT}(\eta)e^{\eta(\hat{X}_1 - \imath\hat{X}_2)}d\eta. \tag{5.44}$$

The characteristic function is related to the Q-function by the relationship:

$$\chi(\eta) = e^{|\eta|^2}\int d^2\alpha e^{\eta\alpha^* - \alpha\eta^*}\mathrm{Q}(\alpha, \alpha^*), \tag{5.45}$$

and to the P-function by the relationship:

$$\chi(\eta) = e^{|\eta|^2}\int d^2\alpha e^{\eta\alpha^* - \alpha\eta^*}\mathrm{P}(\alpha, \alpha^*). \tag{5.46}$$

On the other hand we can express the P-function as the inverse Fourier transform of the characteristic function:

$$\mathrm{P}(\eta, \eta^*) = \frac{1}{\pi^2}\int e^{\hat{a}\eta^* - \hat{a}^\dagger\eta}\chi(\eta)d^2\eta. \tag{5.47}$$

We can also define normally and antinormally ordered characteristic functions:

$$\chi_N(\eta) = \mathrm{Tr}[\hat{\rho}e^{\eta\hat{a}^\dagger}e^{-\eta^*\hat{a}}], \tag{5.48a}$$

$$\chi_A(\eta) = \mathrm{Tr}[\hat{\rho}e^{-\eta^*\hat{a}}e^{\eta\hat{a}^\dagger}]. \tag{5.48b}$$

5.2.4 Wigner Function

The Wigner distribution function[15] (= W-function) — the first quasi-probability distribution introduced into QM [WIGNER 1932] — for a given vector $|\psi\rangle$ [see subsection 5.2.4], in the one-dimensional case, is:

$$W(q, p) = \frac{1}{\pi\hbar}\int_{\Re} dx\langle q + x|\hat{\rho}|q - x\rangle e^{2\imath\frac{\hat{p}x}{\hbar}}. \tag{5.49}$$

[13] Exposition in [*GARDINER* 1991, 109, 122–24]. See also [*BUSCH et al.* 1995a, 150] .

[14] See [*HOLEVO* 1982, 220–26] [*GARDINER* 1991, 121, 123].

[15] For an extensive exposition see [DRAGOMAN 1998].

If the W-function is integrated with respect to p, it gives the correct probability distribution (marginal distribution) of q and *viceversa*, but it can also assume negative values and it does not always yield the same expectation value as QM does[16].

The relationship between W-function and P-function is[17]:

$$W(\alpha) = \frac{2}{\pi} \int d^2\beta e^{-2|\beta-\alpha|^2} \mathrm{P}(\beta, \beta^*). \tag{5.50}$$

That is, the W-function is a Gaussian convolution of the P-function, just like the Q-function – but with a different Gaussian weight [see Eq. (5.42)]. In other terms the Q-function is a less detailed average (it is smoother) than the W-function (because of the factor 2 in the exponential), and hence, unlike the W-function, it is never negative. But for this reason the W-function is better for distinguishing a quantum case from a classical one.

We now calculate the W function for coherent states of form $|\alpha\rangle = \frac{1}{2}|X_1 + \imath X_2\rangle$:

$$W(x'_1, x'_2) = \frac{2}{\pi} e^{-\frac{1}{2}(x'^2_1 + x'^2_2)}, \tag{5.51}$$

where $x'_j = x_j - X_j$, $(j = 1, 2)$. The contour of the W-function is: $x'^2_1 + x'^2_2 = 1$. Thus the error area is a circle with radius 1 centered on the point (X_1, X_2).

5.2.5 Other Problems of Quantum Mechanics in Phase Space

Expectation of observables As we have already seen, the first attempt at developing a QM in phase space was the Wigner function [see subsection 5.2.4]. Weyl [*WEYL* 1931] posed the problem of finding a correspondence between states and distribution functions, and between observables and phase space functions[18] such that the QM expectation values of the form $\mathrm{Tr}(\hat{\rho}\hat{O})$ can be expressed as classical-like averages of the form:

$$\int_\Gamma \rho_C(q,p)\xi(q,p) dq dp, \tag{5.52}$$

where $\xi(q,p)$ is the classical correspondent of the QM operator $\hat{O}$.[19]

Now we can express the expectation value of each observable in terms of the W-function[20] (5.49):

$$\left\langle \hat{O} \right\rangle := (\pi\hbar)^{-1} \int_{-\infty}^{+\infty} dq \int_{-\infty}^{+\infty} dp W(q,p) \hat{O}(q,p). \tag{5.53}$$

We observe that Eq. (5.53) is identical in form to the classical expression (3.90); hence the Wigner function corresponds to the classical distribution function.

Continuity equation Wigner also introduced [WIGNER 1932, 751–52] a definition of the evolution of the distribution[21], which is similar to the classical Eq. of continuity — Liouville Eq. (1.12) — :

$$\frac{\partial W(\mathbf{q}, \mathbf{p}; t)}{\partial t} = -\frac{\mathbf{p}}{m} \frac{\partial W(\mathbf{q}, \mathbf{p}; t)}{\partial \mathbf{q}} + \frac{\partial V(\mathbf{q})}{\partial \mathbf{q}} \frac{\partial W(\mathbf{q}, \mathbf{p}; t)}{\partial \mathbf{p}}. \tag{5.54}$$

---

[16] For further examination see [L. COHEN 1966b, 781] [*BUSCH et al.* 1995a, 146–47].

[17] See [*WALLS/MILBURN* 1994, 58–63].

[18] See [MOYAL 1949] [POOL 1966].

[19] See [*BUSCH et al.* 1995a, 145].

[20] See [O'CONNELL 1983, 85].

[21] See also [*DAVIES* 1976, 44–46] [HILLERY *et al.* 1984] [CINI/SERVA 1990, 105–106].

Other distributions Note that also other distributions can be introduced: Margenau and Hill [MARGENAU/HILL 1961] elaborated a distribution later used by Chandler and co-workers so as to also calculate distribution functions and quasi-probabilities for spin [CHANDLER *et al.* 1992]. Cohen showed a class of quasi-probabilities of which the quasi-probabilities introduced by Margenau and Hill and by Cohen are particular cases [L. COHEN 1966b, 782].

Quadratures and Characteristic Function Let us now return to the problem of quadrature [see Eq.(5.3)], i.e. approximate position and momentum [*DAVIES* 1976, 39–46]. Let us first define a vector $|f\rangle$ of norm one and whose expectation is zero with respect to $\hat{p}$ and $\hat{q}$ as follows:

$$f_{xk}(q) = e^{\imath kq} f(q-x), \tag{5.55}$$

with the following properties:

$$\langle f_{xk}|\hat{p}|f_{xk}\rangle = k, \quad \langle f_{xk}|\hat{q}|f_{xk}\rangle = x, \tag{5.56}$$

and, for some state vector $|\psi\rangle$,

$$\int_{\Re^2} |(2\pi)^{-\frac{1}{2}}\langle\psi|f_{xk}\rangle|^2 dkdx = \int_{\Re^2} |\psi(q)\overline{f(q-x)}|^2 dqdx = \int_{\Re^2} |\psi(q)|^2|f(x)|^2 dqdx. \tag{5.57}$$

Then, for each state vector $|\psi\rangle$, we can write:

$$\begin{aligned} \langle\psi|\hat{X}_1|\psi\rangle &= (2\pi\hbar)^{-1}\int_{x\in\mathcal{B}}\int_{k\in\Re} |\langle\psi|f_{xk}\rangle|^2 dkdx \\ &= \int_{\Re^2} |\psi(q)|^2 \chi_x^{\mathcal{B}} |f(q-x)|^2 dqdx \\ &= \int_{\Re} |\psi(q)|^2 (\chi_q^{\mathcal{B}} \circ |f(q)|^2) dq, \end{aligned} \tag{5.58}$$

where $\mathcal{B}$ means a Borel set [see section 11.3] and $(\chi_q^{\mathcal{B}} \circ |f(q)|^2)$ means the convolution $\int_{-\infty}^{+\infty} \chi_{q'}^{\mathcal{B}} |f(q - q')|^2 dq'$. Finally, from Eq. (5.58), the quadratures can be deduced:

$$\hat{X}_1(\mathcal{B}) = \int_{\Re} (\chi_q^{\mathcal{B}} \circ |f(q)|^2)\hat{q}(dq); \tag{5.59a}$$

$$\hat{X}_2(\mathcal{B}') = \int_{\Re} (\chi_p^{\mathcal{B}'} \circ |\tilde{f}(p)|^2)\hat{p}(dp). \tag{5.59b}$$

5.2.6 Weyl Group

In order to handle unitary transformations, the group correspondent to QM in phase space is the Weil group [*WEYL* 1931].[22] The Weyl pair follows from the Fourier transforms (3.82).

If $\mathbf{p}$ is transformed by $\hat{U}_{\mathbf{p}} = e^{\imath\mathbf{p}\cdot\hat{\mathbf{q}}}$ and $\mathbf{q}$ by $\hat{U}_{\mathbf{q}} = e^{-\imath\mathbf{q}\cdot\hat{\mathbf{p}}}$ [see Eqs. (3.138) and (3.139)], we can write:

$$\hat{U}_{\mathbf{p}}\hat{U}_{\mathbf{q}} = e^{\imath\mathbf{q}\cdot\mathbf{p}}\hat{U}_{\mathbf{q}}\hat{U}_{\mathbf{p}}. \tag{5.60}$$

The Weyl unitary operators then have the form[23]:

$$\hat{U}_{\mathbf{qp}} := e^{\frac{1}{2}\imath\mathbf{q}\cdot\mathbf{p}}\hat{U}_{\mathbf{q}}\hat{U}_{\mathbf{p}}, \tag{5.61}$$

[22] Formalism and problems in [*BUSCH et al.* 1995a, 59–61, 145–150] [*HOLEVO* 1982, 107–114].

[23] In [*HOLEVO* 1982, 110–113] one can find the formula of the Weyl displacement operator, from which position and velocity are deduced. The Weyl transform and its Fourier transform are in [*HOLEVO* 1982, 223–27].

satisfying:

$$\hat{U}_{\mathbf{qp}}\hat{U}_{\mathbf{q'p'}} = e^{\frac{i}{2}(\hat{\mathbf{p}}\cdot\hat{\mathbf{q}}'-\hat{\mathbf{q}}\cdot\hat{\mathbf{p}}')}\hat{U}_{\mathbf{q+q'},\mathbf{p+p'}}. \tag{5.62}$$

which is the Weyl form of the CCR [see Eq. (3.14)]. According to the Stone theorem [p. 47], the two irreducible representations of CCR are equivalent [*HOLEVO* 1982, 114, 224].

In conclusion we can treat CCR and uncertainty relationships in the Weyl, in the Heisenberg and in the Schrödinger representation[24].

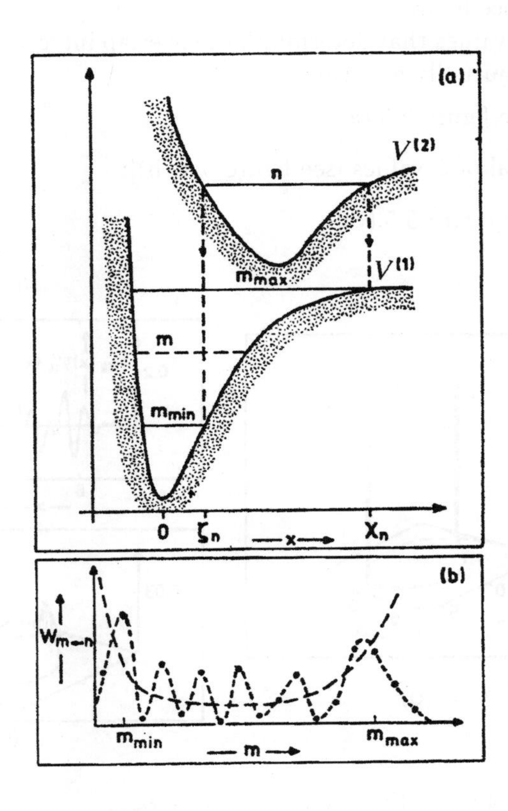

Figure 5.4: (a) Franck/Condon transitions from the n–th vibrational level of the upper electronic state, with intermolecular potential $V^{(2)}$, to vibrational states m in the lower electronic potential $V^{(1)}$. The two vibrational levels of the lower electronic state (shown as solid lines) have the turning points ζ_n, χ_n in common with the nth vibrational motion of the upper electronic state.
(b) This can be seen from the maxima $m = m_{\min}, m_{\mathrm{Max}}$ — from [DOWLING *et al.* 1991, 428–29].

5.2.7 Interference in Phase Space

Formalism

In the standard formalism the probability of transition occurring from the n–th vibrational state of a diatomic molecule in a certain electronic state to the m–th vibrational level of a different

[24]Emch [EMCH 1983] has shown that in the classical limit $\hbar \longrightarrow 0$ of the algebra of observables obtained from QM via the Weyl transformation, one obtains the algebra of classical observables and the Weyl transform of the density operator becomes a true probability density [SCHROECK 1985b].

electronic state is [WHEELER 1985]:[25]

$$\wp_{m,n} := |\vartheta_{m,n}|^2, \tag{5.63}$$

which is the square modulus of the Franck/Condon amplitudes:

$$\vartheta_{m,n} = \int_{-\infty}^{+\infty} dx \psi_m(x)^* \psi_n^{'}(x) \tag{5.64}$$

in one-dimensional form [see figure 5.4].

What is the range of x values that determines the overlap integral (5.64)? The following three possible answers are not mutually exclusive:

- the whole x–axis [see figure 5.5.(a)];
- a well-defined interval of x values [see figure 5.5.(b)];
- a single point x_c [see figure 5.5.(c)].

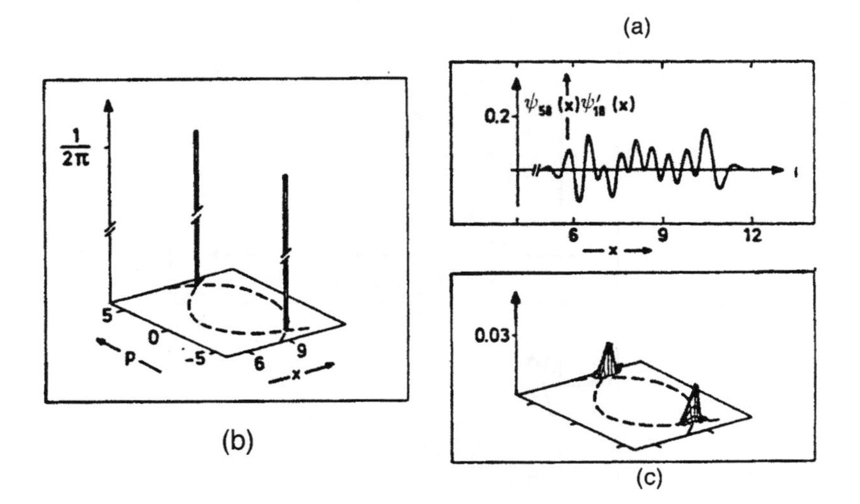

Figure 5.5: Three methods of evaluating jump probability $\wp_{m,n}$.
(a) We can integrate the product of the two vibrational wave functions $\psi_m, \psi_n^{'}$ $(m = 58, n = 18)$ over all space positions (along the entire x axis).
(b) The area of overlap indicated relates the jump probability to the two diamond-shaped phase space zones of cross-over between the two Planck/Bohr/Sommerfeld bands.
(c) The W-function technique determines the jump probability via well-localized, finite domains in the neighborhood of the phase points $(x_c \pm p_m(x_c))$ — from [DOWLING *et al.* 1991, 428–29].

Now it is possible to show that areas in phase space are determined by transition amplitudes: as a consequence of this, interference in phase space is to be understood as the most general way of treating interference. If we take W-functions $W_m(x,p), W_n(x,p)$ [see comments to Eq. (5.50)], probability (5.63) can be written as follows:

$$\wp_{m,n} = 2\pi \int_{-\infty}^{+\infty} dx \int_{-\infty}^{+\infty} dp W_m(x,p) W_n(x,p). \tag{5.65}$$

Hence interference effects must originate from phase space domains where the W-function product takes negative values [see figure 5.6].

[25]For a complete exposition of what follows see [DOWLING *et al.* 1991].

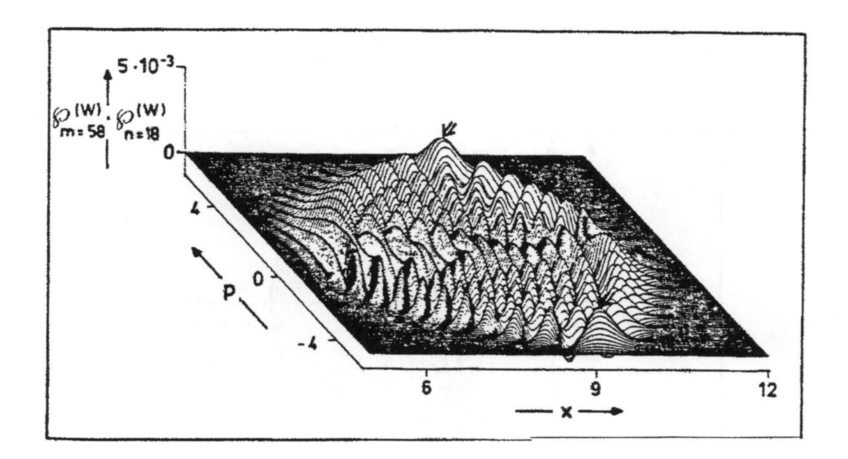

Figure 5.6: Product of two W-functions $W_{m=58}(x,p)$ and $W_{n=18}(x,p)$ as a tool for determining the probability $\wp_{m=58,n=18}$ of a sudden transition from the $n = 18$–th state of the harmonic oscillator about $x_0 = 12$ to the $m = 58$–th state of oscillations about the origin. Two peaks, or Wigner crests, identified by the arrows and located near $x_c = 9.33$ and $\pm p_m(x_c) = \pm 5.5$ dominate the landscape of the product function.The 'wavy sea' between the two peaks, with all its wave crests and troughs where this product assumes negative as well as positive values, contains the interference between these two peaks — see [DOWLING *et al.* 1991, 436].

The phase space area below the wavy sea, caught between the two positive-valued peaks yields the pseudo-probability which follows from Eq. (5.65), where:

$$\wp_{m,n} = 2\wp^{\text{peak}}_{m,n} + \wp^{\text{sea}}_{m,n}. \tag{5.66}$$

Hence we are able to establish a strict connection between interference and area of overlap in phase space [figure 5.7].

Short Commentary*

As we shall see, QM in phase space is the most pregnant demonstration that Hilbert spaces are not necessary (even though they are very useful) to QM [see theorem 9.6: p. 166]. Also stochastic QM [see section 33.2] has shown this possibility. If no sharp observables are used, it is possible to handle non-commuting observables simultaneously. On the other hand, a C* algebra with symmetry conditions and without self-adjointness of operators makes it possible to avoid Hilbert spaces[26]. It is true that self-adjointness is required in Basic-QM [see postulate 3.2: p. 33], but the Naimark theorem [18.5: p. 301] and the Holevo corollary [18.2: p. 301] guarantee the use of symmetric operators.

[26]See for example [*SCHROECK* 1996, 290].

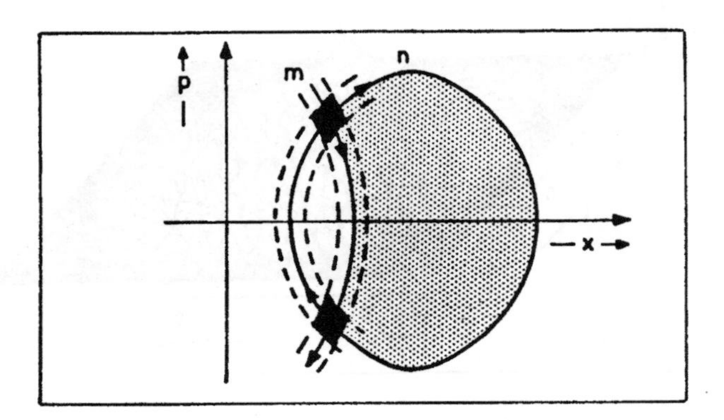

Figure 5.7: The area-of-overlap concept. In semiclassical approximation — Bohr's correspondence principle [see section 3.2] — the QM scalar product between stationary states $|\psi_m\rangle$, $|\psi_n'\rangle$ can be visualized as the overlap between two states depicted in x-p oscillator phase space as occupied bands which are passed through in a clockwise direction. In the case of two or more overlaps (two in the figure) the two diamond shaped black areas differ in their momentum: in one of them both oscillators move to the right, in the other to the left. The phase is fixed by the shaded area — from [DOWLING *et al.* 1991, 435].

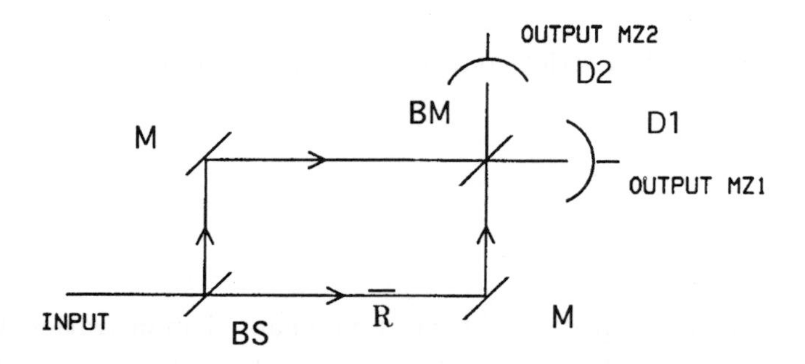

Figure 5.8: MZ Interferometer scheme. BS is the beam splitter while BM is the beam merger and splitter. While the BSs are half-silvered mirrors, the M's are fully reflecting mirrors. R is a delay device (light pipe for photons, a block of aluminum for neutrons, and so on). For other experiments R may be a phase shifter. D1 and D2 are recording devices. Generally BM is prepared so that one device records constructive interference, the other destructive interference (dark output).

5.3 Q-Optics Experimental Apparatus and Methods

5.3.1 Interferometry

An interferometer is an apparatus consisting of Beam splitters (= BS), i.e. half-silvered mirrors which partly transmit the beam and partly reflect it[27]; the BSs separate the beam into two paths, a lower and an upper one; at the end the beams pass through a Beam merger (= BM) and splitter and end up in either one of two detectors D_1, D_2. If the BS and the BM are different and there are two detectors we have a Mach/Zehnder (= MZ) [see figure 5.8]: generally, of the two detectors, one detects constructive interference, and the other destructive interference (it does not detect).[28].

5.3.2 Downconversions

A two-photon decay process from one photon is known as spontaneous parametric fluorescence [LOUISELL *et al.* 1961] [HARRIS *et al.* 1967] [see figure 5.9]. It is subjected to conservation laws of momentum and energy:

$$\mathbf{k}_0 = \mathbf{k}_1 + \mathbf{k}_2, \qquad \omega_0 = \omega_1 + \omega_2, \tag{5.67}$$

where $\mathbf{k}_0$ and ω_0 are initial wave vector and frequency, respectively. $\omega_1 = \omega_2 = \omega_0/2$ is the degenerate case. A specific type of two-photon decay is the Parametric Down Conversion (= PDC) [BURNHAM/D. WEINBERG 1970],[29], where a pumped beam of photons passes through a Non-Linear (= NL) crystal and is separated into two photons, conventionally named the idler photon (= i-photon) and the signal photon (s-photon) [see figure 5.10]. Note that the BS of an interferometer is a linear device[30], such that the same number of photons go in as come out, and this is important for interaction-free measurement [particularly see section 20.2].

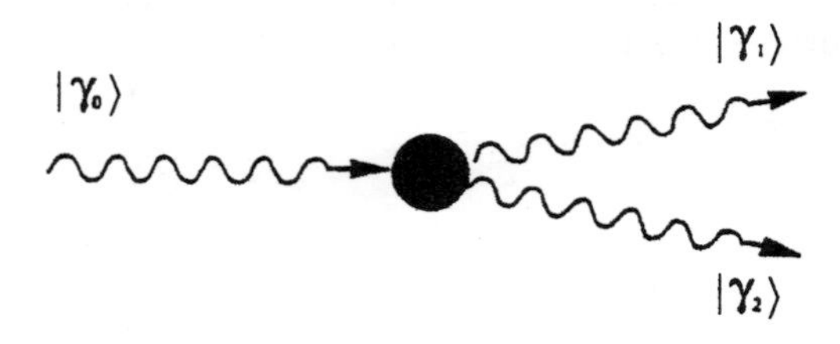

Figure 5.9: Two-photon decay process — from [CHIAO *et al.* 1995b, 261].

Some interesting developments of Q-Optics are to be found in chapter 22.

[27]For a wide account concerning beam splitting, with relative matrix transformations, joint photon-number distributions and variance, can be found in [CAMPOS *et al.* 1989].

[28]For neutron interferometry see [GREENBERGER 1983].

[29]Further developments in [YURKE 1985b] [SCHUMAKER/CAVES 1985] [CAVES/SCHUMAKER 1985].

[30]Anyway some doubts about the absolute linearity of the BSs arose [FRANSON 1997].

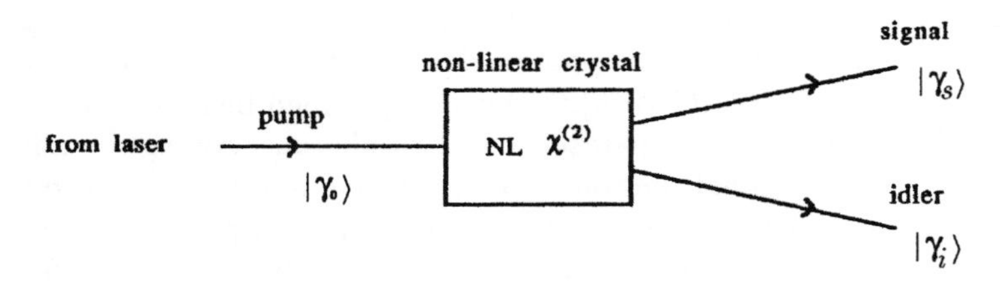

Figure 5.10: Setup of PDC: an entangled photon pair ($|\gamma_s\rangle$ = signal, $|\gamma_i\rangle$ = idler) may be emitted into the two outgoing modes shown — from [MANDEL 1995b, 2].

Part II

THE COPENHAGEN INTERPRETATION

Introduction to Part II This part is centered on the first, and the most important, global interpretation of the theory, the Copenhagen interpretation.

- In *chapter 6*, we first discuss Schrödinger's and Heisenberg's first attempts; then we examine the Ignorance interpretation (from Einstein promoted) and the Statistical interpretation.
- Then, in *chapter 7*, we examine the fundamental Uncertainty Principle: as we shall see, the principle's place in the theory and its character have been clarified in the last years by overcoming some traditional interpretations based on a supposed centrality of the disturbance or of the standard deviation.
- The part ends with the Complementarity Principle [*chapter 8*], different aspects of which are shown, and which are then partly discussed on the basis of recent experiments.

Chapter 6

FIRST INTERPRETATIONS OF QUANTUM MECHANICS

Contents In this chapter we give account of the first interpretations of QM: Schrödinger's interpretation [section 6.1] and the first interpretation of Heisenberg [section 6.2]. Some remarks about the problem of interpretation briefly follow [section 6.3]. Then we shall describe an alternative attempt to Schrödinger's one at reducing QM to a classical case: the Ignorance interpretation. It stems from a basic attitude which is quite the opposite to the Schrödinger's Wave-Packet interpretation: Born's corpuscular interpretation [section 6.4]; then we give a report of the Ignorance interpretation itself with an important variant, the Statistical interpretation [section 6.5]. Finally [section 6.6] an examination of these last two interpretations follows, where some contemporary experiments are also considered.

6.1 Schrödinger's Interpretation

6.1.1 Exposition of Schrödinger's Interpretation

Schrödinger's unsatisfaction In 1926 Wave Mechanics [see section 3.4] appeared to Schrödinger as a solution to microphysical problems that stood in a frame which was compatible with classical physics, i.e. in a wavelike, continuous frame. The model of real mechanical processes could be represented by wave-like evolutions in the coordinate configuration space and not by the evolution of parceled points in that same space [SCHRÖDINGER 1926a, 115]. Hence Schrödinger attacked Bohr's view on quantum jumps [postulate 2.2: p. 20], and in two discussions — July 1926 in München and in September 1926 in Copenhagen, the latter on the occasion of Schrödinger's visit — it was clear that the two points of view were irreconcilable[1].

Eigenfrequencies Successively Schrödinger tried to explain the appearance of jumps and of discrete energy levels in QM by interpreting the latter as eigenfrequencies of a single wave — by numerical agreement between eigenfrequencies and energy levels following Eq. (2.11) —, hence without assuming that — **for example**, as happens for Planck's solution of the black-body problem [see subsection 2.1.1] — we have a distinct resonator for each allowed level [SCHRÖDINGER 1927c, 272–73].[2] By analogy with resonance phenomena, where, if we strike

[1] Historical reconstruction in [*JAMMER* 1966, 344].

[2] See also [SCHRÖDINGER 1953, 694, 696, 704].

a material body, we have a succession of vibrations at different frequencies (proper frequencies), Schrödinger made the Eigenfrequencies Assumption [SCHRÖDINGER 1952, 482, 490–91], which we can synthesize as follows:

Postulate 6.1 (Eigenfrequency Assumption) *QM waves can be understood as a superposition of discrete series of vibrations, each one with a proper frequency.*

Explanation of jumps Now we can explain the jumps as correlations between frequencies and amplitudes of connected subsystems [SCHRÖDINGER 1927c, 271–72].[3] In particular, let us consider two systems, the first defined by the following wave Eq. [see Eq. (3.18)]:

$$\imath\hbar\frac{\partial}{\partial t}|\psi_1\rangle = \left[-\frac{\hbar^2}{2m}\nabla^2 + V_1\right]|\psi_1\rangle, \tag{6.1}$$

with eigenvectors $|\psi_{k_1}\rangle$ and eigenvalues E_{k_1}; and the second one similarly characterized by the following Eq.:

$$\imath\hbar\frac{\partial}{\partial t}|\psi_2\rangle = \left[-\frac{\hbar^2}{2m}\nabla^2 + V_2\right]|\psi_2\rangle, \tag{6.2}$$

with eigenvectors $|\psi_{l_2}\rangle$ and eigenvalues E_{l_2}, and let them interact as a compound system:

$$\imath\hbar|\dot{\Psi}\rangle + \frac{\hbar^2}{2m}\nabla^2|\Psi\rangle - (V_1 + V_2)|\Psi\rangle = 0, \tag{6.3}$$

with respectively eigenvectors $|\psi_{k_1}\rangle|\psi_{l_2}\rangle$ and respectively eigenvalues $E_{k_1} + E_{l_2}$, and add to $V_1 + V_2$ a small interaction potential V_{12}. Consider now four values such that: $E_{k_1} + E_{l'_2} = E_{k'_1} + E_{l_2}$, to which the eigenvectors: $|\psi_{k_1}\rangle|\psi_{l'_2}\rangle$ and $|\psi_{k'_1}\rangle|\psi_{l_2}\rangle$ correspond; so that we have the constant energy difference values (levels): $E_{k_1} - E_{k'_1} = E_{l_2} - E_{l'_2}$. For their amplitudes ϑ_1, ϑ_2 we have Eqs.:

$$\dot{\vartheta}_1 = \frac{1}{\imath\hbar}(c_{11}\vartheta_1 + c_{12}\vartheta_2), \tag{6.4a}$$

$$\dot{\vartheta}_2 = \frac{1}{\imath\hbar}(c_{12}\vartheta_1 + c_{22}\vartheta_2), \tag{6.4b}$$

where c_{ij} $(i, j = 1, 2)$ are appropriate coefficients. Now if one has an increase of the amplitude pertaining to $|\psi_{k_1}\rangle|\psi_{l'_2}\rangle$, we can interpret the thing as an increase of the amplitude of $|\psi_{k_1}\rangle$ at the expenses of $|\psi_{k'_1}\rangle$'s amplitude in the first system, or as an increase of $|\psi_{l'_2}\rangle$ at the expenses of $|\psi_{l_2}\rangle$ in the second system. So that, without using QP [see section 2.1], the 'jumps' are accounted for in terms of a resonance phenomenon, analogous to acoustical bats or to the behaviour of two pendulum of equal proper frequencies, connected with a weak spring. Note that this last assumption is certainly mathematically correct, because we can treat some compound quantum systems in terms of coupled oscillators [on this point see section 27.3].

Wave-packets But though we can equate frequencies and energy levels, in this frame it is always necessary to account for the corpuscular manifestations of microphysical entities — as is clear, for example, from Two-Slit-Exp.s 1–3 [see section 3.1]. Schrödinger tried to do this with the Wave-Packet interpretation, which can be formulated in the following manner[4]:

[3] See also [SCHRÖDINGER 1952, 484–85] . For an analysis see [*JAMMER* 1974, 27–28].

[4] See for example [SCHRÖDINGER 1953, 698, 702].

Proposition 6.1 (Wave-Packet Interpretation) *The apparent corpuscular nature of microentities is a manifestation of a group of waves localized in a extremely concentrated space, i.e. a manifestation of a wave packet.*

In conclusion the Wave-Packet interpretation is an attempt to completely eliminate the ontological reality of particles by affirming the existence only of continuous wave-processes. Sometimes Schrödinger thought of eliminating the ontological status of particles by means of the second quantization [see section 4.3] [SCHRÖDINGER 1953, 706].

6.1.2 Examination of Schrödinger's Interpretation

Criticisms

No ordinary space There are a lot of problems and difficulties regarding the Eigenfrequencies Assumption and the Wave-Packet interpretation. One general difficulty is that the waves of Wave Mechanics [see section 3.4] 'move' in a Hilbert space and not in a ordinary coordinate configuration space, that makes the task of a realistic interpretation of the waves a very difficult one, as noted already by Lorentz in a letter of May 1926 to Schrödinger and as expressed by Bohr at the international congress of physicists of September 1927 [BOHR 1927b, 142] [BOHR 1928, 154]. Schrödinger himself was aware of the difficulty[5].

Dispersion of wave-packets Then the Wave-Packet interpretation is not generalisable. We know that the wave packets disperse in space in the course of time[6] and thus the Wave-Packet interpretation cannot account for the corpuscular behaviour of microentities, in particular, of entities as electrons and atoms that evidently have a long life as corpuscle-like individuals — as Lorentz also remarked in the quoted letter[7].

No integer multiples Another objection to the Eigenfrequency assumption comes from Heisenberg [HEISENBERG 1927, 491]: if we acknowledge the Eigenfrequency Assumption, the radiation emitted by an atom could be expanded into a Fourier series in which the frequencies of the overtones are integer multiples of a fundamental frequency; but the frequencies of atomic spectral lines are never such integer multiples of a fundamental frequency, apart from the case of the harmonic oscillator [see section 3.3].

Other problems Jammer [*JAMMER* 1974, 33] assesses three additional difficulties: 1. ψ is a complex function, which seems to be in contradiction with real Euclidian space; 2. ψ undergoes a discontinuous change during a process of measurement; 3. ψ depends on the set of observables chosen for its representation. Difficulty 1. is solvable because every complex function is equivalent to a pair of real functions. We shall return to 2. by analyzing the measurement problem [see chapters 14 and 17]. 3. is a natural consequence of the transformation theory [see section 3.6].

The Problem of transitions

Schrödinger posed the problem of transition in microevents in which the transition needs a time comparable to the average interval between transitions, so that the concept of single stationary

[5] For reconstruction and analysis see [*JAMMER* 1974, 32–33].

[6] For contemporary research on the problem see [KETZMERICK *et al.* 1997] and articles quoted in it.

[7] See again [*JAMMER* 1974, 31]; see also [SCHRÖDINGER 1953, 698].

states does not explain the whole process [SCHRÖDINGER 1953, 696].[8] This is a very important question.

It is true that at Schrödinger's time it was experimentally demonstrated by Franck and Hertz (using electron collisions) that the energy values of states lying between quantum states do not exist [EINSTEIN 1936, 89]. But, on the other hand, the problem remains: it depends on the great difficulty in analyzing the time intervals of the transitions in such a way that the transition does not appear instantaneous any more. But this is still outside the actual theoretical and experimental means [see also chapter 22].

6.2 Heisenberg's First Interpretation

A Formal Interpretation Heisenberg and Born were not at all convinced of the Wave-Packet interpretation. We shall return below [in section 6.4] to Born's interpretation. Here it is sufficient to say that the Wilson chamber experiments showed a clear particle-like nature of microentities. We know that the discussion between Bohr, Heisenberg and Schrödinger in September 1926 precipitated the crisis[9]: the Wave-Packet interpretation was untenable but, on the other hand, there was no evident substitute for it. Therefore, perhaps as a consequence of his enormous mathematical discoveries, Heisenberg, who already believed that it was impossible to interpret QM observables by using the classical concepts of position, momentum and so on, proposed to consider the formalism F as self-sufficient. We can express Heisenberg's tenet [HEISENBERG 1927, 478] by formulating the following postulate:

Postulate 6.2 (Strong Formal Interpretation) *The meaning of quantum observables coincides with the place that they have in an experimentally tested* F *and consists in their mutual formal relationships.*

A clear example of this interpretation was the new definition of **p** an **q** through the non-commutative multiplication [see subsection 3.3.2] introduced in 1925 by Heisenberg himself [HEISENBERG 1925]. It is an interpretation which is not so different from Planck's Formal interpretation [p. 24], which justifies its name. We leave aside, for the moment, the question regarding the relationship between QM concepts and classical ones.

Ontological Commitment To examine postulate 6.2 we need some meta-theoretical discussions. Firstly we assume following metaprinciple[10]:

---

[8]See also [*DE BROGLIE* 1956, 265–67] and [VIGIER 1985, 655].

[9]Historical reconstruction in [*JAMMER* 1966, 344–45]

[10]See [QUINE 1948, 8–9] for discussion. See also [*LAUDAN* 1977, 15] [IRANI 1984, 310].

Principle 6.1 (Ontological Commitment) *Each — scientific or philosophical — theory must define and delimitate the ontological status of the entities it introduces.*

The reason for assuming principle 6.1 is the following: We cannot have a well-founded theory — a theory which clearly points out what its material is — about the world or one of its aspects without respecting this principle. And we surely cannot have a well-founded physical theory, because respecting principle 6.1 it consists in indicating what type of physical events or individuals or machinery, if any, are beyond the formalism we use, and in the end justify it. We return here to the problem of CRs [p. 3]: without CRs no formalism is in itself a physical theory. It is a fact that the Principle of ontological commitment has always been — at least implicitly — acknowledged by almost all scientists of all times.

For example in classical electrodynamics nobody thought that the potential vector was anything more than a formal device, but surely almost all people thought that electromagnetic waves were something real.

In general, hypothesizing the ontological reality of something has been a great thrust in the progress of physics[11].

Later we shall assume an operational point of view [see chapter 18]. Pragmatism is one of the best philosophical and epistemological expression of such an approach. However, we stress here that a pragmatic or operational approach cannot underestimate the principle of ontological commitment, because without it such an approach becomes too close to a formalistic interpretation. Hence we support an 'ontologically-correct' pragmatism.

Examination of Heisenberg's First Interpretation Now it is clear that the Strong Formal Interpretation does not satisfy principle 6.1. In fact, if we accept the Strong Formal Interpretation, then there is no difference at all between entities that the physicists introduce for the sake of calculation and entities that are believed to be real.

In conclusion the Strong Formal Interpretation cannot be accepted and Heisenberg himself changed his opinions first in the debate with Bohr [chapter 7], and later when proposing a new interpretation [subsection 29.2.2].

6.3 Some Observations Concerning the Structure of the Theory

In the introduction [see p. 3] we assumed the necessity and centrality of an interpretation without justifying this opinion with any intrinsic reasons. Now, on the basis of the discussion of the two previous sections, we can say that the necessity of an *Interpretation of QM* completely derives from the necessity of observing principle 6.1 — it is worth mentioning that also historically, the Copenhagen interpretation starts here. In particular, QM needs a physically coherent interpretation, and this must be a constituent part of QM itself. So that we can say that

$$\text{QM} = \{\text{Basic-QM, QM-Extensions, QM-Foundations, QM-Interpretation}\}$$

[11] For some examples see [BUNGE 1956, 274] . And we shall see [part VIII] the importance of such assumptions, even if wrong, also for QM.

Basic-QM consist of all formalism and theorems or corollaries between horizontal lines, developed in chapter 3. By *QM-Extensions* we understand all the formal extensions of the theory, i.e. the relativistic aspects, quantum optics and several areas which are object of the analysis in parts IV–X. Both Basic-QM and QM-Extensions build the formalism of the theory.

By *QM-Interpretation* we mean the *physical* interpretation of Basic-QM. But it is important to notice that, in order to obtain a physical interpretation, it is necessary to use theorems and principles that are not themselves part of QM, but only of a Meta-theory of QM (= Meta-QM).[12] Part of Meta-QM is, naturally, principle 6.1 itself. Obviously this point is strictly related with the problem of the *foundations* of the theory: which are the basic assumptions of QM? And is there the possibility of formulating a synthetic model from which all aspects of the theory can be derived?

We shall discuss the interpretation in the next two chapters and then in chapter 46. We shall discuss the foundations in the part III of the book and finally in chapter 45.

6.4 Born's Probabilistic Interpretation

As said, from the beginning Born proposed his interpretation [BORN 1926a, 233] in open contraposition to Heisenberg's Strong Formal Interpretation [postulate 6.2: p. 102] and to Schrödinger's Wave Packet interpretation [proposition 6.1: p. 101]. Born's antagonism to the Wave Packet interpretation depends on the experiments related to the Compton effect [p. 26] and on those with the Wilson chamber, where traces of particles were photographed, so that at that time nobody could doubt of the corpuscular nature of microentities[13]. Born accepted the important results of Wave Mechanics [section 3.4], but he believed that the ψ–function was only a mathematical device. Therefore, taking into account of the Probabilistic assumption [postulate 3.5: p. 38], he proposed an interpretation of QM [BORN 1926a, 234] [see also subsection 23.1.3], which can be almost literarily expressed as follows:

Proposition 6.2 (Probabilistic Interpretation) *The wave function determines only the probability that a particle — which brings with itself energy and momentum — takes a path; but no energy and no momentum pertains to the wave.*

In short the Probabilistic interpretation assigns ontological reality only to the particles and not to the waves[14]. But, since there is no way of defining particles' trajectories independently from the probabilistic distribution of positions dictated by the Probabilistic assumption [postulate 3.5: p. 38],[15] QM had, for Born, a *fundamental probabilistic character*. Hence the consequence of the Probabilistic interpretation was the affirmation of the fundamental *indeterminism* of QM [BORN 1926a, 256], which we can synthesize as follows[16]:

Corollary 6.1 (Indeterminism of QM) *QM probabilities do not depend on a partial knowledge but are an intrinsic feature of the QM entities themselves.*

[12]See [*REICHENBACH* 1944, 141–42].

[13]See Born's later reconstruction [BORN 1954, 435]. On that which follows see also [BUNGE 1956, 276].

[14]It is true that Born accepts the language of Bohr/Kramers/Slater regarding the 'Ghost field' [see section 3.2]; but only in the sense of a 'Path-indicator'.

[15]In fact, as we have seen [subsection 3.11.2], path integrals are a means of calculation between different experimental results and not 'paths' in the classical sense. For the conditions of the classicality of trajectories see Eqs. (16.4) and (16.5).

[16]It could also be asked if classical mechanics is as deterministic as it was believed to be [*EARMAN* 1986, 23–54].

The last corollary is fundamental because the Probabilistic interpretation as such does not specify how probabilities are to be understood in QM. One might think that the probabilistic character of QM depends on subjective ignorance (as the supporters of the statistical interpretation, to be analysed below, believe): in this case it would be possible to give QM the structure of classical probability calculus — which, if no additional assumptions are made, is of course contradicted by the Superposition principle [p. 39]. Though the corollary is correct, we do not take it as part of QM because later we shall prove a more powerful theorem which includes this result [theorem 14.6: p. 235; see also chapter 32].

But, note that proposition 6.2 is only a sufficient condition of corollary 6.1, so that it is possible to support the latter (together with the Probabilistic assumption) without assuming the first one.

6.5 The Ignorance and Statistical Interpretations

The two following interpretations, the Ignorance and the Statistical interpretations, stem from a common source but are totally different in character and purpose: in fact from the first the Hidden-Variables theory is derived [see part VIII], and from the second the Many-World interpretation and its variants [see chapters 15 and 25].

6.5.1 Ignorance Interpretation

Einstein's postulates From the Probabilistic assumption above [postulate 3.5] Einstein derived an interpretation exactly opposed the Probabilistic interpretation, in the sense that it rejects corollary 6.1.[17] The assumptions upon which Einstein's interpretation is founded can be formulated as[18]:

Postulate 6.3 (Determined Value Assumption) *Each physical variable has a determinate value in every state of a physical system $\mathcal{S}$ and all indetermination is only due to our approximate knowledge of the state of $\mathcal{S}$;*

and as:

Postulate 6.4 (Faithful Measurement Assumption) *It is always possible to measure a physical variable so that the registered value is an exact counterpart of the real value.*

It is clear that the Faithful Measurement assumption is a consequence of the classical postulate 2.3 [p. 22; see also definition 2.1: p. 22]: in fact only on the assumption of continuity and of an error correction such that the measurement error can be taken down to zero, postulate 6.4 is justifiable. Hence it should be more a corollary than a postulate. But, since the problems of continuity and of error correction do not enter into this discussion as such, we prefer to analyse postulate 6.4 separately.

---

[17]For the history of the problem see [*JAMMER* 1974, 440–43].

[18]For formulations and discussions of the following points see [R. HEALEY 1979] [*HUGHES* 1989, 163] [*REDHEAD* 1987b, 89–90].

Action-at-a-distance Einstein felt particular unsatisfaction with the action-at-a-distance which seems to follow from QM [EINSTEIN 1928].[19] We can imagine the following experiment: a gun fires a particle on a semicircular screen. Before the localization we have a probability distribution (due to the wave function) which assigns values different from zero for all points on the screen. But as soon as the particle is localized in one point of the screen, the probability of reaching the other points falls to zero. So that, if QM is a description of individual processes, it seems that we are faced with action-at-a-distance [for further details see subsection 7.2.1]. Since for Einstein this could not be so, he believed in the possibility of finding a theory which was a description of individual systems, which showed a completely deterministic character and which would have the same relationship with Basic-QM as classical mechanics has with statistical mechanics. Hence Einstein thought that Basic-QM was only an incomplete and hence unsatisfactory theory of microentities and that probabilistic features of QM (which included action-at-a-distance) were to be understood as a subjective imperfect knowledge of an underlying microreality.

Einstein's interpretation Hence he postulated an interpretation which can be synthesized as follows[20]:

Proposition 6.3 (Ignorance Interpretation) *The wave function (or quantum state) provides a description of certain statistical properties of an ensemble of similarly prepared systems, but is not a complete description of an individual system, description which is possible in principle.*

We shall see later the far-reaching consequences of this interpretation [see section 23.1 and chapter 31]. But first we synthesize the Statistical interpretation and in the next section we shall discuss both.

6.5.2 Statistical Interpretation

Later supporters of a Statistical interpretation of QM refuted the Ignorance interpretation, i.e. they did not assume postulate 6.3 and postulate 6.4, and considered the statistical character of the theory as irreducible and not as an approximation to a more fundamental theory — antecedents can be found in Bohr/Slater/Kramers' article of 1924 [see section 3.2]. Proponent of this approach were Blokhintev [*BLOKHINTSEV* 1965] [BLOKHINTSEV 1976] and Ballentine [BALLENTINE 1970].[21] The Statistical interpretation can be formulated as follows:

Proposition 6.4 (Statistical Interpretation) *QM is a statistical theory, i.e. a theory whose objects are statistical ensembles (and not individual entities), but no sound reduction to a deterministic theory is possible.*

The basis of the Statistical interpretation is to be understood from the difficulties originating from the measurement problem: the Projection postulate [postulate 14.1: p. 223] and the so-called reduction of the wave packet proposed by the Standard theory of von Neumann — i.e. the interpretation commonly accepted by almost all physicists [see chapter 14] — seemed untenable

[19] For the history see [*JAMMER* 1974, 115–17]; see also Einstein's reconstruction [EINSTEIN 1953a, 10].

[20] See [EINSTEIN 1949b, 671–72]. For more of Einstein's quotations about the problem of the incompleteness of QM see [PATY 1995, 188–92]. For formulations and analysis see also [BALLENTINE 1970, 360] [BUNGE 1956, 277–78].

[21] For historical reconstruction see [*JAMMER* 1974, 445–46], and [HOME/WHITAKER 1992] for an extensive and recent account.

if applied to individual events and entities[22]. We shall later return to the problem [see chapters 15 and 16].

6.6 Discussion of the Ignorance and Statistical Interpretations

We are here faced with two different questions:

- Is Einstein's Ignorance interpretation correct?
- Is QM concerned only with statistical ensembles even if the Ignorance interpretation fails?

6.6.1 Examination of the Ignorance Interpretation

We begin with the first question. The Ignorance interpretation is not tenable for all mixed states because we know that they are not uniquely decomposable if degenerate or non-orthogonal [see corollary 3.3: p. 46]. It is also possible to put forward a specific **counterexample** [*JAUCH* 1968] [*BUSCH et al.* 1991, 22–24] :

Take a biorthogonal (or polar) (Schmidt) decomposition [see Eq. (14.9) and relative comments] of a system $\mathcal{S} = \mathcal{S}_1 + \mathcal{S}_2$:

$$|\Psi\rangle = \sum \sqrt{w_j}|b_j^1\rangle \otimes |k_j^2\rangle \tag{6.5}$$

Assume now that the reduced density matrix [see Eq. (3.103)] on $\mathcal{H}_2$ gives the mixture: $\sum_j w_j \hat{P}_{b_j^1}$. If we accept the Ignorance Interpretation, then the subsystem $\mathcal{S}_1$ must be in one of the (unknown) states $\hat{P}_{b_j^1}$ with probability w_j. But then, $\mathcal{S}$ is in a state of the form $\hat{P}_{b_j^1 \otimes l_j}$ with probability w_j for some l_j. But then the state $\hat{P}_\Psi$ must represent a *Gemenge* $\{w_j, \hat{P}_{b_j^1 \otimes l_j}\}$, from which it follows that $\hat{P}_{k_j^2} = \hat{P}_{l_j}$, which in general is not the case if $|\Psi\rangle$ is a pure state.

6.6.2 Examination of Statistical Interpretation

A counterexample Turning now to the other question — if QM is concerned only with statistical ensembles even if the Ignorance interpretation fails —, already Schrödinger showed that the Statistical interpretation is inadequate in QM [SCHRÖDINGER 1935a, 487–88].

For example we know that in order to calculate the angular momentum of a particle we must calculate the vectorial product between the vector from a fixed reference point 0 to the origin of the momentum vector and the momentum vector itself. But according to QP [postulates 2.1: p. 19, and 2.2: p. 20], angular momentum is quantized, i.e. it can only have discrete values. So that we must have vector lengths that allow these values. But if we calculate the angular momentum of the same particle from another reference point 0′, we would get different other values which would be (partly) surely impossible.

In conclusion considering statistical ensembles does not help in any way.

---

[22] On this point see also [LANDSBERG/HOME 1987].

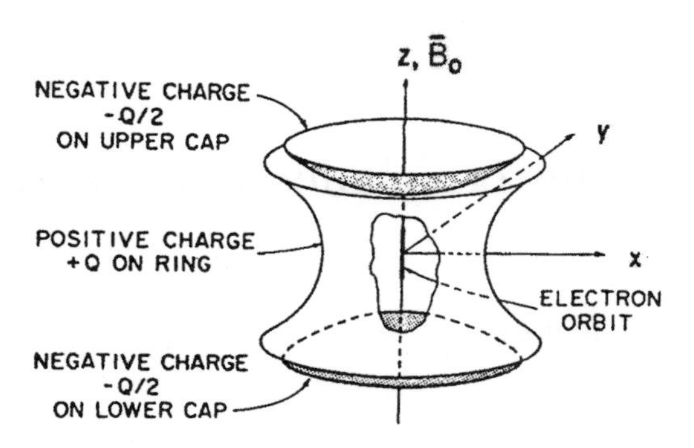

Figure 6.1: The figure shows an ultrahigh vacuum Penning trap, used by Dehmelt and co-workers (in 1973). The simplest motion of an electron in the trap is along its symmetry axis, along a magnetic field (of up to 5 T) line. The axial electric potential well is 5 eV. Each time it comes too close to one of the negatively charged caps, it turns around. The resulting harmonic oscillation took place at 60 MHz in the experiment performed. The electron remains confined in a small cylinder about 30 μm in diameter and 60 μm long. Later (in 1984) van Dyck, Schwinberg and Dehmelt continuously observed an electron so confined for three months — from [DEHMELT 1990, 17].

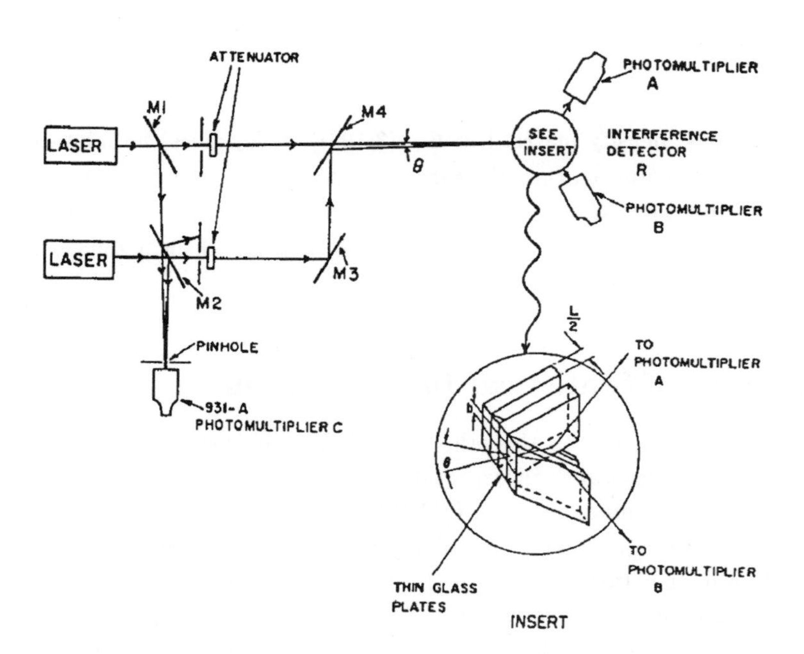

Figure 6.2: Interfer. Exp. 1. The experimental setup. The insert shows the principle of the interference detector. Many sets of correlation coefficients were measured for various angles θ corresponding to various fringe spacings l. The thickness of the glass plates is $L/2$ — from [PFLEEGOR/MANDEL 1967b, 1085].

Confinement of single particles Many of experiments which have been performed in these last years are very useful for judging the Statistical interpretation. From the results of these experiments it is evident that QM has to do also with individual systems and not necessarily with ensembles — the problem will be analysed in more details when we shall speak of quantum non-demolition measurements [chapter 19].

Firstly there are many of experiments, by means of which physicists have successfully confined in some trap an individual electron [WINELAND *et al.* 1973] [see figure 6.1], an individual positron [SCHWINBERG *et al.* 1981], an individual proton [VAN DYCK *et al.* 1986] and antiprotrons [GABRIELSE *et al.* 1986]. [23]

Two-Slit-Exp. 4 But also for photons, there are many of experiments which try to produce interference patterns with a single light quantum. As a conceptual starting point of the analysis of following experiments, we execute another *Gedankenexperiment* (Two-Slit-Exp. 4), which is a variation of the already discussed Feynman's experiments [see p.s 27–28]. We find that, if we decrease the intensity of the source until an average of only one photon at a time in transit between source and screen, we see that, after we have sent some of them, the diffraction pattern still appears, so that we must conclude that diffraction is a property of single photons and does not necessarily involve interference between different photons[24].

Rates	Experimental data	Mean	Standard deviation	Correlation Coefficient
R_1	9 4 9 2 7 5 4 4 5 9 9 2 5 8 3 3 9 4 6 4 4 1 4 6 3	5.08	2.42	- 0.120
R_2	3 5 1 6 3 1 3 2 4 7 4 3 9 5 2 3 6 6 5 4 5 6 6 2 9	4.40	2.13	

Table 6.1: Some examples of measured values of R_1, R_2 in Interfer. Exp. 1 — from [PFLEEGOR/MANDEL 1967b, 1086].

Interfer. Exp. 1 More formally we now discuss a performed Interferometry Experiment (= Interfer. Exp. 1) [PFLEEGOR/MANDEL 1967a] [PFLEEGOR/MANDEL 1967b] [see subsection 5.3.1]. We have [see figure 6.2] two coherent light beams derived from two independent single-mode lasers with aligned polarizations; after having passed through two attenuators, they are superposed with help of (half-silvered) mirrors M1 and M2. A small portion of the unattenuated beams is split off and passes to a phototube C, whose function is to register the difference frequency of the two lasers in form of a beat note, and to activate a gate for the interference detector whenever the beat frequency falls below 50 Kc/s. In this last case, the interference pattern is examined for a 20 μs interval. Since the number of photons registered in 20 μs was about 10, the presence of interference fringes was established by statistical methods (light falling on odd-numbered glass plates is directed toward photomultiplier A, light falling on even-numbered glass plates is directed toward photomultiplier B). When the half fringe spacing coincides with the plate thickness, and the fringe maxima fall on the odd-numbered plates, say, one phototube

[23]For an interesting review see [DEHMELT 1990].

[24]Already Dirac had said that each photon interferes only with itself [*DIRAC* 1930c, 9]. For a review see [GHOSE/HOME 1992]. Also see subsection 41.1.1.

is expected to register nearly all photons and the other one almost none. It is clear that if the number of photons registered in one phototube increases, the number registered by the other one is expected to decrease, provided that the fringe spacing is right for the plates. Thus, if R_1, R_2 are the mean rates at which photons are registered in the two channels, there should be a negative correlation (anticorrelation) between R_1 and R_2 [see table 6.1]. If interference fringes are present, the anticorrelation should be greater when the half-fringe spacing $l/2$ nearly coincides with the thickness of the glass plate $L/2$ [see figure 6.3].

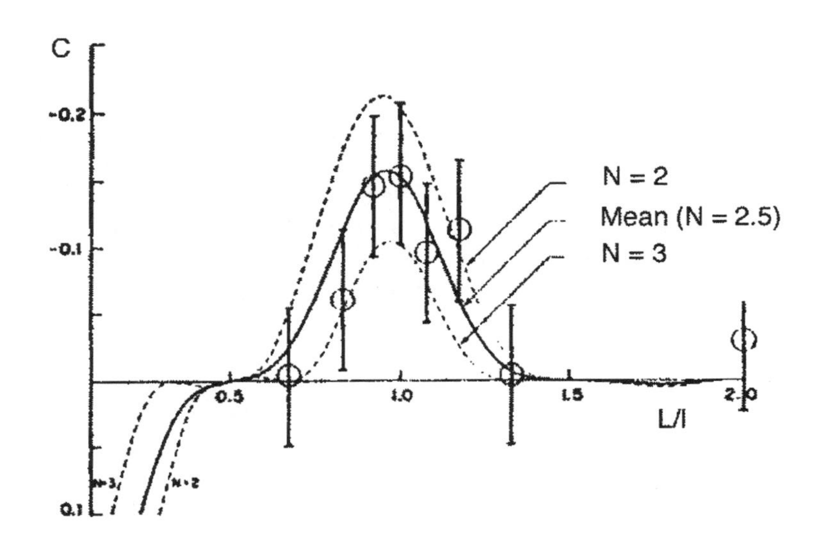

Figure 6.3: Experimental results for the correlation coefficient C, together with theoretical curves for $N = 2, N = 3$ (where N is the number of pairs of illuminated glass plates at the detector) and their mean. One can see that the negative correlation is greatest when $L = l$ — from [PFLEEGOR/MANDEL 1967a, 766].

Interfer. Exp. 2 The interference originated from a single photon was also experimentally tested (Interfer. Exp. 2) by Grangier, Roger and Aspect [GRANGIER *et al.* 1986a] [GRANGIER *et al.* 1986b]. It was shown that a single photon obeys the anticorrelation prediction of QM [see figure 6.4] — it was also shown that the *same* QM system behaves sometimes as a particle and sometimes as a wave [see chapter 30].

Interfer. Exp. 3 Grangier , Roger and Aspect's experiment was further developed by Franson and Potocki [FRANSON/POTOCKI 1988] (= Interfer. Exp. 3). It was a confirmation not only of the correctness of the preceding experiment but also of the non-local nature of a single photon [see part IX and particularly subsection 41.1.1]: they utilized a very long interferometer (45 m) by means of which the absorption or detection of the photon on one arm and the collapse of the wave function on the other are space-like separated events — with no exchange of information — [see figure 6.5]. The results [see figure 6.6] exclude the possibility of solitons, which was advanced by Franson [FRANSON 1982] in the spirit of HV theories [see chapter 32, especially p. 562].

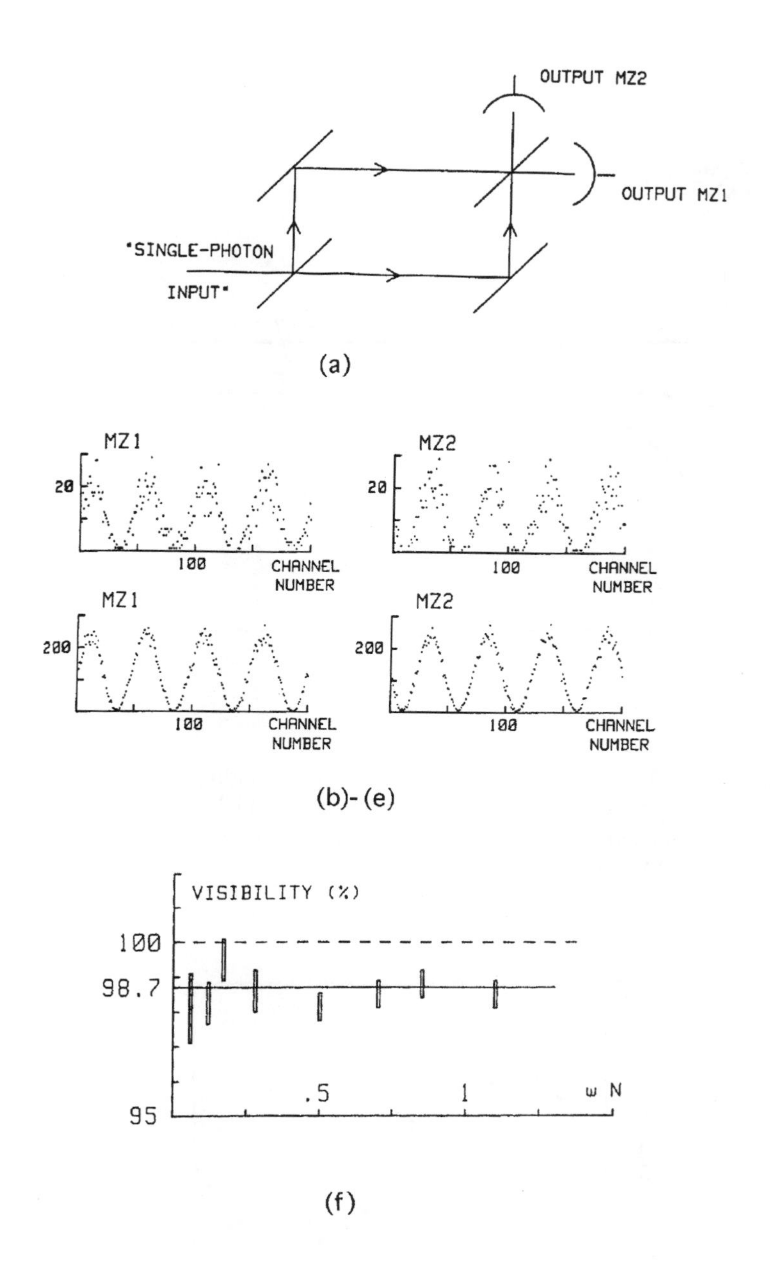

Figure 6.4: Performed Interfer. Exp. 2.
(a) MZ-interferometer for single photon input. Outputs are MZ1 and MZ2.
(b)–(e) Number of counts in outputs MZ1, MZ2 as a function of the path difference δ (one channel corresponds to a $\lambda/50$ variation of δ). (b)–(c) with 1 s counting time for channel. (d)–(e) with 15 s counting time per channel.
(f) Visibility of the fringes as a function of the number of cascades emitted during the gate wN (correction smaller than 0.3%) — from [GRANGIER *et al.* 1986b, 104–105].

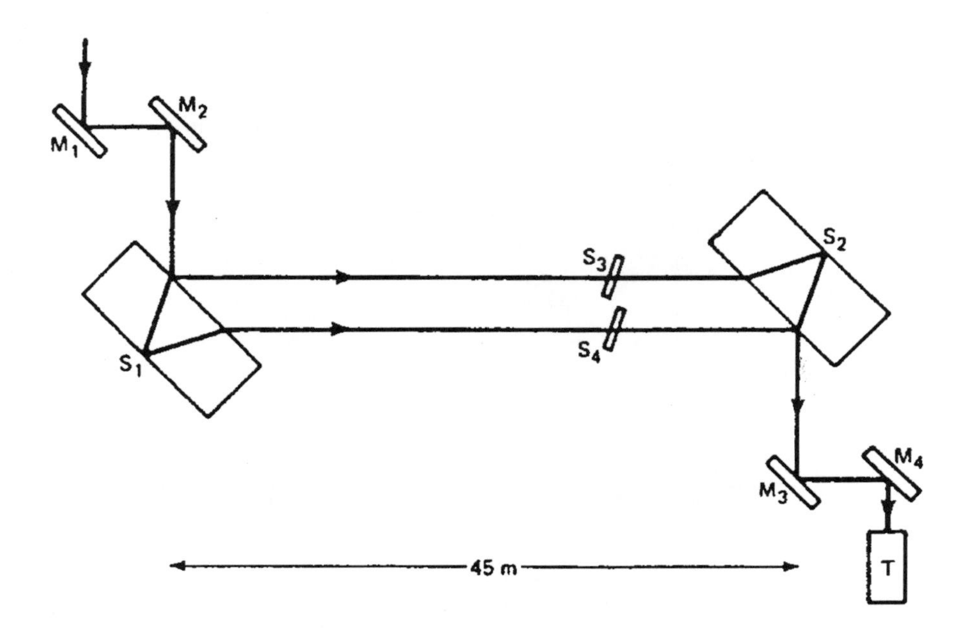

Figure 6.5: Jamin Interferometer and mirrors used to aim the light beams (not drawn to scale). Silica plates S_3, S_4 are used to vary the phase of the two beams, which are recombined by S_2 in order to produce interference. Front surfaces of fused-silica plates S_1, S_2 are half-silvered in order to produce beam splitting — from [FRANSON/POTOCKI 1988, 2512].

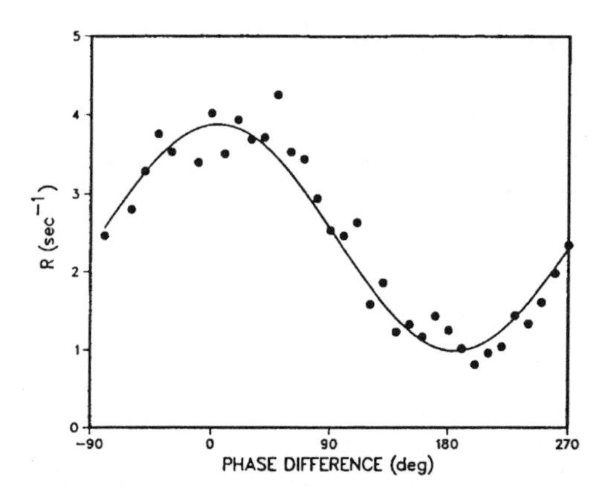

Figure 6.6: Single-photon interference pattern at an average counting rate of 2.4 $\sec^{-1}$. The photon-counting rate R is plotted as a function of the phase difference between the two optical paths — from [FRANSON/POTOCKI 1988, 2513–14].

SPDC Exp. 1 A more complex experiment that has been performed is the following. It is possible to suppress or enhance the emission of entangled photon pairs by means of a Spontaneous Parametric Down Conversion (SPDC Exp. 1) [see figure 6.7; on the SPDC see subsection 5.3.2], which really is a one-photon interference effect. The experimental setup distinguishes itself from normal PDC by mirrors which reflect the pump, the 'idler' photon (i-photon) and the 'signal' photon (s-photon). The experiment consists in seeing the effects on the created i-photon (or on the s-photon) by moving the mirrors (separately or together) [HERZOG *et al.* 1994] [see figures 6.8– 6.10].

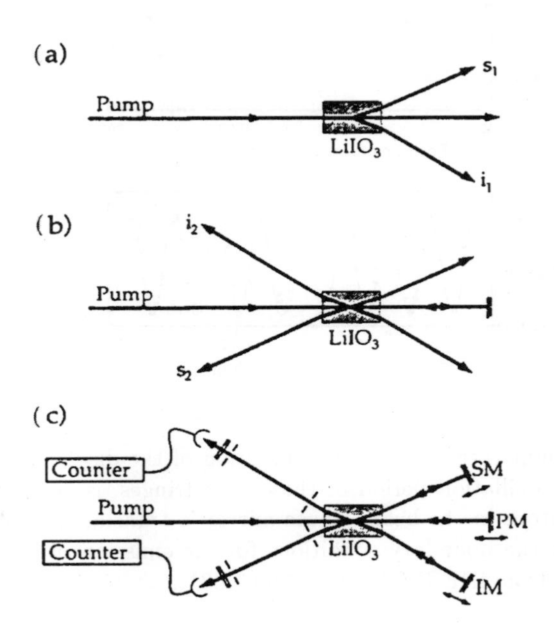

Figure 6.7: Principle of SPDC Exp. 1 setup.
(a) In the standard setup of PDC an entangled photon pair (s= signal, i = idler) may be emitted into the two outgoing modes shown.
(b) The pump is reflected back onto itself in such a way that another possibility for creating the photon pairs arises.
(c) The first mode of the down-converted photons are also reflected back. Thus the two possible ways of creating the photon pair may now interfere — from [HERZOG *et al.* 1994, 629].

Another interesting experiment (which combines the methods for confining particles with the methods for creating interference), was performed by [EICHMANN *et al.* 1993] , by using scattered light from two trapped atoms.

6.6.3 Concluding Remarks

In conclusion, while the Ignorance interpretation is disproved by some formal aspects of the theory (particularly the non-uniqueness of the decomposition of density matrices), the Statistical interpretation is excluded by many of experimental results of these last years.

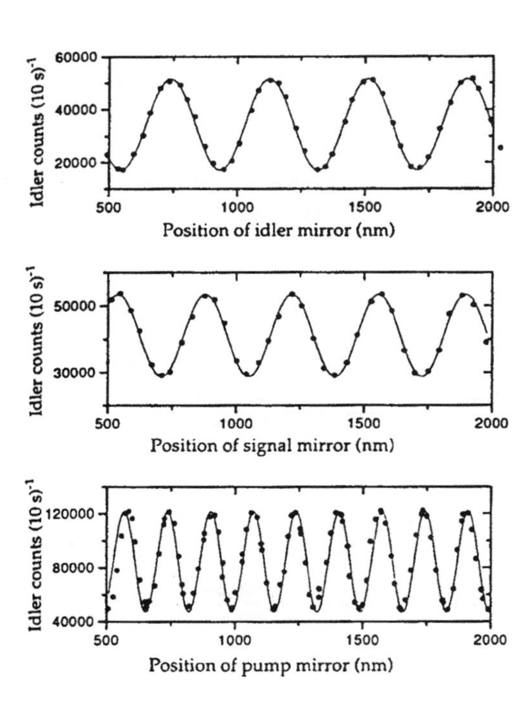

Figure 6.8: The i-photon count rate is shown as a function of the displacement of idler (a), signal (b) and pump mirrors (c). The oscillation period of these idler fringes, is given in all three cases, by half of the wavelength of the photons in the beam whose mirror is translated. The result is clearly that all three mirrors together define the boundary conditions for the emission of i-photons and hence also for the emission of s-photons — from [HERZOG *et al.* 1994, 631].

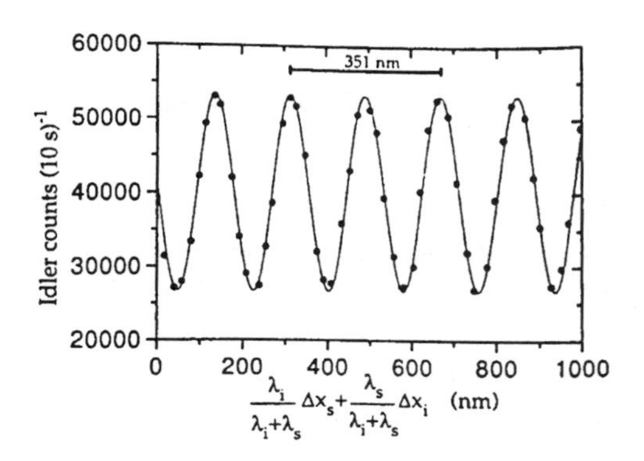

Figure 6.9: Measurement of i-photon count rate when both signal and idler mirrors are translated simultaneously. The abscissa represents a weighted sum of these displacements and it correspond to an equivalent displacement of the pump mirror [see figure 6.8.(c)] — from [HERZOG *et al.* 1994, 631].

Though we have had only recently the means to disprove such interpretations, Heisenberg and Bohr never felt satisfied with them. In fact, as we shall see in the next two chapters, partly as a reaction to such interpretations [see for example section 7.2], they tried another interpretation, the Copenhagen interpretation.

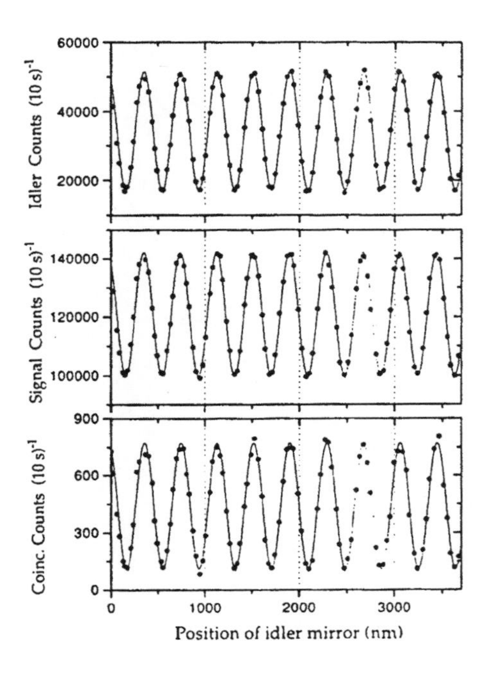

Figure 6.10: Simultaneous measurements of i-photon count rate (a), s-photon count rate (b), and coincidence count rate (c) as a function of the displacement of the idler mirror. The phase agreement of these oscillations clearly shows that the boundary condition controlling the emission of both photons in the entangled state [see definition 3.3: p. 43] can be controlled by changing the position of just one of the mirrors — from [HERZOG *et al.* 1994, 631].

Chapter 7

UNCERTAINTY PRINCIPLE

Contents In this chapter we discuss the first component of the Copenhagen interpretation: the Uncertainty Principle. In section 7.1 we give an account of Heisenberg's formulation with the basic formalism of the uncertainty relations. An initial discussion of the principle follows. In section 7.2 we report Bohr's interpretation of the principle — mainly developed in several discussions with Einstein — which will appear quite different from that of Heisenberg himself. To the latter is devoted section 7.3, which is mainly a discussion of Heisenberg's microscope experiment. Strictly related to the previous point are the different statistical interpretations of the Uncertainty Principle [section 7.4]. In section 7.5 some interpretational problems are discussed, and as a consequence of this examination a new general formulation of the principle is presented [section 7.6].

7.1 Heisenberg's formulation

7.1.1 The Origins of the Problem

Until 1927, excluding Schrödinger's attempt, there was no interpretation at all of QM. Since Heisenberg was unsatisfied with this interpretation, he proposed his Strong Formal interpretation [postulate 6.2: p. 102] based on the belief that it is the theory which decides what we can observe. He thought of Einstein's *dictum* 'it is the theory which decides what can be observed' [HEISENBERG 1977, 5]. But perhaps Einstein intended this in the weaker sense of the Principle of ontological commitment [principle 6.1: p. 103], while, as we have seen, Heisenberg believed that, if the theory denies the strict observability of the trajectory of the particles (position and momentum) and instead regards the 'observed' phenomenon, say in the Wilson chamber, only as a discrete sequence of positions, then some connection could be established between theory and experience. In short Heisenberg hoped that an operational analysis of the concepts of position and momentum could substitute their interpretation[1], similarly to what Einstein already did for the notion of simultaneity in the context of relativity theory[2].

If Heisenberg's article of 1927 is the ending point of the Strong Formal interpretation, it is also the starting point of the Uncertainty Principle (= UP), and hence also the beginning of the Copenhagen interpretation: the new possibilities opened by the UP were understood very quickly by Bohr, so that the Principle was integrated into a more comprehensive interpretation.

[1]And in this sense the Uncertainty Principle can also be understood as the prehistory of an operational approach to QM [see chapter 18].

[2]On this point see [*JAMMER* 1974, 57–58].

Note, nevertheless, that there is a tension here (reflected, as we shall see in the next chapter, in the Complementarity principle itself) between Heisenberg's UP (centered on the problem of uncertainty between conjugate variables), and Bohr' Complementarity (which was mainly meant by Bohr himself as an answer to WP-Dualism).

Perhaps a first stimulus came from Pauli who thought that an electron's orbit could not be exactly described; Pauli was probably also the first to think that the UP and Bohr's Complementarity Principle were two sides of the same interpretation[3]. Heisenberg himself accepted Bohr's interpretation not later than 1930 [*HEISENBERG* 1930b, 135–36].

7.1.2 General formalism and UR1

Heisenberg's argument can be stated in the following form [HEISENBERG 1927, 486–87, 492–94].[4] Take a probability amplitude $\psi(x)$ of the position (in the the one–dimensional case) of Gaussian form:

$$\psi(x) = \frac{1}{\sqrt{2\pi}} e^{-\frac{x^2}{\Delta x^2}}, \tag{7.1}$$

where the *uncertainty* $\Delta\hat{x}$ in respect to $\psi(x)$ was defined by Heisenberg[5] as the *standard deviation* — the square root of the variance [see Eqs. (3.93) and (3.110)] —:

$$\Delta\hat{x} = \langle\psi|[\hat{x} - \langle\psi|\hat{x}|\psi\rangle]^2|\psi\rangle^{\frac{1}{2}}. \tag{7.2}$$

If we now take the Fourier transform of $\psi(x)$ [see Eq. (3.82a)]:

$$\tilde{\psi}(k_x) = \int \psi(x) e^{\imath k_x x} dx, \tag{7.3}$$

which is also of Gaussian form

$$\tilde{\psi}(k_x) = \frac{1}{\sqrt{2\pi}} e^{-\frac{k_x^2}{\Delta k_x^2} - \frac{\imath p'(q-q')}{\hbar}}, \tag{7.4}$$

we finally obtain the product

$$\Delta\hat{x} \cdot \Delta k_x = \frac{1}{2}, \tag{7.5}$$

which, by $\hat{p}_x = \hbar k_x$, becomes the first *Uncertainty Relation* (= UR1):

$$\Delta\hat{x}\Delta\hat{p}_x = \frac{\hbar}{2}, \tag{7.6}$$

which represents the maximum attainable certainty product, i.e. the *minimum uncertainty* product allowed by the UP. Hence we rewrite it as an inequality:

$$\boxed{\Delta\hat{x}\Delta\hat{p}_x \geq \frac{\hbar}{2}} \tag{7.7}$$

Similarly, we can derive the other URs.

[3] See for example [LAURIKAINEN 1985, 274–76].

[4] For further analysis see [RAYSKI/RAYSKI 1977, 16] [*BRAGINSKY/KHALILI* 1992, 5–7, 19–21] [*BUSCH et al.* 1995a, 53–64].

[5] See also [KENNARD 1927].

7.1.3 Other Uncertainty Relations

UR2 The UR for the energy E and time t (= UR2) can be written in a form identical to UR1[6]:

$$\boxed{\Delta t \Delta E \geq \frac{\hbar}{2}} \tag{7.8}$$

But, as we shall see below [subsection 7.5.1], there is a fundamental difference between UR1 and UR2, that is in the fact that the time and the energy (occurring in UR2) are not operators.

UR3 The analysis of the UR between the angular momentum $\hat{\mathbf{J}}$ and the angle ϕ (= UR3) was performed by Judge [JUDGE 1963].[7] A formulation which is isomorphic to UR1–2 is the following:

$$\Delta\phi \Delta\hat{J}_z \geq \frac{\hbar}{2}; \tag{7.9}$$

or, by limiting our analysis to the orbital angular momentum part of $\hat{\mathbf{J}}$:

$$\Delta\hat{L}_z \Delta\phi \geq \frac{\hbar}{2}. \tag{7.10}$$

But since it is possible to show that the maximum value of $\Delta\phi$ is $\pi/3$, this formulation cannot be valid.

Assuming the following function:

$$f(\eta) = \int_{-\pi}^{+\pi} \psi^*(\phi + \eta)\phi^2\psi(\phi + \eta)d\phi, \tag{7.11}$$

one sees that $(\Delta\phi)^2$ is the minimum value of $f(\eta)$. Using the Schwartz inequality one obtains:

$$(\Delta\hat{L}_z)^2 f(\eta) \geq \frac{1}{4}\hbar^2[1 - 2\pi\psi^*(\eta + \pi)\psi(\eta + \pi)]^2. \tag{7.12}$$

From this it follows that:

$$\boxed{\Delta\hat{L}_z \frac{\Delta\phi}{1 - 3(\Delta\phi)^2/\pi^2} \geq 0.15\hbar} \tag{7.13}$$

The factor 0.15 is due to the particular deduction, but with a more general one it can be replaced by $\frac{1}{2}$. Note that the angle as well is not represented by an operator.

UR4 Another UR, of particular importance in Q-Optics [see particularly subsection 5.1.1], is between the number of photons, expressed by the number operator $\hat{N}$ [see Eq. (4.28)], and the phase φ (= UR4) [LOUISELL 1963].[8] Note that the phase as well is not represented by an operator. The difficulty is the following: we can derive the following formula (where $|n\rangle, |m\rangle$ are some states)

$$\langle m|\varphi|n\rangle = \frac{\delta_{mn}}{(n - m)}, \tag{7.14}$$

[6] For more details see [*BUSCH et al.* 1995a, 77–82].

[7] See also [JUDGE/J. LEWIS 1963][LÉVY-LEBLOND 1976]. For further analysis see [*BUSCH et al.* 1995a, 65–70].

[8] For further analysis see [*HOLEVO* 1982, 137–42] [*BUSCH et al.* 1995a, 82–94].

from which one can see that, if the eigenvalues of $\hat{N}$ are integers, then the RHS of the Eq. (7.14) is undefined.

So one can propose the following form:

$$[\hat{N}, f(\varphi)] = -\frac{\partial f(\varphi)}{\partial \varphi}, \tag{7.15}$$

and restrict $f(\varphi)$ to periodic functions of period 2π. Hence one can take $\cos\varphi, \sin\varphi$ as Hermitian operators and write:

$$[\hat{N}, \cos\varphi] = \imath \sin\varphi, \qquad [\hat{N}, \sin\varphi] = -\imath \cos\varphi, \tag{7.16}$$

so that UR4 becomes:

$$\boxed{\Delta\hat{N}\Delta\cos\varphi \geq \frac{1}{2}\left|\langle\sin\varphi\rangle\right|, \quad \Delta\hat{N}\Delta\sin\varphi \geq \frac{1}{2}\left|\langle\cos\varphi\rangle\right|} \tag{7.17}$$

UR5 It is possible to study another uncertainty relation, that between spin and spin phase (= UR5), which makes use of POVM [see section 18.1.2] [*BUSCH et al.* 1995a, 70–73]. But we shall not enter into the problem here.

7.1.4 Formulation of the Uncertainty Principle and First Problems

In conclusion it is possible to formulate the UP itself as follows:

Principle 7.1 (UP) *For each pair of conjugate variables, the product of the uncertainties of both cannot be less than a fixed value given by the respective UR.*

Stated in this general form UP is absolutely correct. However, the greatest problem, as we shall see in the following analysis, is how to interpret the *uncertainty*, which Heisenberg identified with the standard deviation [see Eq. (7.2)].

Another problem is the following: it is possible to reduce UP to the Superposition principle [3.3: p. 39] or *vice versa*, and what are the relationships between the two? The two principles are not the same[9]: The Superposition principle says that, if two states are possible states of a system, then also each linear combination of both is possible, while the content of UP is given by the URs, which establish a certain minimum uncertainty value, which is not mentioned in the Superposition principle. There are also situations in which there is no superposition but in which the URs are certainly valid. Take, for example, the ground state of a harmonic oscillator: it is an eigenstate of the energy but for it the uncertainty $\Delta\hat{\mathbf{q}}\Delta\hat{\mathbf{p}}$ is exactly at the minimum value $\hbar/2$; in the excited states the uncertainty is greater. But it can be shown that in certain superpositions of all states of a harmonic oscillator, in the coherent states [see subsection 5.1.5], the uncertainty has the minimum value again, though only one of the components of the superposition is in a state of minimal uncertainty (these are the states which best resemble classical states). In general

[9]The following examination owes very much to a discussion with Dr. Fortunato.

we shall see that some form of uncertainty between dynamic and kinematic variables can also be assumed in the classical case [see section 33.1] but not so with superposition.

Therefore the two are independent principles, in the sense that they have different contents and different ranges (UP is more extended than the superposition principle and it presupposes this one in its formulation). Below we will also discuss the relationship between UP and CCRs [see section 7.5].

7.2 Bohr's Interpretation of the Principle

Before we enter into some technical difficulties, let us briefly report the discussion between Bohr and Einstein[10], which on one hand is the first attempt at integrating UP in what will become the Copenhagen interpretation [chapter 8], and which, on the other hand, will be particularly useful for later discussion [see subsection 7.5.2].

7.2.1 First discussion with Einstein

Einstein's objections In october 1927 at the Fifth Solvay Conference in Bruxelles, Einstein [EINSTEIN 1928] objected that, if a particle goes through a slit in a wall and then is registered on a photographic plate in a point A, instantaneously the probability that it is in another point B is zero, and this is against the laws of wave propagation [see figure 7.1].

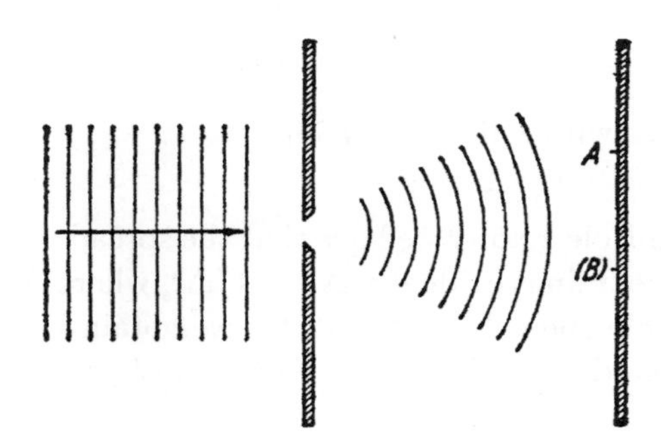

Figure 7.1: The action-at-distance thought experiment — from [BOHR 1949, 212].

We should then suppose some action-at-a-distance. As we have seen, this was the origin of the Ignorance interpretation [subsection 6.5.1] and it also is, indirectly, of the second Heisenberg interpretation [see subsection 29.2.2]. Moreover, the principle of contact forces — i.e. that forces act only over small distances — cannot be adequately formulated in configuration space. The second problem is that of the reduction of the wave packet, and this will be analysed further [see chapters 14 and 17]. To the first problem, Bohr answered by deducing the URs and hence by demonstrating the impossibility of applying a classical deterministic theory to the microworld[11].

[10]For Einstein's positions see [PAIS 1972]. See also [SHIMONY 1988].

[11]We mainly follow Bohr's later reconstruction [BOHR 1949, 212–14]; see also [*JAMMER* 1974, 115–17, 127–29] [*MURDOCH* 1987, 155–57][HILGEVOORD/UFFINK 1988, 93] for further analysis.

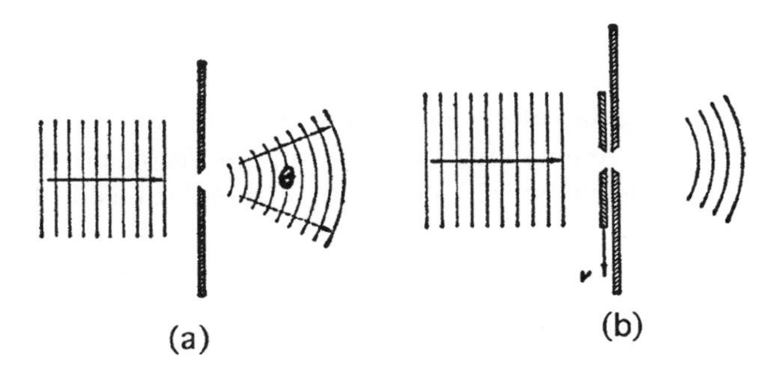

Figure 7.2: (a) One-Hole-Gedanken-Exp. 1.
(b) One-Hole-Gedanken-Exp. 2 — from [BOHR 1949, 214].

One-Hole-Exp. 1 Bohr imagined two experiments, one with a hole without a shutter (One-Hole-Exp. 1) and another with a hole with a shutter (One-Hole-Exp. 2) [see figure 7.2]:

In One-Hole-Exp. 1 we produce a spherical train of waves with a definite angular aperture θ and in One-Hole-Exp. 2 with a limited radial extension. Hence the description of One-Hole-Exp. 1 involves a certain latitude Δp in the momentum component of the particle parallel to the diaphragm, and, in that of One-Hole-Exp. 2 an additional latitude ΔE of kinetic energy. Since a measure of $\Delta\hat{q}$ (the location of the particle in the plane of the diaphragm) is given by the radius l of the hole and since $\theta \simeq (1/lk)$, we obtain, on the basis of $\hat{p} = \hbar k$ [see Eq. (3.2)]:

$$\Delta\hat{p} \simeq \theta\hat{p} \simeq \frac{\hbar}{\Delta\hat{q}}, \tag{7.18}$$

and this result is in accordance with UR1 [Eq. (7.7)].

One-Hole-Exp. 2 For One-Hole-Exp. 2 we have that the spread of frequencies of the harmonic components in the limited wave train is evidently $\Delta\nu \simeq 1/\Delta t$, where Δt is the time interval during which the shutter leaves the hole open. From Eq. (2.11) we obtain UR2 [Eq. (7.8)]. In conclusion the URs are completely respected, and QM presents an intrinsic uncertainty as stated by the UP.

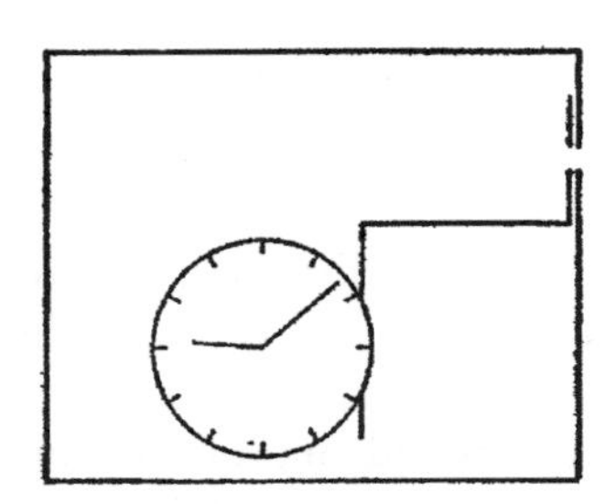

Figure 7.3: Einstein's thought experiment for violating the limits imposed by UR2 — from [BOHR 1949, 225].

7.2.2 Second discussion

At the Sixth Solvay Conference in 1930 Bohr had a new discussion with Einstein [BOHR 1949, 224–28].[12]

Einstein's Experiment Einstein proposed a device [see figure 7.3] consisting of a box with a hole in one of its sides with a shutter moved by means of a clock within the box — Einstein's Experiment[13].

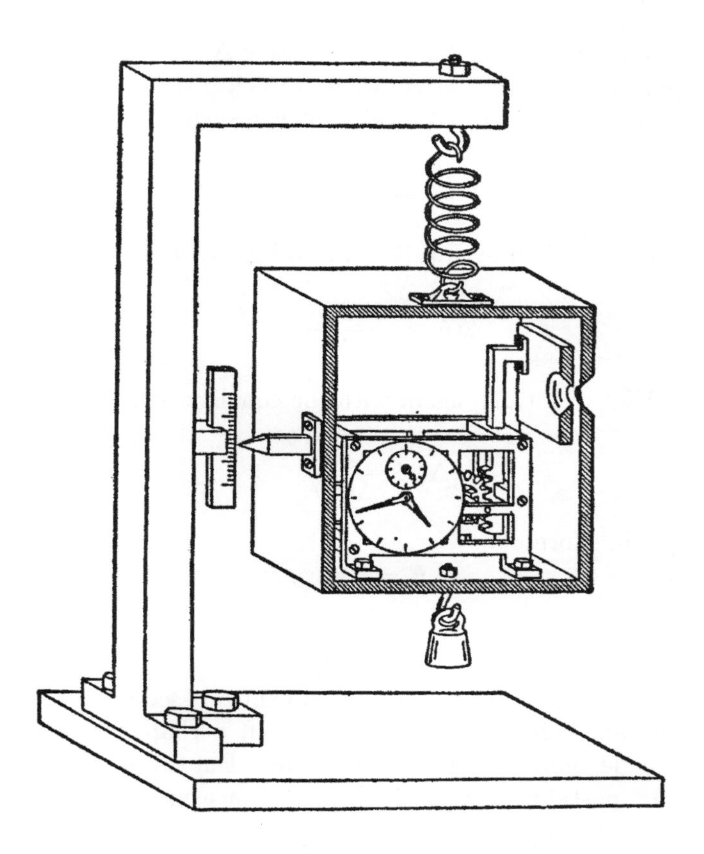

Figure 7.4: A graphical representation of the apparatus proposed by Bohr to test UR2 — from [BOHR 1949, 227].

If in its initial state the box contains a certain amount of radiation and the clock is set to open the shutter after a chosen short interval of time, it could be achieved that a single photon is released through the hole at a moment which is known as exactly as desired. Moreover, if we weigh the box before and after this event, we could measure the energy of the photon as exactly as we want, and this is against UR2.

[12]For historical reconstruction and further analysis see [*JAMMER* 1974, 132–36] [*MURDOCH* 1987, 157–61].
[13]See [TREDER 1970].

Bohr's answer The essential point of Bohr's answer [see figure 7.4] was that any determination of the position of the balance's pointer is given with an accuracy $\Delta\hat{x}$, which will involve a latitude $\Delta\hat{p}_x$ in the control of box' momentum according to UR1 — hence it is clear that Bohr already assumes UR1, by means of which he tries to refute Einstein's criticism to UR2.

This latitude must be smaller than the total momentum which, during the whole interval δt of the balancing procedure, can be imparted by the gravitational field to a body with mass Δm:

$$\Delta\hat{p}_x < \delta t \cdot g \cdot \Delta m, \tag{7.19}$$

where g is the gravity constant. The greater the accuracy of the reading $\hat{x}$ of the pointer, the longer must the balancing interval δt be if a given accuracy Δm of the weight is to be obtained. But according to the general relativity theory, when a clock is displaced in the direction of the gravitational force by an amount $\Delta\hat{x}$, its rate will change in such a way that its reading in δt will differ by an amount Δt given by:

$$\frac{\Delta t}{\delta t} = \frac{1}{c^2} g \Delta\hat{x}. \tag{7.20}$$

By substituting the value of δt given by Eq. (7.20) into Eq. (7.19) we obtain:

$$\Delta\hat{p}_x < \frac{\Delta t \Delta m c^2}{\Delta\hat{x}}; \tag{7.21}$$

and finally, by applying UR1 [Eq. (7.7)] again with the equality sign we obtain:

$$\Delta t > \frac{\hbar}{\Delta m c^2}. \tag{7.22}$$

This, together with Einstein's formula

$$E = mc^2, \tag{7.23}$$

gives UR2 [Eq. (7.8)].

Some considerations It is evident that Bohr's argument is based upon a quantum mechanical interpretation of the pointer. This is not only possible but perhaps necessary because a device suitable to measuring a single photon must obey QM's laws. But this is in clear contradiction with what Bohr always asserts [postulate 8.2: p. 146]: that a measurement must involve apparatus that obey classical physics [see subsection 8.5.3].

It is also possible to ask if there is no way of transforming the quantum mechanical apparatus into a macroscopic one with the effect (as we shall see) of eliminating the uncertainty of the first. In that case one does not use UR1 and hence one cannot derive UR2 following Bohr's proof[14].

7.3 Heisenberg's Interpretation: The Microscope Experiment

Between 1927 and 1930 the URs were mostly proved. Nevertheless their *status* was still very problematic. In 1930 Heisenberg tried a first interpretational answer.

[14]Nevertheless Popper and Agassi's criticism, that QM contradicts Newton's gravitation theory, is surely unsound [*JAMMER* 1974, 137]. See also [*DRIESCHNER* 1979, 150–51].

Heisenberg's assumption Heisenberg considered the URs as experimental consequences of the measuring processes. If this were right, UP would express a type of disturbance originated from the interaction between some apparatus $\mathcal{A}$ and some system $\mathcal{S}$. This is the Heisenberg Assumption [HEISENBERG 1927, 483, 503], which can be formulated:

Postulate 7.1 (Heisenberg Assumption) *The statistical character of QM comes from the observation and not from the definition.*

Microscope experiment We shall show now that the observation or the measuring process has nothing to do with the foundation of UP as such. The idea that UP is fundamentally related to the measuring process — which became a general belief which can be found in almost all textbooks and articles[15] — derives form the original proposal of the Heisenberg microscope experiment [*HEISENBERG* 1930a, 15–19] [*HEISENBERG* 1930b, 125–26].[16] The idea of the experiment can be stated as follows[17]. We wish to measure the position x_1 of a body of mass m. To do so one can attach to this body a stick with a diameter of the order of the wavelength of the light used to detect the body[18]. If we know in advance the approximate position of m, we can arrange a lens and a photographic plate [see figure 7.5].

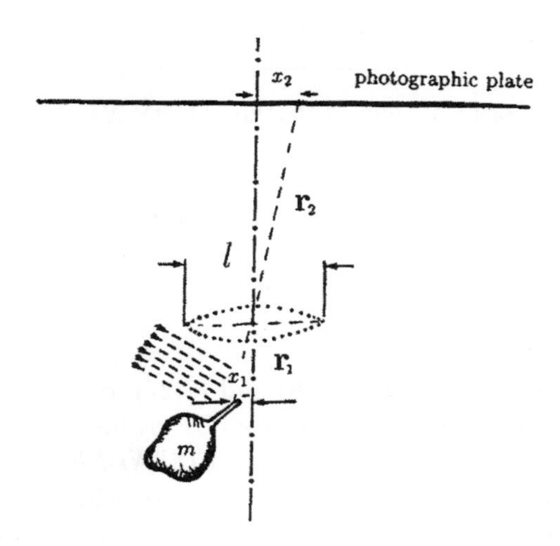

Figure 7.5: The Heisenberg microscope *Gedankenexperiment* — from [*BRAGINSKY/KHALILI* 1992, 9].

The stick must be close to the focal plane of lens, and the optical amplification factor will be approximately r_2/r_1, where r_1 is the focal length. We send a stream of photons from the side and wait for an individual photon to be scattered by the stick and pass through the lens' aperture l, impinge on the photographic plate and produce a small seed of silver in x_2. The transverse

[15]Take for example [*SCHIFF* 1955, 9–11].

[16]See also [BOHR 1928, 151–52].

[17]Formalism and analysis in [*BRAGINSKY/KHALILI* 1992, 8–10].

[18]This can be done with the actual technology.

position of the silver seed, relative to the lens' optical axis can be determined to an accuracy much better than an optical wavelength. From x_2 — due to the geometry $x_1/r_1 = -x_2/r_2$ — we infer the transverse position $x_1 = -x_2 r_1/r_2$ of the stick. But the photon has wave properties [see subsection 6.6.2], and so we cannot claim that it was scattered exactly in x_1. Hence scattering may have occurred everywhere with equal probabilities within a distance

$$(\Delta x)_{\mathrm{M}} \simeq \frac{1}{6}\lambda\frac{r_1}{l} \tag{7.24}$$

of the localization x_1, where $(\Delta x)_{\mathrm{M}}$ is the error of the measured position (M stands for 'measured') and λ is the wavelength of the photon. Now because the photon has mechanical momentum $p_\lambda = \hbar\omega/c$, and because it passed through l, it must have given the stick a random momentum in the x–direction with unknown sign and magnitude of an order such that we have an uncertainty $(\Delta p_x)_{\mathrm{P}}$ (where subscript 'P' stay for 'perturbation')

$$(\Delta p_x)_{\mathrm{P}} \geq \frac{\hbar\omega}{c}\frac{l}{2r_1} \tag{7.25}$$

for $l/r_1 \ll 1$. From Eqs. (7.24) and (7.25) we obtain the following Error/Perturbation UR:

$$\boxed{(\Delta x)_{\mathrm{M}}(\Delta p_x)_{\mathrm{P}} \geq \frac{\hbar}{2}} \tag{7.26}$$

Von Neumann [*VON NEUMANN* 1932, 125–130] [*VON NEUMANN* 1955, 237-47] inferred the inverse relation from a similar thought experiment:

$$\boxed{(\Delta x)_{\mathrm{P}}(\Delta p_x)_{\mathrm{M}} \geq \frac{\hbar}{2}} \tag{7.27}$$

Examination of the experiment Eqs. (7.26) and (7.27) appear to be the same as Eq. (7.7) and Heisenberg thought to have experimentally proved UP by means of his proposed experiment[19]. Later Heisenberg returned to the problem claiming [*HEISENBERG* 1952, 16] that the only place containing uncertainties is the 'dividing line' between the field of classical laws and the field of the laws of quantum theory, where we must understand the effect of macroscopic means of observation upon a microsystem[20]. But Eq. (7.7) expresses the fact that a quantum system — independently from its prehistory, i.e. from its preparation [see chapter 13] — cannot have precise defined values for its position and momentum simultaneously, whereas in Eq. (7.26) and (7.27) an error figures which is due to a perturbation caused by measurement interaction.

Uncertainty and definition That this is the case, is clear from the possibility of gaining information without interaction [RENNINGER 1960, 419] [see section 20.1], which, however, does not jeopardize UP's validity. One can in general assert that not only does UP not derive from the disturbance in measurement, but also that it is UP itself which determines the minimal perturbation which is admitted [*BRAGINSKY/KHALILI* 1992, 26]. This is exactly what a

[19]A belief which is sometimes shared also today: see for example [MARTENS/DE MUYNCK 1992].
[20]See also [*DIRAC* 1930c, 3].

statistical formulation of UP is about [see section 7.4].[21] We can therefore prove the following corollary of UP [see subsection 2.2.1]:

Corollary 7.1 (Uncertainty and Definition) *The UP not only affects the observation of a QM system but also its definition, in the sense that, unlike the classical case, no complete definition is possible.*

The above corollary represents with respect to URs or UP what corollary 3.1 [p. 32] represents with respect to the CCRs.[22]

7.4 Statistical Formulations of the Uncertainty Principle

From the beginning of the theory one has tried to extend, on a formal level, the original formulation of UP with statistical means. But the first attempts were conditioned by the identification, done by Heisenberg himself, between uncertainty and standard deviation [see subsection 7.1.2]. Therefore, in the end they show an interpretational problem which underlies the whole discussion.

I Statistical UR The first essay with a pure statistical formulation of UP was written by Robertson [ROBERTSON 1929].[23] He defined the *uncertainty* $\Delta\hat{O}$ of an arbitrary observable $\hat{O}$ as the square root of the statistical deviation from $\hat{O}$ [Eq. (3.110)], and for two observables $\hat{O}, \hat{O}'$ in state $|\psi\rangle$ he wrote following a Statistical Uncertainty Relation (I Statistical UR):

$$(\Delta_\psi \hat{O}) \cdot (\Delta_\psi \hat{O}') \geq \frac{1}{2} |\langle \psi | [\hat{O}, \hat{O}'] | \psi \rangle|, \tag{7.28}$$

where by Δ_ψ we mean the standard deviation in the state $|\psi\rangle$.

II Statistical UR Schrödinger [SCHRÖDINGER 1930a, 349–51]formulated another statistical expression of UP[24]. He realized that I Statistical UR can be strengthened in (II Statistical UR):

$$\left(\Delta_\psi \hat{O}\right)^2 \left(\Delta_\psi \hat{O}'\right)^2 \geq \left| \left\langle \frac{1}{2} [\hat{O}, \hat{O}'] \right\rangle \right|_\psi^2 + \left(\frac{1}{2} \left\langle \hat{O}\hat{O}' + \hat{O}'\hat{O} \right\rangle - \left\langle \hat{O} \right\rangle \left\langle \hat{O}' \right\rangle \right)_\psi^2 . \tag{7.29}$$

The last squared term normally vanishes in all optimal simultaneous measurements.

---

[21]On this point see [*HOLEVO* 1982, 70]. Heisenberg's experiment was later reexamined by Bohr [BOHR 1949, 202] and Wigner [WIGNER 1983a, 263–64].

[22]On this point also see corollary 26.1 [p. 446] and comments.

[23]There is an extensive literature on the subject: See for example [DITCHBURN 1930] [*JAMMER* 1974, 72] [WIGNER 1983a, 265–66] [*PITOWSKY* 1989b, 56] [GUDDER 1977, 249] [GUDDER 1979, 326] [*BUSCH et al.* 1995a, 134–39]. Popper tried [POPPER 1967] [*POPPER* 1982, 53–60] to interpret the URs as essentially statistical, which will be disproved by the following.

[24]See also [*JAMMER* 1974, 72–73] and [BASS 1992, 4–5].

III Statistical UR Another formulation using variance [*BELTRAMETTI/CASSINELLI* 1981, 24–26] can be the following:

$$\left(\Delta_\psi \hat{O}\right)^2 \left(\Delta_\psi \hat{O}'\right)^2 \geq \frac{1}{4} \left|\langle \psi | [\hat{O}, \hat{O}'] \psi \rangle\right|^2 . \tag{7.30}$$

But the problem with this Eq. is that while the LHS is well defined, the RHS need not be. Hence the following formulation is better (III Statistical UR):

$$\left(\Delta_\psi \hat{O}\right)^2 \left(\Delta_\psi \hat{O}'\right)^2 \geq \frac{1}{4} \left|\langle \hat{O}\psi | \hat{O}'\psi \rangle - \langle \hat{O}'\psi | \hat{O}\psi \rangle\right|^2 . \tag{7.31}$$

However, Eq. (7.31) holds only under some general conditions:

- the two observables do not commute;
- neither is bounded[25]: this is exactly the point which creates asymmetry between CCRs [Eq. (3.14)] and URs;
- neither has a point spectrum or, if one does, its eigenvectors lie outside the domain of the other[26].

7.5 Some Problems

7.5.1 The Problem of Operators

General statement We have seen that the relationships between the UP and the CCRs [Eq. (3.14)] are not evident at all. In the CCRs the observables are treated as operators. The problem is more evident in the analysed statistical formulations because they are about generic non commuting operators. And — as we shall see — it was known from the beginning that there are observables in Basic-QM which are not associated with operators: energy, time and angle observables; and, in Relativistic-QM a position operator for the photon does not exist. The first attempts at solving the problem were inevitably affected by the confusion between UP and the measurement process, a consequence of Heisenberg's interpretation of UP. The object of discussions was, in particular, UR2.

Discussion of UR2 Landau and Peierls [Lev LANDAU/PEIERLS 1931, 467] pointed out[27] that in UR2 must be taken in account the difference between the value of energy obtained as the result of a (predictable) measurement and the value of the energy of the state of the system after the measurement: the state of the system after the measurement is not necessarily identical with the state associated to the measurement's result, and the difference causes an uncertainty of the order of $\hbar/\Delta t$, so that in time Δt no measurement can be performed for which the energy uncertainty in *both* states is less than $\hbar/\Delta t$. Practically, after a time Δt only transitions for which we have

$$|E_{\mathcal{S}_i} - E_{\mathcal{S}_f} + E_{\mathcal{A}_i} - E_{\mathcal{A}_f}| \leq \frac{\hbar}{\Delta t} \tag{7.32}$$

[25]On this point see also [*HOLEVO* 1982, 106–107].

[26]For other formulations see also [LAHTI/MACZYŃSKI 1987] [LÉVY-LEBLOND 1986].

[27]For reconstruction and analysis see also [*JAMMER* 1974, 143–44] [D. PARK 1984].

are of importance (where the four terms on the left of inequality are, respectively, the energy of the observed system $\mathcal{S}$ before and after measurement, and the energy of the apparatus $\mathcal{A}$ before and after measurement). Another way of stating the problem is to say [AHARONOV/BOHM 1961, 716] that energy measurements carried out in short periods of time are not reproducible.

The problem here is that the concept of 'simultaneous measurability' of time and energy has no sense[28]. In fact in order to perform suitable substitutions we should speak of the impossibility of determining the time when a particle passes a fixed point X with an uncertainty smaller than $\hbar/\Delta E$.

Mandelstam/Tamm inequality Mandelstam and Tamm [MANDELSTAM/TAMM 1945] showed that, assuming that the energy of an isolated QM system in a *nonstationary state* has no definite and constant value [QP2: postulate 2.2: p. 20], then only the probability of obtaining in a measurement a specified energy value is constant in time. In a *stationary state* the energy is exactly determined but, as a consequence, the distribution functions of all dynamic variables are constant in time[29].

Fock and Krylov [FOCK/KRYLOV 1947] objected that Mandelstamm/Tamm's calculus had only statistical significance and cannot be applied to individual measurements. We should therefore distinguish between time intervals defined by the wave function — which only have *statistical* significance — and time intervals representing the actual duration of an *individual* measurement. Though with a certain disagreement with Landau and Peierls, Mandelstamm and Tamm also thought that the energy cannot be measured in a time which violates UP. From all this discussion a new statistical formulation of UP can be derived which overcomes the restriction resulting by employing operators and which expresses the spirit of Mandelstamm/Tamm without forgetting the criticism of Krylov and Fock[30]. We name following inequality the *Mandelstamm/Tamm inequality*:

$$\boxed{(\Delta_t\hat{H})^2(\Delta_t\xi)^2 \geq \frac{1}{4}\left|\frac{d}{dt}\langle\xi\rangle_t\right|^2} \tag{7.33}$$

where ξ is an arbitrary observable (not necessarily represented by an operator), Δ_t is the uncertainty at time (interval) t and where we have

$$\frac{d}{dt}\langle\xi\rangle_t = 2\mathrm{Im}\langle\hat{H}e^{-\imath\hat{H}t}\psi|\xi e^{-\imath\hat{H}t}\psi\rangle \tag{7.34}$$

for some state $|\psi\rangle$.[31] Obviously, another problem is if it is possible to construct a time operator in QM. We shall discuss the point later [see section 10.2].

7.5.2 The Problem of the Standard Deviation

One-Hole-Exp. 3 But, assuming this reformulation, are we authorized to establish an equivalence between uncertainty and standard deviation? — and we know that this is a characteristic feature of the statistical interpretation as such [HOME/WHITAKER 1992, 269–70] [see subsection 6.5.2]. The problem, as pointed out by Hilgevoord and Uffink [UFFINK/HILGEVOORD 1985, 925–29] [HILGEVOORD/UFFINK 1988, 95–97, 101], is that the standard deviation is a good

[28]See [RAYSKI/RAYSKI 1977, 13–17] for analysis.

[29]See also [*JAMMER* 1974, 146–47] [*MESSIAH* 1961] [AHARONOV/BOHM 1961, 715].

[30]Formulations and analysis in [*HOLEVO* 1982, 69–70, 105–106, 160] [*DAVIES* 1976, 43].

[31]Further discussions of this point can be found in [PARTOVI/BLANKENBECLER 1986a].

measure of the spread in Gaussian distributions but not of spreads or width of other distributions. To see this consider the following One-Hole Experiment (= One-Hole-Exp. 3), a theoretical reformulation of Bohr's One-Hole-Exp. 1 [p. 122]. Suppose we have a slit of width $2l$. We take a one-dimensional wave function

$$\psi(x) = (2l)^{-\frac{1}{2}} \tag{7.35}$$

where $|x| \leq l$ and $\psi(x) = 0$ elsewhere. The Fourier transform is: $\tilde{\psi}(k) = (2\pi)^{-1/2} \int \psi(x)e^{-ikx}dx$. By inserting Eq. (7.35) into this, we get:

$$\tilde{\psi}(k) = \left(\frac{l}{\pi}\right)^{\frac{1}{2}} \frac{\sin(lk)}{lk} \tag{7.36}$$

Now, by calculating the standard deviation of x, k, we find that Δx is of order l, but Δk is not of order l^{-1}. From the definition of standard deviation we obtain:

$$(\Delta k)^2 = \int k^2 |\tilde{\psi}(k)|^2 dk \tag{7.37}$$

which diverges. This is due to the fact that, because of the factor k^2, the standard deviation assigns an increasingly heavy weight to distant parts of the probability distribution: the standard deviation measures the width of probability distributions by its 'tails' and not by its 'bulk' [see figure 7.6]. In conclusion it is absurd to restrict UP to a formulation in which only standard deviation or variance appear[32].

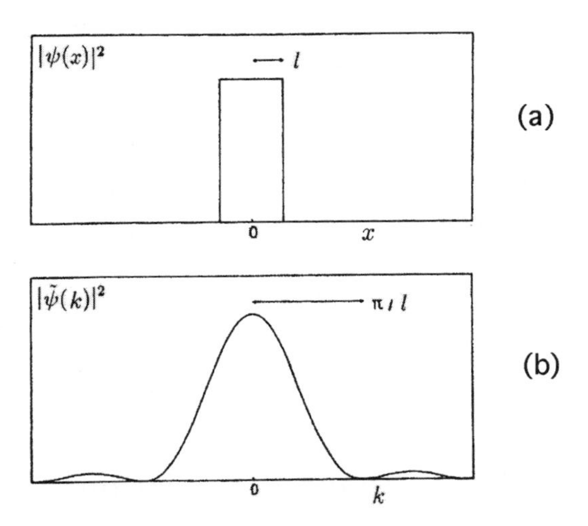

Figure 7.6: The (a) position and (b) momentum distributions for the diffraction at a slit (One-Hole-Exp. 3) — from [HILGEVOORD/UFFINK 1988, 97].

What is very interesting in all Bohr's *Gedankenexperimenten* [see section 7.2] is that there is absolutely no mention of the standard deviation or of the variance [HILGEVOORD/UFFINK 1988, 101, 103, 109].

[32] A very interesting concrete model which proves the distinction between uncertainty and standard deviation is to be found in [LANDSBERG 1990].

Statistical and quantum uncertainties Similar theoretical considerations have been formulated by Cini and Serva [CINI/SERVA 1990] [CINI/SERVA 1992, 323] [CINI 1999]. They distinguish between a statistical or subjective uncertainty — which is best expressed by the standard deviation — and a quantum uncertainty. Hence they distinguish also between a type of measurement which only changes our knowledge of the properties of a system $\mathcal{S}$, and the type which changes the state of $\mathcal{S}$, with a given uncertainty, to another state, with another uncertainty.

7.6 A New More General Formulation

Overall width We have seen that in the standard deviation most probability weights are not in the bulk, so that, searching for a better formulation of the uncertainty we must find a measure of the bulk width [HILGEVOORD/UFFINK 1983] [HILGEVOORD/UFFINK 1988, 98–99]. We define it as the smallest interval W_Δ in which a large fraction $N \leq 1$ of the total probability is situated, i.e. the smallest number W_Δ satisfying (for one-dimensional problems):

$$\int_{r_0 - W_\Delta/2}^{r_0 + W_\Delta/2} |\psi(r)|^2 dr = N, \tag{7.38}$$

where r_0 is some reference point. This measure is always finite. We name it *overall width*. By introducing the Fourier pair $\psi(r), \tilde{\psi}(k)$, we can write:

$$W_\Delta(\psi) W_\Delta(\tilde{\psi}) \geq C(N), \text{ if } N > \frac{1}{2}, \tag{7.39}$$

where $C(N)$ is some constant which depends on the Fourier pair. The first attempt of a similar formulation is to be found in the signal theory [H. LANDAU/POLLAK 1961].

Bohr's Two-Slit Experiment Bohr proposed [BOHR 1949, 216–17] a two-slit experiment (Two-Slit-Exp. 5) which has some importance for this examination. One sends a beam of bosons through a hole in wall I and, before reaching a photographic plate, through two slits in wall II. The problem is to see if the detection of particles' paths destroys the interference fringes — which we know to be typical of QM entities [see p. 27] — and if the presence of interference fringes makes a precise localization of particles impossible. If θ is the small angle between the conjectured paths of a particle passing through slit 1 or 2, the difference of momentum transfer in these two cases will be, according to Eq. (3.2), $\hbar k\theta$, and any control of the momentum of the walls with an accuracy sufficient to measure the difference will, due to UR1 [Eq. (7.7)], involve a minimum latitude of the position of the wall comparable with $1/k\theta$. The number of fringes per unit length is equal to $k\theta$, and since the uncertainty $1/k\theta$ in the position of wall I will cause an equal uncertainty in the position of the fringes, it follows that no interference aspect can appear. Bohr concluded that we face a choice: We either trace the path of a particle, or we observe interference effects. Later we shall discuss the problem of WP-dualism in more detail [see section 8.4 and chapter 30].

A reformulation of Bohr's ideas Now we look for a mathematical formalism allowing us to express these ideas [UFFINK/HILGEVOORD 1985, 933–38] [HILGEVOORD/UFFINK 1988, 100–103] [see figure 7.7].[33]

[33] See also [*BUSCH et al.* 1995a, 137–39] [*SCHROECK* 1996, 203–207, 271–79].

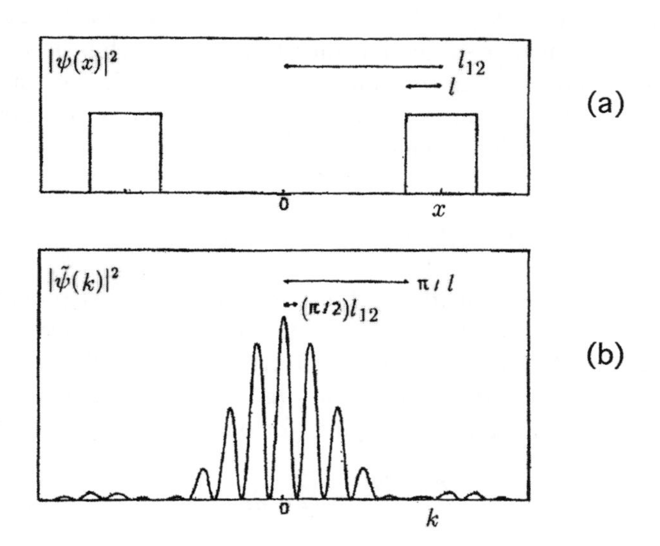

Figure 7.7: The (a) position and (b) momentum distributions for Two-Slit-Exp. 5 — from [HILGEVOORD/UFFINK 1988, 100].

The two slits have a width $2l$ and are separated by a distance $2l_{12}$, where $l \ll l_{12}$. We take as wave functions:

$$\psi(r) = (4l)^{-\frac{1}{2}} \text{ if } l_{12} - l < |r| < l_{12} + l; \quad \psi(r) = 0 \text{ elsewhere} \tag{7.40a}$$

and [see Eq. (7.36)]

$$\tilde{\psi}(k) = \left(\frac{2l}{\pi}\right)^{\frac{1}{2}} \cos(l_{12}k)\frac{\sin(lk)}{lk} \tag{7.40b}$$

Measure of the interference pattern Now we see that the bulk width is the envelope of the interference pattern, so that we encounter two different kinds of width: the bulk width itself of $\psi(r)$, which is of order l_{12} and which is inversely proportional to the fine structure of $\tilde{\psi}(k)$, i. e. the width of the interference lines. *Viceversa* we note that the fine structure of $\psi(r)$, which is of order l, is inversely proportional to the bulk width (the width of the envelope) of $\tilde{\psi}(k)$. We symbolize the measure of the fine structure as w_Δ and name it *Measure of the interference pattern*. Now it is clear what Bohr means about Two-Slit-Exp. 5 [p. 131], i.e. that the determination of a particle's path is inversely proportional to the number of fringes per unit length. Now we look for a UR between w_Δ and W_Δ. Take the overlap integral:

$$w_\Delta(r') = \int \psi^*(r)\psi(r - r')dr \tag{7.41}$$

Clearly this integral is sensitive to the fine structure of ψ. We may define a mean peak width w_Δ as the smallest number w_Δ for which

$$|w_\Delta(\psi)| = \eta \tag{7.42}$$

where $1/2 \leq \eta \leq 1$.

Hilgevoord/Uffink UR Introducing the Fourier transform $\tilde{\psi}(k)$ of $\psi(r)$, we can now write Eq. (7.41) as follows:

$$\int \psi(r)^* \psi(r - r')dr = \int e^{-\imath r'k} |\tilde{\psi}(k)|^2 dk, \tag{7.43}$$

and write the *Hilgevoord/Uffink UR* between the mean peak width of $\psi(r)$ and the bulk width of $\tilde{\psi}(k)$:

$$\boxed{w_\Delta(\psi) W_\Delta(\tilde{\psi}) \geq C(\eta, N) \text{ if } \eta \leq 2N - 1} \tag{7.44}$$

where $C(\eta, N) = 2\arccos(1 + \eta - N)/N$. This uncertainty relation can be extended to every Fourier pair. This UR, when applied to Two-Slit-Exp. 5, says that if l_{12} is reduced, so that $W_\Delta(\psi)$ gets smaller, $w_\Delta(\tilde{\psi})$ must increase, so that the interference lines must broaden.

Derivation of UR1–4 In order to derive UR1–4 from Eq. (7.44), one should use the Galilei group [HILGEVOORD/UFFINK 1988, 104–110] [see subsection 3.10.3]. This is more than a mathematical trick because if observables do not arise from the same symmetry group (domain requisite) URs loose all meaning[34]. By using definition (3.138), we define the spatial translation width $\hat{U}(w_{\Delta_q})$ of state $|\psi\rangle$ — i.e. the probability amplitude of finding a translated state $\hat{U}(w_{\Delta_q})|\psi\rangle$ if the initial state is $|\psi\rangle$ — as the smallest number w_{Δ_q} satisfying:

$$|\langle\psi|\hat{U}(w_{\Delta_q})|\psi\rangle| = \eta \tag{7.45}$$

By using the Fourier transform given by $\tilde{\psi}(k) = \langle k|\psi\rangle$, we obtain [see Eq. (3.138)]:

$$\langle\psi|\hat{U}_a|\psi\rangle = \int e^{-\imath ak} |\tilde{\psi}(k)|^2 dk, \tag{7.46}$$

which is the mathematical equivalent of Eq. (7.43). Hence the translation width w_{Δ_q} of the state and the overall width $W_\Delta(\tilde{\psi})$ of the momentum distribution $|\tilde{\psi}(k)|^2$ satisfy the Hilgevoord/Uffink UR, so that, by writing $W_\Delta(\tilde{\psi}) = W_{\Delta_p}(\psi)$, the equivalent of UR1 is:

$$w_{\Delta_q}(\psi) W_{\Delta_p}(\psi) \geq C(\eta, N), \tag{7.47}$$

under the same conditions of Eq. (7.43). The same can be stated for UR2–4.[35]

[34] See on this point [*SCHROECK* 1996, 49–51].

[35] Explicit calculations of variance and uncertainty are made in [PESLAK 1979]: here again is showed the impossibility of interpreting the uncertainty as a standard deviation. See also [J. PRICE 1983], and [*SCHROECK* 1996, 49–52, 274–79] for a review, and [DE MUYNCK 1999] for a recent study employing POVM [see chapter 18 methods.

Chapter 8

THE COMPLEMENTARITY PRINCIPLE

Contents After a preliminary definition of Complementarity [section 8.1], we analyse in this chapter different forms of it: in section 8.2 the Complementarity between conjugate observables; in section 8.3 the Complementarity between wave-like and corpuscular behaviour; in section 8.4 the Complementarity between 'localization' and dynamic laws; and finally in section 8.5 the Complementarity between separability and phenomenon. Brief conclusions [section 8.6] follow.

8.1 Definition of Complementarity

The Complementarity Principle (= CP) was formulated for the first time by Bohr at the Come Conference in 1927 and communicated to a large audience in an article in *Nature* in 1928 [BOHR 1928].

Some difficulties CP is the main content of the *Copenhagen interpretation.* However it is not a single and unambiguously defined set of ideas but rather a common denominator for a variety of related viewpoints, a variety which is already to be found in Bohr himself.

For example Bohr went very far away when he applied Complementarity to domains such as the relation between body and mind.[1]

Surely these extensions are not only outside of our investigation, but also questionable in themselves. Therefore, it is understandable what in 1971 von Weizsäcker [*VON WEIZSÄCKER* 1971a 225] said: that the Copenhagen interpretation needed itself an interpretation. So in the following we try to give a systematic reconstruction of CP.

A preliminary definition First of all we must understand what Bohr means by *Complementarity.* Complementarity was applied by Bohr himself to the following physical concepts: experimental arrangements, conjugate variables, modes of descriptions — between wave-like and corpuscular modes of description[2] —, but also to the pairs definition/observation and dynamics/kinematics. However, Bohr believed that the wave-like and corpuscular modes of description represent the basic type of Complementarity — because this could eventually provide an an-

[1]See [*JAMMER* 1974, 87–89] on this point.

[2]See [*JAMMER* 1974, 96–97].

swer to WP-Dualism [see subsection 8.3.1]. Hence, summarizing Bohr's ideas, we can give the following definition:

Definition 8.1 (Complementarity) *In QM two modes of description are said to be complementary if they are incompatible at the same time but both classically necessary in order to describe a physical system.*

Two features are characteristic of this definition:

- the exclusiveness of each mode of a complementary pair, so that Complementarity is a yes/no, either/or alternative (exclusive disjunction) [BOHR 1949, 217–18];
- the necessity of both for obtaining a more complete description of the system [BOHR 1929, 205], though, as Bohr acknowledged, this is not possible in the frame of QM.

It is interesting to see what the product uncertainty of a pair of conjugate observables teaches us about the two characters of Bohr's definition of complementarity.

Exclusiveness About the *yes/no character*, it is worth mentioning that until the sixties it was a generally believed that it is impossible to measure two conjugate observables together. That this is not so, has already been shown by the Hilgevoord/Uffink UR [Eq. (7.44)] and it will be shown later to a greater extent [see subsection 18.2.1 and chapter 30].[3] What is important to see here is that it is this belief which probably is the ultimate reason for the attribution of an exclusiveness character to complementarity. Anyway, to a certain extent, conjugate observables are exclusive, in the sense that we can never fully determine two conjugate observables together, and that a minor degree of uncertainty for one of them is equivalent to a major degree of uncertainty for the other. On the other hand, we must not think that complementary observables are always a pair. In reality for every observable it is possible to find many operators which do not commute with it [*VON NEUMANN* 1932][4] — the eventual physical significance of these and whether they correspond to physical observables is another matter.

Necessity About the other character — the *necessity of both complementary aspects* —, it is true that, in order to know a state, we need a knowledge of both observables of a conjugate pair. On the other side there can be situations in which what we need is the best possible knowledge of only one observable, with the consequence of a complete lack of knowledge about the other. Or it can be that we need to prepare some state where only one observable of a conjugate pair is perfectly determined, for example the number of photons[5], i.e. we need a Fock state [see subsection 5.1.1]. One can say that in classical mechanics the perfect knowledge of both variables of a conjugate pair is a sufficient condition for the perfect knowledge of the state of a system [see Eq. (1.1)]. So one could then say that QM is incomplete or in some way imperfect in relation to classical mechanics. But the incompleteness of QM can be rejected [see part VIII] and there are good reasons for assuming that also classical mechanics presents a situation which, to a certain extent, is similar to that of QM [see section 33.1].

In conclusion, we can accept the properties which Bohr established for Complementarity only to a certain extent and under certain conditions. The best we can do is to improve the concept of Complementarity by discussing some concrete proposals put forward by Bohr and by other physicists.

[3]See also [BUSCH/LAHTI 1984, 1638, 1639].

[4]See [*JAMMER* 1974, 94] on the point.

[5]For a discussion see [CAVES *et al.* 1980, 342].

8.2 Complementarity Between Conjugate Observables

8.2.1 Analysis

Formulation and criticism At first sight CP seems to answer in a more general form the problem of conjugate variables than UP, because it can be formulated[6] as follows:

Principle 8.1 (Complementarity of Conjugate Variables (= CP1)) *For each degree of freedom two conjugate variables are a pair of complementary observables, which means that a precise knowledge of one of them implies that all possible outcomes of the measurement of the other are equally probable.*

Is this principle something really new respect to the UP? The reason why some physicists answer yes[7] is that it is assumed that the URs presuppose a measuring process, while CP only asserts that, as such, there can be no determinate values for both of a pair of conjugate observables. But we know that the process of measurement has nothing to do with the URs or with UP [see section 7.3]; hence this formulation of CP is nothing more than UR1–5 or UP itself. The tension regarding this point is due to the fact that, while Bohr tried to assimilate URs and UP in the context of CP2 — which is not correct —, Heisenberg, on the other hand, in the twenties saw the URs as a formal deduction of the theory with no relationship with WP-dualism[8].

Note finally that this form of Complementarity is expressed very well by the Hilgevoord/Uffink UR (7.44) [HILGEVOORD/UFFINK 1988, 107].

8.2.2 Some Consequences

What about the values of an observable? As a consequence of principle 8.1 it was thought that in QM only equations that connect observables are invariant, and that the values of observables are only a result of observation [ROSENFELD 1953, 60] [see postulate 3.3: p. 37], so that the following corollary seems to follow[9]:

Corollary 8.1 (Values of Observables) *Any statement about the value of an observable is meaningless if not preceded by a measurement of the same.*

It is evident that this corollary is semantic. But it is certainly dubious, because, even if we cannot make a statement about individual values without having made measurements, a statistical approach, which is meaningful and sometimes necessary, is always possible [see section 36.2].

A subjectivist interpretation Strengthening corollary 8.1, we arrive to what can be named the *subjectivist interpretation of CP*, which postulates:

Postulate 8.1 (Subjectivist CP) *The properties of QM systems are created by the act of measurement.*

---

[6]See [*MURDOCH* 1987, 58, 80–81].
[7]As happens in [SCULLY *et al.* 1991, 111].
[8]See [COMBOURIEU/RAUCH 1992, 1415].
[9]Formulation in [*REICHENBACH* 1944, 141].

The first support for Subjectivist CP came from Pascual Jordan [P. JORDAN 1934]. Jensen [JENSEN 1934] tried to extend postulate 8.1 also to macrostates but Jordan never followed him[10]. But postulate 8.1 is far from evidence — it will be discussed in chapter 29 — and Complementarity as such does not imply a form of subjectivism[11] though sometimes Bohr means that not only we must ascribe an experimental context to systems for speaking of properties, but also that we need an actual measurement, which is surely a Subjectivist CP[12]. In an other situation Bohr spoke[13] of the subjective character of relativity, forgetting that Einstein's theory presupposes invariants.

Perhaps these positions of Bohr's stem from his philosophical basis and in particular from the mixture of kantism and pragmatism[14] which gave to his interpretation an epistemological 'colour'. However, we have nothing against pragmatism in general, which in the context of an operational approach is certainly appropriate [see chapters 18 and 26]. We remember only that a pragmatist or an operational approach cannot underestimate the fundamental principle of ontological commitment [6.1: p. 103].

8.3 Analysis of the Wave/Particle Dualism

8.3.1 First Formulation

Many of Bohr's formulations give the impression that the two fundamental complementary QM modes of description are the wave-like and the corpuscular ones, so that CP seems to be an answer to the problem of WP-Dualism[15] [see p. 26]. Hence it was said[16] that the ontological basis of CP is the Wave/Particle Dualism. Following Bohr's writings, we can formulate such a principle as follows:

Principle 8.2 (Wave/Particle Complementarity (= CP2)) . *The corpuscular and the wave pictures are two attempts at an interpretation of experimental evidence in which the limitation of classical concepts is expressed in complementary ways.*

The problem is that the terms *wave* and *particle* in the context of Basic-QM are only images because, after Bohr, they are classical terms which do not rigorously adapt to the new situation determined by QM itself. Hence one can conclude that CP2 is not a physical principle but a semantic one: it is about the meaning to be attributed to the two terms. And also a pragmatic one: when talking about microphysical things the use of two terms must be complementary. In this sense it cannot be part of a physical interpretation of QM because it does not individuate physical processes or objects which could justify the complementary use of the two terms.

Later we shall develop an analysis which is able to grasp the essential tenet behind CP2 [see section 30.4]; but for this purpose we need some refinement of quantum concepts.

[10] For historical reconstruction see [*JAMMER* 1974, 161–63]. See also [*REICHENBACH* 1944, 143–44] [CLAUSER/SHIMONY 1978, 1885].

[11] See for example [BOHR 1958a, 7],

[12] See [*MURDOCH* 1987, 147–46, 150, 154] for reconstruction and analysis.

[13] As reported by [*JAMMER* 1974, 132].

[14] As pointed out by Murdoch [*MURDOCH* 1987, 222–25].

[15] See for example [*BOHR* 1961d, 56]. Analysis and comments in [*FOLSE* 1985, 116, 119–21]. This thesis was also named 'symmetry thesis' [HANSON 1959, 2–3] [*MURDOCH* 1987, 68–70, 77–79]

[16] By Jammer [*JAMMER* 1974, 160].

8.3.2 Second Formulation

We could also try to give a new formulation by saying[17]:

Principle 8.3 (Complementarity Continuous/Discontinuous (= CP3)) *The terms* wave *and* particle *are two images of the two aspects inherent to every problem of QM: the continuous and discontinuous, the first more connected with definition (Schrödinger Eq.) and the second more with measurement.*

Such a formulation is too generic and devoid of specific physical contents and meaning. Instead of, the spirit of the Complementarity Wave/particle can be caught by the formulation in section 8.4. Bohr never did explicitly this step because of his epistemological point of view and because of the importance he assigned to macroscopical experience [see postulate 8.2: p. 146].

8.4 Complementarity Between Localization and Dynamic Laws

8.4.1 Formulation of the Principle

A completely different formulation of CP is centered on the Complementarity between localization and dynamic laws of a system [BOHR 1928, 148] [BOHR 1929, 204]. Sometimes Bohr speaks of the dynamic laws as 'causality' as such. If this were right, the consequence would be that no classical causality would be possible in the frame of QM because the dynamic laws of QM are subject to the Superposition principle [principle 3.3: p. 39]. No doubt that this other form of CP was individuated very early by von Weizsäcker, who spoke of *circular* Complementarity, distinct from the Complementarity between conjugate variables, which he named *parallel* Complementarity [VON WEIZSÄCKER 1955, 525–26].[18] It is interesting that for von Weizsäcker the space-time localization is always a classical feature, though only approximate or fragmentary; while the dynamical laws, which in QM are based on the Schrödinger Eq., are strictly associated with the 'anomaly' of superposition. Hence, though 'deterministic', the wave function is not classical.

As said, Bohr never fully accepted this interpretation of von Weizsäcker, because he saw it as an abstract technicisms (and perhaps as an interpretation too much 'ontologically characterized') which did not fit in well with the need of an every-day language [BOHR 1948, 317] [see subsection 8.5.3].[19] The principle can be formulated in this form:

Principle 8.4 (Superposition/Localization Complementarity (= CP4)) *Localization and superposition are complementary.*

The principle is correct (as the following thought-experiment will show) and, in general, goes in the direction of a Complementarity between microphenomena and macrophenomena. Later

[17]Formulation in [*FOLSE* 1985, 97].
[18]See also [*SCHEIBE* 1973].
[19]See also [*JAMMER* 1974, 102–104, 378–79].

we shall see [in chapter 30] that this form of Complementarity must be smooth and that it is a better formulation of a fundamental form of Complementarity than CP2.

Note that by 'Localization' one should not only intend spatial localization, but all types of determinateness of the value of some observable. Otherwise one could think that 'localization' here means 'kinematics' (and this against von Weizsäcker's tenet), which in turn would mix up CP4 with the Complementarity between conjugate observables. That this is not so, can also be seen by considering that superposition can affect every observable of a conjugate pair, a dynamic observable as well as a kinematic one. Hence the best expression of CP4 is to say that it is the Complementarity between wave-like behaviour and corpuscular behaviour, and its mathematical content is then given by inequality (30.19) — see also relative comments. But CP4 should not be confused with CP2, which is an epistemic treatment of the same problem, while, as we shall see in chapter 30, here we are in front of an ontological problem.

8.4.2 An Experiment

Two-Slit-Exp. 6 As has been already said, by means of a thought-experiment, we wish here to verify CP4, i.e. the Complementarity between the *welcher Weg* (which path) — corpuscular behaviour — and diffraction — wave-like behaviour, postponing to part VII an in depth analysis of the problem.

The experiment was developed by Scully, Englert and Walther [SCULLY *et al.* 1991] (Two-Slit-Exp. 6). We have a plane atom wave which goes through two slits of wall I [see figure 8.1]; behind this wall there is a series of wider slits which are used as collimators to define two atomic beams that reach the narrow slits of wall II where the interference originates. But between wall I and wall II and after the collimators the beams receive a laser beam which brings the internal state $|\iota\rangle$ of the two-level atoms from an unexcited (ground) state $|g\rangle$ into an excited state $|e\rangle$, and, after that, they pass through the maser cavities. At the end we have a screen on wall III.

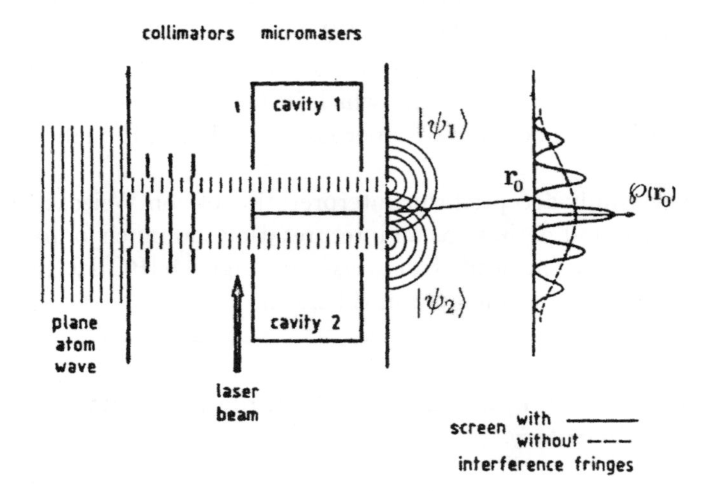

Figure 8.1: Two-Slit-Exp. 6 (with atoms). A set of wider slits collimates two atom beams which illuminate the narrow slits where the interference pattern originates. The collimation of the atomic beams would actually be done using atomic optics. This set-up is supplemented by two high-quality micromaser cavities and a laser beam to provide which-path information — from [SCULLY *et al.* 1991, 112].

In the interference region, the wave function describing the centre-of-mass motion of the atoms is the sum of the two terms referring to slit 1 and slit 2 ($\mathbf{r}$ indicates the centre-of-mass coordinate):

$$\Psi(\mathbf{r}) = \frac{1}{\sqrt{2}} [\psi_1(\mathbf{r}) + \psi_2(\mathbf{r})] ; \tag{8.1}$$

and the probability density of particles falling on the screen, denoted by $\wp(\mathbf{r}_0)$, where $\mathbf{r} = \mathbf{r}_0$, will be given by Eq. (3.40):

$$\wp(\mathbf{r}_0) = \frac{1}{2} \left[|\psi_1(\mathbf{r}_0)|^2 + |\psi_2(\mathbf{r}_0)|^2 + \psi_1^*(\mathbf{r}_0)\psi_2(\mathbf{r}_0) + \psi_2^*(\mathbf{r}_0)\psi_1(\mathbf{r}_0) \right] . \tag{8.2}$$

After being prepared in an excited state by the laser beam, on passing through either one of the maser cavities each atom will emit a microwave photon and could leave which-path information in the cavity. After the atom has passed through the cavity it is again in force-free space and its momentum has the initial value.

Two-Slit-Exp. 7 Let us now discuss another experimental arrangement without the laser-cavities device (Two-Slit-Exp. 7). Then the atom beam, after passing wall II, is described by the state:

$$\Psi(\mathbf{r}) = \frac{1}{\sqrt{2}} [\psi_1(\mathbf{r}) + \psi_2(\mathbf{r})] \, |\iota\rangle. \tag{8.3}$$

The probability density $\wp(\mathbf{r}_0)$ for particles to be on the screen is now:

$$\wp(\mathbf{r}_0) = \frac{1}{2} \left[|\psi_1(\mathbf{r}_0)|^2 + |\psi_2(\mathbf{r}_0)|^2 + \psi_1^*(\mathbf{r}_0)\psi_2(\mathbf{r}_0) + \psi_2^*(\mathbf{r}_0)\psi_1(\mathbf{r}_0) \right] \langle \iota | \iota \rangle, \tag{8.4}$$

which agrees with Eq. (8.2).

No fringes more We now return to Two-Slit-Exp. 6. Passing through the cavities and making transition from $|g\rangle$ to $|e\rangle$, the state of the correlated system (atomic beam plus maser cavity) is given by:

$$\Psi(\mathbf{r}) = \frac{1}{\sqrt{2}} [\psi_1(\mathbf{r}) |1_1 0_2\rangle + \psi_2(\mathbf{r}) |0_1 1_2\rangle] \, |e\rangle, \tag{8.5}$$

where $|1_1 0_2\rangle$ denotes the state in which there is one photon in cavity 1 and none in cavity 2. Note that this Eq. is not the same as eq (8.3). In contrast to Eq. (8.2) the probability density on the screen is now [eq (3.4)]:

$$\begin{aligned} \wp(\mathbf{r}_0) \; &= \; \frac{1}{2} \left[|\psi_1(\mathbf{r}_0)|^2 + |\psi_2(\mathbf{r}_0)|^2 + \psi_1^*(\mathbf{r}_0)\psi_2(\mathbf{r}_0) \langle 1_1 0_2 | 0_1 1_2 \rangle \right. \\ & \left. + \psi_2^*(\mathbf{r}_0)\psi_1(\mathbf{r}_0) \langle 0_1 1_2 | 1_1 0_2 \rangle \right] \langle e | e \rangle. \end{aligned} \tag{8.6}$$

But because $\langle 1_1 0_2 | 0_1 1_2 \rangle$ and $\langle 0_1 1_2 | 1_1 0_2 \rangle$ vanish, the interference terms disappear so that Eq. (8.6) reduces to:

$$\wp(\mathbf{r}_0) = \frac{1}{2} \left[|\psi_1(\mathbf{r}_0)|^2 + |\psi_2(\mathbf{r}_0)|^2 \right] , \tag{8.7}$$

which shows no diffraction fringes [see also Eq. (3.3)].

Discussion The micromaser *welcher Weg* detectors are recoil-free; hence there is no significant change in the spatial wave function of the atoms. It is only the correlation between the centre-of-mass function and the photon's degrees of freedom in the cavities that is responsible for the loss of interference. Therefore, it is a confirmation of CP4. But the authors thought that Two-Slit-Exp. 6 is a violation of UP, because they thought that UP presupposes — as Heisenberg in his article of 1925 also did [see section 7.3] — a disturbance of $\mathcal{A}$ on $\mathcal{S}$ such that it destroys the interference of the spatial wave function [see also p. 28]. But we know that the problem of disturbance has nothing to do with UP as such.

Storey and co-workers [STOREY *et al.* 1994a] pointed out that in Two-Slit-Exp. 6 the which-path determination would be not possible without a double momentum transfer between detector and photon (when a photon is emitted and then reabsorbed from the opposite direction), whose magnitude is in the limits of UR1.[20] But in [ENGLERT *et al.* 1994b] and [ENGLERT *et al.* 1995] the correctness of the fundamental deduction of Scully and co-workers was confirmed.

8.5 Complementarity Between Separability and Phenomenon

8.5.1 General Features

Bohr also thought that there was some form of Complementarity between an apparatus $\mathcal{A}$ and a system $\mathcal{S}$, because we cannot simultaneously solve[21]

- the problem of the distinction between $\mathcal{A}$ and $\mathcal{S}$,
- and the problem of the causal relation of one on the other.

Bohr said that this Complementarity comes from the impossibility of subdividing the quantum phenomena further than the level of energy-quanta [QP1: p. 19] From this Complementarity derives, following Bohr, the fundamentally ambiguity of ascribing traditional (classical) physical properties to quantum objects [BOHR 1949, 210]. However, it is this situation which prevents us from speaking about a system $\mathcal{S}$ independently and separately from an apparatus $\mathcal{A}$ [BOHR 1928, 152].

Terminologically we call (after Bohr) *phenomenon* the whole observational interaction $\mathcal{A}+\mathcal{S}$.[22] The formulation of this Complementarity could be the following [BOHR 1955a, 72]:

Principle 8.5 (Complementarity Separability/Phenomenon (= CP5)) *Any attempt at a well-defined subdivision of a QM phenomenon would require a change in the experimental arrangement incompatible with the appearance of the phenomenon itself.*

Hence the distinction between $\mathcal{A}$ and $\mathcal{S}$ cannot be effected beyond a certain limit. This way of thinking was surely a shared belief among many of QM's fathers. For example Heisenberg says [*HEISENBERG* 1952, 16] [see section 7.3] that the only place of uncertainties is the boundary line dividing microphysics from macrophysics, i.e. the line between $\mathcal{S}$ and $\mathcal{A}$.

[20] See also [STOREY *et al.* 1994b] [STOREY *et al.* 1995].
[21] For an analysis of the problem see [*FOLSE* 1985, 122].
[22] See also [*FOLSE* 1985, 112, 157–62] [*JAMMER* 1974, 210] [VIGIER 1985, 654].

8.5.2 Other Experiments

Two-Slit-Exp. 8

Arrangement That we are generally faced with a coupling between $\mathcal{S}$ and $\mathcal{A}$ such that it is not possible to save the same phenomena when separating them, is a fact which can be shown by the following experimental arrangement (Two-Slit-Exp. 8) [SCULLY *et al.* 1991] [see figure 8.2.(a)].

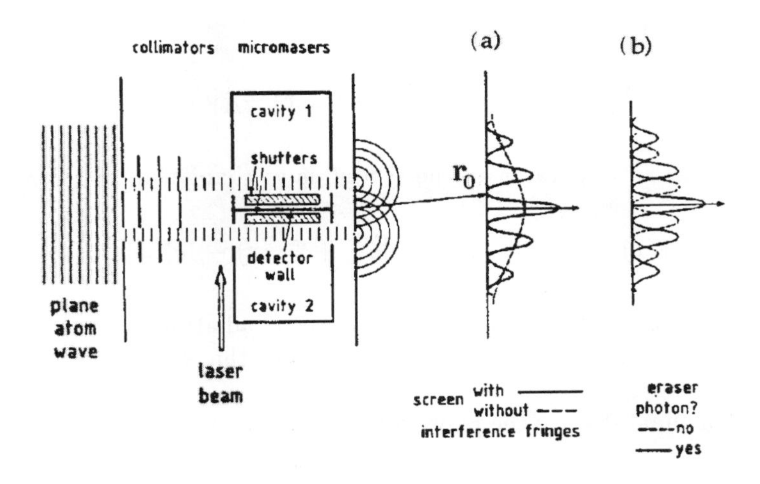

Figure 8.2: Proposed Two-Slit-Exp. 8 (with atoms).
(a) Quantum erasure configuration in which electro-optic shutters separate microwave photons in two cavities form the thin-film semiconductor (detector wall) which absorbs microwave photons and acts as a photodetector.
(b) Density of particles on the screen depending upon whether a photocount is observed in the detector wall ('yes') or not ('no'), demonstrating that correlation between the event on the screen and the eraser photocount is necessary to retrieve the interference pattern — from [SCULLY *et al.* 1991, 115].

We have the same arrangement of Two-Slit-Exp. 6 [p. 140] but now the detectors in the maser cavity are separated by a shutter-detector combination, so that, when the shutters are closed, the photons are forced to remain either in the upper or in the lower cavity. But if the shutters are opened, light will be allowed to interact with the photodetector wall and in this way the radiation will be absorbed and the memory of the passage erased (we call such an operation *quantum eraser*). After the erasure, will we obtain the interference fringes which were eliminated in Two-Slit-Exp. 6? The answer is yes, so that interference effects can be restored by manipulating the *welcher Weg* detectors long after the atoms have passed.

We begin with a description which includes the photodetector, initially in ground state $|g\rangle_{\rm D}$ [see Eq. (8.5)]:

$$\Psi(\mathbf{r}) = \frac{1}{\sqrt{2}}\left[\psi_1(\mathbf{r})|1_1 0_2\rangle + \psi_2(\mathbf{r})|0_1 1_2\rangle\right] |e\rangle_{\rm A}|g\rangle_{\rm D}. \tag{8.8}$$

After absorbing a photon the photodetector passes to an excited state $|e\rangle_{\rm D}$. If we introduce symmetric and antisymmetric atomic states:

$$\Psi_{\pm}(\mathbf{r}) = \frac{1}{\sqrt{2}} [\psi_1(\mathbf{r}) \pm \psi_2(\mathbf{r})], \tag{8.9}$$

and introduce symmetric and antisymmetric states of the radiation fields contained in the cavities:

$$|\pm\rangle = \frac{1}{\sqrt{2}} [|1_1 0_2\rangle \pm |0_1 1_2\rangle], \tag{8.10}$$

we can express Eq. (8.8) as:

$$\Psi(\mathbf{r}) = \frac{1}{\sqrt{2}} [\Psi_+(\mathbf{r})|+\rangle + \Psi_-(\mathbf{r})|-\rangle] \, |e\rangle_{\mathrm{A}} |g\rangle_{\mathrm{D}}. \tag{8.11}$$

The action of the Quantum eraser on the system is to change eq. (8.11) onto:

$$\Psi(\mathbf{r}) = \frac{1}{\sqrt{2}} [\Psi_+(\mathbf{r})|0_1 0_2\rangle |e\rangle_{\mathrm{D}} + \Psi_-(\mathbf{r})|-\rangle |g\rangle_{\mathrm{D}}] \, |e\rangle_{\mathrm{A}}. \tag{8.12}$$

The reason is that the interaction Hamiltonian between radiation and photodetectors only depends on symmetric combinations of radiation variables so that the antisymmetric state remains unchanged.

Now, as long as the final state of the photodetector is unknown, the atomic probability density at the screen is:

$$\wp(\mathbf{r}_{\mathrm{O}}) = \frac{1}{2} \left[\Psi_+^*(\mathbf{r}_{\mathrm{O}})\Psi_+(\mathbf{r}_{\mathrm{O}}) + \Psi_-^*(\mathbf{r}_{\mathrm{O}})\Psi_-(\mathbf{r}_{\mathrm{O}})\right] = \frac{1}{2} [\psi_1^*(\mathbf{r}_{\mathrm{O}})\psi_1(\mathbf{r}_{\mathrm{O}}) + \psi_2^*(\mathbf{r}_{\mathrm{O}})\psi_2(\mathbf{r}_{\mathrm{O}})], \tag{8.13}$$

which does not show any interference fringes. But if we compute the probability density for finding both the photodetector excited and the atom at $\mathbf{R}$ on the screen, we have:

$$\wp_{e_D}(\mathbf{r}_{\mathrm{O}}) = |\Psi_+(\mathbf{r}_{\mathrm{O}})|^2 = \frac{1}{2} \left[|\psi_1(\mathbf{r}_{\mathrm{O}})|^2 + |\psi_2(\mathbf{r}_{\mathrm{O}})|^2\right] + \mathrm{Re}\psi_1^*(\mathbf{r}_{\mathrm{O}})\psi_2(\mathbf{r}_{\mathrm{O}}), \tag{8.14}$$

which exhibits the same fringes of Eq. (8.2). But in contrast, the probability of finding both the photodetector de-excited and the atom at $\mathbf{r}_{\mathrm{O}}$ on the screen is:

$$\wp_{g_D}(\mathbf{r}_{\mathrm{O}}) = |\Psi_-(\mathbf{r}_{\mathrm{O}})|^2 = \frac{1}{2} \left[|\psi_1(\mathbf{r}_{\mathrm{O}})|^2 + |\psi_2(\mathbf{r}_{\mathrm{O}})|^2\right] - \mathrm{Re}\psi_1^*(\mathbf{r}_{\mathrm{O}})\psi_2(\mathbf{r}_{\mathrm{O}}), \tag{8.15}$$

giving rise to the antifringes indicated in figure 8.2.(b).

If one does not consider the Quantum eraser, one obtains the superposition of Eq. (8.13).

Discussion Therefore the question is the following:

- do we want to know whether we registered a 'slit 1'–atom or a 'slit 2'–one (= Two-Slit-Exp. 6),

- or are we interested in the incompatible property of having either excited the microwave-photon sensor ($|e\rangle_{\mathrm{D}}$) or not ($|g\rangle_{\mathrm{D}}$) (= Two-Slit-Exp. 8)?

Both at the same time are not available [ENGLERT *et al.* 1994a, 61]. In other words: we either know the *atom path* (and slit) without eraser and hence without knowing anything about the photodetector and its state (first alternative), or we desire to know the latter (second alternative) and we must reproduce the *interference* by losing all information about the particles' path[23]. In case 1 the absence of interference [Eqs. (8.7) and (8.13)] is due to the sum of probabilities (8.14) and (8.15). It is important not to forget that the photon is a QM object: when the shutters are opened, the two cavities become a single larger one. Now the photon's wave is a union of two initial and partial waves, but such that the two waves can reinforce each other so that the photosensor can detect the photon; or they can mutually extinguish each other with the consequence that the photosensor detects no photon (destructive interference). In other words, there is only a 50% probability of detecting the photon.

What we see here is that in the cases 1 and 2 we have two different global systems $(\mathcal{A}+\mathcal{S})_{\rm I}$ and $(\mathcal{A}+\mathcal{S})_{\rm II}$ that can give information only in incompatible ways. But, since in last analysis we are dealing with the alternative which -path/interference, we see that this form of Complementarity is, in this respect, completely reducible to the Circular Complementarity (CP4) [p. 139], so that we do not need to assume CP5 apart. On the other hand, what seems to be a special feature of the Complementarity separability/phenomenon — the assumption that we cannot subdivide the unity between $\mathcal{A}$ and $\mathcal{S}$ beyond a certain level —, is what we shall later show to be essentially wrong [see section 19.9].

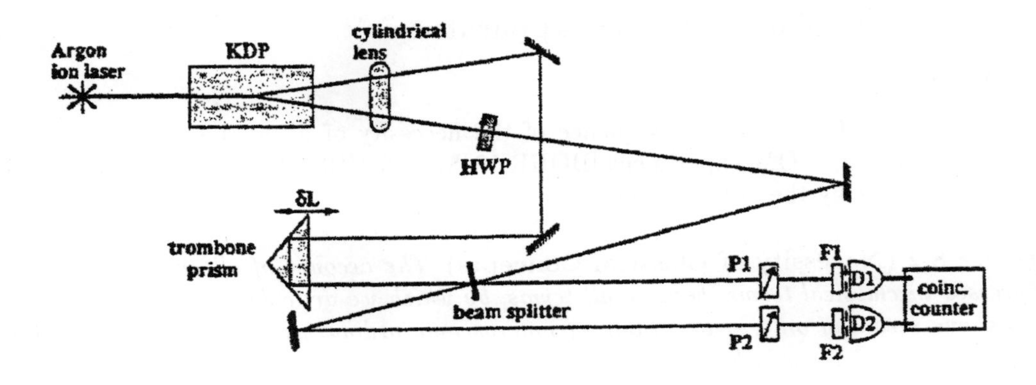

Figure 8.3: Scheme of an experiment (Interfer. Exp. 4) to observe the quantum eraser. $\mathrm{D}_1, \mathrm{D}_2$ are avalanche photodiodes; $\mathrm{P}_1, \mathrm{P}_2$ are polarisers; $\mathrm{F}_1, \mathrm{F}_2$ are bandpass filters, and HWP is a half-wave plate whose optic axis is at an angle $\phi/2$ to the horizontal polarisation of the down-converted beams — from [KWIAT *et al.* 1992, 7730].

Interfer. Exp. 4

It could be said that it is possible to account for the quantum eraser in terms of classical-wave explanations [on the whole matter see also section 28.6]. But in [T. JORDAN 1993] was put forward a similar Mach-Zehnder version of the experiment which can clearly be interpreted as

[23]On this point see also [PERES 1980a, 693].

a quantum eraser. And Kwiat and co-workers [KWIAT *et al.* 1992],[24] performed a two-photon interferometer (Interfer. Exp. 4) [see figure 8.3]: photons pass through a beam splitter and arrive to detectors D_1 and D_2 and we finally have a coincidence counter. If both photons are transmitted or reflected at the same time we have coincidence, so that photons are indistinguishable and hence interfere. But if in one arm of the interferometer we insert a labeller, i.e. a half-wave-plate (HWP) which can rotate the photon polarisation to vertical (or to 45°), we have a way of distinguishing the path and hence of destroying the interference. But, after the HWP, we can put the erasers in the form of two linear polarisers P_1 and P_2 which both polarise photons to 45° to the horizontal, thus erasing the which-path information.

Now the proposed experiment is an intrinsically second-order quantum effect, and cannot be accounted for in classical terms. Another interesting conclusion is that the paths are made distinguishable by correlating them to the environment (wholeness feature). Then we are free to reduce the resulting enlarged Hilbert space either to obtain loss of interference or to re-establish coherence. It is also possible to insert a 'delayed choice' feature [see chapter 26] in the experiment performed [see figures 8.4–8.5].[25]

The reported experiments are also discussed in [KWIAT *et al.* 1992] [SHIH/ALLEY 1988], [PITTMAN *et al.* 1996]. It can be shown that the interference between two photons cannot be reduced to the overlap of the two photons at the BSs. In fact there can be interference even if the photons arrive to the BSs at very much different times. Hence the phenomenon cannot be explained classically[26]. Note that on the basis of this examination it is not possible to assume a classical wave-like mode of description as CP2 [principle 8.2: p. 138] does.

8.5.3 Classical Concepts and Apparatus

A postulate

Bohr thought that CP5 was a consequence of the necessity of using the concepts of classical physics in describing a QM experiment [BOHR 1948, 327] [BOHR 1949, 209–210].[27] We can formulate the tenet by the following postulate:

Postulate 8.2 (Necessity of Classical Concepts) *The account of all evidence in QM must be expressed in classical terms, because all terms, by which we articulate an experience, have an unambiguous meaning only in the frame of macroscopic ordinary experience, which is the only one that can unambiguously be named 'experience'.*

If we understand the postulate only as a general statement about the necessity of a common language in order to speak of physical phenomena[28], almost all people agree.

But it seems that Bohr intended something stronger[29]. But if so, we presuppose the ultimate validity of a 'classical physics' also in relation with QM.[30] But 'classical physics' is more mythological than real. Einstein said in several places [EINSTEIN 1936, 77] [EINSTEIN 1949b] [*EINSTEIN* MW, 159–62] that it is non-sense to speak indifferently of 'classical physics', already

[24]Who follow [HONG *et al.* 1987]. See also [ZAJONC *et al.* 1991] and [CHIAO *et al.* 1995b, 266–69].
[25]On proposed experiments with these features see especially [OU 1997].
[26]For a more comprehensive article on the problem of quantum eraser see [HERZOG *et al.* 1995] .
[27]On this point see also [ROSENFELD 1961], where this view is developed into a new form of anthropocentrism.
[28]As Haag does [*HAAG* 1992, 6].
[29]See also [GHOSE/HOME 1992].
[30]As pointed out in critical form by Bohm [BOHM 1952]. See also [*BOHM/HILEY* 1993, 4].

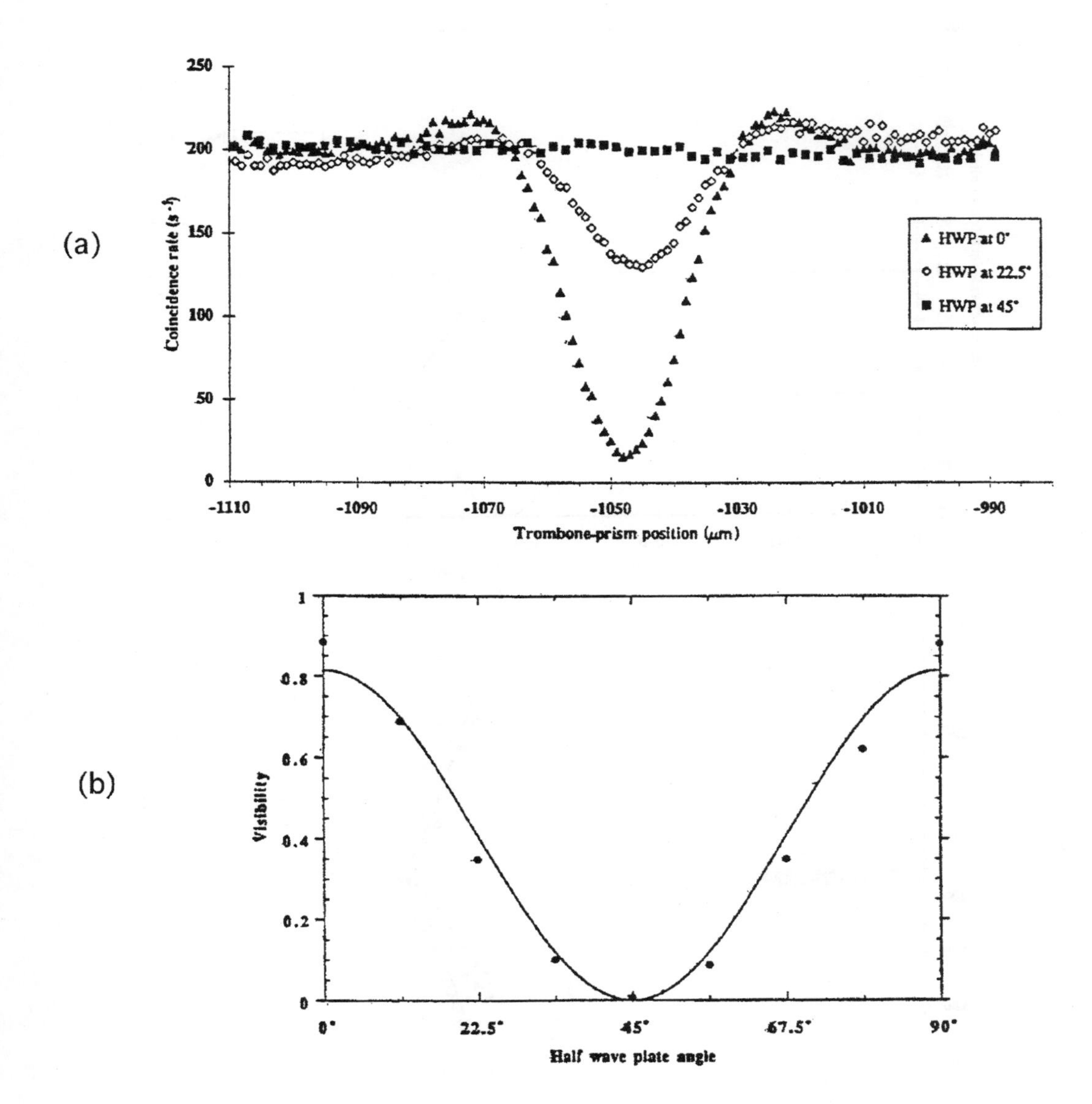

Figure 8.4: (a) Profile of an interference dip in coincidence rate for three wave plate orientations (accidental coincidences have been subtracted, and rates far from dip have been normalized to the same value). The interference effect is seen to vanish when the wave plate is at 45°, i.e. when the input ports of the BS are made distinguishable.
(b) Visibility as a function of HWP angle. The solid line is a fit to theory, with maximum visibility as the free parameter. The experimental points do not lie exactly on the same curve because slight fluctuations in alignment affect the visibility — from [KWIAT *et al.* 1992, 7732].

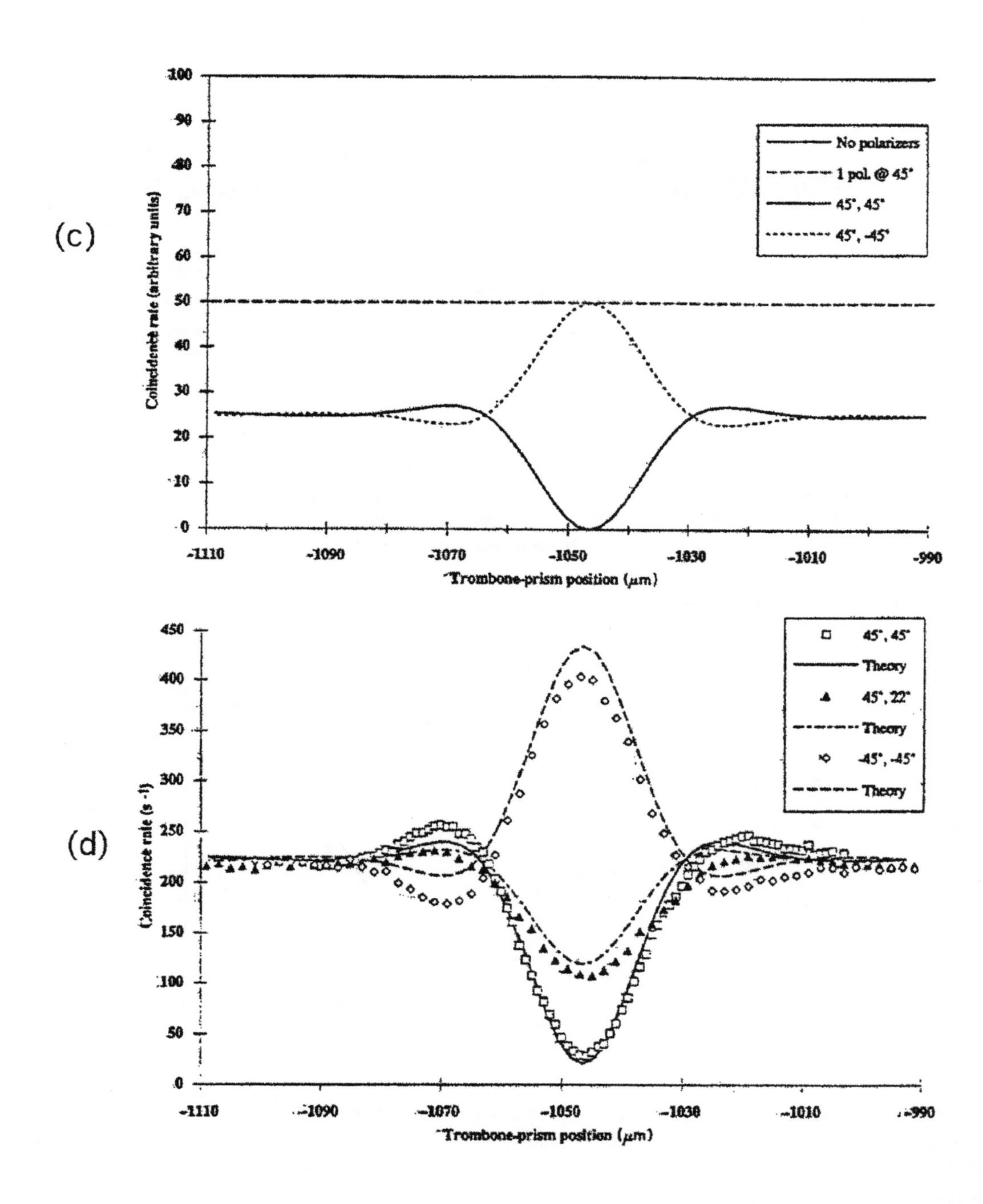

Figure 8.5: (c) Theoretical curves showing how two polarisers at appropriately chosen angles can erase distinguishability, restoring an interference pattern.
(d) Experimental data and scaled theoretical curves (adjusted to fit observed visibility of 91%) with polariser 1 at 45° and polariser 2 at various angles. Far from the dip there is no interference pattern and the angle is irrelevant. At the dip, the non-local collapsing of the polarisation of photon 2 causes us to observe sinusoidal variations as predicted (same normalization as in figure 8.4.(a)) — from [KWIAT *et al.* 1992, 7734].

on the ground of some incompatibility between classical mechanics and Maxwell's electrodynamics [see also section 46.3]. Why did then Bohr assign such a function to 'classical physics'? The point is that Bohr fully agreed with Einstein: he thought that there was in fact such incompatibility, and that exactly this incompatibility is represented in QM by the Complementarity between wave-like and corpuscular modes of description [see principle 8.2 (CP2): p. 138]. In other words, according to Bohr, QM is characterized by the fact that one uses two classical modes of description which are already complementary in themselves. As we have already said [but see especially chapters 30 and 33], this assumption need some correction.

On the other hand, Bohr's thesis of the necessity for all physics of the space-time frame of everyday experience [BOHR 1929, 204][31] and of everyday language [BOHR 1955a, 72] is open to the criticism that the concepts of classical mechanics or of classical electrodynamics are not the same as those of ordinary experience. Nevertheless, he sometimes acknowledges that mathematics is a refinement of ordinary language [BOHR 1955a, 68] and that it is the basis of natural science [BOHR 1958a, 1]. But generally he does not see another problem: That experience is not universally the same, not only in the sense in which, at an individual level, it is culturally and technically contextual[32] but also in the more general sense of the evolution in time of all culture[33].

And science may, on the other hand, have a deep influence on the evolution of everyday beliefs[34]. In conclusion, the everyday experience is a genus, a 'family' of different experiences with certain similarities, which as such is certainly unable to be the basis of scientific knowledge. Probably Bohr's attitude is again to be understood as a result of his Kantian and pragmatistic *forma mentis*, which was generally shared by thinkers of that time[35]. We have nothing against the Pragmatism as such — which fits in very well with an operational approach [see chapter 18] — but only against the consequences derived by Bohr, which are questionable.

And its corollaries

Postulate 8.2 immediately implies the necessity of classical apparatus in order to account for QM experiments:

Corollary 8.2 (Macroscopical apparatus I) *Without macroscopical apparatus (which obey to classical physics) one cannot account for QM experience.*

In fact, without such apparatus, it would be impossible to account for QM phenomena in classical terms. But, it must be said that there are no fixed boundaries between apparatus and systems. Peierls [PEIERLS 1985] has shown that the boundary between $\mathcal{A}$ and $\mathcal{S}$ can be moved in $\mathcal{A}$'s direction, contrary to Bohr's opinion. But there is always a limit where the measurement is already happened. Later [see section 19.9] we shall report a very important proposal in order to analyse the problem in more detail.

Another problem is the necessity to use macroscopic instruments in order to observe QM systems; this seems to derive from the non-commutability or Complementarity of most microscopic observables, and in particular between kinematic and dynamic ones [WIGNER 1963, 338]. We remember here that kinematic observables are necessarily bound to an external localization;

[31]See also [*MURDOCH* 1987, 46, 71–74].

[32]As pointed out, for example, by Khalfin and Tsirelson [KHALFIN/TSIRELSON 1992, 904],.

[33]See [VIGIER 1985, 655] [*FINKELSTEIN* 1996, 34].

[34]Very well known and studied is the passage from the middle-age to the modern age, with the development of a 'scientific' vision of the world; but we may also think the contemporary transformation of society and culture (virtual reality, cloning, and so on), one of the most dramatic in history.

[35]See for example [*DEWEY* 1929, 5] but also [*HUSSERL* 1936, 52–58].

and since they fail to commute with the observables which are conserved (the dynamic ones), in general an external microscopic apparatus is not useful. Hence, following Wigner's intuition, we could try to formulate a corollary which is 'softer' than the preceding one [8.2]:

Corollary 8.3 (Macroscopical Apparatus II) *The distinction between $\mathcal{A}$ and $\mathcal{S}$ can be introduced only at a macroscopic level, but the necessity of macrophysics stems from QM itself.*

However, a more careful analysis is needed in order to arrive to a clear conclusion [see especially theorem 14.3: p. 230, and theorem 16.2: p. 256] on this matter. In a general form, note that not all apparatus which one uses in QM are macroscopic. As we shall see, the amplifier must necessarily be macroscopic [see section 13.2] but the instrument itself could also be a particle or an atom. And surely, not all 'big apparatus' always obey to 'classical' physics.

That it is not so, it can be seen in the following **examples**. Macroscopic objects such as gravity-wave detectors — which can weight a ton —, are to be treated as quantum harmonic oscillators [CAVES *et al.* 1980] [BRAGINSKY *et al.* 1980].[36] Non-classical squeezed states [see subsection 5.1.1] can describe oscillations of suitably prepared electromagnetic fields with macroscopic number of photons [M. TEICH/SALEH 1990]. Superconducting Josephson junctions [JOSEPHSON 1962] have quantum states associated with currents involving macroscopic numbers of electrons, and yet they can tunnel between the minima of effective potential [MARTINIS *et al.* 1987].

8.6 Conclusions on the Copenhagen Interpretation

In conclusion the Copenhagen interpretation consists in a many propositions and corollaries that are partly questionable. We cannot take it *in toto* as an integral part of the theory.

In fact, as we have seen, the barycentre of the Copenhagen interpretation is a last defence of the validity of classical and macroscopical notions in order to account for quantum phenomena [see specially principle 8.2 (CP2): p. 138, and postulate 8.2: p. 146] and an epistemological 'style'[37] which is not completely adequate for the formulation of a sound interpretation of QM.

However, some of its fundamental ideas must come into every subsequent interpretation: certainly, for example, a 'corrected' notion of Complementarity and CP4, UP [principle 7.1: p. 120] and the Hilgevoord/Uffink UR (7.44) as its mathematical content. For other aspects we shall try new formulations which are more adequate to the theory.

[36]See also [ZUREK 1991, 36].

[37]On this point see [SHIMONY 1988, 183].

Part III

FIRST FOUNDATIONS OF QUANTUM MECHANICS

Introduction to Part III The aim of this part is to better develop the foundations of the formalism treated in the first two parts.

Four problems are considered:

- General axiomatic of the theory [*chapter 9*], where different attempts are shown to reduce the complexity of Basic-QM, particularly the 'dualism' between states (vectors) and observables (operators on vectors).
- The structure of Hilbert space [*chapter 10*], where particular attention is devoted to the problem of the observables which in Basic-QM are not represented by operators (time, energy, angle and phase), or which pose some problems in Relativistic-QM (position).
- The quantum probability [*chapter 11*], where particular attention is dedicated to the important Gleason theorem.
- Finally the geometric phase [*chapter 12*], which is a new — very abstract but also very promising — domain of the theory, particularly suitable for explaining phenomena such as the Aharonov/Bohm effect.

Chapter 9

QM AXIOMATICS

Introduction and contents Before we begin the discussion of the problems, we need some preliminary notions [section 9.1]. Then we discuss two order of problems[1]:

- The first is the problem of finding a model with the same logical structure as Basic-QM but a different epistemological structure, in the sense that in the model the propositions which are logically prior determine the meaning of the terms or propositions which are logically posterior. In sections 9.2 and 9.3 some models are presented. Then in section 9.4 we discuss the theoretical problem of whether a model is possible and to what extent it is possible.
- Another problem is that of the logical structures of reasoning in the theory T itself [section 9.5].

9.1 Boolean Algebra and Some Other Technical Notions

Here we shortly synthesize what Boolean algebra is [*BOOLE* 1854]. The first notion is that of *partially ordered* set (= POSet):[2]

Definition 9.1 (POSet) *The set $\mathcal{L}$ is a POSet if there is a reflexive antisymmetric transitive relation $\preceq$ between some elements $a, b, c \in \mathcal{L}$*

$$(\forall a)(a \preceq a) \text{ (Reflexivity)}, \tag{9.1a}$$

$$[(a \preceq b) \wedge (b \preceq a)] \rightarrow (a = b) \text{ (Antisymmetry)}, \tag{9.1b}$$

$$[(a \preceq b) \wedge (b \preceq c)] \rightarrow (a \preceq c) \text{ (Transitivity)}, \tag{9.1c}$$

where $\wedge$ is the logical symbol for conjunction (and $\vee$ for inclusive disjunction). Instead of $a \preceq b$ it is possible to write $a \leq b$ or $a \subseteq b$, but we prefer to use $a \leq b$ as an order relation among numbers and functions of numbers, and $a \subseteq b$ as an order relation between sets. However, many authors prefer to handle the subject of this section in terms of sets. Formally $a \preceq b$

[1]See [HARDEGREE 1979, 50] [*JAMMER* 1974, 11, 341].

[2]Expositions in [*DRIESCHNER* 1979, 209–210] [*BELTRAMETTI/CASSINELLI* 1981, 96] .

is a meta-language implication, the equivalent of $a \vdash b$, which is to be kept distinct from the object-language implication $a \to b$.[3]

For example, by *Modus ponendo ponens*, if we have the assumptions 'a' and '$a \to b$', then we can deduce 'b', and we would write: $a, a \to b \vdash b$.

Hence we define

$$a \preceq b := (a = 1) \to (b = 1) \tag{9.2}$$

where 1 is the truth value.

A POSet can be symbolized by $\mathcal{L}_{\preceq}$ and is characterized by the following features:

- there is at most one element called the null or zero element and denoted by $\mathbf{0}$ (for sets: $\emptyset$) such that:

$$(\forall a)(\mathbf{0} \preceq a); \tag{9.3a}$$

- there is at most one element called the unit element and denoted by $\mathbf{1}$ (symbolized also by the identity operator $\hat{I}$) such that:

$$(\forall a)(a \preceq \mathbf{1}). \tag{9.3b}$$

Another very important definition is that of *lattice*[4]:

Definition 9.2 (Lattice) *The POSet $\mathcal{L}_{\preceq}$ is called a lattice if*

- $\mathbf{1} \neq \mathbf{0}$;

- *for any non-empty finite set $\mathcal{X} \subset \mathcal{L}$ there exists an element $\bigvee_{a \in \mathcal{X}}$ (called* supremum*) such that*

$$(\forall b \in \mathcal{X})(b \preceq \bigvee_{a \in \mathcal{X}}); \tag{9.4a}$$

- *for any non-empty finite set $\mathcal{X} \subset \mathcal{L}$ there exists an element $\bigwedge_{a \in \mathcal{X}}$ (called* infimum*) such that*

$$(\forall b \in \mathcal{X})(\bigwedge_{a \in \mathcal{X}} \preceq b). \tag{9.4b}$$

We symbolize a lattice by $\mathcal{L}_{\preceq,\vee\wedge}$. In the lattice $\mathcal{L}_{\preceq,\vee\wedge}$ a *complement* of a is an element a' such that $a \wedge a' = \mathbf{0}$ and $a \vee a' = \mathbf{1}$. A subset of a lattice $\mathcal{L}_{\preceq,\vee\wedge}$ is called *sublattice* of $\mathcal{L}_{\preceq,\vee\wedge}$ if it is again a lattice with respect to the operations which make $\mathcal{L}_{\preceq,\vee\wedge}$ a lattice.

A complemented lattice is defined by the following:

Definition 9.3 (Complemented Lattice) *$\mathcal{L}_{\preceq,\vee\wedge}$ is said to be* complemented *if*

$$(\forall a \in \mathcal{L}_{\preceq,\vee\wedge})(\exists b)(b = a'). \tag{9.5}$$

Now we give the very important definition of a *distributive lattice*:

[3] As pointed out by Hardegree [HARDEGREE 1979, 58]. See also [*DRIESCHNER* 1979, 9].

[4] The following definitions can be found in [VARADARAJAN 1962, 178] [*VARADARAJAN* 1968, 10, 20] [*BELTRAMETTI/CASSINELLI* 1981, 96–97].

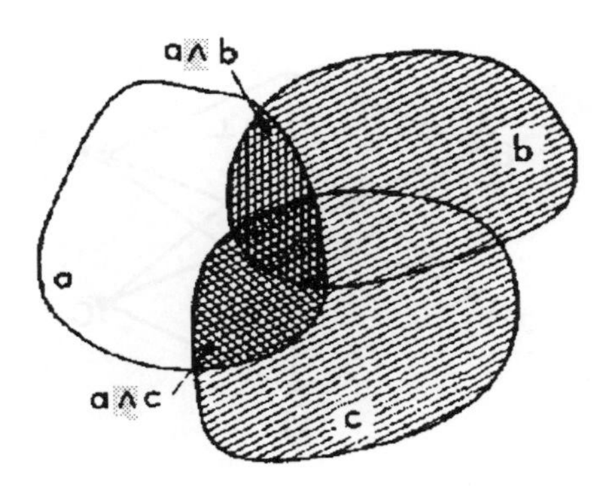

Figure 9.1: Illustration of the distributive law $a \wedge (b \vee c) = (a \wedge b) \vee (a \wedge c)$ — from [*JAUCH* 1968, 79].

Definition 9.4 (Distributive Lattice) *A lattice* $\mathcal{L}_{\preceq,\vee\wedge}$ *is* distributive *if* $\exists a, b, c \in \mathcal{L}_{\preceq,\vee\wedge}$ *such that* [see figure 9.1]*:*

$$a \wedge (b \vee c) = (a \wedge b) \vee (a \wedge c), \tag{9.6a}$$

$$a \vee (b \wedge c) = (a \vee b) \wedge (a \vee c); \tag{9.6b}$$

and of a *modular lattice*:

Definition 9.5 (Modular Lattice) *A Lattice* $\mathcal{L}_{\preceq,\vee\wedge}$ *is said to be* modular *if for any* $a, b, c \in \mathcal{L}_{\preceq,\vee\wedge}$ *for which* $c \preceq a$*, one has:*

$$a \wedge (b \vee c) = (a \wedge b) \vee c. \tag{9.7}$$

The last preliminary definition is that of orthocomplementation[5]:

Definition 9.6 (Orthocomplementation) *An* orthocomplementation *of* $\mathcal{L}_{\preceq,\vee\wedge}$ *is a mapping* $a \mapsto a^{\perp}$ *of* $\mathcal{L}_{\preceq,\vee\wedge}$ *onto itself such that:*

$$a^{\perp\perp} = a, \tag{9.8a}$$

$$(a \preceq b) \rightarrow (b^{\perp} \preceq a^{\perp}), \tag{9.8b}$$

$$a \wedge a^{\perp} = \mathbf{0}, \tag{9.8c}$$

$$a \vee a^{\perp} = \mathbf{1}. \tag{9.8d}$$

A lattice with orthocomplementation is called *orthocomplemented.* We symbolize it by $\mathcal{L}_{\preceq,\vee\wedge,\perp}$. We speak of a *σ–orthocomplemented* lattice if it exists in it the greatest lower bound and the least upper bound of every one of its countable subsets.

With the preceding premisses we can define the very important concept of *Boolean Algebra*[6]:

Definition 9.7 (Boolean Algebra) *A Boolean Algebra is a orthocomplemented distributive lattice* [see figure 9.2] *in which exist* $\bigvee_{a\in\mathcal{X}}$ *and* $\bigwedge_{a\in\mathcal{X}}$ *for every countable subset* $\mathcal{X}$*. In Boolean algebra each element has an unique complement.*

[5] See [*VARADARAJAN* 1968, 105] [*BELTRAMETTI/CASSINELLI* 1981, 96] .

[6] Definition in [*VARADARAJAN* 1968, 10].

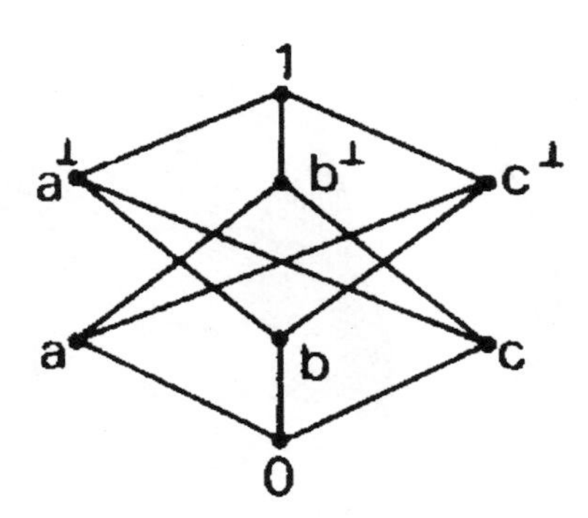

Figure 9.2: A distributive orthocomplemented lattice (Boolean algebra) by means of a Hasse diagram. Order relations are represented by lines: an element which is 'equal or less than' another (for example $a \preceq b^{\perp}$), is placed at the lower end of the line connecting the two. If two elements are connected with a third one, we have two possibilities: either the third element is higher than the other two, which means that it is a sum or disjunction of these two (for example $b^{\perp} = a \vee c$); or the third element is lower, and then it is a multiplication or intersection of the other two (for example $b = a^{\perp} \wedge c^{\perp}$). In fact the unit element (the biggest sum) is always the vertex (*supremum*) of such graphs, as the null element (the smallest intersection) always occupies the lowest position (*infimum*). We see that the figure represents an orthocomplemented lattice by the fact that, for example, we have $a \preceq b^{\perp}$, but also $b \preceq a^{\perp}$ — from [*BELTRAMETTI/CASSINELLI* 1981, 100].

The algebra on a classical phase space [p. 9] is a Boolean algebra. We come now to what is (for our inquiry) the most interesting alternative to Boolean algebra[7]:

Definition 9.8 (Orthomodular Lattice) *A σ-orthocomplemented lattice is* orthomodular *if* [see figure 9.3]*:*

$$(a \preceq b) \rightarrow \{b = [a \vee (b \wedge a^{\perp})]\}. \tag{9.9}$$

Now it can be shown that:

$$(\text{Distributivity} \rightarrow \text{Modularity}) \wedge (\text{Modularity} \rightarrow \text{Orthomodularity}) \tag{9.10}$$

i.e. Distributivity is stronger than Modularity and this is stronger than Orthomodularity.

We define Logic as follows[8]:

Definition 9.9 (Logic) *An orthocomplemented POSet $\mathcal{L}_{\succeq,\perp}$ (it need not to be a lattice: i.e. the greatest lower bound and the least upper bound need not exist for every finite non-empty subset) is said to be a* logic set *(symbolized as $\mathcal{L}_{\succeq,\perp,a_n}$)) if*

- *for any countably infinite sequence $a_1, a_2, \ldots$ of elements of $\mathcal{L}_{\succeq,\perp}$ we have:*

$$\bigvee_n a_n, \bigwedge_n a_n \in \mathcal{L}_{\succeq,\perp}, \tag{9.11a}$$

- *if $a_1, a_2 \in \mathcal{L}_{\succeq,\perp}$ and $a_1 \succeq a_2$, there exists an element $b \in \mathcal{L}_{\succeq,\perp}$ such that:*

$$(b \succeq a_1^{\perp}) \wedge [(b \vee a_1) = a_2]. \tag{9.11b}$$

[7]Definitions and analysis in [*BELTRAMETTI/CASSINELLI* 1981, 97–98] [GUDDER 1979, 341].
[8]Definitions in [*VARADARAJAN* 1968, 105–106].

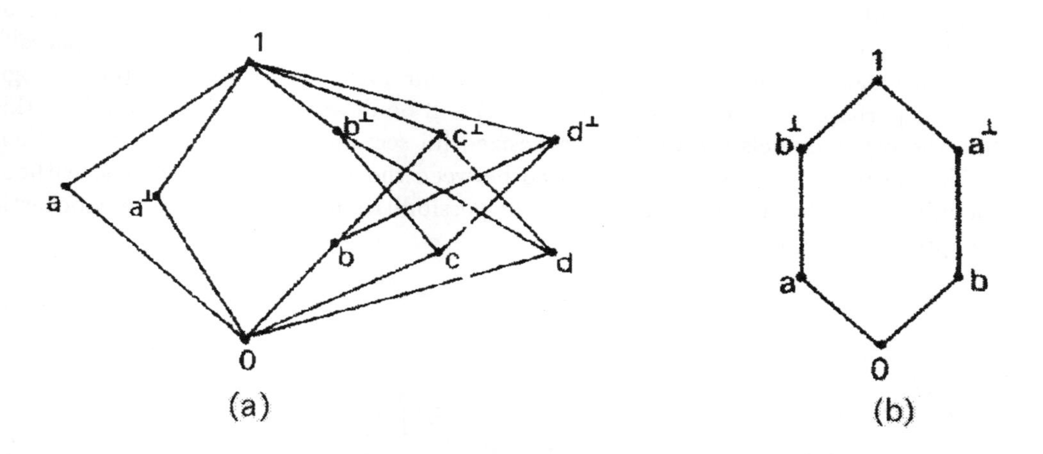

Figure 9.3: (a) A genuine orthomodular lattice. In fact take for example elements b and $d^{\perp}$. They are related in the following way: $b \preceq d^{\perp}$, and we also have $d^{\perp} = b \vee c$, and $c = d^{\perp} \wedge b^{\perp}$. In conclusion Eq. (9.9) is respected.
(b) a non-orthomodular lattice, because we have $a \preceq b^{\perp}$ but $b^{\perp} \neq [a \vee (b^{\perp} \wedge a^{\perp})]$ — from [*BELTRAMETTI/CASSINELLI* 1981, 100].

In a logic the greatest lower bound and the least upper bound, if they exist, are interdependent (De Morgan Law):

$$(\bigwedge_n a_n)^{\perp} = \bigvee_n a_n^{\perp},\ (\bigvee_n a_n)^{\perp} = \bigwedge_n a_n^{\perp}. \tag{9.12}$$

All preceding definitions are correct, but none of them can be part of QM because they are all mathematical notions which mostly refer to classical cases or to cases which are not compatible with QM, as we shall see in that which follows.

9.2 QM Axiomatic and Quantum Logic

9.2.1 Traditional Approach

As we have seen, Basic-QM [chapter 3] is centered on Hilbert spaces [subsection 3.4.3], it associates self-adjoint operators to observables [subsection 3.3.2], density operators to states [subsection 3.5.1], projectors to propositions [subsection 3.5.3], unitary operators to dynamics [subsection 3.6.3]. But the following problems arise[9]:

- it is possible to specify a determinate Hilbert space only for a finite number of degrees of freedom, while in QM there can also be an infinite number (in the case of continuous observables);
- it is not specified where Hilbert spaces come from;
- why is there such duality between observables and states?

[9]See [GUDDER 1979, 330–32].

From these problems arose the search for a more stringent axiomatic approach to QM. Theoretically we have only three elementary possibilities: to take propositions, or observables, or states as primitive elements and try to deduce, in some form, the necessity of Hilbert spaces[10]. The first possibility was developed very early on by quantum logic. Quantum logic (chronologically the first and perhaps the most influential model) will be developed in more detail [in this section], while the other models are briefly synthesized [in section 9.3] in order to grasp their main ideas. This discussion was mainly developed between the 1930's and the 1970's, without arriving to definitive results (as we shall see). Other possibilities may stem from the stochastic theory [see chapters 18 and 33].

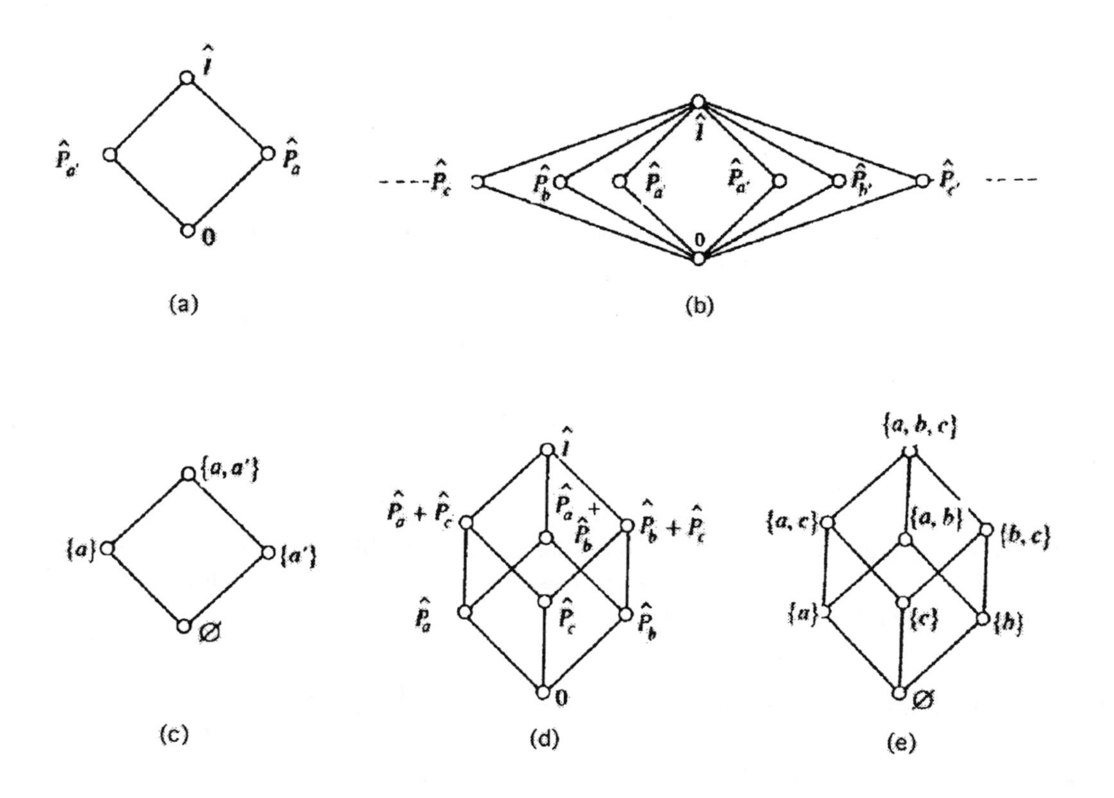

Figure 9.4: (a) Hasse diagram for a Boolean lattice with elements $\mathbf{0}, \hat{P}_a, \hat{P}_{a'}, \hat{I}$;
(b) a Boolean lattice wich is a collection of Boolean sublattices;
(c) a genuine Boolean lattice, isomorfic to (a);
(d) Boolean sublattice for three-dimensional Hilbert space, where a, b, c are orthogonal unit vectors;
(e) Hasse diagram for the Boolean lattice of a three element set $\{a, b, c\}$ — from [*REDHEAD* 1987b, 24–25].

[10]On the classification see [GUDDER 1979, 323] [GUDDER 1981, 126–28] [see also chapter 10]. For another approach see [LUDWIG/NEUMANN 1981].

9.2.2 Quantum Logic

The primitive elements in Quantum logic are the *propositions*. This is made possible thanks to Projectors, which, on one hand, represent the properties of a physical system, and, on the other hand, since they take only $0, 1$ values, they can be seen as propositions ('0' stands then for 'false' and '1' for 'true') [*VON NEUMANN* 1932, 134] [*VON NEUMANN* 1955, 253]. The first attempt was developed in 1936 by von Neumann himself co-working with Birkhoff [BIRKHOFF/VON NEUMANN 1936]. Other representatives of this school are, among others, Jauch, Piron and Varadarajan.

Birkhoff and Von Neumann What distinguishes classical calculus (i.e. a *Boolean algebra*) from the QM one is that *distributivity* [Eq.(9.6)] is valid in the first but not in the second [BIRKHOFF/VON NEUMANN 1936, 9–10]. This is a consequence of the fact that not all observables of QM commute [see Eq. (3.14)].

For example the proposition[11]: 'The electron has a determined momentum component value $p_x = \eta$ and position inside or outside of this box' — formally: $\hat{P}_{p_x=\eta} \wedge (\hat{P}_{q_B} \vee \hat{P}_{q_{B'}})$ — is not equivalent to the proposition 'The electron has momentum component $p_x = \eta$ and position inside of this box, or it has momentum component $p_x = \eta$ and position outside of this box' — formally: $(\hat{P}_{p_x=\eta} \wedge \hat{P}_{q_B}) \vee (\hat{P}_{p_x=\eta} \wedge \hat{P}_{q_{B'}})$. In fact the first proposition is equivalent to the proposition 'The electron has momentum component $p_x = \eta$', because logically 'it has position inside or outside of this box' $(\hat{P}_{q_B} \vee \hat{P}_{q_{B'}})$ is always true. The second proposition is always false in QM because it violates UP and the CCRs ($\hat{p}$ and $\hat{q}$ would be jointly measurable). Hence $[\hat{P}_{p_x=\eta} \wedge (\hat{P}_{q_B} \vee \hat{P}_{q_{B'}})] \neq [(\hat{P}_{p_x=\eta} \wedge \hat{P}_{q_B}) \vee (\hat{P}_{p_x=\eta} \wedge \hat{P}_{q_{B'}})]$.

In conclusion we state following Theorem[12] [see figure 9.4]:

Theorem 9.1 (Birkhoff/von Neumann) *The QM algebra is not Boolean.*

A concrete QM **example** is a spin–1/2 system [see figure 9.5].

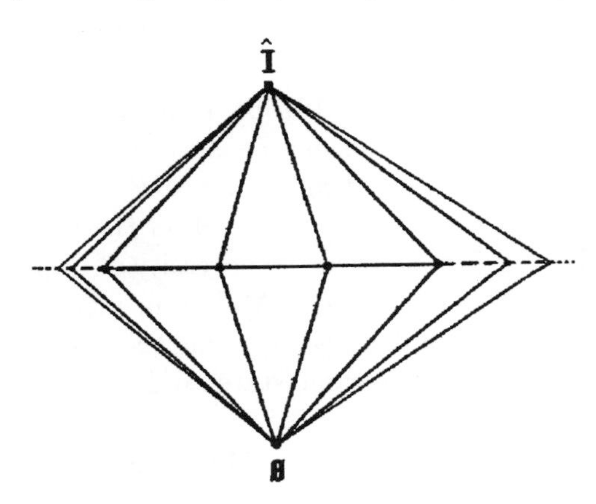

Figure 9.5: The polarisation lattice of a spin–1/2 system — from [BELTRAMETTI/MACZYŃSKI 1991, 1291].

Though correct, this theorem is not integrated into the theory because it is contained by a more general theorem now to be discussed [theorem 9.2]. At first Birkhoff/von Neumann's

[11]The example is taken from [*DRIESCHNER* 1979, 21].

[12]See also [HUSIMI 1937], where one tries to derive the non-Boolean logic of QM from empirical contents.

investigations had some echoes only on Jordan [P. JORDAN 1949b] [P. JORDAN 1952]. Later, in the sixties the problem was tackled by Jauch.

Jauch Birkhoff and von Neumann thought that modularity [Eq. (9.7)] — which is weaker than distributivity [see expression (9.10)] — was still valid for QM. Very early on there were attempts to criticize this approach by pointing out [POPPER 1968][13] the fact that in this manner the lattice proposed by the authors was in fact Boolean[14].

The same was pointed out by Jauch. While projective geometries are always modular lattices, for proposition systems in which the maximal element is not a finite union of points, modularity is no longer compatible with other axioms [*JAUCH* 1968, 122].[15] It is also possible to refute atomicity (see below), but Jauch does not make this step.

Jauch proved that in QM only orthomodularity [Eq. (9.9)] and not modularity is valid[16]:

Theorem 9.2 (Orthomodularity of QM) *QM algebra is a non-modular orthomodular lattice.*

Jauch refined quantum logic further. His starting point are filters: a *filter* that allows only certain types of particles to pass is a proposition, since the particle either passes or does not pass; and a proposition is a special type of observable that has two values, 0 and 1[17]. A question is an experiment leading to alternatives yes/no. It can also be called a test [*HOLEVO* 1982, 15]. A test allows to decide about a property of a system, i.e. if it has this property or not. Hence this approach is a realistic one[18]. The proposition $a \wedge b$ is true if the system passes an infinite sequence of alternating filters for a and b respectively and is otherwise false [*JAUCH* 1968, 75] [see figure 9.6].

The objection, made by Heelan [HEELAN 1970b], that one can never *de facto* produce an infinite sequence of experiments is not a good one because it is possible to work with approximations. The problem is whether the alternating sequence $\hat{P}_a, \hat{P}_b\hat{P}_a, \hat{P}_a\hat{P}_b\hat{P}_a, \ldots$ approaches the same limit as the sequence $\hat{P}_b, \hat{P}_a\hat{P}_b, \hat{P}_b\hat{P}_a\hat{P}_b, \ldots$, and whether the limit itself is a projection operator. Hence Shimony [SHIMONY 1971a] showed the necessity of imposing additional restrictions on the structure of the class of filters to be used. McLaren, inspired by Birkhoff, proposed to measure $\hat{P}_a$ on one system in the state in discussion, then measure $\hat{P}_b$ on a second system in the same state, then $\hat{P}_a$ on a third system in the same state, and so on, and to define $a \wedge b$ as true if all these measurements yield the value 1[19].

Now we state another important property of the QM lattice by means of the following theorem [JAUCH/PIRON 1969, 428] :[20]

[13]See [*JAMMER* 1974, 351–53] for historical reconstruction.

[14]This is a consequence of the combination of the following assumptions: modularity, finiteness, orthocomplementarity, measurability, isotony of measures.

[15]In particular the problem of localizability leads one to reject modularity [*JAUCH* 1968, 83–84, 123, 219–21].

[16]On this point see also [HARDEGREE 1979] [*BELTRAMETTI/CASSINELLI* 1981, 105–109].

[17]See also [GUDDER 1979, 327].

[18]As pointed out in [LUDWIG/NEUMANN 1981, 134].

[19]Historical reconstruction in [*JAMMER* 1974, 354–55].

[20]See also [*BELTRAMETTI/CASSINELLI* 1981, 98, 133–36].

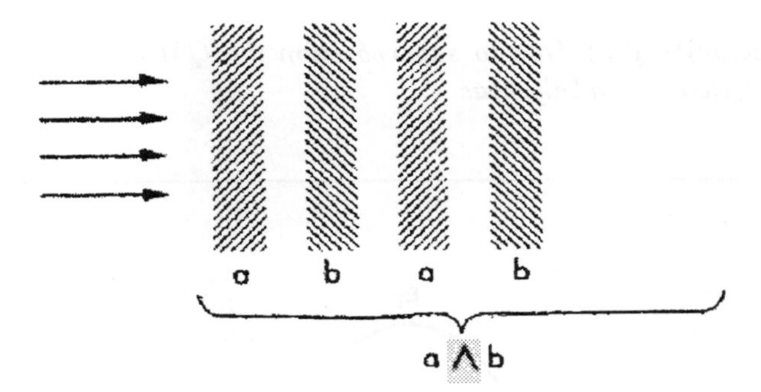

Figure 9.6: Construction of a filter for the propositions $a \wedge b$ in QM — [*JAUCH* 1968, 75].

Theorem 9.3 (Jauch) *The QM lattice also is atomic and fulfills the covering property.*

An element a of a lattice is an *atom* if

$$(\forall b)(\mathbf{0} \preceq b \preceq a) \rightarrow [(b = \mathbf{0}) \vee (b = a)]. \tag{9.13}$$

A lattice is *atomic* if

$$(\forall b \neq \mathbf{0})(\exists a)(a \preceq b), \tag{9.14}$$

where a is an atom. a *covers* b if $a \prec b$ and if

$$(b \preceq c \preceq a) \rightarrow [(c = a) \vee (c = b)]. \tag{9.15}$$

A lattice has the *covering property* if for every atom a and every b such that $b \wedge a = \mathbf{0}$, the element $b \vee a$ covers b.

The *center*[21] of a proposition system $\mathcal{L}_{\succeq,\perp,a_n}$ is constituted by all propositions which are compatible with all propositions of $\mathcal{L}_{\succeq,\perp,a_n}$ — for more details about the concept of compatibility see examination in subsection 18.2.1. An $\mathcal{L}_{\succeq,\perp,a_n}$ is coherent if its center is trivial (it contains only $\hat{I}, \emptyset$). If it is non-trivial, it is always reducible (i.e. any elemnt non pertaining to the center can be decomposed by means of elements pertaining to the center). Now we state following theorem (which, though true, we will not incorporate into the theory because not valid in QM):

Theorem 9.4 (Reducibility) *A Boolean algebra is totally reducible, i.e. down to the last elements $\hat{I}, \emptyset$.*

Therefore irreducibility signifies [*JAUCH* 1968, 124–27] [PIRON 1972, 523–24] the unrestricted validity of the Superposition principle[22]. Each propositional system can be univocally reduced to irreducible ones, so that Boolean algebra is only a limiting class of a more general theory, i.e. classical mechanics can be understood as a limiting case of QM. On the other hand, QM, though not a Boolean algebra as such, as we have seen, can be understood as a collection of Boolean subalgebras [see figure 9.7]:

[21] Definition in [*VARADARAJAN* 1968, 27] [*MACKEY* 1963, 71].

[22] See also [*BELTRAMETTI/CASSINELLI* 1981, 128–29, 165–67].

Theorem 9.5 (Irreducibility) *Due to superposition the QM lattice is irreducible, i.e. it is a collection of maximal Boolean subalgebras*

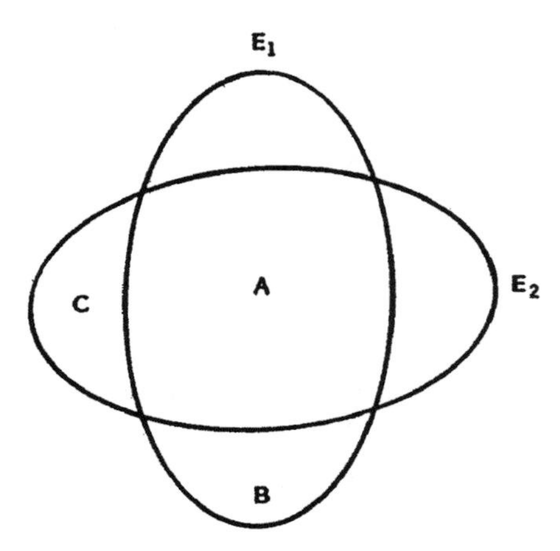

Figure 9.7: Example of QM Boolean subalgebras: the vector space E is the union of E_1 and E_2, and spaces for observables A, B, C are represented. Now the binary relation $C \vee A$ (both belonging to E_2) and $A \vee B$ (both belonging to E_1) does not imply $C \vee B$ (C belongs to E_2 and B belongs to E_1) — from [BASS 1992, 8].

A concrete **example** — of the type already analysed [see Interfer. Exp. 4: p. 146] — is given by Quadt [QUADT 1989, 1030] [see also section 34.3]: it is an example of two Boolean sub-algebras: 1. $\mathcal{L}(\mathcal{A}_{\mathrm{A}}) = \{\mathbf{1}, \mathbf{0}, \mathrm{D_A}, \mathrm{D_A}'\}$, where $\mathrm{D_A}, \mathrm{D_A}'$ are settings for interference–detecting instrument $\mathcal{A}_{\mathrm{A}}$, and 2. $\mathcal{L}(\mathcal{A}_{\mathrm{B}}) = \{\mathbf{1}, \mathbf{0}, \mathrm{D_B}, \mathrm{D_B}'\}$, where $\mathrm{D_B}, \mathrm{D_B}'$ are settings for path instrument $\mathcal{A}_{\mathrm{B}}$ [see figure 9.8].

Jauch derived the notion of lattice directly from propositional calculus (but the propositions of a physical system are a complete, orthocomplemented lattice and not necessarily a logic, even if in the following we assume them to be a logic) and takes the relevant lattice as being isomorphic to that of the subspaces of Hilbert space [*JAUCH* 1968, 74–84].

Every propositional system can be embedded into a projective geometry which is algebraically isomorphic to a linear manifold of some linear vector space with coefficients from a field[*JAUCH* 1968, 127–29]. QM uses coefficients from a complex field. Now, orthocomplementation [Eqs. (9.8)] imposes restrictions on the field of numbers [*JAUCH* 1968, 129–30]. From here we can derive a definite metric. If the field must contain real numbers there are only three possibility: real, complex and quaternionic numbers. We shall later examine these possibilities [see section 10.1]. Jauch adopts the second [*JAUCH* 1968, 131–32], from which complex Hilbert spaces. From this it can be derived that the propositional system is $\mathcal{P}(\mathcal{H})$ (the ensemble of projectors on a given Hilbert space $\mathcal{H}$).[23]

[23]See [*BELTRAMETTI/CASSINELLI* 1981, 87–88, 105–108].

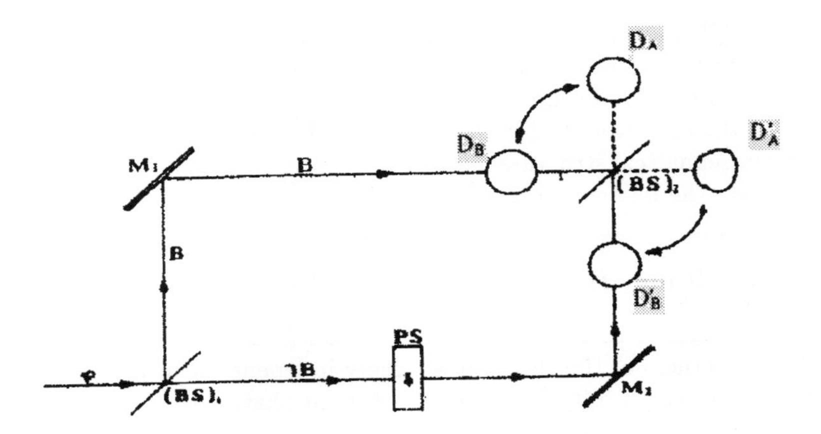

Figure 9.8: Experimental arrangement (= Interfer. Exp. 5) proposed by Quadt — from [QUADT 1989, 1029, 1031].

Later [see p. 199] we shall introduce states as σ–additive positive functionals on projection operators, i.e. as a probability function μ_p on all propositions pertaining to $\mathcal{L}_{\succeq,\perp,a_n}$ [*JAUCH* 1968, 93–94] and on all observables as a σ–homomorphism on $\mathcal{L}_{\succeq,\perp,a_n}$ [*JAUCH* 1968, 98]. The probability measure on a σ–orthocomplete POSet is a mapping to $[0,1]$ which satisfies[24]:

- $\mu_p(1) = 1$,
- $0 \leq \mu_p \leq 1$,
- if $a_1, a_2, \ldots$ is a sequence of mutually orthogonal elements then: $\mu_p(\vee a_i) = \sum \mu_p(a_i)$.

9.2.3 Conclusions on Quantum Logic

The great problem of quantum logic is that it is so general that it is not useful as model of QM. We need more axioms than the assumptions made so as to build a bridge to Hilbert spaces. This was the step taken by Varadarajan [*VARADARAJAN* 1968]. But many of the additional axioms have no convincing physical justification[25].

On the other hand, some theorems of Quantum logic represent general results whose validity is not confined to this area.

9.3 Other Approaches

9.3.1 Algebraic Approach

The beginnings of the algebraic approach are credited to the joint work of Pascual Jordan together with Wigner and von Neumann [P. JORDAN *et al.* 1934]. Other contributions are represented by the work of Segal, Haag and Kastler, Araki.

[24]See also [*BELTRAMETTI/CASSINELLI* 1981, 111] [*SCHROECK* 1996, 118].

[25]On this point see [GUDDER 1979, 345].

Segal [SEGAL 1947] assumes *observables* to be primitive[26] and an algebra with natural physical assumptions of linearity, positivity and normalization is constructed. As a consequence there are states which have a place here though not in any classical theory. The aim was to prove the possibility of a QM theory without Hilbert spaces and to show the possibility of deducing UP from an abstract mathematical structure.

Hence we can state the following theorem:

Theorem 9.6 (QM without Hilbert spaces) *Hilbert spaces are not necessary for QM.*

Note that the correctness of the theorem is largely independent form the algebraic approach as such: in fact one has successfully constructed a QM in phase spaces [see subsection 5.2.7].

On a general level, the enterprise of Haag and Kastler [HAAG/KASTLER 1964] of constructing an algebra of observables (which are not intended as operators on Hilbert spaces) can be seen as an attempt to construct a Relativistic-QM which presents no non-local effects. The first hint in this direction is again thanks to Jordan [P. JORDAN 1932]. The starting point can be following principle (for the concept of compatibility see subsection 18.2.1):

Principle 9.1 (Algebraic Locality) *Observables in causally disjoint space-time regions are always compatible.*

The problem can be examined as follows [HAAG/KASTLER 1964, 852]: we define a *partial state* with respect to a region V_1 as a positive linear form over the algebra $\mathcal{L}(V_1)$, such that we call the partial states of two regions V_1 and V_2 dynamically uncoupled if, by choosing an arbitrary pair of partial states pertaining the first to V_1 and the second to V_2, one can find a global state which is an extension of both (there is complete coupling if each partial state in V_1 determines uniquely a partial state in V_2). Hence we can restate the Algebraic principle of locality by saying that if V_1 and V_2 are causally disjoint, then the partial states of the two regions are uncoupled. On the other hand, we say that there is a causal relation between V_1 and V_2 if V_2 is contained in the causal shadow (future light cone) of V_1, so that the partial states in V_2 are uniquely determined by those of V_1.

Attention must be paid to this concept of *disjointedness*: It can in no way be interpreted in the sense that, without dynamical interaction, two systems are completely uncoupled[27], because, as we know, in QM there can be entanglement [see definition 3.3: p. 43, and discussion].[28] On this point see especially section 40.7.

Another problem is that of *representations*. One one hand, one should avoid the multiplication of representations in different Hilbert spaces of the same algebra. On the other hand, one should have all the required representations. Quantum logic builds an orthomodular lattice [see Eq. (9.9) and theorem 9.2: 162] by a set of yes/no propositions [see p. 162], which turns out to be exactly the lattice which we construct with the set of projectors (trace-one, idem-potent self-adjoint operators) onto the closed subspaces of a Hilbert space. But there are orthomodular lattices which do not have representations with projectors. In order to avoid such problems, and following here the work of Emch [*EMCH* 1972, 70],[29] the necessity arose to use C* algebras with observables as primitive elements. We first give a general mathematical definition of a C* algebra:

---

[26] See also [GUDDER 1979, 324, 327, 332–39, 338] [ARAKI 1964] [*JAMMER* 1974, 380–83].
[27] As some formulations of the authors seem to suggest.
[28] In other words, algebraic locality must be understood only in the sense of theorem 36.3 [p. 634].
[29] See also [*DAVIES* 1976, 6–7, 30–33] [*HAAG* 1992, 5] [*JAUCH* 1968, 142–49].

Definition 9.10 (C* Algebra) *A C* algebra is an involutive Banach (a complex or real associative, complete with respect to the topology induced by the norm and such that* $\| \hat{O} \| = \| \hat{O}^* \|$ *) algebra such that* $\| \hat{O} \|^2 = \| \hat{O}^* \hat{O} \|$ *for all observables* $\hat{O}$*'s.*

One assumes that there are sufficiently many dispersion-free states in order to distinguish the observables using expected values[30]. By the linearity of expected values one equips the set of observables with the structure of a real linear space. By assuming associativity one obtains a C* algebra. And this turns out to be an algebra of bounded operators on a Hilbert space. Thus we have answered not only the question of why Hilbert spaces are necessary [theorem 9.6: p. 166] but also the question about which Hilbert space is appropriate: different Hilbert spaces can be constructed depending on the states used in the construction[31].

Then the central theorem can be formulated as follows:

Theorem 9.7 (Representations) *Two representations are physically equivalent if every neighbourhood of any element of one subset of states corresponding to one representation contains an element of the other and vice versa.*

Now the concept of physical equivalence corresponds to that of weak equivalence, and hence we can use the Fell Theorem [HAAG/KASTLER 1964, 851]:

Theorem 9.8 (Fell) *Two representations are weakly equivalent iff they have the same kernel.*

In consequence of theorem 9.7, the relevant object becomes the algebra and not the representation. The problem here, as already said [see again subsection 5.2.7], is that C* algebras works with symmetry conditions and not with self-adjointness. But the problem can be overcome by the Naimark theorem [18.5: p. 301] and the Holevo corollary [18.2: p. 301]. For further developments of this approach see also subsection 17.1.1].

9.3.2 Convexity Approach

Representatives of this school are Stone, Ludwig, Mielnik, Davies, Edwards, Rüttimann. *States* are here taken as primitive.

The principal concept is that of *face*[32]. A subset $\mathcal{X}$ pertaining to the convex set of density matrices is said to be a face iff for some weights w_1, w_2 and states $\hat{\rho}_1, \hat{\rho}_2$ we have:

$$(\hat{\rho}_1, \hat{\rho}_2 \in \mathcal{X}) \rightarrow (w_1\hat{\rho}_1 + w_2\hat{\rho}_2 \in \mathcal{X}), \tag{9.16}$$

$$(w_1\hat{\rho}_1 + w_2\hat{\rho}_2 \in \mathcal{X}) \rightarrow (\hat{\rho}_1, \hat{\rho}_2 \in \mathcal{X}). \tag{9.17}$$

The first condition is the same as (3.47). It says that faces are convex subsets; the latter that faces are convex in the strong way that, if a face contains an internal point of a 'segment', then it must contain the whole 'segment'. If the convex sets of density matrices have extreme points (i.e. pure states), then each extreme point is a face.

This approach has very important consequences for the problem of open systems [section 21.1] and for a stochastic operational point of view [see section 18.1].

---

[30] See the Gleason Theorem [p. 200].

[31] As we shall see in section 10.1. On this problem see [*SCHROECK* 1996, 26–27, 112–16] [*EMCH* 1972].

[32] Exposition in [*BELTRAMETTI/CASSINELLI* 1981, 208–216] .

9.3.3 States/Observables Approach

Representants of this school are Mackey and Maczyński[33]. *Both states and observables* are taken as primitive. Mackey [*MACKEY* 1963, 75–76] developed a parallelism between QM and classical probability calculus: observables are the random variables [see also section 11.3] and the states are probability measures. It was tried to extend the discussion by inserting joint probability measures and conditional expectations: but the first exist only if the observables commute[34], and the second only if observables have discrete spectrum [DAVIES/J. LEWIS 1970, 239].

The starting point is then the probability of obtaining in the state $\hat{\rho}$ a value in a set $\mathcal{X}$ of real number when an observable $\hat{O}$ is measured: $\wp(\hat{O}, \hat{\rho}, \mathcal{X})$. Now we can identify propositions with probabilities and order them by degrees of probability. Hence we write:

$$a \leq b \text{ whenever } a(\hat{\rho}) \leq b(\hat{\rho}),\ \forall \hat{\rho} \in \mathcal{H}^{\mathrm{Tr}=1;\geq 0}, \tag{9.18}$$

which corresponds to the ordering relation (9.1). The central axiom, which makes $\mathcal{L}$ orthomodular and allows us to see $\mathcal{H}^{\mathrm{Tr}=1;\geq 0}$ as a set of probability measures on $\mathcal{L}$, is the following:

Postulate 9.1 (Mackey) *For every pairwise orthogonal sequence $a_1, a_2, \ldots, \in \mathcal{L}$ (set of propositions), $\exists b \in \mathcal{L}$ such that for every $\hat{\rho} \in \mathcal{H}^{\mathrm{Tr}=1;\geq 0}$, we have $b(\hat{\rho}) + a_1(\hat{\rho}) + a_2(\hat{\rho}) + \ldots = 1$.*

However, Mackey assumed that the POSet $\mathcal{L}$ of all propositions of QM is isomorphic to the POSet of all (closed) subspaces of a separable, infinite-dimensional complex Hilbert space $\mathcal{H}$[35] in order to create a bridge with ordinary QM. But, as pointed out[36], this axiom is completely *ad hoc* [see methodological principle M4: p. 6]. For further developments see sections 11.4 and 39.3.

9.4 Concluding Remarks on the Axiomatic

9.4.1 Relation Between State and Observables

Now the question is: is a model of QM possible? And, eventually, which one of the models just analysed is preferable? We have seen that Quantum logic is not very suitable as a model of QM: the propositions are too abstract a tool, and they do not cover the most interesting developments of these last years, the effects [see section 18.1]. On the other hand, the importance of UP or of the superposition principle can bring to the conclusion that in QM the observables are central. In this sense the algebraic approach would be the best one, and one would affirm a fundamental dependence of states on observables[37]. But there is at least a property of the states which in no way depends from observables: the factorizability [see definition 3.2: p. 42]. Therefore we state following theorem[38]:

[33]Exposition in [*BELTRAMETTI/CASSINELLI* 1981, 146–49] .

[34]Or are POVM [see chapter 18].

[35]On this point see [*JAMMER* 1974, 387].

[36]See [*BELTRAMETTI/CASSINELLI* 1981, 150–51].

[37]As Finkelstein recently does [*FINKELSTEIN* 1996].

[38]The following examination owes very much to several discussions with Dr. Fortunato.

Theorem 9.9 (Factorizability) *If a system is factorisable in one basis, it is also in all other basis.*

We prove it for a two dimension system (generalization can follow by mathematical induction):

Proof

Suppose $|\Psi\rangle = |\psi\rangle_1 |\psi'\rangle_2$. Let $|b_j\rangle_1$ and $|b'_k\rangle_2$ be two arbitrary basis of subsystems $\mathcal{S}_1, \mathcal{S}_2$, respectively; then we have a result of the following form $|\Psi\rangle = [\sum_j c_j |b_j\rangle_1][\sum_k c'_k |b'_k\rangle_2]$. It is easy to see that each other basis must respect this factorization. Q.E.D.

The immediate consequence of the theorem is that factorizability is a property of the state, independently from the basis (observable).[39].

Note that the impossibility of measuring the density matrix of a single system — a recently developed result [see section 28.5] — is a further confirmation of the above conclusion. It is an evidence that the state cannot be measured and hence treated as an observable. Which, on the other hand, also implies that QM cannot only be a theory of states, as proponents of the convenxity approach wish. In general, quantum states are not classical states, i.e. they are not complete collections of all properties one may assign to a system (in a time interval) [see corollaries 3.1: p. 32, and 7.1: p. 127].

This analysis is further confirmed by the fundamental principles upon which the theory is founded. The CCRs [see Eq. (3.14)] are basic relations which are valid for observables independently from the states. The URs (and UP) are also to a certain extent independent from the states: the only condition is that the conjugate observables be taken on the same state, a condition which can be partly avoided by information-theory formulations of the URs [see section 42.2]. It is true that in Basic-QM not all observables are associated to operators [see section 7.4], but, as we shall see, the problem can be partly avoided [see chapter 10]. On the other hand, the entanglement, which in first analysis is a mathematical tool of the theory [see definition 3.3: p. 43 and discussion], is the non factorizability of the probability distributions of compound systems, and this (to a certain extent) independently from the choice of observables. Finally the Superposition principle [p. 39] partly connects observables and states [see also subsection 7.1.4]. In conclusion we have at least two primitive physical entities: observables and states.

9.4.2 Models and Contributions

Models and Completeness In conclusion, up to now, we have found no model which is satisfactory or complete (so as to cover all QM) from all points of view. If this conclusion is sound and it is confirmed by future developments, then QM cannot be an axiomatic system of theorems. We are therefore forced to develop a concept of science which is more adequate to this situation. The new image of science suggested by this examination is that of a *collectio* of partly independent postulates and theorems and not of a system[40]. One may recall here Gödel's theorem [GÖDEL 1931][41] or our choice of an Open Logic [see p. 2], the last justification of which is to be found later in the assumption of an operational point of view [chapter 18].

---

[39] Note that this distinction is not coincident with that between pure states and mixtures, because one can have factorisable pure states and mixtures which are not factorisable [see section 38.2].

[40] As historically pointed out for the fist time by William of Ockham [*OCKHAM* SL, 537] [*OCKHAM* ELP].

[41] On this point see [*T. BREUER* 1997] [*MITTELSTAEDT* 1998, 122–24].

Does this mean that QM is an incomplete theory? The problem is understanding what we mean by 'complete'. If by 'complete' we mean a closed system, then QM is certainly not complete (anyway classical mechanics too was probably not a closed system). If we mean a theory which mirrors every 'element of reality' (*semantical completeness*) [42], then this is only a mythology which conflicts with the reality of theories' pluralism — not only between different disciplines but also in the framework of only one, say physics, and to a certain extent also in QM itself (the plurality of different approaches can be considered as an expression of this situation). Note that semantical completeness is granted for a closed theory. Note also that there are surely very important problems which are not yet solved — for example the problem of quantum transitions [see subsection 6.1.2].

But, on the other hand, QM cannot be an 'incomplete theory' in the sense that there are some physical entities which are in open conflict with the predictions of the theory or which are not at all analysable in the frame of QM, because in that case QM would be a fallacious theory — this point will be object of a later discussion [see part VIII]. And surely not in the sense that there are internal contradictions, what would make QM absurd. All recent developments have confirmed up to now the predictions of the theory.

Hence we are forced to choose a middle solution between both extremes. We shall return later on the point of the applicability of the theory to domains different from microreality [see particularly chapter 24].

Contributions On the other side note that each one of the examined approaches has given many specific contributions. Quantum logic: the theorems on the Orthomodularity [9.2: p. 162], the theorem of Jauch [9.3: p. 163]. The algebraic approach: the theorem on the possibility of QM without Hilbert spaces [9.6: p. 166] and that on Representations [9.7: p. 167]. Concerning the convexity approach, as we have already said, we shall return later to its very important results.

9.5 QM Logical Calculus

Logic can be understood in a more concrete way: as a calculus which reflects and helps different empirical domains. So understood, it is clear that the problem now is: what are the logical connectives and rules which best reflect QM?[43].

The initiator of this approach was Mittelstaedt, whose work began in the late fifties and sixties. While all studies in QM logic presupposed the work of Birkhoff/von Neumann, Mittelstaedt [MITTELSTAEDT 1959] [MITTELSTAEDT 1960] [MITTELSTAEDT 1961] [*MITTELSTAEDT* 1963] based is work on that of Lorenzen [*LORENZEN* 1955].

The work of Mittelstaedt was further developed by Hardegree[44]. If we want a principle of objectification [see definition 14.1: p. 235], we must renounce to the unrestricted availability[45].

---

[42]On this point see [*T. BREUER* 1997, 22–26].

[43]On this point see [FINKELSTEIN 1969] [FINKELSTEIN 1972, 143–46] [*JAMMER* 1974, 404]. Garola [GAROLA 1989] [GAROLA 1992] [GAROLA/SOLOMBRINO 1996] proposes to consider a classical logic, from which, by means of specific constraints, an orthomodular lattice and a QM logical calculus can be derived. To a certain extent (if one does not wish to wipe out the specific quantum features) such a distinction can fit in with our analysis.

[44]See [HARDEGREE 1979, 55].

[45]See [*JAMMER* 1974, 394–98].

Choice disjunction and negation The QM concrete logic $\mathcal{L}_{QM}$ — on a semantical level — is closed under the infinitary conjunction but it is not closed under the *exclusion disjunction* (a disjunction which is true iff only one disjoint is true). In fact, due to the Superposition principle, $\mathcal{L}_{QM}$ obeys to a *choice disjunction*, such that, if we have closed subsets $\mathcal{X}, \mathcal{X}'$ and we consider their supremum, i.e. the smallest subset which includes both, $\mathcal{X} \cup \mathcal{X}'$, this consists in all linear combinations of $\mathcal{X}$ and $\mathcal{X}'$. Generally, it strictly includes $\mathcal{X} \cup \mathcal{X}'$, i.e. there are propositions true in $\mathcal{X} \cup \mathcal{X}'$ but neither in $\mathcal{X}$ nor in $\mathcal{X}'$, which means that the disjunction of some propositions a, b can be true for some state $\hat{\rho}$ but it may be that neither a nor b is true in $\hat{\rho}$ — exactly what happens when we have a superposition or an entanglement.

As a consequence of the choice disjunction, in QM an exclusion negation cannot always be valid. While the formal analogue of the law of excluded middle $(a \vee \neg a)$ is still valid in $\mathcal{L}_{QM}$, i.e. every state satisfies $a \vee a^{\perp}$, it is not closed under *exclusion negation*: i.e. we have not necessarily a or $a^{\perp}$ (which is a particular realization of the choice disjunction). Hence we have a *choice negation.* Now, although both intuitionistic logic and three-valued logic [see chapter 29] employ a choice negation, they nevertheless employ a classical choice disjunction [HARDEGREE 1979, 56].

Truth-value gaps Thus $\mathcal{L}_{QM}$ is not bivalent, i. e. some projectors need not to have a (classical) truth value. But it is not necessary to suppose that $\mathcal{L}_{QM}$ admits third values: there are simply propositions which are truth-valueless, i.e. $\mathcal{L}_{QM}$ admits truth-value gaps[46]. It is exactly the situation showed by the Kochen/Specker Theorem or by the Bell Corollary of the Gleason Theorem [see chapter 32].[47]

In conclusion we state following theorem:

Theorem 9.10 (Hardegree) *QM logical calculus is characterized by choice disjunction and choice negation, and consequently it presents truth-value gaps.*

The above theorem is in direct contradiction with the classical assumption that observables always have a determinate value [see postulate 6.3: p. 105].

We can translate it into a propositional calculus. Propositions are represented by projection operators. If the state $\hat{\rho}$ is an eigenvector of some projector $\hat{P}$, then there are only $0, 1$ as possible values. But if not, $\hat{P}$ may be regarded as neither true, nor false at $\hat{\rho}$. In fact, we can see this by analysing the case when two observables commute. Let the two observables be expressed in terms of projectors $\hat{P}, \hat{P}'$; then we define a commutativity projection as follows[48]:

$$\hat{C}(\hat{P}, \hat{P}') := (\hat{P} \wedge \hat{P}') \vee (\hat{P} \wedge \hat{P}'^{\perp}) \vee (\hat{P}^{\perp} \wedge \hat{P}') \vee (\hat{P}^{\perp} \wedge \hat{P}'^{\perp}). \tag{9.19}$$

The operator $\hat{C}$ is the projector onto the subspaces of vectors on which $\hat{P}$ and $\hat{P}'$ commute. In other words, if two observables commute, then we have no value gaps, but both always have a determined value, 1 or 0.[49] But this is normally excluded by the non-commutability of observables.

---

[46]On this problem see [VAN FRAASSEN 1966] [VAN FRAASSEN 1969] [THOMASON 1970] .

[47]From which A. Fine [A. FINE 1976, 252–53] deduces the incompleteness of QM.

[48]Definitions in [*DRIESCHNER* 1979, 24] [BUSCH 1987b].

[49]A similar reasoning is in [*PITOWSKY* 1989b, 58–61], where the correspondence between the structure of a Hilbert space and a logical calculus is shown to be difficult because the latter cannot be closed under negation and disjunction [*PITOWSKY* 1989b, 59–60, 107, 146]. Hence it is proposed to treat propositions and not equivalence classes [*PITOWSKY* 1989b, 108].

In conclusion a logic can be a Boolean algebra only if one assumes that all non-atomic propositions[50] a, b can be split into disjoint propositions, such that there exist propositions a_1, b_1, c by which: $a = a_1 \vee c$ and $b = b_1 \vee c$. Only on this assumption the implication becomes an inclusion and the negation a complementation. But such an assumption is not valid in $\mathcal{L}_{QM}$[51].

QM implication Theorem 9.10 also has a consequence for the structure of QM implication [HARDEGREE 1979, 58]. We have distinguished between the metalanguaged $a \vdash b$ from the objectlanguaged $a \rightarrow b$ [see p. 156] and said that the lattice order relation $a \preceq b$ corresponds to $a \vdash b$. From the transitivity property derives a weakening law which is not satisfied by counterfactuals. But — as we shall see — QM is to a certain extent counterfactual[52], so that we must reject transitivity [HARDEGREE 1979, 60–62]. The conditional which has been proposed for QM is the Sasaki arrow $\Rightarrow$, which is defined[53] by:

$$a \Rightarrow b := a^{\perp} \vee (a \wedge b) \tag{9.20}$$

and which clearly respects the property of orthomodularity [see definition 9.8], which in turn characterizes QM [see theorem 9.2].

The Sasaki arrow is both residual and locally Boolean, where the last signifies that it agrees with the classical conditional on all Boolean subortholattices of the orthomodular lattice in question — and, as we shall see, the QM orthomodular lattice has Boolean subalgebras. In other words, $a \Rightarrow b = a^{\perp} \vee b$ if we have $a \rightarrow b$. Obviously the Sasaki arrow is different from the classical conditional only under the assumption (characteristic of QM) that distributivity [Eqs. (9.6)] is not valid; because if it were, then the RHS of formulation (9.20) would be equivalent to $(a^{\perp} \vee a) \wedge (a^{\perp} \vee b)$, which, given that $(a^{\perp} \vee a)$ is always true (it is the law of *tertium non datur*), in turn would be equivalent to $a^{\perp} \vee b$, which is equivalent to the classical implication. The Sasaki arrow can be interpreted as a Stalnaker conditional [HARDEGREE 1979, 64–67].

[50]Which are analyzable in further propositions.

[51]See [VARADARAJAN 1962, 181–83].

[52]On the problem of counterfactuals see [*D. LEWIS* 1973] [PIZZI 1978b]. Also the yes/no experiments of Jauch and Piron imply a counterfactual character of QM, as pointed out in [HARDEGREE 1979, 72–75].

[53]See [*MITTELSTAEDT* 1963, 187–88] [*MITTELSTAEDT* 1978] [*DRIESCHNER* 1979, 23].

Chapter 10

HILBERT SPACES AND OPERATORS

Contents In the last chapter we saw a possibility of avoiding Hilbert spaces. Now we discuss what kinds of Hilbert spaces are possible [section 10.1]; superselection rules are also discussed. Then the main part of the chapter is devoted to the a possibility of constructing operators for those observables which in Basic-QM are not represented by operators. Section 10.2 is devoted to the time and energy operators, while section 10.3 to the position operator in a relativistic frame. Finally, section 10.4 discusses the problem of the angle operator.

10.1 Is a Complex Hilbert Space Necessary?

As we have seen, the Quantum Logic [subsection 9.2.2] and the Algebraic approach [subsection 9.3.1] try to deduce the necessity of Hilbert spaces in QM.

Now we wish to investigate how many possibilities there are for constituting a field for an algebra suitable to assure a metric. If the field must contain real number — and it must, because the values of observables and probabilities are real numbers [see p. 32, and postulate 3.3: p. 37] —, we have only three possibilities: real numbers, complex numbers and quaternions [BIRKHOFF/VON NEUMANN 1936]. In the following we synthetically develop these three possibilities in order to see if a complex field is a necessity[1].

10.1.1 Real and Quaternionic Hilbert Spaces

Real Hilbert spaces The possibility of a field of real numbers was developed by Stueckelberg [STUECKELBERG 1960] [STUECKELBERG/GUENIN 1961]:[2] it is possible by introducing a superselection rule, by means of an operator which is commutable with all others[3]. The importance of the superselection rules requires separate treatment [see subsection 10.1.4]. Here we limit ourselves to remarking that such an introduction of superselection rules at a fundamental level of the theory is not very convincing, if one takes into account the theoretical problems which depend on their introduction [see again subsection 10.1.4].

[1]On this point see also [*BELTRAMETTI/CASSINELLI* 1981, 230–41].

[2]See also [MACZYŃSKI 1973].

[3]On this point see also [*JAMMER* 1974, 358–59].

Quaternionic Hilbert spaces The possibility of a field of quaternions (number with three different imaginary units which are anticommuting) was developed by Finkelstein in collaboration with Jauch, Schiminovich and Speiser [FINKELSTEIN *et al.* 1959] [FINKELSTEIN *et al.* 1962] [FINKELSTEIN *et al.* 1963a].[4]

Finkelstein, Jauch and Speiser [FINKELSTEIN *et al.* 1959, 368] justified the introduction of quaternions with the fact that mathematically there can be four different eigenvalue problems for each operator $\hat{O}$ which are not distinguished in the frame of Basic-QM:

$$\hat{O}|\psi\rangle = o|\psi\rangle, \tag{10.1a}$$
$$\hat{O}|\psi\rangle = |\psi\rangle o, \tag{10.1b}$$
$$|\psi\rangle\hat{O} = o|\psi\rangle, \tag{10.1c}$$
$$|\psi\rangle\hat{O} = |\psi\rangle o, \tag{10.1d}$$

and which are distinguished in the frame of quaternionic QM.

But, on the other hand, the price is that the usual machinery connected with eigenvalue problems — for example that of characteristic polynomials and determinants — does not work well on quaternions [FINKELSTEIN *et al.* 1959, 384–85]. Also, the introduction of the concept of co-unitarity [FINKELSTEIN *et al.* 1959, 381–82], typical for quaternionic Hilbert spaces, does not seem very significant for concrete QM problems[5].

10.1.2 Deductions of Complex Hilbert Space

A good analysis of the significance of complex Hilbert spaces has been performed by Strauss We start with the typical feature of QM probability — the non-existence of a joint probability for incompatible operators, which is a consequence of the non-commutability —, by considering the expression $\mathrm{Tr}[\hat{P}_1\hat{P}_2\hat{P}_3]$, which is the expectation of a finale state represented by $\hat{P}_3$, given middle state $\hat{P}_2$, expected on a given initial state $\hat{P}_1$. Now one can choose between a representation in real (Euclidian) space or in one a complex space (Strauss does not consider the possibility of quaternions). But the first alternative is impossible because, this expression is real-valued even if none of the three projectors commutes with the other two, *i.e.* if none of the threee expressions — by permutation — have any physical meaning. Therefore the space must be complex and he metric unitary: iff the expression is complex-valued, none of the operators commutes with any of the other two [STRAUSS 1936, 34–39].[6]

10.1.3 Wootters' Proposal

A very interesting approach was developed by Wootters [WOOTTERS 1981]. He searched for a 'statistical distance' between different quantum preparation procedures [see section 13.4] in a series of N trials, where statistical fluctuations are responsible for some disagreement between the probabilities and the frequencies. As N goes to infinity, Wootters deduced following distance relating the probabilities of the occurrence of the two events:

$$\delta_{\mathrm{S}}(\wp^{(1)}, \wp^{(2)}) = \cos^{-1}\left[\sum_{j=1}^{\infty}(\wp_j^{(1)})^{\frac{1}{2}}(\wp_j^{(2)})^{\frac{1}{2}}\right]. \tag{10.2}$$

[4] Results which were later confirmed by Emch [EMCH 1963].

[5] For further analysis see also [*FINKELSTEIN* 1996, 76–78] and [*BELTRAMETTI/CASSINELLI* 1981, 243–48].

[6] See also [*JAMMER* 1974, 355–57].

Then, in the case of photon polarization, he could deduce a statistical distance between states $|\psi_1\rangle$ and $|\psi_2\rangle$, relative to an apparatus $\mathcal{A}$, having the form

$$\delta_{\mathcal{A}}(\psi_1, \psi_2) = \cos^{-1} |\langle\psi_1|\psi_2\rangle|, \tag{10.3}$$

where the (absolute) statistical distance between two preparations is equal to the angle in Hilbert space between the corresponding rays. The similarity between Eqs. (10.2) and (10.3) is very important because it shows that one can convert a probability space into a section of the unit sphere in a real N-dimensional vector space, with probabilities being the squares of real amplitudes. Then Wootters conjectured that the Hilbert-space structure of QM can derive from the statistical fluctuations in the outcomes of measurements. This provided also one of the first bridges between traditional QM and quantum information [see part X].

In a later study, Wootters showed [WOOTTERS 1990] that complex Hilbert spaces are necessary in order to determine completely the state of a compound system through joint local measurements of the states of its subsystems, which would be not the case if use of real amplitudes only were allowed. For instance, supposing a system consisting of two photons flying in opposite directions, in order to determine the polarization of the two photons we need to determine (with a set, for each 'side', of 3 filters, selecting vertical, circular and at 45° polarizations, respectively) 15 different probabilities (3 for the cases when the photon passes through the right filter, three for when it passes through the left filter, and 9 for the joint passage), which corresponds to the numbers which one needs in order for determining the density matrix of the compound system (since it has four orthogonal states and, for determining the state of a system with n orthogonal states, one needs $N^2 - 1$ numbers). But, if we would allow only real amplitudes, we only had linear and no elliptical polarization, and then we could not determine the density matrix using only local measurements.

In conclusion we can say that there is not an absolute necessity of a complex Hilbert space but it is the best choice in order to fully express the formalism of QM and also for practical purposes.

10.1.4 Superselection Rules

As we have said, a very interesting (possible) feature of Hilbert spaces — especially interesting for, but not restricted to, real Hilbert spaces — are the Superselection Rules (= SSR).

Exposition of the SSRs

Superselection Rules were introduced by Wick, Wigner and Wightman [WICK *et al.* 1952] in connection with the problem of the elimination of superposition[7]. The basic idea of the authors was that they are implicitly contained in the formalism of the relativistic field theory. Fundamental limitations on superposition which seem to lead to SSRs are also the impossibility of a superposition of states with different electric charge or with integer and half-integer spin [ZEH 1996, 8].

First we define a Selection rule:

Definition 10.1 (Selection Rule) *A* Selection rule *is said to operate between subspaces of a given Hilbert space $\mathcal{H}$ if the state vectors of each subspace remain orthogonal to all state vectors of the other subspaces of $\mathcal{H}$ as long as the system remains isolated (i.e. there is no spontaneous transition between them).*

[7]On this problem see also [*BELTRAMETTI/CASSINELLI* 1981, 47–51]. .

It is a Selection rule which prevents any state of an isolated system from changing its total linear momentum. Now we can define a Superselection Rule (= SSR):

Definition 10.2 (Superselection Rule) *A SSR is said to operate between subspaces of a given $\mathcal{H}$ if*

- *there are selection rules between them,*
- *there are no measurable quantities with finite matrix elements between their state vectors (i.e. there are unobservable phase factors).*

SSRs generally situate a physical system between Boolean systems (classical ones) and coherent systems [*JAUCH* 1968, 109]. A classical system is one with a maximal number of SSRs — i.e. with no superposition.

Piron [PIRON 1969] introduced continuous SSRs. But the implantation of continuous SSRs can cause the Hamiltonian generator of the dynamic group not to be an observable[8].

Some Difficulties Regarding SSRs

There many difficulties regarding the SSRs. Through indiscriminate use of SSRs the Hilbert space breaks into little pieces and the door is opened for several exceptions to the Superposition principle [PIRON 1985, 207], with the consequence of a drastic reduction in the applicability of QM laws. The reason for their introduction would be then a classical or semiclassical solution to the measurement problem [see section 16.1]. But we shall see that the measurement problem can be solved also without using SSRs [see chapter 17]. And all recent experiments [see chapter 24] exclude the necessity of such a drastic reduction in the applicability of QM laws. On the other hand, if we employ continuous SSRs, the Hilbert space itself seems to disappear.

Further, SSRs are never strictly necessary, because, under the supposition that a Hilbert space $\mathcal{H}$ is broken by a SSR into a sum of coherent subspaces, it is always possible to consider these subspaces as different Hilbert spaces, each one representing a different physical system (i.e. there is no entanglement between those systems).

Nevertheless, in a more restricted sense, they can be used if one wishes[9]:

- that it not be the case that each self-adjoint operator on some $\mathcal{H}$ represent a physical quantity;
- that it not be the case that each unidimensional subspace of $\mathcal{H}$ represent a pure state.

On the other hand one can avoid the extreme consequences of SSRs by interpreting them as local effects of the correlation with the environment[10]. If so, they would only be relative to a reference frame. In this way one can interpret SSRs for mass, charge and particles with different spin, i.e. in the context of the symmetries[11].

[8] On this problem see [PIRON 1976].
[9] On this problem see [*BELTRAMETTI/CASSINELLI* 1981, 45–51].
[10] See [AHARONOV/SUSSKIND 1967].
[11] For analysis see [GIULINI *et al.* 1995].

10.2 Time and Energy Operators

10.2.1 General Remarks on the Relationship between Operators and Observables

Von Neumann [*VON NEUMANN* 1932, 163-71] [*VON NEUMANN* 1955, 324-25] supposed that every observable can be represented by a self-adjoint operator [see postulate 3.2: p. 33]. This assumption is very dubious, as we shall see now. Such an assumption was the basis of von Neumann's false proof of the impossibility of Hidden-Variables [see subsection 32.2.1].

At a general level Wigner [WIGNER 1952] [WIGNER 1983a, 297–98] showed that is very difficult to find in QM counterparts to some easy classical expression such as qp or q^2p^2: often there is no univocal expression, so that in general we must begin with the measured quantity and later find an operator to express it.

In that which follows we shall see that it is possible to construct operators for observables such as time, position in Relativistic-QM, and phase, but only by sometimes renouncing to the self-adjointness in favour of symmetry [see p. 32]. But, as we shall see, the Naimark theorem [18.5: p. 301] and the Holevo corollary [18.2: p. 301] guarantee the use of symmetry for more complex problems. Furthermore, as we have seen, symmetry is very important in order to construct a representation in phase space without using Hilbert spaces [see subsection 9.3.1].

10.2.2 A Short History of the Problem

Different Times As we have seen when discussing UP [subsection 7.5.1] time, energy, angle and phase are not operators in Basic-QM. The position operator originates a great problem in Relativistic-QM because relativity seems to place time and position at the same level [SCHRÖDINGER 1931b, 385] [see subsection 4.5.2].

In QM, time can be considered from two points of view:

- it is not an observable solidale with the system, but an extraneous topological ordering parameter[12];
- but it can also be considered as an observable of the system itself.

The first aspect will be treated discussing the measurement theory [part IV] and the 'delayed choice problem' [chapter 26]. The second problem poses a lot of questions, and it is often confused with the first one.

Proofs of the impossibility of a time operator From the beginnings it was showed an impossibility as to construct an operator able to express such an observable. Already Pauli [*PAULI* 1980, 63] [13] meant that it is impossible to find a time operator $\hat{t}$ such as to satisfy the CCR $[\hat{t}, \hat{H}] = \imath\hbar$, so that time and energy would be really conjugate. In fact if the energy is bounded from below, i.e. the Hamiltonian operator does not possess a continuous spectrum from $-\infty$ to $+\infty$ — and it must be so if we only want positive values and a ground state of the energy —, we cannot consider time as an operator because, as a result, $-E$ would follow. Or, formulated in other words [*BUSCH et al.* 1995a, 77–80]: the Hamiltonian with a semi-bounded spectrum

[12] Von Neumann found this situation unsatisfactory [*VON NEUMANN* 1932, 188] [*VON NEUMANN* 1955, 353–54]. See also [*JAMMER* 1974, 150].

[13] See also [*HOLEVO* 1982, 130–131] [*JAMMER* 1974, 141–4].

does not admit a group of shifts generated by some self-adjoint operator, which would then be the canonically conjugate time observable.

Again UR2 The first basic discussion began in the sixties. The starting point of the first discussions about the time operator was always the problem of UR2 [see subsection 7.5.1]. Aharonov and Bohm discussed [AHARONOV/BOHM 1961, 718–19] the proposal of Fock and Krylov [FOCK/KRYLOV 1947] of building a time operator: since this time operator would only be representative of the external order established by an apparatus $\mathcal{A}$ on a QM system $\mathcal{S}$, it should commute with every operator of $\mathcal{S}$, and particularly with its Hamiltonian $\hat{H}_{\mathcal{S}}$, that is, we could measure the energy of $\mathcal{S}$ in as a short time as we like. Fock [FOCK 1962] tried to refute the argument by treating the switching off and on of $\mathcal{A}$ in the Aharonov/Bohm argument as instantaneous.

Fock's answer is not sound because here we are dealing with a parameter and not with a dynamic observable subjected to the QM evolution of the combined system and hence, since UP does not apply here — E and t pertain to different systems, $\mathcal{A}$ and $\mathcal{S}$ respectively — we can measure the energy in an arbitrarily short a time [AHARONOV/BOHM 1964].[14]

Engelmann/Fick's and Paul's proposals Engelmann and Fick [ENGELMANN/FICK 1959] developed one of the first attempts to positively construct a time operator. If the time operator is independent of time parameter (and this is the case only if the Hamiltonian is independent of the time parameter), then we can have a commutation relation. But if both are dependent, we have no commutation relation. The expectation value of the time operator must exactly be the time parameter [ENGELMANN/FICK 1959, 69–71].

Paul's model [PAUL 1962] is the free one-dimensional motion of a particle of mass m and takes the following expression for $\hat{t}$:

$$\hat{t} = \frac{1}{2}m\left(\hat{q}\frac{1}{\hat{p}} + \frac{1}{\hat{p}}\hat{q}\right), \tag{10.4}$$

which in momentum representation is

$$\hat{t} = \frac{1}{2}\imath m\hbar\left(\frac{\partial}{\partial p}\frac{1}{\hat{p}} + \frac{1}{\hat{p}}\frac{\partial}{\partial p}\right) = \imath\hbar\left(\frac{\partial}{\partial E} - \frac{1}{4E}\right), \tag{10.5}$$

and which in the energy representation takes the form [FICK/ENGELMANN 1963a, 274] [see also Eq. (3.85)]:

$$\hat{t} = \imath\hbar\frac{\partial}{\partial E}. \tag{10.6}$$

For $-\infty < E < +\infty$, $\hat{t}$ is hypermaximal [see p. 37] and has a continuous spectrum[15]. Paul [PAUL 1962] and Engelmann and Fick [FICK/ENGELMANN 1963a] [FICK/ENGELMANN 1963b] drew the conclusion that this operator is not Hermitian: in fact Hermiticity demands a set of functions complete in $E > 0$, vanishing at $E = 0$ and closed under $\partial/\partial E$, and no such set exists.

[14] See also Allcock's comments [ALLCOCK 1969, 260]. For historical reconstruction see [*JAMMER* 1974, 148–49]

[15] On this problem see also [*HOLEVO* 1982, 132–34, 136].

Allcock's proposal But it is possible to find an approximate solution [ALLCOCK 1969, 262] of the following form:

$$\psi_\tau(E) := e^{(\imath\tau-\epsilon)\frac{E}{\hbar}}, \tag{10.7}$$

where τ is the approximate eigenvalue of $\hat{t}$ and ϵ is a small positive number introduced to give a high energy cut-off and hence a finite form. The probability flux is overwhelmingly concentrated in a narrow band centred on τ and of width of order ϵ. It is this that makes the Eq. a good approximation of the time-of-arrival. But Allcock did not consider this function as a good solution. Firstly, the form of the function is not the only one obtainable. Secondly, expansion coefficients are not mathematically determined because the set of functions $\{\psi_\tau(E) : -\infty < \tau < +\infty\}$ is very much overcomplete for the analysis of functions $\psi(E) : 0 \leq E < +\infty$. Thirdly, the squared moduli of the coefficients cannot be interpreted as probabilities, since the overcompleteness of the vectors (10.7) implies that they cannot be orthogonal. As a possible solution, free-particle wave functions which have components with both positive and negative energies are constructed. The conclusion is that there cannot be an ideal arrival-time for such functions. It is possible to construct an ideal time for functions with only positive energy values, but it is not possible to base any such theory upon the concept of eigenstate measurement [ALLCOCK 1969, 263, 268–71, 284].[16]

10.2.3 Time-of-Arrival Operator

Spectrum of the time-of-arrival The best recent developments of Allcock's ideas about a time-of-arrival, is due to a joint work by Grot, Rovelli and Tate [GROT *et al.* 1996] . They considered the time-of-arrival of a particle at a detector in a fixed position X.

First, we try to find some analogue of the spectral decomposition given by Eq. (3.63):

$$\hat{t} = \int_{-\infty}^{+\infty} t\hat{P}(t)dt. \tag{10.8}$$

But we encounter a difficulty: for an arbitrary self-adjoint operator we have [see Eq. (3.61)]:

$$\int_{-\infty}^{+\infty} \hat{P}(o)do = \hat{I}, \tag{10.9}$$

while we have no reason for thinking that the same is valid for a 'time' operator defined by Eq. (10.8):

$$\hat{P}_{\hat{t}} := \int_{-\infty}^{+\infty} \hat{P}(t)dt. \tag{10.10}$$

The reason is that it is not true that any state is certainly detected at some time. Hence the spectral family $\hat{P}(t)$ is incomplete. Therefore we should say that $\hat{P}_{\hat{t}}$ only projects in the subspace $\mathcal{H}_{\rm D}$ formed by the states detected at some time at position X of a given detector. In fact if we try to define $\hat{P}_{\hat{t}}$ on the entire state space, we cannot distinguish between the range of $\hat{P}(t=0)$ and the ones in the space $\mathcal{H}_{\rm D'}$ of the never-detected states (both are annihilated by $\hat{t}$).

[16]Other attempts to construct a time operator are in [RANKIN 1965] [ROSENBAUM 1969] [F. SMITH 1960] [WIGNER 1955a] [OLKHOVSKY *et al.* 1974] [RECAMI 1977]. Eberly and Singh [EBERLY/L. SINGH 1973] introduced a stationary time operator which can find several applications. On the problem of time operator one could also see the contribution of Halliwell [HALLIWELL 1995a], who sees the subject from the point of view of coherent histories [see chapter 25]. It is also possible to construct a time operator which is a POVM [*BUSCH et al.* 1995a, 79–82] [see chapter 18].

Time-of-arrival operator Consider now a non-relativistic free particle in one dimension. The time-of-arrival of a particle with initial position q_0 and initial momentum p_0, detected at position X can be written:

$$t(X) = \frac{m(X - q_0)}{p_0} \tag{10.11}$$

as a time/space inversion of the classical Eq. of motion:

$$q(t; q_0, p_0) = \frac{p_0}{m}t + q_0. \tag{10.12}$$

Note that, except for the problem at $p_0 = 0$, the particle is always detected (and hence there are no complex $t(X)$ values). In the Heisenberg picture we write Eq. (10.11) as:

$$\hat{t}(X) = \frac{m(X - \hat{q}_0)}{\hat{p}_0}. \tag{10.13}$$

We try now to construct a symmetric ordering for the operator of Eq. (10.13) in the following manner:

$$\hat{t}(X) := \frac{mX}{\hat{p}_0} - m\frac{1}{\sqrt{\hat{p}_0}}\hat{q}_0\frac{1}{\sqrt{\hat{p}_0}}. \tag{10.14}$$

We need a concrete representation for the Hilbert space (a basis), and we look for it in the momentum representation because it makes the definition of $\frac{1}{\sqrt{\hat{p}_0}}$ easier:

$$\hat{t}(X)\tilde{\psi}(k) = \left[-\imath\frac{m}{\hbar}\frac{1}{\sqrt{k}}\frac{d}{dk}\frac{1}{\sqrt{k}} + \frac{m}{\hbar}\frac{X}{k}\right]\tilde{\psi}(k), \tag{10.15}$$

where $\sqrt{k} = \imath\sqrt{|k|}$ for $k < 0$. Note that the 1-parameter family of operators $\hat{t}(X)$ can be generated unitarily via translations of the form:

$$\hat{t}(X) = e^{\imath kX}\hat{t}(0)e^{-\imath kX}, \tag{10.16}$$

where $e^{\imath kX}$ is the space counterpart of $\hat{U}_t$ [Eqs. (3.70) and (3.73)].

Therefore it suffices to study the operator $\hat{t}(0)$ without loss of generality, namely by assuming the detector to be at the origin. Hence from now on we drop the explicit X dependence of $\hat{t}$ and write [see Eq. (10.14) and Eq. (10.15)]:

$$\hat{t} := -\imath\frac{m}{\hbar}\frac{1}{\sqrt{k}}\frac{d}{dk}\frac{1}{\sqrt{k}}. \tag{10.17}$$

In the momentum representation the eigenvalue Eq. for $\hat{t}$:

$$\hat{t}|t\rangle = t|t\rangle \tag{10.18}$$

becomes [see Eq. (10.15)]:

$$\hat{t}\langle k|t\rangle := \left[-\imath\frac{m}{\hbar}\frac{1}{\sqrt{k}}\frac{d}{dk}\frac{1}{\sqrt{k}}\right]\langle k|t\rangle = t\langle k|t\rangle. \tag{10.19}$$

The eigenvalue Eq. is easily solved (in each half of the real line with $k \neq 0$) by:

$$\langle k|t\rangle = (\eta_+\chi(+k) + \eta_-\chi(-k))\sqrt{\frac{\hbar}{2\pi m}}\sqrt{k}e^{\frac{\imath\hbar t k^2}{2m}}, \tag{10.20}$$

where $\chi(+k)$ (or $\chi(-k)$) is the characteristic function [see subsection 5.2.3] of the positive (or negative) half of the real line and $\eta_\pm$ are constants independent of k. In order to fix a relation between η_- and η_+ we first obtain:

$$\hat{t}\langle k|t\rangle = t\cdot\langle k|t\rangle - \imath\sqrt{\frac{m}{2\pi\hbar}}\frac{\delta(k)}{\sqrt{|k|}}(\eta_+ + \imath\eta_-), \tag{10.21}$$

from which we can see that in order to satisfy the eigenvalue Eq. we must have:

$$\eta_- = \imath\eta_+. \tag{10.22}$$

In this construction the problem is that the time operator does not have a basis of orthogonal eigenstates. In order to overcome this difficulty we introduce a small positive number ϵ. We consider now a 1-parameter family of real bounded continuous odd functions $f_\epsilon(k)$ which approaches $1/k$ pointwise. In this way we can circumvent the singularity at the point $k=0$, and construct a self-adjoint time-operator as a sequence of operators which are not self-adjoint. We may choose:

$$f_\epsilon(k) = \frac{1}{k} \quad \text{for} \quad |k| > \epsilon, \tag{10.23a}$$

$$f_\epsilon(k) = \epsilon^2 k \quad \text{for} \quad |k| < \epsilon. \tag{10.23b}$$

Then the regulated time-of-arrival operator [see Eq. (10.17)] becomes:

$$\boxed{\hat{t}_\epsilon = \imath\frac{m}{\hbar}\sqrt{f_\epsilon(k)}\frac{d}{dk}\sqrt{f_\epsilon(k)}} \tag{10.24}$$

Note that on any state with the support $|k| > \epsilon$ the operators $\hat{t}_\epsilon$ and $\hat{t}$ are equal: their action differs only on components of a state with an arbitrarily low momentum ($k \longrightarrow 0$). Apart from this limit the probability distribution computed from $\hat{t}_\epsilon$ turns out to be independent of ϵ. Now operator $\hat{t}_\epsilon$ commutes with $k/|k|$. Let us study the action of the operator in regions $k>0, k<0$:

$$\hat{t}_\epsilon|t,+\rangle_\epsilon = t|t,+\rangle, \tag{10.25a}$$

$$\hat{t}_\epsilon|t,-\rangle_\epsilon = t|t,-\rangle, \tag{10.25b}$$

where

$$\langle k|t,+\rangle_\epsilon = 0 \quad \text{for} \quad k<0, \tag{10.26a}$$

$$\langle k|t,-\rangle_\epsilon = 0 \quad \text{for} \quad k>0. \tag{10.26b}$$

A calculation shows that we have:

$$\langle k|t,\pm\rangle_\epsilon = \chi(\pm k)\sqrt{\frac{m}{2\pi\hbar}}\frac{1}{\sqrt{f_\epsilon(k)}}\exp\left(\imath\frac{\hbar t}{m}\int_{\pm\epsilon}^{k} dk'(f_\epsilon(k'))^{-1}\right). \tag{10.27}$$

Explicitly for $k \geq 0$ we have:

$$\langle k|t,\pm\rangle_\epsilon = \chi(\pm k)\sqrt{\frac{m}{2\pi\hbar}}\sqrt{k}\exp\left(\imath\frac{\hbar t}{2m}(k^2-\epsilon^2)\right). \tag{10.28}$$

It can be shown that, for this momentum representation of the eigenstates of $\hat{t}_\epsilon$, they have the properties of completeness and of orthogonality.

Probability densities Now we can work out the probability density of the time-of-arrival: it is the modulus square of the projection of the state, say $|\psi\rangle$, on the $\hat{t}$–eigenstates. Since these are doubly degenerate, we have:

$$\wp(t) = |_\epsilon\langle t, +|\psi\rangle|^2 + |_\epsilon\langle t, -|\psi\rangle|^2. \tag{10.29}$$

If we assume that the support of $\tilde{\psi}(k)$ does not contain (an arbitrarily small region $|k| < \delta$ around) the origin, we can choose $\epsilon < \delta$, and using the explicit form of eigenstates (10.28), we obtain:

$$\wp(t) = \frac{m}{2\pi\hbar}\left[\left|\int_0^\infty dk\sqrt{k}\exp\left(\imath\frac{\hbar t}{2m}(k^2-\epsilon^2)\right)\tilde{\psi}\right|^2 + \left|\int_{-\infty}^0 dk\sqrt{k}\exp\left(\imath\frac{\hbar t}{2m}(k^2-\epsilon^2)\right)\tilde{\psi}\right|^2\right]. \tag{10.30}$$

Note that the ϵ dependence only gives a phase that disappears when we take the absolute value squared, namely:

$$\wp(t) = \frac{m}{2\pi\hbar}\left[\left|\int_0^\infty dk\sqrt{k}\exp\left(\imath\frac{\hbar t}{2m}k^2\right)\tilde{\psi}\right|^2 + \left|\int_{-\infty}^0 dk\sqrt{k}\exp\left(\imath\frac{\hbar t}{2m}k^2\right)\tilde{\psi}\right|^2\right]. \tag{10.31}$$

Hence we have the result that, for states that do not include an amplitude for zero velocity, the time-of-arrival distribution computed with operator $\hat{t}_\epsilon$ is independent from ϵ. Since the two terms (each one in modulus) of Eq. (10.31) correspond to the movements of the particle from the left and the right (with respect to zero), we have immediately that the probability $\wp^+(t)$ ($\wp^-(t)$) that the particle is detected at $X = 0$ while moving in the positive (negative) direction is given by the first (second) term of the same Eq.

Final results In conclusion, while we calculate the probability distribution of position [see Eq. (3.41) and figure 10.1.(a)] in this way:

$$\int_{-\infty}^{+\infty} |\psi(\mathbf{r},\tau)|^2 d\mathbf{r}, \tag{10.32}$$

where the time τ is kept fixed, we keep X fixed and calculate a distribution for the time in the following way [see figure 10.1.(b)]:

$$\boxed{\int_{-\infty}^{+\infty} |\psi_\epsilon^\pm(X,t)|^2 dt} \tag{10.33}$$

On this basis we can finally calculate the CCR between the time-of-arrival operator and the Hamiltonian:

$$\boxed{[\hat{t}_\epsilon, \hat{H}] = -\imath\hbar(\hat{I} - h_\epsilon(k))} \tag{10.34}$$

where

$$h_\epsilon(k) = 1 - kf_\epsilon(k). \tag{10.35}$$

The function $h_\epsilon(k)$ vanishes for $|k| > \epsilon$, and in the small interval where it has support, it is bounded — by 1, if we choose $f_\epsilon(k)$ as in Eqs. (10.23). For the particle in the state $\tilde{\psi}(k)$ the UR is:

$$\boxed{(\Delta t_\epsilon)^2(\Delta E)^2 \geq \frac{\hbar^2}{4}(1 - \langle\psi|h_\epsilon(k)|\psi\rangle)^2} \tag{10.36}$$

where, for sufficiently small ϵ and for all states with support away from the origin, we have $\Delta t_\epsilon \Delta E \geq \frac{\hbar}{2}$.

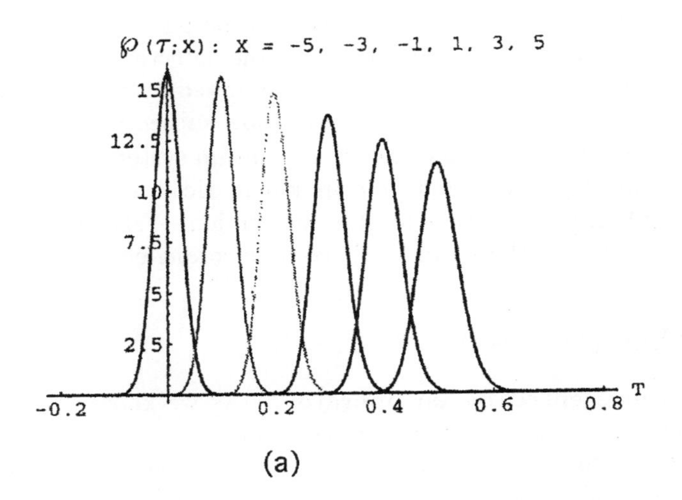

(a)

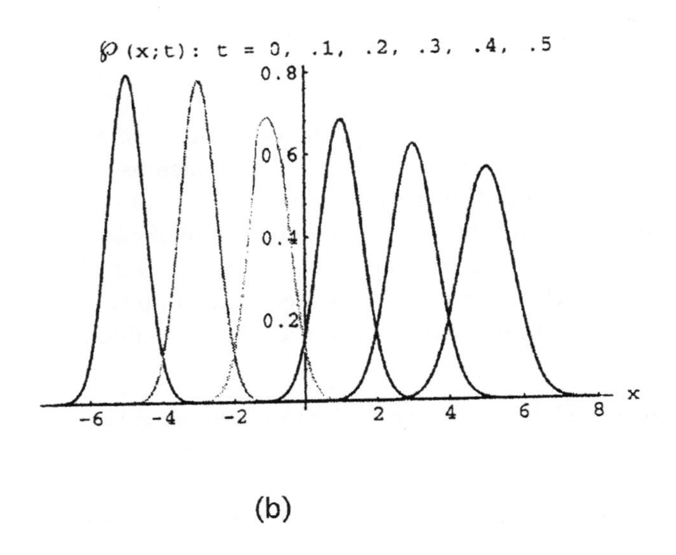

(b)

Figure 10.1: (a) Schrödinger probability densities $\wp(x, \tau)$ for position, plotted at times $t = 0, 1, 2, 3, 4, 5$. (b) Time-of-arrival probability densities $\wp(t; X)$, plotted at the detector positions $X = -5, -3, -1, 1, 3, 5$ — from [GROT *et al.* 1996, 4684, 4685].

Concluding Remarks The greatest interest of the analysed proposal is that it shows a particularity of the wave-function which is not apparent in Basic-QM. In fact in the ordinary (Schrödinger) wave function, as we have said, time is an external parameter while position is an observable of the system. But in Eq. (10.33), we have associated an operator to the time (of arrival) and treated the position as an external parameter: the place (the fixed X) where we have placed a counter in order to detect arrival times. Note that the experimental contexts in the two cases are different and incompatible: if one measures the position (which is then represented by an operator), one does it at a certain time given by the detector (external parameter); but if one wishes to know the arrival time (which can be represented by an operator), the position of the detector is fixed (external parameter) — see also below subsection 10.2.5.

In other words, already in the definition of a quantum system (wave function), a system can be kinematically confined (position and time are kinematic observables), i.e. it can be 'localized' in space (or time) only if there is some form of interaction — shown by the parametric function of time (or space) —, though there can be space-time entanglement [see section 40.1].

10.2.4 Another Proposal

Can energy be represented by an operator? As we know, it was also questioned that energy could be represented by the Hamiltonian operator $\hat{H} = \imath\hbar\partial/\partial t$ [RAYSKI/RAYSKI 1977, 14–15]. But, if we tried to understand the Hamiltonian as a generalized momentum and time as a generalized position, such that $[\hat{H}, \hat{t}] = \imath\hbar$, we would have [see Eq. (10.4)]:

$$\hat{t} = \frac{m}{2}\left(\hat{q}_x \frac{1}{\hat{p}_x} + \frac{1}{\hat{p}_x}\hat{q}_x\right) + f(q_x), \tag{10.37}$$

where f is an arbitrary function. But then, the definition domain of $\hat{H}$ and $\hat{t}$ would not be the same — the inverse of operator $\hat{p}$ does not exist inasmuch as the spectrum of $\hat{p}$ includes zero. Much more: if the spectrum of the energy is discrete, we can construct a stationary solution of the Schrödinger Eq. describing the system in a eigenstate of energy, so that the energy is known exactly at any time instant with arbitrary high precision. But if it is continuous, it is possible to construct a solution such that the energy is determined up to an arbitrarily small uncertainty and this solution remains almost stationary for a very long time interval.

Recent Developments Recently the problem of the singularity at $p = 0$ was circumvented [DELGADO/MUGA 1997, 3427–28] by a method similar to that of Grot and co-workers. Anyway the results are quite different[17].

Take the energy representation defined by the eigenvalue equations [see also Eq. (3.85)]:

$$\hat{H}_0|E, \pm\rangle = E|E, \pm\rangle, \tag{10.38a}$$
$$\hat{p}|E, \pm\rangle = \pm\sqrt{2mE}|E, \pm\rangle, \tag{10.38b}$$

where $\hat{H}_0 = \hat{p}^2/2m$ is the Hamiltonian of the free particle. The set of energy eigenstates $\{|E, \pm\rangle\}$ is orthonormal and complete. Let us now look for a self-adjoint operator $\hat{t}$ with the dimensions of time conjugate to a self-adjoint operator $\hat{H}_{\hat{t}}$, with dimensions of energy and (unlike what happens

[17] A first approach can be found in [KIJOWSKI 1974].

in Basic-QM) a unbounded spectrum such that $[\hat{H}_{\hat{t}}, \hat{t}] = \imath\hbar$. Consider now a projector $\hat{P}(\pm\hat{p})$, onto the subspaces generated by plane waves with positive and/or negative momenta,

$$\hat{P}(\pm\hat{p}) = \int_0^\infty dp| \pm p\rangle\langle\pm p|, \tag{10.39}$$

and define the self-adjoint operator

$$\mathrm{sgn}(\hat{p}) := \hat{P}(+\hat{p}) - \hat{P}(-\hat{p}), \tag{10.40}$$

where $\mathrm{sgn}(\hat{p})$ commutes with the Hamiltonian and satisfies the eigenvalue Eq.:

$$\mathrm{sgn}(\hat{p})|E, \pm\rangle = \pm|E, \pm\rangle. \tag{10.41}$$

Then we can write $\hat{H}_{\hat{t}}$ as follows:

$$\hat{H}_{\hat{t}} := \mathrm{sgn}(\hat{p})\hat{H}_0 \tag{10.42}$$

which, as desired, exhibits an unbounded spectrum:

$$\hat{H}_{\hat{t}}|E, \pm\rangle = \pm E|E, \pm\rangle \quad (E \geq 0). \tag{10.43}$$

Introducing for the energy eigenstates the notation

$$|\eta\rangle = \begin{cases} |+E\rangle := |E, +\rangle & \text{if } \eta \geq 0, \\ |-E\rangle := |E, -\rangle & \text{if } \eta < 0 \end{cases}, \tag{10.44}$$

the above results can be rewritten in terms of the complete and orthonormal set of states $\{|\eta\rangle; \eta \in (-\infty, +\infty)\}$ satisfying the eigenvalue Eqs.:

$$\hat{H}_{\hat{t}}|\eta\rangle = \eta|\eta\rangle, \tag{10.45a}$$

$$\hat{p}|\eta\rangle = \mathrm{sgn}(\eta)\sqrt{2m|\eta|}|\eta\rangle, \tag{10.45b}$$

$$\hat{H}_0|\eta\rangle = |\eta||\eta\rangle. \tag{10.45c}$$

Now we look for a definition of a self-adjoint time operator which is conjugate to the already defined $\hat{H}_{\hat{t}}$. To this purpose let us introduce the states $|\tau\rangle$ defined in the $|\eta\rangle$ basis as:

$$|\tau\rangle = h^{-\frac{1}{2}} \int_{-\infty}^{+\infty} d\eta e^{\frac{\imath}{\hbar}\eta\tau}|\eta\rangle. \tag{10.46}$$

It can be shown that these states constitute a complete and orthonormal set. Then, the spectral decomposition of the time operator $\hat{t}_\tau$, conjugate to $\hat{H}_{\hat{t}}$, is

$$\boxed{\hat{t}_\tau = \int_{-\infty}^{+\infty} d\tau|\tau\rangle\langle\tau|.} \tag{10.47}$$

$\hat{t}_\tau$ turns out to be invariant under time reversal. In order to obtain a consistent physical interpretation of $\hat{t}_\tau$ one can decompose $|\tau\rangle$ as a superposition of negative- and positive-momentum contributions.

Concluding remarks It is evident that Delgado and Muga's proposal can be particularly interesting for relativistic problem — where negative energy solutions are mathematically available and hence the energy is not bounded [see subsection 4.2.3]. On the other hand, while in relativistic frame a time operator can be constructed on these lines, a sharp position operator cannot be defined — as we shall see in the next section. Hence the duality between time and position, which was discussed when concluding the analysis of Grot and co-workers' model, to a certain extent remains.

10.2.5 Interfering Times

If we treat time as a dynamical observable of an object system $\mathcal{S}$ we expect that the probability distribution for the time observable shows an interference pattern as it is the case for the position observable [SOKOLOVSKI/CONNOR 1993]. In the case of a Two-Slit-Exp., we have seen that there are two different probability distributions depending on whether the observable shows an interference pattern [Eq. (3.4)] or not [Eq. (3.3)]:

$$\vartheta' := |\vartheta_1 + \vartheta_2|^2, \tag{10.48a}$$
$$\vartheta'' := |\vartheta_1|^2 + |\vartheta_2|^2. \tag{10.48b}$$

Now, classically, a flight time between positions a, b and times t_1, t_2, can be defined as follows:

$$t^{\mathrm{c}}_{ab}[q(t)] = \int_{t_1}^{t_2} \Theta_{ab}[q(t)]dt, \tag{10.49}$$

where $\Theta_{xz}(y) = 1$ if $x \leq y \leq z$ and 0 otherwise, and $x(t)$ is the trajectory. Using a path-integral approach, a particle starting at t_1 in some state $|\psi_i\rangle$ and reaching at t_2 a final state $|\psi_f\rangle$, travels through a multiplicity of classical paths $t^{\mathrm{c}}_{ab}[q(t)]$, each one equal to a value τ. We assign to each one a traversal time amplitude distribution of the form [see Eqs. (3.93) and (3.155)]:

$$\mathrm{K}^t_{ab}(\psi_f|\tau|\psi_i) = [\vartheta(\psi_f|\psi_i)]^{-1} \int dq_2 \int Dq(\cdot) \int dq_1 \psi^*_f(q_2)\delta\left(t^{\mathrm{c}}_{ab}[q(\cdot)] - \tau\right) e^{\frac{i}{\hbar}\mathrm{S}[q(\cdot)]}\psi_i(q_1), \tag{10.50}$$

where

$$\vartheta(\psi_f|\psi_i) := \int dq_2 \int Dq(\cdot) \int dq_1 \psi^*_f(q_2) e^{\frac{i}{\hbar}\mathrm{S}[q(\cdot)]}\psi_i(q_1), \tag{10.51}$$

δ is the Dirac function and $\psi(q_2) := \langle q_2|\psi_f\rangle$. Now it can be shown that the mean value of τ — i.e. the first moment of $\mathrm{K}^t_{ab}(\psi_f|\tau|\psi_i)$ [see subsection 3.7.3] —, if not disturbed or (potentially) measured, shows an interference pattern:

$$\langle\tau\rangle_I := \int_0^{t_2-t_1} \tau \mathrm{K}^t_{ab}(\psi_f|\tau|\psi_i)d\tau, \tag{10.52}$$

which corresponds to formula (10.48a). If, on the other hand, we perform a measurement such that the interference disappears, we obtain

$$\langle\tau\rangle_{II} := \int_0^{t_2-t_1} \tau \frac{|\mathrm{K}^t_{ab}(\psi_f|\tau|\psi_i)|^2}{\int_0^{t_2-t_1} |\mathrm{K}^t_{ab}(\psi_f|\tau|\psi_i)|^2 d\tau} d\tau, \tag{10.53}$$

which corresponds to Eq. (10.48b). By means of a Larmor clock [PERES 1980b], both behaviours can be simulated. For interesting developments see sections 40.1 and 40.2, concerning time entanglements.

10.3 Position Operator

General considerations In Basic-QM we gave the existence of a position operator for granted. But we have already seen that in Relativistic-QM there are some difficulties, because we cannot find a position operator for bosons [see section 4.5].

On the other hand there are a lot of constraints on the measurability of position which make the matter not easy. A great problem is represented by the limits of measurability — stated by the Wigner/Araki/Yanase theorem [theorem 14.5: p. 234]. When the quantities to be measured do not commute with some additive conserved quantity, we must perform unsharp measurement. Generally, we can state that the main problem is represented by the attribution of sharp values to position.

Hegerfeldt theorem We can formulate the problem better by means of a theorem due to Hegerfeldt [HEGERFELDT 1974].[18] Take a localization operator $\hat{O}(\mathrm{V})$ such that

$$\left\langle \psi_t | \hat{O}(\mathrm{V}) | \psi_t \right\rangle \tag{10.54}$$

gives the probability of finding the particle in region (volume) V at time t. Since probabilities lie between 0 and 1, it follows that $\hat{O}(\mathrm{V})$ is self-adjoint and that its eigenvalues lie between 0 and 1. Now the requirement of relativistic locality (or of relativistic causality) demands that if at time $t_0 = 0$ the particle is localized (probability = 1) in V, at a later time t it is localized in the bounded region

$$\mathrm{V}_t = \{\mathbf{r};\ \text{distance}\,(\mathrm{V}, \mathbf{r}) \leq ct\} \tag{10.55}$$

Hence, if at time t we consider the translated particle state [see Eq. (3.138)]:

$$\hat{U}_{-\mathbf{a}} |\psi_t\rangle, \tag{10.56}$$

with sufficiently large $|\mathbf{a}|$ (larger than a constant η_t), then this state should not be in V. Hence we state following theorem[19]:

Theorem 10.1 (Hegerfeldt) *In Relativistic-QM there is no one-particle state satisfying probability assumption* (10.54) *which is localized in the finite space region* V *satisfying also relativistic locality.*

Instead of proving the theorem directly[20] we wish to discuss a concrete physical model.

[18] See also [HEGERFELDT/RUIJSENAARS 1980] [*PRUGOVEČKI* 1984b, 82] [*HEISENBERG* 1930a, 19–23] [*HEISENBERG* 1930b, 126–27] [WIGNER 1952, 103].

[19] An alternative formulation of the theorem is the following: any kind of sharp QM particle localization that obeys Poincaré or Lorentz covariance necessarily violates Einstein causality [PRUGOVEČKI 1985, 527–28]. As later shown by Hegerfeldt [HEGERFELDT 1985], also systems with exponentially bounded tails at initial time $t = 0$ violate Einstein causality at later times.

[20] For this purpose see the original article.

Wigner's contribution A concrete physical model was developed by Wigner [WIGNER 1983a, 310–12], who showed how Einstein causality is violated because QM obeys to Poincaré covariance [see section 4.1]. In fact, no matter how one defines the position, one has to conclude that the velocity, defined as the difference between two subsequent position measurements divided by the time interval between them, has a finite probability of assuming an arbitrarily large value, exceeding c. One either has to accept this or deny the possibility of a perfectly localized position.

Let us produce out of some initial state $|\psi_i\rangle$ (a one-dimensional system), by means of the operator $\hat{P}_{x_2} - \hat{P}_{x_1}$ (with $x_2 > x_1$), a state $|\psi\rangle$ for which the position is confined to the x_1, x_2 interval. In the momentum representation we write $\tilde{\psi}(p,\sigma)$, where σ is a discrete variable characterizing spin component in one direction. Because of the confinement to interval x_1, x_2, i. e. because of $\tilde{\psi}(p,\sigma) = [\hat{P}_{x_2} - \hat{P}_{x_1}]|\psi_i\rangle$, we have:

$$\hat{P}_x\tilde{\psi}(p,\sigma) = 0, \qquad \forall x \leq x_1, \tag{10.57a}$$

$$\hat{P}_x\tilde{\psi}(p,\sigma) = \tilde{\psi}(p,\sigma), \qquad \forall x \geq x_2. \tag{10.57b}$$

The scalar product of two vectors $|\psi\rangle, |\psi'\rangle$ in the (p,σ)–representation is:

$$\langle\psi|\psi'\rangle = \sum_\sigma \int \tilde{\psi}^*(p,\sigma)\tilde{\psi}'(p,\sigma)\frac{dp}{p_0}, \tag{10.58}$$

where $p_0 = \sqrt{m^2c^2 + p^2}$. The operator for a displacement in space of a and in time of t is simply:

$$e^{-\imath p_0 t + \imath p a}. \tag{10.59}$$

If $|a| > x_2 - x_1$, a space displacement of a moves the original confinement into a new, non-overlapping one, so that $e^{\imath pa}\tilde{\psi}(p,\sigma)$ will be orthogonal to $\tilde{\psi}(p,\sigma)$:

$$e^{\imath pa}\tilde{\psi}^*(p,\sigma)\tilde{\psi}(p,\sigma) = \int e^{-\imath pa}\sum_\sigma |\tilde{\psi}(p,\sigma)|^2\frac{dp}{p_0} = 0. \tag{10.60}$$

This last Eq. expresses the fact that the Fourier transform of

$$\sum_\sigma \frac{\tilde{\psi}(p,\sigma)\tilde{\psi}(p,\sigma)}{p_0} \tag{10.61}$$

is confined to a region of width $2(x_2 - x_1)$. This is a meromorphic function of p, i.e. it has no singularity in the finite complex plane. The scalar product of the displaced $\tilde{\psi}(p,\sigma)$ in space and of the displaced $\tilde{\psi}(p,\sigma)$ in time is:

$$e^{\imath pa}\tilde{\psi}^*(p,\sigma)e^{-\imath p_0 t}\tilde{\psi}(p,\sigma) = \int e^{-\imath pa - \imath p_0 t}\sum_\sigma |\tilde{\psi}(p,\sigma)|^2\frac{dp}{p_0}, \tag{10.62}$$

and is evidently the Fourier coefficient of

$$e^{-\imath p_0 t}\sum_\sigma \frac{\tilde{\psi}^*(p,\sigma)\tilde{\psi}(p,\sigma)}{p_0}. \tag{10.63}$$

This function, for finite t, is not a meromorphic function of p. In fact, the p_0 in the exponent has singularities at $\pm\imath mc$. Hence there is no finite interval of a outside of which its Fourier transform (10.62) will vanish. This means that the time displacement spreads the position, originally confined to the interval x_1, x_2, over all of space. Thus the mentioned conclusion that the velocity has a finite probability of assuming an arbitrarily large value, exceeding c.

Newton/Wigner operator Newton and Wigner [T. NEWTON/WIGNER 1949] , after considering these problems, proposed what later was called a *Newton/Wigner position operator* by imposing the following symmetry conditions (non-self-adjointness)

$$\hat{\mathbf{q}}_{\rm NW} = \imath\nabla_{\mathbf{p}} - \frac{1}{2}\frac{\imath\mathbf{k}}{\mathbf{k}^2 + m^2c^2}. \tag{10.64}$$

But the Newton/Wigner operator does not lead to relativistically covariant probability amplitudes. As we shall see, the only way of handling the problem of a generalized position operator for Relativistic-QM is the stochastic theory [see subsection 33.2.2].

Another possibility is to renounce to a position operator as such [AHMAD/WIGNER 1975], i.e. by trying to establish some other connection between velocity and momentum, which is essential for the relativistic conservation laws. But this solution is far from satisfactory.

10.4 Angle Operator*

Lévy-Leblond showed [LÉVY-LEBLOND 1976] that one can find a solution to UR3 by renouncing to the requirement of Hermiticity. We' follow here the analysis of the problem developed by Carruthers and Nieto [CARRUTHERS/NIETO 1968] [*BUSCH et al.* 1995a, 73–74]. We can consider the orbital ($\hat{J}_z = \hat{L}_z$) and the spin ($\hat{J}_z = \hat{\sigma}_z$) cases separately (we occupy ourselves only with the first case).

Now the $\hat{L}_z$ operator can be represented as a differential operator on the function space $L^2([0, 2\pi), d\phi)$:[21]

$$\hat{L}_z\psi(\phi) = -\imath\hbar\frac{\partial}{\partial\phi}\psi(\phi). \tag{10.65}$$

The spectrum of $\hat{L}_z$ is the set of integers $\mathbf{Z}$, and the eigenvectors $|m\rangle$ are given by the functions $\phi \mapsto \psi_m(\phi) = e^{\imath m\phi}, m \in \mathbf{Z}$. We can introduce an additive unitary group of shift operators:

$$\hat{U}_k : |m\rangle \mapsto |m + k\rangle. \tag{10.66}$$

We now look for the generator $\hat{\phi}$ of this group: $\hat{U}_k = e^{\imath k\hat{\phi}}$ [see theorem 3.6: p. 47 and Eq. (3.143)]. Let us introduce the improper eigenvectors of $\hat{U}_k$:

$$|\phi) := (2\pi)^{-\frac{1}{2}} \sum_{m=-\infty}^{+\infty} e^{-\imath m\phi}|m\rangle, \quad \hat{U}_k|\phi) = e^{\imath k\phi}|\phi). \tag{10.67}$$

Then the following operator is self-adjoint:

$$\boxed{\hat{\phi} := \int_0^{2\pi} \phi|\phi)(\phi|d\phi = \sum_{m\neq k} \frac{1}{\imath(k-m)}|m\rangle\langle k| + \pi\hat{I}} \tag{10.68}$$

The associated spectral measure is:

$$\mathcal{X} \mapsto \hat{P}^{\hat{\phi}}(\mathcal{X}) = \int_{\mathcal{X}} |\phi)(\phi|d\phi = (2\pi)^{-1}\sum_{k,m}\int_{\mathcal{X}} e^{-\imath(k-m)\phi}d\phi|m\rangle\langle k|. \tag{10.69}$$

[21]See [*SCHIFF* 1955, 78–82].

An application of the Weyl relation [see subsection 5.2.6]:

$$e^{\imath \hat{L}_z \phi} e^{\imath k \hat{\phi}} = e^{\imath k \phi} e^{\imath k \hat{\phi}} e^{\imath \hat{L}_z \phi} \tag{10.70}$$

shows that the PV measure $\hat{P}^{\hat{\phi}}(\mathcal{X})$ possesses the covariance, as required. In analogy with the angle operator, it is also possible to construct a phase operator[22].

[22] Formalism and discussion in [CARRUTHERS/NIETO 1968] [R. NEWTON 1980] [GALINDO 1984] [*BUSCH et al.* 1995a, 83–88] [BUSCH *et al.* 1995c] .

Chapter 11

CLASSICAL AND QUANTUM PROBABILITY: AN OVERVIEW

Contents The aim of this chapter is not to give a wide account of the QM probability theory, but to point out some basic concepts which will turn out to be very important for the following investigation. In section 11.1 we briefly discuss the problem of defining what probability is. In section 11.2 we summarize the essential features of classical probability theory, while in 11.3 we treat the quantum case. In section 11.4 we analyse a powerful generalization, which is very useful for the treatment of some theoretical problems to be discussed later. A large part of the chapter is then devoted to the Gleason theorem, which is very important for the foundations of the theory but also for later developments [section 11.5].

11.1 Introduction to the Concept of Probability

As a preliminary investigation we briefly discuss the concept of probability as such. By handling probability, the problem is to determine the way of counting and of classifying the events (which can be also understood as propositions) whose probability one wishes to calculate. One can understand the problem if one thinks of the different partitions presupposed by Fermi/Dirac statistics and the statistics of Bose/Einstein [see section 3.8].

11.1.1 Three Interpretations

Historically we have three possible interpretations of probability[1]:

- The *classical interpretation* considers the probability as 'equal possibility'. But possibility is a concept exterior to probability, so that one should speak of 'equal probability', which makes the classical interpretation circular. For example Gnedenko says [*GNEDENKO* 1969, 23–24]:

  Definition 11.1 (I Probability) *If event* a *is decomposable into k special cases belonging to a complete group of N mutually exclusive and equally probable events, the probability $\wp(a)$ of* a *is $\wp(a) = k/N$.*

[1]See [*GNEDENKO* 1969, 17].

Physically the problem is not so difficult, because we can speak of equiprobability if the same physical conditions apply to different sets of (possible) events. And the same should be valid for each specialised domain. In this case we should have an operational interpretation of the probability. Note that this problem is not as simple as it seems, again due to the different statistics in QM [section 3.8].[2]

- The *objective interpretation* sees the probability as coincident with the frequency. The problem of the objective interpretation is that frequencies and mathematical probabilities coincide only approximately (by the law of large numbers); as a consequence, they must be conceptually different. Therefore, the statistical ensemble of probabilities is never a concrete set but only an abstract mathematical structure[3].

- The *subjective interpretation* considers the probability as coincident with subjective likelihood. But also this interpretation is untenable because it uses the concept of personal belief, while, though some physical laws are accepted for their likelihood, in general we expect from probability the ability to foresee some events to a certain extent. On the other hand, Jaynes gives a relatively good axiomatic of the subjective interpretation.

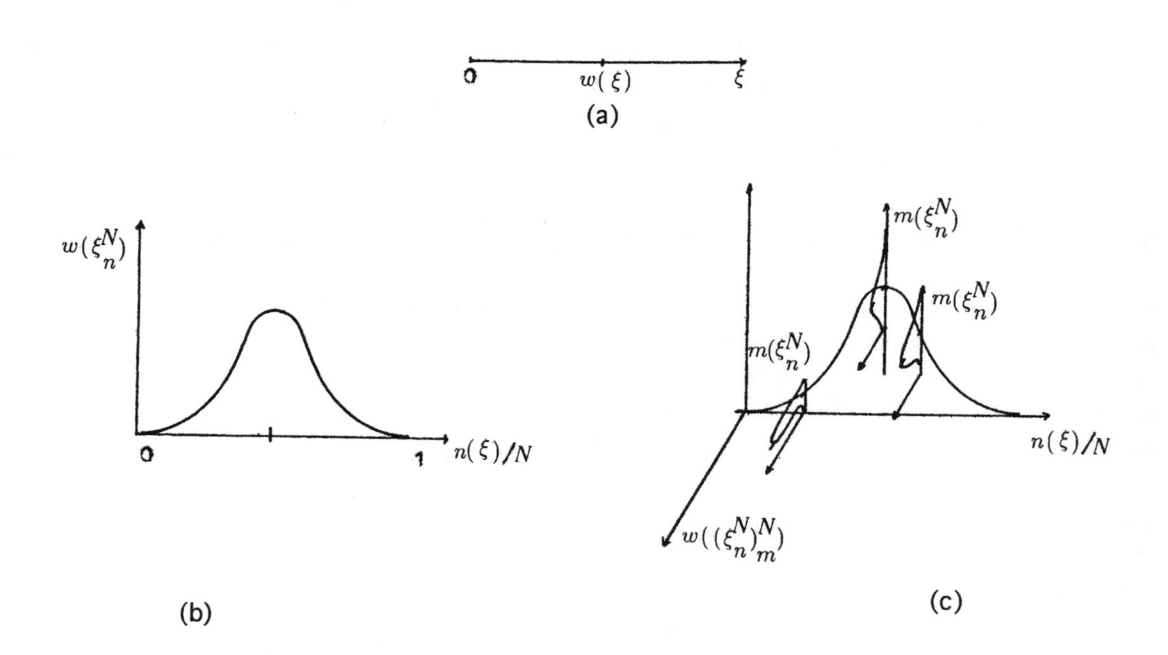

Figure 11.1: (a) predicted relative frequency w of a (classical) variable ξ;
(b) predicted relative frequency of the relative frequency of ξ;
(c) predicted relative frequency of the relative frequency of the relative frequency of ξ — from [*DRIESCHNER* 1979, 84].

[2]On this problem see also [*VAN FRAASSEN* 1991, 57–61].
[3]See also [*POPPER* 1982, 68].

11.1.2 Other Proposals

Hence it seems that the only way out is to synthesize some different points of view. Drieschner tried to synthesize the latter two points of view. His starting point was the concept of *foreseen relative frequency*, which is the most general prediction which is empirically verifiable. He then proposed the following definition [*DRIESCHNER* 1979, 62–64, 69–73, 85]:

Definition 11.2 (II Probability) *Probability is the foreseen (relative) frequency.*

If we try to throw some light on a frequency we always need to go to the meta-level at which we handle the frequency of that frequency [see figure 11.1]. Mathematically this is always possible since, as we have seen, there is the law of large numbers, which guarantees that for infinite series the frequency approaches the probability[4]. Since we can calculate the frequency of different frequencies, Drieschner gives a second definition [*DRIESCHNER* 1979, 79–83]:

Definition 11.3 (III Probability) *Probability is the expectation value of the relative frequency.*

For probability, as we have seen, the most important thing is the distinction between different sets of mutually exclusive events [*GNEDENKO* 1969, 21–22] and their relationship. Now Drieschner tried to deduce the rule of factorization of independent events without using classical definitions. We start with a sequence of trials and investigate the characters of these. Each new single trial consists of a series N of single trials. A possible outcome of such a trial is that the event a occurs k times. We symbolize it by a_k^N. Its probability is given by [*DRIESCHNER* 1979, 81–82, 222]:

$$\wp(a_k^N) = \begin{pmatrix} N \\ k \end{pmatrix} \cdot (\wp(a))^k \cdot (1 - \wp(a))^{N-k} . \tag{11.1}$$

But also this definition is circular because it presupposes the knowledge of the probability $\wp(a)$ and also of the coefficient $\begin{pmatrix} N \\ k \end{pmatrix}$, which one cannot derive without the law of large numbers, which is exactly the content of the objective interpretation.

11.1.3 Some Lessons

Anyway the most important results of this interpretation are:

- The probability cannot be separated from the context of investigation and from the operations which one concretely performs (so we can interpret the concept of 'foreseen' in definition 11.2);
- the values 0, 1 are to be understood more as limit values than as real ones, due to the intrinsic imprecision of probability.

Surely the first point fits in very well with CP, in the sense that it asserts a dependence of probability on the contextuality of experiment. Then, all three interpretations could to a certain extent be reconciled if their statements are interpreted in an operational sense. In fact, on a technical level, we can use each approach without any difficulty. For example, we determine the objective approach with some inequalities which are represented by the supporting *hyperplanes*

[4]On this point see [*GUDDER* 1988b, 23–24, 34–35].

of a polytope — a convex compact subset of an n–dimensional Euclidian space that has finitely many extreme points[5] —; the subjective one by a convex combination of the *vertices* of the same polytope [*PITOWSKY* 1989b, 13, 24, 87, 100] [*DRIESCHNER* 1979, 60, 64–65]. One example is the *Bell/Wigner Polytope*, where the hyperplanes represent Bell inequalities, expressing the distances among the vertices [see figure 11.2; on the subject see sections 39.2 and 39.3]. In conclusion we prefer to renounce to a definition of probability, and to adopt an operational approach.

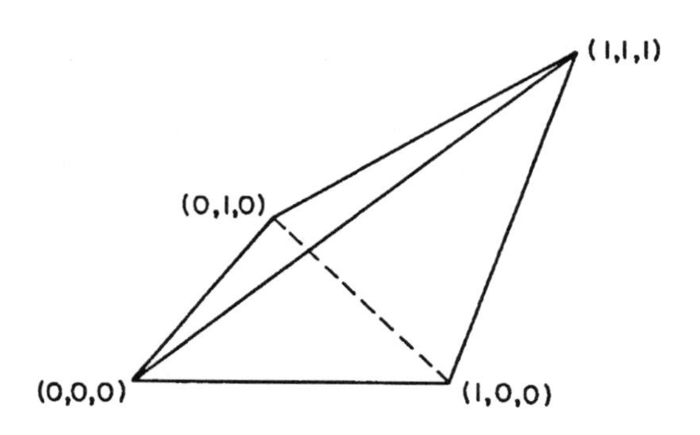

Figure 11.2: Bell/Wigner polytope — from [*PITOWSKY* 1989b, 13].

11.2 Classical Probability

In the following we give a very short account of the classical theory of probability.

Axioms

The classical Kolmogorovian theory of probability is based on the following axiomatic[6]. Let Ω be a set of elements $a, b, c \ldots$, named elementary events, and F a class of subsets of Ω. The elements of F are named random events. F is assumed to be a field (closed under complementation, countable union and intersection).

- F is a set,
- F has Ω as subset,
- for each subset η of F there is a not negative real function $0 \leq \wp(\eta) \leq 1$ named the probability of event η ($\wp : F \mapsto [0, 1]$),
- $\wp(\Omega) = 1$ and $\wp(\emptyset) = 0$,
- If we have N disjoint cases η_i, then: $\wp(\bigvee_{i=1}^{N} \eta_i) = \sum_i \wp(\eta_i)$ (additivity).

We name $(\Omega, F, \wp)$ a classical probability space.

[5] See [*BELTRAMETTI/CASSINELLI* 1981, 211] .

[6] Expositions in [*GNEDENKO* 1969, 25] [VARADARAJAN 1962, 175–77] [*DRIESCHNER* 1979, 75–76] [*GUDDER* 1988b, 1–35]. See also [JAUCH 1976, 126].

Some Features and Laws

When we handle probability, we must respect the following requisites:

- all tests must be done under the same conditions (independence of a test from the precedent ones),
- and the combinations must be linear.

The first point also fixes the boundary between a random sequence and a non random one: if, by a permutation of the order of tests which follow determinate laws, we obtain no changes, then the sequence is non-random [*DRIESCHNER* 1979, 88].

In classical probability calculus — like in a Boolean algebra [see definition 9.7: p. 157] — the commutative law, the associative law and the distributive law are valid [*GNEDENKO* 1969, 23].

Classical probabilities are characterized by the formula:

$$\wp(a \vee b) = \wp(a) + \wp(b) - \wp(a \wedge b). \tag{11.2}$$

The general formula for conditional probability is [*GNEDENKO* 1969, 53]:

$$\wp(a|b) = \frac{\wp(a \wedge b)}{\wp(b)}, \tag{11.3}$$

from which one derives the multiplication theorem:

$$\wp(a \wedge b) = \wp(b)\wp(a|b). \tag{11.4}$$

Bayes' Formula

We saw that the relative frequency corresponds only approximately to the probability. A similar condition is shared by the measurement theory; but the errors which enter probability calculus are specific because each measure of probability supposes already the relative frequency of many tests. Hence we are faced with the problem not only of the passage from a hypothesis to the searched probability, but also with that from some experimented event or series of experimented events to some hypothesis. Thus the importance of Bayes' formula[7]:

$$\wp(a_i|b) = \frac{\wp(a_i)\wp(b|a_i)}{\sum_{j=1}^{N} \wp(a_j)\wp(b|a_j)}, \tag{11.5}$$

where the a_i's are elements of a sequence of which we search the probability under the condition that b happened.

Random Variables

By means of the preceding formalism we can now define a (classical) random variable [JAUCH 1976, 127]:

Definition 11.4 (Classical Random Variable) *In classical probability calculus a random variable ξ is a real-valued function from the space of events Ω to the real line $\Re$, $\xi : \Omega \mapsto \Re$, such that for any subset $\Delta \in \Re$ there is some $\eta \in \Omega$ such that: $\xi^{-1}(\Delta) = \{\xi(\eta) \in \Delta\}$.*

[7]Formulations in [*DRIESCHNER* 1979, 90, 228] [*GNEDENKO* 1969, 58, 298–301] [*HOLEVO* 1982, 170–71] [*GUDDER* 1988b, 10].

11.3 QM Probability

Differences Between Classical and QM Probability

QM probability calculus is characterized by the following: instead of only one set Ω of elementary events, we have several such sets which are mutually irreducible [p. 163 and see theorem 9.4]. This is due to the intrinsic structure of QM probability.

The main differences between classical and QM probability can be synthesized as follows[8]:

- in QM there is no distributivity — it is not a Boolean algebra [see definition 9.8: p. 158, and theorem 9.2: p. 162] —;

- formula (11.2) is not valid in QM.

The reason for the first point, as we know, is that there can be no joint probability of two QM observables if they do not commute [JAUCH 1976, 141–43], i.e. it depends on CCRs. The reason for the second point is that in QM, instead of probabilities, one uses probability amplitudes which are subject to the Superposition principle [Eq. (3.4)]: hence we can have 'interference' between probabilities.

Therefore we can assert the following conclusion [YAMADA/TAGAKI 1991a, 985]: when one considers probabilities for a given exhaustive set of mutually exclusive outcomes, in a classical context the problem is how to determine the value of the probability of each outcome, whereas in QM one must first investigate whether it is really possible to define probabilities for the set [see theorem 14.6: p. 235]. And we return here to the operational approach to probability[9].

Indeterminism of QM

As a consequence of these features, QM is an intrinsically probabilistic theory — we see here again that corollary 6.1 [p. 104] is independent from proposition 6.2 —, i.e. it cannot be reduced to a theory in which probability is only to be ascribed to subjective ignorance regarding conditions (values of different observables) that are all together fulfilled with certainty[10].

QM 'Random Variables'

The absence of a common set of events Ω seems to make the definition of a *random variable* [see definition 11.4] impossible. But this can be accomplished in QM if we substitute to Ω the set $\mathcal{X}$ of subsets of Ω (due to the irreducibility), and, in place of the real line or space, we consider Borel sets [JAUCH 1976, 130] [JAUCH 1977, 46–48].

One defines the Borel set $\mathcal{B}(\Re)$ as the smallest family of subsets of $\Re$ that includes the open sets and is closed under complement and under countable intersections[11]. Hence we can eliminate each reference to the set Ω by handling only subsets $\mathcal{X}$'s and class $\mathcal{L}$ of subsets $\mathcal{X}$'s of Ω — as we shall see [in section 18.1], it is a σ-algebra. Therefore we have [see also definition 11.4]:

[8]Analysis in [FEYNMAN 1951] [VARADARAJAN 1962, 172] [*MACKEY* 1963, 61–66] [*GNEDENKO* 1969, 49] [*GUDDER* 1988b, 35].

[9]Cohen [L. COHEN 1966a] could show by means of the incorrect probability distributions which can be derived from the Wigner function that quantum probability cannot be classical.

[10]On the point see [*VARADARAJAN* 1968, 111–12]. We shall return on this problem when discussing Hidden-Variable theories [see part VIII].

[11]Definition in [*JAUCH* 1968, 5–6] [*BELTRAMETTI/CASSINELLI* 1981, 4] [VARADARAJAN 1962, 176].

Definition 11.5 (Quantum Observable) *A quantum observable $\hat{O}$ can be probabilistically equate to a classical random variable f by the following formula*

$$\boxed{\forall \Delta \in \mathcal{B}(\Re), \exists \mathcal{L} \in \Omega,\ f(\Delta) = \hat{O}^{-1}(\Delta) \in \mathcal{L}.} \tag{11.6}$$

where $f : \mathcal{B}(\Re) \mapsto \mathcal{L}(\mathcal{H})$ and $\mathcal{L}(\mathcal{H})$ is the set of bounded linear operators on the Hilbert space $\mathcal{H}$.

Lessons and Concluding Remarks

In conclusion, we synthesize the preceding results in the following theorem:

Theorem 11.1 (Probability Structure of QM) *The space of events of QM is constituted by an ensemble of mutually irreducible subsets pertaining to some Borel set.*

Note that this analysis is parallel to the quantum logic analysis of QM [see theorem 9.2: p. 162]: there we had an orthomodular lattice and here a Borel set, there Boolean subalgebras [theorem 9.5: p. 164] and here the subsets of a Borel set.

11.4 Generalization to S–Dimensional Probabilities

A powerful generalization, which to some extent is a development of the state-observable approach [see subsection 9.3.3], has been developed by Mączyński and Beltrametti[12]. We start with the following definition [BELTRAMETTI/MACZYŃSKI 1991, 1280–81] :

Definition 11.6 (S–probability) *Let S be a non-empty set. By an S–probability we mean any function $\wp : S \mapsto [0,1]$, i.e. any $\wp \in [0,1]^S$. If $S = \{a\}$ is a one-element set, we call the S–probability simply a probability. If $S = \{a_1, \ldots, a_n\}$ then an S–probability may be identified with a sequence of values $(\wp^{(1)}, \ldots \wp^{(n)})$, i.e. with a vector in $\Re^n$. Hence we call $\wp$ an n–dimensional probability which we write $P(S) = [0,1]^S$.*

For some $\wp_1, \wp_2 \in P(S)$ we define a partial-order relation as $\wp_1 \leq \wp_2$ iff $\wp_1(x) \leq \wp_2(x)$ for all $x \in S$. With $\mathbf{0}$ and $\mathbf{1}$ we indicate functions which only take values 0 and 1 respectively. If $\wp_1, \wp_2 \in P(S)$, we can define a sum $\wp_1 + \wp_2$, although the sum may not belong to $P(S)$. If $(\wp_1 + \wp_2) \in P(S)$, i.e. if $\wp_1 + \wp_2 \leq 1$ then we say that $\wp_1 \perp \wp_2$. A triple of S–probabilities $\wp_1, \wp_2, \wp_3$ is an orthogonal triangle, symbolized by $\triangle(\wp_1, \wp_2, \wp_3)$, if $\wp_i + \wp_j \leq 1$ for $i \neq j; i,j = 1,2,3$.[13] We also write $\wp_2 := 1 - \wp_1$ or $\wp_2 := \wp_1'$.

[12] It will be very useful for the treatment of Bell inequalities [see section 39.3].

[13] See also section 32.4.

Definition 11.7 (Event System) *By an event system we mean a triple $\mathcal{L} = (\Omega, \leq, ')$ where Ω is a set, such that $(\Omega, \leq, ')$ is an orthomodular orthocomplemented POSet (orthoposet)* [see definition 9.8]. *The elements of Ω are called events. $(\Omega, \leq, ')$ satisfies the following axioms:*

i) $a'' = a$, $\forall a \in \Omega$, which is the double negation of a, the first property of orthocomplementation [see Eq. (9.8a)];

ii) $a \leq b \rightarrow b' \leq a'$, $\forall a, b \in \Omega$, which is the second property of orthocomplementation [see Eq. (9.8b)];

iii) if $a_1, a_2, \ldots, a_n \in \Omega$ and $a_i \leq a_{j'}$ for all $i \neq j$, then the least upper bound $a_1 \vee a_2 \vee \ldots \vee a_n$ exists in $(\Omega, \leq, ')$, which is a feature of lattices [see definition 9.2: p. 156];

iv) $a \vee a' = b \vee b', \forall a, b \in \Omega$ (it is denoted by $\mathbf{1}$), which is the fourth property of the orthocomplementation (9.8d);

v) $(a \leq b) \rightarrow \{b = [a \vee (a \vee b')']\}$, which is the definition of orthomodularity [see definition 9.8], *a characteristic of QM.*

By a classical event system we mean a $\mathcal{L}$ which is a Boolean algebra [see definition 9.7] with respect to order and complementation (i.e. it is a complemented distributive lattice).

Definition 11.8 (S–probability on Ω) *By an S-probability on $\mathcal{L}$ (or on Ω) we mean any mapping:*

$$\wp : \Omega \mapsto P(S) = [0, 1]^S, \tag{11.7}$$

with properties:

$$\wp(\mathbf{0}) = 0, \qquad \wp(\mathbf{1}) = 1, \tag{11.8a}$$

$$\wp(a') = 1 - \wp(a) \qquad \forall a \in \Omega, \tag{11.8b}$$

$$\wp(a_1 \vee a_2 \vee \ldots \vee a_n) = \wp(a_1) + \wp(a_2) + \ldots + \wp(a_n), \tag{11.8c}$$

whenever $a_i \leq a'_j$, for $i \neq j$.

By an S–probability event space we mean a pair $(\mathcal{L}, \wp)$.

Definition 11.9 (Completeness of probability) *We say that $\wp$ is* complete *if $a \leq b \leftrightarrow \wp(a) \leq \wp(b), \forall a, b \in \Omega$,*

where $\leftrightarrow$ means logical equivalence.

Definition 11.10 (Representability of S–probabilities) *Let $\mathcal{X}$ be a set of S–probabilities. We say that $\mathcal{X}$ is* representable *if there is a $(\mathcal{L}, \wp)$ such that*

$$\mathcal{X} \subseteq \{\wp(a) : a \in \Omega\}. \tag{11.9}$$

We denote this embedding mapping as $\tilde{\varphi}$ and call $(\mathcal{L}, \wp, \tilde{\varphi})$ a representation for $\mathcal{X}$.

Definition 11.11 (Classical Representability of S–probabilities) *We say that $\mathcal{X}$ is* classically representable *if there is a representation for $\mathcal{X}$ $(\mathcal{L}, \wp, \tilde{\varphi})$ such that $\mathcal{L}$ is a classical event system.*

Definition 11.12 (Consistent Representability of S–probabilities) *We say that an indexed sequence $\mathcal{X} = (\wp_1, \wp_2, \ldots, \wp_n, \ldots \wp_{jk}, \ldots)$ of S–probabilities (= correlation sequence) is* consistently representable *if there is a representation for $\mathcal{X}$ such that $(\mathcal{L}, \wp, \tilde{\varphi})$ and a sequence of pairwise compatible events in Ω such that $\wp_{jk} = \wp(a_j \wedge a_k)$ whenever the pairs j, k appear in $\mathcal{X}$.*

This definitions will be very useful when we shall study generalized Bell inequalities [see section 39.3].

11.5 Gleason Theorem

11.5.1 Introduction

The Gleason Theorem is one of the most important theoretical results since World War II. We can synthesize its contents and consequences as follows. Until now we have used projectors and density operators to represent states, and particularly to calculate expectation values, but we never justified this. Von Neumann demonstrated it with four assumptions, of which at least one is questionable. Since von Neumann's proof is very important for the examination on Hidden-Variables, we postpone the discussion [to section 32.2.1] and limit ourselves to the positive presentation of the Gleason theorem. Also its consequences for Hidden-Variable theories are to be seen later [see section 32.3]. In general the Gleason theorem has far-reaching consequences for all the problem of QM probability distributions. The central aim of Gleason theorem[14] is to show that $\langle \hat{O} \rangle = \mathrm{Tr}(\hat{\rho}\hat{O})$ is the most general formula compatible with the probability structure of QM [see Eq. (3.91)].

From here one deduces:

- that there are no dispersion-free states in QM (if the dimensions of the Hilbert space are at least three) which satisfy the additivity condition;

- that there are gaps in the truth values [see also section 9.5].

11.5.2 Probability Measure

Since in the following we discuss a probability problem in the frame of Hilbert spaces, we cannot use the probability function $\wp$ which acts on a probability space. We need rather a probability measure μ_p [see p.s 164–165] and a *measure space* $(\Omega, \mathcal{X}, \mu_p)$, which is a set of elements Ω together with the σ–ring $\mathcal{X}$ of the subsets of Ω. A *σ–ring* is a ring $\mathcal{R}$ with the property that, for every countable sequence $a_j (j = 1, 2, \ldots)$ of sets in $\mathcal{R}$, we have: $\bigcup_{j=1}^{\infty} a_j \in \mathcal{R}$. The measure space in question here also includes a non-negative function $\mu_p(a)$, defined for all subsets of the class $\mathcal{X}$, which satisfies the properties[15]:

- $0 \leq \mu_p(a) \leq \infty$ ($< \infty$, if we do not allow infinite measures);

- $\mu_p(\emptyset) = 0$;

- for any disjoint sequence $a_j \in \mathcal{X} (j = 1, 2, \ldots)$: $\mu_p\left(\bigcup_{j=1}^{\infty} a_j\right) = \sum_{j=1}^{\infty} \mu_p(a_j)$.

By a measure on the closed subspaces of a Hilbert space, is meant a function μ_p which assigns to every closed subspace a non-negative real number such that if $\{\mathcal{H}_j\}$ is a countable collection of mutually orthogonal subspaces having a closed linear span d, then

$$\mu_p(d) = \sum \mu_p(\mathcal{H}_j). \tag{11.10}$$

Such a measure can be obtained by selecting a vector and, for each closed subspace $\mathcal{H}_j$, taking $\mu_p(\mathcal{H}_j)$ as the square of the projection of the vector on $\mathcal{H}_j$.

[14] See [*BUSCH et al.* 1995a, 124–25] [*GUDDER* 1988b, 51]. See also [*PERES* 1993, 190–96].

[15] Expositions in [*JAUCH* 1968, 7] [*PRUGOVEČKI* 1971, 67] [*PRUGOVEČKI* 1971, 68–70].

11.5.3 Gleason Theorem

In the following examination we shall prove the *Gleason Theorem* [GLEASON 1957, 123]:

Theorem 11.2 (Gleason) *In a separable Hilbert space of dimension $d \geq 3$, whether real or complex, every measure on the closed subspaces can be written in the form*

$$\mu_p(\mathcal{H}_j) = \mathrm{Tr}(\hat{\rho}\hat{P}_{\mathcal{H}_j}), \tag{11.11}$$

where $\hat{P}_{\mathcal{H}_j}$ denotes the orthogonal projection on $\mathcal{H}_j$ and $\hat{\rho}$ denotes a positive semi-definite self-adjoint operator of trace-class one (a density operator).

11.5.4 Proof of the Gleason Theorem

We give here an easy form of Gleason's proof. First we define a frame function of weight w for a separable Hilbert space $\mathcal{H}$ as a real-valued function f defined on the surface of the unit sphere of $\mathcal{H}$ such that if $\{|b_j\rangle\}$ is an orthonormal basis of $\mathcal{H}$, then:

$$\sum f(b_j) = w. \tag{11.12}$$

Then we say that a frame function f is *regular* iff there exists a self-adjoint operator $\hat{O}$ defined on $\mathcal{H}$ such that

$$f(\psi) = \langle \hat{O}\psi|\psi\rangle \qquad \text{for all unit vectors } |\psi\rangle. \tag{11.13}$$

Now Gleason [GLEASON 1957, 125–28] proves the following Theorem :

Theorem 11.3 (I Regularity) *Every continuous frame function on the unit sphere* **S** *in $\Re^3$ is regular and every non-negative frame function on* **S** *in $\Re^3$ is regular,*

where a frame function is said to be regular iff it is the quadratic form of restriction to the unit sphere.

Then we prove the following lemma

Lemma 11.1 (I Gleason Lemma) *if f is a non-negative regular frame function of weight w on a real Hilbert space $\mathcal{H}$, then, for any unit vector $|b_j\rangle$ and $|b_k\rangle$, we have:*

$$|f(b_j) - f(b_k)| \leq 2w \parallel b_j - b_k \parallel . \tag{11.14}$$

Proof

Since f is regular, we have [see Eq. (11.13)] $f(\psi) = \langle \hat{O}\psi|\psi\rangle$ for some self-adjoint operator $\hat{O}$, and, since f is non-negative, we have

$$0 \leq \langle \hat{O}\psi|\psi\rangle \leq w \tag{11.15}$$

for all unit vectors $|\psi\rangle$, and therefore $\parallel \hat{O} \parallel \leq w$. Now, for any unit vector $|b_j\rangle$ and $|b_k\rangle$ we have: $\langle \hat{O}b_k|b_k\rangle = \langle \hat{O}b_j|b_j\rangle$, so that $f(b_j) - f(b_k) = \langle \hat{O}(b_j + b_k)|b_j - b_k\rangle$, and therefore:

$$|f(b_j) - f(b_k)| \leq \parallel \hat{O} \parallel \parallel b_j + b_k \parallel \parallel b_j - b_k \parallel \leq 2w \parallel b_j - b_k \parallel . \tag{11.16}$$

Q.E.D.

Now, using lemma 11.1, Gleason [GLEASON 1957, 129–31] proves another lemma

Lemma 11.2 (II Gleason Lemma) *If f is a non-negative frame function on a two-dimensional complex Hilbert space which is regular on every completely real subspace (i.e. a subspace of the real-linear space $\mathcal{H}$ such that the inner product takes only real values on $\mathcal{H} \otimes \mathcal{H}$), then f is regular.*

Proof

Suppose w is the weight of f and M is its least upper bound. We can choose the unit vectors $|b_n\rangle$'s such that $f(b_n) \longrightarrow M$ and we can arrange that $|b_n\rangle \longrightarrow |\psi\rangle$ because the unit sphere is compact. Let $\lambda_n = \langle\psi|b_n\rangle/|\langle\psi|b_n\rangle|$; then, for $n \longrightarrow \infty$, we have $\lambda_n \longrightarrow 1$ and $\lambda_n|b_n\rangle \longrightarrow |\psi\rangle$. Since $|\lambda_n| = 1$, $f(\lambda_n b_n) = f(b_n)$. Moreover $\langle\lambda_n b_n|\psi\rangle$ is real, so $\lambda_n|b_n\rangle$ and $|\psi\rangle$ are in a completely real subspace. Then, from lemma 11.1 we have:

$$|f(\psi) - M| \leq |f(\psi) - f(\lambda_n b_n)| + |f(b_n) - M| \leq 2w \parallel \psi - \lambda_n b_n \parallel + |f(b_n) - M|. \qquad (11.17)$$

From Eq. (11.17) we see that $f(\psi) = M$ because for $n \longrightarrow \infty$ we have $\lambda_n|b_n\rangle \longrightarrow |\psi\rangle$, which at the limit makes $f(\psi) - f(\lambda_n b_n) = 0$, and since $|b_n\rangle \longrightarrow |\psi\rangle$ we have also $f(b_n) - M = 0$.

Define now a g on $\mathcal{H}$ by

$$g(\psi) = \parallel \psi \parallel^2 f\left(\frac{\psi}{\parallel \psi \parallel}\right) \text{ if } |\psi\rangle \neq 0; \qquad (11.18a)$$

$$g(0) = 0. \qquad (11.18b)$$

The hypothesis concerning f imply that g becomes quadratic in form when restricted to any completely real subspace. Furthermore, since, as we have seen, $f(\lambda\psi) = f(\psi)$ whenever $|\lambda| = 1$, then $g(\lambda\psi) = |\lambda|^2 g(\psi)$ for all scalars λ and vectors $|\psi\rangle$. Let $|b_k\rangle$ be any unit vector orthogonal to $|b_j\rangle$. Then $g(b_j) = f(b_j) = M$ and $g(b_k) = f(b_k) = W - f(b_j) = w - M$. On the completely real subspace determined by $|b_j\rangle$ and $|b_k\rangle$, g is of quadratic form, whose maximum value on the unit circle is at $|b_j\rangle$; therefore the matrix for g relative to the basis $|b_j\rangle, |b_k\rangle$ is diagonal. Hence, if c_j, c_k are real:

$$g(c_j b_j + c_k b_k) = c_j^2 g(b_j) + c_k^2 g(b_k) = M c_j^2 + (w - M) c_k^2. \qquad (11.19)$$

If λ and λ' are non-zero complex numbers, and $|b'_k\rangle = (\lambda'/|\lambda'|)(|\lambda|/\lambda)|b_k\rangle$, then $|b'_k\rangle$ is also an unit vector orthogonal to $|b_j\rangle$; therefore:

$$g(\lambda b_j + \lambda' b_k) = g((|\lambda|/\lambda)(\lambda b_j + \lambda' b_k)) = g(|\lambda| b_j + |\lambda'| b'_k) = M|\lambda|^2 + (w - M)|\lambda'|^2. \qquad (11.20)$$

So we see that

$$g(\psi) = \langle \hat{O}\psi|\psi\rangle \qquad (11.21)$$

for any vector $|\psi\rangle$, where $\hat{O}$ is a self-adjoint operator whose matrix, relative to $|b_j\rangle, |b_k\rangle$ is

$$\begin{bmatrix} M & 0 \\ O & w - M \end{bmatrix}. \qquad (11.22)$$

This shows that f is regular. Q. E. D.

Now Gleason proves the following Lemma:

Lemma 11.3 (III Gleason Lemma) *if f is a non-negative frame function for a Hilbert space $\mathcal{H}$ (either real or complex) and is regular when restricted to any two-dimensional subspace of $\mathcal{H}$, then f is regular.*

We leave the proof aside. Finally he proves following theorem:

Theorem 11.4 (II Regularity) *Every non-negative frame function on either a real or a complex Hilbert space of dimension $d \geq 3$ is regular.*

Proof

A frame function for $\mathcal{H}$ becomes a frame function for any completely real subspace of $\mathcal{H}$ by restriction. Every completely real two-dimensional subspace of $\mathcal{H}$ can be embedded in a completely real three-dimensional subspace, since the dimensions of $\mathcal{H}$ are ≥ 3. Therefore theorem 11.3 shows that any non-negative frame function f is regular on every completely real two-dimensional subspace of $\mathcal{H}$. But the lemma 11.2 shows that f is regular on every two-dimensional subspace; hence f is regular by lemma 11.3. Q. E. D.

We can now prove the Gleason Theorem [theorem 11.2].

Proof

If $\mathcal{H}_\psi$ is the one-dimensional subspace spanned by the unit vector $|\psi\rangle$, then $f(\psi) = \mu_p(\mathcal{H}_\psi)$ defines a non-negative frame function f [see Eqs. (11.10) and (11.12)]. Then there is a self-adjoint operator $\hat{O}$ such that $\mu_p(\mathcal{H}_\psi) = \langle \hat{O}\psi|\psi\rangle$ for all unit vectors $|\psi\rangle$ [see Eq. (11.13)]. Since $\langle \hat{O}\psi|\psi\rangle \geq 0$ for all $|\psi\rangle$, $\hat{O}$ is positive and semi-definite. If $\{|b_j\rangle\}$ is an orthonormal basis for $\mathcal{H}$ we have:

$$\mu_p(\mathcal{H}) = \sum \mu_p(\mathcal{H}_{b_j}) = \sum \langle \hat{O}b_j|b_j\rangle. \tag{11.23}$$

Since this last sum converges, $\hat{O}$ is in the trace class, so that we can substitute $\hat{O}$ with $\hat{\rho}$ and have $\mathrm{Tr}\hat{\rho} = \mu_p(\mathcal{H})$.

If now $\mathcal{H}_j$ is an arbitrary closed subspace of $\mathcal{H}$, we can choose an orthonormal basis $\{|b_j\rangle\}$ for $\mathcal{H}_j$ and add further vectors $\{|\psi_k\rangle\}$ so that $\{|b_j\rangle, |\psi_k\rangle\}$ is an orthonormal basis for $\mathcal{H}$. Then $\hat{P}_{\mathcal{H}_j}|b_j\rangle = |b_j\rangle$ (because $\hat{P}_{\mathcal{H}_j}$ is the projection operator in $\mathcal{H}_j$) for all j and $P_{\mathcal{H}_j}|\psi_k\rangle = 0$ for all k (for the same reason). So that we have:

$$\begin{aligned}
\mu_p(\mathcal{H}_j) = \sum \mu_p(\mathcal{H}_{b_j}) &= \sum_j \langle \hat{\rho} b_j|b_j\rangle + \sum_k \langle \hat{\rho}\psi_k|\psi_k\rangle \\
&= \sum_j \langle \hat{\rho}\hat{P}_{\mathcal{H}_j} b_j|b_j\rangle + \sum_k \langle \hat{\rho}\hat{P}_{\mathcal{H}_j}\psi_k|\psi_k\rangle \\
&= \mathrm{Tr}(\hat{\rho}\hat{P}_{\mathcal{H}_j}).
\end{aligned} \tag{11.24}$$

From this conclusion and Eq. (11.23) we can easily deduce the quantum expectation $\left\langle \hat{O} \right\rangle = \mathrm{Tr}(\hat{\rho}\hat{O})$. Q. E. D.

Chapter 12

GEOMETRIC PHASE

Contents In this chapter, after having reported briefly on some historical antecedents [section 12.1], we discuss Berry's contribution to the theory of Geometric phase [section 12.2]. Section 12.3 is devoted to some generalizations of the theory. Then in the final section some experiments are quoted.

12.1 Antecedents

Is there a possibility of treating the particularities of the quantum phase in general terms, in the same way that we have discussed those of the quantum wave amplitude with the treatment of QM probability? The answer comes from an area developed in these last years: the theory of Geometric Phase in QM. Its generality and the great quantity of problems to which it is and can be applied justify a separate treatment of the problem.

The development of the theory in the eighties is particularly due to Berry [BERRY 1984] [BERRY 1987], but there were already some developments in the fifties [PANCHARATNAM 1956] and later [HERZBERG/LONGUET-H. 1963] [MEAD/TRUHLAR 1979].[1]

Pancharatnam's work was centred on phase shifts by polarisation, and in particular on the problem of whether the phase is conserved after a cyclic series of changes of polarisation. First we state the following theorem with its corollaries [PANCHARATNAM 1956, 53–54]:

Theorem 12.1 (Pancharatnam) *When a vibration of intensity I in the state of elliptical polarisation represented by point C in the Poincaré sphere is decomposed in two vibrations in the opposite states of polarisation — represented by opposite points A, A' —, the intensities of the A-component and of the A'-component are $I\cos^2\theta/2$ and $I\sin^2\theta/2$ respectively, where θ is the angular separation on the sphere between C and A,*

where we recall [see p. 58] that conventionally the two poles represent circular polarisation, the equatorial points linear polarisation [see also figure 12.1].

From this theorem we have following two corollaries (all these results will be integrated in theorem 12.2):

Corollary 12.1 (I Pancharatnam) *Two opposite polarised beams cannot constructively or destructively interfere;*

and

[1] On this point see [BERRY 1989, 25].

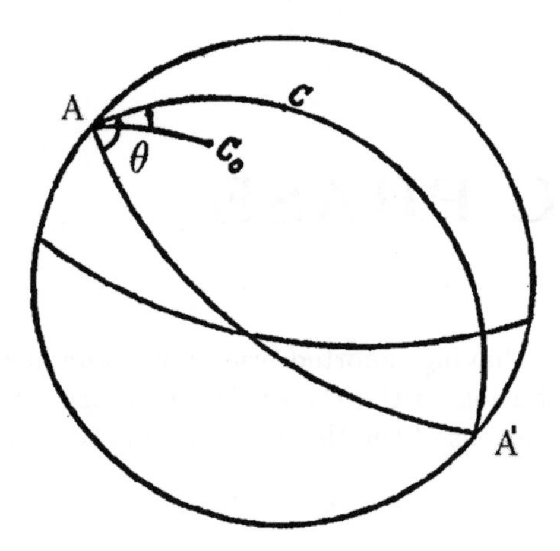

Figure 12.1: Representation on the Poincaré sphere of the orthogonal decomposition of elliptical polarisation — from [PANCHARATNAM 1956, 54].

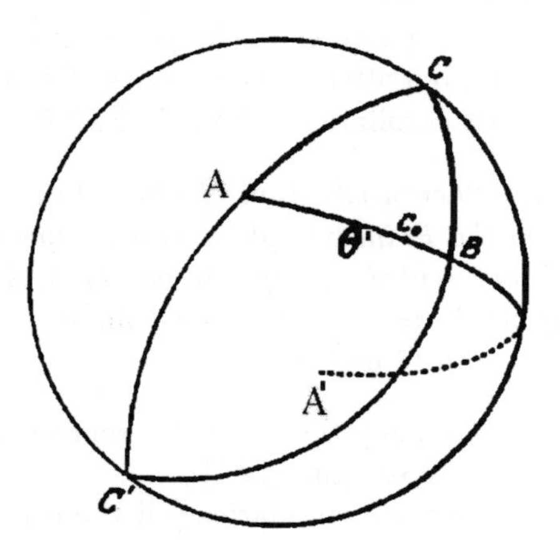

Figure 12.2: Representation on the Poincaré sphere of a non-orthogonal decomposition (A and B) of elliptical polarisation — from [PANCHARATNAM 1956, 55].

Corollary 12.2 (II Pancharatnam) *If the polarised beams are non-orthogonal, the composed intensity of two interfering beams is given by the formula* [see figure 12.2]*:*

$$I = I_1 + I_2 + 2\sqrt{I_1 I_2}\cos\frac{\theta'}{2}\cos\delta\varphi, \tag{12.1}$$

where θ' is the angular separation of the states represented by A, B *on the Poincaré sphere, $\delta\varphi$ is the phase difference and $\cos\theta'/2$ is equal to the visibility of the fringes.*

The formula (12.1) is a generalization for the polarization problem of the interference amplitudes due to the Superposition principle.

Therefore Pancharatnam introduced the idea that an initial state ($|i\rangle$), say before the passage through an analyser, and a final state $|f\rangle$ are 'in phase' if the intensity of their superposition:

$$(\langle i| + \langle f|)(|i\rangle + |f\rangle) = 2 + 2|\langle i|f\rangle|\cos(\delta\varphi\langle i|f\rangle) \tag{12.2}$$

is maximal, a condition equivalent to their overlap $\langle i|f\rangle$ being real and positive. The phase difference between $|i\rangle$ and $|f\rangle$ is simply the phase $\delta\varphi\langle i|f\rangle$ of their scalar product. But this connection is not transitive: suppose an initial state $|\psi_0\rangle$ and three successive measurements I–III such that after I the system is in state $|\psi_1\rangle$, after II $\mathcal{S}$ is in $|\psi_2\rangle$ and after III it returns to $|\psi_0\rangle$. Ignoring the time evolution, the final state is given by:

$$|\psi_0\rangle\langle\psi_0|\psi_2\rangle\langle\psi_2|\psi_1\rangle\langle\psi_1|\psi_0\rangle. \tag{12.3}$$

Final and initial states have a phase difference given by the phase of the complex number $\langle\psi_0|\psi_2\rangle\langle\psi_2|\psi_1\rangle\langle\psi_1|\psi_0\rangle$. All these results are absorbed by Berry's more powerful generalization. Therefore, in conclusion, we can formulate following Theorem by synthesizing both contributions of Pancharatnam and of Berry [PANCHARATNAM 1956, 56–63] [BERRY 1987, 68]:

Theorem 12.2 (Pancharatnam/Berry) *If there is a sequential change of polarisation from state $|i\rangle$ to middle state $|m\rangle$ and from this to state $|f\rangle$, and a return to state $|i'\rangle$ (which occupies the same point on the Poincaré sphere as $|i\rangle$), such that $|f\rangle$ and $|i'\rangle$ are in phase, the phase difference between $|i\rangle$ and $|i'\rangle$ is given by the phase of the factor:*

$$\langle i|i'\rangle = e^{-\frac{\imath\theta_{IMF}}{2}}, \tag{12.4}$$

where θ_{IMF} is the solid angle subtended by the geodesic triangle IMF determined by the points on the Poincaré sphere corresponding to $|i\rangle$, $|m\rangle$ and $|f\rangle$.

12.2 Berry's Contribution

Anholonomy and adiabaticity To better understand the problem one must distinguish a *dynamical phase φ_d:*

$$e^{\imath\varphi_d} := e^{-\frac{\imath}{\hbar}\int E(t)dt} \tag{12.5}$$

from a *geometric phase*, which approaches a finite non-zero limit as the parameters of the Hamiltonian change more and more slowly around a closed path of the Hamiltonian in its parameter space (approach eigenstates of the instantaneous Hamiltonian). The geometric phase is non-integrable (not only does it depend on the endpoints, but also on the geometry of the path) [BERRY 1995, 304]. Two concepts are here important:

- that of *anholonomy*, which is a geometrical phenomenon in which non-integrability causes some variables to fail to return to their original values if others (which drive them) are altered around a cycle;
- and that of *adiabaticity*, which is a slow-change, at the border line between dynamics and static.

The adiabatic theorem guarantees that, if the system is slowly cycling, it returns to its original state. But it acquires a non-trivial phase as a consequence of anholonomy [BERRY 1989, 8]. We speak of *parallel transport* on the surface of a sphere [BERRY 1989, 8–10] if the unit vector $\mathbf{e}$ is transported by rotating the unit radius vector $\mathbf{r}$ [see figure 12.3] at two conditions:

- $\mathbf{e} \cdot \mathbf{r} = 0$ (they are orthogonal),
- the orthogonal frame containing both vectors must not twist about $\mathbf{r}$, i.e. $\mathbf{\Omega} \cdot \mathbf{r} = 0$, where $\mathbf{\Omega}$ is the angular velocity of the triad.

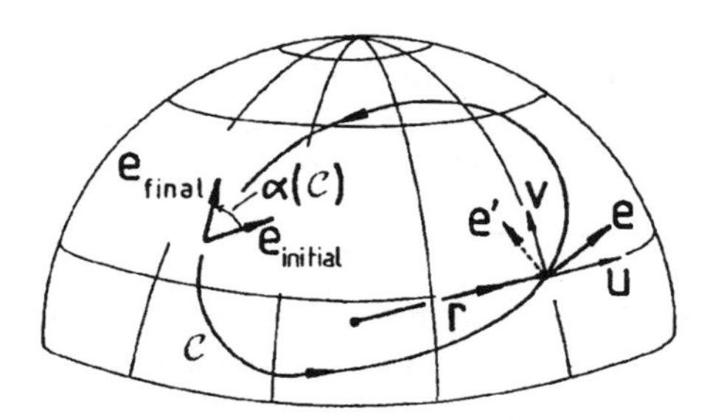

Figure 12.3: Rotation by an angle $\alpha(\mathcal{C})$ after a parallel transport of vector $\mathbf{e}$ around circuit $\mathcal{C}$ on a sphere — from [BERRY 1989, 9].

General case Due to anholonomy, $\mathbf{e}$ after a cycle $\mathcal{C}$ does not necessarily return to the initial position but can present an angle $\alpha(\mathcal{C})$ with the original vector. If $\mathbf{e}'$ constitutes an orthogonal triad with $\mathbf{e}, \mathbf{r}$, in terms of the unit vector

$$\psi := \frac{(\mathbf{e} + \imath \mathbf{e}')}{\sqrt{2}}, \tag{12.6}$$

the transport law can be expressed as:

$$\mathrm{Im} \psi^* \dot{\psi} = 0, \tag{12.7}$$

where $\dot{\psi}$ represents the change in ψ resulting from a change $d\mathbf{r}$. In order to find $\alpha(\mathcal{C})$ we choose a local basis of unit vectors $\mathbf{u}(\mathbf{r}), \mathbf{v}(\mathbf{r})$ at each point of the passage of $\mathbf{e}, \mathbf{e}'$. If we choose $\mathbf{u}, \mathbf{v}$ to lie along parallel of latitude θ and meridian of longitude ϕ, we have $\mathbf{r} = (\sin\theta\cos\phi, \sin\theta\sin\phi, \cos\theta)$, i.e.:

$$\mathbf{u} = (-\sin\phi, \cos\phi, 0), \qquad \mathbf{v} = (-\cos\theta\cos\phi, -\cos\theta\sin\phi, \sin\theta). \tag{12.8}$$

Specifying a local basis is equivalent to specifying a complex unit vector:

$$\mathbf{n}(\mathbf{r}) := \frac{\mathbf{u}(\mathbf{r}) + \imath\mathbf{v}(\mathbf{r})}{\sqrt{2}}. \tag{12.9}$$

If the angle between the transported $\mathbf{e}$ and the local $\mathbf{u}$ is $\alpha(t)$, using Eqs. (12.6) and (12.9) we obtain:

$$\psi = \mathbf{n}e^{-\imath\alpha}. \tag{12.10}$$

Eq. (12.7) now gives the anholonomy as:

$$\alpha(\mathcal{C}) = \oint d\alpha = \mathrm{Im}\int\int_{\partial\mathbf{S}=\mathcal{C}} d\mathbf{n}^* \cdot d\mathbf{n}, \tag{12.11}$$

where the integral in Eq. (12.11) is of the area $\mathbf{S}$ on the sphere bounded by $\partial\mathbf{S} = \mathcal{C}$. It is important that integrand in Eq. (12.11) is independent from the choice of local basis. In terms of arbitrary parameters η_1, η_2 specifying $\mathbf{r}$, Eq. (12.11) can be written:

$$\alpha(\mathcal{C}) = \mathrm{Im}\int\int_{\partial\mathbf{S}=\mathcal{C}} d\eta_1 d\eta_2 (\partial_1\mathbf{n}^* \cdot \partial_2\mathbf{n} - \partial_2\mathbf{n}^* \cdot \partial_1\mathbf{n}). \tag{12.12}$$

Quantum case In QM one needs to substitute state vector $|\psi\rangle$ to unit vector ψ, and position $\mathbf{r}(\eta_1, \eta_2)$ on the sphere with position $\eta = (\eta_1, \eta_2, \ldots)$. The quantum geometric phase $\varphi(\mathcal{C})$ is the change which occurs when returning to initial state

$$\langle\psi_i|\psi_f\rangle = e^{\imath\varphi(\mathcal{C})}, \tag{12.13}$$

which is part of the total phase change transformation:

$$\langle\psi_f|\psi_i\rangle = e^{\imath(\varphi_d + \varphi(\mathcal{C}))}. \tag{12.14}$$

Introducing again a local basis, we can write:

$$|\psi\rangle|n(\eta)\rangle e^{\imath\varphi_d}. \tag{12.15}$$

Rewriting the transport law (12.7) as:

$$\mathrm{Im}\langle\psi|d\psi\rangle = 0, \tag{12.16}$$

we can now calculate the geometric phase:

$$\varphi(\mathcal{C}) = \oint d\varphi = -\mathrm{Im}\oint\langle\mathbf{n}|d\mathbf{n}\rangle = -\mathrm{Im}\int\int_{\partial\mathbf{S}=\mathcal{C}}\langle d\mathbf{n}|\nabla|d\mathbf{n}\rangle, \tag{12.17}$$

by using Stoke's Theorem

$$\oint_{\mathcal{C}} \psi \cdot d\mathcal{C} = \int_{\mathbf{S}} \nabla \times \psi d\mathbf{S}. \tag{12.18}$$

Schrödinger equation In order to obtain more physical concreteness, it is important to find some connection with the Schrödinger Eq. [BERRY 1989, 10–11] [BERRY 1984, 125–26]. The adiabatic theorem guarantees that $|\psi\rangle$ will cycle to one of the eigenstates of $\hat{H}(\eta(t))$ where $\hat{H}(\eta)|\psi\rangle = E_n|\psi\rangle$. The adiabatic condition:

$$|\psi\rangle \simeq |\psi_0\rangle e^{-\frac{\imath}{\hbar}\int_0^t dt' E_n(\eta(t'))} \tag{12.19}$$

provides the connection, where for a cycle which takes time τ we have:

$$\varphi_d = \hbar^{-1}\int_0^\tau dt E_n(\eta(t)). \tag{12.20}$$

If we express the Hamiltonian as a function of a slowly changing parameter $\xi(t)$ [BERRY 1984, 125–26], the system evolves according to Schrödinger Eq.:

$$\hat{H}(\xi(t))|\psi(t)\rangle = \imath\hbar\frac{\partial}{\partial t}|\psi(t)\rangle. \tag{12.21}$$

At any instant we have eigenstates $|n(\xi)\rangle$ which satisfy:

$$\hat{H}(\xi)|n(\xi)\rangle = E_n|n(\xi)\rangle, \tag{12.22}$$

with energies $E_n(t)$. In this situation we have no relation between the phases of the eigenstates $|n(\xi)\rangle$ at different values of ξ.

Taking adiabaticity into account, we can write the state vector as:

$$|\psi(t)\rangle = \exp\left[\frac{\imath}{\hbar}\int_0^t dt' E_n(\xi(t'))\right] e^{\imath\varphi_d^{(n)}(t)}|n(\xi(t))\rangle; \tag{12.23}$$

and substituting Eq. (12.23) into Eq. (12.21) we obtain:

$$\dot{\varphi}(t) = \imath\langle n(\xi(t))|\nabla_\xi n(\xi(t))\rangle \cdot \dot{\xi}(t). \tag{12.24}$$

If the total phase change is now given by [see Eq. (12.20)]:

$$|\psi(\tau)\rangle = e^{\imath\varphi_d^{(n)}(C)} \exp\left[-\frac{\imath}{\hbar}\int_0^\tau dt E_n(\xi(t))\right]|\psi(0)\rangle, \tag{12.25}$$

then the geometrical phase change can be expressed as:

$$\varphi_n(C) = \imath\oint_C \langle n(\xi)|\nabla_\xi n(\xi)\rangle \cdot d\xi. \tag{12.26}$$

12.3 Generalization with Fiber Bundles*

Simon [SIMON 1983] showed that if we have an Hermitian operator $\hat{O}(\eta)$ smoothly depending on a parameter η with isolated non-degenerate eigenvalues $o(\eta)$, then the set

$$\{\langle\eta|\varphi\rangle|\hat{O}(\eta)|\psi\rangle = o(\eta)|\psi\rangle\} \tag{12.27}$$

defines a *line bundle* over the parameter space.

Samuel and Bandhari [SAMUEL/BHANDARI 1988] showed that one does not need cyclic or unitary transformations but only geodesic lines. They used fiber bundles[2]. If the Hamiltonian depends on a slow changing external parameter $\xi(t)$, the phase of wave function can be understood as an internal parameter that depends on $\xi(t)$ through the energies and through the adiabatic phase

$$e^{\int A_n(\xi)d\xi}, \tag{12.28}$$

where $A_n = \langle n(\xi)|\nabla|n(\xi)\rangle$ (we suppose here that the energy $E_n(\xi) = 0$). We use a generalization of the direct product $M \times F$ of two spaces: a base space M and a bundle E with fiber F, such that in a small neighbourhood U of M the bundle E looks like $U \times F$, but globally it may be topologically twisted [figure 12.4].

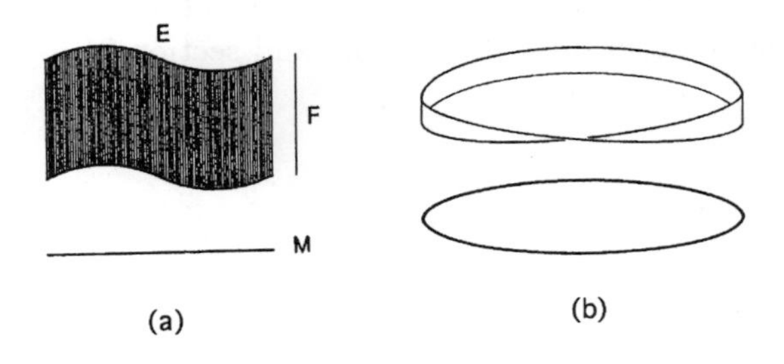

Figure 12.4: (a) A fiber bundle with total space E, fiber F and base space M.
(b) The Möbius strip is one of the simplest non-trivial examples of a fiber bundle — from [*SHAPERE/WILCZEK* 1989, 118].

We are now interested in bundles with group symmetry: G–bundles. If F is G, then we have a mean G–bundle. The action of G may be thought of as a change of coordinate of each fiber. Now a rule for parallel transport is a rule of changing a fiber's coordinate as one moves horizontally over basis space M. It describes a way of lifting a curve in the base manifold to a curve in the bundle. For any two non-orthogonal states the phase difference is given by:

$$\delta\varphi = \int A_s ds, \tag{12.29}$$

where $A_s = \mathrm{Im}\langle\phi_s|u\rangle/\langle\phi_s|\phi_s\rangle$ and ϕ_s is a curve in the projector space $\mathcal{P}(\mathcal{H})$ and $|u\rangle$ denotes the tangent vector to this curve.

In conclusion we have the following theorem[3]:

Theorem 12.3 (Simon/Samuel/Bandhari) *The shifts of phase in QM can be explained in terms of the twisting of a line bundle or in terms of lifting of a curve in the base manifold to a curve in the bundle.*

---

[2]See also [*SHAPERE/WILCZEK* 1989, 117–23]. For mathematics see [HUSEMOLLER 1975].
[3]For further developments in the measurement theory see also [CASSINELLI *et al.* 1994].

12.4 Experiments and Conclusions

Several experiments have measured the geometric phase (often in the Pancharatnam's version). In Q-Optics they can be divided into Interfer. Exp.s [BHANDARI/SAMUEL 1988] and SPDC Exp.s [CHIAO *et al.* 1988] [KWIAT/CHIAO 1991]. Interfer. Exp.s have been made with 1/2–spin (neutrons), 1–spin (photons) and 3/2–spin (chlorine nuclei) particles. The greatest problem is separating the geometric from the dynamical phase. The first proposal for obtaining this result is due to Wagh and Rakhecha [WAGH/RAKHECHA 1990] and an account of the performed experiment is in [WAGH *et al.* 1997].[4]

The geometric phase can be considered a consequence of the superposition between different paths. As we shall see [in section 41.2], the geometric phase is an explication of the Aharonov/Bohm Effect. The geometric phase is a line integral of the Berry connection (the induced vector potential $\mathbf{A}_e$) while the induced magnetic-like field $\mathbf{B}$ is Berry curvature [JACKIW 1988, 29]. For other applications see also subsection 19.8.4 and section 21.6.

[4]In the article there is also a review of preceding attempts.

Part IV

MEASUREMENT PROBLEM

Introduction to Part IV The *fourth part* handles the first thematic area of the book and probably the most important part of the theory: the measurement problem.

- First we examine some basic concepts [*chapter 13*].
- Then we reconstruct von Neumann's original proposal and its refinements [*chapter 14*].

Von Neumann was the first to assume a dichotomy between two different types of evolutions: the unitary one and the measurement process. On this matter, different positions have been developed:

- there are those who have accepted such a dichotomy and tried to assign a fundamental function to the consciousness (after von Neumann: London, Bauer, Wigner): this is the content of *chapter 14*;
- some have tried to eliminate the jump-like features of the measurement by rejecting the projection postulate (Margenau, Ballentine, Everett, deWitt); this is the content of *chapter 15*;
- others have tried to explain measurement in terms of intervention of macrosystems: either as a limiting case of a thermodynamical evolution (Green, van Hove and others) or using SSRs for obtaining the required classical behaviour of the pointer: this is the issue in *chapter 16*;
- there are those who have tried to find some *ad hoc* parameters in order to explain the reduction (Ghirardi/Rimini/Weber): this will be treated later in *chapter 23*, because the proposal will account more generally for spontaneous transitions to localisation, and hence it is more like a theory of the relation between microphysics and macrophysics;
- others have developed a form of non-unitary evolution by using the specific nature of open systems: this is the theory of decoherence (Zurek, Cini, Machida/Namiki, among others), which is treated in *chapter 17*.

It is quit astonishing that measurement theory has led to so many different positions and that, consequently, for fifty years has been a problematic area of QM. The reasons will be briefly discussed in subsection 14.1.1.

The last three chapters of this part are devoted to new developments which partly extend and integrate the theory of decoherence.

- The operational approach [*chapter 18*] represents a formal generalization of the quantum theory of open systems.
- *Chapter 19* is dedicated to the Quantum Non-demolition Measurement.

  The last two chapters contain results which are generally valid and neutral with respect to a particular interpretation. Anyway, their importance for QM (and partly also for some interpretational problems) justifies the place they occupy here.
- Finally we investigate a new field of research which has very interesting interpretational implications and a lot of technological consequences: the so-called interaction-free measurements [*chapter 20*].

In the following we shall use the word 'Measurement' to mean the measurement process in general, while the word 'measurement' always indicates a particular measurement process.

Chapter 13

SOME PRELIMINARY NOTIONS

Contents In section 13.1 we give a definition of Measurement. Then we briefly analyse the different parts [section 13.2] and steps [section 13.3] of Measurement. The first step (Premeasurement) is then analysed [section 13.4] — the other steps will be the object of the next chapter —, and finally different kinds of Measurement are defined [section 13.5].

13.1 Measurement's Definition

We firstly give a preliminary definition of Measurement:

Definition 13.1 (Measurement definition) *It is the temporary insertion into the observed system $\mathcal{S}$ of a certain energy coupling with another system — the apparatus $\mathcal{A}$ — such that it is possible to infer properties of $\mathcal{S}$ from observing properties of $\mathcal{A}$ (reading).*

As we shall see in the next chapter, Measurement supposes an irreversibility such that the result can be registered and later also utilized. The problem is seeing where and how it occurs.

13.2 Elements of the Measurement

We distinguish the following objects in the Measurement: an observed system $\mathcal{S}$, an apparatus $\mathcal{A}$ and sometimes an observer $\mathcal{O}$. $\mathcal{A}$ in turn can be divided into *little measuring devices* (generally characterized by QM laws and, anyway, directly interacting with the microentity to be measured) and *amplifiers* [*JAUCH* 1968, 169]. Traditionally it was supposed that only the amplification is irreversible but Machida and Namiki [MACHIDA/NAMIKI 1980] have shown that the irreversibility can already occur in the little measure device[1].

13.3 Steps of the Measurement

Starting with uncoupled system $\mathcal{S}$ and apparatus $\mathcal{A}$, the Measurement process consists of: *preparation* (premeasurement), *measurement strictu sensu* (detection and amplification), *registration* and/or *reading* and *statistics* [*BUSCH et al.* 1995a, 5–6] [see figure 13.1].[2]

[1]The model will be discussed later [see section 17.3].

[2]See also [*SCHROECK* 1996, 16] [*MITTELSTAEDT* 1998, 22–24].

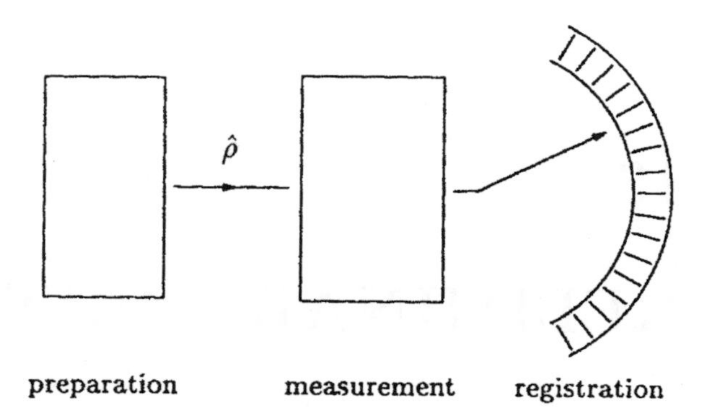

Figure 13.1: Scheme of measurement (we shall speak about the statistics in section 14.3) — from [*BUSCH et al.* 1995a, 6].

We now analyse in more detail some moments of the Measurement process, in particular the premeasurement [in the next section], the Measurement *strictu sensu* [section 14.1], the reading [subsection 14.2.1] and the problems of statistics [section 14.3].

13.4 Premeasurement

Definition The distinction between state preparation and Measurement was made for the first time by Margenau [MARGENAU 1937] [MARGENAU 1958].[3] *Premeasurement* is necessary for posing the adequate constraints on $\mathcal{S}$ which enable us to obtain the desired result [MACHIDA/NAMIKI 1980, 1836]:

Definition 13.2 (Premeasurement) *Premeasurement is the preliminary preparation of the system in order to obtain some desired results, and it usually consists in the confinement of the object system in a small macroscopic space-time region (Localization requirement).*

Classical case We can discuss the point as follows [LAMB 1969, 24–25]:[4] in order to speak of a wave function of a determinate system, one must first be able to confine the system. In classical mechanics one places a particle at an initial position $\mathbf{q}(0)$ and gives it an initial momentum $\mathbf{p}(0)$; then one uses canonical Eqs. (1.1) to calculate the successive evolution. For this purpose we set up a potential well $V_1(\mathbf{q})$ having a minimum at $\mathbf{q} = \mathbf{q}(0)$. Then we catch a particle in the potential well and allow it to settle down to rest. Then we give the particle an impulse derived from a potential $V_2(\mathbf{q})\delta(t)$, where $\delta(t)$ is the Dirac delta-function sharply peaked at initial time $t = 0$. This force can bring the particle momentum to $\mathbf{p}(0)$ without appreciably changing its position. Then we add a potential $V_3(\mathbf{q})$ which corresponds to the problem which interests us here.

[3]For history and further analysis see [*JAMMER* 1974, 228, 486].
[4]See also [ROYER 1989, 18–19] .

Quantum case In QM we desire to prepare a state $\psi(\mathbf{q},t)$. We must distinguish the following cases:

- If this is the lowest energy eigenfunction for a certain potential, we set up such a potential $V_1(\mathbf{q})$ — for example we catch an electron in it, and wait for radiative damping to bring the particle to the ground state. Then we suddenly turn off $V_1(\mathbf{q})$ and turn on $V_3(\mathbf{q})$ for the problem which interests us here (a wave function is not immediately changed by a sudden change in potential if no singular time dependence is involved). Then the future evolution of the system follows the Schrödinger Eq.

- If the initial state is not the known lowest energy eigenfunction of some $V_1(\mathbf{q})$, we can proceed as follows: we may look for a potential $V_1(\mathbf{q})$ for which $\psi(\mathbf{q},0)$ is an eigenfunction with some energy eigenvalue E. Because $\psi(\mathbf{q},0)$ obeys the following Eq. [see also Eq. (3.26)]:

$$[\hat{T}_E + V_1(\mathbf{q})]\psi(\mathbf{q},0) = E\psi(\mathbf{q},0), \tag{13.1}$$

  where $\hat{T}_E$ is the kinetic energy [Eq. (1.6)], we can find that

$$V_1(\mathbf{q}) = E - \frac{\hat{T}_E\psi(\mathbf{q},0)}{\psi(\mathbf{q},0)}. \tag{13.2}$$

  If this potential is experimentally available, we catch the electron in it. The wave function then becomes a linear combination of eigenstates of the Hamiltonian $\hat{H} = \hat{T}_E + V_1(\mathbf{q})$, one of which is the desired $\psi(\mathbf{q},0)$. Then one can carry out a state selection process. The following difficulties may occur: 1. $V_1(\mathbf{q})$ is singular if $\psi(\mathbf{q},0)$ has nodes; 2. the energy spectrum may be not discrete; 3. $V_1(\mathbf{q})$ may be a complex function[5].

13.5 Types of Measurement

Now we treat a specific problem of the statistics: the effects of the Measurement on the statistics before it. Pauli [*PAULI* 1980, 75] distinguishes two types of Measurement: a first kind (= I–Measurement) and a second kind (= II–Measurement).

Definition 13.3 (Type I–Measurement) *We have a type I–Measurement when with the measurement we do not change the value (though unknown) of the observable to be measured (which maintains the same probability distribution before and after the measurement).*

The type I–Measurement is a repeatable measurement. The necessary and sufficient condition of a type I–Measurement of an observable $\hat{O}$, giving a value o_j — represented by a projection operator $\hat{P}_{o_j}$ — is that $\hat{P}_{o_j}$ commutes with the state $\hat{\rho}$ in which the system is[6]. Another way of stating this is to say that the Hamiltonian of the composite system $\mathcal{A}+\mathcal{S}$ commutes with the observable to be measured: $[\hat{H}_{\mathcal{S}+\mathcal{A}}, \hat{O}] = 0$.[7]

We have a type II–Measurement when we change the state of the system — **for example**, allowing the transition from a superposition, relative to the measured observable, to an eigenstate of this observable — and hence the statistics. It is a measurement which is not repeatable.

[5] Lamb discusses the third problem and also gives a solution.

[6] On this point see [*JAUCH* 1968, 167] [*LUDWIG* 1954, 57].

[7] See [WIGNER 1983a, 284]. See also Eq. (19.40).

Chapter 14

VON NEUMANN THEORY AND ITS REFINEMENTS

Contents The chapter begins with the problem as stated by von Neumann and with his solution [section 14.1]. Then we briefly examine some integrations and some criticisms [section 14.2]. Finally we discuss Measurement statistics in some details [section 14.3].

14.1 Von Neumann's Original Exposition

14.1.1 Introduction

Formalization of the Measurement Before the publication of von Neumann's book in 1932, as a whole the Measurement process in QM was not formalized. It is a fact that in classical mechanics Measurement does not pose the problems which characterize QM [*VON NEUMANN* 1932, 212][*VON NEUMANN* 1955, 399] [see section 2.2.2 and subsection 3.3.2]: in fact classical systems are independent from measurements, and predictions made on the basis of past measurements are generally not invalidated by intermediate measurements [ZUREK 1993a, 289] [see also section 2.2].

Hence it is a specific and difficult contribution of QM to have produced a new theory of Measurement. But one cannot account for the delay only in this terms. In fact Basic-QM was essentially developed between 1925 and 1927: so, how is it possible that we arrive to a first formalization of the Measurement process only in 1932?[1] This situation is perhaps a consequence of Bohr's belief that in order to describe a quantum system one needs classical concepts [see postulate 8.2: p. 146], which on the other hand are expressions or refinements of common sense [BOHR 1955a, 68] and hence not formalisable at all[2]. As a matter of fact von Neumann's contributions is characterized by applying to the Measurement process the QM formalism which for Bohr was to be explained in terms of the classical frame. As we shall see it is a form of self-referentiality[3], which, though not necessarily implying semantical completeness [see subsection 9.4.2], is very important for a theory.

[1]And note that von Neumann's book was mostly known only after its english translation in 1955.

[2]On this point see [*JAMMER* 1974, 472].

[3]Discussion of the problem in [*MITTELSTAEDT* 1998, 4–8, 122–24]. The first statement on Measurement theory as a form of 'metatheory' is to be found in [DALLA CHIARA 1977]: the important thing however is that it is a form of *physical* metatheory.

Underestimation of Measurement On the other hand, as we shall see below, von Neumann's formalization was far from being satisfactory. In fact, as we shall see kin the following chapters, there was no rigorous solution to the problem until the early eighties. Again we think that this is so because, among many physicists, the problem was underestimated. The theory worked anyway and with good results: why did one need to waste time for such speculations[4]? If one thinks of the very important theoretical and experimental results which depend on the new solution to the Measurement problem developed in these last years — and to be discussed later [see particularly chapters 17 and 21] —, one has further evidence of the insufficiency of a 'formalistic' interpretation of a theory [see section 6.2].

von Neumann and Copenhagen Returning now to von Neumann, due to some confusion, his Measurement theory was considered to be a fundamental part of the Copenhagen interpretation. As we shall see [subsection 14.1.6], though there are some points in common, von Neumann's theory as such is independent and partly incompatible with the Copenhagen interpretation [see subsection 14.1.6]. Because of the fact that from the 1930's to the 1970's von Neumann's theory was the only one to be accepted by from the community of physicists as the answer to Measurement, we refer to it as the *Standard interpretation.*

14.1.2 General Features of the Problem

Von Neumann assumed [*VON NEUMANN* 1932, 184–91] [*VON NEUMANN* 1955, 347–58] [5] that there are two possible transformations of a state $\hat{\rho}$:

- a continuous, causal, unitary and reversible change $\hat{U}_t : \hat{\rho} \mapsto \hat{\rho}_t$ defined by Eq. (3.71);
- and a discontinuous, non-causal, instantaneously acting, non-unitary and irreversible change (the Measurement process) of the following form:

$$\hat{\rho} \rightsquigarrow \tilde{\hat{\rho}} = \sum_{j=1}^{\infty} \langle \hat{\rho} b_j | b_j \rangle \hat{P}_{b_j}. \tag{14.1}$$

We shall return later to the instantaneous character of Measurement [subsection 14.2.3]. We analyse now the other aspects. Traditionally the second transition is called *Reduction of the Wave-Packet.* The name comes from the fact that, if for example we measure the position, then, firstly, the measurement destroys the phase relations between the components of the Schrödinger wave[6], and then it confines the wave in a restricted region of space by letting it collapse onto one of the possible eigenvalues of the position operator and by loosing all other components which are nevertheless present in the mixture [see figure 14.1].

A Discontinuous and Non-Unitary Change

It is evident that transformation (14.1) begins with a pure state and ends in a mixture of states $|b_k\rangle$ each one with weight $w_k = \langle \hat{\rho} b_j | b_j \rangle$ [see Eq. (3.59)]. Now, there is no continuous evolution able to explain such a change, because of the following theorem [WIGNER 1963, 333]:[7]

[4] Very often unsolved problems counts as genuine problems only when they are already solved [*LAUDAN* 1977, 18].

[5] See also [*JAMMER* 1974, 475].

[6] See for example [*DE BROGLIE* 1956, 62]

[7] See also [A. FINE 1970].

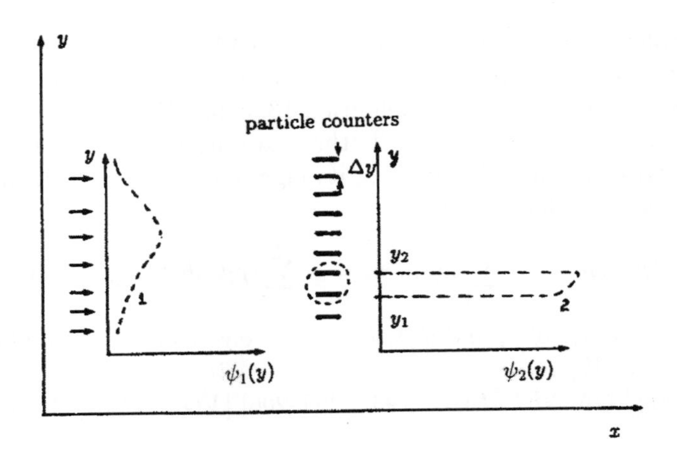

Figure 14.1: An illustration of the reduction of the wave function: before the measurement the probability density of a particle is represented by curve 1. By measuring the y coordinate using a sieve, where each cell contains a device to localize the particle, supposing that the particle has been detected in the circled cell, then the probability distribution is given now by curve 2. The fact that probability 2 is narrower than probability 1 justifies the term *reduction* — from [*BRAGINSKY/KHALILI* 1992, 24].

Theorem 14.1 (Conservation Law of Purity) *The characteristic values of the density matrix are a constant of motion.*

The theorem immediately points out to the non-unitarity of the Measurement, because, according to it, a pure state remains a pure state and a mixture a mixture, which is the case only if the evolution is unitary. The point can be proved in a very simple form [*D'ESPAGNAT* 1976, 46].

Proof

We start with the equality for pure states [see Eq. (3.72)]:

$$\begin{aligned} \hat{\rho}_t^2 &= \hat{U}_t\hat{\rho}_0\hat{U}_t^\dagger\hat{U}_t\hat{\rho}_0\hat{U}_t^\dagger \\ &= \hat{U}_t\hat{\rho}_0\hat{U}_t^\dagger = \hat{\rho}_t \end{aligned} \tag{14.2}$$

so that if $\hat{\rho}_0^2 = \hat{\rho}_0$ [see Eq. (3.45c)], then we also have: $\hat{\rho}_t^2 = \hat{\rho}_t$, i.e., after a unitary and continuous evolution, we again have a pure state and not a mixture. Q.E.D

Later we shall prove the converse, i.e. that the unitarity cannot respect some necessary requirements of the Measurement process [see theorem 14.13: p. 243].

The last consequence seems to be that there is no possible causal explanation of the described evolution. In fact in physics causality is always expressed by a deterministic dynamic evolution — for example the Liouville Eq. [Eq. (1.12)] — of the type described in QM by unitary operators.

Irreversibility

Now we turn to the problem of irreversibility [*VON NEUMANN* 1932, 202] [*VON NEUMANN* 1955, 379–80]: Measurement is generally irreversible, unless the state of $\mathcal{S}$ remains unchanged, i.e. unless we perform a type I–Measurement [see definition 13.3: p. 217]. If $\hat{\rho}_i$ is the state of $\mathcal{S}$ before the measurement, and $\tilde{\hat{\rho}}_f = \sum_{j=1}^{\infty}\langle\hat{\rho}_i b_j|b_j\rangle\hat{P}_{b_j}$ is the state after — where $\langle\hat{\rho}_i b_1|b_1\rangle, \langle\hat{\rho}_i b_2|b_2\rangle, \ldots$ are eigenvalues (weights) of $\hat{\rho}_f$ and $|b_1\rangle, |b_2\rangle, \ldots$ its eigenvectors —, we define the QM or von Neumann entropy of $\hat{\rho}_i$ as [see chapter 42]:

$$-\mathrm{Tr}(\hat{\rho}_i \ln \hat{\rho}_i) = -\sum_{n=1}^{\infty} w_n \ln w_n = -\sum_{n=1}^{\infty}\langle\hat{\rho}_i b'_n|b'_n\rangle \ln\langle\hat{\rho}_i b'_n|b'_n\rangle, \tag{14.3}$$

where $w_1, w_2, \ldots$ are eigenvalues of $\hat{\rho}_i$ and $|b'_1\rangle, |b'_2\rangle, \ldots$ its eigenvectors; and where the entropy of $\tilde{\hat{\rho}}_f$ is $-\sum_{n=1}^{\infty}\langle\hat{\rho}_i b_n|b_n\rangle \ln\langle\hat{\rho}_i b_n|b_n\rangle$.

Von Neumann stated [*VON NEUMANN* 1932, 203–206] [*VON NEUMANN* 1955, 381–87] following theorem:[8]

Theorem 14.2 (II von Neumann) *In the change from a pure state to a mixture, the entropy of $\hat{\rho}_i$ is less than or equal to that of $\tilde{\hat{\rho}}_f$, i.e.:*

$$-\sum_{n=1}^{\infty}\langle\hat{\rho}_i b'_n|b'_n\rangle \ln\langle\hat{\rho}_i b'_n|b'_n\rangle \le -\sum_{n=1}^{\infty}\langle\tilde{\hat{\rho}}_f b_n|b_n\rangle \ln\langle\tilde{\hat{\rho}}_f b_n|b_n\rangle. \tag{14.4}$$

Note that the theorem supposes the already happened transition to a mixture, a transformation which von Neumann could never really explain [see next section]. Later we shall discuss this point and show how a physical model can be constructed in order to prove the theorem [see section 42.3].

On the contrary [*VON NEUMANN* 1932, 206] [*VON NEUMANN* 1955, 388], in the case of the Schrödinger Eq., we have an invariant entropy. This evolution is unitary, since, having: $\hat{\rho}_t = \hat{U}\hat{\rho}\hat{U}^\dagger$, and also: $\hat{\rho}_t \ln \hat{\rho}_t = \hat{U}(\hat{\rho}_i \ln \hat{\rho}_i)U^\dagger$, we also obtain: $\mathrm{Tr}(\hat{\rho}_t \ln \hat{\rho}_t) = \mathrm{Tr}(\hat{\rho}_i \ln \hat{\rho}_i)$.

The asymmetry of the Measurement with respect to the time reversal is also expressed by the probabilistic structure of QM, i.e. by probability distributions for observables which are conditional upon some measurement [*HAAG* 1992, 303]. Suppose we wish to measure two maximal observables [see p. 37] with discrete spectra $\hat{O}_1, \hat{O}_2$, with measurement times respectively t_1 and t_2 ($t_2 > t_1$). The orthonormal systems of eigenstates are respectively $\{|u_j\rangle\}, \{|v_k\rangle\}$; we indicate the number of cases in which both results j and k are registered by n_{jk}. The ensemble consists of n individuals. Conditional probabilities can be expressed as:

$$\wp_{j\rightsquigarrow k} \equiv \frac{n_{jk}}{n_j}; \qquad n_j = \sum_k n_{jk}, \tag{14.5a}$$

$$\wp_{k\rightsquigarrow j} \equiv \frac{n_{jk}}{n_k}; \qquad n_k = \sum_j n_{jk}. \tag{14.5b}$$

But since in QM we have [Eqs. (3.38)– (3.39)] $\wp_{j\rightsquigarrow k} = |\langle v_k|u_j\rangle|^2$ (irrespective of the ensemble studied), then it follows that:

$$\wp_{k\rightsquigarrow j} \equiv \wp_{j\rightsquigarrow k}\frac{n_j}{n_k}, \tag{14.6}$$

which depends on the ensemble measured and considered. In other words a change of the state changes the probability distributions as a consequence.

[8]See also [PERES 1986a] [PERES 1986c].

14.1.3 Von Neumann's Solution

General Features

We divide Measurement into three steps[9] [*BRAGINSKY/KHALILI* 1992, 27–28]:

- the coupling between $\mathcal{A}$ and $\mathcal{S}$ (usually performed in the premeasurement stage);
- the actual Measurement, i.e. the reduction of the initial states of $\mathcal{A}$ and $\mathcal{S}$ to a mixture;
- then, the act of observation.

Von Neumann did not analyse the statistics [see section 13.3]. It is a later development and it will be discussed later [section 14.3]. On the other hand, we will not discuss here the specific content of Premeasurement [on this point see section 13.4] but only the problem of the coupling and of its consequences on the successive steps. Note that the second and third steps are pure information-acquisition, in which we leave unperturbed the observable $\hat{O}$ to be measured, but its probability distribution changes from a wider *a priori* one to a more restricted *a posteriori* one. The last two steps are the content of the Projection postulate — so named by Margenau [MARGENAU 1936, 241], and considered by him as the origin of the EPR paradox [chapter 31]. The postulate is the following [*VON NEUMANN* 1932, 234] [*VON NEUMANN* 1955, 439][10]:

Postulate 14.1 (Projection Postulate) *If the observable $\hat{O}$ is measured on $\mathcal{S}$ in an arbitrary state $|\psi\rangle$, then the latter is projected after the measurement onto one of the basis vectors $|b_j\rangle$ (i.e. $\hat{P}_{b_j}$) of the representation in which $\hat{O}$ is diagonal, i.e. in a eigenstate of $\hat{O}$ for which the probability is $|\langle b_j|\psi\rangle|^2$* [see postulate 3.4: p. 38].

However, it must be observed that, also in the frame of the Standard interpretation, this postulate is not unrestrictly valid. For example it must be modified in the case of degeneracy [LÜDERS 1951]. Later we shall discuss its limits [in subsection 14.2.2].

We assume that before the measurement $\mathcal{S}$ is in state $\hat{\rho}^i_{\mathcal{S}} = |\varsigma\rangle\langle\varsigma|$ with basis $\{|s_j\rangle\}$: $|\varsigma\rangle = \sum c_j|s_j\rangle$ — where the $|s_j\rangle$'s are the eigenvectors of the observable $\hat{O}^S$ to be measured —; and that $\mathcal{A}$ is in state $\hat{\rho}^i_{\mathcal{A}} = |\mathcal{A}\rangle\langle\mathcal{A}|$, with basis $\{|a_k\rangle\}$: $|\mathcal{A}\rangle = \sum |a_k\rangle$ — where the $|a_k\rangle$'s are the eigenvectors of the apparatus' observable $\hat{O}^A$ (the pointer) which we wish to couple with that to be observed. The Hilbert space for the compounded system is $\mathcal{H}_{\mathcal{S}\oplus\mathcal{A}} = \mathcal{H}_{\mathcal{S}} \otimes \mathcal{H}_{\mathcal{A}}$. Initially the states of the two systems are factorised [see definition 3.2: p. 42]; if we let $\mathcal{S}$ and $\mathcal{A}$ interact, normally we obtain an entangled state [see definition 3.3: p. 43] of both.

Entanglement between System and Apparatus

Unitary operator The entanglement between $\mathcal{S}$ and $\mathcal{A}$ can be expressed in terms of the following unitary operator :

$$\hat{U} = e^{\frac{i}{\hbar}\hat{H}_{\mathcal{SA}}\tau}, \tag{14.7}$$

[9] We do not consider here the registration. Obviously such a distinction cannot be applied in the case of continuous measurements [SRINIVAS/DAVIES 1981] [CAVES 1982] [CAVES 1986], where no clear separation between coupling and read-out can be made. In such cases we need more refined instruments, like those given by the operational approach [see chapter 18]. Concerning continuous measurement of photon number see [IMOTO *et al.* 1990] [OGAWA *et al.* 1991] [UEDA *et al.* 1992]. See also [BARCHIELLI/BELAVKIN 1991].

[10] See also [MARGENAU 1958, 29] [MARGENAU 1963, 476] [*BRAGINSKY/KHALILI* 1992, 28–29] .

where τ is the time-interval of the measurement, and the interaction Hamiltonian $\hat{H}_{\mathcal{SA}}$ is given by:

$$\hat{H}_{\mathcal{SA}} = -g_{\mathcal{SA}}(\tau)\hat{\mathbf{p}}_{\mathcal{A}} \otimes \hat{O}^{\mathcal{S}}, \tag{14.8}$$

where $g_{\mathcal{SA}}(\tau)$ is a normalized coupling function between $\mathcal{S}$ and $\mathcal{A}$, with a compact support near the time of measurement, $\hat{O}^{\mathcal{S}}$ is the object observable and $\hat{\mathbf{p}}_{\mathcal{A}}$ is the canonical conjugate of the position $\hat{\mathbf{q}}_{\mathcal{A}}$ of the measuring device (for example of some indicator).

Measure of entanglement If we use the Schmidt decomposition [E. SCHMIDT 1907]:[11]

$$|\Psi\rangle = \sum_j \sqrt{\wp_j}|s_j\rangle|a_j\rangle, \tag{14.9}$$

by starting with an initial state $|s_0\rangle|a_0\rangle$, which has only one component — i.e. $\wp_0(t_0) = 1, \wp_{j\neq 0}(t_0) = 0$ —, the initial value of $\wp_0$ will decrease with the increasing time-delay (the appearance of other components in Eq. (14.9)) so that (to lowest order in t) we can write [JOOS/ZEH 1985, 227–28]:[12]

$$\wp_0(t) = 1 - \mathrm{E}t^2, \tag{14.10}$$

where E is a coefficient representing the measure of entanglement (the rate of de-separation):

$$\begin{aligned} \mathrm{E} &:= \sum_{j\neq 0, k\neq 0} |\langle s_j a_k|\hat{H}_{\mathcal{S}+\mathcal{A}}|s_0 a_0\rangle|^2 \\ &= \langle s_0 a_0|\hat{H}_{\mathcal{S}+\mathcal{A}}(\hat{I} - |s_0\rangle\langle s_0|)(\hat{I} - |a_0\rangle\langle a_0|)\hat{H}_{\mathcal{S}+\mathcal{A}}|s_0 a_0\rangle, \end{aligned} \tag{14.11}$$

where $\hat{H}_{\mathcal{S}+\mathcal{A}}$ is the Hamiltonian of the compound system.

For example, Scully, Schwinger and Englert have shown [SCULLY *et al.* 1989] — by using a Stern/Gerlach Interferometer, with two microcavities in the upper arm — that it is only the correlation between the photons of the cavity and the spin of the particle going through the upper arm which establishes the necessary conditions for a Which-path detection. And the correlation itself can in turn only be established if the photons of the cavity are in a number state and not in a coherent state [see subsection 5.1.1] — see also [SCULLY *et al.* 1989, 1776–78] [SCULLY *et al.* 1978a].

For type I–Measurement E is proportional to the efficiency of the measurement because the system is already in an eigenstate of the measured observable. If not, then an entanglement of the type just defined seems to be in contradiction with the necessity of producing a single and univocal Measurement result (a single value of an observable).

Total and classical rates of change To focalise the problem let us see how a 'classical' evolution can be. The total rate of change Θ of the initial state is given by

$$\Theta := \sum_{j,k\neq 0,0} |\langle s_j a_k|\hat{H}_{\mathcal{S}+\mathcal{A}}|s_0 a_0\rangle|^2, \tag{14.12}$$

while the difference between E and Θ is the classical rate of change:

$$\Theta - \mathrm{E} = \sum_{j\neq 0} |\langle s_j a_0|\hat{H}_{\mathcal{S}+\mathcal{A}}|s_0 a_0\rangle|^2 + \sum_{k\neq 0} |\langle s_0 a_k|\hat{H}_{\mathcal{S}+\mathcal{A}}|s_0 a_0\rangle|^2. \tag{14.13}$$

[11] See also [*VON NEUMANN* 1932, 231–32] [SCHRÖDINGER 1935b]. Note that each state of a system made up of two subsystem can be decomposed in this way [EKERT/KNIGHT 1995, 416–17].

[12] See also [JOOS 1996, 45, 128–31].

Note that here we do not have a correlation anymore, but a factorization.

In conclusion what we need in order to perform a measurement with an univocal result is a final state which is not a pure state, because the entanglement does not allow a 'reading' of a determined value of the observable which we wish to measure[13].

The Mixture Requirement

A mixture What we need instead is a mixture, i. e. a bi-linear correspondence between each state of $\mathcal{S}$ and each state of $\mathcal{A}$ such that we can infer the state of $\mathcal{S}$ from the state of $\mathcal{A}$ or the value of $\hat{O}^{\mathcal{S}}$ from the value of the 'apparatus' observable $\hat{O}^{\mathcal{A}}$. In fact von Neumann supposed that, if $\mathcal{S}$ was already in one of the eigenstates of $\hat{O}^{\mathcal{S}}$ (type I–Measurement), then $\mathcal{A}$ will register the value by a transformation of the following form [see also postulate 3.3: p. 37]:

$$|s_j\rangle \otimes |a\rangle \mapsto |s_j\rangle \otimes |a_j\rangle. \tag{14.14}$$

In other words $\mathcal{A}$ 'registers' a value of $\hat{O}^{\mathcal{S}}$ which is already 'real'[14]. Hence, in the general case (where we do not necessarily perform a type I–Measurement), we need, as a result of the premeasurement, a mixture of the form:

$$\tilde{\hat{\rho}}^{\mathcal{S}+\mathcal{A}} = \sum_j |c_j|^2 |s_j\rangle\langle s_j| \otimes |a_j\rangle\langle a_j|, \tag{14.15}$$

where the states of $\mathcal{A}$, which have not already registered the correspondent j's value, are coupled one-to-one to each corresponding j value of $\hat{O}^{\mathcal{S}}$. It is evident that it is only a mixture of this form to allow a correspondence between respective values $o_j^{\mathcal{A}}$ and $o_j^{\mathcal{S}}$. Without such a connection, it is impossible to speak of a Measurement process.

A Lüders mixture In conclusion not all forms of mixture of states of a system $\mathcal{S}$ and an apparatus $\mathcal{A}$ can be called a Measurement process: the necessary condition is that the density matrix of the apparatus changes during the measurement process — i.e. that $\mathcal{A}$ extracts information from $\mathcal{S}$ — [JOOS 1996, 44]. We can write this requirement in a very short form:

$$\boxed{\hat{\rho}^{\mathcal{S}} \leadsto \sum_j \hat{P}_j \hat{\rho}^{\mathcal{S}} \hat{P}_j} \tag{14.16}$$

for the different eigenvalues j of $\hat{O}^{\mathcal{S}}$. We name such a mixture, which defines a measurement, a *Lüders mixture* [LÜDERS 1951].[15] We usually take this to be a general representation of standard theory. But, as we shall see, such a mixture has various peculiarities. In a more abstract way theorems 14.10 [p. 241] and 14.11 [p. 242] express the same point.

False solutions For this reason it is not useful to try to solve the Measurement paradox by saying that $\mathcal{A}$ before interacting was not in a pure state but already in a mixture $\sum w_j \hat{\rho}_j^{\mathcal{A}}$ [*VON NEUMANN* 1932, 233] [*VON NEUMANN* 1955, 438][16], where we recall that $\hat{\rho}_j^{\mathcal{A}} = |a_j\rangle\langle a_j|$ and w_j are arbitrary weights independent of $|\varsigma\rangle$. After the interaction, $\mathcal{A}+\mathcal{S}$ would be described

[13] On this point see also [KONO *et al.* 1996, 1065].

[14] On this point see [*VAN FRAASSEN* 1991, 252–57].

[15] See also [*BUSCH et al.* 1995a, 38] .

[16] See also [*JAMMER* 1974, 480].

by a density matrix of the form $\sum w_j^f \hat{\rho}_j^{\mathcal{S}+\mathcal{A}}$ where $\hat{\rho}_j^{\mathcal{S}+\mathcal{A}} = \hat{U}_t(\hat{\rho}_j^{\mathcal{A}} \otimes \hat{\rho}^{\mathcal{S}})\hat{U}_t^\dagger$ and initial and final weights would be the same: $w_j = w_j^f$. But QM requires that $w_j^f = |c_j|^2 = |\langle \varsigma | a_j \rangle|^2$ so that w_j would depend on the system's state $|\varsigma\rangle$ contrary to the assumption. Another way of saying the same thing is that [WIGNER 1963, 334–35], even if $\mathcal{A}$ and $\mathcal{S}$ are mixtures before the measurement, it is almost impossible, on the basis of a unitary evolution alone, that the result is a mixture such that each of the states corresponds to a definite final position of the pointer, i.e. a Lüders mixture.

The Act of Observation

Necessary and sufficient conditions We have said that the specific problem of Measurement is to infer the state of $\mathcal{S}$ from the state of $\mathcal{A}$. The Lüders mixture is the necessary condition. But it is not sufficient [FEYERABEND 1957a] [*ZEH* 1989, 84–85]. Von Neumann supposes that as a result of the Measurement, the global system is in a determinate state of the mixture represented by Eq. (14.15), i.e. we have a transition of the following type [see postulate 3.3: p. 37]:

$$\tilde{\hat{\rho}}^{\mathcal{S}+\mathcal{A}} \rightsquigarrow |c_k|^2 |s_k\rangle\langle s_k| \otimes |a_k\rangle\langle a_k|, \tag{14.17}$$

where $|s_k\rangle$ is an eigenstate of the measured observable [see Eq. (3.25)]. This transition is classical and as such presents no specific quantum problems. Anyway, we shall return on the informational aspects of it [see subsection 42.3.3]. It occurs when someone performs an observation or registration act, such that only one component of the mixture is acknowledged to be actual[17].

Power of the observation However, von Neumann gave great importance to the observation and the observer so as to account already for the passage from a pure state to a mixture. In fact he discarded the possibility of using a second measuring device $\mathcal{A}'$: this leads to an infinite regress (*von Neumann chain*) because $\mathcal{A}'$ conceptually stands in the same relation to $\mathcal{A}$ as $\mathcal{A}$ stands to $\mathcal{S}$.[18] From this situation von Neumann led the conclusion that the Measurement could not be explained in terms of a physical apparatus.

Von Neumann's answer was influenced by Szilard [SZILARD 1929], who was developing, at that time, some reflections regarding the Maxwell demon and the possibility of constructing a *perpetuum mobile* — Szilard believed a reduction of entropy (and hence an increase of the information) to be possible by the intervention of an intelligent being (literally a *Deus ex machina*). Following von Neumann, the Measurement process consists in an interaction between a physical system and a being capable of subjective apperception, the observer $\mathcal{O}$. It is $\mathcal{O}$ who, after the interaction between $\mathcal{S}$ and $\mathcal{A}$, is capable of reading the values of the pointer so as to obtain finally a value of the observable $\hat{O}^{\mathcal{S}}$. Hence, as said, von Neumann thought that the observer could be the solution not only of the act of observation but more generally of the apparent acausality of change (14.1) [*VON NEUMANN* 1932, 186, 223–24] [*VON NEUMANN* 1955, 418–20].

It is worth mentioning that von Neumann believed in psychophysical parallelism (a principle which for him was methodologically important in order to keep non-physical explanations out of physical science). Even if it might not be completely sound to defend such a principle, by proposing at the same time the act of observation as an explanation of the reduction, it must be said that it was exactly the psychophysical parallelism that retained von Neumann

[17]Anyway models have recently been developed in which non-linear terms are introduced in order to obtain a result of the form (14.17) and not only a statistical mixture [BLANCHARD *et al.* 1998].

[18]Examination in [*JAMMER* 1974, 479].

from theorizing an intelligent intervention which can violate physical laws (the second law of thermodynamics): in fact while Szilard thought of a decreasing entropy, von Neumann thought of an entropy increase [see theorem 14.2: p. 222]. On the other hand, it was possible for von Neumann to assume together psychophysical parallelism and an act of intelligent intervention because he thought that the limit between $\mathcal{O}$ and $\mathcal{A}+\mathcal{S}$ is arbitrarily fixed and also able to be moved [*VON NEUMANN* 1932, 224, 225–32] [*VON NEUMANN* 1955, 420–21, 422–37].

14.1.4 Poincaré-Sphere Representation of the Measurement

Description of the sphere A very useful representation of all we have analysed is provided by means of a Poincaré sphere [*MITTELSTAEDT* 1998, 37–40] [see also figure 3.7]. Using Pauli operators $\{\hat{I}, \hat{\sigma}_1, \hat{\sigma}_2, \hat{\sigma}_3\}$ we can represent a projector as:

$$\hat{P}(\mathbf{x}) = \frac{1}{2}(\mathbf{1} + x_j\hat{\sigma}_j)\ x_j \in \Re \sum_j x_j^2 = 1 \tag{14.18}$$

and mixtures as:

$$\hat{\tilde{\rho}}(\mathbf{y}) = \frac{1}{2}(\mathbf{1} + y_j\hat{\sigma}_j)\ y_j \in \Re \sum_j y_j^2 = 1 \tag{14.19}$$

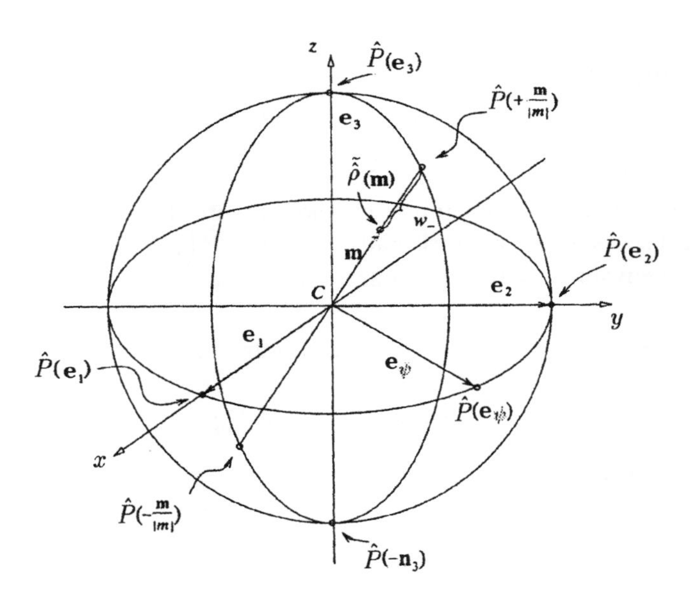

Figure 14.2: Poincaré sphere with centre C, orthogonal coordinates x, y, z and unit vectors $\mathbf{e}_1, \mathbf{e}_2, \mathbf{e}_3$. Any point on the surface, given by an unit vector $\mathbf{e}_\psi$, represents a pure state, while points in the interior of the sphere, with vectors $\mathbf{m}$ such that $|\mathbf{m}| < 1$, represent mixtures — [*MITTELSTAEDT* 1998, 37].

As we know the $\mathbf{x}$'s are points on the surface of the Poincaré sphere and the $\mathbf{y}$'s points in its interior [see figure 14.2]. The projectors $\hat{P}(\mathbf{x})$ and $\hat{P}(-\mathbf{x}) = \mathbf{1} - \hat{P}(\mathbf{x})$ project onto orthogonal states and correspond to diametrically opposite points of the sphere. The unit vectors in the directions of the three coordinate axes x, y, z are symbolized by $\mathbf{e}_1, \mathbf{e}_2, \mathbf{e}_3$. Projectors can be considered either as pure states or as observables with eigenvalues $0, 1$. The projectors $\hat{P}(\mathbf{e}_1)$ and $\hat{P}(\mathbf{e}_3)$ (with orthogonal vectors $\mathbf{e}_1$ and $\mathbf{e}_3$) correspond to complementary observables. A pure state $|\psi\rangle$ is represented by a unit vector $\mathbf{e}_\psi$ and symbolized by a projector $\hat{P}(\mathbf{e}_\psi)$: it is an eigenstate of the observable $\hat{P}(\mathbf{e})$ either if $\mathbf{e} \cdot \mathbf{e}_\psi = 1$ (eigenvalue 1) or if $\mathbf{n} \cdot \mathbf{e}_\psi = -1$ (eigenvalue 0). The respective probabilities for obtaining eigenvalue 1 or 0 are:

$$\wp(\psi, 1) = \frac{1}{2}(\mathbf{1} + \mathbf{e} \cdot \mathbf{e}_\psi) \tag{14.20a}$$

$$\wp(\psi, 0) = \frac{1}{2}(\mathbf{1} - \mathbf{e} \cdot \mathbf{e}_\psi) \tag{14.20b}$$

Decomposition of mixtures The spectral decomposition of mixed states is given by:

$$\tilde{\hat{\rho}}(\mathbf{m}) = w_+ \hat{P}\left(+\frac{\mathbf{m}}{|\mathbf{m}|}\right) + w_- \hat{P}\left(-\frac{\mathbf{m}}{|\mathbf{m}|}\right) \tag{14.21}$$

with $w_+ + w_- = 1$, and projectors given by:

$$\hat{P}\left(\pm\frac{\mathbf{m}}{|\mathbf{m}|}\right) = \frac{1}{2}\left(\mathbf{1} + \frac{m_j}{|\mathbf{m}|}\hat{\sigma}_j\right) \tag{14.22}$$

The eigenvalues of $\tilde{\hat{\rho}}(\mathbf{m})$ are given by

$$w_\pm = \frac{1}{2}(\mathbf{1} \pm |\mathbf{m}|) \tag{14.23}$$

and have an immediate geometrical meaning: on the diameter through point W — which represents $\tilde{\hat{\rho}}(\mathbf{m})$ —, the distances $d_\pm$ between W and the points $\hat{P}\left(\pm\frac{\mathbf{m}}{|\mathbf{m}|}\right)$ on the surface of the sphere are related to these eigenvalues by the relationship $d_\pm = 2w_\mp$.

Measurement Let us now concretely analyse [see figure 14.3] the measurement of the observable:

$$\hat{O} := \hat{P}(\mathbf{e}_1) = \frac{1}{2}(\mathbf{1} + \hat{\sigma}_1) \tag{14.24}$$

pertaining to a system initially in the state:

$$\hat{P}_\psi = \frac{1}{2}(\mathbf{1} + \hat{\sigma}_1|s_1\rangle + \hat{\sigma}_2|s_2\rangle) \tag{14.25}$$

Clearly, vector $\mathbf{e}_\psi$ lies in the equatorial plane, and the angle θ between $\mathbf{e}_1$ and $\mathbf{e}_\psi$ is given by $\cos\theta = \mathbf{e}_1 \cdot \mathbf{e}_\psi = s_1$. The measurement of $\hat{O}$ leads to the mixture:

$$\tilde{\hat{\rho}}(\psi, \hat{O}) = \wp(\psi, 1)\hat{P}(+\mathbf{e}_1) + \wp(\psi, 0)\hat{P}(-\mathbf{e}_1), \tag{14.26}$$

according to Eqs. (14.20):

$$\wp(\psi, 1) = \frac{1}{2}(\mathbf{1} + \mathbf{e}_1 \cdot \mathbf{e}_\psi) = \frac{1}{2}(\mathbf{1} + \cos\theta), \tag{14.27a}$$

$$\wp(\psi, 0) = \frac{1}{2}(\mathbf{1} - \mathbf{e}_1 \cdot \mathbf{e}_\psi) = \frac{1}{2}(\mathbf{1} - \cos\theta). \tag{14.27b}$$

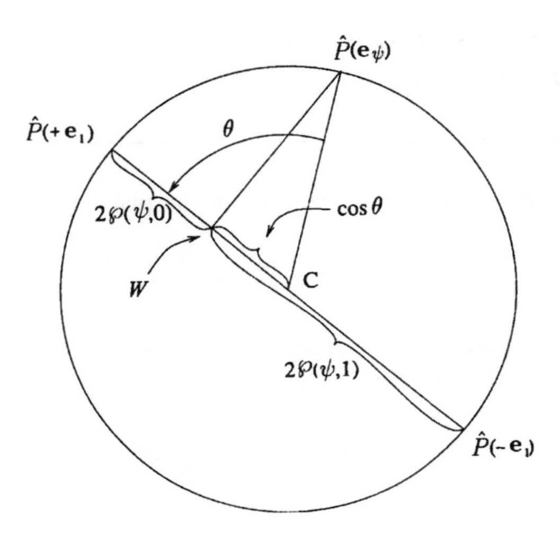

Figure 14.3: Poincaré sphere representation of a measurement: initial state $\hat{P}(\mathbf{e}_\psi)$, measured observable $\hat{P}(\mathbf{e}_1)$, mixed state represented by point W, and probabilities $\wp(\psi, 0), \wp(\psi, 1)$ — [*MITTELSTAEDT* 1998, 40].

Mixed state (14.26) is given by a point W in the interior of the sphere

$$W = \frac{1}{2}(\mathbf{1} + m_1\hat{\sigma}_1). \tag{14.28}$$

Geometrically W can be obtained by orthogonal projection of the point $\hat{P}_{\mathbf{e}_\psi}$ onto the diameter in the direction $\mathbf{e}_1$. The distance between the centre of the sphere and W is given by the length $|\mathbf{m}| = \cos\theta$ and the distances $d_\pm$ between W and $\hat{P}_{+\mathbf{e}_\psi}, \hat{P}_{-\mathbf{e}_\psi}$ are respectively:

$$d_+ = 1 - \cos\theta = 2\wp(\psi, 0), \;\; d_- = 1 + \cos\theta = 2\wp(\psi, 1) \tag{14.29}$$

14.1.5 Necessity and Definition of the Apparatus

In the Standard interpretation the role of an apparatus $\mathcal{A}$ in Measurement is always presupposed and is seen as a necessity in order to read values of the measured observable. In fact, more generally, the connection between observed system and apparatus is only postulated by the Standard interpretation and never argued or demonstrated. The concept of relative state shall give the correct solution [see theorem 15.1: p. 246].

But here one might ask if the apparatus is a necessity. The answer is yes, and we shall show that this is a necessity depending on the internal formalism of QM itself. This is what two theorems, the following one and another which we shall study later [theorem 16.2: p. 256], state [MACHIDA/NAMIKI 1984, 78–79]:[19]

[19]See also [PERES 1984a, 250].

Theorem 14.3 (Watanabe/Machida/Namiki) *A measurement reduction of a system $\mathcal{S}$ leads to contradictions if in the formalism one does not consider an apparatus $\mathcal{A}$.*

Proof

Suppose that, in the case of a type I–Measurement, in the place of Eq. (14.14) we write:

$$\hat{\rho}_i^{\mathcal{S}} = |\varsigma\rangle\langle\varsigma| \leadsto \tilde{\hat{\rho}}_f^{\mathcal{S}} = \sum_j c_j |s_j\rangle\langle s_j|, \tag{14.30}$$

where $c_j = \langle s_j|\varsigma\rangle$. Now the observable $\hat{O}^{\mathcal{S}}$ of which the $|s_j\rangle$ are eigenvectors, is supposed to only have two eigenstates with $c_1 = c_2 = 1/2$. Hence we have:

$$\tilde{\hat{\rho}}_f^{\mathcal{S}} = \frac{1}{2}\left[|s_1\rangle\langle s_1| + |s_2\rangle\langle s_2|\right] \tag{14.31}$$

But setting $|s_\pm\rangle = 2^{-1/2}(|s_1\rangle \pm |s_2\rangle)$, we easily obtain:

$$\tilde{\hat{\rho}}_f^{\mathcal{S}} = \frac{1}{2}\left[|s_+\rangle\langle s_+| + |s_-\rangle\langle s_-|\right], \tag{14.32}$$

since off-diagonal terms wipe out. It is evident that the measurement described by Eq. (14.32) is of another operator, say of $\hat{O}'$, and that in general $[\hat{O}, \hat{O}'] \neq 0$. Hence we are faced with the contradiction of undertaking two incompatible measurements at the same time. Q. E. D.

Theorem 14.3 is a stronger result as corollary 8.3, but anyway both go in the same direction. Note that we cannot define apparatus respecting covariance requirements [see subsection 3.10.3] without introducing unsharp observables[20].

14.1.6 Discussion of the Copenhagen Interpretation and von Neumann's Interpretation

In conclusion we can say that there are very few points of contacts between the Standard interpretation and the Copenhagen interpretation [STAPP 1972, 1100]: in fact von Neumann's quantum mechanical treatment of $\mathcal{A}$ [for example see Eq. (14.8)] is absolutely contrary to the spirit of Bohr's philosophy [see postulate 8.2: p. 146].

The distinction between observing and observed systems is the only point where the Standard interpretation can meet the Copenhagen interpretation; but with the following important difference: for the Copenhagen interpretation the boundaries are between $\mathcal{S}$ and (a macroscopical apparatus) $\mathcal{A}$; for the Standard interpretation — in consequence of the quantum mechanical treatment of $\mathcal{A}$ — between $\mathcal{S} + \mathcal{A}$ and $\mathcal{O}$. In fact, the Copenhagen interpretation is more epistemological[21]: we can account for QM phenomena only in macroscopical and therefore classical terms [see principles 8.2 (CP2): p. 138, 8.3 (CP3): p. 139; postulate 8.2: p. 146 and its corollary 8.2: p. 149].

Note that von Neumann believed himself to agree perfectly with the Copenhagen interpretation: **for example** he said [*VON NEUMANN* 1932, 262] [*VON NEUMANN* 1955, 420] that also Bohr sustained psychophysical parallelism, while it is evident that Bohr never speaks of this concept but only of complementarity [BOHR 1929, 205–206].

---

[20]On this point see subsection 18.3.1.

[21]On this point see [STAPP 1997, 303].

14.2 Criticisms and Problems

The first problem is of a general character: von Neumann's theory contains no prescription about how it is possible to execute the measurement of an operator $\hat{O}$. The general theory of Measurement does not explain how the quantum state of $\mathcal{A}$ is to be understood. In reality, the theory dislocates the question of the differentiability of states from $\mathcal{S}$ to $\mathcal{A}$[22].

Analysing more specific matters, we have four principal classes of problems and contestations:

- the problem of the observer [subsection 14.2.1];
- the limits of the Projection Postulate [subsection 14.2.2];
- the instantaneous character of the Measurement process [subsection 14.2.3];
- the constraints imposed by conservation laws [subsection 14.2.4].

14.2.1 The Observer

Wigner's friend London and Bauer [*LONDON/BAUER* 1939] stressed the role of the observer and also tried to solve by this way the paradox of Schrödinger Cat [see the Introduction to part V]. They understood the activity of the mind as a physical action[23], and hence they were closer to the initial positions of Szilard.

The position of London/Bauer was developed by Wigner with the proposal of a famous paradox, called thereafter the paradox of 'Wigner's friend' [WIGNER 1961, 173, 176–77][24]. Suppose that I have a friend $\mathcal{F}$ that looks at a QM system $\mathcal{S}$, which has a two-state basis: $|s_1\rangle$ and $|s_2\rangle$. I interrogate him about what he saw, and by hypothesis he answers to me only by a *yes* or a *no*. If $\mathcal{S}$ was originally in the state $|s_1\rangle$, the state of $\mathcal{S}+\mathcal{F}$ after observation will be $|s_1\rangle \otimes |f_1\rangle$; if $\mathcal{S}$ was originally in state $|s_2\rangle$, $\mathcal{S}+\mathcal{F}$ will be in state $|s_2\rangle \otimes |f_2\rangle$. The vector states $|f_1\rangle$ and $|f_2\rangle$ are a basis for $\mathcal{F}$: in the first case he is in the state in which he responds to my question about what he saw with a *yes*, and in the second with a *no*. Suppose now that the initial state of $\mathcal{S}$ was a linear superposition $c_1|s_1\rangle + c_2|s_2\rangle$. It follows that after the interaction, the whole system $\mathcal{S}+\mathcal{F}$ is in the state

$$c_1|s_1\rangle|f_1\rangle + c_2|s_2\rangle|f_2\rangle. \tag{14.33}$$

If I ask my friend, as we know he will with probability $|c_1|^2$ answer *yes* and with probability $|c_2|^2$ answer *no*. The probability of other answers is zero because the final wave function has no states of the type $|s_1\rangle \otimes |f_2\rangle$ or $|s_2\rangle \otimes |f_1\rangle$ — i.e. the friend is always faithful. But suppose that, after performing the whole experiment, in case of a *yes* answer, I ask my friend if he had clear in his own mind the answer even before he spoke. $\mathcal{F}$ will answer that he had already decided immediately after his experiment, i.e. immediately after his experiment $\mathcal{S}+\mathcal{F}$ was already in state $|s_1\rangle|f_1\rangle$ (or $|s_2\rangle|f_2\rangle$ if the answer was *no*). But this is in contradiction with Eq. (14.33). Suppose now that, instead of $\mathcal{F}$ we had an apparatus $\mathcal{A}$, say an atom. No doubt that after the interaction the state of $\mathcal{S}+\mathcal{A}$ is defined by Eq. (14.33). In the case of $\mathcal{F}$ our refusal to accept this as a valid definition of the situation, is only because a human mind cannot be in a state of suspended animation before a verbal answer — it is the same situation of the Schrödinger Cat

[22] See [WIGNER 1952, 102–103].

[23] See the historical reconstruction in [*JAMMER* 1974, 484].

[24] On this point see also [*BUSCH et al.* 1991, 101] [*ALBERT* 1992, 112–25].

[see the Introduction to part V]. Hence, after Wigner, in QM a being with consciousness must have a different role than an inanimate being.

Wigner afterwards [WIGNER 1983b] changed opinion and assumed Zeh's point of view [ZEH 1970] regarding the role of the environment [see chapter 17].

Some difficulties The greatest problem with the proposal of an observer as a solution of the reduction problem is intersubjectivity. In fact it has been shown by Shimony [SHIMONY 1963, 17–23] that the interpretation of London/Bauer (and hence to this extent that of von Neumann and Wigner) cannot be reconciled with the intersubjective agreement between several independent observers. We shall return to the problem [chapter 15].

The case of a quantum robot (a mechanical observer) which measures itself was analysed by Albert [ALBERT 1983]. He, firstly, found a new way of formulating UP; secondly, that, in consequence of this, there is something indexical (or subjective) in the QM knowledge; thirdly (as a consequence of the second point), that the description of a QM automata of itself can be different in nature from a description of an external object. But it was shown by Peres [PERES 1984a] that Albert does not discuss Measurement but only Premeasurement

14.2.2 Limits of the Projection Postulate

The greatest difficulty with the Projection Postulate is that it is not valid for continuous observables, i.e. for observables which do not have discrete spectrum [see subsection 3.5.4]. In fact, if we try to make a continuous observable, as $\mathbf{q}$, discrete by using small partitions, we find that measuring requires more than countably many discretized partitions. And by making it discrete, we completely destroy the Euclidian covariance of $\mathbf{q}$ or the translational symmetry of the original problem [*GUDDER* 1988b, 50] [*BUSCH et al.* 1995a, 98].

Generally, all of von Neumann's theory supposes the repeatability of measurement (type I–Measurement) [see Eq. (14.14) and definition 13.3: p. 217],[25] which is not possible for a continuous observable. Hence we state following theorem [OZAWA 1984]:[26]

Theorem 14.4 (I Ozawa) *If an observable $\hat{O}^{\mathcal{S}}$ admits repeatable premeasurement, then $\hat{O}^{\mathcal{S}}$ is discrete.*

The converse can also be proved [*BUSCH et al.* 1995a, 45], so that we have an equivalence between both statements of theorem 14.4.[27]

Another problem [OZAWA 1998] is that in two successive measurements, the reduction which should be obtained after the first one (following the Projection Postulate) cannot, in general, give the joint probability distribution of the outcomes of the consecutive measurements.

[25] See also [MACHIDA/NAMIKI 1980, 1459] for further analysis.

[26] See also [*BUSCH et al.* 1991, 58] [*BUSCH et al.* 1995a, 95] [*SCHROECK* 1996, 25].

[27] Ozawa successively proposed a specific countermodel [OZAWA 1991] which shows that in particular circumstances the position, though a continuous observable, can be measured as precisely as the momentum, i.e. that there are no additional limits other than uncertainties on the measurement of the position which come from linear momentum conservation laws. See also [ARAKI/YANASE 1960] and subsection 18.3.1.

14.2.3 Measurement and Relativity

A problem regarding the objectivity of the reduction of the wave packet is that it is not a relativistic invariant: in QM the Eqs., and not the values which can be obtained through measurements, are invariant[28].

On the other hand the reduction seems to be an instantaneous process, — a problem which had some importance in the discussion between Fock and Bohm/Aharonov [see subsection 10.2.2]. But it is possible to overcome this problem by choosing a formalism — for fields, but which perhaps can be extended — in which the measurement only falls in the future light cone [HELLWIG/KRAUS 1970, 566–67]. We make the following assumptions:

- There exist sets $\mathcal{X}_{\rm V}$ of field observables $\hat{O}_{\rm V}$ for sufficient finite space-time regions V.
- There is a $\hat{U}_L(a,\Lambda)$ such that for each $\hat{O}_{\rm V} \in \mathcal{X}_{\rm V}$ the set of transformations of the type $\hat{U}_L(a,\Lambda)\hat{O}_{\rm V}\hat{U}_L^*(a,\Lambda)$ is the set of observables belonging to the Lorentz transform of V (Lorentz invariance) [see section 4.1, particularly Eq. (4.6)].
- Field observables $\hat{O}_{\rm V} \in \mathcal{X}_{\rm V}, \hat{O}_{\rm V'} \in \mathcal{X}_{\rm V'}$ commute if V and V′ are space-like with respect to each other.

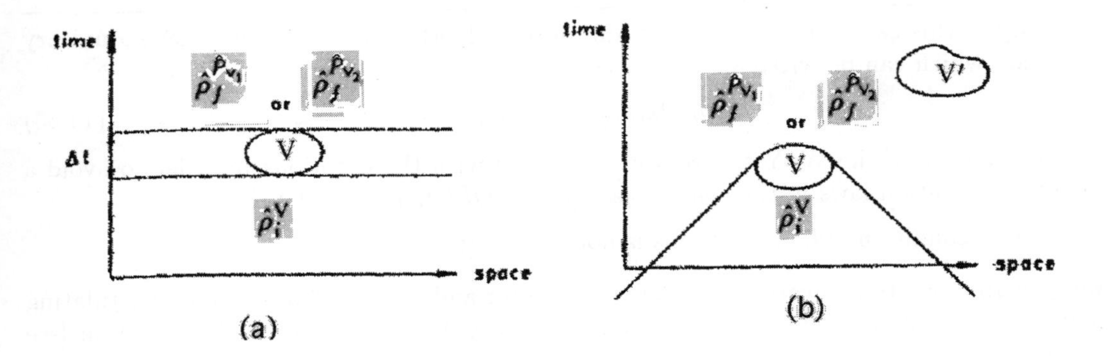

Figure 14.4: (a) Non-relativistic description of a measurement as would follow from the Standard interpretation (V is a region);
(b) proposed relativistic description of a measurement (the field state remains $\hat{\rho}_i^{\rm V}$ in the past cone of V and it changes into $\hat{\rho}_f^{\hat{P}_{{\rm V}_1}}$ or $\hat{\rho}_f^{\hat{P}_{{\rm V}_2}}$ in the future cone of V) — from [HELLWIG/KRAUS 1970, 567].

The field remains $\hat{\rho}_i^{\rm V}$ in the past cone of V and changes into $\hat{\rho}_f^{\hat{P}_{{\rm V}_1}}$ or $\hat{\rho}_f^{\hat{P}_{{\rm V}_2}}$ in the future and side cone of V (the set of all points which are space-like with respect to V). Suppose that the field is initially in state $\hat{\rho}_i$ and that first we measure a property $\hat{P}_{\rm V} \in \mathcal{X}_{\rm V}$ and then another property $\hat{P}_{\rm V'} \in \mathcal{X}_{\rm V'}$ [see figure 14.4].

Suppose that V′ is contained in the union of future and side cones of V. If we have selective premeasurement, the expectation value for V′ is:

$$\mathrm{Tr}(\hat{P}_{\rm V'}\hat{\rho}_f^{\hat{P}_{{\rm V}_2}}) = \frac{\mathrm{Tr}(\hat{P}_{\rm V'}\hat{P}_{\rm V}\hat{\rho}_i^{\rm V}\hat{P}_{\rm V})}{\mathrm{Tr}(\hat{P}_{\rm V}\hat{\rho}_i^{\rm V})} \tag{14.34}$$

[28]Examination in [*D'ESPAGNAT* 1976, 89–90, 264–65]. See also [ROSENFELD 1953, 60] on the invariance.

Another problem was analysed by Gisin [GISIN 1989] [GISIN 1991b]. Different mixtures corresponding to the same density operator — due to the non-uniqueness of the decomposition [see corollary 3.3: p. 46] — would no longer correspond to the same density matrix after the collapse, which could allow at-distance transmission of information, violating relativity. This problem can be lost by a theory, such as decoherence [see chapter 17], which preserves the pureness of the global systems and allows only 'local' reductions[29].

14.2.4 Measurement and Conservation Laws

Wigner proved [WIGNER 1952] that conservation Laws impose some constraints on Measurement which cannot be violated[30]. Wigner's original proof was limited to the case of a measurement of the x–component of the spin of a 1/2–spin particle, where the z–component of angular momentum was the conserved observable. But Araki and Yanase [ARAKI/YANASE 1960] generalized the proof[31]. Hence one speaks of the Wigner/Araki/Yanase (= WAY) Theorem:

Theorem 14.5 (WAY) *In a closed system including $\mathcal{A}$ and $\mathcal{S}$, only quantities which commute with all bounded additive conserved quantities* [Eq. (3.80)] *of the object-apparatus system are measurable exactly.*

Hence only in this case von Neumann's assumption (14.14) is valid. We then have a unitary transformation which can be written:

$$\hat{U}_t : |s_k a_0\rangle \mapsto |s_k a_k\rangle. \tag{14.35}$$

As a consequence of the WAY theorem we are faced with three options in order to avoid a violation of conservation laws in the general case [*SCHROECK* 1996, 79–81]:

- we give up condition (14.35), but this is not crucial[32];
- we give up a one-to-one correlation between pointer and measured observable, postulating only a statistical correlation, but we would obtain only a bad statistical correlation [see theorem 18.4: p. 299];
- we admit unsharp measurements[33], as is proposed by the stochastic theories [see section 18.3].

14.3 Value Reproducibility and Objectification

We now try to develop a more systematic theory of Measurement's statistics. This is the last aspect of the Measurement [see section 13.3] and consists in the relationship between the probability distribution or the expected value of the measured observable before the measurement and the probability distribution or the statistical results after one or more measurements. A pioneer in this analysis was Ludwig [LUDWIG 1964, 239, 250–51] [LUDWIG 1967].

---

[29]See also [AHARONOV/ALBERT 1984] [*D'ESPAGNAT* 1995, 209–214].

[30]See also [BUSCH 1985c]. For historical analysis see [*JAMMER* 1974, 496–98].

[31]For further discussion see also [GHIRARDI *et al.* 1981] [GHIRADI *et al.* 1982] [*BUSCH et al.* 1995a, 111–14] [*SCHROECK* 1996, 74–79].

[32]As proved by Yanase [YANASE 1961].

[33]See [WIGNER 1952] [YANASE 1961] [ARAKI/YANASE 1960].

14.3.1 Objectification Problem

Definition We have already seen [section 11.3] that in QM we do not necessarily have definite probabilities for some observable. In the case of a measurement, this poses the great problem of Objectification [see p. 170].[34]

We can define it (in pure general terms) as follows [*BUSCH et al.* 1991, 1]:

Definition 14.1 (Objectification problem) *It is the problem of whether and how definite measurement outcomes can be obtained.*

Objectification is the necessary condition for the reading and registration of the measurement result. More generally the objectification problem is a preliminary one for the problem of Measurement statistics.

Two hypothesis We now make two hypothesis:

- *A Weak objectification* hypothesis. It is possible to assign a value o_j for the observable $\hat{O}$ to $\mathcal{S}$ in the state $|\psi\rangle$ conditionally upon some other measurement result but such that the property o_j pertains to the system though this value is subjectively unknown to the observer. In other words we suppose that a classical probability distribution for an observable is always possible independently from the state the system is in. Hence by this hypothesis the system possesses the property o_j only in a *potential* sense — see Heisenberg's later interpretation [subsection 29.2.2].
- *A Strong objectification* hypothesis. It is possible to attribute an eigenstate $|\psi^{o_j}\rangle$ of $\hat{O}$ to $\mathcal{S}$ such that $\mathcal{S}$ is actually in state $|\psi^{o_j}\rangle$ but that this state is unknown to the observer who knows only its probability. By this hypothesis $\mathcal{S}$ *actually* has the property o_j [BUSCH/MITTELSTAEDT 1991, 890].

If the state is pure and has the form $\hat{\rho}(\hat{O})$, it allows neither strong nor weak objectification of an arbitrary observable $\hat{O}'$ which does not commute with $\hat{O}$. If the state is mixed we obtain a similar result. Hence, corresponding to what stated in QM probability theory, we say that [BUSCH *et al.* 1992, 949]:[35]

Theorem 14.6 (Busch/Mittelstaedt) *The hypothetical assignment of potential (in the sense of a classical probability distribution) or actual eigenvalues or eigenstates to some observable on a system which is actually in a superposition of eigenstates in relation to that observable, or which is in an eigenstate of another observable, which does not commute with the first one, is incompatible with QM.*

In the following we give explicit proof of this for the weak objectification hypothesis in the case of a pure state.

[34] The problem of objectification was initially investigated by Mittelstaedt [MITTELSTAEDT 1959] [MITTELSTAEDT 1960] [*MITTELSTAEDT* 1963]. See also [VON WEIZSÄCKER 1941, 497–504] [VON WEIZSÄCKER 1955, 522] [*MITTELSTAEDT* 1998, 65–87].

[35] See also [*BUSCH et al.* 1991, 16–20]..

Proof

We only show the incompatibility between QM and the weak objectification hypothesis. We suppose that $\mathcal{S}$ is in an arbitrary pure state $\hat{P}_\psi = |\psi\rangle\langle\psi|$. If $\hat{O}$ is a discrete observable with two eigenvalues, whose spectral decomposition [Eq. (3.64)] is:

$$\hat{O} = o_1 \hat{P}_{o_1} + o_2 \hat{P}_{o_2}, \tag{14.36}$$

and the state $\hat{P}_\psi$ is not an eigenstate of $\hat{O}$, and the probability of having an eigenvalue of $\hat{O}$ is:

$$\wp(\psi, o_j) = \mathrm{Tr}[\hat{P}_\psi \hat{P}_{o_j}]; \tag{14.37}$$

and $\hat{O}' = o'_1 \hat{P}_{o'_1} + o'_2 \hat{P}_{o'_2}$ is another observable which does not commute with $\hat{O}$ and whose $\hat{P}_\psi$ is not an eigenstate, we can write the conditional probability that value o'_k can be obtained assuming that $\mathcal{S}$ has the property o_1 or o_2 as follows:

$$\wp(\psi, o'_k) = \wp(\psi, o_1 \wedge o'_k) + \wp(\psi, o_2 \wedge o'_k). \tag{14.38}$$

But the probability $\wp(\psi, o_j \sqcap o'_k)$ of obtaining o'_k after having obtained o_j, is clearly larger than the joint probability $\wp(\psi, o_j \wedge o'_k)$ to have o_j *and* o'_k — because in the first case we do not suppose that, having obtained o'_k, the system still presents the property o_j (weak objectification assumption). Hence:

$$\wp(\psi, o'_k) \leq \wp(\psi, o_1 \sqcap o'_k) + \wp(\psi, o_2 \sqcap o'_k). \tag{14.39}$$

The sequential probability $\wp(\psi, o_j \sqcap o'_k)$ can be decomposed in:

$$\wp(\psi, o_j \sqcap o'_k) = \wp(\psi, o_j)\wp(o_j, o'_k). \tag{14.40}$$

From Eqs. (14.39) and (14.40) it follows that:

$$\wp(\psi, o'_k) \leq \wp(\psi, o_1)\wp(o_1, o'_k) + \wp(\psi, o_2)\wp(o_2, o'_k). \tag{14.41}$$

By writing the Lüders mixture [see Eq. (14.16)] as follows:

$$\tilde{\hat{\rho}}_L(\psi, \hat{O}) = \wp(\psi, o_1)\hat{P}_{o_1} + \wp(\psi, o_2)\hat{P}_{o_2} = \hat{P}_{o_1}\hat{P}_\psi \hat{P}_{o_1} + \hat{P}_{o_2}\hat{P}_\psi \hat{P}_{o_2}, \tag{14.42}$$

we rewrite the RHS of Eq. (14.41) in the following form:

$$\sum_j \wp(\psi, o_j)\wp(o_j, o'_k) = \mathrm{Tr}[\tilde{\hat{\rho}}_L(\psi, \hat{O}) \cdot \hat{P}_{o'_k}]. \tag{14.43}$$

Hence inequality (14.41) reads:

$$\wp(\psi, o'_k) \leq \mathrm{Tr}[\tilde{\hat{\rho}}_L(\psi, \hat{O}) \cdot \hat{P}_{o'_k}]. \tag{14.44}$$

But in QM we have in general:

$$\wp(\psi, o'_k) = \mathrm{Tr}\left[\tilde{\hat{\rho}}_L(\psi, \hat{O}) \cdot \hat{P}_{o'_k}\right] + 2\mathrm{Re}\left(\hat{P}_{o_1}\hat{P}_\psi \hat{P}_{o_2}\hat{P}_{o'_k}\right). \tag{14.45}$$

Hence inequality (14.44) is violated each time the interference term (the last one of Eq. (14.45)) is > 0. Therefore the weak objectification hypothesis is, in general, not compatible with QM. Q.E.D.

The above proof confirms the non-classical nature of quantum probability [see section 11.3]. It is also an anticipation of a later problem, the proof of the impossibility of Hidden-Variable theories [see chapter 32].

14.3.2 State Transformer and Reduced Final States

Measurement scheme We now discuss the problem of Measurement statistics. In the first instance the value space $(\Upsilon_{\mathcal{A}}, \mathcal{L}_{\mathcal{A}})$ (where $\Upsilon_{\mathcal{A}}$ is some set of values and $\mathcal{L}_{\mathcal{A}}$ some algebra on them) of the pointer observable $\hat{O}^{\mathcal{A}}$ differs from the value space $(\Upsilon_{\mathcal{S}}, \mathcal{L}_{\mathcal{S}})$ of the measured observable $\hat{O}^{\mathcal{S}}$, so that one needs a pointer function which correlates the two values sets: $f : \Upsilon_{\mathcal{A}} \mapsto \Upsilon_{\mathcal{S}}$ such that for $\mathcal{X} \in \Upsilon_{\mathcal{S}}$ we have $f^{-1}(\mathcal{X}) \in \Upsilon_{\mathcal{A}}$. We therefore have a quintuple (named *Measurement scheme*): $< \mathcal{H}_{\mathcal{A}}, \hat{O}^{\mathcal{A}}, \hat{\rho}^{\mathcal{A}}, \mathcal{T}, f >$. Generally one can say that the problem of premeasurement is to find an adequate quintuple [*SCHROECK* 1996, 72].

Coupling is a transformation $\hat{\rho}^{\mathcal{S}} \otimes \hat{\rho}^{\mathcal{A}} \rightsquigarrow \mathcal{T}(\hat{\rho}^{\mathcal{S}} \otimes \hat{\rho}^{\mathcal{A}})$, with complete determination of reduced states and hence of the final reduced state of $\mathcal{A}$ — obtained by tracing out the system [see Eq. (3.103)] —: $\mathrm{Tr}_{\mathcal{S}}[\mathcal{T}(\hat{\rho}^{\mathcal{S}} \otimes \hat{\rho}^{\mathcal{A}})]$ or $\mathcal{R}_{\mathcal{A}}[\mathcal{T}(\hat{\rho}^{\mathcal{S}} \otimes \hat{\rho}^{\mathcal{A}})]$ [*BUSCH et al.* 1991, 29]. We shall not discuss here what type of transformation $\mathcal{T}$ may be. As a result of our analysis it will be shown that it cannot be an unitary transformation, as was already presupposed by von Neumann; and in subsection 18.1.1 it will be shown that it must be an operation.

Probability reproducibility We have two conditions for qualifying a Measurement scheme as a measurement of $\hat{O}^{\mathcal{S}}$, the *Probability reproducibility* and the *Objectification requirement* [*BUSCH et al.* 1995a, 40].[36] The Probability reproducibility is given by following theorem [see also postulate 3.4: p. 38]:

Theorem 14.7 (Probability reproducibility) *A pointer observable $\hat{O}^{\mathcal{A}}$ with $\mathrm{Tr}_{\mathcal{S}}[\mathcal{T}(\hat{\rho}^{\mathcal{S}} \otimes \hat{\rho}^{\mathcal{A}})]$ must reproduce the probability measure of obtaining a result in subset $\mathcal{X}$:*

$$\boxed{\wp^{\hat{O}^{\mathcal{S}}}_{\hat{\rho}^{\mathcal{S}}}(\mathcal{X}) = \wp^{\hat{O}^{\mathcal{A}}}_{\mathcal{R}_{\mathcal{A}}(\mathcal{T}(\hat{\rho}^{\mathcal{S}} \otimes \hat{\rho}^{\mathcal{A}}))}(f^{-1}(\mathcal{X}))} \tag{14.46}$$

for any value set $\mathcal{X} \in \Upsilon_{\mathcal{S}}$ and for all possible initial states $\hat{\rho}^{\mathcal{S}}$. It is this condition which determines whether a Measurement scheme is actually a Premeasurement of $\hat{O}^{\mathcal{S}}$.

The pair $(\hat{O}^{\mathcal{A}}, f)$ defines another pointer observable $\hat{O}^{\mathcal{A}}_f = \hat{O}^{\mathcal{A}}(f^{-1}(\mathcal{X}))$ by means of which we can rewrite condition (14.46) as:

$$\wp^{\hat{O}^{\mathcal{S}}}_{\hat{\rho}^{\mathcal{S}}}(\mathcal{X}) = \wp^{\hat{O}^{\mathcal{A}}_f}_{\mathcal{R}_{\mathcal{A}}(\mathcal{T}(\hat{\rho}^{\mathcal{S}} \otimes \hat{\rho}^{\mathcal{A}}))}(\mathcal{X}) \tag{14.47}$$

This shows that the value spaces of $\hat{O}^{\mathcal{S}}$ and $\hat{O}^{\mathcal{A}}$ can always be identified by an appropriate rescaling of the pointer [*BUSCH et al.* 1991, 29–30].

State transformer Any premeasurement on $\hat{O}^{\mathcal{S}}$ determines a state transformer $\mathcal{I}_{\mathcal{M}}$ according to the relationship [*BUSCH et al.* 1995a, 37] [*BUSCH et al.* 1991, 31]:

$$\mathcal{I}_{\mathcal{M}}(\mathcal{X})\hat{\rho}^{\mathcal{S}} := \mathrm{Tr}_{\mathcal{A}}\left[\hat{I} \otimes \hat{O}^{\mathcal{A}}_f(\mathcal{X})^{\frac{1}{2}} \mathcal{T}(\hat{\rho}^{\mathcal{S}} \otimes \hat{\rho}^{\mathcal{A}}) \hat{I} \otimes \hat{O}^{\mathcal{A}}_f(\mathcal{X})^{\frac{1}{2}}\right] \tag{14.48}$$

[36]See also [*SCHROECK* 1996, 72] [BELTRAMETTI *et al.* 1990, 92, 96–97].

for all $\mathcal{X} \in \Upsilon, \hat{\rho}^{\mathcal{S}} \in \mathcal{H}^{\mathrm{Tr}=1;\geq 0}$. Note that Eq. (14.48) is written in analogy with a Lüders mixture (14.16): in fact the pointer observable can be understood as a projector. The state transformer reproduces $\hat{O}^{\mathcal{S}}$ via the following Eq. [see also Eq. (18.2)]:

$$\wp^{\hat{O}^{\mathcal{S}}}_{\hat{\rho}^{\mathcal{S}}}(\mathcal{X}) = \mathrm{Tr}[\mathcal{I}_{\mathcal{M}}(\mathcal{X})\hat{\rho}^{\mathcal{S}}] \tag{14.49}$$

Note that the relationship between Measurement scheme and $\mathcal{I}_{\mathcal{M}}$ is a many-to-one relationship: different apparatus may effect the 'same' measurement. Also the relationship between $\mathcal{I}_{\mathcal{M}}$ and $\hat{\rho}^{\mathcal{S}}$ is many-to-one: different measurements of $\hat{\rho}^{\mathcal{S}}$ may induce different effects on $\mathcal{S}$. No premeasurement leaves all states of $\mathcal{S}$ unchanged unless the measured observable is trivial (no more available information).

To interpret the probabilities $\wp^{\hat{O}^{\mathcal{S}}}_{\hat{\rho}^{\mathcal{S}}}(X)$ as relative frequencies of pointer values requires that such pointer values become real at the end of measuring process.

Objectification The second condition of a measurement is the *Objectification requirement* [see subsection 14.3.1], the condition of the Measurement as such [*BUSCH et al.* 1991, 30]:

Theorem 14.8 (Objectification requirement) *A measurement must lead to a definite result.*

This requirement entails firstly the *pointer objectification* [see theorem 14.9 in the next subsection] and secondly the *value objectification* [see theorems 14.10, 14.11, and 14.12]. Anyway, in a general form, the requirement entails that, for each initial state $\hat{\rho}^{\mathcal{S}}$ of $\mathcal{S}$, $\mathcal{I}_{\mathcal{M}}$ is the state of $\mathcal{S}$ after the measurement on the condition of reading a value in $f^{-1}(\mathcal{X})$. The probability for this transition is given by Eq. (14.49). The normalized final component states of $\mathcal{S}, \mathcal{A}$ are the following [see Eq. (14.48)] [*BUSCH et al.* 1991, 35]:

$$\begin{aligned}
\hat{\rho}^{\mathcal{S}}(\mathcal{X}, \hat{\rho}^{\mathcal{S}}) &:= [\wp^{\hat{O}^{\mathcal{S}}}_{\hat{\rho}^{\mathcal{S}}}(\mathcal{X})]^{-1}\mathrm{Tr}_{\mathcal{A}}\left[\hat{I} \otimes \hat{O}^{\mathcal{A}}_f(\mathcal{X})^{\frac{1}{2}}\mathcal{T}(\hat{\rho}^{\mathcal{S}} \otimes \hat{\rho}^{\mathcal{A}})\hat{I} \otimes \hat{O}^{\mathcal{A}}_f(\mathcal{X})^{\frac{1}{2}}\right] && \text{(14.50a)}\\
&= [\wp^{\hat{O}^{\mathcal{S}}}_{\hat{\rho}^{\mathcal{S}}}(\mathcal{X})]^{-1}\mathcal{I}_{\mathcal{M}}(\mathcal{X})\hat{\rho}^{\mathcal{S}}\\
\hat{\rho}^{\mathcal{A}}(\mathcal{X}, \hat{\rho}^{\mathcal{S}}) &:= [\wp^{\hat{O}^{\mathcal{S}}}_{\hat{\rho}^{\mathcal{S}}}(\mathcal{X})]^{-1}\mathrm{Tr}_{\mathcal{S}}\left[\hat{I} \otimes \hat{O}^{\mathcal{A}}_f(\mathcal{X})^{\frac{1}{2}}\mathcal{T}(\hat{\rho}^{\mathcal{S}} \otimes \hat{\rho}^{\mathcal{A}})\hat{I} \otimes \hat{O}^{\mathcal{A}}_f(\mathcal{X})^{\frac{1}{2}}\right] && \text{(14.50b)}\\
&= [\wp^{\hat{O}^{\mathcal{S}}}_{\hat{\rho}^{\mathcal{S}}}(\mathcal{X})]^{-1}\hat{O}^{\mathcal{A}}_f(\mathcal{X})^{\frac{1}{2}}\mathrm{Tr}_{\mathcal{S}}\left[\mathcal{T}(\hat{\rho}^{\mathcal{S}} \otimes \hat{\rho}^{\mathcal{A}})\right]\hat{O}^{\mathcal{A}}_f(\mathcal{X})^{\frac{1}{2}}
\end{aligned}$$

whenever $\wp^{\hat{O}^{\mathcal{S}}}_{\hat{\rho}^{\mathcal{S}}}(\mathcal{X})^{-1} \neq 0$.

14.3.3 Pointer Objectification

We introduce now the notion of a reading scale as a partition of the value space of the pointer observable [*BUSCH et al.* 1991, 36]. The finite reading scale $\mathbf{R}_{\mathcal{M}}$ such that $\Upsilon_{\mathcal{A}} = \bigcup f^{-1}(\mathcal{X}_j), \mathcal{X}_j \cap \mathcal{X}_k = \emptyset$ for $j \neq k$, defines a premeasurement $\mathcal{M}^{\mathbf{R}}$, and both define a new discrete observable $\hat{O}^{\mathbf{R}_{\mathcal{M}}}$ [see Eqs. (18.9c) and (18.10c)]. Hence we denote the final states of Eqs. (14.50) in the following form:

$$\begin{aligned}
\hat{\rho}^{\mathcal{S}}_f(j, \hat{\rho}^{\mathcal{S}}) &:= \hat{\rho}^{\mathcal{S}}_f(\mathcal{X}_j, \hat{\rho}^{\mathcal{S}}) && \text{(14.51a)}\\
\hat{\rho}^{\mathcal{A}}_f(j, \hat{\rho}^{\mathcal{S}}) &:= \hat{\rho}^{\mathcal{A}}_f(\mathcal{X}_j, \hat{\rho}^{\mathcal{S}}) && \text{(14.51b)}
\end{aligned}$$

The interpretation that $\hat{\rho}^{\mathcal{A}}_f(j, \hat{\rho}^{\mathcal{S}})$ is the state assumed by $\mathcal{A}$ on the condition that the pointer has its value in $f^{-1}(\mathcal{X}_j)$ follows if *pointer objectification* is postulated [*BUSCH et al.* 1991, 36–37] [see theorem 14.8]:

Theorem 14.9 (Pointer Objectification) *The Pointer Objectification requires*

- the pointer value-definiteness *condition with respect to a* $\mathbf{R}_{\mathcal{M}}$, *written in either of the two following forms:*

$$\mathrm{Tr}[\hat{\rho}_f^{\mathcal{A}}(j,\hat{\rho}^{\mathcal{S}})\hat{O}_j^{\mathcal{A}}] = 1; \ (\text{whenever } \wp_{\hat{\rho}^{\mathcal{S}}}^{\hat{O}^{\mathcal{S}}}(\mathcal{X}_j) \neq 0) \tag{14.52a}$$

$$\hat{O}_j^{\mathcal{A}}\hat{\rho}_f^{\mathcal{A}}(j,\hat{\rho}^{\mathcal{S}}) = \hat{\rho}_f^{\mathcal{A}}(j,\hat{\rho}^{\mathcal{S}}) \tag{14.52b}$$

where Eq. (14.52a) *means that with certainty the pointer has the appropriate value in the final state of the apparatus, and* Eq. (14.52b) *that this final state is an eigenstate of the pointer;*

- *and that the state* $\hat{\rho}^{\mathcal{A}}(\Upsilon_{\mathcal{S}},\hat{\rho}^{\mathcal{S}})$ *is a mixture* [see Eq. (14.16)] *of the pointer eigenstates* $\hat{\rho}_f^{\mathcal{A}}(j,\hat{\rho}^{\mathcal{S}})$*:*

$$\mathrm{Tr}_{\mathcal{S}}\left[\mathcal{T}(\hat{\rho}^{\mathcal{S}}\otimes\hat{\rho}^{\mathcal{A}})\right] = \sum_j \wp_{\hat{\rho}^{\mathcal{S}}}^{\hat{O}^{\mathcal{S}}}(\mathcal{X}_j)\hat{\rho}^{\mathcal{A}}(j,\hat{\rho}^{\mathcal{S}}) \tag{14.53}$$

for all initial states $\hat{\rho}^{\mathcal{S}}$*.*

In order to obtain the pointer objectification see corollary 17.1 [p. 269].

14.3.4 Value Objectification

As we have said, a premeasurement is a measurement if it satisfies the objectification requirement, i.e. pointer and value objectification [theorem 14.8]. We have already spoken of pointer objectification [theorem 14.9]. We can achieve the value objectification through the pointer objectification via strong correlations. A measurement is essentially a transfer of information from a system to an apparatus [see section 19.2]. Now the possibility of transferring information from $\mathcal{A}$ to $\mathcal{S}$ rests on the fact that the state $\mathcal{T}(\hat{\rho}^{\mathcal{S}}\otimes\hat{\rho}^{\mathcal{A}})$ entails statistical dependencies between quantities pertaining to the two subsystems. We consider here three types of dependency:

- strong correlations between observables $\hat{O}^{\mathcal{S}}$ and $\hat{O}^{\mathcal{A}}$ [Eqs. (14.63) and theorem 14.10];

- strong correlations between corresponding values of $\hat{O}^{\mathcal{S}}$ and $\hat{O}^{\mathcal{A}}$ [Eqs. (14.65) and theorem 14.11];

- strong correlations between the final component states of the two subsystems [theorem 14.12].

The first two theorems express in a more abstract and analytical way what is formulated by the Lüders mixture [Eq. (14.16)].

Types of Correlations

Let μ_p be a probability measure on $(\Re^2, \mathcal{B}(\Re^2))$ and μ_p^1, μ_p^2 marginals such that for $\mathcal{X}, \mathcal{Y} \in \mathcal{B}(\Re^2))$ [*BUSCH et al.* 1991, 49–51] [BUSCH/LAHTI 1996b, 2591–92]:

$$\mu_p^1(\mathcal{X}) = \mu_p(\mathcal{X} \times \Re), \qquad \mu_p^2(\mathcal{Y}) = \mu_p(\Re \times \mathcal{Y}) \tag{14.54}$$

The marginal measures correspond to the coordinate projections (on x, y respectively). We assume well defined expectations and variances [Eq. (3.91) and (3.110)], for example: $\langle 1 \rangle = \int x d\mu_p^1(x)$, $(\Delta(1))^2 = \int (x - \langle 1 \rangle)^2 d\mu_p^1(x)$. Let $\langle 12 \rangle = \int xy d\mu_p(x, y)$. Then the correlation of marginals is so defined:

$$\mathrm{C}(\mu_p^1, \mu_p^2; \mu_p) := \int \frac{(x - \langle 1 \rangle)(y - \langle 2 \rangle)}{\Delta(1)\Delta(2)} d\mu_p(x, y) = \frac{\langle 2 \rangle - \langle 1 \rangle \langle 2 \rangle}{\Delta(1)\Delta(2)} \tag{14.55}$$

whenever $\Delta(1) \neq 0 \neq \Delta(2)$. From the Schwarz inequality we obtain:

$$-1 \leq \mathrm{C}(\mu_p^1, \mu_p^2; \mu_p) \leq 1 \tag{14.56}$$

Hence marginals μ_p^1, μ_p^2 are

- *uncorrelated* if $\mathrm{C}(\mu_p^1, \mu_p^2; \mu_p) = 0$;
- *strongly correlated* if $\mathrm{C}(\mu_p^1, \mu_p^2; \mu_p) = 1$;
- and *strongly anticorrelated* if $\mathrm{C}(\mu_p^1, \mu_p^2; \mu_p) = -1$

On the other hand, μ_p^1, μ_p^2 are:

- *independent* if we have a factorization: $\mu_p = \mu_p^1 \times \mu_p^2$, otherwise they are dependent;
- they are *completely dependent* if there is a measurable function $h : \Re \mapsto \Re$ such that $\mu_p(\mathcal{X} \times \mathcal{Y}) = \mu_p^2(h^{-1}(\mathcal{X}) \cap \mathcal{Y})$ for $\mathcal{X}, \mathcal{Y} \in \mathcal{B}(\Re)$; i.e. that μ_p^2 suffices to determine all of measure μ_p.

Strong (anti)correlation entails complete dependence:

$$\mathrm{C}(\mu_p^1, \mu_p^2; \mu_p) = +1 \text{ iff } \mu_p^1, \mu_p^2 \text{ are completely dependent with } h(y) = ay + b, a > 0 \tag{14.57a}$$
$$\mathrm{C}(\mu_p^1, \mu_p^2; \mu_p) = -1 \text{ iff } \mu_p^1, \mu_p^2 \text{ are completely dependent with } h(y) = ay + b, a < 0 \tag{14.57b}$$

with $a = \pm\Delta(1)/\Delta(2), b = \langle 1 \rangle - a \langle 2 \rangle$.

Strong Correlation Between Observables

Now we can handle Value objectification by requiring that the observables $\hat{O}^{\mathcal{S}}$ and $\hat{O}_f^{\mathcal{A}}$ be strongly correlated in the final state $\mathcal{T}(\hat{\rho}^{\mathcal{S}} \otimes \hat{\rho}^{\mathcal{A}})$ [*BUSCH et al.* 1991, 51–55] [BUSCH/LAHTI 1996b, 2592–94]. To avoid complications we assume that the value space of the system is $(\Re, \mathcal{B}(\Re))$. So we have the mapping [DAVIES/J. LEWIS 1970]:

$$\begin{aligned} \mu_p : \mathcal{X} \times \mathcal{Y} \;\mapsto\; & \mathrm{Tr}[\mathcal{T}(\hat{\rho}^{\mathcal{S}} \otimes \hat{\rho}^{\mathcal{A}})\hat{O}^{\mathcal{S}}(\mathcal{X}) \otimes \hat{O}_f^{\mathcal{A}}(\mathcal{Y})] \\ & = \mathrm{Tr}[(\mathcal{I}_{\mathcal{M}}(\mathcal{Y})\hat{\rho}^{\mathcal{S}})\hat{O}^{\mathcal{S}}(\mathcal{X})] \\ & = \mathrm{Tr}[\mathcal{I}_{\mathcal{M}}(\mathcal{X})(\mathcal{I}_{\mathcal{M}}(\mathcal{Y})\hat{\rho}^{\mathcal{S}})] \end{aligned} \tag{14.58}$$

whose marginal distribution are given by:

$$\mu_p^1: \quad \mapsto \text{Tr}[(\mathcal{I}_{\mathcal{M}}(\Re)\hat{\rho}^{\mathcal{S}})\hat{O}^{\mathcal{S}}(\mathcal{X})] \tag{14.59a}$$

$$\mu_p^2: \quad \mapsto \text{Tr}[\hat{\rho}^{\mathcal{S}})\hat{O}^{\mathcal{S}}(\mathcal{Y})] \tag{14.59b}$$

Now we can state:

Theorem 14.10 (Observable Strong Correlation) *A premeasurement $\mathcal{M}$ produces value objectification, i.e. a strong (anti)correlation between observables when the probability measures* (14.59) *are completely dependent* [Eq. (14.57)]*:*

$$\boxed{\text{Tr}[\mathcal{I}_{\mathcal{M}}(\mathcal{X})(\mathcal{I}_{\mathcal{M}}(\mathcal{Y})\hat{\rho}^{\mathcal{S}})] = \text{Tr}[\mathcal{I}_{\mathcal{M}}(h_{\pm}^{-1}(\mathcal{X}) \cap (\mathcal{Y}))\hat{\rho}^{\mathcal{S}}]} \tag{14.60}$$

We have a special case when $h_{+}(y) = y$:

$$\text{Tr}[\mathcal{I}_{\mathcal{M}}(\mathcal{X})(\mathcal{I}_{\mathcal{M}}(\mathcal{Y})\hat{\rho}^{\mathcal{S}})] = \text{Tr}[\mathcal{I}_{\mathcal{M}}((\mathcal{X}) \cap (\mathcal{Y}))\hat{\rho}^{\mathcal{S}}] \tag{14.61}$$

which is the condition for the repeatability of measurements (the Standard interpretation assumption). But this is not a necessary condition for obtaining a strong observable correlation. Eq. (14.60) with $h_{\pm} = \pm(\Delta(1)/\Delta(2))(y - \langle 2\rangle) + \langle 1\rangle)$ is sufficient. Condition (14.61) implies also the equality of marginal measures (14.59) for all $\mathcal{X}$:

$$\wp_{\hat{\rho}^{\mathcal{S}}}^{\hat{O}^{\mathcal{S}}}(\mathcal{X}) = \wp_{\text{Tr}_{\mathcal{A}}}^{\hat{O}^{\mathcal{S}}}(\mathcal{T}(\hat{\rho}^{\mathcal{S}} \otimes \hat{\rho}^{\mathcal{A}})(\mathcal{X}) \tag{14.62}$$

which is a type I–Measurement [see definition 13.3: p. 217] [*BUSCH et al.* 1991, 52].

Necessary and sufficient conditions for strong observable correlations are (for the sake of simplicity we write $\mathcal{I}_{\mathcal{M}} = \mathcal{I}$):

$$\langle 12\rangle = \sum_{jk} jk\mu_p(j,k) = \sum_{jk} \text{Tr}[\mathcal{I}_j(\mathcal{I}_k(\hat{\rho}^{\mathcal{S}}))] \tag{14.63a}$$

$$\langle 1\rangle = \sum_{j} j\mu_p^1(j) = \sum_{j} j\text{Tr}[\mathcal{I}_j(\mathbf{1})(\hat{\rho}^{\mathcal{S}})\hat{O}_j^{\mathcal{S}}] \tag{14.63b}$$

$$\langle 2\rangle = \sum_{k} k\mu_p^2(k) = \sum_{k} k\text{Tr}[\hat{\rho}^{\mathcal{S}}\hat{O}_j^{\mathcal{S}}] \tag{14.63c}$$

$$(\Delta(1))^2 = \sum_{j} j^2\text{Tr}[\mathcal{I}(\mathbf{1})(\hat{\rho}^{\mathcal{S}})(\hat{O}_j^{\mathcal{S}})] - \left(\sum_{j} j\text{Tr}[\mathcal{I}(\mathbf{1})(\hat{\rho}^{\mathcal{S}})\hat{O}_j^{\mathcal{S}}]\right)^2 \tag{14.63d}$$

$$(\Delta(2))^2 = \sum_{k} k^2\text{Tr}[\hat{\rho}^{\mathcal{S}}\hat{O}_j^{\mathcal{S}}] - \left(\sum_{k} k\text{Tr}[\hat{\rho}^{\mathcal{S}}\hat{O}_j^{\mathcal{S}}]\right)^2 \tag{14.63e}$$

where, here, **1** means all the reading scale of $\mathcal{A}$. As we shall see, strong correlation between observables is guaranteed by theorem 17.4 [p. 269].

[37]See also [*BUSCH et al.* 1995a, 45, 96].

Strong Correlations Between Values

If we ask to what degree the values of $\hat{O}^{\mathcal{S}}_{\mathbf{R}_{\mathcal{M}}}$ are correlated to those of $\hat{O}^{\mathcal{A}}_{\mathbf{R}_{\mathcal{M}}}$ [*BUSCH et al.* 1991, 55–57] [BUSCH/LAHTI 1996b, 2594–95] we must study the correlation function $\mathrm{C}(\hat{O}^{\mathcal{S}}_{i}, \hat{O}^{\mathcal{A}}_{i}, \mathcal{T}(\hat{\rho}^{\mathcal{S}} \otimes \hat{\rho}^{\mathcal{A}}))$:

$$\mathrm{C}(\hat{O}^{\mathcal{S}}_{i}, \hat{O}^{\mathcal{A}}_{i}, \mathcal{T}(\hat{\rho}^{\mathcal{S}} \otimes \hat{\rho}^{\mathcal{A}})) = \frac{\langle 12 \rangle - \langle 1 \rangle \langle 2 \rangle}{\Delta(1)\Delta(2)} \tag{14.64}$$

The respective numbers are:

$$\begin{aligned}
\langle 12 \rangle &= \mathrm{Tr}[\mathcal{I}^2_j(\hat{\rho}^{\mathcal{S}})] && (14.65a)\\
\langle 1 \rangle &= \mathrm{Tr}[\mathcal{I}(\mathbf{1})(\hat{\rho}^{\mathcal{S}})\hat{O}^{\mathcal{S}}_j] && (14.65b)\\
\langle 2 \rangle &= \mathrm{Tr}[\hat{\rho}^{\mathcal{S}}\hat{O}^{\mathcal{S}}_j] && (14.65c)\\
(\Delta(1))^2 &= \mathrm{Tr}[\mathcal{I}(\mathbf{1})(\hat{\rho}^{\mathcal{S}})(\hat{O}^{\mathcal{S}}_j)^2] - \mathrm{Tr}[\mathcal{I}(\mathbf{1})(\hat{\rho}^{\mathcal{S}})\hat{O}^{\mathcal{S}}_j]^2 && (14.65d)\\
(\Delta(2))^2 &= \mathrm{Tr}[\mathrm{Tr}_{\mathcal{S}}(\mathcal{T}(\hat{\rho}^{\mathcal{S}} \otimes \hat{\rho}^{\mathcal{A}}))(\hat{O}^{\mathcal{A}}_j)^2] - \mathrm{Tr}[\hat{\rho}^{\mathcal{S}}\hat{O}^{\mathcal{S}}_j]^2 && (14.65e)
\end{aligned}$$

Strong correlation is then equivalent to

$$\langle 12 \rangle - \langle 1 \rangle \langle 2 \rangle = \Delta(1)\Delta(2) \tag{14.66}$$

whenever the RHS is nonzero. We now state the following theorem:

Theorem 14.11 (Strong Value Correlation) *We have a strong-value-correlation measurement if for every j of the chosen partition of the reading scale, we have:*

$$\boxed{\hat{O}^{\mathcal{S}}_j \hat{\rho}^{\mathcal{S}}_f(j, \hat{\rho}^{\mathcal{S}}_i) = \hat{\rho}^{\mathcal{S}}_f(j, \hat{\rho}^{\mathcal{S}}_i)} \tag{14.67}$$

Eq. (14.67) is parallel to Eq. (14.52a). As we shall see, strong value correlation is guaranteed by theorem 17.6 [p. 271].

Strong Correlations Between Final Component States

We now discuss [*BUSCH et al.* 1991, 57–58] [BUSCH/LAHTI 1996b, 2595–96] the correlation between final component states. The following theorem states the necessary and sufficient conditions for strong correlation between final states:

Theorem 14.12 (Strong Correlation between States) Eq. (14.68a) *(condition of orthogonality between systems' final states) is equivalent to the conjunction of* Eq. (14.68b) *and* Eq. (14.68c)*:*

$$\begin{aligned}
\hat{\rho}^{\mathcal{S}}_f(j, \hat{\rho}^{\mathcal{S}}) \cdot \hat{\rho}^{\mathcal{S}}_f(k, \hat{\rho}^{\mathcal{S}}) &= 0, \text{ for } j \neq k && (14.68a)\\
\mathrm{C}(\hat{\rho}^{\mathcal{S}}_f(j, \hat{\rho}^{\mathcal{S}}), \hat{\rho}^{\mathcal{A}}_f(j, \hat{\rho}^{\mathcal{S}}); \mathcal{T}(\hat{\rho}^{\mathcal{S}} \otimes \hat{\rho}^{\mathcal{A}})) &= 1, \text{ for each } j \text{ with } 0 \neq \wp^{\hat{O}^{\mathcal{S}}}_{\hat{\rho}^{\mathcal{S}}}(\mathcal{X}_j) \neq 1 && (14.68b)\\
\hat{\rho}^{\mathcal{A}}_f(j, \hat{\rho}^{\mathcal{S}}) & \text{ is a } 1-\text{eigenstate of } \hat{O}^{\mathcal{A}}_j \text{ for each } j\\
& \text{ with } 0 \neq \wp^{\hat{O}^{\mathcal{S}}}_{\hat{\rho}^{\mathcal{S}}}(\mathcal{X}_j) \neq 1. && (14.68c)
\end{aligned}$$

As we shall see, strong correlation between final states, a necessary but not sufficient condition for Measurement, is provided by theorem 15.1 [p. 246] or theorem 18.3 [p. 298].

14.3.5 Examination of the Objectification Condition

We have generally analysed the requirements for obtaining objectification. Now we analyse the consequences and problems of this condition as such [*BUSCH et al.* 1991, 75–77].

A premeasurement is a measurement if it satisfies the objectification requirement [theorem 14.8]. But it cannot do this if the measurement coupling is taken to be unitary and both $\mathcal{S}, \mathcal{A}$ are proper QM systems, i.e. systems such that all observables can in principle be represented by bounded self-adjoint operators. We have already showed the impossibility of obtaining a measurement by unitary evolution [see theorem 14.1: p. 221; see also theorem 14.2: p. 222]. We now state that no unitary evolution can satisfy the requirements of Measurement statistics:

Theorem 14.13 (Busch/Lahti/Mittelstaedt) *If $\hat{O}^{\mathcal{S}}$ is a non-trivial observable of the proper QM system $\mathcal{S}$, there is no premeasurement $< \mathcal{H}_{\mathcal{A}}, \hat{O}^{\mathcal{A}}, \hat{\rho}^{\mathcal{A}}, \hat{U}, f >$ that satisfies the pointer value definiteness and the following* pointer mixture condition *for $\mathcal{S} + \mathcal{A}$* [see also Eq. (14.53)]*:*

$$\hat{U}(\hat{\rho}^{\mathcal{S}} \otimes \hat{\rho}^{\mathcal{A}})\hat{U}^{\dagger} = \sum_{j} \hat{I} \otimes (\hat{O}_{j}^{\mathcal{A}})^{\frac{1}{2}} \hat{U}(\hat{\rho}^{\mathcal{S}} \otimes \hat{\rho}^{\mathcal{A}}) U^{\dagger} \hat{I} \otimes (\hat{O}_{j}^{\mathcal{A}})^{\frac{1}{2}} \tag{14.69}$$

for some reading scale partition.

As a consequence of this situation, the physical properties of a pure state are not necessarily conserved in the transition to a mixture — **for example** under time-reversal — [JAUCH *et al.* 1967, 147, 149].

To solve the objectification problem, since we do not wish to use the questionable theory of the observer, we are faced with following possibilities: we have to reject

- either the objectification requirement as such;
- or the postulate that $\mathcal{A}$ is a proper QM system (i.e. by assuming that the pointer observable is not quantum-mechanical);
- or the unitarity of the evolution.

The three options are the object of the next three chapters, respectively.

Chapter 15

MANY WORLD INTERPRETATION

Contents In this chapter we shall report and discuss the Many World Interpretation. In section 15.1 its two different formulations (due to Everett and DeWitt, respectively) are reported. Then some difficulties and problems with this interpretation are discussed [section 15.2]. Finally brief concluding remarks follows [section 15.3].

15.1 Presentation of the Interpretation

15.1.1 Antecedents of the Interpretation

The interpretation of Measurement which we shall analyse in this chapter, the Many World Interpretation (= MWI), takes as its starting point a radical refutation of the Projection Postulate [14.1: p. 223]. In this sense, the bridge to this theory is represented by the Statistical interpretation of Measurement [see subsection 6.5.2]. One could say that the MWI is the Statistical interpretation taken literally [PERES/ZUREK 1982, 809] — it is not by chance that Everett is explicitly quoted as a 'statistical supporter' by Pearle [PEARLE 1967, 744].

One of the first criticisms of the Projection Postulate came from Margenau. He claimed [MARGENAU 1936][1] that the Projection Postulate not only violates the dynamic evolution of Schrödinger Eq. and time symmetry [see subsection 14.1.2], but it is also unnecessary and contrary to experience, which shows that, in order to determine a state, one needs a large number of observations and not a single act, and that very often one completely destroys the observed system. Therefore, after this interpretation, in QM is never a question of individual measurements and hence of a reduction of the wave-packet, but only of predictions about ensembles of identical experiments[2]. We have already questioned [see subsection 6.6.2] such an interpretation. We shall return to it when discussing the Quantum Non-demolition Measurement [chapter 19]. Here we only examine the other two interpretations which stem from the statistical one: the MWI [in this chapter] and the thermodynamic approach [in the next chapter].

15.1.2 Everett's Formulation

Relative state The starting point of Everett is constituted by the problematic character of the reduction of the wave packet [EVERETT 1973, 3–9].[3] The aim is to eliminate the choice between

[1] Historical reconstruction in [*JAMMER* 1974, 226–29, 487].

[2] See also [BALLENTINE 1970, 368–70] [BALLENTINE 1984, 74] [PEARLE 1967].

[3] For that which follows see also [*JAMMER* 1974, 501, 510–515].

different components of the wave packet and to propose a conception in which there is no need of macroscopical apparatus, and hence no division between microphysics and macrophysics: QM can be, in an absolute sense, a complete and global description of our world. In other words the MWI is a reductionistic interpretation: one can account for all reality in terms of microentities [see subsection 24.5.3].

The most important concept in Everett's perspective is that of *relative state* [EVERETT 1957, 143]. To understand it, we begin by writing the wave function of the composite system $\mathcal{A}+\mathcal{S}$ in the known form:

$$|\Psi^{\mathcal{S}+\mathcal{A}}\rangle = \sum_{j,k} c_{jk}|s_k\rangle|a_j\rangle, \tag{15.1}$$

where, as usual, $\{|s_k\rangle\}$ is a basis for $\mathcal{S}$ and $\{|a_j\rangle\}$ for $\mathcal{A}$. We can now choose a unique state of one of the two subsystems ($\mathcal{S}$) and assign to it a corresponding relative state of the other subsystem. We choose for example $|s_k\rangle$ and write the following relation for the relative state $|\mathcal{A}(s_k)\rangle$ of $\mathcal{A}$:

$$|\mathcal{A}(s_k)\rangle = N_k \sum_j c_{kj}|a_j\rangle, \tag{15.2}$$

where N_k is a normalization constant. It is evident that the relative state for $|s_k\rangle$ only depends on this vector and is completely independent of the choice of basis $\{|s_n\rangle\}$ $(n \neq k)$ for the orthogonal complements of $|s_k\rangle$. More generally we can write Eq. (15.1) by varying the choice of the states from $\{|s_n\rangle\}$ and writing the corresponding relative states:

$$|\Psi^{\mathcal{S}+\mathcal{A}}\rangle = \sum_n \frac{1}{N_n}|s_n\rangle|\mathcal{A}(s_k)\rangle. \tag{15.3}$$

This is a general analysis which is not peculiar to the problem of Measurement, which will now be discussed.

If we let $\mathcal{S}$ and $\mathcal{A}$ interact, so as to measure some observable of $\mathcal{S}$ from a time $t=0$ to $t=\tau$, and if (for the sake of simplicity) we suppose that the Hamiltonian for $0<t<\tau$ reduces to the interaction one $\hat{H}_{\mathcal{SA}} = -\imath\hbar q(\partial/\partial r)$ (interaction position/momentum), then we could write the Schrödinger Eq. of the whole system as follows:

$$\imath\hbar\frac{\partial}{\partial t}|\Psi_t^{\mathcal{S}+\mathcal{A}}\rangle = \hat{H}_{\mathcal{SA}}|\Psi_t^{\mathcal{S}+\mathcal{A}}\rangle. \tag{15.4}$$

If we explicitly write the dependence of the wave functions of the two subsystems on the position: $\varsigma(\mathbf{q}), \mathcal{A}(\mathbf{r})$ — since measurement with an apparatus can be seen as a position of a pointer on a graduate scale —, then the state:

$$\Psi_t^{\mathcal{S}+\mathcal{A}}(\mathbf{q},\mathbf{r}) = \varsigma(\mathbf{q})\mathcal{A}(\mathbf{r}-\mathbf{q}t) \tag{15.5}$$

is a solution of Eq. (15.4).

Consider now the time $t=\tau$, at which interaction stops, when there is no longer any definite independent apparatus state. Then we can formulate Everett's tenet by the following theorem:

Theorem 15.1 (Everett) *The total wave function of the system + apparatus can be represented after the interaction as a superposition of pairs of subsystem states of the form:*

$$\boxed{\Psi_\tau^{\mathcal{S}+\mathcal{A}}(\mathbf{q},\mathbf{r}) = \int \varsigma(\mathbf{q}')\delta^3(\mathbf{q}-\mathbf{q}')\mathcal{A}(\mathbf{r}-\mathbf{q}\tau)d\mathbf{q}'} \tag{15.6}$$

where the term $\mathcal{A}(\mathbf{r}-\mathbf{q}\tau)$ expresses a shift of $\mathbf{q}\tau$ from the initial $\mathcal{A}(\mathbf{r})$ which allows the storage of the result of a measurement in the memory of $\mathcal{A}$.

Interpretation Superposition (15.6) of states $\Psi_{\mathbf{q}'} = \delta^3(\mathbf{q}-\mathbf{q}')\mathcal{A}(\mathbf{r}-\mathbf{q}\tau)$ represents an ensemble of states of $\mathcal{A}$ relative to the states $\varsigma(\mathbf{q}')$ of $\mathcal{S}$, each one with a definite value $\mathbf{q} = \mathbf{q}'$. The importance of the theorem is due to the fact that, with the notion of *relative state*, it establishes some connection between the observed system and the apparatus, which in general is only presupposed but never analysed by the Standard interpretation [see the beginning of subsection 14.1.5]. Hence the theorem provides the required condition for strong correlation between the states of the system and of the apparatus [see theorem 14.12: p. 242] and therefore it is a necessary but not sufficient condition for Measurement [for this reason we must add theorem 17.4: p. 269]. Anyway, in order to respect covariance conditions, this theorem must be reformulated in terms of unsharp observables [see theorem 18.3: p. 298], and this is why it is not integrated into the theory.

What is important here is that all values of the measured observable $\mathbf{q}$ exist together, as do all readings of $\mathcal{A}$ represented by the elements $\Psi_{\mathbf{q}'} = \delta^3(\mathbf{q}-\mathbf{q}')\mathcal{A}(\mathbf{r}-\mathbf{q}\tau)$ of the superposition (15.6). In conclusion, following Everett there is no reduction at all, no discontinuous change. But what about the fact that, when we actually perform a measurement, we always obtain only one value, i.e. only one relative state and not a superposition of them? Everett believed [EVERETT 1957, 146] that, though there is only a single physical system describing the apparatus — but Everett says: 'describing the observer' —, nevertheless there is no single unique state of the observer himself, but only a superposition, each element of which contains a definite observer state and a corresponding system state (and present, therefore, a definite observation). Thus, in an ensembles of measurements, the observer state branches into a number of different states[4].

The two most important mathematical problems of this theory are

- finding a formalism able to loose the problem of the degeneracy of the eigenvalues of the observable to be measured [EVERETT 1973, 44];

- and finding a unique representation of the superposition of relative states [EVERETT 1973, 47–49, 53, 60].

This two points will be object of Zurek's examination [see following section]. In particular it should be noted that the Relative-state formulation gave a clue toward to Zurek's solution [see subsection 17.2.1].

15.1.3 DeWitt's Formulation

Everett's theory was further developed by DeWitt. DeWitt started with heavy criticism of the Copenhagen interpretation (which for him also included von Neumann's Standard interpretation) and in particular by expressing three refusals [DEWITT 1970, 160–61]:

- no addition to QM formalism is necessary,

- no existence of a separated classical world is to be postulated,

- no collapse of the state vector occurs.

[4]Note that Everett believed to have also lost the EPR paradox [see chapter 31] because observing one of two non-interacting separated but correlated systems cannot have any effect on the other system [EVERETT 1973, 82–83].

The notion of probability The central point of his tenet is the notion of probability. DeWitt started with the intrinsic probabilistic character of QM [see section 6.4 and section 11.3]: QM probability cannot be treated as subjective ignorance, as the Ignorance interpretation believes [see proposition 6.3: p. 106]. From this DeWitt deduced that QM probability is something objective and absolute (not relative to a subject). But if so, then a measurement cannot change the *a priori* probability distribution [see section 19.2].

Let us briefly discuss the point here (and not in the later examination since it is specific to DeWitt's interpretation and does not characterize Everett's one). The questionable point here is the assertion that the probability is something absolute and objective. Certainly in QM the probability is not a measure of our ignorance. But in no way does it follow that it is absolute. We have already discussed the possibility that the probability be context-dependent [see examination in section 11.1]. In this sense, probability has not necessarily to do with subjectivity, but certainly with the experimental context in which it is calculated, so that it must necessarily differ if we ask about the probability of an event before and after some interaction — and specifically, after interaction with some apparatus. It is true that someone could conceive this interaction as a subjective process — as, for example, von Neumann does by introducing the mind of the observer. But it can be shown that, even if no interaction occurs, the probability distribution changes by the sole presence of a detector — hence even without a reading [see theorem 17.2: p. 267] —, and this change has nothing to do with a subjective action. Hence QM probabilities are not subjective but they are also certainly not absolute.

Hence it is evident that DeWitt's position is only tenable under the hypothesis of isolated (non-interacting) systems, which, as we shall see [in subsection 26.4.1] it is not the case when a measurement is performed: in fact, only for an isolated QM system, it is true that its probability distributions never change (the wave function is deterministic).

The universes Everett still had a subjective point of view, while the aim of DeWitt was to completely objectivize this interpretation. Hence DeWitt interpreted the different branches, which are put forward by Everett as a consequence of the superposition of the relative states of the observer, as alternative worlds, in each one of which a piece of apparatus obtains a value which is the component of the above mentioned superposition. Hence our universe always splits into a great number of alternative 'universes', each one not accessible to the others but all as real as 'our' own [DEWITT 1970, 161] [DEWITT 1971, 178–79].[5] In [DEWITT 1971, 179–82] the condition that the split itself is unobservable is treated mathematically and philosophically. Therefore, it is only DeWitt's interpretation which can rightly be named a Many-World interpretation, while often one refers to Everett's one as a *relative state interpretation.* It is also clear that the MWI supposes the existence of a universal wave function for each of the branched universes. And in this sense it is allied with the decoherent-histories approach, particularly in the form given to it by Gell-Mann and Hartle [see section 25.2].

15.2 Difficulties of the Many-World Interpretation

Now we discuss some difficulties which are common to Everett's and DeWitt's proposal.

Multiplication of states The first problem comes from the incredible multiplication of states, of mind states in Everett's proposal [SQUIRES 1987], and of universes in DeWitt's variant —

[5] A philosophical proposal of a theory of a plurality of existing worlds is in [*D. LEWIS* 1986].

see the methodological principle M7 [p. 6].

Interferences Another difficulty is the following. It is supposed that there is no communication and hence no 'interference' at all between the different components of the wave function, observed by different observers or pertaining to different worlds — that it must be so, derives from the fact that nobody has ever experienced such interferences. But if so, then the MWI presupposes the transition from a pure state to a mixture (loss of off-diagonal terms), which is exactly what it denies from the beginning. And in fact supporters of the theory speak only of the 'practical impossibility' of observing such an interference [SQUIRES 1990, 154]. But where does such an impossibility originate from? This is a point which must be explained. Consequentially, a device for detecting such 'interference terms' has been proposed [PLAGA 1997].

Basis Degeneracy Another very important point is the following: who decides which observable is observed — i.e. in which basis of $\mathcal{S}$ we choose to write the relative states of $\mathcal{A}$ [BELL 1976c, 96–97]? In fact not only there is an infinity of terms in a single expansion but perhaps there are also infinitely many possible expansions (each one in terms of eigenvectors of some observable). Now it is impossible to believe that the state of the system is diagonalized for all observables together — this would be, apart from other considerations, a violation of UP.

But that the latter is a consequence of the Relative-state formulation can be seen as follows [ZUREK 1981, 1516].[6] We can rewrite Eq. (15.2) and (15.6) in the following form:

$$|\mathcal{A}_0\rangle \otimes |\varsigma\rangle \mapsto \sum_o c_o |\mathcal{A}_o\rangle \otimes |o\rangle, \tag{15.7}$$

where one could say that we have measured observable $\hat{O}$ whose spectral decomposition is $\hat{O} = \sum_o c'_o |o\rangle\langle o|$. But suppose that we express the state of $\mathcal{A}$ in another basis $\{|\mathcal{A}_{o'}\rangle\}$ composed of superpositions of states $|\mathcal{A}_o\rangle$:

$$|\mathcal{A}_{o'}\rangle = \sum_o \langle \mathcal{A}_o|\mathcal{A}_{o'}\rangle |\mathcal{A}_o\rangle. \tag{15.8}$$

Then we can rewrite state (15.7) of the compound system as:

$$\sum_o c_o |\mathcal{A}_o\rangle \otimes |o\rangle = \sum_{o'} |\mathcal{A}_{o'}\rangle \otimes \sum_o c_o \langle \mathcal{A}_{o'}|\mathcal{A}_o\rangle |o\rangle = \sum_{o'} c''_{o'} |\mathcal{A}_{o'}\rangle \otimes |o'\rangle. \tag{15.9}$$

If the coefficients c_o in Eq. (15.7) have the same magnitude, then [by RHS of Eq. (15.9)] whenever the set $\{|\mathcal{A}_{o'}\rangle\}$ is orthonormal, also the set $\{|o'\rangle\}$ is orthonormal. Then $\mathcal{A}$ contains not only information about the observable $\hat{O}$ but also about another observable $\hat{O}' = c'''_{o'}|o'\rangle\langle o'|$, even if normally $\hat{O}$ and $\hat{O}'$ do not commute. And the same can be said about many other observables [see also theorem 14.3: p. 230, and the relative proof]. We call such a problem the *basis degeneracy problem* [ELBY/BUB 1994] [see also subsection 17.2.1].

Hence, for respecting UP, we are forced to admit that, when measured, the system is in a state whose components are diagonalized only *relatively to one observable*, while the others are indeterminate — contrary to the Relative-state theory.

Choice between basis But suppose that it is possible to maintain MWI by allowing some form of choice between different basis. Then we are faced with other difficulties:

[6] See also [*BUSCH et al.* 1991, 115].

- in this way we have shifted the problem from a choice between the components of a basis to a choice between basis [ZUREK 1991, 37];
- we are now faced with the impossible problem of individuating a subject able to make such a choice;
- we ascribe to the measurement a power greater than that which the Standard interpretation gives it, because it would have the power to choice between which branching universe (choice between basis) we are called to live and to act in.

As a conclusion of the examination of the last two points, we could say [*D'ESPAGNAT* 1976, 277] that MWI does not eliminate subjectivism, as it was supposed to do. Neither does it eliminate the strange nature of Measurement, because, even if we do not consider the problem of the basis degeneracy, MWI always supposes that a measurement causes the universe to branch. Or, in the Everett formulation, we are forced to assume a universal consciousness to avoid that different observers see different components in the same experiment [SQUIRES 1990, 159], which resembles the paradox of Wigner's friend [see subsection 14.2.1].

Branching Also the branching structure as such is not acceptable, because it presupposes some asymmetry between past and future which does not pertain to Basic-QM as such [section 3.10] but only to the QM theory of Measurement itself [subsection 14.1.2], which is exactly what MWI refuses [BELL 1981a, 135]. On the other side without irreversibility we cannot prevent communication between different worlds [*PERES* 1993, 374]. And we return here to the superposition between different components of the wave (worlds).

Another problem with the ramification hypothesis is that we are not granted against branching of measure zero (which is nonsensical) and against fluctuations [*BOHM/HILEY* 1993, 306–307] [BALLENTINE 1973] which cumulate enormously if the sequence approaches infinity, with the consequence that MWI cannot explain QM probabilities for a relatively old world.

Plurality of worlds But apart from the problem of branching, there are problems with the plurality of worlds as such. Take a determinate value o_j and R the number of times o_j occurs in a given sequence $m_1, m_2, \ldots, m_N$ (the number of successes in that sequence). Then every possible value of R between 0 and N will be realized in some world. But then there is no connection between the actual relative frequency R/N in a world and the expected relative frequency $|\langle \Psi_{o_j} | \Psi \rangle|^2$ [GRAHAM 1973, 233–36].[7]

Universal wave function But perhaps also the idea of a universal wave function as such (apart from the problem of the plurality of worlds) is not adequate. If we wish to maintain some statistical value for it, then we cannot use the wave function for very big systems — and certainly not for the whole universe — because it is impossible — even in principle — to reproduce exactly the same conditions (to make identical copies of a state of our universe) [WOO 1986, 924] — we shall return to this point later [see chapter 26].

[7] Anyway, Graham tried to tackle the problem by introducing a statistical frequency operator. But such a hypothesis seems not satisfactory because of its *ad hoc* character.

15.3 Concluding Remarks

In conclusion, on physical and metaphysical grounds, MWI as a whole is not a sound theory because it does not overcome the greatest problems with which the QM theory of Measurement is faced, while posing many unanswerable questions.

Its main weakness is that it does not consider the Measurement theory as a basic aspect of every physical theory, and particularly of QM, due to the peculiar problems that it poses [see section 26.4.1].

Anyway some results [especially theorem 15.1] are already shown to be interesting, particularly in the domain of Measurement theory (decoherence) [see chapter 17] and in the domain of quantum information [se chapter 44].

Chapter 16

SOLUTIONS USING CLASSICAL OR SEMICLASSICAL APPARATA

Introduction and contents We have already shown that the unitary evolution of the system + apparatus cannot solve the Measurement problem. A possible solution could be found if the apparatus were classical or semiclassical: in other words, if it does not strictly obey to QM dynamic laws. But if we do not wish to follow the Copenhagen interpretation, which simply postulates the classicality of the apparatus [see corollary 8.2: p. 149], we need to correct QM in some way.

Two different solutions have been given using the classicality of the pointer: a sharp one by means of SSRs [section 16.1] and a semiclassical one by means of thermodynamics [section 16.2]. In the last section three approaches are analysed: that of Blokhintsev (a supporter of the statistical interpretation), that of van Hove (which partly anticipates some topics to be discussed later) and Daneri/Loinger/Prosperi's approach.

16.1 Classicality as Superselection Rules

16.1.1 Classicality of the Pointer

We discuss now the classicality of the pointer observable as such [*BUSCH et al.* 1991, 77–78]. A system is said to possess a classical observable if there exists an observable which commutes with all other observables of the system — it is a SSR [see subsection 10.1.4]. Hence we state following theorem:

Theorem 16.1 (Classicality of the Pointer) *If $\mathcal{M}_U^m$ is a unitary premeasurement, then pointer objectification is obtained iff the pointer observable $\hat{O}^{\mathcal{A}}$ is a classical observable.*

We shall not prove the theorem itself, but we prefer to study a concrete model of measurement, which is based on the classicality of the pointer, i.e. upon SSRs.

16.1.2 Models of Measurement based on Superselection Rules

Wakita stated for the first time that phase relations could be lost if the system has many degrees of freedom: there could be no observable which has matrix elements between the many subsystems constituting the system [WAKITA 1960] [WAKITA 1962a] [WAKITA 1962b].[1]

[1] See also [KOMAR 1964].

Sherry and Sudarshan [SHERRY/SUDARSHAN 1978, 4581–82] proposed a more concrete model[2]. We do not report all the formalisms developed in the article, but we limit ourselves to deriving a classical apparatus by adding SSRs to QM.

The system They considered a set of commuting (classical) observables $Q := \{Q^1, \ldots, Q^{2n}\} = \{q_1, \ldots, q_n, p_1, \ldots, p_n\}$ and the set P of QM conjugates to Q. Q's observables are superselecting operators, and the P's are unobservable operators. Then we write a 'Hamiltonian' of the form:

$$\hat{H}_{\mathcal{A}} = \frac{\partial H(Q)}{\partial Q^\nu} Q^{\mu\nu} P^\mu, \tag{16.1}$$

where $H(Q)$ is the classical Hamiltonian and $Q^{\mu\nu}$ means Poisson brackets: $Q^{\mu\nu} := \{Q^\mu, Q^\nu\}$. The Hamiltonian (QM) operator of the $\mathcal{A} + \mathcal{S}$ uncoupled system is [SHERRY/SUDARSHAN 1978, 4586–87]:

$$\hat{H}_0 = \hat{H}_{\mathcal{A}} + \hat{H}_{\mathcal{S}}(\eta), \tag{16.2}$$

where η is some subset of the quantum observable P, and $\hat{H}_{\mathcal{A}}$ is given by Eq. (16.1). Suppose now that the (quantum-enlarged) classical apparatus $\mathcal{A}$ is chosen at a fixed moment $t = 0$ to be in an eigenstate of an observable Q^μ. For all times $t > 0$ it will remain in an eigenstate of Q^μ. But if we allow an interaction between $\mathcal{A}$ and $\mathcal{S}$ it will be not so, because we must add to Hamiltonian(16.2) an interaction Hamiltonian $\hat{H}_{\mathcal{SA}}$:

$$\hat{H} = \hat{H}_0 + \hat{H}_{\mathcal{SA}}. \tag{16.3}$$

Conditions of classicality Thus we can claim that the classical system $\mathcal{A}$ does not remain purely classical. Now a system is *classical* if it satisfies both following properties:

- $Q^\mu(t)$ are observables for all times t;
- $Q^\mu(t)$ and $Q^\nu(t')$ are compatible operators for all times t, t'.

The first property says that the trajectories of classical observables are observable; the second that we can measure different trajectories without disturbing the measurable aspects of the system. The first condition mathematically stated is

$$\forall t, \forall \mu, \forall \nu,\ [Q^\mu(t), Q^\nu(0)] = 0; \tag{16.4a}$$

the second is:

$$\forall t, \forall t',\ [Q^\mu(t), Q^\nu(t')] = 0; \tag{16.4b}$$

which are not independent because:

$$[Q^\mu(t), Q^\nu(t')] = e^{\imath \hat{H}_{\mathcal{A}} t'} [Q^\mu(t - t'), Q^\nu(0)] e^{-\imath \hat{H}_{\mathcal{A}} t'} \tag{16.5}$$

Now we require that during the measurement the classical nature of $\mathcal{A}$ must be preserved. If the coupling function between $\mathcal{A}$ and $\mathcal{S}$ is quadratic or higher in the unobservable P^μ, the two properties are lost; if it is linear in P^μ, this is not the case. We desire a coupling function $\hat{f}$ such that the interaction Hamiltonian has the form [SHERRY/SUDARSHAN 1979, 859]:

$$\hat{H}_{\mathcal{SA}} = \hat{f}^\mu(Q; \eta') P^\mu + \hat{g}(Q; \eta''), \tag{16.6}$$

[2]For a different measurement model based on SSRs see [BLANCHARD/JADCZYK 1993].

where $\{\eta'\}, \{\eta''\}$ are subsets of QM observables and $\hat{g}$ is a secondary coupling function. But this condition alone is not sufficient, so that we need further restrictions. In Eq. (16.6) we postulate that $\dot{Q}^\mu(t)$ depends mainly on $\hat{f}^\mu$, so that we can reformulate the two conditions (16.4) or (16.5) in terms of time derivative of $Q^\mu(t)$. We state that

$$\forall t; \forall m, n > 0, \left[\frac{d^m}{dt^m} Q^\mu(t), \frac{d^n}{dt^n} Q^\nu(t)\right] = 0. \tag{16.7}$$

This condition follows from either Eqs. and it is a necessary condition for $\mathcal{A}$ to remain classical. Now we can write Hamiltonian (16.3) in the form [by Eqs. (16.2) and (16.6)]:

$$\hat{H} = \hat{H}_{\mathcal{A}} + \hat{H}_{\mathcal{S}}(\eta) + \hat{f}^\mu(Q,\eta')P^\mu + \hat{g}(Q,\eta'') = \hat{f}_1^\mu(Q,\eta')P^\mu + \hat{g}_1(Q,\eta'''), \tag{16.8}$$

where $\{\eta'''\} = \{\eta''\} \cup \{\eta\}$.

Time derivatives of $Q^\mu(t)$ take the form:

$$\begin{aligned} \dot{Q}^\mu(t) &= -\imath[Q^\mu(t), \hat{H}] = -\hat{f}_1^\mu(Q(t), \eta'(t)), & (16.9a)\\ \ddot{Q}^\mu(t) &= \imath[\hat{f}_1^\mu, \hat{H}] = \imath[\hat{f}_1^\mu, \hat{f}_1^\nu P^\nu + \hat{g}_1], & (16.9b)\\ \frac{d^3}{dt^3} Q^\mu(t) &= [[\hat{f}_1^\mu, \hat{H}], \hat{H}] = [[\hat{f}_1^\mu, \hat{H}], \hat{f}_1^\nu P^\nu + \hat{g}_1], & (16.9c)\\ \frac{d^m}{dt^m} Q^\mu(t) &= -(\imath)^{m-1}[\ldots[[\hat{f}_1^\mu, \hat{H}], \hat{H}]\ldots, \hat{f}_1^\nu P^\nu + \hat{g}_1]. & (16.9d) \end{aligned}$$

Conclusions In conclusion, from Eqs. (16.9) and (16.7), we see that $\hat{f}_1^\mu$ belongs to a commutative algebra for all μ, i.e. the set of $\hat{f}^\mu$ is commutative, which merely tells us that operators $\hat{f}^\mu$ must commute between themselves. If by $\{\hat{\rho}\}$ we denote the maximal algebraically independent subset of operators $\{\hat{f}^\mu\}$ and by $\{\hat{\rho}'\}$ some algebraically operators, which are independent from $\{\hat{\rho}\}$ but commuting with them, then we have the following necessary and sufficient conditions that Eq. (16.7) holds:

- $\hat{f}^\mu$ form a commuting set;
- $\hat{f}_1^\nu[\hat{\rho}_m, P^\nu] + [\hat{\rho}_m, \hat{g}_1]$ commute with $\hat{f}^\mu$ and with each other for all m; let this set be symbolized by $\{\hat{f}_1^\nu\}$;
- $\hat{f}_1^\nu[\hat{\rho}'_m, P^\nu] + [\hat{\rho}'_m, \hat{g}_1]$ commute with each other and with each element of the set $\{\hat{f}_1^\nu\}$ for those $\hat{\rho}'_m$ which occur in the expansion $[\hat{\rho}_m, \hat{H}], [[\hat{\rho}_m, \hat{H}], \hat{H}], \ldots$.

These are necessary and sufficient conditions for $\mathcal{A}$ to remain classical. In conclusion we have shown the possibility, by adding SSRs to QM, of deriving a classical apparatus which interacts with a QM system.

16.1.3 A Difficulty with the Classicality of the Pointer

The assumption that the pointer observable is classical implies [*BUSCH et al.* 1991, 82–83]:[3]:

[3]See also [*MITTELSTAEDT* 1998, 109] [*D'ESPAGNAT* 1995, 176–77] [*T. BREUER* 1997, 90–91].

Corollary 16.1 (Classicality of Pointer) *Let a Measurement scheme* $\left\langle \mathcal{H}_{\mathcal{A}}, \hat{O}^{\mathcal{A}}, |\mathcal{A}\rangle, \hat{U} \right\rangle$ *be a candidate for a* $\mathcal{M}_U$ *of a discrete sharp observable* $\hat{O}^{\mathcal{S}}$. *If* $\hat{O}^{\mathcal{A}}$ *is classical, then the coupling g cannot be generated by an observable of* $\mathcal{S}+\mathcal{A}$*, because in this case* $\left\langle \mathcal{H}_{\mathcal{A}}, \hat{O}^{\mathcal{A}}, |\mathcal{A}\rangle, \hat{U} \right\rangle$ *cannot fulfil the probability reproducibility condition* (14.46).

Proof

Suppose that pointer observable $\hat{O}^{\mathcal{A}}$, referred to a discrete sharp observable $\hat{O}^{\mathcal{S}}$ of $\mathcal{S}$, is classical. Then it can be interpreted as an observable $\hat{I}_{\mathcal{S}} \otimes \hat{O}^{\mathcal{A}}$ of $\mathcal{S}+\mathcal{A}$, i.e. the latter is a classical observable of the compound system.

Writing now the unitary measurement coupling as $\hat{U} = e^{\imath \hat{H}_{\mathcal{S}\mathcal{A}}\tau}$ [see Eq. (14.7)] and assuming that $\hat{H}_{\mathcal{S}\mathcal{A}}$ commutes with $\hat{O}^{\mathcal{A}}$ (a interaction Hamiltonian commutes with all classical observables pertaining to the coupled systems), and hence that it also commutes commutes with $\hat{U}$, it follows that the probability distribution of $\hat{O}^{\mathcal{A}}$ is completely independent from the measured observable; in other words, the apparatus remains uncoupled with the object system. In fact we have:

$$\begin{aligned} \langle \Psi(\tau)|\hat{O}^{\mathcal{A}}\Psi(\tau)\rangle &= \langle \hat{U}\Psi|\hat{O}^{\mathcal{A}}\hat{U}\Psi\rangle = \langle \Psi|\hat{U}^{\dagger}\hat{O}^{\mathcal{A}}\hat{U}\Psi\rangle \\ &= \langle \varsigma \otimes \mathcal{A}|\hat{I}_{\mathcal{S}} \otimes \hat{O}^{\mathcal{A}}\varsigma \otimes \mathcal{A}\rangle = \langle \varsigma|\hat{I}_{\mathcal{S}}\varsigma\rangle\langle \mathcal{A}|\hat{O}^{\mathcal{A}}\mathcal{A}\rangle \\ &= \langle \mathcal{A}|\hat{O}^{\mathcal{A}}\mathcal{A}\rangle, \end{aligned} \tag{16.10}$$

where $|\Psi\rangle = |\varsigma\rangle|\mathcal{A}\rangle$. Such a situation is incompatible with condition (14.46) unless $\hat{O}^{\mathcal{S}}$ is trivial (i.e. constant). Q. E. D.

Therefore the pointer cannot really be classical. This is the point on which the Copenhagen interpretation as a whole really breaks down [see subsection 8.5.3]. But we can find a lesser requirement [weaker than theorem 16.1: p. 253], a kind of semiclassicality [*BUSCH et al.* 1991, 92–93]: this is the matter of the next section

16.2 Semiclassical Apparatus

We have already discussed of the importance of an apparatus for Measurement [theorem 14.3: p. 230]. Another fundamental theorem, which is also the development of what was stated in our analysis of complementarity [see corollary 8.3: p. 150] is the following [MACHIDA/NAMIKI 1980, 1459, 1834–35] [MACHIDA/NAMIKI 1984, 80] [see also subsection 14.3.5]:

Theorem 16.2 (Machida/Namiki) *The value of a macroscopic variable of* $\mathcal{A}$ *read in a measurement is not an eigenvalue of a QM observable but is a kind of average over microscopic observables, i.e. over a large number of Hilbert spaces.*

Hence $\mathcal{A}$ — we mean here not only the little measuring device, but also the amplifiers [see section 13.2] — cannot be considered as a proper quantum system if we consider the final macroscopical result. We shall not prove this theorem as such. It seems very reasonable and is largely confirmed by all the following analysis, particularly that of spontaneous decoherence. We will now look for some concrete model of the theorem, and we shall find it in the thermodynamic approach.

Anyway, notice that this requirement is a necessary but not sufficient condition for obtaining determinate results in a measurement process [see theorem 17.3: p. 268].

16.2.1 Unstable Apparatus

First proposals An antecedent of the thermodynamic approach is surely represented by Jordan's work [P. JORDAN 1949a], which posed the necessity of apparatus subjected to thermodynamical laws in order to obtain the required irreversibility of the Measurement, because, while microphysics is reversible, thermodynamics is irreversible[4].

As we have seen, the Statistical interpretation is the main source of the thermodynamic approach. Ludwig can be considered as a supporter of this approach — he is strongly dependent upon Jordan's theory. Ludwig spoke for the first time of 'channels' in the Measurement process [LUDWIG 1961, 154] and asserted that measurement can be performed only on statistical ensembles. In general we can say that we have macroscopicality when there is a break down of correlations [LUDWIG 1961, 162, 175].[5]

Blokhintsev's proposal One of the first scientists to suppose that the detector must be an *unstable* macroscopic system was Blokhintsev [*BLOKHINTSEV* 1965, 91–98]: otherwise, whenever the detected particle moves a pointer, it would act as a macroscopic particle, which is absurd. Hence only under the instability condition there can be such an effect [BLOKHINTSEV 1976, 154–55]. We can understand Blokhintsev's ideas by means of the following examples.

Suppose we wish to measure the particle's momentum direction [*BLOKHINTSEV* 1965, 85–90] [BLOKHINTSEV 1976, 156–57]. We can perform the measurement by means of a macroscopic ball placed at the top of a cone — a potential trap — [see figure 16.1]; since it is in unstable equilibrium, the impact with the particle determines the falling to the left or to the right of the cone, which gives us the required information.

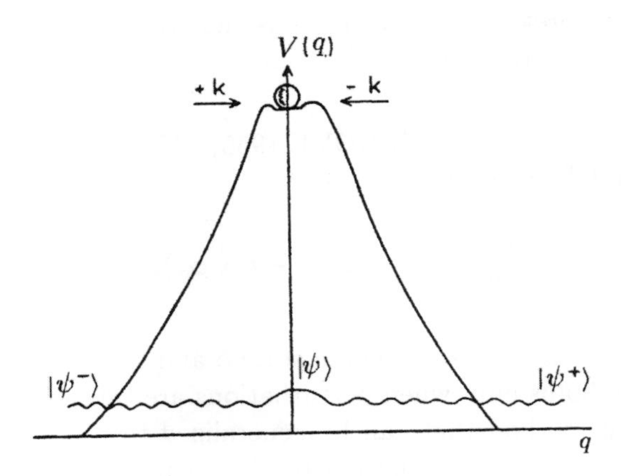

Figure 16.1: Unstable equilibrium of a ball on the top of a 'cone' potential — from [BLOKHINTSEV 1976, 156].

Another example [BLOKHINTSEV 1976, 156–58] is given by a thermodynamically unstable lattice of charged particles [see figure 16.2], with temperature T, near to the equilibrium state with oscillations in the x–y plane along the x-axis. As soon as a particle with magnetic moment penetrates

[4]For historical reconstruction see [*JAMMER* 1974, 413, 488]. See also [*KRAUS* 1983, 1–2].

[5]See also [LUDWIG 1964, 245].

into this medium the temperature falls to $T/2$ because the particle serves as dust particle that provides exchange of energy between two degrees of freedom of the lattice particles[6].

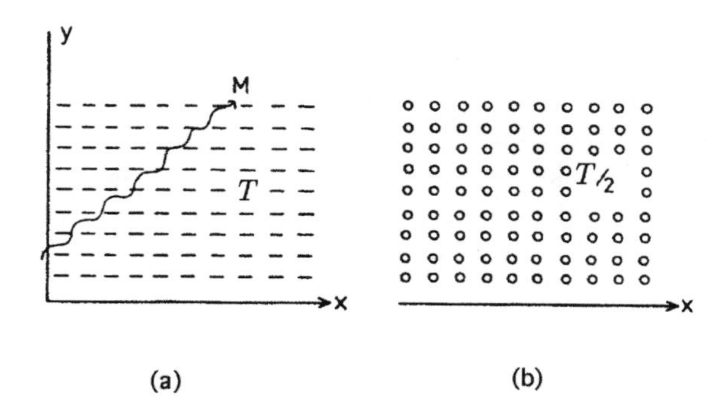

Figure 16.2: (a) Thermodynamically unstable oscillation of the lattice along the x–axis at a temperature T while the oscillation along the y–axis occurs at $T = 0$.
(b) Circular oscillation of the lattice after the interaction with the particle: the temperature is now $T/2$ — from [BLOKHINTSEV 1976, 157].

16.2.2 Van Hove's Master Equation*

The central point of the interpretation of van Hove is that the required Measurement transformation happens already at a microscopic level[7].

Master equation Van Hove starts [VAN HOVE 1955, 517] [VAN HOVE 1957, 441] with a classical master Eq.[8] of the following form:

$$\frac{d\wp_\alpha}{dt} = \sum_\beta (w_{\alpha\beta}\wp_\beta - w_{\beta\alpha}\wp_\alpha), \tag{16.11}$$

where $\wp_\alpha$ is the probability of finding the system in state α and the weights $w_{\alpha\beta}$ are the transition probabilities. This equation holds only when perturbations are taken into account at the lowest non-vanishing order. It is a method that can be generalized to higher orders of perturbation. The long time behaviour of solutions of the master Eq. corresponds to the establishment of microcanonical equilibrium. The quantum equivalent of Eq. (16.11) is the *van Hove Master Eq.*:

$$\frac{d}{dt}\hat{\rho}_{\alpha\alpha} = \sum_\beta w_{\alpha\beta}\hat{\rho}_{\beta\beta}, \tag{16.12}$$

[6]For further discussion see [H. GREEN 1958, 882–83] where the principal function of an apparatus is the separation of microsystems such that each system is characterized by a particular eigenvalue (or range of eigenvalues). Hence it must be diverted into a particular channel by destroying interference terms. See also [MACHIDA/NAMIKI 1980, 1461] where Furry's criticism of this point is reported.

[7]See [*JAMMER* 1974, 491].

[8]See also [JOOS 1996, 109].

where now the $\wp_\alpha$ of Eq. (16.11) are the diagonal elements of the density matrix describing the system. Eq. (16.12) describes an autonomous dynamic for the diagonal part of $\hat{\rho}_{\alpha\beta}$ (in a given basis). This is only the first (and very primitive) example of QM master Eqs. [see section 21.1]. Anyway, the most difficult task is finding a starting point.

Approach to the equilibrium of diagonal operators We have a spontaneous approach to equilibrium [VAN HOVE 1959] not only for diagonal operators in a determinate representation but also for non-diagonal ones (when there is superposition), by means of a convergent series, each term of which is a product of creation and annihilation operators for individual plane wave excitations and where the number of creation and annihilation operators for each term of the series is finite and independent from the large number of the particles in the system. Take an observable $\hat{O}$ [VAN HOVE 1957, 454–58], diagonal in the $|o\rangle$ representation, and the initial state

$$|\psi_i\rangle = \int |o\rangle do \vartheta(o). \tag{16.13}$$

We calculate the expectation value in the course of time to be:

$$\left\langle \hat{O} \right\rangle_t = \langle \psi_i | \hat{U}_{-t} \hat{O} \hat{U}_t | \psi_i \rangle, \tag{16.14}$$

where

$$\hat{U}_t = \exp[-\imath(\hat{H} + \eta V)t] \tag{16.15}$$

and where ηV is a perturbation caused by the interaction. Let now define the resolvent in the following form:

$$\hat{R}_l = (\hat{H} + \eta V - l)^{-1}, \tag{16.16}$$

where l is some complex number; by means of the resolvent we can write:

$$\hat{U}_{-t} \hat{O} \hat{U}_t = -(2\pi)^{-2} \int_{\mathbf{C}} dl \int_{\mathbf{C}} dl' e^{\imath(l-l')t} \hat{R}_l \hat{O} \hat{R}_{l'}, \tag{16.17}$$

where $\mathbf{C}$ is some closed curve in the complex plane. Now we define the function $X_{ll'}$ as follows:

$$\{\hat{R}_l \hat{O} \hat{R}_{l'}\}_d |o\rangle = |o\rangle \int \hat{O}(o') do' X_{ll'}(o'o), \tag{16.18}$$

where d means 'diagonal part'. $X_{ll'}$ approaches zero, for $|l'|^{-1}$ as $l' \longrightarrow \infty$, or for $|ll'|^{-1}$ as $l \longrightarrow \infty$ and $l' \longrightarrow \infty$.

Approach to the equilibrium of non-diagonal operators In analogy with the preceding analysis, take an observable $\hat{O}'$, which is not diagonal in the $|o\rangle$ representation, and a function $Y_{ll'}$ defined as [VAN HOVE 1959, 650–51]:

$$\{\hat{R}_l \hat{O}' \hat{R}_{l'}\}_d |o\rangle = |o\rangle Y_{ll'}(o'o). \tag{16.19}$$

In analogy with Eq. (16.17) we write:

$$\hat{U}_{-t} \hat{O}' \hat{U}_t = -(2\pi)^{-2} \int_{\mathbf{C}} dl \int_{\mathbf{C}} dl' e^{\imath(l-l')t} \hat{R}_l \hat{O}' \hat{R}_{l'}. \tag{16.20}$$

Finally we can calculate the expectation value of $\hat{O}'$ [see Eq. (16.14)]:

$$\left\langle \hat{O}' \right\rangle_t = -(2\pi)^{-2} \int_{\mathbf{C}} dl \int_{\mathbf{C}} dl' e^{\imath(l-l')t} Y_{ll'}(o) do |\vartheta(o)|^2, \tag{16.21}$$

where we have assumed an incoherent phase for amplitudes $\vartheta(o)$.

Now it can be shown that, in the long term, we have as a limit:

$$\left\langle \hat{O}' \right\rangle_{\pm\infty} = \int_{-\infty}^{+\infty} \left\langle \hat{O}' \right\rangle_E p_E dE, \tag{16.22}$$

where $\wp_E dE$ is the probability that the total energy $\hat{H} + \eta V$ be included between E and $E + dE$ for the system in initial state $|\psi_i\rangle$ under the incoherent phase assumption, and where $\left\langle \hat{O}' \right\rangle_E$ is the microcanonical average of $\hat{O}'$ over the energy shell $\hat{H} + \eta V = E$.

Concluding remarks Today van Hove's essay has more of a historical than of a theoretical significance. He influenced later developments in two respects: by proposing a master Eq. as the solution of a class of transitions from coherence to loss of coherence, and by indicating larger systems as those which are able to effect such a transition.

Anyway, his deduction is too specific to be generalized. As we shall see better in the next section, the thermodynamic approach as such cannot be generalized.

16.2.3 Daneri/Loinger/Prosperi's Model

Exposition

The most systematic thermodynamic theory of Measurement was developed by Daneri, Loinger and Prosperi. The ergodic amplification theory gives up the unitarity of the time evolution operator to derive the reduction of wave packet via the dynamic statistical mechanics of thermal irreversible processes[9]. The antecedents are to be found in [BOCCHIERI/LOINGER 1959] [PROSPERI/SCOTTI 1959]: the problem was to find exact ergodicity conditions for the validity in QM of the ergodic theorem — if one waits a sufficiently long time, the ensemble of representative points of a system will cover the entire accessible phase space[10]. Daneri, Loinger and Prosperi [DANERI *et al.* 1962] found that these conditions are satisfied by that class of Hamiltonians for which van Hove [VAN HOVE 1959] derived the master Eq., and, stimulated by Ludwig, tried to establish on this basis a satisfactory theory of Measurement[11].

Suppose a macro-energy shell $\mathbf{C}_a$ to which pertains another macroscopic constant of motion Ω, the spectrum of which can be divided into intervals (Ω_k, Ω_{k+1}), so that each energy shell can be divided in orthogonal sub-manifolds $\mathbf{C}_{ak}$ — where k stands for the k–th channel of the shell. a, k are constant in the case of free evolution of $\mathcal{S}$. We add another index n which denotes macrovariables which change during free evolution of $\mathcal{S}$. The ergodicity condition is satisfied if there is a cell $\mathbf{C}_{ake_k}, (e_k \neq n)$ with a number of dimension which is much larger than $\mathbf{C}_{akn}$ in which $\mathcal{S}$ is (by free evolution) most of the time if it was initially somewhere within $\mathbf{C}_{ak}$. This cell represents the various states of macroscopic equilibrium [DANERI *et al.* 1962, 659]. Therefore, we must have a system possessing many states of macroscopic equilibrium compatible with a given value of macroscopic energy.

If the statistical operator before measurement is:

$$\hat{\rho}_i = \hat{P}^{\mathcal{S}}_{\sum_r c_r|s_r\rangle} \hat{P}^{\mathcal{A}}_{a_0}, \tag{16.23}$$

[9]See [MACHIDA/NAMIKI 1984, 80] for a general statement of the problem.

[10]On this point see [*K. HUANG* 1963, 65].

[11]See [*JAMMER* 1974, 491].

then after the measurement it is [see Eq. (14.14)]:

$$\hat{\rho}_f = \hat{P}^{\mathcal{S}+\mathcal{A}}_{\sum_r c_r|s_r a_r\rangle}. \tag{16.24}$$

But, actually, as far as the macroscopic quantities of $\mathcal{A}$ are concerned, and using the above assumption of a macro-energy shell divided into cells, we have [DANERI *et al.* 1962, 673]:

$$\hat{\tilde{\rho}}_f = \sum |c_r|^2 \hat{P}^{\mathcal{S}}_{\hat{U}_{\mathcal{S}}(t)|s_r\rangle} \frac{1}{N_{rc_r}} \hat{P}^{\mathcal{A}}_{\mathbf{C}_{re_r}}. \tag{16.25}$$

Here the fraction is simply a normalization factor for $\hat{P}^{\mathcal{A}}$. Synthesizing:

- the Hilbert-space vector of $\mathcal{A}$ is actually in an equilibrium cell of a particular channel for a sufficiently long time;
- although this particular channel is unpredictable in principle, QM assigns to it some probability [BUB 1968b, 510–11].

In [BARCHIELLI *et al.* 1982] [BARCHIELLI *et al.* 1983] is developed the idea of choosing a parameter η such that as $\eta \longrightarrow \infty$, we obtain increasingly precise measurements. As $\eta \longrightarrow 0$ the information about the observable decreases.

Criticisms

Bub criticized Daneri/Loinger/Prosperi's proposal [*JAMMER* 1974, 493] on the following grounds:

- Their ideas do not agree with those of Bohr, even if they think so [BUB 1968b, 505–508]; but this criticism has no relevance as such.
- Macrostates do not eliminate micro-superposition. The problem is that $\hat{\tilde{\rho}}_f$ does not follow from QM plus ergodicity. The fact that phase relations between different cells of states are not controllable does not imply their absence: hence each individual measurement will still lead to a superposition of states from different ergodic subspaces. Therefore the authors are forced to admit that this statical operator concerns only the macrolevel. $\hat{\tilde{\rho}}_f$ is derived from the assumption that $\mathcal{A}$ has a specific structure and functions as a macro-instrument, i.e. that the pointer readings correspond to the set of macrostates defined in relation to the set of macrovariables and this ultimately depends on the introduction of a specific observer [BUB 1968b, 514].[12] But, as we shall see in the following chapter, this is not necessary a weakness of the theory.
- They do not distinguish between motion as the result of Measurement (which is problematic) and motion of the free macrosystem [BUB 1968b, 519].

But the best criticism is due to Jauch/Wigner/Yanase [JAUCH *et al.* 1967, 150]: they invoked the fact that in an interaction-free measurement of the type studied by Renninger [see section 20.1] — and defined by the authors as a 'negative-result measurement' —, there is no ergodic amplification[13].

This criticism can be generalized because, as we shall see [theorem 17.2: p. 267; see also SPDC Exp. 2: p. 267], in QM the sole presence of a detector (without even taking a reading) is enough to effect the required transition

[12] See also [ZEH/JOOS 1996, 274].

[13] See also [MACHIDA/NAMIKI 1980, 1462–63].

16.2.4 Conclusion

Many interesting results can be used in more elaborate theories, such as decoherence, and particularly in order to find models of spontaneous transition to decohered states [see chapter 21]. But the negative-measurement model of Renninger excludes the possibility of a generalized theory of Measurement that is formulated in terms of thermodynamic evolution.

Chapter 17

DECOHERENCE

Contents In this chapter we develop the theory of decoherence. In section 17.1 a short historical reconstruction is given. Then the more elaborate form of the theory (Zurek's proposal) is analysed [section 17.2]. Two important alternative (and independent) models are also discussed: that of Machida and Namiki [section 17.3] and that of Cini [section 17.4]. An easy computer simulation of how decoherence may work is then reported [section 17.5] and, finally, brief concluding remarks follow [section 17.6].

17.1 A Recent and Yet Rich History

Many results of the thermodynamic approach were further developed by other schools and by single physicists. Though the idea of a thermodynamical unstable apparatus was abandoned, the central tenet remained because one looked for some effect of the apparatus' macroscopicity able to account for the irreversible dynamics of Measurement.

The prehistory of decoherence consists of some attempts to find (in the classical domain) a model for non-equilibrium processes able to lead naturally to a stationary state through the action of a large reservoir [BERGMANN/LEBOWITZ 1955], and of an approach to the ergodic problem [J. BLATT 1959] which showed that the action of a macroscopic observer upon a small thermodynamical system is not limited to a 'coarse-graining' (such that the resultant lack of complete information about the system gives rise to the irreversible increase of the (coarse-grained) entropy), but rather it produces an increase of entropy due to the collision of external molecules with the outside of the box enclosing the system, such that a truly random and unpredictable perturbation is produced.

With the introduction of the laser [DICKE 1981] [SENITZKY 1960] [SENITZKY 1961] [1] the problem of open systems entered into Q-Optics:[2] in fact sources of light are open systems, where the source can be considered a damped QM system and where the electromagnetic field that the source emits can be considered as the environment into which the source loses energy.

However, the first explicit introduction of the environment in order to overcome the problems of Measurement theory is due to Zeh [ZEH 1970].[3]

[1]See also [*SARGENT et al.* 1973].
[2]On this point see [*CARMICHAEL* 1993, 1, 5]. For a successive model based on laser see [HEPP/LIEB 1973].
[3]See also [KÜBLER/ZEH 1973].

17.1.1 Hepp's Algebraic Approach*

One of the first more organical essays is due to Hepp, and it is developed in full algebraic form. Hence Hepp's approach is to be understood as a development of the local algebra of observables [see subsection 9.3.1].

The starting point is a C* algebra [see definition 9.10: p. 167] with positive linear functionals (states) and a representation $\pi(\hat{O})$ of an arbitrary observable $\hat{O}$ on a Hilbert space $\mathcal{H}_\pi$. First we make a formal distinction [HEPP 1972, 240–42] between the concepts of disjointness and of coherence.

Definition 17.1 (Disjointness) *We say that two states* $\hat{\rho}_1 = |\psi_1\rangle\langle\psi_1|$, $\hat{\rho}_2 = |\psi_2\rangle\langle\psi_2|$ *are* disjoint *iff for every* $\pi(\hat{O})$ *and supposing that* $\psi_1, \psi_2 \in \mathcal{H}_\pi$, *one has that:*

$$\langle\psi_1|\pi(\hat{O})|\psi_2\rangle = 0, \tag{17.1}$$

for every operator $\hat{O}$ *in the algebra.*

If $\hat{\rho}_1$ and $\hat{\rho}_2$ are not disjoint, then they are *coherent*. Even though the definition is correct, we do not integrate it into the theory, because, as we shall see [chapter 24], quantum systems are supposed to never completely lose their coherence.

Now, while coherence cannot be destroyed by automorphic time-evolution, in the case of a sequence $\hat{\rho}_{1,n}, \hat{\rho}_{2,n}$ of coherent states which weakly converge towards disjoint states $\hat{\rho}_1, \hat{\rho}_2$ ($\hat{\rho}_{k,n} \longrightarrow_w \hat{\rho}_k$), all cross terms converge to zero:

$$\langle\psi_{1,n}|\pi_n(\hat{O})|\psi_{2,n}\rangle \longrightarrow 0. \tag{17.2}$$

We now say that [LANFORD/RUELLE 1969]:

Definition 17.2 (Short range correlations) *A state* $\hat{\rho}$ *has short range correlations if for every operator* $\hat{O}$ *pertaining to the algebra and for* $\epsilon > 0$, *there exists a relatively small region (volume)* V *such that:*

$$|\hat{\rho}(\hat{O}\hat{O}') - \hat{\rho}(\hat{O})\hat{\rho}(\hat{O}')| \leq \| \hat{O}' \| \epsilon, \tag{17.3}$$

for all $\hat{O}'$ *pertaining to the algebra* $\mathcal{L}(\mathrm{V})$ *of observables in this region.*

The smallness of the region V is ultimately a practical matter: it depends on the system that we wish to measure and on the measurement that we wish to perform. We can define a classical observable $\hat{O}_C$ as the weak limit of a sequence $\hat{O}_C^n$ with $\hat{O}_C^n$ pertaining to the algebra $\mathcal{L}(\mathrm{V}_n)$ and with $\hat{O}_C^n \leq o'$ (for every o' eigenvalue of $\hat{O}'$):

$$\lim_{N \longrightarrow_w \infty} \frac{1}{N} \sum_{n=1}^{N} \pi(\hat{O}_C^n) = \hat{O}_C. \tag{17.4}$$

Then we state the following lemma:

Lemma 17.1 (Hepp) *If* $\hat{\rho}$ *has short range correlations, we have that if*

$$\lim_{N \longrightarrow \infty} \frac{1}{N} \sum_{n=1}^{N} \hat{\rho}(\hat{O}_C^n) = \hat{\rho}_o \tag{17.5}$$

exists — where o *is an eigenvalue of the observable* $\hat{O}$ *and* $\hat{\rho}_o$ *can informally be understood as the correspondent probability* [see Eq. (42.3)] *—, then in this representation one has:*

$$\lim_{N \longrightarrow_w \infty} \frac{1}{N} \sum_{n=1}^{N} \pi_{\hat{\rho}}(\hat{O}_C^n) = o. \tag{17.6}$$

We omit the proof. Now we are able to formulate the main theorem:

Theorem 17.1 (Hepp) *For primary states $\hat{\rho}_1, \hat{\rho}_2$ and some $\hat{O}^n_C$ we have that if:*

$$\lim_{N \longrightarrow \infty} \frac{1}{N} \sum_{n=1}^{N} \hat{\rho}_j(\hat{O}^n_C) = o_j, \quad j = 1, 2 \tag{17.7}$$

and $o_1 \neq o_2$, then $\hat{\rho}_1, \hat{\rho}_2$ are disjoint.

Hence we can see that different expectation values of macroscopic pointers (classical observables with an infinite number of degrees of freedom) lead to disjointness, i.e. the loss of coherence[4].

17.1.2 Scully/Shea/McCullen's Reduced Density Matrix

SGM-Exp. 1

Partial trace Another antecedent is due to Scully, Shea and McCullen [SCULLY *et al.* 1978a]. They showed that reduction can happen by making a partial trace on the degrees of freedom of the apparatus (here an atom) and that we obtain a mixture even if we do not actively extract the information, i.e. with the sole presence of the detector and without taking a reading, contrary to von Neumann's hypothesis, which needs an observer and an actual observation [see subsection 14.1.3].[5]

Experimental set-up Suppose an arrangement such as that of figure 17.1 — Stern/Gerlach Magnet Experiment 1 (= SGM-Exp. 1).

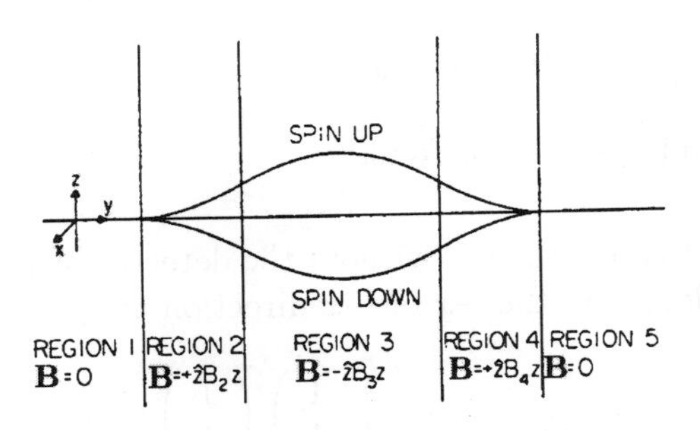

Figure 17.1: Schematic representation of the proposed SGM-Exp. 1 — from [SCULLY *et al.* 1978a, 488].

We begin with a Gaussian wave packet moving in the $+y$ direction (first region), which is then passed (second region) through a magnet with field $\mathbf{B}$ and gradient $\partial\mathbf{B}/\partial z$ in the z–direction. As a result the beam is split into a spatial component with spin up ($|\psi_\uparrow\rangle$) and another one with spin down ($|\psi_\downarrow\rangle$). Then the two partial beams are merged (fourth region). In the fifth region the wave function will be the same as if there were no magnetic field, except for the inevitable spreading.

[4] The model has been recently generalized in [SUN 1993].

[5] See also [NAMIKI/PASCAZIO 1993, 345–47]. Everett and DeWitt reacted exactly against this aspect of von Neumann's interpretation [see section 15.1].

Two alternatives We imagine two alternatives:

- in the described experiment the final density matrix, within the limits of a perfect overlap between $|\psi_\uparrow\rangle$ and $|\psi_\downarrow\rangle$, will be:

$$\hat{\rho}(\mathbf{r}, \mathbf{r}', t) = \begin{pmatrix} 1 & 1 \\ 1 & 1 \end{pmatrix} \langle \psi(\mathbf{r}', t) | \psi(\mathbf{r}, t) \rangle, \tag{17.8}$$

where $|\psi(\mathbf{r}, t)\rangle = |\psi_\uparrow(\mathbf{r}, t)\rangle = |\psi_\downarrow(\mathbf{r}, t)\rangle$;

- and a modified version, in which there is a detector: a two-level atom (ground state $|g\rangle$ and excited state $|e\rangle$), where the final density matrix will be:

$$\hat{\rho} = \begin{bmatrix} \sin^2 \eta_\uparrow |\psi_\uparrow|^2 & \sin \eta_\uparrow \sin \eta_\downarrow \langle \psi_\uparrow | \psi_\downarrow \rangle & \imath \sin \eta_\uparrow \cos \eta_\uparrow |\psi_\uparrow|^2 & \imath \sin \eta_\uparrow \cos \eta_\downarrow \langle \psi_\uparrow | \psi_\downarrow \rangle \\ \sin \eta_\uparrow \sin \eta_\downarrow \langle \psi_\downarrow | \psi_\uparrow \rangle & \sin^2 \eta_\downarrow |\psi_\downarrow|^2 & \imath \sin \eta_\uparrow \cos \eta_\uparrow \langle \psi_\downarrow | \psi_\uparrow \rangle & \imath \sin \eta_\downarrow \cos \eta_\downarrow |\psi_\downarrow|^2 \\ -\imath \sin \eta_\uparrow \cos \eta_\uparrow |\psi_\uparrow|^2 & -\sin \eta_\downarrow \cos \eta_\uparrow \langle \psi_\uparrow | \psi_\downarrow \rangle & \cos^2 \eta_\uparrow |\psi_\uparrow|^2 & \cos \eta_\uparrow \cos \eta_\downarrow \langle \psi_\uparrow | \psi_\downarrow \rangle \\ -\imath \sin \eta_\uparrow \cos \eta_\downarrow \langle \psi_\downarrow | \psi_\uparrow \rangle & -\imath \sin \eta_\downarrow \cos \eta_\downarrow |\psi_\downarrow|^2 & \cos \eta_\uparrow \cos \eta_\downarrow \langle \psi_\downarrow | \psi_\downarrow \rangle & \cos^2 \eta_\downarrow |\psi_\downarrow|^2 \end{bmatrix}, \tag{17.9}$$

where

$$\eta_\uparrow := \frac{g_{\mathcal{SA}}}{\hbar} \int_t |\psi_\uparrow(\mathbf{r}_{\mathcal{A}}, t)|^2 dt, \tag{17.10a}$$

$$\eta_\downarrow := \frac{g_{\mathcal{SA}}}{\hbar} \int_t |\psi_\downarrow(\mathbf{r}_{\mathcal{A}}, t)|^2 dt, \tag{17.10b}$$

where $\mathbf{r}_{\mathcal{A}}$ is the position of the detector and $g_{\mathcal{SA}}$ is the coupling between $\mathcal{A}$ and $\mathcal{S}$.

Now we pass our beam through another SGM apparatus in the $\mathbf{n}$–direction:

$$\mathbf{B} = \mathbf{n} B_x(x, y). \tag{17.11}$$

This device will split the beam into $+x, -x$ packets.

First hypothesis By the first hypothesis, i.e. without the detector [Eq. (17.8)], the probability densities for the beam to be deflected in the $+x$ or $-x$ direction are given by:

$$\wp_{+x}(\mathbf{r}) = \langle +x|\hat{\rho}| + x\rangle = \frac{1}{2}\begin{pmatrix} 1 & 1 \end{pmatrix} \begin{pmatrix} 1 & 1 \\ 1 & 1 \end{pmatrix} \begin{pmatrix} 1 \\ 1 \end{pmatrix} |\psi(\mathbf{r})|^2 = 2|\psi(\mathbf{r})|^2, \tag{17.12a}$$

$$\wp_{-x}(\mathbf{r}) = \langle -x|\hat{\rho}| - x\rangle = \frac{1}{2}\begin{pmatrix} 1 & -1 \end{pmatrix} \begin{pmatrix} 1 & 1 \\ 1 & 1 \end{pmatrix} \begin{pmatrix} 1 \\ -1 \end{pmatrix} |\psi(\mathbf{r})|^2 = 0, \tag{17.12b}$$

where

$$\int |\psi(\mathbf{r})|^2 dr = \frac{1}{2}. \tag{17.13}$$

Hence no particles would be found in the $-x$ direction if a detector was to be placed.

Second hypothesis Now we ask the same question about the second alternative [Eq. (17.9)]. We look for our probabilities irrespective of the state of the detector, i.e. for the quantity

$$\wp_{\pm x}(\mathbf{r}) = \langle \pm x, e_{\mathcal{A}}|\hat{\rho}| \pm x, e_{\mathcal{A}}\rangle + \langle \pm x, g_{\mathcal{A}}|\hat{\rho}| \pm x, g_{\mathcal{A}}\rangle = \langle \pm x|\hat{\rho}_{\mathrm{Tr}_{\mathcal{A}}}| \pm x\rangle, \tag{17.14}$$

where the reduced density matrix $\hat{\rho}_{\mathrm{Tr}_A}$ [see Eq. (3.103)] is:

$$\hat{\rho}_{\mathrm{Tr}_A} = \begin{bmatrix} |\psi_\uparrow|^2 & \cos(\eta_\uparrow - \eta_\downarrow)\langle\psi_\uparrow|\psi_\downarrow\rangle \\ \cos(\eta_\uparrow - \eta_\downarrow)\langle\psi_\downarrow|\psi_\uparrow\rangle & |\psi_\downarrow|^2 \end{bmatrix}. \tag{17.15}$$

For a good measurement — i.e. one such that $\eta_\uparrow = \pi/2, \eta_\downarrow = 0$ — we have:

$$\hat{\rho}_{\mathrm{Tr}_A} = \begin{pmatrix} 1 & 0 \\ 0 & 1 \end{pmatrix} |\psi(\mathbf{r})|^2. \tag{17.16}$$

In conclusion wee find that:

$$\wp_{+x}(\mathbf{r}) = \tfrac{1}{2}(\,1 \;\; 1\,)\begin{pmatrix} 1 & 0 \\ 0 & 1 \end{pmatrix}\begin{pmatrix} 1 \\ 1 \end{pmatrix} |\psi(\mathbf{r})|^2 = \frac{1}{4}|\psi(\mathbf{r})|^2, \tag{17.17a}$$

$$\wp_{-x}(\mathbf{r}) = \tfrac{1}{2}(\,1 \;\; -1\,)\begin{pmatrix} 1 & 0 \\ 0 & 1 \end{pmatrix}\begin{pmatrix} 1 \\ -1 \end{pmatrix} |\psi(\mathbf{r})|^2 = \frac{1}{4}|\psi(\mathbf{r})|^2. \tag{17.17b}$$

Hence in the second case we have 50% of probability of finding the particle in the $+x$ direction and 50% of probability of finding the particle in the $-x$ direction. Therefore a partial trace seems to give an important contribution to the solution of the Measurement problem [see subsection 17.2.2].[6]

Conclusion We have already said that the sole presence of a detector (without reading) is enough for obtaining the transition required by Measurement[7]. We state it as a theorem:

Theorem 17.2 (Scully/Shea/McCullen) *The sole presence of a detector (without reading) is enough to make the measured quantum system undergo the required transition.*

This is very strong evidence of the fact that in QM is important not only what operation is actually performed, but also what is performable, differently from that which is postulated by a generalized thermodynamical solution to the Measurement problem [see subsections 16.2.3 and 16.2.4].

SPDC Exp. 2

Further confirmation can be found in an experiment (SPDC Exp. 2) due to Wang, Zou and Mandel [WANG *et al.* 1991].[8] Two NL crystals effect two SPDCs [see figure 17.2]. A BS merges the two s-photons, one coming from NL1 and the other from NL2, which are then detected by D_s, while the two i-photons are parallel and detected by D_i. Normally neither detector can distinguish between pairs $|\gamma_{s_1}\rangle, |\gamma_{i_1}\rangle$ and $|\gamma_{s_2}\rangle, |\gamma_{i_2}\rangle$ (forth order interference). But if each signal detected by D_s is accompanied by an idler detected by D_i, the detection by the latter becomes superfluous and the counting rate registered by D_s alone suffices to exhibit interference (second order interference), because D_s cannot distinguish between the two paths of the s-photons. The

[6] On this point see also [SCHRÖDINGER 1935b] [BELL/NAUENBERG 1966, 24–25]. For further analysis see also [*JAMMER* 1974, 477–78].

[7] We recall another Scully's experiment [Two-Slit-Exp. 8: p. 143]. See also [DICKE 1981, 99].

[8] See also [MANDEL 1995a].

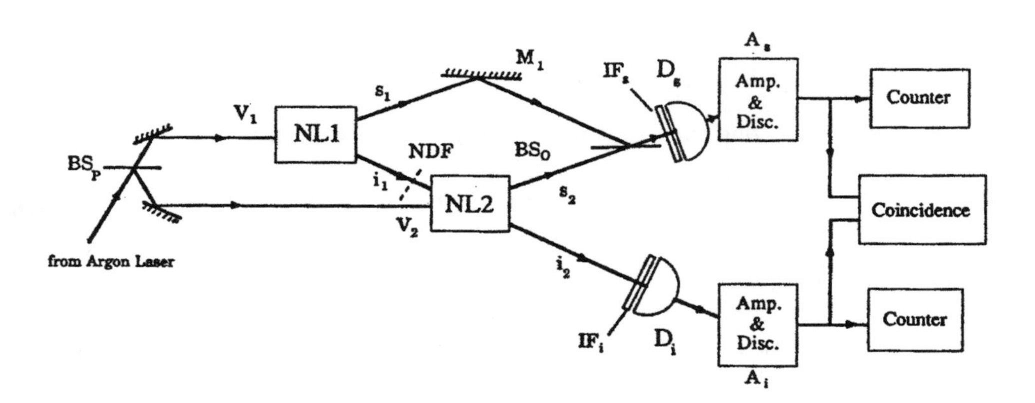

Figure 17.2: Schematic representation of SPDC Exp. 2 ($i = |\gamma_i\rangle, s = |\gamma_s\rangle$ for the sake of simplicity) — from [WANG *et al.* 1991, 318]

question now is: what happens if we insert a beam stop (or attenuator) between the two crystals? If both detectors register, then the s-photon comes from NL2, and if only D_s registers, then the s-photon comes from NL1. Hence the interference pattern has disappeared. But this is not due to some disturbance of the system, but only to the theoretical distinguishability between the two paths, independently from whether D_i is used or not.

17.2 Zurek's Solution

17.2.1 General Features

Zurek's starting point is Everett's theory of Relative states [see theorem 15.1: p. 246]. Zurek showed [section 15.2] that, according to this theory, the apparatus $\mathcal{A}$, due to its being correlated with the state of system $\mathcal{S}$, not only contains all the information about the measured observable, but it must also contain all the information regarding many other observables in spite of the fact that generally the latter ones do not commute with the former, which in QM is certainly impossible (basis degeneracy problem). We can formulate Zurek's conclusion regarding Everett's proposal by means of the following theorem [ZUREK 1981, 1516, 1519]:

Theorem 17.3 (Zurek) *QM, when applied to an isolated composed object consisting of an apparatus and a system, cannot in principle determine which observable has been measured.*

The central problem is finding a pointer basis (a preferred basis) in which the observable of interest, and only this, can be measured[9]. Now what traditionally — only implicitly in the Standard interpretation (because it built no theory of the pointer basis), and explicitly in the MWI — was named a 'measurement' is in reality a premeasurement [see definition 13.2: p. 216],

[9]See also [WALLS *et al.* 1985, 3208–209].

in which we only couple $\mathcal{A}$ and $\mathcal{S}$ [see for example Eq. (15.6)]. The *Measurement* as such is performed when a third factor, the *environment* $\mathcal{E}$, comes into interaction with $\mathcal{A}$ [ZEH 1970, 346] [PERES/ZUREK 1982, 808].[10]

Previously the influence of $\mathcal{E}$ was normally considered a noise factor. Instead, we can understand the importance of this factor by formulating the following theorem (which synthesizes the analysis of both Zeh and Zurek):

Theorem 17.4 (Zeh/Zurek) *The Environment makes all information about the premeasured system unavailable with only one exception: when the Hamiltonian $\hat{H}_{\mathcal{AS}}$, which couples $\mathcal{A}$ and $\mathcal{S}$, commutes with an observable $\hat{O}^{\mathcal{S}}$* [see theorem 14.5: p. 234]:

$$\boxed{[\hat{O}^{\mathcal{S}}, \hat{H}_{\mathcal{AS}}] = 0} \tag{17.18}$$

then this particular observable will not be disturbed, so that the pointer of the apparatus will contain the information about this observable and only this one.

The theorem guarantees the required correlation between the system's and the apparatus' observables [see theorem 14.10: p. 241]. Obviously, such a theorem cannot be understood in the sense of a classicality of the pointer, due to the already analysed problems [corollary 16.1: p. 256]. In fact, as we shall see below, interference terms are never destroyed and other observables which do not commute with the one measured, always enter, to a certain extent, in the result which is never 100% 'determined' — this is the POVM problematic [see chapter 18]. We must also have the following requirement for the pointer observable[11] (which at this level can be treated quantum-mechanically):

Corollary 17.1 (Zurek) *The pointer observable $\hat{O}^{\mathcal{A}}$ commutes with the Hamiltonian $\hat{H}_{\mathcal{AS}}$:*

$$\boxed{[\hat{O}^{\mathcal{A}}, \hat{H}_{\mathcal{AS}}] = 0} \tag{17.19}$$

The combined $\mathcal{A}+\mathcal{S}$ system is now represented by a mixture which is diagonal in a particular product basis consisting of the eigenvectors of the pointer of $\mathcal{A}$ and the corresponding relative states of $\mathcal{S}$ [ZUREK 1981, 1519, 1522] [WALLS *et al.* 1985, 3209]. Therefore, only with the influence of $\mathcal{E}$ do we obtain the required pointer objectification [theorem 14.9: p. 239].

We now prove in a general form that, while biorthogonal decomposition (system + apparatus) suffers the basis degeneracy problem, this is not so for triorthogonal decomposition [ELBY/BUB 1994, 4215–16].[12] First let us state the following lemma:

Lemma 17.2 (Elby/Bub) *Let $\{|a_j\rangle\}$ and $\{|s_j\rangle\}$ be linearly independent sets of vectors, respectively in $\mathcal{H}_1, \mathcal{H}_2$ for two generic systems $\mathcal{S}_1, \mathcal{S}_2$. Let $\{|s'_j\rangle\}$ be a linearly independent set of vectors that differs non trivially from $\{|s_j\rangle\}$. If $|\Psi\rangle = \sum_j c_j|a_j\rangle \otimes |s_j\rangle$, then $|\Psi\rangle = \sum_j c'_j|a'_j\rangle \otimes |s'_j\rangle$ only if at least one of the $\{|a'_j\rangle\}$ vectors is a linear combination of (at least two) $\{|a_j\rangle\}$ vectors.*

[10] For other quotations see [O'CONNELL *et al.* 1986, 267–69, 272]. See also [FEYNMAN/VERNON 1963], where the influence of the environment on the interference of two paths is expressed in terms of a functional of the two paths, which is given in terms of a sum over paths of the environment's degrees of freedom.

[11] Zurek was evidently influenced by the theory of QND-Measurement [see Eq. (19.40)].

[12] See also [*BUB* 1997, 164–67].

We omit the proof of the Lemma[13], but use it to prove *per contradictionem* the uniqueness of triorthogonal decomposition (in order to indirectly prove theorem 17.4 and its corollary, so that the required pointer objectification can be obtained).

Proof

Suppose a vector $|\Psi\rangle = \sum_j c_j|a_j\rangle\otimes|s_j\rangle\otimes|e_j\rangle$, where $\{|a_j\rangle\}, \{|s_j\rangle\}, \{|e_j\rangle\}$ are orthogonal sets of vectors respectively in $\mathcal{H}_1, \mathcal{H}_2, \mathcal{H}_3$ for three generic systems $\mathcal{S}_1, \mathcal{S}_2, \mathcal{S}_3$. Then, we claim that even if some of the $|c_j|$'s are equal, no alternative orthogonal sets $\{|a'_j\rangle\}, \{|s'_j\rangle\}, \{|e'_j\rangle\}$ exist such that $|\Psi\rangle = |a'_j\rangle\otimes|s'_j\rangle\otimes|e'_j\rangle$, unless each alternative set of vectors differs only trivially from the set it replaces. Assume, without loss of generality, that $\{|e_j\rangle\}$ differs not trivially from $\{|e'_j\rangle\}$, and let us write $|\Psi\rangle = \sum_j c_j|f_j\rangle\otimes|e_j\rangle$ (where $|f_j\rangle := |a_j\rangle\otimes|s_j\rangle$). Now supposing $|\Psi\rangle = \sum_j c'_j|f'_j\rangle\otimes|e'_j\rangle$ (where $|f'_j\rangle := |a'_j\rangle\otimes|s'_j\rangle$), we cannot rewrite the factorisable state $|a'_j\rangle\otimes|s'_j\rangle$ as an entangled state [see theorem 3.3: p. 43].

But according to lemma 17.2, since $|\Psi\rangle = \sum_j c_j|f_j\rangle\otimes|e_j\rangle$ and since $\{|e_j\rangle\}$ differs not trivially from $\{|e'_j\rangle\}$, then we have $|\Psi\rangle = \sum_j c'_j|f'_j\rangle\otimes|e'_j\rangle$ only if $|f'_k\rangle = \sum_j g_{jk}|f_j\rangle$, where at least two of the g_{jk}'s are non-zero. But since $|f_j\rangle := |a_j\rangle\otimes|s_j\rangle$, it follows that $|f'_k\rangle$ is an entangled state ($|f'_k\rangle = \sum_j g_{jk}|a_j\rangle\otimes|s_j\rangle$), which is the required contradiction. Q.E.D.

The above proof shows that one should not identify[14] the triorthogonal decomposition presupposed by decoherence with the von Neumann chain [see p. 226]: in the second case, only successive measurements, in which each decomposition is biorthogonal, are considered. Note that, while the partial trace is a mathematical formalism — to be better analysed in the following subsection — which reflects our ignorance (our 'discarding') of the environment and that is therefore relative to the subsystem over which we perform such a partial trace, the triorthogonal decomposition — due to the introduction of a third factor such the environment — is not a relative property (not relative to an observer) but an absolute one[15]. We shall return to this problem later [see subsection 25.2.3].

17.2.2 General Formalism

We can represent Measurement in the following form [ZUREK 1982, 1863]. Let $\mathcal{S}$ be a system of the form: $|\varsigma\rangle = \sum c_n|n\rangle$; then we have:

$$\begin{aligned} |\Psi(t=t_0)\rangle &= |\varsigma\rangle\otimes|a_0\rangle\otimes|\mathcal{E}(t_0)\rangle && (17.20a)\\ &\mapsto |\Psi(t=t_1)\rangle = \left[\sum_n c_n(|n\rangle\otimes|a_n\rangle)\right]\otimes|\mathcal{E}(t_1)\rangle && (17.20b)\\ &\mapsto |\Psi(t>t_2) = \sum_n c_n|n\rangle\otimes|a_n\rangle\otimes|\mathcal{E}_n(t)\rangle. && (17.20c)\end{aligned}$$

The first transition [Eq. (17.20b)] has provided $\mathcal{A}$ with information about $\mathcal{S}$ (it is what we named *Premeasurement*) [see definition 13.2: p. 216] and the second one [eq (17.20c)] is indispensable in order to define the measured observable [see Eq. (17.18)]. If we express the process by using the reduced density matrix, we have:

$$\begin{aligned} \hat{\rho}^{\mathcal{S}+\mathcal{A}+\mathcal{E}}_{\mathrm{Tr}_{\mathcal{E}}} &= \mathrm{Tr}_{\mathcal{E}}\left[|\Psi(t>t_2)\rangle\langle\Psi(t>t_2)|\right]\\ &\simeq \sum_n |c_n|^2|a_n\rangle\langle a_n|\otimes|n\rangle\langle n|\\ &= \tilde{\hat{\rho}}^{\mathcal{S}+\mathcal{A}+\mathcal{E}}_{\mathrm{Tr}_{\mathcal{E}}} \end{aligned} \qquad (17.21)$$

[13]See the original article of Elby and Bub.
[14]As sometimes happens: see [*MITTELSTAEDT* 1998, 112].
[15]See also [*BUB* 1997, 159].

Hence we have a change from an initial (pure) state to a mixture by tracing out the environment.

17.2.3 Characters of the Measurement Process

Proper and improper mixtures Clearly the transition given by Eq. (17.21) [see also Eq. (17.36)] is not a unitary evolution. We have already discussed the impossibility of overcoming the Measurement problem by an unitary evolution [see theorems 14.13: p. 243, and 14.1: p. 221]. Hence we can summarize the results of Shea/Scully/McCullen [see subsection 17.1.2] and of Zurek in the following theorem [see Eqs. (3.103) and (3.104)]:

Theorem 17.5 (Shea/Scully/McCullen/Zurek) *The Measurement-required reduction can only be obtained with a partial trace, which* per definitionem *is not unitary.*

The result of such a process is a mixture, as we have seen. But it is very important to distinguish between a mixture such as that of Eq. (17.21) — which is an 'improper' mixture, because it is generated by a partial trace with respect to a total system which remains in a pure state — and a proper mixture [see p. 52].[16] Note that the Statistical interpretation [see subsections 6.5.2 and 15.1.1] does not distinguish between them.

No destruction of off-diagonal terms Therefore, summarizing the analysis of Zeh [ZEH 1993] and Joos [JOOS 1996, 43–44, 115–24],[17] we can formulate following theorem:

Theorem 17.6 (Joos/Zeh) *Correlations, or off-diagonal terms of a density matrix, cannot be destroyed in the $\mathcal{S}+\mathcal{A}$ system as a whole, but only downloaded into the environment.*

This theorem guarantees the necessary correlation between values [see theorem 14.11: p. 242]. We shall return to the problem in more detail later [section 42.3]. Note also that triorthogonal decomposition together with this theorem has the effect that diagonalization is *objective* but *never fully realized*— this will again be the object of a later discussion [in subsection 25.2.3]. However, at least a short discussion of the point is necessary.

After Zurek, $\mathcal{E}$ acts as an SSR [see subsection 10.1.4] because we have:

- determination (diagonalization) of the observable recorded by $\mathcal{A}$,
- and transition from a pure state to a mixture.

So that Zurek speaks of environment-induced SSRs. But there is also an important difference: according to Wigner and co-workers [WICK *et al.* 1952] , SSRs operate between subspaces of a Hilbert Space of a given system if the phase factors (correlations) between vectors belonging to two distinct subspaces are unobservable, which implies that an operator which has eigenstates

[16] On this point see [*D'ESPAGNAT* 1976, 58–61] [*D'ESPAGNAT* 1995, 101–106] [JOOS 1996, 37, 42].

[17] See also [*BUSCH et al.* 1991, 124, 130].

composed of superpositions of state vectors contained in two or more subspaces cannot be measured. Here, because of decoherence, the phase coherence between two eigenspaces of the pointer observable seems to be destroyed by the interaction or the correlation between $\mathcal{A}$ and $\mathcal{E}$, so that it is this interaction which forces $\mathcal{A}$ to be in one of the eigenstates of the pointer observable and not in some arbitrary superposition of these.

But it is not appropriate to speak of SSRs. SSRs act with sharp effects, whereas decoherence supposes a smooth action of $\mathcal{E}$. If the theorem 17.6 is correct, then we have the consequence that the off-diagonal terms cannot completely vanish even locally (in the reduced state), but that there must always be some connection between the system and and its 'eliminated' part [JOOS 1996, 49–54].[18] In other words we have the following corollary of theorem 17.6:

Corollary 17.2 (Off-diagonal terms) *The off-diagonal terms of the object system tend to zero in a Measurement process but they can never really be zero.*

In fact, if there were SSRs, then the pointer would be completely classical, which has already been excluded [see section 16.1, and particularly corollary 16.1: p. 256]. We shall return to the very great importance of this corollary also later [see chapters 21, especially subsection 21.2.3, and 24].

In general if we use SSRs, we make use of the orthogonality condition for measurements [see subsection 3.5.3 and postulate 14.1], which makes decoherence too close to von Neumann's approach — and hence exposed to the same criticism[19] — and too far away from the best results obtained by Quantum Non-Demolition Measurement and by Stochastic theories [see chapters 18 and 19].

But it is always possible to speak in a special way of *systems obeying SSRs* if we have the following situation: $\hat{H}_{\mathcal{SE}}$ may have the property of commuting with one of the subspaces $\mathcal{H}_n^{\mathcal{S}}$ of the Hilbert space $\mathcal{H}^{\mathcal{S}}$ of $\mathcal{S}$. The result of the interaction of $\mathcal{A}+\mathcal{S}$ with $\mathcal{E}$ is that the state vector of $\mathcal{S}$ is able to remain pure only if it is completely confined to one of the $\mathcal{H}_n^{\mathcal{S}}$. As a consequence, if the correlation between $\mathcal{S}$ and $\mathcal{E}$ is stronger than that between $\mathcal{S}$ and $\mathcal{A}$, the set of observables that can be measured is limited to those which leave subspaces $\mathcal{H}_n^{\mathcal{S}}$ invariant. So $\mathcal{S}$ behaves 'classically' [ZUREK 1982, 1867–68].

Conclusive theorem At the end of this analysis, one can understand why theorems 15.1 [p. 246] and 16.2 [p. 256] represent only necessary but not sufficient conditions for obtaining a determinate measurement result satisfying the probability reproducibility condition (14.46) and the objectification requirement [theorem 14.8: p. 238]. In other words, we obtain a necessary and a sufficient condition by adding theorems 17.4 and 17.5 to these two. Hence we state the following central theorem[20]:

---

[18]See also [*BUSCH et al.* 1991, 133] [*D'ESPAGNAT* 1995, 180–85]. Note that Zurek later [ZUREK 1993a] gave the assumption of SSRs up.

[19]See for example [KONO *et al.* 1996, 1065] [NAMIKI/PASCAZIO 1993, 348].

[20]However, as already said, for covariance requirement, it is better if we substitute theorem 18.3 [p. 298] to theorem 15.1.

Theorem 17.7 (Necessary and Sufficient Conditions for Measurement) *Necessary and sufficient condition in order to satisfy the probability reproducibility condition and the objectification requirement are the following (where* T *stays for 'theorem',* VR *for 'value reproducibility condition' and* O *for 'objectification requirement'):*

$$(\text{T15.1} \wedge \text{T14.3} \wedge \text{T16.2} \wedge \text{T17.4} \wedge \text{T17.5}) \leftrightarrow (\text{VR} \wedge \text{O}). \tag{17.22}$$

Note that the apparatus is a necessary condition *of a measurement* and not of *decoherence* as such. As we shall see [chapter 21 and also subsection 17.2.5], there are models of spontaneous decoherence which consider only the action of the environment on a system, which, therefore, is alone a sufficient condition for decoherence. The apparatus only comes in play if one wishes to detect the system and to record the result [see section 13.3].

17.2.4 Detailed Analysis of the Decoherence Process

Bases for spin We now want to analyse the process in more details. Let us consider [ZUREK 1981, 1517–19] [ZUREK 1982, 1864–67] a pair of two-state systems, a spin-particle (the system $\mathcal{S}$) and one atom (the apparatus $\mathcal{A}$). A possible basis of the spin is $|\uparrow\rangle, |\downarrow\rangle$, parallel or antiparallel to the z axis. Alternative orthonormal basis are for example:

$$|\nearrow\rangle = \frac{1}{\sqrt{2}}(|\uparrow\rangle + |\downarrow\rangle), \tag{17.23a}$$

$$|\searrow\rangle = \frac{1}{\sqrt{2}}(|\uparrow\rangle - |\downarrow\rangle); \tag{17.23b}$$

or

$$|\rightarrow\rangle = \frac{1}{\sqrt{2}}(|\uparrow\rangle + \imath|\downarrow\rangle), \tag{17.24a}$$

$$|\leftarrow\rangle = \frac{1}{\sqrt{2}}(|\uparrow\rangle - \imath|\downarrow\rangle). \tag{17.24b}$$

Bases for the atom The basis of the atom consists in a ground state $|g\rangle$ and an excited state $|e\rangle$. We assume that the energy in each of these states is identical: neither the atom nor the spin possesses self-Hamiltonians. Alternative states of the atom could be, for example:

$$|+\rangle = \frac{1}{\sqrt{2}}(|e\rangle + |g\rangle), \tag{17.25a}$$

$$|-\rangle = \frac{1}{\sqrt{2}}(|e\rangle - |g\rangle); \tag{17.25b}$$

or

$$|\perp\rangle = \frac{1}{\sqrt{2}}(|e\rangle + \imath|g\rangle), \tag{17.26a}$$

$$|\top\rangle = \frac{1}{\sqrt{2}}(|e\rangle - \imath|g\rangle). \tag{17.26b}$$

Coupling Now we can write the interaction Hamiltonian between $\mathcal{A}$ and $\mathcal{S}$ in the following form:

$$\hat{H}_{\mathcal{SA}} = g_{\mathcal{SA}}(|\bot\rangle\langle\bot| - |\top\rangle\langle\top|) \otimes (|\uparrow\rangle\langle\uparrow| - |\downarrow\rangle\langle\downarrow|), \tag{17.27}$$

acting over the time interval $t_i = \pi\hbar/4g$, where $g_{\mathcal{SA}}$ is a coupling constant. It transforms the initial, direct product

$$|\psi_i\rangle = (c_1|\uparrow\rangle + c_2|\downarrow\rangle) \otimes |+\rangle \tag{17.28}$$

into a correlated state vector

$$|\psi_f\rangle = c_1|\uparrow\rangle \otimes |e\rangle + c_2|\downarrow\rangle \otimes |g\rangle, \tag{17.29}$$

which is certainly a pure state. But here $\mathcal{A}$ has not yet 'decided' whether the outcome of the measurement is $|\uparrow\rangle$ or $|\downarrow\rangle$. As we remember from the relative state discussion [section 15.2], the final state can be written, for example, in the following form:

$$|\psi_f\rangle = \frac{1}{\sqrt{2}}(|\nearrow\rangle \otimes |+\rangle + |\searrow\rangle \otimes |-\rangle), \tag{17.30}$$

from which we see that we do not already have the information about what observable of the spin $\mathcal{A}$ was supposed to record.

Environment Now consider an environment $\mathcal{E}$ consisting in N two-level atoms, the k–th of which has a Hilbert space spanned by the basis $\{|g\rangle_k, |e\rangle_k\}$ — to avoid confusion we will indicate the bras and kets of the atom $\mathcal{A}$ with corresponding labels. The apparatus-environment interaction hamiltonian $H_{\mathcal{AE}}$ separates as follows:

$$\hat{H}_{\mathcal{AE}} = \sum_k \hat{H}_{\mathcal{AE}_k}, \tag{17.31}$$

of which each individual component has the form:

$$\hat{H}_{\mathcal{AE}_k} = g_k^{\mathcal{AE}}(|e\rangle\langle e| - |g\rangle\langle g|)_{\mathcal{A}} \otimes (|e\rangle\langle e| - |g\rangle\langle g|)_k \prod_{j\neq k} \otimes \hat{I}_j, \tag{17.32}$$

where, in the following, we write the coupling $g_k^{\mathcal{AE}}$ in the easier form: g_k.

Now [recall Eq. (17.20)], we can rewrite the evolution considering the whole $\mathcal{S}+\mathcal{A}+\mathcal{E}$ system for the initial state of the form:

$$|\Psi(t_1)\rangle = |\psi_f\rangle \prod_{k=1}^{N} \otimes[\eta_k|e\rangle_k + \zeta_k|g\rangle_k], \tag{17.33}$$

where $|\psi_f\rangle$ is given by Eq. (17.29). At an arbitrary time t, substituting Eq. (17.29) into Eq. (17.33), we write the established correlation between $\mathcal{A}$ and $\mathcal{E}$ in the form (where for the sake of simplicity $\hbar = 1$):

$$\begin{aligned} |\Psi(t)\rangle &= c_1|\uparrow\rangle \otimes |e\rangle_{\mathcal{A}} \prod_{k=1}^{N} \otimes[\eta_k e^{ig_k t}|e\rangle_k + \zeta_k e^{-ig_k t}|g\rangle_k] \\ &\quad + c_2|\downarrow\rangle \otimes |g\rangle_{\mathcal{A}} \prod_{k=1}^{N} \otimes[\eta_k e^{-ig_k t}|e\rangle_k + \zeta_k e^{ig_k t}|g\rangle_k]. \end{aligned} \tag{17.34}$$

The measurement The observable $\hat{O}^{\mathcal{A}}$ of $\mathcal{A}$ — the pointer observable —, which is recorded by $\mathcal{E}$, has the form:

$$\hat{O}^{\mathcal{A}} = o_1^{\mathcal{A}}(|e\rangle\langle e|)_{\mathcal{A}} + o_2^{\mathcal{A}}(|g\rangle\langle g|)_{\mathcal{A}}, \tag{17.35}$$

with $o_1^{\mathcal{A}}, o_2^{\mathcal{A}}$ real and different. We can see the measurement process from the following point of view: it can be thought of as a continuous measurement by $\mathcal{E}$ of the pointer observable $\hat{O}^{\mathcal{A}}$ [see corollary 17.1: p. 269]. Now, only observables which commute with $\hat{O}^{\mathcal{A}}$ can successfully be measured or prepared [see figure 17.3].

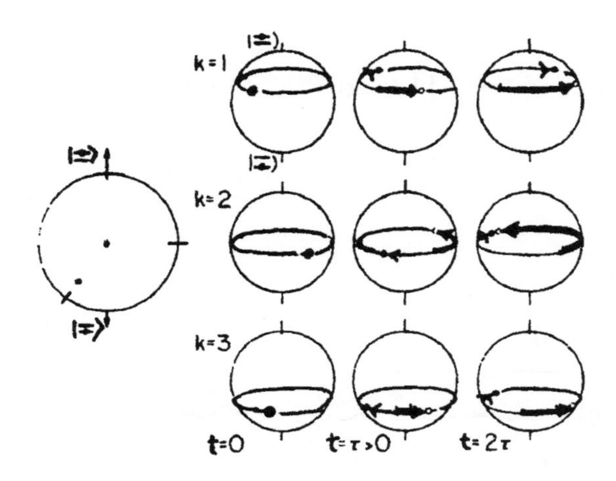

Figure 17.3: Pictorial representation of the environment atoms 'monitoring' the state of the apparatus atom. Resulting apparatus-environment correlations replace some of the correlations between the eigenstate of the pointer observable. The pure states of each of the three ($k = 1, 2, 3$) environment atoms, described by a state vector in a two-dimensional Hilbert space, are represented by points on the surface of the sphere. The initial state of the apparatus atom is represented by a point on the surface of the big sphere on the left. At $t = 0$ (first column of spheres) none of the environment atoms are correlated with the state of the apparatus atom. At $t = \tau > 0$ (second column) the state of each atom is rotated by an angle $\phi_k = \pm g_k t$ around the axis — defined by the k–th eigenstate represented on the large sphere on the left —, which diagonalizes the interaction Hamiltonian. Counterclockwise rotations are correlated with the 'up' state and marked by heavy arrows; clockwise ones are correlated with the 'down' state and marked by light arrows. At $t = 2\tau$ (third column), the states of the environment atoms rotate still further. This evolution can be thought of as a continuous measurement by the environment — from [ZUREK 1982, 1867].

We now calculate the density matrix of $\mathcal{A} + \mathcal{S}$ by performing a partial tracing out of the environment on Eq. (17.34):

$$\begin{aligned}\tilde{\hat{\rho}}^{\mathcal{S}+\mathcal{A}+\mathcal{E}}_{\mathrm{Tr}_{\mathcal{E}}} &= \mathrm{Tr}_{\mathcal{E}}[|\Psi(t)\rangle\langle\Psi(t)|] \\ &= |c_1|^2|\uparrow\rangle\langle\uparrow| \otimes (|e\rangle\langle e|)_{\mathcal{A}} + \vartheta_{\mathcal{SA}}(t)c_1c_2^*|\uparrow\rangle\langle\downarrow| \otimes (|e\rangle\langle g|)_{\mathcal{A}} \\ &\quad + \vartheta^*_{\mathcal{SA}}(t)c_1^*c_2|\downarrow\rangle\langle\uparrow| \otimes (|g\rangle\langle e|)_{\mathcal{A}} + |c_2|^2|\downarrow\rangle\langle\downarrow| \otimes (|g\rangle\langle g|)_{\mathcal{A}},\end{aligned} \tag{17.36}$$

where $\vartheta_{\mathcal{SA}}(t)$ is the following correlation amplitude:

$$\vartheta_{\mathcal{SA}}(t) = \prod_{k=1}^{N}[\cos 2g_k t + \imath(|\eta_k|^2 - |\zeta_k|^2)\sin 2g_k t]. \tag{17.37}$$

The correlation amplitude $\vartheta_{\mathcal{SA}}(t)$ depends on the initial conditions of $\mathcal{E}$ only via the probabilities of finding $\mathcal{S}$ in the eigenstates of the interaction hamiltonian: $\wp(|1\rangle_k) = |\eta_k|^2, \wp(|0\rangle_k) = |\zeta_k|^2$. This property indicates that the ability of $\vartheta_{\mathcal{SA}}(t)$ to damp out correlations will be the same as for a mixture for which only $\wp(|1\rangle_k), \wp(|0\rangle_k)$ can be given. Moreover $\vartheta_{\mathcal{SA}}(t)$ is equal to the scalar product:

$$\vartheta_{\mathcal{SA}}(t) = \langle \mathcal{E}_{|1\rangle}(t)|\mathcal{E}_{|0\rangle}(t)\rangle, \tag{17.38}$$

where we define:

$$|\mathcal{E}_{|1\rangle}(t)\rangle := \prod_{k=1}^{N} \otimes[\eta_k e^{\imath g_k t}|e\rangle_k + \zeta_k e^{-\imath g_k t}|g\rangle_{\mathcal{A}}], \tag{17.39a}$$

$$|\mathcal{E}_{|0\rangle}(t)\rangle := \prod_{k=1}^{N} \otimes[\eta_k e^{-\imath g_k t}|e\rangle_k + \zeta_k e^{\imath g_k t}|g\rangle_{\mathcal{A}}]. \tag{17.39b}$$

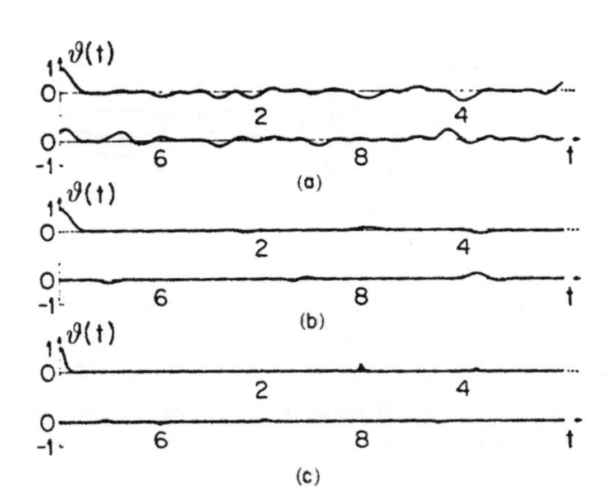

Figure 17.4: Evolution of the correlation-damping factor $\vartheta = \vartheta_{\mathcal{SA}}$, which is calculated as a function of time in three different environment sizes: (a) $N = 5$, (b) $N = 10$, (c) $N = 15$ environment atoms. For simplicity $\vartheta(t) = \prod_{k=1}^{N} \cos 2g_k t$. Constants g_k (proportional to eigenvalues of the interaction Hamiltonian) are chosen at random. Note that the fluctuations of $\vartheta(t)$ decrease their size and frequency with the increase of N — from [ZUREK 1982, 1867].

The states of Eqs. (17.39) represent two distinct records made by the environment of the two alternative outcomes of measurement. The damping of off-diagonal terms by $\vartheta_{\mathcal{SA}}(t)$ is time dependent:

$$\vartheta_{\mathcal{SA}}(t = 0) = 1, \tag{17.40a}$$

$$|\vartheta_{\mathcal{SA}}(t)|^2 \leq 1, \tag{17.40b}$$

$$\langle \vartheta_{\mathcal{SA}}(t)\rangle = \lim_{\mathcal{T}\to\infty} \mathcal{T}^{-1} \int_0^{\mathcal{T}} \vartheta_{\mathcal{SA}}(t)dt = 0, \tag{17.40c}$$

$$\left\langle |\vartheta_{\mathcal{SA}}(t)|^2 \right\rangle = 2^{-N} \prod_{k=1}^{N} [1 + (|\eta_k|^2 - |\zeta_k|^2)^2]. \tag{17.40d}$$

Eq. (17.40) implies that unless the initial state of $\mathcal{E}$ coincides with one of the eigenstates of the interaction hamiltonian, the expected absolute value $|\vartheta_{\mathcal{SA}}(t)|^2$ is much less than the initial value of unity [see figures 17.4 and 17.5].[21]

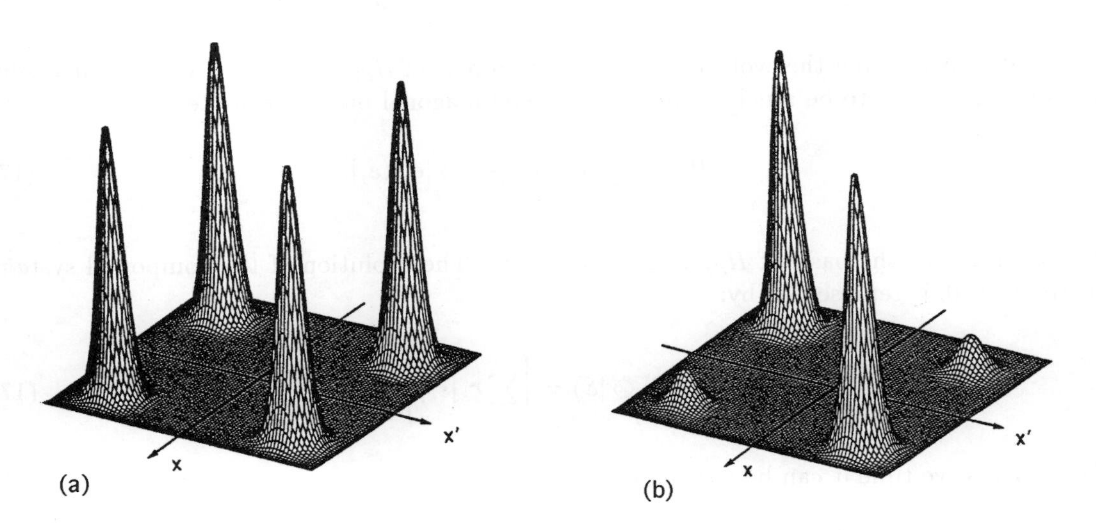

Figure 17.5: (a) Representation of a density matrix $\hat{\rho}(x, x')$. The peaks near the diagonal represent possible locations of the particle, while the right and left peaks (off-diagonal) are due to quantum coherence. Their existence demonstrates that the particle is in none of the two possible positions, but in a superposition of them.
(b) Partial decoherence. Full decoherence (with no off-diagonal peaks) would give us a classical probability distribution with an equal probability of finding the particle in either of the possible locations, corresponding to Gaussian wave-packets — from [ZUREK 1991, 41].

17.2.5 The Action of a Large Environment

The model As we shall see [chapter 21], with the decoherence theory of Measurement we can explain the classical nature of a wide class of macrophysical objects. We shall now very briefly see some general features of the effects of large environments on the decoherence process. Consider the combined Hilbert space of $\mathcal{S}$ and $\mathcal{E}$ [ZUREK 1982, 1868–71]:

$$\mathcal{H}_{\mathcal{S}\oplus\mathcal{E}} = \mathcal{H}_{\mathcal{S}} \otimes \mathcal{H}_{\mathcal{E}}. \tag{17.41}$$

We suppose the Hamiltonian of the compound system to be separable in three distinct components:

$$\hat{H}_{\mathcal{S}} = \sum_k |s_k\rangle\langle s_k|, \tag{17.42a}$$

[21] Easy expositions of the decoherence process are to be found in [JOOS 1996, 40–42, 50–51, 119–24] and in [ZUREK 1993a].

$$\hat{H}_{\mathcal{E}} = \sum_j |e_j\rangle\langle e_j|, \tag{17.42b}$$

$$\hat{H}_{\mathcal{SE}} = \sum_{jk} c_{jk}|s_k\rangle\langle s_k| \otimes |e_j\rangle\langle e_j| + c_{\mathcal{SE}} \sum_k \sum_{k'} \sum_j \sum_{j'} d_{kk'jj'}|s_k\rangle\langle s_{k'}| \otimes |e_j\rangle\langle e_{j'}|. \tag{17.42c}$$

We shall only consider the evolution of the diagonal part of $H_{\mathcal{SE}}$, so that we set $c_{\mathcal{SE}} = 0$, assuming the diagonal terms to be much greater than the off-diagonal ones, and write:

$$\hat{H}^0_{\mathcal{SE}} = \sum_{jk} c_{jk}|s_k\rangle\langle s_k| \otimes |e_j\rangle\langle e_j|. \tag{17.43}$$

We assume that the basis of $H^0_{\mathcal{SE}}$ and $H_{\mathcal{S}}$ coincide. The evolution of the compound system, if, at time $t = 0$, is represented by:

$$|\Psi(t=0)\rangle = |\mathcal{S}\rangle \otimes |\mathcal{E}\rangle = \left[\sum_k c_k|s_k\rangle\right] \otimes \left[\sum_j c'_j|e_j\rangle\right]; \tag{17.44}$$

for a successive time it can be written as:

$$|\Psi(t)\rangle \sum c_k c'_j e^{-\imath t(\delta_k+\epsilon_j+c_{jk})}|s_k\rangle \otimes |e_j\rangle. \tag{17.45}$$

Partial trace To show how decoherence arises we calculate the reduced density matrix:

$$\tilde{\hat{\rho}}^{\mathcal{S}+\mathcal{E}}_{\mathrm{Tr}_{\mathcal{E}}}(t) = \mathrm{Tr}_{\mathcal{E}}|\Psi(t)\rangle\langle\Psi(t)|, \tag{17.46}$$

which can be written as:

$$\tilde{\hat{\rho}}^{\mathcal{S}+\mathcal{E}}_{\mathrm{Tr}_{\mathcal{E}}}(t) = \sum_k \sum_j \hat{\rho}_{jk}|s_k\rangle\langle s_j|. \tag{17.47}$$

The elements of $\tilde{\hat{\rho}}^{\mathcal{S}+\mathcal{E}}_{\mathrm{Tr}_{\mathcal{E}}}$ are:

$$\tilde{\hat{\rho}}_{kk}(t) = |c_k|^2 \sum_j |c'_j|^2 = |c_k|^2, \tag{17.48a}$$

$$\hat{\rho}_{kl}(t) = c_k c^*_l e^{-\imath t(\delta_k-\delta_l)} \sum_j |c'_j|^2 e^{-\imath t(c_{jk}-c_{jl})}. \tag{17.48b}$$

The off-diagonal elements, in the representation in which $\hat{H}^0_{\mathcal{SE}}$ is diagonal, evolve in time in two ways:

- they rotate trivially due to the factor $e^{-\imath t(\delta_k-\delta_j)}$;
- and they decay as a result of the decrease of the correlation amplitude:

$$\vartheta_{kl}(t) = \sum_j |c'_j|^2 e^{-\imath t(c_{jk}-c_{jl})}. \tag{17.49}$$

Correlation amplitude The consequence is that the absolute value of $\vartheta(t)$ is reduced from the initial value $\vartheta(t=0)=1$ to $\vartheta(t) \ll 1$ for large values of t. The average correlation amplitude taken over a sufficiently long time interval approaches zero unless all the differential frequencies:

$$\omega_j^{kl} = c_{jk} - c_{jl} \tag{17.50}$$

are equal to zero; this is equivalent to the statement that $\hat{H}^0_{\mathcal{SE}}$ has the diagonal elements equal to zero. We rewrite the correlation amplitude as

$$\vartheta_{kl}(t) = \sum_j \wp_j e^{-\imath\omega_j^{kl}t}, \tag{17.51}$$

where $\wp_j = |c'_j|^2$ expresses the probability of finding $\mathcal{E}$ in the states corresponding to the distinct eigenvalues of $\hat{H}^0_{\mathcal{SE}}$.

If we now calculate the average absolute value of $\vartheta_{kl}(t)$:

$$\left\langle|\vartheta_{kl}(t)|^2\right\rangle = \frac{1}{\tau}\int_t^{t+\tau} |\vartheta_{kl}(t)|dt, \tag{17.52}$$

we see that as $\tau \longrightarrow \infty$, it tends to:

$$\sum_{j,j'} \wp_j \wp_{j'} \delta(\omega_j^{kl} - \omega_{j'}^{kl}). \tag{17.53}$$

The standard deviation of the correlation amplitude from the average value $\langle\vartheta_{kl}\rangle = 0$ is equal to $\sum_{j=1}^N \wp_j^2$, so that average fluctuations from zero, in the environment of N active states, are approximately equal to $1/\sqrt{N}$. In conclusion, large environments can effectively damp out correlations between those states of the system which diagonalizes $\hat{H}^0_{\mathcal{SE}}$.

17.3 Machida-Namiki Model

General features The influence of the thermodynamic approach on this model is evident. But the authors largely surpass the latter by using the S-matrix formalism [see subsection 3.11.3] to perform the required reduction: due to a phase shift, the cross-correlation parts of the asymptotic total statistical operator have a vanishing integral, from which the reduction. In other words, phase shifts in the interference integral never correlate for $j \neq k$, so that, after integration over state variables on a microscopic scale, all 'cross integrals' in Eq. (17.58) are zero [MACHIDA/NAMIKI 1980, 1464–71] [MACHIDA/NAMIKI 1984, 81, 85–87].

In more details, the model is the following [MACHIDA/NAMIKI 1980, 1837–39].[22] The object system is in an initial state given by:

$$|\varsigma_i\rangle = \left[\sum_j c_j|s_j\rangle\right]|\psi\rangle, \tag{17.54}$$

where $|s_j\rangle$ are eigenvectors of the observable $\hat{O}^{\mathcal{S}}$ to be measured and $|\psi\rangle$ stands for the spatial wave function representing a localized system. The preparation step brings $|\varsigma_i\rangle$ into

$$|\varsigma\rangle = \sum_j c_j|s_j\rangle|\psi_j\rangle, \tag{17.55}$$

[22] See also [NAMIKI/PASCAZIO 1993, 332–41].

where $|\psi_j\rangle$ is the spatial wave function moving toward the j-th detector. Different $|\psi_j\rangle, |\psi_k\rangle$ do not overlap in space. The statistical operator corresponding to Eq. (17.55), which represents the state of $\mathcal{S}$ before detection, is:

$$\hat{\rho}^{\mathcal{S}} = \sum_i |c_j|^2 |j\rangle\langle j| + \sum_{j\neq k} c_j c_k^* |j\rangle\langle k|, \tag{17.56}$$

where $|j\rangle = |s_j\rangle|\psi_j\rangle$. The j-th macroscopic detector is represented by:

$$\hat{\rho}^{\mathcal{A}_j}(X_j) = \int d\zeta_j w^j(\zeta_j, X_j)\hat{\rho}^{(j)}(\zeta_j), \tag{17.57}$$

where ζ_j, X_j are, respectively, variables at microscopic and macroscopic scales, and where $w^j(\zeta_j, X_j)$ is a normalized weight function. The statistical operator for $\mathcal{A}+\mathcal{S}$ before detection is given by:

$$\begin{aligned}
\hat{\rho}^{\mathcal{S}+\mathcal{A}} &= \hat{\rho}^{\mathcal{S}} \otimes \prod_{j=1}^{n} \hat{\rho}^{\mathcal{A}_j}(X_j) \\
&= \sum_j |c_j|^2 \left[\int d\zeta_j w^j(\zeta_j, X_j)[|j\rangle\langle j| \otimes \hat{\rho}^{(j)}(\zeta_j)]\right] \otimes \prod_{l\neq j} \hat{\rho}^{\mathcal{A}_l}(X_l) \\
&\quad + \sum_{j\neq k} c_j c_k^* \left[\int d\zeta_j d\zeta_k w^j(\zeta_j, X_j) w^k(\zeta_k, X_k)[|j\rangle\langle k| \otimes \hat{\rho}^{(j)}(\zeta_j) \otimes \hat{\rho}^{(k)}(\zeta_k)]\right] \\
&\quad \otimes \prod_{l\neq j,k} \hat{\rho}^{\mathcal{A}_l}(X_l).
\end{aligned} \tag{17.58}$$

Overall detection can be represented by the asymptotic time evolution ($t \longrightarrow \infty$):

$$[|j\rangle\langle j| \otimes \hat{\rho}^{(j)}] \leadsto e^{-\frac{\imath}{\hbar}\hat{H}_0 t}\hat{S}_j[|j\rangle\langle j| \otimes \hat{\rho}^{(j)}]\hat{S}_j^\dagger e^{\frac{\imath}{\hbar}\hat{H}_0 t}, \tag{17.59a}$$

$$[|j\rangle\langle k| \otimes \hat{\rho}^{(j)} \otimes \hat{\rho}^{(k)}] \leadsto e^{-\frac{\imath}{\hbar}\hat{H}_0 t}\hat{S}_j[|j\rangle\langle k| \otimes \hat{\rho}^{(j)} \otimes \hat{\rho}^{(k)}]\hat{S}_j^\dagger e^{\frac{\imath}{\hbar}\hat{H}_0 t}, \tag{17.59b}$$

of first and second terms in the RHS of Eq. (17.58), where the S-matrix $\hat{S}_j$ stands for the reaction in the j-th detector and $\hat{H}_0$ is the free Hamiltonian. The S-matrix can be written as:

$$\hat{S}_j = e^{\imath\delta_j}\frac{1+\imath\hat{K}_j}{1-\hat{K}_j}e^{\imath\delta_j}. \tag{17.60}$$

$\hat{K}_j$ gives rise to the formation of resonances and channel-coupling reactions (off-diagonal) and δ_j is a diagonal phase shift matrix which depends on the path length of the particle. Phase shift factors in Eq. (17.59a) cancel each other out, whereas those in Eq. (17.59b) do not since δ_j, δ_k never correlate for $j \neq k$. Hence Eq. (17.59b) becomes zero after integration over ζ. This integration is an averaging [see theorem 16.2: p. 256] which represents the macroscopical action of the apparatus. From here the so-called 'reduction', which in reality is a transformation due to $\mathcal{S}$ and $\mathcal{A}$'s character of open systems (i.e. due to their interaction with the environment).

Discussion What is very interesting in the Machida/Namiki model is that it acknowledges the decoherence in general as imperfect, i.e. such that we never have a complete elimination of off-diagonal terms [NAMIKI/PASCAZIO 1993, 321]. But, on the other hand, there is the risk of interpreting the Measurement process exclusively in statistical terms (not only the measuring apparatus, especially the amplifying device, but also the object system $\mathcal{S}$ itself), by excluding a description of individual events[23].

Furthermore, the statistical character of the model presents some difficulties also for the pointer. In fact, the essential aspect of a measurement is the change in the pointer's position in the apparatus, whereas the scattering process of the Machida/Namiki model would leave the scatterer practically unchanged (if measured by the inner product) — dephasing is not a transformation. Hence the model can only account for ensembles and not for individual measurements[24]. It is for these reasons that the Machida/Namiki model has been questioned [ZEH/JOOS 1996, 270–71].

We know that all new developments in Measurement theory and experiments — see for example Quantum Non-Demolition Measurement [chapter 19] — are centered on individual measurement. This weakness must be overcome, if the model is to be reconciled with Zurek's to a certain extent, which we consider not only possible but also suitable. Anyway Namiki [NAMIKI 1988a] proposed an *experimentum crucis* (an SGM one) in order to decide between Zurek's model and Machida/Namiki's one: a divider separates a wave function into beams I and II, preserving the spin; then a magnetic field performs a spectral decomposition of I and II: one part of each decomposition (I_1 and II_1) goes to a different detector, while the other two parts (I_2 and II_2) mix together. If there is an environmental influence, Namiki says, then no coherence can be found between the latter two, which is not the case according to the Namiki/Machida model. However, there are situations in which there is still coherence and the experiment can be accounted for in terms of Zurek's model: for example by partial isolation of the device.

Another (independent) model based on scattering is to be found in [JOOS/ZEH 1985, 229–31].[25]

17.4 Cini's Model

Cini's article [CINI 1983] presents another interesting model which shows how the action of large systems can cause a transition from a pure state to an improper mixture. The main idea here is that of a large number of particles in the apparatus $\mathcal{A}$, hence no explicit use of the environment concept is made (as we have said the triorthogonal decomposition is important for Measurement only and not for the decoherence as such). In this sense it is an independent model from Zurek's one. Its main result [CINI 1983, 29] is that, for its experimental arrangement, the density $\hat{\rho}^{\mathcal{A}+\mathcal{S}}$ is indistinguishable from a mixture formed by the different states representing the possible outcomes of the measurement, each one consisting in the particle in a given state + the apparatus in the corresponding macroscopically definite state. It is the macroscopic distance between the different spatial locations of the measuring devices which leads to this equivalence, justified by the very small magnitude of interference terms. Another model based on the same general features is to be found in [CINI *et al.* 1979],.

We assume [CINI 1983, 32–36] $\mathcal{S}$ to be a two-level quantum system in initial state:

$$|\mathcal{S}\rangle = c_+|\uparrow\rangle + c_-|\downarrow\rangle, \tag{17.61}$$

where $|\uparrow\rangle, |\downarrow\rangle$ are states of spin-up and spin-down, respectively. $\mathcal{A}$ is composed of N particles which can assume only two possible states: $|a_0\rangle, |a_1\rangle$. The mechanism involved in the interaction

[23] As is acknowledged by Namiki and Pascazio [NAMIKI/PASCAZIO 1993, 325].

[24] However, a recent correction due to Nakazato/Pascazio [NAKAZATO/PASCAZIO 1993] takes into account the energy exchange between the system and the apparatus.

[25] See also [RAUCH *et al.* 1990].

between $\mathcal{S}$ and $\mathcal{A}$ is a direct ionisation of the latter by the former. We suppose a type I–Measurement, i.e. the interaction does not change the state. The Hamiltonian is

$$\hat{H}_{\mathcal{S}+\mathcal{A}}(\hat{a}) = \frac{1}{2} g'_{\mathcal{SA}}(1+\hat{\sigma}_z)(\hat{a}_0^\dagger \hat{a}_1 + \hat{a}_0 \hat{a}_1^\dagger), \tag{17.62}$$

where $\hat{a}_0^\dagger, \hat{a}_0, \hat{a}_1^\dagger, \hat{a}_1$ are creation and annihilation operators on $|a_0\rangle, |a_1\rangle$ respectively and $g'_{\mathcal{SA}}$ is the coupling. A given state of the counter will be defined through the number n of particles in the neutral state $|a_0\rangle$. Hence the number $N-n$ will be the number of particles in the ionised $|a_1\rangle$–state, which is defined by:

$$|n, N-n\rangle = \frac{1}{\sqrt{n!}} \frac{1}{\sqrt{N-n!}} (\hat{a}_0^\dagger)^n (\hat{a}_1^\dagger)^{N-n} |0\rangle. \tag{17.63}$$

The only matrix elements different from zero will be the following:

$$\langle n, N-n | \langle \uparrow | \hat{H} \uparrow \rangle | n+1, N-n-1\rangle = g'_{\mathcal{SA}} \sqrt{n+1}\sqrt{N-n}, \tag{17.64a}$$

$$\langle n, N-n | \langle \uparrow | \hat{H} \uparrow \rangle | n-1, N-n+1\rangle = g'_{\mathcal{SA}} \sqrt{n}\sqrt{N-n+1}. \tag{17.64b}$$

From Eq. (17.64b) it follows that for $n = N$

$$\langle N, 0 | \langle e | \hat{H} e \rangle | N-1, 1\rangle = g'_{\mathcal{SA}} \sqrt{N}. \tag{17.65}$$

This means that, when $\mathcal{A}$'s initial state is chosen to be the neutral state $n = N$, the time τ_0 required to ionise the first particle is of the order

$$\tau_0 \simeq \frac{\hbar}{g'_{\mathcal{SA}}\sqrt{N}}. \tag{17.66}$$

Hence we redefine the coupling constant as $g'_{\mathcal{SA}} = g_{\mathcal{SA}}/\sqrt{N}$. Starting with initial state of $\mathcal{S}+\mathcal{A}$ defined by

$$|\Psi_{\mathcal{A}+\mathcal{S}}(t_0)\rangle = |\mathcal{S}\rangle \otimes |N, 0\rangle; \tag{17.67}$$

and after the following evolution

$$|\Psi_{\mathcal{S}+\mathcal{A}}(t)\rangle = e^{\frac{i}{\hbar}\hat{H}_{\mathcal{S}+\mathcal{A}}t} |\Psi_{\mathcal{S}+\mathcal{A}}(t_0)\rangle, \tag{17.68}$$

we obtain the diagonalized Hamiltonian:

$$\hat{H}_{\mathcal{S}+\mathcal{A}}(\hat{b}) = \frac{1}{2}\frac{g_{\mathcal{SA}}}{\sqrt{N}}(1+\hat{\sigma}_z)\left[\hat{b}_0^\dagger \hat{b}_0 - \hat{b}_1^\dagger \hat{b}_1\right], \tag{17.69}$$

where the following substitutions have been made in Eq. (17.62):

$$\hat{a}_0 = \frac{1}{\sqrt{2}}(\hat{b}_0 - \hat{b}_1), \qquad \hat{b}_0 = \frac{1}{\sqrt{2}}(\hat{a}_0 + \hat{a}_1), \tag{17.70a}$$

$$\hat{a}_1 = \frac{1}{\sqrt{2}}(\hat{b}_0 + \hat{b}_1), \qquad \hat{b}_1 = \frac{1}{\sqrt{2}}(\hat{a}_1 - \hat{a}_0). \tag{17.70b}$$

If we define the eigenstates of the Hamiltonian of Eq. (17.69) in terms of operators $\hat{b}_0, \hat{b}_1$ in analogy with Eq. (17.64):

$$|\eta, N-\eta\rangle|\uparrow\rangle = \frac{1}{\sqrt{\eta!}}\frac{1}{\sqrt{N-\eta!}}(\hat{b}_0^\dagger)^\eta (\hat{b}_1^\dagger)^{N-\eta}|0\rangle|\uparrow\rangle, \tag{17.71}$$

we have that

$$\hat{H}_{\mathcal{S}+\mathcal{A}}|\eta, N-\eta\rangle|\uparrow\rangle = \frac{g_{\mathcal{SA}}}{\sqrt{N}}(2\eta - N)|\eta, N-\eta\rangle|\uparrow\rangle. \tag{17.72}$$

The initial state can now be written as:

$$|\Psi_{\mathcal{S}+\mathcal{A}}(t_0)\rangle = \left(\frac{1}{\sqrt{2}}\right)^N \sum_{\eta=0}^{N} \frac{\sqrt{N!}(-1)^{N-\eta}}{\sqrt{\eta!}\sqrt{N-\eta!}}|\eta, N-\eta\rangle|\mathcal{S}\rangle. \tag{17.73}$$

Inserting Eqs. (17.72) and (17.73) in Eq. (17.68) we obtain:

$$|\Psi_{\mathcal{S}+\mathcal{A}}(t)\rangle = c_+ \left[\left(\frac{1}{\sqrt{2}}\right)^N \sum_{\eta=0}^{N} \frac{\sqrt{N!}(-1)^{N-\eta}}{\sqrt{\eta!}\sqrt{N-\eta!}} \exp\left(\frac{\imath}{\hbar}\frac{g_{\mathcal{SA}}}{\sqrt{N}}(2\eta - N)t\right)|\eta, N-\eta\rangle\right]|\uparrow\rangle + c_-|N,0\rangle|\downarrow\rangle. \tag{17.74}$$

We now go back to the physical states $|a_0\rangle, |a_1\rangle$ defined through $\hat{a}_0, \hat{a}_1$. The result is

$$|\Psi_{\mathcal{S}+\mathcal{A}}(t)\rangle = c_+|\uparrow\rangle\left[\sum_{n=0}^{N} \hat{a}_n(t)|n, N-n\rangle\right] + c_-|\downarrow\rangle|N,0\rangle, \tag{17.75}$$

where

$$\hat{a}_n(t) = \frac{\sqrt{N!}\imath^{N-n}}{\sqrt{n!}\sqrt{N-n!}}\left[\cos\left(\frac{g_{\mathcal{SA}}t}{\hbar\sqrt{N}}\right)\right]^n \left[\sin\left(\frac{g_{\mathcal{SA}}t}{\hbar\sqrt{N}}\right)\right]^{N-n} \tag{17.76}$$

The probability $\wp_n$ of finding n neutral particles at time t is given by

$$\wp_n = \binom{N}{n} p(t)^n q(t)^{N-n}, \tag{17.77}$$

where $p(t) = \cos^2(gt/\hbar\sqrt{N})$ and $q(t) = 1 - p(t) = \sin^2(gt/\hbar\sqrt{N})$. For large values of N the distribution of $\wp_n$ is strongly peaked around its maximum. In fact, at a given t when n is $\overline{n}(t) = Np(t)$ we have:

$$\wp_{\overline{n}} = \frac{N!}{\overline{n}!N-\overline{n}!}\left(\frac{\overline{n}}{N}\right)^{\overline{n}}\left(\frac{N-\overline{n}}{N}\right)^{N-\overline{n}}, \tag{17.78}$$

which for the Stirling formula becomes $\wp_n \simeq 1$. Hence the probability of finding $n \neq \overline{n}(t)$ becomes negligible, i.e. all particles are ionised. The time for the complete discharge of the counter is proportional to $N^{\frac{1}{2}}$. In the limit of very large values of N one can write Eq. (17.75) approximately in the form:

$$|\Psi_{\mathcal{S}+\mathcal{A}}(t)\rangle = c_+|\overline{n}(t), N-\overline{n}(t)\rangle|\uparrow\rangle + c_-|N,0\rangle|\downarrow\rangle. \tag{17.79}$$

Eq. (17.79) shows a one-to-one correspondence between states of $\mathcal{S}$ and states of $\mathcal{A}$. It is not really a mixture (it is an improper one) but to see some difference between such a pure state and a mixture one must perform two different measurements with two different apparatus and compare them. The fact that generally we do not see any difference is due to the partial trace on the spin variables which 'eliminates' interference terms [CINI 1983, 42–43].

If the interaction between $\mathcal{S}$ and $\mathcal{A}$ is reversible, we return to the state before the measurement, but time is practically infinite, due to the second law of thermodynamics [CINI 1983, 46].

A generalization of Cini's model in the spirit of Zurek's decoherence or of Machida/Namichi's model is to be found in [LIU/SUN 1995, 375–76]; for a possible connection between Cini's model and Machida/Namiki's one see [NAMIKI/PASCAZIO 1993, 366–72].

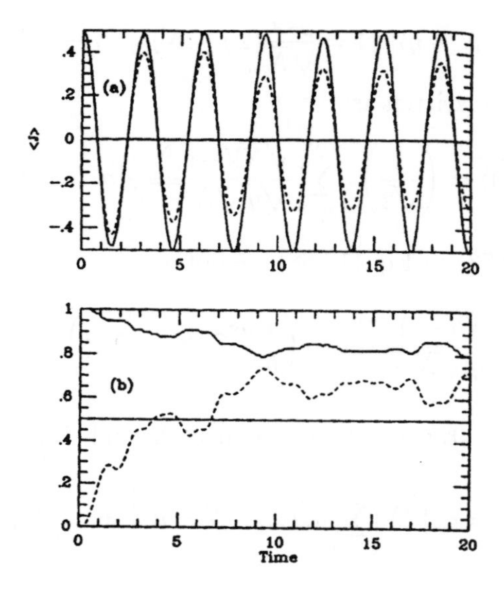

Figure 17.6: Weak coupling ($E_{\mathcal{SE}} = 0.3$) and zero initial entropy.
(a) The solid line is $\left\langle \hat{\mathbf{J}}_{\mathcal{S}} \right\rangle$ for an eigenstate of $\hat{\rho}_{\mathcal{S}}$. The dashed line is $\left\langle \hat{I}_{\mathcal{E}} \otimes \mathbf{J}_{\mathcal{S}} \right\rangle$.
(b) The dashed line is the entropy, the solid line the largest eigenvalue of $\hat{\rho}_{\mathcal{S}}$ — from [ALBRECHT 1992, 5508].

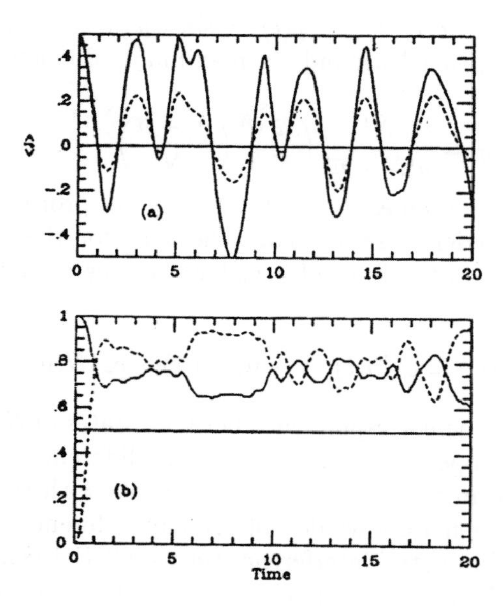

Figure 17.7: Medium coupling ($E_{\mathcal{SE}} = 1$) and zero initial entropy.
Plots as in figure 17.6 — from [ALBRECHT 1992, 5509].

17.5 An Easy Computer Model

The model Let us now discuss a very easy model of decoherence using the interaction between a system and a large environment (we omit the apparatus) [ALBRECHT 1992, 5507–512].[26] Take the full Hamiltonian to be:

$$\hat{H} = \hat{H}_{\mathcal{E}} \otimes \hat{I}_{\mathcal{S}} + \hat{I}_{\mathcal{E}} \otimes \hat{H}_{\mathcal{S}} + \hat{H}_{\mathcal{S}\mathcal{E}}, \tag{17.80}$$

where the self-Hamiltonian of system $\mathcal{S}$ is:

$$\hat{H}_{\mathcal{S}} = E_{\mathcal{S}} \left(| \uparrow\rangle\langle\downarrow | + | \downarrow\rangle\langle\uparrow | \right). \tag{17.81}$$

This causes the spin to rotate from the point of view of the $\{| \uparrow\rangle, | \downarrow\rangle\}$ basis, with frequency proportional to $E_{\mathcal{S}}$. The self-Hamiltonian of the environment $\mathcal{E}$ is taken to be:

$$\hat{H}_{\mathcal{E}} = E_{\mathcal{E}} \times \hat{R}, \tag{17.82}$$

where $\hat{R}$ is a Hermitian matrix with the real and imaginary parts of each independent matrix element initially chosen randomly in the interval $[-0.5, 0.5)$ (by a computer which makes use of a random number generator). The interaction Hamiltonian is:

$$\hat{H}_{\mathcal{S}\mathcal{E}} = E_{\mathcal{S}\mathcal{E}} \left(| \uparrow\rangle\langle\downarrow | \otimes \hat{H}^{\uparrow}_{\mathcal{E}} + | \downarrow\rangle\langle\uparrow | \otimes \hat{H}^{\downarrow}_{\mathcal{E}} \right). \tag{17.83}$$

The matrices $\hat{H}^{\uparrow}_{\mathcal{E}}, \hat{H}^{\downarrow}_{\mathcal{E}}$ are different random matrices, each one constructed like $\hat{R}$.

Zero entropy and weak coupling Now take first an initial state with *zero entropy.* We distinguish between a weak and a strong coupling. As a *weak coupling* we choose:

$$E_{\mathcal{E}} = 1, \quad E_{\mathcal{S}} = 1, \quad E_{\mathcal{S}\mathcal{E}} = 0.3. \tag{17.84}$$

The initial state of the whole system is:

$$|\Psi\rangle = | \uparrow\rangle \otimes |r\rangle, \tag{17.85}$$

where $|r\rangle$ is a randomly chosen state in the $\mathcal{E}$ subspace. The evolution is shown in figure 17.6: the dashed line in (a) gives the expectation value of the operator $\hat{I}_{\mathcal{E}} \otimes \hat{\mathbf{J}}_{\mathcal{S}}$, where:

$$\hat{\mathbf{J}}_{\mathcal{S}} := | \uparrow\rangle \left(+\frac{1}{2}\right) \langle\uparrow | + | \downarrow\rangle \left(-\frac{1}{2}\right) \langle\downarrow |. \tag{17.86}$$

This curve starts at 1/2 (given the initial state $| \uparrow\rangle$), and oscillates with period π, corresponding to the self-Hamiltonian (17.81). However the amplitude of the oscillation decreases with time. This reflects an increasing entropy [see figure 17.6.(b)]. Now if an apparatus is coupled to $\mathcal{S}$ (and not to $\mathcal{E}$), one can treat the spin as being in two different states — shown by a Schmidt decomposition [see Eq. (14.9)] —, each of which leads to its own independent interaction outcome. The oscillating state of the spin is determined by the dynamics, while the only property which is determined by the state of the whole system is the relative probability of finding the spin with one phase or the other.

[26] For other simulations see also [KELLER/MAHLER 1996].

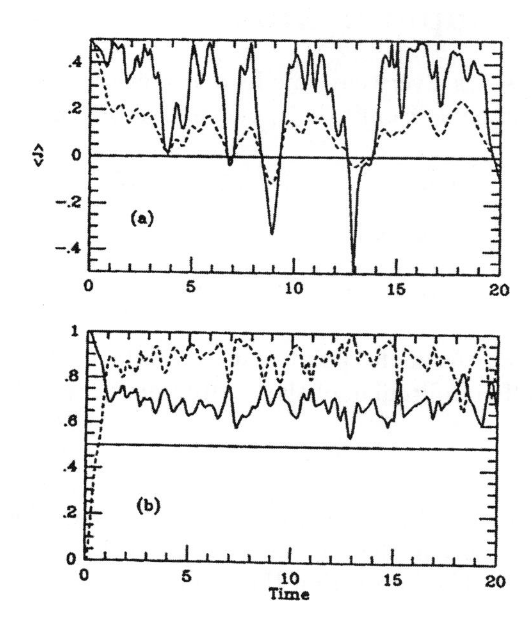

Figure 17.8: Strong coupling ($E_{\mathcal{SE}} = 3$) and zero initial entropy. Plots as in figure 17.6 — from [ALBRECHT 1992, 5509].

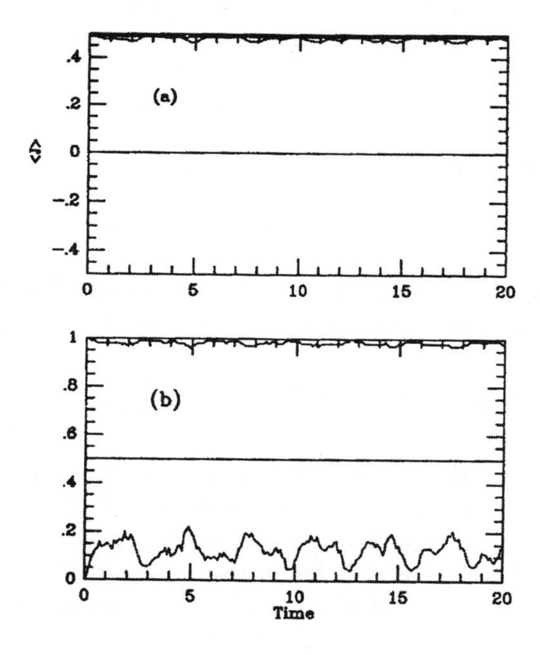

Figure 17.9: Very strong coupling ($E_{\mathcal{SE}} = 50$) and zero initial entropy. Plots as in figure 17.6 — from [ALBRECHT 1992, 5509].

Zero entropy and medium coupling In the case of a *medium coupling* we choose:

$$E_{\mathcal{E}} = 1, \quad E_{\mathcal{S}} = 1, \quad E_{\mathcal{S}\mathcal{E}} = 1. \tag{17.87}$$

Here the periodicity is completely lost [see figure 17.7]: there is no clear pointer basis: whatever states are correlated with $\mathcal{E}$ at a certain time, interactions always decohere them. On the other hand the entropy increases much more rapidly than in the weak coupling case [figure 17.7.(b)].

Zero entropy and strong coupling By increasing the interaction strength to $E_{\mathcal{S}\mathcal{E}} = 3$ (*strong coupling*), we see that the 'Schmidt path' is now more likely to be spin up than spin down, i.e. $\mathcal{S}$ remains in its initial state [see figure 17.8]. This feature becomes more pronounced if we choose $E_{\mathcal{S}\mathcal{E}} = 50$ [figure 17.9].

High entropy By choosing a *high entropy* model, and leaving the same interaction strength ($E_{\mathcal{S}\mathcal{E}} = 50$), we see that the entropy is initially close to maximal and that the Schmidt paths do not reproduce the simple constant spin up evolution as in the low entropy case [figure 17.10].

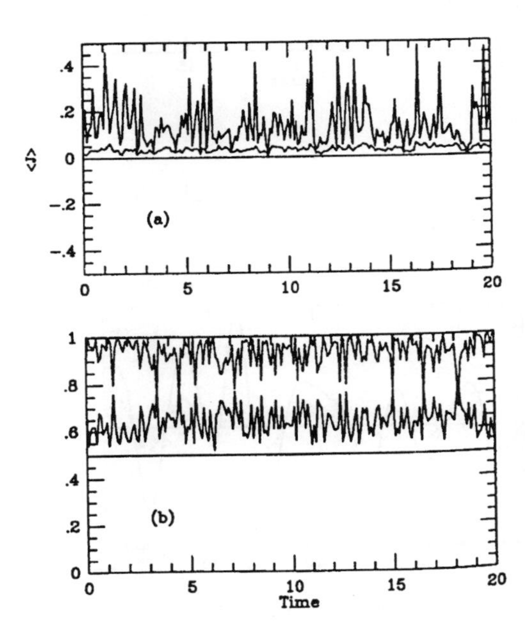

Figure 17.10: Very strong coupling ($E_{\mathcal{S}\mathcal{E}} = 50$) and high initial entropy.
Plots as in figure 17.6 — from [ALBRECHT 1992, 5510].

Increasing the interaction strength very much ($E_{\mathcal{S}\mathcal{E}} = 10.000$) does not change the situation [see figure 17.11].

In the weak coupling case we see vestiges of oscillating behaviour but the results are not as clean as in the zero entropy case [figure 17.12].

Results In conclusion we have seen that, depending on the coupling strength, there is either a constant pointer basis (*strong coupling*), a simply oscillating pointer basis (*weak coupling*), or a noisy behaviour, with no pointer basis (*medium coupling*). The results are very clean if the

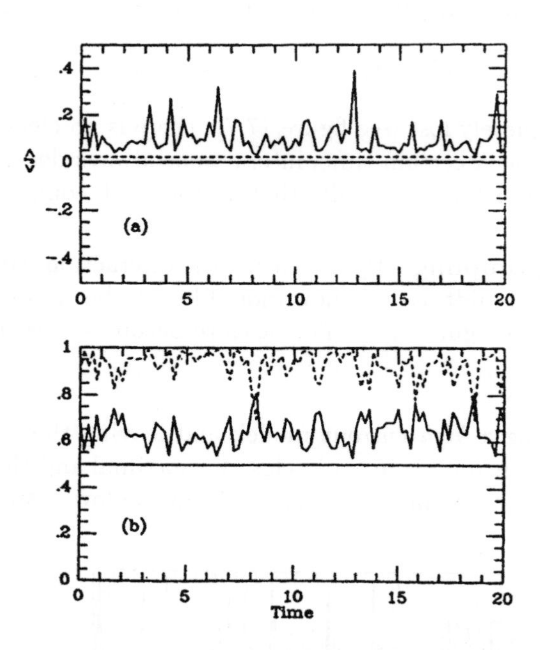

Figure 17.11: Ultra-strong coupling ($E_{\mathcal{SE}} = 10,000$) and high initial entropy. Plots as in figure 17.6 — from [ALBRECHT 1992, 5511].

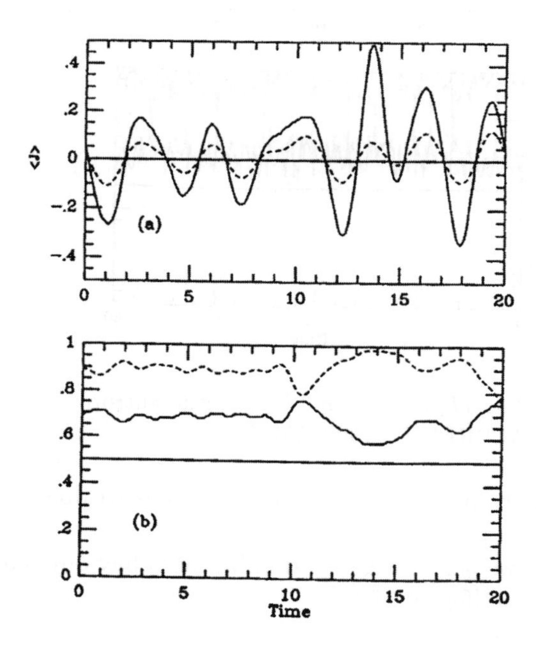

Figure 17.12: Weak coupling ($E_{\mathcal{SE}} = 0.3$) and high initial entropy. Plots as in figure 17.6 — from [ALBRECHT 1992, 5511].

initial state is a zero entropy state (a pure state). In other words, starting with a pure state, a strong action of the environment can produce a better decoherence effect.

17.6 Concluding Remarks on Decoherence

Decoherence is able to solve practically all the problems of Measurement which have been discussed in the preceding chapters.

Bell criticized [BELL 1975, 47–48] this approach by pointing out that decoherence supposes an infinite time in order to bring the off-diagonal factors down to zero. Hence, according to Bell, we never actually have decohered systems. We are only authorized to say that there is a small amount of coherence for the measured observable but not that there is a reduction of the wave packet. We have stated this point in a theorem [17.6: p. 271]; and, as we shall see [in chapter 24], this is a strong point of the theory — common to all models studied in this chapter.

The reported computer model is also a very easy picture of how decoherence can work: we shall extensively return to concrete models of decoherence [chapter 21]. Direct experimental verification of decoherence is another thing. But also on this level decoherence has shown its prediction power [see particularly subsection 24.4.2]. The reason for postponing all this material is that decoherence, as already said, is not only a very good theory for Measurement but also accounts more generally for spontaneous transitions where there is some loss of coherence.

Chapter 18

OPERATIONAL STOCHASTIC QM

Contents Many different results from different schools and domains have converged, in these last years, in a philosophy which can be named Operational Stochastic QM. In this chapter we try to generalize, on a formal level, the solution of the Measurement problem by the introduction of open systems [see section 18.1]. In section 18.2 these instruments are applied to the theoretical problem of non-commutability and complementarity: as we shall see a new formulation of the latter stems from this analysis. Then specific aspects of the Measurement process are treated: covariance and noise operators [section 18.3], the relationship between Projection Valued Measure and Positive Operator Valued Measure [section 18.4]. The relationship between self-adjointness and symmetry is also discussed [section 18.5]. A very important aspect of the Measurement theory is the problem of informational (in-)completeness [section 18.6]. In section 18.7 we treat two formal aspects: decision theory and estimation theory. Strictly related with the previous problems is weak Measurement [section 18.8]. Finally, the stochastic approach is used to solve the problem of the distinguishability of non-orthogonal states [section 18.9] and of the analysis of spin [section 18.10].

We shall later return [chapter 33] to some theoretical generalizations of the stochastic approach: since they concern not only the Measurement problem, but more generally the problem of sharp localization as such, it seems right to postpone the discussion.

18.1 Fundamentals of Operational Stochastic Quantum Mechanics

It is an approach which comes from convexity theory [subsection 9.3.2] and has states as the most important notion – especially interesting are here Davies' results [DAVIES 1969] [*DAVIES* 1976]. But the algebraic approach has also contributed [HAAG/KASTLER 1964]. On the other hand, it is an operational approach due to the character of many basic concepts[1]. Finally it is a stochastic theory because its principal intention is the introduction of unsharp measurements [PRUGOVEČKI 1967].

The starting point of this approach is the problem of open systems, which are not necessarily subjected to the unitary dynamics of Basic-QM. In this sense, this approach is parallel to that of decoherence [chapter 17] because of the role played in the latter by the environment.

[1]Therefore, the epistemology which fits in the best way with this approach is the pragmatistic one. See also [*LAUDAN* 1977, 13]

18.1.1 Operations and Effects

We begin with the concepts of *operation* and *effect* [*DAVIES* 1976, 17–18]:[2]

Definition 18.1 (Operation) *An operation* $\mathcal{T}$ [see Eqs. (14.12) and (14.13)] *is the positive linear mapping from a state space to another, which satisfies following irreversible state transition* [see Eq. (14.4)]*:*

$$\boxed{0 \leq \mathrm{Tr}(\mathcal{T}\hat{\rho}) \leq \mathrm{Tr}(\hat{\rho})} \tag{18.1}$$

and $\parallel \mathcal{T} \parallel = \mathrm{Sup}\,\{\mathrm{Tr}\,[\mathcal{T}(\hat{\rho})]\}$ *for each* $\hat{\rho} \in \mathcal{H}_1^{\mathrm{Tr}=1;\geq 0}$.

We symbolize the operation on $\hat{\rho}$ by $\mathcal{T}(\hat{\rho})$. The probability of the transmission of a state $\hat{\rho}$ by an operation $\mathcal{T}$ is

$$\mathrm{Tr}\,[\mathcal{T}(\hat{\rho})]\,; \tag{18.2}$$

while the output state, dependent upon transmission, is taken to be [see Eq. (3.101)]:

$$\hat{\rho}_f = \frac{\mathcal{T}(\hat{\rho}_i)}{\mathrm{Tr}(\mathcal{T}(\hat{\rho}_i))}. \tag{18.3}$$

Now following theorem can now be proved [*KRAUS* 1983, 42]:

Theorem 18.1 (Kraus) *For an operation* $\mathcal{T}$ *there exist operators* $\hat{O}_k$, $k \in K$ *(a finite or countably infinite index set) on the state Hilbert space satisfying:*

$$\sum_{k \in K_0} \hat{O}_k^\dagger \hat{O}_k \leq 1 \tag{18.4}$$

for all finite subsets $K_0 \leq K$*, such that for an arbitrary state* $\hat{\rho}$ *and an observable* $\hat{O}'$ *the mappings* $\mathcal{T}$ *and* $\mathcal{T}^*$ *are given by*

$$\mathcal{T}\hat{\rho} = \sum_{k \in K_0} \hat{O}_k \hat{\rho} \hat{O}_k^\dagger, \tag{18.5a}$$

$$\mathcal{T}^*\hat{O}' = \sum_{k \in K_0} \hat{O}_k^\dagger \hat{O}' \hat{O}_k, \tag{18.5b}$$

respectively.

Theorem 18.1 describes two types of transformations (passive and active). In fact, associated with an operation is its effect [*DAVIES* 1976, 19],[3] defined as the operator $\hat{E} = \mathcal{T}^*(\mathbf{1})$. The operation (18.5b) performed by $\mathcal{T}^*$ is the Heisenberg picture of this non-unitary dynamic. Since we have:

$$\boxed{\mathrm{Tr}[\mathcal{T}(\hat{\rho})] = \mathrm{Tr}(\hat{\rho}\hat{E})} \tag{18.6}$$

[2] See also [*BUSCH et al.* 1995a, 36]. The concept of operation is introduced for the first time in [HAAG/KASTLER 1964].

[3] First statements in [LUDWIG 1964, 239] [LUDWIG 1968] [LUDWIG 1981, 53].

then $\forall\hat{\rho}$ the effect $\hat{E}$ determines the probability of transmission but not the form of the transmitted state, i.e. it determines the probability of obtaining a determinate outcome o_j, given a state:

$$\hat{\rho} \mapsto \wp_{\hat{\rho}}(o_j) = \text{Tr}(\hat{\rho}\hat{E}_j). \tag{18.7}$$

Kraus uses the following formulation for the two above transformations [*KRAUS* 1983, 13–17, 39–40, 71]: in the case of Eq. (18.2) or (18.6) he speaks of *non-selective operations* (depending on the equivalence class of the initial state only). By keeping the effect apparatus fixed and by varying the states, we obtain all possible state transformations induced by the apparatus — which is a determinative measurement. In the case of Eq. (18.3) he speaks of *selective operations*, which lead to a preparation procedure (but it always depends on the equivalence class of the initial state only) — and it is a preparatory measurement.

We summarize the properties of the effect by the following definition:

Definition 18.2 (Effect) *The effect is an equivalence class of operations.*

Now we are also able to give an operational definition of *state*[4] [*KRAUS* 1983, 3, 6, 13–14]:

Definition 18.3 (State) *Since different preparation procedures may be statistically equivalent (they yield the same statistics for all possible measurements), the state corresponds to an equivalence class of premeasurements.*

Given the specific problems of QM [see theorem 3.1: p. 32, and its corollary, and also corollary 7.1: p. 127], a state can only be defined operationally. On the other hand, it is also possible to give an operational definition of *observable*:

Definition 18.4 (Observable) *Since different registration procedures may be statistically equivalent (they yield the same probability in every state), the observable is an equivalence class of measurements.*

But since the observable is a mapping on the outcome, assigning to it associated effects, we can define it as an *effect valued measure* [*BUSCH et al.* 1995a, 6].

In conclusion unitary transformation is a special case when Eq. (18.5a) or Eq. (18.5b) have only one term $\hat{O}_k = \hat{U}$ of sum (18.4) [see Eqs. (3.72) and (3.73)], i.e. the case in which we have [*DAVIES* 1976, 25–26]:

$$\text{Tr}[\mathcal{T}(\hat{\rho})] = \text{Tr}(\hat{\rho}). \tag{18.8}$$

In other words, while unitary transformations are entropy-preserving, operations in general (18.1) are not [on the whole matter see section 42.3].

---

[4]First statements in [LUDWIG 1964, 239] [LUDWIG 1967] [LUDWIG 1968]. See also [LUDWIG 1981, 53] [*BUSCH et al.* 1995a, 5] [*JAUCH* 1968, 92, 170–72].

18.1.2 Measurement Theory

Now we show that in this frame it is possible to express on a formal level the decoherence analysis.

PVM The standard quantum Measurement theory was based on the Projection Postulate [14.1: p. 223]. A mapping from the Borel subset to the set of bounded linear operators $\mu_p : \mathcal{B}(\Re) \mapsto \mathcal{L}(\mathcal{H})$ [see definition 11.5: p. 197] — where $\mathcal{L}(\mathcal{H})$ is the set of bounded linear operators on the Hilbert space $\mathcal{H}$ — is a *Projection Valued Measure* (= PVM) if [*DAVIES* 1976, 49] [*BUSCH et al.* 1995a, 21]:

$$\mu_p(\mathcal{X}) = \mu_p(\mathcal{X})^* = \mu_p(\mathcal{X})^2, \ \forall \mathcal{X}, \mathcal{X} \in \mathcal{B}(\Re); \tag{18.9a}$$

$$\mu_p(\Re) = \hat{I}; \tag{18.9b}$$

$$\mu_p(\bigcup \mathcal{X}_j) = \sum \mu_p(\mathcal{X}_j), \text{ for all disjoint sequences } (\mathcal{X}_j) \subset \mathcal{B}(\Re). \tag{18.9c}$$

Note that $\mathcal{L}(\mathcal{H})$ is a *Banach space*, i. e. a normed complete space with respect to the norm of the operator $\hat{O}$, i. e. $\| \hat{O} \| = \text{Sup}\{\| \hat{O}\psi \| \ \psi \in \mathcal{H}, \| \psi \| = 1\}$ [see definition 18.1]. Clearly the first requirement [Eq (18.9a)] defines the PVM in terms of projectors [see definition 3.4: p. 44]. Therefore, for each observable-value there is a unique PVM and *vice versa* — due to the spectral theorem [subsection 3.5.4]. In this case we write $\hat{O}_{\text{PV}}$.

Some problems We have already seen the difficulties of the Standard interpretation. Among others, there is the problem of the measurability of continuous observable, where the division presupposed by the Projection Postulate destroys the translational and rotational symmetry of the original problem [see subsection 14.2.2]. Hence the only possibility is an approximate position operator by means of which we define an apparatus which is covariant with respect to the translational group[5].

Other problems are that of the limits imposed by conservation laws [see subsection 14.2.4], from which one could deduce that spin and position can never be observed [*BUSCH et al.* 1995a, 15], and that of the definition of a joint distribution for two non-commuting discrete observables [*DAVIES* 1976, 16–17].

POVM It is from these problems that a theory of a generalized Measurement was born, the *Positive Operator Valued Measure* (= POVM). The first scientist to develop such a powerful generalization was Prugovečki [PRUGOVEČKI 1967].

A POVM is given by the following [*DAVIES* 1976, 35–38] [ALI/PRUGOVEČKI 1977a].[6] Let Υ be some non empty set and $\mathcal{L}_\sigma$ some σ–algebra on subsets of Υ so that $(\Upsilon, \mathcal{L}_\sigma)$ is a measurable space — while the space of a PVM is a Borel space $(\Re, \mathcal{B}(\Re))$ [see section 11.3]. Note that a σ–algebra is a σ–Ring [see p. 199] with an identity element[7]. Then the mapping $\mu_p : \mathcal{L}_\sigma \mapsto \mathcal{L}(\mathcal{H})$ is a POVM if it is positive, normalized, and σ–additive:

[5] For more details see subsection 18.3.1.

[6] Exposition in [*BUSCH et al.* 1991, 7–8, 23]. See also [*HELSTROM* 1976, 54–56] [*SCHROECK* 1996, 121–33, 139–40]. First applications in [PRUGOVEČKI 1977b].

[7] Mathematics in [*KOLMOGOROV/FOMIN* 1980, 44–45, 48].

$$\mu_p(\mathcal{X}) \geq 0, (\forall \mathcal{X})(\mathcal{X} \in \mathcal{L}_\sigma); \tag{18.10a}$$

$$\mu_p(\Upsilon) = \hat{I}; \tag{18.10b}$$

$$\mu_p(\bigcup \mathcal{X}_j) = \sum \mu_p(\mathcal{X}_j), \text{ for all disjoint sequences } (\mathcal{X}_j) \subset \mathcal{L}_\sigma. \tag{18.10c}$$

We can say that projectors are extreme points of the class of effects, i.e. points which only have values $0, 1$ and are orthogonal [see subsection 3.5.3], while:

- an effect $\hat{E}$ can have all values between 0 and 1 though it cannot be represented as a mixture or convex combination of $\{0,1\}$–valued measure [*DAVIES* 1976, 19]:[8] for any state vector $\hat{P}_\psi$ there exists an effect $\hat{E}$ (for example a projection $\hat{E} = \hat{P}_{\psi'}$ with $|\psi'\rangle$ neither collinear nor perpendicular to $|\psi\rangle$) such that $0 \neq \langle\psi|\hat{E}\rangle \neq 1$;

- and an effect $\hat{E}$ is *unsharp* because $\hat{E} \neq \hat{E}^2$, i.e. it is such that $\hat{E}$ and its complement are not orthogonal: $\hat{E}\hat{E}^\perp \neq 0$ [*BUSCH et al.* 1991, 10].

But note that effects respect the completeness condition [Eq. (18.10b)]. Also Quantum Non-Demolition measurements do not respect the orthogonality condition but respect completeness [*BRAGINSKY/KHALILI* 1992, 34] [see chapter 19]. In this way we have distinguished between *sharp properties* (sharp values of an observable) which are defined by projectors and *unsharp properties* (unsharp values of an observable), which are defined by general effects[9], and between *sharp observables*, defined by PVMs, and *unsharp observables* defined by POVMs [*BUSCH et al.* 1995a, 25–26].

This rules out an epistemic interpretation [*BUSCH et al.* 1991, 16] [CASSINELLI/LAHTI 1989, 877], i.e. an objective but unknown 0 or 1–value. In fact, unsharpness must not be confused with inaccuracy: the latter is only due to subjective ignorance, the former expresses the objective uncertainty of some observables. For example, it is possible to construct intermediate situations between a perfect determination of the path of a particle and the diffraction pattern [see section 8.4 and chapter 30]: this intermediate situation is an unsharp property [BUSCH 1987a] [MITTELSTAEDT *et al.* 1987, 893]. Here we can see a consequence of the Gleason Theorem [theorem 11.2: p. 200] or of the Kochen/Specker one [theorem 32.4: p. 551] — there are no QM dispersion-free states or observables [see definition 3.5: p. 54].

From this short analysis — which will be confirmed by later developments [see for example proposition 30.1: p. 526] — the POVMs are ontologically the basic observables of the theory, while the PVMs can be seen more as limiting case of the former. In conclusion we identify an unsharp observable with a POVM and write $\hat{O}_{\text{POV}}$, and identify the set $(\Upsilon, \mathcal{L}_\sigma)$ with the value space of such an observable [*BUSCH et al.* 1995a, 24].

18.2 Stochastic Analysis of Non-commutativity and Complementarity

The main interest of POVMs, as already pointed out, is that they allow the joint measurement of non-commuting observables, as we shall see extensively in this section.

---

[8] See also [*KRAUS* 1983, 11] [BUSCH 1984, 1794].

[9] It is also possible to build an effect algebra [FOULIS/M. BENNETT 1994].

18.2.1 Uncertainty, Commutativity, and Coexistence

With the above formalism it is possible to distinguish between the concepts of *commutativity* and *coexistence*, which is not the case in the Standard interpretation. In fact von Neumann tried to prove the following proposition [*VON NEUMANN* 1932, 89–90] [*VON NEUMANN* 1955, 172–73]:[10]

Proposition 18.1 (I von Neumann) *If two observables $\hat{O}, \hat{O}'$ (self-adjoint operators) are compatible — formally $\hat{O} \odot \hat{O}'$ — then there is a third observable $\hat{O}''$ of which both are functions. Formally:*

$$(\hat{O} \odot \hat{O}') \rightarrow (\exists \hat{O}'')[(\hat{O} = \hat{O}''(\mathcal{B})) \wedge (\hat{O}' = \hat{O}''(\mathcal{B}'))], \tag{18.11}$$

where $\mathcal{B}, \mathcal{B}'$ are Borel sets.

We define compatibility as follows[11]:

$$\hat{O} \odot \hat{O}' := (\hat{O} \wedge \hat{O}') \vee (\neg\hat{O} \wedge \hat{O}') \vee (\hat{O} \wedge \neg\hat{O}') \vee (\neg\hat{O} \wedge \neg\hat{O}'). \tag{18.12}$$

In other words the compatibility guarantees that there are no value gaps in the joint measurement of two observables [see theorem 9.10: p. 171]. A further development in this direction is due to Jauch. First we state the following lemma[12]:

Lemma 18.1 (Jauch) *Two self-adjoint operators commute if their respective spectral measures commute;*

by means of which (for sharp observables) it can be proved that [JAUCH 1974]:

Proposition 18.2 (Jauch) *For arbitrary measurable sets $\mathcal{X}, \mathcal{Y}$ with finite Lesbegues measures, and observables $\hat{\mathbf{p}}, \hat{\mathbf{q}}$ we have:*

$$\hat{\mathbf{q}}\left(\frac{\Re}{\mathcal{X}}\right) \wedge \hat{\mathbf{p}}\left(\frac{\Re}{\mathcal{Y}}\right) \neq 0, \tag{18.13a}$$

$$\hat{\mathbf{q}}(\mathcal{X}) \wedge \hat{\mathbf{p}}(\mathcal{Y}) = 0. \tag{18.13b}$$

Due to the intersection of the ranges of the spectral projections of two conjugate observables, a well-behaved joint probability distribution does not exist. But the result is independent from UP; hence the proposition does not imply an incompatibility. Therefore it is possible to prove [BUSCH 1984] (we omit the proof) a generalized theorem by working with unsharp observables[13]:

Theorem 18.2 (Busch) *There exists a positive lower bound $\hat{\mathbf{q}}_{\mathrm{POV}}(\mathcal{X}) \wedge \hat{\mathbf{p}}_{\mathrm{POV}}(\mathcal{Y}) \neq 0$ of unsharp observables $\hat{\mathbf{q}}_{\mathrm{POV}}, \hat{\mathbf{p}}_{\mathrm{POV}}$ for given measurable sets $\mathcal{X}, \mathcal{Y}$, iff the ranges of the respective square root operators possess non-trivial intersection. Hence operators $\hat{\mathbf{q}}_{\mathrm{POV}}(\mathcal{X})$ and $\hat{\mathbf{p}}_{\mathrm{POV}}(\mathcal{Y})$ can be compatible and therefore coexist even if they do not commute.*

As a consequence of the Busch theorem, we state the following:

[10] See also [GUDDER 1979, 345] [PIRON 1972, 520–21].

[11] On this definition see [VARADARAJAN 1962, 181–83, 188–91] [*DRIESCHNER* 1979, 24].

[12] On this point see [*PRUGOVEČKI* 1971, 261].

[13] See also [BUSCH 1987a] for experimental proposals; see also [*PRUGOVEČKI* 1971, 260] [*HOLEVO* 1982, 70] [A. FINE 1982b] [KRUSZYŃSKI/DE MUYNCK 1987] [SCHROECK 1985d, 680].

Corollary 18.1 (Joint Measurability) *A pair of unsharp position and momentum observables are jointly measurable by means of a continuous phase space observable if they are a Fourier pair.*

In this direction, it can also be shown that the (non-)commutativity has degrees and that there is no sharp yes/no alternative between commutativity and non-commutativity [BUSCH 1987b]. We have already seen how it is possible to define the commutability of observables in terms of projectors (yes/no alternatives) [see Eq. (9.19)]. The formula can be extended to non-orthogonal effects $\hat{E}, \hat{E}'$:

$$\hat{C}_{\mathcal{H}}(\hat{E}, \hat{E}') := \{|\psi\rangle \in \mathcal{H}|(\hat{E}\hat{E}' - \hat{E}'\hat{E})|\psi\rangle = [\hat{E}, \hat{E}']|\psi\rangle = 0\}. \tag{18.14}$$

Now we can call $\hat{E}, \hat{E}'$ commutative if $\hat{C}(\hat{E}, \hat{E}') = \mathbf{1}$, unsharp (non-)commutative if $0 \neq \hat{C}(\hat{E}, \hat{E}') \neq \mathbf{1}$ and sharp (totally) non-commutative if $\hat{C}(\hat{E}, \hat{E}') = 0$ [see also subsection 18.6.1 below].

18.2.2 A Probabilistic Formulation of the Complementarity

By handling measurement outcomes and probabilistic aspects, we can formulate a probabilistic CP — of CP1 or of UP in this case [see principle 8.1: p. 137] — based on the possible the coexistence of Probabilistic Complementary Observables [*BUSCH et al.* 1995a, 104–108]. For some state $\hat{\rho}$ we write:

$$[\wp_{\hat{\rho}}^{\hat{O}_1}(\mathcal{X}) = 1] \rightarrow [0 < \wp_{\hat{\rho}}^{\hat{O}_2}(\mathcal{Y}) < 1] \tag{18.15}$$

and *vice versa* — where $\hat{O}_1$ is a Probabilistic Complementary Observable with respect to $\hat{O}_2$ and $\mathcal{X} \in \mathcal{L}_\sigma^1, \mathcal{Y} \in \mathcal{L}_\sigma^2$. Formula (18.15) means that if $\hat{O}_1$ is determined in the given state, then its complementary observable can be coexistent with some non-zero and non-1 probability. Probabilistic CP is implied by the disjointness of the spectral projections of the two observables, which is a definition of complementarity [proposition 18.1 or proposition 18.2].

In the case of sharp observables, CP and the probabilistic formulation coincide, but it is not so in the case of unsharp observables, where CP is stronger then the Probabilistic version: while Probabilistic Complementary Observables do not admit any repeatable joint Measurement, Complementary Observables do not admit any joint Measurements at all. Neither do they have any order independent sequential Measurements.

But, even if a Fourier pair is jointly measurable, the two observables remain Probabilistic Complementary Observables, so that their predictions are mutually exclusive.

18.3 Covariance, Conservation Laws and Measurement

18.3.1 Davies Theorem

We have seen that if one assumes to measure a continuous observable, such as the position, sharply, then this destroys the translational symmetry of the original problem [see subsection 14.2.2]. On the other hand, on the grounds of the WAY theorem [14.5: p.234], i.e. of a possible violation of the conservation laws, we have already assessed the necessity of unsharp observables. We now develop this point by means of following theorem [*DAVIES* 1976, 66–67]:[14]

[14]See also [*BUSCH et al.* 1995a, 14]. For further generalization see [OZAWA 1993a], where the case of successive approximate measurements is also analyzed.

Theorem 18.3 (Davies) *Let a single spinless particle on space $\mathcal{H}$, on which acts translation $\hat{U}_a\varsigma(x) = \varsigma(r+a)$, be defined by $\varsigma(x)$, and suppose a bounded and normalized wave fucntion $\psi(x)$ describing an apparatus $\mathcal{A}$, and an operator $\hat{O}_a$ such that:*

$$\hat{O}_a\varsigma(x) = \psi(x-a)\varsigma(x). \tag{18.16}$$

Then the formula

$$\boxed{\mathcal{A}(\mathcal{X}_\mathcal{S}, \hat{\rho}_\mathcal{S}) = \int_{\mathcal{X}_\mathcal{S}} \hat{O}^\mathcal{A}_x \hat{\rho}_\mathcal{S} (\hat{O}^\mathcal{A}_x)^\dagger d^3x} \tag{18.17}$$

defines an instrument on $\Re^3$ which is covariant with respect to the translation group. The observable $\hat{O}^\mathcal{A}_{\mathrm{POV}}$ is the approximate covariant pointer observable associated with $\mathcal{A}(\mathcal{X}_\mathcal{S}, \hat{\rho}_\mathcal{S})$ and given by [see Eq. (5.59a)]*:*

$$\hat{O}^\mathcal{A}_{\mathrm{POV}}(\mathcal{X}_\mathcal{S})\varsigma(x) = (\chi^{\mathcal{X}_\mathcal{S}}_x \circ |\psi(x)|^2)\varsigma(x). \tag{18.18}$$

Note that the state of the apparatus is defined in practice as a state relative [see subsection 15.1.2] to the object system, i.e. in this way — as again discussed by Zurek [see subsection 17.2.1] — we confirm our assumption of Relative-state analysis in a general theory of Measurement.

18.3.2 Noise Operators*

We now wish to discuss some aspects that derive from the WAY theorem and the first Ozawa theorem [see subsection 14.2.4] [OZAWA 1993b].[15] We suppose that POVM $\hat{O}^t_{\mathrm{POV}}(\mathcal{X})$ is the observable of a system $\mathcal{S}$ to be measured and that $\hat{O}^\mathcal{A}_{\mathrm{POV}}(\mathcal{X})$ is the pointer observable. We have:

$$\begin{aligned}
\mathrm{Tr}_\mathcal{S}[\hat{\rho}^\mathcal{S}_i \hat{O}^t_{\mathrm{POV}}(\mathcal{X})] &= \left\langle \hat{\rho}^\mathcal{S}_i, \hat{O}^t_{\mathrm{POV}}(\mathcal{X})\right\rangle \\
&= \mathrm{Tr}_{\mathcal{S}+\mathcal{A}}\left([\hat{U}_t(\hat{\rho}^\mathcal{S}_i \otimes \hat{\rho}^\mathcal{A}_i)\hat{U}^\dagger_t][\hat{I} \otimes \hat{O}^\mathcal{A}_{\mathrm{POV}}(\mathcal{X})]\right) \\
&= \mathrm{Tr}_{\mathcal{S}+\mathcal{A}}\left((\hat{\rho}^\mathcal{S}_i \otimes \hat{\rho}^\mathcal{A}_i)\hat{U}^\dagger_t[\hat{I} \otimes \hat{O}^\mathcal{A}_{\mathrm{POV}}(\mathcal{X})]\hat{U}_t\right).
\end{aligned} \tag{18.19}$$

Now we ask if, given $\hat{O}^\mathcal{A}_{\mathrm{POV}}(\mathcal{X})$ and some $\hat{O}^t_{\mathrm{POV}}(\mathcal{X})$, the expected value of

$$\hat{U}^\dagger_t[\hat{I} \otimes \hat{O}^\mathcal{A}_{\mathrm{POV}}(\mathcal{X})]\hat{U}_t - \hat{O}^t_{\mathrm{POV}}(\mathcal{X}) \otimes \hat{I} \tag{18.20}$$

vanishes for the given initial state $\hat{\rho}^\mathcal{A}_i$ and for all initial states $\hat{\rho}^\mathcal{S}_i$. Difference (18.20) is the *first Noise operator*, introduced by Ozawa:

$$\hat{\mathcal{N}}_1(t) = \hat{U}^\dagger_t[\hat{I} \otimes \hat{O}^\mathcal{A}_{\mathrm{POV}}(\mathcal{X})]\hat{U}_t - \hat{O}^t_{\mathrm{POV}}(\mathcal{X}) \otimes \hat{I}, \tag{18.21}$$

and it is useful in order to calculate fluctuations and back-actions on the pointer observable during a measurement process. If Eq. (18.19) holds for all $\hat{\rho}^\mathcal{A}_i$ as well as for all $\hat{\rho}^\mathcal{S}_i$, then $\hat{\mathcal{N}}_1(t)$ vanishes identically.

If, after the interaction, $\hat{O}^t_{\mathrm{POV}}(\mathcal{X})$ is independent from t, then the *second Noise operator*, given by

$$\hat{\mathcal{N}}_2(t) = \hat{U}^\dagger_t[\hat{O}^t_{\mathrm{POV}}(\mathcal{X}) \otimes \hat{I}]\hat{U}_t - \hat{O}^t_{\mathrm{POV}}(\mathcal{X}) \otimes \hat{I}, \tag{18.22}$$

also vanishes from the expected value. Now we are able to give the following definition:

[15]See also [*SCHROECK* 1996, 71–74] [*BUSCH et al.* 1995a, 114–15].

Definition 18.5 (Ozawa) *A unitary Operator $\hat{U}$ is said to give an exact measurement of $\hat{O}^{\mathcal{S}}$ with $\hat{O}^{\mathcal{A}}$ if, for any ϵ, there is some normalized $|\mathcal{A}\rangle \in \mathcal{H}_{\mathcal{A}}$, such that, for any normalized $|\varsigma\rangle \in \mathcal{H}_{\mathcal{S}}$ we have:*

$$\left\langle \varsigma \otimes \mathcal{A} | \hat{\mathcal{N}}_j \varsigma \otimes \mathcal{A} \right\rangle = 0, \tag{18.23a}$$

$$\| \hat{\mathcal{N}}_j \varsigma \otimes \mathcal{A} \| < \epsilon, \tag{18.23b}$$

where $j = 1, 2$.

Now we consider a pair of observables $\hat{O}_1$ acting on $\mathcal{H}_{\mathcal{S}}$ and $\hat{O}_2$ acting on $\mathcal{H}_{\mathcal{A}}$ such that $\hat{O}_1 \otimes \hat{I} + \hat{I} \otimes \hat{O}_2$ is an additive conserved quantity:

$$[\hat{U}_t, (\hat{O}_1 \otimes \hat{I} + \hat{I} \otimes \hat{O}_2)] = 0. \tag{18.24}$$

Using Eqs. (18.21)–(18.24) we compute:

$$[\hat{\mathcal{N}}_1, \hat{U}_t^\dagger (\hat{O}_1 \otimes \hat{I})\hat{U}_t] + [\hat{\mathcal{N}}_2, \hat{U}_t^\dagger (\hat{I} \otimes \hat{O}_2)\hat{U}_t] = [\hat{O}_1, \hat{O}^t_{\mathrm{POV}}(\mathcal{X})] \otimes \hat{I}. \tag{18.25}$$

In the terms already discussed we can state the following theorem:

Theorem 18.4 (II Ozawa) *If $[\hat{O}^{\mathcal{S}}, \hat{O}_1] \neq 0$, and if $\hat{O}_1, \hat{O}_2$ are bounded operators defined by* Eq. (18.24)], *then $\hat{O}^{\mathcal{S}}$ cannot be measured exactly using a pointer observable.*

Proof

Suppose that it is possible; then from Eq. (18.25), Eq. (18.23) and the Cauchy/Schwarz/Buniakowski inequality we have:

$$\left| \left\langle \varsigma | [\hat{O}^{\mathcal{S}}, \hat{O}_1] \varsigma \right\rangle \right| \leq 2(\| \hat{\mathcal{N}}_1 |\varsigma\rangle \otimes |\mathcal{A}\rangle \| \| \hat{O}_1 \| + \| \hat{\mathcal{N}}_2 |\varsigma\rangle \otimes |\mathcal{A}\rangle \| \| \hat{O}_2 \|) < 2\epsilon(\| \hat{O}_1 \| + \| \hat{O}_2 \|) \tag{18.26}$$

for any normalized $|\varsigma\rangle \in \mathcal{H}_{\mathcal{S}}$ and any $\epsilon > 0$, from which it follows that $[\hat{O}^{\mathcal{S}}, \hat{O}_1] = 0$, contrary to the assumption. Q.E.D.

If the conserved observable $\hat{O}$ is additive — i.e. $\hat{O} = \hat{O}_1 \otimes \hat{I} \oplus \hat{I} \otimes \hat{O}_2$ (where $\hat{O}_1 \in \mathcal{L}(\mathcal{H}_{\mathcal{S}}), \hat{O}_2 \in \mathcal{L}(\mathcal{H}_{\mathcal{A}})$), then it satisfies:

$$\hat{U}\hat{O}^{\mathcal{S}} - \hat{O}^{\mathcal{S}}\hat{U} = 0, \tag{18.27}$$

which can be reduced to the condition:

$$[\hat{O}_1, \hat{O}^{\mathcal{S}}] = 0, \tag{18.28}$$

where $\hat{O}^{\mathcal{S}}$ is the observable to be measured.

The interest of theorem 18.4 is that also with POVMs we are forced to take measurements of (discrete or continuous) observables that commute with $\hat{O}_1$. But this condition is not stringent if $\mathcal{A}$ is very big, i.e. it contains many quanta: in [ARAKI/YANASE 1960] it was shown that measurement is possible even if condition (18.24) is not satisfied[16].

[16] A further generalization is in [KUDAKA/KAKAZU 1992].

18.4 Sharpness and Unsharpness: Optimal Measurements

A question strictly related to the precedent one is that of how to optimize a POVM. In fact, a POVM can also contain 'bad' observables' which mix information about the object system with non-information coming from the apparatus. Hence we can establish two extreme limits for each POVM [MARTENS/DE MUYNCK 1990a, 256]:

- an *optimal result*, such that the POVM gives us the subclass of observables which are maximally free from non-information,
- and an *uninformative result*, where the POVM is unable to extract information from the system.

A POVM can be understood as a family of bounded linear operators $\{\hat{O}_k\}$ [see for example figure 30.6 and relative comments] satisfying the completeness and non-negativity conditions. Let us now define the subset K of the countable infinite outcome set Υ of $\{\hat{O}_k\}$ a *support* [MARTENS/DE MUYNCK 1990a, 258–61] [MARTENS/DE MUYNCK 1990b, 361–64] iff:

$$\mathrm{K} := \{k \in \Upsilon | \hat{O}_k \neq 0\}. \tag{18.29}$$

Defining two POVMs, one for the apparatus observable, the other for the system one, $\hat{O}^{\mathcal{A}} = \{\hat{O}^{\mathcal{A}}_k\}_{k\in\mathrm{K}}, \hat{O}^{\mathcal{S}} = \{\hat{O}^{\mathcal{S}}_l\}_{l\in\mathrm{L}}$ (where K and L are respective supports of the first and of the second observable) and supposing that there exists a stochastic matrix $\lambda_{kl} \in \Re^{\mathrm{K}\times\mathrm{L}}$ satisfying the completeness condition:

$$\sum_{k\in\mathrm{K}} \lambda_{kl} = 1, \tag{18.30a}$$

and the non-negativity condition:

$$\lambda_{kl} \geq 0, \tag{18.30b}$$

then $\mathcal{M}^{\mathrm{POV}} : \hat{O}^{\mathcal{S}} \mapsto^{\mathrm{POV}} \hat{O}^{\mathcal{A}}$ is a mapping between the two observables iff the following condition is satisfied:

$$\hat{O}^{\mathcal{A}}_k = \sum_{l\in\mathrm{L}} \lambda_{kl}.\hat{O}^{\mathcal{S}}_l \tag{18.31}$$

Now, if we define the equivalence relationship as:

$$\hat{O}^{\mathcal{A}} \leftrightarrow^{\mathrm{POV}} \hat{O}^{\mathcal{S}} := \left(\hat{O}^{\mathcal{A}} \mapsto^{\mathrm{POV}} \hat{O}^{\mathcal{S}}\right) \wedge \left(\hat{O}^{\mathcal{S}} \mapsto^{\mathrm{POV}} \hat{O}^{\mathcal{A}}\right), \tag{18.32}$$

we have that $\hat{O}^{\mathcal{A}}$ is *maximal* iff:

$$\forall \hat{O}^{\mathcal{S}}_{\mathrm{POV}} : \left(\hat{O}^{\mathcal{S}}_{\mathrm{POV}} \mapsto^{\mathrm{POV}} \hat{O}^{\mathcal{A}}\right) \rightarrow \left(\hat{O}^{\mathcal{A}} \leftrightarrow^{\mathrm{POV}} \hat{O}^{\mathcal{S}}_{\mathrm{POV}}\right), \tag{18.33}$$

i.e., for each system observable, a one-to-one correspondence with a pointer observable can be found; and it is *minimal* iff:

$$\forall \hat{O}^{\mathcal{S}}_{\mathrm{POV}} : \left(\hat{O}^{\mathcal{A}} \mapsto^{\mathrm{POV}} \hat{O}^{\mathcal{S}}_{\mathrm{POV}}\right) \rightarrow \left(\hat{O}^{\mathcal{A}} \leftrightarrow^{\mathrm{POV}} \hat{O}^{\mathcal{S}}_{\mathrm{POV}}\right), \tag{18.34}$$

i.e. given a pointer observable, it is unable to 'decide' among different system observables.

Hence it can be shown that if a POVM is maximal, it is a projector, and if it is minimal, it is uninformative. Note that for every POVM there is a maximal one. In conclusion, an *optimal* POVM can be found between such upper and lower limits, i.e. a measurement which, by renouncing to the requisite of the orthogonality, is able to extract the maximum information.

18.5 Self-Adjointness and Symmetry

As we have seen an unsharp observable can always be obtained from a sharp one via some smearing operation [*BUSCH et al.* 1995a, 26–30].[17] Let $(\Upsilon_1, \mathcal{L}^1_\sigma), (\Upsilon_2, \mathcal{L}^2_\sigma)$ be two value spaces. Consider the mapping $\wp : \mathcal{L}^2_\sigma \times \Upsilon_1 \mapsto [0,1]$ such that for each $\mathcal{X} \in \mathcal{L}^2_\sigma$, $\wp(\mathcal{X}, \cdot)$ is a measurable function on Υ_1 and for every $o \in \Upsilon_1$, $\wp(\cdot, o)$ is a probability measure on $\mathcal{L}^2_\sigma$. If

$$\hat{O}_{\text{PV}} : \mathcal{L}^1_\sigma \mapsto \mathcal{L}(\mathcal{H}) \tag{18.35}$$

is a PVM, then we define a POVM $\hat{O}_{\text{POV}}$ on $\mathcal{L}^2_\sigma$ as [see also Eq. (3.65)]:

$$\hat{O}_{\text{POV}}(\mathcal{X}) := \int_{\Upsilon_1} \wp(\mathcal{X}, o) d\hat{O}_{\text{PV}}(o), \tag{18.36}$$

i.e. it reflects some source of uncertainty in the reading of a measurement. A mapping $\mathcal{T}^*$ (an effect) which allows the passage from PVM to POVM, is very important for the dynamics of open systems, which do not evolve by unitary transformations end hence are described by a non-unitary (but trace-preserving) state transformation $\mathcal{T}$.[18]

We can now express in more general terms the relationship between PVM and POVM by means of the Naimark Theorem [*HOLEVO* 1982, 64–68] [*DAVIES* 1976, 131, 142]:[19]

Theorem 18.5 (Naimark) *If* $\hat{O}_{\text{POV}} : \mathcal{L}_\sigma \mapsto \mathcal{L}(\mathcal{H})$ *is a* POVM, *there exists a Hilbert space* $\mathcal{H}_0 \supset \mathcal{H}$ *and a* PVM $\hat{O}_{\text{PV}} : \mathcal{L}_\sigma \mapsto \mathcal{L}(\mathcal{H})_0$ *such that the equality:*

$$\hat{O}_{\text{POV}}(\mathcal{X})|\psi\rangle = \hat{P}_{\mathcal{H}} \hat{O}_{\text{PV}} |\psi\rangle \tag{18.37}$$

holds for all $|\psi\rangle \in \mathcal{H}$ *and for all* $\mathcal{X} \in \mathcal{L}_\sigma$, *where* $\hat{P}_{\mathcal{H}}$ *is the orthogonal projection of* $\mathcal{H}_0$ *onto* $\mathcal{H}$.

We can express the mathematical content of the Naimark theorem in terms of projectors and effects:

$$\hat{E}(\mathcal{X}) = \hat{P}_{\mathcal{H}} \hat{P}(\mathcal{X}) \hat{P}_{\mathcal{H}}, \tag{18.38}$$

where the effect $\hat{E}(\mathcal{X})$ acts on $\mathcal{H}$ and the projector $\hat{P}(\mathcal{X})$ on $\mathcal{H}_0$ and, as before, $\hat{P}_{\mathcal{H}}$ is the orthogonal projection of $\mathcal{H}_0$ onto $\mathcal{H}$. As an important corollary we have the following:

Corollary 18.2 (Holevo) *For every Effect* $\hat{E}$ *acting on* $\mathcal{H}$ *there exists a Hilbert space* $\mathcal{H}_0$, *a state* $\hat{P}_{\psi_0}$ *in it, and a projector* $\hat{P}$ *on* $\mathcal{H} \otimes \mathcal{H}_0$, *such that:*

$$\boxed{\text{Tr}_{\mathcal{H}\otimes\mathcal{H}_0}[\hat{\rho} \otimes \hat{P}_{\psi_0} \hat{P}(\mathcal{X})] = \text{Tr}_{\mathcal{H}}[\hat{\rho}\hat{E}(\mathcal{X})]} \tag{18.39}$$

for any $\hat{\rho} \in \mathcal{H}^{\text{Tr}=1;\geq 0}$. *Moreover* $\hat{P}$ *can always be chosen to be of either of the form* $\hat{U}^\dagger \hat{P}(\cdot) \otimes \hat{I} \hat{U}$ *or* $\hat{U}^\dagger \hat{I} \otimes \hat{P}(\cdot) \hat{U}$ *for some* $\hat{U}$.

---

[17] UP serves as measure of the degree of this passage [*BUSCH et al.* 1995a, 13].

[18] On this point see [ZEH 1970, 346].

[19] See also [*HELSTROM* 1976, 78–80] [*BUSCH et al.* 1995a, 31–32] [*PERES* 1993, 285].

Hence the POVM acts on a smaller Hilbert space than the corresponding PVM, and from here the smearing operation (coarse-graining) from a PVM to a POVM.

The Naimark Theorem assures us of the measurability of each observable, even if not self-adjoint. Therefore the usefulness of Naimark Theorem and of its corollary can be seen if we handle problems where one reaches only symmetry and not self-adjointness, so that one cannot invoke the spectral theorem [3.5: p. 46] to ensure the existence of probability distributions. One might think, for example, to the time delay in the scattering theory, or to the problem of representation of operators such position (in Relativistic-QM) or the angle where only symmetry can be obtained and not self-adjointness [see sections 10.3 and 10.4]. Now, the Naimark theorem guarantees that any maximal symmetric operator (without extension in $\mathcal{H}$) can be extended to a self-adjoint operator $\hat{O}_{\rm PV}$ acting on $\mathcal{H}_0$, which contains $\mathcal{H}$ [*BUSCH et al.* 1995a, 32–33].

18.6 Informational Completeness and Measurement

18.6.1 Informational Completeness

States We can define *Informational completeness* as follows [PRUGOVEČKI 1977a]:[20]

Definition 18.6 (Informational Completeness) *A family of self-adjoint operators* $\{\hat{O}_k\}$ *(where* $\{k\} = \hat{I}$*) is said to be informationally complete if* $\mathrm{Tr}[\hat{\rho} \cdot \hat{O}_k] = \mathrm{Tr}[\hat{\rho}' \cdot \hat{O}_k]$ *implies that* $\hat{\rho} = \hat{\rho}'$ *for any pair* $\hat{\rho}, \hat{\rho}'$ *of density operators on a Hilbert space* $\mathcal{H}$.

We now say that a projector $\hat{P}$ is commutative if $\hat{P}(\mathcal{X})\hat{P}(\mathcal{Y}) = \hat{P}(\mathcal{Y})\hat{P}(\mathcal{X})$, by means of which we state [BUSCH/LAHTI 1989, 639–40] the following:

Theorem 18.6 (I Busch/Lahti) *No commutative semispectral measure is informationally complete if* $\dim(\mathcal{H}) \geq 2$.

from which it follows:

Corollary 18.3 (Informational Incompleteness) *No family of mutually commuting spectral measures is informationally complete.*

Therefore, if there is an operator (which is not $\hat{I}$) which commutes with all operators of a given set — a requirement which is equivalent to an SSR [see subsection 10.1.4] —, then the set cannot be informationally complete — because there can be states which are not eigenstates of this operator but in which this has the same expectation values as would have have with eigenstates. Hence, in general, a set can be informationally complete iff it is maximal (the set of operators commuting with every operator of the given set consists in multiples of the identity operator $\hat{I}$); and specifically a vector set is informationally complete if all vector states are pure.

[20]See also [BUSCH/SCHROECK 1989, 815] [BUSCH/LAHTI 1989, 638] [*BUSCH et al.* 1991, 89–90] [*BUSCH et al.* 1995a, 120] [*SCHROECK* 1996, 53–54, 116–17].

Therefore, pure states represent maximal collections of properties which a physical system may possess at one time [BUSCH/LAHTI 1989, 633, 659], or, in other words, pure states are in one-to-one correspondence with the maximal set of properties (projectors) which the system may possess at one time:

$$\hat{P}_{\psi} = \bigwedge\{\hat{P} \in \mathcal{P}(\mathcal{H}) | \langle \psi | \hat{P}\psi \rangle = 1\}, \tag{18.40}$$

where $\mathcal{P}(\mathcal{H})$ denotes the set of projectors acting on $\mathcal{H}$, and where $|\psi\rangle \in \mathcal{H}$.

PVM We now prove the following theorem indirectly by means of a specific example:

Theorem 18.7 (Informational Incompleteness of Sharp Observables) *No set of sharp observables* (PVM) *can be informationally complete.*

An observable is informationally complete if its classical embedding is injective, i.e., if its measurement statistics entails unique state determination [*BUSCH et al.* 1991, 90].

Take as an **example** the relationships between $\hat{\mathbf{p}}, \hat{\mathbf{q}}$. The *Pauli problem* [*PAULI* 1980, 17] is[21]: under what conditions do the position and momentum distributions define the state function uniquely? We can now reformulate it in the question of informational completeness of the canonically conjugate position and momentum observables. In fact sharp $\hat{\mathbf{p}}, \hat{\mathbf{q}}$ are totally non-commutative [see comments to Eq. (18.14)], a requirement which only satisfies the necessary condition of informational completeness [theorem 18.6]. But they are not complete (which means that they are not able to distinguish any state). We prove it briefly [PRUGOVEČKI 1977a] for a one-dimensional problem[22].

Proof

Take the unit vector $|\psi\rangle \in L^2(\Re, dx)$ with $\psi(x) = |\psi(x)|e^{\imath\eta(x)}$, where the parameter η satisfies $0 \leq \eta(x) \leq 2\pi$, and $|\psi(x)| = |\psi(-x)|$ ($x \in \Re$ and $\eta(x) + \eta(-x) \neq \text{constant}(\text{mod}2\pi)$). Then $\psi'(x) := |\psi(x)|e^{-\imath\eta(-x)}$ represents a state different from $\psi(x)$ (i.e. $\hat{P}_{\psi'(x)} \neq \hat{P}_{\psi(x)}$ but $|\psi'(x)|^2 = |\psi(x)|^2$ and $|\tilde{\psi}'(k)|^2 = |\tilde{\psi}(k)|^2$, where $\tilde{\psi}$ is the Fourier transform of ψ), then $\hat{p}, \hat{q}$ do not distinguish between these states. Q. E. D.

Now we can state the following:

Theorem 18.8 (II Busch/Lahti) *Let $\hat{P} : \mathcal{L}_\sigma \rightarrow \mathcal{H}^{\mathrm{Tr}=1;\geq 0}$ be a spectral measure; then $[\hat{\rho}]^{\hat{P}} = \{\hat{\rho}\}$ iff $\hat{\rho}$ is in the set of one dimensional spectral projections of $\hat{P}$. By $[\hat{\rho}]^{\hat{P}}$ we denote the set of all states which are equivalent with respect to $\hat{P}$: $\{\hat{\rho}' \in \mathcal{H}^{\mathrm{Tr}=1;\geq 0} : \hat{\rho}' \sim_{\hat{P}} \hat{\rho}\}$.*

On the basis of this theorem (which, though correct, we do not integrate into the theory because interpretationally we assume POVMs to be the basic ontological entities) we have the stronger result that neither $\hat{\mathbf{p}}$ nor $\hat{\mathbf{q}}$ is informationally complete with respect to any state, i.e. for any $\hat{\rho} \in \mathcal{H}_{\mathcal{S}}^{\mathrm{Tr}=1;\geq 0}$:

$$[\hat{\rho}]^q \neq \{\hat{\rho}\}, \qquad [\hat{\rho}]^p \neq \{\hat{\rho}\}. \tag{18.41}$$

---

[21]See also [BUSCH/LAHTI 1989, 645–48] [*BUSCH et al.* 1995a, 123–24] [GALE *et al.* 1968] and sections 28.4 and 28.5.

[22]See also [VON WEIZSÄCKER 1987].

POVM But we know that unsharp observables are a smearing operation on sharp ones [see theorem 18.5: p. 301]. Now this smearing operation on sharp observables, say $\mathbf{p}, \mathbf{q}$, can be understood as coarse-graining operation on a single fine observable. This operation can now have an informationally complete refinement as a result [again see comments to Eq. (18.14)] [*BUSCH et al.* 1995a, 123–24] [*SCHROECK* 1996, 55–59]:

Theorem 18.9 (Informational Completeness of Unsharp Observables) *A Set of (partially) non-commuting unsharp observables can be informationally complete.*

Note that the two aspects of measurement (preparatory and determinative) are mutually exclusive: *determinative measurements*, by means of which we aim to obtain a determined value of a chosen observable, require measurements of strongly non-commutative quantities, whereas *preparatory measurements* are optimal via measurements of complete sets of commuting observables [BUSCH/LAHTI 1989, 634, 669, 672–76].[23] We shall discuss the point in the next section. For further developments see also subsection 33.2.3.

18.7 Statistical Treatment of Definition and Observation*

To acquire information about a system can signify two different things [*HELSTROM* 1976, 1]: either to *decide* what hypothesis describes the system (= definition) or to *estimate* the value of some parameters related to the dynamics of the system (= measurement or observation). This section is devoted to these two topics: the decision theory in subsection 18.7.1 and the estimation theory in subsection 18.7.2.

The importance of the following approach is this: it represents a rigorous framework for an operational approach to the extent in which, on the one hand, it treats the definition of a system in terms of strategies aiming to acquire information and measurements in terms of estimates, and, on the other hand, it treats both problems in terms of cost and risk functions. In fact, epistemologically, for an operational approach theories and concepts are instruments[24]. However, the following formalism is not dependent on an operational approach as such, and can be used for a wide class of problems.

18.7.1 Decision Theory

Classical Decision Theory

The statistical decision theory is called detection theory. First we expose some elementary notions of the Classical detection theory [*HELSTROM* 1976, 8–10]. We call a strategy or a hypothesis a *decision procedure*. If it involves a random element, then there is a probability $\wp_{\mathrm{H}_j}(\mathbf{d})$ that the Hypothesis H_j $(j = 1, 2, \ldots, N)$ is chosen when the set $\mathbf{d} = \{d_1, d_2, \ldots, d_n\}$ of data is observed. The probability that the hypothesis H_j is chosen, when the hypothesis H_k is true, is:

$$\wp(\mathrm{H}_j|\mathrm{H}_k) = \int_{\Re^n} \wp_{\mathrm{H}_j}(\mathbf{d})\wp_k(\mathbf{d})d^n d, \tag{18.42}$$

[23] Examples of determinative and preparatory measurements of spin are in [*SCHROECK* 1996, 185–92, 196].
[24] Wide examination in [*DEWEY* 1929].

where $\wp_k(\mathbf{d})$ is the probability density function when the system is in state k. Upon this event the cost C_{jk} is paid — the numbers C_{jk} assign relative weights to the various possible errors and correct decisions. As H_k is true with *a priori* probability $\wp_k^{\mathrm{AP}}$, the average cost of the strategy is

$$\left\langle C[\wp_{\mathrm{H}_j}]\right\rangle = \sum_{j=1}^{N}\sum_{k=1}^{N} \wp_k^{\mathrm{AP}} C_{jk}\wp(\mathrm{H}_j|\mathrm{H}_k). \tag{18.43}$$

Defining for each H_j a risk function:

$$R_j(\mathbf{d}) = \sum_{k=1}^{N} \wp_k^{\mathrm{AP}} C_{jk}\wp_k(\mathbf{d}), \tag{18.44}$$

we write the average cost as:

$$\left\langle C[\wp_{\mathrm{H}_j}]\right\rangle = \int_{\Re}\sum_{j=1}^{N} R_j(\mathbf{d})\wp_{\mathrm{H}_j}(\mathbf{d})d^n\mathbf{d}, \tag{18.45}$$

and we now seek the N functions $\wp_{\mathrm{H}_j}(\mathbf{d})$ that satisfy general probability conditions and make the average cost as small as possible.

The risk functions $R_j(\mathbf{d})$ are proportional to the posterior risks $R_j^{\mathrm{P}}(\mathbf{d})$ of the hypothesis in view of the data observed:

$$R_j^{\mathrm{P}}(\mathbf{d}) = \sum_{i=1}^{N} C_{ji}\wp(\mathrm{H}_i|\mathbf{d}) = \frac{R_j(\mathbf{d})}{\wp(\mathbf{d})}, \tag{18.46}$$

where

$$\wp(\mathrm{H}_i|\mathbf{d}) = \wp_i^{\mathrm{AP}}\frac{\wp_i(\mathbf{d})}{\wp(\mathbf{d})} \tag{18.47}$$

is the *a posteriori* probability of Hypothesis H_i, while

$$\wp(\mathbf{d}) = \sum_{k=1}^{N} \wp_k^{\mathrm{AP}}\wp_k(\mathbf{d}) \tag{18.48}$$

is the overall joint probability density function of the data. Eq. (18.46) is Bayes' rule [Eq. (11.5)] in the context of decision theory. Bayes' rule results from an assignment of costs which penalizes all errors equally

$$C_{ij} := 1 \text{ for } i \neq j;\ C_{ii} := 0, \tag{18.49a}$$

or from one that equally rewards all correct decisions:

$$C_{ii} := -1;\ C_{ij} := 0 \text{ for } i \neq j. \tag{18.49b}$$

Quantum Decision Theory

POVM Decision Theory In the QM case [*HELSTROM* 1976, 90–93], the hypothesis H_j means that the system is in state $\hat{\rho}_j$.[25] In the case of POVM we associate with hypothesis effects $\hat{E}_{\mathrm{H}_1},\ldots,\hat{E}_{\mathrm{H}_N}$ [see subsection 18.1.1], which we name *detection operators*, non-negative definite Hermitian operators which sum to identity (completeness condition). Then we express

[25]On the quantum decision theory see also [BENIOFF 1972b].

the conditional probability that apparatus $\mathcal{A}$ chooses hypothesis H_j when H_k is true by [see Eq. (18.6)]

$$\wp(\mathrm{H}_j|\mathrm{H}_k) = \mathrm{Tr}(\hat{\rho}_k \hat{E}_{\mathrm{H}_j}). \tag{18.50}$$

The average cost of $\hat{E}_{\mathrm{H}_j}$ is given by [see Eq. (18.43)]:

$$\langle C \rangle = \sum_{j=1}^{N} \sum_{k=1}^{N} \wp_k^{\mathrm{AP}} C_{jk} \mathrm{Tr}(\hat{\rho}_k \hat{E}_{\mathrm{H}_j}) = \mathrm{Tr}\left(\sum_{j=1}^{N} \hat{R}_j \hat{E}_{\mathrm{H}_j} \right), \tag{18.51}$$

where the Hermitian detection risk operator $\hat{R}_j$ is defined by [see Eq. (18.44)]:

$$\hat{R}_j = \sum_{k=1}^{N} \wp_k^{\mathrm{AP}} C_{jk} \hat{\rho}_k. \tag{18.52}$$

If we define a new Hermitian operator $\hat{m}_R$ as follows

$$\hat{m}_R := \sum_{k=1}^{N} \hat{E}_{\mathrm{H}_k} \hat{R}_k = \sum_{k=1}^{N} \hat{R}_k \hat{E}_{\mathrm{H}_k}, \tag{18.53}$$

we can express the requirement of a cost-minimizing strategy by the following conditions:

$$(\hat{R}_j - \hat{m}_R)\hat{E}_{\mathrm{H}_j} = \mathbf{0}, \tag{18.54a}$$

$$\hat{R}_j - \hat{m}_R \geq \mathbf{0}. \tag{18.54b}$$

The role of $\hat{m}_R$ is that of a Lagrangian multiplier, and we name it *Lagrange operator*. Hence the minimum Bayes cost is the following:

$$\langle C \rangle_{\hat{m}} = \mathrm{Tr}(\hat{m}_R), \tag{18.55}$$

and the difference between the cost incurred by using the POVM $\hat{E}_{\mathrm{H}_j}$ [Eq. (18.51)] and the cost arising from the optimal strategy [Eq. (18.55)] is given by:

$$\langle C \rangle - \langle C \rangle_{\hat{m}} = \mathrm{Tr}\left(\sum_{j=1}^{N} (\hat{R}_j - \hat{m}_R) \hat{E}_{\mathrm{H}_j} \right), \tag{18.56}$$

which by Eq. (18.54b) implies

$$\langle C \rangle - \langle C \rangle_{\hat{m}} \geq 0. \tag{18.57}$$

The optimum POVM's form a convex set [Eq. (3.47)]:

$$\hat{E}_{\mathrm{H}_j} = w \hat{E}'_{\mathrm{H}_j} + (1-w) \hat{E}''_{\mathrm{H}_j} \tag{18.58}$$

for $0 < w < 1$. The corresponding Lagrange operator is:

$$\langle \hat{m} \rangle = w\hat{m}' + (1-w)\hat{m}'', \tag{18.59}$$

by means of which we rewrite Eq. (18.54b) in the following form:

$$\hat{R}_j - \langle \hat{m} \rangle = w(\hat{R}_j - \hat{m}') + (w-1)(\hat{R}_j - \hat{m}'') \geq 0. \tag{18.60}$$

The Lagrange operator can be eliminated from the optimization Eqs. (18.54) by writing them [by definition (18.53)] in the form:

$$\hat{E}_{\mathrm{H}_j}(\hat{R}_k - \hat{R}_j)\hat{E}_{\mathrm{H}_k} = \mathbf{0} \tag{18.61}$$

for all pairs j, k. Holevo has derived Eq. (18.61) by considering infinitesimal transformations of operator $\hat{E}_{\mathrm{H}_j}$ that preserves its POVM character.

PVM Decision Theory We analyse now specifically the case of pure states [*HELSTROM* 1976, 97–100]. Let $\mathcal{S}$ be in a pure state $|\psi_k\rangle$ under each hypothesis H_k $(k = 1, 2, \ldots, N)$. The optimum POVM can be confined in an N–dimensional subspace $\mathcal{H}_N$ (spanned by the N vectors $|\psi_k\rangle$) of $\mathcal{H}_\mathcal{S}$ and each component outside $\mathcal{H}_N$ does not contribute to the decision probability $\wp(\mathrm{H}_j|\mathrm{H}_k)$. In order to verify it, let $\hat{P}_N$ project arbitrary vectors $\in \mathcal{H}_\mathcal{S}$ onto $\mathcal{H}_N$ and write each POVM in the following simple identity:

$$\hat{E}_{\mathrm{H}_j} = \hat{P}_N \hat{E}_{\mathrm{H}_j} \hat{P}_N + (\mathbf{1} - \hat{P}_N)\hat{E}_{\mathrm{H}_j}\hat{P}_N + \hat{P}_N \hat{E}_{\mathrm{H}_j}(\mathbf{1} - \hat{P}_N) + (\mathbf{1} - \hat{P}_N)\hat{E}_{\mathrm{H}_j}(\mathbf{1} - \hat{P}_N). \tag{18.62}$$

Then Eq. (18.50) can be written:

$$\wp(\mathrm{H}_j|\mathrm{H}_k) = \mathrm{Tr}(\hat{\rho}_k \hat{E}_{\mathrm{H}_j}) = \langle\psi_k|\hat{E}_{\mathrm{H}_j}|\psi_k\rangle = \mathrm{Tr}(\hat{\rho}_k \hat{E}'_{\mathrm{H}_j}), \tag{18.63}$$

where

$$\hat{E}'_{\mathrm{H}_j} = \hat{P}_N \hat{E}_{\mathrm{H}_j} \hat{P}_N, \tag{18.64}$$

since $(\mathbf{1} - \hat{P}_N)|\psi_k\rangle = 0$ [in Eq. (18.62)]. The optimum POVM can be taken as the generalized resolution of the identity:

$$\sum_{j=1}^{N} \hat{E}'_{\mathrm{H}_j} + (\mathbf{1} - \hat{P}_N) = \mathbf{1}; \tag{18.65}$$

and since the term $(\mathbf{1} - \hat{P}_N)$ has no effect, we can drop it and confine ourselves to subspace $\mathcal{H}_N$. Now we try to find a solution of the optimization Eq. by means of projectors

$$\hat{P}_{\mathrm{H}_j} = |w_j\rangle\langle w_j| \tag{18.66}$$

which project onto N orthonormal vectors $|w_j\rangle$ spanning $\mathcal{H}_N$, called measurement states. Then Eq. (18.61) will be satisfied for all pairs j, k:

$$\langle w_k|(\hat{R}_j - \hat{R}_k)|w_j\rangle = 0, \tag{18.67a}$$

or by [see Eq. (18.52)]:

$$\sum_{i=1}^{N} \wp_i^{\mathrm{AP}}(C_{ji} - C_{ki})\langle w_k|\psi_i\rangle\langle\psi_i|w_j\rangle = 0, \tag{18.67b}$$

which provides $\frac{1}{2}N(N-1)$ Eqs.:

$$\sum_{i=1}^{N} \wp_i^{\mathrm{AP}}(C_{ji} - C_{ki})\eta_{ki}\eta^*_{ji} = 0 \tag{18.68}$$

for N^2 unknowns

$$\eta_{ij} = \langle w_i|\psi_j\rangle. \tag{18.69}$$

These are the components of the N state vectors $|\psi_j\rangle$ along the axes $|w_i\rangle$ of $\mathcal{H}_N$. The inequality (18.54b) requires matrices $T^{(i)}$ whose elements are:

$$T^{(i)}_{mn} = \sum_{j=1}^{N} \wp_j^{\mathrm{AP}}(C_{ij} - C_{mj})\eta_{mj}\eta^*_{nj} \tag{18.70}$$

to be nonnegative definite so that the Bayes cost can be at a true minimum. The special cost function Eq. (18.49b)

$$C_{ij} := -\delta_{ij} \tag{18.71}$$

corresponds to minimizing the average probability error $\wp^e = \langle C \rangle + 1$. Then Eq. (18.68) reduces to

$$\wp_m^{\rm AP} \eta_{km} \eta_{mm}^* = \wp_k^{\rm AP} \eta_{kk} \eta_{mk}^*. \tag{18.72}$$

The minimum attainable error is:

$$\wp_m^e = 1 - \sum_{j=1}^{N} \wp_j^{\rm AP} |\eta_{ii}|^2. \tag{18.73}$$

For this cost function and for N linearly independent states $|\psi_j\rangle$ the optimum POVM is indeed a PVM. But if the states are not linearly independent, the N $|\psi_j\rangle$ span a subspace of $\mathcal{H}_\mathcal{S}$ of dimension less than N and this subspace cannot accommodate N different orthogonal projectors of type (18.66). Then we cannot have PVMs.

In case of binary decision [*HELSTROM* 1976, 107–108], detection operators commute. The Lagrange operator is [Eq. (18.53)]:

$$\hat{m} = \hat{R}_0 \hat{E}_{\rm H_0} + \hat{R}_1 \hat{E}_{\rm H_1}, \tag{18.74}$$

and we have

$$\hat{R}_0 - \hat{m} = \hat{R}_0 - \hat{R}_0 \hat{E}_{\rm H_0} - \hat{R}_1 \hat{E}_{\rm H_1} = (\hat{R}_0 - \hat{R}_1) \hat{E}_{\rm H_1}. \tag{18.75}$$

Eq. (18.54a) is now:

$$(\hat{R}_0 - \hat{R}_1) \hat{E}_{\rm H_1} \hat{E}_{\rm H_0} = \mathbf{0}. \tag{18.76}$$

Since in general the operator $\hat{R}_0 - \hat{R}_1$ does not vanish we again have projectors because $\hat{E}_{\rm H_1} \hat{E}_{\rm H_0}$ are orthogonal.

18.7.2 Estimation Theory

Classical Estimation Theory

We synthesize now the classical estimation theory [*HELSTROM* 1976, 25–31]. The data $\mathbf{d} = \{d_1, d_2, \ldots d_n\}$ acquired by observing $\mathcal{S}$ are random variables whose joint probability density function $\wp(\mathbf{d}|\zeta)$ depends on certain parameters $\zeta_1, \zeta_2, \ldots \zeta_m$. The values of these parameters are unknown and are to be estimated from the data. The estimanda $\zeta_1, \zeta_2, \ldots \zeta_m$ are represented by an m-dimensional point ζ in parameter space $\mathcal{Z}$. The estimates are functions $\tilde{\zeta}_j(\mathbf{d})$ which describe a strategy for calculating the estimates from the data, and are called *estimators*. A common cost function for the estimate of a single parameter is the squared error:

$$C(\tilde{\zeta}, \zeta) = (\tilde{\zeta} - \zeta)^2. \tag{18.77}$$

For a number m of parameters we have

$$C(\tilde{\zeta}, \zeta) = \sum_{j=1}^{m} \sum_{k=1}^{m} f_{jk} (\tilde{\zeta}_j - \zeta_j)(\tilde{\zeta}_k - \zeta_k), \tag{18.78}$$

with the f_{ij} elements of a positive matrix.

The cost function

$$C(\tilde{\zeta}, \zeta) = -\prod_{j=1}^{m} \delta(\tilde{\zeta}_j - \zeta_j) \tag{18.79}$$

is the continuous counterpart of Eq. (18.49b). The data **d** are put into a chance device that selects a set ζ of estimates in such a way that it defines a probability density function which satisfies:

$$\int_{\mathcal{Z}} \wp(\tilde{\zeta}|\mathbf{d})d^m\tilde{\zeta} = 1. \tag{18.80}$$

If we assign to the parameters a joint *a priori* probability density function $\wp^{\mathrm{AP}}(\zeta)$, we can introduce a risk function in the following form:

$$R(\tilde{\zeta};\mathbf{d}) = \int_{\mathcal{Z}} \wp^{\mathrm{AP}}(\zeta)C(\tilde{\zeta},\zeta)\wp(\mathbf{d}|\zeta)d^m\zeta, \tag{18.81}$$

while the average cost is given by

$$\langle C\rangle = \int_{\mathcal{Z}}\int_{\mathcal{Z}} \wp^{\mathrm{AP}}(\zeta)C(\tilde{\zeta},\zeta)\wp^{\mathrm{P}}(\tilde{\zeta}|\zeta)d^m\zeta d^m\tilde{\zeta}, \tag{18.82}$$

where $\wp^{\mathrm{P}}(\tilde{\zeta}|\zeta)$ is the posterior probability density function of the estimate, which in order to express the probability that the estimate lies in a volume element $d^m\tilde{\zeta}$, is written as:

$$\wp^{\mathrm{P}}(\tilde{\zeta}|\zeta)d^m\tilde{\zeta} = \int_{\Re} \wp(\tilde{\zeta}|\mathbf{d})\wp(\mathbf{d}|\zeta)d^n\mathbf{d}d^m\tilde{\zeta}. \tag{18.83}$$

Quantum Estimation Theory

General Introduction The QM estimation theory works with the density operator $\hat{\rho}(\zeta)$ [*HELSTROM* 1976, 235–38].[26] The observational strategy for estimating the parameters $\{\zeta\}$ is a POVM. The resulting estimates are random variables, and the probability that they lie in a value region Υ of $\mathcal{Z}$ is given by:

$$\wp(\tilde{\zeta}\in\Upsilon|\zeta) = \mathrm{Tr}[\hat{\rho}(\zeta)\hat{E}(\Upsilon)]. \tag{18.84}$$

We suppose that the operators for finite regions can be formed as integrals of infinitesimal operators:

$$\hat{E}(\Upsilon) = \int_{\Upsilon} d\hat{E}(\tilde{\zeta}). \tag{18.85}$$

The joint conditional probability density function $\wp^{\mathrm{P}}(\tilde{\zeta}|\zeta)$ [see Eq. (18.83)] is given by:

$$\wp^{\mathrm{P}}(\tilde{\zeta}|\zeta)d^m\tilde{\zeta} = \mathrm{Tr}[\hat{\rho}(\zeta)\hat{E}(\tilde{\zeta})]. \tag{18.86}$$

The Hermitian risk operator is given by

$$\hat{R}(\tilde{\zeta}) = \int_{\mathcal{Z}} \wp^{\mathrm{AP}}(\zeta)C(\tilde{\zeta},\zeta)\hat{\rho}(\zeta)d^m\zeta, \tag{18.87}$$

in terms of which the average cost is

$$\langle C\rangle = \mathrm{Tr}\left(\int_{\mathcal{Z}} \hat{R}(\tilde{\zeta})d\hat{E}(\tilde{\zeta})\right). \tag{18.88}$$

The Eqs. for the optimum observational strategy [see Eqs. (18.54)] are given by:

$$[\hat{R}(\tilde{\zeta}) - \hat{m}_R(\tilde{\zeta})]d\hat{E}(\tilde{\zeta}) = \mathbf{0}, \tag{18.89a}$$

$$\hat{R}(\tilde{\zeta}) - \hat{m}_R(\tilde{\zeta}) \geq \mathbf{0}, \tag{18.89b}$$

[26] See also [*HOLEVO* 1982, 106, 169–74].

where

$$\hat{m}_R(\tilde{\zeta}) = \int_{\mathcal{Z}} \hat{R}(\tilde{\zeta}) d\hat{E}(\tilde{\zeta}); \tag{18.90}$$

and the minimum cost of error is again [Eq. (18.55)]:

$$\langle C \rangle_{\hat{m}} = \mathrm{Tr}[\hat{m}_R(\tilde{\zeta})]. \tag{18.91}$$

For the δ–function cost

$$C(\tilde{\zeta}, \zeta) = -\prod_{k=1}^{m} \delta(\tilde{\zeta}_k - \zeta_k) \tag{18.92}$$

it is convenient to replace the risk function by its negative and look for a POVM for which

$$[\hat{m}^j_R(\tilde{\zeta}) - \hat{R}_j(\tilde{\zeta})] d\hat{E}_j(\tilde{\zeta}) = \mathbf{0}, \tag{18.93a}$$

$$\hat{m}^j_R(\tilde{\zeta}) - \hat{R}(\tilde{\zeta})_j \geq \mathbf{0}, \tag{18.93b}$$

where

$$\hat{m}^j_R(\tilde{\zeta}) = \int_{\mathcal{Z}} \hat{R}_j(\tilde{\zeta}) d\hat{E}_j(\tilde{\zeta}). \tag{18.94}$$

Estimate of the Wave Function A specific problem of great interest is the estimate of a wave function [see also section 28.5]. A wave function can be estimated [*HELSTROM* 1976, 241–43], provided that the corresponding state vector $|\psi\rangle$ is known to lie in a Hilbert space of finite dimensions, and a residual uncertainty is accepted. Select an orthonormal basis $|b_k\rangle$ and consider an estimate of the complex coefficient $\mathbf{c} = (c_1, c_2, \ldots, c_n)$ — where only $2n - 2$ coefficients are independent —:

$$|\psi(\mathbf{c})\rangle = \sum_{k=1}^{n} c_k |b_k\rangle. \tag{18.95}$$

The n complex numbers $\tilde{c}_k = c_{kx} + \imath c_{ky}$ are parameters of the density operator $\hat{\rho}(\mathbf{c})$. Because the vector $|\psi(\mathbf{c})\rangle$ must have unit length, the points $\tilde{\mathbf{c}} = (c_{1x}, c_{1y}, \ldots, c_{nx}, c_{ny})$ must lie on a $(2n)$–dimensional hypersphere $\mathbf{S}_{2n}$ of radius 1. Nothing being known in advance about the state of $\mathcal{S}$, we suppose that the point $\tilde{\mathbf{c}}$ may be anywhere on the surface, assigning to it an *a priori* probability density function

$$\wp^{\mathrm{AP}}(\tilde{\mathbf{c}}) = A_{2n}^{-1} = \frac{1}{2}\Gamma(n)\pi^{-n}, \tag{18.96}$$

where A is the area of the sphere and Γ is the gamma-function. The maximum likelihood estimator is the POVM

$$d\hat{E}(\tilde{\mathbf{c}}) = nA_{2n}^{-1} |\psi(\tilde{\mathbf{c}})\rangle\langle\psi(\tilde{\mathbf{c}})| d\mathbf{S} = nA_{2n}^{-1} \sum_{k=1}^{n}\sum_{j=1}^{n} \tilde{c}_k \tilde{c}^*_j |b_k\rangle\langle b_j| d\mathbf{S}. \tag{18.97}$$

When $d\hat{E}(\tilde{\mathbf{c}})$ is integrated over the hypersphere the terms $k \neq j$ vanish. The Lagrange operator defined by Eq. (18.94) is:

$$\hat{m}^j_R(\tilde{\zeta}) = nA_{2n}^{-2} \int_{\mathbf{S}_{2n}} |\psi(\tilde{\mathbf{c}})\rangle\langle\psi(\tilde{\mathbf{c}})| d\mathbf{S} = nA_{2n}^{-1}\mathbf{1}, \tag{18.98}$$

so that

$$\hat{m}^j_R(\tilde{\zeta}) - \hat{R}_j(\mathbf{c}) = A_{2n}^{-1}[\mathbf{1} - \hat{\rho}(\tilde{c})]. \tag{18.99}$$

The joint conditional probability density function of the estimate $\tilde{c}$ is:

$$\wp^{\mathrm{P}}(\tilde{\mathbf{c}}|\mathbf{c}) = nA_{2n}^{-2}|\langle\psi(\tilde{\mathbf{c}})|\psi(\mathbf{c})\rangle|^2. \tag{18.100}$$

It is independent of the basis vectors $|b_k\rangle$. We can consider

$$|\psi(E)\rangle = \sum_{k=1}^{n} \tilde{c}_k |b_k\rangle \tag{18.101}$$

as an estimate of the state vector $|\psi\rangle$. Then the following absolute value $|\langle\psi(E)|\psi\rangle|$ measures how close the estimated $|\psi(E)\rangle$ lies to the true $|\psi\rangle$ (it is $= 1$ if the estimate is exact).[27]

By similar procedures it is possible to furnish an estimate of the position[28] and of the time-of-arrival[29].

18.8 Weak Measurement

18.8.1 Generals Remarks on Weak Measurement

A *weakening* of the measurement of an observable $\hat{O}$ consists in the sacrifice of the accuracy of measurement of $\hat{O}$ in order to gain some control of the perturbation caused by such measurement to observables which fail to commute with $\hat{O}$ [AHARONOV *et al.* 1987, 199–202]. Briefly a weak measurement preserves a certain commutability, in consequence of which the order of the measurement has no relevance — contrary to what happens ordinarily in QM [see corollary 3.1: p. 32]. Hence weak observable is a special case of unsharp observables [see theorem 18.2: p. 296, and its corollary].

But it is true that, in the case of a succession of weak measurements, they can build up interference with the consequence of a final result far away from the expected values. Clearly this is not the case by orthogonal measurements, where the orthogonality assures that there can be no overlap at all[30].

Hence we give following definition of a *weak value* [AHARONOV/VAIDMAN 1990, 11]:

Definition 18.7 (Weak Value) *A weak value is a physical property of a quantum system between two measurements, i.e. a property of a system belonging to an ensemble that is both preselected and postselected.*

On the problem of postselection and preselection see also [RAUCH 1995]. The weak value of a variable may differ significantly from the eigenvalues of an associated operator. This weakens considerably the relation between observables and associated operators.

18.8.2 Weak Measurement and Time Order

Two measurements We know that the measurement is irreversible. We consider here two successive measurements [AHARONOV/VAIDMAN 1990, 12], say a preparatory one followed by

[27] Recently an universal algorithm for optimal estimation of quantum states has been developed [DERKA *et al.* 1998].

[28] See [*HELSTROM* 1976, 250–52].

[29] See [*HELSTROM* 1976, 277].

[30] Exactly this point was criticized by Busch [BUSCH 1988].

a determinative one. The proposal is to describe the system with two wave functions evolving in opposite directions of time. At initial time t_i, $\hat{O}_i$ is measured and a non-degenerate eigenvalue o_i is found; at t_f, $\hat{O}_f$ is measured and a non-degenerate eigenvalue o_f is found. At intermediate time t the system is described by $|\psi_i\rangle$ (toward future) and $\langle\psi_f|$ (toward past):[31]

$$|\psi_i\rangle = e^{-\imath\int_{t_i}^{t}\hat{H}d\tau}|\hat{O}_i = o_i\rangle, \tag{18.102a}$$

$$\langle\psi_f| = \langle\hat{O}_f = o_f|e^{-\imath\int_{t}^{t_f}\hat{H}d\tau}. \tag{18.102b}$$

We consider for simplicity the free Hamiltonian of the system to be zero. Now the description suggests that at t such that $t_i \leq t \leq t_f$ both $\hat{O}_i = o_i$ and $\hat{O}_f = o_f$. Surely if at t, $\hat{O}_i$ is measured, then we obtain o_i, and if $\hat{O}_f$ is measured, then we obtain o_f. But it may not be the case that at t we obtain a result $o_i \wedge o_f$, because $\hat{O}_i, \hat{O}_f$ may be non-commuting. So both $\mathrm{v}[\hat{O}_i] = o_i$ and $\mathrm{v}[\hat{O}_f] = o_f$ (where 'v' stays for 'value of') are correct if only one measurement is performed. If both were measured successively, in the order $\hat{O}_f$ after $\hat{O}_i$, then the outcome of both may be different.

Weakening Therefore, to apply our description, we must use measurements which do not significantly change either of the two wave functions. Weakening a measurement will necessarily decrease the accuracy of a single measurement such that at the end it will provide almost no information. A weak measurement of $\hat{O}_i$ now has the form [a generalization of Eq. (3.93)]:

$$\hat{O}_i^w = \frac{\langle\psi_f|\hat{O}_i|\psi_i\rangle}{\langle\psi_f|\psi_i\rangle}, \tag{18.103}$$

where the superscript w stands to indicate 'weak'.

A model We consider now [AHARONOV/VAIDMAN 1990, 13–14] an ensemble of one-dimensional systems all members of which are described by $|\psi_i\rangle$ and $\langle\psi_f|$, and a measuring apparatus for each system. The initial state of each $\mathcal{A}$ is assumed to be the Gaussian:

$$\frac{1}{\sqrt{\Delta q}(2\pi)^{\frac{1}{4}}}e^{-\frac{q^2}{4(\Delta q)^2}}. \tag{18.104}$$

We measure the momentum $\hat{p}_j$ of each apparatus after the interaction. Subsequently we perform the final, postselection measurements on the systems of our ensemble. We collect the outcomes p_j only of those systems for which the final state is $|\psi_f\rangle$. The state of each measuring device that has been postselected is:

$$\begin{aligned}\langle\psi_f|e^{-\imath\int\hat{H}dt}|\psi_i\rangle e^{-\frac{q^2}{4(\Delta q)^2}} &= \langle\psi_f|e^{-\imath q\hat{O}_i^w}|\psi_i\rangle e^{-\frac{q^2}{4(\Delta q)^2}} \\ &= \sum_{n=0}^{\infty}\frac{(\imath q)^n}{n!}\langle\psi_f|(\hat{O}_i^n)_w|\psi_i\rangle e^{-\frac{q^2}{4(\Delta q)^2}} \\ &= \langle\psi_f|\psi_i\rangle\sum_{n=0}^{\infty}\frac{(\imath q)^n}{n!}(\hat{O}_i^n)_w e^{-\frac{q^2}{4(\Delta q)^2}}. \end{aligned} \tag{18.105}$$

[31] See also [AHARONOV *et al.* 1964] .

The last expression can be rewritten as the initial wave function of $\mathcal{A}$ multiplied by $e^{\imath q\hat{O}_i^w}$ plus a correction term, which is negligible for small $\Delta\hat{q}$:

$$\langle\psi_f|\psi_i\rangle\sum_{n=0}^{\infty}\frac{(\imath q)^n}{n!}(\hat{O}_i^n)_w e^{-\frac{q^2}{4(\Delta q)^2}} = \langle\psi_f|\psi_i\rangle e^{\imath q\frac{\langle\psi_f|\hat{O}_i|\psi_i\rangle}{\langle\psi_f|\psi_i\rangle}}e^{-\frac{q^2}{4(\Delta q)^2}} + \langle\psi_f|\psi_i\rangle\sum_{n=0}^{\infty}\frac{(\imath q)^n}{n!}\left[(\hat{O}_i^n)_w-(\hat{O}_i^w)^n\right]e^{-\frac{q^2}{4(\Delta q)^2}}. \quad (18.106)$$

By taking $\Delta\hat{q}$ such that for all $n\geq 2$

$$(2\Delta\hat{q})^n\frac{\Gamma(n/2)}{(n-2)!}\left|(\hat{O}_i^n)_w-(\hat{O}_i^w)^n\right|\ll 1, \quad (18.107)$$

where Γ is the gamma-function[32], we can neglect the second term of Eq. (18.106) and, by Fourier transforming, express the final wave function of $\mathcal{A}$ in momentum representation as:

$$\exp\left[-(\Delta\hat{q})^2\left(\hat{p}-\frac{\langle\psi_f|\hat{O}_i|\psi_i\rangle}{\langle\psi_f|\psi_i\rangle}\right)^2\right]. \quad (18.108)$$

The probability distribution of $\hat{p}$ is a Gaussian with spread $\Delta\hat{p}=(2\Delta\hat{q})^{-1}$ centered at $\hat{p}=Re(\hat{O}_i^w)$. $\hat{O}_i^w$ can also have an imaginary part that affects the distribution of $\hat{q}$:

$$e^{\imath q\mathrm{Re}(\hat{O}_i^w)}e^{-\frac{[q+2(\Delta q)^2\mathrm{Im}(\hat{O}_i^w)]^2}{4(\Delta q)^2}}. \quad (18.109)$$

Hence the probability distribution of $\hat{q}$ is also a Gaussian with spread $\Delta\hat{q}$ centered at $\hat{q}=-2(\Delta\hat{q})^2\mathrm{Im}(\hat{O}_i^w)$. Performing measurements on N systems prepared in the same way will reduce the uncertainty by a factor of $1/\sqrt{N}$. Property (18.103) implies that if $\hat{O}_{if}=\hat{O}_i+\hat{O}_f$, then $\hat{O}_{if}^w=\hat{O}_i^w+\hat{O}_f^w$. We call this last property the linearity of weak values. But the value of $\hat{O}_{if}^w$ is o_i+o_f, even if $\hat{O}_i,\hat{O}_f$ are not commuting. The weak values can differ very much from the eigenvalues. In particular $\mathrm{Re}(\hat{O}_i^w)$ can be much bigger (smaller) than the maximum (minimum) eigenvalue of $\hat{O}_i$.

18.8.3 An Example

In [AHARONOV *et al.* 1988] an example of weak measurement is considered. Let us consider an ensemble of particles with initial state $|\psi_i\rangle$ and final state $|\psi_f\rangle$. At the time in between we switch on the measurement interaction. After the postselection the state of $\mathcal{A}$ is:

$$\langle\psi_f|e^{-\imath\int\hat{H}dt}|\psi_i\rangle e^{-\frac{q_{\mathcal{A}}^2}{4(\Delta q)^2}}\simeq\langle\psi_f|\psi_i\rangle e^{\imath q\frac{\langle\psi_f|\hat{O}|\psi_i\rangle}{\langle\psi_f|\psi_i\rangle}}e^{-\frac{q^2}{4(\Delta q)^2}}, \quad (18.110)$$

where $\hat{O}$ is the observable to be measured. This formula is valid if the spread is small:

$$\Delta\hat{q}\ll\max_n\frac{|\langle\psi_f|\psi_i\rangle|}{|\langle\psi_f|\hat{O}^n|\psi_i\rangle|^{\frac{1}{n}}}. \quad (18.111)$$

[32]For each real $s>0$ and each natural number n, the gamma-function is defined as $\Gamma(s+n)=(s+n-1)\cdots(s+1)s\Gamma(s)$, where $\Gamma(s)=\int_{0+}^{\infty}\eta^{s-1}e^{-\eta}d\eta$. If $s=1$ then $\Gamma(n+1)=n!$.

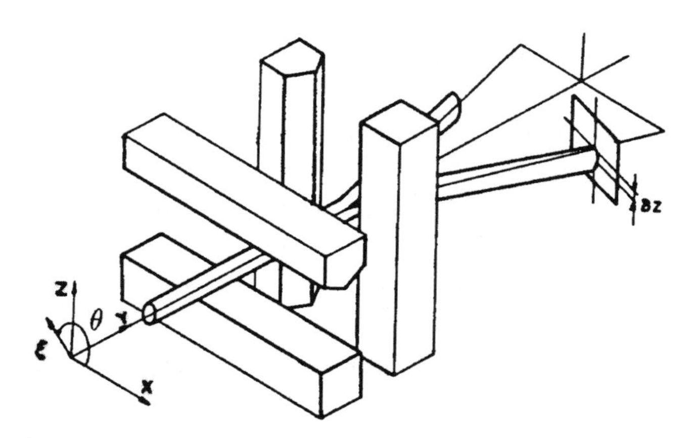

Figure 18.1: The experimental device for measurement of the weak value of $\hat{\sigma}_z$. The beam of particles with the spin pointed in the direction ξ passes through an inhomogeneous (in the z direction) magnetic field and is split by the strong magnet in the x direction. The beam of particles with $\hat{\sigma}_x = 1$ comes toward the screen and the deflection of the spot on the screen in the z direction is proportional to the weak value $\hat{\sigma}_{z_w} = (\delta z p_0 \mu / l)(\partial B_z / \partial z)^{-1}$ — from [AHARONOV *et al.* 1988, 1352].

The particles go in the y–direction and pass through a I SGM [see figure 18.1] where there is a weak interaction (weak magnetic field), with the consequence of a small shift p_z correlated to the two values of $\hat{\sigma}_z$. Then they pass through a II SGM (a ordinary one) which separates the beam into two subbeams which correspond to the two values of $\hat{\sigma}_x$ (SGM-Exp. 2).

We are interested only in $\hat{\sigma}_x = +1$. Then we obtain the weak value:

$$\hat{\sigma}_{z,w} = \frac{\langle \uparrow_x | \hat{\sigma}_z | \uparrow_\xi \rangle}{\langle \uparrow_x \mid \uparrow_\xi \rangle} = \tan \frac{1}{2}\theta, \tag{18.112}$$

where ξ is spin direction before the I SGM, θ the angle between x and ξ and $\langle \uparrow_x \mid$ is the postselected state and $\mid \uparrow_\xi \rangle$ the preselected one.

18.8.4 A Generalization of the Example

Let us now return to the case of two observables $\hat{O}_i$ and $\hat{O}_f$ [AHARONOV/VAIDMAN 1990, 14–15] as discussed before, and let $\hat{O}_i = \hat{\sigma}_x$ and $\hat{O}_f = \hat{\sigma}_\xi$. Now the operator $\hat{O}^w_{if} := \hat{O}_i + \hat{O}_f$ is proportional to $\hat{\sigma}_{\mathbf{e}}$ where $\mathbf{e}$ is the unit vector bisecting θ [see figure 18.2]:

$$\hat{O}^w_{if} = \hat{\sigma}_x + \hat{\sigma}_\xi = 2\cos\frac{\theta}{2}\hat{\sigma}_{\mathbf{e}}. \tag{18.113}$$

If the particles were initially in the state $|\hat{\sigma}_x = 1\rangle$ and found finally in the state $|\hat{\sigma}_\xi = 1\rangle$, the weak values of $\hat{\sigma}_x$ and $\hat{\sigma}_\xi$ will also be 1 for intermediate times. Now we find that the weak value of $\hat{\sigma}_{\mathbf{e}}$ is:

$$(\hat{\sigma}_{\mathbf{e}})_w = \frac{(\hat{\sigma}_x)_w + (\hat{\sigma}_\xi)_w}{2\cos\frac{\theta}{2}} = \frac{1}{\cos\frac{\theta}{2}}. \tag{18.114}$$

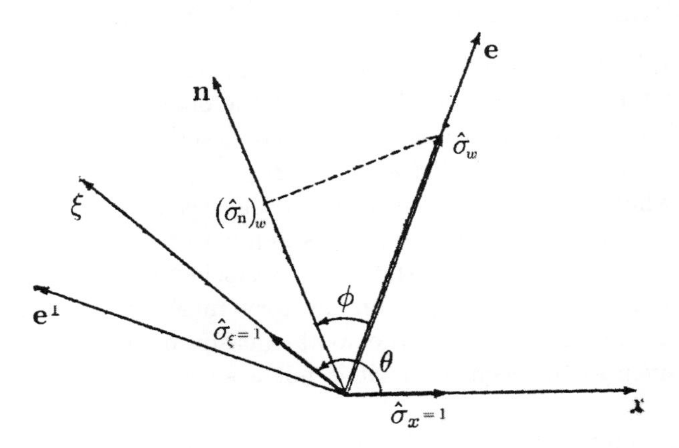

Figure 18.2: Geometry of weak values of spin components of a spin–1/2 particle in the xy plane — from [AHARONOV/VAIDMAN 1990, 14].

We draw in the x–y plane a vector of size $1/\cos(\theta/2)$ pointing in the **e** direction. Its projection on any axis in the x–y plane yields the weak value of this component. The spin component in any direction **n** in the x–y plane can be decomposed along the orthogonal directions **e** and $\mathbf{e}^{\perp}$ where:

$$\mathbf{e} := \frac{\xi + x}{2\cos(\theta/2)}; \quad \mathbf{e}^{\perp} := \frac{\xi - x}{2\cos(\theta/2)}. \tag{18.115}$$

The decomposition is:

$$\hat{\sigma}_{\mathbf{n}} = \hat{\sigma}_{\mathbf{e}} \cos\phi + \hat{\sigma}_{\mathbf{e}^{\perp}} \sin\phi. \tag{18.116}$$

In the case discussed $(\hat{\sigma}_x = 1, \hat{\sigma}_\xi = 1)$ we have:

$$(\hat{\sigma}_{\mathbf{e}^{\perp}})_w = \frac{(\hat{\sigma}_\xi)_w - (\hat{\sigma}_x)_w}{2\cos\theta/2} = 0, \tag{18.117}$$

and therefore

$$(\hat{\sigma}_{\mathbf{n}})_w = (\hat{\sigma}_{\mathbf{e}})_w \cos\phi. \tag{18.118}$$

The weak value of the spin in the z direction is the imaginary:

$$(\hat{\sigma}_z)_w = \frac{\langle \hat{\sigma}_\xi = 1 | \hat{\sigma}_z | \hat{\sigma}_x = 1 \rangle}{\langle \hat{\sigma}_\xi = 1 | \hat{\sigma}_x = 1 \rangle} = \imath \tan(\theta/2). \tag{18.119}$$

In conclusion the weak vector $\hat{\sigma}_w$ with axis $\mathbf{e}, \mathbf{e}^{\perp}, z$ will be [see Eqs. (18.114), (18.117) and (18.119)]:

$$\hat{\sigma}_w = \left[\frac{1}{\cos(\theta/2)}, 0, \imath \tan(\theta/2)\right], \tag{18.120}$$

which apart from its complexity, as a consequence of the linearity property of weak values, behaves similar to a classical vector because it is equivalent to the measurement of two weak observables corresponding to sharp non commuting ones.

18.8.5 Concluding Remarks on the Weak Measurement

As we have already said, the weak measurement is a special case of unsharp measurement — see also [BUSCH/SCHROECK 1989, 854–55]. In fact the operator $\hat{O}^{w}_{if} = \hat{O}_i + \hat{O}_f$, even if both operators on the RHS do not commute, is constructed exactly as a joint measurability of two unsharp observables [see subsections 18.2.1 and 18.2.2].

But if this is so, what is the specific interest of the weak measurement? The interest is that here we do not perform a contemporary measurement of two non commuting observables but two successive ones, one preselected, the other postselected. Hence, as we have said, the first measurement can be interpreted as a preparatory measurement and the second one as a determinative measurement. Therefore the weak measurement can show a very important property of the exclusiveness between both: it is not a sharp, yes/no alternative but a gentle one.

Another interesting point is that a weak measurement can help to find boundaries between classical and QM systems (the limit where both observables commute completely), and also between reversible QM systems (subjected to unitary evolution) and irreversible ones (subjected to some form of measurement) [see also chapter 27 and sections 28.4 and 28.5].

18.9 Discrimination Between Non-Orthogonal States*

The problem The assumption of non-orthogonality made by stochastic theories poses the problem of the strategies for discriminating between non-orthogonal states. We can find the answer by using an indirect measurement [IVANOVIĆ 1987] [DIEKS 1988] [see section 19.3 for details].[33] The problem is that, for two initial non-orthogonal states $|\psi_i\rangle$ and $|n_i\rangle$, there is always a finite 'overlap'. We can choose two different procedures in order to discriminate them:

- we can try to find in the average, a maximum number of correct classifications and a minimum number of incorrect ones;
- or we can try to find a procedure which in the maximum number of cases enables one to infer with certainty whether the system $\mathcal{S}$ is in $|\psi_i\rangle$ or $|n_i\rangle$.

Dieks' procedure The second procedure requires a classification without error. The problem now is that there is no certainty of always arriving at a classification, and there is a finite probability that the procedure will lead to no classification. Hence we must find a method which minimizes the latter probability. Suppose that we have maximal uncertainty about the state of the system before the measurement. Then the probability that the measurement leads to a conclusive result is $\frac{1}{2}(1 - |\langle\psi_i|n_i\rangle|^2)$. Now we use an indirect measurement [more on it in section 19.3] by introducing another system $\mathcal{S}'$ in initial state $|\varsigma_i\rangle$ and letting it interact with $\mathcal{S}$. Consequent upon the possibility that $\mathcal{S}$ can be in $|\psi_i\rangle$ or $|n_i\rangle$ we have two different possible unitary transformations, which can be defined in the following form:

$$|\psi_i\varsigma_i\rangle \mapsto c_1|\psi_1\varsigma_1\rangle + c_2|\psi_2\varsigma_2\rangle, \tag{18.121a}$$
$$|n_i\varsigma_i\rangle \mapsto c'_1|n_1\varsigma_1\rangle + c'_2|n_2\varsigma_2\rangle, \tag{18.121b}$$

where we assume $\langle\varsigma_1|\varsigma_2\rangle = 0$ and $\langle\psi_1|n_1\rangle = 0$. If now a yes–no measurement is performed on $\mathcal{S}'$ such that $|\varsigma_1\rangle$ with probability one gives the answer 'yes' and $|\varsigma_2\rangle$ the answer 'no', we can

[33] See also [PERES 1988a].

then discriminate with certainty whether $|\psi_i\rangle$ or $|n_i\rangle$ was the initial state of $\mathcal{S}$. By the supposed state of maximal uncertainty of the initial state of $\mathcal{S}$ we have following probability for our yes-no experiment:

$$\wp(\varsigma_1) = \frac{1}{2}(|c_1|^2 + |c_1'|^2) = 1 - \frac{1}{2}(|c_2|^2 + |c_2'|^2). \tag{18.122}$$

By the orthogonality $\langle\varsigma_1|\varsigma_2\rangle = 0$ and $\langle\psi_1|n_1\rangle = 0$, and the unitarity of transformations (18.121), we have:

$$|\langle\psi_i|n_i\rangle| = |c_2||c_2'||\langle\psi_2|n_2\rangle|, \tag{18.123}$$

from which it follows:

$$|c_2||c_2'| \geq |\langle\psi_i|n_i\rangle|. \tag{18.124}$$

The minimum value of $|c_2|^2 + |c_2'|^2$ compatible with Eq. (18.124) occurs when $|c_2|^2 = |c_2'|^2 = |\langle\psi_i|n_i\rangle|$. Hence:

$$\wp(\varsigma_1) \leq 1 - |\langle\psi_i|n_i\rangle|, \tag{18.125}$$

where $\wp(\varsigma_1) = 1 - |\langle\psi_i|n_i\rangle|$ holds if $|\psi_2\rangle = |n_2\rangle$ (not considering phase factors), a situation which is always possible to realize. Thus, while the fraction of incorrect classifications by the first average method will be, in the long run, $1/2|\langle\psi_i|n_i\rangle|^2$, by the method just developed, in the long run, the fraction of unclassified systems will be $|\langle\psi_i|n_i\rangle|$. Hence the gain in terms of certainty signifies a price in terms of the probability of a successful operation because

$$|\langle\psi_i|n_i\rangle| > \frac{1}{2}|\langle\psi_i|n_i\rangle|^2, \tag{18.126}$$

unless $\langle\psi_i|n_i\rangle = 0$.

Jaeger and Shimony's generalization Jaeger and Shimony [JAEGER/SHIMONY 1995] extended the preceding analysis: they used an arbitrary proportion of the states of interest (not necessarily 1/2 for both), treated the problem with Hilbert spaces of arbitrary dimensions, and faced in more detail the first question posed at the beginning — how can we find in the average a maximum number of correct classifications and a minimum of incorrect ones. For the case in which there is a proportion μ of states prepared in $|\psi_i\rangle$ and a proportion $\nu = 1 - \mu$ in $|n_i\rangle$ (with $\mu \geq \nu$), they obtained a probability:

$$\wp = \frac{1}{2} + \frac{1}{2}(1 - 4\mu\nu|\langle\psi_i|n_i\rangle|^2)^{\frac{1}{2}}, \tag{18.127}$$

which in the case considered by Dieks ($\mu = \nu = 1/2$) becomes:

$$\wp = \frac{1}{2} + \frac{1}{2}(1 - |\langle\psi_i|n_i\rangle|^2)^{\frac{1}{2}}, \tag{18.128}$$

which is a better probability than that found by Dieks [Eqs. (18.125)–(18.126)].

18.10 Stochastic Analysis of Spin*

We now wish to apply stochastic methods to a concrete analysis of an observable: the spin. The choice is not accidental, because the analysis of sharp spin observable presents a lot of difficulties and unsolved problems[34].

[34] In [*SCHROECK* 1996, 159] [*BUSCH et al.* 1995a, 72] the problems with the traditional analysis of spin are shown. The beginnings of such an analysis are in [PRUGOVEČKI 1977b].

18.10.1 Sharp Treatment

If we analyse a SGM experiment [BUSCH/SCHROECK 1989, 822] we see that the beam separation fails to be perfect even for silver atoms: 1. the vanishing of the divergence of the magnetic field leads to unwanted field components which blur the desired momentum transfer to be produced in the preferred field direction. 2. the spreading of wave packets leaving the magnetic field region causes a partial overlap of spin up and spin down, so that the outcomes are endowed with uncertainty. Hence we cannot speak of the spin of a particle (numerical values) but of unsharp distribution values. Also we cannot assume the Ignorance interpretation [section 6.5] (such that the spin is in a pure but unknown state).

To verify it let us consider the following experiment [BUSCH/SCHROECK 1989, 823]. Let $\mathbf{B} = (B_0 + zb)\mathbf{e}_3$. We assume a Hamiltonian in the field region of the form $\hat{H} = \hat{\mathbf{p}}^2/2M + \mu \cdot \mathbf{B}$ (where $\mu = \mu\hat{\sigma}, \mu = g_s(e/2mc)$). Let us take [see Eq. (3.127)]:

$$\hat{O}_\pm = \hat{I} \otimes \frac{(\hat{I} \pm \hat{\sigma}_3)}{2} = \hat{I} \otimes \hat{O}_{\pm\mathbf{e}_3}. \tag{18.129}$$

We have $\hat{O}_+ + \hat{O}_- = \hat{I}$ and $\hat{O}_+ \cdot \hat{O}_- = 0$. Hence:

$$|\psi\rangle = \hat{O}_+|\psi\rangle + \hat{O}_-|\psi\rangle = c_+|\psi_+\rangle + c_-|\psi_-\rangle, \tag{18.130}$$

where: $c_+ = \parallel \hat{O}_+\psi \parallel, c_- = \parallel \hat{O}_-\psi \parallel$ and $|\psi_+\rangle = c_+^{-1}\hat{O}_+|\psi\rangle$, $|\psi_-\rangle = c_-^{-1}\hat{O}_-|\psi\rangle$ if $c_\pm \neq O$; if $c_\pm = 0$ put $|\psi_\pm\rangle = 0$ too. Putting $\hbar = 1$ we have following transformation by entering in SGM device (at time τ):

$$\psi(\mathbf{r}) \rightsquigarrow c_+e^{-\imath\tau\mu(B_0+bz)}\psi_+(x,y,z) + c_-e^{\imath\tau\mu(B_0+bz)}\psi_-(x,y,z) \equiv \psi_\tau(\mathbf{r}). \tag{18.131}$$

Fourier transforming to momentum space gives:

$$\tilde{\psi}(\mathbf{k}) \rightsquigarrow c_+e^{-\imath\tau\mu B_0}\tilde{\psi}(k_1,k_2,k_3+\tau\mu b) + c_-e^{\imath\tau\mu B_0}\tilde{\psi}(k_1,k_2,k_3-\tau\mu b) \equiv \tilde{\psi}_\tau(\mathbf{k}), \tag{18.132}$$

producing a shift in the third momentum component. It is argued that the two components $|\psi_+\rangle, |\psi_-\rangle$ separate. It can be easily proved that $\langle\psi_+|\psi_-\rangle = 0$. But we understand something else: that we can measure the two components separately. Suppose that each particle exits the SGM at a well-defined port. A classical particle exiting the port with momentum in cone Δ of momentum space will strike the screen in Δ', the intersection of Δ with the screen. Let g be defined by $g(\Delta) = \Delta'$. Intensity at Δ' will be similarly related (with instantaneous intensity of a wave we understand generally the energy that at a given moment flows through a surface orthogonal to the propagation vector):

$$I(\Delta') = \int_\Delta |\tilde{\psi}_\tau(\mathbf{k})|^2 d^3\mathbf{k} = \int_{\Re^3} \tilde{\psi}^*_\tau(\mathbf{k})\chi_\Delta(\mathbf{k})\tilde{\psi}_\tau(\mathbf{k})d^3\mathbf{k} = \langle\psi_\tau|\hat{P}_p(\Delta)|\psi_\tau\rangle, \tag{18.133}$$

where $\hat{P}_p(\Delta)$ is the projection-valued localization observable for sharp localization of momentum. The simplified screen observable is denoted by $\hat{P}'_p(\Delta') = \hat{P}_p(g^{-1}(\Delta'))$. We can pose the relationship between the momentum distribution at port and the screen position distribution at port as $\hat{P}'_p(\Delta') = \hat{P}_q(\Delta') \simeq \hat{P}_p(\Delta)$: it corresponds to the fact that for large flight times in the Heisenberg picture $\hat{q} \simeq \hat{p}t/m$.

18.10.2 Unsharp Treatment

But even with such idealizations the spin can be measured only unsharply [BUSCH/SCHROECK 1989 824–27]. Any observable that is a function of $\mathbf{q}, \mathbf{p}$ only commutes with any pure spin operator. Hence from Eqs. (18.132) and (18.133):

$$\begin{aligned} I(\Delta') &= |c_+|^2\langle\psi_{\tau+}|\hat{P}_p(\Delta)|\psi_{\tau+}\rangle + |c_-|^2\langle\psi_{\tau-}|\hat{P}_p(\Delta)|\psi_{\tau-}\rangle \\ &= \langle\psi|\left[\hat{P}_p(\Delta+\tau\mu b\mathbf{e}_3)\otimes\hat{O}_{\mathbf{e}_3} + \hat{P}_p(\Delta-\tau\mu b\mathbf{e}_3)\otimes\hat{O}_{-\mathbf{e}_3}\right]|\psi\rangle. \end{aligned} \quad (18.134)$$

The actual screen observable is in the expression between {}. Let $|s_\mathbf{q}\rangle$ denote the spatial component and $|s_\sigma\rangle$ the spin component of the wave function (so that $|\psi\rangle = |s_\mathbf{q}\rangle \otimes |s_\sigma\rangle$). We can rewrite Eq. (18.134) in the following form:

$$I(\Delta') = \left\langle s_\sigma \left|\left\{\langle s_\mathbf{q}|\hat{P}_p(\Delta+\tau\mu b\mathbf{e}_3)|s_\mathbf{q}\rangle\hat{O}_{\mathbf{e}_3} + \langle s_\mathbf{q}|\hat{P}_p(\Delta-\tau\mu b\mathbf{e}_3)|s_\mathbf{q}\rangle\hat{O}_{-\mathbf{e}_3}\right\}\right| s_\sigma\right\rangle, \quad (18.135)$$

where the expression between {} := $\hat{E}(\Delta')$ defines a POVM on spin space instead of a PVM: this is so because we obtain a mixing of spin up and spin down of the wave functions impinging in Δ'. If we partition the screen in different horizontal strips Δ' of form $g(a) \leq z \leq g(d)$ then the integral is:

$$\int_{\Delta\pm\tau\mu b\mathbf{e}_3} |\tilde{\psi}_\pm(\mathbf{k})|^2 d^3\mathbf{k} = \int_\Re dk_1 \int_\Re dk_2 \int_{a\pm\tau\mu b}^{b\pm\tau\mu b} |\tilde{\psi}_\pm(\mathbf{k})|^2 dk_3. \quad (18.136)$$

Different strips will have different intensities. Since the spatial component of the wave is well localized in its z–component in order to penetrate the gap between the two magnetic pieces $\sim 1mm$, then the z-component of momentum will spread out. By remembering Eq. (5.19) we can express the visibility as follows:

$$\mathcal{V} = \frac{\mathrm{Max}I(\Delta') - \mathrm{min}I(\Delta')}{\mathrm{Max}I(\Delta') + \mathrm{min}I(\Delta')}. \quad (18.137)$$

The maximum $I(\Delta')$ occurs at strips centered at $\mathbf{k}$ with third component $\pm\tau\mu b$. Then the visibility indicates the purity of the spin composition in the strip.

18.10.3 Proper Screen Observable

Now we analyse the Proper Screen Observables [BUSCH/SCHROECK 1989, 835–36], already studied by Ludwig and by Werner [R. WERNER 1986]. We take into account that detectors or screens do not constitute instantaneous space localization measurements. We assume the screen to be in the plane $y = 0$. Assuming a magnetic field with $\mathbf{B} = (B_0 + bx_3)\mathbf{e}_3$, the wave function can be written:

$$|\psi_\tau\rangle = |s'_{q+}\rangle \otimes |s_{\sigma+}\rangle + |s'_{q-}\rangle \otimes |s_{\sigma-}\rangle. \quad (18.138)$$

It leaves the exit port of the magnetic field undergoing a free evolution until it arrives at the screen. Assuming τ to be in the far remote past (so that we can integrate between $-\infty$ and $+\infty$) and an initial preparation: $|s_q(x,y,z,t_0)\rangle = |s_{13}(x,y,t_0)s_2(y,t_0)\rangle$, the packet will maintain this form for all times. The probability distribution for registering a particle in region Δ' of the screen in time interval $[t_1, t_2] = t_{1-2}$ is given as the expected value of some screen effect $\hat{E}(\Delta'; t_{1-2})$ defined by the following [see Eq. (18.134)]:

$$\langle\psi_\tau|\hat{E}(\Delta';t_{1-2})\otimes\hat{I}|\psi_\tau\rangle = \langle s'_{q+}|\hat{E}(\Delta';t_{1-2})|s'_{q+}\rangle + \langle s'_{q-}|\hat{E}(\Delta';t_{1-2})|s'_{q-}\rangle, \quad (18.139a)$$

$$\langle s^\#_q|\hat{E}(\Delta';t_{1-2})|s^\#_q\rangle = (2\pi)^{-3}\int_{t_{1-2}} f(t)\int_\Delta |s^\#_{13}(x,y,t)|^2 dx dz dt, \quad (18.139b)$$

where $s^{\#}_q = s'_{q+}, s'_{q-}$ and:

$$f(t) = \frac{\hbar}{2\pi M} \left| \int_{-\infty}^{+\infty} |k_2|^{\frac{1}{2}} \tilde{s}^{\#}_2(k_2) e^{-\imath \frac{\hbar}{2M} k_2^2 t} dk_2 \right|^2, \tag{18.140}$$

and

$$|s^{\#}_{13}(x,z,t)|^2 = (2\pi)^{-2} \left| \int\int_{\Re^2} \tilde{s}^{\#}_{13}(k_1,k_3) e^{-\imath(k_1 x + k_2 y)} e^{-\imath \frac{\hbar}{2M}(k_1^2+k_3^2)t} dk_1 dk_3 \right|^2. \tag{18.141}$$

With these equations the operator $\hat{E}(\Delta'; t_{1-2})$ is implicitly defined. It is positive and its expectation values lies between 0 and 1. And we see that Eq. (18.139a) gives rise to a spin POVM $\Delta' \rightsquigarrow \hat{E}''(\Delta')$ — via $\langle \psi_\tau | \hat{E}(\Delta'; t_{1-2}) \otimes \hat{I} | \psi_\tau \rangle = \langle s_\sigma | \hat{E}''(\Delta') | s_\sigma \rangle$ which is of the form obtained before.

18.10.4 Determinative Measurement of Spin

Remembering the distinction between *determinative measurements* and *preparatory measurements* [see section 18.1], we investigate the former with respect to the spin, i.e. how to make a measurement as precise as possible [BUSCH/SCHROECK 1989, 837–39]. Suppose N particles, identically prepared in an eigenstate of spin of the same unknown direction $\mathbf{n}$, which are sent through a SGM with a certain fraction of them deflected up. We investigate the $\mathbf{n}$–orientation in the cases $N = 1$ and N large. We suppose that the process alters $|s_\mathbf{q}\rangle$ but not $|s_\sigma\rangle$. Now $|s_\sigma\rangle$ is an eigenvector of $\hat{O}_\mathbf{n}$. Hence $(\hat{I} \otimes \hat{O}_\mathbf{n})|\psi\rangle = |\psi\rangle$, so that:

$$\begin{aligned} |c_+|^2 &= \| (\hat{I} \otimes \hat{O}_{\mathbf{e}_3})|\psi\rangle \|^2 = \| (\hat{I} \otimes \hat{O}_{\mathbf{e}_3})(\hat{I} \otimes \hat{O}_\mathbf{n})|\psi\rangle \|^2 \\ &= \langle \psi | \hat{I} \otimes \hat{O}_\mathbf{n} \hat{O}_{\mathbf{e}_3} \hat{O}_\mathbf{n} |\psi\rangle = \cos^2 \frac{\theta}{2}, \end{aligned} \tag{18.142}$$

where θ is the angle between $\mathbf{n}$ and $\mathbf{e}_3$. We know from Eq. (18.135) that

$$I(\Delta') = d_+ \cos^2 \frac{\theta}{2} + d_- \sin^2 \frac{\theta}{2}, \tag{18.143}$$

where

$$d_+ = \langle \tilde{s}_\mathbf{q} | \hat{P}_p(\Delta + \tau \mu b \mathbf{e}_3) | \tilde{s}_\mathbf{q} \rangle, \tag{18.144}$$

$$d_- = \langle \tilde{s}_\mathbf{q} | \hat{P}_p(\Delta - \tau \mu b \mathbf{e}_3) | \tilde{s}_\mathbf{q} \rangle. \tag{18.145}$$

It is a useful formula but not very rigorous. Anyway d_+, d_- are measurable simply by turning off the magnet and, for an incoming beam with any spin state, measuring the intensity distribution, which reduces to:

$$I(\Delta') = \langle \psi | \hat{I} \otimes \hat{P}_p(\Delta) | \psi \rangle. \tag{18.146}$$

This is one calibration process. A second one is by supposing the particle to be in a pure and unknown spin state and to rotate the device about the beam axis to obtain a maximum and a minimum for $I(\Delta')$. This will occur for $\theta = 0, \pi$ respectively, so that:

$$I(\Delta')_{\text{Max}} = d_+, \qquad I(\Delta')_{\text{min}} = d_-. \tag{18.147}$$

The visibility is now [Eq. (18.137)]:

$$\mathcal{V}(\Delta') = \frac{d_+ - d_-}{d_+ + d_-}. \tag{18.148}$$

So that by one or another calibration we know experimentally d_-/d_+. We know the limitations on measurability which depend on conservation laws and we know that spin cannot be sharply measured for the Law of Momentum Conservation. With unsharp spin (POVM) the problem is bypassed [BUSCH/SCHROECK 1989, 846–49]. Normally spin effects (unsharp) have sharp counterparts: spin properties. But there are effect observables which have no sharp counterparts, for example the rotation covariant POVM on the Poincaré sphere [BUSCH/SCHROECK 1989, 857].

Chapter 19

NON DEMOLITION MEASUREMENT THEORY

Contents We have already seen that the Statistical interpretation can be questioned on the basis of some experiments which have been performed on individual quantum entities [see subsection 6.6.2]. The theory summarized in this chapter represents the first systematic and general theory of individual Measurement. After a brief history [section 19.1], a more detailed analysis of Measurement is performed [section 19.2]: it will establish the conditions of Quantum Non-Demolition Measurement (= QND-Measurement). The first results of this analysis consist in the theory of indirect Measurement [section 19.3]. After this, the principles of QND-Measurement are summarized [section 19.4]. Then, the problem of the evaluation of minimal Measurement error is analysed [section 19.5], followed by a discussion of UP in this context [section 19.6]. Some experiments are also briefly reported [section 19.7]. The remainder of the chapter is devoted to two special topics: the quantum Zeno paradox [section 19.8] and the problem of the boundary between apparatus and object system [section 19.9].

19.1 Brief History

The first ideas of the QND-Measurement theory were developed by Braginsky and Vorontsov [BRAGINSKY/VORONTSOV 1974] . The aim of the article was the determination of the minimum allowed values of uncertainty in the case of measurement of forces. As we have seen, the problem of the traditional theory and techniques of Measurement was the perturbation of the measured observable, apart from particular cases of type I–Measurement [see definition 13.3: p. 217], where the system is already in an eigenstate of the measured observable. It was shown for the first time by Braginsky and Vorontsov that it is possible to register the n–quantum state of an oscillator without destroying it.

QND-Measurement was improved by the very difficult problem of the measurement of gravitational waves [*BRAGINSKY/KHALILI* 1992, 50, 105], where the necessity arose to overcome the Standard Quantum Limit (of which we shall speak later in this chapter).[1] The first study of a QND-Measurement which takes account of dissipative losses was developed by Imoto and Saito [IMOTO/SAITO 1989]: it contributed to assuring the possibility of a QND measurement in existing media.

[1]See [BRAGINSKY/KHALILI 1996] for an history and also [*WALLS/MILBURN* 1994].

19.2 More on the Measurement Process

Projectors The starting point is the same as the Stochastic QM: an attempt to overcome the traditional Projection postulate [14.1: p. 223, and subsection 14.2.2] [*BRAGINSKY/KHALILI* 1992, 30–33]. Characteristic features of projectors are orthogonality and completeness [see subsection 3.5.3]. Let us express now the analysis of effects in terms of measurability of continuous observables. By measuring a continuous observable (for example $\hat{\mathbf{q}}$, which we take here to be one-dimensional), the probability that it will be found in some arbitrary interval (q_j, q_{j+1}) for initial state $\hat{\rho}_i$ is:

$$\wp_j = \int_{q_j}^{q_{j+1}} \langle q|\hat{\rho}_i|q\rangle dq = \mathrm{Tr}(\hat{P}_j\hat{\rho}_i), \tag{19.1}$$

where $|q\rangle$ is the eigenstate of the measured observable and $\hat{P}_j$ is defined by:

$$\hat{P}_j = \int_{q_j}^{q_{j+1}} |q\rangle\langle q|dq = \int_{-\infty}^{+\infty} |q\rangle\hat{P}_j(q)\langle q|dq, \tag{19.2}$$

where

$$\hat{P}_j(q) = \begin{cases} = 1 & \text{if } q_j \leq \hat{q} \leq q_{j+1} \\ = 0 & \text{if } \hat{q} < q_j \vee \hat{q} \geq q_{j+1} \end{cases}. \tag{19.3}$$

Short examination Now the most positive feature of operators defined by Eqs. (19.2) and (19.3) is that they completely characterize an apparatus $\mathcal{A}$ and that repeated orthogonal measurements leave the state of $\mathcal{S}$ unchanged. In this last sense they are more 'exact' than any other measurement. They can be considered as an exact measurement of the operator [see Eq. (3.63)]

$$\hat{q} = \sum_j q_j\hat{P}_j. \tag{19.4}$$

But the problem is they generate a discontinuous change in the wave function [subsection 14.1.2], as a result of the discontinuous character of the decomposition of the identity operator by the projectors. The consequence is that the observable conjugate to that which is measured, has an infinite variance. On the other hand, even if the orthogonal measurements can be repeated, they do not improve our knowledge of the measured observable. Hence we must understand them as limit cases of measurements.

Effects Therefore, the first thing is to renounce the condition of orthogonality — but not that of completeness, i.e. we must introduce [*BRAGINSKY/KHALILI* 1992, 33–37] effects $\hat{E}$ [see subsection 18.1.1]. We work now with conditional probabilities of the form $\wp(q_{\mathrm{M}}|q)$, which represent the probability of obtaining q_{M} when $\hat{q}$ is in an eigenstate with eigenvalue q — instead of discontinuous operators of the form $\hat{P}_j(q)$. We express the probability of obtaining a result q_{M} by means of these conditional probabilities:

$$\wp(q_{\mathrm{M}}) = \int_{-\infty}^{+\infty} \wp(q_{\mathrm{M}}|q)\hat{\rho}(q)dq, \tag{19.5}$$

where

$$\hat{\rho}(q) = \langle q|\hat{\rho}_i|q\rangle \tag{19.6}$$

is the initial state's *a priori* probability distribution for the values of $\hat{q}$. Now parallel to the standard case we can write the probability (19.5) as:

$$\wp(q_{\mathrm{M}}) = \mathrm{Tr}[\hat{E}(q_{\mathrm{M}})\hat{\rho}_i], \tag{19.7}$$

where the effects in analogy with Eq. (19.2) are:

$$\hat{E}(q_{\rm M}) = \int_{-\infty}^{+\infty} |q\rangle \wp(q_{\rm M}|q)\langle q| dq. \tag{19.8}$$

Two aspects of the measurement To understand the new step we recall [see subsection 14.1.3] [*BRAGINSKY/KHALILI* 1992, 12–13, 41, 54–55, 64–65] the distinction between two different aspects of the Measurement process:

- an unitary evolution, which can be a pre-measurement;
- and in a reduction which allows the passage from a superposition to a Lüders mixture of the form (14.16).

Up to now we demonstrated that the Measurement cannot be unitary [see theorems 14.1: p. 221, and 14.13: p. 243], and this concerns only the second aspect of it. We must also recall [see section 7.3] the distinction between correct measurement error and correct back-action (= perturbation) on $\mathcal{S}$. The back-action is induced only by the unitary evolution, while the reduction to a mixture is completely determined by transition probabilities of the form $\wp(q_{\rm M}|q)$, which is the informational content which we can extract from a measurement (and here is the only source of errors in the measurement).[2] In other words, the unitary evolution (as we have already seen) cannot change the entropy of $\mathcal{S}$ and hence cannot produce any new information. But, and this is the important thing, the strength of the perturbation (back-action) depends on the information which we can obtain in a measurement process, i.e. from $\wp(q_{\rm M}|q)$. Therefore the two aspects are not independent.

Now we write the density operator of the final state of $\mathcal{S}$ in the following form:

$$\hat{\rho}_f(q_{\rm M}) = \frac{1}{\wp(q_{\rm M})}\hat{\vartheta}(q_{\rm M})\hat{\rho}_i\hat{\vartheta}^\dagger(q_{\rm M}), \tag{19.9}$$

where the operator $\hat{\vartheta}(q_{\rm M})$, to be better defined below, is an amplitude operator which completely describes the whole measurement process. From the normalization condition $\mathrm{Tr}[\hat{\rho}_f(q_{\rm M})] = 1$ and from Eq. (19.8) it follows that:

$$\hat{\vartheta}^\dagger(q_{\rm M})\hat{\vartheta}(q_{\rm M}) = \hat{E}(q_{\rm M}). \tag{19.10}$$

Now $\hat{\vartheta}(q_{\rm M})$ can be represented as:

$$\hat{\vartheta}(q_{\rm M}) = \hat{U}(q_{\rm M})\hat{E}^{\frac{1}{2}}(q_{\rm M}), \tag{19.11}$$

where $\hat{E}^{\frac{1}{2}}(q_{\rm M})$ is the same as Eq. (19.8) but with $\wp^{1/2}$ in place of $\wp$. This operator is uniquely determined by $\mathcal{A}$'s conditional probability $\wp(q_{\rm M}|q)$ and commutes with the measure $\hat{\mathbf{q}}$. Hence the measurement can be viewed as a two-step process consisting in the unitary evolution

$$\boxed{\hat{\rho}'(q_{\rm M}) = \hat{U}(q_{\rm M})\hat{\rho}_i(q_{\rm M})\hat{U}^\dagger(q_{\rm M})} \tag{19.12a}$$

followed by the reduction

$$\boxed{\hat{\rho}_f(q_{\rm M}) = \frac{1}{\wp(q_{\rm M})}\hat{E}^{\frac{1}{2}}(q_{\rm M})\hat{\rho}'\hat{E}^{\frac{1}{2}}(q_{\rm M})} \tag{19.13}$$

[2] On this point see also [*HOLEVO* 1982, 164].

19.3 Indirect Measurement

Two steps We can use the already made distinction between the two aspects of measurement by introducing a new form of it, the *indirect measurement*, which we contrast with the *direct measurement*, which is the standard one [*BRAGINSKY/KHALILI* 1992, 40–41]. The latter form of measurement is an interaction between a QM system $\mathcal{S}$ and a macroscopic apparatus $\mathcal{A}$, while the indirect measurement is defined as follows:

Definition 19.1 (Indirect Measurement) *The indirect measurement of a system $\mathcal{S}$ is distinguished by two different steps: first $\mathcal{S}$ interacts with another QM system $\mathcal{S}_{\mathrm{QP}}$, the quantum probe. Here there is no reduction at all, and the evolution is completely unitary — we have back-action. The second step consists in a direct measurement of some chosen observable of $\mathcal{S}_{\mathrm{QP}}$, and here the state of the probe (and of $\mathcal{S}$) is reduced and the information acquired.*

This definition will be integrated below in the definition of QND-Measurement. Now we introduce two conditions:

- the second step of measurement should begin only when the first step has already finished;
- the second step should not contribute significantly to the total error of measurement.

If these conditions are observed, we can infer the magnitudes of the error in the measurement and of the perturbation (back-action) of $\mathcal{S}$ from an analysis of only the first step, i.e. of a unitary evolution, because the only source of error and the perturbation is due to the internal uncertainties of $\mathcal{S}_{\mathrm{QP}}$.

First step We can describe [*BRAGINSKY/KHALILI* 1992, 47–49] the indirect measurement in a formal way as follows. Let the *first step* be represented by the following Eq.:

$$\hat{U}\hat{\rho}_i^{\mathcal{S}}\hat{\rho}^{\mathcal{S}_{\mathrm{QP}}}\hat{U}^{\dagger}. \tag{19.14}$$

The corresponding state of $\mathcal{S}_{\mathrm{QP}}$ alone, after the interaction, is given by the reduced density matrix:

$$\mathrm{Tr}_{\mathcal{S}}(\hat{U}\hat{\rho}_i^{\mathcal{S}}\hat{\rho}^{\mathcal{S}_{\mathrm{QP}}}\hat{U}^{\dagger}). \tag{19.15}$$

Second step Suppose that we have measured the observable $\hat{p}$ on $\mathcal{S}_{\mathrm{QP}}$. Since this measurement contributes negligibly to the experiment's overall error, we can idealize it as arbitrarily accurate. Now we can infer from the value of $\hat{p}$ on $\mathcal{S}_{\mathrm{QP}}$ the value q_{M} of $\hat{q}$ on $\mathcal{S}$. Because of a one-to-one correspondence we can use $\hat{q}$ as a substitute for $\hat{p}$ and hence use it not only as the inferred value of $\hat{q}$ but also as the result of a precise measurement on $\mathcal{S}_{\mathrm{QP}}$ itself. Then we can introduce operator $|q_{\mathrm{M}}\rangle\langle q_{\mathrm{M}}|$ as a definition of the density operator $\hat{\rho}^{\mathcal{S}_{\mathrm{QP}}}$ figuring in Eq. (19.14); just before the second step of the measurement we obtain the following probability:

$$\wp(q_{\mathrm{M}}) = \mathrm{Tr}_{\mathcal{S}_{\mathrm{QP}}}\left[|q_{\mathrm{M}}\rangle\langle q_{\mathrm{M}}|\mathrm{Tr}_{\mathcal{S}}(\hat{U}\hat{\rho}_i^{\mathcal{S}}\hat{\rho}^{\mathcal{S}_{\mathrm{QP}}}\hat{U}^{\dagger})\right], \tag{19.16}$$

which can be rewritten:

$$\wp(q_{\mathrm{M}}) = \mathrm{Tr}_{\mathcal{S}}\left[\hat{E}(q_{\mathrm{M}})\hat{\rho}_i^{\mathcal{S}}\right], \tag{19.17}$$

where

$$\hat{E}(q_{\mathrm{M}}) = \mathrm{Tr}_{\mathcal{S}_{\mathrm{QP}}}\left[\hat{U}^{\dagger}|q_{\mathrm{M}}\rangle\langle q_{\mathrm{M}}|\hat{U}\hat{\rho}^{\mathcal{S}_{\mathrm{QP}}}\right]. \tag{19.18}$$

The back-action of the entire two step measurement on $\mathcal{S}$ is embodied in $\mathcal{S}$'s final state which we have supposed to have the form (19.9). Now we shall prove it. The above considerations imply that $\mathcal{S}$'s final state is:

$$\hat{\rho}_{\mathcal{S}}^{f}(q_{\mathrm{M}}) = \frac{1}{\wp(q_{\mathrm{M}})} \langle q_{\mathrm{M}} | \hat{U} \hat{\rho}_{i}^{\mathcal{S}} \hat{\rho}^{\mathcal{S}_{\mathrm{QP}}} \hat{U}^{\dagger} | q_{\mathrm{M}} \rangle, \tag{19.19}$$

where $\frac{1}{\wp(q_{\mathrm{M}})}$ is the normalization condition.

Amplitude operator Now we are able to better interpret the amplitude operator $\hat{\vartheta}$. If we express $\mathcal{S}_{\mathrm{QP}}$'s initial (premeasured) state by:

$$\hat{\rho}^{\mathcal{S}_{\mathrm{QP}}} = \sum_{k} w_k |\psi_k\rangle\langle\psi_k|, \tag{19.20}$$

by substitution in Eq. (19.19), we obtain:

$$\boxed{\hat{\rho}_{f}^{\mathcal{S}}(q_{\mathrm{M}}) = \frac{1}{\wp(q_{\mathrm{M}})} \sum_{k} \hat{\vartheta}_k(q_{\mathrm{M}}) \hat{\rho}_{i}^{\mathcal{S}} \hat{\vartheta}_k^{\dagger}(q_{\mathrm{M}})} \tag{19.21}$$

where finally we can interpret the amplitude operator as describing all steps of the Measurement: premeasurement ($|\psi_k\rangle$), followed by the unitary evolution ($\hat{U}$) — where they are distinct — and finally by the reduction ($\langle q_{\mathrm{M}}|$):

$$\hat{\vartheta}_k(q_{\mathrm{M}}) = \langle q_{\mathrm{M}} | \hat{U} | \psi_k \rangle, \tag{19.22}$$

or more generally

$$\hat{\vartheta}(q_{\mathrm{M}}) = \langle q_{\mathrm{M}} | \hat{U} | \psi \rangle. \tag{19.23}$$

The amplitude operator gives the probability amplitude for the quantum probe to have evolved from initial state $|\psi\rangle$ to final state $|q_{\mathrm{M}}\rangle$, if the interaction of the first with $\mathcal{S}$ is given by $\hat{U}$. Since an effect is the product of an amplitude operator and its adjoint [see Eq. (19.10)], then we have confirmed its interpretation as an equivalence class of operations [see definition 18.2: p. 293, and Eqs. (18.3) and (18.6)].

Now $\mathcal{S}$'s final state [Eq. (19.19)] is a mixture of states of the type (19.9) with weighting factor w_k. Hence the observable's final state has additional uncertainties not taken into account from the Standard orthogonal theory, uncertainties which derive from the fact that we allowed $\mathcal{S}_{\mathrm{QP}}$ to begin in the mixed state (19.20) and not in a pure one. Hence there are not only QM back actions but also others which derive, for example, from the initial 'temperature' of $\mathcal{S}_{\mathrm{QP}}$ — in the case, for instance, of a bath of harmonic oscillators [see chapter 21].

19.4 Principles of Non-Demolition Measurement

19.4.1 Non-Demolition Measurement

QND-Measurement Now we have all the elements in order to discuss the QND-Measurement. Let us begin [*BRAGINSKY/KHALILI* 1992, 55] [BRAGINSKY/KHALILI 1996, 2]with a formal definition:

Definition 19.2 (QND-Measurement) *A QND-Measurement is a measurement in which an apparatus extracts information only on the single specified observable, i.e. by not perturbing the observable to be measured and by perturbing the other non commuting observables precisely to the minimal extent allowed by UP.*

Some features of the indirect measurement also characterize the QND-Measurement. In general we say that in a QND-Measurement $\mathcal{S}$ interacts only with a probe $\mathcal{S}_{\mathrm{QP}}$, and the interaction between $\mathcal{S}$ and $\mathcal{S}_{\mathrm{QP}}$ is such that only one observable or a set of observables, that are not affected by $\mathcal{S}_{\mathrm{QP}}$'s back-action on $\mathcal{S}$, acts on $\mathcal{S}_{\mathrm{QP}}$.

A QND-Measurement can be performed only on observables that are conserved during the object' free evolution, i. e. on integrals of the motion — for a free mass: energy and momentum; for an oscillator: the energy, number of quanta and the quadrature amplitudes — [see Eq. (19.39)]. In absence of external forces the observable is conserved both during the measurement (absence of back-action) and between consecutive measurements, i.e. during the unitary evolution (because it is an integral of motion) [*BRAGINSKY/KHALILI* 1992, 56].

Variance and SQL From the above definition and the stated features we have the following consequence [BRAGINSKY/KHALILI 1996, 3]:

Corollary 19.1 (Variance and QND-Measurement) *A QND-Measurement does not add any perturbation to the observable to be measured, so that a possible variance is only a consequence of the a priori uncertainty of the value of the observable to be measured.*

To analyse this point, we remember that a QND-Measurement is able to overcome the Standard Quantum Limit (= SQL). To derive the SQL [*BRAGINSKY/KHALILI* 1992, 12–13, 21–22] [BRAGINSKY/KHALILI 1996, 2], we start with an harmonic oscillator with coordinate $x(t)$, mass m and eigenfrequency ω. Expressed by quadrature amplitude [Eq. (5.3)] with the Heisenberg representation:

$$\hat{x}(t) = \hat{X}_1 \cos(\omega t) + \hat{X}_2 \sin(\omega t), \tag{19.24}$$

which satisfy the Quadratures UR:

$$\Delta \hat{X}_1 \Delta \hat{X}_2 \geq \frac{\hbar}{2m\omega}, \tag{19.25}$$

and whose minimum of uncertainty, which we can obtain by continuous monitoring of the oscillator, is exactly the SQL for the coordinate:

$$(\Delta \hat{x})_{\mathrm{SQL}} = \Delta \hat{X}_1 = \Delta \hat{X}_2 = \sqrt{\frac{\hbar}{2m\omega}}. \tag{19.26}$$

The SQL for the energy of the oscillator is:

$$(\Delta E)_{\mathrm{SQL}} = \sqrt{\hbar \omega E}. \tag{19.27}$$

For a coordinate of a free particle the SQL is:

$$(\Delta\hat{\mathbf{q}})_{\mathrm{SQL}} = \sqrt{\frac{\hbar\tau}{2m}}, \tag{19.28}$$

where τ is the time at which the measurement is performed. For the momentum the SQL is:

$$(\Delta\hat{\mathbf{p}})_{\mathrm{SQL}} = \sqrt{\frac{\hbar m}{2\tau}}. \tag{19.29}$$

Error boxes The problem can also be examined by the 'error boxes' method [CAVES *et al.* 1980, 350–351]. We can construct an error box for a harmonic oscillator in the phase space in the following way [see figures 19.1 and 19.2]: the error box is an ellipse, with centroid at expectation value $(\langle\hat{q}\rangle, \langle\hat{p}/m\omega\rangle)$ of the position and momentum.

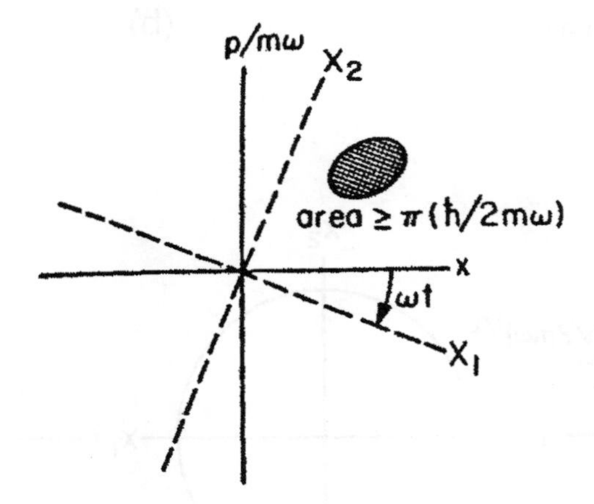

Figure 19.1: Error box in the phase plane for a QM oscillator — from [CAVES *et al.* 1980, 350].

The principal axes of the error ellipse are the eigendirections of the variance matrix:

$$\begin{pmatrix} (\Delta\hat{q})^2 & \Delta_{q,p} \\ \Delta_{q,p} & (\frac{\Delta p}{m\omega})^2 \end{pmatrix} \tag{19.30}$$

and the principal radii are the square roots of the corresponding eigenvalues — here $\Delta_{q,p} := (1/2m\omega)\langle(\hat{q}-\langle\hat{q}\rangle)(\hat{p}-\langle\hat{p}\rangle)+(\hat{p}-\langle\hat{p}\rangle)(\hat{q}-\langle\hat{q}\rangle)\rangle$. The area A of this error box has the properties:

$$\Delta\hat{q}\cdot\frac{\Delta\hat{p}}{m\omega} \geq \frac{1}{\pi}A \geq \frac{\hbar}{2m\omega}, \tag{19.31a}$$

$$\Delta\hat{X}_1\cdot\Delta\hat{X}_2 \geq \frac{1}{\pi}A \geq \frac{\hbar}{2m\omega}, \tag{19.31b}$$

where

$$\hat{x} + \imath\frac{\hat{p}}{m\omega} = (\hat{X}_1 + \imath\hat{X}_2)e^{\imath\omega t}. \tag{19.32}$$

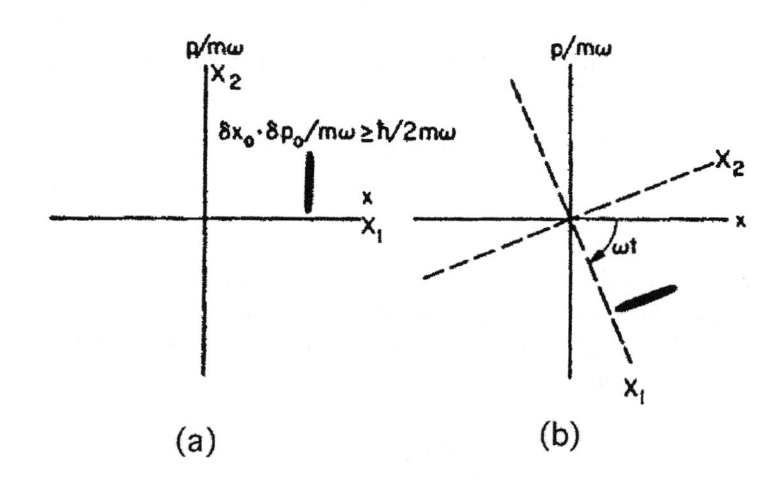

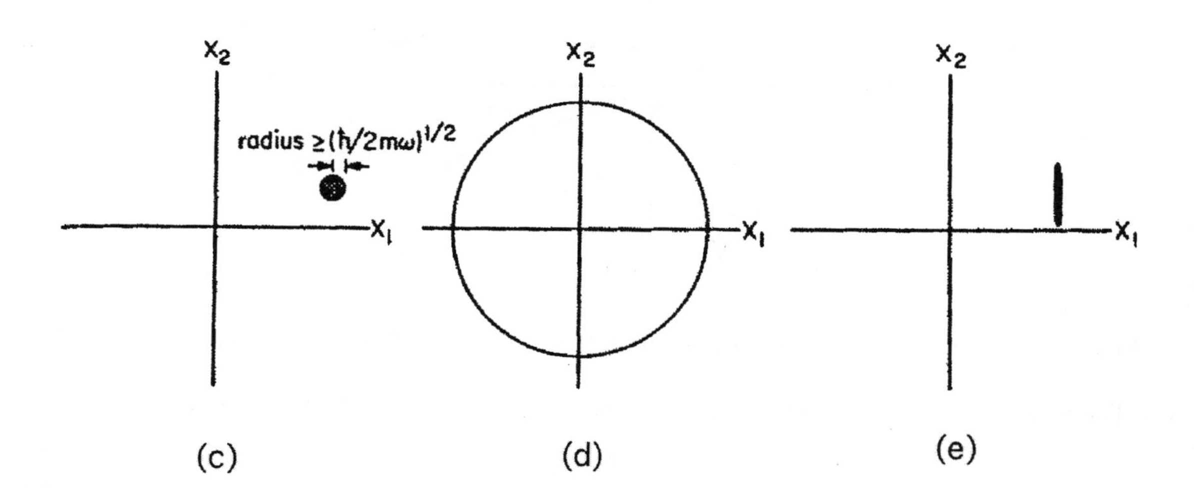

Figure 19.2: Error boxes for various types of measurements of a harmonic oscillator.
(a) The error box characterizing the result of a 'quick measurement' of position. After the measurement the error box rotates clockwise in the phase plane with angular velocity ω, which means that it remains fixed as seen in the rotating (X_1, X_2) coordinates.
(b) The error box for amplitude-and-phase measurement as seen in the (X_1, X_2) coordinate system.
(c) The error anulus ($\delta N \equiv 1$) for quantum-counting measurement.
(d) The error box for a back-action-evading measurement — from [CAVES *et al.* 1980, 351].

Perturbativity and QND-Measurement Now we can give a criterion of non-perturbativity in terms of the SQL [BRAGINSKY/KHALILI 1996, 4–5, 7]:

Definition 19.3 (Non-Perturbativity) *A Measurement is non-perturbative iff it provides both a measurement error and a perturbation of the measured value smaller than the SQL.*

From which we can give an alternative definition of the QND-Measurement:

Definition 19.4 (II QND-Measurement) *A QND-Measurement is a non-perturbative Measurement.*

19.4.2 Non-Demolition Observables

QND-Observable We give a definition of the QND-Observable [*BRAGINSKY/KHALILI* 1992, 60–63, 67] [CAVES *et al.* 1980, 364] :[3]

Definition 19.5 (QND-Observable) *A QND-Observable $\hat{O}_{\rm ND}$ at a given time commutes with itself at a different time:*

$$\boxed{[\hat{O}_{\rm ND}(t), \hat{O}_{\rm ND}(t')] = 0, \quad t < t'} \tag{19.33}$$

and it commutes with the interaction Hamiltonian or the time-displacement unitary operator:

$$\boxed{(\hat{U}^{\dagger}\hat{O}_{\rm ND}\hat{U} - \hat{O}_{\rm ND})|\psi\rangle = 0} \tag{19.34}$$

where the expression between brackets is the Heisenberg-picture change in $\hat{O}_{\rm ND}$ produced by the interaction between $\mathcal{S}$ and $\mathcal{S}_{\rm QP}$. Eq. (19.34) *represents a necessary and sufficient condition of a QND-Observable.*

First condition Equivalently we can express the first condition by saying that a QND-Observable commutes with all its time derivatives, or, by linear measurements [see subsection 19.6.1], having, for example, measured the coordinate and having obtained $q_{\rm M}$, we have:

$$[\hat{\vartheta}(q_{\rm M}), \hat{q}] = 0, \tag{19.35}$$

where $\hat{\vartheta}$ is the amplitude operator in Eq. (19.9).

We can say that a QND-Measurement is characterized by the *repeatability*, so that the first measurement — which determines the values for all subsequent QND ones — is a preparation of $\mathcal{S}$ in the desired state, and the others are the determination of the value. Hence QND-Measurement

[3]See also [BRAGINSKY/KHALILI 1996, 3] [UNRUH 1979].

is a type I–Measurement [see definition 13.3: p. 217] — but not necessarily *vice versa*. Formally [THORNE *et al.* 1978][CAVES *et al.* 1980, 364]:

$$\boxed{\hat{O}_{\rm ND}(t_k) = f_k[\hat{O}_{\rm ND}(t_0)]} \tag{19.36}$$

where t_k is some arbitrary time after the initial t_0 (the time of the first measurement), and f_k is some real-valued function. If one is interested in measurements at arbitrary times (continuously) we have:

$$\hat{O}_{\rm ND}(t) = f[\hat{O}_{\rm ND}(t_0); t, t_0], \tag{19.37}$$

which defines a *continuous* QND-Observable, while, if an observable satisfies Eq. (19.36) only at selected times, it is a *stroboscopic* QND-Observable. Examples of the last one are the position and the momentum of a harmonic oscillator. The simplest way to satisfy Eq. (19.37) is to choose an observable which is conserved in the absence of interactions with the environment. It is interesting that the commutativity of Eq. (19.33) follows from condition (19.37).[4]

Second condition Condition (19.34) can be satisfied

- either by saying that the expression between brackets vanishes, i.e. that $\hat{O}_{\rm ND}$ must return to its initial value after the measurement;
- or, if this does not happen, one can choose the initial state $|\psi\rangle$ to be an eigenstate of that expression.

If the first alternative is adopted, then one can write more simply:

$$[\hat{O}_{\rm ND}, \hat{U}]|\psi\rangle = 0 \tag{19.38}$$

where $|\psi\rangle$ is the initial state of $\mathcal{S}_{\rm QP}$.

Due to the difficulty in determining $\hat{U}$, i.e. the evolution of the coupled system, one can choose alternatively a necessary but not sufficient condition for a QND-Observable: that the measured observable is an integral of the motion for the coupled $\mathcal{S} + \mathcal{S}_{\rm QP}$, i.e. it must remain constant during the interaction, i.e.

$$\imath\hbar\frac{\partial \hat{O}_{\rm ND}}{\partial t} + [\hat{O}_{\rm ND}, \hat{H}_{\mathcal{S}_{\rm QP}+\mathcal{S}}] = 0, \tag{19.39}$$

which, in the case that $\hat{O}_{\rm ND}$ has no specific time-dependence, reduces to the condition [see also Eqs. (17.18) and (17.19)]:

$$[\hat{O}_{\rm ND}, \hat{H}_{\mathcal{S}_{\rm QP}+\mathcal{S}}] = 0, \tag{19.40}$$

which is a more severe condition than (19.38), because, if it is satisfied, the measurement is a QND-Measurement for any time-interval of the interaction between the systems. If the measured observable is an integral of $\mathcal{S}$'s free motion, then it satisfies the following Eq.:

$$\imath\hbar\frac{\partial \hat{O}_{\rm ND}}{\partial t} + [\hat{O}_{\rm ND}, \hat{H}_{\mathcal{S}}] = 0. \tag{19.41}$$

From this Eq. and the obvious $[\hat{O}_{\rm ND}, \hat{H}_{\mathcal{S}_{\rm QP}}] = 0$ we obtain:

$$[\hat{O}_{\rm ND}, \hat{H}_{\mathcal{S}_{\rm QP}\mathcal{S}}] = 0, \tag{19.42}$$

where $\hat{H}_{\mathcal{S}_{\rm QP}\mathcal{S}}$ is the interaction Hamiltonian between $\mathcal{S}$ and $\mathcal{S}_{\rm QP}$.

[4] For the extension to set of observables see [CAVES *et al.* 1980, 365].

19.5 Evaluation of the Minimal Measurement Error

19.5.1 An Example

Now we try to evaluate the minimal measurement error, which is a limit valid also for a QND-Measurement. Let us propose the following QND-Measurement [BRAGINSKY *et al.* 1977]:[5] we wish to measure the electromagnetic energy in a resonator by a ponderomotive pressure on a movable wall [see figure 19.3]. The ponderomotive pressure is given by:

$$F = \frac{E}{l}. \tag{19.43}$$

During time τ (measurement time) the force F changes the momentum of the wall by the

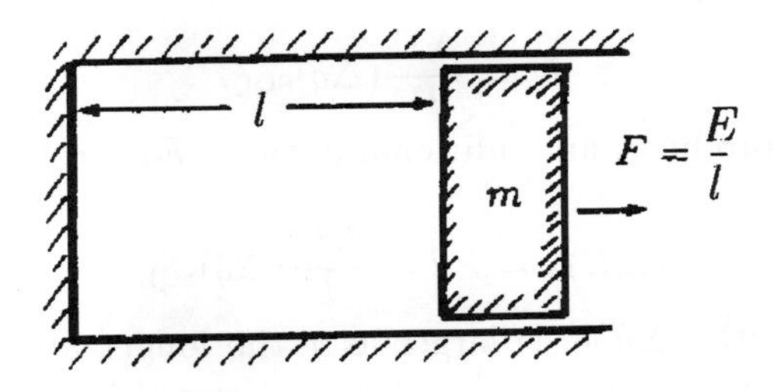

Figure 19.3: The measurement of the electromagnetic energy in a resonator by its ponderomotive pressure on a movable wall [*BRAGINSKY/KHALILI* 1992, 54].

amount:

$$\delta\hat{p} = \frac{E\tau}{l}. \tag{19.44}$$

Thus the more precisely the initial momentum is defined, the more precise will be the measured energy:

$$(\Delta E)_{\mathrm{M}} = \frac{l}{\tau}\Delta\hat{p}. \tag{19.45}$$

On the other hand we have the UR1 [Eq. (7.7)] between the wall's momentum and the wall's position. The latter produces a corresponding uncertainty in the electromagnetic frequency during the measurement:

$$\Delta\omega = \omega\frac{\Delta\hat{q}}{l}, \tag{19.46}$$

and this in turn produces a random change (a perturbation) in the resonator's phase:

$$(\Delta\varphi)_{\mathrm{P}} = \Delta\omega\cdot\tau = \omega\tau\frac{\Delta\hat{q}}{l}, \tag{19.47}$$

from which [see Eq. (7.26)] we obtain the following Error/perturbation UR between energy and phase:

$$(\Delta E)_{\mathrm{M}}(\Delta\varphi)_{\mathrm{P}} \geq \frac{\hbar\omega}{2}. \tag{19.48}$$

[5] Wide exposition in [*BRAGINSKY/KHALILI* 1992, 53–55, 58–60]. See also [BRAGINSKY/KHALILI 1996, 3–4].

By the standard method the minimum value for the error of the measured energy is [see Eq. (19.29)]:

$$(\Delta E)_{\mathrm{S}} = \frac{l}{\tau}(\Delta \hat{p})_{\mathrm{SQL}}. \tag{19.49}$$

Now we must take into account that the electromagnetic field in the resonator is a source of dynamic rigidity for the mass's motion:

$$K = \frac{E}{l^2}. \tag{19.50}$$

The rigidity produces on the wall an unknown force

$$F' = K(\Delta \hat{q})_{\mathrm{SQL}}, \tag{19.51}$$

where the SQL for the position of the wall is given by Eq. (19.28). Hence using Eq. (19.50) and Eq. (19.51) we can estimate the minimum size of the unknown force to be in terms of the initial energy E_i:

$$F' = \frac{\Delta E_i}{l^2}(\Delta \hat{q})_{\mathrm{SQL}}. \tag{19.52}$$

Hence the additional force produces an additional error $(\Delta E)_{\mathrm{A}}$ in the measured value of energy given by:

$$(\Delta E)_{\mathrm{A}} = F'l = \frac{\Delta E_i}{l}(\Delta \hat{q})_{\mathrm{SQL}}. \tag{19.53}$$

Error $(\Delta E)_{\mathrm{A}}$, in contrast with $(\Delta E)_{\mathrm{S}}$, is larger as τ the longer and m and l are smaller. The minimum total error is given when $(\Delta E)_{\mathrm{S}} = (\Delta E)_{\mathrm{A}}$. From Eq. (19.49) and Eq. (19.53) we see that this limit is achieved by

$$ml^2 = \Delta E_i \tau^2, \tag{19.54}$$

with a total error $(\Delta)_{\mathrm{T}}$ given by:

$$(\Delta E)_{\mathrm{T}} = \left[((\Delta E)_{\mathrm{S}})^2 + ((\Delta E)_{\mathrm{A}})^2\right]^{\frac{1}{2}} = \sqrt{\frac{\Delta E_i \hbar}{\tau}}. \tag{19.55}$$

In the worst case, when we know nothing about the energy before the measurement, we have:

$$\Delta E_i \simeq E, \tag{19.56}$$

and correspondingly the minimum total error will be [see Eq. (19.27)]:

$$(\Delta E)_{\mathrm{T}} = \sqrt{\frac{E\hbar}{\tau}} = \frac{(\Delta E)_{\mathrm{SQL}}}{\sqrt{\omega\tau}}. \tag{19.57}$$

Now, if the measurement is made very slowly (with $\omega\tau \gg 1$), then we have $(\Delta E)_{\mathrm{T}} \ll (\Delta E)_{\mathrm{SQL}}$ and we can measure the energy to a precision better than one quantum $(\Delta E_{\mathrm{T}}) \gg \hbar\omega$.

We suppose that before the first measurement we do not know anything about the energy. But by repeating the measurement, the precision of the latter will be determined by Eq. (19.55) with a ΔE_i equal to the error (19.57) of the first measurement. Then the result will be:

$$(\Delta E)_{\mathrm{T}} = E^{\frac{1}{4}}\left(\frac{\hbar}{\tau}\right)^{\frac{3}{4}}. \tag{19.58}$$

By repeating the measurement again and again — compensating each time the electromagnetic rigidity with higher accuracy — we obtain in the limit the QND minimal measurement error:

$$\boxed{(\Delta E)_{\mathrm{ND}} = \frac{\hbar}{\tau} \ll (\Delta E)_{\mathrm{SQL}}} \tag{19.59}$$

19.5.2 Concluding Remarks

In conclusion, the high accuracy of QND-Measurement allows for the first time a measurement on single systems and not only on ensembles [*BRAGINSKY/KHALILI* 1992, 21], so that the traditional Ignorance and Statistical interpretations [see chapter 6] are definitively excluded by recent QM developments.

Another important point — to be discussed later [section 33.1] — is the following: if one can measure the energy to a precision better than one quantum, then one does not need QP [see postulate 2.1: p. 19, and 2.2: p. 20] and has good reason to think that it is a false postulate.

19.6 QND-Measurement and the Uncertainty Principle

We have already stated [see definition 19.2] that the non-commutable observables with respect to the measured ones are to be perturbed to the extent allowed by UP. We wish now to see in more detail the relationship between QND-Measurement and UP. To better discuss the matter we distinguish two cases, the linear and non-linear one.

19.6.1 Linear Case

General Remarks When studying indirect Measurement [section 19.3], we already said that the source of Measurement error and of perturbation are the internal uncertainties of the probe. But, as we know, it is also possible to start with the UR for $\mathcal{S}$ and deduce the UR for the Measurement process [see Eq. (7.26)]. But, since the error and the perturbation do not depend on $\mathcal{S}$'s initial state, we can describe the entire Measurement process by two simple linear formulae [*BRAGINSKY/KHALILI* 1992, 64–66]:

$$\mathbf{q}_{\mathrm{M}} = \hat{\mathbf{q}}_i + (\delta\hat{\mathbf{q}})_{\mathrm{M}} \tag{19.60a}$$

$$\hat{\mathbf{p}}_f = \hat{\mathbf{p}}_i + (\delta\hat{\mathbf{p}})_{\mathrm{P}} \tag{19.60b}$$

where $\mathbf{q}_{\mathrm{M}}$ is the measured value, $\hat{\mathbf{q}}_i, \hat{\mathbf{p}}_i$ are the initial position and momentum, $\hat{\mathbf{p}}_f$ is the final momentum and $(\delta\hat{\mathbf{q}})_{\mathrm{M}}, (\delta\hat{\mathbf{p}})_{\mathrm{P}}$ are random variables that do not depend on the initial state of the system, and thus have no correlation with $\hat{\mathbf{q}}_i, \hat{\mathbf{p}}_i$, and thus have uncertainties $(\Delta\hat{\mathbf{q}})_{\mathrm{M}}, (\Delta\hat{\mathbf{p}})_{\mathrm{P}}$, respectively — measurement error and perturbation which satisfy UR (7.26). A measurement which satisfies this condition is called *linear*. To understand the specificity of a linear measurement we give an example of a non-linear one. Suppose the energy E of a harmonic oscillator is measured and we want to know the perturbation of its generalized coordinate $\hat{\mathbf{q}}$. Then the corresponding UR has the form:

$$(\Delta E)_{\mathrm{M}}(\Delta\hat{\mathbf{q}})_{\mathrm{P}} \geq \frac{\hbar|\langle\hat{\mathbf{p}}\rangle|}{2m}, \tag{19.61}$$

where $\langle\hat{\mathbf{p}}\rangle$ is the expected momentum of the oscillator. Now it is clear that the RHS of Eq. (19.61) is not a constant. Hence, by contrast to the non-linear case, we define a *linear measurement* as a measurement in which in the RHS of measurement uncertainties is constant — as in Eq. (7.26).

Convolutions By a linear QND-Measurement we respect condition (19.35). In the case of a measurement of (one-dimensional) coordinate q we can write the amplitude operator $\hat{\vartheta}$ in the form:

$$\hat{\vartheta}(q_{\mathrm{M}}) = \int_{-\infty}^{+\infty} \vartheta(q_{\mathrm{M}}, q)|q\rangle\langle q|dq, \tag{19.62}$$

where $\vartheta(q_{\rm M}, q)$ is some function normalized to unity $\int_{-\infty}^{+\infty} |\vartheta(q_{\rm M}, q)|^2 dq_{\rm M} = 1$. The probability distribution according to Eqs. (19.7), (19.10) and (19.62) is:

$$\wp(q_{\rm M}) = \int_{-\infty}^{+\infty} \wp_{\rm M}(q_{\rm M}|q)\wp_i(q)dq, \tag{19.63}$$

where the initial probability $\wp_i$ is:

$$\wp_i(q) = \langle q|\hat{\rho}_i|q\rangle, \tag{19.64}$$

and the conditional probability distribution $\wp_{\rm M}(q_{\rm M}|q)$ is:

$$\wp_{\rm M}(q_{\rm M}|q) = |\vartheta(q_{\rm M}, q)|^2, \tag{19.65}$$

and represents the probability that the measured value is $q_{\rm M}$ when the true value of the measured observable is q. The form of this second distribution is determined by the properties of $\mathcal{A}$ and its variance is by definition the measurement error. The assumption of linearity implies that it is independent of q. It is reasonable to assume that the measurement is unbiased, i.e. the mean value of $q_{\rm M}$ is equal to q:

$$\int_{-\infty}^{+\infty} q_{\rm M}\wp_{\rm M}(q_{\rm M}|q)dq_{\rm M} = q. \tag{19.66}$$

From these two conditions follows that $\wp_{\rm M}$ depends only on the difference $\delta q = q_{\rm M} - q$ and that:

$$\int_{-\infty}^{+\infty} \delta q \wp_{\rm M}(\delta q)d(\delta q) = 0, \tag{19.67}$$

so that we can rewrite Eq. (19.63) as a convolution:

$$\wp(q_{\rm M}) = \int_{-\infty}^{+\infty} \wp_{\rm M}(q_{\rm M} - q)\wp_i(q)dq. \tag{19.68}$$

As we know, in statistics such a convolution describes the statistics of the sum of two independent random variables q and δq [see section 5.2]

$$q_{\rm M} = q + \delta q, \tag{19.69}$$

which has the form of Eq. (19.60a). δq is an additional contribution with respect to the true value q produced by $\mathcal{A}$. It is governed by $\wp_{\rm M}(\delta q)$. According to Eq. (19.67) the mean value of δq is zero and the variance — we do not consider here the specific problems born from the identification of the uncertainty with the standard deviation [see subsection 7.5.2 and section 7.6]: for the purposes of this section it is of no relevance — is:

$$(\Delta q)^2_{\rm M} = \int_{-\infty}^{+\infty} (\delta q)^2 \wp_{\rm M}(\delta q)d(\delta q). \tag{19.70}$$

The quantity $(\Delta q)_{\rm M}$ is the rms measurement error.

There is [*BRAGINSKY/KHALILI* 1992, 69–70] a similar situation for the perturbation of a non commutable observable p: we obtain again a convolution, so that the measurement error $(\Delta q)_{\rm M}$ and the perturbation $(\Delta p)_{\rm P}$ are equal to the uncertainties of observable q, p in a state with 'wave function' ϑ (which is the operator product of a reduction and of a linear back action). The minimum of uncertainty is obtained when ϑ has a Gaussian form.

Sequences of measurements In the case of a *sequence of linear measurements* it is often sufficient to calculate for a given j–th measurement the *expectation value* of observable $\hat{O}_j$ (which is the sum of its initial one plus the sum of the perturbations induced by the preceding measurements), its *mean square* (which is the sum of the initial mean square of $\hat{O}_j$, of the measurement error and of the perturbations induced by preceding measurements) and the *cross correlation* of the results between two different measurements [*BRAGINSKY/KHALILI* 1992, 71–75].

Continuous measurements Normally we are faced with discrete measurement, but continuous ones are also possible [CAVES 1982] [CAVES 1986] [CAVES 1987a] [CAVES 1987b].[6] A continuous measurement is the record of some continuous function of time. The dividing line between these two is not the fact that continuous measurements happen in a time-interval because discrete measurements are also not instantaneous. Generally a continuous measurement can be seen as the limiting case of a sequence of discrete measurements when the interval between measurements is much smaller than the characteristic timescale on which the measured quantity changes (smaller than the shortest period present in the spectrum of the observable). In the case of nonlinearity we are faced with enormous technical difficulties, but in the linear case we obtain that the quantity

$$S_{\mathbf{q}} = (\Delta\hat{\mathbf{q}})^2_{\mathrm{M}} \cdot \delta t \tag{19.71}$$

remains constant — here δt represents the interval between measurements, which tends to zero.

The output signal of $\mathcal{A}$ is the input signal plus the noise added by the probe. In our model of linear continuous measurements, the noise is white, i.e. the values at different moments of time are independent. Then the quantity $S_{\mathbf{q}}$ is the spectral density of this noise. On the other side the conjugate variable (say the momentum) undergoes a random process (Brownian motion) because in each measurement the object receives a random kick. The collisions become more frequent and the change of momentum in each collision becomes smaller, so that the mean square deviation of the momentum after some time remains constant. In the limit the separated kicks become a continuous random force $\mathbf{F}$, and after a time τ we have:

$$(\Delta\hat{\mathbf{p}})_{\mathrm{P}(\tau)} = \sqrt{S_{\mathbf{F}}\tau}, \tag{19.72}$$

where $S_{\mathbf{F}}$ is the spectral density of the noise of the force:

$$S_{\mathbf{F}} = \frac{1}{\delta t}(\Delta\hat{\mathbf{p}})^2_{\mathrm{P}}. \tag{19.73}$$

Hence from Eq. (19.71), (19.72) and (19.73) we obtain a UR for continuous linear measurements:

$$\boxed{S_{\mathbf{q}} S_{\mathbf{F}} \geq \frac{\hbar^2}{4}} \tag{19.74}$$

19.6.2 Nonlinear Case

First we must distinguish between two back-actions: the *dynamical* back-action is the influence on the evolution of the expectation value of the measured observable; the *fluctuational* back-action which is a completely random perturbation which changes the dynamic behaviour of $\mathcal{S}$.

[6] See also [*BRAGINSKY/KHALILI* 1992, 76–79].

In linear continuous measurements the two forms are completely independent. It is not so in nonlinear ones [*BRAGINSKY/KHALILI* 1992, 93–100].

We assume as a quantum probe a set of identical particles which do not interact with each other (factorised wave function: white noise) and with vanishing Hamiltonians (when the particle is decoupled from the environment, its quantum state is independent of time, i.e. is conserved). We derive a final form for the equation of the motion of the object's density operator [see Eq. (21.7)]:

$$\frac{\partial}{\partial t}\hat{\rho}_S(t) = \frac{1}{\imath\hbar}[\hat{H}_S, \hat{\rho}_S(t)] - \frac{\sigma_F^2(t)}{2\hbar^2}[\hat{\mathbf{q}}(t), [\hat{\mathbf{q}}(t), \hat{\rho}_S(t)]], \tag{19.75}$$

with the second term, inversely proportional to the precision of monitoring, for particle j as $\Delta t \longrightarrow 0$, given by

$$\sigma_F^2(t_j) := \delta t \left\langle \hat{y}_j^2 \right\rangle, \tag{19.76}$$

where $\left\langle \hat{y}_j^2 \right\rangle$ is the expectation of the coordinate of the j–th particle, and σ_F^2 expresses the *fluctuational back-action.*

We calculate a similar quantity for the conjugate $\hat{\mathbf{p}}_j$, which expresses the *accuracy of monitoring*:

$$\sigma_q^2(t_j) = \frac{\left\langle \hat{\mathbf{p}}_j^2 \right\rangle}{\delta t}, \tag{19.77}$$

and a joint uncertainty:

$$\sigma_{qF}^2(t_j) = |\left\langle \hat{y}_j^2 \hat{\mathbf{p}}_j^2 \right\rangle|^2, \tag{19.78}$$

from which we obtain the following UR for nonlinear measurements:

$$\boxed{\sigma_F^2(t) \cdot \sigma_q^2(t) - \sigma_{qF}^2(t_j) \geq \frac{\hbar^2}{4}} \tag{19.79}$$

which signifies that the accuracy of the monitoring is inversely proportional to the back-action of the monitoring on $\mathcal{S}$.

19.6.3 Classical Forces

In the case of a classical force there are no limits deriving from UP [*BRAGINSKY/KHALILI* 1992, 106–107, 109, 115]. But there are nonetheless two types of limits:

- the more precise the measurement, the more it introduces energy in $\mathcal{S}_{\mathrm{QP}}$, and the more energy, the more $\mathcal{S}_{\mathrm{QP}}$ is perturbed and hence limits the accuracy of the measurement;
- the standard procedures are always based on the monitoring of the coordinate of $\mathcal{S}_{\mathrm{QP}}$.

With the first measurement this interaction perturbs the conjugate momentum, and this reaction produces an uncertainty in the successive measurement of the position (from here the SQL arises). But if the force is classical we can use two methods to overcome these limitations:

- a QND-Measurement on $\mathcal{S}_{\mathrm{QP}}$ by choosing a set of observables which respond to the external force but are not perturbed in the measurement process;
- design $\mathcal{S}_{\mathrm{QP}}$ so that it passes to $\mathcal{A}$ only information about the measured (classical force) and no information about $\mathcal{S}_{\mathrm{QP}}$'s quantum state (arbitrary precision).

But, in the first case there are still limitations connected with thermal dissipation and with the amount of energy that the probe can deal with and still function properly[7].

On the other hand, if one wishes to detect a force by measuring a probe observable, the latter must be well defined, which it cannot be if one does not disturb sufficiently its conjugate by inserting a sufficient quantity of energy in $\mathcal{S}_{\rm QP}$ [*BRAGINSKY/KHALILI* 1992, 125–26]. But this quantity is limited by the limits of $\mathcal{S}_{\rm QP}$ itself, i.e. when its state is destroyed: i. e. we have here an energetic limitation.

19.7 Experimental Context

As we know the point of departure of the QND-Measurement was the measurement of gravitational waves, which requires a precision of the order of 10^{-19} cm, by using aluminum bars with weight between 10 Kg e 10 tons. For times $< 10^{-3}$ sec. the bar behaves as a particle subjected to QM. A way to maximally eliminate the perturbations is that of amplifying the gravitational wave, but then we need two bars some kms away from each other, with an insufficient resolution of 10^{-16}. But we can do the following: we make a measurement of the momentum with a small initial error $\Delta\hat{p} \simeq 10^{-9}$ g cm/sec. We inevitably disturb the bars position by an unknown amount $\Delta\hat{q} \geq 5 \times 10^{-19}$ cm. We wait a while, $t \simeq 10^{-3}$ sec, before making a second momentum measurement: as the bar moves freely between the measurements, its momentum remains fixed. Consequently the second measurement can have as a good accuracy — say 10^{-9} g cm/sec – as the first measurement has, and a momentum change of (a few) $\times 10^{-9}$ g cm/sec, due to a passing gravity wave, can be seen [BRAGINSKY *et al.* 1980, 751–52]. Hence momentum measurements are QND-Measurement, position measurements are not[8]. First experimental proposals and verifications of the QND-Measurement in Q-Optics were performed in the middle of the eighties [YURKE 1985a] [YAMAMOTO *et al.* 1986] [LEVENSON *et al.* 1986].[9]

19.8 Zeno Paradox

19.8.1 General Features

The quantum Zeno effect owes his name to Misra and Sudarshan [MISRA/SUDARSHAN 1977], even if some antecedents had already been developed earlier[10]. It can be defined as follows [ITANO *et al.* 1990, 2295] [PASCAZIO 1996]:

Definition 19.6 (Zeno Effect) *The Zeno effect is the inhibition of spontaneous or induced transitions between quantum states by frequent measurements such that a system remains in its initial state throughout a given time interval.*

---

[7] Examination in [*BRAGINSKY/KHALILI* 1992, 125–35].

[8] See also [UNRUH 1979] [THORNE *et al.* 1978] .

[9] See also [LA PORTA *et al.* 1989] [GRANGIER *et al.* 1991] .

[10] For example, already in 1957 by Khalfin, which recently has returned on the subject [KHALFIN 1982] [KHALFIN 1990b], and in 1961 by Winter [WINTER 1961].

Inhibition of spontaneous transitions — such as decay of an unstable state, originally proposed by Misra and Sudarshan — are very difficult to realize[11].

As a simply model we suppose [*BRAGINSKY/KHALILI* 1992, 95–97, 101–104] that the observable $\hat{\mathbf{q}}$ of $\mathcal{S}$ is being monitored continuously and that $\mathcal{S}$ begins at the n–th eigenstate of $\hat{\mathbf{q}}$. We treat the continuous measurements as limiting cases of discrete measurements separated by small intervals τ. Between two measurements (in the interval τ) the evolution of $\mathcal{S}$ is governed by the Schrödinger Eq. This evolution will cause $\mathcal{S}$ — with some probability — to make a transition from eigenfunction $|\psi_n\rangle$ to some other one. The first interval begins at $t = 0$, and for sufficient small times t produces the following wave Eq., which is a power series of the Schrödinger Eq.:

$$\psi(\mathbf{r}, t) \simeq \left[1 + \frac{\hat{H}t}{\imath\hbar} + \frac{1}{2}\left(\frac{\hat{H}t}{\imath\hbar}\right)^2 + \ldots\right]\psi(\mathbf{r}, 0), \tag{19.80}$$

where $\psi(\mathbf{r}, 0) = \psi_n(\mathbf{r})$. Immediately before the first measurement (at $t = \tau$) the wave function is:

$$\psi(\mathbf{r}, \tau) \simeq \left[1 + \frac{\hat{H}\tau}{\imath\hbar} + \frac{1}{2}\left(\frac{\hat{H}\tau}{\imath\hbar}\right)^2 + \ldots\right]\psi_n(\mathbf{r}). \tag{19.81}$$

The probability that $\mathcal{S}$ will still be in the n–th state (that no transition occurred) is:

$$\begin{aligned} \wp_{nn} &= \left|\int_{-\infty}^{+\infty} \psi_n^*(\mathbf{r})\psi(\mathbf{r}, \tau)d\mathbf{r}\right|^2 \\ &= \left|1 + \frac{\tau}{\imath\hbar}\hat{H}_{nn} - \frac{\tau^2}{2\hbar^2}(\hat{H}^2)_{nn}\right|^2 \\ &\simeq 1 - \frac{\tau^2}{\hbar^2}\left[(\hat{H}^2)_{nn} - (\hat{H}_{nn})^2\right], \end{aligned} \tag{19.82}$$

where $\hat{H}_{nn}, (\hat{H}^2)_{nn}$ are the diagonal matrix elements of the respective operators:

$$\hat{H}_{nn} = \int_{-\infty}^{+\infty} \psi_n^* \hat{H}(\mathbf{r})\psi_n(\mathbf{r})d\mathbf{r}, \tag{19.83a}$$

$$(\hat{H}^2)_{nn} = \int_{-\infty}^{+\infty} \psi_n^* \hat{H}^2(\mathbf{r})\psi_n(\mathbf{r})d\mathbf{r}. \tag{19.83b}$$

Since the quantity in square brackets in the RHS of Eq. (19.82) is the variance of the energy [Eq. (3.110)] in the initial state $|\psi_n\rangle$, we can rewrite it in the following form:

$$\wp_{nn} = 1 - \frac{\tau^2}{\hbar^2}(\Delta E)^2. \tag{19.84}$$

After k steps consisting in free evolution in time-interval τ plus an instantaneous measurement of $\mathbf{q}$, the probability that $\mathcal{S}$ is still in initial eigenstate is the k–th power of the expression (19.84):

$$\wp_{nn}^k(\tau) = \left[1 - \frac{\tau^2}{\hbar^2}(\Delta E)^2\right]^{\frac{t_{\mathrm{T}}}{\tau}}, \tag{19.85}$$

where $t_{\mathrm{T}} = k\tau$ is the total time for these k steps.

[11] On the problem see [GHIRARDI *et al.* 1979].

The continuous limit at $\tau \longrightarrow 0$ is given by[12]:

$$\lim_{\tau \longrightarrow 0} \wp^k_{nn}(\tau) \longrightarrow \exp\left[-\frac{t_{\mathrm{T}}\tau}{\hbar^2}(\Delta E)^2\right], \tag{19.86}$$

and in the ultimate limit of vanishing τ, it becomes:

$$\wp^k_{nn}(t_{\mathrm{T}}) = 1. \tag{19.87}$$

In other words $\mathcal{S}$ is 'frozen' in the initial state $|\psi_n\rangle$. In conclusion when any observable with discrete spectrum is monitored with infinite precision, $\mathcal{S}$ remains in the initial state.

19.8.2 Watchdog Effect

The watchdog effect is a particular case of the Zeno paradox: it is the suppression of the response of a quantum object when the energy of $\mathcal{S}$ has been monitored continuously, and is determined by the relation of the magnitudes of two time-scales: the time scale τ_F for the force to change and the time τ_O required to distinguish the oscillator's energy levels. In other words it is the effect of the fluctuational back-action on the dynamic behaviour, because the response to a fluctuating force is always inferior to that to a precisely harmonic force [*BRAGINSKY/KHALILI* 1992, 94].

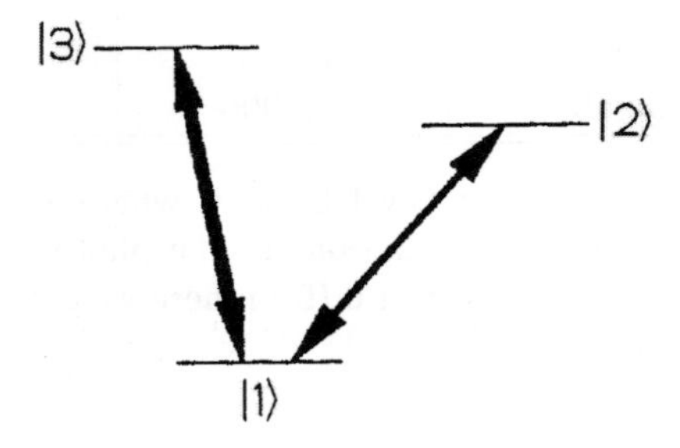

Figure 19.4: Energy-level diagram for Cook's proposed demonstration of the quantum Zeno effect — from [ITANO *et al.* 1990]

19.8.3 Collapse or Not?

The Experiment of Itano and Co-workers

Itano and co-workers [ITANO *et al.* 1990] performed an important experiment based on an idea of Cook closely related to the theory of quantum jumps [see chapter 22]. The idea (= Atomic-Level Exp. 1) is that of a trapped ion which can perform only transitions either from the ground state 1 to metastable excited state 2 (where spontaneous decay from 2 to 1 is negligible) or from 1 to the excited state 3 by a optical pulse [see figure 19.4]. The wave function is projected by the measurement into level 1 or 2: if at the beginning of the pulse the ion is projected into level 1, a cycle occurs between level 1 and 3, and it emits a series of photons until the pulse is turned off, while, if it is projected into level 2, it scatters no photons. If a measurement finds the ion to be in level 1, the ion returns to level 1 after the end of the measurement, within a time approximately

[12]Because $\lim_{n\longrightarrow\infty}\left(1+\frac{x}{n}\right)^n = e^x$.

equal to the lifetime of level 3. If a measurement finds the ion to be in level 2, the ion never leaves that level during the measurement.

Cook now proposed to drive the ion from level 1 to level 2 by a on-resonance π pulse (of duration $\tau = \pi/\omega$, where ω is the *Rabi frequency* , proportional to the amplitude of the applied field), while simultaneously applying a series of short measurement pulses, each one of duration much less than the time between them. Suppose also that the ion is in level 1 at $t = 0$. After N measurements the probability of being in level 2 is given by:

$$\wp_2^{(N)}(\tau) = \frac{1}{2}[1 - \cos^N(\pi/N)]. \tag{19.88}$$

Experimental results for transitions $1 \rightsquigarrow 2$ and $2 \rightsquigarrow 1$, obtained by $^9Be^+$ ions stored in a cylindrical Penning trap[13], are shown in table 19.1. They clearly show monotonic decreasing of $\wp_2(\tau)$ as n grows.

N	$\frac{1}{2}[1-\cos^n(\pi/N)]$	predicted $1 \rightsquigarrow 2$	observed $1 \rightsquigarrow 2$	predicted $2 \rightsquigarrow 1$	observed $2 \rightsquigarrow 1$
1	1.0000	0.995	0.995	0.999	0.998
2	0.5000	0.497	0.500	0.501	0.496
4	0.3750	0.351	0.335	0.365	0.363
8	0.2346	0.201	0.194	0.217	0.209
16	0.1334	0.095	0.103	0.118	0.106
32	0.0716	0.034	0.013	0.073	0.061
64	0.0371	0.006	−0.006	0.080	0.075

Table 19.1: Experimental results of Atomic-Level Exp. 1, where a measurement error of about 0.02 must be taken into account, and random fluctuations in the photon count rate can lead to apparent transition probabilities greater than 1 or less then 0 (from here we get the observed value −0.006 in the last line) — from [ITANO *et al.* 1990, 2298]

Pascazio and Namiki's Analysis

Gaussian behaviour While proponents see the above reported experiment as evidence of a continuous collapse of the wave function provoked by a series of measurements, others questioned this interpretation[14].

Pascazio and Namiki supposed that the Zeno Effect is a process described by a normal unitary evolution [PASCAZIO/NAMIKI 1995, 335–36, 349–50]. It is possible to distinguish [NAMIKI/MUGIBAYASHI 1953] three phases of each dynamical evolution of a QM system $\mathcal{S}$ prepared in an eigenstate of the unperturbed Hamiltonian:

- A Gaussian behaviour at short times,
- a Breit/Wigner exponential decay — where the survival probability of a particle is given by $\exp(-t/\tau_0)$ — at intermediate times[15],

[13] For details of the experiment see the original article.

[14] There is a wide literature on this point. See [BALLENTINE 1990a, 1340–42] [BALLENTINE 1991b] [PERES/RON 1990] [PETROSKY *et al.* 1990] [BLOCK/BERMAN 1991] [FRERICHS/SCHENZLE 1991] [NAKAZATO *et al.* 1995] [PASCAZIO/NAMIKI 1995].

[15] On this point see [NAKAZATO *et al.* 1994].

- a power law at long times[16].

The asymptotic dominance of the second moment is representative of a pure stochastic evolution and can be derived in QM in the van Hove Limit[17]. There is a strict connection between dissipation and exponential decay; the Gaussian behaviour is of particular significance for the Zeno effect.

An experiment Pascazio and Namiki propose a concrete experiment in order to investigate the problem [PASCAZIO/NAMIKI 1995, 339–41]: it consists in inserting detectors $D_j, j = 1, \ldots, N$ and mirrors M, in order to separate the spin-up from spin-down [figure 19.5]. But by increasing N, the experiment is more and more difficult to perform. The authors have shown that, by inserting or removing the lateral detectors (which detect spin-down), the detection probability is the same. Hence projectors are not necessary and the experiment can be considered as a negative-result or interaction-free measurement [see chapter 20].

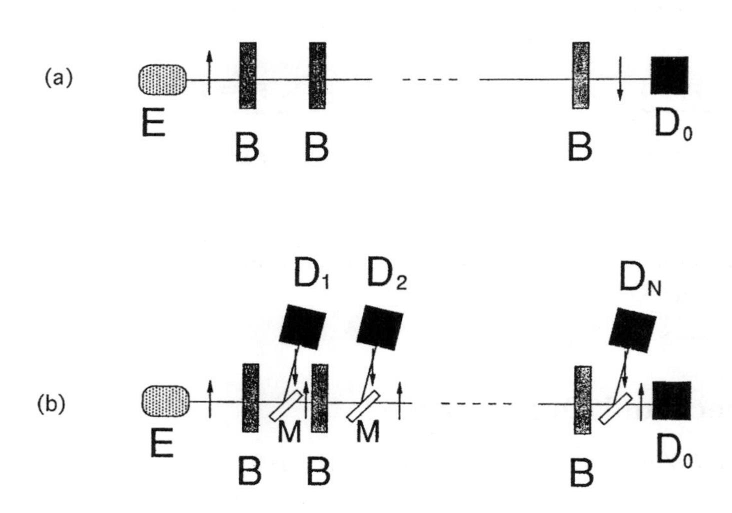

Figure 19.5: (a) 'Free' evolution of a neutron spin under the action of a magnetic field. An emitter E sends a spin-up neutron through several regions where a magnetic field **B** is present. The detector D_0 detect a spin-down neutron: no Zeno effect occurs.
(b) The neutron spin is 'monitored' at every step, by selecting and detecting the spin-down component. D_0 detects a spin-up neutron: the Zeno effect takes place — from [PASCAZIO/NAMIKI 1995].

Another Interpretation of Itano's Experiment Returning now to the experiment proposed by Cook and performed by Itano and co-workers, Pascazio [PASCAZIO 1996, 528–29] objects that the formula (19.88) — or $\wp_1^{(N)}(\tau) = \frac{1}{2}[1 + \cos^N(\pi/N)]$ — is not the quantum Zeno effect [given by definition 19.6] but only expresses the probability that the atom is in level 2 (or 1) at time τ after N measurements, independently from its past history, i. e. it does not only indicate the survival possibility but also includes the possibility that transitions $2 \rightsquigarrow 1 \rightsquigarrow 2$ —

[16]See also [CHO *et al.* 1993, 814].
[17]See [VAN HOVE 1955].

or $1 \rightsquigarrow 2 \rightsquigarrow 1$ take place [see figure 19.6]. In other words, the experiment and the analysis performed by Itano and co-workers, though generally correct, are not about a Zeno effect.

By performing a binomial analysis of formula (19.88) we obtain:

$$\wp_2^{(N)} = 1 - \sum_{n \text{ even}} \binom{N}{n} \sin^{2n}(\pi/2N) \cos^{2(N-n)}(\pi/2N). \tag{19.89}$$

We see that we obtain a Zeno effect if we consider the number of even transitions between level 1 and 2 is zero (i.e. $n = 0$):

$$\wp_2^{(N)} = 1 - \cos^{2N}(\pi/2N). \tag{19.90}$$

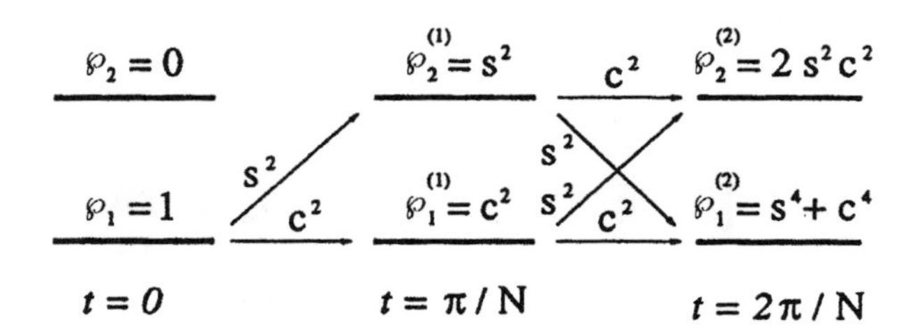

Figure 19.6: Transition probabilities after the first two measurements (with $s = \sin(\pi/2N), c = \cos(\pi/2N)$) — from [PASCAZIO 1996, 530].

19.8.4 Zeno Effect and Decoherence*

Environment Joos and Zeh [JOOS 1984] [JOOS/ZEH 1985, 224] showed that the interaction between a system $\mathcal{S}$ and the environment $\mathcal{E}$ can have different consequences: the freezing of the the internal dynamic of $\mathcal{S}$ (Zeno effect), or the tendential suppression of interference terms (decoherence), or even the respect of Fermi's Golden Rule (independence of transition rates from $\mathcal{E}$ in atom decay). Macroscopical objects are continuously measured by $\mathcal{E}$ and hence the interference terms are always brought near to zero.

A Model A formalization of these ideas was provided by Berry [BERRY 1995, 309–310].[18] The decoherence and the Zeno effect seem very different, because the first happens in very short times, while the second needs a long succession of measurements. But they can be explained with the same two-state model. Take a system governed by the Schrödinger equation:

$$\imath\partial_t|\psi\rangle = \mathbf{H}(\epsilon t) \cdot \hat{\sigma}|\psi\rangle, \tag{19.91}$$

where the state is a 2–spinor:

$$|\psi\rangle = |\psi(t, \epsilon)\rangle = \begin{pmatrix} \psi_1 \\ \psi_2 \end{pmatrix}. \tag{19.92}$$

[18] But on the subject see also [SIMONIUS 1978].

$\hat{\sigma}$ is the vector of spin–1/2 matrices (3.125) and the vector which drives the system will be called the Hamiltonian vector and is:

$$\mathbf{H}(\epsilon t) := [x(\epsilon t), y(\epsilon t), z(\epsilon t)]. \tag{19.93}$$

In Eq. (19.92) ψ_1 and ψ_2 represent the amplitudes for a quantum particle to inhabit each of the two wells constituting our system, separated by a high barrier of a double well potential. Forces within the isolated system are invariant under parity. The energy separation of these two eigenstates is the small parameter ϵ, which is exponentially small in the height and width of the barrier and in $1/\hbar$ — a vanishing ϵ represents the adiabatic limit of slow driving [see section 12.2]. A state initially localized in one of the wells will tunnel resonantly between them, with an oscillation of order $1/\epsilon$. The Hamiltonian of the isolated system is:

$$\hat{H}_0 = \mathbf{H}_0 \cdot \hat{\sigma} = \left(\frac{1}{2}\right) \begin{pmatrix} 0 & \epsilon \\ \epsilon & 0 \end{pmatrix}. \tag{19.94}$$

A crude but adequate representation of the environment $\mathcal{E}$ — for more detailed model see also [BRAY/M. MOORE 1982] — is through time-dependent random forces acting differently on the two wells — for example some collisions with other particles. The environmental forces are supposed to be large compared with ϵ and vary on short time scales compared with $1/\epsilon$. A Hamiltonian describing this situation is:

$$\hat{H} = \mathbf{H}(t) \cdot \hat{\sigma} = \left(\frac{1}{2}\right) \begin{pmatrix} \mathbf{F}(t) & \epsilon \\ \epsilon & -\mathbf{F}(t) \end{pmatrix}, \tag{19.95}$$

where $\mathbf{F}(t)$ is a white noise. The strength of the noise is normalized by:

$$\overline{\left(\int_0^{\mathcal{T}} dt \mathbf{F}(t)\right)^2} = \frac{\mathcal{T}}{\tau_0}, \tag{19.96}$$

where the overline denotes ensemble averaging and τ_0 is a constant — the noise time. Now we calculate the evolution of the ensemble-averaged density matrix:

$$\begin{aligned} \bar{\hat{\rho}} := \overline{|\psi(t)\rangle\langle\psi(t)|} &= \begin{pmatrix} \overline{|\psi_1(t)|^2} & \overline{\psi_1^*(t)\psi_2(t)} \\ \overline{\psi_2^*(t)\psi_1(t)} & \overline{|\psi_2(t)|^2} \end{pmatrix} \\ &\equiv \left(\frac{1}{2}\right) + \hat{\sigma} \cdot \mathbf{r}(t) = \left(\frac{1}{2}\right) \begin{pmatrix} 1 + z(t) & x(t) - \imath y(t) \\ x(t) = \imath y(t) & 1 - z(t) \end{pmatrix}, \end{aligned} \tag{19.97}$$

which expressed in terms of the Bloch vector [BLOCH 1946] is:

$$\mathbf{r}(t) = 2\overline{\langle\psi(t)|\hat{\sigma}|\psi(t)\rangle}. \tag{19.98}$$

For a pure state, $\mathbf{r}$ lies on the surface of a unit sphere; the north pole corresponds to the whole amplitude in well 1 and the south pole to the whole amplitude in well 2. The off-diagonal elements of $\hat{\rho}$ are described by Bloch components x, y. For any given $\mathbf{F}(t)$, the unitarity of quantum evolution ensures that $\mathbf{r}(t)$ remains on the unit sphere. However ensemble averaging, by tracing out $\mathcal{E}$, turns a pure state into a mixture, and makes $\mathbf{r}(t)$ flow into the sphere. Let us now define a parameter:

$$\eta := 4\epsilon\tau_0. \tag{19.99}$$

We are interested now in small ϵ, i.e. in small η. For the initial condition $\mathbf{r}(0) = \mathbf{r}_0$ the evolution is[19]:

$$x(t) = e^{-\frac{t}{2\tau_0}}, \tag{19.100a}$$

$$\begin{pmatrix} y(t) \\ z(t) \end{pmatrix} = \frac{e^{-\zeta_+ t}}{2\sqrt{1-\eta^2}} \begin{bmatrix} (1+\sqrt{1-\eta^2}y_0 + \eta z_0 \\ -\eta y_0 - (1-\sqrt{1-\eta^2})z_0 \end{bmatrix} + \frac{e^{-\zeta_- t}}{2\sqrt{1-\eta^2}} \begin{bmatrix} -(1-\sqrt{1-\eta^2}y_0 - \eta z_0 \\ \eta y_0 + (1+\sqrt{1-\eta^2})z_0 \end{bmatrix}, \tag{19.100b}$$

where

$$\zeta_\pm := \left(\frac{1}{4\tau_0}\right)(1 \pm \sqrt{1-\eta^2}). \tag{19.101}$$

In the limits:

$$\zeta_+ \longrightarrow \frac{1}{2\tau_0}, \quad \zeta_- \longrightarrow 2\epsilon^2\tau_0, \tag{19.102}$$

as $\eta \longrightarrow 0$, we arrive to the following picture of the density matrix flow described by Eq. (19.101)

- on the scale of the noise time τ_0, which is fast compared with the quantum oscillation time $1/\epsilon$, the fast exponential, involving ζ_+, causes the Bloch vector to flow from the surface of the sphere towards the line $x = 0, y = z\eta/(1+\sqrt{1-\eta^2})$, which is close to the z–axis: this fast flow in which off-diagonal elements disappear, is the *decoherence*;

- thereafter, the flow down the z–axis towards $\mathbf{r} = 0$ is dominated by the slow exponential involving ζ_-, whose time scale is of order $1/(\epsilon^2\tau_0)$, and thus is long compared with the quantum oscillation time: this slowing down is the quantum *Zeno effect.*

19.8.5 Zeno Effect of a Single System

We saw that generally the Zeno effect does not represent a collapse of the wave function but is a dynamical effect subject to unitary evolution. However, the above related experiments are centered on a series of measurements performed stepwise on several systems in an ensemble of systems prepared in the same pure state without any type of selection.

Now, a statistical Zeno effect can be interpreted (as Itano and co-workers do) as equivalent to a random decohering process: in fact, for an ensemble of ions undergoing repeated measurements, in order to produce the inhibition it suffices to destroy the coherence by their repeated coupling to a broad decaying state [POWER/KNIGHT 1996]. Instead of, at the individual level, the Zeno effect appears as something different from decoherence [SPILLER 1994]. Here, after each measurement, we choose to retain the measured system only if the measurement result equals the result obtained by measuring the first system, otherwise we discard it. This type of experiment corresponds to the Zeno effect of a *single* system and is truly a measurement effect because the results of such a series of measurement depends statistically on all the previous measurement results, regardless of the precision of the measurement.

As we shall show later [in subsection 28.5.3], the quantum Zeno effect of an individual system is equivalent the indetermination of its state.

[19] For full calculations see [BERRY 1995, 314–17].

19.9 Boundaries Between Apparatus and Object

We have already seen the difficulty of Bohr's position about the boundaries between apparatus and QM systems [section 8.5]. We shall now examine the problem in more detail. Haus and Kärtner showed [HAUS/KÄRTNER 1996] that, with a QND-Measurement, the boundary between A and S can be placed, in a time, after the interaction between them, in agreement with Zurek's decoherence proposal.

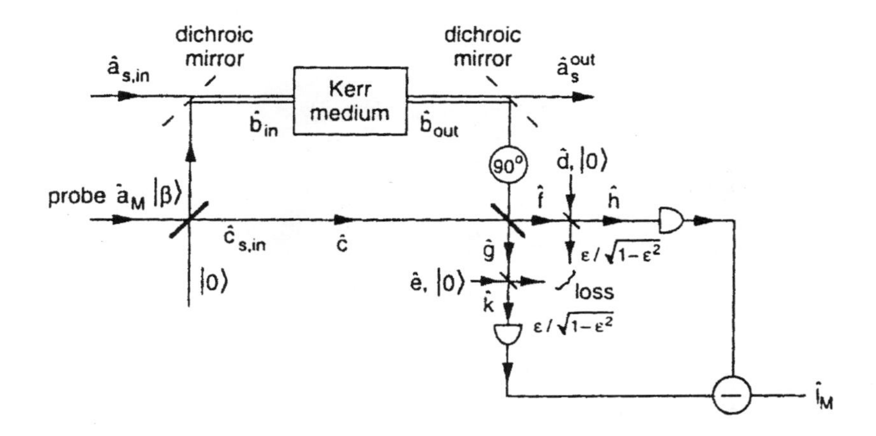

Figure 19.7: Scheme of proposed Interfer. Exp. 6 — from [HAUS/KÄRTNER 1996, 3787].

Experimental set-up Let us consider an optical experimental arrangement (= Interfer. Exp. 6)[20] as shown in figure 19.7 [see subsection 5.3.1 for apparatus] with annihilation operators $\hat{a}_{\rm S}$ for the signal and $\hat{b}_{\rm QP}$ for the probe beam. The Hamiltonian of the field dynamics in the nonlinear arm of the interferometer is:

$$\hat{H} = \hbar g \hat{a}_{\rm S}^{\dagger}\hat{a}_{\rm S}\hat{b}_{\rm QP}^{\dagger}\hat{b}_{\rm QP}. \tag{19.103}$$

The annihilation operators of the signal and probe after passage through the Kerr medium are respectively:

$$\hat{a}_{\rm S,out} = e^{\imath g_l \hat{b}^{\dagger}_{\rm QP,in}\hat{b}_{\rm QP,in}}\hat{a}_{\rm S,in}, \tag{19.104a}$$

$$\hat{b}_{\rm QP,out} = e^{\imath g_l \hat{a}^{\dagger}_{\rm S,in}\hat{a}_{\rm S,in}}\hat{b}_{\rm QP,in}, \tag{19.104b}$$

where $g_l = gl/v_g$, and l is the propagation distance in the Kerr medium and v_g the group velocity [see p. 699]. The photon group $\hat{a}_{\rm S}^{\dagger}\hat{a}_{\rm S}$ is an invariant; hence we can drop the subscript 'in' in the exponent. The Hamiltonian of a BS with inputs represented by annihilation operators $\hat{a}, \hat{b}$ [see figure 19.7] is:

$$\hat{H} = \hbar M(\hat{a}^{\dagger}\hat{b} + \hat{b}^{\dagger}\hat{a}). \tag{19.105}$$

[20] Even if the reported experiment is some pages long, its discussion seems justified by the fact that it is important for the particular problem discussed (which in turn is a fundamental one) and is also interesting for the Q-Optics formalism.

When integrated over a propagation distance $l_{\rm BS}$ or time $l_{\rm BS}/v_g$, so that $\sin(Ml_{\rm BS}/v_g) = \epsilon$ we obtain (where for a 50% beam splitting $\epsilon = \sqrt{1/2}$):

$$\hat{c} = \sqrt{1-\epsilon^2}\hat{a} + \imath\epsilon\hat{b}, \tag{19.106a}$$
$$\hat{d} = \imath\epsilon\hat{a} + \sqrt{1-\epsilon^2}\hat{b}. \tag{19.106b}$$

Now we can evaluate the output operators $\hat{h}, \hat{k}$ that represent the field after the last BS and just before the detection. For intermediate $\hat{f}, \hat{g}$ we find:

$$\hat{f} = \frac{\imath}{\sqrt{2}}(\hat{b}_{\rm out} + \hat{c}_{\rm out}), \tag{19.107a}$$
$$\hat{g} = \frac{1}{\sqrt{2}}(-\hat{b}_{\rm out} + \hat{c}_{\rm out}), \tag{19.107b}$$

where $\hat{c}_{\rm out} = \hat{c}_{\rm in}$. For $\hat{h}, \hat{k}$ we find:

$$\hat{h} = \sqrt{1-\epsilon^2}\hat{f} + \imath\epsilon\hat{d}, \tag{19.108a}$$
$$\hat{k} = \sqrt{1-\epsilon^2}\hat{g} + \imath\epsilon\hat{e}. \tag{19.108b}$$

Measured current We assume 100% quantum efficiency of photodetectors. $\hat{N}_{\hat{h}\hat{k}}$ represents the photon-number difference after the last BS and before the detector, and is given by:

$$\begin{aligned}\hat{N}_{\hat{h}\hat{k}} &:= \hat{h}^\dagger\hat{h} - \hat{k}^\dagger\hat{k} \\ &= (\sqrt{1-\epsilon^2}\hat{f}^\dagger - \imath\epsilon\hat{d}^\dagger)(\sqrt{1-\epsilon^2}\hat{f} + \imath\epsilon\hat{d}) - (\sqrt{1-\epsilon^2}\hat{g}^\dagger - \imath\epsilon\hat{e}^\dagger)(\sqrt{1-\epsilon^2}\hat{g} + \imath\epsilon\hat{e}) \\ &= (1-\epsilon^2)\hat{f}^\dagger\hat{f} - (1-\epsilon^2)\hat{g}^\dagger\hat{g} + \epsilon^2(\hat{d}^\dagger\hat{d} - \hat{e}^\dagger\hat{e}) \\ &+\imath\epsilon\sqrt{1-\epsilon^2}(\hat{f}^\dagger\hat{d} - \hat{d}^\dagger\hat{f}) - \imath\epsilon\sqrt{1-\epsilon^2}(\hat{g}^\dagger\hat{e} - \hat{e}^\dagger\hat{g}).\end{aligned} \tag{19.109}$$

Now we assume that the input probe field of the apparatus $\mathcal{A}$ $\hat{a}_{\rm M}$ is in the coherent state $|\alpha\rangle$ [see subsection 5.1.5] and the other input fields of the interferometer are in the vacuum state $|0\rangle$. Since a BS conserves a coherent state and also power is conserved, $\hat{b}_{\rm in}, \hat{c}_{\rm in}$ are in coherent states $|\alpha/\sqrt{2}\rangle, |\imath\alpha/\sqrt{2}\rangle$. When we take the average over the state of $\mathcal{A}$, i.e. we trace over the probe beams $\hat{b}, \hat{c}$ in Eq. (19.109), we obtain the measured current:

$$\begin{aligned}\left\langle\hat{N}_{\hat{h}\hat{k}}\right\rangle &= {}_{e,d,c,b}\left\langle 0,0,\imath\frac{\alpha}{\sqrt{2}},\frac{\alpha}{\sqrt{2}}|\hat{N}_{\hat{h}\hat{k}}|\frac{\alpha}{\sqrt{2}},\imath\frac{\alpha}{\sqrt{2}},0,0\right\rangle_{b,c,d,e} \\ &= (1-\epsilon^2)|\alpha|^2\sin(g_\imath\hat{a}_{\rm S}^\dagger\hat{a}_{\rm S}).\end{aligned} \tag{19.110}$$

Thus the mean value of the measured current is proportional to the sine of the product of coupling strength and the signal photon number. If we assume further that the signal beam is in the photon number state $|n\rangle$, then the square of the standard deviation of the measured current is:

$$\begin{aligned}(\Delta\hat{N}_{\hat{h}\hat{k}})^2 &= {}_{e,d,c,b,a}\left\langle 0,0,\imath\frac{\alpha}{\sqrt{2}},\frac{\alpha}{\sqrt{2}},n\left|\left(\hat{N}_{\hat{h}\hat{k}} - \left\langle\hat{N}_{\hat{h}\hat{k}}\right\rangle\right)^2\right|n,\frac{\alpha}{\sqrt{2}},\imath\frac{\alpha}{\sqrt{2}},0,0\right\rangle_{a,b,c,d,e} \\ &= (1-\epsilon^2)|\alpha|^2,\end{aligned} \tag{19.111}$$

which is shot noise. Note that by the assumptions made the noise does not depend on the photon-number state measured. If the measurement system has a large amplification — $(1-\epsilon^2)\alpha \gg 1$ —, the measured current can be observed on a macroscopic scale. If the interaction is weak ($g_l \ll 1$), we obtain a linear scale for the photon-number operator. Then, if we restrict the analysis to the case when $\mathcal{S}$ has photon-number states occupied only up to N and assume $g_l \ll 1/N$, we can linearize Eq. (19.110) and obtain:

$$\left\langle \hat{N}_{\hat{h}\hat{k}} \right\rangle = (1-\epsilon^2)|\alpha|^2 g_l \hat{a}_{\mathrm{S}}^\dagger \hat{a}_{\mathrm{S}}. \tag{19.112}$$

If we prepare the signal pulses so that they possess a definite photon number n, then the expectation will be $\left\langle \hat{N}_{\hat{h}\hat{k}} \right\rangle_n = \eta n$, where η is the sensitivity per photon:

$$\eta = (1-\epsilon^2)|\alpha|^2 g_l. \tag{19.113}$$

For a given photon-number state the detector current deviates from the mean value (19.110) from measurement to measurement with a rms deviation $\delta = \sqrt{1-\epsilon^2}|\alpha|$, according to Eq.(19.111), which results from the QM nature of $\mathcal{A}$. Since we use coherent states, the noise, with variance δ, is Gaussian distributed [see figure 19.8].

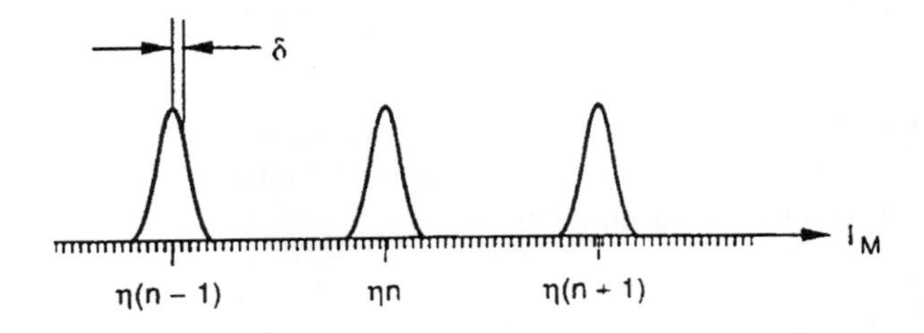

Figure 19.8: Probability distribution of the measured photocurrent difference — from [HAUS/KÄRTNER 1996, 3788].

Thus when $\hat{N}_{\hat{h}\hat{k}}$ lies in the interval $[\eta(n-1/2), \eta(n+1/2)]$, and we decide that the measured photon number is n, this decision has the error probability [see p. 308]:

$$\begin{aligned} \wp^e &= 2\int_{\eta/2}^{\infty} \frac{1}{\sqrt{2\pi}\delta} e^{-\frac{x^2}{2\delta^2}} dx = \operatorname{erfc}\left(\frac{1}{2\sqrt{2(1-\epsilon^2)}|\alpha g_l|}\right) \\ &\simeq 2\left(\frac{2}{\pi(1-\epsilon^2)}\right)^{\frac{1}{2}} \frac{1}{|\alpha g_l|} e^{-(1-\epsilon^2)|\alpha g_l|/8} \end{aligned} \tag{19.114}$$

for $\sqrt{(1-\epsilon^2)}|\alpha g_l| \gg 1$. However, it is possible to reduce drastically the error probability.

Time evolution Now we study the time evolution in the Schrödinger picture of the coupled system. We assume that the signal pulse (the system) is in the state:

$$|\psi\rangle_{\mathrm{S,in}} = \sum_{n=0}^{\infty} c_n |n\rangle_a, \quad \text{with } c_n = 0 \text{ for } n \geq N. \tag{19.115}$$

The total system (signal plus measurement device $\mathcal{A}$ plus environment $\mathcal{E}$ represented by reservoir modes $\hat{d}, \hat{e}$) is:

$$|\Psi\rangle_{\rm I}^{\mathcal{S}+\mathcal{A}+\mathcal{E}} = |\psi\rangle_a^{\rm in} \otimes |\alpha\rangle_b \otimes |0\rangle_c |0\rangle_d \otimes |0\rangle_e. \tag{19.116}$$

After the BS1 we obtain:

$$|\Psi\rangle_{\rm II}^{\mathcal{S}+\mathcal{A}+\mathcal{E}} = |\psi\rangle_a^{\rm in} \otimes \left|\frac{\alpha}{\sqrt{2}}\right\rangle_b \otimes \left|\frac{\imath\alpha}{\sqrt{2}}\right\rangle_c |0\rangle_d \otimes |0\rangle_e. \tag{19.117}$$

After the interaction with the Kerr medium the signal and the measurement system become entangled:

$$\begin{aligned} |\Psi\rangle_{\rm III}^{\mathcal{S}+\mathcal{A}+\mathcal{E}} &= e^{\imath g_I \hat{a}^\dagger \hat{a} \hat{b}^\dagger \hat{b}} |\psi\rangle_a^{\rm in} \otimes \left|\frac{\alpha}{\sqrt{2}}\right\rangle_b \otimes \left|\frac{\imath\alpha}{\sqrt{2}}\right\rangle_c |0\rangle_d \otimes |0\rangle_e \\ &= \sum_{n=0}^{\infty} c_n |n\rangle_a \otimes \left|\frac{\alpha e^{\imath g_I n}}{\sqrt{2}}\right\rangle_b \otimes \left|\frac{\imath\alpha}{\sqrt{2}}\right\rangle_c |0\rangle_d \otimes |0\rangle_e; \end{aligned} \tag{19.118}$$

and after the BS2 the wave function becomes:

$$|\Psi\rangle_{\rm IV}^{\mathcal{S}+\mathcal{A}+\mathcal{E}} = \sum_{n=0}^{\infty} c_n |n\rangle_a \otimes \left|\frac{\alpha}{2}(-1+\imath e^{\imath g_I n})\right\rangle_b \otimes \left|\frac{\alpha}{2}(\imath - e^{\imath g_I n})\right\rangle_c |0\rangle_d \otimes |0\rangle_e. \tag{19.119}$$

Since the measurement pulses are coherent, the splitting off of the output couplers BS3 and BS4 described by Eqs. (19.108) change the amplitude by a factor of $\sqrt{1-\epsilon^2}$ and also entangle $\mathcal{E}$ with $\mathcal{S}+\mathcal{A}$. Hence the state after BS3 and BS4 is:

$$\begin{aligned} |\Psi\rangle_{\rm V}^{\mathcal{S}+\mathcal{A}+\mathcal{E}} = \sum_{n=0}^{\infty} c_n |n\rangle_a &\otimes \left|\frac{\sqrt{1-\epsilon^2}}{2}\alpha(\imath e^{\imath g_I n}-1)\right\rangle_b \otimes \left|\frac{\sqrt{1-\epsilon^2}}{2}\alpha(\imath - e^{\imath g_I n})\right\rangle_c \\ &\otimes \left|\frac{\imath\epsilon}{2}\alpha(\imath e^{\imath g_I n}-1)\right\rangle_d \otimes \left|\frac{\imath\epsilon}{2}\alpha(\imath - e^{\imath g_I n})\right\rangle_e. \end{aligned} \tag{19.120}$$

If ϵ is small and α large enough, detection of $\hat{b}_{\rm out}, \hat{c}_{\rm out}$ still gives complete information about the photon number.

Reduced density matrix If the density matrix of the total system is: $\hat{\rho}^{\mathcal{S}+\mathcal{A}+\mathcal{E}} = |\psi\rangle_{\rm V}^{\mathcal{S}+\mathcal{A}+\mathcal{E}}\,{}_{\rm V}^{\mathcal{S}+\mathcal{A}+\mathcal{E}}\langle\psi|$, by tracing out $\mathcal{E}$ we obtain:

$$\begin{aligned} {\rm Tr}_{\mathcal{E}}(\hat{\rho}^{\mathcal{S}+\mathcal{A}+\mathcal{E}}) &= \sum_{n=0}^{\infty}\sum_{m=0}^{\infty} c_n c_m^* |n\rangle\langle m| \otimes \left|\frac{\sqrt{1-\epsilon^2}}{2}\alpha(\imath e^{\imath g_I n}-1)\right\rangle_b {}_b\left\langle\frac{\sqrt{1-\epsilon^2}}{2}\alpha(\imath e^{\imath g_I m}-1)\right| \\ &\otimes \left|\frac{\sqrt{1-\epsilon^2}}{2}\alpha(\imath - e^{\imath g_I n})\right\rangle_c {}_c\left\langle\frac{\sqrt{1-\epsilon^2}}{2}\alpha(\imath - e^{\imath g_I m})\right| \mathcal{D}_{n,m,d}\mathcal{D}_{n,m,e}, \end{aligned} \tag{19.121}$$

where the decohering factors are given by:

$$\mathcal{D}_{n,m,d} = \left\langle\frac{\imath\epsilon}{2}\alpha(\imath e^{\imath g_I m}-1)\Big|\frac{\imath\epsilon}{2}\alpha(\imath e^{\imath g_I n}-1)\right\rangle, \tag{19.122a}$$

$$\mathcal{D}_{n,m,e} = \left\langle\frac{\imath\epsilon}{2}\alpha(\imath - e^{\imath g_I m})\Big|\frac{\imath\epsilon}{2}\alpha(\imath - e^{\imath g_I n})\right\rangle. \tag{19.122b}$$

For $\epsilon|\alpha|g_l/2 \gg 1$ the decohering factors approach a Kroneker delta $\mathcal{D}_{n,m,d} = \delta_{n,m}$ and the reduced density matrix is given by:

$$\hat{\rho}^{\mathcal{S}+\mathcal{A}+\mathcal{E}}_{\mathrm{Tr}_{\mathcal{E}}} = \sum_{n=0}^{\infty} |c|^2 |n\rangle\langle n| \otimes |\zeta^b_n\rangle\langle\zeta^b_n| \otimes |\zeta^c_n\rangle\langle\zeta^c_n|, \tag{19.123}$$

with

$$|\zeta^b_n\rangle = \left|\frac{\sqrt{1-\epsilon^2}}{2}\alpha(\imath e^{\imath g_l n} - 1)\right\rangle_b, \tag{19.124a}$$

$$|\zeta^c_n\rangle = \left|\frac{\sqrt{1-\epsilon^2}}{2}\imath\alpha(\imath e^{\imath g_l n} - 1)\right\rangle_c. \tag{19.124b}$$

Thus the matrix is diagonal in the photon-number basis of the signal system.

As a result we have that, whether we perform the redout of the information with a photodetector or by sight alone, is completely irrelevant. The decoherence has already occurred in the loss reservoir [HAUS/KÄRTNER 1996, 3789].

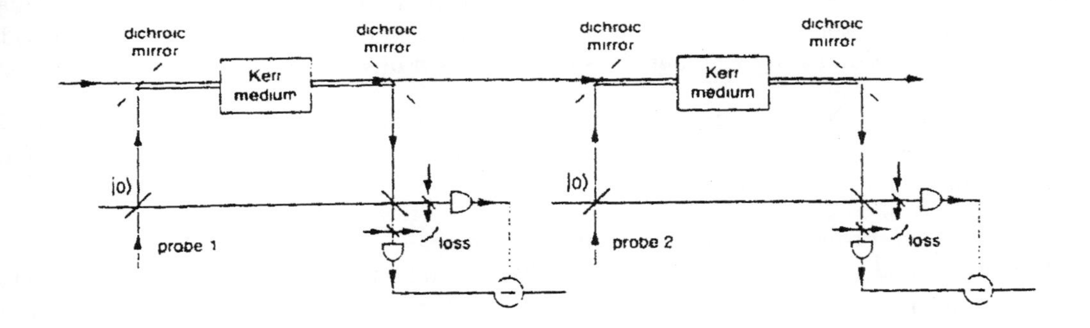

Figure 19.9: Scheme of the two QND measurements — from [HAUS/KÄRTNER 1996, 3790].

QND experimental set-up Now we can show with a second experimental arrangement with an additional QND measurement [see figure 19.9] that the projection postulate is wrong. We calculate the reduced density matrix of signal plus probe-1 plus probe-2:

$$\hat{\rho}^{\mathcal{S}+\mathcal{A}_1+\mathcal{A}_2+\mathcal{E}}_{\mathrm{Tr}_{\mathcal{E}}} = \sum_{n=0}^{\infty}\sum_{m=0}^{\infty} c_n c^*_m |n\rangle\langle m| \otimes |\zeta^{b1}_n\rangle\langle\zeta^{b1}_m| \otimes |\zeta^{c1}_n\rangle\langle\zeta^{c2}_m| \otimes |\zeta^{b2}_n\rangle\langle\zeta^{b2}_m| \otimes |\zeta^{c2}_n\rangle\langle\zeta^{c2}_m|$$
$$\times \mathcal{D}_{n,m,d1}\mathcal{D}_{n,m,e1}\mathcal{D}_{n,m,d2}\mathcal{D}_{n,m,e2}. \tag{19.125}$$

Assuming again approach of decohering factors to Kroneker delta we obtain finally:

$$\mathrm{Tr}_{\mathcal{E}}(\hat{\rho}^{\mathcal{S}+\mathcal{A}_1+\mathcal{A}_2+\mathcal{E}}) = \sum_{n=0}^{\infty} |_n|^2 |n\rangle\langle n| \otimes |\zeta^{b1}_n\rangle\langle\zeta^{b1}_n| \otimes |\zeta^{c1}_n\rangle\langle\zeta^{c2}_n| \otimes |\zeta^{b2}_n\rangle\langle\zeta^{b2}_n| \otimes |\zeta^{c2}_n\rangle\langle\zeta^{c2}_n|. \tag{19.126}$$

Now we ask for the joint probability $\wp_{\mathcal{A}_1,\mathcal{A}_2}(n,m)$ that the first balanced detector will show a value in the interval $[\eta(n-1/2), \eta(n+1/2)]$ and the second in the interval $[\eta(m-1/2), \eta(m+1/2)]$:

$$\wp_{\mathcal{A}_1,\mathcal{A}_2}(n,m) = \mathrm{Tr}\left[\hat{\rho}^{\mathcal{S}+\mathcal{A}_1+\mathcal{A}_2+\mathcal{E}}_{\mathrm{Tr}_{\mathcal{E}}} \otimes m\rangle_{\mathcal{A}_1\mathcal{A}_1}\langle m| \otimes |n\rangle_{\mathcal{A}_2\mathcal{A}_2}\langle n|\right], \tag{19.127}$$

which is equal to:

$$\wp_{\mathcal{A}_1,\mathcal{A}_2}(n,m) = \sum_{k=0}^{\infty} |c_k|^2 \wp_{\mathcal{A}}(m|k)\wp_{\mathcal{A}}(n|k). \tag{19.128}$$

Thus in the limit one measurement becomes precise, i.e. $\wp_{\mathcal{A}}(m|k) = \delta_{m,k}$, we obtain:

$$\wp_{\mathcal{A}_1,\mathcal{A}_2}(n,m) = |c_m|^2 \delta_{n,m}. \tag{19.129}$$

This is the repeatability hypothesis of von Neumann — if the first measurement shows a value, the second one shows the same value [see Eq. (14.14)]. But here it was derived without any discontinuity or Projection postulate.

Possible results Now consider the following situations:

- If the gain in the QND-Measurement is not large enough ($\sqrt{1-\epsilon^2}g_l|\alpha| \ll 1$) — for example a poor resolution —, then $\wp_{\mathcal{A}}(n|m) \neq \delta_{n,m}$;
- if the coupling with the reservoir is still high ($\epsilon g_l|\alpha| \gg 1$), then the entanglement between output signal and output probe is still perfect, and it is meaningful to say that the signal system is in a photon-number state, but the readout of the measurement does not provide enough information to say whether the signal carries n or $n+1$ photons;
- if the gain in the QND measurement is not large enough and the coupling with the reservoir is also poor, so that no perfect decoherence takes place, then no classical interpretation of the measurement is allowed.

In all three cases the Projection postulate is not applicable.

On the other hand, since the Schrödinger Eq. is reversible, the vanishing of the off-diagonal elements does not guarantee that a measurement has in fact been performed: hence restoration of the original state is always possible [see chapter 27]. Hence in conclusion, contrary to Bohr's assumption [see corollary 8.2: p. 149, and relative discussion], the boundaries between system and apparatus are to a certain extent movable[21].

[21] Recent developments such as the calculation of the time in which the measurement happens [ROVELLI 1998b], do not contradict such a result: i) one can discuss the problem only on a probability level, ii) one can discuss the problem only for the compound system: object-system + apparatus.

Chapter 20

INFORMATION WITHOUT INTERACTION

Introduction and Contents We have already quoted [see subsection 16.2.3 and subsection 28.3.1] some experiments related to interaction-free measurements. Now we shall see that, although correlations normally represent useless information, yet they can be utilized to extract information. However, this is only possible if there is such a 'reduction' that we can extract a determinate amount of information about the object we wish to observe. On a theoretical level interaction-free Measurement is a special case of 'reduction' with the sole presence of a detector [see theorem 17.2: p. 267, and related experiments: SGM-Exp. 1: p. 265, and SPDC 2: p. 267].

After a short examination of Renninger's original proposal [section 20.1], an interferometer experiment [section 20.2] and some recent experimental developments [section 20.3] are analysed. Finally, a short theoretical discussion follows [section 20.4].

20.1 Renninger's Experiment

Renninger was the first to have proposed an experiment in which one can measure something without interacting with it. We have already spoken of his pioneering article showing the impossibility of generalizing a thermodynamic model of measurement [see subsection 16.2.3]. In [RENNINGER 1960] this type of measurement is called the Negative-Result Measurement, but today *Interaction-free measurement* is commonly used.

Let us discuss the following experimental arrangement (Renninger Interaction-free Exp.) At a certain distance from a photon source there is a spherical screen (1) with a window with solid angle θ. At a much greater distance from the source there is a second spherical screen (2) without windows. The probabilities that photons are detected by the one or the other screen are, respectively:

$$\wp_1 = \frac{4\pi - \theta}{4\pi}, \tag{20.1a}$$

$$\wp_2 = \frac{\theta}{4\pi}. \tag{20.1b}$$

It is evident that, if a photon is detected by screen 1, the probability that it will be detected by screen 2 becomes immediately zero. But this reduction happens even if the photon is *not* detected by screen 1, because in that case we have $\wp_2 = 1$ [RENNINGER 1960, 418]. To better understand the problem we analyse some more recent experiments.

20.2 Interaction-Free Interferometry

A proposed interaction-free reformulation of Heisenberg's microscope experiment is in [DICKE 1981]. But a more refined experimental proposal (= Interfer. Exp. 7) is based on the use of a Mach-Zehnder interferometer [see subsection 5.3.1] [ELITZUR/VAIDMAN 1993, 988–91] [see figure 20.1].

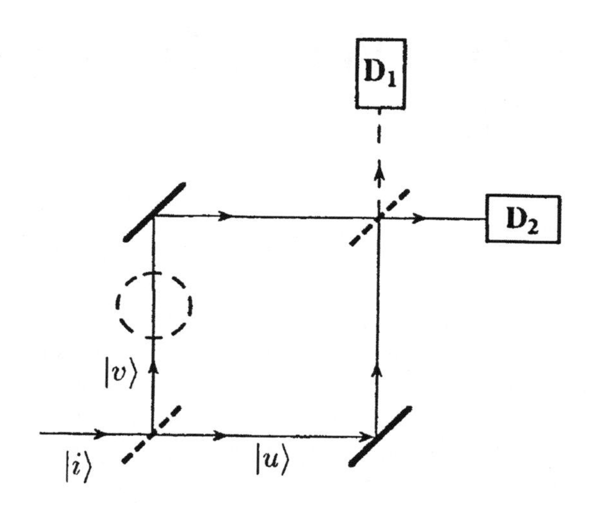

Figure 20.1: Proposed Interfer. Exp. 7. Detector D_2 clicks only if one of the arms is blocked by an object — from [ELITZUR/VAIDMAN 1993, 989].

The BS has a transmission coefficient of 1/2, and, after a symmetric BM, two detectors are arranged in such a way that D_2 detects all photons (due to constructive interference) while D_1 is the dark-output (due to destructive interference). At any time, there is at most one particle in the interferometer. Now we suppose that an object is located in way 2 (upper way). Then there are three possible outcomes of the experiment:

- no detector clicks;
- D_2 clicks;
- D_1 clicks.

In the first case the photon has been absorbed or scattered by the object: the probability is 1/2. In the second case (probability 1/4) the photon could have reached D_1 in both cases: when the object is present and when it is not present in the upper arm. In the third case (probability 1/4) we know that the object is located in the upper way without having interacted with it. To see the QM mechanism let the evolution of the system *in absence of the object* be the following:

$$\begin{aligned} |i\rangle &\mapsto \frac{1}{\sqrt{2}}(|u\rangle + \imath|v\rangle) \mapsto \frac{1}{\sqrt{2}}(\imath|u\rangle - |v\rangle) \\ &\mapsto \frac{1}{2}(\imath|u\rangle - |v\rangle) - \frac{1}{2}(\imath|u\rangle + |v\rangle) = -|v\rangle. \end{aligned} \tag{20.2}$$

Then the photon leaves the interferometer moving to the right, and it is detected by D_2. If *the object is present* the evolution is described by:

$$\begin{aligned}|i\rangle &\mapsto \frac{1}{\sqrt{2}}(|u\rangle + \imath|v\rangle) \mapsto \frac{1}{\sqrt{2}}(|u\rangle + \imath|s\rangle) \\ &\mapsto \frac{1}{\sqrt{2}}(\imath|u\rangle + \imath|s\rangle) \mapsto \frac{1}{2}(\imath|u\rangle - |v\rangle) + \frac{\imath}{\sqrt{2}}|s\rangle, \end{aligned} \tag{20.3}$$

where $|s\rangle$ represents the state of the photon when it is scattered by the object. From Eq. (20.3) we have the three possible outcomes described before (scattering by the object, detection by D_1, detection by D_2):

$$\frac{1}{2}(\imath|v\rangle - |u\rangle) + \frac{\imath}{\sqrt{2}}|s\rangle \rightsquigarrow |s\rangle \tag{20.4a}$$

$$\rightsquigarrow |u\rangle \tag{20.4b}$$

$$\rightsquigarrow |v\rangle. \tag{20.4c}$$

Using the above formalism we can now formulate the experiment (a modification of an experiment originally proposed by Dicke) [ELITZUR/VAIDMAN 1993, 992–93] : suppose that we wish to know if a particle is located in a region V by which it passes by the way v of the interferometer (the other way is free). If the particle is in the superposition $|\psi\rangle = c|\mathrm{A}\rangle + c'|\mathrm{B}\rangle$ (where B is some location disjointed from A), we have the following evolution of the photon plus the particle:

$$|u\rangle|\psi\rangle \mapsto c\left[\frac{1}{2}(\imath|v\rangle - |u\rangle) + \frac{\imath}{\sqrt{2}}|s\rangle\right]|\mathrm{A}\rangle + c'|u\rangle|\mathrm{B}\rangle, \tag{20.5}$$

with the following three possible outcomes:

$$\rightsquigarrow |s\rangle|\mathrm{A}\rangle; \tag{20.6a}$$

$$\rightsquigarrow |u\rangle\left[\frac{2c'}{\sqrt{|c|^2+4|c'|^2}}|\mathrm{B}\rangle - \frac{c}{\sqrt{|c|^2+4|c'|^2}}|\mathrm{A}\rangle\right]; \tag{20.6b}$$

$$\rightsquigarrow |v\rangle|\mathrm{A}\rangle. \tag{20.6c}$$

Result (20.6a) has a probability of $|c|^2/2$, result (20.6b) a probability of $|c|^2/4 + |c'|^2$, result (20.6c) a probability of $|c|^2/4$.

Another experiment has been proposed [ELITZUR/VAIDMAN 1993, 993–95] which underlies the puzzling features of interaction-free Measurement. Suppose there is a stock of bombs. Each bomb can explode if a photon hits a sensor. We wish to know if some bombs are out of control without destroying all the good ones. We now place each bomb in such a way that its sensor is located in the way v of the interferometer. We send photons one at a time until either the bomb explodes or a photon is detected by D_2. If after a large number of tests neither of these two events occurs, then we can conclude that the bomb is not good. By this procedure we can detect about half of the good bombs without destroying them.

20.3 More Recent Experiments

The question is now: is it possible to improve the probability to check the presence of something without interacting with it? We can construct [KWIAT *et al.* 1995a] an apparatus consisting of a series of N interferometers (= Interfer. Exp. 8). The reflectivity of each of the N BSs is chosen to be $\cos^2(\pi/2N)$ and the relative phases between corresponding paths in the upper and lower halves to be zero. The result is that the amplitude of the photon undergoes a gradual transfer from the lower to the upper halves of the interferometers [see figure 20.2a].

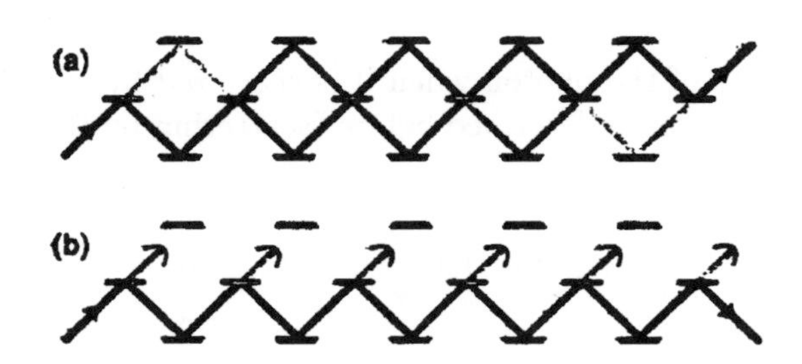

Figure 20.2: (a) The principle of coherently repeated interrogation.
(a) A single photon incident from the lower left gradually transfers to the upper right half of the system. After N stages, where N depends on the BS reflectivities, the photon will certainly exit via the upper port of the last BS.
(b) Introduction of detectors prevents the interference. At each stage the state is projected back into the bottom half of the system if the respective detector does not fire. After all stages there is a good chance (2/3 for the example shown) that the photon now exits via the lower port of the last BS, indicating the presence of the detectors — from [KWIAT *et al.* 1995a, 4764].

After all N stages, the photon will certainly leave by the upper exit. Now, we insert a series of detectors (which represent here the 'obstacle') in the upper half of $\mathcal{A}$ which prevent the interference [see figure 20.2.(b)]. There is a small chance that the photon takes the upper path and triggers a detector, and a large probability $\wp = \cos^2(\pi/2N)$ that it continues to travel on the lower path. The non-firing of each detector projects the state onto the lower half. The probability [see figure 20.3] that the photon will be found in the lower exit after N stages is:

$$\wp = \left[\cos^2\left(\frac{\pi}{2N}\right)\right]^N . \tag{20.7}$$

As the number of stages becomes very large, we may approach an efficiency of 85%.[1]

Recently a combination of the interaction-free device with that of a Zeno experiment [see section 19.8] has been proposed [see figure 20.4] (Interfer. Exp. 9): if a photon enters a commutable mirror from the lower side and runs the optical paths six times before going out through the same commutable mirror, then its final polarisation will be always horizontal if there is an obstacle on one of the paths; otherwise it will be rotated to vertical polarisation.

Other interesting experiments have been recently performed by making use of neutron interferometry [HAFNER/SUMMHAMMER 1997]. By such experiments it has been shown that an interaction-free separation is possible between absorbing objects and transparent or semitransparent ones.

[1] See also [KWIAT *et al.* 1995b].

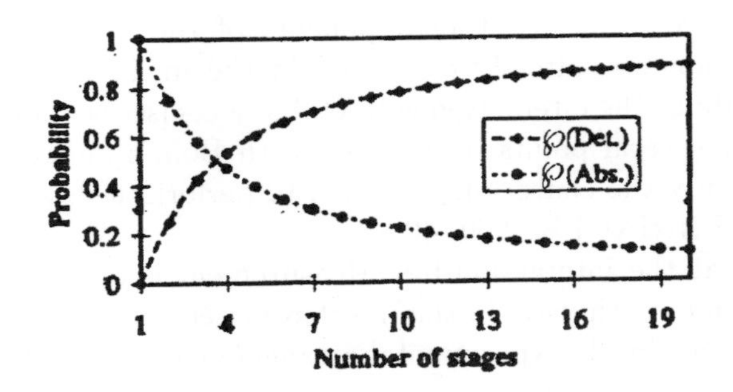

Figure 20.3: The probability ℘(Det) of an interaction-free measurement and the probability ℘(Abs) that the photon is absorbed, i.e. that any of the internal detectors fire, for an incident photon in the set up of figure 20.2b, as a function of the total number of stages, assuming ideal optical components — from [KWIAT *et al.* 1995a, 4764].

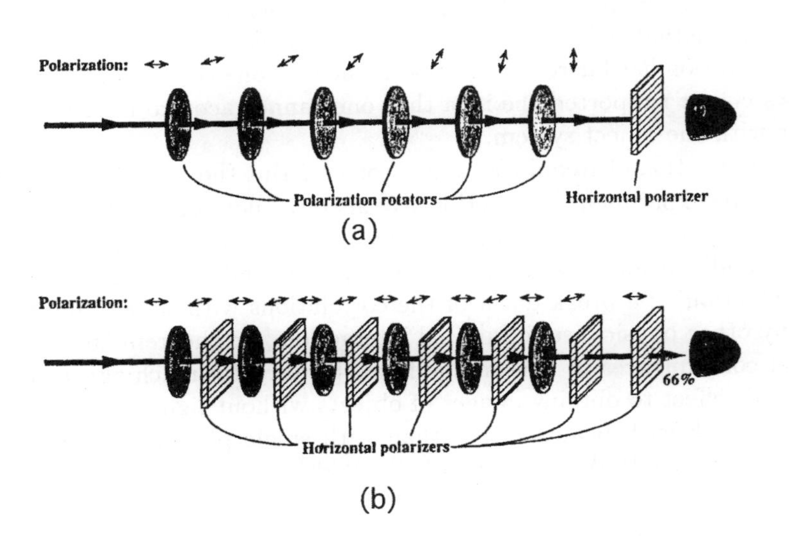

Figure 20.4: Simple optical version of the Zeno effect.
(a) A series of polarisation rotators is used to rotate the polarisation of the input photon from horizontal to vertical. No light is observed in the final detector.
(b) When a series of horizontal polarisers is interspersed between the rotators, the light is projected back into a state of horizontal polarisation at every stage, resulting in light at the final detector. If the number of stages is ≥ 4, more than 50% of the input light will be transmitted. In particular for the case shown (N = 6), the chance of transmission is very nearly twice the chance of absorption — from [KWIAT *et al.* 1995b, 388].

20.4 Discussion

What can the interpretation of the interaction-free Measurement be? It is interesting first to discuss what conclusions were drawn by the proponents of the method. Renninger thought that the obtained information was already contained in the initial wave function, and hence Heisenberg's whole theory about the interaction with QM objects [see section 7.3] would be wrong. Heisenberg's answer shared several points in common with Bohr's answer to EPR [see section 34.2]: he said that it is already the conceptual decision to perform a determinate experiment to determine the reduction [RENNINGER 1960, 420].[2]

Now it appears clear that the interaction-free Measurement is due to an instantaneous 'reduction' of the superposition of both ways in the interferometer due to the simple presence of an obstacle. Let us return to the simple experimental arrangement proposed by Elitzur and Vaidmann. As a consequence of this 'reduction' the photon is now localized: with a probability of 1/2 that it will be in the arm where the object is, and in this case it will be absorbed or scattered, and with a probability of 1/2 that it will be in the other arm (then we have a probability of 1/4 that it will reach the dark output and of 1/4 that it will reach the 'normal' detector). In this sense it is a measurement which uses the QM correlations, but with the condition of the 'reduction'.

Therefore, interaction-free Measurement pertains to the same class of experiments as those with 'reductions' with the sole presence of the detector [see section 8.4, subsection 17.1.2 and section 34.3; see specifically theorem 17.2: p. 267]. Another confirmation of the same results come from the fact that it is the mere possibility of obtaining which-path information that destroys the interference while no actual polarisation measurements need to be made [CHIAO *et al.* 1995b, 269].[3] This interpretation has far-reaching consequences, if one thinks that the 1962 Nobel prize-winner in physics Gabor supported the idea that one cannot acquire information if at least one photon interacts with the object system.

Hence the interpretation of Renninger is not correct. But the young Heisenberg's interpretation of his microscope experiment [see section 7.3] also was not correct, because we see that QM allows some acquisition of information without interacting with the object.

On the other hand, the action-at-a-distance [see p. 106 and p. 121] is also excluded, because such 'distance-reduction' is a break down of the correlations without any information exchange at distance or any other physical superluminal interaction [see theorem 36.3: p. 634].

The technical consequences of this method can be very far-reaching: recently it has been proposed to use this effect to obtains images of objects without light (interaction-free imaging): it would suffice to work with beams rather than with single photons and to associate to each photon a pixel of the image [KWIAT *et al.* 1996] [WHITE *et al.* 1998].

---

[2] See also [*JAMMER* 1974, 495–96].

[3] In [A. STERN *et al.* 1990] it is shown that the presence of a which-path detector in an interferometer always leads to random phase shift, thus washing out interference fringes. See also [CHIAO *et al.* 1995b, 266].

Part V

MICROPHYSICS/MACROPHYSICS

Introduction to Part V

The *fifth part* is centred on the relation between the microworld and the macroworld. The central question is if the transition from a superposition to some eigenstate of a reference observable is due only to a process of measurement, or if it is a more general and spontaneous process of nature. In more general terms, the problem to investigate now is if QM is a specific science of the microscopic world with no consequence on the macroscopic one and if — as it is presumed by the Copenhagen interpretation — there is a sharp boundary line between them. The traditional starting point of the problem is the Schrödinger's cat paradox, which by Schrödinger himself was intended as a paradox in order to show the unsoundness of the theory.

Schrödinger's Analysis

In the following we give in a few words the whole content of the famous article in which, in 1935, Schrödinger drew some conclusions about QM's development. Then we shall analyse it. The interest of the article is in the fact that, after the failure of Schrödinger's interpretative effort [see section 6.1], it is intended as a final balance of the situation of QM with all its problems, open questions and problematic interpretations: in this sense it represents a general synthesis of a lot of material already discussed.

General considerations Schrödinger began the article [SCHRÖDINGER 1935a, 484–87, 490] with the problem of interpretation (which he named 'model') and thought that the dominant opinion (presumably the Copenhagen interpretation) was that there was no interpretation that could be adequate to the microreality. As we know, the problem is that we have no concept of state if we can determine only one half of the canonical conjugates which — classically — together determine the state [see subsection 2.2.1]. Also, the future evolution seems undetermined — due to breaks in the causality, by the complementarity between 'localization' and dynamic laws [section 8.4] and more generally by QP [postulates 2.1: p. 19, and 2.2: p. 20]. Following Schrödinger, while physicists thought that it is impossible to use interpretations in the classical sense, they believed it possible to indicate what the objects of measurement are, as if the elements of their theory (which are ultimately elements of the old classical interpretations) were ontologically real as such. It is so because, following Schrödinger, CP proposed that only what it is measured is real [see postulate 8.1: p. 137]; and what can be measured depends on our apparatus. But it is evident that in QM we have new properties, which have nothing to do with classical ones, for example one can exclude *a priori* some values of an observable without a preceding measurement [see p. 107].

The S-Cat Now Schrödinger criticized the idea that reality is as indeterminate as it seems in QM [SCHRÖDINGER 1935a, 489–95, 500–501]. The problem here is the superposition: We can think of absurd macroscopical situations if QM should be an universal applicable theory. This is expressed paradigmatically by the so named Schrödinger's Cat (= S-Cat) paradox. The *Gedankenexperiment* is the following. A cat is in a box with a very small quantity of radioactive material — the decay probability is, say, one atom in one hour. The decayed atom actives a Geiger counter which is linked to a hammer which releases some poison from an ampulla after having broken it. Now, consequent to the probabilistic character of the radioactive decay, the ψ–function describing the system is a superposition of *cat-alive* and *cat-dead*, which is surely an impossibility: nobody observed such a situation at macroscopic level — it is a very similar situation to the one of the Wigner's friend paradox [see subsection 14.2.1]. On the other hand, we may think that a measurement — the opening of the box to see the cat — will determine the state (*either* alive *or* dead) by eliminating the superposition — by the Projection postulate [14.1: p. 223]. And here we ascribe an enormous power to the measurement.

The ψ–function What in reality is the ψ–function? Schrödinger answered: it is a catalogue of expectations [see postulate 3.4: p. 38] and the link between two measurements. From the point of view of the Standard theory, we have an acausal and abrupt change after a measurement. Hence, according to Schrödinger, the ψ–function cannot be a real thing or an adequate interpretation of a real thing — here Schrödinger's analysis resembles Einstein's one [see subsection 6.5.1].

Suppose that it represents a complete definition of a system: then each function describes a state and two identical ψ–functions describe the same state. Hence no ψ–function can be included in one another. Then it is very difficult to explain how is possible that, when two systems interact, they lose their individuality and their specific ψ–functions to acquire another one which now describes the inseparable whole constituted of both; just as it is very difficult to explain how is possible that a wave function breaks down by a wave-packet reduction.

From the point of view of information, where there is some information gain, we must simultaneously lose some other information (by a completeness hypothesis). But, since the knowledge cannot be lost (it can be recorded), then it must be the system itself which is changed in such a way as to falsify some features of the initial ψ–function. And here we are faced with many difficulties emanating from the superposition principle when we interpret the wave function as a description of a real thing.

Non-separability In addition to the aforementioned difficulties, the superposition principle has the consequence that the described system can be in a determinate state, but this is not necessarily true for its subsystems, which cannot be separated from each other: hence in QM the best possible knowledge of the whole does not necessarily include the best possible knowledge of its parts [SCHRÖDINGER 1936, 424] — we shall return on the problem in parts VIII and IX. And, applied to the measurement, this signifies that we cannot speak of a separated object; and we cannot know *a priori* which value of a given observable is the right one. It is only the observation which provides it. In conclusion, QM's propositions are valid only for some moments and not for dynamic evolutions — see the previously discussed interpretation of Schrödinger [section 6.1]. And this is surely difficult to think in relativistic contexts.

Schrödinger's conclusion is that QM is only a mathematical device and not a physical theory.

A Short Discussion

Theoretically the interest of Schrödinger's analysis is that it poses a sharp alternative[4]: or QM is complete, and then apparently there is no way out of the paradoxes the theory causes; or it is incomplete. In this sense Schrödinger agrees with the Statistical interpretation [section 6.5]. We shall return to the problem of completeness later [see chapter 31].

By examining the details of Schrödinger's arguments, we see that some apparent problems have already been lost, above all the acausality of Measurement. Others are to be treated later: the apparent ontological indetermination [see part VII], the problem of information [see section 42.3] and that of the inseparability of different QM subsystems [see specifically the Schrödinger

[4] As stated by Bell [BELL 1987a, 201].

arguments in section 34.1]. Anyway, the central tenet of Schrödinger remains: if QM is not a regional physical theory — which in turn almost means that it is only a mathematical device if the reasons of this regionality are not shown —, then how it is possible to have a paradox such as that of the S-Cat? Hence, in the following, we must turn ourselves to this central question.

Contents of Part V

- Recently, following the proponents of decoherence, we began to understand that one can account for the 'paradoxical' features of QM by employing QM itself. *Chapter 21* is devoted to these developments, in particularly in connection with some thermodynamic tools: some concrete models are analysed and some estimate of the time of decoherence furnished.

- A further development in this direction is to be found in *chapter 22*, where a formalism of quantum trajectories is studied which is parallel to the Master Eq. one, used in the preceding chapter.

- The first way-out of the S-Cat paradox that one thought of was some form of reduction of QM to a classical case by means of some stochastic theory able to reproduce its probability distributions and to reduce the specific 'quantic' effects of the theory by assimilating QM to some classical (stochastic) theory. Another possible 'classical' way-out makes use of some spontaneous transition at the boundary between microworld and macroworld, such that the latter one does not reproduce the 'strange' behaviour of quantum entities. All these subjects will be discussed in *chapter 23*.

- *Chapter 24* is finally devoted to one of the most important experimental enterprises of recent years: the realisation of a 'Schrödinger cat' in order to show the existence of quantum effects also on a mesoscopic and perhaps a macroscopic level. Hence we can show that what was a paradox for Schrödinger is in reality an impressive prediction power of the theory.

Chapter 21

DECOHERENCE AND THERMODYNAMICS

Introduction and Contents The question is whether there is some smooth transition between microphysics and macrophysics, such that the QM laws are not applicable as such to macrophysics, but the latter is to understand as a specifical case of QM, as the complexity and the size of systems grows. This is the proposal of decoherence. As we have seen in the decoherence theory the off-diagonal terms of the density matrix of a system are never really zero — they only *tend* to zero [see corollary 17.2: p. 272]. In the following the smooth transition between micro- and macrophysical world is based on this property.

As we have already said [subsection 16.2.4], the following developments are strictly dependent upon advancements in the thermodynamic approach. In fact the smallness of off-diagonal terms in a concrete decoherence process is justified by the thermodynamic analysis [VAN HOVE 1955] [VAN HOVE 1957]: interactions occurring in real physical systems 'destroy' phase coherence between the states of the system on a much shorter time scale than that of the relaxation of the thermal equilibrium. Anyway, attention must be paid to the fact that 'thermodynamic decoherence' is not the only form of decoherence (as we shall see in the following sections). It should be kept in mind that there was also a great contribution on an abstract algebraic level: In particular we recall the work of Araki and Lieb [ARAKI/LIEB 1970] [see also subsections 42.3.2 and 17.1.1]. However it is only in the eighties, depending especially in Zurek's proposal of decoherence [see section 17.2], that ripe models of spontaneous decoherence could be developed.

In this chapter we shall first treat [section 21.1] the general problem of the Master equation. Then [section 21.2] we shall apply this formalism to the decoherence problem ('baths' of oscillators). A model (damping), independent of the last one is also discussed [section 21.3]. Then some theoretical problems about these two models are discussed [section 21.4]. A more complex model is that developed by Unruh and Zurek [section 21.5], while a model by Stern and co-workers is interesting for parallelizing two different treatments of the interaction environment/system [section 21.6]. Finally the problem of decoherence parameters is also discussed [section 21.7].

21.1 Heat Baths and Master Equation

The master Eq. treatment is due principally to the work of Davies and Lindblad — but, as we have said, precedents are to be found in van Hove's work [see subsection 16.2.2]. We will briefly introduce the concept of heat baths [*LINDBLAD* 1983, 18; 32–38]. They imply an initial irreversibility based only on statistical mechanics without introducing any form of measurement, and hence are very useful for the problem discussed here. They are prepared in an initial equilibrium (Gibbs) state, hence the memory of the distant past has been wiped out (it is an irreversible process) — which is not so obvious, if we think of the existence of longlived metastable states which preserve some memory of the past. This is a Markovian assumption for the initial state. Heath baths must be infinite systems in order to be able to absorb energy transferred from other systems without changing their equilibrium properties; for the same reason they must always be able to return to an equilibrium state. Normally, for the problem of interest, they are infinite quasi-free boson or fermion systems. In the weak coupling limit Davies [DAVIES 1974] [*DAVIES* 1976, 149–56] showed that the interaction between a reservoir and a small system is governed by a Master Eq. For any transformation $\mathcal{T}$ [see subsection 18.1.1], by assuming complete positivity, factorization, norm continuity and conservation of probability, we can write a generator $\hat{\mathcal{L}}$ of the semigroup [*LINDBLAD* 1983, 17, 28–31, 133–39] [see theorem 3.6: p. 47]:

$$\mathcal{T}(t) = e^{t\hat{\mathcal{L}}}, \quad \hat{\mathcal{L}} = \frac{d}{dt}\mathcal{T}(t)\bigg|_{t=0}, \tag{21.1}$$

where ab belongs to a *semigroup* if both a and b belong to it. A semigroup, which preserves linear maps, is distinct from a group to the extent to which it is characterized by a temporal direction (presupposed in a irreversible dynamics) [KOSSAKOVSKI 1972] [GORINI *et al.* 1976] [LINDBLAD 1976].

Now we can write the Lindblad Master Eq. [see also subsection 16.2.2] for the evolution of the density operator of the system:

$$\frac{d}{dt}\hat{\rho}(t) = \hat{\mathcal{L}}\hat{\rho}(t), \tag{21.2}$$

where (dropping the t dependence) $\hat{\mathcal{L}}$ is the superoperator (*Lindblad superoperator*) for the generalized Liouville transformation given by [see also Eq. (3.72)]:

$$\boxed{\hat{\mathcal{L}}\hat{\rho} = \hat{\mathcal{L}}_d\hat{\rho} - \frac{\imath}{\hbar}[\hat{H}, \hat{\rho}]} \tag{21.3a}$$

Here the second term in the RHS is the 'normal' reversible evolution given by the quantum Liouville Eq. [see Eq. (3.80)], while the term $\hat{\mathcal{L}}_d$ represents the dissipative part given by [see Eq. (18.5a)]:

$$\boxed{\hat{\mathcal{L}}_d\hat{\rho} = \frac{1}{2}\sum_j \left([\hat{V}_j\hat{\rho}, \hat{V}_j^\dagger] + [\hat{V}_j, \hat{\rho}\hat{V}_j^\dagger]\right)} \tag{21.3b}$$

where $\{\hat{V}_j\} \in \mathcal{L}(\mathcal{H})_\mathcal{S}$ is a set of self-adjoint operators. Each $\hat{V}_j$ expresses the action of the reservoir on the j-th component of $\mathcal{S}$ — they are dissipation operators, which, in the easiest case, can be conceived as annihilation and creation operators. We can note that, although dissipation and decoherence have the same physical origin, they are different phenomena [PAZ *et al.* 1993] . It is evident that a QM master Eq. is the quantum analogous of the classical Fokker/Planck Eq. [see Eq. (23.3)].

For an arbitrary observable $\hat{O}$ pertaining to $\mathcal{S}$ we have the active dual $\hat{\mathcal{L}}^\dagger$ (expressing the active or Heisenberg transformation):

$$\frac{d}{dt}\hat{O}(t) = \hat{\mathcal{L}}^\dagger\hat{O}(t), \tag{21.4}$$

which has the form [see Eq. (3.73)]:

$$\hat{\mathcal{L}}^{\dagger}\hat{O} = \hat{\mathcal{L}}^{\dagger}_{d}\hat{O} + \frac{\imath}{\hbar}[\hat{H}, \hat{O}], \tag{21.5a}$$

where the dissipative part [see Eq. (18.5b)] is given by:

$$\hat{\mathcal{L}}^{\dagger}_{d}\hat{O} = \frac{1}{2}\sum_{j}[\hat{V}_{j}^{\dagger}, [\hat{O}, \hat{V}_{j}]]. \tag{21.5b}$$

Since the second term in the RHS of Eq. (21.3a) and of Eq. (21.5a) is the Hamiltonian part, then $\hat{\mathcal{L}}^{\dagger}$ and $\hat{\mathcal{L}}$ are Hamiltonian iff the dissipation is zero. It is clear that $\hat{\mathcal{L}}_{d}$, and hence also $\hat{\mathcal{L}}$, annihilates a Gibbs state.

As we have said, decoherence is generally studied in the Markovian approximation. In [IL KIM *et al.* 1996] [FONSECA R./NEMES 1997] non-Markovian processes are studied. Differently from the first ones, a non-explicitly Hamiltonian correction is present which also allows for forms of decoherence where the interference terms of the density matrix of a system do not immediately fall to zero, which is not possible in the context of Markovian approximations. The importance of this result is that decoherence can be interpreted as a smooth process and not a sharp one[1].

21.1.1 Different Master Equations: General Formal Discussion

Different Master Eqs. Though Lindblad's theory is very neat and clear, it remains more as a general formal structure than as a practical means to solve problems of open systems. Hence different master Eqs. have been developed for practical purposes. Agarwal and Caldeira/Leggett developed the first master Eqs. On Agarwal's proposal, we return often in the following [see Eq. (21.16)]: it is useful for baths of harmonic oscillators. More interesting for its simplicity is the one proposed by Caldeira/Leggett [CALDEIRA/LEGGETT 1983a] [CALDEIRA/LEGGETT 1983b]:

$$\frac{d}{dt}\hat{\rho} = -\frac{\imath}{\hbar}[\hat{H}, \hat{\rho}] - \frac{\kappa}{\hbar}\left[\frac{2mkT}{\hbar}[\hat{\mathbf{q}}, [\hat{\mathbf{q}}, \hat{\rho}]] + \imath[\hat{\mathbf{q}}, [\hat{\mathbf{p}}, \hat{\rho}]_{+}]\right], \tag{21.6}$$

where $[,]_{+}$ means an anticommutator, the reduced density matrix $\hat{\rho}$ is that of a harmonic oscillator with mass m and Hamiltonian $\hat{H}$, coupled to an Ohmic reservoir (with linear dissipation) at temperature T, and where $\kappa = \zeta/2m$ is the damping rate of the oscillator (ζ is the friction coefficient).

From here a general form of the Master Eq. can be derived to describe spontaneous localization processes. Joos and Zeh elaborated such a model for short-length scales [JOOS/ZEH 1985, 229, 232–34] [JOOS 1996, 61–62]:[2]

$$\frac{\partial}{\partial t}\hat{\rho} = \frac{1}{\imath\hbar}[\hat{H}, \hat{\rho}] - \eta[\hat{\mathbf{q}}, [\hat{\mathbf{q}}, \hat{\rho}]], \tag{21.7}$$

which can also be extended to all length scales [GALLIS/FLEMING 1991, 5779].

---

[1] See also [ROYER 1996], where a model is developed for an initial correlated state of the bath and the system.

[2] The Eq. was proposed for the first time by Barchielli/Lanz/Prosperi [BARCHIELLI *et al.* 1982]. Again we see the importance of the thermodynamic approach. On this point see also Eq. (19.75).

Increase of energy A general objection has been directed against the Master Eq. of this form [BALLENTINE 1991a]: by calculating the Ehrenfest relation, there is a constant increase of the average energy. But Gallis and Fleming showed that such an increase depends on the momentum and energy exchange between system and environment presupposed by all the following models [GALLIS/FLEMING 1991, 5781–83].[3]

Violation of positivity However, Master Eqs. such as those of Caldeira/Leggett or Agarwal — differently from that of Lindblad — present the major problem that they violate the positivity requirement of the density operator [see subsection 3.5.1], particularly on short time scales. Hence efforts have been made to come to some synthesis between the Lindblad master Eq. (21.3b) and Eq.(21.6) [DIÓSI 1993a] [DIÓSI 1993b] [TAMESHTIT/SIPE 1996] [GAO 1997]. As we have already said the operators figuring in Eq. (21.3b) can be interpreted as annihilation and creation operators. Hence on the basis of Eq. (5.5) we can write:

$$\hat{V} = \mu\hat{\mathbf{q}} + \imath\nu\hat{\mathbf{p}}, \quad (21.8a)$$

$$\hat{V}^\dagger = \mu\hat{\mathbf{q}} - \imath\nu\hat{\mathbf{p}}, \quad (21.8b)$$

where μ, ν are c-numbers to be determined. By such a substitution we can write Eq. (21.3b) as follows:

$$\frac{d}{dt}\hat{\rho} = -\frac{\imath}{\hbar}[\hat{H}', \hat{\rho}] - \mu^2[\hat{\mathbf{q}}, [\hat{\mathbf{q}}, \hat{\rho}]] - 2\imath\mu\nu[\hat{\mathbf{q}}, [\hat{\mathbf{p}}, \hat{\rho}]_+] - \nu^2[\hat{\mathbf{p}}, [\hat{\mathbf{p}}, \hat{\rho}]], \quad (21.9)$$

where $\hat{H}' = \hat{H} - 2\mu\nu\hbar\hat{\mathbf{q}}\hat{\mathbf{p}}$ and CCRs have been used. Using the model of Caldeira and Leggett — whose Master Eq. is not so far away from the (21.9) — one can calculate that in the general case (by high and low temperature) the coefficients have following values:

$$\mu^2(T) = \frac{\kappa m\Omega}{2\hbar}\coth\left(\frac{\hbar\Omega}{4kT}\right), \quad (21.10a)$$

$$\nu^2(T) = \frac{\kappa}{2\hbar m\Omega}\tanh\left(\frac{\hbar\Omega}{4kT}\right), \quad (21.10b)$$

$$2\mu\nu = \frac{\kappa}{\hbar}, \quad (21.10c)$$

where $\nu^2(T) = 0$ in the high-temperature regime [see figure 21.1]

21.1.2 Indistinguishability of States

We can use the preceding formalism to discuss the problem of indistinguishability between a pure state and a mixture [ALICKI 1984] [see also section 18.9]. Take a time interval $[t, t+\tau]$ and a small quantity ϵ. We say that states $\hat{\rho}, \hat{\rho}'$ are (τ, ϵ)–indistinguishable if for all observables $\hat{O}$:

$$|(\hat{O}, \hat{\rho})_\tau - (\hat{O}, \hat{\rho}')_\tau| \leq \epsilon \parallel \hat{O} \parallel, \quad (21.11)$$

where

$$(\hat{O}, \hat{\rho})_\tau = \tau^{-1}\int_0^\tau (\hat{O}, e^{t\hat{\mathcal{L}}}\hat{\rho})dt. \quad (21.12)$$

[3] See also [JOOS 1996, 78–79].

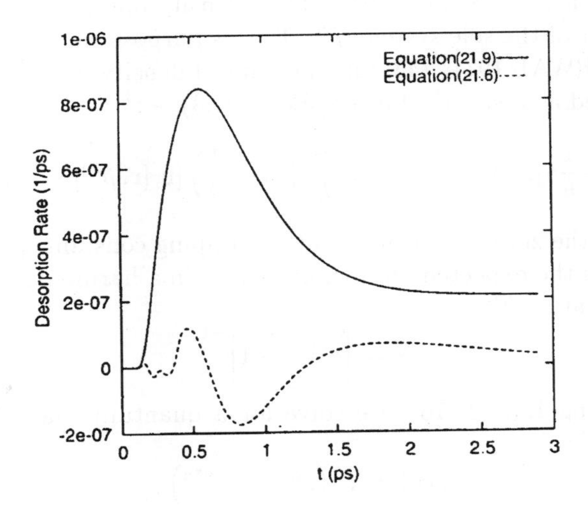

Figure 21.1: Desorption rate as a function of time before absorbing boundary at $T = 200K$. The rate given by Eq. (21.7) — curve below — does not have the right sign in the subpicosecond regime due to the unphysical negative density distribution. This problem does not exist in the rate given by the Gao Master Eq. (21.9) — from [GAO 1997, 3104].

Now take two orthogonal pure states $|\psi\rangle, |\psi'\rangle$. We say that $|\psi\rangle, |\psi'\rangle$ are (τ, ϵ)–disjoint if any coherent superposition of them is (τ, ϵ)–indistinguishable from a statistical mixture of both. As the criterion of (τ, ϵ)–disjointness we give the following:

$$\left\| \int_0^\tau [e^{t\hat{\mathcal{L}}}(|\psi\rangle\langle\psi'|)] dt \right\| \leq \tau\epsilon. \tag{21.13}$$

21.2 Baths of Harmonic Oscillators

21.2.1 First Model of Decoherent Measurement

We now use the Master Eq. Approach. Precedents are in the quoted article of Alicki [ALICKI 1984]. The first model uses a large number of harmonic oscillators [SAVAGE/WALLS 1985a, 2316–17], the k–th of which has mass m_k and frequency ω_k. We choose the zeroth as system $\mathcal{S}$ and the others as a bath (= environment $\mathcal{E}$) to which the system is weakly coordinate-coordinate coupled. The Hamiltonian of $\mathcal{S} + \mathcal{E}$ is [see Eq. (4.25)]:

$$\hat{H} = \hbar\omega_0 \hat{a}_0^\dagger \hat{a}_0 + \hbar \sum_{k(\neq 0)} \omega_k \hat{a}_k^\dagger \hat{a}_k + \mathbf{r}_0 \sum_{k(\neq 0)} g_k \mathbf{r}_k, \tag{21.14}$$

where the first two terms are the free motion of system + bath and the third is the coupling, g_k are coupling strengths and $\mathbf{r}_k$ are coordinates given by:

$$\mathbf{r}_k = \left[\frac{\hbar}{2m_k\omega_k}\right]^{\frac{1}{2}} (\hat{a}_k^\dagger + \hat{a}_k). \tag{21.15}$$

We now seek an equation for the trace of the total density operator over the bath, i.e. for the reduced density operator of the sole system, $\hat{\rho}^{\mathcal{S}}$. To this purpose we can make use of the Agarwal Master Equation [AGARWAL 1971, 743] for the reduced density operator — which, to a certain extent, can be considered as a specific form of Eqs. (21.3) —:[4]

$$\partial_t \hat{\rho}^{\mathcal{S}} = -\frac{1}{\hbar}\gamma[\mathbf{r}, \hat{\mathbf{p}}\hat{\rho}^{\mathcal{S}} + \hat{\rho}^{\mathcal{S}}\hat{\mathbf{p}}] - \frac{2\gamma}{\hbar}\left(\overline{N} + \frac{1}{2}\right)[\mathbf{r}, [\mathbf{r}, \hat{\rho}^{\mathcal{S}}]] - \imath\omega[\hat{a}^\dagger \hat{a}, \hat{\rho}^{\mathcal{S}}], \tag{21.16}$$

where we have dropped the zero subscript. γ is the damping constant, $\hat{\mathbf{p}}$ is the system's momentum observable and $\overline{N}$ is the expected number of quanta in a harmonic oscillator with frequency ω at equilibrium temperature T:

$$\overline{N} = \left[e^{\frac{\hbar\omega}{k_B T}} - 1\right]^{-1}. \tag{21.17}$$

Instead of directly solving Eq.(21.16), we solve for a quantum characteristic function χ [see Eq. (5.43)] in terms of $\hat{\rho}^{\mathcal{S}}$:

$$\chi(\eta) = \mathrm{Tr}\left(\hat{\rho}^{\mathcal{S}} e^{\eta \hat{a}^\dagger} e^{-\eta^* \hat{a}}\right), \tag{21.18}$$

where η is a complex variable. From Eq. (21.16) we find for χ:

$$\{\partial_t + [(-\imath\omega + \gamma)\eta + \gamma\eta^*]\partial_\eta + [\gamma\eta + (\imath\omega + \gamma)\eta^*]\partial_{\eta^*}\}\chi = -\gamma\overline{N}(\eta + \eta^*)^2\chi. \tag{21.19}$$

We can consider now two possible cases:

- zero temperature case,
- finite temperature case.

Zero Temperature Case

At zero temperature [SAVAGE/WALLS 1985a, 2317–19] we have $\overline{N} = 0$ [see Eq. (21.17)] and the solution of Eq. (21.19) is an arbitrary function $I(f, f^*)$, where $f = u\eta + v\eta^*$ and:

$$u = \frac{1}{2}\left[e^{-\mu_- t} + e^{-\mu_+ t} + 2\imath\frac{\omega}{\mu_- - \mu_+}(e^{-\mu_+ t} - e^{-\mu_- t})\right], \tag{21.20a}$$

$$v = \frac{\gamma}{\mu_- - \mu_+}(e^{-\mu_- t} - e^{-\mu_+ t}). \tag{21.20b}$$

The eigenvalues $\mu_\pm$ are given by:

$$\mu_\pm = \gamma \pm (\gamma^2 - \omega^2)^{\frac{1}{2}}. \tag{21.21}$$

If the initial density operator is a superposition of coherent states [see subsection 5.1.1] given by:

$$\hat{\rho}_i = \sum_{\alpha,\beta} N_{\alpha\beta}|\alpha\rangle\langle\beta|, \tag{21.22}$$

the QM characteristic function corresponding to operator $|\alpha\rangle\langle\beta|$ is [see Eqs. (5.43) and (21.18)]:

$$\begin{aligned}\chi &= \mathrm{Tr}(|\alpha\rangle\langle\beta|e^{\eta\hat{a}^\dagger}e^{-\eta^*\hat{a}}) \\ &= \langle\beta|\alpha\rangle e^{\eta\beta^* - \eta^*\alpha},\end{aligned} \tag{21.23}$$

[4] See also [*GARDINER* 1991, 82].

and the zero-temperature solution of Eq. (21.19) has the form:

$$\begin{aligned}\chi(\eta,t) &= \langle\beta|\alpha\rangle e^{\beta^* f-\alpha f^*} \\ &= \langle\beta|\alpha\rangle e^{\eta(u\beta^*-v\alpha)-\eta^*(u^*\alpha-v\beta^*)}.\end{aligned} \tag{21.24}$$

Comparing with Eq. (21.23) we find that the density operator (21.22) evolves according to:

$$\hat{\rho}(t) = \sum_{\alpha,\beta} N_{\alpha\beta}\langle\beta|\alpha\rangle \frac{|u^*\alpha - v\beta^*\rangle\langle u^*\beta - v\alpha^*|}{\langle u^*\beta - v\alpha^*|u^*\alpha - v\beta^*\rangle}. \tag{21.25}$$

We must now distinguish two cases: the underdamped case ($\gamma < \omega$) and the overdamped one ($\gamma > \omega$).

- In the first case we have a mixture of all coherent states after evolution of the system.
- In the second case the eigenvalues of Eq. (21.21) are real. As γ increases, μ_+ increases toward 2γ and μ_- decreases toward zero. Hence we have a fast and a slow eigenvalue.

  In the overdamped case ($\gamma > \omega$) a diagonal (in the coherent state basis) initial density operator $\hat{\rho} = |\alpha\rangle\langle\alpha|$ evolves as:

$$\hat{\rho}(t) \simeq |\mathrm{Re}(\alpha)e^{-\frac{\omega^2}{2\gamma}t} + \imath\mathrm{Im}(\alpha)e^{2\gamma t}\rangle\langle\mathrm{Re}(\alpha)e^{-\frac{\omega^2}{2\gamma}t} + \imath\mathrm{Im}(\alpha)e^{2\gamma t}|, \tag{21.26}$$

  i.e. a coherent state remains coherent.

Finite Temperature Case

In the finite temperature cases, coherent states are no longer preserved [SAVAGE/WALLS 1985a, 2319–22]. We take our initial density operator to be a superposition of coherent states of the form Eq. (21.22). Solving Eq. (21.19) for the corresponding initial characteristic function, we have [see Eq. (21.24)]:

$$\chi(\eta,t) = \sum_{\alpha,\beta} N_{\alpha\beta}\langle\beta|\alpha\rangle e^{\eta(u\beta^*-v\alpha)-\eta^*(u^*\alpha-v\beta^*)} e^{\overline{N}[\eta^2 uv+\eta^{*2}u^*v+|\eta|^2(|u|^2+v^2-1)]}. \tag{21.27}$$

The diagonal coherent-state density-matrix elements and the characteristic function are related by:

$$\langle z|\hat{\rho}|z\rangle = \frac{1}{\pi}\int \chi(\eta)e^{-|\eta|^2-\eta z^*+\eta^* z}d^2\eta, \tag{21.28}$$

where $d^2\eta = d[\mathrm{Re}(\eta)]d[\mathrm{Im}(\eta)]$.

Now the density operator can be expressed as

$$\hat{\rho} = \frac{1}{\pi^2}\int\int \frac{|z+\delta\rangle\langle z-\delta|}{|z-\delta\rangle\langle z+\delta|}e^{-|\delta|^2}\langle z|\hat{\rho}|z\rangle d^2z d^2\delta. \tag{21.29}$$

In the heavily overdamped case $\gamma \gg \omega$ there is a negligible associated spreading of the diagonal coordinate-basis density matrix elements. From Eqs. (21.20) we obtain:

$$\begin{aligned}u &\simeq \frac{1}{2}\left(e^{-2\gamma t} + e^{-\frac{\omega^2}{2\gamma}t}\right), \\ v &\simeq \frac{1}{2}\left(e^{-\frac{\omega^2}{2\gamma}t} - e^{-2\gamma t}\right).\end{aligned} \tag{21.30a}$$

Using these expressions in Eq. (21.27) and (21.28) we find the diagonal coherent-state matrix elements to be:

$$\begin{aligned}\langle z|\hat{\rho}|z\rangle &= \sum_{\alpha,\beta}\langle\beta|\alpha\rangle\frac{1}{\sqrt{jk}}\exp\left[\frac{1}{4j}[(\beta^* - \alpha)e^{-2\gamma t} + z - z^*]^2\right] \\ &\times\exp\left[\frac{1}{4k}[(\beta^* + \alpha)e^{-(\omega^2/2\gamma)t} - z - z^*]^2\right], \end{aligned} \tag{21.31}$$

where $j = 1 + \overline{N}(1 - e^{-4\gamma t})$ and $k = 1 + \overline{N}(1 - e^{-(\omega^2/\gamma)t})$. Substituting this result in Eq. (21.29) taking coordinate-basis matrix elements and evaluating the integrals we finally obtain:

$$\begin{aligned}\langle x - y|\hat{\rho}|x + y\rangle &= (2\pi(\Delta(x))^2)^{\frac{1}{2}}\sum_{\alpha,\beta}\langle\beta|\alpha\rangle\exp\left[\frac{1}{2}(\Delta(y))^2\left(\frac{2m\omega}{\hbar}\right)(\beta^* - \alpha)^2e^{-4\gamma t}\right] \\ &\times\exp\left\{-\frac{1}{2}(\Delta(x))^{-2}\left[x - \left(\frac{2m\omega}{\hbar}\right)^{\frac{1}{2}}(\beta^* + \alpha)e^{-(\omega/2\gamma)t}\right]^2\right\} \\ &\times\exp\left[-\frac{1}{2}(\Delta(y))^{-2}y^2 + \left(\frac{2m\omega}{\hbar}\right)^{\frac{1}{2}}(\beta^* - \alpha)e^{-2\gamma t}y\right]. \end{aligned} \tag{21.32}$$

The variances of the diagonal part $(\Delta(x))^2$ and of the off-diagonal part $(\Delta(y))^2$ are respectively:

$$(\Delta(x))^2 = \frac{\hbar}{2m\omega}[1 + 2\overline{N}(1 - e^{-\omega^2 t/\gamma})], \tag{21.33a}$$

$$(\Delta(y))^2 = \frac{\hbar}{2m\omega}[1 + 2\overline{N}(1 - e^{-4\gamma})]^{-1}. \tag{21.33b}$$

At a high temperature and after a time of few γ^{-1} the off diagonal variance becomes $\hbar^2/4mk_BT = (\lambda/4\pi)^2$ where λ is the wavelength associated with the oscillator's mean kinetic energy at temperature T. However, the spreading of diagonal part occurs at a much slower rate, determined by the quantity ω^2/γ. Thus for time t such that $\gamma/\omega^2 \gg t \gg 1/4\gamma$, the density matrix has substantially diagonalized in the coordinate basis. In conclusion, the coordinate basis is an example of Zurek's pointer basis [see subsection 17.2.1].

The authors also develop another model of spontaneous decoherence in the coordinate basis — whose magnitude increases exponentially with the particle mass —, which is able to explain why, by the decaying of a nucleus in a bubble chamber, although the emitted particle can be represented by an outwardly propagating spherically symmetric wave, a linear track is observed [SAVAGE/WALLS 1985b].

21.2.2 Walls-Milburn Model

We now consider a superposition of two coherent states in a harmonic oscillator, which is here the object system $\mathcal{S}$ [WALLS/MILBURN 1985].[5] We initially prepare $\mathcal{S}$ in the state:

$$|\psi_{\mathcal{S}}\rangle = |\alpha_1\rangle + |\alpha_2\rangle. \tag{21.34}$$

The density operator at time t is

$$\hat{\rho}^{\mathcal{S}}(t) = \sum_{i,j=1}^{2}|\alpha_i(t)\rangle\langle\alpha_j(t)|, \tag{21.35}$$

[5]See also [*FORTUNATO* 1993, 4–6].

where $|\alpha_i(t)\rangle = |\alpha_i\rangle e^{-\imath\omega t}$. The probability density at position x may be evaluated using:

$$\langle x|\alpha\rangle = \left(\frac{\omega}{\pi\hbar}\right)^{\frac{1}{4}} \exp\left[-\frac{\omega}{2\hbar}x^2 + \left(\frac{2\omega}{\hbar}\right)^{\frac{1}{2}} \alpha x - \frac{1}{2}|\alpha|^2 - \frac{1}{2}\alpha^2\right], \tag{21.36}$$

so that we can write:

$$\langle x|\hat{\rho}^{\mathcal{S}}(t)|x\rangle = I_+^2 + I_-^2 + 2I_+I_- \cos\theta(t), \tag{21.37}$$

where, if the initial coherent states have opposite phase ($\alpha_1 = -\alpha_2 = \alpha$, with α real), we have:

$$I_\pm = \left[\frac{\omega}{\pi\hbar}\right]^{\frac{1}{4}} \exp\left[-\left(\left(\frac{\omega}{\hbar}\right)^2 x \pm \sqrt{2}\alpha\cos(\omega t)\right)^2\right] \tag{21.38a}$$

and

$$\theta(t) = 2\left(\frac{2\omega}{\hbar}\right)^{\frac{1}{2}} \sin(\omega t)\alpha x. \tag{21.38b}$$

Again we consider the coupling with a reservoir. We first must show that the amplitude of the harmonic oscillator will be damped. After a lengthy calculation it can be shown that:

$$\langle x|\hat{\rho}(t)|x\rangle = I_+ + I_- + 2e^{-|\alpha|^2(1-e^{\lambda t})} I_+I_- \cos\theta(t), \tag{21.39}$$

where I_+, I_- are given by Eq. (21.38a) with $\alpha \longrightarrow \alpha e^{-\lambda t/2}$. The oscillation amplitude is damped by the interaction with $\mathcal{E}$ at rate $\lambda/2$. But the interference term is damped at a rate $e^{-|\alpha|^2(1-e^{\lambda t})}$, i.e. the greater the initial separation of the two wave packets, the more rapidly the coherence between them is damped. In the same article the authors also show the existence of phase damping in the present model[6].

21.2.3 A Non-Demolition Model of Decoherence

The importance of the following model is due to the fact that QND-Measurement renounces to the orthogonality condition, which shows that decoherence can work without using SSRs — which normally act on the presupposition of orthogonality [see subsection 17.2.3].

A QND-Measurement does not alter the results of subsequent measurements [see section 19.4]. Now the pointer observable of the decoherence theory must be a QND-Observable of $\mathcal{A}$. This ensures that an initial eigenstate of $\hat{O}^{\mathcal{A}}$ evolves entirely within the pointer basis set. In the QND theory it is required that the QND-Observable maintains its QND property in the presence of a subsequently coupled system, in our case the environment $\mathcal{E}$ [see Eq. (19.40)]:

$$[\hat{O}^{\mathcal{A}}(t), \hat{H}_{\mathcal{A}+\mathcal{E}}] = 0. \tag{21.40}$$

The model We now see that by choosing the following model [WALLS *et al.* 1985, 3209–3211]:[7] $\mathcal{S}$ is a harmonic oscillator and $\mathcal{E}$ is a bath of harmonic oscillators. Coupling between $\mathcal{S}$ e $\mathcal{A}$ (both harmonic oscillators) is taken to be quadratic for the $\mathcal{S}$ amplitude in the interaction picture:

$$\hat{H}_{\mathcal{S}\mathcal{A}} = \frac{\hbar}{2}\hat{a}^\dagger_{\mathcal{S}}\hat{a}_{\mathcal{S}}(\hat{a}_{\mathcal{A}}\varphi^* + \hat{a}^\dagger_{\mathcal{A}}\varphi), \tag{21.41}$$

[6] For another model see also [HAAKE/WALLS 1987]. From here Haake and Żukowski [HAAKE/ŻUKOWSKI 1993] established the limits in which an apparatus behaves classically in the sense that its reading is uniquely related to a definite eigenvalue of the measured observable.

[7] See also [ALICKI 1984]. .

where φ is a classical driving field $\varphi(t) = \varphi e^{-\imath\omega_{\mathcal{A}}t}$ oscillating in resonance with the meter oscillator. This interaction represents a back-action which evades coupling (QND) of the $\mathcal{S}$ QND-Observable $\hat{a}_{\mathcal{S}}^{\dagger}\hat{a}_{\mathcal{S}}$. We now assume the following coupling between $\mathcal{A}$ and $\mathcal{E}$:

$$\hat{H}_{\mathcal{A}\mathcal{E}} = \sum_j \hat{a}_{\mathcal{A}} c_j (\hat{a}_j^{\dagger})_{\mathcal{E}} + \hat{a}_{\mathcal{A}}^{\dagger}(\hat{a}_j)_{\mathcal{E}}. \tag{21.42}$$

The reason to choose such coupling is that it leads to a coherent state pointer basis, which has well-defined semiclassical limits[8]. The reduced density operator for $\mathcal{S}+\mathcal{A}$ obeys the Agarwal Master Eq. [AGARWAL 1971] [see Eq. (21.16)]:

$$\frac{d\hat{\rho}_{\mathrm{Tr}_{\mathcal{E}}}^{\mathcal{S}+\mathcal{A}+\mathcal{E}}}{dt} = \frac{1}{2}[(\varphi\hat{a}_{\mathcal{A}}^{\dagger} - \varphi^{*}\hat{a}_{\mathcal{A}})\hat{a}_{\mathcal{S}}^{\dagger}\hat{a}_{\mathcal{S}}, \hat{\rho}_{\mathrm{Tr}_{\mathcal{E}}}^{\mathcal{S}+\mathcal{A}+\mathcal{E}}] + \frac{\gamma}{2}\left(2\hat{a}_{\mathcal{A}}\hat{\rho}_{\mathrm{Tr}_{\mathcal{E}}}^{\mathcal{S}+\mathcal{A}+\mathcal{E}}\hat{a}_{\mathcal{A}}^{\dagger} - \hat{a}_{\mathcal{A}}^{\dagger}\hat{a}_{\mathcal{A}}\hat{\rho}_{\mathrm{Tr}_{\mathcal{E}}}^{\mathcal{S}+\mathcal{A}+\mathcal{E}} - \hat{\rho}_{\mathrm{Tr}_{\mathcal{E}}}^{\mathcal{S}+\mathcal{A}+\mathcal{E}}\hat{a}_{\mathcal{A}}^{\dagger}\hat{a}_{\mathcal{A}}\right), \tag{21.43}$$

where $\mathcal{E}$ is at zero temperature. The initial state of $\mathcal{S}$ is arbitrary, while $\mathcal{A}$ is in the ground state:

$$\hat{\rho}_{\mathrm{Tr}_{\mathcal{E}}}^{\mathcal{S}+\mathcal{A}+\mathcal{E}}(t_0) = \sum_{n,m}(\hat{\rho}_{\mathcal{S}}^{nm}|n\rangle\langle m|)_{\mathcal{S}} \otimes (|0\rangle\langle 0|)_{\mathcal{A}_m}, \tag{21.44}$$

where $\hat{\rho}_{\mathcal{S}}^{nm} = \langle n|\rho_{\mathcal{S}}(0)|m\rangle$. We now find:

$$\begin{aligned}\hat{\rho}_{\mathrm{Tr}_{\mathcal{E}}}^{\mathcal{S}+\mathcal{A}+\mathcal{E}}(t) &= \sum_{n,m}\hat{\rho}_{\mathcal{S}}^{nm}\exp\left[\frac{|\varphi|^2}{\gamma^2}(n-m)^2\left(1-\frac{\gamma t}{2}-e^{-\gamma t/2}\right)\right] \\ &\times(|n\rangle\langle m|)_{\mathcal{S}} \otimes \left[\frac{|\alpha_n(t)\rangle\langle\alpha_m(t)|}{\langle\alpha_m(t)|\alpha_n(t)\rangle}\right]_{\mathcal{A}},\end{aligned} \tag{21.45}$$

where $|\alpha_n(t)\rangle$ designates coherent states of $\mathcal{A}$ with

$$\alpha_n(t) = \frac{\varphi n}{\gamma}(1 - e^{-\frac{\gamma t}{2}}). \tag{21.46}$$

In the long-time limit we have:

$$\hat{\rho}_{\mathrm{Tr}_{\mathcal{E}}}^{\mathcal{S}+\mathcal{A}+\mathcal{E}} \leadsto \sum_n \hat{N}_0(n)(|n\rangle\langle n|)_{\mathcal{S}} \otimes (|\alpha_n\rangle\langle\alpha_n|)_{\mathcal{A}}, \tag{21.47}$$

where $\hat{N}_0(n)$ is the initial number distribution for $\mathcal{S}$. This is a mixture of number states of $\mathcal{S}$ perfectly correlated with a mixture of coherent states of $\mathcal{A}$.

Photon count We now wish to investigate the state of $\mathcal{S}$ once a measured value is known [WALLS *et al.* 1985, 3211–12]. We will consider a photon counting measurement. Photons are absorbed by the counter, and so it is a Quantum Demolition (= QD) Measurement from the point of view of $\mathcal{A}$. The mean number of quanta counted in time t can be shown to be:

$$\langle m\rangle_t = A\left\langle(\hat{a}_{\mathcal{S}}^{\dagger}\hat{a}_{\mathcal{S}}^{\dagger})^2\right\rangle, \tag{21.48}$$

where $A = (\gamma\chi^2)t^3/3$. One may use such an Eq. to infer the value s for $\left\langle(\hat{a}_{\mathcal{S}}^{\dagger}\hat{a}_{\mathcal{S}}^{\dagger})^2\right\rangle$ given by:

$$s = \frac{m}{A}. \tag{21.49}$$

[8]On the point see [YAFFE 1982].

In the case where $\mathcal{S}$ is in an eigenstate of $\hat{a}_{\mathcal{S}}^{\dagger}\hat{a}_{\mathcal{S}}^{\dagger}$, this inference is certain. $\left\langle(\hat{a}_{\mathcal{S}}^{\dagger}\hat{a}_{\mathcal{S}}^{\dagger})^2\right\rangle$ is a fixed but unknown system quantity. Thus, as A increases the most likely values of m also increase.

Hence, as $|\varphi|t$ becomes larger, $\mathcal{S}$'s state, after the measurement of value s for $(\hat{a}_{\mathcal{S}}^{\dagger}\hat{a}_{\mathcal{S}}^{\dagger})^2$, is concentrated on an arbitrarily small nonempty set containing $|\sqrt{s}\rangle\langle\sqrt{s}|$ with all off-diagonal elements vanishing. With the limits $\gamma t \ll 1$, $|\varphi|t \longrightarrow \infty$, the state of $\mathcal{S}+\mathcal{A}$ is:

$$(\hat{\rho}_{\mathrm{Tr}_{\mathcal{E}}}^{\mathcal{S}+\mathcal{A}+\mathcal{E}})^m(t) = (|\sqrt{s}\rangle\langle\sqrt{s}|)_{\mathcal{S}} \otimes (|\alpha_{\sqrt{s}}\rangle\langle\alpha_{\sqrt{s}}|)_{\mathcal{A}}. \tag{21.50}$$

The system at the end of an arbitrarily accurate ($|\varphi|t \longrightarrow \infty$) and instantaneous ($\gamma t \ll 1$) measurement is found to be in an eigenstate of the measured quantity $(\hat{a}_{\mathcal{S}}^{\dagger}\hat{a}_{\mathcal{S}}^{\dagger})^2 = \sum_n n^2|n\rangle\langle n|$, with the eigenvalue equal to the measured result. $\mathcal{A}$ is left in the corresponding correlated highly excited coherent state. A QD measurement (photon counting) on $\mathcal{A}$ has resulted in a QND measurement on $\mathcal{S}$. The form of $\hat{H}_{\mathcal{A}\mathcal{E}}$ determines a preferred basis of the apparatus' states, the pointer basis, which in this model are oscillator's coherent states, and in the limit of an arbitrarily accurate measurement they become increasingly classical, i.e. with large amplitude and vanishingly small overlap.

21.3 Damping

21.3.1 The Model

Instead of considering baths of harmonic oscillators, here we can analyse the influence that a damping factor has on interference terms of two separated systems of which one is moving. The density matrix can be written as follows [CALDEIRA/LEGGETT 1985, 1061]:

$$\hat{\rho}(x,y,t) = \hat{\rho}_1 + \hat{\rho}_2 + \hat{\rho}_{12}. \tag{21.51}$$

The time evolution can be expressed by a propagator G [see Eq. (3.144)]:

$$\hat{\rho}(x,y,t) = \int \mathrm{G}(x,y,t;x',y',0)\hat{\rho}(x',y',0)dx'dy'. \tag{21.52}$$

In order to study the effect of damping on quantum interference, one has to follow the time evolution of $\hat{\rho}(x,x,t)$ given by Eq. (21.51) when $y=x$ and $\hat{\rho}(x',y',0)$ has the appropriate form: a superposition of two well-localized Gaussian wave packets with centres apart from each other. Then the composite system is initially prepared in the two-Gaussian state:

$$\psi(x) = N\left(\exp\left[-\frac{x^2}{4W_\Delta^2}\right] + \exp\left[-\frac{(x-z)^2}{4W_\Delta^2}\right]\right), \tag{21.53}$$

where N is the normalization constant, z is the initial distance between the centres of the packets and W_Δ is the width of each packet (supposed to be equal). The density operator $\hat{\rho}(x,y,0)$ for the wave function $\psi(x)$, which is given by the Eq. (21.51) with the third (interaction) element of its RHS, is given by:

$$\hat{\rho}_{12}(x,y,0) = N^2\left(\exp\left[-\frac{x^2+(y-z)^2}{4W_\Delta^2}\right] + \exp\left[-\frac{y^2+(x-z)^2}{4W_\Delta^2}\right]\right). \tag{21.54}$$

In the case of an underdamped motion — i.e. $\omega_R > \gamma$ [CALDEIRA/LEGGETT 1985, 1062] — by substitutions, we can rewrite the density function in the following form:

$$\begin{aligned}\hat{\rho}(x,\omega_R t) &= \hat{\rho}_1(x,\omega_R t) + \hat{\rho}_2(x,\omega_R t) \\ &\quad +2[\hat{\rho}_1(x,\omega_R t)]^{\frac{1}{2}}[\hat{\rho}_2(x,\omega_R t)]^{\frac{1}{2}} \cos\phi(x,\omega_R t)\exp\left[-\frac{z^2 RI_R(\omega_R t)}{8W_\Delta^2 Q(\omega_R t)}\right], \end{aligned} \quad (21.55)$$

where the third term corresponds to ρ_{12}, $\omega_R^2 := \omega^2 - \gamma^2$ and $R := \gamma/\omega_R$, I represents the intensity of interference fringes [see subsection 5.1.2], $Q(\omega_R t) = 1 + RI_R(\omega_R t) + (R + (\omega/\omega_R)\cot\omega_R t)^2$. The only term which modulates the intensity of the interference fringes is the last exponential in the last expression. It can now be shown that $RI_R(\omega_R t)/Q(\omega_R t) = 0$ when $\omega_R t = 0$, and that $RI_R(\omega_R t)/Q(\omega_R t) \longrightarrow 1$ when $\omega_R t \longrightarrow \infty$. In the last case the attenuation factor tends to $\exp[-\frac{z^2}{8W_\Delta^2}]$ (which is a measure of the initial mean energy of the system in units of $\hbar\omega_R$). As one is interested in a situation where the two packets are quite far apart ($z \gg W_\Delta$), the mean number of quanta present at this moment is certainly much greater than one. Therefore, for long durations, one can say that the attenuation factor is practically zero. Posing $\tilde{N} := z^2/4W_\Delta^2$, we have that, if, for example $z/W_\Delta \sim 10$, then $e^{-\tilde{N}/2} < 10^{-5}$. This signifies that, when two packets are macroscopically distinguishable at $t = 0$, then the number $\tilde{N}$ is huge. The question is: how long does it take for the attenuation factor to become negligible?

We now study the problem in the high-temperature and low-temperature cases.

21.3.2 High-Temperature Case

In the case of high temperature ($\hbar\omega_R \ll \hbar\Omega < 2KT$), we have [CALDEIRA/LEGGETT 1985, 1062–1063]:

$$RI_R(\omega_R t) = \frac{R\tilde{S}e^{R\omega_R t}}{\tilde{K}\sin^2\tilde{S}\omega_R t}\left[\frac{2\tilde{S}}{R}\sinh(R\omega_R t) - e^{-R\omega_R t}\left(\frac{2R}{\tilde{S}}\sin^2\tilde{S}\omega_R t + 2\sin(\tilde{S}\omega_R t)\cos(\tilde{S}\omega_R t)\right)\right], \quad (21.56)$$

where $\tilde{K} := (\hbar\omega_R)/(2k_B T)$ and $\tilde{S} := \omega/\omega_R$.

We study two situations: a weakly damped limit (when $R \longrightarrow 0$ ($\tilde{S} \longrightarrow 1$)) and a strongly damped limit ($R \longrightarrow \infty$ or $\tilde{S} \longrightarrow \imath R$).

Weakly Damped Limit

First we analyse $R \to 0$. For times such that $2\omega_R t \gg \sin\omega_R t$ one can write:

$$\exp\left[-\frac{z^2 RI_R(\omega_R t)}{8W_\Delta^2 Q(\omega_R t)}\right] \simeq e^{-\Gamma_1 t}, \quad (21.57)$$

where $\Gamma_1 := (\tilde{N}R/\tilde{K})\omega_R = (2\tilde{N}k_B T/\hbar\omega_R)\gamma$. As $R/\tilde{K} \ll 1$ the time it takes for the interference pattern to be washed out (Γ_1^{-1}) depends on the ratio between $\tilde{N}$ and $\tilde{K}R^{-1}$. If $\tilde{N} \gg 1$ but $\tilde{N}/R^{-1}\tilde{K} \ll 1$, then $\Gamma_1^{-1} \gg \omega_R^{-1}$, i.e. it will take many cycles before the interference pattern disappears. But if $\tilde{N} \gg 1$ and $\tilde{N}/\tilde{K}R^{-1} > 1$, we have $\Gamma_1^{-1} < \omega_R^{-1}$ and there would be no interference even at the time when the two packets would first overlap. Nevertheless since $\tilde{N}\tilde{K}^{-1} \gg 1$, we have always $\Gamma_1 \gg \gamma$, no matter whether $\Gamma_1 \ll \omega_R$ or $\Gamma_1 \gg \omega_R$. The other two cases ($\tilde{K}/R \simeq 1$ and $\tilde{K}/R \ll 1$) are, in reality, a single situation: $\tilde{K}/R < 1$. When $\omega_R t \ll 1$ we can write:

$$\exp\left[-\frac{z^2 RI_R(\omega_R t)}{8W_\Delta^2 Q(\omega_R t)}\right] \simeq e^{-(\Gamma_2 t)^3}, \quad (21.58)$$

where $\Gamma_2 := (2\tilde{N}R/3\tilde{K})^{1/3}\omega_R$ and one has always $\Gamma_2^{-1} < \omega_R^{-1}$. The interference pattern is destroyed in a typical duration which is shorter than γ^{-1}, possibly being of the order of (or shorter than) ω_R^{-1}.

Strongly Damped Limit

For the strongly damped limit ($R \longrightarrow \infty$) we have:

$$\exp\left[-\frac{z^2 R I_R(\omega_R t)}{8W_\Delta^2 Q(\omega_R t)}\right] \simeq e^{-\Gamma_3 t}, \tag{21.59}$$

where $\Gamma_3 = \tilde{N}\omega_R/2\tilde{K}R = (2\tilde{N}k_BT/\hbar\omega_R)\omega_R^2/2\gamma$. For the overdamped motion we arrive at equilibrium at the origin without oscillating behaviour; then the interference pattern is destroyed in a much smaller time than the relaxation time of the system. With an overdamped system there is no interference to be washed out.

21.3.3 Low-Temperature Case

In the low-temperature case (when $\hbar\omega_R \gg 2KT, \tilde{K} \gg 1$) we have a different approximation [CALDEIRA/LEGGETT 1985, 1063–1064]. As before we distinguish between a weakly damped and a strongly damped limit.

Weakly Damped Limit

In the the weakly damped limit ($R < 1$), we can perform an expansion in the power series of R, which we write here to the first order:

$$\exp\left[-\frac{z^2 R I_R(\omega_R t)}{8W_\Delta^2 Q(\omega_R t)}\right] \simeq e^{-\tilde{N}R+\dots}. \tag{21.60}$$

The time for the destruction of the interference pattern is approximately $\tilde{\Gamma}_1^{-1} := (\tilde{N}\gamma)^{-1}$ which is much shorter than γ^{-1}.

Strongly Damped Limit

In the strong damped limit ($R \longrightarrow \infty$), we have:

$$\exp\left[-\frac{z^2 R I_R(\omega_R t)}{8W_\Delta^2 Q(\omega_R t)}\right] \simeq e^{-\tilde{\Gamma}_2 t}, \tag{21.61}$$

where $\tilde{\Gamma}_2 := \tilde{N}\omega_R^2/2\gamma$ and, as before, the time for the destruction of the interference is shorter (by a factor $\tilde{N}^{-1}$) than the relaxation time of the finite energy wave packet.

Although only a specific case was studied, there is no apparent reason why those results could not be applied to the general case.

21.3.4 Physical Interpretation

We now develop an interpretation of the preceding results [CALDEIRA/LEGGETT 1985, 1064–1065]. At $t = 0$ $\mathcal{S}$ is prepared in the state: $|\psi\rangle = |\psi_z\rangle + |\psi_0\rangle$, where $|\psi_z\rangle$ is a Gaussian centered at $x = z$ and $|\psi_0\rangle$ is the ground state of the oscillator. This state initially contains a mean number n of energy quanta $\hbar\omega_R$. We desire to deal with a zero-temperature case, so that the environment $\mathcal{E}$ is in the ground state $|0\rangle$. Therefore the initial state of the 'universe' is:

$$|\Psi_i\rangle = (|\psi_z\rangle + |\psi_0\rangle) \otimes |0\rangle. \tag{21.62}$$

After time τ (the relaxation time) the final state can be written as follows:

$$|\Psi_f\rangle = |\psi_0\rangle \otimes |n\rangle, \tag{21.63}$$

where $|n\rangle$ is the state of $\mathcal{E}$ containing n quanta of energy. We are assuming that the coupling between $\mathcal{S}$ and $\mathcal{E}$ is vanishingly small. Now we wish to investigate the situation after the oscillator has emitted one quantum of energy to $\mathcal{E}$. It must occur at time τ/n:

$$|\Psi_1\rangle = |\psi_z^R\rangle \otimes |1\rangle + |\psi_0\rangle \otimes |0\rangle, \tag{21.64}$$

but before we investigated

$$\hat{\rho}^{\mathcal{S}+\mathcal{E}}_{\mathrm{Tr}_\mathcal{E}} = \mathrm{Tr}_\mathcal{E}|\Psi\rangle\langle\Psi|. \tag{21.65}$$

Thus at τ/n:

$$\tilde{\hat{\rho}}^{\mathcal{S}+\mathcal{E}}_{\mathrm{Tr}_\mathcal{E}}\left(\frac{\tau}{n}\right) = |\psi_z^R\rangle\langle\psi_z^R| + |\psi_0\rangle\langle\psi_0|, \tag{21.66}$$

where we used orthogonality between $|0\rangle$ and $|1\rangle$. Now we see that eq (21.66) represents a mixture where every interference has disappeared. It is in agreement with quantum theory of measurement because $\mathcal{E}$ acts as if it were measuring the position of the oscillator through the coordinate-coordinate coupling[9].

21.4 Ignorance and Dynamics: Baths and Damping

The usual method of the trace over the environment (thermal bath) does not allow us to distinguish between decoherence as a physical process originated from the presence of large systems and as depending on our insufficient knowledge (inobservability of coherence) of these systems due to their large size [COLLETT 1988, 2233]. In fact, in the reduced density matrix model (phase damping) developed by Walls/Milburn [see subsection 21.2.2], the factor giving the decay of the coefficients of off-diagonal elements arises from the ignorance of the bath rather than from the physical dynamics [COLLETT 1988, 2238].

The proposal of Walls/Collett/Milburn [see subsection 21.2.3] was based on the coupling of three parts: a system coupled to a meter coupled to the environment (heat bath). The coupling to the system is back-action evading. The amplification (necessary to any measurement) is built in by the driven $\mathcal{S}$-$\mathcal{A}$ coupling. The amplitude coupling $\mathcal{A}$-$\mathcal{E}$ means that after the bath has been traced out $\mathcal{A}$ is diagonalized in a basis of coherent states, which have a well-defined classical limit [COLLETT 1988, 2242].

[9] A very detailed study, which resumes the results of Leggett and co-workers is in [LEGGETT *et al.* 1987]: There, the dynamics of a two-state system coupled to a dissipative environment is considered.

A different situation is presented by an amplitude damping model — Collett studied a harmonic oscillator coupled linearly to a heat bath of oscillators in the rotating-wave approximation[10]. By calculating a reduced density matrix we see that the decay has two aspects [COLLETT 1988, 2239–40]: exactly as in the preceding case, the coefficients of off-diagonal elements are reduced, but at the same time we have, for finite temperature, a reduction of any off-diagonal component of the amplitude. The last one is a process arising from the motion of the system, even if, for exhibiting the loss of coherence, ignorance of the state of the bath is still required.

The case of the Raman amplification (the coupling between bath and system is no more direct but via a high-frequency driving field, usually with the pump frequency ω_p which is twice the system frequency $\omega_{\mathcal{S}}$) is similar to the preceding but with the additional complexity of a mixing of annihilation and creation operators. Here we also have two aspects of the decay, but the second one (the physical process) is much more important [COLLETT 1988, 2242].

The physical process of amplitude damping is of the same nature as classical damping or amplification, hence it shows a real irreversible process, even if we need the limit of a continuous number of bath modes in order to obtain a true irreversibility. On the contrary, the first process (tracing out) is typical quantum mechanical but is only based on the ignorance of the bath. The differences are clearer at zero temperature than at finite temperature [COLLETT 1988, 2244–45].

21.5 Unruh/Zurek Model*

The Model Instead of considering a reservoir composed of many harmonic oscillators we use a free field φ. The model supposes a single particle ($= \mathcal{S}$) described by the position observable q and a heat bath ($= \mathcal{E}$) modeled by the field φ. The Lagrangian action [see Eq. (1.7) and section 4.3] for $\mathcal{S} + \mathcal{E}$ is:

$$\mathrm{S} = \int \left[\frac{1}{2} \left[\dot{\varphi}^2 - \left(\frac{\partial \varphi}{\partial x} \right)^2 \right] + \delta(x)(\dot{q}^2 - \omega_0^2 q^2 - \epsilon q \dot{\varphi}) \right] dt dx, \tag{21.67}$$

where ω_0 is the frequency of the undamped harmonic oscillator and ϵ is the strength of the coupling. By writing $\delta(x)$ we distinguish between the space q of $\mathcal{S}$ and the space x in which the field propagates. The energy is a positive-definite quantity. In the Heisenberg picture the evolution of the harmonic oscillator is generated by:

$$m(\ddot{q} + \omega_0^2 q) = -\epsilon \dot{\varphi}, \tag{21.68}$$

while the field obeys:

$$\ddot{\varphi} - \partial_x^2 \varphi = \epsilon \dot{q} \delta(x). \tag{21.69}$$

The scalar field φ interacts with the harmonic oscillator at $x = 0$. q, x may be regarded as distinct 'internal spaces'. In the following we set $\epsilon' = \epsilon/\sqrt{m}, q' = q/\sqrt{m}$ and $m = 1$. Eq. (21.69) is solved by

$$\varphi = \varphi_0 + \frac{\epsilon}{2} \left[q(t-x)\Psi(x) + q(t+x)\Psi(-x) \right], \tag{21.70}$$

where φ_0 is the unperturbed solution. The effect of perturbation is restricted to the light cone $t \leq |x|$. If we insert Eq. (21.70) in (21.68), one obtains:

$$\ddot{q} + \frac{\epsilon^2}{2} \dot{q} + \omega_0^2 q = -\epsilon \dot{\varphi}_0. \tag{21.71}$$

[10]On this point see [*GARDINER* 1991, 69-71].

It is the assumption of retarded interaction in Eq. (21.70) which leads to the damping term $\epsilon^2/2$ rather than to antidamping with the opposite sign. Eq. (21.71) can be seen as a generalized Langevin Equation for q, i.e. the Eq. for a Brownian particle [see subsection 23.1.4] moving in a viscous fluid under the influence of a potential[11]:

$$m\ddot{x} = -V(x) - \gamma\dot{x} + \sqrt{2\gamma k_B T}\xi(t). \tag{21.72}$$

In the high-temperature limit the fluctuation force of the RHS has time dependence on the white noise. Eq. (21.71) is solved by:

$$q = \left[q_0 \cos(\omega t) + (p_0 + \gamma q_0)\frac{\sin(\omega t)}{\omega}\right] e^{-\gamma t} - \frac{\epsilon}{\omega}\int_0^t [\sin\omega(t-t')e^{-\gamma(t-t')}\dot{\varphi}_0(t')]dt', \tag{21.73}$$

where the damping coefficient is

$$\gamma = \frac{\epsilon^2}{4} \tag{21.74}$$

and the angular frequency of the damped oscillator is

$$\omega = \sqrt{\omega_0^2 - \gamma^2}. \tag{21.75}$$

The momentum of $\mathcal{S}$ is given by:

$$p = \dot{q}m = \left[p_0 \cos(\omega t) - \left[\frac{\gamma}{\omega}(p_0 + \gamma q_0) + \omega q_0\right] \sin(\omega t)\right] e^{-\gamma t} - \frac{\epsilon}{\omega}\frac{d}{dt}\int_0^t [\sin\omega(t-t')e^{-\gamma(t-t')}\dot{\varphi}_0(t')]dt'. \tag{21.76}$$

The harmonic oscillator is undamped when ω is real. An imaginary ω corresponds to the case of an overdamped oscillator as long as ω_0 is real [UNRUH/ZUREK 1989, 1072] .

Quantum case The above set of equations applies to classical as well as QM cases. The corresponding quantum system is described by a density operator $\hat{\rho}^{\mathcal{S}+\mathcal{E}}$. But we are only interested in the evolution of $\hat{\rho}^{\mathcal{S}}$:

$$\hat{\rho}^{\mathcal{S}} := \mathrm{Tr}_{\mathcal{E}}(\hat{\rho}^{\mathcal{S}+\mathcal{E}}). \tag{21.77}$$

Now we give the density matrix, not in the position representation ($\hat{\rho}^{\mathcal{S}}(q,q')$, always for a one-dimensional problem), but in the *Unruh-Zurek* representation $(k, \Delta_{q,q'})$ [UNRUH/ZUREK 1989, 1073]:

$$\hat{\rho}^{\mathcal{S}}(k, \Delta_{q,q'}) = \int e^{\imath kq'}\hat{\rho}^{\mathcal{S}}(q' - \Delta_{q,q'}/2, q' + \Delta_{q,q'}/2)dq'. \tag{21.78}$$

$\Delta_{q,q'}$ measures the distance from the diagonal in the position representation ($\Delta_{q,q'} = q - q'$), while k is the wave number in the direction parallel to the diagonal. It is the double Fourier transform of the usual W-Function [see Eq. (5.49)]:

$$W(p,q) = \frac{1}{2\pi}\int \hat{\rho}^{\mathcal{S}}(q - \Delta_{q,q'}/2, q + \Delta_{q,q'}/2)e^{\imath p\Delta_{q,q'}}d\Delta_{q,q'}, \tag{21.79}$$

where $\hbar = 1$.

The density operator of Eq. (21.78) can then be expressed as:

$$\hat{\rho}^{\mathcal{S}}(q - \Delta_{q,q'}/2, q + \Delta_{q,q'}/2) = \mathrm{Tr}_{\mathcal{E}}\left[\delta(\hat{q} - q)e^{\imath\hat{p}\frac{\Delta_{q,q'}}{2}}\hat{\rho}^{\mathcal{S}}e^{\imath\hat{p}\frac{\Delta_{q,q'}}{2}}\right], \tag{21.80}$$

[11] On the point see [*GARDINER* 1991, 42] .

where $\hat{q}, \hat{p}$ are given by Eq. (21.73) and (21.76). Using the commutation relations and Eqs. (21.78) and (21.80) we obtain:

$$\hat{\rho}^{\mathcal{S}}(k, \Delta_{q,q'}) = \mathrm{Tr}_{\mathcal{E}}[e^{\imath(k\hat{q}+\Delta_{q,q'}\hat{p})}\hat{\rho}^{\mathcal{S}}]. \tag{21.81}$$

To evaluate the time derivative of $\hat{\rho}^{\mathcal{S}}(k, \Delta_{q,q'})$ [UNRUH/ZUREK 1989, 1075–77] we make use of the formula:

$$\frac{de^{\hat{O}(t)}}{dt} = \int_0^1 e^{\lambda \hat{O}(t)} \dot{\hat{O}} e^{(1-\lambda)\hat{O}(t)} d\lambda, \tag{21.82}$$

where $\hat{O}$ is an operator which depends on t. Taking $\hat{O} = k\hat{q} + \Delta_{q,q'}\hat{q}$ and using Eq. (21.81), the time derivative is:

$$\dot{\hat{\rho}}^{\mathcal{S}}(k, \Delta_{q,q'}) = \mathrm{Tr}\left[\hat{\rho}_i^{\mathcal{S}} \int_0^1 e^{\imath\lambda(k\hat{q}+\Delta_{q,q'}\hat{p})}[\imath(k\dot{\hat{q}} + \Delta_{q,q'}\dot{\hat{p}})]e^{\imath(1-\lambda)(k\hat{q}+\Delta_{q,q'}\hat{p})}d\lambda\right]. \tag{21.83}$$

As final result we obtain a Master Eq. in the $(k, \Delta_{q,q'})$ representation:

$$\begin{aligned}\dot{\hat{\rho}}^{\mathcal{S}}(k, \Delta_{q,q'}) = & \left[k\frac{\partial}{\partial \Delta_{q,q'}} - \omega_0^2 \Delta_{q,q'}\frac{\partial}{\partial \Delta_{q,q'}} - 2\gamma\Delta_{q,q'}\frac{\partial}{\partial \Delta_{q,q'}} + 4\gamma\Delta_{q,q'}^2 h(t,\Gamma,\eta) - 4\gamma\Delta_{q,q'}kf(t,\Gamma,\eta)\right] \\ & \times \hat{\rho}^{\mathcal{S}}(k, \Delta_{q,q'}),\end{aligned} \tag{21.84}$$

where:

$$f(t,\Gamma,\eta) = \frac{1}{\omega\pi}\int_0^{\Gamma}\left[\int_0^t \cos\omega\tau \sin\omega\tau e^{-\gamma\tau}d\tau\right]\omega\coth(\eta\omega/2)d\omega, \tag{21.85a}$$

$$h(t,\Gamma,\eta) = g - \gamma f, \tag{21.85b}$$

$$g(t,\Gamma,\eta) = \frac{1}{\pi}\int_0^{\Gamma}\left[\int_0^t \cos\omega\tau \cos\omega\tau e^{-\gamma\tau}d\tau\right]\omega\coth(\eta\omega/2)d\omega. \tag{21.85c}$$

Eq. (21.84) can be transformed in the position representation:

$$\begin{aligned}\dot{\hat{\rho}}^{\mathcal{S}} = & \left[\imath\left(\frac{\partial^2}{\partial q^2} - \frac{\partial^2}{\partial q'^2} - \omega_0^2(q^2 - q'^2)\right) - 2\gamma(q-q')\left(\frac{\partial}{\partial q} - \frac{\partial}{\partial q'}\right) - 4\gamma(q-q')^2 h\right]\hat{\rho}^{\mathcal{S}} \\ & + 4\imath\gamma(q-q')\left(\frac{\partial}{\partial q} + \frac{\partial}{\partial q'}\right) f\hat{\rho}^{\mathcal{S}}\end{aligned} \tag{21.86}$$

by substitutions:

$$\begin{aligned}\Delta_{q,q'} = q - q' \qquad & k = \imath\left(\frac{\partial}{\partial q} + \frac{\partial}{\partial q'}\right), \\ \frac{\partial}{\partial k} = \imath(q+q') \qquad & \frac{\partial}{\partial \Delta_{q,q'}} = \frac{\partial}{\partial q} - \frac{\partial}{\partial q'}.\end{aligned} \tag{21.87}$$

In the Wigner representation eq (21.86) takes the Fokker/Planck form [see Eq. (23.3)]:

$$\dot{W} = \left[-\frac{\partial}{\partial q}\hat{\rho}^{\mathcal{S}} + \omega_0^2\frac{\partial}{\partial p}q + 2\gamma\frac{\partial}{\partial p}p + 4\gamma h\frac{\partial^2}{\partial p} - 4\gamma f\frac{\partial}{\partial q}\frac{\partial}{\partial p}\right]W. \tag{21.88}$$

Final considerations The correlation function of the noise induced by the field in its vacuum state is [UNRUH/ZUREK 1989, 1077–80] [see also Eqs. (4.29)]:

$$\left\langle [\dot{\varphi}_0(t), \dot{\varphi}_0(t')]_+ \right\rangle \simeq \frac{1}{t - t'}, \tag{21.89}$$

where $[,]_+$ is the anticommutator of the expression in brackets. The divergence at $t = t'$ forces the introduction of a cutoff Γ (in the plots used in exponential form). The temporal behaviour can be divided into three regimes:

- Initially, over the cutoff time scales of the order of $1/\Gamma$ the diffusion term h becomes very large (of order $\Gamma/2\pi$) and then falls off as t^{-1}. This transient state can be well approximated by the function:

  $$\zeta(t) = \frac{\Gamma}{\pi} \frac{\Gamma t}{\Gamma^2 t^2 + 1}. \tag{21.90}$$

  The initial evolution caused by this transient state is nearly instantaneous and has an enormous impact on the coherence of the quantum state of $\mathcal{S}$. It wipes out all off-diagonal part of density matrix in the position representation [see figures 21.2–21.4]. This stage has not been studied in detail until Unruh/Zurek's article.

- The second stage occurs on the dynamic and dissipative time scales, and especially at high temperature.

- The third stage is the asymptotic ($t \longrightarrow \infty$) regime. Coefficients of the diffusive (h) and correction (f) terms behave differently in the high temperature and vacuum ($T = 0$) environments. f is negligible when $T \ll \Gamma$. In a vacuum it is no longer small compared with h.

21.6 Another Model

In [A. STERN *et al.* 1990] another model is proposed. The authors make use of path integration[12]. The loss of interference can be explained either by the lack of overlap between two states of the environment coupled to the two interfering partial waves, or by the width of the probability distribution of the particle's phase being comparable to 2π: they are equivalent. They propose an AB-interference experiment — with left (= L) and right (= R) paths around a sphere. We shall return later to the AB-effect [see section 41.2]. For the time being it suffices to consider such an effect as a geometric-phase shift [see chapter 12]. The initial wave function of $\mathcal{S} + \mathcal{E}$ is:

$$|\Psi(t=0)\rangle = [|\mathrm{L}\rangle + |\mathrm{R}\rangle] \otimes |e_0\rangle. \tag{21.91}$$

At time t, when the interference is examined the wave function is in general:

$$|\Psi(t)\rangle = |\mathrm{L}\rangle \otimes |e_\mathrm{L}\rangle + |\mathrm{R}\rangle \otimes |e_\mathrm{R}\rangle, \tag{21.92}$$

where the interference term is:

$$2\mathrm{Re}\left[\langle \mathrm{L}|\mathrm{R}\rangle \int dq_\mathcal{E} e^*_\mathrm{L}(q_\mathcal{E}) e^*_\mathrm{R}(q_\mathcal{E})\right]. \tag{21.93}$$

We have two possible explanations.

[12] Developed in [FEYNMAN/VERNON 1963].

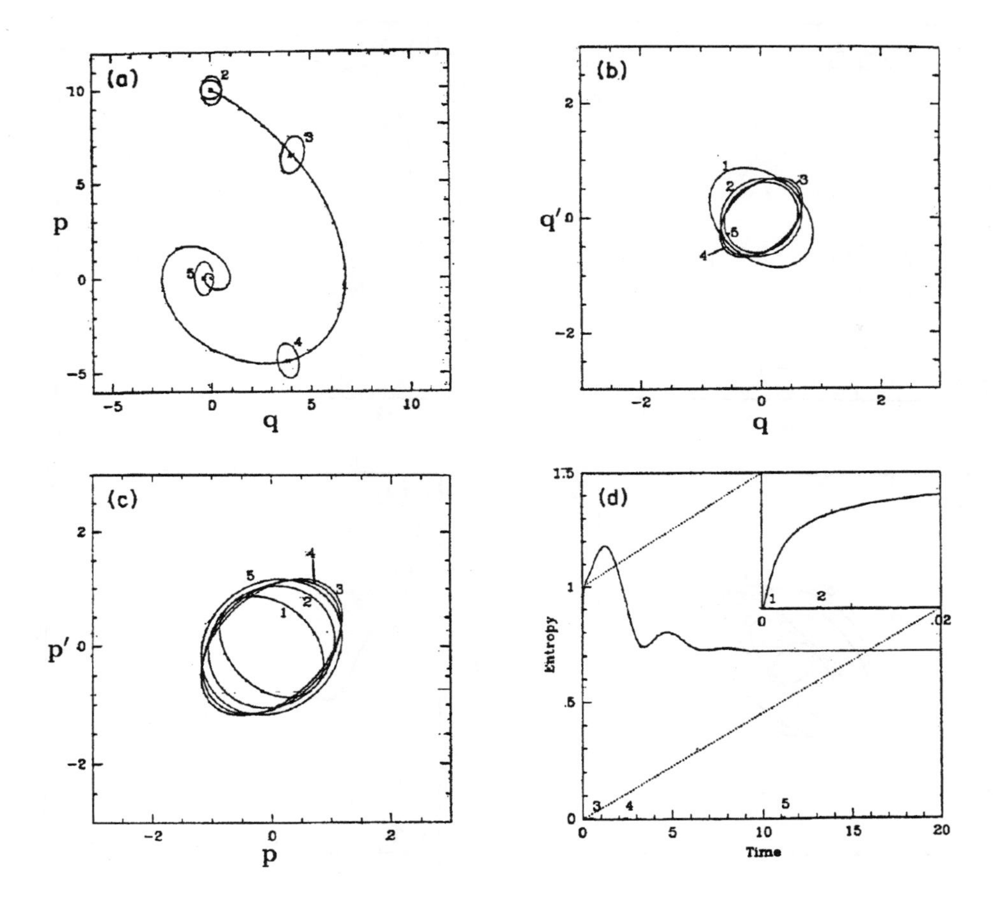

Figure 21.2: Evolution of (a) a $1-\sigma$ contour of the W-function;
(b) density matrix in $q-q'$ representation;
(c) density matrix in the $p-p'$ representation;
(d) entropy for a damped harmonic oscillator ($\omega = 1, \gamma = 0.3, \Gamma = 1000$). The initial state is a Gaussian coherent state ($\Delta q = \Delta p$) — from [UNRUH/ZUREK 1989, 1082].

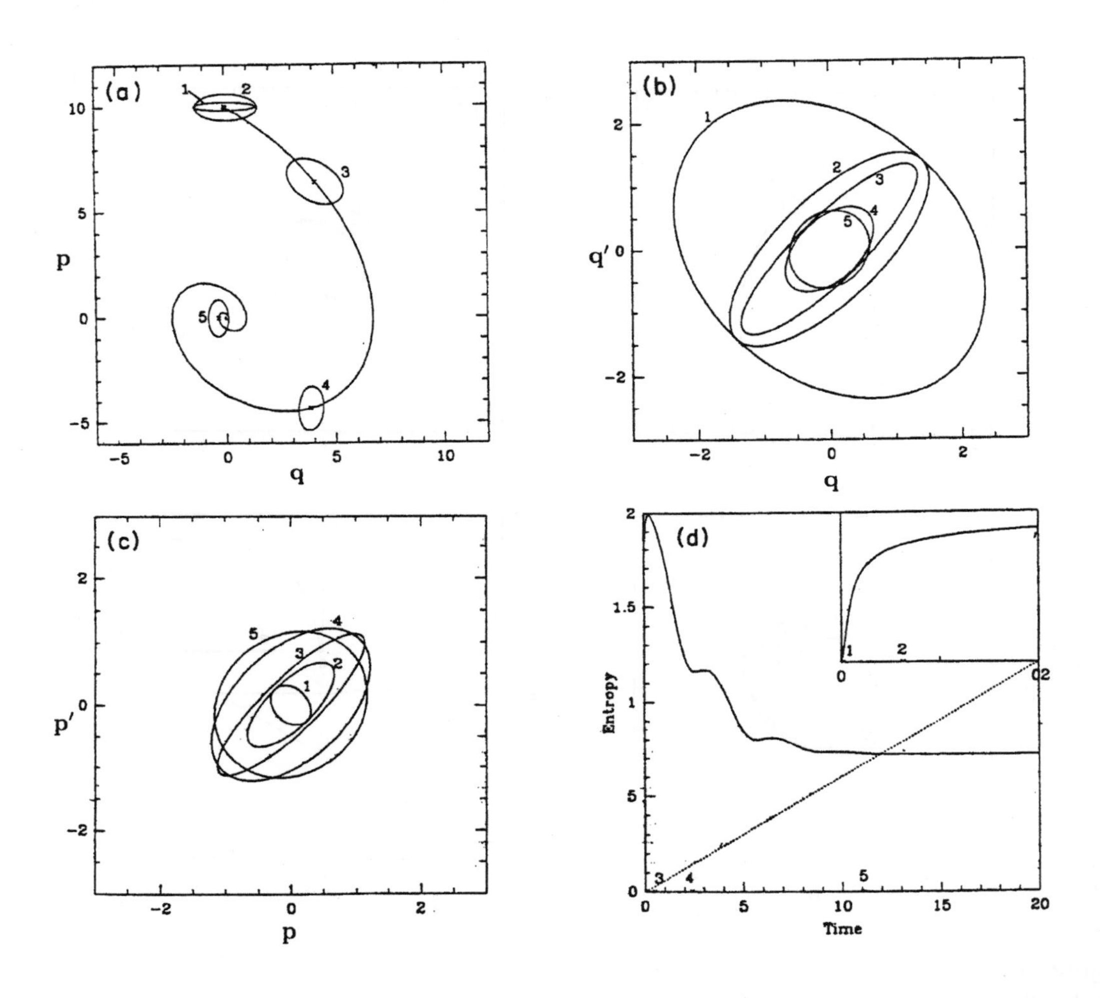

Figure 21.3: The same as figure 21.2, but for the initial state squeezed in momentum ($\Delta q = 8\Delta p$). Note the rapid initial decoherence, manifested in the change of shape of the W-function (a) and by decay of the off-diagonal elements of the density matrix in the position representation (b) — from [UNRUH/ZUREK 1989, 1083].

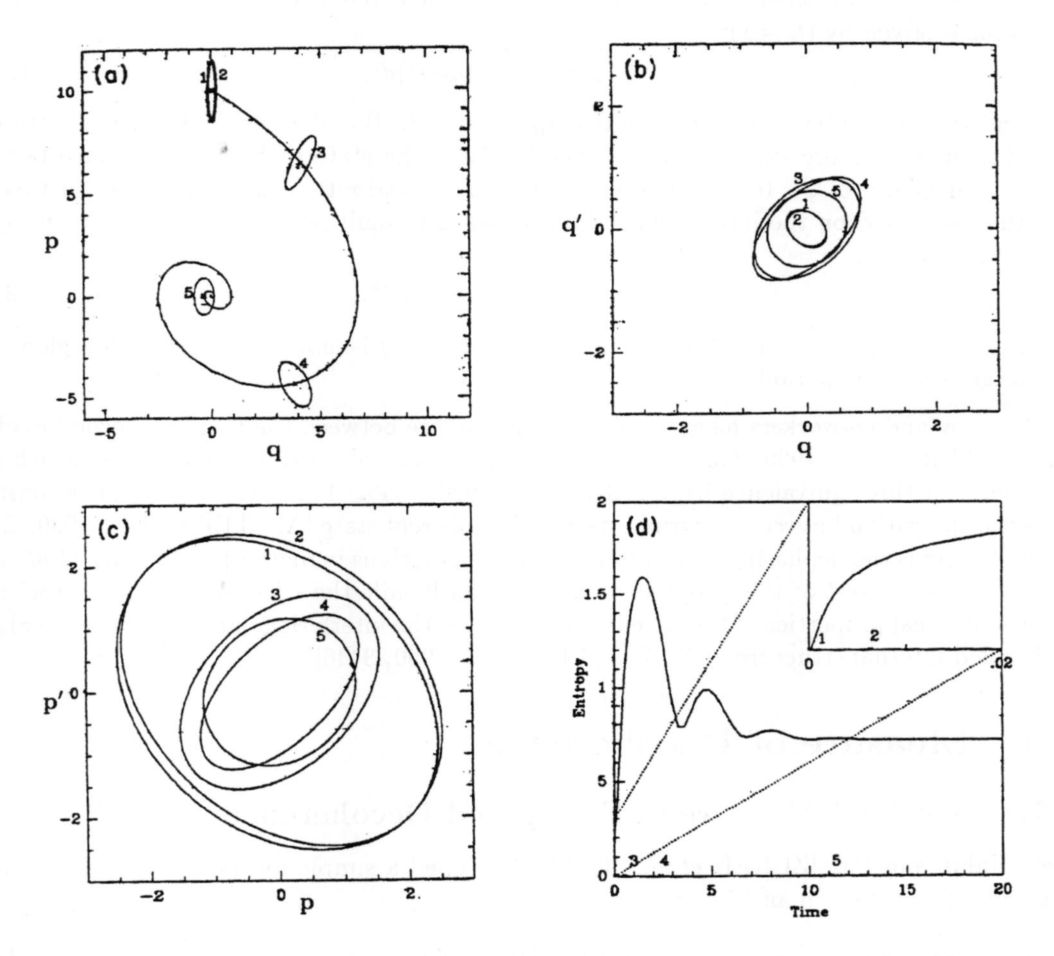

Figure 21.4: The same as figure 21.3, but for the initial state squeezed in position ($\Delta q = \Delta p/8$). Now the initial wave packet is a much better approximation of the position eigenstate. As the system-environment interaction Hamiltonian commutes with q (and hence it measures it) decoherence is rather inefficient — from [UNRUH/ZUREK 1989, 1084].

- The first one is that the interference is lost when the two interfering partial waves shift the environment into states orthogonal to each other. Characteristic of this explanation is that the coherence is lost by an entanglement between the system wave function and that of the environment (or of the apparatus).
- We see now the problem from the opposite point of view, i.e. how the environment affects the waves. When a static potential $V(q_S)$ (for example a function only of a particle's coordinate and momentum) is exerted on one of the two partial waves, the last accumulates a phase given by ($\hbar = 1$):
$$\varphi = -\int V(q_S(t))dt, \tag{21.94}$$
and then the interference term is multiplied by $e^{\imath\varphi}$. But if V depends on the state of $\mathcal{E}$, then it is no more static, and the uncertainty in the state of $\mathcal{E}$ causes an uncertainty of V and finally of φ. In this case we have only a distribution function $\wp(\varphi)$. In this case the effect of $\mathcal{E}$ on the interference of the waves is to multiply the interference term by the average of $e^{\imath\varphi}$:
$$\langle e^{\imath\varphi}\rangle = \int d\varphi \wp(\varphi) e^{\imath\varphi}. \tag{21.95}$$
Since $e^{\imath\varphi}$ is periodic in φ, $\langle e^{\imath\varphi}\rangle$ tends to zero as $\wp(\varphi)$ is slowly varying on a region much larger than one period (2π).

Aharonov and co-workers have proved the equivalence between the first and second explanation [A. STERN *et al.* 1990, 3436–37]. By the same authors also examples are given which show the validity of this equivalence [A. STERN *et al.* 1990, 3437–43], with some applications: the first is the opposition between a thermal state and a coherent state [A. STERN *et al.* 1990, 3443–45], the second is the dephasing by electromagnetic fluctuations in metals [A. STERN *et al.* 1990, 3445–46]. The interest of this proposal is that we can handle the effect of $\mathcal{E}$ on the interference via the statistical properties of the potential $\mathcal{E}$ exerts on the interfering particle without analysing the detailed internal structure of $\mathcal{E}$ [A. STERN *et al.* 1990, 3446].

21.7 Measure of Decoherence

21.7.1 General Measure of Purity and Decoherence

Zurek, Habib and Paz [ZUREK *et al.* 1993] [13] developed a simple formula in order to measure the purity of a system in an arbitrary state $\hat{\rho}$:
$$\kappa(\hat{\rho}) = \mathrm{Tr}(\hat{\rho} - \hat{\rho}^2). \tag{21.96}$$
In the same article it was shown that the rate of linear entropy increase is proportional to the dispersion in the observable of interest, as expected.

Recently Nemes and co-workers [IL KIM *et al.* 1996] have developed such a point by considering two interacting systems $\mathcal{S}_1$ and $\mathcal{S}_2$. Given the following evolution of the compound system $\hat{\rho}(t) = e^{-\imath\hat{H}t}\hat{\rho}(0)e^{\imath\hat{H}t}$, the reduced density matrix of subsystem $\mathcal{S}_2$ is $\mathrm{Tr}_1[\hat{\rho}(t)]$. Then we can introduce the measure of the decoherence of subsystem $\mathcal{S}_2$ by following the 'idempotency defect' of the reduced density matrix for the same subsystem:
$$\delta(t) := \mathrm{Tr}_2\left[\hat{\rho}_2(t) - \hat{\rho}_2^2(t)\right], \tag{21.97}$$

[13] See also [LICHTNER/GRIFFIN 1976] [NEMES/DE TOLEDO P. 1986].

which can be expanded as follows:

$$\begin{aligned}\delta(t) &= 1 - \mathrm{Tr}_2\left[\hat{\rho}_2^2(0) + t\frac{d\hat{\rho}_2^2(0)}{dt} + \frac{t^2}{2!}\frac{d^2\hat{\rho}_2^2(0)}{dt^2} + \cdots\right] \\ &\equiv \delta_0 - \frac{t}{\tau_1} - \frac{t^2}{\tau_2^2} - \cdots \end{aligned} \tag{21.98}$$

Up to order t^2 the coefficients of this expansion are:

$$\delta_0 = 1 - \mathrm{Tr}_2\left[\hat{\rho}_2^2(0)\right]; \tag{21.99}$$

$$\begin{aligned}\frac{1}{\tau_1} &= \mathrm{Tr}_2\left[\hat{\rho}_2(0)\dot{\hat{\rho}}_2(0)\right] \\ &= 2\imath\mathrm{Tr}_2\left[\hat{\rho}_2(0)\mathrm{Tr}_1[\hat{\rho}(0), \hat{H}]\right]; \end{aligned} \tag{21.100}$$

$$\begin{aligned}\frac{1}{\tau_2^2} &= \mathrm{Tr}_2\left[\dot{\hat{\rho}}_2^2(0) + \hat{\rho}_2(0)\ddot{\hat{\rho}}_2(0)\right] \\ &= -\mathrm{Tr}_2\left\{\left(\mathrm{Tr}_1[\hat{\rho}(0), \hat{H}]\right)^2 + \hat{\rho}_2(0)\mathrm{Tr}_1[[\hat{\rho}(0), \hat{H}], \hat{H}]\right\}. \end{aligned} \tag{21.101}$$

The calculus is simplified if one considers systems which are initially uncorrelated. The coefficients of the expansion furnish characteristic times τ_n associated with the correlation processes involving interaction Hamiltonians to order n. Note that double commutators such as those in Eq. (21.101) also appear in the Master Eqs. discussed in this chapter [for example in Eq. (21.7)].[14]

21.7.2 Decoherence Time-Scale After Zurek's Model

For the first time the problem of the time-scale of decoherence was studied by Zurek [ZUREK 1986b], but already Caldeira and Leggett [CALDEIRA/LEGGETT 1985] [see section 21.3] had proposed the introduction of such a parameter, even though it was in the context of a specific model. Zurek proposed the following decoherence time parameter:

$$\tau_D \simeq \gamma\left(\frac{\lambda_{DB}}{\Delta q}\right)^2, \tag{21.102}$$

where γ is the relaxation rate and λ_{DB} is the thermal de Broglie wave length:

$$\lambda_{DB} = \frac{\hbar}{\sqrt{2mk_BT}} \tag{21.103}$$

and Δq is the separation between two Gaussian peaks. A more detailed study, developed by Paz/Habib/Zurek [PAZ *et al.* 1993, 490–91, 493–94][15], confirms such a derivation.

In a recent study [SINHA 1997] it was shown that, differently form the high temperature case, there is no characteristic decoherence time scale at absolute zero. The slow power-law loss of the coherence at absolute zero is qualitatively different from its rapid suppression at high temperatures because of the finite memory in a system at absolute zero[16].

[14]In the same article the authors apply this formalism to the case of a harmonic oscillator coupled to a heat bath, deriving their results in accordance with the usual formalism.

[15]See also [ZUREK 1991, 41–42] and [BRUNE *et al.* 1992, 5205] .

[16]Other calculations of the time of decoherence, depending on more concrete and restricted models are to be found in [PEGG/KNIGHT 1988a] [PEGG/KNIGHT 1988b] [PEGG 1991]. See also [JOOS 1996, 117–18]. For further application of this problematic to quantum computation see subsection 44.5.1.

21.7.3 Decoherence Parameter After Machida/Namiki's Model

The search for a decoherence parameter is a very interesting convergence point between Zurek's approach [see section 17.2] and Machida/Namiki's one [section 17.3]. The model which has been developed in [NAMIKI/PASCAZIO 1990] [NAMIKI/PASCAZIO 1991] [17] is a very easy one. Take an interferometer with an apparatus $\mathcal{A}$ placed in arm 2 so that:

$$|\Psi\rangle = |\psi_1\rangle + c_{\mathrm{T}}|\psi_2\rangle, \tag{21.104}$$

where c_{T} is the transmission coefficient (we suppose that the apparatus is not a sharp one, in other words, it leaves some particle passing) and coefficients are made to equal 1 for the sake of simplicity. After the recombination of the beams, for the j–th particle we have the following density or intensity [see Eq. (3.4)]:

$$\wp_j := ||\psi_1^j\rangle + c_{\mathrm{T}}|\psi_2^j\rangle|^2 = |\psi_1^j|^2 + |c_{\mathrm{T}}^j|^2|\psi_2^j|^2 + 2\mathrm{Re}\langle\psi_1^j|c_{\mathrm{T}}^j\psi_2^j\rangle. \tag{21.105}$$

The event average probability is given by:

$$\overline{\wp} := \frac{1}{N_\wp}\sum_{j=1}^{N_\wp}\wp_j = |\psi_1|^2 + \overline{|c_{\mathrm{T}}|^2}|\psi_2|^2 + 2\mathrm{Re}\langle\psi_1|\overline{c_{\mathrm{T}}}\psi_2\rangle. \tag{21.106}$$

Note that $|\overline{c_{\mathrm{T}}}|^2 \leq \overline{|c_{\mathrm{T}}|^2}$ and that a necessary and sufficient condition for observing no interference is $\overline{c_{\mathrm{T}}} = 0$. Hence we can define a transmission decoherence parameter by:

$$\mathcal{D} = 1 - \frac{|\overline{c_{\mathrm{T}}}|^2}{\overline{|c_{\mathrm{T}}|^2}}. \tag{21.107}$$

It is clear that $0 \leq \mathcal{D} \leq 1$, where $\mathcal{D} = 0$ is the case of null detection, and $\mathcal{D} = 1$ is the case of full decoherence. By means of Eq. (21.107) we will rewrite Eq. (21.106) as follows:

$$\overline{\wp} = |\psi_1|^2 + \overline{|c_{\mathrm{T}}|^2}|\psi_2|^2 + 2\sqrt{\overline{|c_{\mathrm{T}}|^2}}\sqrt{1-\mathcal{D}}\mathrm{Re}\langle\psi_1|e^{\imath\beta}\psi_2\rangle, \tag{21.108}$$

where $\overline{c_{\mathrm{T}}} := |\overline{c_{\mathrm{T}}}|^2 e^{\imath\beta}$. Notice that $\overline{|c_{\mathrm{T}}|^2}$ is the experimentally measured value of the transmission probability. A nonvanishing $\mathcal{D}$ is a consequence of the statistical fluctuations in $\mathcal{A}$. We can also use the visibility of interference given by [see Eq. (5.19)]:

$$\mathcal{V} = \frac{\overline{\wp}_{\mathrm{Max}} - \overline{\wp}_{\mathrm{min}}}{\overline{\wp}_{\mathrm{Max}} + \overline{\wp}_{\mathrm{min}}} = \frac{2\sqrt{\overline{|c_{\mathrm{T}}|^2}(1-\mathcal{D})}}{1+\overline{|c_{\mathrm{T}}|^2}}. \tag{21.109}$$

21.7.4 Coherence Range

The following model is based on interactions by long range forces [JOOS/ZEH 1985, 228–29]. Take an initial factorised two-particle state:

$$\Psi(\mathbf{r}_1, \mathbf{r}_2) = m(\mathbf{r}_1)n(\mathbf{r}_2). \tag{21.110}$$

[17] See also [NAMIKI/PASCAZIO 1993, 326–28] [KONO *et al.* 1996, 1065–66].

The width of wave packets is b_j and they are localized at $\mathbf{r}_j^0$, with $b_j \ll |\mathbf{r}_1^0 - \mathbf{r}_2^0|$. The Hamiltonian is

$$\hat{H} = \frac{\hat{p}_1^2}{2m_1} + \frac{\hat{p}_2^2}{2m_2} + V(\mathbf{r}_1 - \mathbf{r}_2), \tag{21.111}$$

and the interaction part V can be expanded around $\mathbf{r}_1^0 - \mathbf{r}_2^0$:

$$\begin{aligned} V(\mathbf{r}_1 - \mathbf{r}_2) &\simeq V(\mathbf{r}_1^0 - \mathbf{r}_2^0) + \sum_k (r_1 - r_2 - r_1^0 + r_2^0)_k \partial_k V(\mathbf{r}_1^0 - \mathbf{r}_2^0) \\ &+ \frac{1}{2}\sum_{k,l}(r_1 - r_2 - r_1^0 + r_2^0)_k (r_1 - r_2 - r_1^0 + r_2^0)_l \partial_k \partial_l V(\mathbf{r}_1^0 - \mathbf{r}_2^0). \end{aligned} \tag{21.112}$$

For macroscopic objects (with large masses) the effect of the spreading of the wave packet can be neglected. Hence we consider only the interaction $V(\mathbf{r}_1 - \mathbf{r}_2)$ of the Hamiltonian (21.111). The evolution of the total system can be written:

$$\hat{\rho}(\mathbf{r}_1, \mathbf{r}_2, \mathbf{r}_1^0, \mathbf{r}_2^0, t) = \hat{\rho}_0 e^{-\imath t[V(\mathbf{r}_1 - \mathbf{r}_2) - V(\mathbf{r}_1^0 - \mathbf{r}_2^0)]}, \tag{21.113}$$

where $\hat{\rho}_0$ is the density matrix at $t = 0$, corresponding to Eq. (21.110).

The density matrix of particle 1 after time t is:

$$\hat{\rho}_1(\mathbf{r}_1, \mathbf{r}_1^0, t) = \mathrm{Tr}_{\mathbf{r}_2}\hat{\rho} = \hat{\rho}_1(\mathbf{r}_1, \mathbf{r}_1^0, t_0) \int d^3r |n(\mathbf{r})|^2 e^{-\imath t[V(\mathbf{r}_1 - \mathbf{r}) - V(\mathbf{r}_1^0 - \mathbf{r})]}. \tag{21.114}$$

By expanding the interaction part in a similar way to Eq. (21.112), Eq. (21.114) can be rewritten:

$$\begin{aligned} \hat{\rho}_1(\mathbf{r}_1, \mathbf{r}_1^0, t) &= \hat{\rho}_m(\mathbf{r}_1, \mathbf{r}_1^0, t_0) \\ & \exp\left[-\imath t\left(\sum_k (r_1 - r_1^0)_k \partial_k V + \sum_{k,l} V_{kl}\left(\frac{r_{1k}r_{1l} - r_{1k}^0 r_{1l}^0}{2} + (r_1 - r_1^0)_k (r_1 - r_1^0)_l\right)\right)\right] \\ & \times \int d^3r |n(\mathbf{r})|^2 \exp\left[+\imath t \sum_{k,l}(r_1 - r_1^0)_k r_l V_{kl}\right]. \end{aligned} \tag{21.115}$$

The last factor describes a damping of interference terms. Inserting a Gaussian of width b for particle 2, the integral of Eq. (21.115) is:

$$\int d^3r |\varsigma(\mathbf{r})|^2 \exp\left[+\imath t \sum_{k,l}(r_1 - r_1^0)_k r_l V_{kl}\right] = \exp\left[-t^2 b^2 \sum_l \left(\sum_k V_{kl}(r_1 - r_1^0)_k\right)^2\right]. \tag{21.116}$$

For small arguments the integral is $\simeq 1 - t^2 b^2 \sum_l [\sum_k V_{kl}(r_1 - r_1^0)_k]^2$. For a Coulomb-like potential with a coupling constant of g_{12} $V(\mathbf{r}) = g_{12}/|\mathbf{r}|$ one has:

$$V_{kl} = g_{12}\frac{3r_k r_l - r^2 \delta_{kl}}{\mathbf{r}^5}. \tag{21.117}$$

Considering the special case where $\mathbf{r}_1 - \mathbf{r}_1^0$ is parallel to $\mathbf{r}_2^0 - \mathbf{r}_1^0$ we obtain, for the interference term, $\simeq 1 - (4|\mathbf{r}_1 - \mathbf{r}_1^0|b^2 g_{12}^2)t^2/a^6$, where $a := |\mathbf{r}_2^0 - \mathbf{r}_1^0|$. Now it is possible to estimate the coherence range $l = |\mathbf{r}_1 - \mathbf{r}_1^0|$. The coherence under the influence of the second particle is appreciable if the second term of Eq. (21.117) is non-negative. For simplicity we consider the symmetrical situation $l = b$ [see tables 21.1 and 21.2].

	$t = 10^{-5}$s	$t = 1$s	$t = 1$ year
$a = 1$ km	10^6 cm	10^3 cm	10^{-1} cm
$a = 1$ m	10^1 cm	10^{-1} cm	10^{-5} cm
$a = 1$ cm	10^{-2} cm	10^{-4} cm	10^{-8} cm

Table 21.1: Coherence range for two elementary charges after various times t and at various distances a, as resulting from their Coulomb interaction and calculated in short-time approximation. — from [JOOS/ZEH 1985, 229].

	$a = 10^{-3}$ cm dust	$a = 10^{-5}$ cm dust	$a = 10^{-6}$ cm large molecules
Cosmic background radiation	10^6	10^{-6}	10^{-12}
Room temperature	10^{19}	10^{12}	10^6
Sunlight on the Earth	10^{21}	10^{17}	10^{13}
Air	10^{36}	10^{32}	10^{30}
Laboratory vacuum (10^6 particles/cm^3)	10^{23}	10^{19}	10^{17}

Table 21.2: Localization rate in cm$^{-2}s^{-1}$ for three sizes of 'dust particles' and various scattering processes — from [JOOS/ZEH 1985, 234].

21.7.5 Decoherence and Probability*

Breuer and Petruccione [H. BREUER/PETRUCCIONE 1996a, 1151–53] [18] calculated the transformation from a quantum probability to a classical one on the basis of the Walls/Collett/Milburn Model [see subsection 21.2.3]. First we define a QM statistical variance of the operator $\hat{O} = \hat{N}_S^2 = (\hat{a}_S^\dagger \hat{a}_S)^2$:

$$(\Delta\hat{O})^2 = \text{Tr}(\hat{O}^2\hat{\rho}) - [\text{Tr}(\hat{O}\hat{\rho})]^2 \tag{21.118}$$

as the sum of *the quantum statistical variance* $(\Delta\hat{O})^2_{\text{QM}}$ averaged over the pure states contained in the ensemble — it is a measure of the average intrinsic QM fluctuations, i.e. of the distance to an ensemble consisting of eigenstates of $\hat{O}$ — on one side, and of the *dispersion* $(\Delta\hat{O})^2_{\text{C}}$ of the pure state quantum expectation value — i.e. it is a measure of the distance to an ensemble in a pure state — on the other:

$$(\Delta\hat{O})^2 = (\Delta\hat{O})^2_{\text{QM}} + (\Delta\hat{O})^2_{\text{C}}. \tag{21.119}$$

Now the authors prove that $(\Delta\hat{O})^2$ is constant in time — due to Eq. (21.43) —, and that at the beginnings of the measurement process $(\Delta\hat{O})^2$ is practically identical to $(\Delta\hat{O})^2_{\text{QM}}$, while at the end of the measurement $(\Delta\hat{O})^2_{\text{QM}}$ vanishes (because we obtain only eigenstates of the measured observable) and $(\Delta\hat{O})^2$ becomes identical to $(\Delta\hat{O})^2_{\text{C}}$ [see figure 21.5]. Hence the information about the potential outcomes of the measurement contained in the initial pure state are objectivized during the process of measurement.

[18] See also [H. BREUER/PETRUCCIONE 1995a] [H. BREUER/PETRUCCIONE 1995b] [H. BREUER/PETRUCCIONE 1996b].

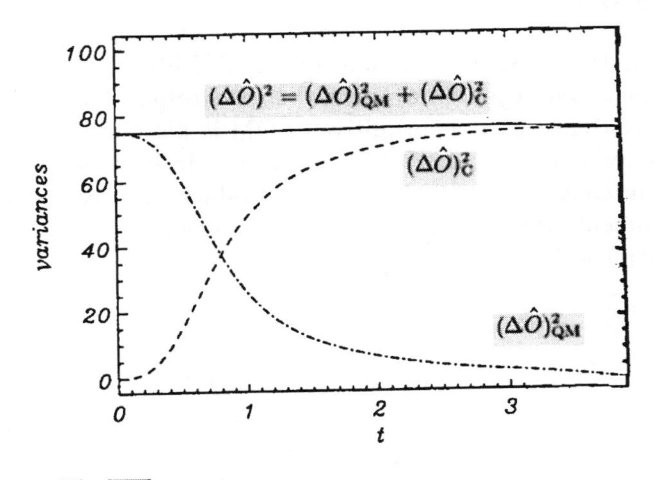

Figure 21.5: The variances $(\Delta\hat{O})^2, (\Delta\hat{O})^2_{\mathrm{QM}}, (\Delta\hat{O})^2_{\mathrm{C}}$ were estimated from a stochastic simulation of the Liouville-master Eq. by averaging over 10^4 realizations — from [H. BREUER/PETRUCCIONE 1996a, 1152].

Suppose an initial state given by

$$|\Psi^i_{\mathcal{S}+\mathcal{A}}\rangle = |\psi_{\mathcal{S}}\rangle \otimes |0\rangle, \tag{21.120}$$

where $|\psi_{\mathcal{S}}\rangle = \sum_n c_n(0)|n\rangle$. The quantum statistical variance can be written as follows:

$$(\Delta\hat{O})^2_{\mathrm{QM}} = \int D\Psi^{\mathcal{S}+\mathcal{A}+\mathcal{E}}_{\mathrm{Tr}_\mathcal{E}} D(\Psi^{\mathcal{S}+\mathcal{A}+\mathcal{E}}_{\mathrm{Tr}_\mathcal{E}})^*(\Delta\hat{O})^2_{\Psi^{\mathcal{S}+\mathcal{A}+\mathcal{E}}_{\mathrm{Tr}_\mathcal{E}}}\wp_{\mathcal{S}\mathcal{A}}(\Psi^{\mathcal{S}+\mathcal{A}+\mathcal{E}}_{\mathrm{Tr}_\mathcal{E}}, t), \tag{21.121a}$$

where $\wp_{\mathcal{S}\mathcal{A}}(\Psi^{\mathcal{S}+\mathcal{A}+\mathcal{E}}_{\mathrm{Tr}_\mathcal{E}}, t)$ is the time-dependent probability distribution of apparatus + system, $|\Psi^{\mathcal{S}+\mathcal{A}+\mathcal{E}}_{\mathrm{Tr}_\mathcal{E}}\rangle$ is the reduced state of apparatus + system (by tracing out the environment) and $D\Psi^{\mathcal{S}+\mathcal{A}+\mathcal{E}}_{\mathrm{Tr}_\mathcal{E}} D(\Psi^{\mathcal{S}+\mathcal{A}+\mathcal{E}}_{\mathrm{Tr}_\mathcal{E}})^*$ denote the unitarily invariant volume elements in the state space of apparatus + system. The dispersion can be written as follows:

$$\begin{aligned}(\Delta\hat{O})^2_{\mathrm{C}} \;=\; & \int D\Psi^{\mathcal{S}+\mathcal{A}+\mathcal{E}}_{\mathrm{Tr}_\mathcal{E}} D(\Psi^{\mathcal{S}+\mathcal{A}+\mathcal{E}}_{\mathrm{Tr}_\mathcal{E}})^* \left\langle\hat{O}\right\rangle^2_{\Psi^{\mathcal{S}+\mathcal{A}+\mathcal{E}}_{\mathrm{Tr}_\mathcal{E}}} \wp_{\mathcal{S}\mathcal{A}}(\Psi^{\mathcal{S}+\mathcal{A}+\mathcal{E}}_{\mathrm{Tr}_\mathcal{E}}, t) \\ & - \left[\int D\Psi^{\mathcal{S}+\mathcal{A}+\mathcal{E}}_{\mathrm{Tr}_\mathcal{E}} D(\Psi^{\mathcal{S}+\mathcal{A}+\mathcal{E}}_{\mathrm{Tr}_\mathcal{E}})^* \left\langle\hat{O}\right\rangle_{\Psi^{\mathcal{S}+\mathcal{A}+\mathcal{E}}_{\mathrm{Tr}_\mathcal{E}}} \wp_{\mathcal{S}\mathcal{A}}(\Psi^{\mathcal{S}+\mathcal{A}+\mathcal{E}}_{\mathrm{Tr}_\mathcal{E}}, t)\right]^2 \end{aligned} \tag{21.121b}$$

The measurement can be understood as a transition from a correspondent initial QM probability distribution

$$\wp^i_{\mathcal{S}\mathcal{A}}[\Psi^{\mathcal{S}+\mathcal{A}+\mathcal{E}}_{\mathrm{Tr}_\mathcal{E}}] = \int_0^{2\pi} \frac{d\varphi}{2\pi}\delta[e^{\imath\varphi}|\Psi^i_{\mathcal{S}+\mathcal{A}}\rangle - |\Psi^{\mathcal{S}+\mathcal{A}+\mathcal{E}}_{\mathrm{Tr}_\mathcal{E}}\rangle], \tag{21.122}$$

where δ is a Dirac measure on the Fock space and the integration over the phase φ ensures the phase invariance, to the classical probability

$$\wp^\infty_{\mathcal{S}\mathcal{A}}[\Psi^{\mathcal{S}+\mathcal{A}+\mathcal{E}}_{\mathrm{Tr}_\mathcal{E}}] = \sum_n |\langle n|\psi_{\mathcal{S}}\rangle|^2 \int_0^{2\pi} \frac{d\varphi}{2\pi}\delta[e^{\imath\varphi}|\psi_n\rangle - |\Psi^{\mathcal{S}+\mathcal{A}+\mathcal{E}}_{\mathrm{Tr}_\mathcal{E}}\rangle], \tag{21.123}$$

where $|\psi_n\rangle = |n\rangle \otimes |\beta_n\rangle$, and $|\beta_n\rangle$ is a coherent meter state, i.e. the state vector $|\Psi^{\mathcal{S}+\mathcal{A}+\mathcal{E}}_{\mathrm{Tr}_\mathcal{E}}\rangle$ is found with probability $|\langle n|\psi_{\mathcal{S}}\rangle|^2$ in one of the states $|\psi_n\rangle$ which appear as alternatives in the sense of the classical theory.

21.8 Conclusion

This chapter has shown the possibility of considering individual systems not as data but as results: they become *individual* through a process of *localization* due to their interaction with relatively large systems. Hence what is primitive in QM is not the individuals themselves but their 'correlations', from whose break-down they acquire their existence [see also section 42.3].

Consequently we can conceive of the particles as products of a continuous measurement of position by the environment [JOOS 1986b, 9, 12], which is what will be discussed later as the concept of 'extended particles' [see section 33.1].

Chapter 22

QUANTUM JUMPS

Introduction and contents In this chapter we describe a quantum theory of open systems and spontaneous transitions by making use of a formalism based on QM trajectories — it is equivalent to the Master Eq. approach. Q-Optics is the area in which such techniques have been developed. In this sense it is also a further development of the formalism treated of in chapter 5. This quantum-jump theory is also known as quantum 'Montecarlo' theory. Its roots can be found in the pioneering work of Carmichael and Walls [CARMICHAEL/WALLS 1975] [CARMICHAEL/WALLS 1976a] [CARMICHAEL/WALLS 1976a] and of Kimble and Mandel [KIMBLE/MANDEL 1976], later developed by Dalibard and co-workers [DALIBARD *et al.* 1992]. In a certain sense the following approach is a theoretical and practical realization of some ideas of Bohr about quantum jumps [see for example postulate 2.2: p. 20].

In what follows we shall first [section 22.1] develop jump-formalism by analysing in more detail the photon counting [see also subsection 5.1.4]. We shall then discuss interpretational problems [section 22.2] and finally [section 22.3] some concrete models and related experiments. Later we shall analyse [chapter 23] some theories which have tried to use some Quantum-jump formalism in order to account for the reduction of the wave packet and spontaneous localizations.

22.1 General Theory

First [subsection 22.1.1] we analyse the case for a field which, apart from the discreteness of the energy, is not characterized by specific quantum features. Later [subsection 22.1.2] we shall discuss the quantum case in full extent.

22.1.1 Photon Counting for 'Classical' Fields

We can express the incident power at a photodetector in terms of the photon energy $\hbar\omega$ multiplied by the average number of photons entering the detector per second ($\overline{N}_D$), which gives the relationship [*CARMICHAEL* 1993, 79–84]:

$$\overline{N}_D = \frac{A\overline{I}}{\hbar\omega}, \tag{22.1}$$

where A is the cross-sectional area of the beam over which the cycle-averaged intensity $\overline{I}$ is uniform. If the detector counts for a time τ, we expect that $\overline{N}_D(\tau) = \frac{A\overline{I}}{\hbar\omega}\tau$. But since we must consider the quantum efficiency s_D of the detector, we finally obtain: $\overline{N}_D(\tau) = s_D\frac{A\overline{I}}{\hbar\omega}\tau = \zeta\overline{I}\tau$, where $\zeta = s_{\rm D}A/\hbar\omega$. The photon flux is given by $\zeta\overline{I}$. Let us now divide the total interval τ into $n = \tau/\Delta t \gg 1$ (sufficiently small) subintervals. Let us write $\wp_+$ for the probability that in a given subinterval Δt one photoelectron is emitted, and $\wp_-$ for the probability that none is emitted. Then

$$\wp_+ = \zeta\overline{I}\Delta t, \qquad \wp_- = 1 - \zeta\overline{I}\Delta t. \tag{22.2}$$

Now $\wp(s,t,\tau)$ is the probability of obtaining s detections and $n-s$ no-detections in the interval $(t, t+\tau]$:

$$\begin{aligned}\wp(s,t,\tau) &= \frac{n(n-1)\cdots(n-s+1)}{s!}(\zeta\overline{I}\Delta t)^s(1-\zeta\overline{I}\Delta t)^{n-s} \\ &= \left(1-\frac{1}{n}\right)\cdots\left(1-\frac{s-1}{n}\right)\frac{(\zeta\overline{I}n\Delta t)^s}{s!}(1-\zeta\overline{I}\Delta t)^{n-s}.\end{aligned} \tag{22.3}$$

As we have said, we want Δt to be very small, so that in the limit $n \longrightarrow \infty$ with $n\Delta t = \tau$ constant, we have $(1-\zeta\overline{I}\Delta t)^{n-s} \longrightarrow e^{-\zeta\overline{I}\tau}$, from which we obtain:

$$\wp(s,t,\tau) = \frac{(\zeta\overline{I}\tau)^s}{s!}e^{-\zeta\overline{I}\tau}, \tag{22.4}$$

which is a Poisson distribution. This is also called the *Mandel photoelectron counting formula* [MANDEL 1958a] [MANDEL 1959] — which was further developed by Kelley and Kleiner [KELLEY/KLEINER 1964].[1]

Until now we have considered only uncertainties due to the quantum nature of the emission process. Now we consider also fluctuations of the field. In that case we can generalize Eq. (22.4) to a situation in which $\overline{I}(t)$ is constant over each counting interval $(t, t+\tau]$, but changes (slowly) between the intervals. Hence we can write:

$$\wp(s,t,\tau) = \left\langle \frac{(\zeta\overline{I}\tau)^s}{s!}e^{-\zeta\overline{I}\tau}\right\rangle = \int_0^\infty d\overline{I}\mathcal{F}(\overline{I})\frac{(\zeta\overline{I}\tau)^s}{s!}e^{-\zeta\overline{I}\tau}, \tag{22.5}$$

where $\mathcal{F}(\overline{I})$ is the probability distribution for the sampled light intensities. Since in general the photoelectron counting time τ will not be much less than intensity correlation time, we can replace $\zeta\overline{I}\tau$ by $\zeta\int_t^{t+\tau}dt'\overline{I}(t')$ — which, for each sampling interval $(t, t+\tau]$, calculates the accumulated number of photons (multiplied by $s_{\rm D}$) incident on the detector —, and write:

$$\wp(s,t,\tau) = \left\langle \frac{(\zeta\int_t^{t+\tau}dt'\overline{I}(t'))^s}{s!}e^{-\zeta\int_t^{t+\tau}dt'\overline{I}(t')}\right\rangle. \tag{22.6}$$

If we use the substitution $\Omega(t,\tau) \equiv \zeta\int_t^{t+\tau}dt'\overline{I}(t')$, we can write Eq. (22.6) in the form:

$$\wp(s,t,\tau) = \left\langle \frac{\Omega(t,\tau)^s}{s!}e^{-\Omega(t,\tau)}\right\rangle. \tag{22.7}$$

[1] In [SRINIVAS/DAVIES 1981] it was shown that, if, by performing a measurement, one does not take into account the modifications of the field distribution produced by the presence of a detector, negative probabilities could result. The result was later confirmed by Imoto, Ueda and Ogawa [IMOTO *et al.* 1990].

Performing a procedure similar to the moments calculated by means of the characteristic function [see subsection 3.7.3], we obtain for the factorial moments:

$$\begin{aligned} s^{(q)} &= \sum_{s=0}^{\infty} s(s-1)\cdots(s-q+1)\wp(s,t,\tau) \\ &= \sum_{s=0}^{\infty} s(s-1)\cdots(s-q+1)\left\langle \frac{\Omega(t,\tau)^s}{s!} e^{-\Omega(t,\tau)} \right\rangle \\ &= (-1)^q \sum_{s=q}^{\infty} \left\langle \frac{\Omega(t,\tau)^{s-q}}{(s-q)!} \frac{d^q}{dx^q} e^{-x\Omega(t,\tau)} \right\rangle \Bigg|_{x=1} \\ &= (-1)^q \frac{d^q}{dx^q} \sum_{k=0}^{\infty} \left\langle \frac{\Omega(t,\tau)^k}{k!} e^{-x\Omega(t,\tau)} \right\rangle \Bigg|_{x=1} \\ &= (-1)^q \frac{d^q}{dx^q} \left\langle e^{(1-x)\Omega(t,\tau)} \right\rangle \Bigg|_{x=1} . \end{aligned} \quad (22.8)$$

By calculating explicitly the two lower moments, we obtain for the mean of photoelectron counting distribution (as expected):

$$\langle s \rangle = \left\langle \zeta \int_t^{t+\tau} dt' \bar{I}(t') \right\rangle ; \quad (22.9)$$

and for the variance [see Eq. (3.110)]:

$$(\Delta s)^2 = \langle s \rangle + \zeta^2 \int_t^{t+\tau} dt' \int_t^{t+\tau} dt'' [\langle \bar{I}(t')\bar{I}(t'') \rangle - \langle \bar{I}(t') \rangle \langle \bar{I}(t'') \rangle]. \quad (22.10)$$

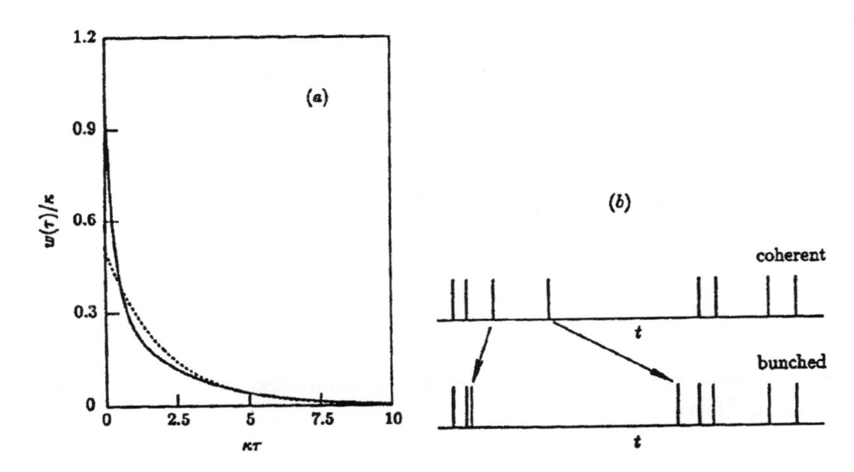

Figure 22.1: (a) Comparison between the waiting-time distributions for filtered thermal light (bunched light) — solid curve — and coherent light of the same intensity — dashed curve. Broadband thermal light ($\langle s \rangle = 1$) is filtered by a cavity with linewidth κ (half-width at half-maximum) and equal transmission coefficients at the input and output mirrors.
(b) Rearrangement of a typical random photoelectron emission sequence to account for the change in the waiting-time distribution shown in (a) — from [*CARMICHAEL* 1993, 88].

Now we define [*CARMICHAEL* 1993, 86–87] the *waiting-time distribution* $\wp_w(\tau_w)$ as the probability, given a photoelectric emission has just occurred, that there are no photoelectric emissions during an interval of length τ_w, followed by a photoelectric emission during next $d\tau_w$. Assuming that the process is stationary and the limit $n \longrightarrow \infty$, we have:

$$\wp_w(\tau_w)d\tau_w = (1 - \zeta\overline{I}\Delta t)^n \zeta\overline{I}\Delta t \longrightarrow \zeta\overline{I}e^{-\zeta\overline{I}\tau_w}d\tau_w. \tag{22.11}$$

The mean waiting time is given by:

$$\langle\tau_w\rangle = \int_0^\infty d\tau_w \, \tau_w \zeta\overline{I}e^{-\zeta\overline{I}\tau_w} = (\zeta\overline{I})^{-1}. \tag{22.12}$$

When the light intensity is a stochastic quantity, the shape of the waiting time distribution changes: the photoelectrons produced at either end of a waiting time interval that is smaller than the intensity correlation time are correlated [see figure 22.1].

22.1.2 Photoelectron Counting for Quantized Fields

In the QM case [*CARMICHAEL* 1993, 88–90] we must replace $\overline{I}$ by $2\epsilon_0 c\hat{E}^{(-)}\hat{E}^{(+)}$ [see Eqs. (5.18) and (5.27)], where the factor $2\epsilon_0 c$ is needed to give the units of intensity. The photoelectron counting distribution [KELLEY/KLEINER 1964] is given by [corresponding to Eq. (22.7)]:

$$\wp(s,t,\tau) = \left\langle : \frac{\hat{\Omega}^s}{s!}e^{-\hat{\Omega}} : \right\rangle, \tag{22.13}$$

where :: indicates that operators have to be written in normal and time order and the integrated intensity is given by the operator:

$$\hat{\Omega} = \zeta \int_t^{t+\tau} dt' \hat{E}^{(-)}(t')\hat{E}^{(+)}(t'), \tag{22.14}$$

with $\zeta = s_{\mathrm{D}} A \frac{2\epsilon_0 c}{\hbar\omega}$.

Similar to the classical case, we write the mean number of the photoelectron counts as:

$$\langle s\rangle = \zeta \int_t^{t+\tau} dt' \langle\hat{E}^{(-)}(t')\hat{E}^{(+)}(t')\rangle \tag{22.15}$$

and its variance [corresponding to Eq. (22.10)] as:

$$\begin{aligned}(\Delta s)^2 &= \langle s\rangle + \zeta^2 \int_t^{t+\tau} dt' \int_t^{t+\tau} dt'' \left[\langle : \hat{E}^{(-)}(t')\hat{E}^{(+)}(t')\hat{E}^{(-)}(t'')\hat{E}^{(+)}(t'') : \rangle\right. \\ &\quad \left. -\langle\hat{E}^{(-)}(t')\hat{E}^{(+)}(t')\rangle\hat{E}^{(-)}(t'')\hat{E}^{(+)}(t'')\rangle\right]\end{aligned} \tag{22.16}$$

Note that, unlike the classical case, it is possible for a quantized field to produce subpoissonian photoelectron counting distribution.

22.1.3 Quantum Trajectories

Exclusive Probability Density If we return to expressions (22.13) and (22.14), we can expand in time the exponential present in the first one, obtaining a series of integrals over the probability densities [*CARMICHAEL* 1993, 114–17]:

$$\wp_m(t_1, t_2, \ldots, t_m) = \zeta_m \langle\hat{E}^{(-)}(t_1)\cdots\hat{E}^{(-)}(t_m)\hat{E}^{(+)}(t_m)\cdots\hat{E}^{(+)}(t_1)\rangle. \tag{22.17}$$

The $\wp_m$ are called *non-exclusive probability densities* or multicoincidence rates for photon counting. The nonexclusive probability $\wp_m(t_1, t_2, \ldots, t_m)\Delta t_1 \Delta t_2 \cdots \Delta t_m$ is the probability that one photoelectron is emitted in each of the non-overlapping intervals $[t_1, t_1+\Delta t_1), [t_2, t_2+\Delta t_2), \ldots, [t_m, t_m+\Delta t_m)$, where $t_1 < t_2 < \ldots < t_m$. The nonexclusive character of this probability comes from the fact that it places no conditions on what may happen at times in between the m infinitesimal intervals during which the specified photoelectron emissions occur. But there is a much simpler expression for the photoelectron counting distribution in terms of the *exclusive probability densities* $\wp_n^e$:[2]

$$\wp(s, t, t+\tau) = \int_t^{t+\tau} dt_s \int_t^{t_s} dt_{s-1} \cdots \int_t^{t_2} dt_1 \wp_s^e(t_1, t_2, \ldots, t_s; [t, t+\tau]). \tag{22.18}$$

$\wp_s^e(t_1, t_2, \ldots, t_s; [t, t+\tau])\Delta t_1 \Delta t_2 \cdots \Delta t_s$ is the probability that s photoelectrons are emitted in the observation interval $[t, t+\tau]$, one in each of the non-overlapping subintervals $[t_1, t_1+\Delta t_1), [t_2, t_2+\Delta t_2), \ldots, [t_s, t_s+\Delta t_s)$, where $t_1 < t_2 < \ldots < t_s$. Hence those events which occur between these subintervals are excluded.

The relationship between exclusive and nonexclusive probability distributions is rather complicated. In fact we can write the first one as follows:

$$\begin{aligned} \wp_s^e(t_1, t_2, \ldots, t_s; [t, t+\tau]) &= \sum_{q=0}^{\infty} \frac{(-1)^q}{q!} \int_t^{t+\tau} dt'_q \int_t^{t+\tau} d'_{q-1} \cdots \int_t^{t+\tau} dt'_1 \langle : \hat{I}(t'_q)\hat{I}(t'_{q-1}) \\ &\quad \cdots \hat{I}(t'_1)\hat{I}(t_m) \cdots \hat{I}(t_1) : \rangle \\ &= \left\langle : e^{-\hat{\Omega}(t,t+\tau)} \hat{I}(t_m) \cdots \hat{I}(t_1) : \right\rangle \\ &= \left\langle : e^{-\hat{\Omega}(t+\tau,t_m)} \hat{I}(t_m) \cdots e^{-\hat{\Omega}(t_2,t_1)} \hat{I}(t_1) e^{-\hat{\Omega}(t_1,t)} : \right\rangle, \end{aligned} \tag{22.19}$$

where $\hat{\Omega}(t_j, t_k) = \int_{t_k}^{t_j} dt' \hat{I}(t')$ and $\hat{I}(t') = \zeta \hat{E}^{(-)}(t')\hat{E}^{(+)}(t')$ is the photon flux operator.

Waiting Time Distribution Note now that, while the waiting-time distribution $\wp_w(\tau_w)$ requires that there be no additional emission during the interval τ_w, the correlation function [see subsection 5.1.2], say of the second order $\mathrm{C}^{(2)}(\tau_w)$ [see Eq. (5.12)], does not make such an assumption. In other words, we have here the simplest form of distinction between exclusive and nonexclusive probability. Hence we can express the normalized second-order correlation function in terms of the non-exclusive probability $\wp_2(t, t+\tau_w)$ and of the waiting-time distributions $\wp_w^1(t)$, $\wp_w^1(t+\tau_w)$:

$$\mathrm{C}^{(2)}(t, t+\tau_w) = \frac{\wp_2(t, t+\tau_w)}{\wp_w^1(t)\wp_w^1(t+\tau_w)}; \tag{22.20}$$

and the waiting-time distribution in terms of the conditional exclusive probability densities:

$$\wp_w^1(t_0) = \frac{\wp_{m+1}^e(t_1, t_2, \ldots, t_m; [t_0, t_m])}{\wp_m^e(t_1, t_2, \ldots, t_m | t_0)}. \tag{22.21}$$

Therefore, the distribution of waiting times τ_w between a photoelectric emission at time t, and the next at time $t+\tau$ is given by:

$$\wp_w(\tau_w | t) := \wp_1^e(t+\tau_w | t) = \frac{\wp_2^e(t, t+\tau_w; [t, t+\tau_w])}{\wp_w^1(t)}. \tag{22.22}$$

[2]See also [ZOLLER *et al.* 1987].

Evaluation of Eq. (22.19) Now we try to evaluate Eq. (22.19) [CARMICHAEL *et al.* 1989] [*CARMICHAEL* 1993, 117–21]. The difficulties are the following:

- the average is taken over the system source + reservoir;
- the average is taken with the operators written in normal and time order, and they do not appear naturally ordered in this way in the same Eq.;
- it is a multi-time average.

To begin with we decompose the field at the detector into free field and source field components:

$$\hat{I}(t) = s_{\mathrm{D}}[\hat{a}_{\mathrm{F}}^{\dagger}(t) + \hat{a}_{\mathrm{S}}^{\dagger}(t)][\hat{a}_{\mathrm{F}}(t)\hat{a}_{\mathrm{S}}(t)], \tag{22.23}$$

where here $\hat{a}_{\mathrm{F}}(t) \propto \hat{E}_{\mathrm{F}}^{(+)}(t), \hat{a}_{\mathrm{S}}(t) \propto \hat{E}_{\mathrm{S}}^{(+)}(t)$ are field operators written in photon flux units. If the reservoir is in the vacuum state, the free field operators will contribute nothing to the average (22.19) because of the normal and time ordering. Thus in Eq. (22.19) we can substitute $s_{\mathrm{D}}\hat{a}_{\mathrm{S}}^{\dagger}(t)\hat{a}_{\mathrm{S}}(t)$ to $\hat{I}(t)$. However, although evaluation is made at different times, the trace remains over the initial state of the source and the reservoir.

Now we take care of the operator ordering. First we consider a waiting-time distribution (22.22) which involves only one exponential and then we make a generalization to arbitrary cases. The steps are:

- We expand the exponentials and write the field operators in explicit normal and time order.
- We write the resulting averages as a trace over an expression written in superoperator form as follows:

$$\mathrm{Tr}\left[\hat{\mathcal{J}} e^{\hat{\mathcal{L}}_{nd}(\tau_w - \tau_w^k)} \hat{\mathcal{J}} e^{\hat{\mathcal{L}}_{nd}(\tau_w^k - \tau_w^{k-1})} \ldots e^{\hat{\mathcal{L}}_{nd}(\tau_w^1)} \hat{\mathcal{J}} \hat{\rho}^{\mathcal{S}+\mathcal{E}}(t - r/c)\right], \tag{22.24}$$

  where the superoperator $\hat{\mathcal{L}}_{nd}$ is the nondissipative part of the superoperator defined by Eq. (21.3a), which acts on an arbitrary operator $\hat{O}$ as follows:

$$\hat{\mathcal{L}}_{nd}\hat{O} = \frac{\imath}{\hbar}[\hat{H}, \hat{O}], \tag{22.25a}$$

  where $\hat{H} = \hat{H}_{\mathcal{S}} + \hat{H}_{\mathcal{E}} + \hat{H}_{\mathcal{ES}}$, while the superoperator (the jump operator) $\hat{\mathcal{J}}$ acts as follows [see Eq. (18.5a)]:

$$\hat{\mathcal{J}}\hat{O} = \hat{a}_{\mathrm{S}}(r/c)\hat{O}\hat{a}_{\mathrm{S}}^{\dagger}(r/c). \tag{22.25b}$$

  Note that the source field is evaluated at time $t = r/c$ so that the source operators themselves will be evaluated at $t = 0$.

Later [see section 22.2] we shall discuss the physical interpretation of such a formalism. Note however that it is the quantum correspondent to the classical Montecarlo method. Numerically two techniques are available in order to integrate a function: either we subdivide the interval in N subintervals and calculate each one separately and then the sum, or we find at random N points and see if they fall under or above the curve and consider finally only the first ones. For two-dimensional spaces the two methods are equivalent and for $N \longrightarrow \infty$ both converge with high accuracy to the integral of the function. But for n–dimensional space, while the first method requires N^n calculations, the second one only requires something more than N itself.

- Then we resum the sums of integrals that came from the exponentials using the identity:

$$e^{(\hat{\mathcal{L}}_{nd}+a\hat{\mathcal{J}})\xi} = \sum_{k=0}^{\infty} a^k \int_0^{\xi} d\xi_k \int_0^{\xi_k} d\xi_{k-1} \cdots \int_0^{\xi_2} e^{\hat{\mathcal{L}}_{nd}(\xi-\xi^k)} \hat{\mathcal{J}} e^{\hat{\mathcal{L}}_{nd}(\xi^k-\xi^{k-1})} \hat{\mathcal{J}} \ldots e^{\hat{\mathcal{L}}_{nd}(\xi^1)} \hat{\mathcal{J}}, \quad (22.26)$$

and replace Eq. (22.19) by:

$$\begin{aligned} \wp_m^e(t_1, t_2, \ldots, t_m; [t, t+\tau]) &= s_{\mathrm{D}}^m \mathrm{Tr} \left[e^{(\hat{\mathcal{L}}_{nd}-s_{\mathrm{D}}\hat{\mathcal{J}})(t+\tau-t_m)} \hat{\mathcal{J}} \cdots \hat{\mathcal{J}} e^{(\hat{\mathcal{L}}_{nd}-s_{\mathrm{D}}\hat{\mathcal{J}})(t_2-t_1)} \hat{\mathcal{J}} \right. \\ & \left. \times e^{(\hat{\mathcal{L}}_{nd}-s_{\mathrm{D}}\hat{\mathcal{J}})(t_1-t)} \hat{\mathcal{J}} \hat{\rho}^{\mathcal{S}+\mathcal{E}}(t - r/c) \right]. \end{aligned} \quad (22.27)$$

- Now we wish to evaluate a trace over the source alone (the system $\mathcal{S}$). To this end, under the Markovian assumption that the future states depend only on the present and not on the past ones, we replace the superoperator $\hat{\mathcal{L}}_{nd}$ by the superoperator $\hat{\mathcal{L}}$, given by Eq. (21.3a), and the density operator by the reduced one $\hat{\rho}^{\mathcal{S}+\mathcal{E}}_{\mathrm{Tr}_{\mathcal{E}}}(t - r/c)$, obtaining finally:

$$\begin{aligned} \wp_m^e(t_1, t_2, \ldots, t_m; [t, t+\tau]) &= s_{\mathrm{D}}^m \mathrm{Tr} \left[e^{(\hat{\mathcal{L}}-s_{\mathrm{D}}\hat{\mathcal{J}})(t+\tau-t_m)} \hat{\mathcal{J}} \cdots \hat{\mathcal{J}} e^{(\hat{\mathcal{L}}-s_{\mathrm{D}}\hat{\mathcal{J}})(t_2-t_1)} \hat{\mathcal{J}} \right. \\ & \left. \times e^{(\hat{\mathcal{L}}-s_{\mathrm{D}}\hat{\mathcal{J}})(t_1-t)} \hat{\mathcal{J}} \hat{\rho}^{\mathcal{S}+\mathcal{E}}_{\mathrm{Tr}_{\mathcal{E}}}(t - r/c) \right] \end{aligned} \quad (22.28)$$

  This average is over source operators alone.

Hence parallel to $\hat{\rho}(t) = e^{\hat{\mathcal{L}}t}\hat{\rho}_0$ for an arbitrary density operator [see Eq. (21.1)], we have the result:

$$\begin{aligned} \hat{\rho}(t) &= e^{[(\hat{\mathcal{L}}-\hat{\mathcal{J}})+\hat{\mathcal{J}}]t} \hat{\rho}_0 \\ &= \sum_{m=0}^{\infty} \int_0^t dt_m \int_0^{t_m} dt_{m-1} \cdots \int_0^{t_2} dt_1 e^{(\hat{\mathcal{L}}-\hat{\mathcal{J}})(t-t_m)} \hat{\mathcal{J}} e^{(\hat{\mathcal{L}}-\hat{\mathcal{J}})(t_m-t_{m-1})} \hat{\mathcal{J}} \cdots e^{(\hat{\mathcal{L}}-\hat{\mathcal{J}})t_1} \hat{\rho}_0 \end{aligned}$$

22.1.4 Wave Function Approach

Some developments by using a wave function approach are to be found in [DALIBARD *et al.* 1992]. Take the simplest case of a two-level atom — ground ($|g\rangle$) and excited ($|e\rangle$) state — coupled to a monochromatic laser field in its ground state. The field is described by a classical function $E(t)\cos\omega_{\mathrm{F}}t$ (the subscript F stands for 'field'). The coupling can be written in the rotating wave approximation ($\hbar = 1$):

$$\hat{H}_0 = -\omega_{\mathrm{AF}}|e\rangle\langle g|g\rangle\langle e| + (\Omega/2)(|e\rangle\langle g| + |g\rangle\langle e|), \quad (22.29)$$

where $\omega_{\mathrm{AF}} = \omega_{\mathrm{L}} - \omega_A$ and the subscript A stays for 'atom'. Suppose that the compound system is described at time t by the wave function

$$|\Psi(t)\rangle = |\psi(t)\rangle \otimes |0\rangle = (c_g|g\rangle + c_e|e\rangle) \otimes |0\rangle, \quad (22.30)$$

where $|0\rangle$ is the ground state (no photon) of the field. At time $t+\tau$ photons may have been spontaneously emitted. We have assumed that at most one is emitted between t and $t+\tau$. The function at $t+\tau$ can then be written as

$$|\Psi(t+\tau)\rangle = |\Psi^{(0)}(t+\tau)\rangle + |\Psi^{(1)}(t+\tau)\rangle, \quad (22.31)$$

where:

$$|\Psi^{(0)}(t+\tau)\rangle = (c_g^{'}|g\rangle + c_e^{'}|e\rangle) \otimes |0\rangle, \tag{22.32a}$$

$$|\Psi^{(1)}(t+\tau)\rangle = |g\rangle \otimes \sum_{\mathbf{k},\epsilon} \vartheta_{\mathbf{k},\epsilon}|\mathbf{k},\epsilon\rangle. \tag{22.32b}$$

$\vartheta_{\mathbf{k},\epsilon}$ is the probability amplitude that a given mode $\mathbf{k}, \epsilon$ is populated after τ. ϵ is a pseudorandom number between 0 and 1 which reproduces the randomness of the results of the measurement. Depending on the result 0 or 1 of the measurement, the compound system is projected either onto (22.32a) or onto (22.32b). Assuming that the detected photon has been destroyed, we get for 0 result, i.e. for $\epsilon > \wp$ (where $\wp$ is the probability of spontaneous emission during τ):

$$|\Psi(t+\tau)\rangle = \mu(c_g^{'}|g\rangle + c_e^{'}|e\rangle) \otimes |0\rangle = \mu(1 - \imath\tau\hat{H})|s(t)\rangle \otimes |0\rangle, \tag{22.33a}$$

and for 1 result, i.e. for $\epsilon < \wp$:

$$|\Psi(t+\tau)\rangle = |g\rangle \otimes |0\rangle, \tag{22.33b}$$

with $\mu = (1-\wp)^{-\frac{1}{2}}$. In both cases we return to a wave function with the same form as (22.30), i.e. an atomic part times the vacuum, and the whole sequence can be repeated to determine the (random) time evolution of the atomic part of the wave function.

The advantage of such a method relatively to the other previously analysed, is that, by using the density operators, we normally need $N^2/2$ elements in order to describe the evolution of a system, while here we only need something more than N.

22.2 Interpretation of Quantum Jumps and Trajectories

But now we can ask: is there a physical interpretation of Eq. (22.28)? The density operator evolves during the interval $t_1 - t$, when there are no photon emissions, under the propagator $e^{(\hat{\mathcal{L}} - s_{\mathrm{D}}\hat{\mathcal{J}})(t_1-t)}$, then collapses under the action of $\hat{\mathcal{J}}$ at the time of the first emission, evolves the next interval without photon emission under the propagator $e^{(\hat{\mathcal{L}} - s_{\mathrm{D}}\hat{\mathcal{J}})(t_2-t_1)}$ and collapses again under the action of $\hat{\mathcal{J}}$, and so on [*CARMICHAEL* 1993, 113–14, 120–21].

It is interesting that this method has analogies with the path-integral method developed by Feynman [see subsection 3.11.2]. In fact, what we are doing is to generate a *trajectory* for the reduced density operator from $\hat{\rho}^{\mathcal{S}+\mathcal{E}}_{\mathrm{Tr}_{\mathcal{E}}}(t - r/c)$ to $\hat{\rho}^{\mathcal{S}+\mathcal{E}}_{\mathrm{Tr}_{\mathcal{E}}}(t+\tau-r/c)$. The interest of such a method is that we can parallel the quantum and the classical cases: the latter one can in fact be described either by a probability distribution and a Fokker/Planck Eq. (23.3) or by an ensemble of noisy trajectories and a set of stochastic Eqs.. In the QM case, instead of the classical probability distribution, we have the Master Eq.; what we have shown is the equivalence between the master Eq. and the quantum trajectory approaches[3].

Now, some physicists[4] saw in the theory of quantum Jumps or in some analogy of it a new possibility to repropose a stochastic dynamics with spontaneous localisations or, more generally, transitions to eigenstates. But the QM-jumps formalism has been shown to be rather a numerical technique. In fact in QM, due to the Superposition principle, one cannot have real 'trajectories' as in the classical case. Also Feynman graphs are more diagrammatic descriptions

[3]Concerning the relationship between Master Eq. and Fokker/Planck Eq. for quasi-probability distributions see [ISAR *et al.* 1991].

[4]For example Shimony [SHIMONY 1991] or Breuer and Petruccione [H. BREUER/PETRUCCIONE 1995a] [H. BREUER/PETRUCCIONE 1995b].

of calculation procedures in order to predict measurement results[5]. Hence one cannot ascribe to the jumps the character which the aforementioned physicists wish to ascribe without some additional physical assumptions [see methodological principle M5: p. 6]. As we shall see [chapter 23] such assumptions are discussible for many reasons.

On the other hand, the idea that a quantum jump theory can be used in order to explain some more general effects[6] will be the object of a later analysis [see postulate 33.1: p. 567]. Let us now discuss some concrete physical model of the preceding formalism.

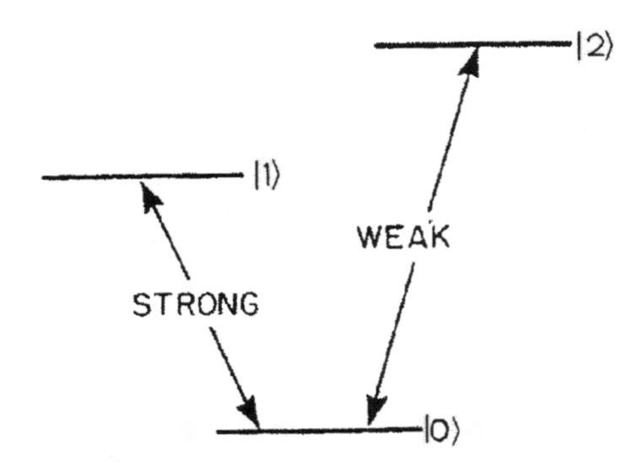

Figure 22.2: Energy-level scheme for single-atom double-resonance experiment — from [COOK/KIMBLE 1985, 1023].

22.3 Models and Experiments: Jumps as Telegraph Signals

First concrete models and numerical simulations of the quantum jumps were presented in[7] [COOK/KIMBLE 1985] [JAVANAINEN 1986] [SCHENZLE/BREWER 1986] and in [COHEN-T./DALIBARD 1986]. In the following we study in particular the model proposed in [COOK/KIMBLE 1985]: Cook and Kimble understood quantum jumps as some type of telegraph signals. The *Gedankenexperiment* (Atomic-Level Exp. 2) is originally due to Dehmelt [DEHMELT 1975]:[8] an atom can perform a strong transition from level $|0\rangle$ to level $|1\rangle$ ($\simeq 10^8$ photons/sec) or a weak transition (1 photon/sec) from the same level $|0\rangle$ to a level $|2\rangle$ [see figure 22.2].

If the atom is excited to level $|1\rangle$, we will detect a fluorescence intensity (irradiance) I_0, which, due to the high emission of photons, can be approximately considered constant in time. Now we let radiation be applied to the weak transition and let the saturating field continue to act on the strong one. Hence the detector will continue to register the intensity I_0 until the atom makes a transition to level $|2\rangle$. This turns off the strong fluorescence because the atom is no longer available for transition between $|0\rangle$ and $|1\rangle$. In consequence of this situation the atomic fluorescence $I(t)$ has the form of a random telegraph signal [see figures 22.3 and

[5]See [*D'ESPAGNAT* 1995, 316–18].
[6]As in [BLANCHARD/JADCZYK 1995].
[7]See also [KIMBLE *et al.* 1986].
[8]See also [SCHENZLE *et al.* 1986] [PEGG/KNIGHT 1988a].

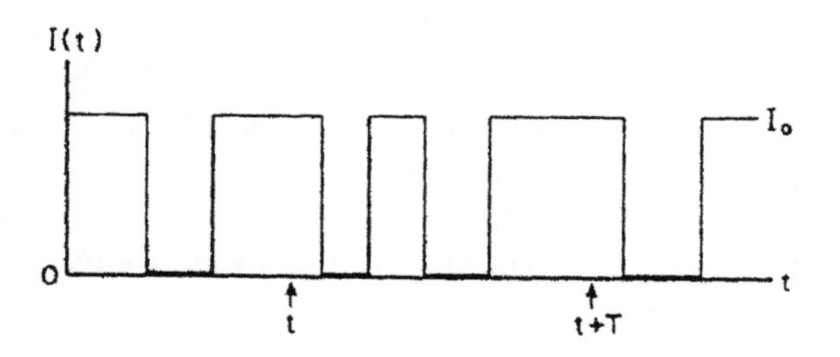

Figure 22.3: Single-atom fluorescent intensity versus time. Interruptions of fluorescence are due to excitations of the weak transition $|0\rangle \rightsquigarrow |2\rangle$ — from [COOK/KIMBLE 1985, 1023].

22.4]. An experimental realization of these ideas can be found in [BERGQUIST *et al.* 1986] [ITANO *et al.* 1987] and in [SAUTER *et al.* 1986].[9]

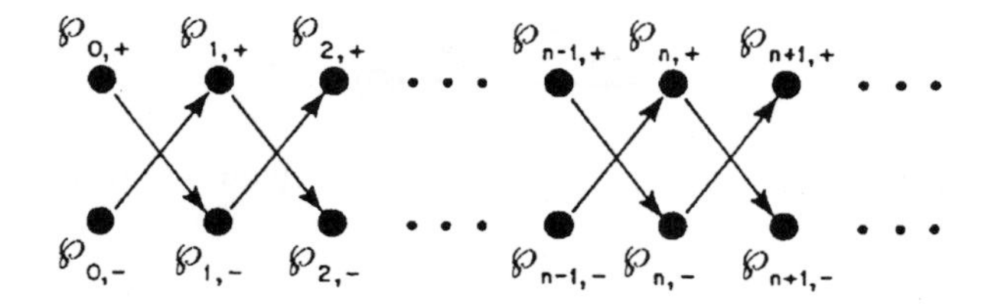

Figure 22.4: Solid dots represent probabilities $\wp_{n,\pm}(\tau)$ that n transitions occur in time τ and the atom is left in state $\pm$ (allowed states are $-,+$). Arrows indicate the flow of probability as τ increases — from [COOK/KIMBLE 1985, 1024].

[9]See also an interesting photon antibunching effect reported in [GRANGIER *et al.* 1986c]. Antibunching cannot be accounted for in terms of classical fields [KIMBLE *et al.* 1977]. Several examples of applications of the quantum-jump formalism can be found in Carmichael's textbook [*CARMICHAEL* 1993]: by means of such examples the power of this formalism can be better understood.

Chapter 23

REDUCTIONS TO CLASSICAL CASE

Introduction and contents In this chapter we collect all interpretations which tried and try to reduce quantum behaviour to some classical behaviour. The first proposal to solve the apparent conflict between QM and macrophysics was to consider QM as an incomplete theory which constitutes a part of some classical stochastic theory. The accent on the incompleteness of QM makes this interpretation a further development of the Ignorance interpretation [proposition 6.3] and an historical antecedent of Hidden-Variables theories [chapter 32]. To distinguish the stochastic theories examined in section 23.1 from those analysed in chapter 18 [see also chapter 33] we call (in our QM context) the first ones *Classical Stochastic Theories*: they try to lead QM to some classical stochastic theory or to derive QM from such theory.

There also are a lot of theoretical constructions which partly parallels the Master equation approach [chapter 21] and the Montecarlo theory [chapter 22], but whose nature is quite different from both. We name them collectively *Reduction Jump theories*. We place them into the following two families of theories whose aim is to obtain the required spontaneous localization-jumps:

- theories which add some parameters to QM [section 23.2];
- theories which propose a stochastic correction to QM unitary dynamics (normally a form of non-linearity) [section 23.3].

23.1 Classical Stochastic Theories

23.1.1 Hydrodynamics and QM

The first proposal of a hydrodynamic model of QM is due to Madelung [MADELUNG 1926]. The proposal was also developed in [KORN 1927].[1] This approach is an antecedent of the Pilot wave theory [see section 28.1] and, as we have said, has some further developments in the Hidden-Variable context [see subsection 32.6.2]; hence we do not discuss it in detail, but limit ourselves to a pair of fundamental questions.

The main conceptual difficulty of such an approach derives from the fact that it supposes an incompressible fluid, while for QM only a compressible one would be adequate.

It is also difficult to think of the density operator as a hydraulic density: hydrodynamics is a theory which works without the assumption of particles, while here one uses it to explain the corpuscular microreality.

[1] For history see [*JAMMER* 1974, 35–38].

23.1.2 The Fokker Equation

In the following subsections we shall study some models centered on Brownian motion: one supposes here that QM is intrinsically stochastic. As a general feature all the following proposals make use of a model with some sort of Brownian 'particles' in a fluid. The similarity between microrealities and Brownian motion was already acknowledged by Schrödinger [SCHRÖDINGER 1931a] [SCHRÖDINGER 1932], and developed by Fürth [FÜRTH 1933].[2]

Fényes' proposal Fényes showed [FÉNYES 1952a, 89–90] that the Fokker Eq.:

$$\frac{\partial \varrho}{\partial t} = \nabla \cdot \varrho \mathbf{v} - D\nabla^2 \varrho, \tag{23.1}$$

where $\mathbf{v}$ is the velocity of the macroscopic current, D is the diffusion coefficient of the fluid, and ϱ is the probability density of the distribution of Brownian particles, though formally different from the Schrödinger Eq., it is always the valid in QM but by taking $\psi^*\psi$ and not ψ. The similarity is particularly evident if we eliminate the exterior force and in both cases obtain [see Eq. (3.16)]:

$$\imath\hbar\frac{\partial}{\partial t}|\psi\rangle = -\frac{\hbar^2}{2m}\frac{\nabla^2}{\partial x^2}|\psi\rangle, \tag{23.2a}$$

$$\frac{\partial \varrho}{\partial t} = D\nabla^2 \varrho. \tag{23.2b}$$

Thus the argument of Schrödinger, after which his Eq. is reversible but the Fokker Eq. is not, fails because, after Fényes, both in QM and in classical mechanics $\psi^*\psi$ is irreversible and ψ is not.

Criticisms The proposal was contested by Weizel [WEIZEL 1953a] on the grounds of an analysis of a particular equilibrium problem in which forces balance the diffusion velocity, and produce a stationary distribution in contradiction with Fényes' result.

Another criticism is to be found in [NICHOLSON 1954] where it is shown that the Lagrangian used by Fényes is *ad hoc*, and the assumed Eq. does not lead to QM without supplementary assumptions[3].

The Fokker/Planck Eq. However, the efforts of Fényes are very important for having contributed to stressing the importance of the Fokker/Planck Eq. (which, as we have seen, is useful in quantum thermodynamics and Q-optics), which in terms of the P-function [see subsection 5.2.2] can be written[4]:

$$\frac{\partial \mathrm{P}(\alpha)}{\partial t} = \left(\frac{\partial}{\partial \alpha_j}A_j\alpha_j + \frac{1}{2}\frac{\partial^2}{\partial \alpha_j \alpha_k}D_{jk}\right)\mathrm{P}(\alpha), \tag{23.3}$$

where A is the drift matrix (the first derivative term represents the deterministic motion), and D is the diffusion matrix (the second derivative term represents the broadening of $\mathrm{P}(\alpha)$).

[2] See [*JAMMER* 1974, 419–22].

[3] On this problem see also [*JAMMER* 1974, 426].

[4] See [*WALLS/MILBURN* 1994, 101, 133].

23.1.3 A Development of the Probabilistic Interpretation

Bopp [BOPP 1947] [BOPP 1952] [BOPP 1953] [BOPP 1954] found a correlation probability which could be split once into two factors, obtaining results in agreement with quantum Gibbs ensembles. Hence he conceived of the waves as properties of ensemble of particles.

He proposed to map each QM system in a statistical ensemble of particles in a phase space so that every QM process would correspond to a movement of this ensemble.

There is no doubt that the point of departure is Born's interpretation [proposition 6.2: p. 104], and it is worth mentioning that Born himself approved the proposal[5]. We already know the difficulties of a statistical interpretation [see subsection 6.6.2]. Later we shall return to the problem of corpuscular interpretation of QM [see parts VII and VIII, particularly subsection 32.6.2].

23.1.4 Nelson's Proposal

We now analyse the most developed proposal of this family of theories in more detail.

Brownian-Motion Formalism Nelson's proposal [NELSON 1966] [6] represent a direct and explicit derivation[7] of Bohm's study [BOHM 1952].[8] It is known that stochastic processes (which we symbolize in the following by $x(t)$) are very often not differentiable (as in the case of Wiener process). What we must do is introduce a mean forward and a mean backward derivative:

$$\overline{d}x(t) = \lim_{\delta t \longrightarrow 0+} \left\langle \frac{x(t+\delta t) - x(t)}{\delta t} \right\rangle_t , \tag{23.4a}$$

$$\overline{d}^* x(t) = \lim_{\delta t \longrightarrow 0+} \left\langle \frac{x(t) - x(t-\delta t)}{\delta t} \right\rangle_t , \tag{23.4b}$$

where $\langle\rangle_t$ is the conditional expectation value, given the state of the system $\mathcal{S}$ at time t. We now define the stochastic acceleration by:

$$a(t) = \frac{1}{2}\overline{d}\overline{d}^* x(t) + \frac{1}{2}\overline{d}^*\overline{d}x(t). \tag{23.5}$$

Using this formalism we can write a forward and a backward Fokker Eq. [see Eq. (23.1)]:

$$\frac{\partial \varrho}{\partial t} = -\nabla \cdot \varrho \mathbf{b} + D\nabla^2 \varrho, \tag{23.6a}$$

$$\frac{\partial \varrho}{\partial t} = -\nabla \cdot \varrho \mathbf{b}^* - D\nabla^2 \varrho, \tag{23.6b}$$

where $\mathbf{b}$ is a function such that the current velocity is expressed:

$$\mathbf{v} = \frac{1}{2}(\mathbf{b} + \mathbf{b}^*), \tag{23.7}$$

[5] See also [BOPP 1961b] [*JAMMER* 1974, 428].

[6] See also [*PARISI* 1988, 334–36].

[7] As it is acknowledged by Nelson himself [NELSON 1966, 1085].

[8] In [BOHM/VIGIER 1954, :209] a theory of fluid is proposed where the particles are inhomogeneities sharp localized: see also [BOHM/HILEY 1981, 532]. See also subsection 32.6.2

so that the mean between both Eqs. (23.6) gives:

$$\frac{\partial \varrho}{\partial t} = -\nabla \cdot \varrho \mathbf{v}. \tag{23.8}$$

On the other hand, we define the osmotic velocity $\mathbf{v_o}$ as the velocity acquired by a Brownian particle in equilibrium with respect to an external force, to balance the osmotic force. It is given by:

$$\mathbf{v_o} = \frac{1}{2}(\mathbf{b} - \mathbf{b}^*) = D\frac{\nabla \varrho}{\varrho} = D\nabla \ln \varrho, \tag{23.9}$$

in the light of which we write the following evolution Eq.:

$$\frac{\mathbf{v_o}}{\partial t} = -D\nabla(\nabla \cdot \mathbf{v}) - \nabla(\mathbf{v} \cdot \mathbf{v_o}). \tag{23.10}$$

By using Eq. (23.5) we write:

$$\mathbf{a} = \frac{1}{2}\frac{\partial}{\partial t}(\mathbf{b} + \mathbf{b}^*) + \frac{1}{2}(\mathbf{b} \cdot \nabla)\mathbf{b}^* + \frac{1}{2}(\mathbf{b}^* \cdot \nabla)\mathbf{b} - \frac{1}{2}D\nabla^2(\mathbf{b} - \mathbf{b}^*). \tag{23.11}$$

Since by Eq. (23.7) and Eq. (23.9) we have:

$$\mathbf{b} = \mathbf{v} + \mathbf{v_o}, \quad \mathbf{b}^* = \mathbf{v} - \mathbf{v_o}, \tag{23.12}$$

then Eq. (23.11) can be rewritten as follows:

$$\frac{\partial \mathbf{v}}{\partial t} = \mathbf{a} - (\mathbf{v} \cdot \nabla)\mathbf{v} + (\mathbf{v_o} \cdot \nabla)\mathbf{v_o} + D\nabla^2\mathbf{v_o}. \tag{23.13}$$

Quantum Case By treating the quantum case one assumes:

$$D = \frac{\hbar}{2m}. \tag{23.14}$$

Note that this is an *ad hoc* assumption, which weakens the whole deduction [see methodological principle M4: p. 6]. By means of assumption (23.14) we can rewrite Eqs. (23.10) and (23.13) as follows:

$$\begin{aligned} \frac{\mathbf{v_o}}{\partial t} &= -\frac{\hbar}{2m}\nabla(\nabla \cdot \mathbf{v}) - \nabla(\mathbf{v} \cdot \mathbf{v_o}), & (23.15\text{a}) \\ \frac{\partial \mathbf{v}}{\partial t} &= \frac{\mathbf{F}}{m} - (\mathbf{v} \cdot \nabla)\mathbf{v} + (\mathbf{v_o} \cdot \nabla)\mathbf{v_o} + \frac{\hbar}{2m}\nabla^2\mathbf{v_o}. & (23.15\text{b}) \end{aligned}$$

Suppose that force is given by some potential: $\mathbf{F} = -\nabla V$, and that $\mathbf{v} = 0$, so that [by Eqs. (23.8) and (23.9)] ϱ and $\mathbf{v_o}$ are independent of t and Eq. (23.15a) says that $\mathbf{v_o}/\partial t = 0$ while Eq. (23.15b) becomes:

$$\mathbf{v_o} \cdot \nabla\mathbf{v_o} + \frac{\hbar}{2m}\nabla^2\mathbf{v_o} = \frac{\nabla V}{m}. \tag{23.16}$$

By Eq. (23.9) $\mathbf{v}_o$ is a gradient, so that $\mathbf{v_o} \cdot \nabla\mathbf{v_o} = 1/2\nabla^2\mathbf{v_o}$ and $\nabla^2 = \nabla(\nabla \cdot \mathbf{v_o})$. Thus Eq. (23.16) becomes:

$$\nabla\left(\frac{1}{2}\mathbf{v_o}^2 + \frac{\hbar}{2m}\nabla \cdot \mathbf{v_o}\right) = \frac{\nabla V}{m} \tag{23.17}$$

or

$$\frac{1}{2}\mathbf{v_o}^2 + \frac{\hbar}{2m}\nabla \cdot \mathbf{v_o} = \frac{V}{m} - \frac{E}{m}, \tag{23.18}$$

where E is a constant with the dimension of energy. Since it can be shown — by multiplying Eq. (23.18) by $m\varrho$ and integrating — that $E = 1/2m\mathbf{v_o}^2 + V$, it follows that E can be interpreted as the mean energy of the particle. In this case $\mathbf{b} = -\mathbf{b}^* = \mathbf{v_o}$, so that we could replace $\mathbf{v_o}$ by $\mathbf{b}$ or $\mathbf{b}^*$. Eq. (23.18) is non-linear, but it is equivalent to a linear one by a change of the dependent variable. By Eq. (23.9) we pose $\varphi = \frac{1}{2}\ln\varrho$, which is the potential of $m\mathbf{v_o}/\hbar$. Let [see Eq. (3.21)]:

$$\psi = e^{\varphi}. \tag{23.19}$$

Then ψ is real and $\varrho = |\psi|^2$. In conclusion Eq. (23.18) is equivalent to the time dependent Schrödinger Eq. [see Eq. (3.18)]:

$$\left[-\frac{\hbar^2}{2m}\nabla^2 + V - E\right]\psi = 0 \tag{23.20}$$

for a real ψ.

Nelson [*NELSON* 1967, 139–40] acknowledged that some aspects of its Brownian theory are incompatible with QM, but said that there are no contradictions between the measurable predictions of the one and of the other. Other similar developments are thanks to De La Peña-Auerbach [DE LA PEÑA-A. 1969].[9]

23.1.5 General Criticism

There are some results which stem from the Classical Stochastic Theories (the Fokker/Planck Eq.) which can be used in specific experimental or theoretical contexts. There also are many problems in respect to the proposed interpretation.

The greatest problem was posed by Bohm himself: it is impossible to construct a theory which reproduces QM and which is classical in all respects [*BOHM/HILEY* 1993, 200–202].

In fact the osmotic velocity of Nelson's model already presupposes a quantum field. Anyway, in the case of a multiparticle system, we have a typical quantum non-separability phenomenon [see part IX]: the osmotic velocities of different particles are interrelated, and this is impossible to explain by classical means.

More generally, as was already proved [see p. 107 and also part IX], there is never an ensemble of classical states that is adequate to describe the totality of properties of a proper quantum system [SCHRÖDINGER 1935a, 488].

In any case, it is the very idea of Brownian 'particles' moving in a current which cannot be completely adequate. In fact, as we shall see:

- even in the macroscopic world the quantum effects never cease acting [see chapter 24], so that one cannot use a 'macro'-model in order to explain QM without exhibiting specific physical reasons which are able to justify it;
- the relationship between corpuscular and wave-like behaviour cannot be accounted for in terms of Brownian motion [see chapter 30].

[9] On this point see also [*JAMMER* 1974, 433–35].

As we have said, although classical stochastic theories are not generalisable, they can be used in specific contexts: the Madelung approach, or similar ones, can be interesting for some semi-classical situations [GUERRA 1981], while, more recently, following the Fokker formalism, other proposals have been made [MANCINI *et al.* 1997b] based on simplectic tomography procedures in order to obtain an estimate of the initial quantum state [see subsection 28.5.2].

23.2 Ghirardi/Rimini/Weber

Among the theories which try to explain the reduction of the wave packet by spontaneous jumps which are produced by adding some parameters to QM, perhaps the best known one is the model of Ghirardi/Rimini/Weber (= GRW).

23.2.1 Exposition of the Model

The central idea of GRW is the introduction of some nature constants in consequence of which QM systems sometimes jump spontaneously to a 'localized' state.

Authors start with the dynamical Eq. proposed in 1982 by Barchielli, Lanz and Prosperi [BARCHIELLI *et al.* 1982] [see Eq. (21.7)]:

$$\frac{d}{dt}\hat{\rho}(t) = -\frac{\imath}{\hbar}\left([\hat{H}, \hat{\rho}(t)] - \frac{\eta}{4}[\hat{q}, [\hat{q}, \hat{\rho}(t)]]\right). \tag{23.21}$$

For a N–particles compound system Eq. (23.21) is rewritten by GRW [GHIRARDI *et al.* 1986, 477–80] as follows[10]:

$$\frac{d}{dt}\hat{\rho} = -\frac{\imath}{\hbar}[\hat{H}, \hat{\rho}] - \sum_{j=1}^{N} \eta_j \left(\hat{\rho} - \mathcal{T}_j[\hat{\rho}_j]\right), \tag{23.22}$$

where $\hat{\rho}, \hat{\rho}_j$ are, respectively, the density operators for the total system and for the j–th particle, η_j is the frequency of the process undergone by constituent j and where the transformation $\mathcal{T}_j$ [see subsection 18.1.1] acts as follows:

$$\mathcal{T}_j[\hat{\rho}_j] = \sqrt{\frac{\zeta}{\pi}} \int_{-\infty}^{+\infty} dx \; e^{-\frac{\zeta}{2}(\hat{q}_j - x)^2} \hat{\rho} e^{-\frac{\zeta}{2}(\hat{q}_j - x)^2}, \tag{23.23}$$

where ζ is a localization constant to be specified.

If we introduce the center of mass position operator $\hat{Q}$ and the position operator $\hat{r}_k, (k = 1, 2, \ldots, N-1)$ related to $\hat{q}_j$ by:

$$\hat{q}_j = \hat{Q} + \sum_{k=1}^{N-1} c_{jk}\hat{r}_k, \tag{23.24}$$

then we can write the operation $\mathcal{T}_j[\hat{\rho}_j]$ as follows:

$$\mathcal{T}_j[\hat{\rho}_j] = \sqrt{\frac{\zeta}{\pi}} \int_{-\infty}^{+\infty} dx \; \exp\left[-\frac{\zeta}{2}\left(\hat{Q} + \sum_{k=1}^{N-1} c_{jk}\hat{r}_k - x\right)^2\right] \hat{\rho} \exp\left[-\frac{\zeta}{2}\left(\hat{Q} + \sum_{k=1}^{N-1} c_{jk}\hat{r}_k - x\right)^2\right]; \tag{23.25}$$

[10] See also [BELL 1987a, 202–204] [*BOHM/HILEY* 1993, 326–29].

and Eq. (23.22) as follows:

$$\frac{d}{dt}\hat{\rho} = -\frac{\imath}{\hbar}[\hat{H}_{\hat{Q}}, \hat{\rho}] - \frac{\imath}{\hbar}[\hat{H}_{\hat{r}}, \hat{\rho}] - \sum_{j=1}^{N} \eta_j(\hat{\rho} - \mathcal{T}_j[\hat{\rho}_j]). \tag{23.26}$$

The dynamical evolution of the center of mass of the system is described by

$$\hat{\rho}_Q = \mathrm{Tr}_r(\hat{\rho}), \tag{23.27}$$

obtained by taking the partial trace over the internal degrees of freedom of $\hat{\rho}$ for the complete N–particle system. By shifting the integration variable x by an amount $\sum_{k=1}^{N-1} c_{jk}\hat{r}_k$ one finds:

$$\mathrm{Tr}_r(\mathcal{T}_j[\hat{\rho}_j]) = \mathcal{T}_Q[\mathrm{Tr}_r\hat{\rho}], \tag{23.28}$$

where

$$\mathcal{T}_Q[\mathrm{Tr}_r\hat{\rho}] = \sqrt{\frac{\zeta}{\pi}} \int_{-\infty}^{+\infty} dx e^{-\frac{\zeta}{2}(\hat{Q}-x)^2} \hat{\rho} e^{-\frac{\zeta}{2}(\hat{Q}-x)^2}. \tag{23.29}$$

By having already performed the partial trace over r on Eq. (23.26), one obtains the following evolution of the reduced density matrix:

$$\frac{d}{dt}\hat{\rho}_{\hat{Q}} = -\frac{\imath}{\hbar}[\hat{H}_{\hat{Q}}, \hat{\rho}_{\hat{Q}}] - \sum_{j=1} \eta_j(\hat{\rho}_{\hat{Q}} - \mathcal{T}_{\hat{Q}}[\hat{\rho}_{\hat{Q}}]). \tag{23.30}$$

Now we assume that the matrix elements $\langle Q', r'|\mathcal{T}_j[\hat{\rho}]|Q'', r''\rangle$ are non-negligible only when the following conditions are satisfied for all $j = 1, \ldots, N$:

$$\left|\sum_{k=1}^{N-1} c_{jk}\hat{r}'_k - a_j\right| \ll \frac{1}{\sqrt{\zeta}}, \tag{23.31a}$$

$$\left|\sum_{k=1}^{N-1} c_{jk}\hat{r}''_k - a_j\right| \ll \frac{1}{\sqrt{\zeta}}, \tag{23.31b}$$

where a_j is the equilibrium position of particle i relative to the center of mass. The conditions (23.31) imply:

$$\left|\sum_{k=1}^{N-1} c_{jk}(\hat{r}'_k - r''_k)\right| \ll \frac{1}{\sqrt{\zeta}}, \tag{23.32}$$

from which it follows that $\langle Q', r'|\hat{\rho}|Q'', r''\rangle$ is negligibly small unless condition (23.32) is satisfied. From Eq. (23.25) one gets:

$$\begin{aligned}\langle Q', r'|\mathcal{T}_j[\hat{\rho}]|Q'', r''\rangle &= \sqrt{\frac{\zeta}{\pi}} \int_{-\infty}^{+\infty} dx \exp\left[-\frac{\zeta}{2}\left(\hat{Q}' + \sum_{k=1}^{N-1} c_{jk}\hat{r}'_k - x\right)^2\right] \langle Q', r'|\mathcal{T}_j[\hat{\rho}]|Q'', r''\rangle \\ &\times \exp\left[-\frac{\zeta}{2}\left(\hat{Q}'' + \sum_{k=1}^{N-1} c_{jk}\hat{r}''_k - x\right)^2\right] \\ &= \exp\left[-\frac{\zeta}{2}\left(\hat{Q}' - \hat{Q}'' + \sum_{k=1}^{N-1} c_{jk}(\hat{r}'_k - \hat{r}''_k)\right)^2\right] \langle Q', r'|\mathcal{T}_j[\hat{\rho}]|Q'', r''\rangle. \end{aligned} \tag{23.33}$$

Because of Eq. (23.32) the displacement of the Gaussian in the variables $\hat{Q}' - \hat{Q}''$ (the exponential factor in the RHS) can be neglected with respect to its width, so that we obtain in this approximation:

$$\mathcal{T}_j[\hat{\rho}] = \mathcal{T}_Q[\hat{\rho}], \tag{23.34}$$

which signifies that the localization of a single component is equivalent to the localization of the center of mass. If the system is initially in a product state $\hat{\rho}_r\hat{\rho}_Q$, it remains in a state of the same type, and, while $\hat{\rho}_Q$ obeys Eq. (23.30), $\hat{\rho}_r$ obeys the following:

$$\frac{d}{dt}\hat{\rho}_r = -\frac{\imath}{\hbar}[\hat{H}_r, \hat{\rho}_r], \tag{23.35}$$

so that even though the internal motion of the system remains unaffected by the localization process, the center-of-mass motion is affected by it with a characteristic frequency equal to the sum of the frequencies of all constituents.

The authors give the following values to the introduced constants:

$$\eta \simeq 10^{-16}\text{sec}^{-1}, \tag{23.36a}$$

$$\frac{1}{\zeta} \simeq 10^{-5}\text{cm}. \tag{23.36b}$$

Condition (23.36a) means that the localization happens spontaneously every $10^8 - 10^9$ years for a single particle. Obviously the localization time-rate is shorter for greater systems and is shorter the larger the system is. On the other hand, the value of $1/\zeta$ is much larger than the length of a de Broglie wave.

This means that both constants are very important only for big systems: hence they leave almost unaffected the microworld of QM, but allow description of how localized and 'reduced' macrobjects are possible.

23.2.2 Discussion and Criticisms

The value of such a model has been to show that it is possible to account for macrophenomena in terms of the formalism of QM. On the other hand, there the specific contents and methodology of the model are not entirely satisfactory. Let us briefly discuss, in that which follows, some problems related to the model.

- The first problem of such a proposal is that the hypothesis is *ad hoc* [see methodological principle M4: p. 6]: in other words the authors are unable to explain where the values of η and ζ should come from, and why. As a consequence authors exhibit no physical mechanism at all able to explain such a (GRW) jump.

- Another problem [JOOS 1987, 3285] is that the starting point of GRW [Eq. (23.21)] is formally similar to a decoherence Master Eq. [see Eq. (21.7)]. Hence we do not need a new dynamic in order to explain how localization processes are produced.

- Finally [JOOS 1987, 3286], due to the non-unique decomposition of the density matrix [see corollary 3.3: p. 46], the suppression of off-diagonal terms in the density matrix of the center-of-mass is not a sufficient condition to extract single particle trajectories.

The above criticisms do not exclude the use of the above formalism for some specific situations. They only exclude its use as a general interpretation of the localization process.

23.3 Stochastic Non-Linear Jump-Theories

23.3.1 Brief Exposition

Generally such a family of theories follows a basic idea of Pearle [PEARLE 1976] [PEARLE 1984], which has been further developed by Gisin [GISIN 1984] [GISIN 1989].[11] The basic idea is to change the unitary Schrödinger dynamics by adding a non-Hamiltonian term which accounts for spontaneous transitions to localisations and more generally to eigenstates. Gisin is motivated [GISIN/PERCIVAL 1993b] by the irrelevancy of de Broglie waves in a lot of experimental situations — such as Two-Slit-Exp.s, where the detected objects are always sharply confined and hence show a corpuscular behaviour[12].

It is assumed to be a stochastic diffusion of quantum state $|\psi\rangle$, by using a Master Eq. [see section 21.1]. Then one writes the master Eq. in the following stochastic non-linear form [GISIN/PERCIVAL 1992a, 316] [GISIN/PERCIVAL 1992b, 5678] [GISIN/PERCIVAL 1993a, 2245] [see Eqs. (21.3)] — the Langevin/Itô Eq.[13] —:

$$d|\psi\rangle = -\frac{\imath}{\hbar}\hat{H}|\psi\rangle dt + \sum_m \left(\left\langle \hat{V}_m^\dagger \right\rangle_\psi \hat{V}_m - \frac{1}{2}\hat{V}_m^\dagger \hat{V}_m - \frac{1}{2}\left\langle \hat{V}_m^\dagger \right\rangle_\psi \left\langle \hat{V}_m \right\rangle_\psi \right)|\psi\rangle dt + \sum_m \left(\hat{V}_m - \left\langle \hat{V}_m \right\rangle_\psi \right)|\psi\rangle d\xi_m, \tag{23.37}$$

where $\left\langle \hat{V}_m \right\rangle_\psi = \left\langle \psi \left| \hat{V}_m \right| \psi \right\rangle$ is the expectation of $\left\langle \hat{V}_m \right\rangle$ for the state $|\psi\rangle$, $d\xi_m$ represents the Itô form of the complex normalized Wiener increment that satisfies[14]:

$$\mathrm{Re}(d\xi_m)\mathrm{Re}(d\xi_n) = \mathrm{Im}(d\xi_m)\mathrm{Im}(d\xi_n) = \delta_{nm} dt, \tag{23.38a}$$

$$\mathrm{Re}(d\xi_m)\mathrm{Im}(d\xi_n) = 0, \tag{23.38b}$$

23.3.2 Difficulties

We do not wish to go into a deep analysis of these models but only to show that they present several difficulties.

- We have already discussed the theoretical question concerning whether the quantum jumps can be interpreted as reductions of wave-packet as such. On the basis of the models discussed in section 22.3, it can be shown [JOOS 1996, 115] that such an interpretation is not sound: in fact, while a collapse happens at every measurement, a jump is only a change of the measurement result and occurs at random times [see figure 23.1].
- In particular the non-linearity can be criticized [see section 28.2].
- One the other hand, calculations have been made [S. BOSE *et al.* 1998b], based on a *Gedankenexperiment* of field in a cavity with a movable mirror (which can be considered here as a harmonic oscillator).[15] By a direct confrontation of a decoherence — Zurek-like

[11] See also [GISIN 1991b]. For other proposals and further analysis see also [DIÓSI 1987] [DIÓSI 1988] [DIÓSI 1989] [*DIÓSI/LUKÁCZ* 1994] [GOETSCH *et al.* 1996].

[12] See also for a more complete exposition [PERCIVAL 1994].

[13] For the mathematics see [ITÔ 1951]; for physical applications see [HASEGAWA/EZAWA 1980].

[14] For a non-Markovian generalization of such stochastic Eqs. See also [DIÓSI/STRUNZ 1997].

[15] For the experiment see [S. BOSE *et al.* 1997] and also [MANCINI *et al.* 1997a] .

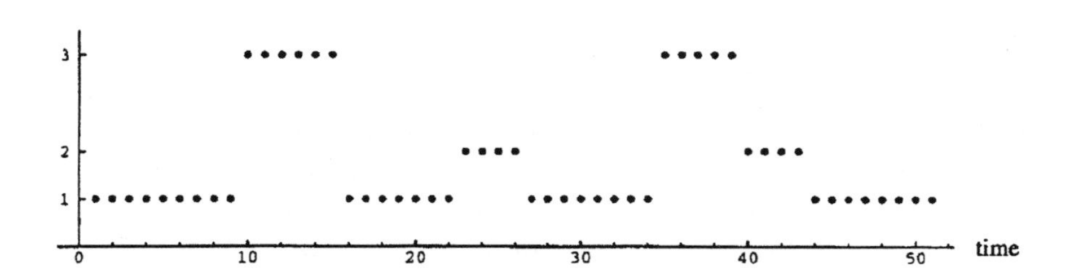

Figure 23.1: While a measurement collapse (following stochastic jump theories) corresponds to a single dot, a jump is characterized by the abrupt change in the sequence of aligned dots (the first one happens at $t = 10$) — from [JOOS 1996, 116].

— model and a Diósi-like jump model [DIÓSI 1989], it turns out that in the latter case one needs an improvement of some parameters by sixteen orders of magnitude, which, if it does not exclude the latter model, makes it very improbable, particularly if one considers how efficiently decoherence parameters work[16].

- Finally, if the point of Gisin and Percival is to treat of quantum systems subjected to measurement as open systems, then their theory seems hardly to say more than decoherence and the quantum-jumps theory already do. If, on the other hand, the aim is the introduction of a real collapse[17] — and not a 'local' reduction (partial trace) as in the decoherence approach —, then a real collapse is obviously contradicted by QM laws. The GRW model was surely intended as such 'real collapse', and also recently Pearle, who, as we have said, gave the first hint towards stochastic non-linear theories, in collaboration with Squires has proposed another version of real collapse [PEARLE/SQUIRES 1994] . It is true that the model agrees with QM for systems containing a small number of particles, but, as we shall see in the next chapter, we have to exclude a real loss of non-diagonal terms.

On the other hand, the formalism developed by Gisin and Percival can have interesting applications[18].

[16]A previous study [GARRAWAY/KNIGHT 1994] could show no appreciable difference between the two approaches.

[17]Following [STAMATESCU 1996, 254–55, 264–67].

[18]See [SPILLER 1994] [subsection 19.8.5].

Chapter 24

GENERATION OF THE CAT

Contents The experiments discussed in this chapter are a direct confirmation of the hypothesis of decoherence and show that, at a mesoscopic level, it is possible to obtain a S-Cat, i.e. a quantum state which is a superposition of macroscopically (mesoscopically) distinguishable states. Though S-Cats were introduced in the thirties into the theory, only in the 1990's an experimental generation became possible. Firstly, we shall summarize briefly some features of the generation of a S-Cat on the basis of a concrete model [section 24.1]. We shall then recount briefly some historical developments in this area [section 24.2]. The main part of the chapter is devoted to the detailed description of two performed experiments [section 24.3 and 24.4]. Finally a theoretical and interpretational discussion of some features of QM generalizes these results [section 24.5].

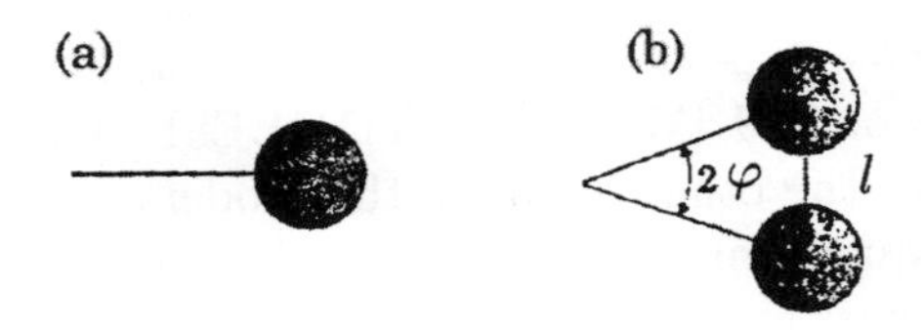

Figure 24.1: (a) Pictorial representation in phase space of a coherent state of a quantum oscillator. (b) The two components separated by a distance l of a S-Cat — from [BRUNE *et al.* 1996b, 4887].

24.1 How to Construct a Schrödinger Cat

Before we discuss in some detail the different experiments, let us firstly see how it is possible to conceive of a S-Cat in a very simple way [BRUNE *et al.* 1996b]. Consider a two-level atom (ground and excited state: $|g\rangle, |e\rangle$) coupled to an apparatus $\mathcal{A}$ represented by a quantum oscillator in a coherent state. The state vector $|\alpha\rangle$ defining it [see Eq. (5.31)] has a circular Gaussian distribution of radius unity due to quantum fluctuations which make the tip uncertain [see figure 24.1.(a)]. Consider an ideal measurement in which the atom-oscillator interaction entangles the phase of the oscillator $\pm\varphi$ to the internal state of the atom leading to:

$$|\Psi\rangle = \frac{1}{\sqrt{2}}(|e, \alpha e^{\imath\varphi}\rangle + |g, \alpha e^{-\imath\varphi}\rangle). \tag{24.1}$$

When the distance $l = 2\sqrt{N}\sin\varphi$ (where N is the photon number) is larger than 1, a S-cat is obtained [see figure 24.1.(b)].

Practically the biggest difficulty to generate a S-Cat is the preservation of the quantum coherence when enlarging sufficiently the 'distance' between the component states in order to obtain a quantum behaviour on a mesoscopic scale.

24.2 Short History

We see briefly some steps of the research until the factual generation of a S-Cat[1].

24.2.1 Non-Linear Media Arrangements

Free evolution A pioneering work of Yurke and Stoler [YURKE/STOLER 1986] [2] showed for the first time the possibility of constructing a superposition — which, as we know, is not necessarily an entanglement [see subsection 3.5.2] — between states which are macroscopically distinguishable. The superposition between two coherent states 180° out of phase with each other is created in the evolution of a coherent state propagated through an amplitude-dispersive medium (NL Kerr medium). Practically, we have at the same time two Gaussians separated from each other and interference fringes between them. Naturally the problem is that they are very sensitive to dissipation (decoherence).

Thermal and squeezed baths An article of Milburn and Holmes [MILBURN/HOLMES 1986] studied the coupling between an anharmonic oscillator with a thermal bath (at zero temperature). In order to make the effect of dissipation less destructive, a squeezed bath instead of a thermal one was later proposed [MECOZZI/TOMBESI 1987] [KENNEDY/WALLS 1988] [D'ARIANO *et al.* 1995a]. On the other hand, in the model of Yurke and Stoler there can be inhibition of the detection of quantum interference even by zero dissipation, due to vacuum fluctuations which enter the nonlinear medium. Instead of, if a squeezed vacuum enters the nonlinear medium, one can control the influence of vacuum fluctuations. It has also been proved that this influence may be less destructive.

A generalization of S-Cat generation to thermal, Fock, coherent, and squeezed states, by using another method, i.e. the vibrational motion of a trapped ion is also possible [MEEKHOF *et al.* 1996] [LEIBFRIED *et al.* 1996] — later we shall discuss in great details this method [section 24.3].

QND-Measurement A QND-Measurement scheme [see chapter 19] was used by Song, Caves and Yurke [SONG *et al.* 1990].[3] A different but equivalent approach is used by Yurke, Schleich and Walls [YURKE *et al.* 1990].

24.2.2 Amplification Methods

Very early it was considered possible to use amplification in order to create a mesoscopic or macroscopic S-Cat[4]. But, in a detailed study, Daniel and Milburn [DANIEL/MILBURN 1989,

[1] For an account and a history of the problem see [BUŽEK/KNIGHT 1995].
[2] Some antecedents can also be found in [MILBURN 1986].
[3] Based on the techniques proposed in [LA PORTA *et al.* 1989]
[4] For the formal treatment of the amplification, which is the inverse of damping, see [*GARDINER* 1991, 188–91].

4637] showed that, in the case of a NL oscillator, the amplification destroys the quantum coherence earlier than damping, because the latter process consists in a random loss of quanta to the heat bath, which depends on the number of quanta remaining in the system, while the amplification is a gain of quanta, which is possible also for a zero-temperature bath.

24.2.3 Microcavity Experiments

Cavity exp. 1 As we have seen, the greatest problem of the generation of a S-Cat is that of dissipation due to a rapidly increasing decoherence. In a *Gedankenexperiment* Savage, Braunstein and Walls [SAVAGE *et al.* 1990], in order to avoid dissipation, proposed the use of a single atom in a vacuum.

In the same year Slosser, Meystre and Wright [SLOSSER *et al.* 1990] [5] showed that at low temperatures a steady-state field in a micromaser cavity (Cavity exp. 1) pumped by atoms injected in a coherent superposition of their upper and lower states undergoes a transition as the parameter $\eta = (n_k + 1)/k^2\pi^2$ is increased — where the number states $|n_k\rangle$ are seen as $2k\pi$ pulses by the successive atoms as they pass through the cavity in a time τ, such that $(n_k + 1)^{1/2} g_{\text{AF}}\tau = k\pi$ (g_{AF} is the atom-field dipole coupling constant and k is an integer). In the high η–limit the micromaser field reaches a pure state that exhibits the characteristics of a macroscopic superposition.

Cavity Exp. 2 Another very interesting experiment has been proposed by Haroche and coworkers [BRUNE *et al.* 1990] [BRUNE *et al.* 1992].[6] It is a QND-Measurement based on the detection of the dispersive phase shift produced by the field on the wave function of nonresonant Rydberg atoms[7] crossing a microcavity (Cavity Exp. 2) [see figure 24.2]. The field is assumed to be in a coherent state [see Eq. (5.31)]. The aim is to count photons without absorbing them.

Before the atoms enter the cavity [see figure 24.3] they are excited in a superposition of level $|g\rangle$ and $|e\rangle$. When each atom emerges from the cavity, the component $|g\rangle$ remains unchanged, but the component $|e\rangle$ has had its phase changed proportionally to the number of photons in the cavity and the time spent there (depending from the velocity). Then it is excited in a second Ramsey zone. The result of the two excitations is a set of interference fringes in the probability of the atoms being found in $|e\rangle$: they are named *Ramsey fringes*, and are due to the interference between the amplitude for excitation in the two Ramsey zones. The detection of an excited atom implies the existence of a superposition of two field states in the cavity, one of which is phase shifted from the original one.

The information acquired by detecting a sequence of atoms modifies the field, until it eventually collapses in a Fock state. At the same time the field undergoes a diffusive process as a result of the back-action of the measurement on the photon-number. Once a Fock state has been generated, its evolution under weak perturbation can be continuously monitored, revealing quantum jumps between various photon numbers. When applied to an initial coherent field, the intermediate steps of the measuring sequence produce and reproduce quantum superpositions of

---

[5] See also [SLOSSER *et al.* 1989].

[6] See also [KNIGHT 1992] for short report.

[7] I.e. atoms which have horizontal electron orbit and which are characterized by a very long radiative decay time, much longer than the transit time across the apparatus. Studies on the Rydberg atoms in [HULET/KLEPPNER 1983] [BRUNE *et al.* 1994, 3339].

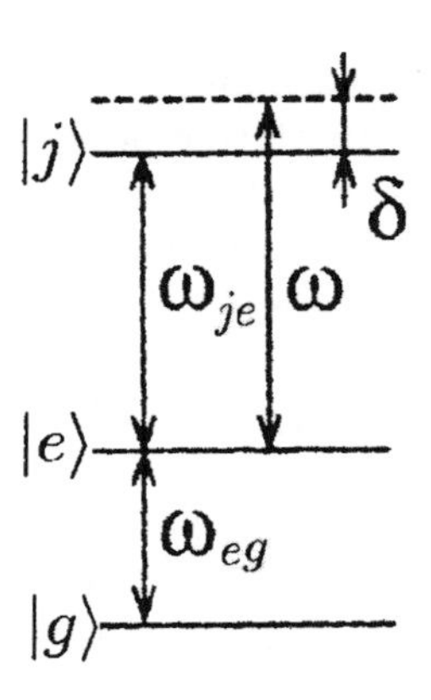

Figure 24.2: Possible configuration of Rydberg levels in proposed Cavity Exp. 2. δ is the frequency mismatch (detuning), i.e. the difference between the frequency of the radiation in the cavity and the resonance of the atomic states: the detuning is large enough to ensure that the atoms cannot absorb any photons and to shift the resonant frequency by a little amount (the shift increases with the number of photons). State $|g\rangle$ is too far away to be affected this way — from [BRUNE *et al.* 1992, 5195].

classical fields (preserved by atoms with constant velocity), the S-cat states of the form:

$$|\psi_{\epsilon}^{(\pm)}(\alpha)\rangle = \frac{|\alpha\rangle \pm |\alpha e^{-\imath\epsilon}\rangle}{\sqrt{2(1 \pm \mathrm{Re}\langle\alpha|\alpha e^{-\imath\epsilon}\rangle)}}, \tag{24.2}$$

where signs $+, -$ correspond respectively to a detection of $|g\rangle$ or $|e\rangle$, ϵ is the shift of the field phase produced by the detection, and the denominator ensures state normalization. We shall return with more detail on a very important experiment performed by the same team, which is to some extent a development of this one [see section 24.4].

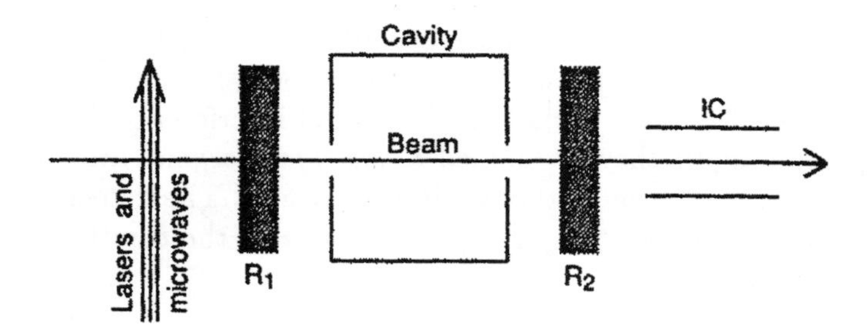

Figure 24.3: Microcavity Exp. 2.: Microwaves in the first Ramsey zone (R_1) creates superposition of states. Then atoms enter in a microcavity, whose field photons are to be counted. After a further excitation in the second Ramsey zone (R_2), atoms are detected in an ionisation counter (IC) — from [KNIGHT 1992, 438].

Cavity Exp. 3 In [BUŽEK *et al.* 1992b], an experiment is proposed which is also very interesting for the problem of reversibility [see chapter 27]. The authors use a Jaynes-Cummings model in a cavity (= Cavity Exp. 3) of optical resonance [see subsection 5.1.3]: a quantum interaction between a two-level atom and a quantized field mode (bosonic system) produces entanglement. But at precisely half of the revival time the atom and the field become asymptotically disentangled — the more as the coherent-state amplitude increases. The entangled state is a superposition state of two distinct parts of the Q-function [see subsection 5.2.1], which leads to oscillations in the photon-number distributions. The Q-function is defined as [see particularly Eq. (5.39)]:

$$Q(\alpha) = \langle \alpha | \hat{\rho}_F | \alpha \rangle, \tag{24.3}$$

where $\hat{\rho}_F$ is the reduced density operator of the field.

Since one expects a very rapid destruction of the oscillations in the photon-number distribution of the cavity field, the idea is to send repeatedly different atoms in order to create a stable superposition[8].

24.2.4 Other Proposals

Many others experiments have been proposed in the last few years. Among them we remember here the generation of a coherent superposition of two spatially localized wave packets (separated by ca. 0.4 μm) at the opposite extremes of a Kepler orbit of an electron around a nucleus [NOEL/STROUD 1996].

We remember also [DAKNA *et al.* 1997], where the generation of a S-Cat is proposed, making use of a simple beam-splitter scheme for a certain type of conditional measurement[9].

24.3 Detailed Exposition of a Performed Experiment

By Wineland and co-workers [MONROE *et al.* 1996] a first experimental generation of a S-Cat (Trapped Ion Exp.) was realized.

24.3.1 Preparation

The experiment makes use of a *laser-cooling* technique on a trapped ion. The technique consists of a resonant exchange of linear momentum between photons and atoms in order to control the external degrees of freedom of the latter ones and hence to reduce their kinetic energy[10].

An atom located in a trap was prepared in a quantum superposition of two spatially separated but localized positions. The following entangled state was created:

$$|\Psi\rangle = \frac{|\psi_1\rangle|\uparrow\rangle + |\psi_2\rangle|\downarrow\rangle}{\sqrt{2}}, \tag{24.4}$$

[8] On the same line we remember also the model proposed by [AGARWAL *et al.* 1997], of two-level atoms interacting with a dispersive cavity.

[9] These measurements are called 'quasi-continuous measurements' and are studied in [BAN 1997a]: the statistics of the measurement of photon number is equivalent to that obtained by continuous measurements described by the quantum Markovian process, even though one assumes, on the contrary, the Projection Postulate.

[10] This technique was proposed for the first time by Hänsch and Schawlow [HÄNSCH/SCHAWLOW 1975] and by Dehmelt and Wineland [WINELAND/DEHMELT 1975]. For reviews see [WINELAND/ITANO 1987] [COHEN-T./PHILLIPS 1990]. Preparatory study about the motion of a trapped ion in [DE MATOS F./W. VOGEL 1996]. The experiment makes use of an apparatus described in [JEFFERTS *et al.* 1995] [MONROE *et al.* 1995a]. See also [BARDROFF *et al.* 1996].

where $|\psi_1\rangle$ and $|\psi_2\rangle$ denote classical-like wave-packet states corresponding to spatially separated positions of the atom, and $|\uparrow\rangle$ and $|\downarrow\rangle$ refer to distinct internal electronic quantum states of the atom. By means of a sequence of laser pulses we form a superposition of two coherent-state wave packets. Each wave packet is correlated to a particular internal state of the atom. To analyse this state we apply an additional laser pulse to couple the internal states of the atom [see figure 24.4].

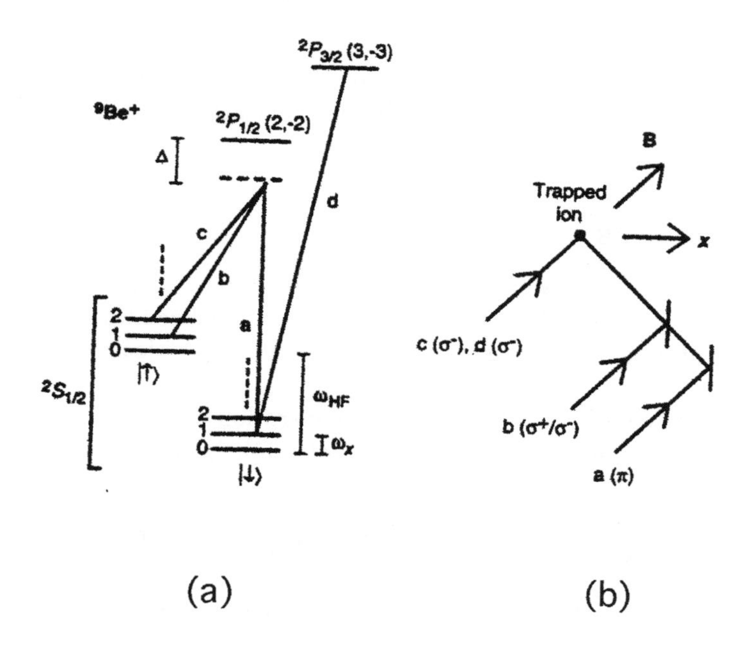

Figure 24.4: (a) Electronic (internal) and motional (external) energy levels (not to scale) of the trapped $^9\mathrm{Be}^+$ ion, coupled by the indicated laser beams a through d. The difference frequency of the 'carrier' Raman beams a, b is set near $\omega_{HF}/2\pi \simeq 1.250$ GHz, providing a two-photon Raman coupling between the $^2S_{1/2}(\mathbf{J}=2, m_\mathbf{J}=-2)$ and $^2S_{1/2}(\mathbf{J}=1, m_\mathbf{J}=-1)$ hyperfine ground states (denoted by $|\uparrow\rangle$ and $|\downarrow\rangle$ respectively). The difference frequency of the displacement Raman beams b and c is set to $\omega_x/2\pi \simeq 11.2$ MHz. Detection of the internal state is accomplished by first illuminating the ion with σ^-–polarised detection beam d, which drives the cycling $^2S_{1/2}(F\mathbf{J}=2, m_\mathbf{J}=-2) \mapsto {}^2P_{3/2}(\mathbf{J}=3, m_\mathbf{J}=-3)$ and then observing the scattered fluorescence.
(b) Geometry of the three Raman laser beams a, b, c (with the indicated polarisations). The quantization axis defined by the applied magnetic field $\mathbf{B}$ is 45° from the x axis of the harmonic trap potential — [MONROE *et al.* 1996, 1132].

The key features of the present method are:

- One controls the harmonic motion of the trapped atom to a high degree by exciting the motion from initial zero-point wave-packets to coherent states of well defined amplitude and phase;
- one does not rely on a conditional measurement to project out the desired S-Cat state;
- wave-packet dispersion of the atomic motion is negligible.

A single ^{9}Be$^+$ ion is confined in a coaxial-resonator radio frequency (= RF) ion trap (Paul Trap) that provides harmonic oscillation frequencies $(\omega_x, \omega_y, \omega_z)/2\pi \simeq (11.2, 18.2, 29.8)$MHz. The authors laser-cooled the ion to the QM ground state of motion and then coherently manipulated its internal (electronic) and external (motional) state by applying pairs of off-resonant laser beams, which drive two-photon stimulated Raman transitions. The two *internal states* of interest are the stable ^{2}S$_{1/2}$($\mathbf{J} = 2, m_\mathbf{J} = -2$) and ^{2}S$_{1/2}$($\mathbf{J} = 1, m_\mathbf{J} = -1$) hyperfine ground states (denoted respectively by $| \downarrow\rangle$ and $| \uparrow\rangle$) separated in frequency by $\omega_{HF}/2\pi \simeq 1.250$MHz[11]. **J** and $m_\mathbf{J}$ are total angular momentum and its projection along a quantization axis. The Raman beams are detuned at $\Delta \simeq -12$GHz from the $^2P_{1/2}$($\mathbf{J} = 2, m_\mathbf{J} = -2$) excited state, which acts as a virtual level, providing the Raman coupling. The external motional states are characterized by quantized vibrational harmonic oscillator states $|n\rangle_e$ in the x dimension, separated in frequency by $\omega_x/2\pi \simeq 11.2$MHz. One first applies 'carrier' beams a and b, such that the ion experiences a coherent Rabi oscillation[12] between states $| \downarrow\rangle$ and $| \uparrow\rangle$; then one applies 'displacement' Raman beams b and c, whose effect is formally equivalent to applying a displacement operator to the state of motion. Hence the initial state $|0\rangle_e$ becomes the coherent one:

$$|\alpha\rangle = e^{-|\alpha|^2/2} \sum_n \frac{\alpha^n}{(n!)^{\frac{1}{2}}} |n\rangle_e, \tag{24.5}$$

where $|\alpha\rangle = \vartheta e^{\imath\varphi}$ is the dimensionless complex number which represents the amplitude and the phase of motion in the harmonic potential. The polarisation of a, b and c produces $\pi, \sigma^+/\sigma^-, \sigma^-$ coupling respectively with respect to a quantization axis. As a result displacement beams (b and c) affect only the motional state correlated with $| \uparrow\rangle$ because the σ^- polarised beam c cannot couple the internal state $| \downarrow\rangle$ to any virtual ^{2}P$_{1/2}$ state. This selectivity provides an entanglement of the internal state with the external motional state. While the motional state can be thought of as 'classical', its entanglement with the internal atomic levels cannot be analysed in a classical or semiclassical way.

24.3.2 Generation of a Schrödinger Cat

The ion is first laser-cooled in the factorised state $| \downarrow\rangle|n_x = 0\rangle_e$. One then applies five sequential pulses of the Raman beams [see table 24.1]:

- In step 1 a $\pi/2$ pulse on the carrier splits the wave function into an equal superposition of $| \uparrow\rangle|0\rangle_e$ and $| \downarrow\rangle|0\rangle_e$ [see figure 24.5.(b)].
- In step 2 the displacement beams excite the motion correlated with the $| \uparrow\rangle$ component to a coherent state $|\vartheta e^{-\imath\varphi/2}\rangle_e$ [see figure 24.5.(c)].
- In step 3 a π–pulse on the carrier swaps the internal states of superposition [see figure 24.5.(d)].
- In step 4 the displacement beams excite the motion correlated with $| \uparrow\rangle$ component to a second coherent state $|\vartheta e^{\imath\varphi/2}\rangle_e$ [see figure 24.5.(e)].
- In step 5 a final $\pi/2$ pulse on the carrier combines the two coherent states [see figure 24.5.(f)].

[11]On this point see [*SCHIFF* 1955, 426–428].
[12]Treatment of the Rabi oscillation in [*GARDINER* 1991, 290].

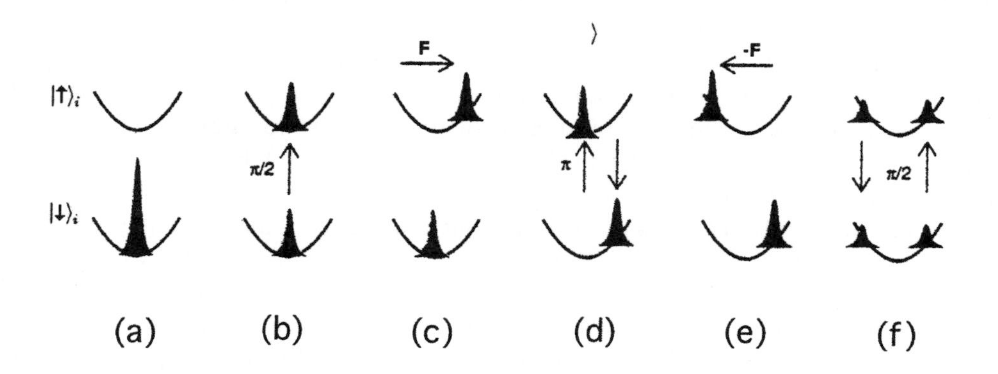

Figure 24.5: Evolution of the position-space atomic wave packet entangled with the internal states $|\uparrow\rangle$ and $|\downarrow\rangle$ during the creation of a S-cat state with amplitude and phase respectively $\vartheta = 3, \varphi = \pi$ [table 24.1]. The wave packets are snapshots in time, taken when the atom is at the extremum of motion in the harmonic trap potential (represented by the parabolas). The area of the wave packets corresponds to the probability of finding the atom in the given internal state.
(a) The initial wave packet corresponds to the quantum ground state of motion after laser-cooling.
(b) The wave packet is split after a $\pi/2$–pulse on the carrier.
(c) The $|\uparrow\rangle$ wave packet is excited to a coherent state by the force $\mathbf{F}$ of the displacement beams. Note that $\mathbf{F}$ acts only on $|\uparrow\rangle$, hence entangling the internal and the motional systems.
(d) The $|\uparrow\rangle$ and $|\downarrow\rangle$ wave packets are interchanged following a π–pulse on the carrier.
(e) The $|\uparrow\rangle$ wave packet is excited to a coherent state by the displacement beam force $-\mathbf{F}$, which is out of the phase with respect to the force in (c). The state in (e) corresponds most closely to a S-Cat state.
(f) The $|\uparrow\rangle$ and $|\downarrow\rangle$ wave packets are finally combined after a $\pi/2$–pulse on the carrier — from [MONROE *et al.* 1996, 1133].

Step	Function	Approximate duration (μs)	Phase	State created (initial state $\vert\downarrow\rangle\vert 0\rangle_e$)
1	Carrier $\pi/2$–pulse	0.5	μ	$[\vert\downarrow\rangle\vert 0\rangle_e - \imath e^{-\imath\mu}\vert\uparrow\rangle\vert 0\rangle_e]/\sqrt{2}$
2	Displacement	$\tau \simeq 10.0$	$-\varphi/2$	$[\vert\downarrow\rangle\vert 0\rangle_e - \imath e^{-\imath\mu}\vert\uparrow\rangle\vert\vartheta e^{-\imath\varphi/2}\rangle_e]/\sqrt{2}$
3	Carrier π–pulse	1.0	ν	$[e^{-\imath(\nu-\mu)}\vert\downarrow\rangle\vert\vartheta e^{-\imath\varphi/2}\rangle_e + \imath e^{-\imath\nu}\vert\uparrow\rangle\vert 0\rangle_e]/\sqrt{2}$
4	Displacement	$\tau \simeq 10.0$	$\varphi/2$	$[e^{-\imath(\nu-\mu)}\vert\downarrow\rangle\vert\vartheta e^{-\imath\varphi/2}\rangle_e + \imath e^{-\imath\nu}\vert\uparrow\rangle\vert\vartheta e^{\imath\varphi/2}\rangle_e]/\sqrt{2}$
5	Carrier	0.5	0	$1/2\vert\downarrow\rangle[\vert\vartheta e^{-\imath\varphi/2}\rangle_e - e^{\imath\delta}\vert\vartheta e^{\imath\varphi/2}\rangle_e]$ $-\imath/2\vert\uparrow\rangle[\vert\vartheta e^{-\imath\varphi/2}\rangle_e + e^{\imath\delta}\vert\vartheta e^{\imath\varphi/2}\rangle_e]$

Table 24.1: Raman beam pulse sequence for the generation of a S-Cat. The magnitude (phase) of the coherent states is controlled by the duration (phase) of the applied displacement beams in step 2, 4. The phases of internal state carrier operations in steps 1, 3, 5 are relative to step 5. The states created after each step do not include overall phase factors, and the phase appearing in the final state is $\delta \equiv \mu - 2\nu + \pi$ — from [MONROE *et al.* 1996, 1133].

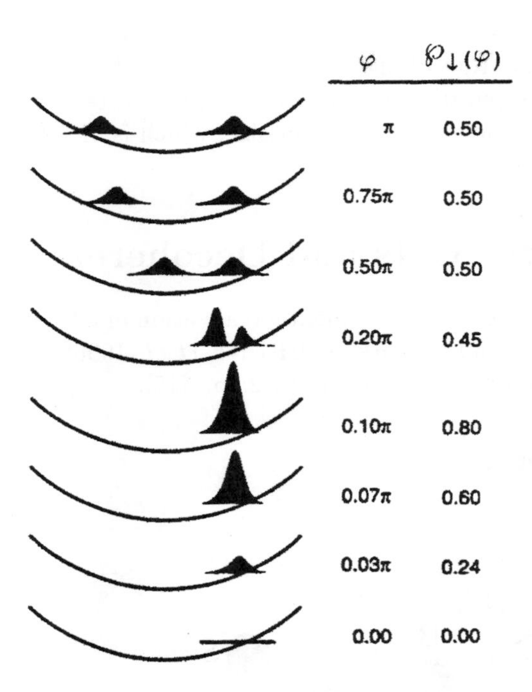

Figure 24.6: The evolution of the position-space wave packet superposition correlated with the $| \downarrow\rangle$ internal state as the phase separation φ of the two coherent states is varied for $\vartheta = 3, \delta = 0$. The expected signal $\wp_\downarrow(\varphi)$ is the integrated area under these wave packets. The wave packets are maximally separated at $\varphi = \pi$ ($\wp_\downarrow(\varphi) \simeq 1/2$) but they begin to overlap as φ gets smaller, and finally at $\wp_\downarrow(\varphi) = 0$ they interfere destructively. This vanishing interference signal is a signature of an odd S-cat associated with $| \downarrow\rangle$, because $\delta = 0$ — from [MONROE *et al.* 1996, 1134].

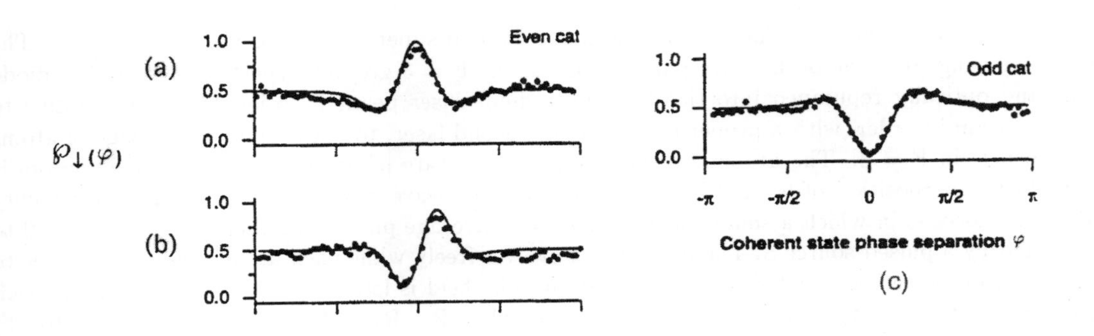

Figure 24.7: The measured interference signal $\wp_\downarrow(\varphi)$ for three values of δ ($\vartheta \simeq 1.5$). The curve (a) corresponds to $\delta = 1.03\pi$ (approximate even cat state correlated with $| \downarrow\rangle$, exhibiting constructive interference);

the curve (b) to $\delta = 0.48\pi$;

the curve (c) to $\delta = 0.06\pi$ (approximate odd cat state exhibiting destructive interference) — from [MONROE *et al.* 1996, 1134].

The state created after step 4 is a superposition of two independent coherent states each correlated with an internal state of the ion, in the spirit of the original proposal of the S-cat. The maximum of spatial separation between component states obtained is 83(3) nm, which is significantly larger than the single wave packet size, which is of 7.1(1) nm [see also figures 24.6–24.7].

24.4 Schrödinger Cats and Decoherence

An experiment which connects the experimental realization of a S-cat with the observation of decoherence is due to Haroche and co-workers [BRUNE *et al.* 1996b].[13] It is a further development of a previously discussed proposal [Cavity Exp. 2: p. 415].

24.4.1 Schrödinger Cat

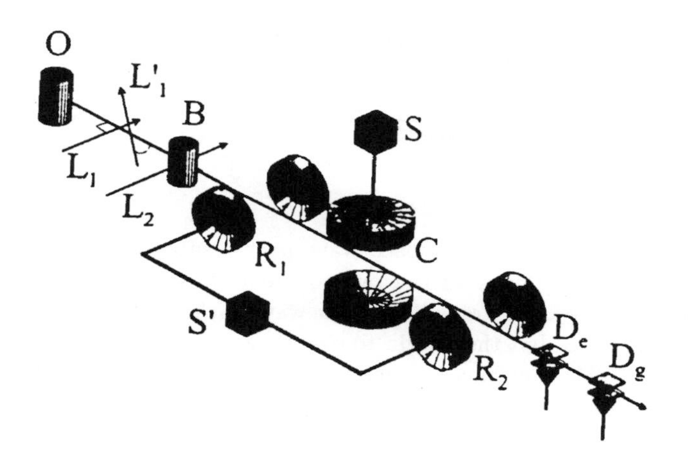

Figure 24.8: Cavity Exp. 4. The cavity C is made by two superconducting niobium mirrors. The Rb atoms effusing from the oven O are pumped out of the F = 3 ground hyperfine level with a diode laser L_1 and optically repumped into this level by a diode laser beam L_1' oriented at 58° relative to the atomic beam in order, with a proper tuning of the second laser, to allow only the passage of atoms at 400 ± 6 m/s in F = 3. The atoms are then excited into state $|e\rangle$ in box B. Each circular atom is prepared in a superposition of $|e\rangle$ and $|g\rangle$ by a resonant microwave $\pi/2$ pulse in a low quantum cavity R_1. It then crosses C in which a small coherent field with average photon number N varying from 0 to 10 is injected by a pulsed source S. The field, which evolves freely while each atom crosses C, relaxes to a vacuum before being regenerated for the next atom. The field is left coherent. After leaving C, each atom undergoes another $\pi/2$ pulse in a cavity R_2 identical to R_1. R_1 and R_2 are fed by a cw source S′ whose frequency ν is swept across ν_0. The atoms are finally counted in $|e\rangle$ and $|g\rangle$ by detectors D_e, D_g — from [BRUNE *et al.* 1994, 4888].

In this case the mesoscopic S-Cat is obtained by sending a rubidium atom, prepared in a superposition of two circular Rydberg states, excited $|e\rangle$ and ground state $|g\rangle$ (quantum numbers 51 and 50 respectively), across a high quantum microwave cavity C (Cavity Exp. 4) storing a

[13]See also [DAVIDOVICH *et al.* 1996].

small coherent field $|\alpha\rangle$. The coupling between the cavity and the atom is measured by the Rabbi frequency ω. The $|e\rangle \rightsquigarrow |g\rangle$ transition (transition frequency $\nu_0 = 51.099$ GHz) and the cavity frequency are slightly off resonance (detuning δ), so that the atom and the field cannot exchange energy but only undergo $1/\delta$ dispersive frequency shifts. The atom-field coupling during time t produces an atomic-level dependent dephasing of the field and generates an entangled state, which, for $\omega/\delta \ll 1$, is given by Eq. (24.1) with $\varphi = \omega^2 t/\delta$ [see figure 24.8].

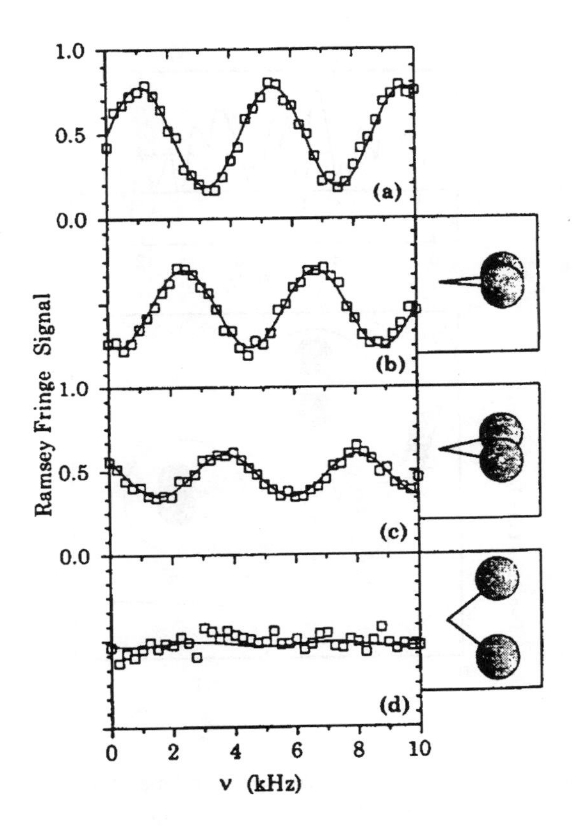

Figure 24.9: $\wp_g^{(1)}(\nu)$ signal exhibits Ramsey fringes.
(a) C empty.
(b)–(d) C stores a coherent field.
Insets show the phase space representation of the field components left in C — from [BRUNE *et al.* 1994, 4888].

Figure 24.9.(a) shows the signal obtained when C is empty ($\delta/2\pi = 712$ kHz): the final probability distribution $\wp_g^{(1)}(\nu)$ (of finding an atom in $|g\rangle$ as a function of ν) exhibits Ramsey fringes (interference) typical of atoms subjected to successive pulses. In fact transitions $|e\rangle \rightsquigarrow |g\rangle$ occur either in R_1 or in R_2: hence the two 'paths' are indistinguishable, leading to an interference term between the corresponding probability amplitudes. The phase difference between these amplitudes is $2\pi(\nu - \nu_0)\tau$ (where $\tau = 230\mu s$ is the time between the two pulses).

Figures 24.9.(b)–(d) show the fringes for $\delta/2\pi = 712.347$, and 104 kHz when there is a field in C ($N = 9.5, |\alpha| = 3.1$). When δ is reduced, the contrast of the fringes decreases (and the 'paths' of the atoms become partially distinguishable) and their phase is shifted. The fringe contrast reduction demonstrates the separation of the field state into two components and provides a measurement of the 'separation' l^2. When an atom leaves C, the system is prepared in state (24.1), so that the field phase 'points' toward $|e\rangle$ and $|g\rangle$ at the same time: we obtain a S-Cat. Finally [figure 24.9.(d)], the field components do not overlap at all and there is no longer any interference. In this sense Cavity Exp. 3 is also a marvellous confirmation of which-paths or wave/particle experiments [see section 8.4 and chapter 30].

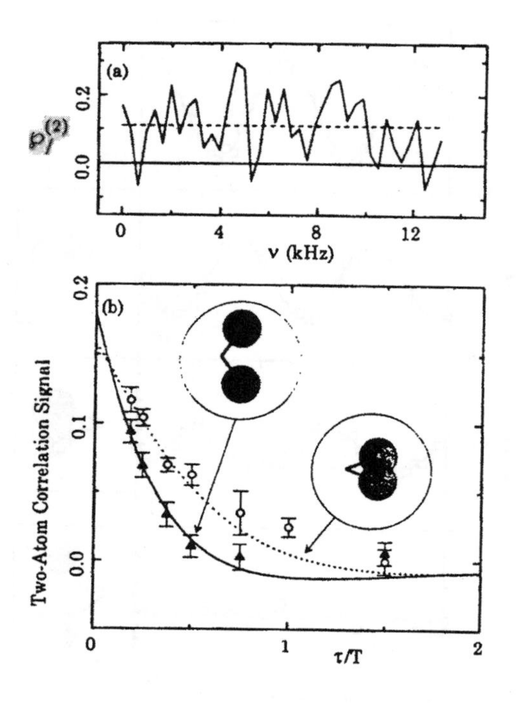

Figure 24.10: (a) The correlation signal $\wp_l^{(2)}$ is shown to be independent from ν.
(b) Decoherence: it proceeds at a faster rate when the distance between field components is increased (shown in the insets) — from [BRUNE *et al.* 1994, 4890].

24.4.2 A Decoherence Experiment

The coherence between the two components of the state and its quantum decoherence were revealed by a subsequent two-atom correlation experiment [BRUNE *et al.* 1996b, 4889–90] .[14] While a first atom creates a superpositional state involving the two field components, a second atom (the probe) crosses C with the same velocity after a short delay τ_2 and dephases the field again by an angle $\pm\varphi$. The two field components turn into three, with phases: $+2\varphi, -2\varphi, 0$. The zero component may be obtained via two different paths, since the atoms may have crossed C either in the (e, g) configuration or in the (g, e) configuration (i.e. the second atom undoes the phase shift of the first one). Since the atomic states are mixed after C in R_2, the (e, g) and (g, e) 'paths' are indistinguishable. As a consequence, there is an interference term in the joint

[14]The experiment follows outlines of [DAVIDOVICH *et al.* 1996].

probabilities $\wp_{ee}^{(2)}, \wp_{eg}^{(2)}, \wp_{ge}^{(2)}, \wp_{gg}^{(2)}$. The difference between conditional probabilities (the correlation between the two atoms)

$$\wp_{I}^{(2)} = \left(\frac{\wp_{ee}^{(2)}}{\wp_{ee}^{(2)} + \wp_{eg}^{(2)}} \right) - \left(\frac{\wp_{ge}^{(2)}}{\wp_{ge}^{(2)} + \wp_{gg}^{(2)}} \right) \tag{24.6}$$

is independent of ν, except around $\varphi = 0, \pi/2$ [see figure 24.10.(a)]. Equal to 0.5 at short times τ_2 when the quantum coherence is fully preserved, $\wp_{I}^{(2)}$ is shown to decay to 0 when the system I atom + field has evolved in a fully incoherent mixture[see figure 24.10.(b)]. Hence we have obtained a decoherence.

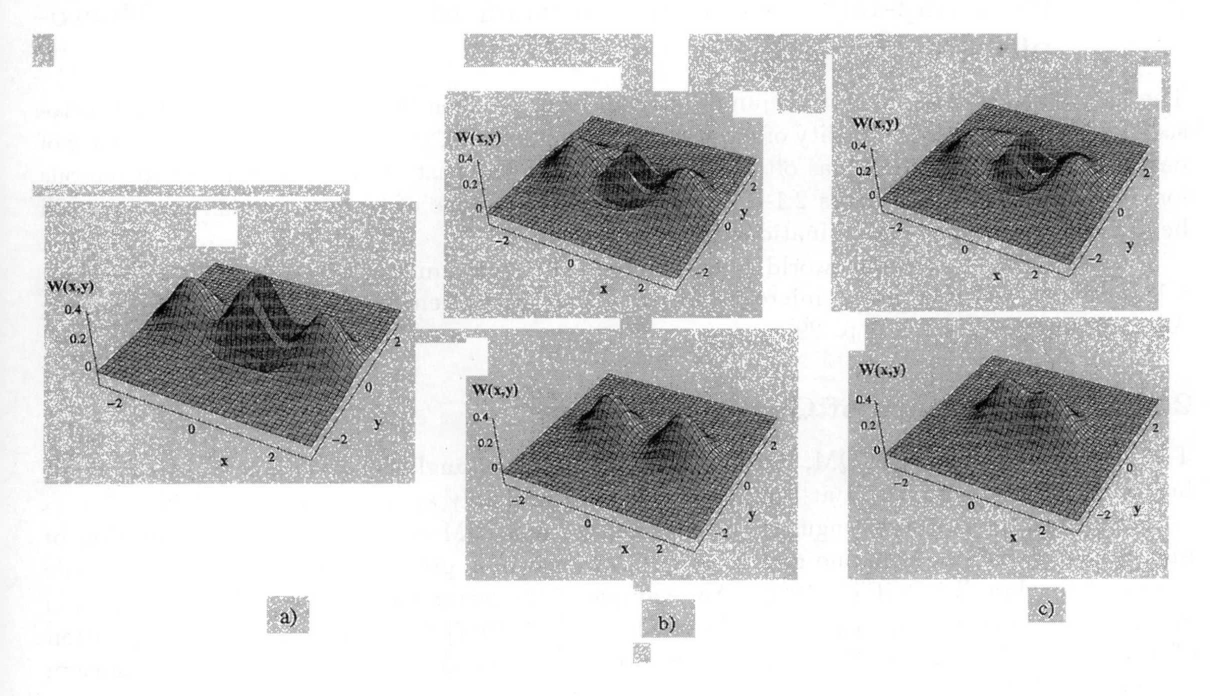

Figure 24.11: (a) Wigner function of the initial Schrödinger-cat state, $|\psi\rangle = N_{-}(|\alpha\rangle - |-\alpha\rangle)$, $|\alpha|^2 = 3.3$ (b) Wigner function of (left) the same cat state after 13 feedback cycles and (right) after one relaxation time in the absence of feedback; (c) Wigner function of (left) the same state after 25 feedback cycles and (right) after two relaxation times in the absence of feedback. The comparison between left and right is striking: in absence of feedback the Wigner function becomes quickly positive definite, while in the presence of feedback the quantum aspects of the state remain well visible for many decoherence times — from [FORTUNATO *et al.* 1999].

A modification of the preceding experiment was introduced by Vitali, Tombesi and Milburn [VITALI *et al.* 1997]: a slowing down of the decoherence obtained by means of stroboscopic feedback makes it possible to control the process[15].

A further improvement of this feedback scheme has been recently proposed by Fortunato, Raimond, Tombesi, and Vitali [FORTUNATO *et al.* 1999]. They have devised an "automatic" mechanism which, thanks to a second auxiliary cavity, is able to significantly increase the deco-

[15] See also [MANCINI *et al.* 1997a], where a cavity with a movable mirror (treated as a quantum harmonic oscillator) can be used in order to generate S-Cats, and [S. BOSE *et al.* 1997] for further developments.

herence time of the superposition state [see Fig. 24.11] thus overcoming the critical role played by the detection efficiency in the previous scheme.

24.5 Conclusion

What are the lessons to be drawn from the preceding experiments? We shall now discuss them in what follows.

24.5.1 No Jump-Like Transition between Microphysics and Macrophysics

The first lesson is that there is no jump-like transition between the microscopic and the macroscopic level because of the reality of mesoscopic entanglements, and that there is no 'difference of nature' between them, as it has often been imagined in the past. Hence there are good reasons for refuting QP [see postulates 2.1–2.2: p.s 19–20]. What type of transitions there can be, will be the matter of further examinations [see chapter 33].

However, the macroscopic world is to be understood as a 'limiting case' of the microworld for a large number of interacting microsystems. To a certain extent one can recover the Limiting Assumption [postulate 3.1: p. 29].

24.5.2 Persistence of Quantum Effects

The fundamental feature of QM, the superposition or the entanglement, is never completely lost, but there are also traces of it at the mesoscopic, and probably at the macroscopic level.

Note that one must distinguish between macroscopic QM effects produced by addition of microscopic variables, each one governed by QM, and those produced by a single macroscopic degree of freedom [LEGGETT 1980]. An example of the latter macroscopic QM phenomena is given by a Josephson tunnel junction [MARTINIS *et al.* 1987].[16] In the case of a superposition of n particles [JOOS 1996, 39], there is a fundamental difference between macroscopic quantum effects such that:

$$|\Psi\rangle = (c_1|\psi_1\rangle + c_2|\psi_2\rangle)^n, \tag{24.7a}$$

and a superposition of macroscopic states such as S-Cats:

$$|\Psi\rangle = (c_1|\psi_1\rangle)^n + (c_2|\psi_2\rangle)^n. \tag{24.7b}$$

Therefore, if the localization of macro-objects can be understood as a continuous scattering process produced by the environment [CAVES/MILBURN 1987] [JOOS 1996, 55–69, 83–84], on the other hand, it is not the number of microsystems which guarantees as such the transition to the macroscopic domain: in fact the entanglement between a system and the apparatus happens very quickly if there is a large number of quanta involved [JOOS 1996, 128–29].

As a consequence of this general situation, the macro world is not so 'determinate' as it seems to our perception. Evidently there is nothing wrong with the senses and perceptions as such: their scope is to serve our struggle for survival, and it is normal that living beings interpret their environment in a selective and instrumental way. What is wrong is the arbitrary

[16] For 'macroscopical experiments' using SQUID (= Superconducting Quantum-Interference Device) see [TESCHE 1990] and also [CARELLI *et al.* 1995].

scientific generalization, which takes the 'determinate' beings of everyday experience as primary and perhaps absolute experience. Hence, also for these reasons, Bohr's postulate 8.2 [p. 146] cannot be accepted. Here is the philosophical limit of a subjectivistic Pragmatism, which partly influenced Bohr himself.

Therefore, all attempts to find *ad hoc* solutions for the measurement problem which try to reproduce zero values for off-diagonal elements of a superposition — for example the GRW model [section 23.2] — are inadequate.

24.5.3 Diagonalization, Dissipation and Reductionism

The existence of an environment in which correlations are discarded and from which quasi-individuals proceed, is a type of duality: the environment is at the same time the condition of the generation of individuals and of their quasi-individuality and finally of their absorption by loss of individuality. This is something absolutely new, which no classical theory has ever foreseen. In fact a classical environment can only be understood as a sum of already perfectly determined individual systems.

The point can be better understood by discussing briefly the problem of reductionism. If we understand *reductionism* as the idea that the world can be explained in some form in terms of its 'components'[17], then it is surely false. Firstly, there is no guarantee at all that one day we shall find the 'last elements'. In so far as we know now, this is a mythology.

Secondly, the 'last elements', as far as we know them, i.e. photons, electrons, quarks, are characterized by the features of entanglement and superposition, which make them completely inadequate to serve as 'elements'. And here we return to the point that the environment cannot be understood as a 'sum' of individuals. It is also just as difficult to consider particles as elements in quantum field theory.

Furthermore, what the experiments reported in this chapter and the models discussed in chapter 21 show, is also that macroscopic behaviour, though not different in nature from microscopic behaviour, emerges as a process of complication and localization[18], which is completely quantic at its starting point, but capable of producing a new form of being in its results[19]. But then the macroscopic world cannot be understood as a '*simulacrum*', i.e. as a phenomenon of the underlying atomic or subatomic structure. Complication, localization and individualisation are not 'illusions' — as the MWI postulates [see chapter 15] —, but depend on physical interactions which are surely 'real' [see chapter 26]: in fact triorthogonal decomposition is an absolute property and not relative to a specific system [see subsection 17.2.1]. In other words: it is a fact of experience that there are (partly fuzzy) individuals and localisations. And there are no grounds to reject such experience. We shall return to problem of localization in part IX.

The history of the pair of concepts *appearance/reality* is very long, stemming from ancient Platonism up until our own times. We certainly cannot discuss the problem fully in this present context. It suffices to remark that, within the framework of modern physics, it also had and partly has some place. It is well known that Galilei thought [*GALILEI* S, 333] that 'qualities' as colour, taste, sound, and flavour were all merely illusory and not 'objective'. After the birth of optics and of acoustics, and after the technological and industrial production of tastes and flavours, such a supposition now seems a little unsound.

---

[17]Examinations in [*PRIMAS* 1983] [*DUMMETT* 1991, 322–24, 330–31].

[18]See also [*D'ESPAGNAT* 1995, 321–24].

[19]See also [BRODSKY/DRELL 1980] [LYONS 1983].

On the other hand, at a philosophical level, the supposition itself of illusory things, or of ontologically inferior realities is surely a theoretical dinosaur. Therefore one can postulate that no realities of the second order exist, but all that which exists exists. Hence we postulate:

Postulate 24.1 (Appearance) *There are not illusory things.*

In conclusion we have a confirmation of CP4 [p. 139], but, on the other side, a smooth form of this complementarity is also suggested by the preceding analysis. We shall return later on the point [see inequality (30.19) and comments].

24.5.4 Self-Referentiality

We have already discarded semantical completeness for QM [see subsection 9.4.2]. A necessary but insufficient condition of semantical completeness is *self-referentiality* [*MITTELSTAEDT* 1998, 4, 122–24], i.e. the capacity of a theory to furnish also the means of its proof. This is what happens with the theory of Measurement, which, on one hand, must account for QM phenomena (as objects) and, on the other hand, is the QM description of the process itself (and especially of the apparatus). This is also what happens to a greater extent with the relationship between microphysics and macrophysics. We interpret the quantum phenomena with our macroscopic devices and instruments (also on a conceptual level). But, on the other hand, QM forces us to reinterpret these means and these instruments in the light of itself (for example the S-Cat). Perhaps, as stressed by Mittelstaedt, one should speak in this case of a certain *feed-back* of the theory.

Note that the self-referentiality is a very important property of every theory, for example of mathematics (Gödel theorem). It is this property alone which guarantees that no alien concepts or instruments are applied to a theory [see also methodological principle M1: p. 6], for example that no psychological theories are applied to physics. The greatest defect of solutions to the Measurement problem, such as that of von Neumann [see subsections 14.1.3 and 14.2.1] is precisely such a methodological violation by the introduction of consciousness.

The failure of self-referentiality is also the main limit of the Copenhagen interpretation. In fact, as we have just said, this interpretation affirms a fundamental need of classical and macroscopical concepts and instruments in order to handle quantum phenomena [see postulate 8.2: p. 146, and corollary 8.2: p. 149]. On the contrary we have shown that the classical theory can be understood as an 'emergence' from the quantum world [see also section 19.9].

Part VI

TIME AND QM

Introduction to Part VI As we have seen [chapter 21], there can be spontaneous decoherence, such that the interference terms fall to zero in a relatively short time. It is natural to ask: Is it possible to generalize that model in order to build a microtheory in which the interference is lost and a deterministic *history* is gained? This is the aim of the consistent or decoherent-histories approach. The question is very interesting on the cosmological level, because, if the answer is affirmative, then we can describe quantum mechanically a consistent history of our universe.

Another problem is the following: von Neumann assumed two types of evolution, an unitary reversible evolution and the reduction of the wave-packet. In part V we have showed how is possible to overcome the gap between micro- and macro-world postulated by von Neumann. But a problem remains: the basic equations of QM seem reversible, while decoherence seems to presuppose some type of reversibility. How to solve such a problem? Summing up:

- In *chapter 25* we analyse different proposals to build cosmological decoherent wave functions.
- Then the 'delayed choice' issue is analysed theoretically and experimentally [*chapter 26*], because of its enormous consequence on the theory, and in particular on the interpretation of QM.
- While in chapter 26 time is treated of as an external parameter, in *chapter 27* we discuss the problem of the reversibility and irreversibility of quantum systems, i.e. treating time as a dynamical observable of the system itself, following the discussion of the time operator [see subsection 10.2.3].

Even if the theme of this part has only taken a central place in scientific debate since the seventies, its connection with the preceding part and its importance for the successive two parts, justify its treatment here.

Chapter 25

CONSISTENT AND DECOHERING HISTORIES

Contents In that which follows, we present [section 25.1] the approach of the first physicist to have proposed a consistent-histories approach: Griffiths. Then we analyse the approach of Gell-Mann and Hartle [section 25.2], which is known as that of 'decohering histories', and is quite a different variant. In the same section we also recount the Wheeler/DeWitt Eq. for its importance in the matter. In section 25.3, the theorem of Yamada/Takagi is then discussed: though the theorem is not intended as a direct criticism of the theories analysed in the preceding sections, it give important insights into the whole matter. Finally, concluding remarks follow in [section25.4].

25.1 Consistent Histories

25.1.1 General Theory

The possibility of speaking of consistent histories is based on the possibility of handling soundly quantum systems independently of any measurement or interaction, in other words isolated (or closed) QM systems [see definition 3.2: p. 42] [GRIFFITHS 1984, 220–26, 272]. Therefore, the main theoretical proposition of the consistent history approach can be formulated as follows [GRIFFITHS 1984, 230–31, 238, 250–51] [see also postulate 6.3: p. 105]:

Proposition 25.1 (Consistent Histories) *If a quantum system has a certain property after a measurement, it had it already shortly before (and shortly after) the measurement.*

We have already seen the importance of open systems [chapters 17, 18 and 21]; but we have not categorically excluded the possibility of isolated systems. The verification of this possibility will be a central issue in the following .

In Griffiths' model, one does not suppose individuals but events as ground entities. We associate with each event a projection operator $\hat{P}^E$. A history on the Hilbert space $\mathcal{H}$ is a sequence which has the following form:

$$\hat{P}^i \mapsto \hat{P}^{E_1} \mapsto \hat{P}^{E_2} \mapsto \ldots \mapsto \hat{P}^{E_n} \mapsto \hat{P}^f, \qquad (25.1)$$

where $\hat{P}^i, \hat{P}^f$ are initial and final event and the mapping is subjected to unitary evolution. The weight associated with a history (in the Heisenberg picture) is defined by [see Eq. (3.99)]:

$$w(\hat{P}^i \wedge \hat{P}^{E_1} \wedge \hat{P}^{E_2} \wedge \ldots \wedge \hat{P}^{E_n} \wedge \hat{P}^f) = \mathrm{Tr}[\hat{P}^{E_n}\hat{P}^{E_{n-1}} \ldots \hat{P}^{E_2}\hat{P}^{E_1}\hat{P}^i\hat{P}^{E_1}\hat{P}^{E_2} \ldots \hat{P}^{E_{n-1}}\hat{P}^{E_n}\hat{P}^f]. \quad (25.2)$$

A conditional weight is defined by

$$w(\hat{P}^{E_1} \wedge \hat{P}^{E_2} \wedge \ldots \wedge \hat{P}^{E_n} | \hat{P}^i \wedge \hat{P}^f) = \frac{w(\hat{P}^i \wedge \hat{P}^{E_1} \wedge \hat{P}^{E_2} \wedge \ldots \wedge \hat{P}^{E_n} \wedge \hat{P}^f)}{w(\hat{P}^i \wedge \hat{P}^f)}, \tag{25.3}$$

provided that $w(\hat{P}^i \wedge \hat{P}^f)$, defined by

$$w(\hat{P}^i \wedge \hat{P}^f) = \mathrm{Tr}[\hat{U}(t_f, t_i)\hat{P}^i \hat{U}(t_i, t_f)\hat{P}^f], \tag{25.4}$$

does not vanish. Hence the formalism supposes determined initial and final events, independently from whatever has happened or will happen. However, it is better to speak of sets of events instead of individual events. In that case, for a time t_k, we write $\{\hat{P}^{E_k^\mu}\}$, assuming that the sum over k is the identity operator:

$$\hat{I} = \sum_{\mu=1}^{N_k} \hat{P}^{E_k^\mu} \tag{25.5}$$

Hence we rewrite Eq. (25.1) in the following form:

$$\hat{P}^i \mapsto \{\hat{P}^{E_1^\mu}\} \mapsto \{\hat{P}^{E_2^\mu}\} \mapsto \ldots \mapsto \{\hat{P}^{E_n^\mu}\} \mapsto \hat{P}^f \tag{25.6}$$

Now we introduce the concept of *consistent histories*. The family F_C of histories described by Eq. (25.6) will be called consistent when, for every time t_k such that $1 \leq k \leq n$ and every history in F_C, we have:

$$w(\hat{P}^i \wedge \hat{P}^{E_1} \wedge \ldots \hat{P}^{E_k} \wedge \ldots \wedge \hat{P}^{E_n} \wedge \hat{P}^f) = \sum_\mu{}' w(\hat{P}^i \wedge \hat{P}^{E_1} \wedge \ldots \hat{P}^{E_k^\mu} \wedge \ldots \wedge \hat{P}^{E_n} \wedge \hat{P}^f), \tag{25.7}$$

where the sum $\sum'$ on the RHS is to be understood as a sum precisely over those projections which make up $\hat{P}^{E_k}$ on the left:

$$\hat{P}^{E_k} = \sum_\mu{}' \hat{P}^{E_k^\mu}. \tag{25.8}$$

The consistent condition (25.7) is equivalent to the following:

$$\mathrm{Re}\left(\mathrm{Tr}\left[\hat{P}^{E_n} \ldots \hat{P}^{E_k^\mu} \ldots \hat{P}^{E_1} \hat{P}^i \hat{P}^{E_1} \ldots \hat{P}^{E_k^{\mu'}} \ldots \hat{P}^{E_n} \hat{P}^f\right]\right) = 0 \tag{25.9}$$

for every pair $\mu \neq \mu'$ and for every $1 \leq k \leq n$, where $\hat{P}^{E_k^\mu} + \hat{P}^{E_k^{\mu'}} = 1$ and with $\hat{P}^{E_j}$ we denote $\hat{P}^{E_j^\mu}$ or $\hat{P}^{E_j^{\mu'}}$ or 1. In short the consistent condition is equivalent to the assertion that the effects of interference are negligible, i.e. that there is no interference in quantum amplitudes propagating from $\hat{P}^i$ to $\hat{P}^f$ along the separate paths passing through $\hat{P}^{E_k^\mu}$ and $\hat{P}^{E_k^{\mu'}}$ [see figure 25.1]. Iff this consistent condition is satisfied, we replace the weight symbol with the probability symbol $\wp$ and we speak of a consistent *individual* history.

25.1.2 Criticism

Note that the 'realistic' and 'classical' tenets of this proposal are very similar to that of Hidden-Variable theories, which are objects of later analysis [in the following part]. Here we wish to point out that counterexamples based on incompatible conditional measurements of properties which, after Griffiths, should pertain objectively to systems, are possible — think of the EPR experiment itself [see chapter 31], which already invalidates the proposal [*D'ESPAGNAT* 1995, 236–38]. Section 25.3 will explicitly show that Griffiths' proposal conflicts with some aspects of quantum theory.

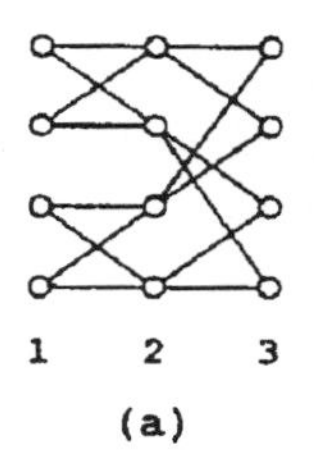

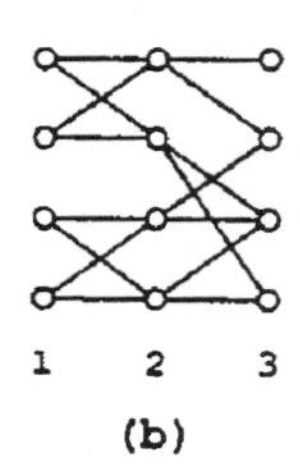

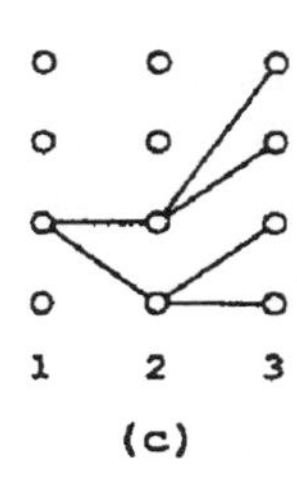

Figure 25.1: The columns in these trajectory graphs correspond to three times $t_1 < t_2 < t_3$. The non-interference condition (25.9) is satisfied for (a) but not for (b): note that the two paths connecting the next-to-lowest nodes at t_1 and t_3: in (a), starting from the third node from the top in t_1, we can reach the third node from the top in t_3 along one path alone (through the fourth node from the top in t_2); whereas in (b), by starting from the third node from the top in t_1, we can reach the third node from the top in t_3 along two paths (through the third node and the latter node from the top in t_2). An elementary family of those trajectories in (a) passing through one of the nodes at time t_1 in (a) is shown in (c) — from [GRIFFITHS 1993, 2202].

25.1.3 Omnès' Proposal

In [GRIFFITHS 1993] the consistent-history approach is developed by using the concept of trajectory in phase space as a proposal which aims to connect deterministic unitary evolution and the Probabilistic Assumption [see postulate 3.5: p. 38]. The formalism is that of cells in the phase space as suggested by Omnès.

Omnès openly criticized the Copenhagen interpretation for its dualism [OMNÈS 1988, 925] [OMNÈS 1990, 359–60, 370, 387, 421]. He proposed to associate a macroscopic cell of phase space with a projector in the Hilbert space, with the aim of overcoming the 'jump' between microstates and macrostates [OMNÈS 1990, 355, 421] [OMNÈS 1992, 352–53] [OMNÈS 1988, 925]. Such a proposal could be defined as a form of 'reductionism'.

Another feature of Omnès' proposal is the construction of a consistent-history model by means of a Boolean logic [OMNÈS 1990, 358, 361, 371, 423, 430].[1] But there seems to be some inconsistency here, because, on one hand, Boolean logic and the consistent-history approach presuppose some form of determinateness, and hence the classicality of reality; but, on the other hand, Omnès, especially in later works [*OMNÈS* 1994, 270–77], seems to acknowledge a form of decoherence similar to the Gell-Mann/Hartle's model — to be analysed in the next section —, and here the quantum character of reality is evidently in open conflict with Boolean logic (there is interference!).

25.2 Gell-Mann/Hartle's Model

25.2.1 Wheeler/DeWitt Equation

The application of Griffiths' ideas on a cosmological level came principally from the work of Gell-Mann and Hartle. The fundamental aim is to construct a universal wave function — in the

[1]See also [OMNÈS 1990, 410] [OMNÈS 1992, 352].

spirit of the MWI [see chapter 15] — such that macroscopically it does not present the typical QM superposition.

We do not wish here to enter in a difficult and controversial matter as quantum gravity, which is a basic tenet the of Hartle/Hawking model. We limit ourselves to saying that such a model presupposes the following Wheeler/DeWitt Eq. [a generalization of Eq. (4.10)] plus boundary conditions (Hermiticity), such that, starting with a ground state of the universe, all its possible types of expansion from zero volume (initial singularity) are studied [HARTLE/HAWKING 1983] [HAWKING 1984]:[2]

$$\left\{-G_{ijkl}\frac{\delta^2}{\delta h_{ij}\delta h_{kl}}+h^{\frac{1}{2}}\left[-{}^3R(h)+2\eta+l^2T_{nn}\left(-\imath\frac{\delta}{\delta\hat{\varphi}},\hat{\varphi}\right)\right]\right\}\Psi(h_{ij},\hat{\varphi})=0, \tag{25.10}$$

where h_{ij}, h_{kl} are 3–metric on a space-like surface, $h^{\frac{1}{2}}$ can be thought of as the time coordinate, G_{ijkl} can be viewed as a metric on a superspace, i.e. the space of all three-geometries:

$$G_{ijkl}=\frac{1}{2}h^{\frac{1}{2}}(h_{ik}h_{jl}+h_{il}h_{jk}-h_{ij}h_{kl}), \tag{25.11}$$

while ${}^3R$ denotes the covariant derivative and scalar curvature constructed from the three metric h_{ij}, η is a cosmological constant, l is the Planck length $(16\pi g)^{1/2}$ in units $\hbar=c=1$, T_{nn} is the stress energy tensor of the matter field projected in the direction normal to the surface, $\hat{\varphi}$ is the field and the operator $-\imath\frac{\delta}{\delta\hat{\varphi}}$ the momentum conjugate to the field [see Eq. (4.19)], and we have finally:

$$\Psi(h_{ij},\hat{\varphi})=\int_C \delta g\delta\hat{\varphi}e^{\imath S(g,\hat{\varphi})}, \tag{25.12}$$

where the path integral is over the class C of space-times and g is the gravitational constant.

25.2.2 General Formalism

Gell-Mann and Hartle were the first scientists to have understood spontaneous decoherence [see chapter 21] as a possibility for giving a concrete physical model to Griffiths' consistent histories. Naturally, the closed system is here the universe itself (only in this case we prefer to use the term 'closed' instead of 'isolated'). Hence the role played by the environment in Zurek's model, is played here by the 'rest' of the universe. In general the formalism is the same as Griffiths' one, except that, instead of speaking of events, the subjects here are various scalar fields at different points in space [GELL-MANN/HARTLE 1990, 432–39] [GELL-MANN/HARTLE 1993] [HARTLE 1991]. A *history* is a particular sequence of alternatives $\{\hat{P}_\mu\}=(\hat{P}^1_{\mu_1}(t_1),\hat{P}^2_{\mu_2}(t_2),\ldots \hat{P}^n_{\mu_n}(t_n))$. A complete *fine-grained* history is specified by giving the values of a complete set of operators at all times. One history is a *coarse-graining* of another one if the set $\{\hat{P}_\mu\}$ of the first consists of sums of the set $\{\hat{P}_\mu\}$ of the second. The completely coarse grained history is the unit operator. The reciprocal relationships of coarse and fine graining histories constitute only a partial ordering of sets of alternative histories [see figure 25.2].

The formalism is based on an application of Feynman's path integrals, by specifying the amplitude for a completely fine-grained history in a particular basis of generalized coordinates $\hat{\mathbf{q}}_j(t)$, say all fundamental field variables at all points in space. This amplitude is proportional to [see Eq. (3.151)]:

$$\vartheta[\hat{\mathbf{q}}_j(t)]\propto e^{\frac{\imath}{\hbar}S[\hat{\mathbf{q}}_j(t)]} \tag{25.13}$$

[2]See also [ZEH 1986].

The operators $\hat{\mathbf{q}}_j(t)$ are the various scalar fields at different points in space. There are at least three types of coarse-graining:

- not specifying observables at all times, but only at certain times;
- specifying a non-complete set of observables at any time;
- specifying ranges of values for only some observables.

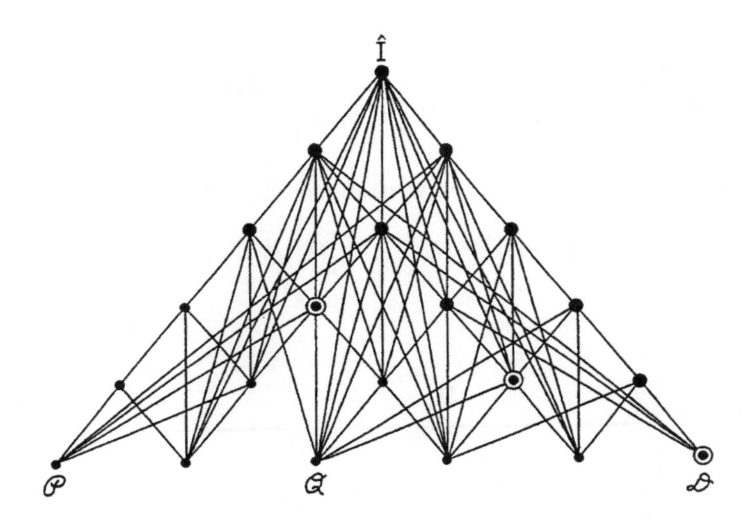

Figure 25.2: The schematic structure of the space of *sets* of possible histories for the universe. Each dot in the diagram represent an exhaustive set of alternative histories. Such sets (denoted by $\{\hat{P}_\mu\}$ in the text), correspond in the Heisenberg picture to time sequences $(\hat{P}^1_{\mu_1}(t_1), \hat{P}^2_{\mu_2}(t_2), \ldots, \hat{P}^n_{\mu_n}(t_n))$ of sets of projectors, such that at each time t_k the alternatives μ_k are an orthogonal and exhaustive set of possibilities for the universe. At the bottom of the diagram are the completely fine-grained sets of histories, each arising from taking projections into eigenstates of a complete set of observables for the universe at every time. For example the set $\mathcal{Q}$ is the set in which all field variables at all points of the space are specified at every moment in time. $\mathcal{P}$ might be the completely fine-grained set in which all field momenta are specified at each time. $\mathcal{D}$ may be a degenerate set, in which the same complete set of operators occurs at every time. The dots above the bottom row are coarse-grained sets of alternative histories. If two dots are connected by a path, the one above is the coarse graining of the one below, i.e. the projections in the set above are sums of those in the set below. At the very top is the degenerate case in which complete sums are taken at every time yielding no projections at all other than the unit operator. The space of sets of alternative histories is thus partially ordered by the operation of coarse-graining. The heavy dots denote decoherent sets of alternative histories. Coarse graining of decoherent sets remain decoherent. Maximal sets (heavy dots surrounded by circles) are those decohering sets for which there is no finer-grained decoherent set — from [GELL-MANN/HARTLE 1990, 433].

To illustrate all three, let us divide $\hat{\mathbf{q}}_j(t)$ up into variables x_j, X_j, and consider only set of ranges $\{\Delta^k_\mu\}$ of x_j at times $t_k, k = 1, \ldots, n$. A set of alternatives at any time consists of ranges Δ^k_μ which exhaust the possible values of x_j as μ ranges of all integers. An *individual* history is specified by a particular Δ^k_μ at particular times $t_k, k = 1, \ldots, n$. We write $[\Delta_\mu] = (\Delta^1_{\mu_1}, \ldots, \Delta^n_{\mu_n})$ for a particular history.

We now introduce a decoherence functional $\mathcal{D}$, which is a complex functional on any pair of histories in a set of alternative histories:

$$\mathcal{D}[\hat{\mathbf{q}}'_j(t), \hat{\mathbf{q}}_j(t)] = \delta\left[\hat{\mathbf{q}}'_j(t_f) - \hat{\mathbf{q}}_j(t_f)\right] \exp\left[\frac{\imath(S[\hat{\mathbf{q}}'_j(t)] - S[\hat{\mathbf{q}}_j(t)])}{\hbar}\right] \hat{\rho}^U_i(\hat{\mathbf{q}}'_j(t_j), \hat{\mathbf{q}}_j(t_j)), \tag{25.14}$$

where $\hat{\rho}^U_i$ is the initial density matrix of the universe in the $\hat{\mathbf{q}}_j$ representation. The decoherence functional for coarse-grained histories is obtained from Eq. (25.14) according to the superposition principle by summing over all terms which are not specified by the coarse graining [see figure 25.3]:

$$\mathcal{D}([\Delta_{\mu'}], [\Delta_\mu]) = \int_{[\Delta_{\mu'}]} \delta\hat{\mathbf{q}}' \int_{[\Delta_\mu]} \delta\hat{\mathbf{q}}\delta\left[\hat{\mathbf{q}}'_j(t_f) - \hat{\mathbf{q}}_j(t_f)\right] \exp\left[\frac{\imath(S[\hat{\mathbf{q}}'_j(t)] - S[\hat{\mathbf{q}}_j(t)])}{\hbar}\right] \hat{\rho}^U_i(\hat{\mathbf{q}}'_j(t_j), \hat{\mathbf{q}}_j(t_j)). \tag{25.15}$$

In the Heisenberg picture Eq. (25.15) can be written [see Eq. (25.2)]:

$$\mathcal{D}\left([\hat{P}_{\mu'}, \hat{P}_\mu]\right) = \mathrm{Tr}[\hat{P}^n_{\mu'_n}(t_n)\ldots\hat{P}^1_{\mu'_1}(t_1)\hat{\rho}^U_i\hat{P}^1_{\mu_1}(t_1)\hat{P}^n_{\mu_n}(t_n)] \tag{25.16}$$

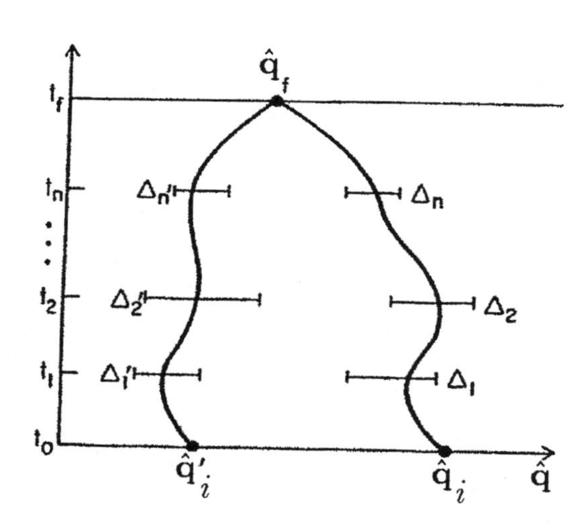

Figure 25.3: The sum-over-histories construction of the decoherence functional — from [GELL-MANN/HARTLE 1990, 435].

A set of coarse grained alternative histories is said to *decohere* when the off-diagonal terms of $\mathcal{D}$ are sufficiently small:

$$\mathcal{D}\left([\hat{P}_{\mu'}, \hat{P}_\mu]\right) \simeq 0, \ \forall \mu'_k \neq \mu_k \tag{25.17}$$

Hence decoherence is an interesting notion only for strings of projectors that involve more than one point in time. We can write the probability for *predicting* alternatives $\mu_{k+1}, \ldots, \mu_n$ given that alternatives $\mu_1 \ldots \mu_k$ have already happened in the form:

$$\wp(\mu_n t_n, \ldots, \mu_{k+1} t_{k+1} | \mu_k t_k, \ldots, \mu_1 t_1) = \frac{\wp(\mu_n t_n, \ldots \mu_1 t_1)}{\wp(\mu_k t_k, \ldots \mu_1 t_1)}. \tag{25.18}$$

By *prediction*, we mean here a certain (probability equal to 1) statement about events which have not yet occurred.

On the other hand, the probability that $\mu_{n-1}, \ldots, \mu_1 t_1$ happened in the past (*retrodiction*) given present data summarized by the alternative μ_n is:

$$\wp(\mu_{n-1}t_{n-1}, \ldots, \mu_1 t_1 | \mu_n t_n) = \frac{\wp(\mu_n t_n, \ldots \mu_1 t_1)}{\wp(\mu_n t_n)}, \tag{25.19}$$

where with *retrodiction* we understand a certain statement about what we suppose to be already-happened events

But, while future predictions can be obtained from an effective density matrix summarizing information about what has happened, there is no effective density matrix representing present information from which probability for the past can be derived. In fact the actual state of universe $\hat{\rho}_U^A$ is given by:

$$\hat{\rho}_U^A = \frac{\hat{P}^k_{\mu_k}(t_k) \cdots \hat{P}^1_{\mu_1}(t_1) \hat{\rho}_i^U \hat{P}^1_{\mu_1}(t_1) \cdots \hat{P}^k_{\mu_k}(t_k)}{\mathrm{Tr}[\hat{P}^k_{\mu_k}(t_k) \cdots \hat{P}^1_{\mu_1}(t_1) \hat{\rho}_i^U \hat{P}^1_{\mu_1}(t_1) \hat{P}^k_{\mu_k}(t_k)]}. \tag{25.20}$$

From Eq. (25.20) we obtain the following prediction probability, i.e. reformulation of Eq. (25.18):

$$\wp(\mu_n t_n, \ldots, \mu_{k+1} t_{k+1} | \mu_k t_k, \ldots, \mu_1 t_1) = \mathrm{Tr}[\hat{P}^n_{\mu_n}(t_n) \cdots \hat{P}^{k+1}_{\mu_{k+1}}(t_{k+1}) \hat{\rho}_U^A \hat{P}^{k+1}_{\mu_{k+1}}(t_{k+1}) \cdots \hat{P}^n_{\mu_n}(t_n)]. \tag{25.21}$$

As is shown by Eq. (25.21), history requires knowledge both of present data and the initial condition of the universe. Due to the cyclic property of trace, any final alternatives decohere and an event can be predicted; but we expect certain variables only to have decohered in the past.

25.2.3 A Symmetric Proposal

But this asymmetry of the treatment is overcome in [GELL-MANN/HARTLE 1993, 3369–70] — probably after Page's criticism [PAGE 1993a] —, so that future and past are treated similarly. The first to propose a completely time-symmetric treatment was Page himself [PAGE 1993a]. Practically, even though each history is not time-symmetric, the whole set of decohering histories is. In this case we can rewrite the decoherence functional (25.15) in the form:

$$\mathcal{D}(\mu', \mu) = N \int_{\mu'} \delta\hat{\mathbf{q}}' \int_{\mu} \delta\hat{\mathbf{q}} \hat{\rho}_U^f(\hat{\mathbf{q}}(t_f), \hat{\mathbf{q}}'(t_f)) \exp\left[\frac{\imath(\mathrm{S}[\hat{\mathbf{q}}'(\tau)] - \mathrm{S}[\hat{\mathbf{q}}(\tau)])}{\hbar}\right] \hat{\rho}_i^U(\hat{\mathbf{q}}'(t_i), \hat{\mathbf{q}}(t_i)), \tag{25.22}$$

where

$$N^{-1} = \int d\hat{\mathbf{q}}' \int d\hat{\mathbf{q}} \hat{\rho}_U^f(\hat{\mathbf{q}}', \hat{\mathbf{q}}) \hat{\rho}_i^U(\hat{\mathbf{q}}, \hat{\mathbf{q}}'). \tag{25.23}$$

This idea is further developed in [GELL-MANN/HARTLE 1994] and [KIEFER 1996b]. We write the decoherence functional as a function of variables for a chosen system and variables for the environment:

$$\begin{aligned} \mathcal{D}[\xi^{\mathcal{E}}(t), \xi^{\mathcal{S}}(t); \xi_i^{\mathcal{E}}, \xi_i^{\mathcal{S}}] &= \int d\xi_f^{\mathcal{E}} d\xi_f^{\mathcal{S}} \delta(\xi_f^{\mathcal{S}}) \exp\left[\imath \mathrm{S}_{\mathcal{S}}\left(\xi^{\mathcal{E}}(t) + \frac{\xi^{\mathcal{S}}(t)}{2}\right) - \imath \mathrm{S}_{\mathcal{S}}\left(\xi^{\mathcal{E}}(t) - \frac{\xi^{\mathcal{S}}(t)}{2}\right)\right] \\ &\times \exp\left[\imath\varphi(\xi^{\mathcal{E}}(t), \xi^{\mathcal{S}}(t))\right] \hat{\rho}_{\mathcal{S}}\left(\xi_i^{\mathcal{E}} + \frac{\xi^{\mathcal{S}}(t)}{2}, \xi^{\mathcal{E}}(t) - \frac{\xi^{\mathcal{S}}(t)}{2}\right) \end{aligned} \tag{25.24}$$

where φ is the phase. This functional is completely fine-grained. By performing a coarse-graining it reads:

$$\mathcal{D}(\mu, \mu') = \int_{(\mu,\mu')} \mathcal{D}(\xi^{\mathcal{E}})\mathcal{D}(\xi^{\mathcal{S}})\delta(\xi_f^{\mathcal{S}})e^{\imath \mathcal{S}}\hat{\rho}^{\mathcal{S}}. \tag{25.25}$$

The fundamental idea is taken from Zeh and Joos [JOOS 1996, 39, 41, 120, 123], and is a reformulation of the concept of relative state (MWI).

25.2.4 Discussion

The symmetry of the decoherence functional makes the proposal of Gell-Mann/Hartle model more similar to that of Griffiths [see for example Eq. (25.2)]. However, we are faced with the following alternatives:

- either we maintain decoherence, and then we must accept some form of irreversibility;
- or we wish to treat of time symmetrically.

If we wish to maintain decoherence, we are obliged to take two steps: first we must introduce an operation of tracing out the environment [see theorem 17.5: p. 271]; and note here that, while the decoherence theory supposes the action of an external environment relative to the considered (object) system, it seems quite difficult to apply such an operation to the evolution of the whole universe — we shall return later on the point [see subsection 26.4.3]. We must then introduce a form of triorthogonal decomposition [see theorem 17.4: p. 269]: but, as we have seen, this is not a relative property but an absolute one, because, once a state of a system is decomposed in the basis diagonal for the measured observable, every other decomposition is excluded. Note that we do not necessarily need to perform a measurement: each spontaneous interaction between three systems effects a similar result. It is from here that the irreversibility stems [see also chapter 27]. And, to a certain extent, i.e. for large systems, where the quantum features are partly lost, it seems reasonable to assert the possibility of predictions (in other words, given that a partial decoherence has already happened).

But it is possible to argue that a reversibility is excluded only in the system + apparatus + environment reference 'frame', and that, by taking larger systems or other ones, one could obtain other decompositions. Apart from the problem of what 'larger' or 'other' systems could be, if so, i.e. if we accept the second alternative (symmetry of time), then we are faced with another alternative: either we follow Griffiths, and suppose that there is a real loss of diagonal terms, or we suppose that we never have a real loss of coherence. And here, while Griffiths' proposal assumes that alternative histories pertain to a statistical mixture (without interference terms between them),[3] the proponents of the decohering history approach seem to suppose that the alternative histories are in superposition. But if so, the original tenet of Griffiths vanishes and then the existence of 'decohered' or macroscopical objects becomes only illusory [on this point see postulate 24.1: p. 428] while the decohered histories should be more imaginary than real.

But, if so, then the decohering-histories interpretation is nothing more than the MWI itself [see chapter 15] and, as the MWI, is subjected to the fallacy of the basis degeneracy. Moreover, the decoherent histories theory also overcomes the problem of the choice of one observable, by also making the latter illusory.

Such contradictions are more evident in other variants of the preceding theories which try to synthesize the decoherent-histories theory with a stochastic reduction-jump model such as

[3] See [PAZ/ZUREK 1993] for analysis of this problem.

that of Gisin [see section 23.3] [DIÓSI *et al.* 1994] or with a 'traditional' jump model such as that of Carmichael [see section 22.1] [BRUN 1997]. Actually, it seems difficult to integrate such different perspectives because stochastic reduction theories generally acknowledge the collapse of a wave function as a real fact, which is not the case with decoherence or decohering histories [STAMATESCU 1996, 254–55, 264–67].

25.3 Yamada/Takagi Theorem

An interesting model, again based on path integrals, has been developed by Yamada and Takagi. In Basic-QM it seems that, for particles, Newtonian time plays an important role in selecting a family of sets for which QM probabilities can be defined. From path integrals it follows that probabilities for momenta at a fixed time are reducible to those for positions by a time-of-flight [see subsection 10.2.3]. Hence a probability-assignable set has been limited to a set of possible positions of the particle on a hypersurface of constant time in the space-time [YAMADA/TAGAKI 1991a, 985] [see figure 25.4].

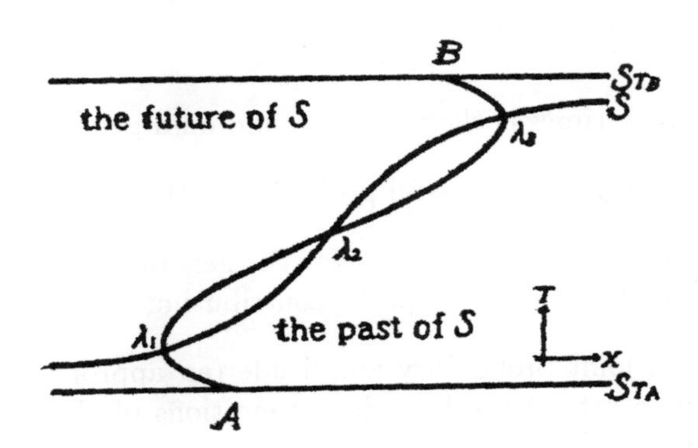

Figure 25.4: Intersection of a classical path with a surface **S** in a $(1+1)$–dimensional Newtonian space-time. A classical path is shown which connects two points A and B lying in the opposite sides of **S** and intersects S three times at $\lambda_1, \lambda_2, \lambda_3$ — from [YAMADA/TAGAKI 1991a, 986].

The problem is whether we can assign probabilities to numbers and places of intersection between possible paths of a particle and a general hypersurface connecting space-lines containing points of departure and arrival. Their reference model is a proposal stemming from Hartle, who argued that the role of time in sum-over-paths formulation is not central and looked for a wave function on a general hypersurface as a natural extension of Schrödinger wave [HARTLE 1988a]. Yamada/Takagi considered probabilities instead of wave function; however, the discussion is very interesting in view of the symmetric model proposed by Gell-Mann/Hartle [see Eq. (25.22)].

Now there are assignable probability sets only on the hypersurfaces of constant time and not on general hypersurfaces. Hence Newtonian time is understood as a selector of assignable probability sets [YAMADA/TAGAKI 1991a, 988]. The questions (by means of which we can distinguish between different sets) are [YAMADA/TAGAKI 1991a, 989] [see figure 25.5]:

- where and how many times does the particle intersect a general surface?

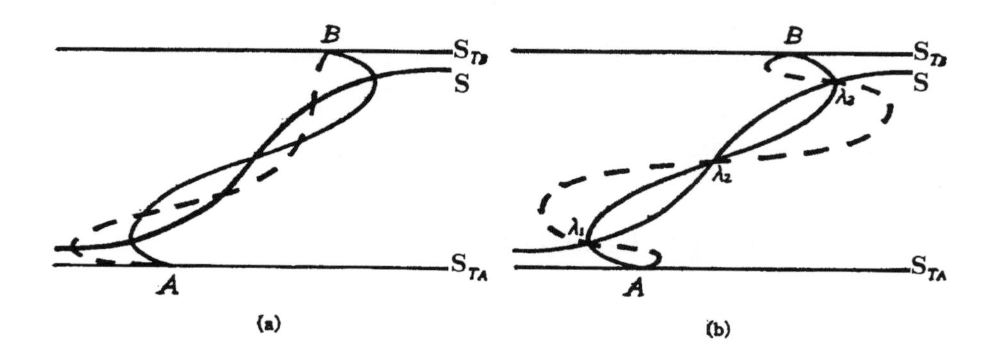

Figure 25.5: Intersection of virtual paths with a surface **S**.
(a) Two examples of virtual paths which connect A and B and intersect **S** three times. In QM many virtual paths contribute to the motion from A to B; they intersect **S** an odd number of times, and the number and the places of intersection vary from one virtual path to another.
(b) Examples of virtual paths which contribute to $\vartheta(\mathrm{B}; \lambda_3, \lambda_2, \lambda_1; \mathbf{S}; \mathrm{A})$ for the present model. There are many paths which intersect **S** three times at $\lambda_1 < \mathrm{l}_\lambda < \lambda_3$. The sum of $e^{\imath \mathrm{S}}$ over all such paths defines $\vartheta(\mathrm{B}; \lambda_3, \lambda_2, \lambda_1; \mathbf{S}; \mathrm{A})$ — from [YAMADA/TAGAKI 1991a, 986].

- and where and how many times is the particle found on a general surface?

Two conditions are imposed for a successful probabilistic theory:

- The possible places where the particle intersects the general surface are the points of intersection of such a surface **S** with virtual paths contributing to the sum.
- First one calculates the joint probability amplitude (an appropriately defined sum of $e^{\imath \mathrm{S}}$ over virtual paths which satisfy the boundary conditions of the problem), and then one forms the absolute square of the amplitude.

A QM probability is obtained by summing the absolutely squared amplitudes over those observables which are irrelevant to the probability in question [YAMADA/TAGAKI 1991a, 990–91]. Hence we obtain the following expression for probability to be assigned to the sample (subset) $\Delta(\lambda_n)$ (where n = number of intersections, λ_n = places of intersection) of the λ set on S:

$$\wp(\Delta(\lambda_n), \mathbf{S}) = \int dX_\mathrm{B} \left| \int dX_\mathrm{A} \int_{\Delta(\mathrm{l}_n)} d\mathbf{e}\mathrm{K}_n(\mathrm{B}; \mathbf{e}, \mathbf{S}; \mathrm{A})\vartheta(\mathrm{A}) \right|^2, \tag{25.26}$$

where A, B are departure and arrival points, respectively, K is a propagator [see subsection 3.11.2], $\vartheta(\mathrm{A})$ an initial normalized amplitude, **e** an n–components vector.
Now we impose two conditions:

- the first condition, on the propagator, is the *classifiable condition*, which means that K is decomposable into components $\{\mathrm{K}_n(\mathrm{B}; \mathbf{e}, \mathbf{S}; \mathrm{A})\}$, each of which is associated with each of the samples of the set in question:

$$\mathrm{K}(\mathrm{B}, \mathrm{A}) = \sum_n \int d\mathbf{e}\mathrm{K}_n(\mathrm{B}; \mathbf{e}, \mathbf{S}; \mathrm{A}). \tag{25.27}$$

- The second condition is the *non-interference condition* [see also Eq. (25.9)]:

$$\mathrm{Re}\int dX_{\mathrm{B}}\int\int dX_{\mathrm{A}}dX_{\mathrm{A}'}\mathrm{K}_n^*(\mathrm{B};\mathbf{e};\mathrm{A})\mathrm{K}_m(\mathrm{B};\mathbf{e}';\mathrm{A}')\vartheta^*(\mathrm{A})\vartheta(\mathrm{A}')\propto\delta_{nm}\delta(\mathbf{e}-\mathbf{e}'), \qquad (25.28)$$

where:

$$\delta_{nm}\delta(\mathbf{e}-\mathbf{e}'):=\delta_{nm}\prod_{j=1}^{n}\delta(e_j-e_j'). \qquad (25.29)$$

This framework can be adapted to the other sets. We do not have any longer more intersection points but only points that are common (which includes both intersection and touching) to **S** and virtual paths [YAMADA/TAGAKI 1991a, 992–993]. The interesting feature of Yamada/Takagi's proposal is that it is a more concrete physical model than Griffiths' ideas, and hence it is particularly useful in testing all matter.

Furthermore, by using the random-walk method, the authors prove [YAMADA/TAGAKI 1991a, 995–1008] the following theorem:

Theorem 25.1 (Yamada/Takagi) *Given* (25.26) *probability-expression, condition* (25.27) *does not hold.*

As a consequence of theorem 25.1, we have the following corollary:

Corollary 25.1 (Yamada/Takagi) *A probabilistic description is theoretically impossible either for the question: Where and how many times the particle intersects the hypersurface?, or for the question: Where and how many times the particle is found on the hypersurface?*

In [YAMADA/TAGAKI 1991b] one makes an attempt to find nontrivial examples of a set of non-interfering alternatives which are not restricted to an instant of time and for which QM probabilities are defined. The authors found a set corresponding to yes/no experiments. The absence of interference in this example depends on a special choice of the symmetry of an initial amplitude, so that in general we return to constant-time hypersurfaces[4].

25.4 Conclusions

The importance of theorem 25.1 and its corollary is evident: it forbids a symmetric construction [Eq. (25.22)] such as that of Gell-Mann/Hartle if intended as a decohering-history model: in QM retrodictions or predictions (or both) are impossible. In fact, we are faced with the following alternative: either we discard the classifiable condition (25.27) and maintain the non-interference condition (25.28), and then we are unable to build histories by failing a well-defined probability (i.e. disjoint sets); or we reject the non-interference condition (25.28) and maintain the classifiable condition (25.27).

If we accept the latter solution, i.e. if we assign unrestricted value to the Superposition principle, then the predictions or the retrodictions (or both) are surely impossible. That it is so for

[4]See also [YAMADA/TAGAKI 1992].

the *predictions* — a question which concerns Griffiths' model especially —, is confirmed by the analysis of chapter 21 and chapter 24 about the permanence of (even if locally vanishing) interference terms [see also theorem 17.6: p. 271]. Hence nothing forbids new forms of superposition and entanglement in a long time delay — for example by field oscillations, as in Cavity Exp. 3 [p. 416] — even in the case where they are locally vanishing. Again the conclusion is that histories are not possible if we do not introduce some form of decoherence and hence of irreversibility.

In conclusion, it is quite strange that Gell-Mann/Hartle have tried to apply the decoherence to predictions or to future histories of our universe, which would be possible only under the hypothesis of a total destruction of off-diagonal terms, which is openly questioned by the proponents of decoherence.

The only question then to be analysed is if *retrodictions*, in a proper quantum context, are possible. This is the matter of the next chapter.

Chapter 26

DELAYED CHOICE

Contents In that which follows we discuss a new issue of the quantum theory proposed for the first time by Wheeler in 1978[1]. In section 26.1 we present the general theoretical problem. In section 26.2 we report Wheeler's proposal of some experiments in order to prove this theoretical assumption, while in section 26.3 we study some experimental realizations. Finally [in section 26.4] we try to generalize this matter at an interpretational level by returning to some of the issues discussed in the previous chapter.

26.1 The Theoretical Problem

Until the 1970's, the common belief was that, for proper quantum systems, only predictions (and not retrodictions) are impossible [*JAMMER* 1974, 208–209]. This is due to the Superposition principle. The theorem 25.1 [p. 443] and its corollary [25.1], proved by Yamada and Tagaki, show that predictions or retrodictions (or both) are impossible. *Predictions* seems to be excluded [see section 25.4]. Wheeler shows now by the Delayed-choice paradox that *retrodictions* are also impossible. The paradox can be better understood on a cosmological level [WHEELER 1983, 190–99] — the importance of the point is due to the theory of the universal wave function [see section 25.2].

Suppose a quasar, whose light, before reaching the Earth, encounters a lenticular galaxy. As one knows, such a galaxy can act as a gravitational lens[2], such that the light can take two paths, one of which, say , passes 500,000 light years apart from the 'straight line' path. Now the problem is the following: we can choose here on Earth if we wish to observe an interference phenomenon, hence receiving the light from both paths, or to detect a determinate path (with a corpuscular behaviour of the photons). Hence it is as we could decide a determinate behaviour (wave-like or corpuscular) about an event which seems *to have already happened* millions of years ago.

How can we solve the problem? It is clear that the possibility of retrocausation can be excluded [WHEELER 1978b, 41] [WHEELER 1983, 183–84]. As we shall see[3], in fact one can exclude the possibility that a particle (or an antiparticle) inverts its 'time direction'. It is true that mathematical models are thinkable by which, by preserving always the proper 'time direction' at local level, one ends in the past, as proposed by Gödel [GÖDEL 1949].[4] But models such as

[1]The hint was given by Einstein and co-workers [EINSTEIN *et al.* 1931] and von Weizsäcker [VON WEIZSÄCKER 1931]. See also [BOHR 1949] [HELLMUTH *et al.* 1985].

[2]On this point see [TURNER 1988].

[3]The problem will be shortly discussed in the final conclusions [see especially section 46.4].

[4]On this point a good and easy exposition is in [*HORWICH* 1987, 111–14].

this one are no more than mathematical games devoid, for all we know, of any physical content[5]. Hence there is no concrete element for hypothesizing such a possibility.

The only reasonable possibility is then that the assumption 'the event has already occurred' is false. To deny this assumption is equivalent to deny that the past exists. But in what sense? Surely not in the sense that it has never existed, which would be, in the context of physical science, an absurd and extreme form of subjective idealism. Hence, we methodologically reject this possibility here [see M1: p. 6] — the ontological aspects will be discussed in part VII.

But we can also understand this point in the sense that the past has no reality at all *now*. In other terms, what we call 'past' (the light of million years ago) factually is only a present thing, i.e. the light, *not* of million years ago, but the light which we detect now, which, in order to arrive at us, has run all the way (and all the time) up to us. Hence, by performing the quoted experiment, we are not at all acting on the past, but only on the *effects which past events have transmitted to our present*. In other words we can postulate what follows:

Postulate 26.1 (Wheeler) *We cannot speak of or handle the past without interacting with its effects on the present, and whenever we do that, we are also free to choose the form to receive them, i.e. to manipulate them.*

In conclusion the delayed choice shows on a chronological scale the fundamental quantum result that we cannot speak (neither on the level of the definition) of a phenomenon without interacting in some form with it [WHEELER 1990, 13–14] [see also principle 3.1: p. 28, and corollary 7.1: p. 127]:

Corollary 26.1 (QM Phenomenon) *In QM a phenomenon can be defined iff we interact in some form with it.*

With *phenomenon* we understand here 'what is manifest' (by measurement, for example) and not 'appearance'. In consequence of the postulate 26.1 we can better understand the distinction between proper time of a quantum system — **for example** the time-of-arrival, to which an operator can be associated [see subsection 10.2.3] — and the time as an external (respectively to the same system) parameter. The delayed choice, as we shall see in that which follows, has nothing to do with the proper time of the system, but only with the time of its interaction with us from our frame of reference — which can be viewed in a nice way by a proposed experiment [see subsection 26.2.4]. Again, note that speaking of 'external parameter' has nothing to do with a form of subjectivism: we only mean, in general terms, a time frame of reference of a system external to a given (object) system, which is the case in every form of interaction, not only in the case of measurements.

Note that 'quantum eraser' experiments [see subsection 8.5.2] have some relationship with the 'delayed choice'. In fact recently, one [HERZOG *et al.* 1995, 3034] proposed considering the quantum eraser a stronger form of delayed choice to the extent to which one can postpone the decision to read or to erase some information (typically a 'which-path' one) until after detection. Only, the quantum eraser has more to do with the reversible behaviour of the (quantum) detector [see chapter 27], and in this sense it is partly a different aspect of the theory.

[5]For some philosophical problems of time travels see [D. LEWIS 1976].

26.2 Proposed Delayed Choice Experiments

Following Wheeler's original exposition in [WHEELER 1978b], we now propose some experiments which can throw more light on the matter. If the postulate 26.1 is correct, then we can 'decide' how to measure or interact with a system $\mathcal{S}$ while $\mathcal{S}$ is 'in flight' toward us. Or, in other terms, we can change the experimental arrangement while the system is approaching the present. Wheeler himself proposed a lot of different experimental arrangements. Due to the importance of the matter for the interpretation of the theory we report most of them briefly.

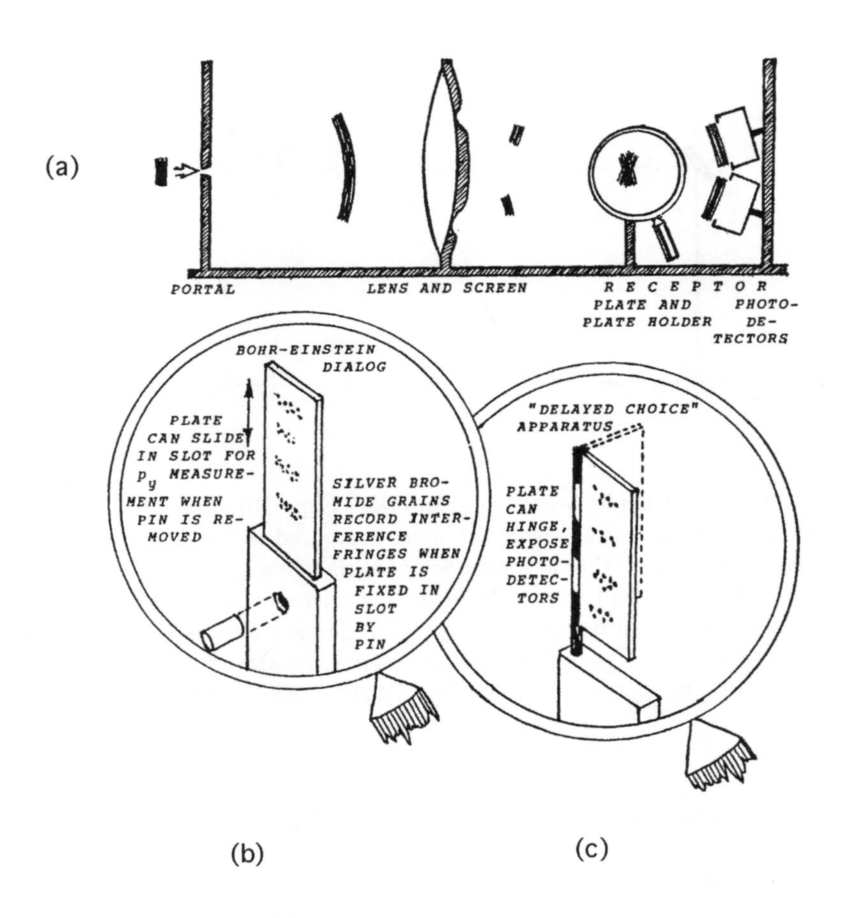

Figure 26.1: (a) Idealized Two-Slit-Exp.
(b) The version of Bohr-Einstein dialog. The plate catches every photon. It registers precisely either the y coordinate of impact or the y–component of impulse delivered — but does not and cannot do both. Omit the photodetectors.
(c) The delayed choice version. It includes the photodetectors — from [WHEELER 1978b, 11].

26.2.1 Another Two-Slit Experiment

The first one [WHEELER 1978b, 10–11] is a modification of Bohr's experiment Two-Slit-Exp. 5 [p.131], say Two-Slit-Exp. 9 [see figure 26.1]. We can slide a plate inside an envelope by removing a pin or remove it by inserting the pin in the hole: in the first case the receptor has a definite localization and will register the component p_y of momentum; in the second case silver bromide grains will record the interference pattern. Suppose now that we perform a delayed choice experiment in the following form: we place some photodetectors behind the envelope plus plate, and, when the plate is out, we can swing it aside, in order to expose photodetectors and to register a quantum of energy. If the plate is in place, we register the interference (wave-like behaviour), if it is aside, we infer through what slit the quantum of energy comes (corpuscular behaviour). Now we can decide the one or the other arrangement when the particle has *already* passed through the wall with slits.

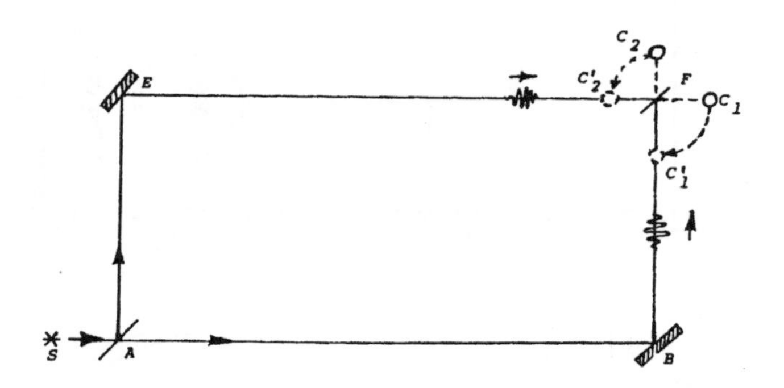

Figure 26.2: The delayed choice split-beam exp. (Interfer. Exp. 10). S is the pulsed source, giving a wave train long compared to one wave length but short compared to the total distance of travel. It is operated at an intensity level so low, that at most one quantum of energy is transported per pulse. A and F are BS and BM respectively. B and E are fully reflecting mirrors. C_1, C_2 are photodetectors in order to determine, by their relative counting rates, the phase difference between the two alternative routes from the point of observation. C'_1, C'_2 are the same photodetectors swung into position to determine the route by which the quantum of energy arrives — from [WHEELER 1978b, 32].

26.2.2 Again on the Heisenberg Microscope

A second proposed experiment [WHEELER 1978b, 29–31] is another version of the gamma ray microscope of Heisenberg (Delayed Microscope Exp.) [see section 7.3]. When the lens is used to fix the position, the quantum of energy scattered into the lens gives the electron a lateral kick, the amount of which is subjected to UR1 [see Eq. (7.7)]. But the uncertainty in the lateral kick can be reduced to a very small fraction by placing a sufficiently large collection of small photodetectors at a little distance above the lens. Whichever one of them goes off, signals the direction of the scattered photon and thus the momentum imparted to the electron. In this way the lens is used in an improper way because we have lost the possibility of knowing the position of the electron by the limits imposed again by UR1.

Now the delayed choice experiment consists of the arrangement of alternative lattices of photodetectors: only after the quantum of energy has *already* passed through the lens do we decide which of different lattices of photodetectors to swing into action. Say that we swing a lattice located in the focal plane of the lens. If one of them goes off, then this irreversible act of amplification tells us where the electron was when it scattered the radiation within a latitude Δp. But, alternatively, we can also swing another lattice located in a plane some small fraction of the way up from the lens to the focal plane. This record will tell us — within a very small range — what lateral kick was given to the electron in the Compton process: it is deduced from the direction of the scattered photon, and this direction is revealed by the coordinates of the photocounter that registered it, because it responds to a quantum of energy only if the energy goes through a highly restricted portion of the aperture of the lens. Hence the alternative 'the whole aperture transmitted the energy' or 'only a small fraction of it did' can be decided after the lens has already finished transmitting energy.

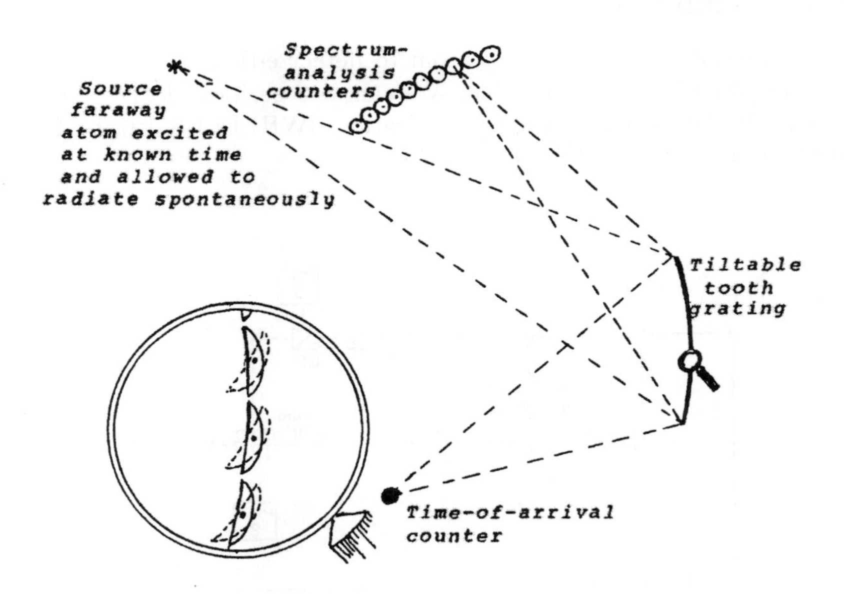

Figure 26.3: Use of tiltable tooth grating to make a delayed choice between alternative pictures (wave train versus pulse) of the radiation that the atom has already emitted. The magnified view shows a few typical teeth of the grating depicted by dashed lines when the teeth are oriented so as to reflect or 'blaze' the light to the spectrum analyzing counters, by full lines when the teeth are aligned. Then the grating becomes a spherical mirror, and the time-of-arrival counter alone is operative — from [WHEELER 1978b, 34].

26.2.3 Beam Splitting

Another experiment is constituted by an Interferometer (Interfer. Exp. 10) [WHEELER 1978b, 31–32]. Normally the detectors are placed after a Beam merger, but we can move them so as to place both before the BM [see figure 26.2]. Hence, after the beam has already passed through the BS, we can decide if we want to look at the interference (first arrangement) or to detect the which-path (second arrangement). It is this experiment in particular which has been realized [see section 26.3].

26.2.4 Energy and Time

Another proposed experiment concerns the complementarity energy/time [WHEELER 1978b, 33], i.e. UR2. An atom placed far away is excited so that it radiates energy. We place a tiltable tooth grating (Tooth Grating Exp.), so that when teeth are aligned we obtain a spherical mirror which reflects the radiation to a time-of-arrival counter. If teeth are oriented differently, they reflect the radiation to a spectrum analyzer counter which determines the energy. Hence we decide after the radiation has been already emitted if it has to manifest itself as pulselike or as nearly monochromatic [see figure 26.3].

26.2.5 Other Examples

We can also use radiation from an excited atom to detect either the angle of emission or the angular momentum [WHEELER 1978b, 33]. We can also perform delayed choice experiments on the complementarity between alternative polarisations [WHEELER 1978b, 34–35] or an EPR coincidence rate experiment [WHEELER 1978b, 35–39].

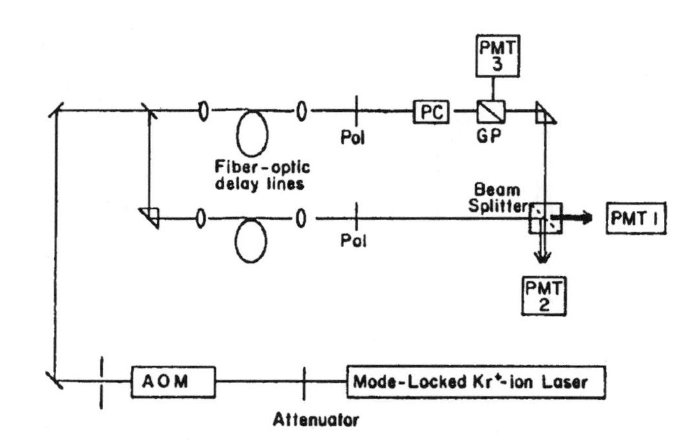

Figure 26.4: Realized Interfer. Exp. 10 — from [HELLMUTH *et al.* 1985, 419].

26.3 A Delayed Choice Performed Experiment

26.3.1 The Experiment

Walther and co-workers [HELLMUTH *et al.* 1985] [HELLMUTH *et al.* 1986] realized an experiment (Interfer. Exp. 10) [see figure 26.4] following Wheeler's original proposal [see subsection 26.2.3]. A mode-locked, krypton-ion laser provides 150 psec pulses of 647 nm radiation at a repetition rate of 80 MHz. The pulses of light pass through an optical attenuator (10^{-9}) and then through an acousto-optic modulator which reduces the repetition rate to 10 MHz. In addition the modulator has a measured cw attenuation of 1.6×10^{-2}. This implies that the 300 MWatt average power of initial beam is reduced to 6×10^{-16} Watt or 0.4 eV/pulse: therefore the average number of photons per pulse is less than 0.2. One can say that 95% of the pulses have one photon or less (80% zero and 15% one).

The incident beam passes through the first BS and the two resulting beams are then directed and focused into a single-mode optical fiber 5 ms in length. Transit time is approximately 24 nsec. A Pockel cell (PC) is placed in one arm. By means of the appropriate voltage, a phase shift of one-half wave is introduced between the two orthogonal polarisation components of the incident light. The polarisation is thereby rotated of 90°. The PC is followed by a Galn polarising prism (GP) which reflects light polarised in the plane of the figure and transmits the orthogonal polarisation. The switching time of the PC is 4–5 nsec (= 4–5 $\times 10^{-9}$ sec). The two beams are recombined by the second BS, a cube having BM function, before going to the cooled photomultipliers (PMT1 and PMT2).

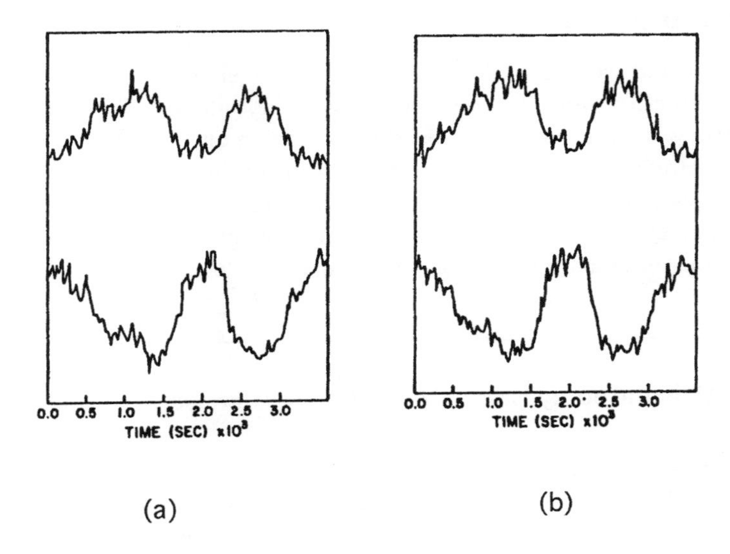

Figure 26.5: Results of (a) the 'ordinary' experiment and of (b) the delayed-choice one. On the y axis the number of photodetections — from [HELLMUTH *et al.* 1985, 421].

In the ordinary, non-delayed-choice experiment, the optical switch is 'opened' 5 nsec before the light pulse reaches the first BS. Thus the experiment has a fixed configuration. In the delayed-choice mode, the optical switch is normally closed (reflection to PMT3), and only opened 5 nsec after the pulse passes through the first BS. The 40 MHz driver for the mode-locker is used as a master clock. The data of the delayed-choice experiment are then confronted with those of the 'ordinary' one [see figure 26.5]: as one see there is no difference between 'normal' experimental set up and delayed choice one. In conclusion we are totally free to perform delayed choice experiments without altering QM predictions.

A mathematically more refined description has been given by Hellmuth, Walther, Zajonc and Schleich [HELLMUTH *et al.* 1987].[6]

[6]On the point see also [WICKES *et al.* 1983, 459].

26.3.2 Conclusion

Therefore, on the basis of the experiments discussed, we have proved that, for a proper quantum system, retrodiction is impossible:

Theorem 26.1 (Retrodiction) *Quantum-mechanically retrodiction is not possible.*

26.4 Lessons of the Delayed Choice

What are the principal lessons which we can draw from the above analysis? Very schematically we sum up the principal ones.

26.4.1 Definition and Measurement

We have already stated that we cannot speak, even on the level of Definition, of QM systems independently of some interaction (measurement) with them [corollary 26.1: p. 446]. In other words, the Measurement theory is an integral part of QM because, in order to know something of a quantum system, we must already have some preliminary information about it, i.e. we must know the wave function of the system, which is due to a premeasurement [*BRAGINSKY/KHALILI* 1992, 21] [see also subsection 28.4.3]. In other words we cannot speak of closed systems as Griffiths wishes. Hence we can assume a second corollary of postulate 26.1 [p. 446; see also definition 3.2: p. 42]:[7]

Corollary 26.2 (QM isolated systems) *In QM there is no isolated physical system.*

The corollary is of great importance, because in classical mechanics it was always possible to assume an isolated system. Hence, rigorously speaking, there are isolated objects only within the limit of a classical environment [see discussion in section 24.5], even if SSRs [see definition 10.2: p. 176] are valid to a high degree of accuracy [*BUSCH et al.* 1991, 27–28].

This does not signify that the microsystems have no reality independently from us (or from the consciousness). In fact, as we have already said, Measurement is only a particular form of interaction [*HAAG* 1992, 309–312]: we know that there are also spontaneous transitions to eigenstates — spontaneous decoherence [see chapter 21]. And, on the other side, we know that experiments are performable in which the 'reduction' also happens with the sole presence of a detector, and therefore without reading [see subsection 17.1.2, especially theorem 17.2, and chapter 20].

But the problem is that we can only speak of physical microsystems in the context of a determinate experimental arrangement or more generally in the context of some physical interaction [see also principle 3.1: p. 28]. Here we have the most important theoretical justification of an Operational approach [see chapter 18]. It is evident then that we can encounter here some fundamentals tenets of Pragmatism [*DEWEY* 1929] without following some subjective formulations

[7] Formulations in [ROSENFELD 1953, 48] [*FOLSE* 1985, 113].

of it, especially in the frame of the Copenhagen interpretation, or without supposing some break between macrophysics and microphysics, again as the Copenhagen Interpretation does.

In fact, as Peirce pointed out [*PEIRCE* 1872-73, 30], we can speak of a property of an object, even if macroscopical, say of the hardness of a diamond, only if we know that there is no stone able to scratch it. This means in no way that the object does not have this property if it is not actually scratched by some stone — or, following Berkeley [*BERKELEY* 1710], that it ceases to have it if a stone does not scratch it or it does not yet have this property if some stone has not already scratched it —, but only that we cannot speak of this property without imagining some circumstances under which such a property can manifest itself — it is a counterfactual reasoning [see also section 36.2] —, which, on the other side, implies that someone has at one time or another experienced such a factual presence of the property. In other words, something exists only by virtue of some conditions, under which it can happen (e. g. some other stones which can be scratched against the diamond), and can be thought of only under the condition that someone can experience such a phenomenon.

Obviously the difference between the classical and the quantum case is due to the fact that some properties in the quantum domain begin to exist only when and if we interact with them, due to the superposition — we shall return to this problem later [chapter 29].

If so, then the discussion of the last years[8], in order to see whether QM is realistic or not, is a little bit displaced. In fact under *Realism* we can understand the doctrine stating that systems have properties independently from the observational process [BALLENTINE 1987, 1493] [*DUMMETT* 1978] [*DUMMETT* 1991, 4]. In other words the realistic tenet is in exact opposition to the subjectivistic interpretation of QM [see postulate 8.1: p. 137]. But here we are claiming that physical properties can be given only through some form of physical interaction, which in turn presupposes a type of reality completely different from that imagined by classically realistic positions. In other words, proponents of Realism apply to QM an ontology which is fully inadequate to it — we shall return to the problem in part VII. D'Espagnat, in order to express the point of view supported here, uses the term 'weak objectivity' [*D'ESPAGNAT* 1995, 23, 310–11, 325–30].

26.4.2 The Present Determines the 'Past'

And here we come to the second, more radical lesson: the present experimental arrangement has the power to define how a system will be manifested to us. As a consequence of this fact, i.e. of the impossibility of a prediction or of a retrodiction [theorems 25.1 and 26.1], we cannot speak quantum-mechanically of only *one* history of our universe or of a branching of alternative paths: each path of a system coming from the 'past' can be in superposition with others paths of the same system and in entanglement with other systems' paths, and whether or not it is depends also on the choice made in what, each time, is the present, which is the only way to decohere to a certain extent such a superposition of different paths [see figure 26.6].

26.4.3 Can the Universe be Treated as a Closed System?

We have seen that the decoherent-histories approach [see subsection 25.2.3] can be developed into a new version of the MWI, where there is never a 'reduction' to one alternative branch of the universe (it is only a local perspective depending on a reduced density matrix). We would then have a universal wave function which is ever in superposition (between different possible

[8]See for example [*REDHEAD* 1987b].

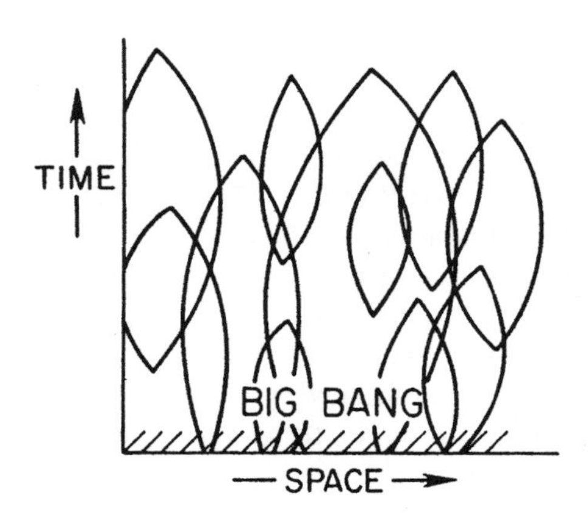

Figure 26.6: Symbolic description of how all that 'has happened' in the past is influenced by choices made in the present as to what to observe. The upper tip of each 'leaf' stands for the elementary act of registration. The lower end of each 'leaf' stands for the beginning of the elementary phenomenon being investigated — from [WHEELER 1983, 198].

'paths') and decoheres only locally, while from the universal point of view it is only an illusory phenomenon.

We have also seen the problems in taking this step. We know [see theorem 17.4: p. 269, and its corollary] that the trio-orthogonal decomposition is something more than the partial trace: in fact, even by spontaneous transitions (not necessarily by measurements), each time that two open systems (open to the environment) interact, we have a diagonalization, hence a spontaneous localization and 'individualisation'. But let us leave aside such a problem and consider only the operation of partial tracing. It seems then, that in this way one can distinguish the decoherent-history approach from the MWI. And in this way Joos and Zeh, Omnès in later works [*OMNÈS* 1994] and Kiefer [KIEFER 1996b, 177] seem to understand the subject[9]. But how can a history be decohering if it is a history of the whole universe, and hence without an 'external environment' able to produce such a *local* and *partial* effect? Naturally it should happen relatively to us, i.e. to some particular and partial tracing out. But here the situation is inverted: we have not one or two small systems ($\mathcal{S} + \mathcal{A}$) which interact with a large environment, but a very big environment (the whole universe!) which is 'measured' in some way by small systems *within it*.

But, apart from the problems posed by the above inversion as such, is it sound to suppose that it is possible to 'measure' the universe as such or some of its observables? How could it be possible to 'measure' in one way or another the wave function of the universe? By which technical and theoretical means? In fact there is no way to know directly the wave function or the state of the universe, since our universe is unique [see sections 28.4 and 28.5].

And, as a consequence, since partial tracing depends from the point of view of an observer or of an apparatus 'internal' with respect to the observed system (the universe), this signifies that there can be reduced states (by tracing out the rest of the world) of different states of the

[9]See also [ALBRECHT 1992, 5512–18] [ALBRECHT 1993, 3771–76].

universe which cannot be distinguished by this 'internal' observer or apparatus[10].

For example [*T. BREUER* 1997, 109] there are several 'states of the universe' which are different from one other only through EPR-correlations [see chapter 31] between several observers of the same event: these observers then 'see' the same event though the states of the whole are different.

Generalizing, since a Measurement is a bilinear correspondence between the states of a system $\mathcal{S}$ (here the environment) and the states of an apparatus $\mathcal{A}$ from which the properties of $\mathcal{S}$ can be inferred [see theorem 15.1: p. 246], it follows that $\mathcal{S}$ determines $\mathcal{A}$. But, as we know [see introduction to part V and also part IX], in QM a whole (the universe) in entanglement does not necessarily determines the parts (the 'measuring' apparatus). And, if we are not able to measure some observables, how can we speak in any form of the wave function of the universe?

Another problem is that, following such a proposal, the world should be quantum-mechanical as such and the macroscopic phenomena only an appearance. We have already discussed the unsoundness of such an interpretation [see postulate 24.1: p. 428, and comments].

Therefore, we have sufficient grounds to exclude the possibility of a universal wave function as a description of our universe and hence also to renounce the possibility of describing the universe as a whole as a closed system: we can describe systems only if we can interact (directly or indirectly) with them[11], which is always a local phenomenon, and we cannot think of a useful universal wave function which avoids the basis degeneracy problem. Therefore, concluding this examination, we can state the following corollary:

Corollary 26.3 (Universal Wave function) *A universal wave function cannot be defined.*

The only remaining possibility is that the wave function of the universe *exists*, even if nobody or nothing in our universe can ever know it — which would be anyway an exception to corollary 26.1. But this choice shifts the problem to that of the existence of some observer outside of our universe. And here we have two possible cases:

- this observer is God;
- this observer is in another universe outside of our own.

If we choose the first answer, we are confronted with a theological problem which is of no relevance in the frame of QM and of our analysis[12]. If we choose the second, we are moving into a unknown domain in which everything is possible but completely beyond all means of verification for what we know at present. Epistemologically the situation is not very different from that in which one insists on maintaining the notion of 'contemporaneity' even after Einstein's theory of Relativity.

26.4.4 Participation

In conclusion, after Wheeler [WHEELER 1990], QM ontology must be founded on the category of *participation*, or in other words of interaction. It is a consequence of the fact that all physical processes are local and context-dependent[13].

[10] As can be deduced from [*MITTELSTAEDT* 1998, 120–21]. T. Breuer has shown [*T. BREUER* 1997, 53–75] that in general an 'internal' apparatus cannot distinguish all states of a system, and this result is valid in both classical and quantum mechanics.

[11] See also [HAAG 1990] [MARGENAU/WIGNER 1962].

[12] However, it could be asked if one can reach the God's level of knowledge.

[13] See also [BASTIN 1976, 198]. .

Obviously in this context we cannot use the concept of causality in the sense of a one-directional relation between *one* condition and *one* effect. Hence it is right to say with Bohr that QM represents a break in causality [see subsection 8.4.1]. But for different reasons: a quantum-mechanical world is a cluster of interdependent things where each local interaction represents a new reformulation of such interdependence and a partial decoherence of some paths.

26.4.5 Universality of Quantum Mechanics

As a final consequence of the preceding examination, we discuss the problem of the universality of the theory. We have already said that QM is self-referential [see subsection 24.5.4]. But we have also excluded the semantical completeness of the theory [see subsection 9.4.2]: In that discussion we have chosen a middle way between a 'strong' completeness and a strong 'incompleteness'.

All of the preceding discussion moves us toward a similar way out. QM is surely universal, because, as far as we know, it applies to all domains of 'reality'. However, this cannot be understood as a static mirroring of a global reality, but only as an effort to extend the theory ever more. Therefore, it is surely not universal, if we understand by 'universal' a total explanation of reality. In conclusion: no metaphysics of a 'universal wave function' but only an operational approach is able to account for the dynamics of microsystems.

26.4.6 Epistemological Aspects

The fact that QM is based on phenomena such as entanglement and superposition, can be viewed as a fundamental limitation of the theory relative to classical mechanics, which can conceive isolated systems. But it can also be shown that what an 'isolated system' is, also in classical mechanics is context-dependent. This has been also showed to be valid for thermodynamical descriptions [J. BLATT 1959] [see section 17.1].

For example [SHARP 1961, 231] take the gravitational influence of Syrius on the system Sun/Earth: it is negligible. But, suppose that Syrius and the Earth are a compound isolated system: then the Earth sooner or later will experience a deviation from the straight line.

Already in 1914 Borel pointed out that even the gravitational effect resulting form the shifting of a small piece of rock as distant as Syrius by a few centimeters would completely change the microscopic state of a gas in a vessel here on Earth [ZEH 1996, 27] [JOOS 1996, 35–36].

Chapter 27

REVERSIBILITY AND IRREVERSIBILITY

Contents The general problem arises as to whether quantum systems can show irreversible behaviour. In fact, on one hand, the laws of QM seems reversible [see also section 46.4]; on the other, we suppose some form of irreversibility by measurements and interactions between three or more involved systems [see chapters 17, 21 and 26]. This apparent contradiction must be overcome. A number of concrete experiments have been proposed and performed in order to solve the problem. In section 27.1 we show a proposed spin experiment. In section 27.2 we report an interesting proposal of Greenberger and Yasin (Haunted measurements). In section 27.3 we return to the problem of revival by S-Cat experiments. In section 27.4 we report what is perhaps the most interesting experiment proposed (by Mabuchi and Zoller): it makes use of Montecarlo methods. Finally [in section 27.5] some conclusions are drawn.

27.1 A Spin Experiment

We wish to know if Measurement is really irreversible as postulated by von Neumann [see subsection 14.1.2]. As an example take, the reconstruction of the coherence of spin in pairs of 1/2–spin particles originated by beam splitting through a SGM (SGM Exp. 3) [ENGLERT *et al.* 1988].[1] The proposed experiment is very similar to the previously analysed use of a quantum eraser [see Two-Slit-Exp. 8: p. 143 and Interfer. Exp. 4: p. 146]. The set-up is shown in figure 27.1. Our interest in such a device is that it represents an attempt to make the measurement reversible.

Given a Hamiltonian for the spin and the z motion like the following:

$$\hat{H} = \frac{1}{2m}p_z^2 - E(t)\hat{\sigma}_z - F(t)z\hat{\sigma}_z, \tag{27.1}$$

where the energy parameter $E(t) := \mu B(t)$ (μ is the magnetic moment of the particle and B the magnetic field) and the force parameter $F(t) := \mu\partial B/\partial z$ are non-zero only while the particle is inside the SGM, we find the following Heisenberg Eqs. of motion for the spin:

$$\hat{\sigma}_z(t) = \hat{\sigma}_z(0), \tag{27.2a}$$

$$\hat{\sigma}_+(t) = \exp\left[-\imath\left(\varphi(t) + 2z(0)\frac{\Delta p_z(t)}{\hbar} - 2p_z(0)\frac{\Delta z(t)}{\hbar}\right)\right]\hat{\sigma}_+(0), \tag{27.2b}$$

[1] The first proposal of a so-called 'quantum eraser' in [SCULLY/DRÜHL 1982].

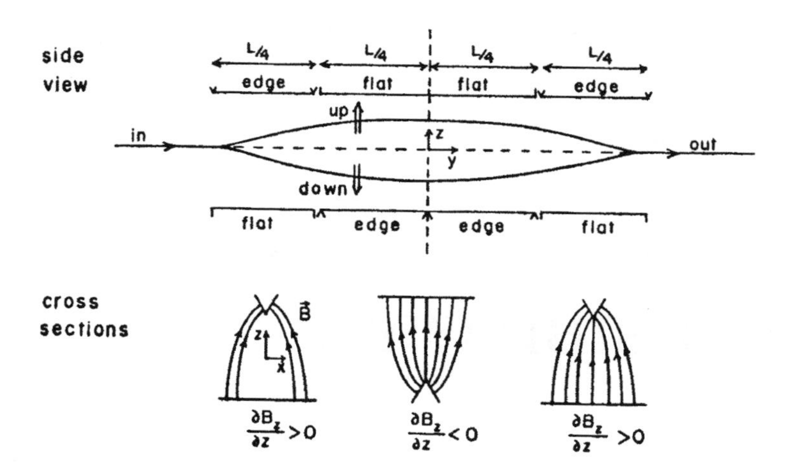

Figure 27.1: Proposed SGM Exp. 3 — from [ENGLERT *et al.* 1988, 1046].

where $\hat{\sigma}_+ = \hat{\sigma}_x(t) + \imath\hat{\sigma}_y(t)$ and:

$$\Delta p_z(t) = \int_0^t dt' F(t'), \quad \Delta z(t) = -\int_0^t dt' t' \tfrac{F(t')}{m}, \quad \varphi(t) = \frac{2}{\hbar}\int_0^t dt' E(t'). \tag{27.3}$$

Consider that the incoming particle is in a eigenstate of $\hat{\sigma}_x$ so that:

$$\langle \hat{\sigma}_x(0) \rangle = 1, \quad \langle \hat{\sigma}_y(0) \rangle = 0, \quad \langle \hat{\sigma}_z(0) \rangle = 0. \tag{27.4}$$

Then the expectation that after a $\hat{\sigma}_z$ measurement, the particle will be in an eigenstate of $\hat{\sigma}_x$ is given by:

$$\begin{aligned} \langle \hat{\sigma}_x(t) \rangle &= \mathrm{Re}\,\langle \hat{\sigma}_+(t) \rangle & \text{(27.5a)} \\ &= \mathrm{Re}\left\{ e^{-\imath\varphi(t)} \left\langle \psi \left| \exp\left[\frac{2\imath}{\hbar}\left(p_z(0)\Delta z(t) - z(0)\Delta p_z(t)\right)\right] \right| \psi \right\rangle \right\}, & \text{(27.5b)} \end{aligned}$$

where $|\psi\rangle$ means the normalized spatial state of the incoming particle. By using z' or p' integration, we find:

$$\begin{aligned} &\left\langle \psi \left| \exp\left[\frac{2\imath}{\hbar}(p_z(0)\Delta z(t) - z(0)\Delta p_z(t))\right] \right| \psi \right\rangle \\ &= \int dz' \psi^*(z' - \Delta z(t)) e^{-2\imath z' \Delta p_z(t)/\hbar} \psi(z' + \Delta z(t)) \\ &= \int dp'_z \tilde{\psi}^*(p'_z - \Delta p_z(t)) e^{2\imath p'_z \Delta z(t)/\hbar} \tilde{\psi}(p'_z + \Delta p(t)). \end{aligned}$$

The integrals will tend to vanish unless both $\Delta z(t)$ and $\Delta p_z(t)$ are small on the scale set by the initial spreads of position δz and momentum δp_z. Hence a macroscopic splitting either in the position or in the momentum produces $\langle \hat{\sigma}_x \rangle = 0$, indicating a total loss of coherence. Hence by beam merging (at time τ), one would need ideally $\Delta z(\tau) = 0$, $\Delta p_z(\tau) = 0$ and an integer multiple of 2π for $\varphi(\tau)$. More precise calculations allow (where the subscript s' means 'spatial' and the s means 'spin'):

$$|\delta\varphi(\tau)| \leq \epsilon_s, \quad \left|\frac{\Delta z(\tau)}{\delta z}\right| \leq \frac{\epsilon_{s'}}{\sqrt{2}}, \quad \left|\frac{\Delta p_z(\tau)}{\delta p_z}\right| \leq \frac{\epsilon_{s'}}{\sqrt{2}}, \tag{27.6}$$

with $\epsilon_{s'}, \epsilon_s \ll 1$. Then the ability of the SGM to conserve spin coherence is given by:

$$|1 - \langle \hat{\sigma}_x(\tau) \rangle| \leq \frac{1}{2}|\delta\varphi(\tau)|^2 + \frac{1}{2}\left|\frac{\Delta z(\tau)}{\delta z}\right|^2 + \frac{1}{2}\left|\frac{\Delta p_z(\tau)}{\delta p_z}\right|^2 \geq \frac{1}{2}(\epsilon_s^2 + \epsilon_{s'}^2). \tag{27.7}$$

For instance, in order to maintain 99% of coherence we need $\epsilon_s = \epsilon_{s'} = 1/10$. But the final result is that there is always a certain loss of coherence: Even if it is largely reconstructed, it can never happen totally, and therefore there is some form of irreversibility [SCULLY *et al.* 1989, 1776].[2]

27.2 Haunted Measurements

Similar techniques have been developed by Greenberger and Yasin under the name of haunted Measurement. Take the example of a spin echo experiment [GREENBERGER/YASIN 1989]: a magnetic pulse apparently disorders a system but a second pulse reorders the system so that all spins are coherently aligned for a moment, which also proves that between the I and the II pulse the system was already ordered. Now imagine a system $\mathcal{S}$ which interacts with a macroscopic body $\mathcal{A}$ inducing a macroscopic change in $\mathcal{A}$ and a loss of order in $\mathcal{S}$. $\mathcal{S}$ then interacts again undoing the macroscopic effect and completely regaining its coherence so that the measurement has been 'haunted'. Normally we do not consider this the case for QM. The problem is when an irreversible recording process has taken place [GREENBERGER/YASIN 1989, 680].

Definition 27.1 (Latent Order Systems) *Latent Order Systems* (= LOS) *are systems which appear to be incoherent but are in reality still coherent.*

It is difficult to distinguish between incoherent systems and LOSs. Take a heavy particle ($\mathcal{A}$) interacting with a photon ($\mathcal{S}$): the impact scrambles the wave function of both, so that they would appear to be incoherent. But there is a latent order: both are correlated — in an EPR way [see chapter 31]. But if one looks at the pointer, the correlation is lost and we have a true measurement. But if we do not look, both can interact again transforming latent into manifest order. And if the I collision were a little stronger, there would be no need of a second interaction: there was already a true measurement by the first [GREENBERGER/YASIN 1989, 681]. In other words, Greenberger and Yasin mean that the actual measurement theory is incomplete until one is unable to distinguish completely between an already performed measurement (irreversible) and a normal reversible unitary evolution.

In order to demonstrate the problem they propose the following thought-experiment: we have an interferometer (Interfer. Exp. 11) [see figure 27.2] with an incident neutron with narrow momentum spread and wide position spread $a \gg \lambda = 2\pi/k_0$ (k_0 is the average x–momentum):

$$|\psi(x_1)\rangle = Ne^{\imath k_0 x_1}e^{-\frac{x_1^2}{a^2}}, \tag{27.8}$$

[2]See also [SCULLY/WALTHER 1989, 5231] . An experiment with a quantum eraser has been performed by Kwiat and co-workers [KWIAT *et al.* 1992].

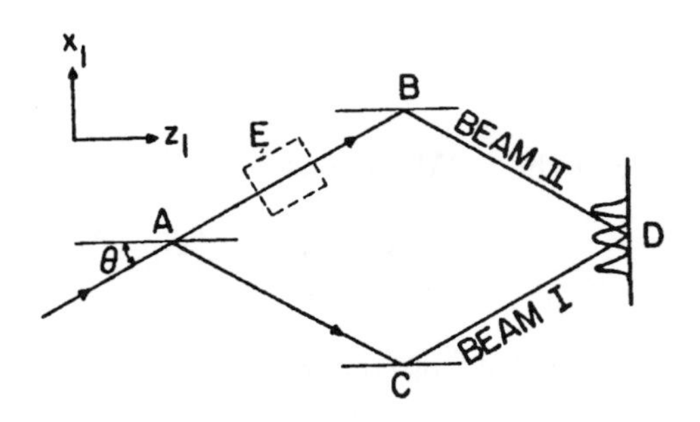

Figure 27.2: Simplified model of the neutron interferometry. The beam is split by a BS at A, bounces off fully reflecting mirrors at B and C, and forms a diffraction pattern at a screen at D. The effects on the phase of some apparatus at E can be monitored by placing it in beam II, and observing the shift in diffraction peaks at D — from [GREENBERGER/YASIN 1989, 683].

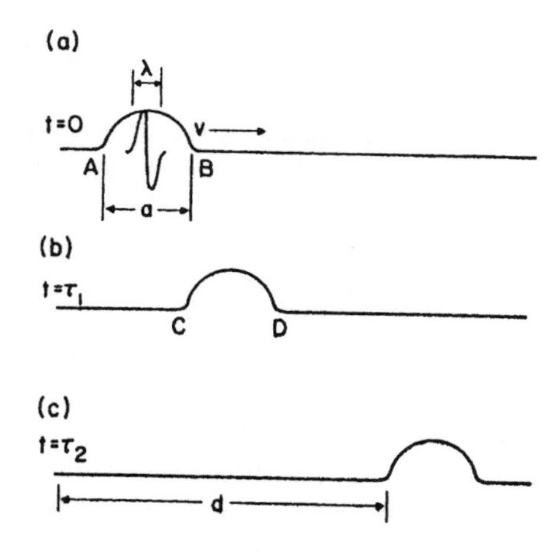

Figure 27.3: (a) The coherent length of the packet is a, and the phases at A and B start to become incoherent with each other.
(b) After a time τ_1, when the beam has travelled distance a, in the group velocity approximation, the phases at A and C are coherent, as are those at B and D, although not those at C and D. This is because the approximation cancels out first-order effects due to δk.
(c) After a time τ_2, when the beam has travelled distance $d \simeq a^2/\lambda$, the second-order effects in $(\delta k)^2$ produce general incoherence, so that $d \gg L$, the size of the experiment — from [GREENBERGER/YASIN 1989, 685].

where N is a normalization constant and x_1 is the neutron coordinate. The distance between mirrors is very large: $l \gg a$. a is the phase coherence length of the packet, so that at τ_0, at points A and B the phases start to become incoherent with each other [see figure 27.3]:

$$e^{\imath kx} \simeq e^{\imath(k_0+\delta k)x}, \qquad e^{\imath \delta kx} = e^{\imath \delta \varphi}, \tag{27.9}$$

where $\delta\varphi \sim 2\pi$ for $x \sim a$. At later time $\tau_1 \sim a/v_g$ (where v_g is the group velocity):

$$e^{\imath(kx-\omega t)} = e^{\imath(k_0 x-\omega_0 t)} e^{\imath \delta k(x-v_g t)} e^{\imath \frac{\hbar(\delta k)^2}{2M_1} t}. \tag{27.10}$$

Now the phase at A is completely correlated with the phase at C and the phase at B with that at D [GREENBERGER/YASIN 1989, 685–86]. The coherence say between B and D will persist until a time τ_2 when the contribution in $(\delta k)^2$ becomes significant, after some distance $d \gg L$.[3]

In conclusion, the hypothesis of Greenberger/Yasin about an incompleteness of the Measurement theory can be turned out into a positive hypothesis about the boundaries between irreversible and reversible behaviour. Hence we state what follows:

Postulate 27.1 (Complementarity Reversibility/Irreversibility (= CP6)) *Reversible and irreversible behaviour of quantum systems are complementary, and this complementarity is a smooth one.*

As we shall see, in the following sections all further experiments of the last years confirm such an assumption. It can be that such a postulate can be reduced in a form or another to CP4. However, we prefer to single out different aspects of the theory separately instead of trying a more compact but less evident construction.

27.3 Schrödinger Cats and Revival Behaviour

Another direction in research has been initiated by Eberly and co-workers [EBERLY *et al.* 1980]: they showed a collapse and revival behaviour in the interaction between an atom and a single mode of a quantized radiation field (JCM).

Similarly, Rempe, Walther and Klein [REMPE *et al.* 1987] showed the collapse and revival behaviour of a single Rydberg atom interacting with a single mode of an electromagnetic field (JCM) in a superconducting cavity.

On the same line Gea-Banacloche [GEA-BANACLOCHE 1990] has later shown that a two-level atom interacting with a quantum field (in the resonant JCM) presents the revival behaviour. This model has been developed further by Bužek and co-workers [BUŽEK *et al.* 1992b] [see Cavity Exp. 3: p. 416]. Another interesting proposal by Raimond, Brune and Haroche [RAIMOND *et al.* 1997] follows the outlines of Cavity Exp. 4 [see especially subsection 24.4.2]. It shows that leakage of information and irreversibility can be partly different things in a model where loss and revival of coherence are dynamical processes[4].

[3] See also [GREENBERGER/YASIN 1989, 696–97]. The proposed experiment can also be viewed as a delayed choice experiment — or an EPR one [GREENBERGER/YASIN 1989, 700–702].

[4] See also [BRUNE *et al.* 1996a].

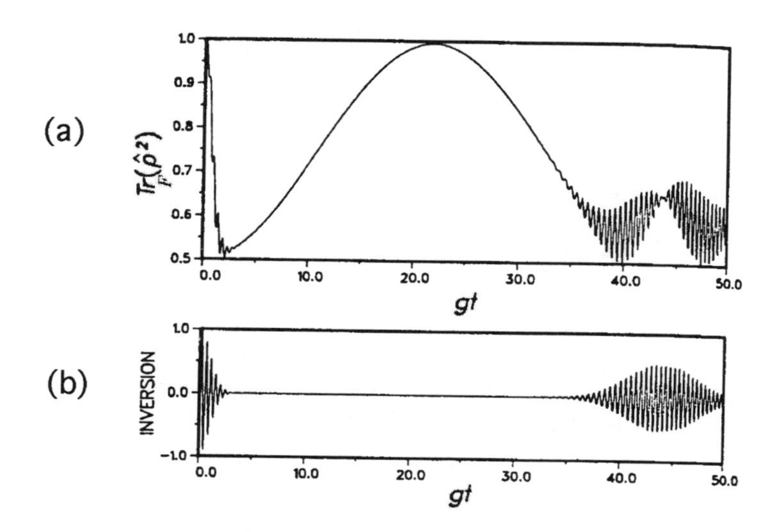

Figure 27.4: (a) The quantity $\mathrm{Tr}_{\mathrm{F}}(\hat{\rho}^2)$ (trace over the field) as a function of time for an atom initially prepared in state $|g\rangle$ and field initially in a coherent state with $\overline{\hat{N}} = 49$.
(b) The average atomic inversion as a function of time for the same initial conditions as in (a). $g = g_{\mathrm{AF}}$ — from [GEA-BANACLOCHE 1990, 3386]

Returning to Gea-Banacloche's proposal, by taking the atom to be in the initial lower state $|g\rangle$ and the field to be in the coherent state $|\alpha\rangle = \overline{\hat{N}}^{\frac{1}{2}} e^{-\imath\varphi}$ (where $\overline{\hat{N}}$ is the average number of photons of the field), we find first of all that the atom remains for a short time of the order of $1/g_{\mathrm{AF}}$ (where g_{AF} is the coupling constant) in a pure state, then follows a collapse. Finally, by $\tau = \overline{\hat{N}}^{\frac{1}{2}} \pi / g_{\mathrm{AF}}$ a revival takes place [see figure 27.4.(a)]. It is worth mentioning that the recreation of the state vector takes place in the middle of what is traditionally considered the JCM collapse region [see figure 27.4.(b)]: in fact the *inversion* (the probability of finding the atom in the upper state minus the probability of finding it in the lower state) would suggest that throughout the whole collapse region the atom has somehow lost coherence.

Experiments or *Gedankenexperimenten* like this one show that 'reversible' QM systems can be treated as pendulum. In fact the coupling of an atom with a single mode of an electromagnetic field (= JCM) can be represented as a pair of coupled pendulum (the atom performs a transition from ground to excited state by absorbing a photon while the field goes from $|1\rangle$ to $|0\rangle$, and *vice versa*). But precisely for this reason, they tend to undergo some dispersion, so that, after a long series of 'oscillations', the irreversible behaviour is evident.

Similar revival behaviour can also be found in wave packets: even if they can refer to different systems (atoms, molecules and cavity QED), the behaviour is essentially the same [LEICHTLE *et al.* 1996]. However, there is always some dynamical dispersion.

27.4 Montecarlo Methods

27.4.1 First Attempts

In a pioneering work Ueda and Kitagawa [UEDA/KITAGAWA 1992] showed that a measurement can be made reversible if

- it is unsharp, and
- it is sensitive to the vacuum field fluctuations.

But the authors do not consider a conserved quantity; therefore the argument is not resolutive.

A conserved quantity, here the photon-number operator, is considered in a theoretical study by Imamoḡlu [IMAMOḠLU 1993], i.e. repetitive (logically reversible) measurements seem possible. Imamoḡlu uses a QND-Measurement scheme with a three-level atom that exhibits coherent population trapping at the two-photon resonance — the frequency of the metastable level $|m\rangle$ plus the frequency difference between the frequency of $|m\rangle$ and that of the excited level $|e\rangle$ is equal to the frequency of the ground level $|g\rangle$ plus the difference of frequencies between $|e\rangle$ and $|g\rangle$. It has been demonstrated that the state of the coupling field before and after the measurement is unchanged.

What Imamoḡlu has shown is that there are conditions under which one can recover information which is apparently lost during a normal (reduction-like) measurement. But he has not shown (and has not pretended to have shown) that this information is normally not actually lost. In fact, recent attempts to make completely reversible quantum measurements in order to measure the density matrix have also failed because of this problem [see subsection 28.4.2].

27.4.2 A 'Unitarily Reversible' Measurement

Recently Mabuchi and Zoller [MABUCHI/ZOLLER 1996] have shown under which conditions reversibility is possible in the case of a system coupled to an environment. In short they prove following theorem:

Theorem 27.1 (Jump operator) *The action of a jump superoperator cannot in general be inverted as such. Inversion is possible if one consider the system as pertaining to a subset of the original Hilbert space and the jump as unitary: but then no new information is obtained.*

The theorem is a further specification of the second theorem of von Neumann [14.2: p. 222]. In fact a jump superoperator can generally account for a measurement, due to its non-unitarity.

Let us discuss the problem on an abstract level (the authors also propose a concrete experiment). Let the evolution of some system be subjected to the action of jump superoperator $\hat{\mathcal{J}}_{j_r}$ [see subsection 22.1.3; particularly Eq. (22.24)]:

$$|\psi_S(t_r + dt)\rangle = \hat{\mathcal{J}}_{j_r}|\psi_S(t_r)\rangle, \tag{27.11}$$

where t_r is the observation time of jumps j_r (j is the j–th channel, for example the j–th harmonic oscillator of a bath reservoir). Between different counts the system wave function obeys:

$$|\psi_S(t)\rangle = e^{-\imath \hat{H}_{\text{eff}}(t-t_r)}|\psi_S(t_r)\rangle, \tag{27.12}$$

where $\hat{H}_{\rm eff} = \hat{H}_{\mathcal{E}+\mathcal{S}} - \imath\frac{1}{2}\sum_j \hat{\mathcal{J}}_j^\dagger \hat{\mathcal{J}}_j$ is an effective non-Hermitian Hamiltonian. Their final result is that the quantum jump superoperator is not invertible on the entire Hilbert space of the system, and that it will eventually be on a subspace of the same assuming a feedback, i.e. the action on the input of the output of the same system[5] described by a unitary operator such that:

$$|\psi_{\mathcal{S}}(t_r + dt)\rangle = \hat{U}_j \hat{\mathcal{J}}_{j_r} |\psi_{\mathcal{S}}(t_r)\rangle \propto |\psi_{\mathcal{S}}(t_r)\rangle. \tag{27.13}$$

Hence $\hat{\mathcal{J}}_j = c_j \hat{U}_j^\dagger$ (where c_j is a complex number). If by two QM jumps the unitary dynamic is not distorted by damping factors:

$$|\psi_{\mathcal{S}}(t)\rangle = e^{-\imath\hat{H}_{\rm eff}t}|\psi_{\mathcal{S}}\rangle = e^{-\frac{1}{2}\sum_j |c_j|^2 t} e^{-\imath\hat{H}_{\mathcal{E}+\mathcal{S}}t}|\psi_{\mathcal{S}}\rangle, \tag{27.14}$$

where the wave functions pertains to the subspace considered, then the feedback can undo the effect of the quantum jump:

$$\begin{aligned} |\Psi_{\mathcal{E}+\mathcal{S}}(t)\rangle &= \frac{1}{N} e^{-\imath\hat{H}_{\rm eff}(t-t_n)} \hat{U}_{j_n}\hat{\mathcal{J}}_{j_n}\cdots\hat{U}_{j_1}\hat{\mathcal{J}}_{j_1} e^{-\imath\hat{H}_{\rm eff}t_1}|\Psi_{\mathcal{E}+\mathcal{S}}\rangle \\ &= e^{-\imath\hat{H}_{\mathcal{E}+\mathcal{S}}t}|\Psi_{\mathcal{E}+\mathcal{S}}\rangle, \end{aligned} \tag{27.15}$$

where N is some normalization constant.

27.4.3 Is New Information Gained?

The model chosen by Mabuchi and Zoller is not a measurement in the more general sense of the word [MENSKY 1996]. In fact they suppose the action of a destruction operator (which causes the jump by photon absorption, for example) $\hat{a}$ which acts as a unitary operator on a subspace $\mathcal{H}_1^{\rm Tr=1;\geq 0}$ ($\mathcal{H}_1$ for the sake of simplicity, where the subscript '1' means here 1 photon) of the original problem [Eq. (27.13)]:

$$\hat{a}|\psi\rangle = \hat{U}|\psi\rangle, \quad \langle\psi|\hat{a}^\dagger = \langle\psi|\hat{U}^\dagger. \tag{27.16}$$

Due to the unitarity of $\hat{U}$ we have:

$$\langle\psi|\hat{N}|\psi\rangle = 1, \tag{27.17}$$

where $\hat{N}$ is the number operator, which means that the expectation value of the photon number is equal to unity for an arbitrary state from the specified subset $|\psi\rangle \in \mathcal{H}_1$. If we expand $|\psi\rangle$ as follows

$$|\psi\rangle = c_0|0\rangle + c_1|1\rangle + \ldots + c_n|n\rangle, \tag{27.18}$$

where $|n\rangle$ is a (normalized) state with n photons, then Eq. (27.17) can be written:

$$\wp_1 + 2\wp_2 + 3\wp_3 + \ldots + n\wp_n + \ldots = 1, \tag{27.19}$$

with positive numbers $\wp_n = |c_n|^2$. For this Eq. to be fulfilled, at least one of the numbers $\wp_1, \wp_2, \ldots$ must be non-zero. And therefore the state $|\psi\rangle \in \mathcal{H}_1$ cannot be the vacuum. Since a jump diminishes here the number of photons by single unity, the fact that a jump occurred gives the information that the initial state was not the vacuum: but we know this already, because the initial state belonged to the subset $\mathcal{H}_1$. But since $|\psi\rangle \in \mathcal{H}_1$ depends on the preparation, no new information is gained by such a reversible quantum jump [see also what is said in subsection 19.5.2]. The same can be proven if we consider a double jump operator (for the absorption of two photons) and so on.

Hence we have proved theorem 27.1, i.e. that, in the general case, the action of a jump operator cannot be inverted.

[5] On this point see [WISEMAN/MILBURN 1993].

27.4.4 Formal Generalization of Unitarily Reversible Measurements

Nielsen and Caves provided [NIELSEN/CAVES 1997, 2247–51] a powerful generalization of the preceding analysis. We remember here the formula (18.5a) for operations: $\mathcal{T}\hat{\rho} = \sum \hat{O}_k \hat{\rho} \hat{O}_k^\dagger$ [see also Eq. (21.3b)]. If we have a single (not necessarily unitary) operator such that $\mathcal{T}\hat{\rho} = \hat{O}_k \hat{\rho} \hat{O}_k^\dagger$, then we speak of an *ideal* measurement (perfect redout of the state of the apparatus). We can then speak of a unitarily reversible measurement — on a subspace $\mathcal{H}_0^{\mathrm{Tr}=1;\geq 0}$ ($\mathcal{H}_0$ for the sake of simplicity) of the state space $\mathcal{H}^{\mathrm{Tr}=1;\geq 0}$ ($\mathcal{H}$ for the sake of simplicity) of the original problem, i.e. the initial state of the system —, if there exists a unitary operator $\hat{U}$ acting on $\mathcal{H}$ such that [see Eq. (18.3)]:

$$\hat{\rho} = \hat{U} \frac{\mathcal{T}\hat{\rho}}{\mathrm{Tr}[\mathcal{T}\hat{\rho}]} \hat{U}^\dagger \tag{27.20}$$

for all $\hat{\rho}$ whose support lies in $\mathcal{H}_0$. Now suppose the following generalized form of measurement [see Eq. (18.4)]:

$$\sum_{jk} \hat{O}_{jk}^\dagger \hat{O}_{jk} = \hat{I}. \tag{27.21}$$

j labels the outcome of the measurement. Supposing that j occurs, then the unnormalized state after the measurement is:

$$\mathcal{T}_j \hat{\rho} := \sum_k \hat{O}_{jk} \hat{\rho} \hat{O}_{jk}^\dagger. \tag{27.22}$$

The probability that outcome j occurs is:

$$\wp(j) = \mathrm{Tr}[\mathcal{T}_j \hat{\rho}] = \mathrm{Tr}\left[\hat{\rho} \sum_k \hat{O}_{jk}^\dagger \hat{O}_{jk}\right]. \tag{27.23}$$

Analogous expressions can be written in terms of state vector. The authors prove that the following conditions are equivalent:

- The ideal quantum operation $\mathcal{T}_j \hat{\rho} = \hat{O}_j \hat{\rho} \hat{O}_j^\dagger$ is unitarily reversible on a subspace $\mathcal{H}_0$ of the total space $\mathcal{H}$.
- The POVM $\hat{O}_{\mathrm{POV}} = \hat{O}_j^\dagger \hat{O}_j$, when restricted to $\mathcal{H}_0$, is a positive multiple of the identity operator on $\mathcal{H}_0$, i.e.:
$$\hat{P}_{\mathcal{H}_0} \hat{O}_{\mathrm{POV}} \hat{P}_{\mathcal{H}_0} = \eta^2 \hat{P}_{\mathcal{H}_0}, \tag{27.24}$$
where η is a real constant satisfying: $0 < \eta \leq 1$ and $\hat{P}_{\mathcal{H}_0}$ is the projector onto $\mathcal{H}_0$.
- We have
$$\mathrm{Tr}\left[\hat{\rho} \hat{O}_j^\dagger \hat{O}_j\right] = \mathrm{Tr}\left[\hat{\rho} \hat{O}_{\mathrm{POV}}\right] = \eta^2 \tag{27.25}$$
for all density operators whose support lies in $\mathcal{H}_0$.
- The operator $\hat{O}_j$ can be written as follows:
$$\hat{O}_j = \eta \hat{U} \hat{P}_{\mathcal{H}_0} + \hat{O}_j \hat{P}_{\mathcal{H}_0^\perp}, \tag{27.26}$$
where $\hat{U}$ is some unitary operator acting on the whole $\mathcal{H}$ and $\hat{P}_{\mathcal{H}_0^\perp}$ projects in the orthogonal complement of $\mathcal{H}_0$, that is $\hat{P}_{\mathcal{H}_0} + \hat{P}_{\mathcal{H}_0^\perp} = \hat{I}$.

Hence an ideal measurement is reversible iff no information about the identity of the prior state is obtained from the measurement, since each state is equally likely, given result j — as stated by Eq. (27.25). Therefore, it is not a measurement.

For further applications of this analysis see subsection 43.2.2. For the general problem of the impossibility of truly reversing measurements see also subsection 28.4.3.

27.5 Concluding Remarks

Practically QM systems are never completely free from some form of dispersion or dissipation. This is another way of saying that quantum systems are open systems, which confirms the main results of chapter 26. This signifies a connatural irreversibility of quantum systems. In other words, although QM laws are reversible, an isolated QM system is theoretically and practically impossible, which shows that factually we always have to some extent an irreversible behaviour [*HAAG* 1992, 312–17]. In conclusion we can postulate that what follows [see also postulate 27.1: p. 461]:

Postulate 27.2 (Irreversibility) *The reversibility of QM systems is only an ideal and limiting case of a series of irreversible behaviours, where a system is thought to be more and more isolated.*

Obviously there can be a partial reversibility only if there is no triorthogonal decomposition — i.e. in the case of a coupling between a system and a reservoir —, while, as has already been said, if three systems are coupled — which is the case when performing measurements [see theorem 17.4: p. 269, and its corollary] but also for some form of spontaneous interactions —, irreversible behaviour is stronger.

Part VII

WAVE/PARTICLE DUALISM

Introduction to Part VII We have touched many times upon the problem of WP-Dualism. Now we shall analyse the problem, by starting with what is chronologically the first discussion about it. An organic answer to the problem came from de Broglie, but, for many reasons yet to be analysed in greater depth, it was not accepted by the majority of physicists. This theory was later included in the theoretical program of the Hidden-Variable theory by David Bohm, so that one normally speaks of the 'de Broglie-Bohm Pilot wave'. However, de Broglie's proposal was directed more toward the problem of WP-Dualism as such, while Bohm's proposal was directed more toward the construction of a deterministic theory by contesting the completeness of QM — and here the point of departure is more Einstein's Ignorance interpretation [see proposition 6.3: p. 106] and the historical article of Einstein/Podolsky/Rosen [EINSTEIN *et al.* 1935]. As a consequence of the Hidden Variable theory the problem of non-locality was posed.

The problem of the WP-Dualism and that of Hidden Variable theories (and non-locality) are theoretically different and hence they require different examinations. Therefore we divide all matter into three different parts: in this part we shall discuss de Broglie's theory and the problem of WP-Dualism; in part VIII, the problem of completeness and determinism (Bohm's Hidden Variable theory); in part IX the problem of separability and non locality.

While part VIII, and especially chapter 33, is devoted to an interpretation of QM particles, we discuss in particular here what quantum waves are; if they are classical entities, if and how they can be ontologically interpreted, and what a quantum state is. We shall confirm our main distinction between states and observables.

- We begin [in *chapter 28*] with the classical discussion stimulated by de Broglie with his proposal of the double solution theory — a proposal following which both particles and waves always exist together. Then more recent proposals concerning the nature of quantum waves and of quantum states are discussed (non-analytical solutions, empty waves, ontological interpretation of waves based on protective measurements).

- Three-valued logic [*chapter 29*] represents to a certain extent the opposite point of view relative to de Broglie's: the microentities are not determined in themselves (neither corpuscular nor wave-like), and, only under very specific circumstances (measurements), can the truth or falsehood of propositions about the values of observables be decided. Even if three-valued logic is chronological precedent to the matter of the precedent chapter, for systematical reasons it seemed better to discuss it after the latter.

- Then in *chapter 30* the relationship between which-path information and superposition is discussed, and very recent (proposed and performed) experiments are reported; finally we begin to discuss an ontological interpretation.

Chapter 28

THE REALITY OF QUANTUM WAVES AND STATES

Contents In this chapter we discuss different and partly unrelated proposals developed around an ontologically realistic meaning (in a classical sense) of the wave function. In section 28.1 we present de Broglie's proposal, the first historical attempt along this line. Then we recall the principal criticisms developed in 1927 and discuss de Broglie's theory in more detail. In section 28.2 we examine (theoretically and experimentally) the non-linear corrections to the Schrödinger Eq. which have been proposed in the previous years. This is a further development of de Broglie's original proposal. In section 28.3 several proposed and realized experiments about the existence of empty-waves (another consequence of de Broglie's assumption about the existence and permanence of classical waves) are reported. Different interpretations of the subject are also discussed. In section 28.4 the interesting proposal of Aharonov and co-workers is discussed, which seems to assign an ontological meaning to the wave function by means of a single measurement, while in section 28.5 the impossibility of performing it is demonstrated (a new and very interesting field). Finally [section 28.6] the nature of quantum waves (classical or specifically quantic?) is discussed and conclusions are drawn [section 28.7].

28.1 De Broglie's Pilot Wave

28.1.1 Pilot Wave and Double Solution

Two problems have to be spelled out here before proceeding further into the analysis:

- that concerning the pilot wave,
- and that of the double solution.

Pilot wave The theory of the *pilot wave* is a proposal in order to understand the basic ontology of the microworld as composed of two different entities given simultaneously: a wave and a particle. The wave ψ is a field which moves wave-like in the space, and which 'pilots' a particle embedded in the field and which is sensible to all wave superpositions of the same field. In the example of Two-Slit-Exp. 1 [see p. 27], the particle, factually, though both slits are open, always passes only through one slit, and the diffraction pattern is only due to the strange and wave-like trajectory impressed by the field. In this perspective there is no complementarity [see CP2: p.

138, and CP3: p. 139] and no 'indeterminacy' at all. The antecedents of the Pilot wave theory have to be found in the hydrodynamic theories [MADELUNG 1926] [KORN 1927] [see subsection 23.1.1].[1]

Double solution The *double solution theory* is a mathematical aspect of the same idea: the correlation between particle and wave is a phase correlation, such that the particle is a singularity of the field, which differs from ψ only in amplitude, and which represents another, non-analytical solution of the wave equation.

Short history De Broglie published his results in a series of articles [DE BROGLIE 1927a] [DE BROGLIE 1927b] [DE BROGLIE 1927c]. But, on occasion of an exposition to a great scientific auditorium at the Fifth Physical Conference of the Solvay Institute in Brussels (October 1927), he presented only the simplified version of the whole: the Pilot wave theory [DE BROGLIE 1955, 38–40]. The many important criticisms to his proposal, pushed de Broglie to abandon the theory. Hence in a public lecture at the university of Hamburg in early 1928 he embraced CP.[2] Later (1955–56) he returned to his old proposal in a more systematic way.

28.1.2 De Broglie's Theory

Some formalism There is little doubt that de Broglie presented his proposal more organically in 1955–56 only after Bohm published his historical article in 1952 and after Vigier showed him the importance of relativistic considerations for the problem.

The model [*DE BROGLIE* 1956, 88–89, 92 107] considers a particle moving in the direction of z–axis, with proper mass m_0, and a point of observation, an observer $\mathcal{O}(x, y, z, t)$. The total energy and momentum of the particle from the point of view $\mathcal{O}$ are given by:

$$E = \frac{m_0 c^2}{\sqrt{1 - \frac{\mathbf{v}^2}{c^2}}}; \quad \mathbf{p} = \frac{m_0 \mathbf{v}}{\sqrt{1 - \frac{\mathbf{v}^2}{c^2}}}, \tag{28.1}$$

where $\mathbf{v}$ is the velocity of the particle in the frame of reference of $\mathcal{O}$. The quantized energy for the particle is:

$$E = m_0 c^2 = h\nu_0. \tag{28.2}$$

The Lorentz transformations from the frame of reference of $P(x_0, y_0, z_0, t_0)$, which is that of the particle, to point $\mathcal{O}(x, y, z, t)$ are given by [see Eqs. (4.1a)]:

$$x_0 = x, \quad y_0 = y, \quad z_0 = \frac{z - \mathbf{v}t}{\sqrt{1 - \frac{\mathbf{v}^2}{c^2}}}, \quad t_0 = \frac{t - \frac{\mathbf{v}}{c^2} z}{\sqrt{1 - \frac{\mathbf{v}^2}{c^2}}}, \tag{28.3a}$$

while the mass is transformed according to [see Eq. (4.4)]:

$$m = \frac{m_0}{\sqrt{1 - \frac{\mathbf{v}^2}{c^2}}}. \tag{28.3b}$$

Trying to find a wave associated with the dynamic evolution of the particle, we can write it in the following form [see Eq. (3.21)]:

$$|\psi_0\rangle = \vartheta_0 e^{2\pi \imath \nu_0 t_0}. \tag{28.4}$$

[1] See also [*DE BROGLIE* 1956, 87] for further evidence.

[2] Historical reconstruction in [*JAMMER* 1974, 110–14].

If we write the wave equation from the point of view of $\mathcal{O}$, we must use the transformation Eq. of the time (28.3a), so that we obtain:

$$|\psi\rangle = \vartheta_0 e^{2\pi\imath\nu_w(t-z/\mathbf{v}_w)}, \tag{28.5}$$

where frequency and velocity are given respectively by:

$$\nu_w = \frac{\nu_0}{\sqrt{1-\frac{\mathbf{v}^2}{c^2}}}, \tag{28.6a}$$

$$\mathbf{v}_w = \frac{c^2}{\mathbf{v}}. \tag{28.6b}$$

Interpretation However, ν_w and $\mathbf{v}_w$ are not at all equal to the frequency and the velocity of the particle from the point of view $\mathcal{O}$:

- The velocity of the particle is $\mathbf{v}$, which is smaller than $\mathbf{v}_w$, as is evident from Eq. (28.6b). We incidentally note that $\mathbf{v}_w$ is greater than c.
- For the frequency we have the following situation: because of the slowing down of the clock associated with P in relation to $\mathcal{O}$ (due to the velocity $\mathbf{v}$ of the particle in relation to $\mathcal{O}$), the frequency of the particle is diminished when we see it in the frame of reference of $\mathcal{O}$ by an amount:

$$\nu_p = \nu_0\sqrt{1-\frac{\mathbf{v}^2}{c^2}}; \tag{28.7}$$

  while the frequency ν_w is augmented in relation to ν_0 by the amount given by Eq. (28.6a).

Hence we must conclude that, from the observation point $\mathcal{O}$ there is a particle with velocity $\mathbf{v}$ and frequency ν_p, and a wave with velocity $\mathbf{v}_w$ and frequency ν_w. Now the particle can be interpreted as a clock which displaces itself with a velocity $\mathbf{v}_w - \mathbf{v}$, remaining in phase with the wave. Hence we can prove that:

Theorem 28.1 (de Broglie) *For any Galilean observer the phase of the internal clock of a moving quantum particle is at each instant equal to the value of the phase of the wave calculated at the same point at which the particle is.*

Proof
The phase *of the particle* from the point of view of $\mathcal{O}$ is:

$$\varphi_p = \nu_p t = \nu_0\sqrt{1-\frac{\mathbf{v}^2}{c^2}}t; \tag{28.8a}$$

and the phase *of the wave* from the point of view of O is:

$$\varphi_w = \nu_w\left(t-\frac{z}{\mathbf{v}_w}\right) \tag{28.8b}$$

Now let us rewrite Eq. (28.8a) in the following form:

$$\varphi_p = \frac{m_0c^2}{h}\sqrt{1-\frac{\mathbf{v}^2}{c^2}}\frac{z}{v}, \tag{28.9}$$

considering Eq. (28.2) and the equality $t = z/\mathbf{v}$. On the other hand, Eq. (28.8b) can be transformed in the following manner:

$$\varphi_w = \frac{m_0 c^2}{h} \frac{1}{\sqrt{1-\frac{\mathbf{v}^2}{c^2}}} \left(\frac{z}{\mathbf{v}} - \frac{z\mathbf{v}}{c^2} \right) = \frac{m_0 c^2}{h} \sqrt{1 - \frac{\mathbf{v}^2}{c^2}} \frac{z}{\mathbf{v}}, \tag{28.10}$$

which is the same as Eq. (28.9). Q. E. D.

Obviously, we could prove the theorem only because we presupposed the existence of a (classical) particle embedded in (and different from) a wave.

Consequences Now the amplitude cannot be represented by a factor ϑ_0 with the same distribution in space: it must fall with increasing distance from the singularity represented by the particle. Hence de Broglie supposed that the wave amplitude is only a statistical mean which does not depict the real trajectory of the particle. We need a more complex wave function than Eq. (28.5), in order to represent the global phenomenon of a wave with an internal singularity (a particle). This wave must be of the following form [see again Eq. (3.21)]:

$$|\Psi_{pw}\rangle = \vartheta(x, y, z, t) e^{\frac{i}{\hbar}\varphi(x,y,z,t)}, \tag{28.11}$$

where $|\Psi_{pw}\rangle$ stays for the global state vector of wave plus particle. In this sense we speak of a 'double solution' for the same system, the Schrödinger wave Eq. (28.5) and the de Broglie wave Eq. (28.11) where the two state vectors have the same phase φ [*DE BROGLIE* 1956, 2–7, 86, 94–95].[3]

Parallels with Classical mechanics De Broglie's theory has to be understood as an attempt to establish a complete equivalence between classical mechanics and QM [*DE BROGLIE* 1956, 9, 96, 100]. In fact, central in his work are the equivalents of the classical Eq. of Jacobi [see Eqs. (1.10) and (4.10)], i.e. a de Broglie/Jacobi Eq.:

$$\frac{1}{c^2}\left(\frac{\partial \varphi}{\partial t} - V\right)^2 - \nabla^2 \varphi = m_0^2 c^2 + \hbar^2 \frac{\Box \vartheta}{\vartheta}, \tag{28.12}$$

and of the classical continuity Eq. [see Eq. (1.12) and also (5.54)], i.e. a de Broglie/Liouville Eq.:

$$\frac{1}{c^2}\left(\frac{\partial \varphi}{\partial t} - V\right)\frac{\partial \vartheta}{\partial t} - \nabla\varphi\nabla\vartheta = -\frac{1}{2}\vartheta\Box\varphi. \tag{28.13}$$

A more concrete model De Broglie [*DE BROGLIE* 1956, 102–103] also tried to develop a more concrete model. Let $\mathbf{S}$ be a very small spherical surface around the particle, in whose interior the amplitude ϑ is very big. We suppose that in the interior of the sphere the wave Eq. is not linear, but it is linear on the surface. We admit that on $\mathbf{S}$, φ and its first derivatives have everywhere the same value — we suppose that $\mathbf{S}$ is very small in relation to the local wave-length which is related to φ, while ϑ has different values in $\mathbf{S}$ [see figure 28.1].

[3]Note however that the phase is not the only identity between the global wave function and the relative Schrödinger $|\psi\rangle$ vector; another one is that of current densities $\mathbf{j}$ [*DE BROGLIE* 1956, 219]. For an exposition see also [LOCHAK 1984, 3–4]

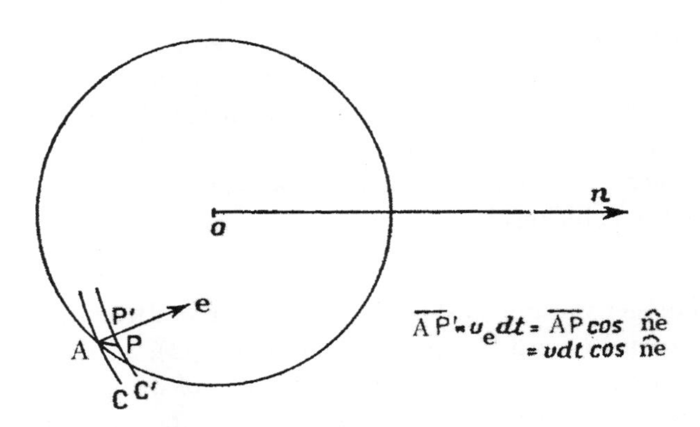

Figure 28.1: The model of double solution proposed by de Broglie — from [*DE BROGLIE* 1956, 102].

In proximity of a point A of the surface we see positions at time t and $t + dt$ of the same surface $\vartheta = C$ (where C is a constant). The direction $\mathbf{e}$ is that of $\nabla\vartheta$ in A, and that of $\mathbf{n}$ of $\nabla\varphi$ on all the sphere. Our hypothesis is that:

$$\left(\frac{\vartheta}{\frac{\partial\vartheta}{\partial\mathbf{e}}}\right)_{\mathrm{A},t} \simeq 0. \tag{28.14}$$

Now Eq. (28.13) allows us to write:

$$\frac{E - V}{c^2}\frac{\partial\vartheta}{\partial t} + \frac{\partial\vartheta}{\partial\mathbf{e}}\frac{\partial\varphi}{\partial\mathbf{n}}\cos(\widehat{\mathbf{ne}}) = \frac{1}{2}\vartheta\Box\varphi \tag{28.15}$$

If we divide by $\partial\vartheta/\partial\mathbf{e}$ and consider Eq. (28.14), and see that the velocity of ϑ in A and at t is

$$v_{\mathrm{e}} = \left(-\frac{\frac{\partial\vartheta}{\partial t}}{\frac{\partial\vartheta}{\partial\mathbf{e}}}\right)_{\mathrm{A},t}, \tag{28.16}$$

we then conclude:

$$\mathbf{v}_{\mathrm{e}} = \frac{c^2}{E - V} \mid \nabla\varphi \mid \cos(\widehat{\mathbf{ne}}). \tag{28.17}$$

Now this result is valid at each point of the sphere's surface. Since $\mathbf{v} = \mathbf{v}_{\mathrm{e}}\cos(\widehat{\mathbf{ne}})$ and since $E - V/c^2 = m_0/\sqrt{1 - v^2/c^2}$ we finally write:

$$\mathbf{p} = \frac{m_0\mathbf{v}}{\sqrt{1 - \frac{\mathbf{v}^2}{c^2}}} = -\nabla\varphi, \tag{28.18}$$

which is the *Pilot-wave equation* — an analogy of Jacobi's classical formula — which expresses the action of the field on the particle. It is very important — we shall return to this point — that this pilot equation makes the particle sensitive to the topological structure of the space.

In conclusion de Broglie assigned a dual role to $|\psi|^2$: it determines the localization and it influences the localization by exerting a force on the orbit [*DE BROGLIE* 1956, 78, 90, 99].[4]

[4]On this point see also [*HOLLAND* 1993, 16]

28.1.3 First Critical Positions

Pauli As we have said, the community of physicists rejected de Broglie's proposal. Remember also that the object of the discussion in the twenties was only the Pilot wave theory. In the Brussels conference of October 1927 Pauli opposed de Broglie in saying that, in case of collision between the particle and a rigidly fixed rotator, the phase would determine a motion of the system in the configuration space incompatible with the final quantized state of the rotator (stationary state before and after the collision).[5]

Schrödinger Schrödinger was opposed to the fact that in de Broglie's wave mechanics a system of N particles is represented by $3N$ separate functions $q_k(t)$, while in multidimensional wave mechanics it is represented by one single state vector $|\Psi\rangle$ of $3N$ variables q_k in addition to the variable t. This implies some correlation between the components of the wave which are not present at all in de Broglie's proposal: the entanglement [see subsection 3.5.2 and also further developments in section 34.1]. In other terms the $3N$ de Broglie's wave functions are factorised, contrary to the principal feature of QM. The space of the waves of QM is not a tridimensional or fourdimensional 'real' space but a Hilbert space, which makes the ontological interpretation of the wave very difficult[6].

De Broglie De Broglie avoided the validity of these criticisms [*DE BROGLIE* 1956, 87]. Notwithstanding, he seemed to believe that his proposal was sound. Perhaps one may think it possible to answer to Pauli's objection on the basis of some correction to dynamics, as later did other physicists [see section 28.2]. But, supposing that de Broglie accepted that the Schrödinger $|\psi\rangle$–vector is in a Hilbert space, in order to maintain the ontological interpretation of the wave he should have shown that the $|\Psi_{pw}\rangle$–vector 'lives' in an ordinary tridimensional space. He never did prove it, and it is also doubtful if it is in general possible.

On the other hand, if the pilot Eq. has to reproduce all typical and experimentally verified quantum correlations (for example the non separability of one particular path in all Two-Slit-Exp.s), we need to develop a more specific model. This is what Bohm tried to do later [section 32.6].

28.1.4 Further Examination of de Broglie's Theory

However, even if these objection were all refuted, it seems that there are some other problems with this theory. We have already seen that the wave velocity can be greater than the speed of light. In that what follows we briefly examine some other difficulties.

The position De Broglie's starting point seems to have been [*DE BROGLIE* 1956, 56–58] the following. According to the Copenhagen interpretation all possible representations of $|\psi\rangle$ [see subsection 3.6.4], which correspond to possible measurements of different observables, are equivalent (before a concrete measurement), because it is the experiment which determines which observable is 'real' [see postulate 8.1: p. 137, and theorem 14.6: p. 235]. According to de Broglie

[5]See [*DE BROGLIE* 1956, 174–76], where the criticism is quoted, and [*JAMMER* 1974, 111–13] for historical reconstruction. For further analysis see also [*HOLLAND* 1993, 16].

[6]See also [*DE BROGLIE* 1956, 95] [*JAMMER* 1974, 117].

this is not the case. While the momentum depends on observation, position is independent from the latter[7].

De Broglie always considered the problem in analogy with classical mechanics. In the case of a collision of a non-monochromatic wave on a prism, before the interaction it is the wave train and not the observed frequencies — i.e. the components of a Fourier decomposition — which has a meaning. The latter are only created by the interaction; hence, following de Broglie, only the amplitude has physical significance, while the frequencies (or the phase) are only the result of an analytical process. Now the specificity of $\mathbf{q}$ came from the fact that in this representation a wave-like behaviour (which, following de Broglie, is an ontological reality) is explicit, which (again following de Broglie) is not present as such in other representations [*DE BROGLIE* 1956, 59, 63–65, 81].

But this point is absolutely dubious because we know that the Hilgevoord/Uffink UR [see Eq. (7.44)], which exists between Fourier pairs, and CP4 [p. 139] do not allow such an interpretation: the superposition is in no way confined to position observable only. The diffraction is a manifestation on a screen of a general phenomenon which concerns all QM observables. And in general if a difference has to be made between position and momentum, then surely it is the position which may be associated more easily with a corpuscular behaviour of microrealities — for example in which-path experiments.

Quantum potential Another very difficult matter is the pilot Eq. as such [Eq. (28.18)]. The potential which can be derived from the Pilot equation was named the *Quantum potential* by Bohm — we shall return to this matter in greater detail later [in section 32.6]. This potential is in strict connection with the proper mass m_0 of the particle — which is a function of position and time — because it is equivalent to the action of a 'quantic field' which can also exist in the absence of classical fields and which is present as soon as interference or diffraction are manifest. In this sense de Broglie could quote Newton or Laplace, who believed in the corpuscular nature of the light: in fact, Newton and Laplace[8] also thought that obstacles could influence the trajectory of a particle and give it the 'strange' behaviour which results as wave-like [*DE BROGLIE* 1956, 111, 115–17]. But this point, which is strictly dependent on Schrödinger's objection on correlation, is a source of many difficulties. In fact, in order to speak of a special quantum field, one must experimentally verify and measure it. And here the work of de Broglie ended. Bohm tried to give this point a better foundation [see subsection 32.6.2].

Hydrogen atom One of the better tests for de Broglie's theory is the *hydrogen atom*, a subject to which de Broglie himself applied his point of view. Generally the Quantum potential is counterbalanced by the Coulomb force, so that for the ground state they are equal. But the consequence is that the electron is at rest, and de Broglie knew that the electron's immobility is in contradiction with the — experimentally verified — isotropy of the ground state, i.e. spherical symmetry of $|\vartheta|^2$; his answer — that the statistical distribution is random — is very unsatisfactory [*DE BROGLIE* 1956, 120–22].

Non-linearity But the greatest problem with the hydrogen atom is that the solution with singularity exists iff the singularity is positioned in a point where the Schrödinger wave function is zero, i.e. the particle is exactly in the place where it has no probability to be. From here the

[7]The same is true for Bohm's theory: see [PAGONIS/CLIFTON 1995].

[8]Examination of Newton's optics in [A. R. HALL 1963].

necessity of non-linearity stems [see following section]. But de Broglie avoids a more detailed formulation of such an equation [*DE BROGLIE* 1956, 214–216].

Galilei invariance Another problem worth mentioning is that de Broglie's waves are not gauge invariant and Galilei invariant [LÉVY-LEBLOND 1974] [LÉVY-LEBLOND 1976] [ZEILINGER 1990, 19–20].

28.2 Non-Linear Solutions

28.2.1 Non-Linear Theory

As we have already seen, de Broglie proposed a new non-linear wave Eq., although he was not able to specify a more concrete form of it [see subsection 28.1.4]. In the seventies, the matter acquired some independence from the original context of de Broglie's proposal.

The work of Bialynicki-Birula and Mycielski [BIALYNICKI-B./MYCIELSKI 1976] represents one of the first attempts to find some non-linear Eq.[9] The proposed non-linear Schrödinger Eq. was the following [BIALYNICKI-B./MYCIELSKI 1976, 64] [see also Eq. (3.18)]:

$$\imath\hbar\frac{\partial\psi(\mathbf{r},t)}{\partial t} = \left[-\frac{\hbar^2}{2m}\nabla^2 + V(\mathbf{r},t) + f\left(|\psi(\mathbf{r},t)|^2\right)\right]\psi(\mathbf{r},t), \tag{28.19}$$

where the non linear factor is represented by $f\left(|\psi|^2\right)$.

In general, by an equation of this form for compound systems, the motion of a subsystem would be influenced by the state of another subsystem — not necessarily interacting or having interacted with it. To avoid the problem, the authors postulated the separability of non-interacting systems, by means of which they deduced the particular form of the non-linear factor [BIALYNICKI-B./MYCIELSKI 1976, 68–69]:

$$f\left(|\psi(\mathbf{r},t)|^2\right) = -\eta \lg\left(|\psi(\mathbf{r},t)|^2\zeta^n\right), \tag{28.20}$$

where η, ζ are some parameters and n is the number of dimensions of the configuration space.

But the great problem is that the completeness, the orthogonality of eigenvectors, the superposition of state vectors, etc. are all properties which derive from the linearity of QM — in particular of QM operators [see definition 3.1: p. 32]. However, the authors intend that all these properties can be derived from the Probabilistic assumption [postulate 3.5: p. 38] plus the non-linear Eq. (28.19) if we not only respect the postulate of the separability of non-interacting systems but also a weak superposition principle: the sum of two solutions whose overlap is negligible is also a solution [BIALYNICKI-B./MYCIELSKI 1976, 64, 82–84]. The only sensible difference is that the non-linearity implies the existence of *gaussons*, i.e. uniformly moving Gaussian wave packets, which do not dissolve in space because their wave-length is inversely proportional to the mass — see also Schrödinger's Wave-Packet interpretation [proposition 6.1: p. 101]. They are also the limit between microbjects with wave-like properties and macrobjects [BIALYNICKI-B./MYCIELSKI 1976, 77–78, 84].

[9] Theories which introduce a non-linear term in order to explain specifically the reduction of the wave-packet, such as that of Pearle's, have already been briefly considered [see section 23.3].

The authors calculate also the upper limit of η [BIALYNICKI-B./MYCIELSKI 1976, 85]: the most restrictive results by precision measurements are obtained from the Lamb shift measurement in Hydrogen from ${}^2S \mapsto {}^2P$ transitions:

$$\eta < 4 \times 10^{-10}\text{eV}. \tag{28.21a}$$

In a similar way, a lower bound was also calculated:

$$\eta > 2.5 \times 10^{-12}\text{eV}. \tag{28.21b}$$

28.2.2 Experimental Proofs

Proposal Shimony [SHIMONY 1979] was the first to outline a possible experiment in order to verify if the wave Eq. of microrealities can be non-linear. One makes use of a neutron interferometer (Interfer. Exp. 12) in which a beam is split into two coherent beams I, II, with negligible overlap which after some centimeters are recombined [see figure 28.2].

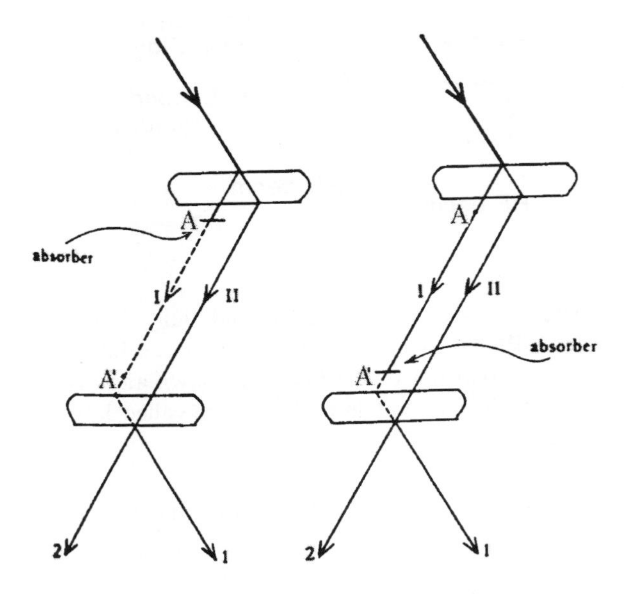

Figure 28.2: Dashed lines indicate the attenuation of the amplitude of a beam by a factor α. Output beams I and II are directed towards the neutron counters — from [SHIMONY 1979, 51].

The experiment consists of observing a phase shift $\delta\varphi$ due to moving a partial absorber in the path I from position A to position A′. The nonlinearity should cause a non-negligible phase shift while QM predicts that it is negligible.

Performance Shull and co-workers [SHULL *et al.* 1980] realized the experiment and proved that there is in fact a phase shift as previewed by Shimony, but the value of η is one order of magnitude smaller than the lower bound resulting from theoretical calculation after Bialynicki-Birula/Mycielski [Eq. (28.21b)]:

$$\eta < 3.4 \times 10^{-13}\text{eV}. \tag{28.22}$$

The experiment follows the outline of Shimony's Interfer. exp. 12, with beams which pass through two perfect-crystal plates and two attenuating plates A, B (of LiF and Cd). A is downstream in path I and B upstream in path II; but during the experiment A begins downstream in path II and B upstream in path I — they are 'interchanged' [see figure 28.3]

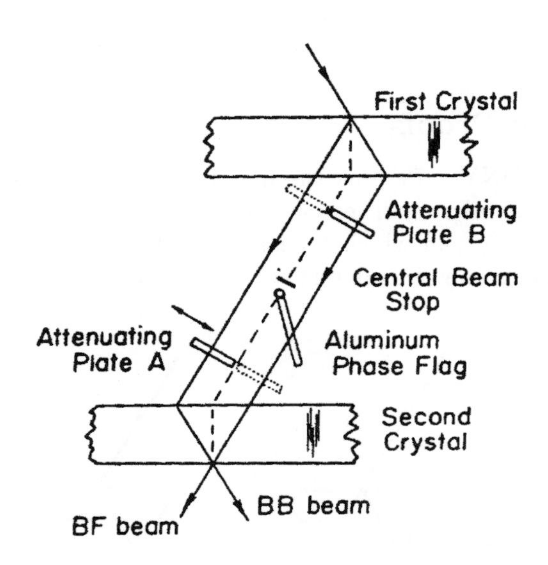

Figure 28.3: Coherence exists between rays on the left and right sides of the beams passing between the two perfect crystal plates. Phase effects are exhibited in the intensities of the outgoing beams, BB and BF. Intensity attenuating plates are positioned as shown and these are shuttled left and right in the experiment. The solid and dotted shuttle positions are called initial and inverted configuration respectively — from [SHULL *et al.* 1980, 766].

They calculated that if α is the intensity attenuation due to partial attenuating plates, the phase shift is:

$$\delta\varphi = \frac{\tau\eta}{\hbar}\lg(|\alpha|)^2, \tag{28.23}$$

where τ is the particle's transit time over distance l. If we define θ_n (n = normal) as the measured phase angle shift in the initial (normal) configuration of attenuating plates, we have

$$\theta_n = 2\delta\varphi + 2(\varphi_\text{A} - \varphi_\text{B}), \tag{28.24a}$$

while, by changing the configuration of the attenuating plates, we have

$$\theta_i = 2\delta\varphi - 2(\varphi_\text{A} - \varphi_\text{B}), \tag{28.24b}$$

where the 'i' in θ_i stands for 'inverted'.

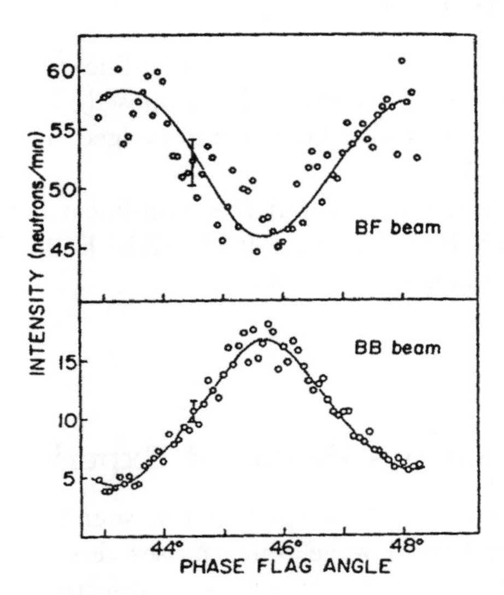

Figure 28.4: Neutron-interferometer-fringe display as relative phase between left and right rays is continuously varied by rotating the aluminium phase flag — from [SHULL *et al.* 1980, 766].

Results Experimentally they found that, by using LiF attenuating plates, we obtain the following values:

$$\delta\varphi = +0.08(\pm 0.30), \qquad (\varphi_{\text{A}} - \varphi_{\text{B}}) = +2.69(\pm 0.30), \tag{28.25}$$

from which it follows:

$$\theta_n = +5.53(\pm 0.79), \qquad \theta_i = -5.22(\pm 0.92). \tag{28.26}$$

By using a pair of cadmium attenuating plates we have:

$$\delta\varphi = -0.22(\pm 0.27), \qquad (\varphi_{\text{A}} - \varphi_{\text{B}}) = -0.52(\pm 0.27), \tag{28.27}$$

from which:

$$\theta_n = -1.49(\pm 0.79), \qquad \theta_i = 0.60(\pm 0.76). \tag{28.28}$$

Combining the two results in Eq. (28.23), with $\tau = 1.123 \times 10^{-5}$ s, we finally obtain:

$$\eta = (-1.0 \pm 3.7) \times 10^{-13}\text{eV}. \tag{28.29}$$

In conclusion the obtained phase shift is smaller than the experimental uncertainty [see figure 28.4] and smaller than the lower bound (28.21b).

Other experiments and studies In another series of experiments performed by Zeilinger and co-workers [GÄHLER *et al.* 1981] and by Klein and Werner [A. KLEIN/WERNER 1983], this value was diminished to $\eta < 3.3 \times 10^{-15}$eV.

A very important and now classical theoretical study on the problem is due to Weinberg [S. WEINBERG 1989b], following which other experimental proposals were performed[10].

[10] See for example [BOLLINGER *et al.* 1989]: the result is in principle the same, i.e. a value which is under the limit of the value predicted.

28.2.3 Concluding Remarks

In conclusion no positive proof of the existence of non-linear factors has been reached, while the smallness of the reached value disproves the proposed theoretical model and practically rules out the hypothesis, as it was also later acknowledged by the proponents of the theory [*BIALYNICKI-B. et al.* 1992, 50].

On the other hand, it has been pointed out that non-linear factors can lead to superluminal forms of communication [GISIN 1989] [GISIN 1990] [PEARLE 1986], which are excluded by other considerations and developments [see section 36.5]

28.3 Empty Waves

28.3.1 Croca's and Hardy's Proposed Experiment

Supposing that de Broglie's ontological interpretation is sound, we face the following situation: in a typical Two-Slit-Exp. we have a wave which passes through both slits and an associated particle which passes only through one (say slit 1) [see subsection 28.1.1]. Hence we have a wave with a particle through slit 1 and a wave without particle through slit 2, i.e. an *empty wave*. The first hint in this direction was probably due to Einstein [EINSTEIN 1917].[11] We see now some *Gedankenexperimenten* with the aim of proving the possibility or the existence of such a wave, or at least of deciding on the matter.

Croca's proposal

The first attempt to handle the matter and propose concrete experiments is due to Selleri [SELLERI 1969] and Tarozzi [TAROZZI 1985].[12] Croca [CROCA 1987] proposed an experiment (Interfer. Exp. 13) to produce empty waves and to detect them.

We have a source which sends N particles one at a time. By means of a Beam Splitter (BS) the beam is split into components $|\psi_1\rangle, |\psi_2\rangle$. If the detector D_R clicks, it means that $|\psi_2\rangle$ carried the particle and that $|\psi_1\rangle$ is by hypothesis an empty wave. Now, once triggered, D_R sends a signal to open the Gate G for a short time t_G allowing the empty wave to pass [see figure 28.5].

Now, to detect it, imagine a second source which sends systems described by wave function $|\varsigma\rangle$. The beam encounters another Beam splitter (BS′), in consequence of which the system has the probability of being reflected (amplitude $|\varsigma_1\rangle$) or transmitted (amplitude $|\varsigma_2\rangle$). The two beams converge in the detector D [see figure 28.6].

Now suppose that we send the empty wave $|\psi_1\rangle$ into the interference region. If the hypothesis is correct, then instead of a particle distribution on D proportional to $|\varsigma_1 + \varsigma_2|^2$ [see Eq. (3.4)] we should have something proportional to $|\psi_1 + \varsigma_1 + \varsigma_2|^2$. Croca then proposes a concrete experiment by means of a neutron interferometer.

Why should such a proposal be of interest? Croca assumed that a empty wave ($|\psi_1\rangle$) can interact with an 'ordinary' quantum wave ($|\varsigma\rangle$). On this hypothesis, the proposed experiment can decide whether such an empty wave exists or not. On a theoretical level, however, we should already exclude the first alternative: in fact, if empty waves did have measurable properties and then could interact with other quantum systems, experiments would be constantly perturbed by background noise caused by extraneous fields, which is not the case [*HOLLAND* 1993, 372–73].

[11] On this point see [SELLERI 1982, 1088–90].

[12] For a review see [SELLERI 1982].

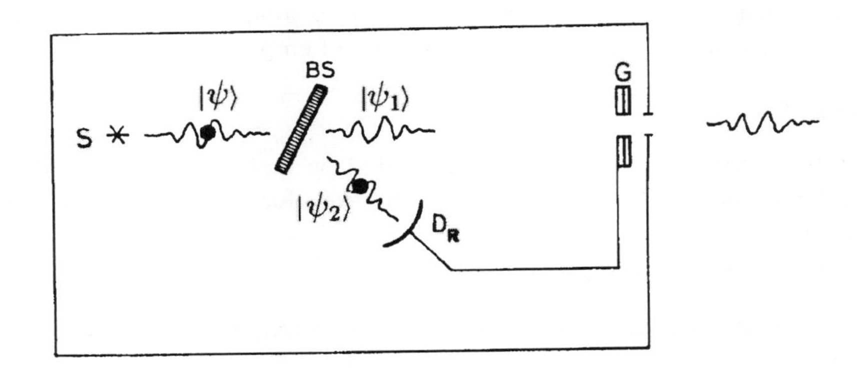

Figure 28.5: Proposed Interfer. Exp. 13. Empty wave generator (EWG). The source S sends particles one at time. On its way, the real singularity carrying $|\psi\rangle$ is split into two real waves $|\psi_1\rangle$ and $|\psi_2\rangle$ by the BS. If detector D_R clicks, this means that the wave $|\psi_2\rangle$ carried the singularity, and $|\psi_1\rangle$ is a real empty wave. Once triggered the detector D_R sends a signal to open the gate G, for a short time t_G, allowing the empty wave to pass — from [CROCA 1987, 973].

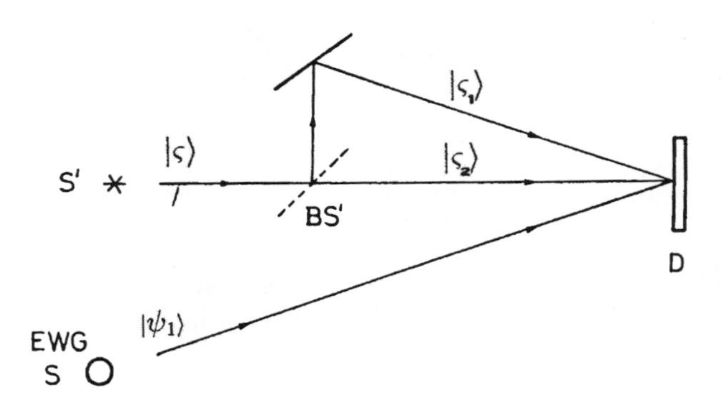

Figure 28.6: Taylor-like experiment with an additional empty wave — from [CROCA 1987, 974].

For this reason, Holland assumed that empty waves can only interact with empty waves — which would invalidate a test having the form of Croca's proposal. But, if so, empty waves could not then be detected at all, and it seems to be something non-physical to postulate the existence of something about which we cannot know something else apart from that it should be 'something' — see also methodological principle M2 [p. 6]. In other words, we face here a dilemma where both ways out seem hardly favorable to the assumption of empty waves.

Hardy's proposal

The experiment Croca's proposal was reformulated by Hardy [HARDY 1992b] in a more complex thought-experiment. The two sufficient conditions for the existence of an empty wave according to Hardy are:

- we know which path the particle takes (*condition 1*);
- a measurable property of a system placed in the other path (of an interferometer) is changed (*condition 2*).

If no particle is detected at the end of the interferometer (due to destructive interference) we have a *dark output.*

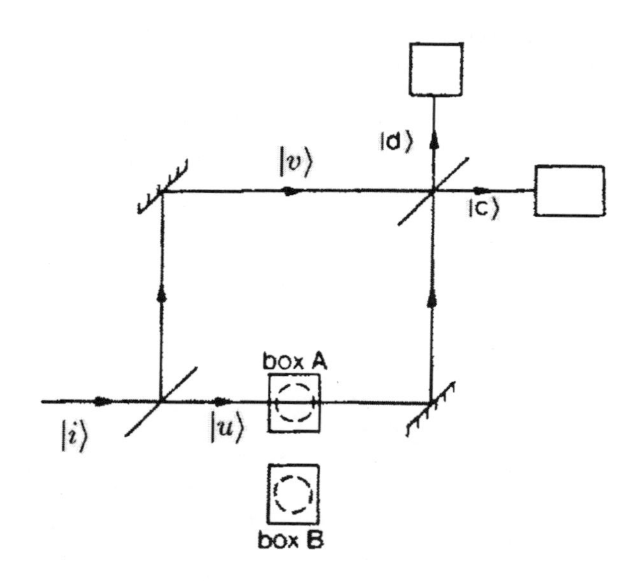

Figure 28.7: Proposed Interfer. Exp. 14. MZ Interferometer is arranged such that, when there are no obstructions in either of the paths, no particles will be detected in output d. Box A is then placed in path u — from [HARDY 1992b , 13].

We now describe the proposed experiment (Interfer. Exp. 14) [see subsection 5.3.1 for apparatus]. In the experiment [see figure 28.7] the input beam $|i\rangle$ is divided by a first BS (BS1) in path $|u\rangle$ and $|v\rangle$, and by means of a second BS (BS2) recombined and split through detectors $|d\rangle$ and $|c\rangle$. Now we imagine an atom with spin 1/2 along the x–axis and with initial state given by:

$$|\psi\rangle = \frac{1}{\sqrt{2}}(|\uparrow\rangle + |\downarrow\rangle), \tag{28.30}$$

where the arrows stand for directions of spin along the z–axis. The atom is in the path u, placed in a box subjected to a non-uniform magnetic field aligned along the z–axis. This acts as an SGM and splits the state of the atom in an upper part of the box with state $|\uparrow\rangle$ and in a lower part with state $|\downarrow\rangle$. If walls are introduced we obtain an upper box A and a lower one B. If the atom is in A, it will absorb the particle with probability of one if the particle goes along the path u — in this case we write $|A,\uparrow\rangle_e$ (where 'e'stands for 'excited'). This is the case which holds no interest for us. After the particle has passed and been detected, A and B are brought together by removing the separating walls. Finally a measurement of the spin along the x axis is made on the atom [see figure 28.8].

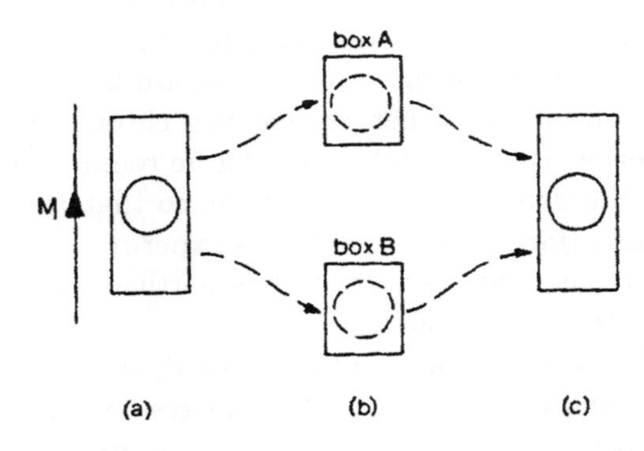

Figure 28.8: (a) The atom with spin+1/2 along the x axis, is placed in a box and a non-uniform magnetic field M is placed along the box.
(b) Dividing walls are placed so as to create two boxes A and B.
(c) The boxes are brought back together by removing the dividing wall — from [HARDY 1992b , 13].

The evolution of the whole system can be described as follows:

$$\frac{1}{\sqrt{2}}|i\rangle\left(|A,\uparrow\rangle+|B,\downarrow\rangle\right) \mapsto \frac{1}{2}\left(|u\rangle+\imath|v\rangle\right)\left(|A,\uparrow\rangle+|B,\downarrow\rangle\right) \tag{28.31a}$$

$$\mapsto \frac{1}{2}|A,\uparrow\rangle_e+\frac{1}{2}\imath|v\rangle|A,\uparrow\rangle+\frac{1}{2}\left(|u\rangle+\imath|v\rangle\right)|B,\downarrow\rangle \tag{28.31b}$$

$$\mapsto \frac{1}{2}|A,\uparrow\rangle_e+\frac{\imath}{2\sqrt{2}}|c\rangle|A,\uparrow\rangle-\frac{1}{2\sqrt{2}}|d\rangle|A,\uparrow\rangle+\frac{1}{\sqrt{2}}|c\rangle|B,\downarrow\rangle \tag{28.31c}$$

With this formalism we reformulate *condition 2* in the following form: If, at any time t, the state of the atom is $|A,\uparrow\rangle$, then the atom is actually in box A even if we do not perform a measurement. This motivates a third assumption:

Proposition 28.1 (Hardy) *If a box is closed at a time t_1 and one establishes at some later time t_2 that an atom is actually in the box, then the atom is there for all the time from t_1 till its opening at t_3.*

We are only interested in those runs for which there is a detection in $|d\rangle$, i.e. when the system is projected on the third term of Eq. (28.31c). But if so, then we know with certainty that the particle took path $|v\rangle$, so that *condition 1* is satisfied. But now, if we bring together boxes A and B, there is a 50% probability that the spin along the x–axis becomes $-1/2$ from the original $+1/2$, i.e. there is a change induced by the empty wave: and, if so, then *condition 2* is also satisfied. Normally the atom should already be in a state of superposition of $|\mathrm{A}\rangle$ and $|\mathrm{B}\rangle$. But, according to Hardy, in the case supposed here, the superposition is destroyed because the atom is certainly in the state $|\mathrm{A}, \uparrow\rangle$, after which we have a 50% probability of the change of spin by putting the boxes together.

Discussion Now Hardy pointed out that the experiment only shows that an 'empty something' is passed but does not positively demonstrate the wave nature of this something. Note[13] that it is not the 'something' alone that effects the result, but this together with detection $|d\rangle$. Another problem seen by Hardy himself is that the whole experiment is frame-dependent: it is necessary that the particle is detected by dark output *before* boxes are brought together: however these events can be space-like apart from each other, so that this chronological succession is not guaranteed in all frames of reference. But the ambiguity can be removed if we bring boxes together in the forward light cone of the detection of the particle, so that the succession is the same in all frames of reference. Finally, Hardy thought that his experiment is in agreement with the de Broglie/Bohm model for fermions (which are treated as particles) but not for de Broglie/Bohm model for bosons (which are treated as fields).

However, even if we do not consider the problems quoted, the experiment is not at all so unquestionable. In fact the first and the third assumptions have been questioned [PAGONIS 1992, 219] on the basis that QM does not acknowledge the locality presupposed by both: the photon is in a superposition after partition of the box and not in a factorised state.

On the other hand, the proposition 28.1 seems to assign the 'empty wave property' also — and in particular — to the atom: if the primitive box is divided into two boxes, in one or the other there is an empty wave. Hence Hardy's argument eventually demonstrates only, that, *if* the atom has the empty wave property, then it passes it to the photon [ŻUKOWSKI 1993]. This point is not meaningless, because in this case we would have a form of interaction between empty waves and 'ordinary' quantum systems, and we know from the preceding discussion of Croca's proposal that this is an unlikely assumption.

But the central question to raise is the following: in which aspects does Hardy's experiment differ from an interaction-free one [see chapter 20]? Also in interaction-free experiments, if we have a click in one brace of the interferometer, we have the possibility of detecting the state of an atom in the other brace without interacting with it: but, as we know, this is due to the correlation between the two subsystems [PAGONIS 1992, 221]. Hence the empty waves are not necessary at all in order to explain the phenomenon: i.e. it can be explained with the formalism of QM without such a further assumption — see methodological principle M7 [p. 6].

The following is another very important problem: the circumstance that the empty waves only produce effects in the cases in which detector d clicks and not in others is highly suspect. Suppose that there is no click in d. This does not mean that there was necessarily a destructive interference, since the atom could be in A — the second term of Eq. (28.31c). This means that, even if there is no destructive interference — i.e. the photon is certainly passed through path v —, the empty wave manifests itself only if the photon is detected in d and not in c. On the other hand, if the photon is detected in c, the atom is projected in the superposition constituted by the second and fourth term of Eq. (28.31c), exactly as it was at the beginning: in this case a

[13]The following examination owes very much to Dr. Fortunato.

measurement of the spin along the x–axis would give no information at all.

It is not by chance that Hardy later abandoned *condition 1* [DEWDNEY *et al.* 1993].[14]

28.3.2 Other Proposals for Detecting the Energy of the Empty Waves

An experiment realized by Badurek and co-workers and described in [BADUREK *et al.* 1983a] has been interpreted by Vigier [VIGIER 1985, 665–75] and co-workers [VIGIER *et al.* 1987, 184–87] as a proof of the possibility of a corpuscular behaviour in one channel and of a wave-like behaviour in both channels of an interferometer[15].

The experiment We begin with a short description of the experiment, which belongs to the neutron interferometry type (Interfer. Exp. 15). In order to create a neutron spin-state superposition, one inverts the spin-state of one of the beams. But instead of doing so by a time-independent static polarisation-turn device — by which the change in the Zeeman potential energy is compensated exactly by a corresponding inverse change of the kinetic energy, i.e. of the neutron wave-length, so that the total energy of the neutron is a constant of the motion —, one uses a time-dependent radio-frequency (= rf) flipper, whereby the total energy of the neutrons is no longer conserved because there is an exchange of photons of energy $\hbar\omega_{\rm rf}$ between the neutrons and the rf field [see figure 28.9].

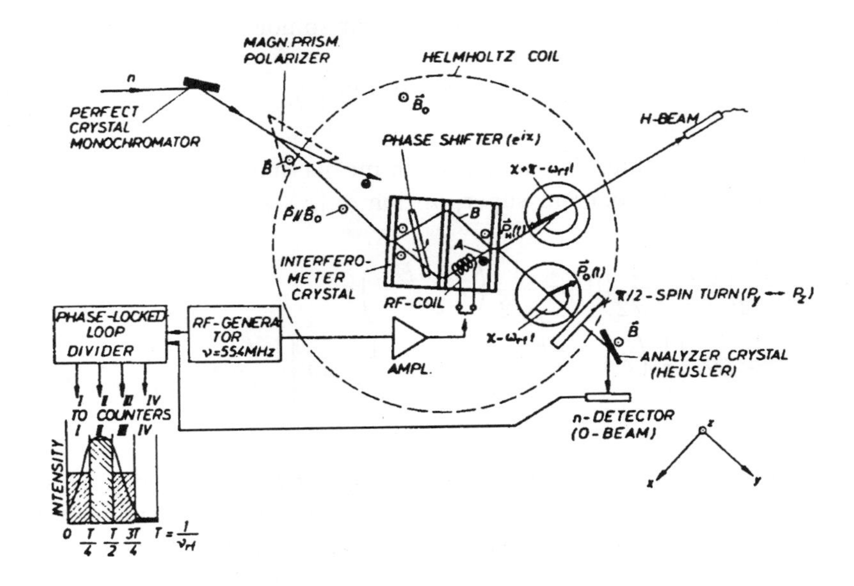

Figure 28.9: Schematic diagram of spin-superposition experiment (Interfer. Exp. 15) and of a stroboscopic neutron registration – from [BADUREK *et al.* 1983a, 1016].

Neutrons initially polarised parallel to direction of the static field component B_0 (z direction) and with energy $E = \hbar^2 k_i^2/2m + |p|B_0$, by flipping in the opposite direction will have lost an amount of energy $\delta E = 2|p|B_0$.

[14]For another proposal of empty-wave experiment see [GRÖSSING 1995].

[15]See also [BADUREK *et al.* 1986].

As long as no rf field acts, their wave function belongs to the same polarisation state but differs for a phase factor $e^{\imath\delta\varphi}$, where $\delta\varphi = k(1 - n_R)\delta l$, where k is the wave number of the neutron, n_R the refraction index, and δl the path difference; whereas they maintain their initial momentum k_i. For simplicity's sake, choosing a single-plane-wave representation for beams A and B, we can describe the two beams by the state vectors

$$
\begin{aligned}
|\mathrm{A}\rangle &= e^{\imath \mathbf{k}_i \cdot \mathbf{r}_A} e^{-\frac{\imath}{\hbar}Et} | \uparrow_z\rangle, & (28.32a)\\
|\mathrm{B}\rangle &= e^{\imath\delta\varphi} e^{-\frac{\imath}{\hbar}Et} | \uparrow_z\rangle. & (28.32b)
\end{aligned}
$$

Omitting all common phase factors and neglecting the propagation-direction difference, the coherent superposition of states $|\mathrm{A}\rangle$ and $|\mathrm{B}\rangle$ yields the final state:

$$
|f\rangle = \frac{1}{\sqrt{2}}|\mathrm{A}\rangle + \frac{1}{\sqrt{2}}|\mathrm{B}\rangle = \frac{1}{\sqrt{2}}(1 + e^{\imath\delta\varphi})| \uparrow_z\rangle, \tag{28.33}
$$

which is polarised in the z–direction and exhibits an oscillatory behaviour of the intensity as a function of the phase shift:

$$
I(\delta\varphi) \propto \langle f|f\rangle = \frac{1}{2}(1 + \cos\delta\varphi). \tag{28.34}
$$

Now, if an rf field acts, say, on $|\mathrm{A}\rangle$, the correspondent spinor is transformed in:

$$
|\mathrm{A}\rangle' = e^{\imath \mathbf{k}_i \cdot \mathbf{r}_A} e^{-\frac{\imath}{\hbar}(E-\delta E)t} | \downarrow_z\rangle. \tag{28.35}
$$

The two interfering beams now differ in energy, which means that they cannot be in a stationary interference pattern. In fact the coherent superposition of $|\mathrm{A}\rangle'$ and $|\mathrm{B}\rangle$ is given by:

$$
\begin{aligned}
|f\rangle &= \frac{1}{\sqrt{2}}|\mathrm{A}\rangle' + \frac{1}{\sqrt{2}}|\mathrm{B}\rangle = \frac{1}{\sqrt{2}}\begin{pmatrix} e^{\frac{\imath}{\hbar}\delta Et} \\ e^{\imath\delta\varphi} \end{pmatrix} \\
&= \frac{1}{\sqrt{2}}\cos\left[\frac{1}{2}\left(\delta\varphi - \frac{\delta E}{\hbar}t\right)\right] | \uparrow_z\rangle - \frac{1}{\sqrt{2}}\imath \sin\left[\frac{1}{2}\left(\delta\varphi - \frac{\delta E}{\hbar}t\right)\right] | \downarrow_z\rangle. \qquad (28.36)
\end{aligned}
$$

Discussion According to Vigier and co-workers the experiment quoted proves that there is an energy transfer to the apparatus of beam $|\mathrm{A}\rangle$ which can only be explained in terms of a corpuscular behaviour in that channel [see proposition 6.2: p. 104]. On the other hand, since we have interference of both beams, then we have a wave which goes both ways. In conclusion, while for the Copenhagen interpretation a particle can be in a superposition of different stationary states of the energy — in which case it does not have a definite level of energy —, for Vigier it always has a definite state.

But Badurek and co-workers analysed the experiment in very different terms. In fact, they acknowledged [BADUREK *et al.* 1986, 2605] that one could measure the energy transfer without destroying the interference pattern. But — as we have already seen for the complementarity between the interference pattern and localization [see section 8.4] and shall see later in more detail [chapter 30] —, this is so because there is no jump-like transition between the corpuscular (exchange of energy) and wave-like (interference) behaviour.

28.4 Meaning of the Wave Function

28.4.1 Protective Measurement applied to a Spatial Wave Function

Another proposal — which is not in the line of de Broglie's proposal[16] — for interpreting the wave function ontologically is the result of the work of Aharonov, Anandan and Vaidman. The main idea is that of protective Measurement, a member of the same family of the weak Measurement [section 18.8]: a measurement of the wave function during which it is prevented from changing noticeably by means of another interaction which it undergoes at the same time [AHARONOV *et al.* 1993a, 4616–19].[17]

Conditions If we wish to measure an observable $\hat{O}$, normally the interaction between an apparatus $\mathcal{A}$ and a system $\mathcal{S}$ creates an entangled state [subsection 14.1.3] which may be written as

$$\Psi(\tau) = \sum_j c_j |s_j\rangle |a_j\rangle. \tag{28.37}$$

If g_0 represents the switching on and off of the interaction [see Eq. (14.8)], i.e. $g_0(t)$ is non zero only in interval $[0, \tau]$ (beginning and end of the interaction), then $|s_j\rangle = e^{-\imath/\hbar g_0 o_j \hat{\mathbf{q}}_\mathcal{A}}|\varsigma\rangle$ (where $\hat{\mathbf{q}}_\mathcal{A}$ is the pointer observable) are states of $\mathcal{A}$ which, for sufficiently large g_0 are orthogonal for distinct o_j's; while $|s_j\rangle$ are eigenvectors of $\hat{O}$. The Standard interpretation of measurement states that the result of a measurement must be a reduction or a collapse of wave-packet [Eq. (14.17)]:

$$\sum_j c_j |s_j\rangle |a_j\rangle \rightsquigarrow |s_k\rangle |a_k\rangle. \tag{28.38}$$

By hypothesis now, no entanglement takes place. Hence, instead of Eq. (28.37), we have the factorised state:

$$|\varsigma(0)\rangle |\mathcal{A}(0)\rangle \rightsquigarrow |\varsigma(t)\rangle |\mathcal{A}(t)\rangle, \ t > 0. \tag{28.39}$$

Neither is there collapse, so that, instead of Eq. (28.38), we have:

$$\frac{d}{dt}\langle \varsigma(t)| \langle \mathcal{A}(t)| \hat{\mathbf{p}}_\mathcal{A} |\varsigma(t)\rangle |\mathcal{A}(t)\rangle = -g_0(t) \langle \varsigma(t)| \hat{O} |\varsigma(t)\rangle, \tag{28.40}$$

where $\hat{\mathbf{p}}_\mathcal{A}$ is the canonical conjugate to $\hat{\mathbf{q}}_\mathcal{A}$ and we have [see Eq. (3.81)]:

$$\frac{d}{dt}\hat{\mathbf{p}}_\mathcal{A} = \frac{\imath}{\hbar}[\hat{\mathbf{H}}_{\mathcal{A}+\mathcal{S}}, \hat{\mathbf{p}}_\mathcal{A}] = -g_0(t)\hat{O}, \tag{28.41}$$

where $\hat{H}_{\mathcal{A}+\mathcal{S}}$ is the Hamiltonian of global system $\mathcal{A}+\mathcal{S}$. Hence Eq. (28.41) shows that $\hat{\mathbf{p}}_\mathcal{A}$ changes by different amounts for distinct eigenvalues o_j, and by eq (28.40) we can determine $\langle \varsigma(t)| \hat{O} |\varsigma(t)\rangle$ by the change in the apparatus' momentum.

A protective measurement can be made in two different ways:

- If $|\varsigma(t)\rangle$ is an eigenstate of the Hamiltonian, then the interaction is assumed to be sufficiently weak so that $\hat{H}_{\mathcal{A}+\mathcal{S}}$ changes slowly and $|\varsigma(t)\rangle$ is nearly equal to $|\varsigma(0)\rangle$ up to a phase factor for $t \in [0, \tau]$. $|\varsigma(t)\rangle$ then remains an eigenstate of the Hamiltonian and no entanglement takes place.

[16] The ontological differences between this model and that of de Broglie are analysed in [BLÄSI/HARDY 1995]: they are a consequence of the time symmetry postulated by the first one, which is also a characteristic of the weak measurement.

[17] See also [AHARONOV/VAIDMAN 1993] [AHARONOV/VAIDMAN 1995].

- If we have an arbitrary evolution, so that $|\varsigma(t)\rangle$ is not necessarily an eigenstate of the Hamiltonian, we can operate in the following manner. If $|\varsigma_0(t)\rangle$ is the evolution of $|\varsigma\rangle$, determined by the unperturbed Hamiltonian $\hat{H}_0$ of the system $\mathcal{S}$ — where $\hat{H}_{\mathcal{A}+\mathcal{S}} = \hat{H}_0 + \hat{H}_{\mathcal{AS}} + \hat{H}_{\mathcal{A}}$ —, then one can measure an observable $\hat{O}(t)$, for which $|\varsigma_0(t)\rangle$ is a non-degenerate eigenstate, a large number of times which are dense in $[0, \tau]$ — say at times $t_n = (n/N)\tau, n = 1, 2, \ldots N$, where N is an arbitrarily large number. Then $|\varsigma(t)\rangle$ does not noticeably depart from $|\varsigma_0(t)\rangle$ — it is a type of Zeno effect [see section 19.8]. Now consider the branch of combined system evolution in which each measurement of $\hat{O}(t_n)$ results in the state $|\varsigma_0(t_n)\rangle$ of $\mathcal{S}$:

$$\begin{aligned} |\Psi(\tau)\rangle_0 &:= |\varsigma_0(t_N)\rangle\langle\varsigma_0(t_N)|e^{-\frac{\imath}{\hbar}\frac{\tau}{N}\hat{H}(t_N)} \ldots |\varsigma_0(t_2)\rangle\langle\varsigma_0(t_2)| & \text{(28.42a)} \\ &\quad \times e^{-\frac{\imath}{\hbar}\frac{\tau}{N}\hat{H}(t_2)}|\varsigma_0(t_1)\rangle\langle\varsigma_0(t_1)|e^{-\frac{\imath}{\hbar}\frac{\tau}{2}\hat{H}(t_1)}|\varsigma(0)\rangle|\mathcal{A}(0)\rangle \\ &= |\varsigma_0(t_N)\rangle\langle\varsigma_0(t_N)|e^{-\frac{\imath}{\hbar}\frac{\tau}{N}g(t_N)\hat{\mathbf{q}}_{\mathcal{A}}\hat{O}} \ldots |\varsigma_0(t_3)\rangle\langle\varsigma_0(t_2)| & \text{(28.42b)} \\ &\quad \times e^{-\frac{\imath}{\hbar}\frac{\tau}{N}g(t_2)\hat{\mathbf{q}}\hat{O}}|\varsigma_0(t_2)\rangle\langle\varsigma_0(t_1)|e^{-\frac{\imath}{\hbar}\frac{\tau}{N}g(t_1)\hat{\mathbf{q}}_{\mathcal{A}}\hat{O}}|\varsigma_0(t_1)\rangle|\mathcal{A}_0(\tau)\rangle, \end{aligned}$$

where $|\mathcal{A}_0(\tau)\rangle$ is the state of $\mathcal{A}$ when it evolves under Hamiltonian $\hat{H}_{\mathcal{A}}$. We now calculate the following expectation value — the last one in Eq. (28.42c) — to the second order in $1/N$:

$$\begin{aligned} \langle\varsigma_0(t_1)|e^{-\frac{\imath}{\hbar}\frac{\tau}{N}g(t_1)\hat{\mathbf{q}}_{\mathcal{A}}\hat{O}}|\varsigma(0)(t_1)\rangle &= 1 - \frac{\imath}{\hbar}\frac{\tau}{N}g(t_1)\hat{\mathbf{q}}_{\mathcal{A}}\left\langle\hat{O}\right\rangle - \frac{1}{2\hbar^2}\frac{\tau^2}{N^2}g(t_1)^2\hat{\mathbf{q}}_{\mathcal{A}}^2\left\langle\hat{O}\right\rangle^2 \\ &\quad - \frac{1}{2\hbar^2}\frac{\tau^2}{N^2}g(t_1)^2\hat{\mathbf{q}}_{\mathcal{A}}^2\Delta\hat{O}^2 \\ &= e^{-\frac{\imath}{\hbar}\frac{\tau}{N}g(t_1)\hat{\mathbf{q}}_{\mathcal{A}}\langle\hat{O}\rangle}\left[1 - \frac{1}{2\hbar^2}\frac{\tau^2}{N^2}g(t_1)^2\hat{\mathbf{q}}_{\mathcal{A}}^2\Delta\hat{O}^2\right]. \quad \text{(28.43)} \end{aligned}$$

In the limit $N \longrightarrow \infty$ — where the product of the factors containing $\Delta\hat{O}^2$ tends towards 1 — Eq. (28.42c) reads:

$$|\Psi(\tau)\rangle_0 = |\varsigma_0(\tau)\rangle e^{-\frac{\imath}{\hbar}\int_0^\tau g_0(t)\hat{\mathbf{q}}_{\mathcal{A}}\langle\hat{O}\rangle dt}|\mathcal{A}_0(\tau)\rangle. \quad \text{(28.44)}$$

In this limit the branch represented by Eqs. (28.42) undergoes an unitary evolution and therefore the contribution from other branches, giving states different from $|\varsigma_0(t)\rangle$ tends towards zero.

A proposed experiment Let us now apply these results to the *measurement of the spatial wave function.* Consider the function of a single charged particle [AHARONOV *et al.* 1993a, 4620–21]:

$$\psi(\mathbf{r}, t) = \frac{1}{\sqrt{2}}\left[\psi_1(\mathbf{r}, t) + \psi_2(\mathbf{r}, t)\right], \quad \text{(28.45)}$$

where the two components are confined within two boxes (Two-Boxes Exp.) in their ground states [see figure 28.10].

The particle has charge q. An electron which is prepared in the state of a small localized wave packet, is shot so that if the charge q were not be present, it would go along a straight line

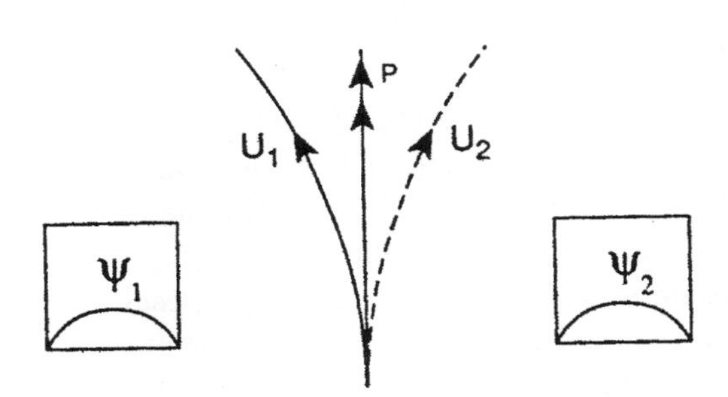

Figure 28.10: A single particle with charge q is in an equal superposition of normalized states $|\psi_1\rangle$ and $|\psi_2\rangle$ localized inside the boxes 1 and 2. If the state is unprotected, then the trajectory of an electron moving midway between the two boxes would be either U_1 or U_2, corresponding to q being in box 1 or 2. But for a protected state, the trajectory would be P as if half of the q is in each box — from [AHARONOV *et al.* 1993a, 4620]

through the midpoint perpendicular to the line of separation between the boxes. In the presence of q it would be influenced by charge. Suppose that $|\psi\rangle$ is protected. It is orthogonal to:

$$\psi'(\mathbf{r},t) = \frac{1}{\sqrt{2}}\left[\psi_1(\mathbf{r},t) - \psi_2(\mathbf{r},t)\right]. \tag{28.46}$$

Therefore, if $|\psi\rangle$ is an eigenfunction of $\hat{H}_0$, so also is $|\psi'\rangle$, degenerate with $|\psi\rangle$. Hence method 1. is not applicable. But we can protect $|\psi\rangle$ by allowing only local interactions by means of a tunnelling between the two boxes, which removes degeneracy. In the case of protective measurement, if each box contains half charge, the trajectory of an electron would be exactly as if there were no charge q, i.e. along the central straight line.

Since there is no collapse, the authors state [AHARONOV *et al.* 1993a, 4619, 4624–25] that two non-orthogonal states can be distinguished by two protective measurements and this enables the measurement of the wave function of a single system. Hence, according to the authors, these are sufficient motivations to dismiss the usual Standard interpretation, which for our purposes also includes the Probabilistic assumption [postulate 3.5: p. 38], and which states that:

- a particle is point-like,
- by measurements there is a collapse,
- it is not possible to distinguish between two nonorthogonal states with probability 1 because this would violate the unitarity of time evolution.

In conclusion then, the meaning of the wave function should not be that of an ensemble average but that of a time average of a single system (the duration of the protective interaction is always very long). In other words, one supposes here — and this must be proven later [see chapters 30, especially its concluding remarks, and 33] — that the basic quantic entity, to which the wave function is referred, is an extended object.

28.4.2 Reversibility and Measurement of the Density Operator of a Single System

Another proposal has been made by Royer [ROYER 1994]. Royer uses the reversibility techniques developed by Imamoḡlu [see section 27.4]. In fact, even if one does not know *a priori* the state of a system, the reversibility of the latter could allow (in the case of a type I-Measurement) for a knowledge of the state before the measurement.

We do not wish to enter here into the details of this proposal. It suffices to say that, as Royer himself acknowledged[18] in a later article [ROYER 1995], the outcome of each basic measurement influences the probability of subsequent reversal, and the price of a successful reversal is that one does not gain any information on the system. On the other hand, if it were possible to gain such information, one could violate the Einstein locality, which is not the case [see chapter 36]. Hence this is not a suitable method for measuring the density operator or the wave function of a single system.

28.4.3 Discussion of Aharonov's Proposal

The two proposals (that of Aharonov and his co-workers and that of Royer) were criticized by many physicists. We mainly discuss the objections to Aharonov's proposal.

Unruh's criticism Unruh [UNRUH 1994] pointed out that Aharonov and his co-workers only showed that, *if* the wave vector of a system $\mathcal{S}$ is known beforehand to be the eigenstate of the unknown Hamiltonian of $\mathcal{S}$, then it is possible to determine the properties of that eigenstate: in other words one can determine some of the properties of an unknown Hamiltonian on $\mathcal{S}$, if one knows that $\mathcal{S}$ is in an eigenstate of that operator. In fact the main condition of their model is a protective measurement: i.e. the wave function of $\mathcal{S}$ interacts with an apparatus $\mathcal{A}$ or with the rest of the world in such a way that it remains unchanged after the measurement but affects $\mathcal{A}$ in such a way that a succession of measurements can completely determine it. This in turn signifies that, if $\mathcal{S}$ is in an energy eigenstate and if the interaction between $\mathcal{S}$ and the rest of the world is adiabatic, then the wave vector after the measurement would still represent the same energy eigenstate. While the wave vector is unchanged, the rest of the world has been changed in a manner dependent on the specific state of $\mathcal{S}$. But, if so, what we have obtained is only the measurement of an observable (the Hamiltonian) and not of the wave function as such. The problem is that we can force the wave function to be the eigenstate of an observable, but we cannot force the observable to have the unknown wave function as its eigenstate.

On the other hand, the articles of Aharonov and his co-workers are significant in having pointed out the difference between the probabilistic distribution of the results of a measurement, which are determined by the Hamiltonian, and the results themselves, which may not be eigenvalues of any operator of direct relevance to the system.

Alter and Yamamoto's criticism Alter and Yamamoto developed a very argumented criticism of Aharonov's approach [ALTER/YAMAMOTO 1995a] [ALTER/YAMAMOTO 1995b] [ALTER/YAMAMOTO 1996a].

Alter and Yamamoto showed that, with a series of weak QND-Measurements, we cannot extract sufficient information in order to measure the wave function of a single system. In fact

[18] After criticisms of Finkelstein, Gisin and Huttner.

there is no one-to-one correspondence between the result of a single measurement and the state of the system before the measurement. Hence we need to perform a statistics on an ensemble of systems prepared in the same state. We need, not only information about the average of the measured observable (the center of the wave function), but also about the variance of the same — which here, in an informal way, represents the width of the wave function [but see section 7.6], as well as of other moments. It is the variance which reveals the wave-like behaviour.

Their formal starting point is represented by Eq. (19.23), which gives the probability amplitude for the quantum probe to have evolved from initial state $|\psi\rangle$ to final $|q_{\mathrm{M}_1}\rangle$, if the interaction of it with $\mathcal{S}$ is given by $\hat{U}$. Now all Measurement's statistics must satisfy two requirements (a one-dimensional problem is considered):

- The first is a general requirement, i.e. that the transition probability of the probe has to be normalized over all possible final states of the probe:

$$\int dq_{\mathrm{M}_1} \hat{E}(q, q_{\mathrm{M}_1}) = 1. \tag{28.47}$$

- The second is the probability reproducibility condition [see theorem 14.7: p. 237], which we can express by saying that q_{M_1} should be equal, on average, to the center of the probability density of $\hat{q}$, i.e. $\langle q_{\mathrm{M}_1}\rangle = \int dq_{\mathrm{M}_1} \wp(q_{\mathrm{M}_1}) q_{\mathrm{M}_1} = \langle q_i \rangle = \int dq \wp_i(q) q$, where q_i is the initial value of $\hat{q}$ and $\wp_i(q)$ is the initial probability density of $\hat{q}$.

From these two conditions it follows:

$$\int dq_{\mathrm{M}_1} \hat{E}(q, q_{\mathrm{M}_1}) q_{\mathrm{M}_1} = q. \tag{28.48}$$

The system $\mathcal{S}$ (the signal) and the probe should be independent of each other; hence the probability error associated with the result of the measurement should equal the sum of the measurement error Δ^2_{M} and of the intrinsic uncertainty of the wave function — i.e. the variance of the measured observable —: $\left\langle \Delta q^2_{\mathrm{M}_1} \right\rangle = \langle \Delta q_i^2 \rangle + \Delta^2_{\mathrm{M}}$, where $\langle \Delta q_i^2 \rangle = \int dq_{\mathrm{M}_1} \wp(q_{\mathrm{M}_1}) q^2_{\mathrm{M}_1}$. From this we obtain:

$$\int q_{\mathrm{M}_1} \hat{E}(q, q_{\mathrm{M}_1}) q^2_{\mathrm{M}_1} = q^2 + \Delta^2_{\mathrm{M}}. \tag{28.49}$$

The measurement is *weak* if $\Delta^2_{\mathrm{M}} \gg \langle \Delta q_i^2 \rangle$. After this measurement the signal $\mathcal{S}$ is described by $\hat{\rho}_{\mathcal{S}} = \wp(q_{\mathrm{M}_1})^{-1} \hat{\vartheta}(\hat{q}, q_{\mathrm{M}_1}) \hat{\rho}_i \hat{\vartheta}^\dagger(\hat{q}, q_{\mathrm{M}_1})$ [see Eq. (19.9)]. The corresponding probability density of $\hat{q}$ is:

$$\wp(q, q_{\mathrm{M}_1}) = \langle q|\hat{\rho}|q\rangle = \wp(q_{\mathrm{M}_1})^{-1} \hat{E}(q, q_{\mathrm{M}_1}) \wp_i(q). \tag{28.50}$$

We then measure the conjugate $\hat{p}$. This measurement changes the density probability of $\hat{q}$ from $\wp(q, q_{\mathrm{M}_1})$ [given by Eq. (28.50)] to $\wp_{\mathrm{M}(p)}(q, q_{\mathrm{M}_1})$. The center remains unchanged, but the width increases due to the back-action noise Δ^2_b [see subsection 19.6.1]:

$$\int dq\ \wp_{\mathrm{M}(p)}(q, q_{\mathrm{M}_1}) q^2 = \int dq\ \wp(q, q_{\mathrm{M}_1}) q^2 + \Delta^2_b. \tag{28.51}$$

Now we measure $\hat{q}$ a second time, and we obtain the following conditional probability:

$$\wp(q_{\mathrm{M}_2}|q_{\mathrm{M}_1}) = \int dq \hat{E}(q, q_{\mathrm{M}_2}) \wp_{\mathrm{M}(p)}(q, q_{\mathrm{M}_1}). \tag{28.52}$$

Each of the measurement results — q_{M_1} or q_{M_2} — estimates the initial center $\langle q_i \rangle$. Hence we can estimate the initial second-order moment using one of them:

$$\langle q_{\mathrm{M}_1} \rangle = \int dq_{\mathrm{M}_1} \wp(q_{\mathrm{M}_1}) q^2_{\mathrm{M}_1} = \langle q_i \rangle + \Delta^2_{\mathrm{M}}, \tag{28.53a}$$

$$\langle q_{\mathrm{M}_2} \rangle = \int dq_{\mathrm{M}_1} \wp(q_{\mathrm{M}_1}) \int dq_{\mathrm{M}_2} \wp(q_{\mathrm{M}_2} | q_{\mathrm{M}_1}) q^2_{\mathrm{M}_2} = \langle q_i \rangle + \Delta^2_{\mathrm{M}} + \Delta^2_{b}. \tag{28.53b}$$

However, one cannot estimate the initial width $\langle q_i^2 \rangle$ using the result of a single measurement because a single measurement does not contain information about $\langle q_i \rangle^2$ [see Eq. (3.110)]. If $q_{\mathrm{M}_1}, q_{\mathrm{M}_2}$ were two independent results — obtained from two different systems prepared in the same initial state —, then their correlation would provide this information. But in our case the second measurement depends conditionally upon the first, so that we have:

$$\langle q_{\mathrm{M}_1} q_{\mathrm{M}_2} \rangle = \int dq_{\mathrm{M}_1} \wp(q_{\mathrm{M}_1}) q_{\mathrm{M}_1} \int dq_{\mathrm{M}_2} \wp(q_{\mathrm{M}_2} | q_{\mathrm{M}_1}) q_{\mathrm{M}_2} = \left\langle q_i^2 \right\rangle. \tag{28.54}$$

This treatment can be extended to an arbitrary succession of measurements. In conclusion no estimate of the initial widths of $\wp_i(q)$ and $\wp_i(p)$ is possible, and hence it is impossible to measure the wave function of a single system using repeated QND measurements without an *a priori* knowledge of the wave function itself. But then we do not need protective measurements [ALTER/YAMAMOTO 1995b].[19] Similar considerations can also be developed for Royer's Model.

D'Ariano and Yuen's criticism D'Ariano and Yuen [D'ARIANO/YUEN 1996, 2832–33] provided a further generalization of the proof about the impossibility of measuring the wave function of a single system — see also the Pauli Problem [subsection 18.6.1]. By using a theorem to be demonstrated later [42.5: p. 723] about the impossibility to clone two non-orthogonal states — because, if one could, then the unitarity of the evolution would be violated —, they showed the impossibility of distinguishing two non-orthogonal states (and hence of measuring a single-system wave function). In fact, if it were possible, then one could easily produce n clones of one of the two states (depending on the measurement result).

This interesting result, which we shall state later as a corollary of the quoted theorem, is complementary to the one shown in subsection 27.4.4: there it was demonstrated [see theorem 27.1: p. 463] that an ideal measurement is reversible iff no information about the identity of the initial state is obtained, and then, if it is reversible, is generally unitary.

In the same article D'Ariano and Yuen also proved [D'ARIANO/YUEN 1996, 2833–34] that a sequence of measurements on the same system yields the same information on the state of a system as an appropriately chosen single measurement. Finally they proved [D'ARIANO/YUEN 1996, 2834–35] concretely that a measurement requires a minimum of nonvanishing efficiency of the detector $= 1/2$, which in turn requires a violation of the weakness of the measurement due to the necessary uncontrollable back-action produced in order to satisfy the efficiency requirement.

[19] In [ALTER/YAMAMOTO 1996a] it is shown that the protective measurement is equivalent to a measurement of an ensemble of systems.

28.5 Measurement of the Density Matrix

28.5.1 General Abstract Formalism

The attempts of Royer and of Aharonov and his co-workers in order to obtain a measurement of the wave function has given a new impulse in developing a more general formalism concerning the measurement of the density matrix (or state) of a system.

The following results confirm the impossibility of measuring the state of a single system. In fact this measurement is only possible if we perform a series of measurements, each one on a single system of a set of identically prepared systems. At the beginning the problem was not posed in its full generality, because only measurements of particles moving in one dimension (pure states and spatial wave functions) were considered [KEMBLE 1951]. Gale and his co-workers [GALE *et al.* 1968] used methods which are good for spin states but not for translational states[20].

Royer [ROYER 1985] [ROYER 1989, 17, 19–23, 27–29] analysed the problem in all its generality by working out the premeasurement techniques developed by Lamb [see section 13.4]. The problem is: given a well-defined preparation procedure, is it possible to determine experimentally (to measure) the state into which such a procedure puts the systems? It is possible if one is able to calculate the W-function, due to a one-to-one correspondence between the W-function and the corresponding density matrix of a system [see subsection 5.2.4].

First take the following displacement operator in phase space (for the sake of simplicity the system is one-dimensional):

$$\hat{D}_{qp} := e^{\frac{i}{\hbar}(p\hat{q}-q\hat{p})}, \tag{28.55}$$

and the following parity operator about the phase space point (q, p):

$$\hat{\Pi}_{qp} = \hat{D}_{qp}\hat{\Pi}\hat{D}_{qp}^{-1}, \tag{28.56}$$

where

$$\hat{\Pi} = \int_{-\infty}^{+\infty} dq|-q\rangle\langle q| = \int_{-\infty}^{+\infty} dp|-p\rangle\langle p|, \tag{28.57}$$

from which it follows:

$$\hat{\Pi}_{qp}(\hat{q}-q)\hat{\Pi}_{qp} = -(\hat{q}-q), \qquad \hat{\Pi}_{qp}(\hat{p}-p)\hat{\Pi}_{qp} = -(\hat{p}-p). \tag{28.58}$$

Then the W-function is the expectation value of the parity operator $\hat{\Pi}_{qp}$ [see Eq. (5.49)]:

$$W_{\hat{\rho}}(q,p,t) = \frac{1}{\pi\hbar}\left\langle \hat{\Pi}_{qp}\right\rangle_{\hat{\rho}(t)}. \tag{28.59}$$

$\hat{\Pi}_{qp}$ is an observable whose eigenvalues are ± 1; a complete set of eigenstates $|\psi^n_{qp}\rangle, n = 1, 2, \ldots$ satisfying $\hat{\Pi}_{qp}|\psi^n_{qp}\rangle = (-1)^n|\psi^n_{qp}\rangle$ may be obtained by displacing in phase space any complete orthogonal set of kets (in the corresponding Hilbert space) $|\psi^n\rangle$ of definite parity about the origin. Thus:

$$\hat{\Pi}_{qp} = \sum_n (-1)^n |\psi^n_{qp}\rangle\langle\psi^n_{qp}|, \tag{28.60a}$$

$$|\psi^n_{qp}\rangle = \hat{D}_{qp}|\psi^n\rangle, \tag{28.60b}$$

$$|\psi^n(-q)\rangle = |(-)^n\psi^n(q)\rangle; \tag{28.60c}$$

[20]See also [ROYER 1989, 4–5] [POOL 1966] [BAND/PARK 1979] [IVANOVIĆ 1981] [IVANOVIĆ 1983] [WOOTTERS 1987] [WOOTTERS/FIELDS 1989].

so that Eq. (28.59) can be rewritten as follows:

$$W_{\hat{\rho}}(q,p,t) = \frac{1}{\pi\hbar}\sum_n (-1)^n \langle \psi^n_{qp}|\hat{\rho}(t)|\psi^n_{qp}\rangle. \tag{28.61}$$

We now try to measure $W_{\hat{\rho}}(q,p,t)$ at some definite time (e.g. $t=0$). This can be done by measuring each transition probability $\langle \psi^n_{qp}|\hat{\rho}(0)|\psi^n_{qp}\rangle$ in the previously quoted manner of Lamb. A very easy approach is possible if we choose the $|\psi^n\rangle$'s as eigenstates of following Hamiltonian:

$$\hat{H} = \frac{\hat{p}^2}{2m} + V(\hat{q}), \tag{28.62}$$

where $V(-q) = V(q)$ is a symmetric potential. Then the $|\psi^n_{qp}\rangle$'s are eigenstates of the phase-space displaced Hamiltonian:

$$\hat{H}_{qp} = \hat{D}_{qp}\hat{H}\hat{D}^{-1}_{qp} = \frac{(\hat{p}-p)^2}{2m} + V(\hat{q}-q). \tag{28.63}$$

A method of measuring $\hat{H}_{qp}$ almost in the strict sense is as follows. First, we place ourselves in a reference frame moving with uniform speed $v = p_{\mathcal{A}}/m$ relative to the preparation apparatus $\mathcal{A}$. By Galilei transformations, the density operator seen (for $t \leq 0$) is:

$$\hat{\rho}'(t) = \hat{D}^{-1}_{vt,p_{\mathcal{A}}}\hat{\rho}(t)\hat{D}_{vt,p_{\mathcal{A}}}. \tag{28.64}$$

At time $t=0$ we turn on the potential $V(q-q_{\mathcal{A}})$ in the moving frame. The eigenstates of

$$\hat{H}^{q_{\mathcal{A}}}_0 = \frac{(\hat{p}')^2}{2m} + V(\hat{q}'-q_{\mathcal{A}}) \tag{28.65}$$

are $\hat{D}^{q_{\mathcal{A}}}_0|\psi^n\rangle = |\psi^n_{q_{\mathcal{A}},0}\rangle$ with corresponding energies E_n. Then at times $t \geq 0$ we obtain:

$$\begin{aligned}
\hat{\rho}'(t) &= e^{-\imath t\hat{H}^{q_{\mathcal{A}}}_0/\hbar}\hat{\rho}'(0)e^{\imath t\hat{H}^{q_{\mathcal{A}}}_0/\hbar} \\
&= \sum_{m,n} e^{-\imath t(E_n-E_m)/\hbar}|\psi^n_{q_{\mathcal{A}},0}\rangle\langle\psi^m_{q_{\mathcal{A}},0}|\langle\psi^n_{q_{\mathcal{A}},0}|\hat{\rho}'(0)|\psi^m_{q_{\mathcal{A}},0}\rangle \\
&= \sum_{m,n} e^{-\imath t(E_n-E_m)/\hbar}|\psi^n_{q_{\mathcal{A}}0}\rangle\langle\psi^m_{q_{\mathcal{A}},0}|\langle\psi^n_{q_{\mathcal{A}}p_{\mathcal{A}}}|\hat{\rho}(0)|\psi^m_{q_{\mathcal{A}}p_{\mathcal{A}}}\rangle.
\end{aligned} \tag{28.66}$$

Now the transition probabilities

$$\langle\psi^n_{q_{\mathcal{A}},0}|\hat{\rho}'(0)|\psi^n_{q_{\mathcal{A}},0}\rangle = \langle\psi^n_{q_{\mathcal{A}}p_{\mathcal{A}}}|\hat{\rho}(0)|\psi^n_{q_{\mathcal{A}}p_{\mathcal{A}}}\rangle \tag{28.67}$$

are time independent, so that we have a long time available to perform a measurement of $\hat{H}^{q_{\mathcal{A}}}_0$ referring to the set $\{|\psi^n_{q_{\mathcal{A}},0}\rangle\}$ and 'find' the particle in one of the states pertaining to this set. Repeating the measurement many times will allow the building of the distribution (28.67), from which $W(q_{\mathcal{A}},p_{\mathcal{A}},0)$ can be deduced by means of Eq. (28.61). What has been done is to measure $W(q_{\mathcal{A}},p_{\mathcal{A}},0)$ by measuring $W(q_{\mathcal{A}},0,t) = W(q_{\mathcal{A}}+vt,p_{\mathcal{A}},t)$ at $t=0$ in the moving frame.

In conclusion we can state the following theorem[21]:

[21] See also [*BUSCH et al.* 1995a, 131].

Theorem 28.2 (Individual State Determination) *Determination of the state of a single (individual) system is in QM not possible.*

Since a single measurement does not suffice, in the general case the method developed by Royer can be employed only for ensembles of systems prepared in the same state. If the determination of the state of a single system were possible, then any two different states would be orthogonal, which is surely wrong: no pair of overlapping states can be uniquely separated by a single measurement. On the other hand, assuming it is possible, we could then exchange signals to superluminal velocity, but this is excluded by the successive analysis [see chapter 36].

Note that such a theorem is a confirmation of the basic duality of QM between states and observables [see subsection 9.4.1]. What we can do, is only to measure observables and infer the state of the system.

28.5.2 Concrete Measurement Proposals in Q-Optics

Harmonic oscillator case The method used here is called *quantum tomography* in analogy with the evaluation of recorded transmission profiles of radiation. Specifically, one can use the tomographic inversion of a set of measured probability distributions of a quadrature amplitude in order to obtain the W-function [BERTRAND/BERTRAND 1987] [K. VOGEL/RISKEN 1989].

The homodyne detector[22] utilizes the interference of the signal field and a local oscillator which has phase φ and measures the probability distributions of the following linear combination of creation and annihilation operators [YUEN/SHAPIRO 1980] [K. VOGEL/RISKEN 1989] [see also Eq. (19.24)]:

$$\hat{x}(\varphi) = \hat{x}^{\dagger}(\varphi) = \hat{X}_1 \cos\varphi - \hat{X}_2 \sin\varphi, \tag{28.68}$$

where the quadratures [see Eq. (5.3)] are $\hat{X}_1 = \hat{x}(0), \hat{X}_2 = \hat{x}(\pi/2)$. Now we can introduce the probability distribution $\wp(x,\varphi)$ as the Fourier transform of the characteristic function [see Eq. (3.111)]:

$$\wp(\xi,\varphi) = \mathrm{Tr}\left[e^{\imath\xi\hat{x}(\varphi)}\hat{\rho}\right]; \tag{28.69}$$

or:

$$\wp(x,\varphi) = \frac{1}{2\pi}\int \tilde{\wp}(\xi,\varphi)e^{-\imath\xi x}d\xi. \tag{28.70}$$

We then establish a one-to-one correspondence between the set of the probability distributions (28.70) for $0 < \varphi < \pi$ and the W-function (we remember that it is a quasi-probability distribution):

$$W(\hat{X}_1,\hat{X}_2) = \frac{1}{4\pi^2}\int_{-\infty}^{+\infty}\int_{-\infty}^{+\infty}\int_{0}^{\pi} \wp(x,\varphi)e^{\imath\xi(x-\hat{X}_1\cos\varphi-\hat{X}_2\sin\varphi)}|\xi|dx d\xi d\varphi. \tag{28.71}$$

[22] For the problem of noise arising in homodyne and heterodyne detection see [YUEN/CHAN 1983].

Other methods Optical homodyne tomography of a single mode of the electromagnetic field has been experimentally realized [SMITHEY *et al.* 1993]: an ensemble of repeated measurements of the quadrature for various phases relative to the local oscillator of the homodyne detector[23].

The method was also extended in invariant form by relating the homodyne output distribution directly to the density matrix (without the intermediate step of reconstructing the Wigner function) in a quadrature-component basis [KÜHN *et al.* 1994] [ZUCCHETTI *et al.* 1996] and in the Fock basis [D'ARIANO *et al.* 1994] [D'ARIANO *et al.* 1995b] [SCHILLER *et al.* 1996].[24] In fact the reconstruction of the W-function can be a difficult task: the knowledge of the W-function is equivalent to the knowledge of all independent moments of the system operators (in the case of the harmonic oscillator the knowledge of all moments of the creation and annihilation operators). But the state under consideration is generally characterized by an infinite number of independent moments, i.e. we would need an infinite time in order to reconstruct the W-function. Hence, in the general case, we can only partially reconstruct the W-function[25].

Recently Schleich and his co-workers [BARDROFF *et al.* 1995] have proposed a new method, named *quantum state endoscopy.* Wallentowitz and Vogel [WALLENTOWITZ/W. VOGEL 1996] have proposed a very easy technique: a mixing of signal and coherent fields with controlled amplitude on the BS may serve the reconstruction of the W-function and other distribution functions by counting photons only. Hradil [HRADIL 1997] has proposed methods deriving from the statistical estimation of the wave function [see subsection 18.7.2].[26]

Generalization However, these techniques are restricted to undamped harmonic oscillators. But, in many situations (as S-Cats), we are faced with anharmonic (modified by damping) oscillators. Hence other methods have recently been found which apply to both cases. Leonhardt and Raymer [LEONHARDT/RAYMER 1996] developed a method by observing a one-dimensional wave packet moving in an arbitrary potential $V(x)$.[27] At time $t = 0$ let us write the density operator of the system in the energy representation:

$$\hat{\rho}_{mn} = \langle m|\hat{\rho}|n\rangle, \tag{28.72}$$

where m, n are energy eigenstates of the system. The position probability distribution $\wp(x,t)$ is spectrally decomposed into terms oscillating at differences of eigenfrequencies:

$$\wp(x,t) = \sum_{mn} \hat{\rho}_{mn}\psi^*_m(x)\psi_n(x)e^{-\imath(\omega_m-\omega_n)t}. \tag{28.73}$$

where $\psi_n(x)$ denotes the state vector corresponding to the energy eigenstate $|n\rangle$. Since none of the discrete levels are degenerate, $\psi_n(x)$ is the only normalized solution of the stationary Schrödinger Eq. with eigenfrequency ω_n [see Eq. (3.18)]:

$$\left[-\frac{\hbar^2}{2m}\frac{\partial^2}{\partial x} + V(x)\right]\psi_n(x) = \omega_n\psi_n(x). \tag{28.74}$$

[23] See also [FREYBERGER/HERKOMMER 1994] [WALLENTOWITZ/W. VOGEL 1995].

[24] See also [LEONHARDT 1995] .

[25] On this point see [BUŽEK *et al.* 1996a], where quantifications of the precision in the reconstruction of the W-function and the minimal bound in order to discriminate between pure states and corresponding mixtures are also given.

[26] For other techniques see [ROYER 1989] [PAUL *et al.* 1996] [STEUERNAGEL/VACCARO 1995].

[27] See also [LEONHARDT/SCHNEIDER 1997]. For other methods see [RICHTER/WÜNSCHE 1996], where time-averaged position distributions instead of phase-averaged quadrature component distributions are used.

Now we Fourier-transform $\wp(x,t)$ in order to obtain a spectrally decomposed probability distribution:

$$\begin{aligned} \tilde{\wp}(x,\omega_m-\omega_n) &:= \lim_{\tau\longrightarrow\infty}\frac{1}{\tau}\int_{-\frac{\tau}{2}}^{+\frac{\tau}{2}}\wp(x,t)e^{-\imath(\omega_m-\omega_n)t}dt \\ &= \sum_{\mu\nu}{}' \hat{\rho}_{\mu\nu}\psi^*_\mu(x)\psi_\nu(x), \end{aligned} \tag{28.75}$$

where $\sum'$ indicates a sum restricted to values of μ,ν such that:

$$\omega_\mu-\omega_\nu=\omega_m-\omega_n. \tag{28.76}$$

In order to extract the density-matrix elements, we try to find functions $f_{mn}(x)$ orthogonal to products of wave functions $\psi^*_\mu(x)\psi_\nu(x)$, provided that constraint (28.76) is given, i.e.:

$$\int_{-\infty}^{+\infty}\psi^*_\mu(x)\psi_\nu(x)f_{mn}(x)dx=\delta_{m\mu}\delta_{n\nu}. \tag{28.77}$$

Sampling the spectrally decomposed probability distributions $\tilde{\wp}(x,\omega_m-\omega_n)$ with respect to the functions $f_{mn}(x)$ would produce the density matrix:

$$\hat{\rho}_{mn}=\int_{-\infty}^{+\infty}\tilde{\wp}(x,\omega_m-\omega_n)f_{mn}(x). \tag{28.78}$$

The sampling functions turn out to be very simple: they are just derivatives of products of regular and irregular state vectors:

$$f_{mn}(x)=\frac{\partial}{\partial x}\left[\psi^*_m(x)\psi^{\rm I}_n(x)\right]. \tag{28.79}$$

Any linear differential Eq. of the second order such as the Schrödinger Eq. must have two linearly independent solutions for a given frequency. Since the stationary states $|n\rangle$ are non-degenerate, the irregular vectors such as $|\psi^{\rm I}_n\rangle$ cannot be normalizable and they must be discarded as representative of physical states. Nevertheless they can be used in order to obtain the required information about the density matrix[28].

28.5.3 Zeno Effect and Indetermination of the Initial Density Matrix*

We have seen that one can arrange an experiment in order to obtain a Zeno effect of a 'single' system [see subsection 19.8.5]. Spiller [SPILLER 1994] and Power/Knight [POWER/KNIGHT 1996] proved that, although, as we have seen, a statistical Zeno effect is generally equivalent to a random decoherence, it is something different in the case of a single system.

Aharonov and Vardi [AHARONOV/VARDI 1980] proposed for the first time an 'inverse Zeno effect', by means of which we induce a a sequence of slight changes of the state by a series of measurements on a system that otherwise would stay at rest. Such a phenomenon has been studied in detail in [ALTENMÜLLER/SCHENZLE 1993].

Alter and Yamamoto [ALTER/YAMAMOTO 1996b] [ALTER/YAMAMOTO 1997, R2501–502] followed this line of research and developed the formalism they used for criticizing Aharonov and his co-workers' proposal [see subsection 28.4.3], showed that the Zeno effect is equivalent to

[28] On this matter see also [OPATRNÝ *et al.* 1997].

the indetermination of the initial state of the system, and these two aspects are respectively the Schrödinger picture and the Heisenberg picture description of the same phenomenon.

Consider a series of N measurements of the observable $\hat{q}$ performed on a single (one-dimensional) quantum system $\mathcal{S}$ during its unitary evolution in the time interval $t \in [0, \tau]$. The initial state of the system is described by the density operator $\hat{\rho}_i$ and the deterministic time evolution of $\mathcal{S}$ between the $(k-1)$–th and the k–th measurements at $t_{k-1} = (k-1)\tau/N$ and $t_k = k\tau/N$, respectively, is described by $\hat{U}_k$ such that $\hat{\rho}^{(k)} = \hat{U}_k \hat{\rho}_{k-1} \hat{U}_k^\dagger$. The k–th measurement at t_k (the preparation of the k–th probe in the pure state $|\psi\rangle_{\rm QP}^{(k)}$), the interaction of the probe with $\mathcal{S}$ (described by $\hat{U}_{\rm M}(\hat{q})$), and the result of the measurement $q_{\rm M}^{(k)}$, which corresponds to the state of the probe after the measurement $|q_{\rm M}^{(k)}\rangle_{\rm QP}$, is described by the probability amplitude operator [see Eq. (19.23)]:

$$\hat{\vartheta}_k := \hat{\vartheta}(\hat{q}, q_{\rm M}^{(k)}) = {}_{\rm QP}\langle q_{\rm M}^{(k)} | \hat{U}_{\rm M}(\hat{q}) | \psi \rangle_{\rm QP}. \tag{28.80}$$

The state of $\mathcal{S}$ after the kth measurement is $\hat{\rho}_{\rm AM}^{(k)} = \wp_{\rm S}(q_{\rm M}^{(k)})^{-1} \hat{\vartheta}_k \hat{\rho}_{\rm BM}^{(k)} \hat{\vartheta}_k^\dagger$, where $\hat{\rho}_{\rm BM}^{(k)}$ is the state of $\mathcal{S}$ before the k–th measurement, and $\wp_{\rm S}(q_{\rm M}^{(k)}) = {\rm Tr}_{\mathcal{S}}[\hat{\vartheta}_k \hat{\rho}_{\rm BM}^{(k)} \hat{\vartheta}_k^\dagger]$ (trace over the operators of the measured system) is the probability of obtaining result $q_{\rm M}^{(k)}$. The density operator which defines the single system at $t = \tau$, after the N–th measurement is:

$$\hat{\rho}_{\rm S} = \wp_{\rm S}(q_{\rm M}^1, \ldots, q_{\rm M}^{(N)})^{-1} \hat{\vartheta}_N \hat{U}_N \cdots \hat{\vartheta}_1 \hat{U}_1 \hat{\rho}_i \hat{U}_1^\dagger \hat{\vartheta}_1^\dagger \cdots \hat{U}_N^\dagger \hat{\vartheta}_N^\dagger, \tag{28.81}$$

and the probability density to obtain the series of results $(q_{\rm M}^1, \ldots, q_{\rm M}^{(N)})$ is

$$\begin{aligned} \wp_{\rm S}(q_{\rm M}^1, \ldots, q_{\rm M}^{(N)}) &= {\rm Tr}_{\mathcal{S}} \left[\hat{\vartheta}_N \hat{U}_N \cdots \hat{\vartheta}_1 \hat{U}_1 \hat{\rho}_i \hat{U}_1^\dagger \hat{\vartheta}_1^\dagger \cdots \hat{U}_N^\dagger \hat{\vartheta}_N^\dagger \right] \\ &= {\rm Tr}_{\mathcal{S}} \left[\hat{O}_N \cdots \hat{O}_1 \hat{\rho}^0 \hat{O}_1^\dagger \cdots \hat{O}_N^\dagger \right] \\ &:= \wp_{\rm H}(q_{\rm M}^1, \ldots, q_{\rm M}^{(N)}), \end{aligned} \tag{28.82}$$

where we used the identity $\hat{U}_k \hat{U}_k^\dagger = \hat{I}$, and where

$$\hat{O}_k := \hat{O}(\hat{q}^{(k)}, q_{\rm M}^{(k)}) = \hat{U}_1^\dagger \cdots \hat{U}_k^\dagger \hat{\vartheta}_k \hat{U}_k \cdots \hat{U}_1. \tag{28.83}$$

Therefore, $\wp_{\rm H}(q_{\rm M}^1, \ldots, q_{\rm M}^{(N)})$ describes the probability density of obtaining a series of results $(q_{\rm M}^1, \ldots, q_{\rm M}^{(N)})$ in the series of measurements $(\hat{O}_1, \ldots, \hat{O}_N)$ of the single system, with no time evolution in between successive measurements. The state of the system after this series of measurements is described by the Heisenberg-picture density operator:

$$\hat{\rho}_{\rm H} = \wp_{\rm H}(q_{\rm M}^1, \ldots, q_{\rm M}^{(N)})^{-1} \hat{O}_N \cdots \hat{O}_1 \hat{\rho}^0 \hat{O}_1^\dagger \cdots \hat{O}_N^\dagger. \tag{28.84}$$

Since the statistics of $\wp_{\rm S}$ and $\wp_{\rm H}$ are the same, $\hat{\rho}_{\rm S}$ and $\hat{\rho}_{\rm H}$ are respectively the density operator of $\mathcal{S}$ in the Schrödinger picture [see Eq. (3.72)] and in the Heisenberg picture [see Eq. (3.73)], where the time evolution is that of the observables associated with $\mathcal{S}$. Now $\hat{\rho}_{\rm S}$ defines the quantum Zeno effect of a single system and $\wp_{\rm S}(q_{\rm M}^1, \ldots, q_{\rm M}^{(N)})$ describes the information about the free time evolution of $\mathcal{S}$ that is contained in the measurement results, while $\hat{\rho}_{\rm H}$ defines the stochastic evolution of a single system due to a series of measurements of time-varying observables, and $\wp_{\rm H}(q_{\rm M}^1, \ldots, q_{\rm M}^{(N)})$ describes the indetermination of the unknown quantum state of the single system.

28.6 Classical Waves or Quantum Waves?

De Broglie — not followed here by other proponents of the ontological interpretation of QM waves — supposed the *coexistence* of 'classical waves' and of 'classical particles'. We shall speak about the concept of coexistence in chapter 30 and about the nature of particles in part VIII. But here we can pose the following question: are the QM waves classical or do they have a different nature?

Again Interfer. Exp. 2 We have already reported an experiment performed by Grangier, Roger and Aspect [GRANGIER *et al.* 1986a] [see Interfer. Exp. 2: p. 110; figure 6.4 and figure 28.11] on the basis of which we suppose that QM waves cannot be reduced to classical ones.

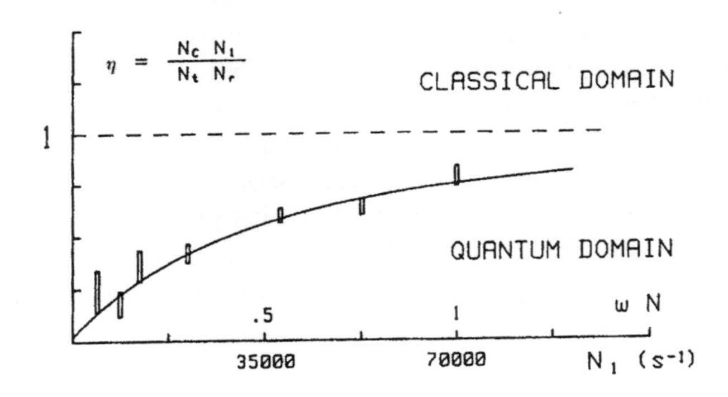

Figure 28.11: Results of Interfer. Exp. 2: Anticorrelation parameter η as a function of wN (number of cascade emitted during the gate) and of N_1 (triggered rate). The inequality $\eta \geq 1$ characterizes the classical domain — from [GRANGIER *et al.* 1986b, 103].

Criticism and correction Santos, Marshall and Selleri tried to make the experiment compatible with a classical wave theory [MARSHALL *et al.* 1983] [MARSHALL/SANTOS 1987]. After the latter the ratio between the coincidence probability and the product of the reflection probability and the transmission probability η is ≥ 1. It is true that Grangier and co-workers showed [GRANGIER *et al.* 1986a] that in QM one can obtain $\eta < 1$. However Santos, Marshall and Selleri [MARSHALL *et al.* 1983] pointed out that, if we use stochastic theories, this value can be allowed . Finally Hardy [HARDY 1991b, 593–95] proved that in this case one must have:

$$\eta \geq \frac{\mathcal{V}_S^2}{4R_0T_0}, \tag{28.85}$$

where $\mathcal{V}_S$ is the fringe visibility by leaving the two beams interfering, $R_0 = \wp'_R/(\wp'_R + \wp_T), T_0 = \wp_T/(\wp'_R + \wp_T)$ and $\wp_R, \wp'_R$ are the probabilities of reflection at $t, t' = t - x/c$ respectively and $\wp_T, \wp'_T$ the probabilities of transmission at at $t, t' = t - x/c$ respectively. One can deduce this inequality only by means of some specific assumptions. However, QM violates it. We can already conclude therefore that the QM waves are not classical[29].

[29]See also [MANDEL 1983].

Fourth-order interference Another typical quantum effect is that Fock states (which normally never carry information about the phase of the electromagnetic field), if in superposition with the vacuum, allow that light carries phase information, which rules out a second-order interference effect because it is always a manifestation of the intrinsic indistinguishability of several possible paths of the detected photon [HONG *et al.* 1988] [OU *et al.* 1990a] [ZOU *et al.* 1991]. In general, any two classical light waves can in principle give rise to second-order interference effects because the phase of a classical field always exists, whereas that of a quantum field does not, even if the field is quasi-monochromatic. And since a single photon does not have a definite phase according to Q-Optics, two single photons cannot exhibit second-order interference but only fourth-order interference [see also subsection 5.1.2].

I-order interference Also very interesting is the following experiment, which shows that it is possible also to perform I-order interference experiments in QM.[30] Weinfurter, Herzog, Kwiat, Rarity and Zeilinger discussed a I-order interference experiment [WEINFURTER *et al.* 1995] which uses virtual photons, that are understood as energetically indistinguishable from the vacuum field and therefore undetectable. We can obtain constructive and destructive interference by means of a mirror. In the latter case the question arises as to whether any SPDC took place at all. An alternative explanation does not use virtual photons but the possibility that a pair of back-reflected photons are spontaneously upconverted with probability equal to that of downconversion (= SPDC Exp. 3). But photons could now be detected in the downconverted mode between the crystal and the mirror even in the case of destructive interference [see figure 28.12].

For the ordinary experiment we have:

$$|\Psi\rangle = c|\gamma_s^1\rangle|\gamma_i^1\rangle, \tag{28.86}$$

where γ_i, γ_s stand for i-photon and s-photon. If we reflect the beam, due to the very low efficiency of DC process, there is a phase correlation with phase shift φ_p between the pumped beam and the reflected one:

$$|\Psi\rangle = c|\gamma_s^1\rangle|\gamma_i^1\rangle + \alpha e^{\imath\varphi_p}|\gamma_s^2\rangle|\gamma_i^2\rangle. \tag{28.87}$$

If we use a mirror IM to overlap mode $|\gamma_i^1\rangle$ (with phase shift φ_i) with $|\gamma_i^2\rangle$ to obtain $|\gamma_i\rangle$, a definite phase relation with φ_p only exists for the whole product state, so that the remaining s-photons could be used for which-path detection of the overlapping i-photons. Interference effects are excluded.

But if we use another mirror SM to overlap s-photons, there is indistinguishability, i.e. I order interference for both i- and s-photons:

$$|\Psi\rangle = c(e^{\imath\varphi_p} + e^{\varphi_p} + e^{\varphi_s+\varphi_i})|\gamma_s\rangle|\gamma_i\rangle. \tag{28.88}$$

28.7 Conclusion

Every attempt here examined to ascribe to the QM entities a classical 'wave' nature together with a classical particle feature has failed — and hence also Bohr's use of classical wave-like and corpuscular modes of descriptions [see principle 8.2 (CP2): p. 138] is without foundations. Does

[30]See [WANG *et al.* 1991] [HERZOG *et al.* 1994].

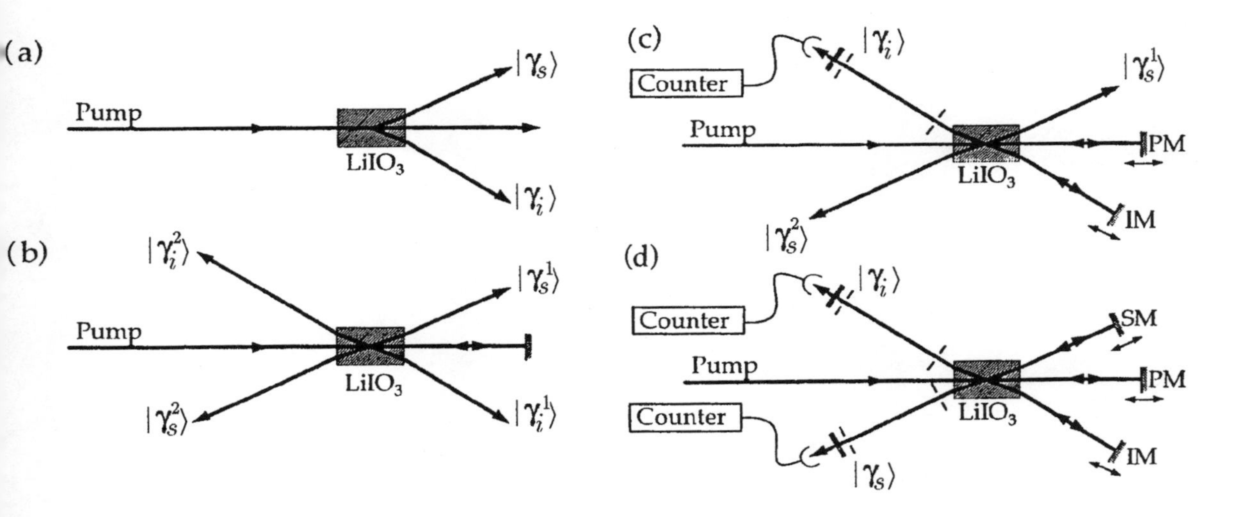

Figure 28.12: Principle of the experimental setup (= SPDC Exp. 3): In addition to the standard setup (a) of PDC, a second possibility for a PDC process is given by back-reflecting the pump beam onto itself (b).
Overlapping both idler modes alone (c) is not sufficient to obtain interference because the signal photons could still be used for which-path analysis. When the signal photons are reflected as well (d), the two PDC processes become indistinguishable and first-order interference can be observed in both the signal and idler intensities — from [WEINFURTER *et al.* 1995, 63].

this mean that the only ontological interpretation of the wave-like behaviour is the Probabilistic interpretation [proposition 6.2]? Not at all. As we shall see more clearly in chapter 30, the wave has an ontological reality, but:

- this reality is smoothly complementary to the corpuscular one,
- it is an intrinsically 'superposed' reality which has no classical analogy.

Anyway the previously observed intrinsically quantum nature of the wave excludes that, in the context of QM, one can support a naive 'realism' in the classical sense of the word [see also subsection 26.4.1]. We remember here our conclusions about the imperfect determination of quantum entities [see section 24.5]. This will be the subject of the next part.

Chapter 29

THE INDETERMINACY OF QUANTUM WAVES AND STATES

Contents Having discussed some recent literature on the problem of WP-Dualism, let us now analyse an answer to the problem diametrically opposed to that of de Broglie: three-valued logic. The following discussion is important due to the fact that we have rejected a form of subjectivism on purely epistemological grounds [see sections 26.1 and 26.4]. We shall now discuss the problem in more detail beginning from within the theory. The chapter is organized in two sections: in section 29.1 we analyse the logical and foundational aspects of the proposal; while in section 29.2 we discuss the ontological and interpretational issues.

29.1 Three-valued Logic

29.1.1 Exposition of Three-valued Logic

Short history From the beginning of QM, some philosophers and physicists considered WP-Dualism [see principle 8.2 (CP2): p. 138] as an ultimate fact, and tried to interpret it in the sense of the Indetermination Hypothesis [see proposition 3.1: p. 28; see also postulate 8.1: p. 137] [*REICHENBACH* 1944, 20–32] [BUNGE 1956, 283–84]. Bohr, Kramers and Slater's proposal [see section 3.2] can be considered as an antecedent of the following interpretation.

Following Łukasiewicz[1], Zawirski [ZAWIRSKI 1932] made the first attempt to apply a logic with three or more values to QM. Reichenbach [REICHENBACH 1932] and Zwicky [ZWICKY 1933] developed more complex examinations[2].

Reichenbach's proposal We have said that Reichenbach's explicit point of departure is the Wave/Particle complementarity (CP2) [principle 8.2: p. 138]. He could therefore interpret CP in the following meta-physical form [*REICHENBACH* 1944, 143]:

Postulate 29.1 (Three-valued-logic Interpretation of CP) *If two statements are complementary, at most one of them is meaningful; the other is meaningless.*

[1]See [LUKASIEWICZ 1973] for a philosophical analysis.

[2]A more complete exposition can be found in a later work [*REICHENBACH* 1944]. Another proposal along the same line can be found in [DESTOUCHES-F. 1937a] [*DESTOUCHES-F.* 1951]. An examination of the latter is in [MCKINSEY/SUPPES 1954]. History of the problem in [*JAMMER* 1974, 344–46]. Recent developments in [DALLA CHIARA 1977].

Such an interpretation needs both meaningful and meaningless expressions: hence Reichenbach was forced to introduce meaningless statements into the language of physics. To deal with this problem he proposed considering QM as a theory which was founded on a new type of logic: a three-valued logic, i.e. a logic which has not only the values 'true' (1) and 'false' (0) but also 'indeterminate' (1/2) (which is the logical expression for 'meaningless'), so that, independently of a corpuscular language and of a wave language [see section 8.3], one also has a neutral language [*REICHENBACH* 1944, 144–46]. Now consider the following two statements:

- the indicator of apparatus $\mathcal{A}$ will show a value $\mathbf{q}_\mathrm{M}$ of the observable $\hat{\mathbf{q}}$;
- indicator of $\mathcal{A}$ will show a value $\mathbf{p}_\mathrm{M}$ of the observable $\hat{\mathbf{p}}$.

Both statements are verifiable, since the indicator will, or will not, show the predicted value, even if no measurement is performed. Therefore, they are either true or false. However, the statements:

- a measurement of $\hat{\mathbf{q}}$ implies a result $\mathbf{q}_\mathrm{M}$;
- a measurement of $\hat{\mathbf{p}}$ implies a result $\mathbf{p}_\mathrm{M}$

are complementary, so that we have an implication which can have an indeterminate value. Reichenbach names it *quasi-implication* (we symbolize it by $\rhd$) and assigns the following values to it — by a parallel comparison with normal (bivalued) implication ($\rightarrow$) [*REICHENBACH* 1944, 151, 167–68] [see table 29.1].

a	b	$a \rightarrow b$	$a \rhd b$
1	1	1	1
1	$\frac{1}{2}$	$\frac{1}{2}$	$\frac{1}{2}$
1	0	0	0
$\frac{1}{2}$	1	1	$\frac{1}{2}$
$\frac{1}{2}$	$\frac{1}{2}$	1	$\frac{1}{2}$
$\frac{1}{2}$	0	$\frac{1}{2}$	$\frac{1}{2}$
0	1	1	$\frac{1}{2}$
0	$\frac{1}{2}$	1	$\frac{1}{2}$
0	0	1	$\frac{1}{2}$

Table 29.1: Truth table of classical implication and of the quasi-implication proposed by Reichenbach — from [*REICHENBACH* 1944, 151].

Note however that the proposed quasi-implication is not defined as the traditional three-valued implication [PRIOR 1953, 318], which in the table is rather defined as the 'ordinary' implication '$\rightarrow$'.

von Weizsäcker's proposal Another proponent of a logic with three truth-values was von Weizsäcker [VON WEIZSÄCKER 1958, 246].[3] He considered [VON WEIZSÄCKER 1955, 528]

[3]See also [*JAMMER* 1974, 377–78].

the relationship of classical mechanics to QM as the same as that of classical two-valued logic to three-valued logic, in the sense that in both cases the first theory is only a limiting case of the second one. Later, von Weizsäcker considered the propositions of QM theory contingent, because they can be true at a certain time and false later on, and this standing in opposition to eternal propositions which always have a determinate truth value [*VON WEIZSÄCKER* 1971a, 243],[4] as seems to be the case, at least partly, for classical mechanics. He therefore proposed considering as *determinate* only those propositions which refer to the present or to the (documented) past, or which one can derive through mathematical arguments; and as *indeterminate* all those propositions which refer to the future [*VON WEIZSÄCKER* 1971a, 247–48] — there are some connections between such an analysis and definition 11.2 [p. 193]. Obviously von Weizsäcker is here supposing the validity of retrodictions and not of predictions, but we know that retrodictions too are difficult for a proper quantum system [see chapter 26]. However, the difficulty can be removed if our statements only refer to the present and not to the past.

29.1.2 An analysis of Three-valued Logic

The main question is now the following[5]: in which aspects does a two-valued logic — for example, that of QM Logical calculus [section 9.5] — differ from three-valued logic? Take the proposition a: 'the value of an observable $\hat{O}$ in some physical system $\mathcal{S}$ in state $|\psi\rangle$ is o — which we express mathematically as $\hat{O}|\psi\rangle = o|\psi\rangle$. If a is true, there is no difference at all between the two logics. But if a is false, then there are two possibilities:

- There exists a value o' different from o such that $\hat{O}|\psi\rangle = o'|\psi\rangle$.
- No value η exists such that $\hat{O}|\psi\rangle = \eta|\psi\rangle$.

The first answer is clearly untenable in QM, at least in general, because the state may well be a superposition for $\hat{O}$ [see also subsection 14.3.1]. We know that this answer presupposes an exclusion negation which we do not accept in the frame of a QM Logical calculus [see section 9.5]. But what does the second answer say? Exactly that *there is no such value*. Now we know that it is possible to acknowledge such truth-gaps in the context of the two-value logic [again section 9.5].[6] In other words, what QM denies is the Determined Value assumption [see postulate 6.3: p. 105], as already acknowledged by Feyerabend [FEYERABEND 1958b, 110–11]. Hence a two-valued logic is completely satisfactory in order to understand the QM complementarity, without introducing a new logical value for propositions. In conclusion three-valued logic is at least superfluous — see the methodological principle M7 [p. 6].

But there are also some more general difficulties, in connection with three-valued logic as such — not only in its application to QM.[7] Hempel [HEMPEL 1945] and Nagel [NAGEL 1945] criticized the fact that Reichenbach did not define the meaning of truth values 'true' and 'false' in the new context of three-valued logic, which cannot be the same as in two-valued logic.

McKinsey and Suppes [MCKINSEY/SUPPES 1954] followed the same line as Hempel: one cannot change the logical structure of QM alone without changing all the mathematical apparatus[8].

---

[4] On this point see also [*FINKELSTEIN* 1996, 59].

[5] See also [*JAMMER* 1974, 414–16].

[6] From another point of view the same is said by Piron [PIRON 1985, 209] and Mittelstaedt. See also [*JAMMER* 1974, 394–404].

[7] For that which follows see also [*JAMMER* 1974, 368–70, 400].

[8] See also [NAGEL 1946], an examination of the problems deriving by introducing meaningless propositions.

Reichenbach was followed by the philosopher Putnam [PUTNAM 1957]. To both, Feyerabend [FEYERABEND 1958b, 110] [FEYERABEND 1967] addressed the following criticism[9]: three-valued logic is a device which presents a theory (QM), which after Reichenbach and Putnam should be defective but in such a way that its defects do not become apparent. Every analogy with non-Euclidian geometries and their application to physics is misleading because the application of the latter to physics led to fruitful new theories and results — see methodological principle M2 [p. 6].[10]

29.1.3 A Possibility of an Infinite-Valued Logic

In general there is no reason to dismiss a bi-valued logic given the centrality of projectors in the formalism of QM. Obviously, the projectors are only extreme cases of the effects: we said that QM has truth-value gaps (i.e. we do not always have projectors). In the context of an Operational stochastic approach, we have developed the theory of effects [see subsection 18.1.1] and have seen the importance of such an approach to define coexistent non-commutable observables [see subsection 18.2.1] and a probabilistic formulation of CP [see subsection 18.2.2]. And here it is perhaps possible to develop the main idea of Reichenbach in a form valid for observables and their values. We can try to find some logical equivalent of formula (18.15). It is possible by constructing an infinite-valued logic which reflects the structure of effects, i.e. a logic which has all values between 0 and 1 (considered as limiting cases). Then 1/2 is only the case of an equally distributed uncertainty for both observables of a non-commutable pair (minimum of uncertainty). Modifying the structure of the Sasaki arrow [formula (9.20)], we can define the following infinite-valued QM implication (for effects and POVM):

$$\boxed{a \Uparrow b := [(0 < a < 1) \wedge (b = 1)] \vee [(a = 1) \wedge (b = 1)],} \tag{29.1}$$

where we have $(0 < a < 1) \wedge (b = 1)$ iff a represents an unsharp observable partly non commuting with the observable represented by b, and $(a = 1) \wedge (b = 1)$ iff the observables represented by a and b are commutable. We are not considering here the case of sharp non-commuting observables (which, as has been said, is to be understood as a limiting case for which the ordinary Sasaki arrow holds).

However, three-valued logic, in the form developed by Reichenbach, is not applicable as such to the POVM and effects. In fact after Reichenbach an observable is maximally undetermined (truth value = 1/2) when it is in a symmetric superposition of its eigenstates. But this situation can always be represented by a projector. On the contrary, an effect represents a situation in which two non-commutable observables are (unsharply) coexistent [see subsection 18.2.1, and for an example see in particular subsection 30.3.1]. And in this case the observables are not maximally undetermined in the sense of Reichenbach.

29.2 Ontological Aspects

There are other proposals which can be defined as the ontological counterparts of three-valued logic. In particular that of Popper and Heisenberg's later interpretation.

[9] See also [*JAMMER* 1974, 373–75] [*HAAK* 1974, 153–67].

[10] A partly different approach can be found in [PUTNAM 1968]. Its examination is in [FRIEDMAN/GLYMOUR 1972], where it is shown that, by formalizing Putnam's proposal, the consequence would be that the value of an observable cannot belong to one of two disjoint Borel sets.

29.2.1 Propensity

Popper [POPPER 1934b] formulated an interpretation of UP — as we have already said, an antecedent of the following interpretation is the proposal by Bohr, Kramers and Slater [see section 3.2]. According to Popper quantum vectors can only be expressed by statistical statements. But, following Popper, the Copenhagen school chose to ascribe statistical distributions to the single elements of statistical ensembles and not to the ensembles as such; if only the last ones are considered, then no WP-Dualism arise, because each time, when we speak of 'wave' or of 'particle', we refer us to different statistical ensembles.

However the problem remains of how it is possible that some ensembles behave as particles and other as waves. Hence Popper [POPPER 1957] [POPPER 1959] [*POPPER* 1982, 68–74, 126–34] formulated a propensity interpretation of probability which he then extended to all form of being:

Postulate 29.2 (Propensity) *Things have in themselves the tendency to realize some situations: the assigned weights are to be understood as measures of the propensity of a possibility to realize itself upon repetition. The situations, on the level of possibility, may also be alternative. But no contradiction follows because a contradiction can only arise between existent realities.*

We do not wish to discuss this postulate as an interpretation of probability as such, but only its ontological implication for QM. A strong criticism was made by Margenau [MARGENAU 1954, 11] and by Feyerabend [FEYERABEND 1968]. Feyerabend showed that an ideal experiment such as the Two-Slit-Exp. 1 [p. 27] cannot be accounted for in the frame of Popper's propensity because he always presupposed definite trajectories of particles. And what is most important — and remains unexplained in Popper's proposal — is why some trajectories, which are classically allowable, are not possible in QM.[11]

On the other hand, it seems contradictory or almost unsound if one presupposes that no contradiction could be possible between possibilities and then, by interpreting them as propensities, one believes at the same time that, in the case of a superposition, they can interact with one other, as would be the case with existing things alone, which, naturally, *can be* in contradiction [POPPER 1959, 28]. As we have said, here we are confronted again with an entity which is between being and non-being such as the ghost-field of Bohr, Slater and Kramers .

That the propensity interpretation is not sound if applied to the specific problem of possible values of observables[12], can be seen in the analysis of the objectification problem [subsection 14.3.1] [BARRETTO B. F./SELLERI 1995].[13] Inevitably we should consider pure states as inhomogeneous ensembles, with the consequence that we have to acknowledge some form of Bell inequality, in violation of the prediction and the experimental results of QM [on this problem see chapter 35]. A better understanding of these and other difficulties can stem from an analysis of the Heisenberg's later interpretation.

29.2.2 Heisenberg's Later Interpretation

Heisenberg's proposal Another variant is represented by Heisenberg's later interpretation [*HEISENBERG* 1959, 175–76]. Already in 1927, shortly after having completed his article on

[11] See also [*JAMMER* 1974, 448–53].

[12] Sneed [SNEED 1966] saw in what he called a 'dispositional interpretation' the interpretation frame of the projection postulate.

[13] On this point see also [*MITTELSTAEDT* 1998, 62–63] .

UP [see chapter 7], Heisenberg felt dissatisfaction with the Strong Formal interpretation [postulate 6.2].[14] Perhaps it was the Probabilistic interpretation [proposition 6.2: p. 104] which spurred Heisenberg on in a new direction: namely it is possible to understand the Probabilistic interpretation in the sense that, before a measurement, a physical observable has no determinate value but only possible values[15]. Note that Heisenberg never sustained a form of dualism. He was particularly engaged, therefore, by WP-Dualism [see principle 8.2 (CP2): p. 138], and defended a reformulation of the Indetermination Hypothesis [see proposition 3.1] [*HEISENBERG* 1959, 35–39, 135–36], a proposition which we can call the Potentiality hypothesis[16]:

Proposition 29.1 (Potentiality hypothesis) *Microentities only have a potentiality to behave in a certain manner or to acquire certain properties. The dispositions become real only by interaction with measurement devices.*

This new ontology (which explicitly refers to that of Aristotle) presupposes that reality is composed of 'factual' (existent) and 'potential' parts: normally microentities are only potential, but their interaction with measuring apparatus causes them to pass from possibility to actuality. The background of this interpretation is represented by the so-called reduction of the wave packet [see postulate 14.1: p. 223].

Here also the word *potentiality* has an ambiguous meaning. Generally we may use the word in two different ways:

- as a type of program or disposition;
- as a pure possibility.

Disposition If we interpret the *potentialities* of proposition 29.1 as *programs*, we think of something like DNA, which, from a single — for example — human cell, has the potentiality to develop the whole organism. It was precisely in this way that Aristotle understood [*ARISTOTLE De Gen. et corr.*, 320a8 ff.] the δύναμις (*potentia*). So we must think by analogy that a microentity, before observation, is neither a particle nor a wave, but has the potentiality to be one or the other.

However, the DNA is an existing thing and biologically determines the development of a living organism completely and in an *univocal* sense. And how is it possible that there is a program which can produce incompatible states of affairs such as wave-like or particle-like behaviour? Is this not a shifting of the contradiction further back to another level, without overcome it? It is evident that we encounter here the same difficulty as in the case of Popper's propensities.

Possibility The only way, therefore, to think of incompatible things together is as pure possibilities[17]. This second interpretation of the Potentiality hypothesis is excluded because ontologically possibilities have no existence at all[18]: only that which exists exists[19]. When we speak of *existent possibilities*, we refer either to projects, ideas, strategies, etc., that are naturally completely

[14]See [*JAMMER* 1974, : 76].

[15]De Broglie understood the Probabilistic interpretation exactly in this way [*DE BROGLIE* 1956, : 48].

[16]See also [STAPP 1979, 18–25] where some references to Whitehead are made.

[17]See also [MITTELSTAEDT 1981, 85–86, 89].

[18]One of the best philosophical analysis of the problem remains [QUINE 1948], even if we do not follow Quine in all his conclusions.

[19]There is a strange precedent, perhaps unknown to Heisenberg, of this second interpretation of the Potentiality Hypothesis: Leibniz' theory of possible beings which come to existence [AULETTA 1994a] .

existent in their specifical manner — i.e. as (for example) writings that are basic for some social groups or organizations —; or we refer to the objective possibilities that offer this or another situation or more generally our world. In this case possibilities are not things but only a *language* that we use to speak about the opportunities which the situation in which we exist offers to us — i.e. it is a specific mode of considering things in relation to some project.

Hence we can assume the following metatheoretical Principle :

Principle 29.1 (Possibility) *There are no such things as possibilities. The relative concepts have no existent referents but are part of a language which one uses to speak of real things.*

In other words, there are no holes in existence, but only an incompleteness of the conditions which are required to execute an action or an operation [see postulate 24.1: p. 428]. And here we return to the Operational approach [see also chapters 26 and section 26.4].

It is also true that Aristotle normally speaks [*ARISTOTLE De Gen. et corr.*, 319a17–21] of the potentiality of matter of being formed in different ways (by some artist) — and here we have no univocal program anymore. But in this sense we would accept — as Heisenberg did — a jump between the QM indeterminacy and a determination through a measurement. As we have seen, this vision is not at all adequate for describing QM [see chapter 24].

QM ontology To see the problem more specifically in the context of QM, it is evident that the only reality which corresponds to the uncertainty of an observable is the superposition — in fact the wave-like behaviour corresponds to a superposition relative to the position 'operator' —, which is not something 'indeterminate' and surely not a possibility. This phenomenon is such that, for a determinate state, it is not possible to make predictions about some observable because the state is an eigenstate of some other (non-commuting) observable — in our case of momentum. The same is valid for entanglement [see subsection 3.5.2]: if two subsystems are entangled, we cannot obtain values for some observables on a subsystem independently from other observables of the other subsystems [see also chapter 35] — if the state is not destroyed. But, in both cases, what we can say is only that *respective to some operation* which we name a 'measurement', some observable is not in eigenstate, or that we cannot separate a subsystem without destroying the state, and hence we can speak of 'indeterminate' not with an ontologically privative meaning but only with an *operational* meaning. As we have seen in the concluding remarks of the last chapter [section 28.7] and we shall see later on [in section 30.4], it is true that quantum entities are not classical and not 'determinate' in a classical sense. However they are not defective in relationship to a supposed perfect *determinatio*, since it is the assumption of classical mechanics itself — the perfect determination of being — which is discussible [see especially sections 24.5 and 33.1]. Obviously, the main difficulty is that we cannot 'grasp' the correlations in the same sense in which we can grasp (by a local and individual measurement) some physical property of a system — we shall return later to the problem [see chapter 30].

Chapter 30

BETWEEN WAVE AND PARTICLE

Introduction and contents That we cannot interpret quantum 'waves' as classical waves is clear from the preceding discussion [see section 28.6]. Now we examine in some detail what this ontological nature can be. We have already seen that Complementarity cannot be a yes/no alternative [section 8.1]. The Hilgevoord/Uffink UR is a good mathematical expression of this fact [section 7.6 and section 8.2]. We have also see — Interfer. Exp. 15 [in subsection 28.3.2] — that there can be a corpuscular behavior even if there is some wave-like behaviour. We now investigate in greater detail the complementarity between which-path (the corpuscular behaviour) and diffraction (the wave-like behaviour) [see section 8.4]. The aim is to give a better interpretation of CP4 [p. 139] once we have examined the Measurement problem and discussed tools like the POVM.

In section 30.1 we discuss the starting point of the following developments: a *Gedankenexperiment* of Wootters and Zurek's by means of which for the first time a 'partial' coexistence between wave-like and corpuscular behaviour was proven. Later, a new inequality between Predictability and Visibility [section 30.2] was established. The stochastic aspects of the problem are discussed in section 30.3. Finally, the first elements of an ontological interpretation are developed [section 30.4].

30.1 Wootters/Zurek Gedankenexperiment

Pioneers of a new interpretation of Complementarity were Wootters and Zurek — who also partly anticipated the work of Hilgevoord/Uffink, though not with the same generality because they used Gaussian distributions and still thought of the problem of uncertainty in analogy with the standard deviation. On the other hand, Wootters and Zurek's *Gedankenexperiment* has strong similarities with the experiment proposed independently by Scully and co-workers [see subsection 8.4.2].

The Experiment

Our experiment (Two-Slit-Exp. 10) [WOOTTERS/ZUREK 1979] consists of a photon which goes through plate 1 (with one slit), then plate 2 (with two slits) and ends on screen 3 [see figure 30.1]. The distance between plates 1 and 2 and between plates 2 and 3 is given by l_1; the distance from slit I and slit II by l_2. Consider a Gaussian wave function of the photon and its Fourier

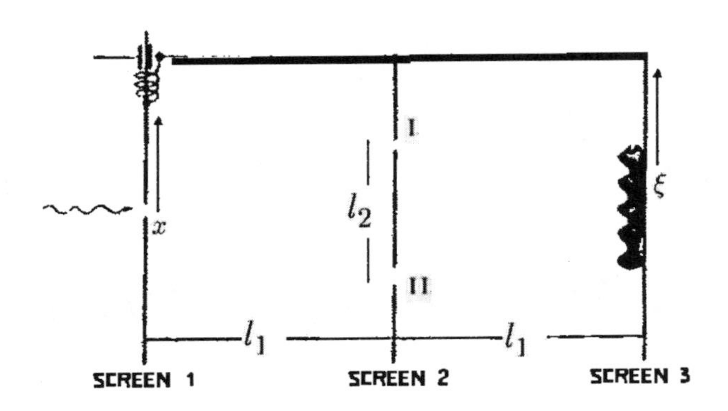

Figure 30.1: Path determination in the Two-Slit-Exp. 5 as proposed by Einstein — from [WOOTTERS/ZUREK 1979, 444].

transform (one-dimensional problem):

$$\psi(x) = \pi^{-\frac{1}{4}}\varphi^{-\frac{1}{2}}e^{-\frac{r^2}{2\varphi^2}}, \ \Delta r^2 = \frac{\varphi^2}{2}, \tag{30.1a}$$

$$\tilde{\psi}(k) = \pi^{-\frac{1}{4}}\varphi^{\frac{1}{2}}e^{-\frac{\varphi^2 k^2}{2}}, \ \Delta k^2 = \frac{1}{2\varphi^2}. \tag{30.1b}$$

Now consider the wave function of photons at screen 3 which have passed through slit I — slit II being closed —:

$$\psi_I(\xi) = f(\xi)e^{i\frac{2\pi}{\lambda}\left[l_1^2+(\xi-\frac{l_2}{2})^2\right]^{\frac{1}{2}}}, \tag{30.2}$$

where f is a slowly varying envelope function due to the fact that the slit is not infinitesimally small. We consider a region $|\xi| < \xi_0$ in which $f(\xi)$ is essentially constant for $\xi \ll l_1$. Then Eq. (30.2) can be approximated as:

$$\psi_I(\xi) = f(0)e^{\eta(\xi)}e^{ik_0\xi}, \tag{30.3a}$$

where $\eta(\xi) = (2\pi/l_1)[l_1^2 + l_2^2/8 + \xi^2/8]$ and $k_0 = \pi l_2/(l_1\lambda)$. Similarly the wave function of photon coming from slit II is given by:

$$\psi_{II}(\xi) = f(0)e^{\eta(\xi)}e^{-ik_0\xi}. \tag{30.3b}$$

If we superimpose these two wave functions, the factor $e^{\eta(\xi)}$ does not contribute to the interference pattern, and therefore it will be neglected.

After a photon is passed through a plate we can measure either the position or the momentum of the plate. Let us assume that we measure the position and let us calculate the resulting interference. All photons starting at the same place x form a subensemble whose contribution $I_x(\xi)$ to the total interference pattern is shifted by an amount $\delta x = -x$:

$$I_x(\xi) = 1 + \cos 2k_0(\xi + x). \tag{30.4}$$

The number of photons characterized by the initial position x is proportional to $|\psi(x)|^2$. Hence the total interference can be described by:

$$I(\xi) \propto \int dx|\psi(x)|^2 I_x(\xi) = 1 + e^{-k_0^2\varphi^2}\cos 2k_0\xi. \tag{30.5}$$

On the other hand, if we decide to measure the momentum of plate 1 there are only two possible values of the wave vector imparted to the screen 1: $\pm k_0$ — where we assume $\varphi \ll l_2 \ll l_1$. But there is some a priori uncertainty due to the fact that the initial momentum of the plate is not well defined but a distribution given by $|\tilde{\psi}(k_i)|^2$, where k_i is the initial vector. The plate's final vector k_f can only have values: $k_f = k_i \pm k_0$. The total recorded distribution $\mathcal{F}(k_f)$ of wave numbers of the plate will be the sum of partial distributions $\mathcal{F}_{k_i}(k_f)$ weighted by $|\tilde{\psi}(k)|^2$:

$$\mathcal{F}_{k_i}(k_f) = \frac{1}{2}[\delta(k_f - k_i + k_0) + \delta(k_f - k_i - k_0)], \tag{30.6a}$$

$$\mathcal{F}(k_f) = \int dk_i |\tilde{\psi}(k_i)|^2 \mathcal{F}_{k_i}(k_f) = \frac{\varphi}{2\pi^{1/2}}[e^{-\varphi^2(k_f+k_0)^2} + e^{-\varphi^2(k_f-k_0)^2}]. \tag{30.6b}$$

It is shown that the interference pattern of Eq. (30.5) and the distribution of Eq. (30.6b) are not mutually exclusive, in the sense that both can have determined values, contrarily to the hypothesis of a yes/no complementarity [see definition 8.1: p. 136].

But one can deny that measurements of position and momentum were performed on the same individual photons. Look now for the distribution of scintillations arising only from photons which have been associated with a definite measured momentum k_f of plate 1. We now obtain the following for the interference pattern on screen 3 after having measured the momentum of plate 1:

$$I_{k_f}(\xi) = 1 + 2\wp_I^{\frac{1}{2}} \wp_{II}^{\frac{1}{2}} \cos 2k_0\xi, \tag{30.7}$$

where $\wp_I$ is the probability that a photon passes through slit I. Now if we take a very large ratio $\wp_I/\wp_{II}$ — e.g. 99 photons out of 100 pass through slit I — the crest to valley ratio of interference

$$R_I = \frac{1 + 2\wp_I^{\frac{1}{2}} \wp_{II}^{\frac{1}{2}}}{\left(1 - 2\wp_I^{\frac{1}{2}} \wp_{II}^{\frac{1}{2}}\right)} \tag{30.8}$$

is approximately 3/2; a relatively high value. Hence photons still have a wave-like behaviour even if the path is predicted almost certainly. The authors then show that we would have arrived at exactly the same result of Eq. (30.5) by taking as a starting point Eq. (30.7) [WOOTTERS/ZUREK 1979 444–47].

Concluding remarks

In conclusion we are faced with the following alternatives:

- either we measure the position of screen 1, and we obtain perfect but shifted partial interference patterns;
- or we measure the momentum of screen 1, and we obtain smeared out but centered partial interference patterns. Hence our choice affects the wave function of the photon

In the first case it arrives at screen 3 in the state:

$$\psi_x(\xi) \propto e^{\imath k_0(\xi+x} + e^{-\imath k_0(\xi+x}. \tag{30.9a}$$

In the second case:

$$\psi_{k_f}(\xi) \propto \wp_I^{\frac{1}{2}}(k_f)e^{\imath k_0\xi} + \wp_{II}^{\frac{1}{2}}(k_f)e^{-\imath k_0\xi}. \tag{30.9b}$$

But our choice has no influence on the global interference pattern of the system constituted by photon + plate 1 — in this non-separability the present authors [WOOTTERS/ZUREK 1979, 446–48] see a type of EPR situation [see chapter 31].

30.2 Inequality Between Predictability and Visibility

30.2.1 Greenberger/Yasin Inequality

The starting point for the study of Greenberger and Yasin [GREENBERGER/YASIN 1988a] was represented by the article of Wootters/Zurek [WOOTTERS/ZUREK 1979].[1] The authors noted first of all that, also in the presence of a small intensity of the beam, say 0.1%, the contrast (visibility) of the interference fringes [see Eq. (5.19)] will be 20%. Hence by an almost certain determination of the path, we always have a significant diffraction pattern. Then they proposed the following thought-experiment (Interfer. Exp. 16) in order to test this hypothesis [see figure 30.2].

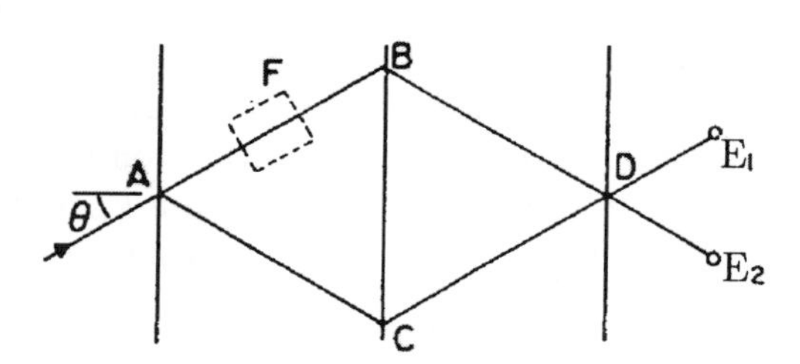

Figure 30.2: Proposed Interfer. Exp. 16: The incident beam enters at A and is split into beam I (ABD) and II (ACD), recombined at D and detected at E1, E2. One can introduce an absorber at F — from [GREENBERGER/YASIN 1988a, 391].

We assume fully coherent beams I, II with respective amplitudes ϑ_1, ϑ_2. Supposing that at point D there is a screen, the wave function there can be written as:

$$\psi = \left(\vartheta_1 e^{+\imath k_x x} + \vartheta_2 e^{\imath\delta\varphi} e^{-\imath k_x x}\right) e^{\imath k_z z}, \tag{30.10}$$

where k_x, k_z are determined by the Bragg scattering condition with ϑ_1, ϑ_2 real, and $\delta\varphi$ represents the phase difference between the beams. Then

$$|\psi|^2 = \vartheta_1^2 + \vartheta_2^2 + 2\vartheta_1\vartheta_2 \cos(2k_x x + \delta\varphi), \tag{30.11}$$

and the visibility [again Eq. (5.19)] is given by

$$\mathcal{V} = \frac{2\vartheta_1\vartheta_2}{\vartheta_1^2 + \vartheta_2^2}. \tag{30.12}$$

If we write [see again figure 30.2]

$$\vartheta_1 = \eta\cos\theta, \quad \vartheta_2 = \eta\sin\theta, \tag{30.13}$$

then the visibility becomes:

$$\mathcal{V} = \sin 2\theta, \tag{30.14}$$

[1]See also [BARTELL 1980] [MANDEL 1991].

and has its maximum at $\theta = 45°$, i.e. when the beams are equal.

Now imagine two extreme situations: one in which we have no knowledge at all about the path of the particles (probability = 1/2), and one in which we predict that all particles will be in beam I: $|\vartheta_1|^2 > |\vartheta_2|^2$. In the second case there will be just a fraction $|\vartheta_1|^2/(|\vartheta_1|^2 + |\vartheta_2|^2)$. Therefore, by comparing both probabilities we obtain:

$$\mathcal{P} = \frac{\frac{|\vartheta_1|^2}{|\vartheta_1|^2+|\vartheta_2|^2} - \frac{1}{2}}{\frac{1}{2}} = \frac{|\vartheta_1|^2 - |\vartheta_2|^2}{|\vartheta_1|^2 + |\vartheta_2|^2} = \cos 2\theta, \tag{30.15}$$

where $\mathcal{P}$ is the *predictability of path*.

Therefore, in the case of fully coherent beams, we obtain [by Eqs. (30.14) and (30.15)] the equality:

$$\mathcal{P}^2 + \mathcal{V}^2 = 1. \tag{30.16}$$

In the case of a non-fully coherent beams, we can separate the amplitudes in a coherent part $\vartheta_{\rm C1}, \vartheta_{\rm C2}$ and an incoherent one $\vartheta_{\rm I1}, \vartheta_{\rm I2}$, in which case the probability density becomes:

$$|\psi|^2 = |\vartheta_{\rm C1}|^2 + |\vartheta_{\rm C2}|^2 + |\vartheta_{\rm I1}|^2 + |\vartheta_{\rm I2}|^2 + 2\vartheta_{\rm C1}\vartheta_{\rm C2}\cos(2k_x x + \delta\varphi); \tag{30.17}$$

and the visibility:

$$\mathcal{V} = \frac{2\vartheta_{\rm C1}\vartheta_{\rm C2}}{|\vartheta_{\rm C1}|^2 + |\vartheta_{\rm C2}|^2 + |\vartheta_{\rm I1}|^2 + |\vartheta_{\rm I2}|^2}. \tag{30.18}$$

Greenberger and Yasin deduced finally the following inequality, which is a generalization of Eq. (30.16):

$$\boxed{\mathcal{P}^2 + \mathcal{V}^2 \leq 1} \tag{30.19}$$

We name such inequality the *Greenberger/Yasin inequality*. This inequality is exactly the mathematical content of CP4 [p. 139].[2] For further developments of the same point in an information theory context see also section 42.6.

30.2.2 Englert's Model

Englert [ENGLERT 1996a] [ENGLERT 1996b] proposed another experiment, by means of a Ramsey interferometer (Interfer. Exp. 17), in order to test what we have already analysed [see figure 30.3].

A two level atom, prepared in its *excited state*, passes successively through two stretches of microwave radiation (beam splitter and merger), with intensity equivalent to a $\pi/2$ pulse for each one, and between both through a classical static electric field which gives a relative phase shift of $\delta\varphi$ between the wave function components of the two atomic states [see figure 30.4]. The initial state of the atom is described by the density matrix

$$\hat{\rho}^a_i = \frac{1}{2}(1 + s\hat{\sigma}_z), \tag{30.20}$$

with inversion $-1 \leq s \leq 1$ (- 1 is the ground state, 1 the excited one, and the values in between represent a superposition of both). The action of each of the two microwave fields is given by:

$$\hat{\rho}^a \mapsto e^{-\imath\frac{\pi}{4}\hat{\sigma}_y}\hat{\rho}^a e^{\imath\frac{\pi}{4}\hat{\sigma}_y}; \tag{30.21}$$

[2] Experiments performed along these lines by using neutron interferometry are reported in [RAUCH *et al.* 1990]

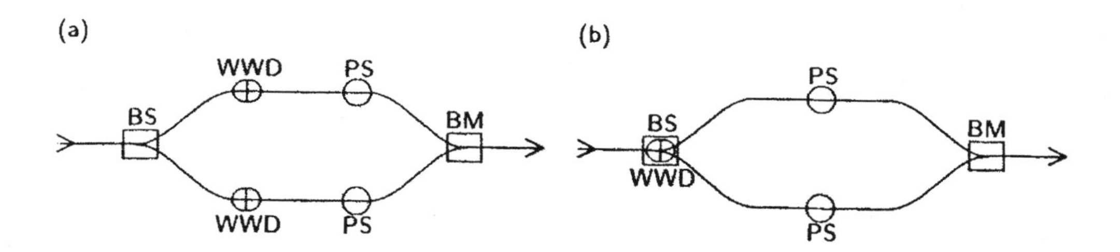

Figure 30.3: Schematic two-way interferometers with which-path detectors. The BS distributes the input among the two ways, the BM recombines the contributions and produces the output. The phase shifter PS introduces a relative phase which eventually modulates the interference pattern. In addition, there is a which way detector (WWD).
(a) Another physical interaction is used for this purpose; this is the situation analysed in [ENGLERT 1996a].
(b) The same mechanism is used both for splitting the beam and for detecting the way – from [ENGLERT 1996b, 250].

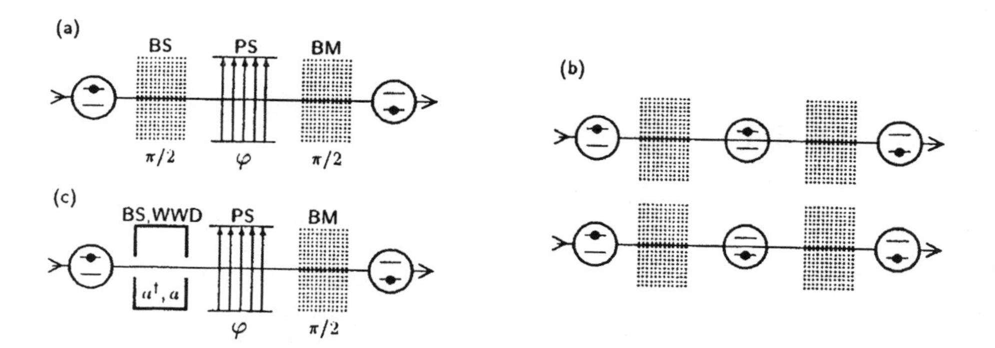

Figure 30.4: Ramsey interferometer.
(a) A two-level atom passes through a first microwave field that plays the role of a beam splitter BS; then it crosses a static field that acts as a phase shifter PS; finally a second microwave field is the BM.
(b) The paths through the interferometer are characterized by the state of the atom at the central stage. In one path the atomic transition happens in the BM, in the other, it occurs in the BS.
(c) Upon replacing the classical microwave field of the BS by the quantized field of a resonator, the BS may also function as a which-path detector WWD. This is a realization of the abstract setup of figure 30.3b – from [ENGLERT 1996b, 250].

and of the static electric field by:

$$\hat{\rho}^a \mapsto e^{-\imath\frac{\delta\varphi}{2}\hat{\sigma}_z}\hat{\rho}^a e^{\imath\frac{\delta\varphi}{2}\hat{\sigma}_z}. \tag{30.22}$$

Hence we have:

$$\hat{\rho}^a_f = \frac{1}{2}[1 + s(\hat{\sigma}_y \sin\delta\varphi - \hat{\sigma}_z \cos\delta\varphi)]. \tag{30.23}$$

The probability of finding the atom in ground state is:

$$\wp_g = \mathrm{Tr}_a\left[\frac{1}{2}(1-\hat{\sigma}_z)\hat{\rho}^a_f\right] = \frac{1}{2}(1 + s\cos\delta\varphi), \tag{30.24}$$

because $\hat{\sigma}_y$ and $\hat{\sigma}_z$ are orthogonal, where $(1-\hat{\sigma}_z)/2 = \hat{P}_g$ is the projection operator on the ground state. This is the interference pattern in the Ramsey interferometer — because the atom can be deexcited either by microwave radiation or by electric field, so that the probability of this result is a superposition of both possibilities. Now if, instead of the beam splitter (first microwave stretches), we set a quantized radiation field (a resonator), the two ways could become distinguishable by detection of the effects on resonator photons. Whether which way information will be stored or not in the state of resonator depends on cavity initial state $\hat{\rho}^\gamma_i$. The initial state of the combined system is:

$$\hat{\rho}_i = \hat{\rho}^a_i \hat{\rho}^\gamma_i. \tag{30.25}$$

The interaction between atom and cavity effects the transition:

$$\hat{\rho} \mapsto e^{\imath\phi\hat{g}}\hat{\rho}e^{\imath\phi\hat{g}} = \hat{U}^\dagger\hat{\rho}\hat{U}, \tag{30.26}$$

where ϕ is the accumulated Rabbi angle and $\hat{g}_c$ is the coupling operator given by:

$$\hat{g}_c = \frac{1}{2\imath}(\hat{\sigma}_+\hat{a} - \hat{\sigma}_-\hat{a}^\dagger), \tag{30.27}$$

with $\hat{\sigma}_\pm = \hat{\sigma}_x \pm \imath\hat{\sigma}_y$. It is a resonant Jaynes/Cummings model [see subsection 5.1.3].

It is possible to write the unitary operator of Eq. (30.26) in terms of operators $\hat{W}$ which only affect the degrees of freedom of the which-path detector:

$$\hat{U} = \frac{1}{\sqrt{8}}[(1+\hat{\sigma}_z)\hat{W}_{++} + \hat{\sigma}_+\hat{W}_{+-} - \hat{\sigma}_-\hat{W}_{-+} + (1-\hat{\sigma}_z)\hat{W}_{--}], \tag{30.28}$$

where we remember that the terms associated with the first and fourth which-path operator correspond respectively to the projector on excited state and the projector on ground state [Eq. (30.24)]. In conclusion the system is transformed from the initial state (30.25) into a final state of the form:

$$\hat{\rho}_f = \frac{1+s}{2}\hat{\rho}^+_f + \frac{1-s}{2}\hat{\rho}^-_f, \tag{30.29}$$

with

$$\begin{aligned}\hat{\rho}^+_f &= \frac{1+\hat{\sigma}_x}{4}\hat{W}^\dagger_{++}\hat{\rho}^\gamma_i\hat{W}_{++} + \frac{1-\hat{\sigma}_x}{4}\hat{W}^\dagger_{+-}\hat{\rho}^\gamma_i\hat{W}_{+-} \\ &\quad -\frac{\hat{\sigma}_z - \imath\hat{\sigma}_y}{4}e^{-\imath\delta\varphi}\hat{W}^\dagger_{++}\hat{\rho}^\gamma_i\hat{W}_{+-} - \frac{\hat{\sigma}_z + \imath\hat{\sigma}_y}{4}e^{\imath\delta\varphi}\hat{W}^\dagger_{+-}\hat{\rho}^\gamma_i\hat{W}_{++}.\end{aligned} \tag{30.30}$$

By replacing $\hat{W}_{++}$ by $-\hat{W}_{-+}$ and $\hat{W}_{+-}$ by $\hat{W}_{--}$ one obtains $\hat{\rho}^-_f$.

Now the probability that the atom will end up in the ground state is calculated by means of the corresponding projector, and is:

$$\wp_g = \mathrm{Tr}_a \mathrm{Tr}_\gamma [\hat{P}_g \hat{\rho}_f] = \frac{1}{2} + \frac{1}{2} \mathrm{Re}(e^{-\imath \delta\varphi} \mathbf{C}), \tag{30.31}$$

where the complex contrast factor $\mathbf{C}$ is given by:

$$\begin{aligned} \mathbf{C} &= \mathrm{Tr}_\gamma \left[\frac{1+s}{2} \hat{W}^\dagger_{++} \hat{\rho}^\gamma_i \hat{W}_{+-} + \frac{1-s}{2} \hat{W}^\dagger_{-+} \hat{\rho}^\gamma_i \hat{W}_{--} \right] \\ &= \frac{1+s}{2} \left\langle \hat{W}_{+-} \hat{W}^\dagger_{++} \right\rangle^\gamma_i - \frac{1-s}{2} \left\langle \hat{W}_{--} \hat{W}^\dagger_{-+} \right\rangle^\gamma_i . \end{aligned} \tag{30.32}$$

The maxima and minima in $\wp_g$ of Eq. (30.31) lead naturally to recognize that the fringe visibility [see Eq. (5.19)] is:

$$\mathcal{V} = |\mathbf{C}|. \tag{30.33}$$

The which-way information is obtained either by measuring $\hat{\sigma}_z$ before the atom passes the beam merger or by measuring $\hat{\sigma}_x$ in state $\hat{\rho}_f$. These two possibilities correspond to calculate the following probabilities $\wp_\pm$ of taking either the upper or the lower way:

$$\begin{aligned} \wp_\pm &= \mathrm{Tr}_a \mathrm{Tr}_\gamma \left[\frac{1}{2}(1 \pm \hat{\sigma}_z) e^{-\imath \frac{\pi}{4} \hat{\sigma}_y} \hat{\rho}_f e^{\imath \frac{\pi}{4} \hat{\sigma}_y} \right] \\ &= \mathrm{Tr}_a \mathrm{Tr}_\gamma \left[\frac{1}{2}(1 \pm \hat{\sigma}_x) \hat{\rho}_f \right], \end{aligned} \tag{30.34}$$

with outcomes:

$$\wp_+ = \frac{1+s}{4} \left\langle \hat{W}_{++} \hat{W}^\dagger_{++} \right\rangle^\gamma_i + \frac{1-s}{4} \left\langle \hat{W}_{-+} \hat{W}^\dagger_{-+} \right\rangle^\gamma_i , \tag{30.35a}$$

$$\wp_- = \frac{1+s}{4} \left\langle \hat{W}_{+-} \hat{W}^\dagger_{+-} \right\rangle^\gamma_i + \frac{1-s}{4} \left\langle \hat{W}_{--} \hat{W}^\dagger_{--} \right\rangle^\gamma_i . \tag{30.35b}$$

This probabilities appear together with the contrast factor in the final state of the atom:

$$\hat{\rho}^a_f = \mathrm{Tr}_\gamma(\hat{\rho}_f) = \frac{1}{2} \left[1 + \hat{\sigma}_x(\wp_+ - \wp_-) - \hat{\sigma}_y \mathrm{Im}(e^{-\imath\delta\varphi}\mathbf{C}) - \hat{\sigma}_z \mathrm{Re}(e^{-\imath\delta\varphi}\mathbf{C}) \right]. \tag{30.36}$$

If the path $\hat{\sigma}_z = +1$ has been taken, the final photon state $\hat{\rho}^{\gamma+}_f$ is identified by:

$$\wp_+ \hat{\rho}^{\gamma+}_f = \mathrm{Tr}_a \left[\frac{1+\hat{\sigma}_x}{2} \hat{\rho}_f \right]; \tag{30.37a}$$

and if we get $\hat{\sigma}_z = -1$:

$$\wp_- \hat{\rho}^{\gamma-}_f = \mathrm{Tr}_a \left[\frac{1-\hat{\sigma}_x}{2} \hat{\rho}_f \right]. \tag{30.37b}$$

The weighted sum of $\hat{\rho}^{\gamma+}_f$ and $\hat{\rho}^{\gamma-}_f$ is the final photon state if no determination of the path has been performed:

$$\hat{\rho}^\gamma_f = \mathrm{Tr}_a[\hat{\rho}_f] = \wp_+ \hat{\rho}^{\gamma+}_f + \wp_- \hat{\rho}^{\gamma-}_f. \tag{30.38}$$

Now we calculate the predictability $\mathcal{P}$ of the path. If both ways are equally probable ($\wp_+ = \wp_- = 1/2$), then $\mathcal{P} = 0$. On the other hand, we have $\mathcal{P} = 1$ if either $\wp_+ = 1$ or $\wp_- = 1$. Hence more generally we have:

$$\mathcal{P} = |\wp_+ - \wp_-|. \tag{30.39}$$

If we use the concept of distinguishability $\mathcal{D}$ (the which-way knowledge stored in the final state of the which-path detector) we have:

$$\mathcal{D} = \mathrm{Tr}_\gamma \left[|\wp_+ \hat{\rho}_f^{\gamma+} - \wp_- \hat{\rho}_f^{\gamma-}| \right]. \tag{30.40}$$

It is clear that the distinguishability is the which way information acquired by an optimized reading of the detector, and therefore we obtain:

$$\boxed{\mathcal{D} \geq \mathcal{P}.} \tag{30.41}$$

By positivity of $\hat{\rho}_f^\gamma$ given by Eq. (30.38), and by Eqs. (30.39) and (30.40) one can finally derive inequality (30.19).

30.2.3 Other Experiments

There are also other experiments which show the same co-presence of corpuscular and wave-like behaviour. An experiment with photons has been proposed by Ghose, Home and Agarwal [GHOSE *et al.* 1991] [GHOSE *et al.* 1992], and recently realized [MIZOBUCHI/OHTAKÉ 1992].[3] It is the QM equivalent of the classical experiment of Ghose. The basic idea is a single photon which tunnels [see section 41.3] and creates anticoincidence between two detectors — one for the transmitted and the other for the reflected part of the initial system (Tunnelling Exp. 1). But now transmission is a wave phenomenon where anticoincidence presupposes a particle-like behaviour [see figure 30.5]. However, it should be stressed that Ghose and Home interpret [GHOSE/HOME 1996] such an experiment as evidence in favour of de Broglie's model and against Complementarity.

30.3 Stochastic Complementarity Visibility/Predictability

30.3.1 Mittelstaedt/Prieur/Schieder Model

Mittelstaedt, Prieur and Schieder [MITTELSTAEDT *et al.* 1987] developed the first stochastic treatment of the problem. First let us introduce a path projector $\hat{P}_\mathcal{P}$ and an interference-pattern projector $\hat{P}_\mathcal{V}$ (both sharp), which are orthogonal on the Poincaré sphere [see figure 30.6].

Then consider all sharp measurements of an arbitrary observable $\hat{O}$ such that $\hat{P}_{\hat{O}}$ lies on the arc going from $\hat{P}_\mathcal{P}$ to $\hat{P}_\mathcal{V}$, where $\hat{P}_{\hat{O}}$ satisfies: $[\hat{P}_{\hat{O}}, \hat{P}_\mathcal{P}] \neq 0$ and $[\hat{P}_{\hat{O}}, \hat{P}_\mathcal{V}] \neq 0$. The outcomes of these $\hat{O}$–experiments can be considered as joint measurements of $\hat{P}_\mathcal{P}$ and $\hat{P}_\mathcal{V}$ [see subsection 18.2.1].

The authors proposed the following interferometry experiment (Interfer. Exp. 18) [see figure 30.7]: an incoming photon $|\psi\rangle$ passes a first BS and is subjected to a phase shift $\delta\varphi$ in the full transmitted component, becoming $1/\sqrt{2}(|\mathcal{P}\rangle + e^{\imath\delta\varphi}|\neg\mathcal{P}\rangle)$, where $\hat{P}_\mathcal{P} = |\mathcal{P}\rangle\langle\mathcal{P}|$ and $\hat{P}_{\neg\mathcal{P}}$ represents the opposite point with respect to $\hat{P}_\mathcal{P}$ in the Poincaré sphere, whereas $\hat{P}_{\neg\mathcal{V}}$ represents the

[3] See also [GHOSE/ROY 1991] [GRANGIER *et al.* 1986a] and subsection 6.6.2, where the latter experiment is reported.

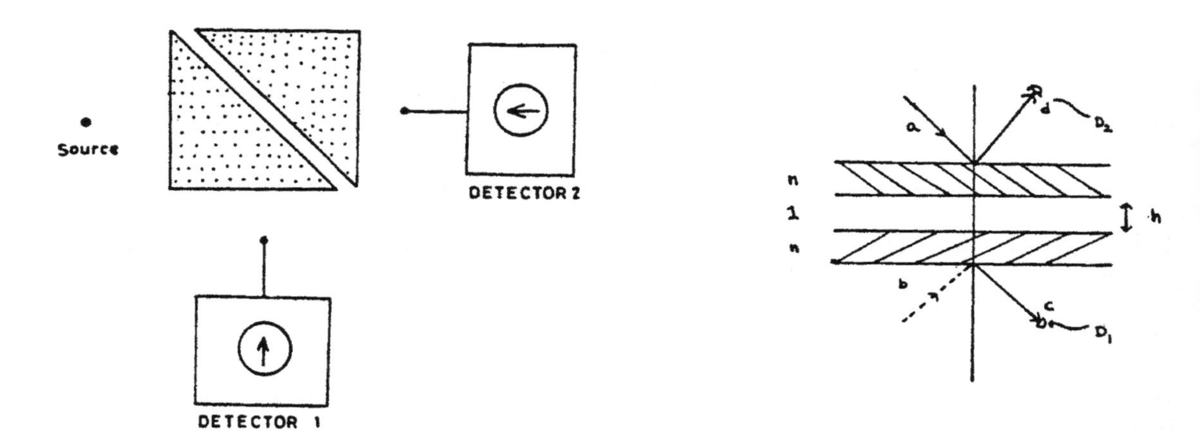

Figure 30.5: Tunnelling Exp. 1 — from [GHOSE *et al.* 1991, 404].

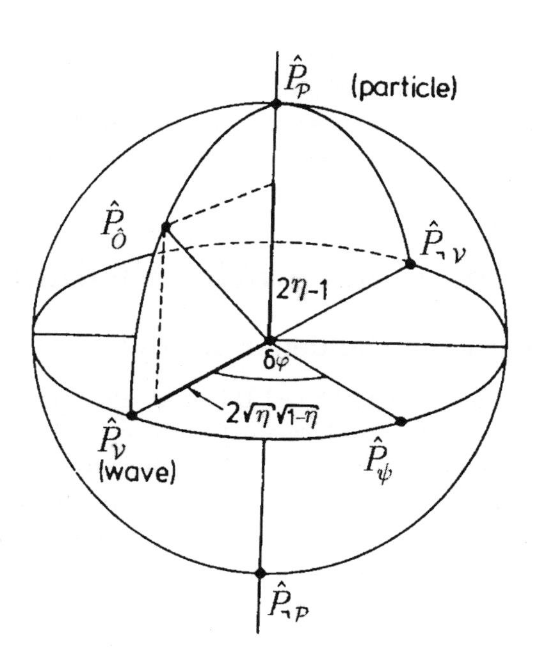

Figure 30.6: Poincaré-sphere representation of a stochastic measurement of which-path/visibility — from [MITTELSTAEDT *et al.* 1987, 893] .

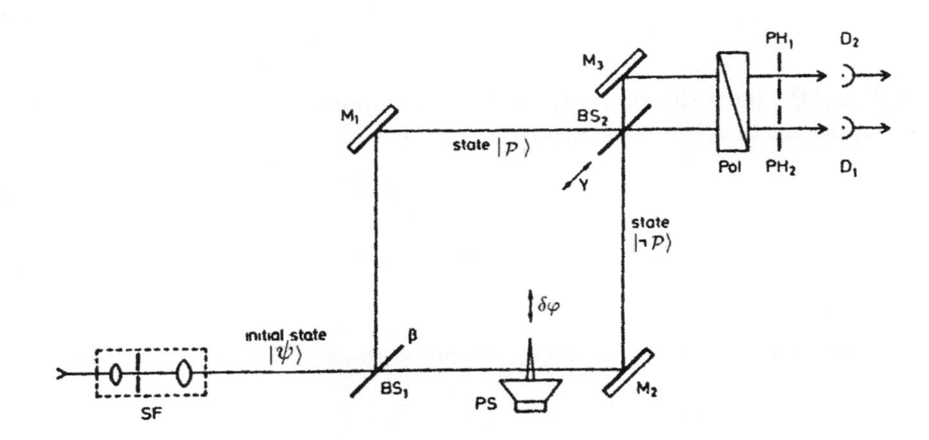

Figure 30.7: Schematic representation of the proposed interferometer experiment — from [MITTELSTAEDT *et al.* 1987, 892] .

antifringes of the diffraction [see also Two-Slit-Exp. 8: p. 143]. The transparency of the second BS is expressed by the variable parameter $\eta \in [0, 1]$, which thus determines the observable $\hat{P}_{\hat{O}}(\eta)$. If $\eta = 0$, then the path is completely determined:

$$\hat{P}_{\hat{O}}(0) = \hat{P}_{\mathcal{P}}; \tag{30.42}$$

and the same for $\eta = 1$ ($\hat{P}_{\hat{O}}(1) = \hat{P}_{\neg\mathcal{P}}$); but if $\eta = \frac{1}{2}$, then one measures the interference observable

$$\hat{P}_{\hat{O}}\left(\frac{1}{2}\right) = \hat{P}_{\mathcal{V}} = \frac{1}{2}\left(\hat{I} + |\mathcal{P}\rangle\langle\neg\mathcal{P}| + |\neg\mathcal{P}\rangle\langle\mathcal{P}|\right). \tag{30.43}$$

In the general case $0 \leq \eta \leq 1$ we have:

$$\begin{aligned} \hat{P}_{\hat{O}}(\eta) &= (\eta - \sqrt{(1-\eta)\eta})\hat{I} + 2\sqrt{(1-\eta)\eta}\hat{P}_{\mathcal{V}} + (1-2\eta)\hat{P}_{\mathcal{P}} \\ &= \left(\eta - \frac{1}{2}\mathcal{V}(\eta)\right)\hat{I} + \mathcal{V}(\eta)\hat{P}_{\mathcal{V}} + (1-2\eta)\hat{P}_{\mathcal{P}}, \end{aligned} \tag{30.44}$$

with the visibility given by $\mathcal{V} = 2\sqrt{(1-\eta)\eta}$.

The study of Mittelstaedt, Schieder and Prieur presents a very easy model which, by means of the Poincaré sphere and POVMs, demonstrates in a direct way the smoothness of the Complementarity which-path/visibility.

30.3.2 de Muynck/Martens/Stoffels Model

De Muynck, Stoffels and Martens [DE MUYNCK *et al.* 1991] [MARTENS/DE MUYNCK 1990a] developed another POVM analysis of the Complementarity between interference fringes and which-path, with results formally similar to that of Greenberger/Yasin or Englert.

If $\{\hat{O}^1_k\}$ and $\{\hat{O}^2_n\}$ are two POVMs — $[\hat{O}^1_k, \hat{O}^2_n] \neq 0$ in general [see subsection 18.2.1] —, we can choose a third POVM $\hat{O}^{12}_{ij}$ such that

$$\sum_j \hat{O}^{12}_{ij} = \sum_n o^2_{in}\hat{O}^2_n, \; o_{in} \geq 0, \; \sum_i o_{in} = 1, \tag{30.45a}$$

$$\sum_i \hat{O}^{12}_{ij} = \sum_k o^1_{jk}\hat{O}^1_k, \; o^1_{jk} \geq 0, \; \sum_j o^1_{jk} = 1. \tag{30.45b}$$

The matrices $\{\hat{O}^1_k\}, \{\hat{O}^2_n\}$ have the property that, if one approaches unity, the other becomes more nonideal. Supposing that there are inverses of both

$$\hat{O}^2_n = \sum_i [(o^2)^{-1}]_{ni} \sum_j \hat{O}^{12}_{ij}, \tag{30.46a}$$

$$\hat{O}^1_k = \sum_j [(o^1)^{-1}]_{kj} \sum_i \hat{O}^{12}_{ij}, \tag{30.46b}$$

we define the Wigner measure (an operator $\hat{\mathcal{W}}_{nk}$ on phase space which need not be positive) as:

$$\hat{\mathcal{W}}_{nk} = \sum_{ij} [(o^2)^{-1}]_{ni}[(o^1)^{-1}]_{kj}\hat{O}^{12}_{ij}, \tag{30.47}$$

satisfying:

$$\sum_k \hat{\mathcal{W}}_{nk} = \hat{O}^2_n, \; \sum_n \hat{\mathcal{W}}_{nk} = \hat{O}^1_k. \tag{30.48}$$

Now let us consider an experimental arrangement (Interfer. Exp. 19) such as that shown in figure 30.8. BS1, BS2 have transparencies = 1/2. A BS3 is inserted with transparency η. In one arm the photon undergoes a phase shift $\delta\varphi$. If $\eta = 1$ and detectors have efficiency 100%, then the detection probabilities of detectors D1, D2 are given by $\wp_k = \langle \hat{\mathcal{V}}_k \rangle, k = 1, 2$:

$$\hat{\mathcal{V}}_1 = \begin{pmatrix} \cos^2 \frac{\delta\varphi}{2} & -\frac{i}{2}\sin\delta\varphi \\ \frac{i}{2}\sin\delta\varphi & \sin^2 \frac{\delta\varphi}{2} \end{pmatrix}, \quad \hat{\mathcal{V}}_2 = \hat{I} - \hat{\mathcal{V}}_1. \tag{30.49}$$

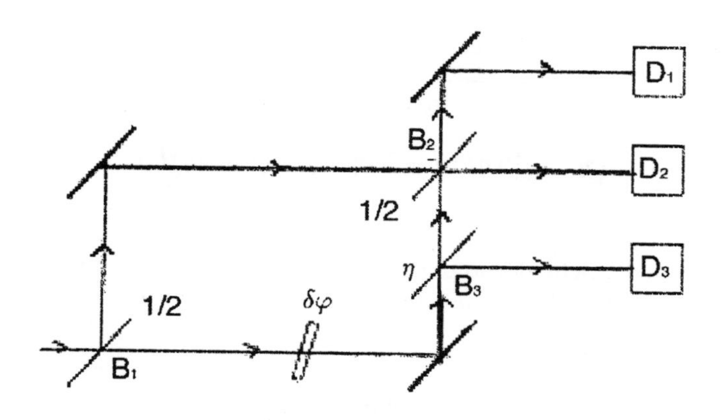

Figure 30.8: Proposed Interfer. Exp. 19 — from [DE MUYNCK *et al.* 1991, 128].

We call the PVM $\{\hat{\mathcal{V}}_1, \hat{\mathcal{V}}_2\}$ the *interference observable*. If $\eta = 0$, the path of the photon is measured and probability for left $\wp_1 + \wp_2$ and for right path $\wp_3$ are given by $\langle \hat{\mathcal{P}}_n \rangle$, with $n = \mathrm{L}, \mathrm{R}$ respectively. The PVM $\{\hat{\mathcal{P}}_\mathrm{L}, \hat{\mathcal{P}}_\mathrm{R}\}$ is called the *path observable*. For $0 < \eta < 1$ (stochastic assumption) we obtain the following detection probabilities $\wp_j = \left\langle \hat{O}^j_{\rm POV} \right\rangle$ of Dj (j = 1, 2, 3):

$$\hat{O}^1_{\rm POV} = \frac{1}{2}\left[\hat{\mathcal{P}}_{\rm L} + \eta\hat{\mathcal{P}}_{\rm R} + \sqrt{\eta}(\hat{\mathcal{V}}_1 - \hat{\mathcal{V}}_2)\right], \tag{30.50a}$$

$$\hat{O}^2_{\rm POV} = \frac{1}{2}\left[\hat{\mathcal{P}}_{\rm L} + \eta\hat{\mathcal{P}}_{\rm R} - \sqrt{\eta}(\hat{\mathcal{V}}_1 - \hat{\mathcal{V}}_2)\right], \tag{30.50b}$$

$$\hat{O}^3_{\rm POV} = (1-\eta)\hat{\mathcal{P}}_{\rm R}. \tag{30.50c}$$

The POVM $\{\hat{O}^1_{\rm POV}, \hat{O}^2_{\rm POV}, \hat{O}^3_{\rm POV}\}$ can be interpreted as a joint measurement of interference and path.

Proof

In order to show this, order the POVM into a bivariate POVM according to:

$$(\hat{O}^{jk}_{\rm POV}) = \begin{pmatrix} \hat{O}^1_{\rm POV} & \hat{O}^2_{\rm POV} \\ \zeta\hat{O}^3_{\rm POV} & (1-\zeta)\hat{O}^3_{\rm POV} \end{pmatrix}. \tag{30.51}$$

Choosing transparency $\zeta = 1/2$, the marginals of this bivariate POVM can be computed as:

$$\begin{pmatrix} \hat{O}^1_{\rm POV} + \hat{O}^2_{\rm POV} \\ \zeta\hat{O}^3_{\rm POV} + (1-\zeta)\hat{O}^3_{\rm POV} \end{pmatrix} = \begin{pmatrix} o_{\rm LL} & o_{\rm LR} \\ o_{\rm RL} & o_{\rm RR} \end{pmatrix}\begin{pmatrix} \hat{\mathcal{P}}_{\rm L} \\ \hat{\mathcal{P}}_{\rm R} \end{pmatrix}, \tag{30.52a}$$

$$\begin{pmatrix} \hat{O}^1_{\rm POV} + \zeta\hat{O}^3_{\rm POV} \\ \hat{O}^2_{\rm POV} + (1-\zeta)\hat{O}^3_{\rm POV} \end{pmatrix} = \begin{pmatrix} o_{11} & o_{12} \\ o_{21} & o_{22} \end{pmatrix}\begin{pmatrix} \hat{\mathcal{V}}_1 \\ \hat{\mathcal{V}}_2 \end{pmatrix}, \tag{30.52b}$$

where the matrices (o_{jn}) — with $j, n = \rm L, R$ — and (o_{kl}) — with $k, l = 1, 2$ — are given by:

$$(o_{jn}) = \begin{pmatrix} 1 & \eta \\ 0 & 1-\eta \end{pmatrix}, \tag{30.53a}$$

$$(o_{kl}) = \frac{1}{2}\begin{pmatrix} 1+\sqrt{\eta} & 1-\sqrt{\eta} \\ 1-\sqrt{\eta} & 1+\sqrt{\eta} \end{pmatrix}. \tag{30.53b}$$

We see that Eqs. (30.52) and (30.53) satisfy conditions (30.45): hence the bivariate POVM can be interpreted as a joint measurement of Visibility and Predictability. Q. E. D.[4]

30.4 Beginning of an Ontological Interpretation

The smooth transition between wave and particle behaviour forbids an interpretation which acknowledges an ontological status of only one of the pair, as, for example, Born's Probabilistic interpretation [proposition 6.2: p. 104]. Either both or none have an ontological status. The second possibility is excluded from the discussed experiments — think of the exchange of energy

[4]Since matrices (30.53) can be inverted, we see that conditions (30.46) too are satisfied, so that we find the following Wigner measure [Eq. (30.47)] for this experiment:

$$\hat{\mathcal{W}}_{11} = \frac{1}{2}(\hat{\mathcal{P}}_{\rm L} + \hat{\mathcal{V}}_1 - \hat{\mathcal{V}}_2), \tag{30.54a}$$

$$\hat{\mathcal{W}}_{12} = \frac{1}{2}(\hat{\mathcal{P}}_{\rm L} - \hat{\mathcal{V}}_1 + \hat{\mathcal{V}}_2), \tag{30.54b}$$

$$\hat{\mathcal{W}}_{21} = \hat{\mathcal{W}}_{22} = \frac{1}{2}\hat{\mathcal{P}}_{\rm R}. \tag{30.54c}$$

between the neutron and the radio-frequency field in Interfer. Exp. 15 [see subsection 28.3.2] — and from the absurd theoretical consequence that QM would have no ontological reference at all. Therefore the superposition — the wave-like behaviour — cannot only be a probabilistic tool: it is the objective non-separability of different possible values of an observable, for example, considering the which-path experiments, the non-separability of different paths. Hence we have here the ultimate grounds on which to reject the interpretations which see in the wave a form of ontological indeterminacy [see chapter 29].

However, one cannot interpret the ontological status of both as a contemporary *presence* of a wave and of a particle, which is not confirmed by the reported experiments about non-linear wave Eqs. [see section 28.2] and excluded by the smooth complementarity between both. The quantum wave is something real exactly as macroscopic things are; but its ontology is quite different from that which we normally suppose the macroworld to have: it is a spread and entangled reality. On the contrary, the corpuscular behaviour is the *emergence* (and not the presence) of a 'singularity' — of a localization [see chapter 24]. In conclusion we can advance the following proposition, which must be discussed and tested in detail in the following examination:

Proposition 30.1 (Wave/Particle Ontology) *The wave-like and the corpuscular behaviour of microentities are only two extreme forms of being of the same ontological entity, which is governed by the Greenberger/Yasin inequality* (30.19).

In other words, the basic observables of the theory are POVMs — as shown especially by the stochastic treatment of the problem [section 30.3] — and not PVMs, which can be understood as limiting cases of the former [on this point see chapter 33].

In the analysis to be developed in the following chapters, we shall examine in greater depth this point when we discuss in more detail the problem of localization as such — the matter of part VIII — and the problem of non-locality — the matter of part IX. What can be said here, is that the correlations are not a reality with which we can 'interact' in the normal sense of the word [see subsection 29.2.2]. In fact we know that a probability distribution can also change with the sole presence of a detector, i.e. without reading [see theorem 17.2: p. 267].

If so, then, as already proposed [see subsection 26.4.1], it is nonsensical to discuss whether QM is realistic or not, if we do not state preliminarily which type of ontology QM implies.

Finally we stress that to some extent this proposition is the correct formulation of a fundamental tenet of the Copenhagen interpretation [see principle 8.2 (CP2): p. 138], but by following a philosophy which could be called an 'ontologically correct Pragmatism' [see also section 6.2].

Part VIII

COMPLETENESS AND DETERMINISM

Introduction to Part VIII The *eighth part* is devoted to the Hidden-Variables theories, an attempt to interpret QM in a deterministic way.

- The starting point of the discussion is most surely represented by the experiment proposed by Einstein/Podolsky/Rosen (= EPR), which aims to show the incompleteness of the theory [*chapter 31*]. As we have seen, from its beginning, the 'strangeness' of QM was intended by a number of physicists and philosophers as evidence of the incompleteness of the theory and of the necessity of finding a more adequate account of microphenomena. In particular, one tried to interpret the statistical aspect of the theory, i.e. the Probabilistic assumption [postulate 3.5: p. 38], as an approximation to a more basic theory with which QM statistics would have the same relationship as classical statistical mechanics has with classical mechanics. It is not by chance that Einstein, the proponent of the Ignorance interpretation [proposition 6.3: p. 106], was also the proponent of an argument which sought to demonstrate the incompleteness of QM.

  In the fifties, many proposals aimed to prove that Hidden-Variable corrections to QM were not possible. In the middle of the sixties, Bell proposed a theorem with a related inequality, by means of which he could quantify the difference between the predictions of Hidden-Variable theories and those of QM, allowing the possibility of experimentally testing such predictions. But, as we shall see, Bell Inequality has consequences which go much further than the problem of Hidden-Variable theories because they pose the problem of non-locality. Although dependent on Hidden-Variable theory, the latter problem presents a lot of new and positive questions. Hence we postpone this aspect until part IX.

- In this part, we limit ourselves to the analysis of the more abstract (logical) confutations of Bohm's proposal [*chapter 32*]. In the seventh part we already examined a certain spread in the position of quantum entities due to their wave-like nature. Here we analyse the complementary problem which is more concerned with their corpuscular nature: is there a contradiction if we add to QM an assumption such as the Determined Value Assumption [postulate 6.3: p. 105]?

  Then we discuss the interpretational issues of Hidden-Variable theories. Bohm began with the proposal of EPR and tried to accomplish this by introducing some variables underlying the quantum observables able to account for latter ones in order to reduce QM to a deterministic theory. As we shall see, the proposal of Bohm, especially his physical interpretation, can also be understood as a partial continuation of de Broglie's theory of double solution.

- Finally, in *chapter 33*, a radical (undeterministic) counterproposal is presented: stochastic theories. The problem of renormalization and that of the position operator is also discussed again.

Chapter 31

EINSTEIN/PODOLSKY/ROSEN ARGUMENT

Contents In this chapter we examine the EPR argument. In section 31.1 we recount the basic definitions used by EPR, while section 31.2 is devoted to the structure of the argument. In section 31.3 Bohm's reformulation is given (a standard quoted by almost all successive works and a great approximation to experimental tests). While in section 31.4 the preliminary problem of whether or not QM is formally inconsistent is discussed, brief concluding remarks follow in section 31.5.

31.1 Basic Definitions

Some antecedents of the following problematic are to be found in an article of Einstein in collaboration with Podolsky and Tolman [EINSTEIN *et al.* 1931] and in Popper's book [*POPPER* 1934a, 243–45].[1] The aim of the EPR paper was to show, by means of a *Gedankenexperiment*, that QM is not a complete physical theory. Of course at the beginning of the article, EPR gave a set of definitions and, among these, that of completeness too.

31.1.1 Theory and reality

EPR distinguished between the *objective reality* and the *physical concepts*, the latter having to correspond to the former; so that, in judging a physical theory (likely made of physical concepts), we must make inquiry as to the *correctness* and *completeness* of the theory [EINSTEIN *et al.* 1935, 138].

31.1.2 Correctness

The *correctness* consists in the degree of agreement between the conclusions of the theory and human experience — the objective reality. Hence we can formulate it as follows:

Definition 31.1 (Correctness) *A theory is totally correct if every element of the theory has a counterpart in reality.*

In other words a totally correct theory is one without superfluous theoretical terms [see p. 3].

[1] See also [HOOKER 1972, 84]. For history see [*JAMMER* 1974, 167–68].

31.1.3 Completeness

Complementarily to correctness, we can formulate the following definition of Completeness[2]:

Definition 31.2 (Completeness) *A theory is complete if every element of reality has a counterpart in it.*

It is evident that correctness together with completeness (in analogy with the theory of completeness of mathematical logic) establish an equivalence between physical theory and reality[3], which is probably the aim of EPR:

$$(\text{Theory} \rightarrow \text{Reality}) \wedge (\text{Reality} \rightarrow \text{Theory}). \tag{31.1}$$

However, some preliminarily questions can be asked concerning the condition of completeness: we may ask if it is possible to have a theory which satisfies the criterion (*semantical completeness*). If the answer were yes, sooner or later we would have a *unique theory* which mirrors every 'element' of reality. But, as we have already said [see subsection 9.4.2], this is only a mythology which is in conflict with the existence of a plurality of theories and with the current evolution of science. Hence we are favorable disposed towards self-referentiality but not to completeness in this strong sense [see subsection 24.5.4].

Therefore, it is better to state a principle of ontological commitment [6.1: p. 103], which presupposes the context of the theory in which the entities have to appear and to be definite. Note that Einstein in other writings is more elastic and refined on the problem and generally acknowledges the theoretical contextuality of the 'reality' [*EINSTEIN* MW, 159].[4] Perhaps the reason for this is to be found in the fact that the material redactor of the EPR article was Podolsky, and that the condition of completeness was likely to have been proposed by himself[5].

The aim of the article, however, is to show that there are possible elements of reality which are in open conflict with QM, and hence we do not need such a strong formulation of completeness. In other words, the aim is to show the *incompleteness* of QM in the sense of the absence in it of a satisfactory explanation of entities which are considered fundamental in order to make QM work, and not to decide positively about its completeness (it is a 'disproof' and not a proof). Hence in the following we limit ourselves to this point and use the term *incompleteness* in this specific sense.

31.1.4 Separability

Following the example of many commentators[6] EPR use what seems a locality principle, something expressing the impossibility of exchanging superluminal signals [see also definition 36.1: p. 633]. But the EPR argument has nothing to do with locality. It is evident that the core is more a *Separability principle*, which we can express as follows:

Principle 31.1 (Separability) *Two spatially separated systems possess their own separate real states.*

---

[2] On this point see also [BALLENTINE 1970, 362] .

[3] As stated by [*JAMMER* 1974, 181].

[4] See examinations in [MARGENAU 1949, 247–50] [*MURDOCH* 1987, 219].

[5] As supposed by Jammer [JAMMER 1985, 142–43].

[6] See [STAPP 1980], but what we say in the following is already evident in some titles of Stapp's articles.

The Separability principle is the principle of individuation for physical systems [HOWARD 1985, 173, 175, 180]. Practically, it identifies the term *isolated* [see definition 3.2: p. 42] and the term *dynamically independent*: in other words it does not acknowledge a form of interdependence between systems other than the dynamically or the causal local form of interaction — i.e. the first point of the same definition 3.2.

Therefore, it is important to distinguish carefully the problem of locality — connected with a passage of information or of physical effects — and that of Separability, which concerns only the impossibility of a correlation (= C) between separated systems *without dynamical and causal connections* [see subsections 3.5.2 and 3.7.2]. Part of the EPR argument is that, in the absence of physical interactions, the systems have no relation at all, and not that physical systems cannot be superluminally connected. In other words, following EPR, we could think that entanglement is only a defective (incomplete) way to describe quantum systems. In conclusion, with EPR, we assume the weaker principle of separability, postponing the examination of locality [see section 36.5].

31.1.5 Reality

On the other hand, EPR state a sufficient condition for the reality of observables, which can be formulated [BALLENTINE 1970, 362] as follows:

Proposition 31.1 (Sufficient Condition of Physical Reality) *If, without in any way disturbing a system, we can predict with probability equal to unity the value of a physical quantity, then there exists an element of the physical reality corresponding to this physical quantity.*

It is evident that the proposition 31.1 is a reformulation — intended as a condition — of the definition of correctness, under the Determined Value assumption [postulate 6.3: p. 105], the Faithful Measurement assumption [postulate 6.4: p. 105] and the Separability principle:

$$(\text{Determined} - \text{Value} \wedge \text{Faithful Measurement} \wedge \text{Separability}) \rightarrow \text{Suff. Cond. Reality.} \quad (31.2)$$

31.1.6 Counterfactuality

One generally thinks that EPR use implicitly a counterfactual principle of value-definiteness of the form [*REDHEAD* 1987b, 92]:

Proposition 31.2 (Counterfactual Values-Definiteness) *A physical theory must be able to predict all possible outcomes of different measurements concerning the same variable which are alternative — by a different setting of the apparatus — to the measurement factually performed.*

In fact EPR also used counterfactual reasoning. But, as we shall see, their thought experiment is more an anticipation of the Relative state discussion [see theorem 15.1: p. 246] and in this respect it is developed at a purely formal and hypothetical level. We can therefore examine the problem without needing such an assumption[7] . A counterfactual argument is on the contrary necessary for proofs of Stapp's type [see section 36.1].[8] In conclusion note that proposition 31.2 is weaker than postulate 6.3 [p. 105].

[7]Which, on the other hand, could already be questioned in its generality on the grounds of theorem 14.6 [p. 235]. An examination of this point can be found in section 36.2.

[8]Where alternative settings of the same measuring apparatus are considered, which cannot be simultaneously the case in any physical theory, neither in classical mechanics.

31.2 The Argument

31.2.1 Structure

The structure of the argument is as follows. If we assume the sufficient condition of reality [proposition 31.1], then we can prove that QM does not satisfy the condition of completeness [proposition 31.2]:

$$\text{Suff. Cond. Reality} \rightarrow \neg\text{QM} - \text{Completness}, \tag{31.3}$$

or in terms of formula (31.2):

$$(\text{Determined} - \text{Value} \wedge \text{Faithful Measurement} \wedge \text{Separability}) \rightarrow \neg\text{QM} - \text{Completeness}, \tag{31.4}$$

so that, if we accept the EPR argument, the only way to refute the conclusion is to refute at least one of the three conditions which constitute the *antecedents*. We leave aside the Faithful Measurement assumption [it will be the subject of chapter 33], because it is of no relevance here, and concentrate our efforts on the Determined-Value assumption and the Separability principle.

31.2.2 The Argument Itself

The aim of EPR is to prove the incompleteness of QM by showing that there are elements of reality that have no counterpart in the theory. In particular, if we take two non-commuting observables, QM assures us that we cannot have simultaneously determined values of both. But if one can show that this simultaneous measurement is possible, then one has demonstrated that there is a reality which has no counterpart in the theory and hence the incompleteness of QM.

The problem Now let us suppose [EINSTEIN *et al.* 1935, 140] [9] we have two systems — $\mathcal{S}_1$ and $\mathcal{S}_2$ —, say, two particles which interact for a time between t_1 and t_2, after which they no longer interact — after t_2 they are separated by a space-like interval. For the sake of simplicity we suppose a one-dimensional case. If $o_1, o_2, \ldots$ are eigenvalues of a variable $\hat{O}$ pertaining to $\mathcal{S}_1$ and $u_1(x_1), u_2(x_1), \ldots$ the corresponding eigenfunctions and $\psi_1(x_2), \psi_2(x_2), \ldots$ eigenfunctions of some observable of $\mathcal{S}_2$, where x_1 is a variable used to describe system $\mathcal{S}_1$, x_2 a variable used to describe system $\mathcal{S}_2$, we have a function Ψ that describes for all times the composite system $\mathcal{S}_1 + \mathcal{S}_2$:

$$\Psi(x_1, x_2) = \sum_{n=1}^{\infty} \psi_n(x_2) u_n(x_1). \tag{31.5}$$

Suppose that quantity $\hat{O}$ is measured and the eigenvalue o_k is found; the composite system reduces to [see Eq. (14.17)]:

$$\psi_k(x_2) u_k(x_1). \tag{31.6}$$

But — and here we remember the Relative-State analysis [see subsection 15.1.2 and section 15.2] — if we had chosen $\hat{O}'$ (on $\mathcal{S}_1$), with eigenvalues $o'_1, o'_2, \ldots$ and eigenfunctions $v_1(x_1), v_2(x_1), \ldots$, we would have, for example, instead of Eq. (31.5), the following:

$$\Psi(x_1, x_2) = \sum_{s=1}^{\infty} \varsigma_s(x_2) v_s(x_1). \tag{31.7}$$

[9] See also [*JAMMER* 1974, 182–83] [JAMMER 1985, 148–49].

If $\hat{O}'$ is measured and eigenvalue o'_r found, we have the reduction:

$$\varsigma_r(x_2)v_r(x_1). \tag{31.8}$$

We see that, as a consequence of two different measurements performed upon $\mathcal{S}_1$, $\mathcal{S}_2$ is left in states with different wave functions and this without any form of causal relation between $\mathcal{S}_1$ and $\mathcal{S}_2$ — admitting the Separability principle [31.1].

Quantum formalism Let us now discuss [EINSTEIN *et al.* 1935, 141] this point more exactly on a formal level. Suppose that the two non-commuting operators (in the one-dimensional case) are $\hat{p}_1$ ($\hat{p}$ for the sake of simplicity) and $\hat{q}_1$ ($\hat{q}$). Since the two particles are separated and hence free, we have a continuous spectrum, so that Eq. (31.5) will now be written:

$$\Psi(x_1, x_2) = \int_{-\infty}^{+\infty} dp\psi_p(x_2)u_p(x_1). \tag{31.9}$$

Now the eigenfunctions of $\hat{O} = \hat{p}$ in the position representation will be[10]:

$$u_p(x_1) = e^{\imath\frac{1}{\hbar}px_1}, \tag{31.10}$$

with eigenvalue p, while for $\mathcal{S}_2$ we have:

$$\psi_p(x_2) = e^{-\imath\frac{1}{\hbar}(x_2 - x_0)p}, \tag{31.11}$$

where x_0 is a constant and $\psi_p(x_2)$ is the eigenfunction in the position representation of the operator [see Eq. (3.86)]

$$\hat{p}_2 = \imath\hbar\frac{\partial}{\partial x_2}, \tag{31.12}$$

corresponding to the eigenvalue $-p$ of the second particle's momentum. It is worth mentioning that this original argument supposes a rigid bar (due to the constant x_0) between the two subsystems; this can be questioned on physical grounds. But, since it is possible to restate the argument without the use of such an assumption, we do not discuss this problem.

Eq. (31.9) can be rewritten:

$$\Psi(x_1, x_2) = \int_{-\infty}^{+\infty} e^{\imath\hbar(x_1 - x_2 + x_0)p}dp. \tag{31.13}$$

Now, if $\hat{O}' = \hat{q}$ for the first particle, we have for its v_x eigenfunctions corresponding to the eigenvalue x:

$$v_x(x_1) = \delta(x_1 - x). \tag{31.14}$$

So that Eq. (31.7) becomes:

$$\Psi(x_1, x_2) = \int_{-\infty}^{+\infty} \varsigma_x(x_2)v_x(x_1)dx. \tag{31.15}$$

But, since this is just another decomposition of the same wave function described by Eq. (31.13) — we still have no reduction — then we must also have:

$$\varsigma_x(x_2) = \int_{-\infty}^{\infty} e^{\imath\hbar(x - x_2 + x_0)p}dp = h\delta(x - x_2 + x_0). \tag{31.16}$$

Now ς_x is the eigenfunction of the position operator $\hat{q}_2$ of $\mathcal{S}_2$.

[10] On the point see [*SCHIFF* 1955, 54].

Conclusions On the ground of the CCRs [see Eq. (3.14)], we conclude that $\psi_p(x_2)$ and $\varsigma_x(x_2)$ must be eigenfunctions of two non-commuting operators corresponding to two physical quantities.

Thus — using now the Sufficient Condition of Reality [proposition 31.1] —, by measuring $\hat{O}$ or $\hat{O}'$ of $\mathcal{S}_1$ we can predict with certainty and without interacting with $\mathcal{S}_2$ (Separability), either the value of $\hat{p}_2$ or the value of $\hat{q}_2$ of $\mathcal{S}_2$. In the first case $\hat{p}_2$ is an element of reality, in the second $\hat{q}_2$ is; note that, since there is no apparent correlation between $\mathcal{S}_1$ and $\mathcal{S}_2$, these realities must exist in $\mathcal{S}_2$ before and independently from any measurement on $\mathcal{S}_1$ [see also proposition 25.1: p. 433]. But we saw that ψ_p and ς_x together cannot have reality for QM because they are incompatible — this being reflected in the multiplicity of reduced wave functions for $\mathcal{S}_2$ [Eqs. (31.6) and (31.8)]: it is another form of the basis degeneracy problem already analysed [see section 15.2]. So that EPR conclude that QM is not complete.

31.3 Bohm's Reformulation

31.3.1 Theory

Before discussing the problems deriving from the EPR argument, we present another form of EPR paradox, elaborated by Bohm. It is the universal point of reference of all subsequent discussion because it is a very easy model and it is testable — in fact one very often speaks of the EPRB argument, and not simply of EPR; and so also do we.

The new formulation [*BOHM* 1951, 614–23] only deals with discrete observables, instead of continuous ones such as position and momentum. This step was originally understood as a further simplification of EPR argument. Actually, by performing local measurement of position and momentum, it can turn out, as shown recently [O. COHEN 1997], that, despite the entanglement between the subsystems, there is no violation of the separability principle.

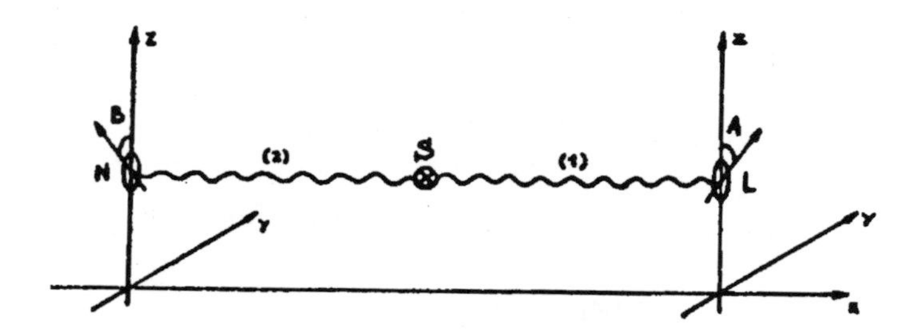

Figure 31.1: Schematic overview of an EPR-Bohm experiment — from [VIGIER 1985, 663].

Consider two particles with spin $\frac{1}{2}$ that are in a state in which the total spin is zero (singlet state). After a time t_0 the two particles begin to separate and at time t_1 they are space-like separated, so that they no longer interact [see figure 31.1]. On the hypothesis that they are not disturbed, they remain in a singlet state. By an initial decay of a particle with spin zero to two particles, the Law of angular momentum conservation (in absence of perturbation) guarantees it. We may write the equation (for z–direction) for the singlet state in the following manner [see

Eq. (3.132d)]:

$$|\Psi_0\rangle = \frac{1}{\sqrt{2}}(|\uparrow\rangle_1 \otimes |\downarrow\rangle_2 - |\downarrow\rangle_1 \otimes |\uparrow\rangle_2). \tag{31.17}$$

Now if a measurement of the spin of particle 1 leads to a result $|\uparrow\rangle = |+\frac{1}{2}\rangle_z$, that of particle 2 must give value $|\downarrow\rangle = |-\frac{1}{2}\rangle_z$. This result seems obvious because of the conservation of the total angular momentum for the total system. But $|\Psi_0\rangle$ is rotationally invariant. Let us write it in terms of the z–component eigenspinors:

$$|\Psi_0\rangle = \frac{1}{\sqrt{2}}\left[\begin{pmatrix}1\\0\end{pmatrix}_1 \otimes \begin{pmatrix}0\\1\end{pmatrix}_2 - \begin{pmatrix}0\\1\end{pmatrix}_1 \otimes \begin{pmatrix}1\\0\end{pmatrix}_2\right]. \tag{31.18}$$

Then $|\Psi_0\rangle$ in terms of the z–component eigenspinors is also an eigenvector, with eigenvalue -1, of both operators: $(\hat{\sigma}_{1x}\hat{\sigma}_{2x}), (\hat{\sigma}_{1y}\hat{\sigma}_{2y})$ [FRY 1995, 231]. For example [see Eq. (3.125)]:

$$\begin{aligned}
(\hat{\sigma}_{1y}\hat{\sigma}_{2y})|\Psi_0\rangle &= \frac{1}{\sqrt{2}}\begin{pmatrix}0 & -\imath\\ \imath & 0\end{pmatrix}_1 \begin{pmatrix}0 & -\imath\\ \imath & 0\end{pmatrix}_2 \left[\begin{pmatrix}1\\0\end{pmatrix}_1 \otimes \begin{pmatrix}0\\1\end{pmatrix}_2 - \begin{pmatrix}0\\1\end{pmatrix}_1 \otimes \begin{pmatrix}1\\0\end{pmatrix}_2\right] \\
&= \frac{1}{\sqrt{2}}\left[\begin{pmatrix}0\\ \imath\end{pmatrix}_1 \otimes \begin{pmatrix}-\imath\\0\end{pmatrix}_2 - \begin{pmatrix}-\imath\\0\end{pmatrix}_1 \otimes \begin{pmatrix}0\\ \imath\end{pmatrix}_2\right] \\
&= (-1)|\Psi_0\rangle.
\end{aligned} \tag{31.19}$$

Note that in the following, for the sake of simplicity, we generally drop the sign $\otimes$.

Hence, if the y–component of particle 1 is measured, after QM a measurement of the y–component of particle 2 must give the opposite result. This consequence clearly violates the Separability principle. As we have already said, it is another form of the degeneracy problem already analysed — see Zurek's criticism of Everett [section 15.2].

Suppose now that, instead of having found value $+\frac{1}{2}$ for particle 1, we had found the value $-\frac{1}{2}$: would we find value $+\frac{1}{2}$ for particle 2? If so, we have this situation: while Eq. (31.17) is neutral concerning the two possibilities, the measurement on particle 1 seems to 'decide' the result of the measurement on the separated particle 2.[11]

31.3.2 Preparing a Singlet State

An interesting question, preliminary to all further investigation is: how can one prepare a singlet state? An article of Cirac and Zoller shows this [CIRAC/ZOLLER 1994] . One can prepare a singlet state by allowing two atoms, — 1 and 2 —, initially in their excited ($= |\uparrow\rangle$) and ground ($= |\downarrow\rangle$) states respectively, to interact with a resonant cavity mode in the vacuum state (Cavity Exp. 5). After the preparation of the entangled states the cavity mode is left in its original state. The preparation takes place in two steps. First we send atom 1, initially in excited state $|\uparrow\rangle_1$ through the cavity (= c) in its vacuum state $|0\rangle_c$ in direction $\mathbf{r}_1$. After 1 has left the cavity, the atom 2, prepared in its ground state $|\downarrow\rangle_2$, is sent through the cavity in a different direction $\mathbf{r}_2$. The interaction Hamiltonian in a rotating frame at the cavity mode frequency and in the rotating wave approximation is

$$\hat{H} = \imath g(t)(\hat{\sigma}_+\hat{a} - \hat{a}^\dagger\hat{\sigma}_-), \tag{31.20}$$

where $\hat{\sigma}_+ = (\hat{\sigma}_-)^\dagger = |\uparrow\rangle\langle\downarrow|$ is the spin–$\frac{1}{2}$ excitation operator for a two level system [see Eq.s (5.25)]. The coupling parameter between c and 1, 2 is $g(t)$. Here we use the resonant

[11] As pointed out by [BELL 1987a, 206].

Jaynes/Cummings model [see Eqs. (5.24)], an important property of which is the conservation of the excitation number $\hat{a}^\dagger\hat{a} + |\uparrow\rangle\langle\uparrow|$. In particular, the state $|\downarrow\rangle|0\rangle_c$ does not change during the interaction, and the state $|\uparrow\rangle|0\rangle_c$ and $|\downarrow\rangle|1\rangle_c$ (where the $|1\rangle$ describes the presence of a photon in the cavity) will experience vacuum Rabi oscillations, which will depend on the interaction time t_j of particle j with the cavity field and therefore on the atomic velocity $v_j = L_c/t_j$ of j — where L_c is the length of the cavity. We suppose that v_1 has been selected in such a way that it undergoes 1/4 of a Rabi oscillation.

Therefore, the state of the combined system (atom 1 + atom 2 + cavity mode), after the first atom has crossed the cavity, will be given by

$$|\Psi\rangle = \frac{1}{\sqrt{2}}(|\uparrow\rangle_1|0\rangle_c - |\downarrow\rangle_1|1\rangle_c)|\downarrow\rangle_2. \tag{31.21}$$

The state of the system after particle 2 crosses c can be calculated if one takes into account that $|0\rangle_c|\downarrow\rangle_2$ remains unchanged during the interaction. By selecting v_2 in such a way that the state $|1\rangle_c|\downarrow\rangle_2$ performs half a Rabi cycle, the final state of the system is:

$$\begin{aligned}|\Psi\rangle_f &= \frac{1}{\sqrt{2}}(|\uparrow\rangle_1|\downarrow\rangle_2|0\rangle_c - |\downarrow\rangle_1|\uparrow\rangle_2|0\rangle_c) \\ &= \frac{1}{\sqrt{2}}(|\uparrow\rangle_1|\downarrow\rangle_2 - |\downarrow\rangle_1|\uparrow\rangle_2)|0\rangle_c \\ &= |\Psi\rangle_0|0\rangle_c;\end{aligned} \tag{31.22}$$

and therefore the singlet state (31.17) has been prepared.

31.4 Wave Function and Density Matrix

Another very important aspect of the problem is the following: EPR insisted that by measuring two different observables on subsystem $\mathcal{S}_1$ one obtains two different reduced wave functions for $\mathcal{S}_2$ [Eq. (31.6) and (31.8)]. Now, since $\mathcal{S}_2$ was already in the state ascribed to it by the measurement on $\mathcal{S}_1$, one could think that the two different reduced wave functions show a contradiction: the reality of two incompatible initial states of $\mathcal{S}_2$ (before the measurement) — a conclusion not drawn by EPR themselves. However, independently from the validity of EPR argument, the question remains: is there perhaps some form of contradiction in the formalism of QM? In no way, as the following proof shows [CANTRELL/SCULLY 1978]. One obtains the correct answer by using the density matrix formalism instead of the wave function formalism. In fact, we know [see subsection 3.5.1] that the wave-function treatment is not fully correct if the final measurement is performed on a subsystem, which is the case here.

Two possible experiments Imagine the source to be the disintegration of a positronium in the singlet ground state in order to emit two particles (a positron and an electron) with spin $\frac{1}{2}$. Two alternative experiments are possible: in the first one (SGM-Exp. 4) we measure the z component [see figure 31.2] and in the second one (SGM-Exp. 5) the x component [see figure 31.3] of $\mathcal{S}_1$'s spin.

SGM-Exp. 4 Suppose we perform a measurement at $t = t_0$. By the first experiment (symbolized by I) suppose that the wave function for $t < t_0$ is [see Eq. (31.17)]:

$$|\Psi_{1+2}^{<}(\mathrm{I})\rangle = \frac{1}{\sqrt{2}}(|\uparrow_1\rangle\otimes|\downarrow_2\rangle - |\downarrow_1\rangle\otimes|\uparrow_2\rangle). \tag{31.23}$$

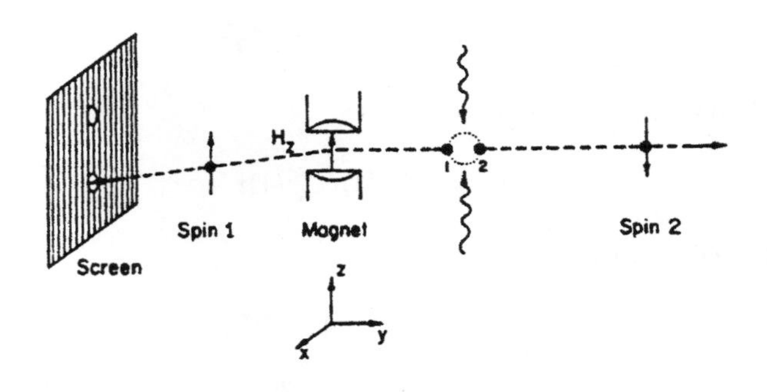

Figure 31.2: Measurement of the z–component of the spin of particle 1 — from [CANTRELL/SCULLY 1978, 502].

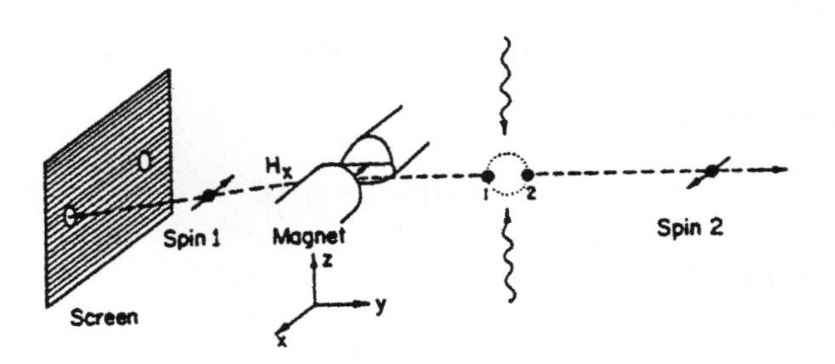

Figure 31.3: Measurement of the x–component of spin of particle 1 — from [CANTRELL/SCULLY 1978, 503].

If after the measurement one finds $| \downarrow_1\rangle$, then the state of the second particle is:

$$|\psi_2^>(\mathrm{I})\rangle = | \uparrow_2\rangle. \tag{31.24}$$

Now if this value were an element of reality in the sense of EPR, then we would have:

$$|\psi_2^<(\mathrm{I})\rangle = |\psi_2^>(\mathrm{I})\rangle = | \uparrow_2\rangle, \tag{31.25}$$

which, expressed in density-matrix language, makes:

$$\hat{\rho}_2^<(\mathrm{I}) = |\psi_2^<(\mathrm{I})\rangle\langle\psi_2^<(\mathrm{I})|, \tag{31.26}$$

and in matrix notation [see Eq. (3.125)]:

$$\hat{\rho}_2^<(\mathrm{I}) = \begin{bmatrix} 1 & 0 \\ 0 & 0 \end{bmatrix}_2 . \tag{31.27}$$

SGM-Exp. 5 By the second experiment (symbolized by II), we expand in terms of $| \pm x\rangle$ states:

$$| \pm x\rangle := |\pm\rangle = \frac{1}{\sqrt{2}}(| \uparrow\rangle \pm | \downarrow\rangle), \tag{31.28}$$

from which we obtain for the initial state:

$$|\Psi_{1+2}^<(\mathrm{II})\rangle = \frac{1}{\sqrt{2}}\left(|+\rangle_1 \otimes |-\rangle_2 - |-\rangle_1 \otimes |+\rangle_2\right). \tag{31.29}$$

Suppose we find the spin of $\mathcal{S}_1$ to be $|-\rangle_1$; then we have:

$$|\psi_2^>(\mathrm{II})\rangle = |+\rangle_2, \tag{31.30}$$

and, again according to EPR:

$$|\psi_2^>(\mathrm{II})\rangle = |\psi_2^<(\mathrm{II})\rangle. \tag{31.31}$$

In the language of density matrix:

$$\hat{\rho}_2^<(\mathrm{I}) = |\psi_2^<(\mathrm{II})\rangle\langle\psi_2^<(\mathrm{II})|. \tag{31.32}$$

Hence EPR would write in matrix notation:

$$\hat{\rho}_2^<(\mathrm{II}) = \frac{1}{2}\begin{bmatrix} 1 & 1 \\ 1 & 1 \end{bmatrix}_2 . \tag{31.33}$$

Concluding remarks We see clearly that, according to EPR, we have:

$$\hat{\rho}_2^<(\mathrm{II}) \neq \hat{\rho}_2^<(\mathrm{I}). \tag{31.34}$$

But suppose now that we have the following expectation value of the spin of $\mathcal{S}_2$:

$$\langle\hat{\sigma}_2\rangle = \mathrm{Tr}[(\hat{I}_1 \otimes \hat{\sigma}_2)\hat{\rho}_{1+2}]. \tag{31.35}$$

Now it is clear that

$$\langle\hat{\sigma}_2\rangle = \mathrm{Tr}_2[\hat{\sigma}_2 \mathrm{Tr}_1(\hat{I}_1\hat{\rho}_{1+2})] = \mathrm{Tr}_2[\hat{\sigma}_2\mathcal{R}_2], \tag{31.36}$$

and that a correct description of the spin of $\mathcal{S}_2$ is given by

$$\hat{\rho}^{<}_{\mathrm{Tr}_1} = \mathrm{Tr}_1(\hat{\rho}^{<}_{1+2}). \tag{31.37}$$

Now we are able to prove the equality of the two density matrices describing the system before measurements I and II.

Proof

We can rewrite the density matrix before experiment I in the following form:

$$\begin{aligned}
\hat{\rho}^{<}_{\mathrm{Tr}_1}(\mathrm{I}) = \mathrm{Tr}_1[\hat{\rho}^{<}_{1+2}(\mathrm{I})] &= \langle\uparrow_1|\Psi^{<}_{1+2}(\mathrm{I})\rangle\langle\Psi^{<}_{1+2}(\mathrm{I})|\uparrow_1\rangle + \langle\downarrow_1|\Psi^{<}_{1+2}(\mathrm{I})\rangle\langle\Psi^{<}_{1+2}(\mathrm{I})|\downarrow_1\rangle \\
&= \langle\uparrow_1|\frac{1}{\sqrt{2}}(|\uparrow_1,\downarrow_2\rangle - |\downarrow_1,\uparrow_2\rangle)\,\frac{1}{\sqrt{2}}(\langle\uparrow_1,\downarrow_2| - \langle\downarrow_1,\uparrow_2|)|\uparrow_1\rangle \\
&\quad +\langle\downarrow_1|\frac{1}{\sqrt{2}}(|\uparrow_1,\downarrow_2\rangle - |\downarrow_1,\uparrow_2\rangle)\,\frac{1}{\sqrt{2}}(\langle\uparrow_1,\downarrow_2| - \langle\downarrow_1,\uparrow_2|)|\downarrow_1\rangle \\
&= \frac{1}{2}(|\downarrow_2\rangle\langle\downarrow_2| + |\uparrow_2\rangle\langle\uparrow_2|) = \frac{1}{2}\begin{bmatrix} 1 & 0 \\ 0 & 1 \end{bmatrix}_2 .
\end{aligned} \tag{31.38}$$

The initial density matrix before the second experiment is:

$$\begin{aligned}
\hat{\rho}^{<}_{\mathrm{Tr}_1}(\mathrm{II}) = \mathrm{Tr}_1[\hat{\rho}^{<}_{1+2}(\mathrm{II})] &= \langle +_1|\Psi^{<}_{1+2}(\mathrm{I})\rangle\langle\Psi^{<}_{1+2}(\mathrm{I})|+_1\rangle + \langle -_1|\Psi^{<}_{1+2}(\mathrm{I})\rangle\langle\Psi^{<}_{1+2}(\mathrm{I})|-_1\rangle \\
&= \frac{1}{2}(|-_2\rangle\langle -_2| + |+_2\rangle\langle +_2|),
\end{aligned} \tag{31.39}$$

which in terms of z–component basis is:

$$\begin{aligned}
\hat{\rho}^{<}_{\mathrm{Tr}_1}(\mathrm{II}) &= \frac{1}{4}[(|\uparrow_2\rangle - |\downarrow_2\rangle)(\langle\uparrow_2| - \langle\downarrow_2|) + (|\uparrow_2\rangle + |\downarrow_2\rangle)(\langle\uparrow_2| + \langle\downarrow_2|)] \\
&= \frac{1}{2}(|\uparrow_2\rangle\langle\uparrow_2| + |\downarrow_2\rangle\langle\downarrow_2|) = \frac{1}{2}\begin{bmatrix} 1 & 0 \\ 0 & 1 \end{bmatrix}_2 ;
\end{aligned} \tag{31.40}$$

and we see now that the density matrix of states before the first experiment (31.38) is the same as that from before the second experiment (31.40). Q. E. D.

31.5 Concluding Remarks

The directions of the debate after EPR's article are now two-fold: the first one is the consequent proposal of a deterministic theory based on Hidden-variables which tries to 'complete' QM in a deterministic sense [postulate 6.3: p. 105] by assuming the separability; the other one is a sharp criticism of the Separability principle [see formula (31.4)], which was criticized at the beginning already by the fathers od the theory. In the following chapters we discuss the aim of completing QM, while the problem of separability will be the subject of part IX.

Chapter 32

EXAMINATION AND INTERPRETATION OF HIDDEN VARIABLE THEORIES

Contents After a preliminary definition of what a Hidden-Variable (= HV) Theory can be [section 32.1], in section 32.2 we analyse the first (and incomplete) proofs against HV-Theories. Section 32.3 is devoted to a very important corollary to the Gleason theorem: Bell's corollary concerning the impossibility of dispersion-free states (in Hilbert spaces with three or more dimensions). In section 32.4 the Kochen/Specker Theorem is analysed and some variants of it reported. Then other proofs which partly parallelize the latter theorem are examined [section 32.5]. Finally, in section 32.6, we analyze the interpretational issues of the HV-Theory (Bohm's interpretation), and, in section 32.7, concluding remarks follows.

32.1 Basic Definitions of the Hidden Variable Problem

In that which follows, we speak of HV-Theories as deterministic[1]. Sometimes one speaks of the 'realism' and not of the 'determinism' of HV-Theories. It is true that there are also other forms of realism, which have nothing to do with HVs. An HV-Theory presupposes a classical form of realism, which we have already excluded for waves [section 28.6] and which in the following analysis we shall exclude also for particles. Firstly, we give a basic definition of HV [*JAMMER* 1974, 256–57, 262]:

Definition 32.1 (HV) *If a given theory* T *contains a set of observables* $\{\hat{O}\}$ *which describe a physical system* $\mathcal{S}$*, and there are some variables* $\{\lambda_{\rm HV}\}$ *about* $\mathcal{S}$ *which are not experimentally detectable within the framework of* T *and the values of each* $\hat{O}$ *can be obtained by some averaging operation on the values of some* $\lambda_{\rm HV}$*, then the* $\{\lambda_{\rm HV}\}$ *are named HVs with respect to* T.

Now we define an HV-Theory (evidently for QM alone, which is the only case of interest here) [*JAUCH* 1968, 116]:[2]

[1]See [PYKACZ/SANTOS 1991] and [R. WERNER 1989, 4278] for a different point of view.

[2]See also [GUDDER 1970, 433].

Definition 32.2 (HV Theory) *If for each quantum system there exists a measure space Ω together with a finite measure μ (normalized, so that $\mu(\Omega) = 1$) on Ω such that every state $\hat{\rho}$ of an arbitrary QM system can be represented as a mixture*

$$\hat{\rho}(\hat{P}) = \int_{\Omega} d\hat{\rho}(\lambda_{\text{HV}})\hat{\rho}_{\lambda_{\text{HV}}}(\hat{P}) \tag{32.1}$$

of dispersion-free states [see definition 3.5] $\hat{\rho}_{\lambda_{\text{HV}}}(\hat{P})$ *for all PVM* $\hat{P}$*, then we call the theory which admits such mixtures a HV-Theory.*

In other words, a HV-Theory supposes that the state of a system and its observables are all — in classical sense — perfectly determined [see postulate 6.3: p. 105], and hence also the whole dynamic evolution of the same [see definition 2.1: p. 22]. In this sense it is a deterministic theory. In other words all HV-Theories examined in the following share the belief that the ontological entities object of the theory are not relative to some form of measurement or more generally of interaction [*D'ESPAGNAT* 1995, 287]; and in this sense the present discussion can be understood as a further development of chapter 26.

Therefore, a HV-Theory should be exactly a complete theory in the sense of EPR [see definition 31.2: p. 532]: it is characterized by the predictive exhaustivity of all possible outcomes — as we shall see [chapter 35] a necessary condition for deterministic HV-Theories and a sufficient and necessary condition for stochastic HV-Theories [JARRETT 1984a] [JARRETT 1984b, 24–25] [ELBY *et al.* 1993, 978].[3]

32.2 First Proofs

32.2.1 Von Neumann's Proof Concerning Hidden Variables

Von Neumann's analysis In 1932, von Neumann gave a demonstration of the impossibility of dispersion-free states in QM, with the evident aim of showing the completeness of QM.[4] It is interesting to note that EPR never took into account von Neumann's proof [SHIMONY 1971b, 77–78]. Von Neumann's proof can be formulated as follows [*VON NEUMANN* 1932, 163–71] [*VON NEUMANN* 1955, 324–25]:[5]

Proposition 32.1 (II von Neumann) *Assuming what follows:*

- *If an observable is represented by an operator* $\hat{O}$*, then a function* f *of this quantity is represented by operator* $\hat{f}(\hat{O})$.
- *If observables are represented by operators* $\hat{O}, \hat{O}', \ldots$*, then the sum of these observables is represented by operator* $\hat{O} + \hat{O}' + \ldots$.
- *If the observable* $\hat{O}$ *is by nature non-negative, then its expectation* $\langle\hat{O}\rangle$ *is also non-negative.*
- *If* $\hat{O}, \hat{O}', \ldots$ *are arbitrary observables and* $a, b, \ldots$ *real numbers, then* $\langle a\hat{O} + b\hat{O}', \ldots\rangle = a\langle\hat{O}\rangle + b\langle\hat{O}'\rangle + \ldots$.

[3] See also [SHIMONY 1984c].
[4] On this point see [JAMMER 1985, 138].
[5] See also [*BELINFANTE* 1973, 24–31] [*JAMMER* 1974, 267–69] [*DE BROGLIE* 1957, 23–25].

One then obtains the following formula: $\left\langle \hat{O} \right\rangle = \mathrm{Tr}[\hat{\rho}\hat{O}]$ [see Eq. (3.91) and section 11.5].

The preceding proposition implies the following corollary:

Corollary 32.1 (von Neumann) *If* Eq. (3.91) *is valid, then there cannot be dispersion-free states.*

We now outline von Neumann's original proof of this corollary.

Proof

If an observable is dispersion free we have [see Eq. (3.110)]:

$$\left\langle \hat{O}^2 \right\rangle = \left\langle \hat{O} \right\rangle^2; \tag{32.2}$$

or for some state $\hat{\rho}$

$$\mathrm{Tr}(\hat{\rho}\hat{O}^2) = [\mathrm{Tr}(\hat{\rho}\hat{O})]^2. \tag{32.3}$$

Now, choosing $\hat{O} = \hat{P}_\psi = |\psi\rangle\langle\psi|$, since $\hat{P}_\psi = \hat{P}_\psi^2$ and by $\mathrm{Tr}(\hat{\rho}\hat{P}_\psi) = \langle\hat{\rho}\psi|\psi\rangle$, we obtain immediately:

$$\langle\hat{\rho}\psi|\psi\rangle = \langle\hat{\rho}\psi|\psi\rangle^2, \tag{32.4}$$

which means that $\langle\hat{\rho}\psi|\psi\rangle$ is equal either to 0 or to 1. But for each $\| \psi \|= 1$ we obtain the same result, which implies that $\hat{\rho}$ is either 0 or 1. The first possibility is excluded, because it gives no information. If one accepts the second one, then $\mathrm{Tr}(\hat{I})$ implies that the observable is not dispersion-free. Q.E.D.

In conclusion, von Neumann excluded the possibility of HV theories and asserted that an homogeneous ensemble describing a pure state cannot be split up into subensembles corresponding to HV states. As Wigner pointed out, there cannot be HVs because, by a sequence of demolition measurements, the statistical distribution remains so unsharp that the outcomes are always as so unpredictable as by the first measurement [WIGNER 1970, 1006].

Discussion Von Neumann's proof was discussed and exemplified by Bell [BELL 1966, 4] [BELL 1971, 31–32]. Bell centered his criticism on von Neumann's fourth assumption in proposition 32.1. First he gave an example of its apparent soundness. Take the QM operator:

$$\frac{(\hat{\sigma}_x + \hat{\sigma}_y)}{\sqrt{2}}, \tag{32.5}$$

whose eigenvalues are certainly not the correspondent linear combination of the eigenvalue of $\hat{\sigma}_x$ and $\hat{\sigma}_y$:

$$\frac{(\pm 1 \pm 1)}{\sqrt{2}}. \tag{32.6}$$

Now a dispersion-free theory supposes that expectation values are identical with eigenvalues [definition 3.5: p. 54]. As Bell pointed out, the problem is that the abstract mathematical possibility of the combination of operators (the II assumption in the same proposition) does not imply the linear combination of expectation values, and hence of eigenvalues (IV assumption), because it may be the case that the two combined operators correspond to observables which are experimentally mutually exclusive. In the example chosen by Bell the three observables $\hat{\sigma}_x, \hat{\sigma}_y, (\hat{\sigma}_x + \hat{\sigma}_y)/\sqrt{2}$ correspond to three different orientations of an SGM and cannot be set simultaneously. Von Neumann's proof fails because, in order to use the formula $\left\langle \hat{O} \right\rangle = \mathrm{Tr}[\hat{\rho}\hat{O}]$ in a generalized way, we need a stronger mathematical theorem: the Gleason Theorem [see section 11.5].

Gudder was also on the same line as Bell [GUDDER 1970, 432]. He said: the dispersion-free states need not be defined on the entire proposition system of the pertinent Hilbert space since HV-Theories only allows predictions concerning single measurements.

32.2.2 Jauch/Piron Theorem

Theorems Jauch and Piron proposed [JAUCH/PIRON 1963] a refinement of von Neumann's proof. They showed that HV-Theories can exist only if every proposition is compatible with every other one, which empirically is not the case [JAUCH/PIRON 1963, 834–35]:

Theorem 32.1 (Jauch/Piron) *If there exists a dispersion-free state $\hat{\rho}$ on a propositional system $\mathcal{L}_{\succeq,\perp,a_n}$, then there exists a point r in the centre* [see p. 163] *of $\mathcal{L}_{\succeq,\perp,a_n}$ for which $\hat{\rho}(r) = 1$.*

From which two corollaries follow:

Corollary 32.2 (I Jauch/Piron) *There exist no dispersion-free states on a coherent proposition system (with trivial centre);*

and

Corollary 32.3 (II Jauch/Piron) *If a proposition system $\mathcal{L}_{\succeq,\perp,a_n}$ admits HVs, then every proposition of $\mathcal{L}_{\succeq,\perp,a_n}$ is compatible with every other proposition of $\mathcal{L}_{\succeq,\perp,a_n}$.*

Corollary 32.3 means [by theorem 9.4: p. 163] that an HV-Theory can only be Boolean, which is not the case for QM [see theorem 9.1: p. 161 or 9.2: p. 162]. In other words, through a generalized introduction of SSRs alone, one can obtain such a result for QM.

A similar proof has been developed by Jauch [*JAUCH* 1968, 117] [see figure 32.1]. He showed that

- every dispersion-free state is pure;
- on non-trivial coherent proposition systems there exist no dispersion-free states.

He then proved that if a proposition of $\mathcal{L}_{\succeq,\perp,a_n}$ admits HVs, then any pair of propositions $\in \mathcal{L}_{\succeq,\perp,a_n}$ is compatible.

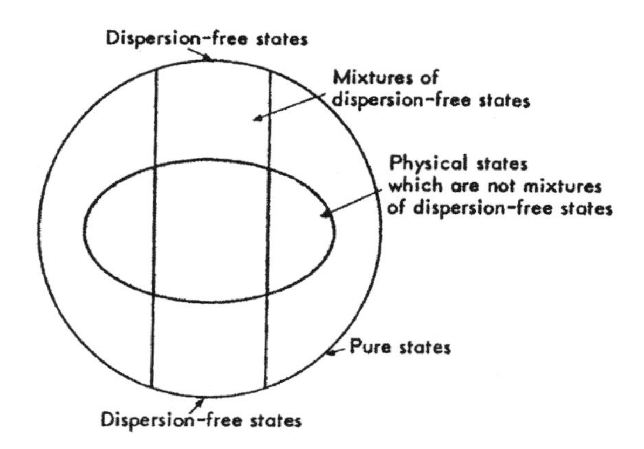

Figure 32.1: The sphere represents all the states. Pure states, as usually, are on the surface, and the dispersion-free states are the indicated subset of the sphere — from [*JAUCH* 1968, 119].

Misra [MISRA 1967] performed a stronger proof in which it was shown that even if one uses structures of a C^* algebra which allows dispersion-free positive linear functionals, the algebra is always Abelian, i.e. commutative [see subsection 3.10.2].

Discussion Bohm and Bub [BOHM/BUB 1966b] answered by pointing out that the proof of Jauch/Piron is circular: they admit that the impossibility of perfectly observing two non-commuting observables at the same time is an empirical fact. But it is so, only if one admits the interpretation to be proven already. The answer of Jauch and Piron [JAUCH/PIRON 1968] was centred on the fact that in an HV-Theory every physically representable state can be expressed as a mixture of dispersion-free states [definition 32.2: p. 543], and this entails the contradiction with the commutation laws (CCRs) of QM.

Bell also criticized the proof of Jauch and Piron [BELL 1966, 5–6] by assimilating it with that of von Neumann: we are dealing with measurements and not with abstract propositional systems. By concrete measurements the orientations of magnets, for example, change[6].

32.2.3 Gudder Theorem

Jauch and Piron [JAUCH/PIRON 1963, 828] showed that in a system with SSRs [see subsection 10.1.4] von Neumann's proof is false. Gudder [GUDDER 1968a] agrees with Bell, Bohm and Bub that there can be mathematical QM models which admit HV, but one must see the consequences of admitting a model with HV. He proves the following theorem[7]:

Theorem 32.2 (Gudder) *A system without SSRs cannot contain HV, and a system which contains them is classical.*

This theorem holds a certain interest. It can be deduced, however, from a theorem of the form of Kochen/Specker. We, therefore, neither prove it, nor integrate it into the theory.

32.2.4 Informational Completeness and Hidden Variable Theories

Before discussing the main proofs concerning HV-Theory, we remember here that sharp observables (PVM) are informationally incomplete [see theorem 18.7: p. 303], while unsharp ones (POVM) can be informationally complete [theorem 18.9: p. 304]. And this is clearly an indirect proof of the impossibility of dispersion-free observables in QM, because, if sharp observables were dispersion-free, they would be informationally complete — one could determine univocally the state of a system by the knowledge of such observables, as happens in classical mechanics. On the other hand, unsharp observables cannot be dispersion-free, because the values of unsharp observables are associated with effects, which are not projectors — whereas each value of a dispersion-free observable is associated with a projector[8].

32.3 A Corollary of the Gleason Theorem

32.3.1 General Features

The I Gleason Theorem [11.2: p. 200] was a confirmation of von Neumann's proof about the QM formula for expectation values [corollary 32.1] without using the additivity postulate, contested by Bell — the fourth assumption of von Neumann [in proposition 32.1]. Hence The I Gleason theorem assures that QM probabilities and expectation values can always be calculated by means

[6]See also [*JAMMER* 1974, 305] [*BELINFANTE* 1973, 54–60] [FREEDMAN *et al.* 1976, 43–44].

[7]See also [GUDDER 1968b] [GUDDER 1971] [*JAMMER* 1974, 319–21, 327–29].

[8]See section 39.4 for further discussion of the informational incompleteness of QM.

of a density matrix and of projectors. Bell now proved the following corollary to the I Gleason Theorem [BELL 1966, 6–8]:

Corollary 32.4 (Bell Dispersion-Free) *For a Hilbert space of dimension of at least three, there cannot be dispersion-free states.*

But we shall prove corollary 32.4 by making use of another theorem, formulated by Bell himself.

32.3.2 Some Lemmas

Let $\hat{P}_\varsigma$ be the projector on the Hilbert space vector $|\varsigma\rangle$, i.e. acting on an arbitrary vector $|\psi\rangle$ in the following manner [see Eq. (3.62)]:

$$\hat{P}_\varsigma|\psi\rangle = \frac{|\varsigma\rangle\langle\varsigma|\psi\rangle}{\langle\varsigma|\varsigma\rangle}. \tag{32.7}$$

If $\{|\varsigma_j\rangle\}$ is a complete set of orthogonal vectors, we have:

$$\sum_j \hat{P}_{\varsigma_j} = 1. \tag{32.8}$$

Since the $\hat{P}_{\varsigma_j}$ commute by hypothesis, we have:

$$\sum_j \left\langle \hat{P}_{\varsigma_j} \right\rangle = 1. \tag{32.9}$$

We know that the expectation value of a projector is non-negative (each measurement allows values 0 or 1). Now let us make the following assumption (the dispersion-free one):

Lemma 32.1 (I Bell) *If some arbitrary vector $|\varsigma\rangle$ is such that $\left\langle \hat{P}_\varsigma \right\rangle = 1$ for a given state, then for that state $\left\langle \hat{P}_\psi \right\rangle = 0$ for any $|\psi\rangle$ orthogonal to $|\varsigma\rangle$.*

Now, if $|\psi_1\rangle$ and $|\psi_2\rangle$ are another orthogonal basis for the subspace spanned by some vectors $|\varsigma_1\rangle$ and $|\varsigma_2\rangle$, then from Eq. (32.9) we have:

$$\left\langle \hat{P}_{\psi_1} \right\rangle + \left\langle \hat{P}_{\psi_2} \right\rangle = 1 - \sum_{j\neq 1, j\neq 2} \left\langle \hat{P}_{\varsigma_j} \right\rangle; \tag{32.10}$$

or

$$\left\langle \hat{P}_{\psi_1} \right\rangle + \left\langle \hat{P}_{\psi_2} \right\rangle = \left\langle \hat{P}_{\varsigma_1} \right\rangle + \left\langle \hat{P}_{\varsigma_2} \right\rangle. \tag{32.11}$$

We now make a second assumption:

Lemma 32.2 (II Bell) *Since $|\psi_1\rangle$ and $|\psi_2\rangle$ can be any linear combination of $|\varsigma_1\rangle$ and $|\varsigma_2\rangle$, we have that, if for a given state*

$$\left\langle \hat{P}_{\varsigma_1} \right\rangle = \left\langle \hat{P}_{\varsigma_2} \right\rangle = 0 \tag{32.12}$$

for some pairs of orthogonal vectors, then:

$$\left\langle \hat{P}_{c'\varsigma_1 + c''\varsigma_2} \right\rangle = 0 \tag{32.13}$$

for all scalars c', c''.

32.3.3 The First Bell Theorem and the Proof of the Corollary

We develop Bell's central idea, i.e. a theorem, by means of which along with the I Gleason theorem [11.2: p. 200] we shall prove corollary 32.4:

Theorem 32.3 (I Bell) *Let $|\varsigma\rangle$ and $|\psi\rangle$ be some vectors in a Hilbert space with dimension $d \geq 3$ such that we have:*

$$\left\langle \hat{P}_\psi \right\rangle = 1, \qquad (32.14a)$$

$$\left\langle \hat{P}_\varsigma \right\rangle = 0. \qquad (32.14b)$$

Then $|\varsigma\rangle$ and $|\psi\rangle$ cannot be arbitrarily chosen. In fact:

$$|\varsigma - \psi| > \frac{1}{2}|\psi|. \qquad (32.15)$$

First, we prove the theorem itself.

Proof

Let us normalize $|\psi\rangle$ and write $|\varsigma\rangle$ in the following form:

$$|\varsigma\rangle = |\psi\rangle + \epsilon|\psi^\perp\rangle, \qquad (32.16)$$

where $|\psi^\perp\rangle$ is orthogonal to $|\psi\rangle$ and normalized, and ϵ a real number. Let $|\psi^{\perp\perp}\rangle$ be a normalized vector orthogonal to both $|\psi\rangle$ and $|\psi^\perp\rangle$ — from here the necessity of dimensions $d \geq 3$. By the orthogonality of $|\psi^\perp\rangle$ and $|\psi^{\perp\perp}\rangle$ to $|\psi\rangle$ [lemma 32.1] and by Eqs. (32.14) we have

$$\left\langle \hat{P}_{\psi^\perp} \right\rangle = 0, \quad \left\langle \hat{P}_{\psi^{\perp\perp}} \right\rangle = 0. \qquad (32.17)$$

But by the Eqs. (32.12) and (32.13) we have:

$$\left\langle \hat{P}_{\varsigma + a^{-1}\epsilon\psi^{\perp\perp}} \right\rangle = 0, \qquad (32.18)$$

where we have posed $c' = 1$ and $c'' = a^{-1}\epsilon$, $|\varsigma_1\rangle = |\varsigma\rangle$ and $|\varsigma_2\rangle = |\psi^{\perp\perp}\rangle$, and a is a real number. Again by the lemma 32.2 we also have:

$$\left\langle \hat{P}_{\epsilon\psi^\perp + a\epsilon\psi^{\perp\perp}} \right\rangle = 0. \qquad (32.19)$$

The vector arguments in (32.18) and (32.19) are mutually orthogonal, so that, combining both, using substitution (32.16) and using again the lemma 32.2, we have:

$$\left\langle \hat{P}_{\psi + \epsilon(a + a^{-1})\psi^{\perp\perp}} \right\rangle = 0. \qquad (32.20)$$

Now, if ϵ is less than $\frac{1}{2}$, there is a real a such that:

$$\epsilon(a + a^{-1}) = \pm 1. \qquad (32.21)$$

Therefore:

$$\left\langle \hat{P}_{\psi + \psi^{\perp\perp}} \right\rangle = \left\langle \hat{P}_{\psi - \psi^{\perp\perp}} \right\rangle = 0. \qquad (32.22)$$

The vectors $|\psi\rangle \pm |\psi^{\perp\perp}\rangle$ are orthogonal to each other; adding them and again using the lemma 32.2 we have:

$$\left\langle \hat{P}_\psi \right\rangle = 0, \qquad (32.23)$$

which contradicts the assumption (32.14a). Therefore $\epsilon > 1/2$. Q. E. D.

Now, using this theorem, we can prove the corollary 32.4.

Proof

In a dispersion-free state each projector has an expectation value which is given by either 0 or 1 [lemma 32.1]. It is clear from Eq. (32.9) that both values must occur, and, since no other values are possible, there must be arbitrarily close vectors $|\psi\rangle, |\varsigma\rangle$ with different expectation values 0 and 1 respectively. But we saw, by the I Bell theorem, that such vectors cannot be arbitrarily close (i.e. ϵ cannot be less than 1/2). Therefore there are no dispersion free states. Q. E. D.

32.3.4 Brief Discussion

It is possible to reject lemma 32.2 on this ground [BELL 1966, 8–9] [MERMIN 1993, 811–12]: $\hat{P}_{c'\varsigma_1+c''\varsigma_2}$ commutes with $\hat{P}_{\varsigma_1}$ and $\hat{P}_{\varsigma_2}$ only if, respectively, $c'' = 0, c' = 0$. Thus in general the measurement of $\hat{P}_{c'\varsigma_1+c''\varsigma_2}$ requires a distinct experimental arrangement (an objection similar to that of Bell directed at von Neumann). In other words it is implicitly assumed that the measurement of an observable must yield the same value independently of whichever other measurements may be made simultaneously, i.e. that measurement of an observable $\hat{O}$ yields the same value if it is measured as part of set $\hat{O}, \hat{O}'_1, \hat{O}''_1, \ldots$ of mutually commuting observables or of a second set $\hat{O}, \hat{O}'_2, \hat{O}''_2, \ldots$ of mutually commuting observables, though in general some observables of the second set fail to commute with some of the first set.

Hence corollary 32.4 supposes a non-contextuality of HV Theories. Is this assumption correct? It should be if, as we shall see [section 36.5], QM yields the same marginal distribution of an observable independently of whichever other measurements may be made simultaneously. On the other side the II Bell theorem [35.1: p. 591] excludes also contextual HV-Theories which satisfy a locality requirement.

Moreover, every local HV-Theory should be non-contextual, because it supposes that the value of each observable is already determined by HVs, independently of measurements on other observables[9]. In conclusion, if the non-contextuality of HV-Theories is accepted, there is no reason to deny the validity of the proof [see also subsection 32.5.2].

32.4 The Kochen/Specker Theorem

32.4.1 First Exposition

Proofs Another important consequence of Gleason Theorem is a theorem proved by Kochen and Specker [KOCHEN/SPECKER 1967]. We follow here a simplified proof given by Pitowsky [*PITOWSKY* 1989b, 109–117].[10] Let $\{a_1, a_2, \ldots, a_n, \ldots\}$ be a set of atomic propositions[11]. Define:

$$b(a_i, a_j) := \neg(a_i \wedge a_j), \tag{32.24a}$$
$$c(a_i, a_j, a_k) := (a_i \vee a_j \vee a_k). \tag{32.24b}$$

[9] This is the reason why Bohm was later forced to assume a fundamental non-locality of his theory [see subsection 32.6.2]. For contextual HV theories see [GUDDER 1970] [WIENER/SIEGEL 1955] [SIEGEL/WIENER 1956] [*BELTRAMETTI/CASSINELLI* 1981, 175].

[10] See also [*BELINFANTE* 1973, 35–39].

[11] Such that they are not analyzable in further propositions. Compound propositions are said to be 'molecular' propositions.

In QM $b(a_i, a_j)$ can be true even if both a_i, a_j are true — the case of non-commutability of the projections (on different subsystems) represented by propositions a_i, a_j. By de Morgan's Law we see that $b(a_i, a_j) = \neg a_i \vee \neg a_j$, which means that the disjunction (or sum) of a_i and a_j can be true even if both are false: this is the content of the choice disjunction which characterizes QM [see theorem 9.10: p. 171].

Then $c(a_i, a_j, a_k)$ can be true even when a_i, a_j, a_k are all false — by de Morgan's Law $\neg(a_i \wedge a_j) \leftrightarrow (\neg a_i \vee \neg a_j)$, suitable substitutions and the addition of an arbitrary element a_k. We now consider a proposition with ten atoms [see figure 32.2]:

$$\begin{aligned} d = d(a_1, a_2, \ldots, a_{10}) \;=\; & b(a_1, a_2) \wedge b(a_1, a_3) \wedge b(a_1, a_9) \wedge b(a_2, a_4) \wedge b(a_2, a_6) \\ & \wedge b(a_3, a_5) \wedge b(a_3, a_7) \wedge b(a_4, a_6) \wedge b(a_4, a_8) \\ & \wedge b(a_5, a_7) \wedge b(a_5, a_8) \wedge b(a_6, a_7) \wedge b(a_8, a_9) \wedge b(a_8, a_{10}) \wedge b(a_9, a_{10}) \\ & \wedge c(a_2, a_4, a_6) \wedge c(a_3, a_5, a_7) \wedge c(a_8, a_9, a_{10}). \end{aligned} \tag{32.25}$$

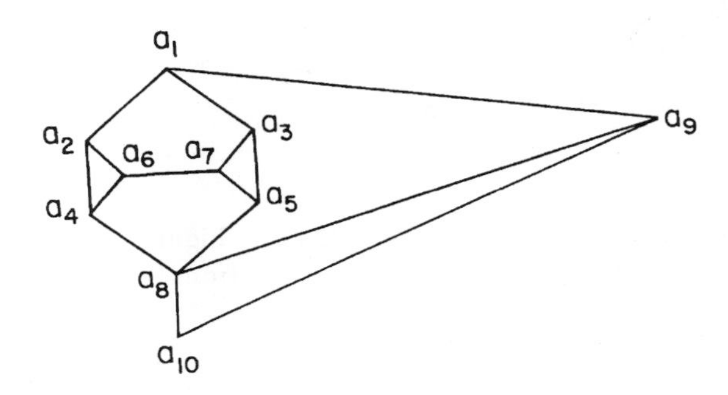

Figure 32.2: Graphic representation of the proposition (32.25) — from [*PITOWSKY* 1989b, 110].

The vertices are the atomic propositions. Two propositions a_i, a_j, such that $1 \leq i \leq j \leq 10$ are connected by an edge iff $b(a_i, a_j)$ is in Eq. (32.25). And three pairs $\{a_i, a_j\}, \{a_i, a_k\}, \{a_j, a_k\}$ such that $1 \leq i \leq j \leq k \leq 10$, are edges (i.e. $\{a_i, a_j, a_k\}$ is a triangle) iff $c(a_i, a_j, a_k)$ is in Eq. (32.25). Let $f : \{a_1, a_2, \ldots, a_{10}\} \mapsto \{0, 1\}$ be a classical truth proposition such that if $f(d) = 1$ then no pair of atomic propositions connected by an edge can have truth value 1, otherwise we would have classically $[f(a_i) = f(a_j) = 1 \rightarrow f(b(a_i, a_j)) = 0] \rightarrow f(d) = 0$. Now we state the following lemma (a simple evidence can be obtained by studying the graph 32.2):

Lemma 32.3 (Pitowsky) *If $f : \{a_1, a_2, \ldots, a_{10}\} \mapsto \{0, 1\}$ is a classical truth function such that $f(d) = 1$ and $f(a_1) = 1$, then $f(a_{10}) = 1$.*

We now prove the following Theorem (= KS theorem):

Theorem 32.4 (Kochen/Specker) *Consider the following proposition*

$$e = \left[\bigwedge b(a_i, a_j)\right] \wedge \left[\bigwedge c(a_i, a_j, a_k)\right], \tag{32.26}$$

composed of 117 *atomic propositions* [see figure 32.3], *where each $\{a_i, a_j\}$ is an edge and each $\{a_i, a_j, a_k\}$ is a triangle. e is a classical logical falsity, but true in QM.*

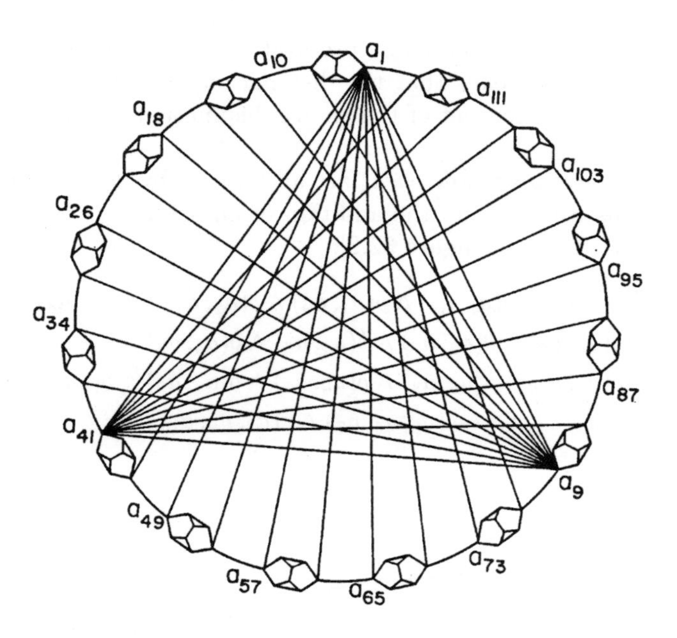

Figure 32.3: Graphic representation of proposition (32.26). Figure 32.2 represents a detail of the present one (consider especially the propositions a_1, a_9, a_{10}) — from [*PITOWSKY* 1989b, 112].

Proof

Suppose that $f(e) = 1$, then $f(c(a_1, a_9, a_{41})) = 1$ because the three propositions constitute a triangle. Then, classically, at least one of the three atomic propositions is true. Suppose now $f(a_1) = 1$ then by the lemma 32.3 also $f(a_{10}) = 1$, and thus by the same lemma we have $f(a_{18}) = 1$ and so on until $f(a_{41}) = 1$. But $\{a_1, a_{41}\}$ is an edge, so that $f(b(a_1, a_{41})) = 0$ and hence $f(e) = 0$: a contradiction.

That, on the other hand, $f(e) = 1$ can be realized in QM with the operators of angular momentum for three components $\hat{J}_x, \hat{J}_y, \hat{J}_z$ — which are the same as the components of spin given by Eq. (3.126). As we know the square of each component commutes with the square of any other, which means that simultaneous values of $\hat{J}_x^2, \hat{J}_y^2, \hat{J}_z^2$ can be measured. But, while the eigenvalues of $\hat{J}_x^2, \hat{J}_y^2, \hat{J}_z^2$ are either 0 or 1, we find that $\hat{J}_x^2 + \hat{J}_y^2 + \hat{J}_z^2 = 2\hat{I}$. Hence we conclude that in one and only one direction we can have a zero value. From here it can be seen that we can construct a situation such that $f(e) = 1$: given directions $r_1, r_2, \ldots, r_{117}$, the 117 propositions a_j correspond to 'the angular momentum is zero in the r_j direction'. Since we cannot then have zero for two orthogonal directions we must accept that $b(a_j, a_k)$ is true for all j, k such that $\{a_j, a_k\}$ is an edge. Moreover since we must have zero in one direction, then $c(a_i, a_j, a_k)$ is true whenever $\{a_i, a_j, a_k\}$ is a triangle. The truth of theorem 32.4 easily follows. Q.E.D.

Discussion In conclusion the KS theorem is a more refined form of corollary 32.4,[12] because acknowledging that in QM there can be value assignments which are not classical means acknowledging that we cannot have, for all states or values of observables, projectors with 0 or 1 value (which is the classical assignment), which in turn signifies that there cannot be dispersion-free

[12]See also [A. FINE/TELLER 1978].

states or observables.

We have here another confirmation of the existence of truth gaps in QM [see section 9.5]. On the other hand the KS-Theorem confirms the non-classical structure of quantum probability [see section 11.3 and theorem 14.6: p. 235; see also chapter 39].

Belinfante [*BELINFANTE* 1973, 40–43] criticized the KS theorem because it supposes that there is an observable for any operator[13] and that one cannot always evaluate a value of an observable independently from the choice of a concrete orthonormal set (the problem of contextuality), and, according to Belinfante, both assumptions are not necessarily acknowledged by an HV-Theory. But the first assumption is not made in the proof already seen — because we have shown a concrete counterexample in which the proposition (32.26) is a QM truth, while it is always a classical falsity —, and the second is reasonable for an HV theory because (as we have seen for the corollary 32.4) the latter should always suppose determined values of the observables independently from the choice of a set or of an experimental context. Anyway in the next section we shall see some proofs which partly overcome the problem of contextuality.

32.4.2 Brown/Svetlichny Theorem*

In [H. BROWN/SVETLICHNY 1990, 1382–84] it is possible to find an interesting direct application to the case of EPRB-Exp. [p. 536]. We mathematically reformulate some principles already analysed by EPR. Consider two separated systems defined by $|\Psi\rangle$ which have interacted in the past and both depend on a parameter λ_{HV}. We reformulate principle 31.1 in the following way: Any observables $\hat{O}_1 \otimes \hat{I}$ and $\hat{I} \otimes \hat{O}_2$ are single-valued functions of λ_{HV} with respective values symbolized by $[O_1 \otimes \hat{I}]$ and $[\hat{I} \otimes O_2]$, which result from separated measurements of $\hat{O}_1$ on the first subsystem and of $\hat{O}_2$ on the second one. Suppose now that we acknowledge the non-contextuality of the theory, i.e. $[O_1 \otimes \hat{I}]$ and $[\hat{I} \otimes O_2]$ are insensitive to the choice of measurements of any mutually incompatible observables on the separated system. Then we may formulate the following proposition:

Proposition 32.2 (Excluded Joint Events) *If $\wp^{\Psi}(\hat{O}_1 = o_1, \hat{O}_2 = o_2)$ denotes the joint probability in QM of obtaining, for a state $|\psi\rangle$ of the system $\mathcal{S}_1 + \mathcal{S}_2$, the pair of values o_1, o_2 in the measurement of $\hat{O}_1, \hat{O}_2$, performed respectively on subsystems $\mathcal{S}_1, \mathcal{S}_2$, then*

$$\{\wp^{\Psi}(\hat{O}_1 = o_1, \hat{O}_2 = o_2) = 0\} \rightarrow \{([O_1 \otimes \hat{I}] \neq o_1) \vee ([\hat{I} \otimes O_2] \neq o_2)\}. \tag{32.27}$$

Now we prove the following theorem:

Theorem 32.5 (Brown/Svetlichny) *A non-contextual theory based on principle* 31.1 *and proposition* 32.2 *is inconsistent with QM.*

Proof

Consider the triad θ of mutually orthogonal directions x, y, z and the locally maximal [see p. 37] observable $\hat{H}^{\theta} \otimes \hat{I}$, where on $\mathcal{H}_1$ we have:

$$\hat{H}^{\theta} = h_x \hat{P}_{|\hat{\sigma}_x = 0\rangle} + h_y \hat{P}_{|\hat{\sigma}_y = 0\rangle} + h_z \hat{P}_{|\hat{\sigma}_z = 0\rangle}, \tag{32.28}$$

and h_j for $j = x, y, z$ are real numbers, $\hat{P}$ is the projection operator and $\hat{\sigma}_i$ for $j = x, y, z$ is the component of spin in the j direction. If we choose $\hat{H}^{\theta}$ to be the spin Hamiltonian operator $a\hat{\sigma}_x^2 + b\hat{\sigma}_y^2 + c\hat{\sigma}_z^2$, then

[13] On this point see [VAN FRAASSEN 1973] [HEYWOOD/REDHEAD 1983] [*REDHEAD* 1987b, 133–37].

$h_x = b + c, h_y = a + c, h_z = a + b$. Consider now the observable $1 \otimes \hat{\sigma}_{\mathbf{n}}$ the component of spin in the direction $\mathbf{n}$ for subsystem $\mathcal{S}_2$. Now for all $\theta, \mathbf{n}$ the predictions $[\hat{H}^\theta \otimes \hat{I}], [\hat{I} \otimes S_{\mathbf{n}}]$ are well defined (for the separability), and $[\hat{H}^\theta \otimes \hat{I}] \in \{h_x, h_y, h_z\}, [\hat{I} \otimes \hat{\sigma}_{\mathbf{n}}] \in \{-1, 0, 1\}$. For the non-contextuality we have that $[\hat{H}^\theta \otimes \hat{I}]$ is unaffected by the choice of measurement on $\mathcal{S}_2$. Suppose that the composite system is in a singlet state:

$$|\Psi\rangle = -\frac{1}{\sqrt{3}}[|\hat{\sigma}_x = 0\rangle \otimes |\hat{\sigma}_x = 0\rangle - |\hat{\sigma}_y = 0\rangle \otimes |\hat{\sigma}_y = 0\rangle + |\hat{\sigma}_z = 0\rangle \otimes |\hat{\sigma}_z = 0\rangle]. \quad (32.29)$$

Suppose furthermore that $[\hat{H}^\theta \otimes \hat{I}] = h_x$. Since we have:

$$\wp^\Psi(\hat{H}^\theta = h_x, \hat{\sigma}_x = \pm 1) = 0 \quad (32.30a)$$

and

$$\wp^\Psi(\hat{H}^\theta = h_x, \hat{\sigma}_{y,z} = 0) = 0, \quad (32.30b)$$

from the proposition 32.2 and letting $\wp'$ represent the joint of the two probabilities (32.30),we obtain:

$$\wp' \to \{([(\hat{H}^\theta \neq h_x]) \vee [(\hat{\sigma}_x = 0) \wedge (\hat{\sigma}_{y,z} = 1, -1)]\}. \quad (32.31)$$

But, since we assumed Eqs. (32.30) and assumed $\hat{H}^\theta = h_x$, we have as a consequence of (32.31) that:

$$[\hat{\sigma}_x = 0] \wedge [\hat{\sigma}_{y,z} = 1, -1]. \quad (32.32)$$

Thus to each of the spatial directions x, y, z in θ we can assign a number $[\hat{I} \otimes \hat{\sigma}_{\mathbf{n}}], \mathbf{n} = x, y, z$ from the set $\{-1, 0, 1\}$ such that zero is assigned to one and only one direction. But now we utilize the KS Theorem to show that we cannot find a classical assignment of this type. Q. E. D.

Though we have proved the theorem, we do not integrate it into the theory, because it brings nothing new, though it is important for the physical concreteness of the matter.

Brown and Svetlichny preferred to use a more visual version of the KS theorem due to Redhead [*REDHEAD* 1987b, 123–31], consisting in coloring a unit hypersphere in Hilbert space. Practically we translate the problem posed by the projectors associating to the value 1 the colour *red* and to the value 0 the colour *blue*. The question now is: can we colour the hypersphere with the colours red and blue in such a way that the following conditions are satisfied?:

- Every point (unit vector) is coloured red or blue.
- For every complete orthogonal set of unit vectors only one is coloured red.
- Unit vectors belonging to the same ray have the same colour.

Redhead proved the following theorem:

Theorem 32.6 (Two-Colour Theorem) *If the dimension of the Hilbert space is greater than two, the colouring of the unit hypersphere in the way described is not possible.*

KS Theorem states that it is impossible to assign determined and independent values to all quantum states and the Brown/Svetlichny theorem demonstrates, by means of KS Theorem or of the Two-Colour theorem, a direct contradiction to principle 31.1.[14]

[14]For other versions of the KS Theorem (which generally simplify it) see also [PERES 1991] (33 vectors instead of 117) [MERMIN 1993, 808–811] [KERNAGHAN 1994] (20 vectors) [CERECEDA 1995] [BUB 1996] [CABELLO *et al.* 1996a] (18 vectors) [*BUB* 1997, 82–95] [CABELLO/GARCÍA-A. 1998] — the last book quoted is also a review of several attempts.

32.5 Other Proofs

The following proofs are less general than the KS-Theorem but very interesting because they allow measurements on distant subsystems only of commutable observables, hence bypassing the problem of non-contextuality.

32.5.1 Non-local Measurement

The first examination of the problem was given by Peres [PERES 1990a].[15] Consider two space-like separated spin-$\frac{1}{2}$ particles in a singlet state. The result of the measurement of $\hat{\sigma}_{1x}$, (± 1) will be called $\mathrm{v}(\hat{\sigma}_{1x})$ (where 'v' stands for 'value of') and similar notations are used for the other components. We now measure a product $\hat{\sigma}_{1x}\hat{\sigma}_{2x}$, either by measuring $\hat{\sigma}_{1x}$ and $\hat{\sigma}_{2x}$ separately, with result $\mathrm{v}(\hat{\sigma}_{1x})\mathrm{v}(\hat{\sigma}_{2x})$, or in a single non-local procedure. The result of the latter must be -1 because $\langle\hat{\sigma}_{1x}\hat{\sigma}_{2x}\rangle = -1$ in a singlet state. It follows that

$$\mathrm{v}(\hat{\sigma}_{1x}\hat{\sigma}_{2x}) = -1; \quad \mathrm{v}(\hat{\sigma}_{1y}\hat{\sigma}_{2y}) = -1; \quad \mathrm{v}(\hat{\sigma}_{1z}\hat{\sigma}_{2z}) = -1. \tag{32.33}$$

Now consider a measurement of $\hat{\sigma}_{1x}\hat{\sigma}_{2y}$. If we measure $\hat{\sigma}_{1x}$ and $\hat{\sigma}_{2y}$ separately, we have results $\mathrm{v}(\hat{\sigma}_{1x})$ and $\mathrm{v}(\hat{\sigma}_{2y})$. In all these situations we assume that a measurement on one particle has no influence at all on the other one (separability). If we measure non-locally $\hat{\sigma}_{1x}\hat{\sigma}_{2y}$ we have result $\mathrm{v}(\hat{\sigma}_{1x}\hat{\sigma}_{2y})$. Remember that

$$[\hat{\sigma}_{1x}\hat{\sigma}_{2y}, \hat{\sigma}_{1y}\hat{\sigma}_{2x}] = 0, \tag{32.34}$$

so that we can make a non-local measurement of $\hat{\sigma}_{1y}\hat{\sigma}_{2x}$ together with a non-local measurement of $\hat{\sigma}_{1x}\hat{\sigma}_{2y}$ without mutual disturbance.

Now we have [see matrices (3.125)]:

$$\hat{\sigma}_{1x}\hat{\sigma}_{2y}\hat{\sigma}_{1y}\hat{\sigma}_{2x} = \hat{\sigma}_{1x}\hat{\sigma}_{1y}\hat{\sigma}_{2y}\hat{\sigma}_{2x} = \hat{\sigma}_{1z}\hat{\sigma}_{2z}, \tag{32.35}$$

whose expectation value is $\langle\hat{\sigma}_{1z}\hat{\sigma}_{2z}\rangle = -1$. On the hypothesis of factorization, it follows that:

$$\mathrm{v}(\hat{\sigma}_{1x}\hat{\sigma}_{2y})\mathrm{v}(\hat{\sigma}_{1y}\hat{\sigma}_{2x}) = \langle\hat{\sigma}_{1x}\hat{\sigma}_{2y}\hat{\sigma}_{1y}\hat{\sigma}_{2x}\rangle = -1. \tag{32.36}$$

Eq. (32.36) is in contradiction with Eq. (32.33) in which $\mathrm{v}(\hat{\sigma}_{1x}\hat{\sigma}_{2x}) = \mathrm{v}(\hat{\sigma}_{1x}\hat{\sigma}_{2x}) = -1$. So that, according to Peres, QM violates the Separability principle. Peres saw this proof only as an exhibition of the non-separability feature of QM — to which we shall return in the next part. But Mermin[16] showed that this proof can be generalized as an optimal KS-like theorem bypassing the difficulty deriving from contextuality. However, the proof has less generality than the KS-Theorem itself. This will be the subject of the next subsection.

32.5.2 Mermin's Generalization

It is possible to obtain a more general proof of Peres' results — without supposing a singlet state — by employing the following nine operators [MERMIN 1990b]:[17]

$$\begin{array}{ccc} \hat{\sigma}_{1x} & \hat{\sigma}_{2x} & \hat{\sigma}_{1x}\hat{\sigma}_{2x} \\ \hat{\sigma}_{2y} & \hat{\sigma}_{1y} & \hat{\sigma}_{1y}\hat{\sigma}_{2y} \\ \hat{\sigma}_{1x}\hat{\sigma}_{2y} & \hat{\sigma}_{1y}\hat{\sigma}_{2x} & \hat{\sigma}_{1z}\hat{\sigma}_{2z}. \end{array} \tag{32.37}$$

[15] See also [*PERES* 1993, 151–52].
[16] As later acknowledged by Peres himself [PERES 1996b].
[17] See also [HARDY 1991a, 3] [STAPP 1980, 774].

The three operators in each row and in each column commute, and the value of their product is 1, except for that of the last column, which is -1. However there is no way to assign numerical values ± 1 to the operators with this multiplicative property.

We work in the eight dimensional space of three spins 1/2, and attempt to assign values to the ten operators: $\hat{\sigma}_{1x}, \hat{\sigma}_{1y}, \hat{\sigma}_{2x}, \hat{\sigma}_{2y}, \hat{\sigma}_{3x}, \hat{\sigma}_{3y}, \hat{\sigma}_{1x}\hat{\sigma}_{2y}\hat{\sigma}_{3y}, \hat{\sigma}_{1y}\hat{\sigma}_{2x}\hat{\sigma}_{3y}, \hat{\sigma}_{1y}\hat{\sigma}_{2y}\hat{\sigma}_{3x}, \hat{\sigma}_{1x}\hat{\sigma}_{2x}\hat{\sigma}_{3x}$. In QM we have the five constraints:

$$\mathrm{v}(\hat{\sigma}_{1x}\hat{\sigma}_{2y}\hat{\sigma}_{3y})\mathrm{v}(\hat{\sigma}_{1x})\mathrm{v}(\hat{\sigma}_{2y})\mathrm{v}(\hat{\sigma}_{3y}) = 1; \tag{32.38a}$$

$$\mathrm{v}(\hat{\sigma}_{1y}\hat{\sigma}_{2x}\hat{\sigma}_{3y})\mathrm{v}(\hat{\sigma}_{1y})\mathrm{v}(\hat{\sigma}_{2x})\mathrm{v}(\hat{\sigma}_{3y}) = 1; \tag{32.38b}$$

$$\mathrm{v}(\hat{\sigma}_{1y}\hat{\sigma}_{2y}\hat{\sigma}_{3x})\mathrm{v}(\hat{\sigma}_{1y})\mathrm{v}(\hat{\sigma}_{2y})\mathrm{v}(\hat{\sigma}_{3x}) = 1; \tag{32.38c}$$

$$\mathrm{v}(\hat{\sigma}_{1x}\hat{\sigma}_{2x}\hat{\sigma}_{3x})\mathrm{v}(\hat{\sigma}_{1x})\mathrm{v}(\hat{\sigma}_{2x})\mathrm{v}(\hat{\sigma}_{3x}) = 1; \tag{32.38d}$$

$$\mathrm{v}(\hat{\sigma}_{1x}\hat{\sigma}_{2x}\hat{\sigma}_{3x})\mathrm{v}(\hat{\sigma}_{1x}\hat{\sigma}_{2y}\hat{\sigma}_{3y})\mathrm{v}(\hat{\sigma}_{1y}\hat{\sigma}_{2x}\hat{\sigma}_{3y})\mathrm{v}(\hat{\sigma}_{1y}\hat{\sigma}_{2y}\hat{\sigma}_{3x}) = -1. \tag{32.38e}$$

Once again, since the eigenvalues of all the operators are ± 1 and since each value appears twice in Eqs. (32.38), the product of all five Eqs. (32.38) must be $+1$, which is impossible due to the value of each of the same Eqs. separately. Hence, the argument shows the impossibility of assigning precise values to all operators in Eq. (32.38). Therefore, we overcome in this way the difficulties of a non-local measurement, which could be questionable in the view of EPR: with the argument of Mermin[18] there is no such problem because the four operators $\hat{\sigma}_{1x}\hat{\sigma}_{2y}\hat{\sigma}_{3y}$, $\hat{\sigma}_{1y}\hat{\sigma}_{2x}\hat{\sigma}_{3y}$, $\hat{\sigma}_{1y}\hat{\sigma}_{2y}\hat{\sigma}_{3x}$, $\hat{\sigma}_{1x}\hat{\sigma}_{2x}\hat{\sigma}_{3x}$ are mutually commuting, so that all can be assigned definite values by an appropriate choice of the state vector[19]. However, Garola [GAROLA 1998b] showed that the proof cannot completely bypass the problem of contextuality — see for example the discussion about the I Bell theorem [subsection 32.3.4] —, because it is generally not the case that the operators of a set commute which all operators of the other sets — where each set is represented by a single Eq. of the (32.38).

32.6 Interpretation of Hidden-Variable Theories

This section is centered on the interpretational issues of the HV-Theory: in fact the last sections presented an HV-Theory only on a mathematical level and no physical model was discussed. As remarked by Pagonis and Clifton [PAGONIS/CLIFTON 1995], many of the proofs and discussions about the HV-Theories make no reference at all to Bohm's theory, a fact which explains a certain gap between the previous sections and the present one. On the other hand, the historical merit of Bohm was to have reformulated in a more concrete physical model the EPR argument [see section 31.3] and to have proposed in 1952 an HV-Theory allowing a new discussion of the problem in the fifties and sixties after von Neumann's first examination in 1932 [see section 32.2 and also subsection 28.1.1].

32.6.1 Bohm's Basic Theory

Copenhagen interpretation Bohm was the first proponent of an HV-Theory for QM. Bohm's proposal [BOHM 1952, 371–72] originated from a sharp criticism — in the sense of EPR — of the Copenhagen interpretation: the wave function cannot be the best description of microreality.

---

[18] And that of GHSZ [see section 36.4]

[19] A simpler result is in [MERMIN 1981] and [MERMIN 1990c]. Another example is in [SVETLICHNY 1987] — based on Leggett's article [LEGGETT 1980].

On the other hand, we have to do with individuals, and so each Statistical interpretation is useless [BOHM/BUB 1966a, 456, 459]. Hence, in standard QM (a hybrid constituted by Basic-QM plus Copenhagen plus Standard interpretations) the relationship between individual systems and ensembles is not at all clear, which gives to QM the aspect of being more a mathematical algorithm for calculating probabilities of experimental results than that of a physical theory[20].

Against the completeness claimed for QM, Bohm recalled that the theory had already some *ad hoc* extensions in the case of high energies and little distances [BOHM/BUB 1966a, 453]. However, we know that this is not a decisive argument against the theory [see subsection 9.4.2].

It is evident from Bohm's analysis that building a deterministic theory of microentities is not the only aim of his proposal, but he also tried to guarantee some form of objectivity independent of the observer[21].

Quantum potential About the co-presence of waves and of particles, Bohm's proposal is substantially the same of de Broglie. But, in order to account for their reciprocal relationships and to face the measurement problems which QM poses, Bohm developed to a great extent the concept of *Quantum potential* [BOHM 1952, 372–75].[22] Bohm started with the usual wave Eq. (3.21) [see also Eq. (23.19)] and wrote the Eqs. for amplitude and phase (both real) in the following form:

$$\frac{\partial \vartheta}{\partial t} = -\frac{1}{2m}[\vartheta \nabla^2 \varphi + 2\nabla\vartheta \cdot \nabla\varphi], \qquad (32.39a)$$

$$\frac{\partial \varphi}{\partial t} = -\left[\frac{(\nabla\varphi)^2}{2m} + V(\mathbf{r}) - \frac{\hbar^2}{2m}\frac{\nabla^2\vartheta}{\vartheta}\right]. \qquad (32.39b)$$

In the classical limit the phase φ is a solution of the Hamilton/Jacobi Eq. In this case we can consider the probability density $\wp(\mathbf{r}) = \vartheta^2(\mathbf{r})$ as a probability density for ensembles of particles (similar to the classical stochastic interpretation) — we shall return to this point in the next subsection.

From Eqs. (32.39) we derive a QM equivalent of the Hamilton/Jacobi Eq. for an ensemble of particles, a Bohm/Jacobi Eq. [see its relativistic counterpart (28.12)]:

$$\frac{\partial \varphi}{\partial t} + \frac{(\nabla\varphi)^2}{2m} + V(\mathbf{r}) - \frac{\hbar^2}{4m}\left[\frac{\nabla^2\wp}{\wp} - \frac{1}{2}\frac{(\nabla\wp)^2}{\wp^2}\right] = 0, \qquad (32.40)$$

where $\wp = \vartheta^2$. Hence, in the case of QM, we assume that the particles are acted upon not only by the 'classical' potential $V(\mathbf{r})$, but also by a Quantum potential — which is the term in square brackets in Eq. (32.40) —:

$$V_Q(\mathbf{r}) = \frac{-\hbar^2}{4m}\left[\frac{\nabla^2\wp}{\wp} - \frac{1}{2}\frac{(\nabla\wp)^2}{\wp^2}\right] = -\frac{\hbar^2}{2m}\frac{\nabla^2\vartheta}{\vartheta}. \qquad (32.41)$$

It is evident that the Quantum potential guarantees that the position of a particle is always 'determined'.

---

[20]See also [BELL 1975, 51] [EINSTEIN 1949b, 671–72] [EINSTEIN 1953a, 12–14].

[21]See [BELL 1982, 160] [BELL 1984b, 173].

[22]See also [*BOHM/HILEY* 1993, 28–30] [*HOLLAND* 1993, 72–74].

Therefore, following de Broglie, Bohm acknowledged a guidance formula [see Eq. (28.18)] for the particle. Its velocity is given by [*BOHM/HILEY* 1993, 215–16, 272]:[23]

$$\mathbf{v} = \frac{\mathbf{j}}{\hat{\rho}}, \tag{32.42}$$

where $\mathbf{j}$ is the Dirac current given by

$$\mathbf{j} = \langle \psi | \alpha \psi \rangle, \tag{32.43}$$

where $\alpha = \{\alpha^k\}$ and $\alpha^k = \gamma^0 \gamma^k, k = 1, 2, 3$ [see section 4.2].

But the most important difference between Bohm and de Broglie is that in Bohm's proposal a non-linear theory is practically absent. He insisted that the field which acts through the Quantum potential has no source, in order to avoid that the wave function becomes non-linear [*BOHM/HILEY* 1993, 30, 345–46].

According to Bohm, the most important evidence for the existence of the Quantum potential is given by the Aharonov/Bohm-effect[24] [see figure 32.4]. Though the AB-Effect can be explained in pure quantum-mechanical terms [see section 41.2], it is one of the most important contributions of Bohm to have discovered it. It is another evidence for that the struggle for interpreting a theory, even if guided by wrong principles, can give rise to very important results — the same is valid for EPR. In the article of 1952 the theory is then applied to the problem of the hydrogen atom — already discussed by de Broglie [*DE BROGLIE* 1956, 121, 181] — and of a potential barrier [see figure 32.5].[25]

32.6.2 Bohm's Later Interpretation

We shall see in the next subsections how Bohm tried to develop a more global interpretation and to give a better conceptual basis to this basic proposal. The *summa* of all these efforts is the book [*BOHM/HILEY* 1993] — but we shall see in short some precedents too.

Hydraulic Model

At the beginning Bohm together with Vigier [BOHM/VIGIER 1954] proposed a hydraulic model, on the line of that of Madelung [see subsection 23.1.1]: the density matrix was interpreted as the density of a fluid; then the Quantum potential can arise from the effect of an internal stress in the fluid. The particle is a highly localized inhomogeneity.

Successively, it was proved that it is possible to deduce the QM probability density from the assumption of a fluid[26]. Note that, independently of this assumption, which Bohm later will fail to make, it is difficult to explain how the ψ–field should act as guidance for a single particle while determining the density of its 'sisters' — and without the standard distribution $|\psi|^2$ one cannot reproduce the statistics predicted in QM for an ensemble of particles [ANANDAN/BROWN 1995, 259].

[23] See also [*HOLLAND* 1993, 503–509].

[24] Analysis in [PHILIPPIDIS *et al.* 1982] [*BOHM/HILEY* 1993, 51–52] .

[25] In [BOHM/SCHILLER 1955] and [BOHM *et al.* 1955] an application to the spin problem (Pauli equation) is discussed.

[26] See also [BOHM 1953a] [*BOHM/HILEY* 1993, 182–84] [DÜRR *et al.* 1992a, 856–60].

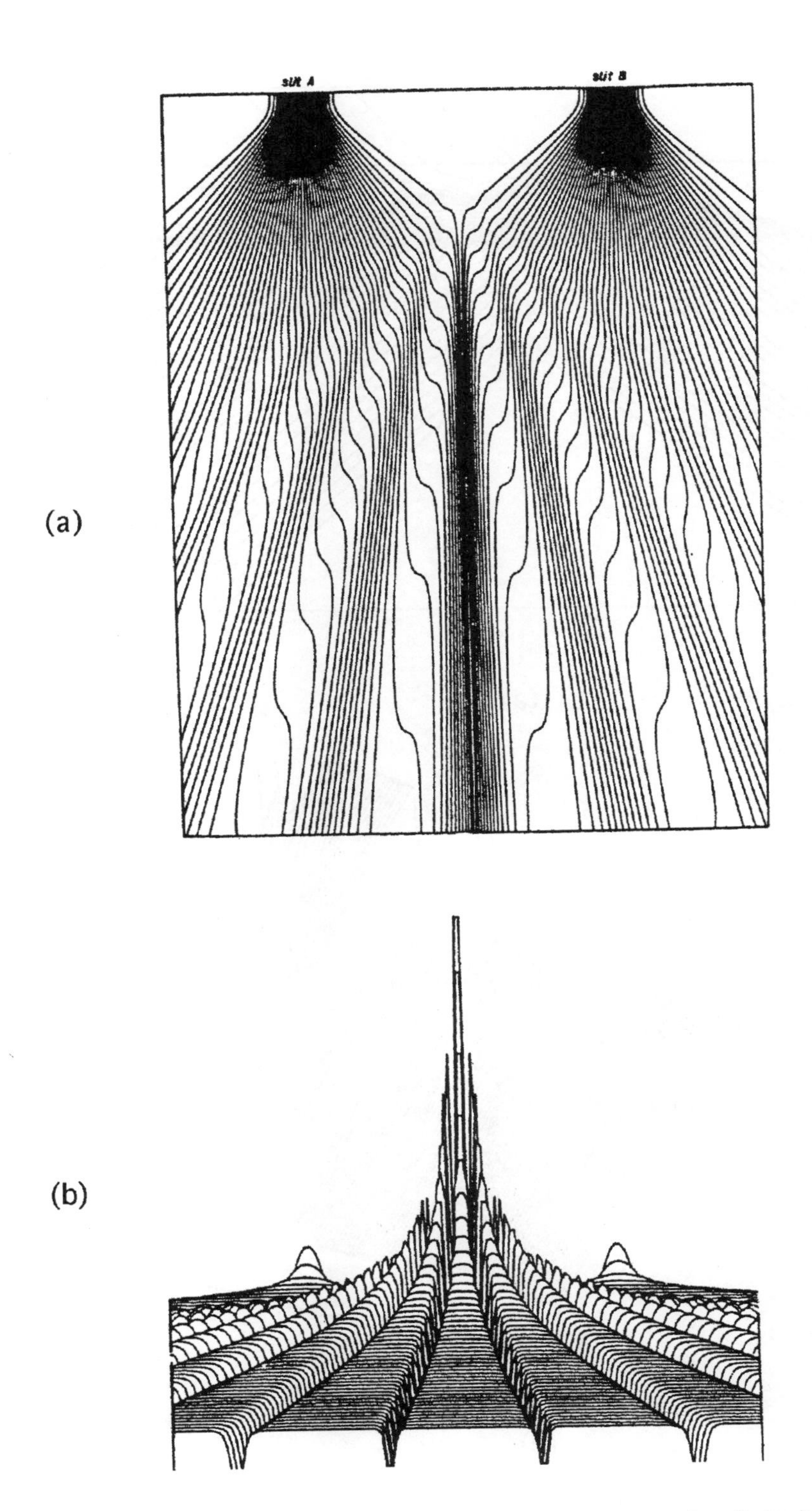

Figure 32.4: (a) Particle trajectories for two Gaussian slit systems after Bohm's model, and (b) relative Quantum potential. It is interesting that the same figures are also utilized by Bohm/Hiley for the AB-effect — from [*BOHM/HILEY* 1993, 33–34].

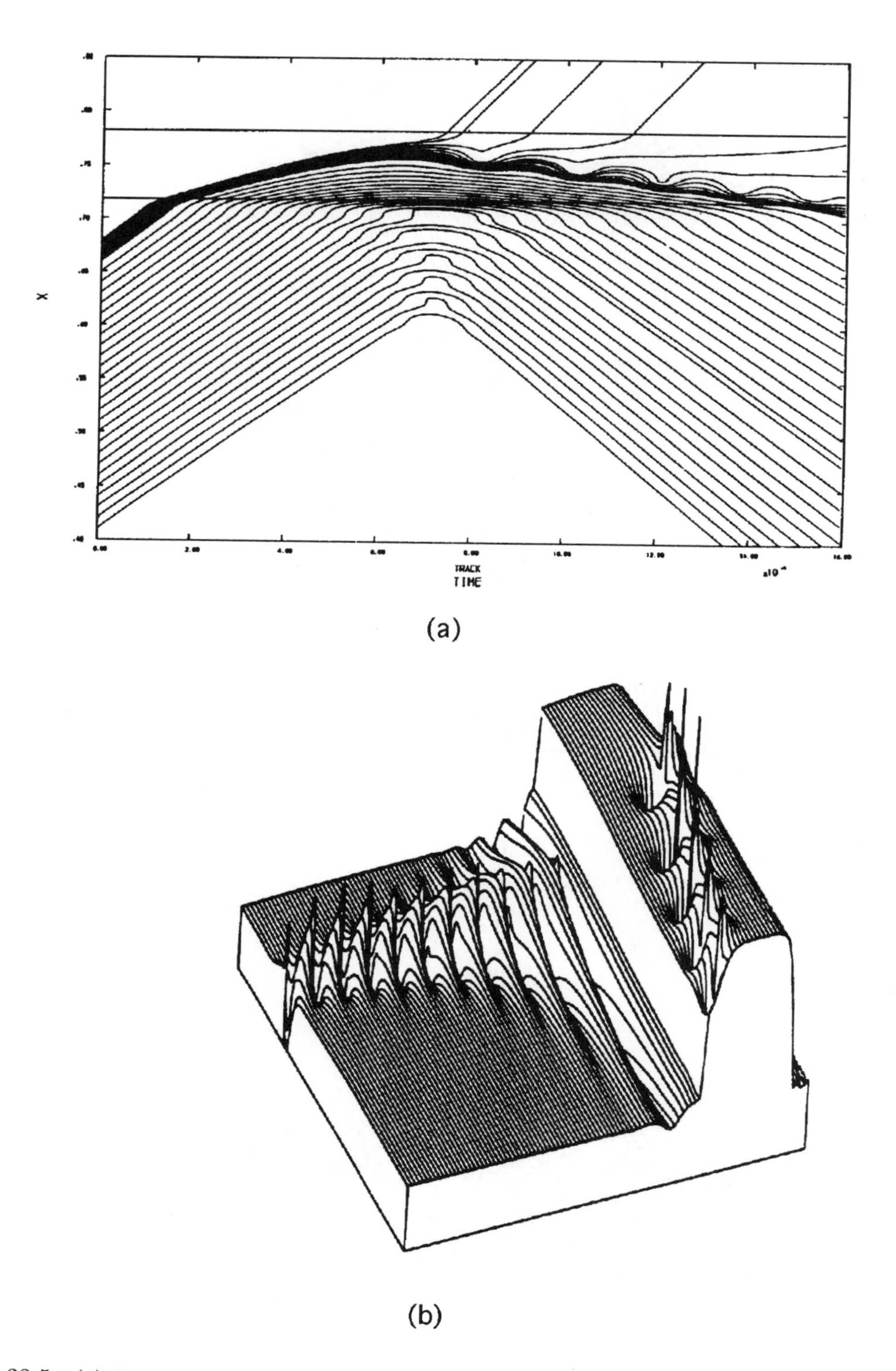

Figure 32.5: (a) Trajectories for a potential barrier ($E = V/2$) after Bohm's model, and (b) corresponding Quantum potential — from [*BOHM/HILEY* 1993, 76–77].

Quantum Potential and Active Information

But the greatest difficulty for the Hydraulic HV model comes from the Quantum potential. In fact by the Quantum potential the wave or field ψ appears both in the numerator and in the denominator of Eq. (32.41), so that the Quantum potential is not changed at all if multiplied by an arbitrary constant, i.e. the effect of the Quantum potential is independent of the strength of the field [*BOHM/HILEY* 1993, 31]. If so, then it is impossible to conceive of the Quantum potential as a form of physical energy. Therefore, in [BOHM/HILEY 1984, 260], and in a more general form in [*BOHM/HILEY* 1993, 35–37], the Quantum potential was interpreted as a type of information, an *active information* — so-called in order to distinguish it from the potential information represented by the entropy [see chapter 42]. It is evident that this interpretation is suggested by de Broglie's concept of the Pilot wave. The active information, according to Bohm, is a form of action where the strength of the signal is not important but only the form, or the structure of the message is[27]. Bohm also thought that the multidimensionality of information corresponds to a certain extent to the structure of Hilbert spaces [*BOHM/HILEY* 1993, 61]. But it is evident that in this way we introduce into the frame of physics an entity which is not physical at all — an action or effect which seems to be unquantifiable —, and, if we respect methodological rule M1 [p. 6], it should not be allowed.

Wholeness

Now, as a consequence of this step, since the strength of the action does not decrease with distance, Bohm was forced to affirm a strong form of non-locality [*BOHM/HILEY* 1993, 57, 62]. The reasons are also different: it was largely proven that an HV-Theory must necessarily be non-local in some form [see chapters 35 and 36]. This position was strengthened in a form of universal Holism, where everything conspires with everything [BOHM/HILEY 1981, 538–40] [BOHM/HILEY 1984, 260] [*BOHM/HILEY* 1993, 38, 58, 350–57].[28] In order to account for such holism in physical terms, the concept of ether was introduced [BOHM/HILEY 1984, 270]: a field of superluminal connections with a privilegiate frame of reference[29].

But even if we do not follow such speculations, the fact is that the non-locality and the holism — as consequences of the theory of active information — imply violation of the Lorentz invariance [*BOHM/HILEY* 1993, 282–85, 289–93, 350].[30] The problem is the ontological interpretation of the multiparticle Dirac Eq. [see Eq. (4.12)]. It is true that Bohm and Hiley said that this violation only happens in one privilegiate frame; but this is no better — see methodological rule M3 [p. 6]. These authors actually propose a new theory which goes further than QM and the relativity: it should be centered on a universal frame of reference where the non-local effects are factually local ones but with superluminal speed [see figure32.6]. It is again evident that the reasons for this step are not only the inner logic of the theory but also that it was proved — as we shall see in the next part — that an HV theory must necessarily violate the requirement of locality.

A variant of the theory has been proposed by Bohm himself [*BOHM* 1980, 80–88] and Vigier [VIGIER 1982]: even if there is direct interaction between elements of this field and space-like

[27]Incidentally a form of Platonism or of Cartesian semiotic [*CHOMSKY* 1966] [AULETTA 1992].

[28]See also [DÜRR *et al.* 1992a, 858].

[29]This has given rise to a number of speculations which seem to have very little to do with science [CUFARO-P./VIGIER 1979] [CUFARO-P./VIGIER 1982] [CUFARO-P. *et al.* 1984b].

[30]See also [BELL 1981a, 133] [*ALBERT* 1992, 160].

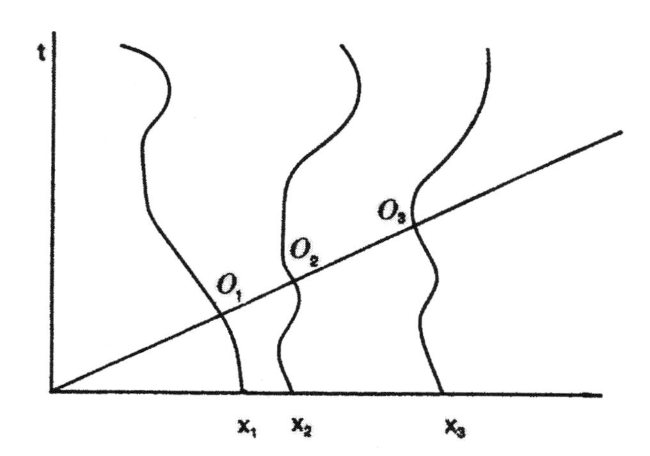

Figure 32.6: Trajectory intersection of hyperplane $t' =$ constant: three successive measurements (at t_1, t_2, t_3 with the three results o_1, o_2, o_3 respectively) are considered contemporary — from [*BOHM/HILEY* 1993, 284].

separated systems, such that there are non-local connections, fluctuations do not allow communication of ordered configurations. Vigier in fact proposed a ripe model [VIGIER 1991], in which particles are treated as joint gravitational, electromagnetic and wave soliton-like singularities, piloted by their surrounding external gravity and electro-magnetic forces and by the associated wave fields (where the latter are also subjected to external gravitational and electromagnetic forces). In this model it is not assumed that the density associated with the state of the particles is proportional to $|\psi|^2$ (as happened in early Bohm's model), but rather that the wave only represents an average behaviour, where each individual particle follows random stochastic Feynman-like paths with positive probability weights. Obviously (again in contrast with preceding Bohm's attempts), Vigier proposed a non-linear equation — a modification of the Klein/Gordon Eq. (4.10).

Later [section 36.6] we shall discuss whether or not it is possible to find some violations of Lorentz invariance implied by HV Theories.

Quantum Potential and Measurement

The Quantum potential was interpreted very early as the solution to the Measurement problem. The model was developed systematically in [PHILIPPIDIS *et al.* 1979] and extended in [*BOHM/HILEY* 1993, 75, 78–82, 99–100, 105–108].[31] The Quantum potential always depends on the whole system. If we start with the model of a beam of particles banged against a potential barrier, they separate after the clash with the potential, each one entering in a particular 'channel' independently from the others, so that the situation is now that of separated systems — transmitted or reflected particles. In order to create new interference between them, this should be produced in the configuration space of all subsystems that had interacted before the split, which in turn would imply allowing all channels interact again, which is normally impossible. Irreversibility derives from this point.

[31]Some antecedents can also be found in [*DE BROGLIE* 1956, 182–88] [*DE BROGLIE* 1957, 81–86, 89–92] [see subsection 28.1.4].

Let us now apply this model to the Measurement problem. The most important point here is the large number of particles of an apparatus $\mathcal{A}$ [see also theorem 16.2: p. 256]. The transition to an eigenstate is understood as the entering of the particle into a particular channel. Hence the potential global information is reduced by a dispersion: the other channels are made inactive. But, on the other hand, there is no way to predict into which individual channel the particle will enter. Hence there is no way to speak of the property of the particle independently of $\mathcal{A}$. This is another aspect of holism. But the most important consequence of this step is to render the theory very near to the original Copenhagen interpretation that Bohm had so strongly criticized in his article of 1952 — in fact in his last book, in collaboration with Hiley, he acknowledged CP [*BOHM/HILEY* 1993, 107]. And it may be asked if this whole theory is necessary if we can obtain the same result with a more economic one — see methodological rule M7 [p. 6].

However, that Bohm's interpretation is not so neutral for QM can be demonstrated by the following example. Recently Leavens [LEAVENS 1995, 247–57] proved in a very detailed study that, due to the fundamental assumption of Bohm's theory that two different trajectories never intersect or touch each other, the decomposition of the total probability density in Reflection (R) and Transmission part (T) by the tunnelling time problem [see section 41.3], is

$$|\psi|^2 = |\psi|^2_{\mathrm{T}} + |\psi|^2_{\mathrm{R}}; \tag{32.44}$$

while after calculations using the Feynman path-integral approach (which gives the experimentally correct formula) is:

$$|\psi|^2 = |\psi_{\mathrm{T}} + \psi_{\mathrm{R}}|^2. \tag{32.45}$$

Trajectories for Particles

One should also mention that some more technical objections can be advanced. In the following we shall examine some of them. Englert, Scully, Süssman and Walther [ENGLERT *et al.* 1993] proved that Bohm's trajectories can differ from observed ones, and Dewdney, Hardy and Squires [DEWDNEY *et al.* 1993] proved that in Bohm's model a particle can excite a detector on one path of an interferometer, even if takes the other path as a consequence of the Quantum potential.

Some properties which should pertain to particles, are not ascribable to the latter: for example in the case of neutrons, mass and magnetic moment cannot be localized at the hypothetical position of the particle [H. BROWN *et al.* 1995]. Surely one may think of a new 'non-classical' type of particle. But if so, then the whole enterprise seems problematic and it is hard to see the necessity of extending or correcting QM [see section 33.1].

32.6.3 Modal Interpretation

A 'soft' realistic interpretation of QM is the *modal* interpretation. Van Fraassen supposed [*VAN FRAASSEN* 1991, 273–337] that states evolve deterministically (following the Schrödinger Eq.) while values of observables are subjected to the statistical constraints which derive mainly from postulate 3.4 [p. 38] — in other words he distinguished between a 'dynamic state' and a 'property state' (or 'value state') [32]. One could avoid the Projection postulate by assuming the

[32]See also [*BUB* 1997, 115–26].

formalism of reduced matrices [KOCHEN 1985] [DIEKS 1989] [DIEKS 1994]: if it is possible to decompose a QM system bi-orthogonally in the form:

$$|\Psi\rangle = \sum_j c_j |m_j\rangle |n_J\rangle, \tag{32.46}$$

then it is possible to assign properties (if no degeneracy occurs) to the reduced states of the two subsystems to observables with eigenvectors respectively $|m_j\rangle$ (state of the system) and $|n_J\rangle$ (state of the apparatus).

Synthesizing, the modal interpretation [VERMAAS/DIEKS 1995] [CLIFTON 1995a] assumes that:

Postulate 32.1 (Modal interpretation) *For every collection of systems $\{\mathcal{S}_1, \mathcal{S}_2, \mathcal{S}_3, \ldots, \mathcal{S}_n\}$ and for every state $\hat{\rho}^T$ with $\mathcal{S}_1, \mathcal{S}_2, \mathcal{S}_3, \ldots, \mathcal{S}_n$ subsystems of $\mathcal{S}_T$, there exists a classical probability measure $\wp(\hat{P}_a^1, \ldots, \hat{P}_d^n)$ on the set $\{\hat{P}_j^1\}_j \times \ldots \times \{\hat{P}_m^n\}_m$ generated by the eigenprojections of $\hat{\rho}^1, \ldots, \hat{\rho}^n$, which is consistent with the probability measures for subsets of $\{\mathcal{S}_1, \mathcal{S}_2, \mathcal{S}_3, \ldots, \mathcal{S}_n\}$.*

However, if postulate 32.1 is intended as a variant of some realistic interpretations [KOCHEN 1985] [DIEKS 1994] [VERMAAS/DIEKS 1995] [*BUB* 1997, 137–44], then it certainly has unsound consequences, because, for a number $N \geq 3$ of subsystems, Vermaas proved [VERMAAS 1997] that it fails for QM, i.e. that one cannot simultaneously ascribe a property to each subsystem, which is something analogous to the KS Theorem [see section 32.4]. Another problem is that the modal interpretation cannot avoid some form of perspectivalism (partial traces) [CLIFTON 1995a], even though efforts have been made in order to overcome this problem [CLIFTON 1996].

A difference between de Broglie/Bohm's proposal and the modal interpretation is that according to the first theory the position is always determinate, while according to the modal interpretation the position has no special *status* [*BUB* 1997, 185].

32.6.4 Cini's Proposal

Recently Cini [CINI 1999] proposed a new corpuscular interpretation of QM aiming at eliminating quantum waves by employing quasi-probability distributions and assigning an ontological meaning to negative probabilities. It seems that the greatest problem here is how to understand probability values in the interval $[-1, 0)$. In fact 0 probability already means impossibility of the event or certainty that the event cannot happen. Then, how could one distinguish between a probability $-\frac{1}{2}$ and a probability $-\frac{1}{3}$? Cini, quoting Feynman, answers that negative probabilities should be understood as a measure of the degrees of unlikelihood of the conditions under which the event must occur, which is a means of bringing the problem one level back. It seems here a case where methodological principle M5 [p. 6] should be applied.

32.7 Concluding Remarks

In conclusion, every attempt to complete QM along the lines of the Determined Value assumption [p. 105] leads to contradictions, while the attempts to develop a corresponding physical model are inadequate. This conclusion justifies a generalization of the stochastic interpretation and formalism, which we shall discuss in the following chapter.

Chapter 33

STOCHASTIC GENERALIZATION

Introduction and contents We have already seen the importance of a stochastic approach to the problem of Measurement. However, the models of a smooth transition from microworld to macroworld and the failure of a deterministic theory for QM pose the more general problem of a global stochastic formulation of QM in order to account for a form of radical imprecision which affects its entities and such that one does not assume violent breaks with macrophysics. Therefore, we must critical examine the Faithful measurement assumption, which enters as premise in formula (31.4). This stochastic approach has been developed in particular by Prugovečki. It is an important integration of the Operational Stochastic QM [see chapter 18]. The present chapter develops in particular an analysis of the position and of the ontological status of the QM particles, and it is therefore a further development of parts VI and VII, respectively.

In section 33.1 the discussion of the problem is developed in ontological and interpretational terms. A non-relativistic and relativistic stochastic theory of QM in phase space is then presented [section 33.2]. Finally, brief concluding remarks follow in section 33.3.

33.1 General Statement of the Problem

In the last chapter we saw the impossibility of a fully deterministic interpretation of QM. However, no general interpretation of the problem has been given. In particular the relations between a supposed determined-value macrophysics and a fuzzy microphysics are not clear at all. QP [postulates 2.1 and 2.2] affirms a total break between them. Is this the correct answer? We know that such a break is excluded [see section 24.5], but no general account of the problem originating from the Determined Value assumption and from the Faithful Measurement assumption [see subsection 6.5.1] has yet been given.

33.1.1 Again Concerning Jumps

Brief history The first physicist who hinted at a new interpretation of the problem was Born — it is not by chance that the corollary 6.1 [p. 104] is due to Born. He asserted [BORN 1954, 438] that speaking of absolutely exact initial conditions makes no sense in classical mechanics too. In the simple application of numbers such as π we have already the problem of where to stop in the infinite series of digits. In conclusion, after Bohr, the determinism of classical physics is only a false image, an idol [BORN 1954, 439] [BORN 1961, 458–59].[1] The conclusion is that the assertion that classical mechanics is deterministic is not correct [BORN 1961, 459–60].[2]

Landé made another important contribution. Not all in his proposal is acceptable, but Landé has the historical merit of having posed a number of important questions. Criticizing the hypothesis of HV, he drew the conclusion that microphysical probability cannot be reduced to a classical determinism. The leading principles of his research were continuity, symmetry and invariance.

The *continuity* principle is initially elaborated as a principle of thermodynamical continuity [LANDÉ 1952, 353–55]: according to Landé, the orthogonality of the QM states is a consequence of the fact that work is zero for two identical gases in their isothermal diffusion and non-zero only if the two gases are not identical. Now, in order to avoid the Gibbs paradox[3], he postulated a continuous function, i.e. a likeness fraction between the gases. Now the likeness fraction signifies a relation between states which is not of total separability (orthogonality) and not of total inseparability (identity). Therefore, if one uses a filter through which only a state ρ_j can pass, there is another state ρ_k which is in a relation of fractional likeness with the first: apparently identical ρ_k's in part pass and in part do not, i.e. some become identical and some orthogonal to ρ_j. The QM jumps may derive from here. But, since initially there are not evident differences between apparently identical ρ_k's, then these jumps are only probabilistic[4]. Landé also deduced the complex amplitudes of QM. But the problem is that we must give determinate values to these amplitudes in order to obtain the distributions of QM [LANDÉ 1952, 356–57].

Jumps There were different criticisms of this proposal [SHIMONY 1966] [ZEILINGER 1986b, 168–71].[5] At any rate, another step against the supposed break between microphysics and macrophysics was made. In fact the importance of the idea of fractional likeness can be seen as follows. We have already seen Jauch's proposal to consider yes/no filters as propositions (projectors) for the case of sharp observables [see figure 9.6 and related comments: p. 163]. But, in the case of unsharp observables, we can use two different consecutive filters: the first represented by an effect and the second by a projector. The states which pass through the first filter generally are not assumed to be sharp (say that only zero-valued states are discarded): as they arrive at the second filter, some will pass (they jump to value 1 of the corresponding projector or to 'yes') and some not (they jump to the value 0 or to 'no').

To see this more concretely, we can think of the stochastic model of the Complementarity between wave and particle — for example Interfer. Exp. 16 [in subsection 30.3.1]. The two extreme projectors here are visibility and path projectors ($\hat{P}_{\mathcal{V}}, \hat{P}_{\mathcal{P}}$). But the projectors can also represent states: with one we conventionally associate the value 1 and with the other the value 0. All intermediate states — represented by intermediate points on the curve connecting $\hat{P}_{\mathcal{V}}$ and $\hat{P}_{\mathcal{P}}$ on the sphere shown in figure 30.6 — are effects. Hence the distance from every position on the curve to, say projector $\hat{P}_{\mathcal{P}}$ gives the probability to pass or not to pass a sharp test as the following: is the system localized in this position? Here we again find the idea of constructing a concrete QM logic for unsharp observables [see for example formula (29.1) and the relative discussion].

[1] Born shows the problem with the example of the Hamiltonian of classical mechanical systems: the elements are not separable, so that one conjugate variable is taken as constant: but it is exactly this which is unacceptable.

[2] See also [*LANDÉ* 1965a] .

[3] On this point see [*K. HUANG* 1963, 141].

[4] Landé interpreted the relation of fractional likeness as a geometrical distance between points on a circle with unity diameter.

[5] See also [*JAMMER* 1974, 460–64].

In conclusion we can legitimately postulate the following:

Postulate 33.1 (Quantum jumps) *Quantum jumps are generally the result of a localization (spontaneous or not), which acts as a sharp filter.*

Of course, in this way we have not resolved all the problems: we have given only a general framework in order to solve the problem of how systems can become localized or jump to almost sharp determined ones, but we have not pretended to offer a concrete solution, for example, to the problem of why electrons occupy only determinate energy levels. However, we think that this problem can be worked out in this direction. Surely we do not intend to destroy the 'boundaries' between the quantum world and the macroscopic world. We only want to say that the transition between them can be seen as a smooth one and what in QM appear as inexplicable transitions are in reality localization processes. We also suppose that such a localization can never be 100% sharp.

33.1.2 Uncertainty and Extended Particles

Now we discuss the other side of the problem: is continuity as such sound? The most important contribution to the problem came from Wheeler [WHEELER 1983]. The fundamental idea is that the continuum, supposed by classical mechanics [see postulate 2.4: p. 23] and denied by QP, is a useful mathematical tool but without the physical consequences which one ascribes to it [WHEELER 1986, 312] [WHEELER 1990, 9]. The supposed continuity of physical macroprocesses is evidently an idealization: factually we experiment in each moment and in each domain of the macroworld breaks, points of crisis, singularities, which more and more become in recent times the matter of new sciences, such as the mathematics of chaos.

More generally the mathematical continuum and the physical one should not be identified [ALI/EMCH 1974]. In fact in the mathematical continuum, for every three points A, B, C, we have:

$$[(\mathrm{A} = \mathrm{B}) \wedge (\mathrm{B} = \mathrm{C})] \rightarrow (\mathrm{A} = \mathrm{C}). \tag{33.1}$$

But in the physical continuum we have no guarantee that the physical indistinguishability is the same as identity. In fact for a sequence $\{\mathrm{A}_1, \mathrm{A}_2, \ldots, \mathrm{A}_{n-1}, \mathrm{A}_n\}$ we can assume that A_1 is the same as A_2, A_2 the same as A_3, and so on. But generally, after n steps, it can also be the case that $\mathrm{A}_1 \neq \mathrm{A}_n$ results. The passage from the physical continuum to the mathematical one requires the idealization of an infinitely precise measurement, which is never actually attainable — and we pass here to the Faithful Measurement assumption. In other words, given a small experimental uncertainty, there is always the possibility of reducing it but never of eliminating it; or, in other words, there is always the possibility of inserting other values between the known ones, gaining more and more subtle subdivisions.

The same results were reached by Margenau [MARGENAU 1949, 259–60]. We can suppose that to execute two identical measurements is impossible because there is always an ineliminable imprecision. Hence, against the Faithful Measurement assumption [p. 105], we can suppose, for example, that the fluctuations increase when we increase our level of subdivision [PRUGOVEČKI 1985, 530–531] [see also subsection 19.6.3]. We remember here that decoherence assumes that there is never a complete diagonalization in the basis of the measured observable [see corollary 17.2: p. 272], and hence never a perfect determined value.

A new insight comes from quantum information [see chapter 44]: the real world is unlikely to supply us with unlimited memory or unlimited Turing machine tapes [LANDAUER 1996]. The individuals of our universe (product of localization processes) are only a special physical realization of an information potentially infinite (given by the superposition or the entanglement) but not available as such [see chapters 42 and 44], i.e. each time only a finite amount of information is really available. Moreover: not all information physically present in our universe is available for one of its components or parts, because there is always some form of dispersion and because each gain in information is a local interaction which partly destroys the input. Therefore, mathematical continuum is not (physically) realizable, which, from an operational point of view, is very unsatisfactory. In other words, in a world with limited memory we cannot distinguish between π and a very close neighbor.

Therefore, at the place of the postulated reduction to zero of the measurement uncertainty [postulate 2.3: p. 22], we can assume the following one[6], which can cover also the QM case without postulating, with QP, a specific break between macrophysics and microphysics:

Postulate 33.2 (Certainty) *Even if the physical processes (micro and macro) can sometimes be mathematically treated as continuous, experimentally we can never reach an absolutely certain value such that the measurement uncertainty is zero.*

Recently it has been shown by Granik [GRANIK 1997] that the postulate 2.3 leads to problems if we consider very small time intervals and abandon the assumption about the independence of energy on the time interval irrespective of the scale. He showed that, in consequence of such a formalism, one can also derive a UR which is the classical counterpart of UR1 [Eq. (7.7)]. On the other hand, there are perhaps means to overcome the minimal quantum: for example, using a QND-Measurement, we have seen [subsection 19.5.2] that we can measure the energy better than one quantum, to the expense of the uncertainty of the conjugate observable.

Again this situation does not signify at all that there is no difference between micro-wold and macroworld. In the former the constant $\hbar$ has a strong meaning, while in the latter — due to the different magnitude scale — it has no relevance. But the transition from one to the other is a smooth passage — remember the models of decoherence discussed in chapters 21 and 24. On the other hand, we are not rejecting (as the word 'smooth' proves) a specific use of continuity as such: we are only claiming that the postulate of reducing the measurement uncertainty to zero is physically unfounded.

Therefore, if we never, also in the classical case, reach a completely sharp measurement, we have grounds for understanding the basic QM entity [see proposition 30.1: p. 526] as an *extended particle* or *quantum particle*. In other words the superposition character is never completely lost also in a corpuscular behaviour, but the 'particle' conserves a form of fuzziness and intrinsic uncertainty. Hence QM becomes intrinsically stochastic and every localization is always imperfect.

But if so, then all the problems relating to sharp localization are unfounded. Prugovečki's work is devoted to this fundamental idea of an extended particle: it will be examined in the next section.

[6]See [GALLAVOTTI 1986a] [GALLAVOTTI 1986b]. Following formulation owes much to Dr. Fortunato.

33.1.3 Complementary Observables

Another interesting point, partly related to the preceding is the following: Are the URs typical of QM or can one also speak of some uncertainty relations in classical mechanics? In fact we have already discussed that there is also a classical measurement problem [see proposition 2.1: p. 23], due to the impossibility of measuring simultaneously both variables of a canonical conjugate pair.

Now, in the light of the preceding discussion, the point can be generalized [LANDSBERG 1988, 970] as follows. In classical mechanics q and $\dot{q}$ are on the same footing, so that both can have determined values at the same time. But it is dubious that one can measure with infinite precision the velocity of a particle at a given point. It can be said only that *in a small space interval* it has a velocity $\dot{q}$, but it is surely conceptually erroneous to assign to it both a perfectly determined velocity $\dot{q}$ and a perfectly determined position q. Landsberg showed with a concrete example that it is possible to deduce some uncertainty relations, similar in structure to those of QM, also in the classical case [LANDSBERG 1988, 974–77]. This result fits very well with the fact that there can be coexistence of unsharp complementary observables [see section 18.2].

33.2 Stochastic Quantum Mechanics in Phase Space

What follows is a mathematical aspect of the preceding examination: the aim is to introduce unsharp operators, especially an unsharp position operator in the frame of Relativistic-QM. The theoretical model makes use of phase spaces [see also section 5.2.7].

33.2.1 Non-Relativistic Treatment

Stochastic space and points The principal concept, as we have already said, is that of extended or *quantum particle*: it is neither a wave nor a particle in the classical sense of word, but rather a spatio-temporal phenomenon with a numerous properties and particularly the proper wave function [*PRUGOVEČKI* 1984b, XIX].

Banishing exact values from all physics [BORN 1955b] [*PRUGOVEČKI* 1984b, 3–7], the Measurement process becomes a process of concurrence between different measurements, which cannot be achieved if the result of each measurement is described by a single number α.[7] A confidence interval must be assigned to each reading, or more generally, a probability distribution $\mathcal{F}_\alpha$ peaked at α and describing the margin of uncertainty in the reading. Hence we introduce the following stochastic probability measure:

$$\mu_\alpha(\Delta) = \int_\Delta \mathcal{F}_\alpha(x)dx, \tag{33.2}$$

where $\mathcal{F}_\alpha(x)$ is a distribution function [see section 3.7]. Eq. (33.2) expresses the idea that, when the result α is obtained, the actually prepared value had been within Δ. If the reading were absolutely accurate, then μ_α would be a δ–measure. We introduce the concept of stochastic point $\underline{\alpha}$. Given a measurable space $(\mathcal{X}, \mathcal{L}_\sigma)$ consisting of a set $\mathcal{X}$ and of an algebra $\mathcal{L}_\sigma$, we define a stochastic space $(\underline{\mathcal{X}}, \mathcal{L}_\sigma)$ by assigning to each $\alpha \in \mathcal{X}$ a μ_α on $(\mathcal{X}, \mathcal{L}_\sigma)$ such that:

$$\underline{\mathcal{X}} = \{\underline{\alpha} = (\alpha, \mu_\alpha) | \alpha \in \mathcal{X}\}, \tag{33.3}$$

where $\underline{\alpha}$ is a stochastic value or point.

[7] Antecedents in [ALI/DOEBNER 1976].

If the set $\mathcal{X}$ is a metric space — with distance $d^2(\alpha_1, \alpha_2)$ between points α_1, α_2 —, the stochastic set $\underline{\mathcal{X}}$ becomes a stochastic metric space allowing for square-metric distribution functions:

$$\mathcal{F}_{\underline{\alpha_1}, \underline{\alpha_2}}(y) = \int_{d^2(\omega_1,\omega_2)\leq y} d\mu_{\alpha_1}(\omega_1) d\mu_{\alpha_2}(\omega_2). \tag{33.4}$$

It is clear that we are dealing here with POVMs [*PRUGOVEČKI* 1984b, 15–19]. But, while the stochastic theory normally treats position and time asymmetrically (the first in terms of unsharp localizability, the second as a sharply defined parameter), here we treat both as unsharp quantities. Spacetime localizability can be consistently defined only in relation to certain spacetime domains $\underline{D}$, i.e. we refer ourselves only to conditional probabilities $\wp(\underline{R}|\underline{D})$ to observe a particle in spacetime region $\underline{R}$ in relation to domain $\underline{D}$ in which $\underline{R}$ is contained:

$$\wp_\psi(\underline{R}|\underline{D}) = \int_{\underline{R}} |\psi(\mathbf{r}, t)|^2 d\mathbf{r}dt \int_{\underline{D}} |\psi(\mathbf{r}, t)|^2 d\mathbf{r}dt. \tag{33.5}$$

The reason why $\underline{D}$ cannot be taken as the whole space (and for us the whole spacetime) is the condition (3.41), from which would result a divergent expression. $\underline{R}$ can be taken to be a Borel set.

Stochastic phase space We now search for a phase space representation Γ of the Galilei group [see subsection 3.10.2] [*PRUGOVEČKI* 1984b, 21–26], i.e. some unitary transformation $\hat{U}^\xi_\Gamma$ that maps the space $L^2(\Re^3)$ onto subspaces $\hat{U}^\xi_\Gamma L^2(\Re^3)$ of $L^2(\Gamma)$:

$$\hat{U}^\xi_\Gamma : \psi(\mathbf{r}) \;\mapsto\; \psi(\mathbf{q}, \mathbf{p}) = \int_{\Re^3} \xi^*_{\mathbf{q},\mathbf{p}}(\mathbf{r})\psi(\mathbf{r})d\mathbf{r}, \tag{33.6a}$$

$$\xi_{\mathbf{q},\mathbf{p}}(\mathbf{r}) \;=\; \hat{U}\left(0, \mathbf{q}, \frac{\mathbf{p}}{m}, \hat{I}\right) \xi(\mathbf{r}) = e^{\frac{i}{\hbar}\mathbf{p}\cdot(\mathbf{r}-\mathbf{q})}\xi(\mathbf{r}-\mathbf{q}). \tag{33.6b}$$

We can also write:

$$\hat{U}^\xi_\Gamma \psi(\mathbf{q}, \mathbf{p}) = \langle \xi_{\mathbf{q},\mathbf{p}} | \psi \rangle. \tag{33.7}$$

The function $\xi(\mathbf{r})$ can be regarded as a parametric quantity. Its properties follow from the isometric property:

$$\langle \hat{U}^\xi_\Gamma \psi_1 | \hat{U}^\xi_\Gamma \psi_2 \rangle_\Gamma = \langle \psi_1 | \psi_2 \rangle. \tag{33.8}$$

Using the unitarity property of Fourier/Plancherel transforms, we obtain:

$$\langle \hat{U}^\xi_\Gamma \psi | \hat{U}^\xi_\Gamma \psi \rangle_\Gamma = (2\pi\hbar)^3 \int_{\Re^3} d\mathbf{r} |\xi(\mathbf{r})|^2 \int_{\Re^3} d\mathbf{q} |\psi(\mathbf{r}+\mathbf{q})|^2, \tag{33.9}$$

from which it can be derived that

$$\parallel \xi \parallel = \langle \xi | \xi \rangle^{\frac{1}{2}} = (2\pi\hbar)^{-\frac{3}{2}}. \tag{33.10}$$

Using Eq. (33.6a) we see that Eq. (33.10) is equivalent to

$$\int_\Gamma d\mathbf{q}d\mathbf{p} |\xi_{\mathbf{q},\mathbf{p}}\rangle\langle\xi_{\mathbf{q},\mathbf{p}}| = \mathbf{1}. \tag{33.11}$$

The inverse transformation can be written as

$$(\hat{U}^\xi_\Gamma)^{-1} : \psi(\mathbf{q}, \mathbf{p}) \mapsto \psi(\mathbf{r}) = \int_\Gamma \xi_{\mathbf{q},\mathbf{p}}(\mathbf{r})\psi(\mathbf{q}, \mathbf{p}) d\mathbf{q}d\mathbf{p}. \tag{33.12}$$

We can now treat of the problem of localizability in Stochastic phase space. The physical meaning of $\psi(\mathbf{q},\mathbf{p})$ is obtained from the marginal properties:

$$\int_{\Re^3} |\psi(\mathbf{q},\mathbf{p})|^2 d\mathbf{p} = \int_{\Re^3} \mathcal{F}^{\xi}_{\mathbf{q}}(\mathbf{r})|\psi(\mathbf{r})|^2 d\mathbf{r}, \tag{33.13a}$$

$$\int_{\Re^3} |\psi(\mathbf{q},\mathbf{p})|^2 d\mathbf{q} = \int_{\Re^3} \tilde{\mathcal{F}}^{\xi}_{\mathbf{p}}(\mathbf{k})|\tilde{\psi}(\mathbf{k})|^2 d\mathbf{k}, \tag{33.13b}$$

with

$$\mathcal{F}^{\xi}_{\mathbf{q}}(\mathbf{r}) = (2\pi\hbar)^3|\xi(\mathbf{r}-\mathbf{q})|^2, \tag{33.14a}$$

$$\tilde{\mathcal{F}}^{\xi}_{\mathbf{p}}(\mathbf{k}) = (2\pi\hbar)^3|\tilde{\xi}(\mathbf{k}-\mathbf{p})|^2. \tag{33.14b}$$

The normalized distributions $\mathcal{F}^{\xi}_{\mathbf{q}}, \tilde{\mathcal{F}}^{\xi}_{\mathbf{p}}$ can be interpreted as confidence functions of the respective stochastic points [see Eqs. (33.2)–(33.4)]:

$$\underline{\mathbf{q}} = (\mathbf{q}, \mathcal{F}^{\xi}_{\mathbf{q}}) \in \underline{\Re}^3_{\mathbf{q}}, \ \underline{\mathbf{p}} = (\mathbf{p}, \tilde{\mathcal{F}}^{\xi}_{\mathbf{p}}) \in \underline{\Re}^3_{\mathbf{p}}, \tag{33.15}$$

where $\underline{\Re}^3_{\mathbf{q}}, \underline{\Re}^3_{\mathbf{p}}$ are stochastic configuration and momentum space respectively. The probability that a position indicator will provide response $\mathbf{q}$ when a particle in the state $|\psi(\mathbf{r})\rangle$ is present, is [see also Eqs. (15.6) and (18.17)]:

$$\int_{\Re^3} \mathcal{F}^{\xi}_{0}(\mathbf{r}-\mathbf{q})|\psi(\mathbf{r})|^2 d\mathbf{r}, \tag{33.16}$$

where $\mathcal{F}^{\xi}_{0}(\mathbf{r}-\mathbf{q}) = \mathcal{F}^{\xi}_{\mathbf{q}}(\mathbf{r})$. In the sharp point limit

$$\mathcal{F}^{\xi}_{\mathbf{q}}(\mathbf{r}) = \mathcal{F}^{\xi}_{0}(\mathbf{r}-\mathbf{q}) \longrightarrow \delta^3(\mathbf{r}-\mathbf{q}) \tag{33.17}$$

of a perfectly accurate position indicator, the RHS of Eq. (33.13a) merges into $|\psi(\mathbf{r})|^2$, i.e. the traditional interpretation is recovered. In conclusion $|\psi(\mathbf{q},\mathbf{p})|^2$ can be interpreted as a probability density on stochastic phase space

$$\mathit{\Gamma}_{\xi} = \underline{\Re}^3_{\mathbf{q}} \times \underline{\Re}^3_{\mathbf{p}} = \{\underline{\zeta} = (\mathbf{q}, \mathcal{F}^{\xi}_{\mathbf{q}}) \times (\mathbf{p}, \tilde{\mathcal{F}}^{\xi}_{\mathbf{p}}) | (\mathbf{q},\mathbf{p}) \in \mathit{\Gamma}\}, \tag{33.18}$$

where $\zeta = (\mathbf{q},\mathbf{p})$.

Proper wave function and effects The operational interpretation of $|\xi_{\mathbf{q},\mathbf{p}}\rangle$ is the following [*PRUGOVEČKI* 1984b, 26–29]: it is obtained from the configuration space resolution generator $\xi(\mathbf{r})$ [see Eq. (33.6b)] by the operation of translating $\xi(\mathbf{r})$ in space by an amount $\mathbf{q}$ and then boosting it to the 3–velocity $\mathbf{p}/m$. Hence, if we think of $\xi(\mathbf{r})$ as representing the configuration space *proper wave function* of an extended test particle of mass m that is used as a microdetector placed at the origin of the laboratory, then $\xi_{\mathbf{q},\mathbf{p}}(\mathbf{r})$ is the proper wave function of the test particle after it has been submitted to the aforementioned operations. It is legitimate to think of $|\xi_{\mathbf{q},\mathbf{p}}\rangle$ as the state of a test particle with position $\underline{\mathbf{q}}$ and momentum $\underline{\mathbf{p}}$. We can interpret the following probability

$$\wp^{\xi}_{\psi}(\mathcal{B}) = \int_{\mathcal{B}} |\hat{U}^{\xi}_{\Gamma}\psi(\mathbf{q},\mathbf{p})|^2 d\mathbf{p}d\mathbf{q} \tag{33.19}$$

as the probability of observing stochastic values $\underline{\mathbf{q}}, \underline{\mathbf{p}}$ within the Borel set $\mathcal{B}$ when a simultaneous measurement of both stochastic observables is performed on a particle with state vector $|\psi(\mathbf{r})\rangle$. The probability (33.19) can be written as an expectation value [Eq. (33.7)]

$$\wp^{\xi}_{\psi}(\mathcal{B}) = \langle\psi|\hat{E}_{\xi}(\mathcal{B})|\psi\rangle, \tag{33.20}$$

where

$$\hat{E}_\xi(\mathcal{B}) = \int_{\mathcal{B}} d\mathbf{q}d\mathbf{p}|\xi_{\mathbf{q},\mathbf{p}}\rangle\langle\xi_{\mathbf{q},\mathbf{p}}|; \tag{33.21a}$$

or, by considering Eq. (33.8):

$$\hat{E}^{\hat{U}}_\xi(\mathcal{B}) = \int_{\mathcal{B}} d\mathbf{q}d\mathbf{p}|\hat{U}^\xi_\Gamma\xi_{\mathbf{q},\mathbf{p}}\rangle\langle\hat{U}^\xi_\Gamma\xi_{\mathbf{q},\mathbf{p}}|. \tag{33.21b}$$

It is evident that $\hat{E}_\xi(\mathcal{B})$ is an effect on $L^2(\Re^3)$ whereas $\hat{E}^{\hat{U}}_\xi(\mathcal{B})$ is an effect on $L^2(\Gamma_\xi)$.

$\hat{E}_\xi(\mathcal{B})$ constitutes a system of covariance [see subsection 3.10.3] for the representation $\hat{U}(G)$ restricted to the isochronous Galilei Group $\mathcal{G}'$ [see subsection 3.10.2]:

$$\hat{U}(G)\hat{E}_\xi(\mathcal{B})\hat{U}^*(G) = \hat{E}_\xi(G\mathcal{B}),\ G \in \mathcal{G}', \tag{33.22}$$

where:

$$\begin{aligned}\hat{U}(G) : \psi(\mathbf{r},t) \mapsto \psi'(\mathbf{r},t) &= \exp\left[\frac{\imath}{\hbar}\left(-\frac{m\mathbf{v}^2}{2}(t-b) + m\mathbf{v}\cdot(\mathbf{r}-\mathbf{a})\right)\right] \\ &\times\psi(\mathbf{R}^{-1}[\mathbf{r}-\mathbf{a}-\mathbf{v}(t-b)], t-b).\end{aligned} \tag{33.23}$$

By Eq. (33.7) and Eq. (33.6a) we can write:

$$\hat{U}^\xi_\Gamma\xi_{\mathbf{q},\mathbf{p}}(\mathbf{q}',\mathbf{p}') = \int_{\Re^3} \xi^*_{\mathbf{q}',\mathbf{p}'}(\mathbf{r})\xi_{\mathbf{q},\mathbf{p}}(\mathbf{r})d\mathbf{r}; \tag{33.24}$$

and for the isometry of $\hat{U}^\xi_\Gamma$ [Eq. (33.8)] we can write:

$$\hat{U}^\xi_\Gamma\xi_{\mathbf{p},\mathbf{q}}(\mathbf{q}',\mathbf{p}') = \langle\hat{U}^\xi_\Gamma\xi_{\mathbf{q}',\mathbf{p}'}|\hat{U}^\xi_\Gamma\xi_{\mathbf{q},\mathbf{p}}\rangle_\Gamma, \tag{33.25}$$

which in turn can be rewritten as

$$\mathrm{K}_\xi(\mathbf{q}',\mathbf{p}';\mathbf{q},\mathbf{p}) = \int_\Gamma \mathrm{K}_\xi(\mathbf{q}',\mathbf{p}';\mathbf{q}'',\mathbf{p}'')\mathrm{K}_\xi(\mathbf{q}'',\mathbf{p}'';\mathbf{q},\mathbf{p})d\mathbf{q}''d\mathbf{p}'' \tag{33.26}$$

in terms of the kernel defined by

$$\mathrm{K}_\xi(\mathbf{q}',\mathbf{p}';\mathbf{q},\mathbf{p}) = \hat{U}^\xi_\Gamma(\mathbf{q}',\mathbf{p}')\xi_{\mathbf{q},\mathbf{p}}(\mathbf{q}',\mathbf{p}') = \langle\xi_{\mathbf{q}',\mathbf{p}'}|\xi_{\mathbf{q},\mathbf{p}}\rangle, \tag{33.27}$$

with the property

$$\hat{E}^{\hat{U}}_\xi(\mathcal{B})\psi(\mathbf{q},\mathbf{p}) = \int_\Gamma \mathrm{K}_\xi(\mathbf{q}',\mathbf{p}';\mathbf{q},\mathbf{p})\psi(\mathbf{q}',\mathbf{p}')d\mathbf{q}'d\mathbf{p}'. \tag{33.28}$$

Since

$$\mathrm{K}^*_\xi(\mathbf{q}',\mathbf{p}';\mathbf{q},\mathbf{p}) = \mathrm{K}_\xi(\mathbf{q},\mathbf{p};\mathbf{q}',\mathbf{p}'), \tag{33.29}$$

then $\hat{E}^{\hat{U}}_\xi(\mathcal{B})$ is self-adjoint, and, by Eq. (33.27), it is also idempotent; therefore it can be considered as a projector acting on $L^2(\Gamma)$ [see corollary 18.2: 301].

Current Another argument [*PRUGOVEČKI* 1984b, 40–42] for the interpretation of $\psi(\mathbf{q},\mathbf{p})$ as a probability amplitude for measurements of stochastic $\mathbf{q},\mathbf{p}$ is the existence for real $\xi(\mathbf{r})$ of a conserved and Galilei covariant current $\mathbf{j}^{\xi}(\mathbf{q})$ which, in the sharp point limit

$$\mathcal{F}^{\xi}(\mathbf{q}) = (2\pi\hbar)^3|\xi(\mathbf{r}-\mathbf{q})|^2 \longrightarrow \delta^3(\mathbf{r}-\mathbf{q}), \tag{33.30}$$

converges to its traditional counterpart. In the stochastic phase space we have:

$$\hat{\rho}^{\xi}(\mathbf{q},t) = \int_{\Re^3} |\psi(\mathbf{q},\mathbf{p},t)|^2 d\mathbf{p}, \tag{33.31a}$$

$$\mathbf{j}^{\xi}(\mathbf{q},t) = \int_{\Re^3} \frac{\mathbf{p}}{m}|\psi(\mathbf{q},\mathbf{p},t)|^2 d\mathbf{p}, \tag{33.31b}$$

which are analogous to the classical expressions

$$\rho_C(\mathbf{q},t) = \int_{\Re^3} \rho_C(\mathbf{q},\mathbf{p},t) d\mathbf{p}, \tag{33.32a}$$

$$\mathbf{j}_C(\mathbf{q},t) = \int_{\Re^3} \frac{\mathbf{p}}{m}\rho_C(\mathbf{q},\mathbf{p},t)|^2 d\mathbf{p}. \tag{33.32b}$$

Stochastic position operator Let us now define [*PRUGOVEČKI* 1984b, 20, 44] a stochastic position operator. If we define a Galilei system of covariance such that $\hat{q}^j_G = \hat{q}^j + \imath\hbar\partial/\partial\hat{p}^j$ we have that

$$\hat{q}^j_G \hat{E}^{\hat{U}}_{\xi} = \hat{U}^{\xi}_{\Gamma}\hat{q}^j(\hat{U}^{\xi}_{\Gamma})^{-1}\hat{E}^{\hat{U}}_{\xi}. \tag{33.33}$$

These relations extend to arbitrary functions and in particular to

$$V(\hat{\mathbf{q}})\hat{E}^{\hat{U}}_{\xi} = \hat{U}^{\xi}_{\Gamma}V(\mathbf{r})(\hat{U}^{\xi}_{\Gamma})^{-1}\hat{E}^{\hat{U}}_{\xi}. \tag{33.34}$$

Hence the Schrödinger Eq. can be rewritten in $L^2(\Gamma_{\xi})$ as follows:

$$\imath\hbar\partial_t\psi(\mathbf{q},\mathbf{p},t) = \left(-\frac{\hbar^2}{2m}\nabla^2_q + V(\mathbf{q})\right)\psi(\mathbf{q},\mathbf{p},t). \tag{33.35}$$

On the other hand we could also consider interaction Hamiltonians which contain functions $V(\underline{\mathbf{q}})$ of a stochastic position operator:

$$\hat{\underline{q}}^j(\mathbf{q},\mathbf{p})\psi(\mathbf{q},\mathbf{p}) = q^j\psi(\mathbf{q},\mathbf{p}). \tag{33.36}$$

33.2.2 Relativistic Treatment*

We know the difficulties of constructing a position operator for Relativistic-QM [see subsection 4.5.2 and section 10.3]. We now try to develop a Relativistic-QM without the sharp localization assumption[8].

[8] One of the first attempts to build such a theory is in [PRUGOVEČKI 1978].

Generals Here we work with the Poincaré group [see section 4.1] [*PRUGOVEČKI* 1984b, 89–99], which gives the following mapping

$$\hat{U}^{\mathbf{p}}_{\mathcal{Q}}(\mathbf{a}, \Lambda) : \tilde{\psi}(\mathbf{k}) \mapsto \tilde{\psi}'(\mathbf{k}) = e^{\frac{i}{\hbar}\mathbf{a}\cdot\mathbf{k}}\tilde{\psi}(\Lambda^{-1}\mathbf{k}). \tag{33.37}$$

The $\tilde{\psi}(\mathbf{k})$ are Fourier transforms of spin-zero vectors $\psi(\mathbf{r})$ subjected to the Klein/Gordon Eq. [Eq. (4.10)], and are elements of a relativistic Hilbert space $L^2(\mathcal{H}_m)$ consisting of functions of the mass hyperboloid

$$\mathcal{H}_M = \mathcal{H}^+_m \cup \mathcal{H}^-_m, \ \mathcal{H}^\pm_m = \{\mathbf{k}|k^2 = \mathbf{k}\cdot\mathbf{k} = m^2c^2, k_0 >< 0\} \tag{33.38}$$

that are square integrable with respect to the relativistically invariant measure

$$d\mathcal{X}_m(k) = \delta(k^2 - m^2c^2)d^4k. \tag{33.39}$$

We find a relativistic counterpart of Eq. (33.6b) [see also p. 68]:

$$\tilde{\eta}_{q,p}(k) = (\hat{U}_{\mathcal{Q}}(q, \Lambda_\nu)\tilde{\eta})(k) = e^{\frac{i}{\hbar}q\cdot k}\tilde{\eta}(\Lambda_\nu^{-1}k); \tag{33.40a}$$

and also of Eq. (33.6a):

$$\hat{U}^\eta_\Gamma : \tilde{\psi}(k) \mapsto \psi(q,p) = \int_{\mathcal{H}^\pm_m} \tilde{\eta}^*_{q,p}(k)\tilde{\psi}(k)d\mathcal{X}_m(k). \tag{33.40b}$$

As for the non-relativistic case $\hat{U}^\eta_\Gamma$ defines an isometric mapping, while $\tilde{\eta}(k)$ is rotationally invariant (it depends exclusively on $|\mathbf{k}|$ and not on the orientation of $\mathbf{k}$), so that

$$\tilde{\eta}(k) = \eta(mck_0), \ k = (k_0, \mathbf{k}) \in \mathcal{H}^\pm_m; \tag{33.41}$$

or also

$$\tilde{\eta}(\Lambda_\nu^{-1}\mathbf{k}) = \eta(mc(\Lambda_\nu^{-1}k)_\circ) = \eta(\mathbf{p}\cdot\mathbf{k}). \tag{33.42}$$

We can also establish a one-to-one correspondence with the non-relativistic resolution generator $\tilde{\xi}(\mathbf{k})$ [see Eq. (33.14b)]:

$$\tilde{\eta}(\mathbf{k}) = (2mc)^{\frac{1}{2}}\tilde{\xi}(\mathbf{k}). \tag{33.43}$$

Parallel to Eq. (33.11) we set:

$$\int_{q_0=\mathrm{const}, p\in\mathcal{H}^\pm_m} d\mathbf{q}d\mathbf{p}|\tilde{\eta}_{q,p}\rangle\langle\tilde{\eta}_{q,p}| = \mathbf{1}_\pm, \tag{33.44}$$

where $\mathbf{1}_\pm$ are identity operators on $\mathcal{H}^\pm_m$.

Relativistic position operator We now discuss the localizability problem in the relativistic context. First we need marginals equivalent to Eqs. (33.13). Consider the following Eq. at $q_0 = \mathrm{const}$:

$$\wp^\eta(p) = \int_{\Re^3} |\hat{U}^\eta_\Gamma(q,p)\psi(q,p)|^2 d\mathbf{q}, \tag{33.45}$$

where [see Eqs. (33.40)]

$$\hat{U}^\eta_\Gamma(q,p)\tilde{\psi}(q,p) = \int_{\mathcal{H}^\pm_m} e^{-\frac{i}{\hbar}q\cdot k}\eta^*(p\cdot k)\tilde{\psi}(k)d\mathcal{X}_m(k). \tag{33.46}$$

Hence we can perform integration in Eq. (33.45) using the unitarity property of Fourier/Plancherel transforms [see Eq. (33.9)]:

$$\wp^\eta(p) = \pm(2\pi\hbar)^3 \int_{\mathcal{H}^\pm_m} |\eta(p\cdot k)|^2 |\tilde{\psi}(k)|^2 (2k_0)^{-1} d\mathcal{X}_m(k). \tag{33.47}$$

Introducing substitution $k_{(p)} := \Lambda^{-1}_{p/m} k$ and collecting Eqs. (33.42), (33.43) and (33.14b), after setting $d\mathcal{X}_m(k) := d\mathcal{X}_m(k_{(p)})$, we obtain:

$$\wp^\eta(p) = \pm mc \int_{\Re^3} \frac{d\mathbf{k}_{(p)}}{k^0_{(p)}} \tilde{\mathcal{F}}^\xi_0(\mathbf{k}_{(p)}) \frac{|\tilde{\psi}(\Lambda_p k_{(p)})|^2}{2(\Lambda_{p/m} k_{(p)})_0}. \tag{33.48}$$

Hence in the sharp momentum limit $\tilde{\mathcal{F}}^\xi_0(\mathbf{k}_{(p)}) \longrightarrow \delta^3(\mathbf{k}_{(p)})$ we obtain (where $k = \Lambda_p k_{(p)} = p$ when $k_{(p)} = (\pm mc, \mathbf{0})$):

$$\wp^\eta(p) \longrightarrow (2p_0)^{-1} |\tilde{\psi}(p)|^2. \tag{33.49}$$

Therefore, in the sharp momentum limit, the traditional relativistic probabilities are valid:

$$\int_{\mathcal{B}} d\mathbf{p} \int_{\Re^3} |\psi(q,p)|^2 d\mathbf{q} \longrightarrow \int_R |\tilde{\psi}(p)|^2 d\mathcal{X}_m(p), \tag{33.50}$$

where $\psi(q,p) = \hat{U}^\eta_\Gamma(q,p)\psi(q,p)$ [see Eq. (33.45)].

If we try to duplicate the same procedure in the case of position

$$\wp^\eta(q) = \int_{\Re^3} |\psi(q,p)|^2 d\mathbf{p},\ p = (p_0, \mathbf{p}) \in \mathcal{H}^\pm_m, \tag{33.51}$$

we discover that the $\mathbf{p}$ integration cannot be performed in any manner that would give rise to a distribution $\mathcal{F}^\xi_0(\mathbf{q})$ utilizing Eqs. (33.14), (33.42) and (33.43). But $\wp^\eta(q)$ is equal to the component $j^0_\eta(q)$ of a relativistically covariant and conserved current $j^\mu_\eta(q)$, which indicates that $|\psi(q,p)|^2$ has a physical meaning despite of the impossibility of localizing particles sharply.

In analogy with Eq. (33.36) we write:

$$\hat{\underline{q}}^\mu(q,p)\psi(q,p) = q^\mu \psi(q,p),\ \mu = 0,1,2,3. \tag{33.52}$$

Let us introduce the operator

$$\hat{q}^\mu_\eta(\sigma) = \hat{P}_\eta(\Sigma^+_m)\hat{\underline{q}}^\mu \hat{P}_\eta(\Sigma^+_m), \tag{33.53}$$

where $\Sigma_m = \Sigma^+_m \cup \Sigma^-_m$, $\Sigma^\pm_m = \sigma \times \mathcal{H}^\pm_m$, σ is an initial-data hypersurface of operational simultaneity which at q_0 is a constant, and $\hat{P}_\eta(\Sigma^+_m)$ are orthogonal projectors onto the subspaces

$$L^2(\Sigma^\pm_{m,\eta}) := \hat{U}^\eta_\Gamma L^2(\mathcal{H}^\pm_m) \subset L^2(\Sigma^\pm_m). \tag{33.54}$$

Under this conditions if η is real, using the Newton/Wigner operator [see Eq. (10.64)], we finally obtain the following stochastic position operator:

$$\boxed{\hat{\underline{q}}^j_\eta(\sigma) = \hat{U}^\eta_\Gamma \hat{q}^j_{\mathrm{NW}} (\hat{U}^\eta_\Gamma)^\dagger} \tag{33.55}$$

with $j = 1,2,3$.

Then the physical significance of the Newton/Wigner operator is that of a mean stochastic position operator but only in stochastic rest frames Σ^+_m of the microdetectors. But even if we have:

$$\langle\tilde{\psi}|\hat{\mathbf{q}}_{\mathrm{NW}}|\tilde{\psi}\rangle_{\mathcal{H}^+_m} = \langle\hat{U}^\eta_\Gamma\tilde{\psi}|\hat{\underline{q}}^\mu|\hat{U}^\eta_\Gamma\tilde{\psi}\rangle_{\Sigma^+_m}, \tag{33.56}$$

this equality does not extend to functions of these operators, except in the approximate sense when the spread of the proper wave function $\xi(\mathbf{r})$, resulting from Eq. (33.43) is sufficiently small.

Covariance An important problem is to establish a system of covariance. The interpretation of $\psi(q,p)$ as a probability amplitude at points (q,p) implies that [see Eq. (33.20): but we have now a PVM]

$$\wp^{\eta}_{\psi}(\mathcal{B}) = \langle\tilde{\psi}|\hat{\tilde{P}}_{\eta}(\mathcal{B})\tilde{\psi}\rangle_{\mathcal{H}^{\pm}_{m}} \tag{33.57}$$

is a probability measure over Borel sets $\mathcal{B} \subset \Sigma_{m,\eta}$, where $\hat{\tilde{P}}_{\eta}(\mathcal{B})$ is the Bochner integral [see Eq. (33.21a)]:

$$\hat{\tilde{P}}_{\eta}(\mathcal{B}) = \int_{\mathcal{B}} d\Sigma_m(q,p)|\tilde{\eta}_{q,p}\rangle\langle\tilde{\eta}_{q,p}|. \tag{33.58}$$

$\hat{\tilde{P}}_{\eta}(\mathcal{B})$ gives rise to a system of covariance such that [see Eq. (33.37)]:

$$\hat{U}^{p}_{\mathcal{Q}}(G)\hat{\tilde{P}}_{\eta}(\mathcal{B})(\hat{U}^{p}_{\mathcal{Q}}(G))^{-1} = \hat{\tilde{P}}_{\eta}(G\mathcal{B}). \tag{33.59}$$

Since proper Lorentz transformations do not preserve simultaneity, we do not have the counterpart of Eq. (33.22), but we can introduce nevertheless a POVM:

$$\hat{\tilde{E}}_{\eta}(\mathcal{B}) = \int_{-\infty}^{+\infty} \hat{\tilde{P}}_{\eta}(\mathcal{B}_{q_0})dq_0, \tag{33.60}$$

with

$$\mathcal{B}_{q_0} = \{(q,p)|(q,p) \in \Sigma^{(q_0)\pm}_{m,\eta} \cap \mathcal{B}\}. \tag{33.61}$$

Now $\hat{\tilde{E}}_{\eta}(\mathcal{B})$ gives rise to a system of covariance such that [see Eq. (33.37)]:

$$\hat{U}^{p}_{\mathcal{Q}}(g_{\mathcal{G}})\hat{\tilde{E}}_{\eta}(\mathcal{B})(\hat{U}^{p}_{\mathcal{Q}}(G))^{-1} = \hat{\tilde{E}}_{\eta}(G\mathcal{B}), \tag{33.62}$$

able to constitute a system of covariance on the relativistic stochastic phase spaces $\Gamma^{\pm}_{m,\eta}$.

Propagators and current By pursuing further the analogies with the nonrelativistic case, we introduce the following relativistic free propagator [see Eq. (33.27)]:

$$\mathrm{K}_{\eta}(q',p';q,p) = \langle\tilde{\eta}_{q',p'}|\tilde{\eta}_{q,p}\rangle_{\mathcal{H}^{\pm}_{m}} = \int_{\mathcal{H}^{\pm}_{m}} e^{\frac{i}{\hbar}(q-q')\cdot k}\eta^{*}(p'\cdot k)\eta(p\cdot k)d\mathcal{X}_m(k), \tag{33.63}$$

where $|\eta_{q,p}\rangle$ constitutes a continuous resolution of identity in $L^2(\Sigma^{\pm}_{m,\eta})$ [see Eq. (33.44)]:

$$\int_{\Sigma^{\pm}_{m}} d\Sigma_m(q,p)|\eta_{q,p}\rangle\langle\eta_{q,p}| = \mathbf{1}^{\pm}_{m,\eta}. \tag{33.64}$$

For $\zeta = (q,p)$ the two properties of the propagator are:

$$\mathrm{K}^{*}_{\eta}(\zeta';\zeta) = \mathrm{K}_{\eta}(\zeta;\zeta'), \tag{33.65a}$$

$$\mathrm{K}^{*}_{\eta}(\zeta';\zeta) = \int_{\Sigma^{\pm}_{m}} \mathrm{K}^{*}_{\eta}(\zeta';\zeta'')\mathrm{K}^{*}_{\eta}(\zeta'';\zeta)d\Sigma_m(\zeta''). \tag{33.65b}$$

In the relativistic case we can also construct an analogy of Eqs. (33.31) [*PRUGOVEČKI* 1984b, 111–12]:

$$j^{\mu}_{\eta}(q) = \pm 2\int_{\mathcal{H}^{\pm}_{m}} \frac{p^{\mu}}{m}|\psi(q,p)|^2 d\mathcal{X}_m(p), \tag{33.66}$$

which is covariant under relativistic transformation (a,Λ). We can also write:

$$j^{0}_{\eta}(q) = \wp_{\eta}(q) = \int_{\Re^3} |\psi(q,p)|^2 d\mathbf{p},\ p_0 = \pm(\mathbf{p}^2 + m^2c^2)^{1/2}, \tag{33.67}$$

where $d\mathcal{X}_m(k) = \frac{dk}{2}(\mathbf{p}^2 + m^2c^2)^{1/2}$. We see that $j^{0}_{\eta}(q) \geq 0$ and hence $j^{\mu}_{\eta}(q)$ is really a probability current.

33.2.3 Statistical Mechanics on Stochastic Phase Spaces*

We need a phase space representation [*PRUGOVEČKI* 1984b, 137–48] in which there is a one-to-one affine map between states and probabilities such that:

$$\wp_{\hat{\rho}}(\mathcal{B}) = \int_{\mathcal{B}} \hat{\rho}(\mathbf{q}, \mathbf{p}) d\mathbf{q} d\mathbf{p}, \tag{33.68}$$

where $\mathcal{B}$ is a measurable space; the probabilities are also Galilei covariant:

$$\wp_{\hat{U}_G^* \hat{\rho} \hat{U}_G}(\mathcal{B}) = \wp_{\hat{\rho}}(G\mathcal{B}), \ G \in \mathcal{G}'. \tag{33.69}$$

The required type of phase-space representation for one-particle systems might be obtained from the (isochronous) Galilei systems of covariance of Eq. (33.22) by setting:

$$\wp_{\hat{\rho}}^{\xi}(\mathcal{B}) = \mathrm{Tr}[\hat{\rho} \hat{E}_{\xi}(\mathcal{B})], \tag{33.70}$$

provided that a one-to-one map $\hat{\rho} \mapsto \wp_{\hat{\rho}}^{\xi}$ can be established. It is evident that Eq. (33.70) is not the most general form of phase-space representation compatible with Eq. (33.69). Hence we define a normalized POVM $\hat{E}(\mathcal{B})$such that

$$\hat{U}(G)\hat{E}(\mathcal{B})\hat{U}^*(G) = \hat{E}(G\mathcal{B}), \ G \in \mathcal{G}', \tag{33.71}$$

from which

$$\wp_{\hat{\rho}}(\mathcal{B}) = \mathrm{Tr}[\hat{\rho} \hat{E}(\mathcal{B})]. \tag{33.72}$$

Now we can say that a phase-space system of covariance is informationally complete [see subsection 18.6.1] iff the Eq.

$$\mathrm{Tr}[\hat{O} \hat{E}(\mathcal{B})] \tag{33.73}$$

can be satisfied for all Borel sets $\mathcal{B}$ in Γ^N only by $\hat{O} = 0$ for an arbitrary operator $\hat{O}$.

Theorem 33.1 (I Prugovečki) *Every phase space representation defined by* Eq. (33.68) *and* Eq. (33.69) *of quantum statistical mechanics determines a unique informationally complete phase-space system of covariance for which* Eq. (33.72) *is true, and conversely to each informationally complete phase-space system of covariance we can assign by* Eq. (33.72) *a unique phase space representation of type* (33.68) *and* (33.69) *of quantum statistical mechanics.*

We say that the POVM $\hat{E}_1(\mathcal{B}), \hat{E}_2(\mathcal{B})$, are informationally equivalent iff, for $\hat{\rho}', \hat{\rho}''$ and for all $\mathcal{B}$, we have:

$$\{\mathrm{Tr}[\hat{\rho}' \hat{E}_1(\mathcal{B})] = \mathrm{Tr}[\hat{\rho}'' \hat{E}_1(\mathcal{B})]\} \rightarrow \{\mathrm{Tr}[\hat{\rho}' \hat{E}_2(\mathcal{B})] = \mathrm{Tr}[\hat{\rho}'' \hat{E}_2(\mathcal{B})]\}. \tag{33.74}$$

Consequently we say that a POVM $\hat{E}_1(\mathcal{B})$ is informationally complete iff the antecedent of Eq. (33.74) holds only when $\hat{\rho}' = \hat{\rho}''$. By starting now with expression (33.21b) we try to generalize it to space $L^2(\Re^N)$ in the following manner:

$$\hat{E}_{\gamma^{\xi}}(\mathcal{B}) = \int_{\mathcal{B}} \gamma_{q,p}^{\xi} dq dp, \tag{33.75}$$

where

$$\gamma_{q,p}^{\xi} = \hat{U}_{q,p} \gamma^{\xi} \hat{U}_{q,p}^{-1}, \tag{33.76a}$$

$$\gamma^{\xi} = |\xi\rangle\langle\xi|, \ \xi = \xi_{(1)} \otimes \cdots \otimes \xi_{(N_0)}; \tag{33.76b}$$

and

$$\hat{U}_{q,p} = e^{\frac{i}{\hbar}q\cdot p}e^{\frac{i}{\hbar}p\cdot\hat{q}}e^{-\frac{i}{\hbar}q\cdot\hat{p}}, \tag{33.77}$$

with

$$q \cdot p = \sum_{\mu=1}^{N} q_\mu p_\mu, \quad q \cdot \hat{p} = \sum_{\mu=1}^{N} q_\mu \hat{p}_\mu, \quad p \cdot \hat{q} = \sum_{\mu=1}^{N} p_\mu \hat{q}_\mu. \tag{33.78}$$

Theorem 33.2 (II Prugovečki) *A POVM on $\mathcal{B} \in \Gamma^N$ of the form*

$$\hat{E}_{\gamma^\xi}(\mathcal{B}) = \int_{\mathcal{B}} \hat{U}^*_{q,p}\gamma^\xi \hat{U}_{q,p} dqdp \tag{33.79}$$

with $\mathrm{Tr}\gamma^\xi = (2\pi\hbar)^{-N}$ *is informationally complete iff the Weyl transform* $\hat{\gamma}^\xi_{WT}$ *[see subsection 5.2.6] given by*

$$\hat{\gamma}^\xi_{WT}(q,p) = \mathrm{Tr}[\gamma^\xi \hat{U}_{q,p}] \tag{33.80}$$

of the generalized resolution operator γ^ξ *is different from zero for almost all* $(q,p) \in \Gamma$.

Note that if $\gamma^\xi = |\xi\rangle\langle\xi|$, with $\xi \in L^2(\Re^N)$ then

$$\hat{\gamma}^\xi_{WT}(q,p) = \langle\xi|\hat{U}_{q,p}\xi\rangle. \tag{33.81}$$

Theorem 33.1 and theorem 33.2 are very important in order to understand some practical implications of the stochastic theory, but are not fundamental to it. Hence we do not need to integrate them into the theory and limit ourselves to consider both as examples of the fertility of the stochastic theory.

33.3 Conclusion

Though many results of this chapter are to be verified and used practically to fully judge the model, which eventually can imply some changes or accommodations of it, it is clear that the path taken by Prugovečki is hopeful. The formalism here developed is able to satisfy the fundamental requirements, on both the general theoretical and the interpretational level, which were advanced in section 33.1. Moreover, for relativistic QM, a stochastic position operator of the form (33.55) can avoid the difficulties depending on the assumption of a sharp position operator.

Part IX

THE PROBLEM OF NON-LOCALITY

Introduction to Part IX The *ninth part* is devoted to a problem originally developed from the HV problematic, but which in previous years has gone much further than the original discussion: the non-locality problem.

- In *chapter 34* we discuss the initial historical criticisms of the EPR argument.
- Bell inequalities are then discussed [*chapter 35*].
- In *chapter 36*] the type of locality which is violated by QM is examined.
- *Chapter 37* is devoted to different theorems in order to determine more precise values which are violated by QM or classical theories.
- In the following chapters different generalizations of the subject are examined. Firstly, we ask if mixtures violate Bell inequalities [*chapter 38*].
- Then a generalization on a probabilistic level is provided and several probabilistic formulations of Bell inequalities are discussed [*chapter 39*].
- Thereafter, Bell inequalities for observables different from the spin [*chapter 40*] are examined.
- Finally, other non-local effects such as entanglements from different sources, AB-Effect and tunnelling time [*chapter 41*] are discussed.

The theoretical and mathematical foundations of the non-local features of QM are especially due to Bell's article of 1964. However, this area was mostly developed in the eighties, especially as the new techniques developed in the Q-Optics domain allowed for the first time an experimental verification and development of the theory.

Chapter 34

INITIAL CRITICISMS OF THE EINSTEIN/PODOLSKY/ROSEN ARGUMENT

Contents EPR's article was originally intended by the authors as a proof of the incompleteness of QM. It therefore implicitly showed the necessity of completing it, which occurred in the fifties with the HV-theories — examined in the previous part. Some answers to EPR (those of Bohr and Schrödinger, for example) were centered on another aspect, which in previous years became a central tenet of the theory: they denied the validity of the Separability principle in QM [see formula 31.4 and section 31.5]. This is the object of this chapter and of the successive analysis. While in section 34.1 we report and discuss Schrödinger's answer, in section 34.2 Bohr's answer is analysed. Bohr's answer will also be formalized [section 34.3], so that some fundamental issues already discussed [see section 9.2] will be applied and developed further.

34.1 Schrödinger's Answer

The question of separability [principle 31.1: p. 532] can be understood in very different ways [*JAMMER* 1974, 195–97]. In fact Schrödinger and Bohr both tried to deny that the Separability principle is valid in QM, but while Schrödinger understood it in the sense of the inseparability between two subsystems, Bohr rather understood it in the sense of inseparability between a system and a measurement apparatus [see principle 8.5: p. 142]. We begin with the first problem, because it is more closely related to the original proposal of EPR, although Bohr's article is chronologically precedent.

As we shall see now, Schrödinger's analysis of the problem posed by EPR is strictly related to the paradox of the S-Cat [see the introduction to the V part]. In fact Schrödinger accepted and developed consequently the EPR argument, and in so doing he discovered entanglement as the basilar character of QM [see definition 3.3: p. 43].

Take two subsystems $\mathcal{S}_1, \mathcal{S}_2$ with observables, respectively, $\hat{q}_1, \hat{p}_1$ and $\hat{q}_2, \hat{p}_2$. Let there be two observables

$$\hat{q} = \hat{q}_1 - \hat{q}_2, \qquad \hat{p} = \hat{p}_1 + \hat{p}_2 \tag{34.1}$$

operating on the state of the compound system and having numerical values q', p', which we assume to know. Then we have:

$$\hat{q}|\Psi_{1+2}\rangle = q'|\Psi_{1+2}\rangle, \qquad \hat{p}|\Psi_{1+2}\rangle = p'|\Psi_{1+2}\rangle, \tag{34.2}$$

where $|\Psi_{1+2}\rangle$ is the state vector of the compound system. This is possible because $[\hat{q}, \hat{p}] = 0$.

Since we have

$$q'_1 - q'_2 = q', \tag{34.3}$$

q'_1 can be predicted from q'_2 and viceversa. A similar argument holds for p'_1, p'_2. So that, knowing the result of a measurement on $\mathcal{S}_2$, we can predict either q'_1 or p'_1 without interfering with $\mathcal{S}_1$. It is as if we could always know the right answer to our question, whatever it be [SCHRÖDINGER 1936, 427–29].[1] In other words Schrödinger discovered a fundamental interdependence between subsystems, the entanglement, which does not depend on the physical interactions between them [see definition 3.2: p. 42]. Note, on the other hand, that he did not accept at all such a result, which for him showed some fundamental error in Basic-QM itself [see also introduction to the V part]. We think, on the contrary, that the problem is not in the formalism itself, but in the physical interpretation of it. The following chapters of this part are devoted to the problem of the interpretation of the entanglement.

34.2 Bohr's Answer

Bohr [BOHR 1935a] [BOHR 1935b] [2] answered EPR by pointing out that, even if the EPR thought-experiment excludes any direct physical interaction of the system with the measuring apparatus, the measurement process has an essential influence on the conditions on which the very definition of the physical observables in question rests [see subsection 26.4.1]. And these conditions must be considered as an inherent element of any phenomenon to which the term 'physical reality' can be unambiguously applied [see principles 3.1: p. 28, and CP5: p. 142]. Bohr acknowledged that it is possible to determine experimental arrangements such that the measurement of the position or of the momentum of one particle automatically determines the position or the momentum of the other. But each time the experimental arrangements for measuring momentum and position are incompatible. Generally the pairs $\hat{q}_1, \hat{p}_1$ and $\hat{q}_2, \hat{p}_2$ obey CCR while the pairs $(\hat{q}_1, \hat{q}_2), (\hat{p}_1, \hat{p}_2), (\hat{q}_1, \hat{p}_2), (\hat{q}_2, \hat{p}_1)$ commute. However, one can also choose new observables $\hat{q}', \hat{p}'$ related to the first ones by rotation of an angle θ in the planes defined by $\hat{q}_1, \hat{q}_2$ and $\hat{p}_1, \hat{p}_2$:

$$\hat{q}_1 = \hat{q}'_1 \cos\theta - \hat{q}'_2 \sin\theta, \tag{34.4a}$$
$$\hat{q}_2 = \hat{q}'_1 \sin\theta + \hat{q}'_2 \cos\theta, \tag{34.4b}$$
$$\hat{p}_1 = \hat{p}'_1 \cos\theta - \hat{p}'_2 \sin\theta, \tag{34.4c}$$
$$\hat{p}_2 = \hat{p}'_1 \sin\theta + \hat{p}'_2 \cos\theta. \tag{34.4d}$$

These new observables obey the same commutation relations as before. It follows that in the description of the state both $\hat{q}'_1$ and $\hat{p}'_1$ cannot be assigned definite values. However, since $\hat{q}'_1$ commutes with $\hat{p}'_2$, it is possible to assign values simultaneously to $\hat{q}'_1$ and $\hat{p}'_2$. Hence we are

[1] Some important developments of these ideas are in Furry [FURRY 1936] Cooper and Sharp [SHARP 1961]. See also [*JAMMER* 1974, 211–16, 221–24, 237].

[2] See also [*JAMMER* 1974, 195–97].

forced to consider a measurement as a whole[3] — such that one subsystem cannot be considered independently from the other and that both cannot be considered independently from the experimental context which has been chosen [see again principle 8.5: p. 142].

Discussion Therefore Bohr, like Schrödinger, has rejected the Separability principle [31.1: p. 532] and in this way solved the EPR paradox[4].

It may be thought that Bohr rejects the Determined Value Assumption [postulate 6.3: p. 105; see also subsections 31.1.5 and 31.2.1], and this from a subjectivistic interpretation of the problem of the values of observables [see postulate 8.1: p. 137].[5] But there is no real ground to think that Bohr did so [*MURDOCH* 1987, 107, 150].

So that in conclusion the problem of the Completeness of QM is not decided at all by EPR *Gedankenexperiment* because we know that, by an implication such as (31.4), if the antecedent is false, the consequent can be whatever (true or false), always keeping the implication as such true.

34.3 Bub's Formulation

Bub [BUB 1989a] gave a very concrete meaning to Bohr's position concerning the conditions of a QM experiment[6]. We have already seen [see section 8.4 and subsection 17.1.2, particularly theorem 17.2] that the sole presence of a detector can suffice to effect that a system undergoes a transition, and that there can be interaction-free measurements [see chapter 20]. We see now some applications of the quoted theorem to the EPR problem.

Bub stressed that we must express the EPR problem in terms of maximal Boolean subalgebras [see theorem 9.5: p. 164, and figure 9.7 with relative comments] so that it cannot be sufficient to indicate only some of the conditions of the EPR problem: we must indicate all of them. Translated into EPR language: we cannot consider only the values of an observable of one subsystem (which would be a non-maximal subalgebra), rather we must always consider the values of this observable and of its 'brothers' (or generally of the commuting observables) in the other subsystem. The measurement of two correlated but space-like separated systems generates a 16–element Boolean algebra $\mathcal{L}_{\mathcal{B}}$ [see figure 34.1].

In fact, following the EPRB model [see section 31.3], if we suppose that the maximal measurement is one of the complete commuting sets of observables $\{\hat{\mathbf{J}}^2(\mathcal{S}_1 + \mathcal{S}_2), \hat{J}_x(\mathcal{S}_1 + \mathcal{S}_2)\}$, then the corresponding complete orthonormal set of eigenvectors in the four-dimensional Hilbert space of $\mathcal{S}_1 + \mathcal{S}_2$ is given [see Eqs. (3.132)] by the three vectors of the triplet state:

$$| \uparrow\rangle| \uparrow\rangle, \tag{34.5a}$$

$$\frac{1}{\sqrt{2}}(| \uparrow\rangle| \downarrow\rangle \ + \ | \downarrow\rangle| \uparrow\rangle), \tag{34.5b}$$

$$| \downarrow\rangle| \downarrow\rangle; \tag{34.5c}$$

and by the vector of the singlet state

$$\frac{1}{\sqrt{2}}(| \uparrow\rangle| \downarrow\rangle - | \downarrow\rangle| \uparrow\rangle). \tag{34.6}$$

[3]See also [BOHR 1939, 20].

[4]On this point see also [HOOKER 1972, 144–45].

[5]See also [SHARP 1961].

[6]See also [*BUB* 1997, 204–211].

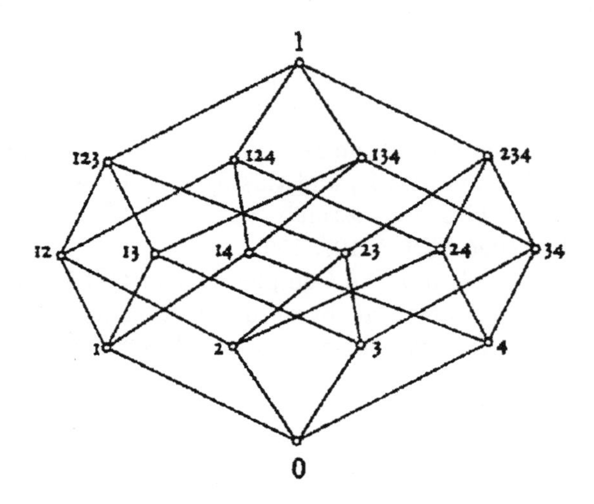

Figure 34.1: Haasse diagram for the 16–element Boolean algebra of the EPRB model (where with 123, for example, is to be understood $1 \vee 2 \vee 3$ — from [*BUB* 1997, 207].

Now the sixteen elements of the Boolean algebra $\mathcal{L}_{\mathcal{B}}$ are the following:

$$
\begin{aligned}
\emptyset: &\quad \mathbf{0} \\
1: &\quad \hat{P}_{\uparrow\uparrow} \\
2: &\quad \hat{P}_{\downarrow\downarrow} \\
3: &\quad \hat{P}_{\uparrow\downarrow+\downarrow\uparrow} \\
4: &\quad \hat{P}_{\uparrow\downarrow-\downarrow\uparrow} \\
1 \vee 2: &\quad \hat{P}_{\uparrow\uparrow} + \hat{P}_{\downarrow\downarrow} \\
1 \vee 3: &\quad \hat{P}_{\uparrow\uparrow} + \hat{P}_{\uparrow\downarrow+\downarrow\uparrow} \\
1 \vee 4: &\quad \hat{P}_{\uparrow\uparrow} + \hat{P}_{\uparrow\downarrow-\downarrow\uparrow} \\
2 \vee 3: &\quad \hat{P}_{\downarrow\downarrow} + \hat{P}_{\uparrow\downarrow+\downarrow\uparrow} \\
2 \vee 4: &\quad \hat{P}_{\downarrow\downarrow} + \hat{P}_{\uparrow\downarrow-\downarrow\uparrow} \\
3 \vee 4: &\quad \hat{P}_{\uparrow\downarrow-\downarrow\uparrow} + \hat{P}_{\uparrow\downarrow+\downarrow\uparrow} = \hat{P}_{\uparrow\downarrow} + \hat{P}_{\downarrow\uparrow} \\
1 \vee 2 \vee 3: &\quad \hat{P}_{\uparrow\uparrow} + \hat{P}_{\downarrow\downarrow} + \hat{P}_{\uparrow\downarrow+\downarrow\uparrow} = \hat{P}^{\perp}_{\uparrow\downarrow-\downarrow\uparrow} \\
1 \vee 2 \vee 4: &\quad \hat{P}_{\uparrow\uparrow} + \hat{P}_{\downarrow\downarrow} + \hat{P}_{\uparrow\downarrow-\downarrow\uparrow} = \hat{P}^{\perp}_{\uparrow\downarrow+\downarrow\uparrow} \\
1 \vee 3 \vee 4: &\quad \hat{P}_{\uparrow\uparrow} + \hat{P}_{\uparrow\downarrow+\downarrow\uparrow} + \hat{P}_{\uparrow\downarrow-\downarrow\uparrow} = \hat{P}_{\uparrow\uparrow} + \hat{P}_{\uparrow\downarrow} + \hat{P}_{\downarrow\uparrow} = \hat{P}^{\perp}_{\downarrow\downarrow} \\
2 \vee 3 \vee 4: &\quad \hat{P}_{\uparrow\downarrow} + \hat{P}_{\downarrow\uparrow} + \hat{P}_{\downarrow\downarrow} = \hat{P}^{\perp}_{\uparrow\uparrow} \\
1 \vee 2 \vee 3 \vee 4: &\quad \hat{I}.
\end{aligned}
\tag{34.7}
$$

The elements 1, 2, 3 are spanned by the triplet, and the element 4 is spanned by the singlet. If we measure, for example, the x–component of the spin of subsystem $\mathcal{S}_2$ we select only four elements within the non-maximal Boolean subalgebra $\mathcal{L}_{\mathcal{B}_2}$ generated by the following projectors:

$$\hat{P}^{(2)}_{\uparrow} = \hat{P}_{\uparrow\uparrow} + \hat{P}_{\downarrow\uparrow}, \qquad \hat{P}^{(2)}_{\downarrow} = \hat{P}_{\uparrow\downarrow} + \hat{P}_{\downarrow\downarrow}, \tag{34.8}$$

where $\hat{P}^{(2)}_{\uparrow}$ covers the case in which the result for $\mathcal{S}_2$ is spin-up, while $\hat{P}^{(2)}_{\downarrow}$ covers the case of spin-down for the same subsystem.

Now we must find the elements of the first subsystem which are compatible with $\mathcal{L}_{\mathcal{B}_2}$ so that we arrive at a maximal subalgebra $\mathcal{L}'_{\mathcal{B}}$ of 16 elements (by different alternatives) in the partial Boolean algebra $\mathcal{L}_{\mathcal{B}}$ of the propositions of $\mathcal{S}_1 + \mathcal{S}_2$:

$$\begin{array}{rl}
\emptyset: & \mathbf{0} \\
1: & \hat{P}_{\uparrow\uparrow} \\
2: & \hat{P}_{\uparrow\downarrow} \\
3: & \hat{P}_{\downarrow\uparrow} \\
4: & \hat{P}_{\downarrow\downarrow} \\
1 \vee 2: & \hat{P}_{\uparrow\uparrow} + \hat{P}_{\uparrow\downarrow} \\
1 \vee 3: & \hat{P}_{\uparrow\uparrow} + \hat{P}_{\downarrow\uparrow} \\
1 \vee 4: & \hat{P}_{\uparrow\uparrow} + \hat{P}_{\downarrow\downarrow} \\
2 \vee 3: & \hat{P}_{\uparrow\downarrow} + \hat{P}_{\downarrow\uparrow} \\
2 \vee 4: & \hat{P}_{\uparrow\downarrow} + \hat{P}_{\downarrow\downarrow} \\
3 \vee 4: & \hat{P}_{\downarrow\uparrow} + \hat{P}_{\downarrow\downarrow} \\
1 \vee 2 \vee 3: & \hat{P}_{\uparrow\uparrow} + \hat{P}_{\uparrow\downarrow} + \hat{P}_{\downarrow\uparrow} \\
1 \vee 2 \vee 4: & \hat{P}_{\uparrow\uparrow} + \hat{P}_{\uparrow\downarrow} + \hat{P}_{\downarrow\downarrow} \\
1 \vee 3 \vee 4: & \hat{P}_{\uparrow\uparrow} + \hat{P}_{\downarrow\uparrow} + \hat{P}_{\downarrow\downarrow} \\
2 \vee 3 \vee 4: & \hat{P}_{\uparrow\downarrow} + \hat{P}_{\downarrow\uparrow} + \hat{P}_{\downarrow\downarrow} \\
1 \vee 2 \vee 3 \vee 4: & \hat{I}.
\end{array} \tag{34.9}$$

The elements labelled $\emptyset, 1, 4, 1 \vee 4, 2 \vee 3, 1 \vee 2 \vee 3, 2 \vee 3 \vee 4, \hat{I}$ are derived from $\mathcal{L}_{\mathcal{B}}$ (they correspond respectively to elements $1, 2, 1 \vee 2, 3 \vee 4, 1 \vee 3 \vee 4, 2 \vee 3 \vee 4$ of the latter algebra): 1 and 4 correspond, respectively, to the first term of the projector $\hat{P}^{(2)}_{\uparrow}$ and to the second term of $\hat{P}^{(2)}_{\downarrow}$ in Eq. (34.8), whereas 2 and 3 correspond to the second term of the projector $\hat{P}^{(2)}_{\uparrow}$ and to the first term of the projector $\hat{P}^{(2)}_{\downarrow}$ in Eq. (34.8). The elements $1 \vee 3, 2 \vee 4$ from $\mathcal{L}_{\mathcal{B}}{}'$ are precisely those of Eq. (34.8). The remaining elements are generated as least upper bounds or greatest lower bounds or orthocomplements of these elements.

Hence $\mathcal{L}'_{\mathcal{B}}$ is simply the maximal Boolean subalgebra generated by the elements labelled $1, 2, 3, 4$, corresponding to the lines spanned by the vectors

$$| \uparrow\rangle | \uparrow\rangle, \tag{34.10a}$$

$$| \uparrow\rangle | \downarrow\rangle, \tag{34.10b}$$

$$| \downarrow\rangle | \uparrow\rangle, \tag{34.10c}$$

$$| \downarrow\rangle | \downarrow\rangle, \tag{34.10d}$$

again with spin-up and spin-down in the x–direction, i.e. to the eigenvectors of the maximal measurement of the commuting set $\{\hat{J}_x(\mathcal{S}_1), \hat{J}_x(\mathcal{S}_2)\}$. In conclusion we have constructed a maximal subalgebra corresponding to the measurement of the x–direction of the spin on the subsystem $\mathcal{S}_2$.

Now we introduce our hypothesis that the total system $\mathcal{S}_1 + \mathcal{S}_2$ is in a singlet state. Under this hypothesis, the elements which are true in $\mathcal{L}_{\mathcal{B}}$ are those generated by the elements which in (34.7) are labelled $4, 1 \vee 4, 2 \vee 4, 3 \vee 4, 1 \vee 2 \vee 4, 1 \vee 3 \vee 4, 2 \vee 3 \vee 4$. These elements, together with those which are true in $\mathcal{L}_{\mathcal{B}_2}$ — i.e. either the ultrafilter $1 \vee 3$ $(\hat{P}_{\uparrow\uparrow} + \hat{P}_{\downarrow\uparrow})$ or the ultrafilter $2 \vee 4$ $(\hat{P}_{\uparrow\downarrow} + \hat{P}_{\downarrow\downarrow}, \hat{I})$ [see Eq. (34.8)] —, determine the elements which are true in $\mathcal{L}'_{\mathcal{B}}$. Then the only elements in $\mathcal{L}_{\mathcal{B}}$ which are transferred by the ultrafilters to $\mathcal{L}'_{\mathcal{B}}$, are at positions labelled $2 \vee 3, 1 \vee 2 \vee 3, 2 \vee 3 \vee 4, \hat{I}$ in expression (34.9) — corresponding respectively to locations $3 \vee 4, 1 \vee 3 \vee 4, 2 \vee 3 \vee 4$ in expressions (34.7).

Therefore, we have to consider the following elements of set (34.9): $1 \vee 3, 2 \vee 4, 2 \vee 3, 1 \vee 2 \vee 3, 2 \vee 3 \vee 4, \hat{I}$; but it suffices to consider only $1 \vee 3, 2 \vee 4, 2 \vee 3$. Now a truth-value assignment to $\mathcal{L}'_{\mathcal{B}}$ is a two-valued homomorphism on $\mathcal{L}'_{\mathcal{B}}$. Since location $2 \vee 3$ is mapped onto truth-value 1 (either 2 or 3 must occur in a singlet state), and either location $1 \vee 3$ or $2 \vee 4$ is also mapped onto 1, it follows that either location 2 is mapped onto 1 — which corresponds to state (34.10b) —, or location 3 is mapped onto 1 — which corresponds to state (34.10c) —, exactly the two possibilities allowed by the singlet state [BUB 1989a, 802–804].[7] In a similar way we can build any maximal subalgebra pertaining to the measurement of the spin of one subsystem along any direction.

In conclusion we see that the intuitive answer of Bohr to EPR is perfectly confirmed by an analysis of the problem on an algebraic level, which is able to show in an abstract form the typical entanglement feature of QM.

[7] Concerning this problem see also [*JAUCH* 1968, 188–89] [HOOKER 1972, 96–98] [AERTS 1982a] and for some concrete models see also subsection 32.5.

Chapter 35

BELL INEQUALITIES

Contents In this chapter we analyse one of the most important contributions to QM in the last fifty years: the second Bell theorem and its refinements. In section 35.1 the original Bell contribution is reported, followed [section 35.2] by some further developments. In section 35.3 we report and discuss many of the experimental realizations in order to (dis-)prove Bell theorem. Other experiments are reported in section 35.4: our aim is to overcome some difficulties (loop-holes) which have arisen in the more recent discussion of the above experiments. Finally a more abstract analysis (Fine's theorems) is developed in order to bridge the gap between deterministic HV theories and stochastic HV theories [section 35.5].

35.1 The Second Bell Theorem

As we have already said in the introduction to this part [see also section 32.3], Bell's examination was conditioned by the problems arisen in the context of HV-theories, so that originally there was a mixing of the problematic of HVs and that of separability or locality. Finally, Bell proved that no deterministic local HV-theory can give the same predictions as QM[1]. Later, there must be a clear cut between the concept of HV-theory and that of separability (or classical correlation) [see particularly section 35.6].

The *Gedankenexperiment* proposed by Bell is the following — the model is that of EPRB [see subsection 31.3.1]. We suppose [BELL 1964, 15–19] a hidden parameter λ_{HV} such that, given λ_{HV}, the result A of measuring the spin of the first particle, along a chosen direction $\mathbf{a}$ of a first polariser, a Stern-Gerlach magnet (SGM1) — $\hat{\sigma}_1 \cdot \mathbf{a}$ —, depends only on λ_{HV} and on $\mathbf{a}$, and the result B of measuring the spin of the second particle by a chosen direction $\mathbf{b}$ of a SGM2 $\hat{\sigma}_2 \cdot \mathbf{b}$ — depends only on $\mathbf{b}$ and λ_{HV} [Separability principle 31.1: p. 532; see p. 58 for a definition of polariser and related formalism]. As we know this hypothesis can be formulated mathematically in a more rigorous way as a factorization rule [see definition 3.2: p. 42]:

$$(A_{\mathbf{a}} \cdot B_{\mathbf{b}})(\lambda_{\mathrm{HV}}) = A_{\mathbf{a}}(\lambda_{\mathrm{HV}}) \cdot B_{\mathbf{b}}(\lambda_{\mathrm{HV}}), \tag{35.1}$$

i.e. the probability distributions for the two particles are mutually independent.

We formalize the results of each measurement (spin up or down) with ± 1:

$$A_{\mathbf{a}}(\lambda_{\mathrm{HV}}) = \pm 1, \quad B_{\mathbf{b}}(\lambda_{\mathrm{HV}}) = \pm 1. \tag{35.2}$$

[1] See [STAPP 1980, 767].

If $\hat{\rho}(\lambda_{\text{HV}})$ is the HV state of the compound system and $\rho(\lambda_{\text{HV}})$ indicates (by integration) the probability distribution of λ_{HV} — the present symbolism is legitimated by Eq. (42.3) —, then the expectation value of the product of the two components is:

$$\mathrm{C}(\mathbf{a}, \mathbf{b}) := \int_{\Lambda_{\text{HV}}} d\lambda_{\text{HV}} \rho(\lambda_{\text{HV}}) A_{\mathbf{a}}(\lambda_{\text{HV}}) B_{\mathbf{b}}(\lambda_{\text{HV}}), \tag{35.3}$$

where, instead of the usual formalism of Eq. (3.90) we prefer to use symbol C to indicate the correlation between the two measurements. In fact the Separability principle denies that there can be a form of interdependence between two systems if they do not interact dynamically. Hence what we must investigate here is whether or not there is such a dependence — an entanglement [see definition 3.3: p. 43] —, in which case the two systems are separated [see again the definition 3.2: p. 42].

Since we do not know the values of the hidden parameter(s) λ_{HV}, we integrate over the possible and unknown values $\lambda_{\text{HV}} \in \Lambda_{\text{HV}}$ (in the following we normally drop the explicit mention of integration over the entire space Λ_{HV}). Because $\rho(\lambda_{\text{HV}})$ is supposed to be a normalized probability distribution, we have:

$$\int d\lambda_{\text{HV}} \rho(\lambda_{\text{HV}}) = 1, \tag{35.4}$$

so that, because of property (35.2), in Eq. (35.3) $\mathrm{C}(\mathbf{a}, \mathbf{b})$ takes the minimum value of -1.

Eq. (35.3) should be equal to the QM expectation value, which for the singlet state is necessarily — one subsystem is in spin up the other in spin down [see Eq. (3.132d) and (31.17)]:

$$[\mathrm{C}(\mathbf{a}, \mathbf{b})]_{\Psi_0} := \langle \Psi_0 | \hat{\sigma}_1 \cdot \mathbf{a} \hat{\sigma}_2 \cdot \mathbf{b} | \Psi_0 \rangle = -\mathbf{a} \cdot \mathbf{b}. \tag{35.5}$$

When SGM1-2 are parallel we have:

$$[\mathrm{C}(\mathbf{a}, \mathbf{a})]_{\Psi_0} = -1, \tag{35.6}$$

and we say that there is a perfect *anticorrelation* between the results of the two measurements.

But on the HV hypothesis, Eq. (35.6) holds iff

$$A_{\mathbf{a}}(\lambda_{\text{HV}}) = -B_{\mathbf{a}}(\lambda_{\text{HV}}), \tag{35.7}$$

and in this case we reach the minimum value of Eq. (35.3). Under this supposition we can rewrite Eq. (35.3) in the following manner:

$$\mathrm{C}(\mathbf{a}, \mathbf{b}) = -\int d\lambda_{\text{HV}} \rho(\lambda_{\text{HV}}) A_{\mathbf{a}}(\lambda_{\text{HV}}) A_{\mathbf{b}}(\lambda_{\text{HV}}). \tag{35.8}$$

Now suppose *three different orientations* $\mathbf{a}, \mathbf{b}, \mathbf{c}$ of the SGMs (two for SGM2); then we may write:

$$\begin{aligned} \mathrm{C}(\mathbf{a}, \mathbf{b}) - \mathrm{C}(\mathbf{a}, \mathbf{c}) &= -\int d\lambda_{\text{HV}} \rho(\lambda_{\text{HV}}) [A_{\mathbf{a}}(\lambda_{\text{HV}}) A_{\mathbf{b}}(\lambda_{\text{HV}}) - A_{\mathbf{a}}(\lambda_{\text{HV}}) A_{\mathbf{c}}(\lambda_{\text{HV}})] \\ &= \int d\lambda_{\text{HV}} \rho(\lambda_{\text{HV}}) A_{\mathbf{a}}(\lambda_{\text{HV}}) A_{\mathbf{b}}(\lambda_{\text{HV}}) [A_{\mathbf{b}}(\lambda_{\text{HV}}) A_{\mathbf{c}}(\lambda_{\text{HV}}) - 1], \end{aligned} \tag{35.9}$$

because $[A_{\mathbf{a}}(\lambda_{\text{HV}})]^2 = 1$. Using property (35.2), Eq. (35.9) can be written:

$$|\mathrm{C}(\mathbf{a}, \mathbf{b}) - \mathrm{C}(\mathbf{a}, \mathbf{c})| \le \int d\lambda_{\text{HV}} \rho(\lambda_{\text{HV}}) [1 - A_{\mathbf{b}}(\lambda_{\text{HV}}) A_{\mathbf{c}}(\lambda_{\text{HV}})], \tag{35.10}$$

so that we finally obtain:

$$|C(\mathbf{a},\mathbf{b}) - C(\mathbf{a},\mathbf{c})| \leq 1 + C(\mathbf{b},\mathbf{c}). \tag{35.11}$$

This formula is the first of a family of inequalities, collectively called *Bell inequalities*. Now we can formulate the second Bell Theorem, which we also call 'Bell theorem' as such for its importance:

Theorem 35.1 (II Bell) *A deterministic HV theory, which acknowledges the Separability principle* [31.1: p. 532] *must satisfy the inequality* (35.11). *The prediction of QM on the contrary violates it.*

Proof

In order to prove the theorem, it suffices to show a contradiction between Eq. (35.11) and Eq. (35.5) by means of a counterexample (= SGM-Exp. 6) [CLAUSER/SHIMONY 1978, 1888–90]. We take **a**, **b** and **c** coplanar, with **c** making an angle ϕ of $2\pi/3$ with **a**, and **b** making an angle θ of $\pi/3$ with both **a** and **c** [see figure 35.1].

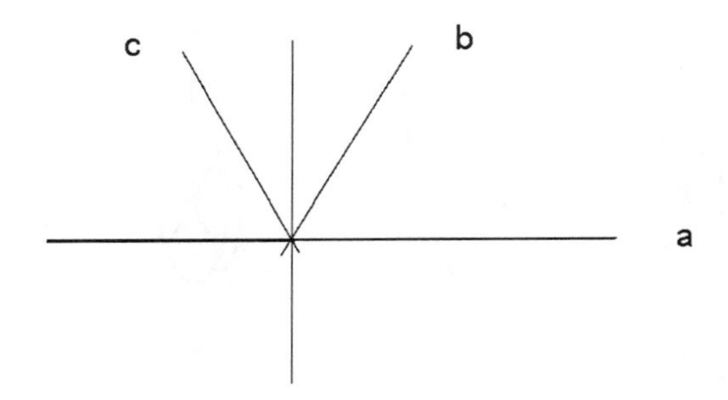

Figure 35.1: SGM-Exp. 6

Then, according to the elementary formula

$$cos\theta = \frac{\mathbf{a}\cdot\mathbf{b}}{|\mathbf{a}||\mathbf{b}|}, \tag{35.12}$$

and by choosing unitary vectors ($\| \mathbf{a} \| = \| \mathbf{b} \| = \| \mathbf{c} \| = 1$), we have:

$$\mathbf{a}\cdot\mathbf{b} = \mathbf{b}\cdot\mathbf{c} = \frac{1}{2}; \quad \mathbf{a}\cdot\mathbf{c} = -\frac{1}{2} \tag{35.13}$$

and

$$|[C(\mathbf{a},\mathbf{b})]_{\Psi_0} - [C(\mathbf{a},\mathbf{c})]_{\Psi_0}| = 1, \tag{35.14}$$

while

$$1 + [C(\mathbf{b},\mathbf{c})]_{\Psi_0} = \frac{1}{2}. \tag{35.15}$$

These values do not satisfy inequality (35.11). Q.E.D..[2]

[2] Wigner [WIGNER 1970, 1007–1008] worked out the Bell theorem with spin measurements for the singlet state.

35.2 Refinements of Bell Theorem

35.2.1 Clauser/Horne/Shimony/Holt Inequality

In a paper by Clauser, Shimony, Horne and Holt (= CHSH) [CLAUSER *et al.* 1969] Bell's Theorem was generalized and determined more accurately in the following way:

- they left aside the assumption of determinism so that they were able to show an incompatibility between QM and stochastic HV theories which acknowledge the Separability principle;

- they conceived a more realistic *Gedankenexperiment* so that a considerable step toward factual experiments was made.

We suppose correlated pairs of particles such that one enters apparatus $1_{\mathbf{a}}$ and the other enters apparatus $2_{\mathbf{b}}$ — where $\mathbf{a}$ and $\mathbf{b}$ are two parameters —, consisting of two analyzers and of two different detectors beyond each analyzer — it is an idealized coincidence count experiment (Cascade Exp. 1) [see figure 35.2].

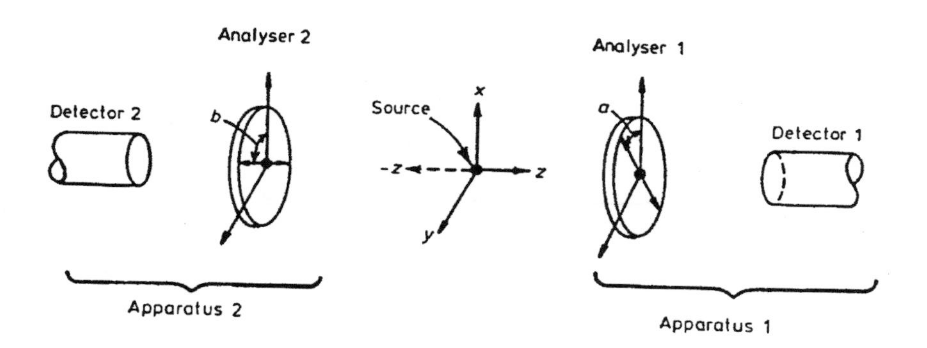

Figure 35.2: Apparatus configuration used in the proof by CHSH and by CH (see below). A source emitting particle pairs is analyzed by two apparatus. Each apparatus consists of an analyzer and an associated detector. The analyzers have parameters $\mathbf{a}, \mathbf{b}$, respectively, which can be adjusted by the experimenter. In the above example $\mathbf{a}$ and $\mathbf{b}$ make some angles with a fixed reference axis — from [CLAUSER/SHIMONY 1978, 1892].

In each apparatus a particle must select one of two channels labelled by +1 and −1 respectively. The results are represented by $A_{\mathbf{a}}$ and $B_{\mathbf{b}}$, each of which equals +1 or −1. Suppose again HVs and independence of $A_{\mathbf{a}}(\lambda_{\mathrm{HV}})$ from $\mathbf{b}$ and of $B_{\mathbf{b}}(\lambda_{\mathrm{HV}})$ from $\mathbf{a}$ [see Eq. (35.1)]. The normalized probability distribution $\rho(\lambda_{\mathrm{HV}})$ is independent of both $\mathbf{a}$ and $\mathbf{b}$. Defining the correlation function

$$\mathrm{C}(\mathbf{a}, \mathbf{b}) := \int_{\Lambda_{\mathrm{HV}}} d\lambda_{\mathrm{HV}} \rho(\lambda_{\mathrm{HV}}) A_{\mathbf{a}}(\lambda_{\mathrm{HV}}) B_{\mathbf{b}}(\lambda_{\mathrm{HV}}), \tag{35.16}$$

we have [see Eqs. (35.9) and (35.10)]:

$$|\mathrm{C}(\mathbf{a}, \mathbf{b}) - \mathrm{C}(\mathbf{a}, \mathbf{c})| \le \int_{\Lambda_{\mathrm{HV}}} d\lambda_{\mathrm{HV}} \rho(\lambda_{\mathrm{HV}}) |A_{\mathbf{a}}(\lambda_{\mathrm{HV}}) B_{\mathbf{b}}(\lambda_{\mathrm{HV}}) - A_{\mathbf{a}}(\lambda_{\mathrm{HV}}) B_{\mathbf{c}}(\lambda_{\mathrm{HV}})|$$

$$= \int_{\Lambda_{\rm HV}} d\lambda_{\rm HV}\rho(\lambda_{\rm HV})|A_{\mathbf{a}}(\lambda_{\rm HV})B_{\mathbf{b}}(\lambda_{\rm HV})|[1 - B_{\mathbf{b}}(\lambda_{\rm HV})B_{\mathbf{c}}(\lambda_{\rm HV})]$$
$$= 1 - \int_{\Lambda_{\rm HV}} d\lambda_{\rm HV}\rho(\lambda_{\rm HV})B_{\mathbf{b}}(\lambda_{\rm HV})B_{\mathbf{c}}(\lambda_{\rm HV}). \tag{35.17}$$

Suppose now that for some $\mathbf{b}'$ and $\mathbf{b}$ we have

$$\mathrm{C}(\mathbf{b}', \mathbf{b}) = 1 - \delta, \tag{35.18}$$

where $0 \leq \delta \leq 1$. In this way we avoid Bell's experimentally unrealistic condition that for some pair $\mathbf{b}$ and $\mathbf{b}'$ there is perfect correlation, i.e. $\delta = 0$. Dividing $\Lambda_{\rm HV}$ into two regions $\Lambda^{+}_{\rm HV}$ and $\Lambda^{-}_{\rm HV}$, such that $\Lambda^{\pm}_{\rm HV} = \{\lambda_{\rm HV}|A_{\mathbf{b}'}(\lambda_{\rm HV}) = \pm B_{\mathbf{b}}(\lambda_{\rm HV})\}$, we have:

$$\int_{\Lambda^{-}_{\rm HV}} d\lambda_{\rm HV}\rho(\lambda_{\rm HV}) = \frac{1}{2}\delta, \tag{35.19}$$

and hence:

$$\begin{aligned}\int_{\Lambda_{\rm HV}} d\lambda_{\rm HV}\rho(\lambda_{\rm HV})B_{\mathbf{b}}(\lambda_{\rm HV})B_{\mathbf{c}}(\lambda_{\rm HV}) &= \int_{\Lambda_{\rm HV}} d\lambda_{\rm HV}\rho(\lambda_{\rm HV})A_{\mathbf{b}'}(\lambda_{\rm HV})B_{\mathbf{c}}(\lambda_{\rm HV}) \\ &\quad -2\int_{\Lambda^{-}_{\rm HV}} d\lambda_{\rm HV}\rho(\lambda_{\rm HV})|A_{\mathbf{b}'}(\lambda_{\rm HV})B_{\mathbf{c}}(\lambda_{\rm HV})| \\ &= \mathrm{C}(\mathbf{b}', \mathbf{c}) - \delta.\end{aligned} \tag{35.20}$$

In conclusion we obtain:

$$|\mathrm{C}(\mathbf{a}, \mathbf{b}) - \mathrm{C}(\mathbf{a}, \mathbf{c})| \leq 2 - \mathrm{C}(\mathbf{b}', \mathbf{b}) - \mathrm{C}(\mathbf{b}', \mathbf{c}), \tag{35.21}$$

which is the Second inequality of the family of Bell's inequalities and is named *CHSH inequality* [CLAUSER *et al.* 1969, 881]. Now CHSH inequality conflicts with QM — we shall prove this in subsection 35.2.2.

We must note that, if we read the parameters $\mathbf{a}$ and $\mathbf{b}$ as polarisers' orientations and suppose a perfect parallelism [as in Eq. (35.6)], inequality (35.21) implies inequality (35.11) as a special case [CLAUSER/SHIMONY 1978, 1890], so that CHSH inequality is a more general case of the I Bell inequality.

CHSH divide the detection possibilities in four cases: a) when both particles of a pair are observed; b) only particle 1 is observed; c) only particle 2; d) neither particle is observed. But, in order to develop their proof, CHSH are obliged to assume that the distribution ρ of the union of these four cases is independent of $\mathbf{a}$ and $\mathbf{b}$ [CLAUSER *et al.* 1969, 881].[3]

35.2.2 Bell in 1971

Stimulated by the progress of the research, Bell returned to the problem in 1971 [BELL 1971, 37]. Differently from CHSH, Bell implicitly proposes a device with an auxiliary apparatus (= event-ready detectors) to measure the number of pairs emitted by the source, and collocated it before the analyzers (= Cascade exp. 2) [see figure 35.3].

However, because instruments themselves could contain HV which could influence the result, he preferred to work with averages, so that in place of Eq. (35.3), we have:

$$\mathrm{C}(\mathbf{a}, \mathbf{b}) = \int d\lambda_{\rm HV}\rho(\lambda_{\rm HV})\overline{A}_{\mathbf{a}}(\lambda_{\rm HV})\overline{B}_{\mathbf{b}}(\lambda_{\rm HV}); \tag{35.22}$$

[3] On this point see also [CLAUSER/SHIMONY 1978, 1890, 1904].

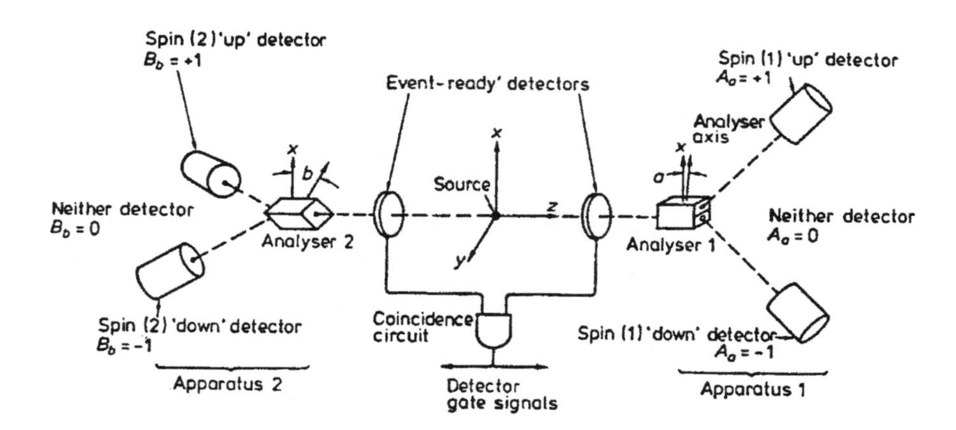

Figure 35.3: Apparatus configuration used by Bell in 1971. Event-ready detectors signal to both arms that a pair of particles has been emitted. For a given gate signal, the result on either arm is assigned the value +1 if the corresponding spin-up detector responds, −1 if the spin-down detector responds, and 0 if neither detector responds — from [CLAUSER/SHIMONY 1978, 1891].,

and in place of conditions (35.2), we have now:

$$|\overline{A}| \leq 1, \quad |\overline{B}| \leq 1. \tag{35.23}$$

In practice there will be some occasions on which one or both detectors fail to register either way. If a' and b' are alternative settings of instruments we have:

$$\begin{aligned} \mathrm{C}(\mathbf{a},\mathbf{b}) - \mathrm{C}(\mathbf{a},\mathbf{b}\prime) &= \int d\lambda_{\mathrm{HV}}\rho(\lambda_{\mathrm{HV}})[\overline{A}_{\mathbf{a}}(\lambda_{\mathrm{HV}})\overline{B}_{\mathbf{b}}(\lambda_{\mathrm{HV}}) - \overline{A}_{\mathbf{a}}(\lambda_{\mathrm{HV}})\overline{B}_{\mathbf{b}'}(\lambda_{\mathrm{HV}})] \\ &= \int d\lambda_{\mathrm{HV}}\rho(\lambda_{\mathrm{HV}})[\overline{A}_{\mathbf{a}}(\lambda_{\mathrm{HV}})\overline{B}_{\mathbf{b}}(\lambda_{\mathrm{HV}})(1 \pm \overline{A}_{\mathbf{a}'}(\lambda_{\mathrm{HV}})\overline{B}_{\mathbf{b}'}(\lambda_{\mathrm{HV}}))] \\ &\quad - \int d\lambda_{\mathrm{HV}}\rho(\lambda_{\mathrm{HV}})[\overline{A}_{\mathbf{a}}(\lambda_{\mathrm{HV}})\overline{B}_{\mathbf{b}'}(\lambda_{\mathrm{HV}})(1 \pm \overline{A}_{\mathbf{a}'}(\lambda_{\mathrm{HV}})\overline{B}_{\mathbf{b}}(\lambda_{\mathrm{HV}}))] \end{aligned} \tag{35.24}$$

Then using inequalities (35.23) we have:

$$|\mathrm{C}(\mathbf{a},\mathbf{b}) - \mathrm{C}(\mathbf{a},\mathbf{b}')| \leq \int d\lambda_{\mathrm{HV}}\rho(\lambda_{\mathrm{HV}})(1 \pm \overline{A}_{\mathbf{a}'}(\lambda_{\mathrm{HV}})\overline{B}_{\mathbf{b}'}(\lambda_{\mathrm{HV}})) + \int d\lambda_{\mathrm{HV}}\rho(\lambda_{\mathrm{HV}})(1 \pm \overline{A}_{\mathbf{a}'}(\lambda_{\mathrm{HV}})\overline{B}_{\mathbf{b}}(\lambda_{\mathrm{HV}})); \tag{35.25}$$

or

$$|\mathrm{C}(\mathbf{a},\mathbf{b}) - \mathrm{C}(\mathbf{a},\mathbf{b}')| \leq 2 \pm (\mathrm{C}(\mathbf{a}',\mathbf{b}') + \mathrm{C}(\mathbf{a}',\mathbf{b})); \tag{35.26}$$

or:

$$|\mathrm{C}(\mathbf{a},\mathbf{b}) - \mathrm{C}(\mathbf{a},\mathbf{b}')| + |\mathrm{C}(\mathbf{a}',\mathbf{b}') + \mathrm{C}(\mathbf{a}',\mathbf{b})| \leq 2, \tag{35.27}$$

which is the *III Bell inequality*.

It is easy to find a QM counterexample to inequality (35.27) [BELL 1981b, 152–53].[4]

Proof

[4]See also [CLAUSER/SHIMONY 1978, 1893–94].

Taking into account the imperfections of instruments, the QM correlations have the form:

$$\mathrm{C}(\mathbf{a},\mathbf{b})_{\mathrm{QM}} = C'\mathbf{a}\cdot\mathbf{b}, \tag{35.28}$$

where C' is some coefficient which is ± 1 only in the idealized case. If we take $\mathbf{a}$, $\mathbf{a}'$, $\mathbf{b}$ and $\mathbf{b}'$ to be coplanar and the angle between $\mathbf{a}$ and $\mathbf{b}$, between $\mathbf{b}$ and $\mathbf{a}'$, between $\mathbf{a}'$ and $\mathbf{b}'$ to be $\theta = \pi/4$, we have [see figure 35.4]:

$$\begin{aligned}[\mathrm{C}(\mathbf{a},\mathbf{b}) - \mathrm{C}(\mathbf{a},\mathbf{b}') + \mathrm{C}(\mathbf{a}',\mathbf{b}') + \mathrm{C}(\mathbf{a}',\mathbf{b})]_{\mathrm{QM}} &= \left[\frac{\sqrt{2}}{2} - (-\frac{\sqrt{2}}{2}) + \frac{\sqrt{2}}{2} + \frac{\sqrt{2}}{2}\right] C' \\ &= 2\sqrt{2}C', \end{aligned} \tag{35.29}$$

with $|\mathbf{a}| = |\mathbf{b}| = |\mathbf{a}'| = |\mathbf{b}'| = 1$ so that $\mathbf{a}\cdot\mathbf{b} = \cos\theta$. The necessary and sufficient condition to see the QM violation is that the 'efficiency' $C' > \sqrt{2}/2$. Q.E.D.

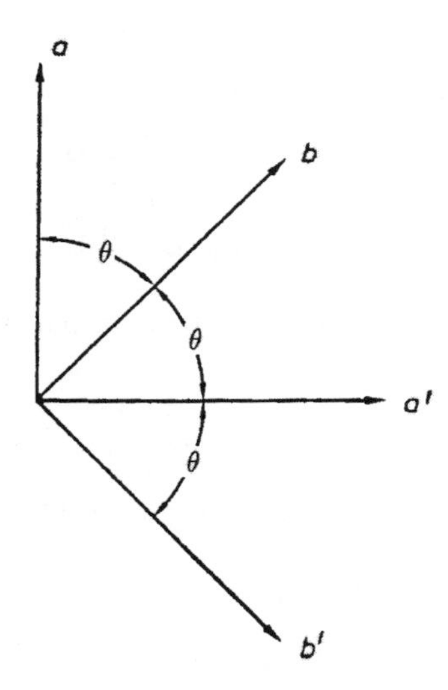

Figure 35.4: Optimal orientation for $\mathbf{a}, \mathbf{a}', \mathbf{b}, \mathbf{b}'$ for Cascade exp. 2 in order to show QM violation of inequality (35.27) — from [CLAUSER/SHIMONY 1978, 1894].,

35.2.3 Clauser/Horne Inequality

The preceding proofs were generalized by Clauser and Horne (= CH) [CLAUSER/HORNE 1974, 527][5] eliminating event-ready detectors of Bell's proof of 1971: another step was made towards the realization of concrete experiments.

Statement of the General Problem

The apparatus is the same as that in Cascade Exp. 1 of CHSH [see subsection 35.2.1]. The source during a fixed time emits N pairs of particles. Let us denote with $N_1(\mathbf{a})$ and $N_2(\mathbf{b})$ the number

[5]See also [CLAUSER/SHIMONY 1978, 1894–95].

of counts at detector 1 and 2, respectively, and with $N_{12}(\mathbf{a}, \mathbf{b})$ the number of simultaneous counts (coincidence counts). When N is sufficiently large, the corresponding probabilities are:

$$\wp_1(\mathbf{a}) = \frac{N_1(\mathbf{a})}{N}; \quad \wp_2(\mathbf{b}) = \frac{N_2(\mathbf{b})}{N}; \quad \wp_{12}(\mathbf{a}, \mathbf{b}) = \frac{N_{12}(\mathbf{a}, \mathbf{b})}{N}. \tag{35.30}$$

Assuming an HV parameter λ_{HV}, as we did before, the requirement of separability is now [Eq. (35.1)]:

$$\wp_{12}(\lambda_{\text{HV}}, \mathbf{a}, \mathbf{b}) = \wp_1(\lambda_{\text{HV}}, \mathbf{a})\wp_2(\lambda_{\text{HV}}, \mathbf{b}). \tag{35.31}$$

The ensemble average probabilities of Eqs. (35.30) are:

$$\wp_i(j) = \int d\rho \; \wp_i(\lambda_{\text{HV}}, j) \;, \quad i = 1, 2, \quad j = \mathbf{a}, \mathbf{b}; \tag{35.32a}$$

$$\wp_{12}(\mathbf{a}, \mathbf{b}) = \int d\rho \wp_1(\lambda_{\text{HV}}, \mathbf{a})\wp_2(\lambda_{\text{HV}}, \mathbf{b}). \tag{35.32b}$$

The probabilities for orientations $\mathbf{a}$ and $\mathbf{a}'$ of analyzer 1 and $\mathbf{b}$ and $\mathbf{b}'$ of analyzer 2, respectively, must satisfy the inequalities:

$$0 \leq \wp_1(\lambda_{\text{HV}}, \mathbf{a}) \leq 1, \tag{35.33a}$$
$$0 \leq \wp_1(\lambda_{\text{HV}}, \mathbf{a}') \leq 1, \tag{35.33b}$$
$$0 \leq \wp_2(\lambda_{\text{HV}}, \mathbf{b}) \leq 1, \tag{35.33c}$$
$$0 \leq \wp_2(\lambda_{\text{HV}}, \mathbf{b}') \leq 1. \tag{35.33d}$$

To proceed further in the derivation we need a mathematical lemma [CLAUSER/HORNE 1974, 532]:

Lemma 35.1 (Clauser/Horne) *Given six numbers x_1, x_2, y_1, y_2, X and Y such that:*

$$0 \leq x_1 \leq X, \quad 0 \leq x_2 \leq X, \quad 0 \leq y_1 \leq Y, \quad 0 \leq y_2 \leq Y \tag{35.34}$$

the function $f = x_1y_1 - x_1y_2 + x_2y_1 + x_2y_2 - Yx_2 - Xy_1$ satisfies the inequalities:

$$-XY \leq f \leq 0. \tag{35.35}$$

So that from Eqs. (35.31), (35.33) and (35.35) we obtain:

$$-1 \leq \wp_{12}(\lambda_{\text{HV}}, \mathbf{a}, \mathbf{b}) - \wp_{12}(\lambda_{\text{HV}}, \mathbf{a}, \mathbf{b}') + \wp_{12}(\lambda_{\text{HV}}, \mathbf{a}', \mathbf{b}) + \wp_{12}(\lambda_{\text{HV}}, \mathbf{a}', \mathbf{b}') - \wp_1(\lambda_{\text{HV}}, \mathbf{a}') - \wp_2(\lambda_{\text{HV}}, \mathbf{b}) \leq 0. \tag{35.36}$$

Integrating Eq. (35.36) over $d\rho$ and using Eqs. (35.31) and (35.32b), the result is:

$$-1 \leq \wp_{12}(\mathbf{a}, \mathbf{b}) - \wp_{12}(\mathbf{a}, \mathbf{b}') + \wp_{12}(\mathbf{a}', \mathbf{b}) + \wp_{12}(\mathbf{a}', \mathbf{b}') - \wp_1(\mathbf{a}') - \wp_2(\mathbf{b}) \leq 0, \tag{35.37}$$

which is the *CH Inequality*, the Fourth of the Bell family.

We can rewrite the RHS of this inequality in the following form:

$$\frac{\wp_{12}(\mathbf{a}, \mathbf{b}) - \wp_{12}(\mathbf{a}, \mathbf{b}') + \wp_{12}(\mathbf{a}', \mathbf{b}) + \wp_{12}(\mathbf{a}', \mathbf{b}')}{\wp_1(\mathbf{a}') + \wp_2(\mathbf{b})} \leq 1, \tag{35.38}$$

which involves only quantities which are independent of N.[6]

[6] Another inequality has recently been proposed in [ARDEHALI 1998] as a generalization of CHSH's and CH's ones.

Different Theoretical Predictions

Assuming now [CLAUSER/HORNE 1974, 528–29][7] **a** and **b** to be orientation angles relative to some reference axis in a fixed plane, cylindrical symmetry about a line normal to the fixed plane and reflection symmetry with respect to the fixed plane, QM makes the following predictions:

$$\wp_i(j)_{\rm QM} \equiv \wp_i, \quad i = 1, 2, \quad j = \mathbf{a}, \mathbf{b}; \tag{35.39a}$$

$$\wp_{12}(\mathbf{a}, \mathbf{b})_{\rm QM} \equiv \wp_{12}(|\,\mathbf{a} - \mathbf{b}|)_{\rm QM}; \tag{35.39b}$$

$$\mathrm{C}(\mathbf{a}, \mathbf{b})_{QM} \equiv \mathrm{C}(|\mathbf{a} - \mathbf{b}|)_{\rm QM}. \tag{35.39c}$$

Stochastic HV-theories which assume the Separability principle exhibit parallel symmetries, but these do not follow from QM:

$$\wp_i(j) \equiv \wp_i, \quad i = 1, 2, \quad j = \mathbf{a}, \mathbf{b}; \tag{35.40a}$$

$$\wp_{12}(\mathbf{a}, \mathbf{b}) \equiv \wp_{12}(|\mathbf{a} - \mathbf{b}|); \tag{35.40b}$$

$$\mathrm{C}(\mathbf{a}, \mathbf{b}) \equiv \mathrm{C}(|\mathbf{a} - \mathbf{b}|). \tag{35.40c}$$

Suppose now that we take **a**, **a**′, **b** and **b**′ so that [see again figure 35.4]:

$$|\mathbf{a} - \mathbf{b}| = |\mathbf{a}' - \mathbf{b}| = |\mathbf{a}' - \mathbf{b}'| = \frac{1}{3}|\mathbf{a} - \mathbf{b}'| = \theta. \tag{35.41}$$

We can reformulate Eq. (35.37) in the following manner:

$$-1 \leq 3\wp_{12}(\theta) - \wp_{12}(3\theta) - \wp_1 - \wp_2 \leq 0; \tag{35.42}$$

or:

$$|3\mathrm{C}(\theta) - \mathrm{C}(3\theta)| \leq 2 \tag{35.43}$$

and

$$\tilde{\mathrm{C}}(\theta) \leq 1, \tag{35.44}$$

where

$$\tilde{\mathrm{C}}(\theta) := \frac{3\wp_{12}(\theta) - \wp_{12}(3\theta)}{\wp_1 + \wp_2}. \tag{35.45}$$

We can now prove that QM predictions can be in conflict with the CH inequality.

Proof

Consider an experiment with the following QM predictions:

$$\wp_{12}(\theta)_{\rm QM} = \frac{1}{4}\eta_1\eta_2\wp_1\wp_{2/1}[\epsilon_+^1\epsilon_+^2 + \epsilon_-^1\epsilon_-^2\,\mathrm{E}\cos n\theta], \tag{35.46a}$$

$$(\wp_1)_{\rm QM} = \frac{1}{2}\eta_1\wp_1\epsilon_+^1, \tag{35.46b}$$

$$(\wp_2)_{\rm QM} = \frac{1}{2}\eta_2\wp_2\epsilon_+^2, \tag{35.46c}$$

where η_j represents the effective quantum efficiency of detector j and $\epsilon_+^j := \epsilon_M^j + \epsilon_m^j$ and $\epsilon_-^j := \epsilon_M^j - \epsilon_m^j$ – where ϵ_M^j and ϵ_m^j are the maximum and the minimum transmissions of the analyzers; the functions

[7] See also [CLAUSER/SHIMONY 1978, 1896–97, 1901–902] .

$\wp_1$ and $\wp_2$ are the probabilities that an emission enters apparatus 1 or 2; and $\wp_{2/1}$ is the conditional probability that, if emission 1 enters apparatus 1 then emission 2 enters apparatus 2. E is a measure of initial-state purity. The values of n are 1 or 2 depending upon wether the experiment is performed with fermions or bosons.

Now the QM predictions (35.46) for the function $\tilde{\mathrm{C}}(\theta)$ [Eq. (35.45)] are given by:

$$\tilde{\mathrm{C}}_{\mathrm{QM}}(\theta) = \frac{1}{4}\eta\wp_{2/1}\left\{2\epsilon_+ + [3\cos n\theta - \cos 3n\theta]\mathrm{E}\left(\frac{\epsilon_-^2}{\epsilon_+}\right)\right\}, \tag{35.47}$$

where for simplicity $\eta \equiv \eta_1 = \eta_2$, $\wp_1 = \wp_2$, $\epsilon_+ = \epsilon_+^1 = \epsilon_+^2$, $\epsilon_- \equiv \epsilon_-^1 = \epsilon_-^2$ Selecting the value $\theta = \pi/4n$, one finds that the condition for violation of inequality (35.44) is:

$$\eta\wp_{2/1}\epsilon_+\left[\sqrt{2}(\frac{\epsilon_-^2}{\epsilon_+})^2\mathrm{E} + 1\right] > 2. \tag{35.48}$$

Q. E. D.

We shall see later [subsection 35.3.2] how it is possible to arrive at a physical test of what has been proved here on a mathematical plane. Before that, we discuss [in the next section] a generalization of Holt and a generalization to N quanta [subsection 35.2.5] and present a short review of the fundamental experiment types [subsection 35.3.1].

35.2.4 Holt's Proof

Holt, following [WIGNER 1970, 1007–1008], obtained an interesting result [CLAUSER/SHIMONY 1978, 1897–98].. Consider the same apparatus as that of CHSH and CH [subsections 35.2.1 and 35.2.3, particularly figure 35.2]. Assume that the detection of the component 1 is completely determined by a parameter a of the first polariser but independent of b of the second one, and the same for component 2. We can exhaustively divide the space λ_{HV} into 16 mutually disjoint subspaces $\lambda_{\mathrm{HV}}(ij, kl)$, where each index can be 0 or 1 depending on non-detection or detection; i and j, respectively, refer to the two possible parameters a or a' of SGM1; and k and l to the two possible parameters b or b' of SGM2. If there is a probability measure μ_p on λ_{HV}, then $\mu_p(ij, kl)$ is defined as the probability that the composite state is in $\Lambda_{\mathrm{HV}}(ij, kl)$. Because the 16 subspaces are disjoint we have:

$$\sum_{ijkl}\mu_p(ij, kl) = 1. \tag{35.49}$$

We now define $\wp_1(a)$ as the probability that the component 1 will be detected with position a, and $\wp_{12}(a, b)$ as the probability of joint detection with 1-2-polariser positions on a and b. Then we have:

$$\wp_{12}(a, b) = \mu_p(11, 11) + \mu_p(11, 10) + \mu_p(10, 11) + \mu_p(10, 10); \tag{35.50a}$$
$$\wp_{12}(a, b') = \mu_p(11, 11) + \mu_p(11, 01) + \mu_p(10, 11) + \mu_p(10, 01); \tag{35.50b}$$
$$\wp_{12}(a', b) = \mu_p(11, 11) + \mu_p(11, 10) + \mu_p(01, 11) + \mu_p(01, 10); \tag{35.50c}$$
$$\wp_{12}(a', b') = \mu_p(11, 11) + \mu_p(11, 01) + \mu_p(01, 11) + \mu_p(01, 01); \tag{35.50d}$$

from which we obtain:

$$\wp_1(a') = \mu_p(11, 11) + \mu_p(11, 10) + \mu_p(11, 01) + \mu_p(11, 00) + \mu_p(01, 11) + \mu_p(01, 10) + \mu_p(01, 01) + \mu_p(01, 00); \tag{35.51a}$$
$$\wp_2(b) = \mu_p(11, 11) + \mu_p(11, 10) + \mu_p(10, 11) + \mu_p(10, 10) + \mu_p(01, 11) + \mu_p(01, 10) + \mu_p(00, 11) + \mu_p(00, 10). \tag{35.51b}$$

From which it follows:

$$\begin{aligned}\wp_{12}(a,b) &- \wp_{12}(a,b\prime)+\wp_{12}(a\prime,b)+\wp_{12}(a\prime,b\prime)-\wp_{1}(a\prime)-\wp_{2}(b) = -\mu_p(11,01)-\mu_p(11,00)\\ &- \mu_p(10,11)-\mu_p(10,01)-\mu_p(01,10)-\mu_p(01,00)-\mu_p(00,11)-\mu_p(00,10).\end{aligned} \quad (35.52)$$

35.2.5 Generalization with an Arbitrary Number of Quanta

Before we discuss experiments on Bell inequalities, we briefly report a powerful generalization with N quanta (bosons) due to Drummond [DRUMMOND 1983] [see also section 40.5]. The quantum-field state of interest is defined to be:

$$|n\rangle = \frac{1}{n!(n+1)^{\frac{1}{2}}}[\hat{a}_1^\dagger \hat{b}_1^\dagger + \hat{a}_2 \hat{b}_2]^N |0\rangle, \quad (35.53)$$

where n is the number of quanta produced at energies E_a, E_b; operators $\hat{a}_1, \hat{a}_2(\hat{a}_1^\dagger, \hat{a}_2^\dagger)$ annihilate (create) orthogonally polarised bosons with energy E_a propagating in the $+x$ direction with polarisation in the y, z directions. Operators $\hat{b}_1, \hat{b}_2(\hat{b}_1^\dagger, \hat{b}_2^\dagger)$ behave identically, except that they annihilate (create) quanta with energy E_b propagating in the $-x$ direction. The intensity correlation [see subsection 5.1.2]has following structure:

$$\mathrm{C}^{IJ}(\gamma_\theta, n) = \langle n|A^I(1,a)A^J(\gamma_\theta, b)|n\rangle, \quad (35.54)$$

where:

$$\begin{aligned}\gamma_\theta &:= \cos^2\theta, & \theta &= \theta_a - \theta_b, & (35.55a)\\ a &:= [\hat{a}_1^\dagger, \hat{a}_2^\dagger, \hat{a}_1, \hat{a}_2], & b &:= [\hat{b}_1^\dagger, \hat{b}_2^\dagger, \hat{b}_1, \hat{b}_2], & (35.55b)\\ A^J(\gamma_\theta, a) &:= & & [\gamma_\theta^{\frac{1}{2}}\hat{a}_1^\dagger + (1-\gamma_\theta^{\frac{1}{2}})\hat{a}_2^\dagger]^J[\gamma_\theta^{\frac{1}{2}}\hat{a}_1 + (1-\gamma_\theta^{\frac{1}{2}})\hat{a}_2]^J, & (35.55c)\\ A^J(\infty, a) &:= & & :[\hat{a}_1^\dagger\hat{a}_1 + \hat{a}_2^\dagger\hat{a}_2]^J: . & (35.55d)\end{aligned}$$

The correlation $\mathrm{C}^{IJ}(\gamma_\theta, n)$ is proportional to the probability of observing I quanta of type 'a' at position X, with polariser angle θ_a; and of J quanta of type 'b' at position $-X$ with polariser angle θ_b. The measurement $\mathrm{C}^{IJ}(\infty, n)$ is identical to $\mathrm{C}^{IJ}(\gamma_\theta, n)$ except that there is no polariser in the detection of type-'b' quanta. Relative probabilities are defined by

$$g_n^J(\theta) := \frac{\mathrm{C}^{IJ}(\gamma_\theta, n)}{\mathrm{C}^{IJ}(\infty, n)}. \quad (35.56)$$

The Bell inequality which can be used in our context is the IV one [see Eq. (35.43)], which now reads:

$$3g(\theta) - g(3\theta) - 2 \leq 0. \quad (35.57)$$

By calculating explicitly the quantum intensity correlations one obtains:

$$\begin{aligned} g_n^1(\theta) &= \frac{[n-1+\gamma_\theta(n+2)]}{3n}, & (35.58a)\\ g_J^J(\theta) &= \gamma_\theta^J, & (35.58b)\end{aligned}$$

for the cases $I = J = 1$ and $I = J = n$, respectively. One sees that the inequality (35.57) is violated if $n = 1$ (this is the standard Bell violation) and when $J = n$ for finite θ.

35.3 Experimental Tests

We now show the passage from the theory in the form of *Gedankenexperimenten* to performed experiments by means of only one inequality and only one experiment, limiting ourselves to remembering the results of other tests.

From the nature of the Bell inequalities (generically understood) it follows that there are two different questions to test [*REDHEAD* 1987b, 107]:

- Are Bell inequalities experimentally violated?
- And, eventually, does the violation conform to the predictions of QM?

35.3.1 A General Presentation of Different Tests

Different tests of the Bell inequalities are possible. First of all we can distinguish, among others, the following correlation types, each one of which requires different experimental setups:

- Two-particle polarisation correlation (spin entanglement).
- Two-particle energy-time entanglement.
- Time entanglement for fields.
- A simultaneous entanglement of the first two types.
- Photon-number correlation.

We shall return later to the second [see section 40.1], the third [section 40.2], the fourth [see section 40.3] and the fifth [see section 40.5] setups. Here we restrict our examination to the first correlation type, and divide all experimental tests of Bell inequalities into three classes [BERTLMANN 1990, 1207–209] [*REDHEAD* 1987b, 107–113]:

- Photon Correlation Experiments
- Proton-Proton Scattering Experiments.
- Positron Annihilation Experiments.

Of all the experiments only two of them — one executed by Holt/Pipkin in 1972 and the other by Faraci, Gutowsky, Notarrigo and Pennisi in 1974 [FARACI *et al.* 1974], the first one with visible photons and the second one with γ-rays —, disagree with the result of a violation of Bell inequalities in respect of QM predictions[8].

The experiment in which we choose to test the theoretical predictions is of the first type: a photon-correlation one on the line of CHSH or CH instead of massive spin$-\frac{1}{2}$ particles, and linear polarisers instead of SGMs [see figure 35.5].[9]

Experiments of this type consist either in atomic cascade or in a Spontaneous parametric downconversion (= SPDC) [see subsection 5.3.2]. In the latter case a sufficiently intense pump beam is incident on a birefringent crystal so that nonlinear (= NL) effects lead to the spontaneous emission of a pair of entangled photons ($\nu_1 + \nu_2 = \nu_0, k_1 + k_2 = k_0$, where ν_0, k_0 are respectively frequency and wave vector of the photon which enters the crystal). The frequency and propagation directions are determined by the orientation of the NL crystal.

[8] See also [FRY 1995] and [CHIAO *et al.* 1995b].

[9] For a review of first experimental attempts in Q-Optics see [REID/WALLS 1986].

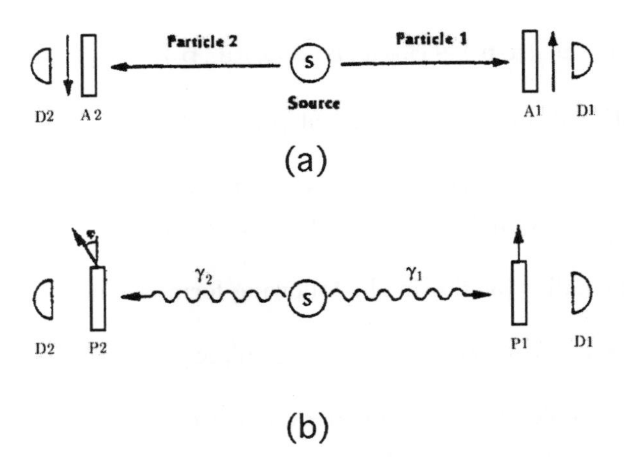

Figure 35.5: (a) Bohm's version of the EPR experiment.
(b) Optical version of the EPR experiment — from [CHIAO *et al.* 1995b, 260].

35.3.2 From the Theory to the Experiment

CH assumption We now return to the testing of Bell inequalities. CHSH assumed that, given that a pair of photons emerges from the polarisers, the probability of their joint detection is independent of the polarisers' orientations **a** and **b** [subsection 35.2.1]. The problem is that there is no way to test this assumption [CLAUSER/SHIMONY 1978, 1912–13]. CH made another assumption which leads to the same results: for every pair emitted (i.e. for each value of $\lambda_{\rm HV}$), the probability of a count with a polariser in place is less than or equal to the corresponding probability with the polariser removed [CLAUSER/HORNE 1974, 530] — in fact very often only a small number of photons is factually detected. This assumption appears reasonable because the insertion of a polarisation analyser imposes an obstacle between the source of the emission and the detector, and it is natural to suppose that an obstacle cannot increase the detection probability. But it is very difficult to prove positively CH's assumption because it requires that the probability be diminished upon the insertion of a polariser for all $\lambda_{\rm HV}$ [CLAUSER/SHIMONY 1978, 1904, 1913]. On the other hand, attempts to invalidate CH's assumption as a means of invalidating experimental proofs for the non-locality or non-separability of QM[10] do not seem as reasonable as the CH's assumption itself [see also subsection 35.4.3].

We therefore assume the CH hypothesis and look for the general conditions in order to test it [CLAUSER/SHIMONY 1978, 1905–906] [FRY 1995, 234–35]; finally, we report an experimental test.

Count rates We denote with ∞ an apparatus configuration in which the analyser is absent. $\wp_i(\lambda_{\rm HV}, \infty)$ denotes the probability of a count from detector i when analyser i is absent and the state of emission is $\lambda_{\rm HV}$. The CH assumption is:

$$0 \leq \wp_i(\lambda_{\rm HV}, j) \leq \wp_i(\lambda_{\rm HV}, \infty) \leq 1, \quad i = 1, 2 \quad j = \mathbf{a}, \mathbf{b}. \tag{35.59}$$

[10]For example by Caser [CASER 1992, 24–25].

From Eq. (35.59) and Eq. (35.37) we have:

$$-\wp_{12}(\infty,\infty) \leq \wp_{12}(\mathbf{a},\mathbf{b}) - \wp_{12}(\mathbf{a},\mathbf{b}') + \wp_{12}(\mathbf{a}',\mathbf{b}) + \wp_{12}(\mathbf{a}',\mathbf{b}') - \wp_{12}(\mathbf{a}',\infty) - \wp_{12}(\infty,\mathbf{b}) \leq 0. \tag{35.60}$$

We again use the rotational invariance argument [Eqs. (35.40a) and (35.41)]:

$$\wp_{12}(j,\infty) := \wp_{12}(\infty), \quad j = \mathbf{a},\mathbf{b}; \tag{35.61a}$$
$$\wp_{12}(\mathbf{a},\mathbf{b}) := \wp_{12}(\theta),\ \theta = |\mathbf{a}-\mathbf{b}|, \tag{35.61b}$$

so that we may rewrite inequality (35.60) in the following form [see inequality (35.42)]:

$$-\wp_{12}(\infty,\infty) \leq 3\wp_{12}(\theta) - \wp_{12}(3\theta) - \wp_{12}(\mathbf{a}',\infty) - \wp_{12}(\infty,\mathbf{b}) \leq 0, \tag{35.62}$$

for all $\mathbf{a}'$ and $\mathbf{b}$. Since the emission rates in all experiments would be held constant, we can write the ratio of probabilities as ratios of count rates:

$$\frac{\wp_{12}(j,\infty)}{\wp_{12}(\infty,\infty)} = \frac{R_i}{R_0}, \quad i = 1,2, \quad j = \mathbf{a},\mathbf{b}; \qquad \frac{\wp_{12}(\theta)}{\wp_{12}(\infty,\infty)} = \frac{R(\theta)}{R_0}, \tag{35.63}$$

so that we can write Eq. (35.62) in the following form:

$$-R_0 \leq 3R(\theta) - R(3\theta) - R_1 - R_2 \leq 0. \tag{35.64}$$

If we take $\theta = \pi/8$ for the maximum violation we have:

$$-R_0 \leq 3R(\pi/8) - R(3\pi/8) - R_1 - R_2 \leq 0. \tag{35.65}$$

If we take for the minimum violation $\theta = 3\pi/8$, using the fact that $9\pi/8$ is the same angle as $\pi/8$, we have:

$$-R_0 \leq 3R(3\pi/8) - R(\pi/8) - R_1 - R_2 \leq 0. \tag{35.66}$$

Dividing by R_0 and subtracting the second inequality from the first we obtain [see also figure 35.6]:

$$\frac{|R(\pi/8) - R(3\pi/8)|}{R_0} \leq \frac{1}{4}. \tag{35.67}$$

Ideal case If we take in the ideal case [SHIMONY 1971b, 82–85][11] pairs of photons propagating in opposite directions from the source along the z axis with total angular momentum 0 and total parity +1, for the polarisation part of the wave function we have:

$$\Psi_0 = \frac{1}{2}\left[\begin{pmatrix}1\\0\\0\end{pmatrix}_1 \otimes \begin{pmatrix}1\\0\\0\end{pmatrix}_2 + \begin{pmatrix}0\\1\\0\end{pmatrix}_1 \otimes \begin{pmatrix}0\\1\\0\end{pmatrix}_2\right], \tag{35.68}$$

where the first two expressions between parentheses represent polarisation vectors along the x–axis and the other two along the y–axis.

[11] See also [CLAUSER/HORNE 1974, 530] [CLAUSER/SHIMONY 1978, 1906–907].

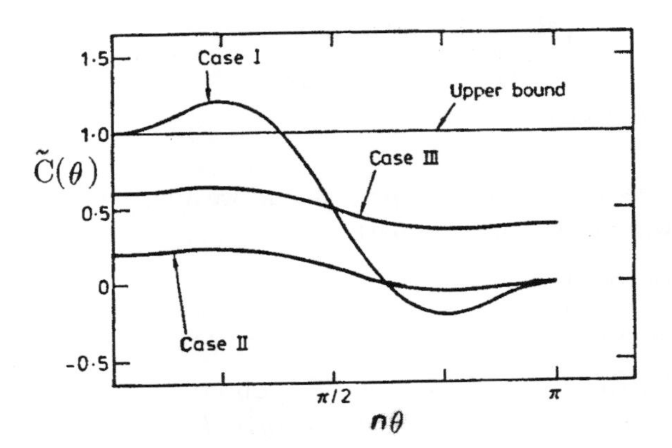

Figure 35.6: Typical dependence of $\tilde{C}(\theta)$ upon $n\theta$ for cases I–III. The upper bound for $\tilde{C}(\theta)$ set by inequality (35.44) is +1.
Case I (nearly ideal): we have: $\eta \simeq \wp_{2/1} \simeq \mathrm{E} \simeq \epsilon_+ \simeq \epsilon_- \simeq 1$ [see Eqs. (35.46)].
Case II: experiments have nearly ideal parameters $\mathrm{E} \simeq \epsilon_+ \simeq \epsilon_- \simeq 1$, but have $\eta \ll 1$ and/or $\wp_{2/1} \ll 1$.
Case III: experiments have nearly ideal parameters $\eta \simeq \wp_{2/1} \simeq 1$ but have $\mathrm{E} \ll 1$ and/or $\epsilon_-/\epsilon_+ \ll 1$ — from [CLAUSER/SHIMONY 1978, 1903].

A projection operator for linear polarisation along an axis lying in the xy plane and making angle ϕ with the x–axis is

$$\hat{P}(\phi) := \begin{bmatrix} \cos^2\phi & \cos\phi\sin\phi & 0 \\ \cos\phi\sin\phi & \sin^2\phi & 0 \\ 0 & 0 & 0 \end{bmatrix}. \tag{35.69}$$

The vector

$$\begin{pmatrix} \cos\phi \\ \sin\phi \\ 0 \end{pmatrix}, \tag{35.70}$$

representing linear polarisation in that chosen direction, is an eigenvector of the projector (35.69) with eigenvalue 1; while vector

$$\begin{pmatrix} -\sin\phi \\ \cos\phi \\ 0 \end{pmatrix}, \tag{35.71}$$

representing linear polarisation perpendicular to the chosen direction, is again an eigenvector of $\hat{P}(\phi)$ with eigenvalue 0. Vector

$$\begin{pmatrix} 0 \\ 0 \\ 1 \end{pmatrix} \tag{35.72}$$

represents polarisation along the z–axis.

The QM predictions (using again $\theta = |\mathbf{a} - \mathbf{b}|$) are:

$$\left[\frac{R(\theta)}{R_0}\right]_{\Psi_0} = \left\langle \Psi_0 | \hat{P}(\mathbf{a}) \otimes \hat{P}(\mathbf{b}) | \Psi_0 \right\rangle = \frac{1}{4}(1 + \cos 2\theta), \tag{35.73}$$

from which we find:

$$\left[\frac{R(\pi/8)}{R_0} - \frac{R(3\pi/8)}{R_0}\right]_{\Psi_0} = \frac{1}{4}\sqrt{2}, \tag{35.74}$$

which obviously violates Eq. (35.67).

Cascade Exp. 3 setup As has already been pointed out, we divide [CLAUSER/SHIMONY 1978, 1907–908, 1909] the following test, into two parts:

- the test for stochastic HV prediction,
- the test for QM predictions.

Freedman/Clauser [FREEDMAN/CLAUSER 1972] observed (= Cascade Exp. 3) the 5513 Å and 4227 Å pairs of photons produced by $4p^2\,{}^1S_0 \rightsquigarrow 4p4s\,{}^1P_1 \rightsquigarrow 4s^2\,{}^1S_0$ cascade in calcium [see figure 35.7].

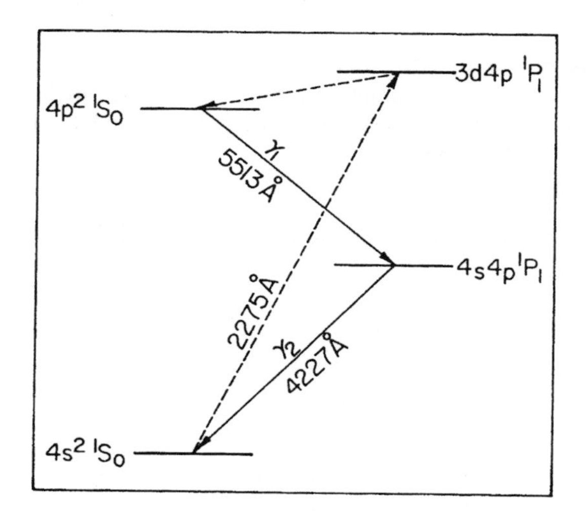

Figure 35.7: Partial Grotrian diagram of atomic calcium for Cascade Exp. 3 — from [FREEDMAN *et al.* 1976, 53].

Calcium atoms in a beam from an oven were excited by resonance absorption to the $4p4s\,{}^1P_1$ level, from which a considerable fraction decayed to the $4p^2\,{}^1S_0$ state at the top of the cascade[12] [see figure 35.8].

Testing HV-Theories The average ratios for approximately 200 h of running time are:

$$\left\langle\frac{R(\pi/8)}{R_0}\right\rangle = 0.400 \pm 0.007, \tag{35.75a}$$

$$\left\langle\frac{R(3\pi/8)}{R_0}\right\rangle = 0.100 \pm 0.003, \tag{35.75b}$$

$$\left\langle\frac{R(\pi/8) - R(3\pi/8)}{R_0}\right\rangle = 0.300 \pm 0.008, \tag{35.75c}$$

[12] More technical details can be found in the literature cited.

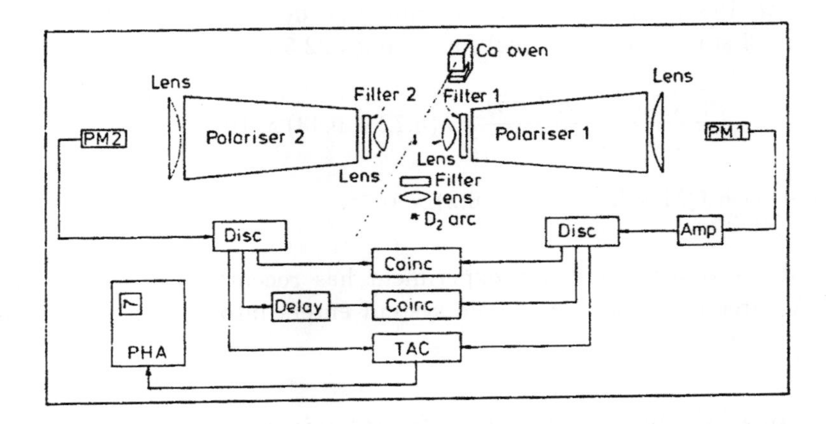

Figure 35.8: Schematic diagram of apparatus and associated electronics of the experiment by Freedman and Clauser. Scalers (not shown) monitored the outputs of the discriminators and coincidence circuits — from [FREEDMAN/CLAUSER 1972].

which clearly violates inequality (35.67) — 6 standard deviations.

Testing QM We now test the *QM predictions*. With all the necessary corrections from an ideal case, QM predictions (35.74) can be written as:

$$\frac{[R(\pi/8) - R(3\pi/8)]}{R_0} = (0.401 \pm 0.005) - (0.100 \pm 0.005) = 0.301 \pm 0.007. \tag{35.76}$$

And we see a clear agreement with experimental results [Eq. (35.75a)–(35.75c)].

35.3.3 Other Atomic Cascade Experiments

Cascade Exp. 4 A very similar experiment (= Cascade Exp. 4) is that of Fry and Thompson [FRY/THOMPSON 1976]:[13] they analysed linear polarisation correlation of photon pairs from $7^3S_1 - 6^3P_1 - 6^1S_0$ cascade of mercury (Hg^{200}). Using the restriction of HV local theories in the form [see Eq. (35.67)]:

$$\delta = \left|\frac{R(67,5)}{R_0} - \frac{R(22,5)}{R_0}\right| - \frac{1}{4} \leq 0, \tag{35.77}$$

we find that experimental results are:

$$\delta_{\text{exp}} = 0.046 \pm 0.014, \tag{35.78}$$

which contradicts inequality (35.77) but not QM predictions, which are:

$$\delta_{\text{QM}} = 0.044 \pm 0.007. \tag{35.79}$$

[13]See also [FRY 1995, 237–55].

Cascade Exp. 5 An experiment (= Cascade Exp. 5) was performed by Aspect, Grangier and Roger [ASPECT *et al.* 1981] with a radiative atomic cascade of calcium, obtaining a maximal violation (more then 13 standard deviations) for θ values 22.5 and 67.5:

$$\frac{[R(22.5) - R(67.5)]}{R_0} = (5.72 \pm 0.43) \times 10^{-2}, \tag{35.80}$$

in perfect agreement with QM values: $(5.8 \pm 0.2) \times 10^{-2}$.

Another proposal Another interesting experiment has recently been proposed by Ansari [ANSARI 1997]. A three-level cascade atomic system enters into a multiwave mixing[14] with two field modes.

35.3.4 Low Energy Proton-Proton Scattering

Lamehi-Rachti and Mittig [LAMEHI-R./MITTIG 1976] performed the 'classical experiment' (SGM-Exp. 7) [described briefly in figures 35.9]. The agreement with QM predictions is again confirmed.

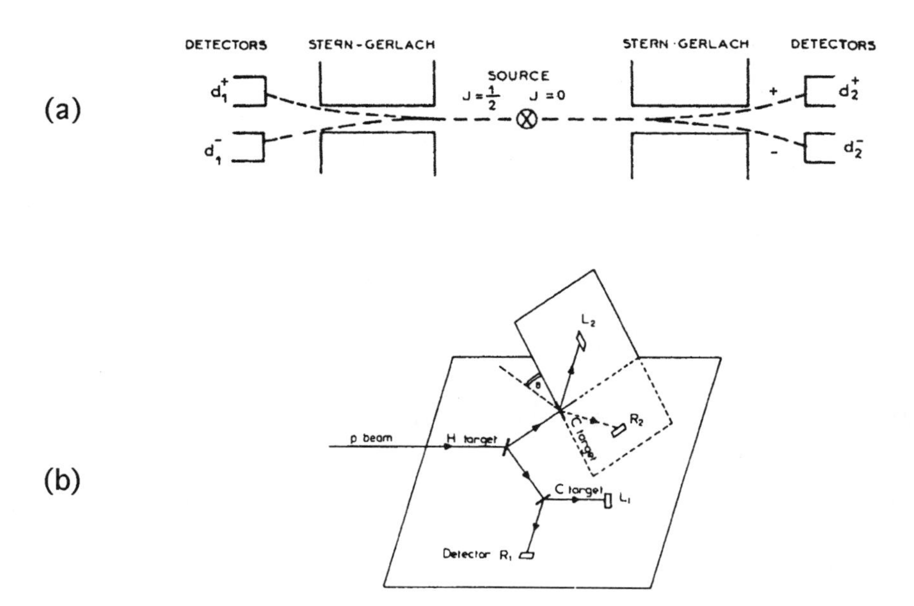

Figure 35.9: (a) A source prepares two particles of spin $J = 1/2$ in an intermediate state $J = 0$. This state disintegrates by emitting the particles.
(b) Schematic experimental setup for the measurement of spin correlation. A beam of protons hits a target containing hydrogen. After the scattering, the two protons enter in kinematical coincidence into analyzers at $\theta = 45°$. In the analyzers the protons are scattered by a carbon foil and the coincidence counted — from [LAMEHI-R./MITTIG 1976, 424–25].

[14]Theory in [ANSARI/ZUBAIRY 1988].

35.3.5 Chained Bell Inequalities

Braunstein and Caves [BRAUNSTEIN/CAVES 1989, 34] [BRAUNSTEIN/CAVES 1990, 33–45] developed an interesting refinement on the conceptual level. They showed that we can reach greater discrepancies between experimental results and Bell inequalities by using chained Bell inequalities[15]. By writing the third Bell inequality (35.27) — where three of the angles between the polarisation analyzers are equal (maximal violation) [see also Eq. (35.57) and figure 35.4] — as follows:

$$-2 \leq 3\mathrm{C}\left(\frac{\theta}{3}\right) - \mathrm{C}(\theta) \leq +2, \tag{35.81}$$

and reformulating its RHS as follows:

$$3\left[1 - \mathrm{C}\left(\frac{\theta}{3}\right)\right] \geq 1 - \mathrm{C}(\theta), \tag{35.82}$$

we can iterate it as follows:

$$9\left[1 - \mathrm{C}\left(\frac{\theta}{9}\right)\right] \geq 3\left[1 - \mathrm{C}\left(\frac{\theta}{3}\right)\right] \geq [1 - \mathrm{C}(\theta)]. \tag{35.83}$$

The effect of the iteration is shown in figure 35.10.

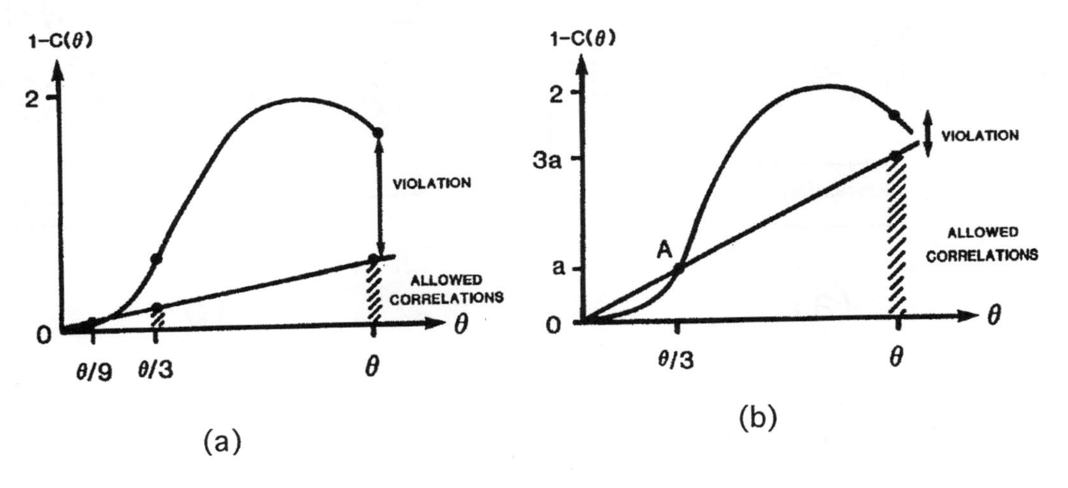

Figure 35.10: (a) A measurement of $\mathrm{C}\left(\frac{\theta}{9}\right)$ fixes $1 - \mathrm{C}\left(\frac{\theta}{9}\right)$ for the LHS of Eq. (35.83), which gives an upper bound for $1 - \left(\frac{\theta}{3}\right)$ and in turn an upper bound for $1 - \mathrm{C}(\theta)$: a measurement at $\theta/9$ (an interpolate detector setting) yields a stronger violation at θ, as compared with (b) violation obtained from a direct measurement at $\theta/3$ — from [BRAUNSTEIN/CAVES 1989, 33–34].

35.4 Loopholes

The test we have discussed so far seem rather convincing and in favour of QM. However, in order to be careful, one has to admit the possibility of some loopholes. In the following five subsections we analyse five possible loopholes, by means of which one could deny the validity of the tests.

[15]See also [Lw. LANDAU 1987a, 54–56] .

35.4.1 Aspect/Dalibard/Grangier/Roger Experiment

First loophole In all the previously discussed experiments, there is always the possibility that the result of a measurement by using a polariser direction depends on the orientation of the other polariser. In other words it could be the case that QM violates not only separability but also locality, which we understand in this context as a superluminal interaction between both polarisers.

We shall discuss this problem in more detail later [see section 36.5]. Here we are interested in excluding not the non-locality as a possibility, but only its possible influence on the experiments' results, in order to test the separability as such.

Cascade Exp. 6 In 1976 Aspect [ASPECT 1976] proposed for the first time an experiment in which, instead of a fixed apparatus as is shown in figure 35.11.(a), an apparatus with optical commutators is used [figure 35.11.(b)]. The experiment (= Cascade Exp. 6) was performed by Aspect, Dalibard, Grangier and Roger [ASPECT *et al.* 1982a] [ASPECT *et al.* 1982b].[16]

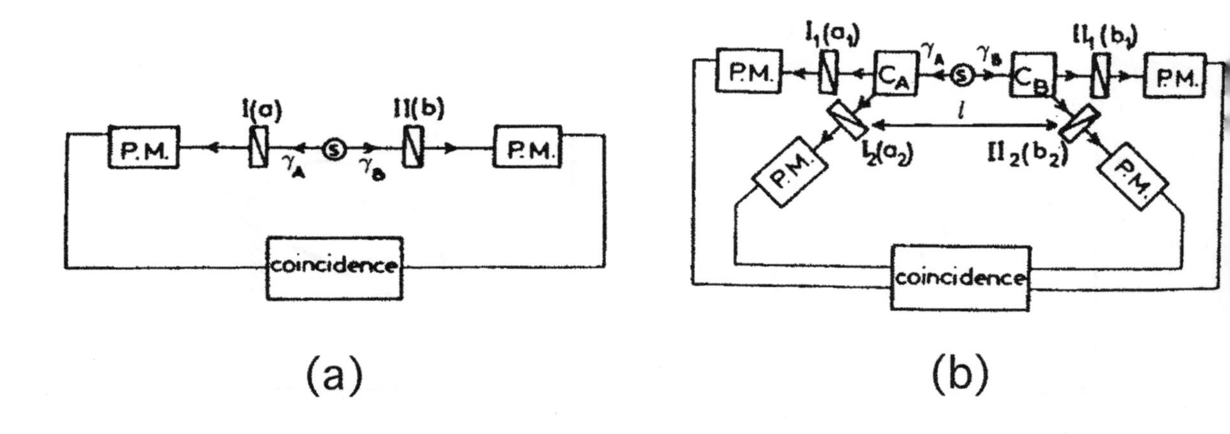

Figure 35.11: (a) Cascade Exp. 3 (Friedman/Clauser experiment): the correlated photons γ_A, γ_B coming from the source S impinge upon the linear polarisers I, II oriented in directions $\mathbf{a}, \mathbf{b}$. The rate of joint detection by the photomultipliers is monitored for various combinations of orientations.
(b) Experiment proposed by Aspect: The optical commutator C_A directs the photon γ_A either towards polariser I_1 with orientation $\mathbf{a}_1$ or to polariser I_2 with orientation $\mathbf{a}_2$. Similarly for C_B. The two commutators work independently (the time intervals between two commutations are taken to be stochastic). The four joint detection rates are monitored and the orientations $\mathbf{a}_1, \mathbf{a}_2, \mathbf{b}_1, \mathbf{b}_2$ are not changed for the whole experiment — from [ASPECT 1976, 436].

Switching between the two channels occurs about every 10 ns. Since this delay, as well as the lifetime of the intermediate level of the cascade (5 ns), is small compared with l/c [see figure 35.11.(b)], a detection event and the corresponding change in the orientation on the other side are separated by space-like intervals.

[16] See also [SHIH/ALLEY 1988] [BUSCH/SCHROECK 1989, 817] [SHIH *et al.* 1995a, 41] [MANDEL 1995b, 1–2].

By testing the CH inequality [Eq. (35.37)] choosing angles between orientation **a** and **b**, **b** and **a**$'$, **a**$'$ and **b**$'$ = 22.5°, and angle between **a** and **b**$'$ = 67.5°, the experimental result for the expression in Eq. (35.37) was 0.101 ± 0.020, clearly violating the corresponding inequality by 5 standard deviations.

Later on, a cascade experiment was performed [PERRIE *et al.* 1985], in which the photons are emitted simultaneously, which provides a space-like separation between detection events and bypasses the difficulty that may be significant absorption and reemission of photons in the source.

Contextuality Note that the second Bell theorem excludes not only non-contextual HV-theories [see subsection 32.3.4], because they respect the requirement of separability, but also contextual ones. In fact, if the two detectors are space-like separated, a supporter of an HV-theory is forced to accept a strong form of non-locality, i.e. an action-at-a-distance, in order to explain how it is possible to obtain the same results with fixed polarisers and with randomly moving ones, if the results are determined by HVs. Hence for the supporter of an HV-theory it would be absurd to reject only the weaker separability and assume the locality [SHIMONY 1984a, 109–116]. It was probably in order to overcome this difficulty that Bohm and others proposed a non-local HV-theory later [see subsection 32.6.2].

35.4.2 Spontaneous Parametric Downconversion Experiments

Second loophole The problem of a possible correlation between polarisers is not the only difficulty in performed experiments which seek to test Bell inequalities. Another difficulty (*second loophole*) concerns the *angular correlation* [SANTOS 1991a] [SANTOS 1992c]: Because of the cosine-squared angular correlation of the directions of photons emitted in an atomic cascade [see Eq. (3.130)], there is an inherent polarisation decorrelation, due to the transversality condition. Hence the very polarisation correlation which could result in a violation of one of the Bell inequalities is reduced for non-collinear photons, so that it seems strictly impossible to disprove a separable HV-theory.

The problem can be overcome [KWIAT *et al.* 1994a, 3210] by using SPDC experiments instead of atomic cascade experiments. These photons can have an angular correlation of better than 1 mrad, although in general they need not be collinear. Initially such experiments could reach only limited efficiencies, but recently efficiencies as high as 90% have been reached.

SPDC Exp. 4 We report here an SPDC experiment (= SPDC Exp. 4) performed for the first time by Alley and Shih and successively improved by Ou/Mandel [OU/MANDEL 1988a]:[17] while Alley/Shih obtained a violation of Bell inequality by three standard deviations, the experiment performed by Ou/Mandel obtained violations as large as six standard deviations [see figure 35.12].

Light from the 351.1–mm line of an argon-ion laser falls on a NL crystal of potassium dihydrogen phosphate, where down-converted photons of wave length about 702 nm are produced. When the condition for degenerate phase matching is satisfied, down-converted, linearly polarised signal and idler photons emerge at angles of about $\pm 2°$ relative to the uv (=ultraviolet) pump beam with the electric vector in the plane of the diagram. The *idler photons* (= i-photons) pass through a 90° polarisation rotator, while the *signal photons* (= s-photon) traverse a compensating glass plate C_1 producing equal time delay. S-photons and i-photons are then directed from opposite sides towards a beam splitter (BS). The light beams emerging from BS, consisting of

[17] See also [GRANGIER *et al.* 1988].

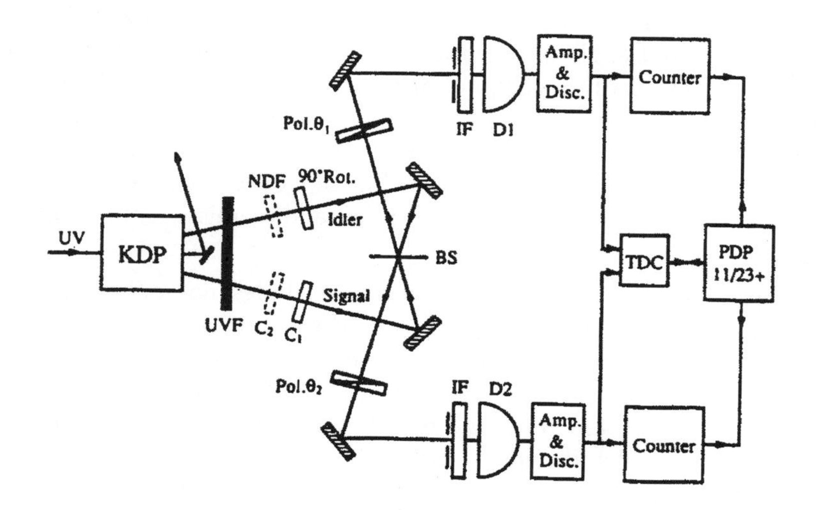

Figure 35.12: Outline of SPDC Exp. 4 — from [OU/MANDEL 1988a, 50].

a mixing of i-photons and s-photons, pass through linear polarisers set at adjustable angles θ_1 and θ_2, through similar interference filters (= IF) and finally fall on two photodetectors D_1 and D_2. The photoelectric pulses from D_1 and D_2 are amplified and shaped and fed to the start and stop inputs of a time-to-digital converter (TDC) under computer control which functions as a coincidence counter. $\wp(\theta_1, \theta_2)$ is the joint probability of detecting two photons for a setting θ_1, θ_2 of two linear polarisers. First we rewrite CH inequality (35.37) as follows:

$$\mathrm{C} = \wp(\theta_1, \theta_2) - \wp(\theta_1, \theta_2') + \wp(\theta_1', \theta_2') + \wp(\theta_1', \theta_2) - \wp(\theta_1', \infty) - \wp(\infty, \theta_2) \leq 0, \tag{35.84}$$

where again ∞ stands for the absence of analyzers.

QM describes the output state as the following linear superposition state:

$$|\Psi\rangle = (\mathrm{T}_x\mathrm{T}_y)^{1/2}|1_{1x}, 1_{2y}\rangle + (\mathrm{R}_x\mathrm{R}_y)^{1/2}|1_{1y}, 1_{2x}\rangle - \imath(\mathrm{R}_y\mathrm{T}_x)^{1/2}|1_{1x}, 1_{1y}\rangle + \imath(\mathrm{R}_x\mathrm{T}_y)^{1/2}|1_{2x}, 1_{2y}\rangle, \tag{35.85}$$

where $\mathrm{R}_x, \mathrm{R}_y, \mathrm{T}_x, \mathrm{T}_y$ are the beam splitter reflectivities and transmissivities with $\mathrm{R}_x + \mathrm{T}_x = 1$ and $\mathrm{R}_y + \mathrm{T}_y = 1$. Using polarised scalar fields at the two detectors, we can calculate that:

$$\wp(\theta_1, \theta_2) = K\left[(\mathrm{T}_x\mathrm{T}_y)^{1/2}\cos\theta_1\sin\theta_2 + (\mathrm{R}_x\mathrm{R}_y)^{1/2}\sin\theta_1\cos\theta_2\right]^2, \tag{35.86}$$

which reduces to

$$\wp(\theta_1, \theta_2) = K\sin^2(\theta_2 + \theta_1) \tag{35.87}$$

if $\mathrm{R}_x = \mathrm{T}_x = 1/2, \mathrm{R}_y = \mathrm{T}_y = 1/2$. If the polariser angles are chosen so that $\theta_1 = \pi/8, \theta_2 = \pi/4, \theta_1' = 3\pi/8, \theta_2' = 0$, one sees that quantum correlation function $[\mathrm{C}]_\psi$ violates the inequality (35.84):[18]

$$[\mathrm{C}]_\psi = \frac{1}{4}K(\sqrt{2} - 1) > 0. \tag{35.88}$$

[18]Such a derivation from Eq. (35.84) has been criticized by Garuccio [GARUCCIO 1995].

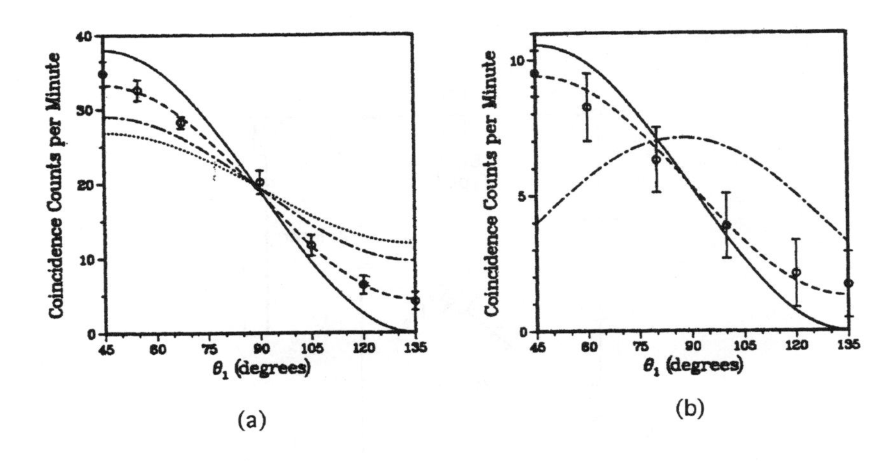

Figure 35.13: (a) Measured coincidence counting rate as a function of the polariser angle θ_1, with θ_2 fixed at 45°. The full curve represents QM prediction and the dash-dotted curve is the classical prediction. The dashed and dotted curve are the preceding curves with some corrections in order to reduce modulation caused by imperfect alignment.
(b) Measured coincidence counting rate as a function of the polariser angle θ_1 with θ_2 fixed at 45°, when a 8:1 attenuator is inserted into the idler beam. The curves are the same as in (a) (only the dotted curve fails) — from [OU/MANDEL 1988a, 52–53].

If we express the correlation C in terms of coincidence rates, we have experimentally the following result:

$$C_{exp} = R(22.5, 45) - R(22.5, 0) + R(67.5, 45) + R(67.5, 0) - R(67.5, \infty) - R(\infty, 45) = (11.5 \pm 2.0). \quad (35.89)$$

Hence C_{exp} is positive with an accuracy of about 6 standard deviations, in violation of inequality (35.84) [see figure 35.13].

SPDC Exp. 5 Another experiment was performed by Kiess and co-workers [KIESS *et al.* 1993], showing a violation of Bell inequality by 22 standard deviations (SPDC Exp. 5). The authors used for the first time a type–II phase matching: While, in the ordinary type–I phase matching, the correlated photons have the same polarisation, in the type–II phase matching, one photon is ordinary polarised and the other extraordinary, with polarisations orthogonal to each other. We report schematically SPDC Exp. 5 in figure 35.14 and analyse in more detail another experiment of the same type, which is in some respects a further generalization.

SPDC Exp. 6 This experiment was performed by Kwiat and co-workers [KWIAT *et al.* 1995c] Our analysis starts with the following observations: while in type–I the two photons emerge with equal wavelength on a cone, centered on the pump beam and whose opening angle depends on the angle θ_{pm} between the crystal optic axis and the pump, here in the type–II, we have two cones, which in the collinear situation are tangential to one another on exactly the line representing the pump beam direction but can also be separated from each other (if θ_{pm} is increased) or intersect (if θ_{pm} is decreased) [see figure 35.15]. Along the two directions (1 and 2), where the cones

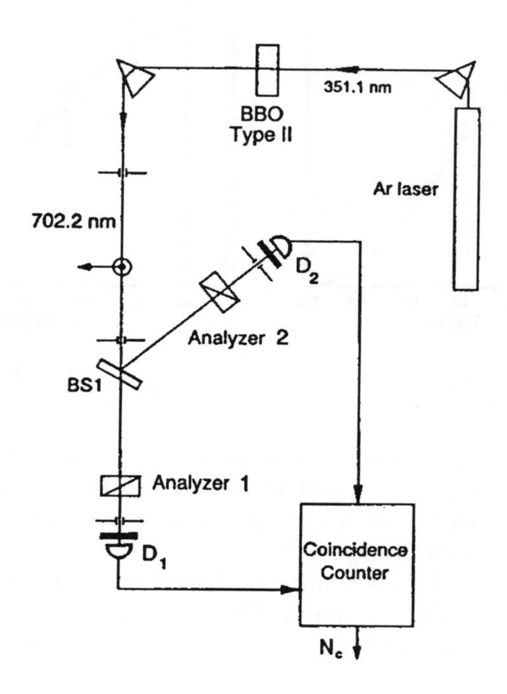

Figure 35.14: SPDC Exp. 5. Pairs from a collinear type–II down-conversion in a BBO ($\beta-\mathrm{Ba_zB_2O_4}$: beta-barium-borate) crystal are separated from the pump at a prism and directed to a 50:50 BS. The coincidence registrations in detectors 1, 2 are recorded as a function of the angles θ_1, θ_2 of the Glan/Thompson analyzers, for each bandwidth filter installed in front of the detectors — from [KIESS *et al.* 1993, 3894].

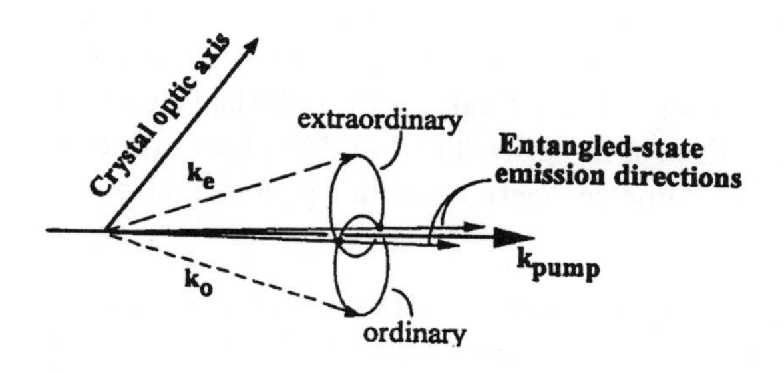

Figure 35.15: SPDC with type–II phase matching — from [KWIAT *et al.* 1995c, 4338].

overlap, the light can be described by the following entangled state:

$$|\Psi\rangle = \frac{1}{\sqrt{2}}(|\leftrightarrow\rangle_1|\updownarrow\rangle_2 + e^{\imath\varphi}|\updownarrow\rangle_1|\leftrightarrow\rangle_2), \tag{35.90}$$

where $\updownarrow, \leftrightarrow$ stand for vertical and horizontal polarisation, respectively.

By using (= SPDC Exp. 6) two extra birefringent crystals in order to compensate the birefringent nature of the down-conversion crystal (photons with different velocities and propagating along different directions), and a combination of a half-wave plate and a quarter-wave plate in order to change the polarisation [see figure 35.16], one can easily produce any of the four EPR states:

$$|\Psi_1^{\pm}\rangle = \frac{1}{\sqrt{2}}(|\leftrightarrow\rangle_1|\updownarrow\rangle_2 \pm |\updownarrow\rangle_1|\leftrightarrow\rangle_2), \tag{35.91a}$$

$$|\Psi_2^{\pm}\rangle = \frac{1}{\sqrt{2}}(|\leftrightarrow\rangle_1|\leftrightarrow\rangle_2 \pm |\updownarrow\rangle_1|\updownarrow\rangle_2). \tag{35.91b}$$

Experimentally, a violation by over 100 standard deviations in less than 5 min was found[19].

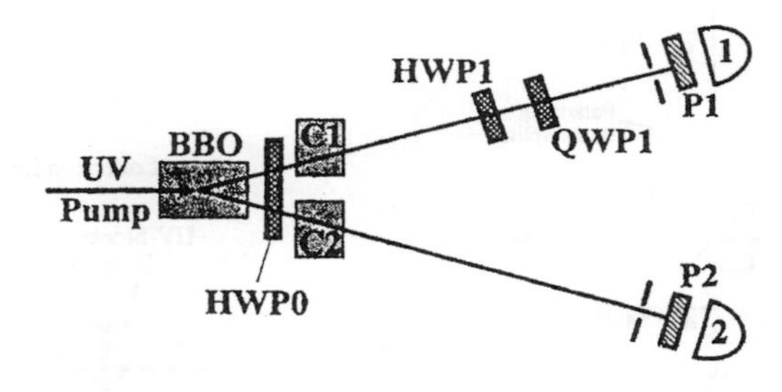

Figure 35.16: Setup of SPDC Exp. 6 — from [KWIAT *et al.* 1995c, 4339].

35.4.3 Detection Loophole

We have already observed the difficulty concerning possible non-local correlations [see subsection 35.4.1] and concerning the angular correlation [see subsection 35.4.2]. Another problem is that of detection [FERRERO *et al.* 1990, 686–87] [KWIAT *et al.* 1994a, 3209].

Efficiency If we pose the question of how high the detection efficiencies — hence the parameter η_i given by Eq. (35.46) — must be for the experimental confirmation of the quantum theoretical predictions, we have: by experiments such as that of Aspect and co-workers a sufficient efficiency is 83%, a value greater than $2(\sqrt{2}-1)$ [MERMIN 1986]. But if the inequality is optimized changing the angle of the settings after the correction for $\eta < 100\%$, then a lower requirement is sufficient, varying from 66.7% to 100% depending on the variation of the background level from 0.00% to 10.36% [EBERHARD 1993]

[19] However, recently, Santos [SANTOS 1996] was not completely satisfied with the quoted results of SPDC Exp.s. In [HUELGA *et al.* 1995a] [HUELGA *et al.* 1995b] another solution is proposed centered on cascade experiments.

Discarding by counting The problem here is that, even with high detection efficiency, one must discard a part of the counts — we have already seen this on a theoretical level in experiments proposed by CH and CHSH [FERRERO/SANTOS 1997, 786]. Hence a more refined solution is required, which is shown and summarized in figure 35.17.[20]

Enhancement A related problem is that Clauser and Horne's assumption — with polarisers in place the probability of a count is not larger than without polarisers — has been questioned [MARSHALL *et al.* 1983] by supposing a form of *enhancement* in the detection process. At the beginning, it was shown that, even in the hypothesis of an enhancement in the detection process, there is a detectable difference between HV theories and QM in the case of experiments with three polarisers [GARUCCIO/SELLERI 1984] [SELLERI 1985a].

But the work of Marshall/Santos [MARSHALL/SANTOS 1985] does not strictly exclude a form of enhancement which invalidates at least a part of the preceding experiments[21]. However, such a form of enhancement seems not very plausible.

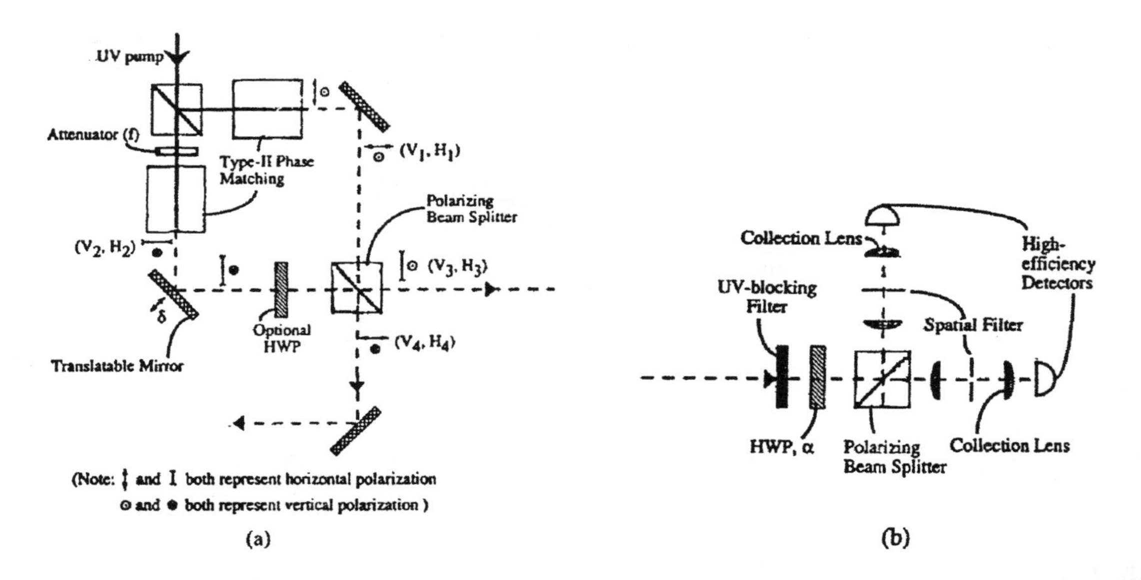

Figure 35.17: (a) An ultraviolet pump photon may be spontaneously down-converted in either of two nonlinear crystals, producing a pair of orthogonally polarised photons at half the frequency. One photon from each pair is directed to each output port of a polarising BS. When the outputs of both crystals are combined with an appropriately relative phase $\delta\varphi$, a true singlet- or triplet-like state may be produced. By using a half-wave plate (HWP) to effectively exchange the polarisations of photons originating in crystal 2, one overcomes several problems arising from nonideal phase matching. An additional mirror is used to direct the photons into opposite direction towards separated analyzers.
(b) A typical analyzer, including an HWP to rotate by θ the polarisation component selected by the analyzing BS, and precision spatial filters to select only conjugate pairs of photons. In an advanced version of the experiment, the HWP could be replaced by an ultrafast polarisation rotator (such as Pockels or Kerr cells) to close another loophole: the space-like-separation — from [KWIAT *et al.* 1994a, 3211].

[20] Another solution to the same problem (using cascade experiments) is proposed in [FRY *et al.* 1995].
[21] See [HAJI-HASSAN *et al.* 1987] for more details.

35.4.4 Radial Correlation

Santos [SANTOS 1995] has stressed a fourth loophole: the radial correlation between the particle positions — in fact he proposed to introduce explicitly in the Eqs. to be tested a spatial part to be added to the spin part. By following Fry, Santos shows that the loophole could be overcome if one made use of large measurement regions for the particles by imposing space-like separation and very short measurement times. The final result is that one cannot prove a violation of a Bell inequality if the particles have not travelled form the source during some relatively long time interval. An experiment has also been proposed [R. JONES/ADELBERGER 1994] which takes into account such a problem by showing no violation of Bell inequalities.

35.4.5 Duration of the Pump Pulse

The practical realisability of an entanglement of three or more photons through SPDC requires that the duration of the pump pulse must be much shorter than the coherence length of the signal and idler photons generated by each SPDC crystal, which in turn signifies generally the use of femtosecond (10^{-15} s) pump pulse [ŻUKOWSKI *et al.* 1993] (*V loophole*). But this condition is also responsible for a strong decoherence of the output signal from each of the SPDC crystals. This leads to the loss of stable phase relationships between the optical field amplitudes in the signal and idler photons and hence destroys the interference visibility. One can overcome this problem by the spectral postselection of the entangled photons by placing narrowband interference filters in front of detectors [DI GIUSEPPE *et al.* 1997c].

35.5 Fine's Theorems

The theorem Until now we have developed two different types of proofs: one concerning deterministic HV theories (Bell Theorem), the other concerning stochastic HV-theories (for example the CHSH inequality). But what is the relationship between these two approaches? We have seen that it is possible to derive the I Bell inequality (deterministic case) from the CHSH Inequality (stochastic case) [see Eq. (35.21) and relative comments]. But what is the relation between these two cases in more general and systematic terms? An answer to this question can be found in an important theorem proved in [A. FINE 1982a]:[22]

Theorem 35.2 (Fine) *The necessary and sufficient condition for the existence of a deterministic factorisable HV model is the existence of a stochastic factorisable HV model for the same experiment.*

To prove the theorem one needs a long succession of proofs.

Generals Consider two space-like separated regions V_1 and V_2 with non commuting observables $\hat{O}^a$ and $\hat{O}^{a'}$ in V_1, and non-commuting $\hat{O}^b$ and $\hat{O}^{b'}$ in V_2 (each couple can be represented, as before, as alternative polarisers' orientations). The values are ± 1. We write $\wp(\hat{O}^a)$ to express the probability that $A = +1$ and $\wp(\tilde{\hat{O}}^a)$ to express that the complementary of $\hat{O}^a$ takes value $+1$ and

[22] See also [GARG/MERMIN 1984, 15, 34–35].

hence $\hat{O}^a = -1$. Now if a variable η ranges over the four preceding variables plus their complements, and η, ζ ranges over the compatible pairs of preceding variables and their complements, we have:

$$\wp(\eta) = \sum d\lambda_{\mathrm{HV}} \tilde{\eta}(\lambda_{\mathrm{HV}}) \rho(\lambda_{\mathrm{HV}}), \tag{35.92a}$$

$$\wp(\eta\zeta) = \sum d\lambda_{\mathrm{HV}} \tilde{\eta}(\lambda_{\mathrm{HV}}) \tilde{\zeta}(\lambda_{\mathrm{HV}}) \rho(\lambda_{\mathrm{HV}}), \tag{35.92b}$$

where $\tilde{\eta}(\lambda_{\mathrm{HV}}) = 1$ if $\eta(\lambda_{\mathrm{HV}}) = 1$ and $\tilde{\eta}(\lambda_{\mathrm{HV}}) = 0$ if $\eta(\lambda_{\mathrm{HV}}) = -1$; similarly for ζ.

Deterministic HV-theories and and probability distributions If we have a *deterministic factorisable HV model*, the following Eq. defines a joint distribution for the four observables:

$$\wp(\hat{O}^a \hat{O}^{a'} \hat{O}^b \hat{O}^{b'}) = \sum_{\lambda_{\mathrm{HV}}} d\lambda_{\mathrm{HV}} \tilde{\hat{O}}^a(\lambda_{\mathrm{HV}}) \tilde{\hat{O}}^{a'}(\lambda_{\mathrm{HV}}) \tilde{\hat{O}}^b(\lambda_{\mathrm{HV}}) \tilde{\hat{O}}^{b'}(\lambda_{\mathrm{HV}}) \rho(\lambda_{\mathrm{HV}}). \tag{35.93}$$

It can be shown that the marginals yield the probabilities of experiments (for example that the sum of all alternative non overlapping possibility subensembles are 1). The consequence is that the idea of a deterministic HV model is just the idea of a suitable joint probability function. In fact if there were an HV model (i.e. such joint distribution), then there would be distributions for triples of observables $\hat{O}^a \hat{O}^b \hat{O}^{b'}$ satisfying $\wp(\hat{O}^a \hat{O}^b \hat{O}^{b'}) = \sum d\lambda_{\mathrm{HV}} \tilde{\hat{O}}^a(\lambda_{\mathrm{HV}}) \tilde{\hat{O}}^b(\lambda_{\mathrm{HV}}) \tilde{\hat{O}}^{b'}(\lambda_{\mathrm{HV}}) \rho(\lambda_{\mathrm{HV}})$ that return the correct experimental probabilities $\wp(\hat{O}^a), \wp(\hat{O}^b), \wp(\hat{O}^{b'}), \wp(\hat{O}^a \hat{O}^b), \wp(\hat{O}^a \hat{O}^{b'})$ as marginals. Similarly there would be a distribution $\wp(\hat{O}^{a'} \hat{O}^b \hat{O}^{b'})$ for triple $\hat{O}^{a'}, \hat{O}^b, \hat{O}^{b'}$. Moreover each of these triple distribution would give rise to one and the same distribution for the non-commuting pair $\hat{O}^b, \hat{O}^{b'}$ — the same obviously for triple distributions $\wp(\hat{O}^a \hat{O}^{a'} \hat{O}^b)$ and $\wp(\hat{O}^a \hat{O}^{a'} \hat{O}^{b'})$.

Now we state the following lemma:

Lemma 35.2 (I Fine) *The necessary and sufficient condition for the existence of a deterministic HV model is the existence of a distribution $\wp(\hat{O}^a \hat{O}^b \hat{O}^{b'})$ and a distribution $\wp(\hat{O}^{a'} \hat{O}^b \hat{O}^{b'})$ whose marginals yield the experimental probabilities and which in addition yield one and the same joint distribution $\wp(\hat{O}^b \hat{O}^{b'})$.*

Proof

The *necessary* condition of lemma 35.2 (Deterministic HV $\rightarrow$ joint distribution) is trivial, because it is what defines an HV-theory [see definition 32.2: p. 543].

To show the *sufficiency* (joint distribution $\rightarrow$ Deterministic HV), it is sufficient to build a distribution $\wp(\hat{O}^a \hat{O}^{a'} \hat{O}^b \hat{O}^{b'})$ from $\wp(\hat{O}^a \hat{O}^b \hat{O}^{b'})$, $\wp(\hat{O}^{a'} \hat{O}^b \hat{O}^{b'})$ and $\wp(\hat{O}^b \hat{O}^{b'})$. If we set

$$\wp(\hat{O}^a \hat{O}^{a'} \hat{O}^b \hat{O}^{b'}) = \frac{\wp(\hat{O}^a \hat{O}^b \hat{O}^{b'}) \wp(\hat{O}^{a'} \hat{O}^b \hat{O}^{b'})}{\wp(\hat{O}^b \hat{O}^{b'})}, \tag{35.94}$$

we can show that this is a proper joint distribution with the required marginals (if $\wp(\hat{O}^b \hat{O}^{b'}) = 0$, set $\wp(\hat{O}^a \hat{O}^{a'} \hat{O}^b \hat{O}^{b'}) = 0$). Q. E. D.

Deterministic HV-theories and CH inequality Now we also prove the following lemma.

Lemma 35.3 (II Fine) *The necessary and sufficient condition for a deterministic factorisable HV model for an experiment is that CH inequality holds for the same experiment.*

First we prove the *necessity* condition, i.e. that a deterministic factorisable HV model implies CH inequality (Deterministic HV → CH) — it is possible to extend the argument to the other Bell inequalities.

Proof

We search restrictions on probability distributions. Using the marginals, we have:

$$\begin{aligned} \wp(\hat{O}^a\hat{O}^b\hat{O}^{b'}) &= \wp(\hat{O}^a\hat{O}^{a'}\hat{O}^b\hat{O}^{b'}) + \wp(\hat{O}^a\tilde{\hat{O}}^{a'}\hat{O}^b\hat{O}^{b'}) \\ &\leq \wp(\hat{O}^{a'}\hat{O}^b) + \wp(\tilde{\hat{O}}^{a'}\hat{O}^{b'}) \\ &= \wp(\hat{O}^{a'}\hat{O}^b) + \wp(\hat{O}^{b'}) - \wp(\hat{O}^{a'}\hat{O}^{b'}); \end{aligned} \quad (35.95a)$$

$$\begin{aligned} \wp(\tilde{\hat{O}}^a\hat{O}^b\hat{O}^{b'}) &= \wp(\tilde{\hat{O}}^a\hat{O}^{a'}\hat{O}^b\hat{O}^{b'}) + \wp(\tilde{\hat{O}}^a\tilde{\hat{O}}^{a'}\hat{O}^b\hat{O}^{b'}) \\ &\leq \wp(\hat{O}^{a'}\hat{O}^{b'}) + \wp(\tilde{\hat{O}}^{a'}\hat{O}^b) \\ &= \wp(\hat{O}^{a'}\hat{O}^{b'}) + \wp(\hat{O}^b) - \wp(\hat{O}^{a'}\hat{O}^b). \end{aligned} \quad (35.95b)$$

Then we can derive:

$$0 \leq \wp(\hat{O}^a\tilde{\hat{O}}^b\tilde{\hat{O}}^{b'}) = \wp(\hat{O}^a) - \wp(\hat{O}^a\hat{O}^b) - \wp(\hat{O}^a\hat{O}^{b'}) + \wp(\hat{O}^a\hat{O}^b\hat{O}^{b'}); \quad (35.96a)$$

$$0 \leq \wp(\tilde{\hat{O}}^a\tilde{\hat{O}}^b\tilde{\hat{O}}^{b'}) = 1 - \wp(\hat{O}^a) - \wp(\hat{O}^b) - \wp(\hat{O}^{b'}) + \wp(\hat{O}^a\hat{O}^b) + \wp(\hat{O}^a\hat{O}^{b'}) + \wp(\tilde{\hat{O}}^a\hat{O}^b\hat{O}^{b'}) \quad (35.96b)$$

Using the inequality (35.95a) for $\wp(\hat{O}^a\hat{O}^b\hat{O}^{b'})$ in Eq. (35.96), and the inequality (35.95b) for $\wp(\tilde{\hat{O}}^a\hat{O}^b\hat{O}^{b'})$ in Eq. (35.96), we obtain:

$$-1 \leq \wp(\hat{O}^a\hat{O}^b) + \wp(\hat{O}^a\hat{O}^{b'}) + \wp(\hat{O}^{a'}\hat{O}^{b'}) - \wp(\hat{O}^{a'}\hat{O}^b) - \wp(\hat{O}^a) - \wp(\hat{O}^{b'}) \leq 0, \quad (35.97)$$

which is the IV Bell inequality — CH inequality [Eq. (35.37)], where $\wp(\mathbf{a}, \mathbf{b}')$ is substituted by $\wp(\mathbf{a}', \mathbf{b})$, and consequently $\wp(\mathbf{a}')$ by $\wp(\mathbf{a})$ and $\wp(\mathbf{b})$ by $\wp(\mathbf{b}')$ [see also Eq. (39.19a)]. Similar calculations are possible for other triple distributions. Q.E.D.

We have proved the necessity condition of the lemma. Now we report the proof of the *sufficient condition* of the lemma 35.3 (CH → Deterministic HV) — by using lemma 35.2 [A. FINE 1982b, 1307].[23]

Proof Let β be the minimum of the following (non-negative) terms:

$$\begin{aligned} &\wp(\hat{O}^b),\ \wp(\hat{O}^{b'}),\ \wp(\hat{O}^a\hat{O}^b) - \wp(\hat{O}^{b'}) - \wp(\hat{O}^a\hat{O}^{b'}),\ \wp(\hat{O}^a\hat{O}^{b'}) + \wp(\hat{O}^b) - \wp(\hat{O}^a\hat{O}^b), \\ &\wp(\hat{O}^{a'}\hat{O}^b) - \wp(\hat{O}^{b'}) - \wp(\hat{O}^{a'}\hat{O}^{b'}),\ \wp(\hat{O}^{a'}\hat{O}^{b'}) + \wp(\hat{O}^b) - \wp(\hat{O}^{a'}\hat{O}^b). \end{aligned} \quad (35.98)$$

Then we will set $\beta = \wp(\hat{O}^b\hat{O}^{b'})$ and define the rest of the distribution for $\hat{O}^b, \hat{O}^{b'}$ by:

$$\wp(\hat{O}^b\tilde{\hat{O}}^{b'}) = \wp(\hat{O}^b) - \beta,\ \wp(\tilde{\hat{O}}^b\hat{O}^{b'}) = \wp(\hat{O}^{b'}) - \beta,\ \wp(\tilde{\hat{O}}^b\tilde{\hat{O}}^{b'}) = 1 - \wp(\hat{O}^b) - \wp(\hat{O}^{b'}) - \beta. \quad (35.99)$$

This is well defined by a correct choice of β. Then let α be the minimum of

$$\beta,\ \beta - [\wp(\hat{O}^a) + \wp(\hat{O}^b) + \wp(\hat{O}^{b'}) - \wp(\hat{O}^a\hat{O}^{b'}) - \wp(\hat{O}^a\hat{O}^b) - 1],\ \wp(\hat{O}^a\hat{O}^b),\ \wp(\hat{O}^a\hat{O}^{b'}), \quad (35.100)$$

and let be α' the minimum of

$$\beta,\ \beta - [\wp(\hat{O}^{a'}) + \wp(\hat{O}^b) + \wp(\hat{O}^{b'}) - \wp(\hat{O}^{a'}\hat{O}^{b'}) - \wp(\hat{O}^{a'}\hat{O}^b) - 1],\ \wp(\hat{O}^{a'}\hat{O}^b),\ \wp(\hat{O}^{a'}\hat{O}^{b'}). \quad (35.101)$$

[23] An attempt to question the following proof can be found in [SVETLICHNY *et al.* 1988].

The CH inequality guarantees that α and α' are non-negative. Then we have:

$$0 \leq \alpha \leq \beta \leq 1, \quad 0 \leq \alpha' \leq \beta \leq 1. \tag{35.102}$$

We now set $\wp(\hat{O}^a\hat{O}^b\hat{O}^{b'}) = \alpha$ and fill out the remainders of the distribution for $\hat{O}^a, \hat{O}^b, \hat{O}^{b'}$ as follows:

$$\begin{aligned}
\wp(\hat{O}^a\hat{O}^b\tilde{\hat{O}}^{b'}) &= \wp(\hat{O}^a\hat{O}^b) - \alpha; && (35.103a)\\
\wp(\hat{O}^a\tilde{\hat{O}}^{b}\hat{O}^{b'}) &= \wp(\hat{O}^a\hat{O}^{b'}) - \alpha; && (35.103b)\\
\wp(\hat{O}^a\tilde{\hat{O}}^{b}\tilde{\hat{O}}^{b'}) &= \wp(\hat{O}^a) - \wp(\hat{O}^a\hat{O}^b) - \wp(\hat{O}^a\hat{O}^{b'}) + \alpha; && (35.103c)\\
\wp(\tilde{\hat{O}}^{a}\hat{O}^b\hat{O}^{b'}) &= \beta - \alpha; && (35.103d)\\
\wp(\tilde{\hat{O}}^{a}\hat{O}^b\tilde{\hat{O}}^{b'}) &= \wp(\hat{O}^b) - \wp(\hat{O}^a\hat{O}^b) - (\beta - \alpha); && (35.103e)\\
\wp(\tilde{\hat{O}}^{a}\tilde{\hat{O}}^{b}\hat{O}^{b'}) &= \wp(\hat{O}^{b'}) - \wp(\hat{O}^a\hat{O}^{b'}) - (\beta - \alpha); && (35.103f)\\
\wp(\tilde{\hat{O}}^{a}\tilde{\hat{O}}^{b}\tilde{\hat{O}}^{b'}) &= 1 - \wp(\hat{O}^a) - \wp(\hat{O}^b) - \wp(\hat{O}^{b'}) + \wp(\hat{O}^a\hat{O}^b) + \wp(\hat{O}^a\hat{O}^{b'}) + (\beta - \alpha). && (35.103g)
\end{aligned}$$

In just the same way we set $\wp(\hat{O}^{a'}\hat{O}^b\hat{O}^{b'}) = \alpha'$ and fill out in the same way distributions for $\hat{O}^{a'}, \hat{O}^b, \hat{O}^{b'}$. Now one can check that

$$\wp(\hat{O}^a\hat{O}^b\hat{O}^{b'}) + \wp(\tilde{\hat{O}}^{a}\hat{O}^b\hat{O}^{b'}) = \wp(\hat{O}^{a'}\hat{O}^b\hat{O}^{b'}) + \wp(\tilde{\hat{O}}^{a'}\hat{O}^b\hat{O}^{b'}) = \beta, \tag{35.104}$$

and that the other terms in the distribution of $\hat{O}^b, \hat{O}^{b'}$ come out the same, whether calculated from the triple $\hat{O}^a, \hat{O}^b, \hat{O}^{b'}$ or from the triple $\hat{O}^{a'}\hat{O}^b\hat{O}^{b'}$. Similarly the experimental distributions for the observables and compatible pairs also come out correctly. But this is exactly the case of a deterministic factorisable HV model. Q.E.D.

Thus we have also proved the sufficient condition of lemma 35.3. But, on the other hand, the sufficient and necessary conditions for deterministic HV models of lemma 35.2 hold if the CH inequality holds. Now we are in position to prove the Fine Theorem [theorem 35.2].

Proof

Consider a factorisable HV stochastic model: we relax the requirement that each λ_{HV} determines a unique measurement response ± 1 but require only a probability of each result. Now, as the lemma 35.3 states, from a stochastic Bell inequality, such as the CH one, one can derive a deterministic factorisable HV model. Intuitively each deterministic factorisable HV model is a stochastic one with all probability at λ_{HV} either 0 or 1 — and this is already shown by the II Bell inequality, the CHSH inequality [see Eq. (35.21) and comments]. Hence we have shown that a stochastic factorisable HV model is a sufficient condition of a deterministic one.

Now we show the *necessity*, i.e. that the deterministic model implies the stochastic model: we know that a stochastic factorisable HV model must obey the CH inequality and we have already derived the CH inequality from a deterministic HV model [see the derivation of Eq. (35.97)]. Hence, again by lemma 35.3, we have the result. Q.E.D.

The importance of Fine's Theorem is that, in principle, it makes all formulations analysed in this chapter equivalent.

35.6 Hidden-Variable Theories and Classicality

We have discussed often the determinism of HV-theories. Therefore, an interesting question may be posed: are the concepts of HV-theories and classicality equivalent (since, though perhaps not correctly, it is supposed that classical mechanics is deterministic)? In no way, as shown by Garg and Mermin [GARG/MERMIN 1982c]: in fact the Bell Inequalities are necessary conditions of HV-theories but not sufficient ones. Each set of correlations admitting HV-theories is a convex set, and as such it is completely determined by some set of linear inequalities: we call such a set *Generalized Bell Inequality* [R. WERNER 1989, 4278] [see chapter 39]. But a general procedure for obtaining all generalized inequalities is not known.

Each HV-theory can be stated in the form

$$\int \mathcal{M}(d\omega) f_{\Omega_1}(\nu, \lambda_{HV}) f_{\Omega_2}(\mu, \lambda_{HV}), \tag{35.105}$$

where λ_{HV} is an HV variable, $\mathcal{M}$ is a measure, μ, ν are outcomes, f is a response function such that $\lambda_{HV} \mapsto f_\Omega(\mu, \lambda_{HV}) \in \Re$. Now formula (3.106) implies formula (35.105), i.e. all C-Correlated states admit HV-Models and hence satisfy Bell inequalities. Hence each state which violates some generalized Bell inequality is EPR-Correlated. But the converse is not true: it is not true that every state admitting a HV-Model is classically correlated[24].

[24]In the article cited, Garg and Mermin offer a concrete example. The interesting point of the conditions posed on C-Correlations and QM-Correlations will be the object of chapter 37.

Chapter 36

THE PROBLEM OF SEPARABILITY AS SUCH

Introduction and contents We have seen that QM cannot be an HV-theory which acknowledges the Separability principle. Now we shall see that QM itself violates the separability. Hence the matter of this chapter is a more stringent formalization of the answers given by Schrödinger [section 34.1] and Bohr [see section 34.2] to EPR. On the other hand, all proofs which we report in this chapter have strong analogies with the KS theorem. The reason is that the non-separability is a logico-algebraic property of QM.

Section 36.1 is devoted to Stapp theorem. This theorem poses the great problem of the status of counterfactuals in QM: this is the object of section 36.2. In section 36.3 the relationship between Stapp theorem and Bell theorem is discussed. Section 36.4 is devoted to other proofs which follow the main line of Stapp's proof. In section 36.5 we analyse the important Eberhard theorem, the first attempt to prove that QM does not violate locality. The problem of whether HV theories violate Lorentz invariance is also discussed [section 36.6]. Finally 36.7 is devoted to the contextuality of entanglement.

36.1 Stapp Theorem

Stimulated by Bell's results but in an independent way, Stapp tried to prove that QM as such is incompatible with at least separability [see principle 31.1: p. 532] or with a stronger form of non-locality — we limit ourselves to separability, postponing the general discussion about non-separability and non-locality [see section 36.5]. Hence Stapp was able to develop a demonstration which leaves aside the problem of HVs — and hence that of the completeness of QM —, overcoming the confusion between different problems which was still present at the beginning of Bell's work[1]. On the other hand, all proofs which derive directly or indirectly from Stapp's proof seem to concern individual events: Hence they are more in the spirit of EPR's original proof; but then we lose the statistical correlations which were central in CHSH and CH and in the third Bell inequality [see section 35.2]. Generally this important theoretical step has a price: the simplification — as always — is paid in terms of verifiability. In other words, formulations without inequality, as with most of the followings, pose stronger constraints and hence are more

[1]Note that the first proposal of a proof without using inequalities is due to Wigner [WIGNER 1970], but it is always centered on HV theories.

difficult to test[2].

We begin with the original formulation of the Stapp theorem [STAPP 1971, 1306–307]:[3]

Theorem 36.1 (Stapp) *No theory can*

- *give contingent general predictions of the individual results of measurements,*
- *be compatible with the statistical predictions of QM (to within, say,* 5%*),*
- *satisfy* principle 31.1.

36.1.1 First Requirement

The *first requirement* must be understood in terms of the following *Gedankenexperiment* (SGM-Exp. 8). Take two SGM apparatus, the axis of which can be rotated, allowing different alternative settings. The word 'contingent' signifies that the theory gives predictions for various possible *alternative* settings. We denote the axis of SGM1 and SGM2 by D_1 and D_2, respectively. They are both normal to the line of flight and $\theta(\mathrm{D}_1, \mathrm{D}_2)$ is the angle between them. Two different settings D_{1a} and $\mathrm{D}_{1a'}$ of D_1, D_{2b} and $\mathrm{D}_{2b'}$ of D_2, respectively, are considered. Let j label the individual experiment and $n_{ij}(\mathrm{D}_1, \mathrm{D}_2) = \pm 1$ according to whether the theory predicts that the particle from the j pair that passes through the SGM is deflected up or down when the settings are D_1 and D_2. Hence the first condition of the theorem means that for each individual pair j the numbers $n_{1j}(\mathrm{D}_1, \mathrm{D}_2)$ and $n_{2j}(\mathrm{D}_1, \mathrm{D}_2)$ are perfectly defined for all four combinations of arguments D_1 and D_2.

36.1.2 Second Requirement

According to QM the following relation holds with increasing accuracy as the number N of tests increases:

$$\frac{1}{N}\sum_{j=1}^{N} n_{1j}(\mathrm{D}_1, \mathrm{D}_2) n_{2j}(\mathrm{D}_1, \mathrm{D}_2) = -\cos\theta(\mathrm{D}_1, \mathrm{D}_2), \tag{36.1}$$

so that the second requirement is simply that Eq. (36.1) holds. Now let us choose the directions D_{1a}, $\mathrm{D}_{1a'}$, D_{2b} and $\mathrm{D}_{2b'}$ so that:

$$\cos\theta(\mathrm{D}_{1a}, \mathrm{D}_{2b}) = 1; \tag{36.2a}$$
$$\cos\theta(\mathrm{D}_{1a}, \mathrm{D}_{2b'}) = 0; \tag{36.2b}$$
$$\cos\theta(\mathrm{D}_{1a'}, \mathrm{D}_{2b}) = -\frac{1}{\sqrt{2}}; \tag{36.2c}$$
$$\cos\theta(\mathrm{D}_{1a'}, \mathrm{D}_{2b'}) = \frac{1}{\sqrt{2}}. \tag{36.2d}$$

In other words what Stapp assumes is a Counterfactual Values-Definiteness [proposition 31.2: p. 533], differently from the mathematical structure of EPR's original argumentation. We shall return later on this point [section 36.2]: here we assume it as a premise of the proof.

[2]On this point see [KWIAT *et al.* 1994a] .

[3]See also [STAPP 1979, 3–7] [STAPP 1980, 767–68] [BELL 1981a, 132] [BELL 1981b].

36.1.3 Third Requirement

The third condition (of Separability) can be expressed:

$$n_{1j}(\mathrm{D}_{1a}, \mathrm{D}_{2b}) = n_{1j}(\mathrm{D}_{1a}, \mathrm{D}_{2b'}) \equiv n_{1ja}; \tag{36.3a}$$
$$n_{1j}(\mathrm{D}_{1a'}, \mathrm{D}_{2b}) = n_{1j}(\mathrm{D}_{1a'}, \mathrm{D}_{2b'}) \equiv n_{1ja'}; \tag{36.3b}$$
$$n_{2j}(\mathrm{D}_{1a}, \mathrm{D}_{2b}) = n_{2j}(\mathrm{D}_{1a'}, \mathrm{D}_{2b}) \equiv n_{2jb}; \tag{36.3c}$$
$$n_{2j}(\mathrm{D}_{1a}, \mathrm{D}_{2b'}) = n_{2j}(\mathrm{D}_{1a'}, \mathrm{D}_{2b'}) \equiv n_{2jb'}. \tag{36.3d}$$

36.1.4 Proof of the Stapp Theorem

Proof
Inserting Eqs. (36.2) and Eqs. (36.3) into Eq. (36.1) we obtain:

$$\frac{1}{N}\sum n_{1ja}n_{2jb} = -1; \tag{36.4a}$$
$$\frac{1}{N}\sum n_{1ja}n_{2jb'} = 0; \tag{36.4b}$$
$$\frac{1}{N}\sum n_{1ja'}n_{2jb} = \frac{1}{\sqrt{2}}; \tag{36.4c}$$
$$\frac{1}{N}\sum n_{1ja'}n_{2jb'} = -\frac{1}{\sqrt{2}}. \tag{36.4d}$$

From Eq. (36.4a) we have:

$$n_{1ja} = -n_{2jb}, \tag{36.5}$$

which combined with Eq. (36.4b) gives:

$$\frac{1}{N}\sum n_{2jc}n_{2jd} = 0. \tag{36.6}$$

Subtraction of Eq. (36.4d) from the (36.4c):

$$\frac{1}{N}\sum n_{1jb}(n_{2jb} - n_{2jd}) = \sqrt{2}. \tag{36.7}$$

Using the fact that $n_{2jc}n_{2jb'} = 1$ (because the allowed values are only ± 1), one obtains:

$$\sqrt{2} = \frac{1}{N}\sum n_{1ja'}n_{2jb'}(n_{2jb}n_{2jb'} - 1), \tag{36.8a}$$
$$\leq \frac{1}{N}\sum |n_{2jb}n_{2jb'} - 1|, \tag{36.8b}$$
$$\leq \frac{1}{N}\sum (1 - n_{2jb}n_{2jb'}), \tag{36.8c}$$
$$\leq 1 - \frac{1}{N}\sum n_{2jb}n_{2jb'}, \tag{36.8d}$$
$$= 1 \tag{36.8e}$$

which is impossible, so that the QM predictions are not compatible with a theory which acknowledges separability or the condition 1. Q.E.D. [4]

[4] Another proof of Stapp Theorem is to be found in [STAPP 1977a, 193–95]. Eberhard, as we shall see, has further developed and refined the theorem: he always used a counterfactual logic but based his proof upon the CHSH inequality. See also [PERES 1978a] and [BRODY/DE LA PEÑA-A. 1979].

36.2 The Problem of Counterfactuals

The Problem It is evident that the whole of Stapp's argumentation presupposes a counterfactuality [see proposition 31.2 [p. 533] about individual events. Stapp thought that Bohr did not acknowledge it [STAPP 1971, 1314] and felt the necessity of justifying it by using the instruments of modal logic and particularly the device of possible worlds [STAPP 1980, 791]. This is not convincing or it is far from universal consensus [CLAUSER/SHIMONY 1978, 1899]. Stapp also said [STAPP 1980, 780] that one has the right to consider, in a logical analysis, the information that could be obtained from alternative, mutually exclusive experiments, if he carefully treats this information exclusively as information that could be obtained from alternative, mutually exclusive experiments. But the argument is too generic in order to be accepted as resolutive.

Therefore, although Stapp has the historical merit of having focused attention upon the question of non-separability as such, he was unable to find a more solid justification of counterfactuality. Our opinion here is that, even if counterfactuality has a large application in physics, it can be employed in QM only in a restricted form.

Bohr In order to better understand this point, let us return to Bohr's answer to EPR[5]. Bohr said [see section 34.2]: our freedom of handling the measuring instruments is characteristic of the very idea of experiment; we have a completely free choice whether we want to determine one or the other of these quantities [BOHR 1935b]. What is the meaning of this assertion? Is it a rejection of counterfactuality? We know that complementarity was very often interpreted in the sense of an absence of meaning of propositions about not-observed observables [see corollary 8.1: p. 137; see also chapter 29] and sometimes also in a subjective sense [see postulate 8.1: p. 137]. It is evident on the other side that Bohr was influenced here by some neo-positivistic epistemology[6].

Now Bohr's statement is surely a rejection of proposition 31.2, if intended as a general statement about measurements of individuals: i.e. we cannot in general say of the *same* system: 'we have performed a measurement of the observable $\hat{O}$ and obtained result o_j, but, if we had performed a measurement of the observable $\hat{O}'$, we would have obtained o'_k as result' — generally in QM one cannot assign values to all observables even hypothetically [see theorem 14.6: p. 235]. The reason is that the state in which the system is before one or the other measurement can have different effects on one or the other measurement (for example if $\hat{O}, \hat{O}'$ do not commute, that state can be an eigenstate only of $\hat{O}$ or only of $\hat{O}'$ or of none). And we know that normally if the system was not in an eigenstate of the measured observable (i.e. it is not a type I–Measurement), the state before the observation is destroyed by the measurement process. On the other hand, the only way to assure to ourselves some state is a good preparation, but we also know that preparatory and determinative measurement are exclusive. And there is also no method of knowing the *a-priori* state of a single system [see section 28.5].

Unpredictability Hence in general nothing can be said *a priori* about the results of alternative measurements. In conclusion the QM results are normally unpredictable and the chosen experimental context determines the definition of the phenomenon itself — see for example the delayed-choice experiments [chapter 26]. Here we return to the problem that in QM one cannot describe phenomena without and independently from some form of interaction, contrary to what

[5] See also [HARDEGREE 1979, 77–79].
[6] As an example see [*HEMPEL* 1953].

presupposed by the HV-theories [see section 32.1]. In general, counterfactuality cannot imply in one way or another that an observable necessarily has some value [see proposition 31.1: p. 533, and postulate 6.3: p. 105].

Statistical counterfactuals Surely counterfactuality has always a *statistical validity*, in the sense that we are sure that a determined ensemble of particles, prepared in some way, will give a determined result in an experimental arrangement and another one in a different experimental arrangement: however, it is possible to test such results after such preparations and therefore prediction is also possible to a certain extent. However, this form of statistical counterfactuality may not always suffice.

Note that in measurement we always use such counterfactual reasonings [UNRUH 1998]. Generally *Gedankenexperimenten* are based on counterfactual reasoning [SCHLESINGER 1996], and we have seen the importance of thought experiments from the beginning of QM (the whole debate between Bohr and Einstein, and particularly the EPR experiment itself).

Correlations Counterfactuality has a larger application than the one described: The correlations are constraints on the individual particles and not statistical relations: *each pair* of QM particles violates separability [ELITZUR *et al.* 1992].[7] Hence, returning to the EPRB model, we can affirm that of each pair we have an *a priori* probability such that we can say: if we obtain a spin-up on one subsystem, we have spin-down on the other, and if we obtain spin-down on the first we obtain spin-up on the other. Practically, in a system of two particles in singlet state, measuring the spin on one system or on the other is an equivalent operation [*D'ESPAGNAT* 1995, 219–20].

Conclusion In this case the counterfactuality has a direct individual meaning, and in this sense Stapp's assumption is completely correct.

But, on the other hand, if counterfactuality is used together with an HV-theory or with some other form of classical realism, then it is no longer a valid conceptual mean in QM [UNRUH 1998].[8]

However, if we wish to parallelize Stapp's proof with Bell-type proofs, we also need some statistical formulation of the first. And, as we shall see in the next section, Stapp achieved it.

36.3 The Relationship Between Stapp Theorem and Bell Theorem

Cascade Exp. 7 In [STAPP 1980, 776–77] the content of Stapp Theorem changed: it is no more a question of individuals but rather of statistical ensembles of n particles. Given this new approach, Stapp could prove by means of a new *Gedankenexperiment*, centered on coincidence rates (Cascade Exp. 7), an equivalence between his Theorem and the family of Bell-like Theorems and inequalities. If we take the polarisers A and B, and suppose they have rotation possibility so that there are two different possible settings represented by two possible values of two independent stochastic variables $\xi_A = 1, 2$ and $\xi_B = 1, 2$, for the four values of (ξ_A, ξ_B) there are 4^n conceivable

[7] See also [HARDY 1994a] [GERRY 1996].

[8] This is the error which Stapp makes in his essay [STAPP 1997] aiming to generalize a proof of Hardy [see subsection 36.6.2].

observations. In the case $(\xi_A, \xi_B) = (1,1)$ we label the 4^n conceivable observations with index j (from 1 to 4^2). Let 4^n other conceivable observations in the case $(\xi_A, \xi_B) = (2,2)$ be labelled with index k (always running from 1 to 4^n). The j–th conceivable observation (coincidence) is identified by a sequence of n pairs of numbers

$$R_i(A,1,1,j), R_i(B,1,1,j); \quad i = 1,2,\ldots n, \tag{36.9a}$$

while for the k–th conceivable observation we have the corresponding sequence:

$$R_i(A,2,2,k), R_i(B,2,2,k); \quad i = 1,2,\ldots n, \tag{36.9b}$$

where $R_i = \pm 1$ depending on whether the individual detected is a deflection up or down. The factorization (separability) gives [see Eqs. (36.3)]:

$$\begin{aligned} R_i(A,1,2,j,k) &= R_i(A,1,1,j,k) = R_i(A,1,1,j); & (36.10a)\\ R_i(B,1,2,j,k) &= R_i(B,2,2,j,k) = R_i(B,2,2,k); & (36.10b)\\ R_i(A,2,1,j,k) &= R_i(A,2,2,j,k) = R_i(A,2,2,k); & (36.10c)\\ R_i(B,2,1,j,k) &= R_i(B,1,1,j,k) = R_i(B,1,1,j). & (36.10d) \end{aligned}$$

The Separability principle generates $4^n \times 4^n$ possible quadruples. We can reformulate Eqs. (36.10) by means of a variable q which runs from 1 to 4^{2n}:

$$\begin{aligned} R_i(A,1,2,q) &= R_i(A,1,1,q); & (36.11a)\\ R_i(B,1,2,q) &= R_i(B,2,2,q); & (36.11b)\\ R_i(A,2,1,q) &= R_i(A,2,2,q); & (36.11c)\\ R_i(B,2,1,q) &= R_i(B,1,1,q). & (36.11d) \end{aligned}$$

Let $\mathcal{X}$ denote the set of all conceivable quadruples and $\mathcal{Y}$ denote the subset of $\mathcal{X}$ which satisfies conditions (36.11). This set $\mathcal{Y}$ consists precisely of the quadruple (36.10). On the other hand, let us define the subset Q of $\mathcal{X}$ by the condition that, for each of the four observations comprising any q in Q, the observed value of the correlation parameter lies (with a reasonable statistic) in the limits predicted by QM. Now Stapp Theorem [36.1: p. 622] says that:

$$\mathcal{Y} \cap Q = \emptyset. \tag{36.12}$$

Bell inequalities Let us now discuss this matter in terms of Bell inequalities. Consider the conditional probability of a pair of results (R_A, R_B) subject to the condition that the two setting parameters have the values ξ_A and ξ_B:

$$\langle R_A, R_B | \xi_A, \xi_B \rangle = \sum_{\lambda_{\rm HV}=1}^{N} \rho(\lambda_{\rm HV}) \wp_A(\lambda_{\rm HV}, R_A, \xi_A) \wp_B(\lambda_{\rm HV}, R_B, \xi_B), \tag{36.13}$$

where the factorization is expressed in the following form:

$$\begin{aligned} \langle R_A | \xi_A \rangle &= \sum_{\lambda_{\rm HV}=1}^{N} \rho(\lambda_{\rm HV}) \wp_A(\lambda_{\rm HV}, R_A, \xi_A); & (36.14a)\\ \langle R_B | \xi_B \rangle &= \sum_{\lambda_{\rm HV}=1}^{N} \rho(\lambda_{\rm HV}) \wp_B(\lambda_{\rm HV}, R_B, \xi_B). & (36.14b) \end{aligned}$$

The distribution functions satisfy:

$$\sum_{\lambda_{\rm HV}=1}^{N} \rho(\lambda_{\rm HV}) = 1, \tag{36.15a}$$

$$\sum_{R_A=\pm 1} \wp_A(\lambda_{\rm HV}, R_A, \xi_A) = 1, \tag{36.15b}$$

$$\sum_{R_B=\pm 1} \wp_B(\lambda_{\rm HV}, R_B, \xi_B) = 1, \tag{36.15c}$$

and that ρ, $\wp_A$ and $\wp_B$ are all non negative.

Let us rewrite the separability condition as follows:

$$R_i(A, \xi_A, \xi_B) = R_i(A, \xi_A), \quad i = 1, \ldots, n, \tag{36.16a}$$

$$R_i(B, \xi_A, \xi_B) = R_i(B, \xi_B), \quad i = 1, \ldots, n, \tag{36.16b}$$

and suppose that there is a function f such that

$$[f_i(A, R_A, \xi_A) = 1] \rightarrow [R_i(A, \xi_A) = R_A]; \tag{36.17a}$$

$$[f_i(A, R_A, \xi_A) = 0] \rightarrow [R_i(A, \xi_A) \neq R_A]; \tag{36.17b}$$

$$[f_i(B, R_B, \xi_B) = 1] \rightarrow [R_i(B, \xi_B) = R_B]; \tag{36.17c}$$

$$[f_i(B, R_B, \xi_B) = 0] \rightarrow [R_i(B, \xi_B) \neq R_B]; \tag{36.17d}$$

then one can prove the following theorem:

Theorem 36.2 (Equivalence) *If a quadruple of individual results satisfies* principle 31.1 *(factorization), then the average values defined by*

$$\langle R_A, R_B | \xi_A, \xi_B \rangle = \frac{1}{n}\sum_{i=1}^{n} f_i(A, R_A, \xi_A) f_i(B, R_B, \xi_B), \tag{36.18a}$$

$$\langle R_A | \xi_A \rangle = \frac{1}{n}\sum_{i=1}^{n} f_i(A, R_A, \xi_A), \tag{36.18b}$$

$$\langle R_B | \xi_B \rangle = \frac{1}{n}\sum_{i=1}^{n} f_i(B, R_B, \xi_B) \tag{36.18c}$$

can be expressed in the form of Eqs. (36.13)–(36.14) — *trivially by identifying* $\lambda_{\rm HV}$ *with the* i*'s. Conversely probabilities satisfying* Eqs. (36.13)–(36.14) *can be reproduced — up to terms that vanish as* n *tends to infinity as averages* (36.18) *over individual results that satisfy the conditions* (36.11)

This result is very important because it is a definitive legitimation for leaving aside the problem of HV-theories and discussing only separability or locality for QM.

36.4 Other Proofs on the Same General Lines as Stapp's Proof

36.4.1 Greenberger/Horne/Zeilinger Proof

Stimulated by Stapp's results, Greenberger/Horne/Zeilinger (= GHZ) proved the violation of separability in the spirit of Bell proofs but without using inequalities at all and with a more direct demonstration by assuming the counterfactuality [proposition 31.2: p. 533]. The argument of GHZ [GREENBERGER *et al.* 1989] uses only perfect anticorrelation, i.e. the case in which the directions of two different analyzers are opposite to each other[9].

We can construct a simple model (SGM-Exp. 9): A particle has spin 1, it is in the state $m = 0$ in a magnetic field along the z axis. It decays into two particles each one of spin 1, one along $+z$ and the other along $-z$. Both particles then decay into two spin 1/2 particles, the first two (1 and 2) also moving along $+z$, the other two (3 and 4) along $-z$. The orientation of SGM1–4 are respectively $\mathbf{n}_1$, $\mathbf{n}_2$, $\mathbf{n}_3$ and $\mathbf{n}_4$ [see figure 36.1].

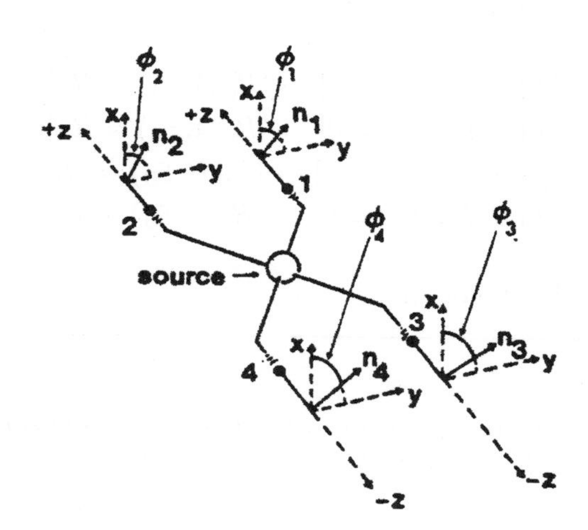

Figure 36.1: A four-particle Gedankenexperiment (SGM-Exp. 9). The source emits a quadruple of spin–1/2 particles: 1, 2, 3, and 4, in the state of Eq.(36.19). Particle j ($j = 1, 2, 3, 4$) enters its own SGM oriented along direction $\mathbf{n}_j$. We emphasize that the four SGMs can be separated by arbitrarily large distances. Behind each SGM two detectors, not shown, record whether the result is up or down — from [GREENBERGER *et al.* 1990, 1134].

The QM spin state (GHZ state) can be written:

$$|\Psi\rangle = \frac{1}{\sqrt{2}}(|\uparrow\rangle_1|\uparrow\rangle_2|\downarrow\rangle_3|\downarrow\rangle_4 - |\downarrow\rangle_1|\downarrow\rangle_2|\uparrow\rangle_3|\uparrow\rangle_4). \tag{36.19}$$

The expectation value of the product of outcomes depending on the indicated conditions is:

$$C^{\Psi}(\mathbf{n}_1, \mathbf{n}_2, \mathbf{n}_3, \mathbf{n}_4) = \langle\Psi|(\mathbf{n}_1\cdot\hat{\sigma}_1)(\mathbf{n}_2\cdot\hat{\sigma}_2)(\mathbf{n}_3\cdot\hat{\sigma}_3)(\mathbf{n}_4\cdot\hat{\sigma}_4)|\Psi\rangle$$

[9] See also [GREENBERGER *et al.* 1990, 1134–35, 1140–41].

$$= \frac{1}{2}[\langle\uparrow\uparrow\downarrow\downarrow|(\mathbf{n}_1\cdot\hat{\sigma}_1)(\mathbf{n}_2\cdot\hat{\sigma}_2)(\mathbf{n}_3\cdot\hat{\sigma}_3)(\mathbf{n}_4\cdot\hat{\sigma}_4)|\uparrow\uparrow\downarrow\downarrow\rangle$$
$$-\langle\uparrow\uparrow\downarrow\downarrow|(\mathbf{n}_1\cdot\hat{\sigma}_1)(\mathbf{n}_2\cdot\hat{\sigma}_2)(\mathbf{n}_3\cdot\hat{\sigma}_3)(\mathbf{n}_4\cdot\hat{\sigma}_4)|\downarrow\downarrow\uparrow\uparrow\rangle$$
$$-\langle\downarrow\downarrow\uparrow\uparrow|(\mathbf{n}_1\cdot\hat{\sigma}_1)(\mathbf{n}_2\cdot\hat{\sigma}_2)(\mathbf{n}_3\cdot\hat{\sigma}_3)(\mathbf{n}_4\cdot\hat{\sigma}_4)|\uparrow\uparrow\downarrow\downarrow\rangle$$
$$+\langle\downarrow\downarrow\uparrow\uparrow|(\mathbf{n}_1\cdot\hat{\sigma}_1)(\mathbf{n}_2\cdot\hat{\sigma}_2)(\mathbf{n}_3\cdot\hat{\sigma}_3)(\mathbf{n}_4\cdot\hat{\sigma}_4)|\downarrow\downarrow\uparrow\uparrow\rangle]. \tag{36.20}$$

Now:

$$\langle\uparrow|(\mathbf{n}\cdot\hat{\sigma})|\uparrow\rangle = \begin{pmatrix} 1 & 0 \end{pmatrix}\begin{pmatrix} \cos\theta & \sin\theta e^{-i\phi} \\ \sin\theta e^{-i\phi} & -\cos\theta \end{pmatrix}\begin{pmatrix} 1 \\ 0 \end{pmatrix} = \cos\theta; \tag{36.21a}$$

$$\langle\uparrow|(\mathbf{n}\cdot\hat{\sigma})|\downarrow\rangle = (1 \;\; 0)\begin{pmatrix} \cos\theta & \sin\theta e^{-i\phi} \\ \sin\theta e^{i\phi} & -\cos\theta \end{pmatrix}\begin{pmatrix} 0 \\ 1 \end{pmatrix} = \sin\theta e^{-i\phi}; \tag{36.21b}$$

$$\langle\downarrow|(\mathbf{n}\cdot\hat{\sigma})|\uparrow\rangle = \sin\theta e^{i\phi}; \tag{36.21c}$$
$$\langle\downarrow|(\mathbf{n}\cdot\hat{\sigma})|\downarrow\rangle = -\cos\theta; \tag{36.21d}$$

where θ and ϕ are polar and azimuthal angles of a vector $\mathbf{n}$, respectively. Of the four chains in brackets in Eq. (36.20), the first and fourth are products of terms similar to Eq. (36.21a) and Eq. (36.21d) with result $\cos\theta_1\cos\theta_2\cos\theta_3\cos\theta_4$ (where the subscript $j = 1, 2, 3, 4$ indicates the $\mathbf{n}_j$ vector). The second and third chains of Eq. (36.20) add analogously to: $\sin\theta_1\sin\theta_2\sin\theta_3\sin\theta_4\cos(\phi_1 + \phi_2 - \phi_3 - \phi_4)$, so that we obtain:

$$\mathrm{C}_\Psi(\mathbf{n}_1, \mathbf{n}_2, \mathbf{n}_3, \mathbf{n}_4) = \cos\theta_1\cos\theta_2\cos\theta_3\cos\theta_4 - \sin\theta_1\sin\theta_2\sin\theta_3\sin\theta_4\cos(\phi_1 + \phi_2 - \phi_3 - \phi_4). \tag{36.22}$$

If we limit ourselves to study the vectors $\mathbf{n_J}$ in the $x - y$ plane, we have:

$$\mathrm{C}_\Psi(\mathbf{n}_1, \mathbf{n}_2, \mathbf{n}_3, \mathbf{n}_4) = -\cos(\phi_1 + \phi_2 - \phi_3 - \phi_4). \tag{36.23}$$

We choose particular values (perfect anticorrelation) so that:

$$[(\phi_1 + \phi_2 - \phi_3 - \phi_4) = 0] \rightarrow (\mathrm{C}_\Psi = -1), \tag{36.24a}$$

and

$$[(\phi_1 + \phi_2 - \phi_3 - \phi_4) = \pi] \rightarrow (\mathrm{C}_\Psi = +1). \tag{36.24b}$$

If the complete state of the four particles is symbolized by λ (not necessary an hidden parameter), we have four functions $A_\lambda(\phi_1)$, $B_\lambda(\phi_2)$, $C_\lambda(\phi_3)$, $D_\lambda(\phi_4)$ which are the outcomes of spin measurements along the respective directions and have value $+1$ or -1. This signifies that, if we know three angles we can predict with certainty the fourth.

On the hypothesis of separability and hence of factorization of the possible outcomes, we can rewrite the consequent of implication (36.24a) in the following form:

$$A_\lambda(\phi_1)B_\lambda(\phi_2)C_\lambda(\phi_3)D_\lambda(\phi_4) = -1; \tag{36.25a}$$

and the consequent of implication (36.24b) in the following form:

$$A_\lambda(\phi_1)B_\lambda(\phi_2)C_\lambda(\phi_3)D_\lambda(\phi_4) = +1. \tag{36.25b}$$

But there is no way to satisfy these conditions.

Proof
Four instances of Eq. (36.25a) are:

$$A_\lambda(0)B_\lambda(0)C_\lambda(0)D_\lambda(0) = -1, \quad (36.26a)$$
$$A_\lambda(\phi)B_\lambda(0)C_\lambda(\phi)D_\lambda(0) = -1, \quad (36.26b)$$
$$A_\lambda(\phi)B_\lambda(0)C_\lambda(0)D_\lambda(\phi) = -1, \quad (36.26c)$$
$$A_\lambda(2\phi)B_\lambda(0)C_\lambda(\phi)D_\lambda(\phi) = -1, \quad (36.26d)$$

whatever ϕ is. From Eq. (36.26a) and Eq. (36.26b) we obtain:

$$A_\lambda(0)C_\lambda(0) = A_\lambda(\phi)C_\lambda(\phi); \quad (36.27a)$$

and from Eq. (36.26a) and Eq. (36.26c) we obtain:

$$A_\lambda(0)D_\lambda(0) = A_\lambda(\phi)D_\lambda(\phi). \quad (36.27b)$$

From Eq. (36.27a) and Eq. (36.27b):

$$\frac{C_\lambda(0)}{D_\lambda(0)} = \frac{C_\lambda(\phi)}{D_\lambda(\phi)}, \quad (36.28)$$

which can be rewritten as

$$C_\lambda(0)D_\lambda(0) = C_\lambda(\phi)D_\lambda(\phi), \quad (36.29)$$

because $D_\lambda(\phi) = \pm 1$ and $D_\lambda(0) = \pm 1$ and hence equal their inverse. From the last Eq. and Eq. (36.26d) we have:

$$A_\lambda(2\phi)B_\lambda(0)C_\lambda(0)D_\lambda(0) = -1, \quad (36.30)$$

which, together with Eq. (36.26a), gives:

$$A_\lambda(2\phi) = A_\lambda(0) = C', \quad (36.31)$$

where C' is a constant. But now we use Eq. (36.25b), for which, for whatever ϕ', we have:

$$A_\lambda(\phi' + \pi)B_\lambda(0)C_\lambda(\phi')D_\lambda(0) = 1, \quad (36.32)$$

which, together with Eq. (36.26b), gives:

$$A_\lambda(\phi' + \pi) = -A_\lambda(\phi'), \quad (36.33)$$

which confirms the anticorrelation but also contradicts the result of Eq. (36.31) — **for example** for $\phi = \pi/2, \phi' = 0$. So we must conclude that a theory based upon perfect correlations between four particles, separability and Counterfactuality is contradictory. Q.E.D.

36.4.2 Greenberger/Horne/Shimony/Zeilinger Proof

The proof of GHZ can be generalized to each system with three or more particles without using spin correlations — hence the following proof is important because it is the bridge to Bell inequalities with other observables [chapter 40]. Using an argument found in [HORNE/ZEILINGER 1985], Greenberger/Horne/Shimony/Zeilinger (= GHSZ) demonstrate the contradiction between separability and QM [GREENBERGER *et al.* 1990, 1135–36, 1141]

Suppose (Interfer. Exp. 20) a particle with (mean) momentum 0 that decays into three photons. If all three particles have the same energy, then, by momentum conservation they must be emitted 120° apart from each other. The central source is surrounded by an array of six

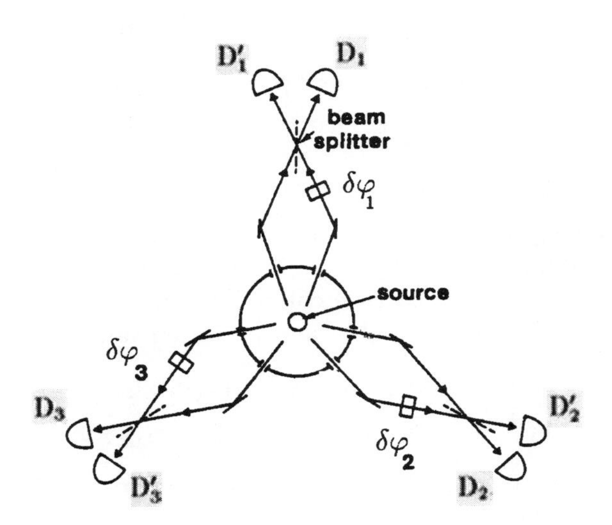

Figure 36.2: A *Gedankenexperiment* with a three-particle interferometer. The source emits a triple of particles, 1, 2, and 3, in six beams, with the state given by Eq.(36.34). A phase shift $\delta\varphi_1$ is imparted to beam a' of particle 1, and beams a, a' are brought together on a BS before illuminating detectors D_1, D'_1. Likewise for particles 2 and 3 — from [GREENBERGER *et al.* 1990, 1135].

apertures: a, b and c at 120° separation, and a', b' and c' also at 120° separation. Because of the placement of apertures, the three particle 1, 2 and 3 must emerge either through a, b and c or through a', b' and c' [see figure 36.2].

Thus the state of the three particles beyond the apertures will be the superposition (GHSZ state):

$$|\Psi\rangle = \frac{1}{\sqrt{2}}(|a\rangle_1|b\rangle_2|c\rangle_3 + |a'\rangle_1|b'\rangle_2|c'\rangle_3), \tag{36.34}$$

where $|a\rangle_1$ denotes the particle 1 in beam a, and so on.

Beyond the apertures beams $|a\rangle_1$ and $|a'\rangle_1$ are totally reflected so as to overlap at a 50/50 (half possibility of transmission and half of reflection) BS, and the two outgoing beams are monitored by detectors D_1 and D'_1. Suppose that on the way $|a'\rangle_1$ passes through a phase plate which causes a phase shift φ_1. Consequently we have the evolutions

$$|a\rangle_1 \mapsto \frac{1}{\sqrt{2}}(|D_1\rangle + \imath|D'_1\rangle); \tag{36.35a}$$

$$|a'\rangle_1 \mapsto \frac{1}{\sqrt{2}}e^{\imath\varphi_1}(|D'_1\rangle + \imath|D_1\rangle). \tag{36.35b}$$

Particles 2 and 3 are subjected to similar treatment with detectors D_2 and D'_2 for particle 2 and detectors D_3 and D'_3 for 3.

The evolution of all three particles is given by:

$$\begin{aligned} |\Psi\rangle \mapsto\ & \frac{1}{4}[(1 - \imath e^{\imath(\varphi_1+\varphi_2+\varphi_3)})|D_1\rangle|D_2\rangle|D_3\rangle + (\imath - e^{\imath(\varphi_1+\varphi_2+\varphi_3)})|D_1\rangle|D_2\rangle|D'_3\rangle \\ & +(\imath - e^{\imath(\varphi_1+\varphi_2+\varphi_3)})|D_1\rangle|D'_2\rangle|D_3\rangle + (-1 + \imath e^{\imath(\varphi_1+\varphi_2+\varphi_3)})|D_1\rangle|D'_2\rangle|D'_3\rangle \\ & +(\imath - e^{\imath(\varphi_1+\varphi_2+\varphi_3)})|D'_1\rangle|D_2\rangle|D_3\rangle + (-1 + \imath e^{\imath(\varphi_1+\varphi_2+\varphi_3)})|D'_1\rangle|D_2\rangle|D'_3\rangle \\ & +(-1 + \imath e^{\imath(\varphi_1+\varphi_2+\varphi_3)})|D'_1\rangle|D'_2\rangle|D_3\rangle + (-\imath + e^{\imath(\varphi_1+\varphi_2+\varphi_3)})|D'_1\rangle|D'_2\rangle|D'_3\rangle. \end{aligned} \tag{36.36}$$

The probability for detection of the three particles by respective detectors D_1, D_2 and D_3 is:

$$\wp^{\Psi}_{D_1D_2D_3}(\varphi_1,\varphi_2,\varphi_3) = \frac{1}{16}|(1 - \imath e^{\imath(\varphi_1+\varphi_2+\varphi_3)}|^2 = \frac{1}{8}[1 - \sin(\varphi_1+\varphi_2+\varphi_3)]; \qquad (36.37a)$$

and likewise:

$$\wp^{\Psi}_{D'_1D_2D_3}(\varphi_1,\varphi_2,\varphi_3) = \frac{1}{8}[1 + \sin(\varphi_1+\varphi_2+\varphi_3)]; \qquad (36.37b)$$

and so on for the remaining six possible outcomes.

The sum of the probabilities for all eight possible outcomes is of course 1. Assuming the GHZ argument we give value +1 when a particle enters into an unprimed detector and −1 when it enters into a primed one. Now we calculate the expectation value:

$$\begin{aligned} C_{\Psi}(\varphi_1,\varphi_2,\varphi_3) &= \wp^{\Psi}_{D_1D_2D_3}(\varphi_1,\varphi_2,\varphi_3) - \wp^{\Psi}_{D'_1D'_2D'_3}(\varphi_1,\varphi_2,\varphi_3) \\ &\quad + \wp^{\Psi}_{D_1D'_2D'_3}(\varphi_1,\varphi_2,\varphi_3) + \wp^{\Psi}_{D'_1D_2D'_3}(\varphi_1,\varphi_2,\varphi_3) + \wp^{\Psi}_{D'_1D'_2D_3}(\varphi_1,\varphi_2,\varphi_3) \\ &\quad - \wp^{\Psi}_{D'_1D_2D_3}(\varphi_1,\varphi_2,\varphi_3) - \wp^{\Psi}_{D_1D'_2D_3}(\varphi_1,\varphi_2,\varphi_3) - \wp^{\Psi}_{D_1D_2D'_3}(\varphi_1,\varphi_2,\varphi_3) \\ &= \sin(\varphi_1+\varphi_2+\varphi_3). \end{aligned} \qquad (36.38)$$

In the following we synthesize very briefly how to deduce an important inequality which in principle can be tested.

For $\varphi_1+\varphi_2+\varphi_3 = \pi/2$ we obtain $C_{\Psi} = +1$ and for $\varphi_1+\varphi_2+\varphi_3 = 3\pi/2$ we obtain $C_{\Psi} = -1$ [see Eqs. (36.24)]. Hence we can now apply the same demonstration that we utilized for the GHZ theorem. Q.E.D.

We can show a possible test of the theorem by deducing a strong constraint on the probability measure [GREENBERGER *et al.* 1990, 1136–37]. Consider four different choices of the phase angles: $(\pi/2,0,0)$, $(0,\pi/2,0)$, $(0,0,\pi/2)$ and $(\pi/2,\pi/2,\pi/2)$. As in the GHZ proof we have three functions $A_\lambda(\varphi_1)$, $B_\lambda(\varphi_2)$, $C_\lambda(\varphi_3)$. Since the three particles are emitted before encountering the phase plate, one can suppose that the same probability measure μ_p governs the four possible alternatives. Consider now the following statements:

$$A_\lambda(\pi/2)B_\lambda(0)C_\lambda(0) = 1; \qquad (36.39a)$$
$$A_\lambda(0)B_\lambda(\pi/2)C_\lambda(0) = 1; \qquad (36.39b)$$
$$A_\lambda(0)B_\lambda(0)C_\lambda(\pi/2) = 1. \qquad (36.39c)$$

Multiplying the three Eqs. (36.39) we have:

$$A_\lambda(\pi/2)B_\lambda(\pi/2)C_\lambda(\pi/2) = 1, \qquad (36.40)$$

because the other factors are equal to one, since $A^2_\lambda(\phi) = B^2_\lambda(\phi) = C^2_\lambda(\phi) = 1$ for any ϕ. Consequently the statement:

$$A_\lambda(\pi/2)B_\lambda(\pi/2)C_\lambda(\pi/2) = -1 \qquad (36.41)$$

implies that at least one of Eqs. (36.39) is false. Expressing the implication in the language of set theory:

$$\Lambda_4 \subseteq (\Lambda'_1 \bigcup \Lambda'_2 \bigcup \Lambda'_3), \qquad (36.42)$$

where Λ_j $(j = 1,2,3)$ is the set of all parameters λ such that respectively the j eq of the (36.39) holds, Λ_4 is the set of all parameters λ such that Eq. (36.41) holds, and Λ' is the complement of Λ. Using the probability measure we have:

$$\mu_p(\Lambda_4) \subseteq \mu_p(\Lambda'_1 \bigcup \Lambda'_2 \bigcup \Lambda'_3). \qquad (36.43)$$

Using the formalism of classical probability theory [see Eq. (11.2)] we obtain:

$$\mu_p(\Lambda'_1 \bigcup \Lambda'_2 \bigcup \Lambda'_3) \leq \mu_p(\Lambda'_1) + \mu_p(\Lambda'_2) + \mu_p(\Lambda'_3). \quad (36.44)$$

From the last two Eqs. we have finally:

$$\mu_p(\Lambda_4) \leq \mu_p(\Lambda'_1) + \mu_p(\Lambda'_2) + \mu_p(\Lambda'_3). \quad (36.45)$$

This inequality is the desired constraint and, under auxiliary technical assumptions, it is testable [10]. Recently, a proposal[11] has been made [ZEILINGER *et al.* 1997] in order to generate a three particle entanglement from two entangled pairs by erasure of the information about the source of one of the four particles[12]. Note that inequality (36.45) has the same formal structure as inequalities (39.3) and (42.94). On a formal level it has been also confirmed by the analysis developed in section 39.2. Moreover, GHZ or GHSZ states are very useful for multiparticle communication schemes, such as teleportation [see section 43.2].

36.5 Eberhard Theorem

Eberhard' proof Eberhard's work [EBERHARD 1978] focused on the problem of separability or locality as such[13]. Let $\hat{O}_1$ and $\hat{O}_2$ be [EBERHARD 1978, 394, 406, 413–414] two observables on subsystems $\mathcal{S}_1$ and $\mathcal{S}_2$ of a system $\mathcal{S}$, respectively, and $\wp(o_a, \mathbf{a}; o_b, \mathbf{b})$ be the probability that the results of a measurement on each one give respectively $\hat{O}_1 = o_a$ and $\hat{O}_2 = o_b$ when the knob settings are respectively $\mathbf{a}$ and $\mathbf{b}$.

We now give the following definition of locality [EBERHARD 1978, 396, 403, 416–17]:

Definition 36.1 (Locality) *The probability distribution of $\hat{O}_1$ ($\hat{O}_2$), integrated over o_a (o_b) is independent of the other setting* $\mathbf{b}$ ($\mathbf{a}$)*:*

$$\sum_{o_b} \wp(o_a, \mathbf{a}, o_b, \mathbf{b}) = \hat{\rho}(o_a, \mathbf{a}); \quad \sum_{o_b} \wp(o_a, \mathbf{a}, o_b, \mathbf{b}) = \hat{\rho}(o_b, \mathbf{b}). \quad (36.46)$$

If locality were violated, we would have a causal interdependence between the two subsystems, because, by changing the setting $\mathbf{a}$ ($\mathbf{b}$), we would be able to act on the result of the other measurement, and hence, if we perform experiments on subsystems that are space-like separated, we would be able to transmit a message with superluminal or perhaps instantaneous velocity. Hence a violation of the locality is surely a violation of special relativity. We can now state the following theorem:

[10] But for some difficulties in order to perform experiments see [GREENBERGER 1995].

[11] Based on a preceding proposal of Yurke/Stoler [YURKE/STOLER 1992a].

[12] For a further generalization see also [CLIFTON *et al.* 1991]. The experiment has also been recently performed [WEINFURTER *et al.* 1999].

[13] Eberhard distinguishes 4 types of Locality, of which the first one is the same as 'separability'. We consider here only the fourth form.

Theorem 36.3 (Eberhard) *QM does not violate locality as expressed by* Eq. (36.46).

Proof

Let $\hat{P}_{o_a}$ and $\hat{P}_{o_b}$ be two projectors and $\hat{\rho}$ a density matrix which represents the compound state. . The probability $\wp_a(o_a)$ that $\hat{O}_1 = o_a$ is

$$\wp_a(o_a) = \mathrm{Tr}[\hat{P}_{o_a}\hat{\rho}]. \tag{36.47}$$

After a measurement of $\hat{O}_1$ with result $\hat{O}_1 = o_a$ we obtain the transformation:

$$\hat{\rho} \leadsto \hat{\rho}' = \frac{\hat{P}_{o_a}\hat{\rho}\hat{P}_{o_a}}{\wp_a(o_a)}. \tag{36.48}$$

Now we perform a second measurement on the second subsystem, and calculate the conditional probability of obtaining $\hat{O}_2 = o_b$ by $\mathbf{b}$ setting:

$$\wp_b'(o_b|o_a) := \wp_b'(o_b) = \mathrm{Tr}\left[\frac{\hat{P}_{o_b}\hat{P}_{o_a}\hat{\rho}\hat{P}_{o_a}}{\wp_a(o_a)}\right]. \tag{36.49}$$

We compute now the global probability of obtaining the two results:

$$\begin{aligned}\wp(o_a, \mathbf{a}, o_b, \mathbf{b}) &= \wp_a(o_a)\wp_b'(o_b) \\ &= \wp_a(o_a)\frac{\mathrm{Tr}[\hat{P}_{o_b}\hat{P}_{o_a}\hat{\rho}\hat{P}_{o_a}]}{\wp_a(o_a)} \\ &= \mathrm{Tr}[\hat{P}_{o_b}\hat{P}_{o_a}\hat{\rho}\hat{P}_{o_a}].\end{aligned} \tag{36.50}$$

By the property $\sum_{o_a}\hat{P}_{o_a} = \hat{I}$ for each projector $\hat{P}_{o_a}$, we have [see Eq. (36.46)]:

$$\sum_{o_a}\wp(o_a, \mathbf{a}, o_b, \mathbf{b}) = \mathrm{Tr}\sum_{o_a}\left(\hat{P}_{o_a}\hat{P}_{o_b}\hat{\rho}\hat{P}_{o_a}\right) = \mathrm{Tr}[\hat{P}_{o_b}\hat{\rho}] = \wp(o_b, \mathbf{b}). \tag{36.51}$$

Similarly for the conditional probability of having $\hat{O}_1 = o_a$ having obtained $\hat{O}_2 = o_b$. In conclusion Locality is not violated by QM. Q.E.D.

Other proofs A very interesting proof along the same lines is due to Jordan [T. JORDAN 1983].[14] Let $\mathcal{S}$ be a system, $\hat{P}$ be a projection operator whose expectation value $\langle \hat{P} \rangle$ is the probability of a particular result for a measurement on the subsystem $\mathcal{S}_1$ of some system $\mathcal{S}$. Let $\{\hat{P}_k'\}$ be a set of mutually orthogonal projection operators whose expectation values $\langle \hat{P}_k' \rangle$ are the probabilities for a complete set of different results of a measurement on a subsystem $\mathcal{S}_2$. Since the measurements are on different subsystems, each $\hat{P}_k'$ commutes with $\hat{P}$. Then $\langle \hat{P}_k'\hat{P} \rangle$ is the joint probability of finding the result corresponding to $\hat{P}$ in $\mathcal{S}_1$ together with the result corresponding to $\hat{P}_k'$ in $\mathcal{S}_2$. But for the property $\sum_k \hat{P}_k' = \hat{I}$ we obtain:

$$\sum_k \langle \hat{P}_k'\hat{P} \rangle = \left\langle \sum_k \hat{P}_k'\hat{P} \right\rangle = \langle \hat{P} \rangle. \tag{36.52}$$

[14] See also [BUSSEY 1982] [GHIRARDI *et al.* 1980] . For other interesting demonstrations of the same result see also [JARRETT 1984b, 24–25] [BALLENTINE/JARRETT 1987].

More recently it was shown that, even by supposing an instantaneous reduction of the wave packet, it is always possible to measure certain non-local properties without violating locality [AHARONOV/ALBERT 1981], which is another confirmation of Eberhard's result. On the other hand, it was shown that it is impossible to account properly for all experimental results with a single covariant state history — see the delayed choice problem [chapter 26 and subsection 36.6.2] — because then each distant observer could cause a wave collapse with contradictory results [*PERES* 1993, 154]. Dieks [DIEKS 1982] has examined Herbert's proposal [HERBERT 1982] in order to perform superluminal communication and confirmed the precedent results by pointing out that the linearity of quantum formalism prevent such a possibility.

Conclusion The above proofs are the clear confirmation that in QM it is possible to distinguish (and one must do so too) between probability distributions and physical interactions, in the sense that we can have physical systems which are physically separated but not probabilistically independent [see definition 3.2: p. 42]. In the following we use the term *non-locality* as a short-hand way to understand globally all QM correlations based on entanglement. But we should never forget that in QM there is no violation of locality *stricto sensu.*

36.6 Hidden Variables and Lorentz Invariance

We have already seen in the preceding section that QM does not violate Lorentz invariance, while we know [see subsection 32.6.2] that Bohm argued for some violation of Lorentz invariance. The question now is: is it possible to demonstrate positively some violation of Lorentz invariance by HV theories? This is the aim of the following proof. First we analyse the general proof about the separability and then consider the specific problem of Lorentz-invariance.

36.6.1 Hardy's Proof of the Non-separability of Quantum Mechanics

Hardy's Proof

Before we discuss the problem of Lorentz invariance in HV theories, let us examine the following experiment (Interfer. Exp. 21) which makes use of GHSZ' results [HARDY 1992a , 2981–82]. We consider two Mach-Zehnder interferometers (= MZ) [see subsection 5.3.1], one for positrons (MZ^+) and one for electrons (MZ^-), arranged so that their paths overlap [see figure 36.3].

Each $MZ^\pm$ has an input mode $|i^\pm\rangle$, two paths inside the interferometer, $u^\pm$ and $v^\pm$, due to beam splitters $BS1^\pm$, and two outputs modes $c^\pm$ and $d^\pm$, preceded by beam splitters $BS2^\pm$, which end in respective detectors $C^\pm$ and $D^\pm$. Each interferometer is so arranged that no positron or electron will be detected by $D^\pm$. $BS2^\pm$ are removable. Now if e^+ takes path u^+ and e^- takes path u^-, particles meet at point A and annihilate one other. Hence we have

$$|u^+\rangle|u^-\rangle \mapsto |\gamma\rangle. \tag{36.53}$$

We shall see that, as a consequence of this possible interaction, it becomes possible for e^+ and e^- to arrive at detectors $D^\pm$.

The operation of $BS1^\pm$ is:

$$|i^\pm\rangle \mapsto \frac{1}{\sqrt{2}}(\imath|u^\pm\rangle + |v^\pm\rangle). \tag{36.54}$$

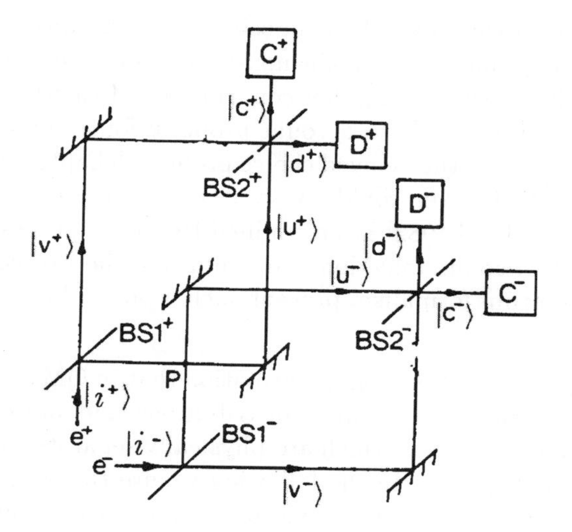

Figure 36.3: Proposed Interfer. Exp. 21. Two MZ-type interferometers, one for positrons and one for electrons, arranged so that, if a positron takes the path u^+ and an electron the path u^-, then they will meet at point A and annihilate one another — from [HARDY 1992a , 2981].

The operation effected by BS2$^\pm$ is given by:

$$|u^\pm\rangle \mapsto \frac{1}{\sqrt{2}}(|c^\pm\rangle + \imath|d^\pm\rangle) \tag{36.55a}$$

and

$$|v^\pm\rangle \mapsto \frac{1}{\sqrt{2}}(\imath|c^\pm\rangle + |d^\pm\rangle). \tag{36.55b}$$

If BS2$^\pm$ is removed, then:

$$|u^\pm\rangle \mapsto |c^\pm\rangle; \tag{36.56a}$$
$$|v^\pm\rangle \mapsto |d^\pm\rangle. \tag{36.56b}$$

After passing BS1$^\pm$ this state evolves according to eq. (36.54):

$$\frac{1}{2}(\imath|u^+\rangle + |v^+\rangle)(\imath|u^-\rangle + |v^-\rangle). \tag{36.57}$$

After passing point A the state becomes [using Eq. (36.53)]:

$$\frac{1}{2}(-|\gamma\rangle + \imath|u^+\rangle|v^-\rangle + \imath|v^+\rangle|u^-\rangle + |v^+\rangle|v^-\rangle). \tag{36.58}$$

Now we discuss four alternatives:

- If both BS2$^+$ and BS2$^-$ are removed, then using Eqs. (36.56), we see that the state represented by Eq. (36.58) evolves to the final state (*case I*):

$$\frac{1}{2}(-|\gamma\rangle + \imath|c^+\rangle|d^-\rangle + \imath|d^+\rangle|c^-\rangle + |d^+\rangle|d^-\rangle). \tag{36.59}$$

- If BS2$^+$ is in place and BS2$^-$ is removed, using Eqs. (36.55) and Eqs. (36.56), we see that the state represented by Eq. (36.58) evolves to the final state (*case II*):

$$\frac{1}{2\sqrt{2}}(-\sqrt{2}|\gamma\rangle - |c^+\rangle|c^-\rangle + 2\imath|c^+\rangle|d^-\rangle + \imath|d^+\rangle|c^-\rangle). \tag{36.60}$$

- If BS2$^+$ is removed and BS2$^-$ in place, the state represented by Eq. (36.58) evolves to the final state (III case):

$$\frac{1}{2\sqrt{2}}(-\sqrt{2}|\gamma\rangle - |c^+\rangle|c^-\rangle + \imath|c^+\rangle|d^-\rangle + 2\imath|d^+\rangle|c^-\rangle). \tag{36.61}$$

- If both BS2$^+$ and BS2$^-$ are in place, using Eqs. (36.55), we find that the state represented by Eq. (36.58) evolves to the final state (*IV case*):

$$\frac{1}{4}(-2|\gamma\rangle - 3|c^+\rangle|c^-\rangle + \imath|c^+\rangle|d^-\rangle + \imath|d^+\rangle|c^-\rangle - |d^+\rangle|d^-\rangle). \tag{36.62}$$

Proof

Now we develop the proof as follows. We can make two measurements on each particle, with BS in place (0) and with BS removed (∞). We suppose a parameter λ (not necessarily hidden) such that it determines perfectly the results. If a particle is detected, for example by detector C we write $C^+_\lambda(\infty) = 1$ if BS removed and $C^-_\lambda(0) = 1$ if BS is not removed, and if we do not detect any particle we write $C^+_\lambda(\infty) = 0$ if BS removed and $C^-_\lambda(0) = 0$ if BS is not removed. Similarly with detectors D$^\pm$.

- In the first case [Eq. (36.59)] we have:

$$C^+_\lambda(\infty)C^-_\lambda(\infty) = 0 \tag{36.63}$$

for every experiment, because we have no term $|c^+\rangle|c^-\rangle$.

- In the second case [Eq. (36.60)] we have:

$$[D^+_\lambda(0) = 1] \rightarrow [C^-_\lambda(\infty) = 1] \tag{36.64}$$

for every experiment, because if e$^+$ is detected the state is projected in the last term ($|d^+\rangle|c^-\rangle$).

- In the third case [Eq. (36.61)] we have:

$$[D^-_\lambda(0) = 1] \rightarrow [C^+_\lambda(\infty) = 1] \tag{36.65}$$

for every experiment, because if e$^-$ is detected the state is projected in the third term ($|d^-\rangle|c^+\rangle$).

- In the fourth case [Eq. (36.62)] we have for 1/16th of experiments (see in particular the the last term: $|d^+\rangle|d^-\rangle$):

$$D^+_\lambda(0)D^-_\lambda(0) = 1. \tag{36.66}$$

Now consider the following ideal experiment. We have the situation of Eq. (36.66). From Eq. (36.63) and Eq. (36.64), and from the factorization (Separability) we see that this implies that:

$$C^+_\lambda(\infty)C^-_\lambda(\infty) = 1, \tag{36.67}$$

which contradicts result (36.63), which is valid for all experiments. Q.E.D.[15]

[15] Another more direct proof is in [HARDY 1993a]. See also [S. GOLDSTEIN 1994] [T. JORDAN 1994].

Experiments

In order to perform factual experiments, the main idea is that it is possible to use two spin-$\frac{1}{2}$ (instead of three as by GHZ and GHSZ) in a non-maximally entangled state to develop non-locality proofs without inequalities. The importance of this simplification is due to the previously discussed difficulty to test Eqs. which make no use of inequalities.

Mandel and co-workers [TORGESON *et al.* 1995] developed an experimental demonstration by using 'postselected' entangled states with a fixed degree of entanglement. They obtained about 45 standard deviations.

More sophisticated is the experiment performed by De Martini, Di Giuseppe and Boschi [DI GIUSEPPE *et al.* 1997a] [DI GIUSEPPE *et al.* 1997b] (SPDC Exp. 7).[16] Two photons, 1 and 2, with identical wavelengths and the same polarisation are generated by SPDC. Then their polarisations are rotated by angles ϕ_1, ϕ_2 and both impinge a polarising beam splitter (PBS). On emerging they are now in entangles state. Each one is then subjected first to another rotation (by angle $\theta_j, j = 1, 2$), then it passes through another polarising beam splitter PBSj, and finally it reaches the detectors [see figure 36.4]. It is a clear confirmation of the theoretical results developed by Hardy.

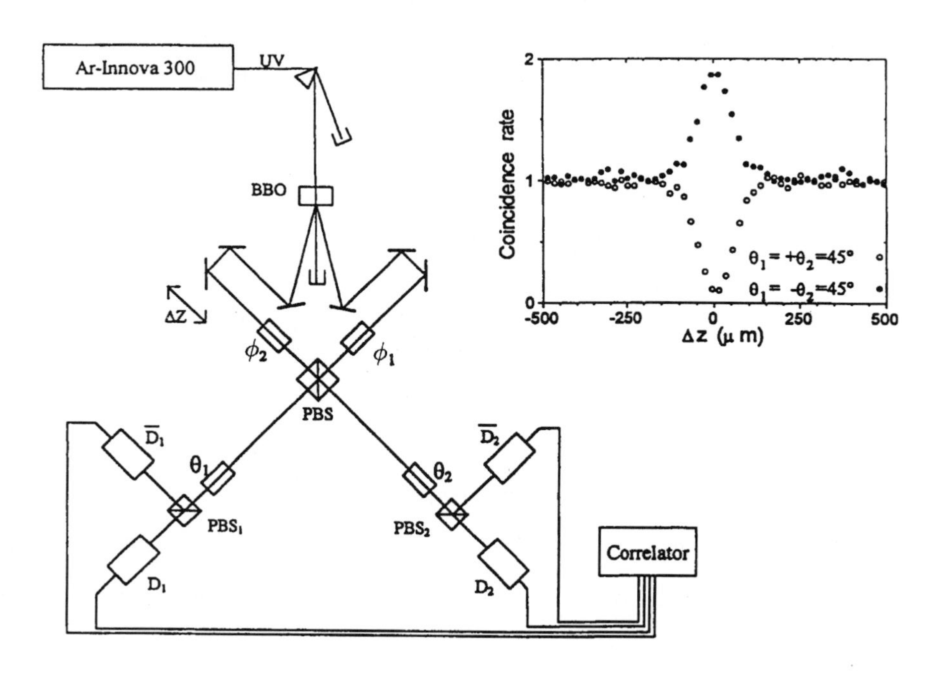

Figure 36.4: Scheme and results of SPDC Exp. 7 — from [DI GIUSEPPE *et al.* 1997a, 228].

[16]The experiment is mathematically and experimentally developed in [BOSCHI *et al.* 1997b] .

36.6.2 Hardy's Theorem

Exposition

Hardy tried to prove [HARDY 1992a , 2982–84] that theories which acknowledge separability and the Determined Value Assumption [postulate 6.3: p. 105] necessarily violate Lorentz-invariance. Now we apply the latter postulate (completeness condition) to the problem of Lorentz-invariance:

Proposition 36.1 (Lorentz-invariance necessary condition) *If we suppose that an element of physical reality is Lorentz-invariant, then his value is Lorentz-invariant.*

We have therefore the following theorem:

Theorem 36.4 (Hardy) *Each theory based on QM correlations, postulate* 6.3, *separability and proposition* 36.1 *is contradictory.*

Proof
In the framework of the experiment discussed previously, we introduce a projection operator:

$$\hat{P}^{\pm} = |u^{\pm}\rangle\langle u^{\pm}|. \tag{36.68}$$

Using the Determined Value Assumption [postulate 6.3: p. 105] we have:

$$\left(\hat{P}^{\pm}|u^{\pm}\rangle = |u^{\pm}\rangle\right) \quad \rightarrow \quad \left(\mathrm{v}[\hat{P}^{\pm}] = 1\right); \tag{36.69}$$

$$\left(\hat{P}^{+}\hat{P}^{-}|u^{+}\rangle|u^{-}\rangle = |u^{+}\rangle|u^{-}\rangle\right) \quad \rightarrow \quad \left(\mathrm{v}[\hat{P}^{+}\hat{P}^{-}] = 1\right); \tag{36.70}$$

and, if $|u^{+}, u^{-}\rangle_{\perp}$ is any state vector orthogonal to $|u^{+}\rangle|u^{-}\rangle$, we have:

$$\left(\hat{P}^{+}\hat{P}^{-}|u^{+}, u^{-}\rangle_{\perp} = 0\right) \rightarrow \left(\mathrm{v}[\hat{P}^{+}\hat{P}^{-}] = 0\right). \tag{36.71}$$

From proposition 36.1 we obtain:

$$\left(\mathrm{v}[\hat{P}^{+}]\mathrm{v}[\hat{P}^{-}] = 1\right) \rightarrow \left(\mathrm{v}[\hat{P}^{+}\hat{P}^{-}] = 1\right). \tag{36.72}$$

We arrange the apparatus so that measurements of e^{-} and e^{+} are simultaneous. But we can consider a frame of reference F^{+} in which the measurement on e^{+} is made when it passed $\mathrm{BS2}^{+}$ and before e^{-} arrives at $\mathrm{BS2}^{-}$. Using Eqs. (36.55) we see that the state represented by Eq. (36.58) evolves to the final state:

$$\frac{1}{2\sqrt{2}}\left(-\sqrt{2}|\gamma\rangle - |c^{+}\rangle|u^{-}\rangle + 2\imath|c^{+}\rangle|v^{-}\rangle + \imath|d^{+}\rangle|u^{-}\rangle\right). \tag{36.73}$$

If e^{+} is detected by D^{+} still before e^{-} arrives at $\mathrm{BS2}^{-}$, the state is projected in the last term of Eq. (36.73) and the state of the electron becomes $|u^{-}\rangle$. Therefore:

$$(\mathrm{D}^{+} = 1) \rightarrow \left(\mathrm{v}[\hat{P}^{-}] = 1\right). \tag{36.74}$$

Now consider the frame F^{-} in which the measurement of e^{-} occurs before e^{+} reaches $\mathrm{BS2}^{+}$. By the symmetry of apparatus we have:

$$(\mathrm{D}^{-} = 1) \rightarrow \left(\mathrm{v}[\hat{P}^{+}] = 1\right). \tag{36.75}$$

Finally consider the rest frame in which both measurements appear simultaneous. Before the particles arrive at $\mathrm{BS2}^{\pm}$ then the state is given by Eq. (36.58). However this state is orthogonal to $|u^{+}\rangle|u^{-}\rangle$ — there is no therm $|c^{+}\rangle|c^{-}\rangle$ [see Eq. (36.63)]. Therefore we have:

$$\mathrm{v}[\hat{P}^{+}\hat{P}^{-}] = 0 \tag{36.76}$$

for all experiments. But now observables $\hat{P}^{\pm}$ and $\hat{P}^{+}\hat{P}^{-}$ are Lorentz-invariant and therefore, if we adopt the Lorentz-invariance condition, we must also say that results (36.74), (36.75) and (36.76) are Lorentz-invariant and hence independent of the frame of reference. Consider now an experiment in which there are detections at D^{+} and D^{-}. From the Eq. (36.66) we know that this is the case only in 1/16th of all experiments. For such experiments we have

$$v[\hat{P}^{+}\hat{P}^{-}] = 1, \tag{36.77}$$

which contradicts Eq. (36.76), which is valid in all experiments. Q.E.D.[17]

Discussion

The proof makes use of the dubious completeness condition [see Eqs. (36.69)–(36.71)], as it is shown by [BERNDL/S. GOLDSTEIN 1994].[18] In fact, as we know by KS theorem [32.4: p. 551], in QM there can also be truth-value gaps. Hardy partially accepted the criticism [HARDY 1994b].

In conclusion it can be said that Hardy's proof is not resolutive, but that on a general plane, in a theory in which (hidden) parameters are added to QM to determine the results of individual measurements without changing the statistical predictions, there must be a mechanism whereby the setting of one measuring device can influence the reading of another instrument however remote. In other words, each 'quantum' HV theory should share both assumptions, of completeness and of determinateness of values. As we have seen, this is the reason why David Bohm in latter years became a supporter of a non-local HV-theory [see subsection 32.6.2].[19]

36.7 Context-Dependent Entanglement

Another interesting aspect of our examination is the following: in some cases entanglement can be context-dependent [KRENN/ZEILINGER 1995] [KRENN/ZEILINGER 1996] [20]. In the case of three particles with spin-$\frac{1}{2}$ (GHSZ state) emitted by a common source (z-direction is considered), defined by:

$$|\Psi\rangle = \frac{1}{\sqrt{2}}(|\uparrow\rangle|\uparrow\rangle|\uparrow\rangle + |\downarrow\rangle|\downarrow\rangle|\downarrow\rangle), \tag{36.78}$$

we consider the special case in which observers 1 and 2 perform spin measurements within the x–y plane — in spherical coordinates (θ, ϕ) we have $\theta_1 = \theta_2 = \frac{\pi}{2}$. Now we generate two subensembles depending on the result (up or down) of the measurement performed by observer 3. Then the correlation functions for these subensembles are [see Eqs. (36.22)–(36.23)]:

$$C^{\pm}_{12} = \pm\sin(\theta_3)\cos(\phi_1 + \phi_2 + \phi_3). \tag{36.79}$$

Generally, after having measured the third particle, the other two are entangled unless the third is measured along the z-direction ($\theta_3 = 0$), because in this case we obtain: $C^{+}_{12} = C^{-}_{12} = 0$.

[17] Hardy's argument can also be translated into an experimental arrangement using an SGM instead of an interferometer. In this way it is generalized to a two-particle system with arbitrary spin s [CLIFTON/NIEMANN 1992]. Hardy's proof has been developed further by Stapp [STAPP 1997]. But Stapp's proof is defective because, as has been already said, it uses counterfactual reasoning together with a form of classical realism [see section 36.2].

[18] See also [HILLERY/YURKE 1995, 219–20] and [CLIFTON/NIEMANN 1992, 181–82].

[19] See also [BELL 1964, 20] [BELL 1981b, 149–53] [BELL 1987a, 206].

[20] See also [CERECEDA 1997], where maximum-violation parameters are shown.

What does this examination prove? That we cannot speak of entanglement in general without specifying some conditions under which we are able to define it (which in general are the measurement conditions). In other words, we cannot handle entanglement without limiting it, because the measurement conditions are necessarily a limitation of the correlations[21].

[21] On the subject see again [CERECEDA 1997].

Chapter 37

MORE EXACT BOUNDS FOR THE INEQUALITIES

Contents The object of this chapter is the search for more exact values for the violation of the separability principle in QM. After an introduction to the problem [section 37.1], several theorems are discussed: in section 37.2 Tsirelson theorem, in section 37.3 Landau theorem and in section 37.4 Hillery/Yurke theorem. Finally in section 37.5 some more stringent conditions for separability are proposed.

37.1 Introduction to the Problem

We know that the Bell inequalities [see for example inequality (35.27)] determine a bound on the possible values that a factorisable HV deterministic or stochastic theory [see theorem 35.2: p. 615] allows for joint probabilities:

$$|C(\mathbf{a},\mathbf{b}) - C(\mathbf{a},\mathbf{b}')| + |C(\mathbf{a}',\mathbf{b}') + C(\mathbf{a}',\mathbf{b})| \leq 2. \tag{37.1}$$

On the other hand, it has been calculated that a theory satisfying only the constraint of locality [see theorem 36.3: p. 634] has a weaker condition on possible correlations:

$$|C(\mathbf{a},\mathbf{b}) - C(\mathbf{a},\mathbf{b}') + C(\mathbf{a}',\mathbf{b}') + C(\mathbf{a}',\mathbf{b})| \leq 4. \tag{37.2}$$

However, we have already demonstrated that QM requires an intermediate value between them [see Eq. (35.29)]:

$$[C(\mathbf{a},\mathbf{b}) - C(\mathbf{a},\mathbf{b}') + C(\mathbf{a}',\mathbf{b}') + C(\mathbf{a}',\mathbf{b})]_{\mathrm{QM}} = 2\sqrt{2}. \tag{37.3}$$

Now we wish to investigate the problem in a more systematic fashion.

37.2 Tsirelson Theorem

37.2.1 The Theorem and the Proofs

Tsirelson proved an important theorem [TSIRELSON 1980, 94–95]:[1]

[1] See also [KHALFIN/TSIRELSON 1992, 885–87] [HILLERY/YURKE 1995, 221–22] [*PERES* 1993, 174].

Theorem 37.1 (Tsirelson) *The QM expectation values for the expression* (35.27) *for correlated systems lie between* 2 *and* $2\sqrt{2}$.

Even though this theorem is sound, as we shall prove in the following, it is contained in the stronger theorem of Landau [37.2: p. 645], so that we need not integrate it into the theory.

The first part of the theorem, that the values are larger than 2, was proved by the Bell Theorem and CHSH inequality. We prove now that they are smaller than $2\sqrt{2}$.

To this end, note first that all Bell inequalities can be represented in the form of inequalities linear in observables and holding for all states. Thus they can be treated as inequalities for observables, while reference to states can be eliminated [KHALFIN/TSIRELSON 1985, 441].

Proof

Let $\hat{O}^a, \hat{O}^{a'}, \hat{O}^b, \hat{O}^{b'}$ be Hermitian operators on a Hilbert space $\mathcal{H}$ — we have conserved the symbolism a, a', b, b' as for polarisation problems in order to mark some continuity, but note that the analysis here is completely general and in no way restricted to the particular problems analysed in the last two chapters. Each operator has eigenvalues 1 and -1 and they satisfy the condition $[\hat{O}^a, \hat{O}^b] = 0$ and so on for the other couples $(a, b'), (a', b), (a', b')$. If we calculate the correlation functions from the density matrix: $\langle \hat{O}^a\hat{O}^b\rangle = \mathrm{Tr}(\hat{O}^a\hat{O}^b\hat{\rho})$, then there are four unit vectors in a four dimensional Euclidian space m, m', n, n' such that:

$$\left\langle \hat{O}^a\hat{O}^b \right\rangle = \langle m|n\rangle; \tag{37.4}$$

and so on for the other couples.

Since we have:

$$|\langle m|n\rangle + \langle m'|n\rangle + \langle m|n'\rangle - \langle m'|n'\rangle| \;\le\; |\langle m|n+n'\rangle| + |\langle m'|n-n'\rangle|, \tag{37.5}$$

which follows from Schwartz inequality, and the RHS is always equal to 2, so that

$$|\langle m|n+n'\rangle| + |\langle m'|n-n'\rangle| \;\; \le 2\sqrt{2}, \tag{37.6}$$

then we can write down the Tsirelson inequality:

$$\left|\left\langle \hat{O}^a\hat{O}^b \right\rangle + \left\langle \hat{O}^{a'}\hat{O}^b \right\rangle + \left\langle \hat{O}^a\hat{O}^{b'} \right\rangle - \left\langle \hat{O}^{a'}\hat{O}^{b'} \right\rangle\right| \;\le\; 2\sqrt{2}, \tag{37.7}$$

which is analogous to the II Bell inequality [Eq. (35.21)]. Q. E. D.

Tsirelson also gave another and more direct proof of this result[2], which we outline in the following.

Proof

We define the operator (we are again considering the II Bell inequality)

$$\hat{\mathrm{C}} = \hat{O}^a\hat{O}^b + \hat{O}^{a'}\hat{O}^b + \hat{O}^a\hat{O}^{b'} - \hat{O}^{a'}\hat{O}^{b'}. \tag{37.8}$$

We know that the square of each operator is equal to the identity, and this implies:

$$2\sqrt{2} - \hat{\mathrm{C}} = \frac{1}{\sqrt{2}}\left[(\hat{O}^a)^2 + (\hat{O}^{a'})^2 + (\hat{O}^b)^2 + (\hat{O}^{b'})^2\right] - \hat{\mathrm{C}} = \frac{1}{\sqrt{2}}\left(\hat{O}^a - \frac{\hat{O}^b + \hat{O}^{b'}}{\sqrt{2}}\right)^2 + \frac{1}{\sqrt{2}}\left(\hat{O}^{a'} - \frac{\hat{O}^b + \hat{O}^{b'}}{\sqrt{2}}\right)^2, \tag{37.9}$$

since $\frac{1}{\sqrt{2}}\left[(\hat{O}^a)^2 + (\hat{O}^{a'})^2 + (\hat{O}^b)^2 + (\hat{O}^{b'})^2\right] = 2^2 \cdot 2^{-1/2} = 2 \cdot 2^{1/2}$. Since the expression in the RHS consists of the sum of squares of Hermitian operators, it is clearly an operator with expectation value greater than or equal to zero. This leads to the conclusion:

$$\left\langle \hat{\mathrm{C}} \right\rangle \;\le\; 2\sqrt{2}. \tag{37.10}$$

[2] An analogous proof is developed by Landau [Lw. LANDAU 1987a, 54].

A similar argument leads to

$$\left\langle \hat{\mathrm{C}} \right\rangle \geq -2\sqrt{2}. \tag{37.11}$$

Now we see that Eq. (37.7) follows immediately. Q. E. D.

37.2.2 Khalfin/Tsirelson Inequality

In [KHALFIN/TSIRELSON 1985, 442–43] we can also find a generalized inequality from which one can derive all Bell inequalities [for example Eq. (37.1)] and the QM ones: the Khalfin/Tsirelson inequality:

$$\left\| \hat{O}^a\hat{O}^b + \hat{O}^a\hat{O}^{b'} + \hat{O}^{a'}\hat{O}^b - \hat{O}^{a'}\hat{O}^{b'} \right\| \leq \sqrt{4 + \| [\hat{O}^a, \hat{O}^{a'}] \| \cdot \| [\hat{O}^b, \hat{O}^{b'}] \|}. \tag{37.12}$$

If the commutators vanish we have the classical case; if not, since norms of commutators are at most 2, the right-hand side is at most $\sqrt{4+2\cdot 2} = 2\sqrt{2}$, which is the Tsirelson inequality.

Palatnik proposed the following correlation function for three subsystems[3]:

$$\hat{\mathrm{C}}_3 = \left\langle \hat{O}^a\hat{O}^b\hat{O}^{c'} \right\rangle + \left\langle \hat{O}^a\hat{O}^{b'}\hat{O}^c \right\rangle + \left\langle \hat{O}^{a'}\hat{O}^b\hat{O}^c \right\rangle - \left\langle \hat{O}^{a'}\hat{O}^{b'}\hat{O}^{c'} \right\rangle, \tag{37.13}$$

for which we have the following inequality:

$$\hat{\mathrm{C}}_3^2 \leq 8 - 2\left(\left\langle \hat{O}^b_j \hat{O}^c_k \right\rangle \right)^2, \tag{37.14}$$

for any j, k, which shows that an extremal correlation between two subsystems becomes impossible if at least one of them is correlated with a third subsystem.

37.3 Landau Theorem

Lawrence Landau [Lw. LANDAU 1988] showed that there are some sets of probabilities which are not allowed in QM and which violate the Landau condition and not Tsirelson inequality. Thus Landau's work is a step further on the way to fix bounds for QM.

Theorem 37.2 (I Landau) *QM correlations must satisfy the following Landau inequality:*

$$\begin{aligned} \left| \left\langle \hat{O}^a\hat{O}^b \right\rangle \left\langle \hat{O}^a\hat{O}^{b'} \right\rangle \right. & - \left. \left\langle \hat{O}^{a'}\hat{O}^b \right\rangle \left\langle \hat{O}^{a'}\hat{O}^{b'} \right\rangle \right| \\ & \leq \left(1 - \left\langle \hat{O}^a\hat{O}^{b'} \right\rangle^2 \right)^{\frac{1}{2}} + \left(1 - \left\langle \hat{O}^{a'}\hat{O}^b \right\rangle^2 \right)^{\frac{1}{2}} \left(1 - \left\langle \hat{O}^{a'}\hat{O}^{b'} \right\rangle^2 \right)^{\frac{1}{2}} \end{aligned} \tag{37.15}$$

One can prove constraints on QM that inequality (37.15) implies which are easier to prove than the theorem itself and represent also a good introduction to the following theorem. In fact, if the values of two of the correlation functions are known, we have constraints on the values of the other two [HILLERY/YURKE 1995, 222–24] . Consider first the following situation:

$$\left\langle \hat{O}^a\hat{O}^{b'} \right\rangle = \left\langle \hat{O}^{a'}\hat{O}^b \right\rangle = 1. \tag{37.16}$$

[3]See [KHALFIN/TSIRELSON 1992, 887].

We can rewrite Eq. (37.1) in the following form:

$$\left|\left\langle \hat{O}^{a'}\hat{O}^{b'}\right\rangle + \left\langle \hat{O}^{a'}\hat{O}^{b}\right\rangle + \left\langle \hat{O}^{a}\hat{O}^{b'}\right\rangle - \left\langle \hat{O}^{a}\hat{O}^{b}\right\rangle\right| \leq 2. \tag{37.17}$$

If we substitute now Eq. (37.16) in Eq. (37.17) we find:

$$\left\langle \hat{O}^{a}\hat{O}^{b}\right\rangle = \left\langle \hat{O}^{a'}\hat{O}^{b'}\right\rangle. \tag{37.18}$$

Now we obtain in QM the same result. Suppose the usual Hilbert space with the commutation relation: $[\hat{O}^a, \hat{O}^b] = \hat{I}$ $(\hbar = 1)$ and with allowed eigenvalues of the operators being $1, -1$, which implies that the square of each operator is 1:

$$(\hat{O}^a)^2 = 1; \; (\hat{O}^b)^2 = 1; \; (\hat{O}^{a'})^2 = 1; \; (\hat{O}^{b'})^2 = 1. \tag{37.19}$$

If $|\psi_1\rangle$ is a vector such that $\langle\psi_1|\hat{O}^a_1\hat{O}^{a'}|\psi_1\rangle = 1$, then our assumptions imply that:

$$\hat{O}^a_1\hat{O}^{b'}|\psi_1\rangle = |\psi_1\rangle. \tag{37.20}$$

Multiplying both sides by $\hat{O}^a$ and taking into account Eq. (37.19), we obtain:

$$\hat{O}^{b'}|\psi_1\rangle = \hat{O}^a_1|\psi_1\rangle. \tag{37.21}$$

Similarly if $\langle\psi_2|\hat{O}^{a'}\hat{O}^b|\psi_2\rangle = 1$, then:

$$\hat{O}^b|\psi_2\rangle = \hat{O}^{a'}|\psi_2\rangle. \tag{37.22}$$

Now if $|\psi\rangle$ is a vector such that $\langle\psi|\hat{O}^a\hat{O}^{b'}|\psi\rangle = \langle\psi|\hat{O}^{a'}\hat{O}^b|\psi\rangle = 1$ then we can use Eqs. (37.21) and (37.22) to obtain respectively:

$$\hat{O}^{a'}\hat{O}^{b'}|\psi\rangle = \hat{O}^a\hat{O}^{a'}|\psi\rangle; \tag{37.23a}$$

$$\hat{O}^a\hat{O}^b|\psi\rangle = \hat{O}^a\hat{O}^{a'}|\psi\rangle, \tag{37.23b}$$

from which we have:

$$\langle\psi|\hat{O}^a\hat{O}^b|\psi\rangle = \langle\psi|\hat{O}^{a'}\hat{O}^{b'}|\psi\rangle. \tag{37.24}$$

Finally if $\hat{\rho}$ is a density matrix such that:

$$\mathrm{Tr}\left(\hat{O}^a\hat{O}^{b'}\hat{\rho}\right) = \mathrm{Tr}\left(\hat{O}^{a'}\hat{O}^b\hat{\rho}\right) = 1, \tag{37.25}$$

then it must be of the form:

$$\hat{\rho} = \sum_n c_n|\psi_n\rangle\langle\psi_n|, \tag{37.26}$$

where $\langle\psi_n|\hat{O}^a_1\hat{O}^{b'}|\psi_n\rangle = \langle\psi_n|\hat{O}^{a'}\hat{O}^b|\psi_n\rangle = 1$. Eq. (37.24) implies:

$$\left\langle \hat{O}^a\hat{O}^b\right\rangle = \mathrm{Tr}(\hat{O}^a\hat{O}^b\hat{\rho}) = \mathrm{Tr}(\hat{O}^{a'}\hat{O}^{b'}\hat{\rho}) = \left\langle \hat{O}^{a'}\hat{O}^{b'}\right\rangle, \tag{37.27}$$

which is Eq. (37.18).

We present now a set of probabilities [see table 37.1] which satisfy the laws of probability and the requirement of locality [see Eq. (37.2)] and also $\left\langle \hat{O}^a\hat{O}^{b'}\right\rangle = \left\langle \hat{O}^{a'}\hat{O}^b\right\rangle$ but for which $\left\langle \hat{O}^a\hat{O}^b\right\rangle \neq \left\langle \hat{O}^{a'}\hat{O}^{b'}\right\rangle$. This set yields the following values for correlation functions:

$$\left\langle \hat{O}^a\hat{O}^{b'}\right\rangle = 1; \tag{37.28a}$$

$$\left\langle \hat{O}^a\hat{O}^b\right\rangle = 1 - 4\epsilon; \tag{37.28b}$$

$$\left\langle \hat{O}^{a'}\hat{O}^b\right\rangle = 1; \tag{37.28c}$$

$$\left\langle \hat{O}^{a'}\hat{O}^{b'}\right\rangle = 1. \tag{37.28d}$$

Now for $\epsilon > 0$ this set of probabilities is not allowed quantum-mechanically. If $0 < \epsilon < (\sqrt{2} - 1)/2$ then our set allows Tsirelson inequality but violates QM predictions.

	$\hat{O}^a = 1$	$\hat{O}^a = -1$	$\hat{O}^{a'} = 1$	$\hat{O}^{a'} = -1$
$\hat{O}^b = 1$	x	ϵ	$x+\epsilon$	0
$\hat{O}^b = -1$	ϵ	$1-x-2\epsilon$	0	$1-x-\epsilon$
$\hat{O}^{b'} = 1$	$x+\epsilon$	0	$x+\epsilon$	0
$\hat{O}^{b'} = -1$	0	$1-x-\epsilon$	0	$1-x-\epsilon$

Table 37.1: QM-violating set of probabilities. Each entry is a probability $\wp_j(\hat{O}^a_j = m, \hat{O}^b_k = n)$, where $j, k = 1, 2$ and m, n are either 1 or -1. The variables x and ϵ are chosen to be greater than or equal to zero so that all entries in the table are between 0 and 1. This set of probabilities satisfies the causal communication constraint, but cannot be derived from a density matrix — from [HILLERY/YURKE 1995, 224].

37.4 Hillery/Yurke Theorem

Stronger constraints were proved by Hillery and Yurke [HILLERY/YURKE 1995, 224–26] . Let us suppose that:

$$\left\langle \hat{O}^a \hat{O}^{b'} \right\rangle = 1 - \epsilon_1, \tag{37.29a}$$

$$\left\langle \hat{O}^{a'} \hat{O}^{b} \right\rangle = 1 - \epsilon_2, \tag{37.29b}$$

where both ϵ_1 and ϵ_2 are between zero and two. The case considered in Eq. (37.28a) and Eq. (37.28c) is when $\epsilon_1 = \epsilon_2 = 0$. If we calculate the correlation function from a Bell-like inequality, from Eqs. (37.29) we have:

$$\left| \left\langle \hat{O}^a \hat{O}^b \right\rangle - \left\langle \hat{O}^{a'} \hat{O}^{b'} \right\rangle \right| \leq \epsilon_1 + \epsilon_2. \tag{37.30}$$

Thus the extent to which $\left\langle \hat{O}^a \hat{O}^b \right\rangle$ differs from $\left\langle \hat{O}^{a'} \hat{O}^{b'} \right\rangle$ is limited by the deviations of $\left\langle \hat{O}^a \hat{O}^{b'} \right\rangle$ and $\left\langle \hat{O}^{a'} \hat{O}^b \right\rangle$ from 1. In QM there are also restrictions on how much $\left\langle \hat{O}^a \hat{O}^b \right\rangle$ and $\left\langle \hat{O}^{a'} \hat{O}^{b'} \right\rangle$ can differ but the bound is different. The importance of the proof lies in the fact that it is normally very difficult to guarantee that $\left\langle \hat{O}^a \hat{O}^{b'} \right\rangle$ and $\left\langle \hat{O}^{a'} \hat{O}^b \right\rangle$ are exactly 1, so that a statistical result of this type is preferable. Hence we can prove the following theorem

---

Theorem 37.3 (Hillery/Yurke) *Assuming* Eqs. (37.29) *the QM difference between* $\left\langle \hat{O}^a \hat{O}^b \right\rangle$ *and* $\left\langle \hat{O}^{a'} \hat{O}^{b'} \right\rangle$ *is given by*

$$\boxed{\left| \left\langle \hat{O}^a \hat{O}^b \right\rangle - \left\langle \hat{O}^{a'} \hat{O}^{b'} \right\rangle \right| \leq \sqrt{2\epsilon_1} + \sqrt{2\epsilon_2} + 2\sqrt{\epsilon_1 \epsilon_2}} \tag{37.31}$$

The importance of the Hillery/Yurke theorem is not only due to its statistical nature. But firstly it is due to the fact that it gives a concrete quantitative nature to the correlations between observables: it is a measure of correlation or of Entanglement [see sections 38.3, 42.7 and 42.8].[4]

[4]See also the pioneering work of Margenau/Hill [MARGENAU/HILL 1961, 725–26].

We prove the theorem for the case of a pure state, but the generalization to mixed ones is straightforward.

Proof

We make the same assumption on the values of operators as before, from which Eqs. (37.29) become:

$$\langle\psi|\hat{O}^a\hat{O}^{b'}|\psi\rangle = 1-\epsilon_1, \tag{37.32a}$$
$$\langle\psi|\hat{O}^{a'}\hat{O}^{b}|\psi\rangle = 1-\epsilon_2. \tag{37.32b}$$

The operator $\hat{O}^a\hat{O}^{b'}$ has eigenvalues 1 and -1 and its eigenvectors form a complete set. Hence each vector can be represented as the sum of two orthogonal vectors both of which are eigenvectors of $\hat{O}^a\hat{O}^{b'}$, one of which corresponds to eigenvalue 1 and the other to -1. In particular we can express $|\psi\rangle$ in this form:

$$|\psi\rangle = |\psi_{1+}\rangle + |\psi_{1-}\rangle, \tag{37.33}$$

where

$$\hat{O}^a\hat{O}^{b'}|\psi_{1+}\rangle = |\psi_{1+}\rangle, \tag{37.34a}$$
$$\hat{O}^a\hat{O}^{b'}|\psi_{1-}\rangle = -|\psi_{1-}\rangle. \tag{37.34b}$$

If $|\psi\rangle$ is normalized to one, we have:

$$\parallel \psi_{1+} \parallel^2 + \parallel \psi_{1-} \parallel^2 = 1, \tag{37.35}$$

which, together with Eq. (37.32b), gives:

$$\parallel \psi_{1+} \parallel^2 - \parallel \psi_{1-} \parallel^2 = 1-\epsilon_1. \tag{37.36}$$

From Eq. (37.35) and Eq. (37.36) we obtain:

$$\parallel \psi_{1+} \parallel^2 = 1-\frac{1}{2}\epsilon_1; \tag{37.37a}$$
$$\parallel \psi_{1-} \parallel^2 = \frac{1}{2}\epsilon_1. \tag{37.37b}$$

If we now decompose $|\psi\rangle$ in the states:

$$|\psi\rangle = |\psi_{2+}\rangle + |\psi_{2-}\rangle, \tag{37.38}$$

with similar considerations we arrive at:

$$\parallel \psi_{2+} \parallel^2 = 1-\frac{1}{2}\epsilon_2; \tag{37.39a}$$
$$-\parallel \psi_{2-} \parallel^2 = \frac{1}{2}\epsilon_2. \tag{37.39b}$$

These decompositions can be used to express $\langle\psi|\hat{O}^a\hat{O}^b|\psi\rangle$ in terms of $\langle\psi|\hat{O}^{a'}\hat{O}^{b'}|\psi\rangle$. We employ Eq. (37.21) and Eq. (37.22) to show that:

$$\hat{O}^a|\psi_{1+}\rangle = \hat{O}^{b'}|\psi_{1+}\rangle; \tag{37.40a}$$
$$\hat{O}^{a'}|\psi_{2+}\rangle = \hat{O}^{b}|\psi_{2+}\rangle; \tag{37.40b}$$

and that:

$$\hat{O}^a|\psi_{1-}\rangle = -\hat{O}^{b'}|\psi_{1-}\rangle; \tag{37.41a}$$
$$\hat{O}^{a'}|\psi_{2-}\rangle = -\hat{O}^{b}|\psi_{2-}\rangle; \tag{37.41b}$$

so that we can write:

$$\hat{O}^a|\psi\rangle = \hat{O}^{b'}|\psi_{1+}\rangle - \hat{O}^{b'}|\psi_{1-}\rangle = \hat{O}^{b'}|\psi\rangle - 2\hat{O}^{b'}|\psi_{1-}\rangle, \quad (37.42a)$$

$$\hat{O}^b|\psi\rangle = \hat{O}^{a'}|\psi_{2+}\rangle - \hat{O}^{a'}|\psi_{2-}\rangle = \hat{O}^{a'}|\psi\rangle - 2\hat{O}^{a'}|\psi_{2-}\rangle, \quad (37.42b)$$

which gives:

$$\langle\psi|\hat{O}^a\hat{O}^b|\psi\rangle = \langle\psi|\hat{O}^{a'}\hat{O}^{b'}|\psi\rangle - 2\langle\psi|\hat{O}^{a'}\hat{O}^{b'}|\psi_{2-}\rangle - 2\langle\psi_{1-}|\hat{O}^{a'}\hat{O}^{b'}|\psi\rangle + 4\langle\psi_{1-}|\hat{O}^{a'}\hat{O}^{b'}|\psi_{2-}\rangle. \quad (37.43)$$

Moving the term $\langle\psi|\hat{O}^{a'}\hat{O}^{b'}|\psi\rangle$ from the RHS to the LHS, taking the absolute values of both sides, applying the Schwartz inequality and using Eq. (37.19), Eq. (37.37) and Eq. (37.39) for the remaining terms on the RHS, we obtain:

$$|\langle\psi|\hat{O}^{a'}\hat{O}^{b'}|\psi_{2-}\rangle| \leq \sqrt{\frac{\epsilon_2}{2}}, \quad (37.44a)$$

$$|\langle\psi_{1-}|\hat{O}^{a'}\hat{O}^{b'}|\psi\rangle| \leq \sqrt{\frac{\epsilon_1}{2}}, \quad (37.44b)$$

$$|\langle\psi_{1-}|\hat{O}^{a'}\hat{O}^{b'}|\psi_{2-}\rangle| \leq \frac{\sqrt{\epsilon_2\epsilon_1}}{2}, \quad (37.44c)$$

which yield the final result of Eq. (37.31). The importance of the result is that it holds for mixed states as well. Q.E.D.

37.5 Conditions for Separability

As we have seen [section 35.6] all classically correlated states admit HV-Models and hence satisfy Bell inequalities; but it is not true that every state admitting an HV-Model is classically correlated. In other words while a separable [see definition 3.2: p. 42] system always satisfies Bell inequalities, the converse is not necessarily true. We seek now a necessary condition (more stringent than that of Bell) for separability [PERES 1996a].[5] Let us write a separable system as follows:

$$\hat{\rho}_{m\mu,n\nu} = \sum_j w_j(\hat{\rho}'_j)_{mn} \otimes (\hat{\rho}''_j)_{\mu\nu}, \quad (37.45)$$

where latin indices refer to the first subsystem and greek indices to the second one (the dimensions can be different). Note that this Eq. can be always be satisfied if we replace density matrices by Liouville functions, which have to be non-negative. In the case of QM we require the non-negativity of the eigenvalues [see subsection 3.4.5]. In order to find such a criterion let us define a new density matrix:

$$\hat{\rho}_{n\mu,m\nu} = \sum_j w_j(\hat{\rho}'_j)^{\mathrm{T}} \otimes (\hat{\rho}''_j) \quad (37.46)$$

where only latin indices have been transposed. It is a non-unitary transformation, but $\hat{\rho}_{n\mu,m\nu}$ is still Hermitian. Since a transposed matrix is a non-negative matrix with unit trace, it follows that none of the eigenvalues of $\hat{\rho}_{n\mu,m\nu}$ is negative: this is the necessary condition for separability.

[5]See also [LEWENSTEIN/SANPERA 1998].

For example take a pair of spin-$\frac{1}{2}$ particles in a Werner state (impure singlet) [see Eq. (38.12)] consisting of a fraction w and a random fraction $1-w$ [see Eq. (3.47)]. The latter includes singlets, mixed in equal proportion with the three triplet components [see Eqs. (3.132)]. We have:

$$\hat{\rho}_{m\mu,n\nu} = w\hat{\rho}^0_{m\mu,n\nu} + (1-w)\frac{\delta_{mn}\delta_{\mu\nu}}{4}, \tag{37.47}$$

where the density matrix for a pure singlet state is given by:

$$\hat{\rho}^0_{\downarrow\uparrow,\downarrow\uparrow} = \hat{\rho}^0_{\uparrow\downarrow,\uparrow\downarrow} = -\hat{\rho}^0_{\downarrow\uparrow,\uparrow\downarrow} = -\hat{\rho}^0_{\uparrow\downarrow,\downarrow\uparrow} = \frac{1}{2}, \tag{37.48}$$

and all other components of $\hat{\rho}^0$ vanish. Now a calculation shows that $\hat{\rho}_{n\mu,m\nu}$ (the density matrix of the I system has been transposed) has three eigenvalues equal to $(1+w)/4$ and the fourth one $(1-3w)/4$. The condition for the lower one to be positive (necessarily a requirement for separability) is $w < \frac{1}{3}$. It is a stricter criterion than those of Bell inequality, which hold for $w < \frac{1}{\sqrt{2}}$, and of the entropic inequality [R. HORODECKI *et al.* 1996a], which is $w < \frac{1}{\sqrt{3}}$.

In the particular case chosen, this criterion is also sufficient for separability — and it is for composite systems having dimensions 2×2 and 2×3. But for higher dimensions it is not a sufficient condition, as demonstrated in [R. HORODECKI *et al.* 1996b] .

For further interesting developments see also section 42.8.

Chapter 38

ENTANGLEMENT WITH PURE STATES AND WITH MIXTURES

Contents Until now we have proved that QM violates Bell inequalities and hence the separability principle, but not positively that each quantum entangled state — also a mixture — violates Bell inequalities. This is the subject of the present chapter. In section 38.1 we discuss again the problem of pure states in order to find some generalization. The remainder of the chapter is devoted to the violation of separability by mixed states [section 38.2]. Some recent formal generalizations of the latter results follow [section 38.3].

38.1 Pure States

38.1.1 Gisin Theorem

Gisin [GISIN 1991a] proved for the first time the important result that all entangled pure states violate a CHSH type inequality — i.e., for the Fine Theorem [p. 615], all the family of Bell inequalities, and for the Equivalence Theorem [p. 627], they respect also Stapp Theorem [p. 622] and similar formulations [see chapter 36].

Theorem 38.1 (Gisin) *Let $|\Psi\rangle \in \mathcal{H}_1 \otimes \mathcal{H}_2$. If $|\Psi\rangle$ is entangled, then $|\Psi\rangle$ violates CHSH inequality.*

Proof
Let $\{|m_j\rangle\}$ and $\{|n_j\rangle\}$ be orthonormal bases of $\mathcal{H}_1$ and $\mathcal{H}_2$, respectively, such that

$$|\Psi\rangle = \sum_j c_j |m_j\rangle \otimes |n_j\rangle, \tag{38.1}$$

for some real c_j with $c_1 \neq 0 \neq c_2$. Suppose one has:

$$|\Psi\rangle = |\psi\rangle + |\psi_\perp\rangle, \tag{38.2}$$

where $|\psi\rangle \perp |\psi_\perp\rangle$ and

$$|\psi\rangle = c_1|m_1\rangle \otimes |n_1\rangle + c_2|m_2\rangle \otimes |n_2\rangle \in C^2 \otimes C^2. \tag{38.3}$$

For simplicity we write:

$$|\psi\rangle = c_1|\uparrow\downarrow\rangle + c_2|\downarrow\uparrow\rangle. \tag{38.4}$$

Now we have:

$$\langle \mathbf{a}\hat{\sigma} \otimes \mathbf{b}\hat{\sigma}\rangle_\psi = -2c_1c_2(a_xb_x + a_yb_y) - a_zb_z, \tag{38.5}$$

for all vectors $\mathbf{a}, \mathbf{b} \in \Re^3$. Now let $a_x = \sin\theta, b_x = \sin\phi, a_y = 0, b_y = 0, a_z = \cos\theta, b_z = \cos\phi$, and similarly for vectors $\mathbf{a}', \mathbf{b}'$. Furthermore let $\theta = 0, \theta' = \pm\pi/2$, where the sign is the opposite to that of the product c_1c_2. We now have:

$$|\wp(\mathbf{a},\mathbf{b}) - \wp(\mathbf{a},\mathbf{b}')| + \wp(\mathbf{a}',\mathbf{b}) + \wp(\mathbf{a}',\mathbf{b}') = |\cos\phi - \cos\phi'| + 2\,|c_1c_2|\,[\sin\phi + \sin\phi']\,. \tag{38.6}$$

This is maximum for $\cos\phi = -\cos\phi' = (1+4|c_1c_2|)^{-\frac{1}{2}}, \sin\phi > 0, \sin\phi' > 0$. For this values one has:

$$|\wp(\mathbf{a},\mathbf{b}) - \wp(\mathbf{a},\mathbf{b}')| + \wp(\mathbf{a}',\mathbf{b}) + \wp(\mathbf{a}',\mathbf{b}') = 2(1+4|c_1c_2|)^{-\frac{1}{2}}. \tag{38.7}$$

This is greater than 2 and hence violates CHSH inequality [see Eq. (35.21)] for all non-zero c_1, c_2, that is for all entangled $|\Psi\rangle$. Q.E.D.

38.1.2 Popescu/Rohrlich Theorem

In [GISIN/PERES 1992] Gisin Theorem was extended: from the model of two particles with spin-$\frac{1}{2}$, to a pair of entangled systems with effective Hilbert spaces of arbitrary dimensions — see also [POPESCU/D. ROHRLICH 1992a, 294–95] [*PERES* 1993, 175].

Theorem 38.2 (Gisin/Peres) *For any entangled pure state of two systems, a set of measurements may be prescribed for which the predictions of QM are inconsistent with the separability.*

Since this and the following theorem are only extensions of Gisin theorem, we do not need to integrate them in the theory.

In [POPESCU/D. ROHRLICH 1992a, 296–97] Gisin Theorem has been extended further to any number of entangled systems (pure states) of any kind.

Theorem 38.3 (Popescu/Rohrlich) *Any n–system entangled pure state violates the separability principle.*

To prove the theorem we state the following Lemma:

Lemma 38.1 (Popescu/Rohrlich) *If Ψ is a n–system entangled state, for any two of the n–systems there exists a projection onto a direct product of states of the other $n-2$ systems that leaves the two systems in an entangled state.*

Proof

We assume that it is not so, i.e. that a projection of this kind leaves the remaining two systems in a product state. Let now $|b^i\rangle_j$ be a given basis in the subspace of the j–th system and consider projections onto the direct products of these basis vectors:

$$\langle b^{i_3}|_3\langle b^{i_4}|_4 \ldots \langle b^{i_n}|_n|\Psi\rangle = |\varsigma\rangle_1|\varsigma'\rangle_2, \tag{38.8}$$

where $|\varsigma\rangle_1$ and $|\varsigma'\rangle_2$ are the states of the two remaining subsystems. By assumption they are in a product state for any choice of indices $i_3, i_4, \ldots i_n$, although the product state will in general depend on the indices. Hence we show this dependence explicitly:

$$|\varsigma\rangle_1 = |\varsigma(i_3, i_4, \ldots i_n)\rangle_1, \quad |\varsigma'\rangle_2 = |\varsigma'(i_3, i_4, \ldots i_n)\rangle_2. \tag{38.9}$$

Now suppose that in the direct product of basis vectors we choose a different basis vector $|b^{i'_3}\rangle$ from the subspace 3. It is clear that either $|\varsigma\rangle_1$ or $|\varsigma'\rangle_2$ must remain unchanged (up to a phase factor). In fact, suppose that both change. Then, by substituting a linear combination of $\langle b^{i_3}|_3$ and $\langle b^{i'_3}|_3$ for $\langle b^{i_3}|_3$ in the direct product (38.8), we produce an entangled state of the two remaining systems, contrarily to the assumption. Thus either $|\varsigma\rangle_1$ or $|\varsigma'\rangle_2$ are independent of i_3. Repeating the argument for the other subspaces we conclude that each index actually appears either in $|\varsigma\rangle_1$ or $|\varsigma'\rangle_2$, but not in both. So that we can write:

$$|\varsigma\rangle_1 = |\varsigma(i_3, i_4, \ldots i_k)\rangle_1, \quad |\varsigma'\rangle_2 = |\varsigma'(i_{k+1}, i_{k+2}, \ldots i_n)\rangle_2. \tag{38.10}$$

Now we have [see Eq. (38.8)]

$$\begin{aligned} |\Psi\rangle = \hat{I} \cdot |\Psi\rangle &= \sum_{i_3, i_4, \ldots i_n} |b^{i_n}\rangle_n \ldots |b^{i_4}\rangle_4 |b^{i_3}\rangle_3 \langle b^{i_3}|_3 \langle b^{i_4}|_4 \ldots \langle b^{i_n}|_n |\Psi\rangle \\ &= \sum_{i_3, i_4, \ldots i_k} |\varsigma(i_3, i_4, \ldots i_k)\rangle_1 |b^{i_3}\rangle_3 \ldots |b^{i_k}\rangle_k \\ &\times \sum_{i_{k+1}, \ldots i_k} |\varsigma'(i_{k+1}, i_{k+2}, \ldots i_n)\rangle_2 |b^{i_{k+1}}\rangle_{k+1} \ldots |b^{i_n}\rangle_n; \end{aligned} \tag{38.11}$$

and we see that state $|\Psi\rangle$ factorizes into a product contradicting the assumption that it was an entangled state. Q.E.D.

Once proved the lemma, also Popescu/Rohrlich Theorem is easily proved by using Gisin/Peres Theorem.

38.2 Mixed States

38.2.1 Werner States

But what about mixed states? At the beginning the problem of which states would not violate any Bell inequality was considered. Mixtures (convex combinations) of direct products are classically correlated states, because they can always be reproduced by a classical random generator. It was also thought that the following proposition was sound [R. WERNER 1989]:[1]

Proposition 38.1 (Werner) *There are mixtures of entangled states (and not only of direct products) — known as 'Werner states' — which do not violate any of the family of Bell inequalities.*

An example of Werner state is the following [see Eq. (43.23)]:

$$\hat{\rho}_{\mathrm{W}} = w\hat{P}_{\Psi_0} + (1-w)\hat{I}, \tag{38.12}$$

where $|\Psi_0\rangle$ is a singlet state.

But, as we shall see in the following, there also exist Werner states which are not product states and which can show non-local features.

38.2.2 Andås Theorem

Another important step is represented by the following proposition [ANDÅS 1992]:[2]

[1] See also [POPESCU 1994, 798].

[2] Partial antecedents can be found in [GARUCCIO *et al.* 1990] [SUPPES/ZANOTTI 1991].

Proposition 38.2 (Andås) • *If the state $\hat{\rho}$ allows a description of its properties by classical probability concepts, then*

$$\left|\left\langle \hat{\mathrm{C}} \right\rangle_{\mathrm{CL}}^{\hat{\rho}}\right| \leq M_0, \tag{38.13}$$

where $M_0 = \mathrm{Max}(\sum_{\mu\nu} c_{\mu\nu}\xi_\mu\eta_\nu)$, with $\xi_\mu, \eta_\nu = \pm 1, \forall \mu, \nu$.

- *If $\hat{\rho}$ is a state described by QM, then*

$$\left|\left\langle \hat{\mathrm{C}} \right\rangle_{\mathrm{QM}}^{\hat{\rho}}\right| \leq \left(\sum_{\mu\nu} c_{\mu\nu}^2 + \sum_{\nu\nu'}\left|\sum_{\mu} c_{\mu\nu}c_{\mu\nu'}\right| + \sum_{\mu\mu'}\left|\sum_{\nu} c_{\mu\nu}c_{\mu'\nu}\right| + \sum_{\mu\nu}|c_{\mu\nu}|\sum_{\mu'\nu'}|c_{\mu'\nu'}|\right)^{\frac{1}{2}}, \tag{38.14}$$

where $\nu' \neq \nu, \mu' \neq \mu$ and where the upper limit can be reached by pure states only.

We do not prove the proposition[3]. Though sound, we do not integrate it in the theory because some more general theorems will be proven in the following. However, there are important applications of inequality (38.14) which we demonstrate now. In fact it is possible to derive different inequalities from inequality (38.14). Posing

$$\hat{\mathrm{C}}_{\mathrm{B}} = \hat{O}_1^a\hat{O}_1^b + \hat{O}_1^a\hat{O}_2^b + \hat{O}_2^a\hat{O}_2^b - \hat{O}_2^a\hat{O}_1^b, \tag{38.15}$$

we obtain for $\left|\left\langle \hat{\mathrm{C}} \right\rangle_{\mathrm{B}}^{\mathrm{CL}}\right| \leq 2$ the first Bell inequality (35.11), and from $\left|\left\langle \hat{\mathrm{C}} \right\rangle_{\mathrm{B}}^{\mathrm{QM}}\right| \leq 2\sqrt{2}$ we obtain Tsirelson inequality (37.10). On the other hand, if we consider

$$\begin{aligned} \hat{\mathrm{C}}_{\mathrm{K}} &= \hat{O}_1^a\hat{O}_1^b + \hat{O}_1^a\hat{O}_2^b + \hat{O}_2^a\hat{O}_2^b - \hat{O}_2^a\hat{O}_1^b + \hat{O}_1^a\hat{O}_3^b - \hat{O}_1^a\hat{O}_4^b + \hat{O}_2^a\hat{O}_3^b - \hat{O}_2^a\hat{O}_4^b \\ &\quad + 2\hat{O}_3^a\hat{O}_2^b - \hat{O}_3^a\hat{O}_3^b + \hat{O}_3^a\hat{O}_4^b + \hat{O}_4^a\hat{O}_3^b + \hat{O}_4^a\hat{O}_4^b, \end{aligned} \tag{38.16}$$

we obtain, for $\left|\left\langle \hat{\mathrm{C}} \right\rangle_{\mathrm{K}}^{\mathrm{CL}}\right| \leq 6$, Kempermann inequality [SUPPES/ZANOTTI 1991] and another QM inequality $\left|\left\langle \hat{\mathrm{C}} \right\rangle_{\mathrm{K}}^{\mathrm{QM}}\right| \leq 2\sqrt{33}$ [see also section 37.2].

38.2.3 Succession of Measurements of Werner States

We have seen that Werner states do not seem to violate Bell inequalities. But if we study a succession of measurements of Werner states — which can be taken as a single POVM —, the matter is quite different. In fact Werner states have also 'hidden' non-classical features because they can be used for teleportation [see section 43.2] [POPESCU 1995]. It can be shown that consecutive measurements on two Werner subsystems produce a violation of CHSH inequality.

Suppose a Werner state of the following type:

$$\hat{\rho}_{\mathrm{W}} = \frac{1}{q^2}\left(\frac{1}{d}\hat{I}^{d\times d} + 2\sum_{j<k;j,k=1}^{d} |\Psi_0^{jk}\rangle\langle\Psi_0^{jk}|\right), \tag{38.17}$$

where $\hat{I}^{d\times d}$ is the identity matrix in the $(d \times d)$ Hilbert space of the two subsystems and $|\Psi_0^{jk}\rangle$ is the spin-$\frac{1}{2}$ singlet state

$$|\Psi_0^{jk}\rangle = \frac{1}{\sqrt{2}}(|j\rangle_1|k\rangle_2 - |k\rangle_1|j\rangle_2). \tag{38.18}$$

[3] See the original article.

We consider now Werner matrices with dimension $\geq (5 \times 5)$. We first subject each particle to a measurement of a twodimensional projector, $\hat{P}_1$ for the first particle:

$$\hat{P}_1 = |1\rangle_1\langle 1|_1 + |2\rangle_1\langle 2|_1; \tag{38.19a}$$

and $\hat{P}_2$ for the second particle:

$$\hat{P}_2 = |1\rangle_2\langle 1|_2 + |2\rangle_2\langle 2|_2. \tag{38.19b}$$

Now, only after measurement (38.19a) has been performed, an observer near to the first particle decides at random whether to measure an observable $\hat{O}_1$ or another one $\hat{O}'_1$, and an observer near to particle two only after measurement (38.19b) has been performed, decides at random to measure either an observable $\hat{O}_2$ or another one $\hat{O}'_2$. The four operators have eigenvalues $1, -1, 0$. $1, -1$ are non degenerate and the corresponding eigenstates belong to subspaces $\{|1\rangle_1, |2\rangle_1\}, \{|1\rangle_2, |2\rangle_2\}$, respectively. Eigenvalue 0 is degenerate and corresponds to the rest $\{|3\rangle_1, \ldots |d\rangle_1\}, \{|3\rangle_2, \ldots |d\rangle_2\}$ of the two bases, respectively. The non-degenerate part of these operators is chosen so that there is maximal violation of the III Bell inequality for the singlet state $|\Psi_0^{12}\rangle$ [see Eq. (38.24a)]:

$$\langle\Psi_0^{12}|\hat{O}_1\hat{O}_2 + \hat{O}_1\hat{O}'_2 + \hat{O}'_1\hat{O}_2 - \hat{O}'_1\hat{O}'_2|\Psi_0^{12}\rangle = 2\sqrt{2}. \tag{38.20}$$

Then the ensemble corresponding to $\hat{P}_1 = 1, \hat{P}_2 = 1$ is:

$$\hat{\rho}'_{\mathrm{W}} = \frac{1}{N}\hat{P}_1\hat{P}_2\hat{\rho}_{\mathrm{W}}\hat{P}_2\hat{P}_1 = \frac{2d}{2d+4}\left(\frac{1}{2d}\hat{I}^{2\times 2} + 2|\Psi_0^{12}\rangle\langle\Psi_0^{12}|\right); \tag{38.21}$$

from which we have the following violation of the III Bell inequality:

$$\mathrm{Tr}\left[\hat{\rho}'_{\mathrm{W}}(\hat{O}_1\hat{O}_2 + \hat{O}_1\hat{O}'_2 + \hat{O}'_1\hat{O}_2 - \hat{O}'_1\hat{O}'_2)\right] = \frac{2d}{2d+4}2\sqrt{2} \geq 2, \tag{38.22}$$

for $d \geq 5$. The reason for this situation is the following. In general, in an HV model we can have:

$$\int d\lambda_{HV}\hat{\rho}(\lambda_{HV})\wp(\hat{O}_1 = 0, \lambda_{HV}) = \int d\lambda_{HV}\hat{\rho}(\lambda_{HV})\wp(\hat{O}'_1 = 0, \lambda_{HV}); \tag{38.23a}$$

but it does not follow that

$$\wp(\hat{O}_1 = 0, \lambda_{HV}) = \wp(\hat{O}'_1 = 0, \lambda_{HV}). \tag{38.23b}$$

Since the case in question consists of a succession of measurements, the probabilities (38.23b) have to be generally understood as conditional probabilities.

38.2.4 The Bell Operator

The operator itself In [BRAUNSTEIN *et al.* 1992a] it was shown that there can also be a maximal violation of Bell inequality for mixed states[4]. Braunstein and co-workers first defined the Bell operator for the third Bell inequality [the (35.27)] as follows:

$$\hat{\mathcal{B}}_{\mathrm{III\,Bell}} := \mathbf{a}(\mathbf{b} + \mathbf{b}') + \mathbf{a}'(\mathbf{b} - \mathbf{b}'), \tag{38.24a}$$

$$\hat{\mathcal{B}}^2_{\mathrm{III\,Bell}} = 4\hat{I} - [\mathbf{a}, \mathbf{a}'][\mathbf{b}, \mathbf{b}']. \tag{38.24b}$$

[4]See also [MANN *et al.* 1992] [*PERES* 1993, 174].

It can have four vectors as basis with the form:

$$|\Psi^{\pm}\rangle = \frac{1}{\sqrt{2}}(|\uparrow\rangle \otimes |\downarrow\rangle \pm |\downarrow\rangle \otimes |\uparrow\rangle), \quad (38.25a)$$

$$|\Phi^{\pm}\rangle = \frac{1}{\sqrt{2}}(|\uparrow\rangle \otimes |\uparrow\rangle \pm |\downarrow\rangle \otimes |\downarrow\rangle). \quad (38.25b)$$

Now it is clear that any mixture of the $|\Psi^{+}\rangle$ states will yield a maximal violation of $+2\sqrt{2}$, and any mixture of the $|\Phi^{-}\rangle$ yields a violation of $-2\sqrt{2}$. The importance of the above operator will be evident in a successive analysis [see section 43.2].

Some applications In [R. HORODECKI *et al.* 1995] the necessary and sufficient condition for violating the III Bell inequality by an arbitrary mixed spin–1/2 state is shown. In the Hilbert space $\mathcal{H} = C^2 \otimes C^2$ we choose the following state:

$$\hat{\rho} = \frac{1}{4}\left(\hat{I} \otimes \hat{I} + \mathbf{r} \cdot \hat{\sigma} \otimes \hat{I} + \hat{I} \otimes \mathbf{n} \cdot \hat{\sigma} + \sum_{j,k=1}^{3} c_{jk}\hat{\sigma}_j \otimes \hat{\sigma}_k\right), \quad (38.26)$$

where $\mathbf{r}, \mathbf{n}$ are some vectors and $\mathbf{r} \cdot \hat{\sigma} = \sum_{j=1}^{3} r_j \hat{\sigma}_j$. The coefficients $c_{jk} = \mathrm{Tr}(\hat{\rho}\hat{\sigma}_j \otimes \hat{\sigma}_k)$ form a real matrix T_ρ. The condition $\mathrm{Tr}(\hat{\rho}^2) \leq 1$ (not considering the non-negativity condition) implies:

$$\sum_{m=1}^{3}(r_m^2 + n_m^2) + \sum_{j,k=1}^{3} c_{jk}^2 \leq 3. \quad (38.27)$$

We now take the Bell operator [Eq. (38.24a)] in the following form:

$$\hat{\mathcal{B}}_{\mathrm{III\,Bell}} = \mathbf{a} \cdot \sigma \otimes (\mathbf{b} + \mathbf{b}') \cdot \sigma + \mathbf{a}' \cdot \sigma \otimes (\mathbf{b} - \mathbf{b}') \cdot \sigma, \quad (38.28)$$

subjected to condition

$$\left|\left\langle \hat{\mathcal{B}}_{\mathrm{III\,Bell}} \right\rangle_{\hat{\rho}}\right| \leq 2, \quad (38.29)$$

where $\left\langle \hat{\mathcal{B}}_{\mathrm{III\,Bell}} \right\rangle_{\hat{\rho}} = \mathrm{Tr}(\hat{\mathcal{B}}_{\mathrm{III\,Bell}}\hat{\rho})$. Now it is possible to prove the following theorem:

Theorem 38.4 (Horodecki I) *The density matrix* (38.26) *violates inequality* (38.29) *iff* $\eta(\hat{\rho}) > 1$, *where*

$$\eta(\hat{\rho}) := \mathrm{Max}_{\mathbf{c},\mathbf{c}'}(\| T_\rho \mathbf{c} \|^2 + \| T_\rho \mathbf{c}' \|^2) = u + u', \quad (38.30)$$

where u, u' *are eigenvalues of the matrix* $T_\rho^{\mathrm{T}} T_\rho$ *(*T^{T} *is the transposed of* T*) and* $\mathbf{c}, \mathbf{c}'$ *are the following orthogonal vectors:*

$$\mathbf{c} = \frac{\mathbf{b} + \mathbf{b}'}{2\cos\theta}, \quad \mathbf{c}' = \frac{\mathbf{b} - \mathbf{b}'}{2\sin\theta}, \quad (38.31)$$

where $\theta \in [0, 1/2\pi]$.

We do not integrate in the theory this useful theorem because of other results of greater generality [see next section].

38.3 Purification Procedures

What the preceding subsection has shown, is that there are states which apparently seem separable, and which, by some procedure, show quantum non-local features. In the following we try to generalize such a method by finding some purification procedure by means of which the correlations become more evident and we can finally define what states are not quantum-mechanically entangled.

Gisin [GISIN 1996a] showed that there can be generalized measurements which can violate Bell inequalities even for mixed states which are local. This is a *purification* procedure: one extracts pure quantum entanglement from a partially entangled state — it is an information *compression* [see subsection 44.3.1].

A maximally entangled state is for example a singlet state of the form (31.17). A partially entangled state can be written as follows [BENNETT *et al.* 1996b] :

$$\cos\theta | \uparrow\rangle_1 \otimes | \downarrow\rangle_2 - \sin\theta | \downarrow\rangle_1 \otimes | \uparrow\rangle_2, \tag{38.32}$$

where we have a singlet if $\theta = \pi/4$ and a product state (zero entanglement) if $\theta = 0$.

Interesting procedures of purification have been developed in [BENNETT *et al.* 1996b] and [DEUTSCH *et al.* 1996]. Here we wish to stress only some general features of these procedures. Any purification has three ingredients [VEDRAL *et al.* 1997a, 2275–76]:

- *local measurement* (= LM) such that they are performed on the two subsystems $\mathcal{S}_1, \mathcal{S}_2$ (1, 2 shortly) separately, which are described by operators satisfying the conditions of completeness: $\sum_j (\hat{O}^1_j)^\dagger \hat{O}^2_j = \hat{I}$, $\sum_k (\hat{O}^2_k)^\dagger \hat{O}^2_k = \hat{I}$;
- *classical communication* (= CC) such that the actions on I and II are classically correlated so that the transformation involving LM + CC would look like:

$$\hat{\rho}_{1+2} \mapsto \sum_j \left[\hat{O}^1_j \otimes \hat{O}^2_j\right] \hat{\rho}_{1+2} \left[(\hat{O}^1_j)^\dagger \otimes (\hat{O}^2_j)^\dagger\right]. \tag{38.33}$$

  This map is completely positive; to assure that it is also trace preserving it suffices to require: $\sum_j (\hat{O}^1_j)^\dagger \hat{O}^1_j \otimes (\hat{O}^2_j)^\dagger \hat{O}^2_j = \hat{I}$;
- *postselection* such that one rejects part of the original ensemble, making the whole transformation nonlinear:

$$\hat{\rho}_{1+2} \rightsquigarrow \frac{\left[\hat{O}^1_j \otimes \hat{O}^2_j\right] \hat{\rho}_{1+2} \left[(\hat{O}^1_j)^\dagger \otimes (\hat{O}^2_j)^\dagger\right]}{\mathrm{Tr}\left\{\left[\hat{O}^1_j \otimes \hat{O}^2_j\right] \hat{\rho}_{1+2} \left[(\hat{O}^1_j)^\dagger \otimes (\hat{O}^2_j)^\dagger\right]\right\}}. \tag{38.34}$$

For example [VEDRAL/PLENIO 1998, 1620]: Let the initial ensemble contain states of the form $| \downarrow_1\rangle \otimes (| \downarrow_2\rangle + | \uparrow_2\rangle)/\sqrt{2}$: the correlations (measured by the mutual information) between 1 and 2 are zero. Now let 1 perform a measurement on his particles in the $\downarrow, \uparrow$ basis. If $\uparrow$ is obtained, 1 communicates to 2 the result and 2 'rotates' the state in order to obtain $| \uparrow_1\rangle$; otherwise they do nothing. The final state will be $\hat{\rho} = \frac{1}{2}(| \downarrow_1\rangle\langle \downarrow_1 | \otimes | \downarrow_2\rangle\langle \downarrow_2 | + | \uparrow_1\rangle\langle \uparrow_1 | \otimes | \uparrow_2\rangle\langle \uparrow_2 |$, whose correlations are now $\ln 2$.[5]

In conclusion we state the following theorem [R. HORODECKI *et al.* 1998] (proof in the original article):[6]

[5] On the problem of 'hidden' entanglement see also [O. COHEN 1998].

[6] See also [R. HORODECKI *et al.* 1996b] and [BARNETT/PHOENIX 1992], where use of the Schmidt decomposition is made in order to prove following result.

Theorem 38.5 (Horodecki II) *Only states of two two-level systems that can be written as a sum over density operators which are direct product states of the two subsystems (i.e.* $\hat{\rho}_{1+2} = \sum_j \hat{\rho}_1^j \otimes \hat{\rho}_2^j$*) cannot be purified, and hence only states with this feature cannot show quantum non-separability (correlations).*

Chapter 39

GENERALIZED BELL INEQUALITIES

Introduction and contents All of the EPR problematic can be synthesized by saying that the Hilbert space formalism of QM relaxes considerably the constraints on correlations which exist in classical mechanics [*PITOWSKY* 1989b, 52]. In fact we remember the KS theorem where it is evident that there are truth values which are admissible in QM and are not so in classical mechanics. We shall now analyse the relationship between classical distributions which satisfy Bell-type inequalities and quantum probability distributions in a more systematic way.

In section 39.1 we discuss Pykacz/Santos theorems — a further development of the problem of the QM lattice [see section 9.2]. A probabilistic analysis at the level of polyhedron and relative surfaces [see figure 11.2: p. 194] is performed in section 39.2. A different approach to the same problem has been developed by Beltrametti and Maczyński [section 39.3]. A stochastic approach to Bell inequalities (on the Poincaré sphere) is the object of section 39.4. Then, again on a probabilistic level, an extension to arbitrary numbers of spin is developed [section 39.5]. Finally [section 39.6], a brief concluding remark follows.

39.1 Pykacz/Santos Theorems

Until now we have spoken of separability and non-separability as if they were absolute concepts. However, the purification procedures [see section 38.3] have already shown different degrees of entanglement, and hence of non-separability. Therefore it is very useful to introduce a measure of the 'separation'.

We can define the concept of separation between two propositions a, b for a probability (state) $\wp$ [SANTOS 1986] [PYKACZ/SANTOS 1991, 1289–91]:

Definition 39.1 (Separation) *The separation between a and b in state $\wp$ is given by the real number:*

$$\Xi(a,b) := \wp(a) + \wp(b) - 2\wp(a \wedge b). \tag{39.1}$$

Now it is possible to prove (we omit the proof) the following theorem:

Theorem 39.1 (I Pykacz/Santos) *For a Boolean algebra* [see definition 9.7] $\mathcal{L}$, *and for any $a, b, c \in \mathcal{L}$ and any $\wp$ on $\mathcal{L}$ the separations of the form* (39.1) *fulfill the triangle inequality:*

$$\Xi(a,b) + \Xi(b,c) \geq \Xi(a,c). \tag{39.2}$$

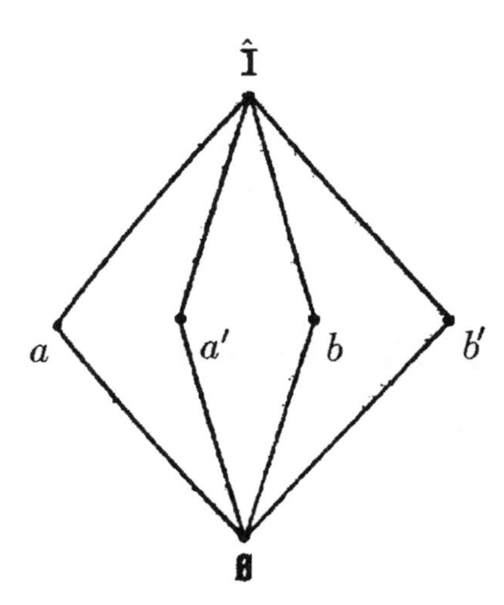

Figure 39.1: Representation of the Hasse diagram of the $\mathcal{L}_6$ logic. Clearly $\mathcal{L}_6$ is not Boolean: see for example figures 9.3 [p. 159] and 9.5 [p. 161] — from [PYKACZ/SANTOS 1991, 1291].

Though sound, such a theorem does not concern QM. Therefore, it cannot be integrated directly into the theory. Since Eq. (39.2) only holds if some additional assumptions are made (for example, in the original model of Bell, that the spin projections along the same axis of two spin-$\frac{1}{2}$ particles, which are in a singlet state, are opposite), we can express the same issue without any additional assumption by the following quadrilateral inequality — which is equivalent to the triangular one if all mentioned conditions of the latter are satisfied —: which happens for the GHSZ model [see subsection 36.4.2 and especially inequality (36.45)]:

$$\Xi(a_1, a_2) + \Xi(b_1, a_2) + \Xi(b_1, b_2) \geq \Xi(a_1, b_2). \tag{39.3}$$

If the same inequality is written for coincidence probabilities, i.e. if we make use of $\wp(a_1 \wedge a_2)$, instead of $\Xi(a_1, a_2)$, we obtain the IV Bell inequality [see Eq. (35.37)].

But it is possible to give a general condition under which Eq. (39.2) holds. Let us define a Jauch/Piron state by the following condition:

$$(\forall a, b \in \mathcal{L})[\wp(a) = \wp(b) = 1] \rightarrow [\wp(a \wedge b) = 1], \tag{39.4}$$

by means of which we formulate the following theorem:

Theorem 39.2 (II Pykacz/Santos) *If $\mathcal{L}$ is an orthocomplemented orthomodular lattice and $\wp$ is a dispersion-free state on $\mathcal{L}$ which fulfills condition* (39.4), *then, $\forall a, b, c \in \mathcal{L}$, inequality* (39.2) *holds.*

We can now generalize the Bell inequalities. First we define a Logic [see definition 9.9] by the following condition: $\mathcal{L}_{\succeq,\perp,a_n}$ is a logic iff, whenever $a, b \in \mathcal{L}_{\succeq,\perp,a_n}$ and a is compatible with b $(a \odot b)$, we have:

$$\Xi(a, b) = \wp(a \vee b) - \wp(a \wedge b). \tag{39.5}$$

Then let us state the following definition of circular compatibility:

Definition 39.2 (Circular Compatibility) *A sequence* $a_1, a_2, \ldots, a_n$ *of elements of the logic* $\mathcal{L}_{\succeq,\perp,a_n}$ *is circular compatible iff* $a_1 \odot a_2 \odot a_3 \ldots \odot a_n \odot a_1$.

Now we state the following theorem[1]:

Theorem 39.3 (III Pykacz/Santos) *If* $\mathcal{L}_{\succeq,\perp,a_n}$ *is a logic, the sequence* $a_1, a_2, \ldots, a_n$ *is circular compatible, and* $\wp$ *is dispersion-free on a pair* (a_j, a_{j+1}) *then the following generalized Bell inequality holds:*

$$\left[\sum_{k=1;k\neq i}^{n} \Xi(a_k, a_{k+1})\right] \geq \Xi(a_i, a_{i+1}), \tag{39.6}$$

where $a_{n+1} = a_1$.

In every HV-Theory inequality (39.6) holds but never in QM. It is possible to study a logic [see definition 9.9] $\mathcal{L}_6 = \{\emptyset, a, a' = 1 - a, b, b' = 1 - g, \hat{I}\}$, consisting of affine functionals that map a square $pqrt$ into the interval $[0, 1]$ and which take on the vertices p, q, r, t of the square only values 0 or 1 [see figure 39.1 and table 39.1]. The square represents a set of states; pure states are vertices of the square and all other points are convex combinations of the vertices, i.e. they are mixed states; a, a', b, b' can be roughly understood as alternative settings for polarisers, as usually.

As we shall see the fundamental results of this section can be reproduced in the context of information theory [see subsection 42.7.2].

	$\emptyset$	a	a'	b	b'	$\hat{I}$
p	0	1	0	1	0	1
q	0	1	0	0	1	1
r	0	0	1	1	0	1
t	0	0	1	0	1	1

Table 39.1: Values taken by elements of $\mathcal{L}_6$ on vertices of the square $pqrt$ (pure states) — from [PYKACZ/SANTOS 1991, 1290].

39.2 A Probabilistic Treatment of GHSZ States

The factorised probabilities in the Bell model, with two polarimeters and two different settings [see section 35.1], can be expressed mathematically by 10 different 2×2 matrices [FROISSART 1981], given by all permutations of rows and columns of the following extremal matrices:

$$\begin{bmatrix} 0 & 0 \\ 0 & 0 \end{bmatrix}; \begin{bmatrix} 1 & 0 \\ 0 & 0 \end{bmatrix}; \begin{bmatrix} 1 & 0 \\ 1 & 0 \end{bmatrix}; \begin{bmatrix} 1 & 1 \\ 1 & 1 \end{bmatrix}. \tag{39.7}$$

These represent 10 points in the 4–dimensional space of 2 × 2 matrices and define a polyhedron with a convex hull [see also figure 11.2: p. 194 and relative discussion]. As we know, we can look

[1] See also [BRAUNSTEIN/CAVES 1988].

at the polyhedron as the intersection between surfaces: each surface is defined by an inequality. The inequalities are as follows:

$$0 \leq \wp_{ab} \leq 1; \tag{39.8a}$$

$$\wp_{ab} - \wp_{ab'} - \wp_{a'b} + 1 \geq 0; \tag{39.8b}$$

and all those deduced from them by interchanging a and a' and b and b'. We can write inequalities (39.8) in matricial form:

$$\mathrm{Tr}\wp \begin{bmatrix} 1 & 0 \\ 0 & 0 \end{bmatrix} \geq 0 \quad \text{for } f = 1, \tag{39.9a}$$

$$\mathrm{Tr}\wp \begin{bmatrix} -1 & 0 \\ 0 & 0 \end{bmatrix} + 1 \geq 0 \quad \text{for } f = 0.5, \tag{39.9b}$$

$$\mathrm{Tr}\wp \begin{bmatrix} 1 & -1 \\ -1 & 0 \end{bmatrix} + 1 \geq 0 \quad \text{for } f = 1, \tag{39.9c}$$

for a quality factor f which we define below. Here inequalities (39.9a) and (39.9b) correspond to inequality (39.8a), and inequality (39.9c) to inequality (39.8b), where the matrices are:

$$\begin{bmatrix} \wp_{ab} & \wp_{a'b} \\ \wp_{ab'} & \wp_{a'b'} \end{bmatrix} \tag{39.10}$$

For transmission coincidence experiments we have correlations of the form [see also Eq. (3.130)] $\frac{1}{2}\cos^2\frac{(\theta-\phi)}{2}$ for azimuthal angles $\frac{\theta}{2}, \frac{\phi}{2}$ of the analyzers — the factor $\frac{1}{2}$ comes from the normalization condition for the two polarisation states. Then Eq. (39.8b) becomes:

$$\frac{1}{2}\cos^2\frac{\theta-\phi}{2} - \frac{1}{2}\cos^2\frac{\theta'-\phi}{2} - \frac{1}{2}\cos^2\frac{\theta-\phi'}{2} + 1 \geq 0; \tag{39.11a}$$

or

$$\cos(\theta-\phi) - \cos(\theta'-\phi) - \cos(\theta-\phi') + 3 \geq 0, \tag{39.11b}$$

which is satisfied for any settings $\theta, \phi, \theta', \phi'$. Now we define f as the ratio of the maximum value of the angle-dependent part of the inequalities (39.11) to the angle-averaged LHS. Wherever $f > 1$, then the LHS may have the wrong sign. Then the excess of f over 1 is a measure of the violation of inequalities by QM. Note that, on this abstract level, inequalities (39.9) — for two subsystems — are not violated in QM.

If we have three settings for three polarimeters (for three subsystems), we obtain a polyhedron with 204 faces, of which 9 are of the type (39.9a), 9 of the type (39.9b), 36 of the type (39.9c), and the remaining are the following:

$$\mathrm{Tr}\wp \begin{bmatrix} 1 & 1 & -1 \\ 1 & -1 & 0 \\ -1 & 0 & 0 \end{bmatrix} + 1 \geq 0 \,(\times 36) \quad \text{for } f = 1.21; \tag{39.12a}$$

$$\mathrm{Tr}\wp \begin{bmatrix} -2 & 0 & 1 \\ 0 & 1 & -1 \\ 1 & -1 & -1 \end{bmatrix} + 2 \geq 0 \,(\times 36) \quad \text{for } f = 1.12; \tag{39.12b}$$

$$\mathrm{Tr}\wp \begin{bmatrix} 1 & 0 & -1 \\ 0 & -1 & 1 \\ -1 & 1 & 0 \end{bmatrix} + 1 \geq 0 \,(\times 12) \quad \text{for } f = 1.30; \tag{39.12c}$$

$$\mathrm{Tr}\wp\begin{bmatrix} 2 & -1 & -1 \\ -1 & 1 & -1 \\ -1 & -1 & 1 \end{bmatrix} + 2 \geq 0\ (\times 18) \qquad \text{for } f = 1.17; \tag{39.12d}$$

$$\mathrm{Tr}\wp\begin{bmatrix} 1 & -1 & 2 \\ -1 & -2 & 2 \\ -2 & 2 & 0 \end{bmatrix} + 3 \geq 0\ (\times 36) \qquad \text{for } f = 1.22; \tag{39.12e}$$

$$\mathrm{Tr}\wp\begin{bmatrix} -2 & -1 & 2 \\ -1 & 2 & -2 \\ 2 & -2 & -1 \end{bmatrix} + 3 \geq 0\ (\times 12) \qquad \text{for } f = 1.04; \tag{39.12f}$$

where $\times 36$ signifies that we can obtain 36 inequalities from the preceding one by permutation of all rows and columns. Through inequalities (39.12) we have evident violations in the frame of QM. This result corresponds to that of GHSZ [see subsection 36.4.2], that it is possible to obtain a conflict between the separability and QM by three particles with a direct formal proof.

39.3 Beltrametti/Maczyński Theorems

Now we try to use the instruments exposed in section 11.4 in order to generalize the analysis of Bell inequalities by stating some important theorems [BELTRAMETTI/MACZYŃSKI 1991, 1282–84]. First we formulate the following theorem, which is a generalization of that proved for the first time by Pitowsky [*PITOWSKY* 1989b, 22]:

Theorem 39.4 (I Pitowsky/Beltrametti/Maczyński) *Let $\mathcal{X} = (\wp_1, \wp_2, \wp_{12})$ be a correlation sequence of S–probabilities. The following conditions are equivalent:*
i) K is consistently representable [see definition 11.12: p. 198];
ii) the following inequalities hold:

$$0 \leq \wp_{12} \leq \wp_1 \leq 1, \quad 0 \leq \wp_{12} \leq \wp_2 \leq 1, \quad \wp_1 + \wp_2 - \wp_{12} \leq 1; \tag{39.13}$$

iii) there exist numbers $q_1, q_2, q_3 \in P(S)$ with $q_1 + q_2 + q_3 \leq 1$ such that[2]*:*

$$\wp_1 = q_1 + q_3, \quad \wp_2 = q_2 + q_3, \quad \wp_{12} = q_3; \tag{39.14}$$

iv) $\mathcal{X}$ is consistently representable in a classical event space.

Though sound, it is evident that the theorem does not concern QM. A further extension of the preceding theorem to the case of three events is the following:

Theorem 39.5 (II Pitowsky/Beltrametti/Maczyński) *Let $\mathcal{X} = (\wp_1, \wp_2, \wp_3, \wp_{12}, \wp_{13}, \wp_{23})$ be a correlation sequence of S–probabilities. The following conditions are equivalent:*
i) $\mathcal{X}$ is classically representable;
ii) the following inequalities hold for $1 \leq i \leq j \leq 3$:[3]

$$0 \leq \wp_{ij} \leq \wp_i \leq 1, \qquad 0 \leq \wp_{ij} \leq \wp_j \leq 1 \tag{39.15}$$

[2]See also [*PITOWSKY* 1989b, 15–16].
[3]See also [*PITOWSKY* 1989b, 25–26].

$$\wp_i + \wp_j - \wp_{ij} \leq 1, \tag{39.16a}$$
$$\wp_1 - \wp_{12} - \wp_{13} + \wp_{23} \geq 0, \tag{39.16b}$$
$$\wp_2 - \wp_{12} - \wp_{23} + \wp_{13} \geq 0, \tag{39.16c}$$
$$\wp_3 - \wp_{13} - \wp_{23} + \wp_{12} \geq 0, \tag{39.16d}$$
$$\wp_1 + \wp_2 + \wp_3 - \wp_{12} - \wp_{13} - \wp_{23} \leq 1; \tag{39.16e}$$

iii) there exist numbers $q_1, q_2, q_3, q_4, q_5, q_6, q_7 \in P(S)$, *with property* $\sum_{i=1}^{7} q_i \leq 1$, *such that:*

$$\wp_1 = q_1 + q_2 + q_4 + q_7, \tag{39.17a}$$
$$\wp_2 = q_2 + q_5 + q_6 + q_7, \tag{39.17b}$$
$$\wp_3 = q_3 + q_4 + q_6 + q_7, \tag{39.17c}$$
$$\wp_{12} = q_2 + q_7, \tag{39.17d}$$
$$\wp_{13} = q_4 + q_7, \tag{39.17e}$$
$$\wp_{23} = q_6 + q_7. \tag{39.17f}$$

Inequalities (39.16a)–(39.16e) are the Bell inequalities: For example the I Bell inequality can be derived by taking inequality (39.16a) and rewriting it as follows [using the properties (39.15)]: $|\wp_{12} - \wp_{13}| - \wp_{23} \leq 1$ — which is formally equivalent to inequality (35.11). We see that when these inequalities do not hold, the system is not classically representable, and hence not Boolean, as we have already seen. Bell inequalities are represented by Bell/Wigner polytope [see figure 11.2: p. 194], by means of which we can demonstrate the equivalence between conditions i) and ii) [*PITOWSKY* 1989b, 25].[4]

We have also CH inequalities which are the following [*PITOWSKY* 1989b, 27–30]:

$$0 \leq \wp_{ij} \leq \wp_i \leq 1, \tag{39.18a}$$
$$0 \leq \wp_{ij} \leq \wp_j \leq 1, \tag{39.18b}$$
$$\wp_i + \wp_j - \wp_{ij} \leq 1, \tag{39.18c}$$

for $i = 1, 2; j = 3, 4$, and[5]:

$$-1 \leq \wp_{13} + \wp_{14} + \wp_{24} - \wp_{23} - \wp_1 - \wp_4 \leq 0, \tag{39.19a}$$
$$-1 \leq \wp_{23} + \wp_{24} + \wp_{14} - \wp_{13} - \wp_2 - \wp_4 \leq 0, \tag{39.19b}$$
$$-1 \leq \wp_{14} + \wp_{13} + \wp_{23} - \wp_{24} - \wp_1 - \wp_3 \leq 0, \tag{39.19c}$$
$$-1 \leq \wp_{24} + \wp_{23} + \wp_{13} - \wp_{14} - \wp_2 - \wp_3 \leq 0, \tag{39.19d}$$

of which inequality (39.19d) is the CH inequality — or the IV Bell inequality [the (35.37)] [6] — and (39.19a) was derived by A. Fine [see Eq. (35.97)].

A generalization of theorem 39.5 to correlation of more than three events is impossible, as proved by Pitowsky [*PITOWSKY* 1989b, 33–45]. On the other side Beltrametti and Maczyński proved the following theorem:

[4] This result was already proved in [A. FINE 1982a] [see section 35.5]. In [*PITOWSKY* 1989b, 63–76] some QM correlation polytopes are shown.

[5] See again [A. FINE 1982a].

[6] Take '1' as polarisation direction $\mathbf{a}$, '2' as $\mathbf{a}'$, '3' as $\mathbf{b}$ and finally '4' as b'.

Theorem 39.6 (I Beltrametti/Maczyński) *The following conditions are equivalent:*
i) $\mathcal{X}$ *is in the range of a complete S–probability measure on some* $\mathcal{L}$ [see definition 11.9: p. 198];
ii) $\mathcal{X}$ *has the properties* [see p. 197]:

$$0 \in \mathcal{X}, \quad (\wp \in \mathcal{X}) \to (1 - \wp \in \mathcal{X}), \quad (\triangle(\wp_1, \wp_2, \wp_3) \in \mathcal{X}) \to (\wp_1 + \wp_2 + \wp_3 \in \mathcal{X}). \quad (39.20)$$

This theorem provides a criterion for verifying that $\mathcal{X}$ is in the range of a complete probability measure on an arbitrary (possibly non-classical) event system. The following theorem gives a criterion for the non-classicality of systems:

Theorem 39.7 (II Beltrametti/Maczyński) *The following properties are equivalent:*
i) $\mathcal{X}$ *is in the range of a complete S–probability measure on some classical event system;*
ii) $\mathcal{X}$ *has the three properties* (39.20), *i.e. it is in the range of a complete probability measure, plus the following: for any* $\wp_1 + \wp_2 \in \mathcal{X}$ *there exists a triangle* $\triangle(q_1, q_2, q_3) \in \mathcal{X}$ *such that* $\wp_1 = q_1 + q_2$ *and* $\wp_2 = q_2 + q_3$.

From the latter two theorems, this corollary follows which gives a necessary and sufficient condition for a set of S–probabilities in order to be non-classical (a case which includes QM too):

Corollary 39.1 (Beltrametti/Maczyński) *Let there be* $\mathcal{X}$ *in the range of a complete S–probability measure on some event system* $(\Omega, \leq, ')$. $\mathcal{X}$ *is non classical iff there exists a pair* $\wp_1, \wp_2 \in \mathcal{X}$ *such that whenever* $\exists a, c \in \mathcal{X}$ *such that* $a \leq c$ *and* $\wp_1 - \wp_2 \neq a + c$*, then* $\wp_1 - \wp_2 \neq a + c - 1$.

Proof
Let us write $a = q_1$ and $1 - c = q_3$. We know that the classicality condition, i.e. condition ii) of theorem 39.7, poses the constraint $q_2 = \wp_1 - q_1 = \wp_1 - a$, so that we can define the triangle $\triangle(a, \wp_1 - a, 1 - c)$ if we have following constraints [see p. 197]: i) $a + (\wp_1 - a) \leq 1$, ii) $a + (1 - c) \leq 1$, iii) $(\wp_1 - a) + (1 - c) \leq 1$, which can be reduced respectively to i) $\wp_1 \leq 1$ (which always holds), ii) $a \leq c$, iii) $(\wp_1 \leq a + c$. But supposing that $\mathcal{X}$ is non classical implies that $\wp_2 \neq (q_2 + q_3)$, which is equivalent to $\wp_2 \neq [(\wp_1 - a) + (1 - c)]$, which in turn can be rewritten as $\wp_1 - \wp_2 \neq a + c - 1$. Q.E.D.

39.4 Stochastic Quantum Mechanics and Bell Inequalities

39.4.1 Informational Completeness and Bell inequalities

In [BUSCH/LAHTI 1989] and [BUSCH/SCHROECK 1989, 814–15] it is shown that QM is not informationally complete [see subsections 18.6.1 and 32.2.4] by performing an experiment on four possible alternative outcomes on the spin of a single photon. Now, if we would have here informational completeness, there would be a Hilbert space with dimension 2 with four projectors adding to the unit operator; and there is none.

In other words: simultaneous objectification of incompatible spin components leads to a contradiction with the boundedness of the spin spectrum: if it could be possible to measure simultaneously σ_x and σ_y of a spin–$\frac{1}{2}$ system, both would have maximal value $\frac{1}{2}$. But as these components are projections of component $\sigma_{\mathbf{n}} = (\sigma_x + \sigma_y)/\sqrt{2}$, it follows that $\sigma_{\mathbf{n}}$ must have the value $\sqrt{2}(1/2)$, which is outside of the spectrum of $\sigma_{\mathbf{n}}$.[7]

[7] Corrections to the theoretical prediction given by Busch and Schroeck are to be found in [KAR/ROY 1995].

39.4.2 Other Developments

A further development of the theorems of Beltrametti and Maczyński is due to Mittelstaedt and co-workers [BUSCH *et al.* 1993] . They introduced the pair correlation function $\mathrm{C}(a,b)$ as follows

$$\mathrm{C}(a,b) := \wp(a,b) + \wp(\bar{a},\bar{b}) - \wp(\bar{a},b) - \wp(a,\bar{b}), \tag{39.21}$$

where $\wp(\bar{\jmath}) := 1-\wp(j)$ $(j=a,b)$, $\wp(\bar{a},\bar{b}) := 1-\wp(a)-\wp(b)+\wp(a,b)$, $\wp(\bar{a},b) := \wp(b)-\wp(a,b)$ and $\wp(a,\bar{b}) := \wp(a)-\wp(a,b)$. By means of these definitions for 4 polarisation directions we obtain the following inequalities — which are equivalent to inequalities (39.19) —:

$$-2 \leq \ +\mathrm{C}(a,b) + \mathrm{C}(a,b') + \mathrm{C}(a',b) - \mathrm{C}(a',b') \ \leq +2, \tag{39.22a}$$
$$-2 \leq \ +\mathrm{C}(a,b) + \mathrm{C}(a,b') - \mathrm{C}(a',b) + \mathrm{C}(a',b') \ \leq +2, \tag{39.22b}$$
$$-2 \leq \ +\mathrm{C}(a,b) - \mathrm{C}(a,b') + \mathrm{C}(a',b) + \mathrm{C}(a',b') \ \leq +2, \tag{39.22c}$$
$$-2 \leq \ -\mathrm{C}(a,b) + \mathrm{C}(a,b') + \mathrm{C}(a',b) + \mathrm{C}(a',b') \ \leq +2, \tag{39.22d}$$

which under the substitution $\mathrm{C}(a,b) = \mathbf{a}\cdot\mathbf{b}$ and after some algebra can be reduced to[8]:

$$|\mathbf{a}\cdot(\mathbf{b}+\mathbf{b}')| + |\mathbf{a}'\cdot(\mathbf{b}-\mathbf{b}')| \ \leq \ 2, \tag{39.23a}$$
$$|\mathbf{a}\cdot(\mathbf{b}-\mathbf{b}')| + |\mathbf{a}'\cdot(\mathbf{b}+\mathbf{b}')| \ \leq \ 2. \tag{39.23b}$$

For stochastic analysis of Bell inequalities for the triple case see [BUSCH *et al.* 1992, 953–54, 958].

39.5 Garg/Mermin Model

The following treatment, due to Mermin and Garg, though confined to the original problem of the spin, represents another powerful generalization.

39.5.1 General Statement of the Problem

Point of departure of Mermin's analysis was the possibility of extending previous proofs [see chapter 35] to an arbitrary large number s of spins [MERMIN 1980]. But it seemed that there was a classical limit for $s \longrightarrow \infty$. In [MERMIN/SCHWARZ 1982] it was conjectured for the first time that this belief could depend on particular values chosen in [MERMIN 1980]. Garg and Mermin [GARG/MERMIN 1982b] showed for the first time that, for any spin number s, for any EPR correlations for spin measurements along all pair of directions from a set of N axes, Bell inequalities are violated for any set of distinct coplanar axes when $N=3$ and for any set of distinct axes whatever when $N=4$.

The whole subject was generalized in a successive article [GARG/MERMIN 1984]. The problem can be stated in the following form [GARG/MERMIN 1984, 3–9]: given a set $\mathcal{X}_1$ of N discrete random variables, the i-th of which has domain containing n_i values and a distribution $\wp_i$ assigning probabilities to each of those values, and given a set $\mathcal{X}_2$ of pair distributions $\wp_{ij}$ for some but not necessarily all pairs of the variables, such that each one occurs at least in one of pair

[8] For details and geometrical representations see the original article.

distributions $\wp_{ij}$, what conditions have to be respected if we wish to construct sets of third or higher-order distributions $\wp_{ijk...}$?[9] The requirements are then the following three:

- *Non-negativity*: none of the high-order distributions can be negative.
- *Compatibility*: The two-variable marginals of higher-order distributions agree with the specified distributions in $\mathcal{X}_2$, i.e. the sum of $\wp_{ijk...}$ over all its variables except the ith and the jth must be $\wp_{ij}$.
- *Consistency*: if any two of the higher-order distributions have n of their variables in common, then the two nth order distributions obtained by adding each $\wp_{ijk...}$ over its remaining variables must be the same.

Suppose now that $\{\alpha\}$ is a set of discrete parameters large enough to specify each $\wp_{ij}$ in $\mathcal{X}_2$, and $\{\mu\}$ a set of discrete parameters that specifies a particular one of the higher-order distributions. The *compatibility condition* can be expressed by

$$\wp_\alpha = \sum_\mu \wp_\mu v_{\mu\alpha}, \tag{39.24}$$

where $v_{\mu\alpha}$ is a matrix to be specified. The *consistency condition* can be expressed as a set of Eqs. of the form:

$$\sum u_\mu \wp_\mu = \sum w_\mu \wp_\mu. \tag{39.25}$$

The solution of the problem posed is given by the Farkas' Lemma: the necessary and sufficient conditions for the existence of non-negative real numbers $\wp_\mu$ satisfying Eqs. (39.24)–(39.25) for given $\wp_\alpha, v_{\mu\alpha}$ is that every set of real numbers c_α that satisfies

$$\sum_\alpha v_{\mu\alpha} c_\alpha \geq 0, \ \forall \mu, \tag{39.26}$$

should also satisfy

$$\sum_\alpha c_\alpha \wp_\alpha \geq 0. \tag{39.27}$$

$v_{\mu\alpha}$ can be regarded as a rectangular matrix, whose columns are labeled by the parameter α — specifying points in the domain of all $\wp_{ij} \in S_2$ — and whose rows are labeled by the parameter μ — specifying points in the required high-order set distributions. While the column vectors have dimension D, the row vectors have dimension d. We can now restate conditions (39.26)–(39.27) defining the row vector c with components c_α and writing:

$$c \cdot \wp \geq 0, \tag{39.28}$$

for all row vectors c satisfying

$$c \cdot v^{(\mu)} \geq 0, \ \forall \mu, \tag{39.29}$$

where $v^{(\mu)}$ are the row vectors of matrix $v_{\mu\alpha}$. The set of c satisfying inequality (39.29) is a convex polyhedral cone, bounded by hyperplanes through the origin and normal to the vectors $v^{(\mu)}$. Now if the column vectors of matrix $v_{\mu\alpha}$ are linearly independent, then the convex polyhedral cone determined by inequality (39.29) is generated by a finite number of extreme rays $c^{(q)}$ each

[9] We remember here that to obtain a violation of classical probability distributions by QM, we require at least a third order distribution, as is evident in the I Bell inequality [Eq. (35.11)] or in inequalities (39.16e) or by corollary 39.1 [p. 665].

of which is orthogonal to $d-1$ linearly independent vectors from the full set of d–dimensional row vectors $v^{(\mu)}$, $\mu = 1, \ldots, D \geq d$. Any vector in the cone (39.29) can be expressed as a linear combination of the $c^{(q)}$ with non-negative coefficients. Hence any vector c satisfying inequality (39.29), will also satisfy inequality (39.28) iff inequality (39.28) is satisfied by all the extreme rays $c^{(q)}$. Hence the Farkas conditions (39.26)–(39.27) or (39.28)–(39.29) reduce to the finite set of conditions:

$$c^{(q)} \cdot \wp \geq 0, \tag{39.30}$$

i.e. to the problem of finding the extreme rays of a specified polyhedral cone.

By the orthogonality of the $c^{(q)}$'s to the $v^{(\mu)}$'s,, it is clear that $d-1$ of Eqs. (39.30) have to be equalities. Since the probabilities are non-negative, it is evident that the condition (39.30) is trivially satisfied for all non-negative $c^{(q)}$ (which we call *irrelevant*). In other words it suffices to find the negative $c^{(q)}$ (the *relevant* ones) in order to solve the problem.

We now apply the method to some spin numbers s and verify the possible conflict with QM (i.e. we show that such high-order distributions are incompatible with QM) [GARG/MERMIN 1984, 10–28].

39.5.2 Spin 1/2

First we analyse the case $s = \frac{1}{2}$. We have three random variables $\mathbf{a}, \mathbf{b}, \mathbf{c}$ with pair distributions $\wp_{\mathbf{ab}}, \wp_{\mathbf{ac}}, \wp_{\mathbf{bc}}$ and possible values $-1, +1$. We state that $\wp_{\mathbf{ab}} = 0$ if $\mathbf{a}, \mathbf{b}$ have the same value, and $= 1/2$ if the values are different. We can also generalize the latter condition by writing:

$$\begin{aligned} \wp_{\mathbf{ab}} &= \frac{1}{4}(1+\eta),\ a = b, & (39.31a) \\ &= \frac{1}{4}(1-\eta),\ a \neq b, & (39.31b) \end{aligned}$$

where η is a parameter satisfying $-1 \leq \eta \leq +1$. We now search for a third-order distribution. Using symmetry and permutation equivalences, we see that the system of 12 Eqs. for the eight numbers $\wp_{\mathbf{abc}}$ (two possible values for each of the three variables) reduces to the existence of a solution for the following system:

$$\begin{aligned} \wp(11) &= \wp(111) + \wp(11\tilde{1}), & (39.32a) \\ \wp(1\tilde{1}) &= 2\wp(11\tilde{1}), & (39.32b) \end{aligned}$$

where $\tilde{1} = -1$. Hence the matrix $v_{\mu\alpha}$ [see inequality (39.27)] is:

$$\begin{array}{ccc} \alpha \rightarrow, \mu \downarrow & \wp(11) & \wp(1\tilde{1}) \\ \wp(111) & 1 & 0 \\ \wp(11\tilde{1}) & 1 & 2; \end{array} \tag{39.33}$$

by means of which we can write inequality (39.30) as a system of two Eqs.:

$$\begin{aligned} c(11) &\geq 0, & (39.34a) \\ c(11) + 2c(1\tilde{1}) &\geq 0. & (39.34b) \end{aligned}$$

Since inequality (39.34a) requires that $c(11)$ be non-negative, then a relevant extreme ray, if it exists, can only be $c(1\tilde{1})$. But if $c(1\tilde{1})$ is negative, then inequality (39.34b) requires $c(11)$ to be positive, which in turn prevents the (39.34a) being an equality. Since $d-1 = 1$ of the inequalities must hold as equalities, the remaining condition (39.34b) must be an equality. Hence there is a unique (up to a scaling) relevant extreme ray given by $c(11) = 2, c(1\tilde{1}) = -1$ and there is only one Farkas condition (39.30)

$$2\wp(11) - \wp(1\tilde{1}) \geq 0, \tag{39.35}$$

which for Eqs. (39.31) is:

$$\eta \geq -\frac{1}{3}. \tag{39.36}$$

To see the conflict with QM, we choose (spin case) η to be $\cos\theta$, where θ is the angle between $\mathbf{a}, \mathbf{b}, \mathbf{c}$ (understood here as detection directions of polarisers, as usually). Since the angle that three symmetrically oriented axes make with each other cannot exceed 120°, we have $-1/2 \leq \eta \leq 1$. The condition (39.36) is then violated for $-1/2 \leq \eta \leq -1/3$ or

$$120° \geq \theta > \cos^{-1}\left(-\frac{1}{3}\right) = 109.47°. \tag{39.37}$$

39.5.3 Spin 1

For the case of spin $s = 1$ ($m_{\hat{\sigma}} = -1, 0, 1$) we have the following matrix $v_{\mu\alpha}$ (where $d = 4, D = 6$):

$\alpha \rightarrow, \mu \downarrow$	$\wp(11)$	$\wp(00)$	$\wp(1\tilde{1})$	$\wp(10)$
$\wp(111)$	1	0	0	0
$\wp(000)$	0	1	0	0
$\wp(11\tilde{1})$	1	0	2	0
$\wp(110)$	1	0	0	1
$\wp(100)$	0	2	0	1
$\wp(10\tilde{1})$	0	0	1	1

(39.38)

We remember that $d - 1 = 3$ of the conditions (39.30) must be equalities. Now the first two row vectors of the table require any extreme ray to have its first two components as non-negative: $c_1 \geq 0, c_2 \geq 0$. The last one requires that c_3 and c_4 cannot both be negative. Hence there are at most two types of relevant extreme rays: with only c_3 negative and with only c_4 negative. In the first case we find the $c^{(q)} = c^{(1)} = (2, 0, -1, 1)$. In the second one $c^{(2)} = (2, 1, 2, -2)$. Hence the conditions (39.30) are:

$$c^{(1)} \cdot \wp = 2\wp(11) - \wp(1\tilde{1}) + \wp(10) \geq 0, \tag{39.39a}$$
$$c^{(2)} \cdot \wp = 2\wp(11) + \wp(00) + 2\wp(1\tilde{1}) - 2\wp(10) \geq 0, \tag{39.39b}$$

which after some manipulation can be reduced to

$$\wp(00) \leq 1 + 6\left|\wp(11) - \wp(1\tilde{1})\right|, \tag{39.40a}$$
$$\wp(00) \geq \wp(0) - \frac{1}{3}. \tag{39.40b}$$

Since $\wp(0) = 1/3$, Eq. (39.40b) requires $\wp(00)$ to be non-negative. Hence the only non-trivial condition is Eq. (39.40a), which becomes:

$$4 + 9P^{(1)}(\cos\theta) - P^{(2)}(\cos\theta) \geq 0, \tag{39.41}$$

where $P^{(l)}(\cos\theta)$ is the l–th Legendre polynomial. Then condition (39.41) is violated for

$$120° \geq \theta > \cos^{-1}(3 - 2\sqrt{3}) = 117.65°. \tag{39.42}$$

39.5.4 Spin 3/2

For a spin $s = 3/2$ ($m_{\hat{\sigma}} = -3/2, -1/2, 1/2, 3/2$) we have $d = 6, D = 10$:

$$\begin{array}{ccccccc} \alpha \rightarrow, \mu \downarrow & \wp(33) & \wp(11) & \wp(3\tilde{3}) & \wp(1\tilde{1}) & \wp(31) & \wp(3\tilde{1}) \\ \wp(333) & 1 & 0 & 0 & 0 & 0 & 0 \\ \wp(111) & 0 & 1 & 0 & 0 & 0 & 0 \\ \wp(33\tilde{3}) & 1 & 0 & 2 & 0 & 0 & 0 \\ \wp(11\tilde{1}) & 0 & 1 & 0 & 2 & 0 & 0 \\ \wp(331) & 1 & 0 & 0 & 0 & 1 & 0 \\ \wp(113) & 0 & 1 & 0 & 0 & 1 & 0 \\ \wp(33\tilde{1}) & 1 & 0 & 0 & 0 & 0 & 1 \\ \wp(11\tilde{3}) & 0 & 1 & 0 & 0 & 0 & 1 \\ \wp(31\tilde{3}) & 0 & 0 & 2 & 0 & 1 & 1 \\ \wp((13\tilde{1}) & 0 & 0 & 0 & 2 & 1 & 1 \end{array} \tag{39.43}$$

where ± 3 stands for $\pm 3/2$ and ± 1 for $\pm 1/2$.

After calculations similar to those of the case $s = 1$ we obtain the following extreme rays:

$$c^{(1)} = (2, 0, -1, 0, 2, 0), \tag{39.44a}$$
$$c^{(2)} = (1, 1, 1, 1, -1, -1), \tag{39.44b}$$
$$c^{(3)} = (2, 2, -1, -1, -2, 4). \tag{39.44c}$$

Three other extreme rays can be obtained from Eq. (39.44a) and one more from Eq. (39.44c) by using symmetries $c_1 = c_2, c_3 = c_4, c_5 = c_6$. All these conditions can be further simplified in:

$$\wp(31) + \wp(3\tilde{1}) \geq \frac{1}{8} + \frac{1}{2}|\wp(31) - \wp(3\tilde{1})| - \frac{3}{4}[\wp(33) + \wp(11)] + \frac{1}{4}|3[\wp(33) - \wp(11)] + 4\wp(1) - 1|; \tag{39.45a}$$
$$\wp(31) + \wp(3\tilde{1}) \leq \frac{1}{6}; \tag{39.45b}$$
$$\wp(31) + \wp(3\tilde{1}) \geq \frac{1}{6} + |\wp(31) - \wp(3\tilde{1})| - [\wp(33) + \wp(11)]; \tag{39.45c}$$

which can be rewritten in terms of Legendre polynomials (we omit the common argument $\cos\theta$) as:

$$15(1 + P^{(1)} + P^{(3)}) - 5P^{(2)} - 18|P^{(1)} - P^{(3)}| \geq 0; \tag{39.46a}$$
$$1 + 3P^{(2)} \geq 0; \tag{39.46b}$$
$$10 + 15(P^{(1)} + P^{(3)}) - 9|P^{(1)} - P^{(3)}| \geq 0. \tag{39.46c}$$

Inequalities (39.46) are violated by QM for the following values of the angle θ:

$$120° \geq \theta > 116.40°, \tag{39.47a}$$
$$109.47° > \theta > 70.53°, \tag{39.47b}$$
$$64.03° > \theta > 57.06°. \tag{39.47c}$$

39.5.5 Conclusion

As we can see, though the range of the violation reduces when one passes from $s = \frac{1}{2}$ to $s = 1$, it increases from $s = 1$ to $s = \frac{3}{2}$, behaviour which is also confirmed in the cases $s = 2$ and $s = \frac{5}{2}$.[10] For further interesting developments see subsection 42.7.1 and on the problem of the extensibility of Bell inequalities to the macroscopic domain see section 40.6.

39.6 Concluding Remark

From the results of this chapter and of chapter 32 it is evident that classical mechanics pose stronger constraints upon the observables and the states than QM does. Notwithstanding, the non-locality is a positive property (not a defective property) which can also used, as we shall see below [see part X].

[10] For the corresponding proofs see the original article. More complicated is the treatment of CH inequality [GARG/MERMIN 1984, 28–33].

Chapter 40

BELL INEQUALITIES FOR OTHER OBSERVABLES

Introduction and contents The following entanglement experiments have been proposed and performed [see subsection 35.3.1] :

- Two (or more) particle polarisation correlation (spin entanglement), discussed in sections 35.3 and 35.4.
- Two particle space-time entanglement [this will be discussed in section 40.1].
- Time entanglement for fields [section 40.2].
- A simultaneous entanglement of the first two types [section 40.3].
- Phase-momentum entanglement [section 40.4].
- Photon-number correlation [section 40.5].

Finally, the violations of Bell inequalities in the macroscopic domain [section 40.6] and in the vacuum [section 40.7] are also discussed. The first one to propose an extension of Bell inequalities to other observables was Franson in a pioneering work [FRANSON 1982]. He showed on a theoretical level that in the case of a single system, for example a photon which is absorbed by an atom in some finite time interval, there are QM predictions which conflict with some form of local requirement.

40.1 Space-Time Entanglement

40.1.1 The Theory

Space-time-entanglement experiments were proposed by Franson [FRANSON 1989].[1] As a first example (= Atomic-Level Exp. 3) take a three-level system [see figure 40.1]. At time $t = 0$ an atom is assumed to be excited into the upper state $|\psi_1\rangle$ which has a relatively long lifetime τ_1. After emission of a photon in state $|\gamma_1\rangle$ and with wavelength λ_1, the atom will be in the intermediate state $|\psi_2\rangle$, which has a relatively short lifetime $\tau_2 \ll \tau_1$. Thus a second photon in state $|\gamma_2\rangle$ and with wavelength λ_2 will be emitted very soon after γ_1, and a coincidence counting

[1] See also [FRANSON 1986] [FRANSON 1994a].

experiment will show a very narrow peak with a width $\simeq \tau_2$. The final state $|\psi_3\rangle$ can be assumed to be the ground state.

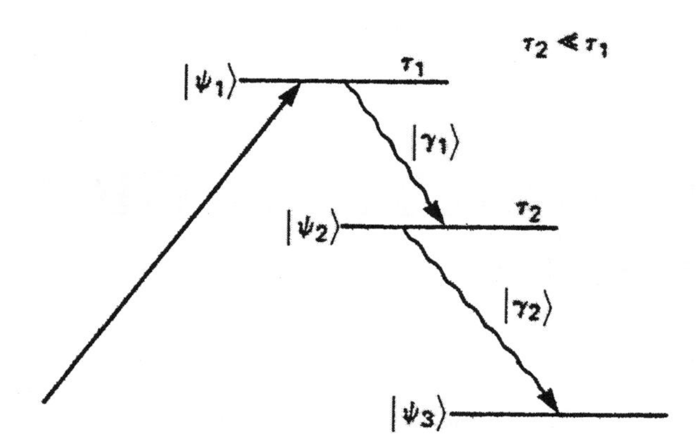

Figure 40.1: Three-level atomic system for Atomic-Level Exp. 3, proposed by Franson — from [FRANSON 1989, 2205].

Now we collimate photons in states $|\gamma_1\rangle, |\gamma_2\rangle$ [see figure 40.2] by lenses L1 and L2 into beams which propagate towards distant detectors D1 and D2 respectively. Spectral filters F1 and F2 transmit only wavelength λ_1, λ_2. Half-silvered mirrors M_1, M_1', M_2, M_2' can be inserted into the beams.

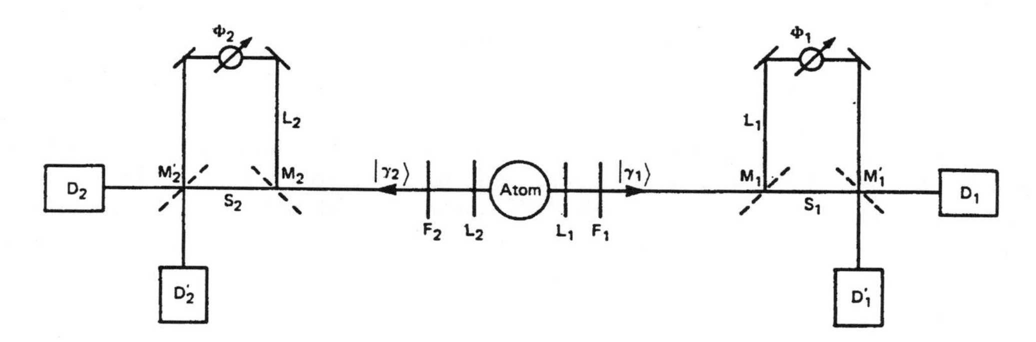

Figure 40.2: Photon coincidence measurements including interference between the amplitudes along the shorter paths S1 and S2 and the longer ones L1 and L2 — from [FRANSON 1989, 2206].

In the absence of half-silvered mirrors the coincidence counting rate will show a narrow peak indicating that $|\gamma_1\rangle, |\gamma_2\rangle$ were emitted at times which were the same to within a small uncertainty $\simeq \tau_2$. But the quantum-mechanical description is highly non-local since the time at which either photon was emitted was initially uncertain over a much larger time interval $\simeq \tau_1$. Hence the time and position of both photons are relatively uncertain. However, the detection of one photon, say $|\gamma_1\rangle$, immediately determines the time of emission of the other one and thus its position to within a small uncertainty, which must be reflected by a non-local change of its wave function. This 'reduction' is analogous to that which occurs in the polarisation measurement in traditional Bell inequalities. We can write the wave function of the two photons as follows $|\gamma\rangle = |\gamma_1(\mathrm{L}_1)\rangle|\gamma_2(\mathrm{S}_2)\rangle + |\gamma_1(\mathrm{S}_1)\rangle|\gamma_2(\mathrm{L}_2)\rangle$, which expresses the fact that when one photon is measured to follow the L path, the other one must have followed the S path.

The fact that single photons produce interference patterns over distances larger than $c\tau_1$ suggests, however, that the wave-like nature of the photons cannot be neglected — and hence there is no perfect determination of time and position all along the paths, as is presupposed by classical mechanics and HV-theories. This will be shown by inserting the half-silvered mirrors. M_1 and M_2 split the beams into equal components which travel along either the short paths S1 and S2 (to the detectors) or the longer ones L1 and L2. The time difference between the two paths δt is assumed to be the same for both photons and to satisfy

$$\tau_2 \ll \delta t \ll \tau_1. \tag{40.1}$$

Phase shift plates ϕ_1, ϕ_2 are introduced in order to create shifts $\delta\varphi_1, \delta\varphi_2$. The BM M_1' and M_2' recombine the components. An efficiency of 100% is assumed. As a final result it is possible to obtain the following Eq. [FRANSON 1989, 2207]:

$$R_c = \frac{1}{4}R_0 \cos^2\left(\delta\varphi_1' - \delta\varphi_2'\right), \tag{40.2}$$

where R_c is the coincidence rate with mirrors inserted, R_0 with mirrors removed, and where:

$$\delta\varphi_1' = \frac{\delta\varphi_1}{2}, \tag{40.3a}$$

$$\delta\varphi_2' = -\frac{\delta\varphi_2 + \Delta E\delta t/\hbar}{2}; \tag{40.3b}$$

and $\Delta E = E_1 - E_3$. Eq. (40.2) is formally equivalent to a Bell inequality[2].

40.1.2 The Experiments

The first realization of the experiment proposed by Franson was performed by Ou, Zou and Mandel [OU *et al.* 1990b]. However, such an experiment presented the problem of too little a visibility in order to prove space-time entanglements: in fact it should be not less than 71%. The first experiments which have overcome such a limit are due to Brendel, Martienssen and Mohler [BRENDEL *et al.* 1992a] [BRENDEL *et al.* 1992b].[3]

The experiment we consider here is an SPDC realization (= SPDC Exp. 8) of Franson's original idea [KWIAT *et al.* 1993] [CHIAO *et al.* 1995b]. For the realization one uses two interferometers with longer and shorter paths and a coincidence counter [see figure 40.3]. For appropriate choice of the phases $\varphi_1 = 45°$ and $135°$ and $\varphi_2 = 0°$ and $-90°$, using the CHSH inequality [inequality (35.21)] in the form $|\mathrm{C}| \leq 2$ we obtain a violation of the order: $\mathrm{C} = -2.63 \pm 0.08$ [CHIAO *et al.* 1995b, 263–65] [see figure 40.4].

Strekalov and co-workers [STREKALOV *et al.* 1996] have proposed an experiment (SPDC Exp. 9) without using BSs [see figure 40.5]: it has the advantage that it does not involve any time discrimination — the only upper limit is constituted by the requirement that only one couple of photons must be in the apparatus —: hence one can use much slower detectors and a much large coincidence window; another advantage is the greater stability of the interference fringes.

[2]See also [FRANSON 1993].

[3]See also [RARITY *et al.* 1990] [LARCHUK *et al.* 1993].

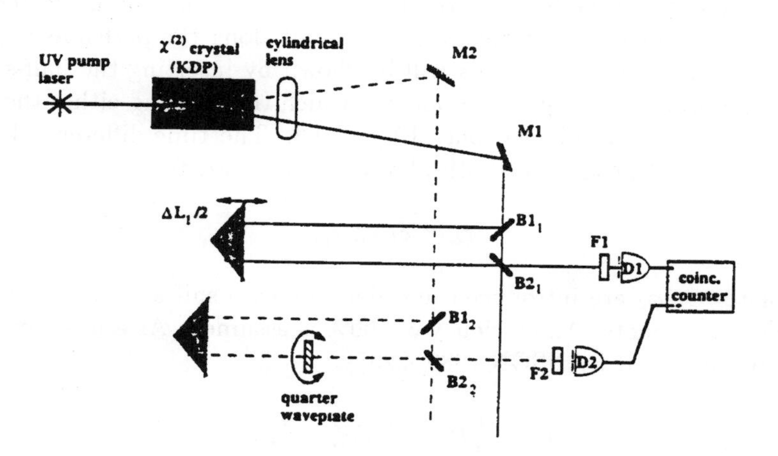

Figure 40.3: Apparatus used at Berkeley to realize the Franson experiment (SPDC Exp. 8) — from [CHIAO *et al.* 1995b, 263].

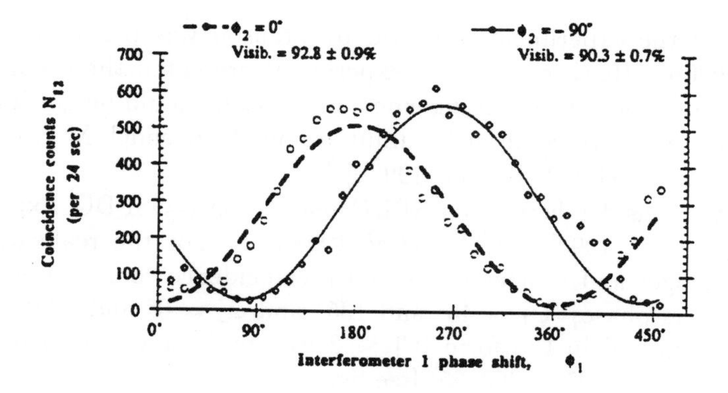

Figure 40.4: Interference fringes obtained in SPDC Exp. 8, for different phase settings of the quarter-wave plate: $\varphi_2 = 0°$ and $-90°$ — from [CHIAO *et al.* 1995b, 264].

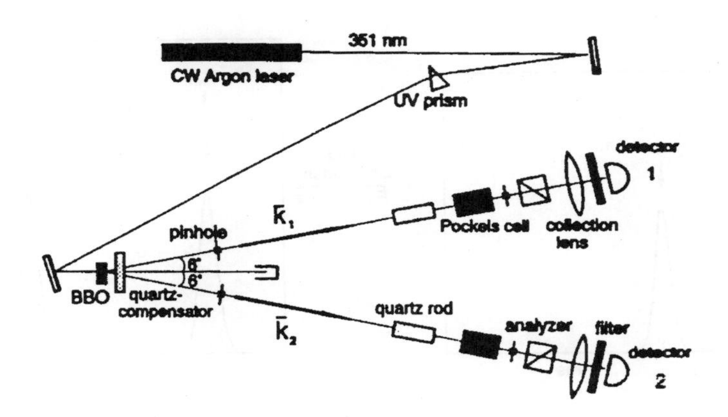

Figure 40.5: Scheme of SPDC Exp. 9: The UV laser beam pumps the crystal (BBO). The signal and idler beams propagate at 6° relative to the pump in the same plane. Each of them passes through a quartz compensator, quartz rod, Pockel cell, and analyzer. Both beams are then focused by lenses to detectors 1 and 2, passing through similar interference filters — from [STREKALOV *et al.* 1996, R2].

40.2 Time Entanglement for Fields

Franson has proposed another interesting inequality [FRANSON 1991]. Take the experimental arrangement shown in figure 40.2 — where only the paths and the detectors are here of interest and the source emits at the same time photons by PDC. Classical fields must respect the following inequality:

$$\langle E_1^*(t)E_2^*(t)E_2(t-\delta\tau)E_1(t-\delta\tau)\rangle \leq \frac{\langle E_1^*(t)E_2^*(t-\delta\tau)E_2(t-\delta\tau)E_1(t)\rangle}{2} + \frac{\langle E_2^*(t)E_1^*(t-\delta\tau)E_1(t-\delta\tau)E_2(t)\rangle}{2}, \quad (40.4)$$

where $\delta\tau$ is the time difference between longer and shorter path, and E_1, E_2 refer to fields at detector positions 1, 2. The physical significance is the following: if E_1 is evaluated at time t and E_2 at time $t \pm \delta\tau$, then one or other of the fields must be zero and their product vanishes [see figure 40.6.(a)]. The RHS of inequality (40.4) is then zero, which requires that the LHS also vanishes. This is a consequence of the fact that classically, the fields are well-defined (complex) numbers. But the inequality is violated in QM.

In QM the intensity is expressed by $I(t) = E^{(-)}(t)E^{(+)}(t)$ [see Eq. (5.18)]. Hence the QM equivalent of inequality (40.4) would be:

$$\langle E_1^{(-)}(t)E_2^{(-)}(t)E_2^{(+)}(t-\delta\tau)E_1^{(+)}(t-\delta\tau)\rangle \leq \frac{\langle E_1^{(-)}(t)E_2^{(-)}(t-\delta\tau)E_2^{(+)}(t-\delta\tau)E_1^{(+)}(t)\rangle}{2} + \frac{\langle E_2^{(-)}(t)E_1^{(-)}(t-\delta\tau)E_1^{(+)}(t-\delta\tau)E_2^{(+)}(t)\rangle}{2}.$$

Now, *ex hypothesi* we require that:

$$E_2^{(+)}(t \pm \delta\tau)E_1^{(+)}(t) = 0; \quad (40.5)$$

while the conservation of the energy in the SPDC requires that:

$$E_1^{(+)}(t-\delta\tau)E_2^{(+)}(t-\delta\tau) = e^{\imath(\omega_1+\omega_2)\delta\tau}E_1^{(+)}(t)E_2^{(+)}(t), \quad (40.6)$$

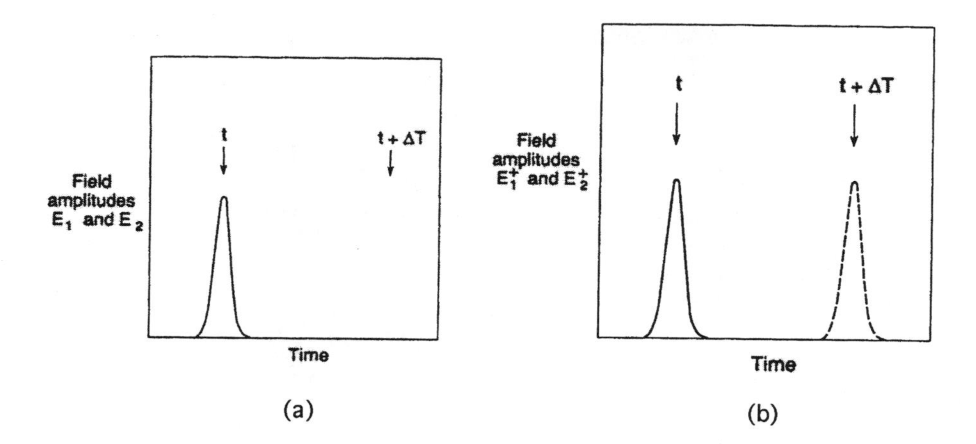

Figure 40.6: (a) A pair of coincident classical pulses.
(b) QM field corresponding to a pair of entangled pairs of coincident photons, with a superposition of times at which the pair may have been emitted — from [FRANSON 1991, 291, 293].

where $\omega_1 + \omega_2 = \omega_0$. Inserting Eq. (40.5) into the RHS of inequality (40.5) gives zero, whereas inserting Eq. (40.6) into the LHS of the same inequality gives $\langle E_1^{(-)}(t)E_2^{(-)}(t)E_2^{(+)}(t)E_1^{(+)}(t)\rangle$, which is a non-zero quantity. Hence the inequality is violated in Q-Optics [see figure 40.6.(b)].

40.3 Polarisation and Space-Time Entanglements

It is also possible to perform experiments in which both a space-time-entanglement and a spin entanglement are used. An example is given by the following experiment, proposed by Shih, Sergienko, Pittman and Rubin (Interfer. Exp. 22). Let us discuss the following Eq. [SHIH/SERGIENKO 1994a] [SHIH *et al.* 1995a, 42–44]:

$$|\Psi\rangle = \int d\nu_1 \delta(\nu_1 + \nu_2 - \nu_0)\psi(k_1 + k_2 - k_0)\hat{a}_p^\dagger[\nu_1(k_1)]\hat{a}_o^\dagger[\nu_2(k_2)]|0\rangle, \tag{40.7}$$

where $|0\rangle$ indicates the state before the interaction with the NL crystal, $|1\rangle$ describes the s-photon and $|2\rangle$ the i-photon, p, o indicates polarisation directions parallel or orthogonal to the incident beam (x, y are along these two directions, respectively) [see figure 40.7]. It is a type–II SPDC [see SPDC Exp. 5 and 6: pp. 611 and 613]. Phase match conditions and conservation laws $\nu_1 + \nu_2 = \nu_0, k_1 + k_2 = k_0$ [see Eqs. (5.67)] are satisfied (where ν_0 represents the pump).

Taking the origin of the coordinate z at the output surface of the NL crystal:

$$\Psi(\delta k) = \frac{[1 - e^{\imath\delta k\cdot l}]}{\imath\delta k \cdot l}, \tag{40.8}$$

where l is the length of the crystal and $\delta k = k_1 + k_2 - k_0$. Suppose now that phase match conditions are satisfied by a set $\{\nu_o, \nu_p, k_o, k_p\}$. Because of the finite spectral bandwidth of the two-photon state we assume $\nu_1 = \nu_p + \nu, \nu_2 = \nu_o - \nu$, where $|\nu| \ll \nu_{p,o}$. Now we have:

$$k_1 = k_p + \frac{\nu}{u_p}, \tag{40.9a}$$

$$k_2 = k_o + \frac{\nu}{u_o}, \tag{40.9b}$$

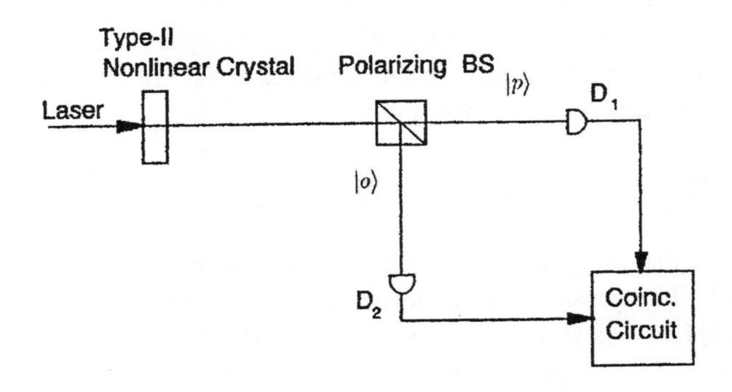

Figure 40.7: Performed Interfer. Exp. 22. The BS transmits the p-ray and reflects the o-ray — from [SHIH *et al.* 1995a, 42].

where u_o, u_p are the group velocities for the two beams. Now Eq. (40.8) can be written:

$$\Psi(\nu) = \frac{1 - e^{\imath\nu\delta u l}}{\imath\nu\delta u}, \tag{40.10}$$

where $\delta u = 1/u_p - 1/u_o$ and we assume that the crystal is negative ($u_o > u_p$). Fields at detectors $\mathrm{D}_1, \mathrm{D}_2$ are:

$$E_1^{(+)}(t) = c_{\mathrm{T}} \int d\nu_p e^{\imath\nu_p(t-\tau_1)} \hat{a}_p(\nu_p), \tag{40.11a}$$

$$E_2^{(+)}(t) = c_{\mathrm{R}} \int d\nu_o e^{\imath\nu_o(t-\tau_2)} \hat{a}_o(\nu_o), \tag{40.11b}$$

where $c_{\mathrm{T}}, c_{\mathrm{R}}$ are transmission and reflection coefficients, $\tau_i := s_i/c$ (s is the optical path length from the output surface) and we assume $\tau_1 = \tau_2$. The average coincidence rate is given by

$$\begin{aligned} R_c &= \frac{1}{t^{\mathrm{D}}} \int\int_0^{t^{\mathrm{D}}} dt_1^{\mathrm{D}} dt_2^{\mathrm{D}} \langle \Psi | E_1^{(-)} E_2^{(-)} E_2^{(+)} E_1^{(+)} | \Psi \rangle \\ &= \frac{1}{t^{\mathrm{D}}} \int\int_0^{t^{\mathrm{D}}} dt_1^{\mathrm{D}} dt_2^{\mathrm{D}} |\Psi(t_1, t_2)|^2, \end{aligned} \tag{40.12}$$

where $t_j := t_j^{\mathrm{D}} - \tau_j$ and t_j^{D} is the detection time of the j–th detector. We now have:

$$|\Psi(t_1, t_2)\rangle = \langle 0 | E_1^{(+)} E_2^{(+)} | \Psi \rangle. \tag{40.13}$$

Substituting Eqs. (40.7) and (40.11) into (40.13) we obtain:

$$|\Psi(t_1, t_2)\rangle = v(t_1 + t_2) u(t_1 - t_2), \tag{40.14}$$

where

$$v(t) = v_{\mathrm{C}} e^{\frac{-\imath\nu_0 t}{2}}, \tag{40.15a}$$

$$\begin{aligned} u(t) &= u_{\mathrm{C}} e^{-\frac{\imath\nu_d t}{2}} \int_{-\infty}^{+\infty} d\nu \frac{1 - e^{\imath\nu\delta u l}}{\imath\nu\delta u} e^{-\imath\nu t} \\ &= e^{\frac{-\imath\nu_d t}{2}} \Pi(t), \end{aligned} \tag{40.15b}$$

where v_C is a normalization constant, $\Pi(t)$ has possible values $u_C, 0$ (u_C being another normalization constant) and $\nu_d := \nu_p - \nu_o$. In the calculation we approximate the pump to be a plane wave; but it can be taken to be a Gaussian with bandwidth σ_0. In this case a Gaussian function $v_C e^{-\sigma_0 t^2/8}$ replaces the constant v_C, and we have:

$$|\Psi(t_1, t_2)\rangle = v_C e^{-\frac{\sigma_0^2(t_1+t_2)^2}{8}} \Pi(t_1 - t_2) e^{-\imath \nu_p t_1} e^{-\imath \nu_o t_2}. \tag{40.16}$$

Eq. (40.16) defines a biphoton (two dimensional wave packet referred to as the two photon effective wave function). The p-ray goes to detector 1, the o-ray to detector 2; the rectangular shaped function $\Pi(t_1 - t_2)$ means that if D_2 is triggered at t_2^D, D_1 will be triggered at a later time, but not later than $t_2^D + \delta ul$. The joint triggering probability at t_1^D, t_2^D is a constant during this period and zero otherwise. The function $v(t_1 + t_2)$ has a finite width along the axis of $t_1 + t_2$. It is clear that the biphoton is entangled in space-time because the wave function cannot factorise in a function of time t_1 and a function of time t_2.[4]

40.4 Phase-Momentum Entanglement

Similarly to the case of space-time entanglement, Rarity and Tapster [RARITY/TAPSTER 1990] proposed and realized an experiment as shown in figure 40.8 (SPDC Exp. 10).

A measure of the distribution of coincidences between detectors on the same side and on the opposite sides of the BS for a particular phase setting is given by the correlation coefficient:

$$\mathrm{C}(\varphi_a, \varphi_b) = \frac{\overline{R}_{a3,b3}(\varphi_a, \varphi_b) + \overline{R}_{a4,b4}(\varphi_a, \varphi_b) - \overline{R}_{a3,b4}(\varphi_a, \varphi_b) - \overline{R}_{a4,b3}(\varphi_a, \varphi_b)}{\overline{R}_{a3,b3}(\varphi_a, \varphi_b) + \overline{R}_{a4,b4}(\varphi_a, \varphi_b) + \overline{R}_{a3,b4}(\varphi_a, \varphi_b) + \overline{R}_{a4,b3}(\varphi_a, \varphi_b)}, \tag{40.17}$$

where $\overline{R}_{aj,bk}(\varphi_a, \varphi_b)$ $(j, k = 3, 4)$ is the mean coincidence rate between respective detectors. The Bell inequality is in our case:

$$-2 \leq \mathrm{C} \leq +2, \tag{40.18}$$

where C is the following combination of four measurements at various phase angles [see III Bell inequality (35.27)]:

$$\mathrm{C} = \mathrm{C}(\varphi_a, \varphi_b) - \mathrm{C}(\varphi_a, \varphi_b') + \mathrm{C}(\varphi_a', \varphi_b) + \mathrm{C}(\varphi_a', \varphi_b'). \tag{40.19}$$

On setting at the minimum position corresponding to a difference of path length $\delta x = 0$ and tilting either phase plate, one modulates the correlation coefficient $\mathrm{C}(\varphi_a, \varphi_b)$. Phase plate thickness varies quadratically with tilt angle θ when the tilt (from orthogonal to the beam) θ_0 is small. Figure 40.9 shows a plot of the coefficient versus the square of the tilt angle of phase plate Pa showing the expected cosinusoidal variation. A least-squares fit to the data estimates the visibility $\mathcal{V} = 0.78$ and provides a calibration of the phase plate allowing settings nominally corresponding to $\varphi_a = 0, \varphi_a' = \pi/2$ to be chosen. A similar measurement made using plate Pb with $\varphi_a = 0$ fixed the phase delays $\varphi_b = \pi/4, \varphi_b' = 3\pi/4$. Experimentally a value of

$$\mathrm{C} = 2.21 \pm 0.022 \tag{40.20}$$

was obtained from five measurement pairs, which is a clear violation of inequality (40.18).

[4]In the same article a performed test is also discussed (with additional conditions) [SHIH *et al.* 1995a, 44–52]. See also [SHIH/SERGIENKO 1994a].

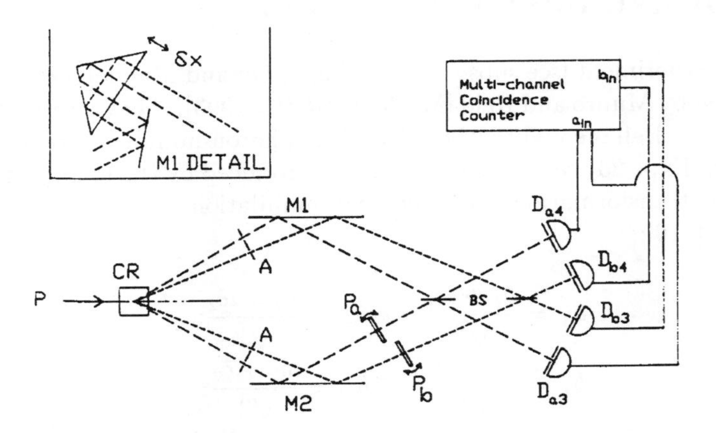

Figure 40.8: Scheme of SPDC Exp. 10. Light from a krypton-ion laser operating at 413.4 nm wavelength is weakly focused in a crystal (CR) of deuterate potassium dihydrogen phosphate (KD*P) with a crystal axis cut at 90° to the incident beam. As a result a small fraction of the vertically polarised incident photons is down-converted to pair of horizontally polarised photons satisfying energy conservation and propagating in directions set by momentum conservation within the birefringent crystal. In this case the symmetric pairs of 826.8 nm wavelength photons are emitted in directions subtending an angle of 28.6°. Conjugate colors a and b with wavelength above and below 826.8 nm, respectively, are selected using double apertures (A) in each arm. Pairs of photons are now detected in beams of different wavelength. Mirrors M1 and M2 reflect the beams onto a BS where recombinations of the different colours occur at points separated by several millimeters. The four outputs are detected by photon counting using a modified photon correlator with a gate time $\delta t = 10$ ns. The path difference can be changed by adjusting the mirror M1 (see insert: 'optical trombone'). Two tiltable glass plates Pa and Pb allow independent adjustment of the relative phases φ_a, φ_b at recombination of colours a and b — from [RARITY/TAPSTER 1990, 2495].

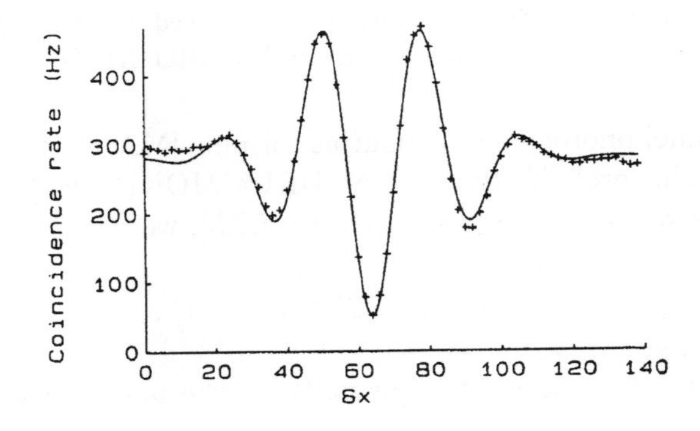

Figure 40.9: Results of SPDC Exp. 10. Coincidence rate as a function of gross path length difference δx between arms 1 and 2 of the apparatus. Experimental points are shown by '+' — from [RARITY/TAPSTER 1990, 2498].

40.5 Photon-Number Correlation

The already related experiment [see subsection 35.4.2] of Ou and Mandel [OU/MANDEL 1988a], was further developed by Munro and Reid [MUNRO/REID 1993]. They wrote down an inequality corresponding to the IV Bell inequality [the (35.37)] for photon-number correlations. The original experiment (Interfer. Exp. 23) can be schematically introduced by the following equations, which describe the relevant transformations between the annihilation operators of the various modes involved [see figure 40.10]:

$$\hat{a}_+ = \frac{\hat{a}_1 + \imath\hat{c}_1}{\sqrt{2}}, \qquad \hat{a}_- = \frac{\hat{a}_2 + \imath\hat{c}_2}{\sqrt{2}}; \tag{40.21a}$$

$$\hat{b}_+ = \frac{\imath\hat{a}_1 + \hat{c}_1}{\sqrt{2}}, \qquad \hat{b}_- = \frac{\imath\hat{a}_2 + \hat{c}_2}{\sqrt{2}}; \tag{40.21b}$$

$$\hat{c}_+ = \hat{a}_+ \cos\theta + \hat{a}_- \sin\theta, \qquad \hat{c}_- = -\hat{a}_+ \sin\theta + \hat{a}_- \cos\theta; \tag{40.21c}$$

$$\hat{d}_+ = \hat{b}_+ \cos\phi + \hat{b}_- \sin\phi, \qquad \hat{d}_- = -\hat{b}_+ \sin\phi + \hat{b}_- \cos\phi, \tag{40.21d}$$

where $\hat{a}_1, \hat{a}_2$ are orthogonal input modes and $\hat{c}_1, \hat{c}_2$ are modes for the input vacuum states.

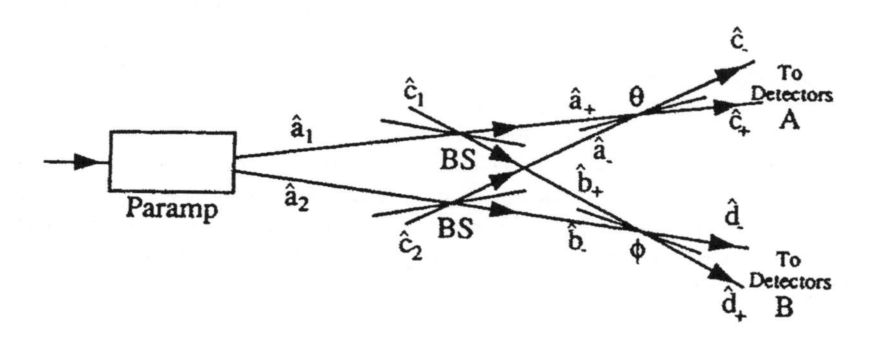

Figure 40.10: Schematic diagram of the transformation involved in the experimental arrangement (Interfer. Exp. 23) proposed by Munro and Reid — from [MUNRO/REID 1993, 4413].

Making use of the Mandel photoelectron counting formula [MANDEL 1958a] [see Eq. (22.4)], in order to calculate the joint probability $\wp_{\hat{N}}(\theta, \phi)$ [DRUMMOND 1983] of detecting N photons at point A and N photons at point B [see subsection 35.2.5], we obtain:

$$\wp_{\hat{N}}(\theta, \phi) = \eta \left\langle : \hat{c}_+^{\dagger N} \hat{c}_+^N \hat{d}_+^{\dagger N} \hat{d}_+^N : \right\rangle, \tag{40.22}$$

where η is an efficiency factor (assumed to be small). If the polariser preceding B is removed QM predicts:

$$\wp_{\hat{N}}(\theta, -) = \eta \left\langle : \hat{c}_+^{\dagger N} \hat{c}_+^N (\hat{d}_+^\dagger \hat{d}_+ + \hat{d}_-^\dagger \hat{d}_-)^N : \right\rangle; \tag{40.23a}$$

and if the polariser preceding A is removed QM predicts:

$$\wp_{\hat{N}}(-, \phi) = \eta \left\langle : (\hat{c}_+^\dagger \hat{c}_+ + \hat{c}_-^\dagger \hat{c}_-)^N \hat{d}_+^{\dagger N} \hat{d}_+^N : \right\rangle. \tag{40.23b}$$

Munro and Reid have shown that — classically — the following inequality — which is equivalent to inequality (35.38) — must be respected:

$$C_{\hat{N}} = \frac{\wp_{\hat{N}}(\theta, \phi) - \wp_{\hat{N}}(\theta, \phi') + \wp_{\hat{N}}(\theta', \phi) + \wp_{\hat{N}}(\theta', \phi')}{\wp_{\hat{N}}(\theta', -) + \wp_{\hat{N}}(-, \phi)} \leq 1. \tag{40.24}$$

$n = N$	θ	θ'	ϕ	ϕ'	$\hat{C}_{\hat{N}}$ (Max)
1	0.393	1.178	0.785	0.000	2.414
2	0.365	3.050	- 1.662	1.955	2.617
3	0.263	- 0.062	- 1.615	- 1.260	2.717
4	0.235	- 0.033	- 1.604	1.800	2.780
5	0.194	- 0.023	- 1.548	- 1.375	2.820
6	0.170	- 0.017	- 1.590	- 1.400	2.846
7	0.140	- 0.014	- 1.580	- 1.420	2.863
8	0.140	- 0.009	- 1.580	- 1.450	2.880
9.	0.140	- 0.007	- 1.580	- 1.480	2.890
10	0.130	- 0.005	- 1.580	- 1.482	2.898
30	0.017	- 0.002	- 1.571	- 1.487	2.917
100	0.007	0.000	- 1.571	- 1.550	2.922

Table 40.1: Results of Interfer. Exp. 23. The angle $\theta, \theta', \phi, \phi'$ are chosen in order to obtain a maximal violation of inequality (40.24) — from [MUNRO/REID 1993, 4414].

That this inequality (40.24) is violated by QM is shown by the experimental results reported in table 40.1. Two interesting points should be noted concerning this experiment:

- it is a direct confirmation of the proof developed already by Mermin and Garg about the violation of Bell inequalities by higher spin states [see section 39.5 and also subsection 35.2.5].

- It could be a confirmation that quantum effects (in this case EPR states) can extend also to mesoscopic and macroscopic domains for high number of photons (by increasing of N — see particularly [REID/MUNRO 1992] [REID/DRUMMOND 1988].

In the following section, we analyse the second point[5].

40.6 Violation of Bell Inequalities in Macroscopic Domain

We have already analysed quantum effects at the macroscopic or the mesoscopic level [see chapter 24]. The following discussion is a particular extension centered on Bell inequalities.

[5] Franson [FRANSON 1994a] has shown that it is possible that the phase measurement on one component leaves the other in a coherent state with the corresponding phase. For other related experiments see [GHOSH/MANDEL 1987] [OU/MANDEL 1988b] [SHIH *et al.* 1993].

For experiments such as that of Munro/Reid (*boson fields*) there is a problem in creating violations of Bell inequalities in macroscopic domain: in fact the measurement process must project the two orthogonal states which are initially superposed onto a common final state in order to produce the interference signaling non-local effects, which is very difficult when the two states are macroscopically different. In fact Sanders [SANDERS 1992] has shown that the correlation coefficient decreases exponentially with increasing the photon number.

Leggett and Garg [LEGGETT/GARG 1985] showed that the macroscopic realism plus non-invasive measurability at the macroscopic level (no interference) are in contradiction with QM at the macroscopic level. We can derive some Bell-like inequalities — using time at the place of polariser settings — which are violated by QM [LEGGETT/GARG 1985, 858] . The experiment proposed by Leggett and Garg was later criticized by Ballentine [BALLENTINE 1987] and Peres [PERES 1988b].[6] In fact, though Leggett and Garg proposed a SQUID experiment, the basic model remained a *polarisation experiment* such as that conceived by Garg and Mermin [see section 39.5]. Now there is also a general problem in using polarisation experiments: experiments must be conducted on individual particles, which makes an n–spin experiment more similar to a complicated quantum system than to a macroscopic phenomenon.

Mermin [MERMIN 1990a] generalized the GHZ experiment [see subsection 36.4.1] to models with n particles: violation of Bell inequalities grows exponentially with n. Again the experiments are increasingly difficult to realize with increasing n and it is not at all sure that they are free from loopholes.

But a partial theoretical confirmation of Mermin's proposal came from Braunstein and Mann [BRAUNSTEIN/MANN 1993] . They showed that, by calculating the ratio between signal and noise in presence of three forms of noise (loopholes): source-noise (variation in the quantum states used as source of correlations), noise from data-flipping errors (noise from the flipping of the spin polarisations due the read-out), and noise from counting statistics (one measures only a finite number of N-particle coincidences), the violation of Bell inequalities grows exponentially faster than the noise.

For experiments such as those performed by Franson the situation is very different: it is possible to violate the Bell inequalities also in the presence of macroscopic fields [FRANSON 1993].

40.7 Violation in Vacuum State

Lawrence Landau [Lw. LANDAU 1987c, 115] showed that there are observables associated with three causally separated space-time regions which do not possess a joint distribution in the vacuum state [see also subsection 9.3.1]:

- in two space-like separated regions there are vector states, say $|\psi\rangle$, which violate Bell inequality, and thus there is no joint distribution for arbitrary observables $\hat{O}_a, \hat{O}_b \in \mathrm{V}_1; \hat{O}_{a'}, \hat{O}_{b'} \in \mathrm{V}_2$ (here two-valued observables).

- If $\hat{O}_3 \in \mathrm{V}_3$ and can take only values $+1, -1$, we find the conditional Bell inequality [Lw. LANDAU 1987c, 116] — analogously to the III Bell inequality [the (35.27)] by substituting b to a, a' to b', b' to b, a to a' (each taking only ± 1 values):

$$\left|\mathrm{C}_{\hat{O}_3}\right| := \left|\left\langle \hat{O}_a \hat{O}_{a'} \right\rangle_{\hat{O}_3} + \left\langle \hat{O}_a \hat{O}_{b'} \right\rangle_{\hat{O}_3} + \left\langle \hat{O}_b \hat{O}_{b'} \right\rangle_{\hat{O}_3} - \left\langle \hat{O}_b \hat{O}_{a'} \right\rangle_{\hat{O}_3}\right| \leq 2, \tag{40.25}$$

[6] See also [BRAUNSTEIN/CAVES 1990, 53].

where $\left\langle \hat{O}_a \hat{O}_{a'} \right\rangle_{\hat{O}_3}$ represents the conditional expectation of the product of $\hat{O}_a$ and $\hat{O}_{a'}$ given that $\mathrm{v}[\hat{O}_3] = 1$.

Then we state the following theorem[7]:

Theorem 40.1 (II Landau) *For any given $\epsilon > 0$ and space-like separated* $\mathrm{V}_1, \mathrm{V}_2, \mathrm{V}_3$ *there are* $\hat{O}_a, \hat{O}_b \in \mathrm{V}_1; \hat{O}_{a'}, \hat{O}_{b'} \in \mathrm{V}_2; \hat{O}_3 \in \mathrm{V}_3$ *(two-valued observables) such that in vacuum state*

$$|\mathrm{C}_{\hat{O}_3}| > 2\sqrt{2} - \epsilon. \tag{40.26}$$

It is possible to generalize for observables which are not two-valued and to all state vectors and density matrices in the vacuum sector[Lw. LANDAU 1987c, 117].

In [SUMMERS/R. WERNER 1985] it is shown that the vacuum state in any Bose or Fermi free QM-field Theory maximally violates Bell inequalities: using suitable detectors, maximal violations may be obtained without setting up a source. But they are very difficult to measure[8].

---

[7]For the proof see original article.

[8]Generalizations are to be found in [SUMMERS/R. WERNER 1987a].

Chapter 41

OTHER NON-LOCAL EFFECTS

Contents The subject of this chapter is other forms of quantum non-locality. In section 41.1 we discuss the non-locality of a single photon and with independent sources. Section 41.2 is devoted to one of the most interesting aspects of QM: the Aharonov/Bohm Effect. Finally section 41.3 is devoted to tunnelling experiments. Another non-local effect is teleportation. However, since we need to make an examination within the frame of the information theory, we shall discuss this subject later [in section 43.2].

41.1 Non-Locality by Single Particles and by Independent Sources

41.1.1 Non-Locality of Single Photons

A first but not wholly satisfactory attempt to prove experimentally the self-interference of a photon is due to Jánossy and Naray [JÁNOSSY/NARAY 1958].[1] Already the reported [see p. 110 and section 28.6] experiment (Interfer. Exp. 2), developed in 1986 by Grangier, Roger and Aspect [GRANGIER *et al.* 1986a], but especially its development (Interfer. Exp. 3) by Franson/Potocki [FRANSON/POTOCKI 1988] [p. 110] clearly show the non-local nature of a single photon[2].

The experiment of Tan, Walls, and Collett More recently the problem has been discussed by Tan, Walls and Collett [TAN *et al.* 1991]. We consider (Interfer. Exp. 24) a pair of homodyne detectors, each of which consists in a 50%–50% BS, a coherent local oscillator with amplitude $\vartheta_k = \vartheta e^{\imath\varphi_k}$ (k = 1, 2) and two photodetectors. The inputs to these homodyne detectors are themselves derived by a third BS [see figure 41.1]. The detectors may be regarded as making a measurement on mode b_k, with local parameter φ_k, which is analogous to the angle of the analyzer by EPRB type experiments. The relevant transformations between the annihilation operators of the various modes involved are given by:

$$\begin{pmatrix} \hat{c}_k \\ \hat{d}_k \end{pmatrix} = \frac{1}{\sqrt{2}} \begin{pmatrix} 1 & \imath \\ \imath & 1 \end{pmatrix} \begin{pmatrix} \hat{a}_k \\ \hat{b}_k \end{pmatrix}; \tag{41.1a}$$

[1] For reviews see [GHOSE/HOME 1992] [COMBOURIEU/RAUCH 1992].
[2] See also [HONG/MANDEL 1986].

$$\begin{pmatrix} \hat{b}_1 \\ \hat{b}_2 \end{pmatrix} = \frac{1}{\sqrt{2}} \begin{pmatrix} 1 & \imath \\ \imath & 1 \end{pmatrix} \begin{pmatrix} \hat{f} \\ \hat{e} \end{pmatrix}. \tag{41.1b}$$

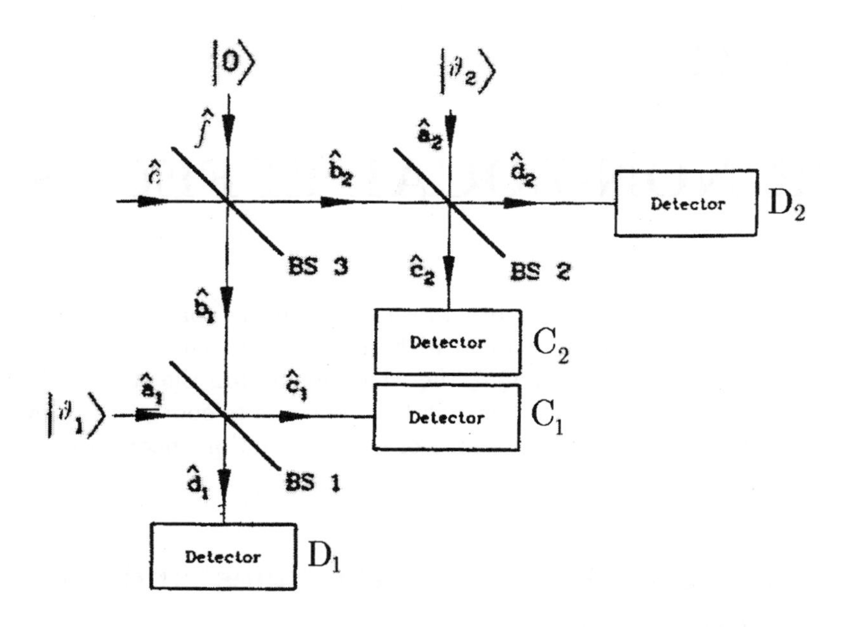

Figure 41.1: Setup of Interfer. Exp. 24 — from [TAN *et al.* 1991, 252].

Now we calculate the corresponding coincidence probabilities. First consider the case of vacuum inputs to the modes $\hat{f}, \hat{e}$. The local oscillators are assumed to be in coherent states: $|\vartheta e^{\imath\varphi_1}\rangle, |\vartheta e^{\imath\varphi_2}\rangle$. The intensities at all detectors are equal:

$$\langle I_{C_1}\rangle = \langle I_{C_2}\rangle = \langle I_{D_1}\rangle = \langle I_{D_2}\rangle = \frac{1}{2}\vartheta^2. \tag{41.2a}$$

The two-photon coincidence rates are also equal:

$$\langle I_{C_1} I_{C_2}\rangle = \langle I_{D_1} I_{D_2}\rangle = \langle I_{C_1} I_{D_2}\rangle = \langle I_{D_1} I_{C_2}\rangle = \frac{1}{4}\vartheta^4. \tag{41.2b}$$

Consider now the input of a single photon in mode $\hat{e}$ while the mode $\hat{f}$ is in the vacuum state. The states of $\hat{b}_1, \hat{b}_2$ after the first BS are an entangled state of one-photon and the vacuum:

$$|\Psi\rangle = \frac{1}{\sqrt{2}}\left(\imath|1\rangle_{b_1}|0\rangle_{b_2} + |0\rangle_{b_1}|1\rangle_{b_2}\right). \tag{41.3}$$

The photon count probabilities are respectively

$$\langle I_{C_1}\rangle = \langle I_{C_2}\rangle = \langle I_{D_1}\rangle = \langle I_{D_2}\rangle = \frac{1}{2}\vartheta^2 + \frac{1}{4}, \tag{41.4a}$$

which is increased by $\frac{1}{4}$ relatively to calculation (41.2a), and:

$$\langle I_{C_1} I_{C_2}\rangle = \langle I_{D_1} I_{D_2}\rangle = \frac{1}{4}\left\{\vartheta^4 + \vartheta^2[1 + \sin(\varphi_1 - \varphi_2)]\right\}, \tag{41.4b}$$

$$\langle I_{C_1} I_{D_2}\rangle = \langle I_{D_1} I_{C_2}\rangle = \frac{1}{4}\left\{\vartheta^4 + \vartheta^2[1 - \sin(\varphi_1 - \varphi_2)]\right\}, \tag{41.4c}$$

where, if we set $\varphi_1 - \varphi_2 = -\frac{\pi}{2}$, we get the minimum of coincidence rate $\frac{1}{4}\vartheta^4$ for detector pairs $(\mathrm{C}_1, \mathrm{C}_2)$ and $(\mathrm{D}_1, \mathrm{D}_2)$, and the maximum value $\frac{1}{4}\vartheta^4 + \frac{1}{2}\vartheta^2$ for pairs $(\mathrm{C}_1, \mathrm{D}_2)$ and $(\mathrm{D}_1, \mathrm{C}_2)$. The classical wave description of light also shows a similar non-local behaviour. However, it is possible to distinguish between them [see figure 41.2].

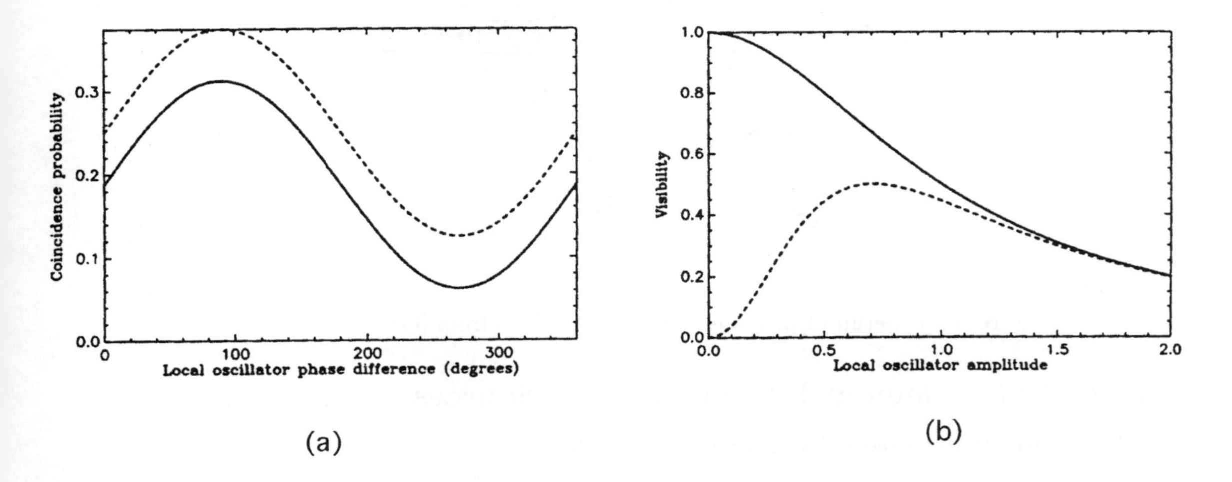

Figure 41.2: Experimental results of Interfer Exp. 22.
(a) Coincidence probabilities for QM model (solid line) and classical wave model (dashed line).
(b) Variation of visibility with local oscillator amplitude for QM model (solid line) and classical wave model (dashed line) — from [TAN *et al.* 1991, 254].

Now we calculate an intensity correlation coefficient as follows [see Eq. (5.19)]:

$$\mathrm{C}(\varphi_1, \varphi_2) = \frac{\langle (I_{\mathrm{D}_1} - I_{\mathrm{C}_1})(I_{\mathrm{D}_2} - I_{\mathrm{C}_2}) \rangle}{\langle (I_{\mathrm{D}_1} + I_{\mathrm{C}_1})(I_{\mathrm{D}_2} + I_{\mathrm{C}_2}) \rangle}. \tag{41.5}$$

By evaluating it in the case of a single photon we obtain:

$$\mathrm{C}(\varphi_1, \varphi_2) = \left[\frac{1}{\vartheta^2 + 1}\right] \sin(\varphi_1 - \varphi_2). \tag{41.6}$$

If the coefficient of $\sin(\varphi_1 - \varphi_2)$ is greater of $1/\sqrt{2}$, there is a violation of a Bell-type inequality.

Santos [SANTOS 1992a] tried to invalidate the experiment by arguments similar to those previously discussed [see the beginning of subsection 35.4.2]. However, the author answered by reaffirming the non-enhancement hypothesis [TAN *et al.* 1992].

Gerry's proposed experiment Another experiment in order to throw light on the non-locality of a single photon[3] has been proposed by Gerry [GERRY 1996] (Cavity Exp. 6): two separated cavities are prepared in an entangled state containing only one photon by means of an excited atom traversing both. If the atom is found to be in ground state (photon emitted), then two atoms in ground state are sent, each one traversing a cavity toward opposite directions. This establishes the non-locality. Then the atoms are analysed by classical microwave fields and selective ionisation, with the consequent violation of Bell-type inequality [see figure 41.3].

[3]For a more abstract treatment see also [REVZEN/MANN 1996].

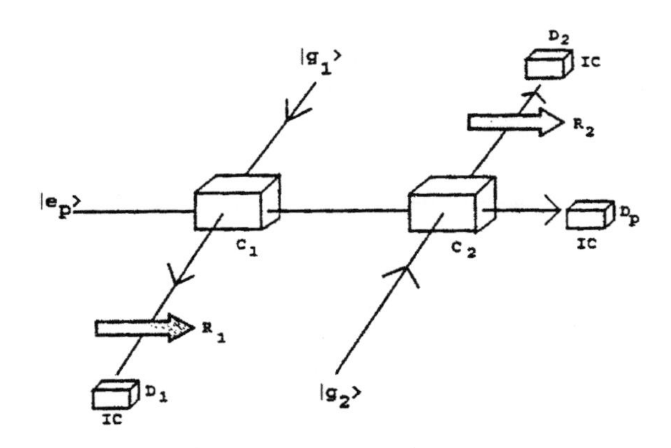

Figure 41.3: Setup of proposed Cavity Exp. 6 — from [GERRY 1996, 45844].

41.1.2 Entanglement by Independent Sources

An Experiment proposed by Yurke and Stoler

Until the end of the eighties a general belief was that entanglement is a consequence of the fact that the particles involved originated from the same source. But Yurke and Stoler, by means of a thought experiment, showed [YURKE/STOLER 1992a] [YURKE/STOLER 1992b] that the entanglement can also originate for two or three particles deriving from different sources and combined by BSs (Interfer. Exp. 25). It is a variant of the proposed GHSZ experiment [see subsection 36.4.2] and of one proposed by Mermin [MERMIN 1990c]. We give here the proof for the first Bell inequality using two particle sources [YURKE/STOLER 1992b].[4] By the geometry of the experiment [see figure 41.4] we have — R stands for 'red' and G for 'green', terms originally employed by Mermin, and which here mean particle input and vacuum input, respectively —:

$$\wp(R, G, \varphi_a, \varphi_b) = \wp(RGR; RGR) + \wp(RGG; RGG), \tag{41.7a}$$
$$\wp(R, G, \varphi_a, \varphi_c) = \wp(RRG; RRG) + \wp(RGG; RGG), \tag{41.7b}$$
$$\wp(G, R, \varphi_b, \varphi_c) = \wp(RGR; RGR) + \wp(GGR; GGR), \tag{41.7c}$$

where $\wp(R, G, \varphi_a, \varphi_b)$ is the probability that detector 1 reports the event R and detector 2 the event G, given that the detector phase φ_1 of detector 1 (given by $\varphi_1 = \varphi_{R1} - \varphi_{G1}$) is set to φ_a and the detector phase φ_2 (given by $-\varphi_{R2} + \varphi_{G2}$ for bosons) of detector 2 is set to φ_b; it is presupposed that both φ_1 and φ_2 can assume only one of the three values: $\varphi_a, \varphi_b, \varphi_c$, and $\wp(RGR; RGR)$ is the probability that instruction set $(RGR; RGR)$ is sent.

From Eq. (41.7a) and (41.7c) one obtains:

$$\wp(R, G, \varphi_a, \varphi_b) \geq \wp(RGG; RGG), \tag{41.8a}$$
$$\wp(G, R, \varphi_b, \varphi_c) \geq \wp(RGR; RGR). \tag{41.8b}$$

From the (41.8) and Eq. (41.7b) one obtains the following Bell inequality [see the (35.11)]:

$$\wp(R, G, \varphi_a, \varphi_b) \leq \wp(R, G, \varphi_a, \varphi_c) + \wp(R, G, \varphi_b, \varphi_c), \tag{41.9}$$

[4] In [YURKE/STOLER 1992a] three sources are used. On entanglement by independent sources, see also [*PERES* 1993, 169].

which can be rewritten as follows:

$$\sin^2(\varphi_a - \varphi_b) \leq \sin^2(\varphi_a - \varphi_c) + \sin^2(\varphi_b - \varphi_c). \tag{41.10}$$

Taking $\theta = \varphi_a - \varphi_c = \varphi_c - \varphi_b$ we obtain:

$$\sin^2(2\theta) \leq 2\sin^2\theta, \tag{41.11}$$

which is violated when $0 < |\theta| < \pi/4$.[5]

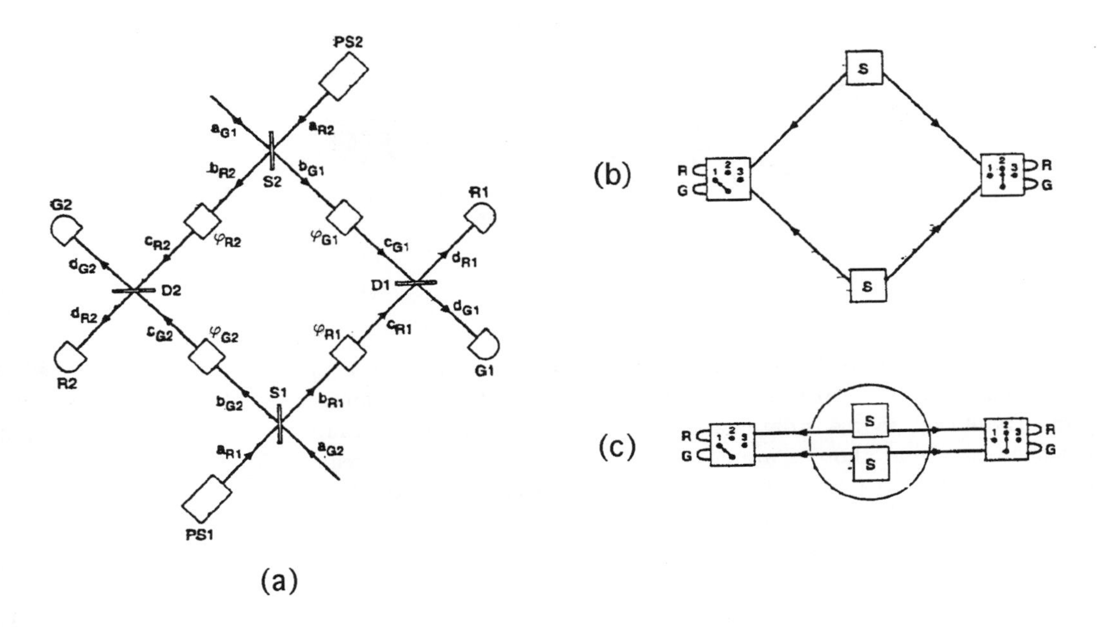

Figure 41.4: (a) Schematics of Interfer. Exp. 25, proposed by Yurke/Stoler: particles enter by pumps PS1 and PS2 (red), vacuum by the other input port (green) before beam splitters S1 and S2. Each arm consists of a phase shifter φ_{Gj} (j = 1, 2) or φ_{Rj} (j = 1, 2), a BM (D1 or D2) and a particle counter Gj ($j = 1, 2$) or Rj ($j = 1, 2$). Detector j consists of phase shifters φ_{Gj} and φ_{Rj}, of beam splitter Dj and of particle counters Gj and Rj
(b) An abstraction of the apparatus shown in (a). The detectors are represented as boxes with red R and green G lights. The switches (which can point to 1, 2, or 3) select the detector phases.
(c) The (b) configuration can be topologically distorted in order to obtain the conventional EPR experiment, where particles are emitted by a central source — from [YURKE/STOLER 1992b, 2229].

Another Proposed Experiment Zeilinger and co-workers employed a technique named *entanglement swapping*. They [ŻUKOWSKI *et al.* 1993] stated that the choice of coincidence timing enables one

[5]Similar results were obtained by using four harmonic oscillators [YURKE/STOLER 1995], pairs of which interact to allow the interchange of energy. It has been shown that the energy becomes distributed between the four harmonic oscillators in a manner that is incompatible with the separability assumption after which information is exchanged only when oscillators interact with each other. On this point see also [HACYAN 1997].

- to monitor the emission events, from the sources;
- to erase *Welcher-Weg* information.

The proposed experiment (SPDC Exp. 11) employs PDC sources pumped by cw lasers. One of the requirements is the narrow filtering of i-photons and detecting them in ultracoincidence (= coincidence window narrower than the filter bandwidth time). In [ŻUKOWSKI *et al.* 1995] these restrictions are abandoned [see figure 41.5].

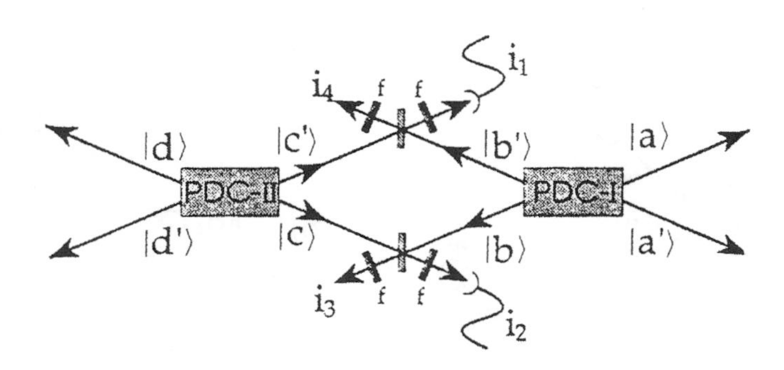

Figure 41.5: Proposed SPDC Exp. 11. Principle of entanglement swapping. Two cw pumped down-conversion sources — SPDC-I and SPDC-II [see p. 611] — emit each a photon pair. The specific geometry of each source is obtainable by a suitable arrangement of mirrors and apertures. The initially independent signal photons are entangled by an ultracoincident registration of the idlers (for sake of simplicity $\gamma_i = i$) at i_1, i_2 — from [ŻUKOWSKI *et al.* 1995, 94].

Consider the factorisable two-particle state:

$$|\Psi\rangle = |\psi\rangle_1 |\varsigma\rangle_2; \tag{41.12}$$

and the entangled state:

$$|\Psi^+\rangle = \frac{1}{\sqrt{2}} \left(|a\rangle_1 |b\rangle_2 + |c\rangle_1 |d\rangle_2 \right), \tag{41.13}$$

where for simplicity $\langle a|c\rangle = 0, \langle b|d\rangle = 0$. It is always possible to obtain an entanglement from Eq. (41.12) by means of a projection operator $\hat{P} = |\Psi^+\rangle\langle\Psi^+|$ acting on $|\Psi\rangle$:

$$\hat{P}|\Psi\rangle = \eta|\Psi^+\rangle. \tag{41.14}$$

Let there be now the two pumps of figure 41.5, with two pairs of particles represented respectively by $\frac{1}{\sqrt{2}}(|a\rangle_1|b\rangle_2 + |a'\rangle_1|b'\rangle_2)$, $\frac{1}{\sqrt{2}}(|c\rangle_3|d\rangle_4 + |c'\rangle_3|d'\rangle_4)$, we have for the initial four-photons state:

$$|\Psi_i\rangle = \frac{1}{\sqrt{2}} \left(|a\rangle_1 |b\rangle_2 + |a'\rangle_1 |b'\rangle_2 \right) \left(|c\rangle_3 |d\rangle_4 + |c'\rangle_3 |d'\rangle_4 \right). \tag{41.15}$$

where 1, 4 are s-photons, the 2, 3 i-photons, and 1 is entangled with 2 and 3 is entangled with 4. In order to entangle uncorrelated s-photons (1 and 4) and to obtain

$$|\Psi^+\rangle_{\gamma_s} = \frac{1}{\sqrt{2}} \left(|a\rangle_1 |d'\rangle_4 + |a'\rangle_1 |d\rangle_4 \right), \tag{41.16}$$

we project the i-photons into an entangled state such as that of Eq. (41.13). This projection can be done (destructively) after overlapping their modes at two BS and by observing the i-photons at detectors i_1, i_2. If the two i-photons are indistinguishable, the joint detection projects the i-state into:

$$|\Psi^+\rangle_{\gamma_i} = \frac{1}{\sqrt{2}}\left(|b\rangle_2|c'\rangle_3 + |b'\rangle_2|c\rangle_3\right). \tag{41.17}$$

A consequence is that the registration of photons 2, 3 can operationally define at distance the now entangled pair 1, 4. But the joint detection of i-photons must be in coincidence, and this poses the experimental requirements for this purpose. In the case of the coincidence between i- and s-photon of the same source, the bandwidth determines the relation between registration time [HONG/MANDEL 1986]. We have $t_c \simeq 1/\Delta\omega$. In the case of two sources the ultracoincidence $\tau_i \ll t_c$ erases welcher-Weg Information. It is also possible to obtain the same result without ultracoincidence: it is a scheme with only one source, but one can also perform an experiment with two independent sources [ŻUKOWSKI *et al.* 1995, 96–97].[6]

41.2 Aharonov/Bohm Effect

41.2.1 Theory

The Aharonov/Bohm Effect (= AB-Effect) was introduced in QM in [AHARONOV/BOHM 1959].[7] It consists in the effect of an isolated electromagnetic field on the surrounding environment. The most easy way to detect it is to let two twins particles pass to the right and left of the isolated field: they present a relative phase shift [see figure 41.6].

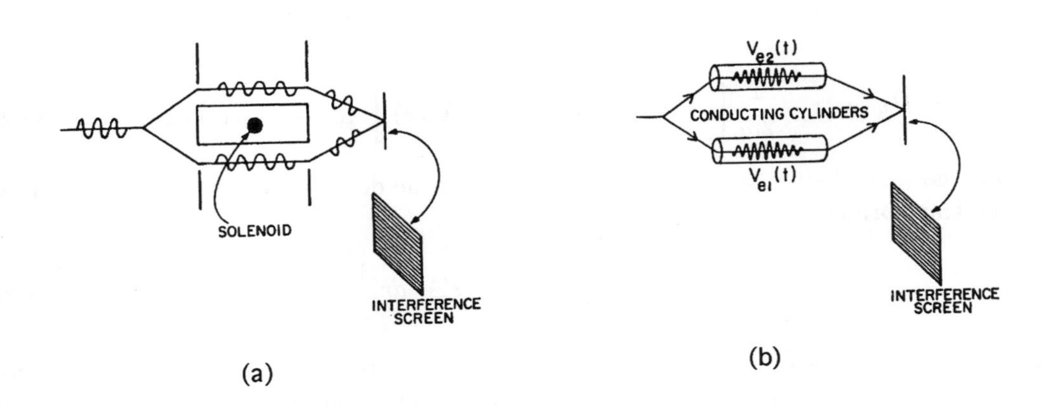

Figure 41.6: (a) Magnetic AB-effect. The axis of the solenoid is perpendicular to the page. The wave function is a split plane wave.
(b) Electric AB-effect. $V_{e_1} = V_{e_2} = 0$ except when the wave packet is shielded from the electric field — from [*PESHKIN/TONOMURA* 1989, 5].

In the classical theory there is no AB-Effect: it enters into QM by electromagnetic potentials V_e and $\mathbf{A}_e$ in the Schrödinger Eq. The information of AB-Effect is recorded in the action $\hbar S$

[6] For a generalization to multiparticle systems of entanglement swapping see [S. BOSE *et al.* 1998a].

[7] See also [*HOLLAND* 1993, 190–97].

[*PESHKIN/TONOMURA* 1989, 55]. Hence the AB-Effect can be analysed in two forms: the magnetic and the electric one.

The phase shift provided by the *magnetic* one (vector potential) [*PESHKIN/TONOMURA* 1989, 4–6, 41–42] is given by:

$$\delta\varphi_M = -\frac{\mathrm{e}}{\hbar}\int \mathbf{A}_e ds, \tag{41.18}$$

where e is the electrical charge and the integral is carried out along a closed curve connecting the two paths.

The *electric* form (scalar potential) of the AB-Effect [*PESHKIN/TONOMURA* 1989, 6–7, 40–41] consists in the following phase shift:

$$\delta\varphi_E = \frac{\mathrm{e}}{\hbar}\int [V_1(t) - V_2(t)]dt, \tag{41.19}$$

where $V_1(t), V_2(t)$ are time dependent potentials of the two sub-beams.

In the case of a magnetic field confined in a torus, which is isolated with respect to an electron, the vector potential cannot vanish because we have:

$$\int \mathbf{A}_e(\mathbf{r}) \cdot d\mathbf{r} = \phi_M, \tag{41.20}$$

where ϕ_M is the magnetic flux through the torus. In the absence of the isolated magnetic field we have:

$$\imath\hbar\frac{\partial}{\partial t}|\psi_0\rangle = \hat{H}_0\psi_0(\mathbf{r},t) = \frac{1}{2m}\left[-\imath\hbar\nabla + \frac{e}{c}\mathbf{A}_0(\mathbf{r},t))\right]^2|\psi_0\rangle - eV_0(\mathbf{r},t)|\psi_0\rangle, \tag{41.21}$$

where $V_0, \mathbf{A}_0$ are potentials due to ordinary electromagnetic field. By adding the isolated field we have:

$$\imath\hbar\frac{\partial}{\partial t}|\psi\rangle = \hat{H}\psi(\mathbf{r},t) = \frac{1}{2m}\left[-\imath\hbar\nabla + \frac{\mathrm{e}}{c}(\mathbf{A}_0(\mathbf{r},t) + \mathbf{A}_e(\mathbf{r}))\right]^2|\psi\rangle - eV_0(\mathbf{r},t)|\psi\rangle. \tag{41.22}$$

Consider for the sake of simplicity only coordinate r_x. If the domain r_x is *simply connected*, we have a gauge transformation from $\hat{H}_0$ to $\hat{H}$:

$$\hat{U}(r_x) = \exp\left[-\frac{\imath e}{\hbar c}\int_{r_x} A_e(r'_x)\cdot dr'_x\right], \tag{41.23}$$

with $\phi_M = \hat{U}\phi_M^0$ and $\hat{H} = \hat{U}\hat{H}_0\hat{U}^\dagger$. In this case we have no observable effects on the electron. But if the domain of the electron is *multiply connected*, there is no gauge transformations and an AB-Effect on the electron results [*PESHKIN/TONOMURA* 1989, 9–10]. As has already been said, the AB-Effect is due to the fact that the Hamiltonian contains not only a kinetic energy component but also a potential one. In other words, the presence of a potential suffices for producing a phase shift.

The AB-Effect does not represent a violation of the locality. If a beam of electrons is split by an impenetrable solenoid before impinging on a detector screen, one may ask whether the changed diffraction pattern receives a superluminal signal [see figure 41.7].

The Maxwell equations forbid arbitrary fluxes and they forbid vanishing electromagnetic fields outside a cylinder containing a time-dependent magnetic flux. Hence the resulting potential moves outwith with velocity c [*PESHKIN/TONOMURA* 1989, 18–19] [see figure 41.8].

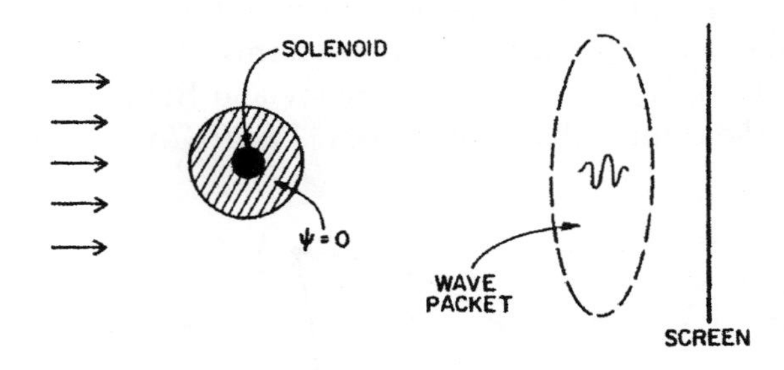

Figure 41.7: From [*PESHKIN/TONOMURA* 1989, 18].

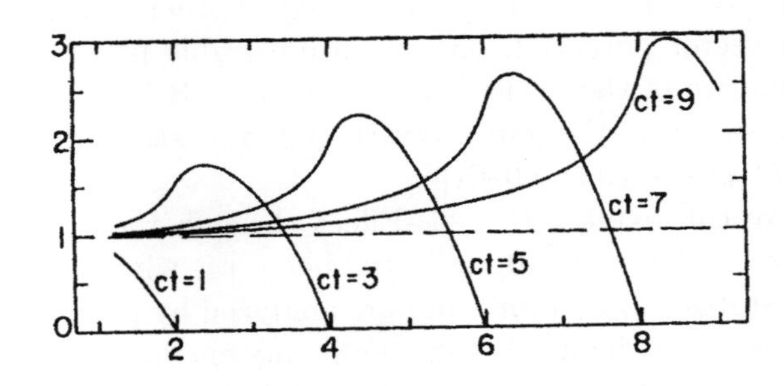

Figure 41.8: There is no violation of locality in the AB-Effect (the y–axis represents a scaled potential) — from [*PESHKIN/TONOMURA* 1989, 14].

41.2.2 Interpretation of Aharonov/Bohm-Effect

How can we interpret the AB-Effect now? Aharonov and Bohm [AHARONOV/BOHM 1959] supposed the physical reality of vector potentials and this reality was understood as more fundamental then that of the field strength. The reason for this interpretation is to be found in the theory of Quantum potential, which Bohm was developing in those years [see section 32.6.1 and subsection 32.6.2 for criticism]. In other words Bohm saw in the AB-Effect the manifestation of a potential which cannot be accounted for in classical terms. In 1964 Feynman, Leighton and Sands [*FEYNMAN et al.* 1965, II, 15] followed Aharonov and Bohm's interpretation. But there was no great agreement in the community of physicists [*PESHKIN/TONOMURA* 1989, 53–55].

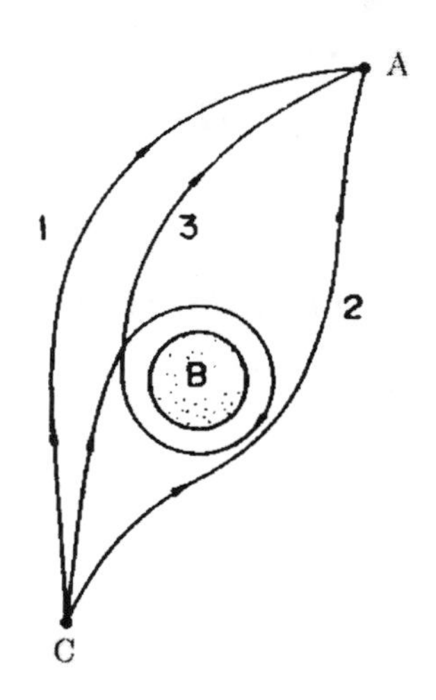

Figure 41.9: Phase factors have different values along paths 1, 2, and 3 — from [*PESHKIN/TONOMURA* 1989, 60].

Later, Wu and Yang [WU/YANG 1975] introduced another quantity, which was neither a field strength, neither a vector potential, but a non-integrable phase factor. The phase factor reflects the topological nature of the paths involved in the AB-Effect [see figure 41.9]. Electromagnetism is then understood as the gauge-invariant manifestation of a non-integrable phase factor [*PESHKIN/TONOMURA* 1989, 60–61].[8]

Following this line, we can explain the AB-Effect through the Geometric Phase formalism [see chapter 12]. Suppose [see figure 41.10] an initial beam $| i\rangle$ which is divided in two subbeams which pass intermediate states $|m\rangle$ and $|m'\rangle$ and are analyzed by polarisers: their final states $|f\rangle$ and $|f'\rangle$ will differ in phase: what here is the magnetic flux enclosed by beams, in the phase shift for polarised light is represented [see theorem 12.2: p. 205] by the solide angle of the polygon on the Poincaré sphere [BERRY 1987, 70].

41.2.3 Experiments

Among the first experiments there was Chamber's experiment, Möllenstedt/Bayh's experiment [*PESHKIN/TONOMURA* 1989, 46–50]. But they were contested by different physicists (Bocchieri and Loinger, Roy under others).

[8] On the topology see [*PESHKIN/TONOMURA* 1989, 84–88].

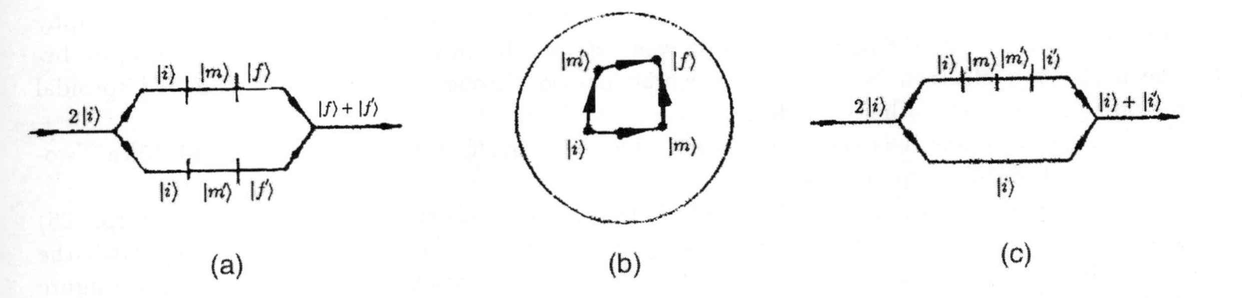

Figure 41.10: AB-effect with polarised light.
(a) Different histories of two beams.
(b) Equivalent circuit on Poincaré sphere.
(c) Alternative scheme — from [BERRY 1987, 70].

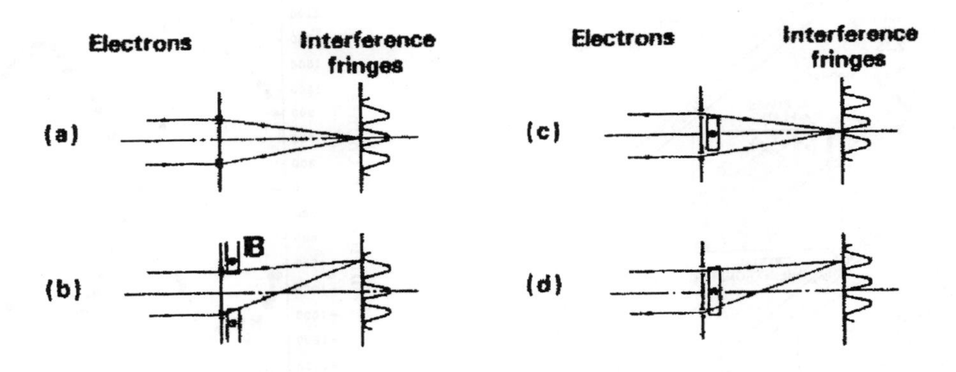

Figure 41.11: Two-Slit-Exp. 11 with electrons: (a) without and (b)–(d) with different positions of the solenoid(s) — from [*PESHKIN/TONOMURA* 1989, 44].

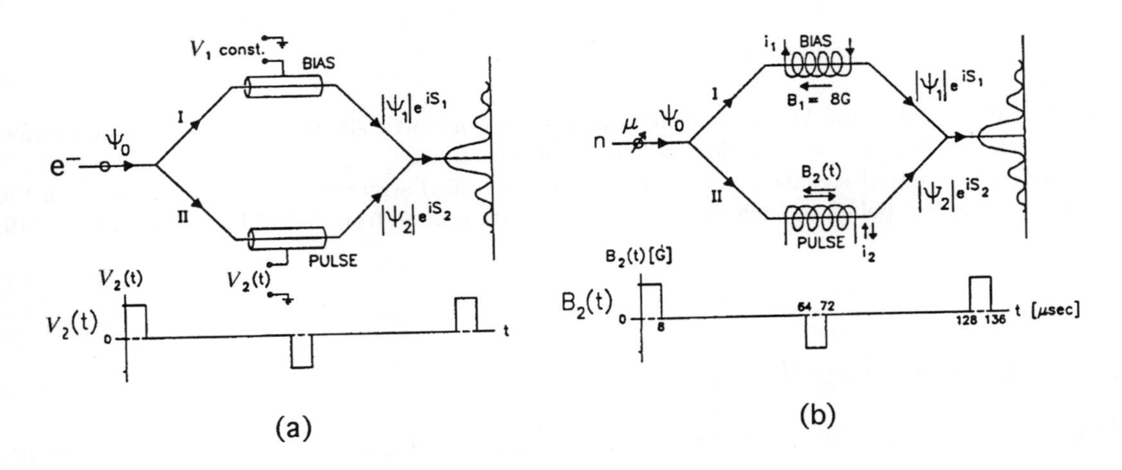

Figure 41.12: Schematic parallel between (a) the traditional scalar AB-experiment for electrons, and (b) the scalar AB-experiment for neutrons — from [ALLMAN *et al.* 1992, 2410].

In recent years the experimental evidence has been provided by the method of holography [TONOMURA *et al.* 1986]: a phase shift was detected by means of an electron holography between the two electron beams, one of which passed through the hole of a shielded toroidal ferromagnet, and the other outside[9].

On an ideal plane it is very easy to consider [*PESHKIN/TONOMURA* 1989, 44–45] a Two-Slit-Exp. (Two-Slit-Exp. 11) [see figure 41.11].

Allman and co-workers [ALLMAN *et al.* 1992] have recently obtained (Interfer. Exp. 26) a scalar effect with neutrons [see figure 41.12]:[10] The electromagnetic field interacts with the magnetic moment of the neutron and not with the electrical charge of the electron [see figure 41.13]. Our interest in the latter experiment lies in that it becomes more difficult to interpret the AB-Effect as a manifestation of a physical potential.

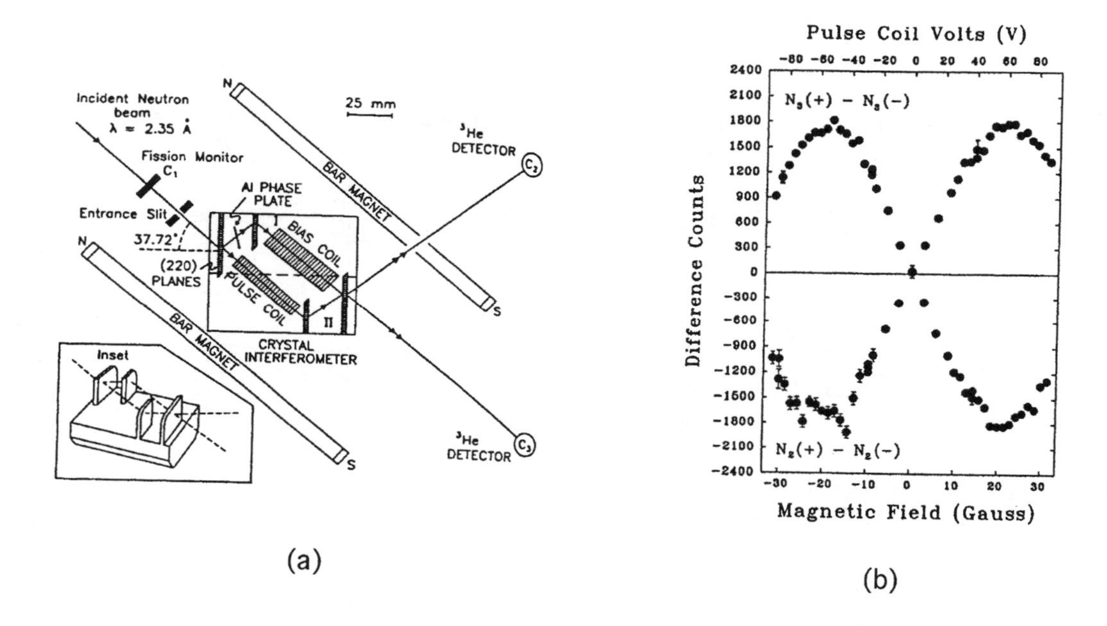

Figure 41.13: (a) Layout of AB-experiment using a skew-symmetric single-Si-crystal neutron interferometer. Inset: An isometric view of the interferometer crystal.
(b) Interferometer output signals as a function of pulse coil field strengths. This is obtained from the average of the central points in each plateau region of data sets — from [ALLMAN *et al.* 1992, 2410, 2412].

41.3 Tunnelling Time

What is Tunnelling Non-local effects arise normally in the context of multiparticle systems, or of multipath systems. But we know that non-local effects are also possible for a single particle [see subsection 41.1.1]. One may pose the problem as to whether there are non-local effects by

[9] For other recent experiments see [*PESHKIN/TONOMURA* 1989, 101–117] and [TONOMURA 1984].
[10] See also [A. KLEIN 1995] [PESHKIN 1995].

the tunnelling (an abrupt passage of a QM entity through a barrier of potential or through an obstacle of other type[11]) of a single particle. In fact, by tunnelling, the peak of the wave can move with a speed greater of the light.

To understand this phenomenon better, we distinguish [*BRILLOUIN* 1960, 1–2] between[12]:

- Phase velocity: ω/k, at which the zero-crossing of the carrier wave would move.
- Group velocity: $v_g = \partial\omega/\partial k$, at which the peak (or the valley) of a wave packet would move [see figure 41.14].
- Energy velocity: at which energy would be transported by the wave.
- Signal velocity: at which the half-maximum wave amplitude would move.
- Front velocity: at which the first appearance of a discontinuity would move.

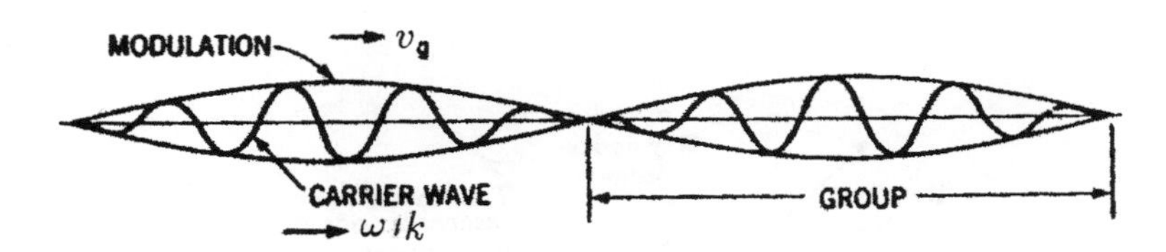

Figure 41.14: Group velocity — from [*BRILLOUIN* 1960, 2].

Classically the first and the second type may be superluminal, the third and the fourth ones are always subluminal, the fifth type is equal to the vacuum speed of light. But in QM there can be media in which also energy and signal velocity may be superluminal. In a media with inverted atomic populations the signal velocity can be equal to the group velocity and hence be superluminal [CHIAO *et al.* 1995a, 401]. But there is never violation of the locality because the front velocity remains always subluminal[13], and hence there is no information contained in the peak of an analytic wave packet which is not contained already in its forward tail. Hence only the front velocity is a genuine information velocity [CHIAO *et al.* 1995b, 272–77]. The difference between the peak velocity and that of the forward tail can be explained by the fact that, by passing a barrier, the wave packet is divided in the reflected and transmitted part. The effect arises only for the transmitted part (which has tunneled) [see figure 41.15]. Now the transmitted part is smaller than the original wave packet (the probability of a tunnelling is always < 1), and hence it has a tail which is superposed to that of the original packet while the peak is nearer to the preceding tail than before. Hence the peak has moved with a speed greater than that of the tail.

The theoretical turning point concerning tunnelling time was represented by a study of Caldeira and Leggett [CALDEIRA/LEGGETT 1981] and another study of Büttiker and Landauer [BÜTTIKER/LANDAUER 1982]. Then the theory and the experiments were developed

[11]For a detailed exposition of the following see [CHIAO/STEINBERG 1997].

[12]See also [CHIAO *et al.* 1995a].

[13]Mathematical proof in [HASS/BUSCH 1994] [JAPHA/KURIZKI 1996] [CHIAO/STEINBERG 1997, 397–98]. The last quoted article is also a very good review of the subject of this section. For other detailed studies see [HAUGE/STØVNENG 1989] [OLKHOVSKY/RECAMI 1992] [LANDAUER/MARTIN 1994] [LEAVENS 1995].

particularly in a series of studies by Enders and Nimtz [ENDERS/NIMTZ 1993] [NIMTZ *et al.* 1994]: see figures 41.16–41.18.[14]

Some mathematics Let us now treat of a one-dimensional case [STEINBERG/CHIAO 1994, 3287–89] . Let a barrier of potential extend from $x = 0$ to x_d [see figure 41.19]. The incident part of the electron wave function is:

$$|\psi_{k_x}(x < 0)\rangle = e^{\imath k_x x - \imath Et/\hbar}. \tag{41.24}$$

The field at x_d is $\vartheta_T(k_x)e^{\imath k_x x_d - \imath Et/\hbar}$, where $\vartheta_T(k_x)$ is the transmission amplitude and the energy is $E = (\hbar k_x)^2/2m$. The total phase at x_d is

$$\begin{aligned} \varphi_T &= \arg\vartheta_T(k_x) + k_x x_d - \frac{Et}{\hbar} \\ &:= \varphi_d(k_x) - \frac{Et}{\hbar}. \end{aligned} \tag{41.25}$$

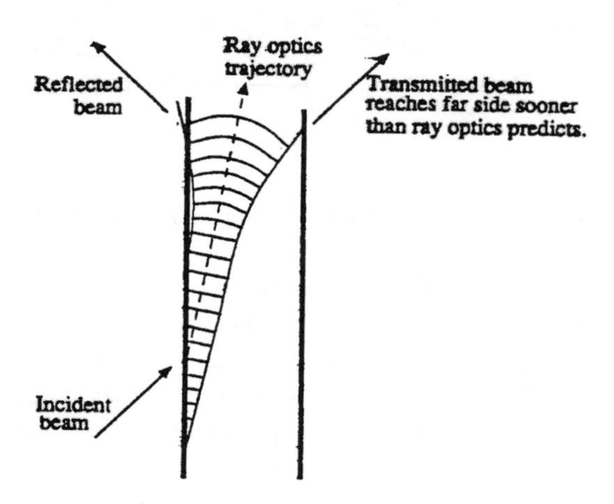

Figure 41.15: A simplified picture of the wave nature of transmission through an air gap. When light is incident just below the critical angle, the ray inside the barrier region is nearly parallel to the interfaces; at the critical angle, the ray optics traversal time thus diverges. The wave-mechanical group delay, on the other hand, remains finite: the transversely bounded beam undergoes diffraction, and begins to couple out of the barrier sooner than the predicted ray optics. Depending on the various length scales (i.e. how much diffraction occurs), the transmitted portion may preferentially consist of wave vectors closer to the normal, or of those parts of the beam which originate closer to the RHS and thus reached the barrier earliest. The former case simply corresponds to the preferential transmission of those components which were incident furthest from the critical angle, while the latter may offer some insight into anomalously small group delay — from [STEINBERG/CHIAO 1994, 3286].

Consider an electron wave packet:

$$|\Psi(x)\rangle = \int c(k_x)|\psi_{k_x}(x)\rangle dk_x, \tag{41.26}$$

[14] See also [RANFAGNI *et al.* 1993].

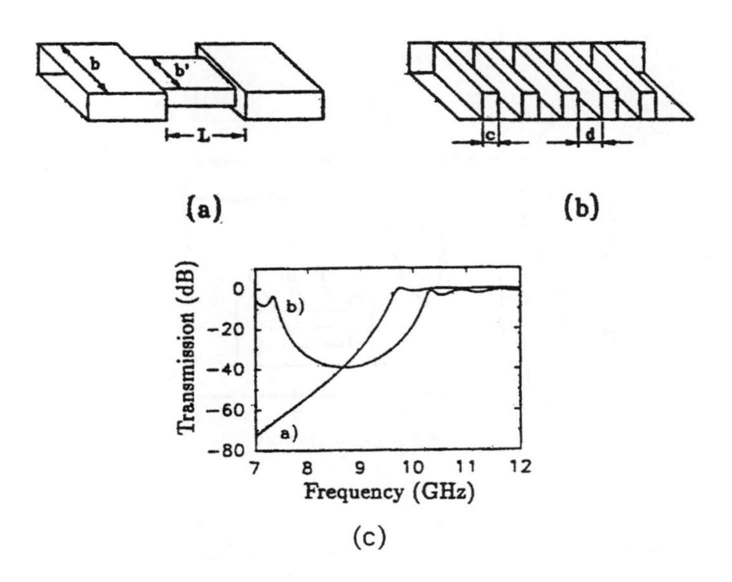

Figure 41.16: (a)–(b) Two investigated barrier types.
(a) Rectangular waveguide, the center part of the wave guide is operated below, the two adjacent waveguides are operated above the cut-off frequency. The longest transverse extension of the guide b or b' of such a rectangular waveguide determines the cut-off frequency, i.e. the transmission from normal wave propagation to tunnelling.
(b) Rectangular waveguide periodically loaded with dielectric quarter wave-length layers.
(c) Transmission of the two barriers (a) and (b) as a function of the frequency — from [NIMTZ *et al.* 1994, 566–67].

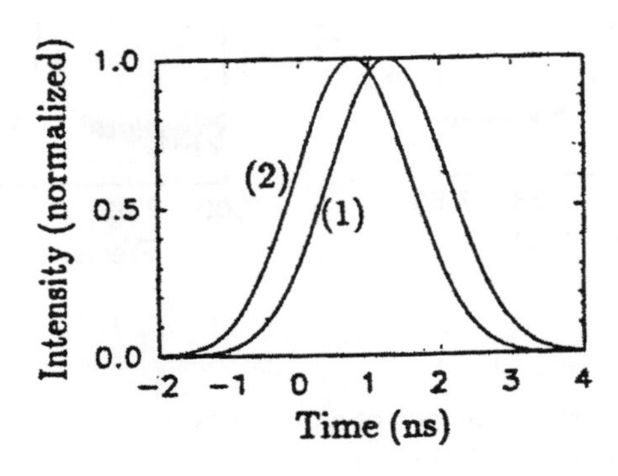

Figure 41.17: Measured propagation time of a Gaussian wave-packet (1) through an empty waveguide of 249.5 mm length [figure 41.16a] and (2) through the same guide with seven plexiglas layers each 6 mm thick and separated by 12 mm air-layers [figure 41.16.(b)], the refractive index of the plexiglas has been 1.6 — from [NIMTZ *et al.* 1994, 567].

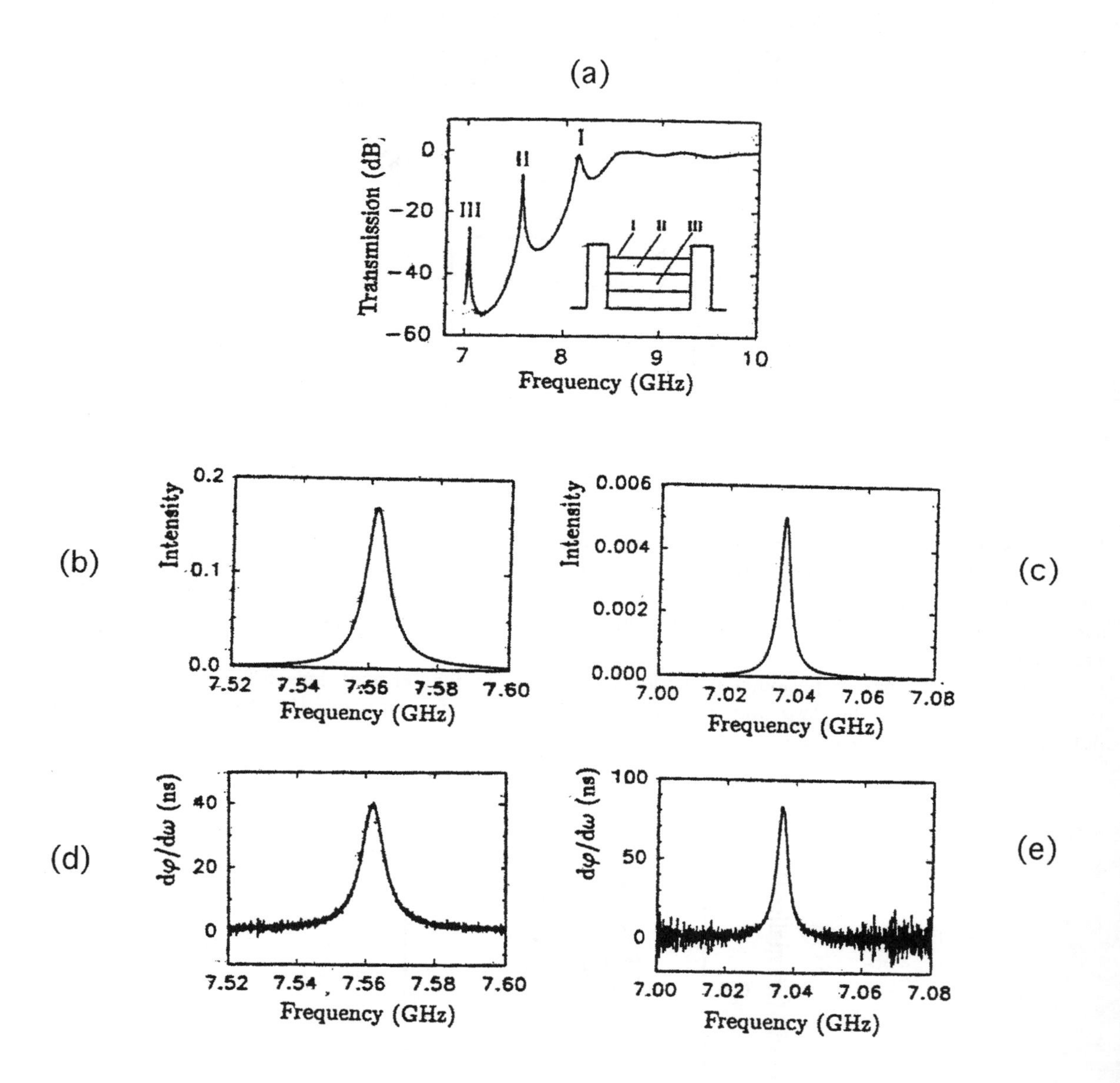

Figure 41.18: (a) Transmission *versus* frequency of a double-barrier structure with three resonant transition lines below the barriers' cut-off frequency (insert).
(b)–(c) Intensity *versus* frequency of the resonant lines.
(d)–(e) The phase time $\partial\varphi/\partial\omega$ of lines II and III of (a) as a function of frequency — from [NIMTZ *et al.* 1994, 568].

with coefficients $c(k_x)$ all real. For c sufficiently smooth and narrow band, we may use the stationary-phase approximation. The incident wave packet is peaked at $x = 0$ at $t = 0$, since the phase of each component vanishes there (we do not consider here the complications due to interference between incident and reflected waves). By the same reasoning the transmitted wave packet arrives at x_d at a time $t = \tau_e$ such that:

$$0 = \frac{\partial \varphi_T}{\partial E} = \frac{\partial \varphi_d(k_x)}{\partial E} - \frac{\tau_e}{\hbar}, \tag{41.27}$$

implying

$$\tau_e = \hbar \frac{\partial \varphi_d(k_x)}{\partial E}. \tag{41.28}$$

For further experimental evidence see also [STEINBERG *et al.* 1993].

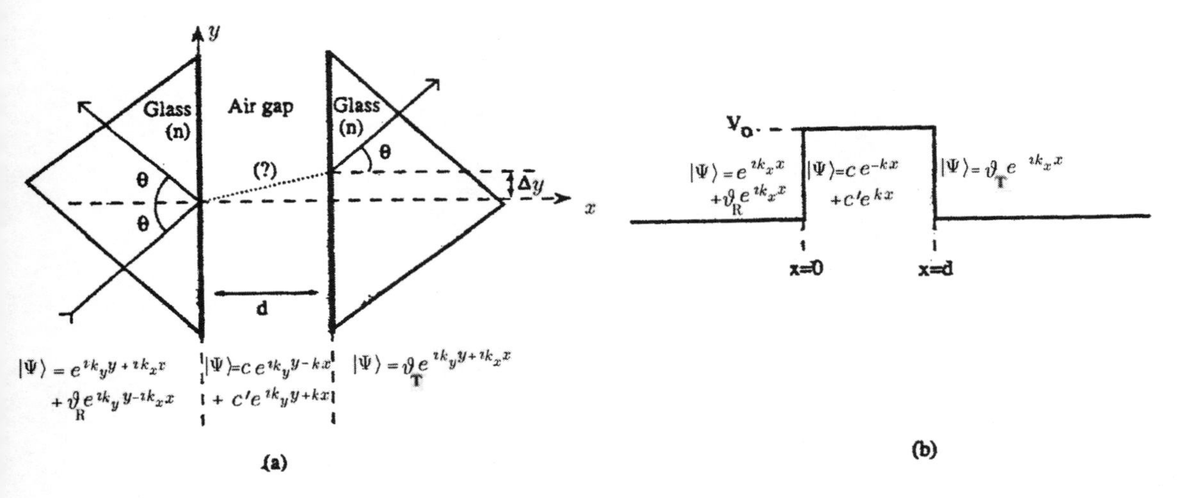

Figure 41.19: (a) Two dimensional tunnelling through an air gap when $\theta > \theta_c = \sin^{-1}(1/n)$. (b) One-dimensional tunnelling for $E < V_0$. If k_x and κ are arranged to be the same as in (a), then the transmission and reflection coefficients T and R will be the same in the two problems — from [STEINBERG/CHIAO 1994, 3284].

Conclusions In tunnelling Exp.s a more general phenomenon is evident: QM allows values which are not allowed by classical mechanics. Using the Fourier transform of the Feynman averaged characteristic function for the classical traversal time functional, negative probabilities in the distributions for the dwell, reflection and traversal time can result [SOKOLOVSKI/BASKIN 1987] [SOKOLOVSKI/CONNOR 1991] [FERTIG 1990].

Sokolovski and Connor [SOKOLOVSKI/CONNOR 1990] pointed out that no general method of solving the tunnelling time by using wave-packet simulation can be found because based on an the arbitrary generalization of the assumption of well-localized transmitted wave-packets. They developed an alternative method based on path integrals [see subsection 3.11.2]. Finally, their result is that it does not supply a universal time parameter, analogous to a classical time; therefore, the hope of finding well-defined real tunnelling time should be erroneous[15]

[15]Concerning this point see also [DELGADO/MUGA 1996] and section 10.2.

By using a pointer in order to detect how much time the particle spends in barrier before being transmitted, one can interpret the real part for transmission time as representing a shift in the pointer position, whereas the imaginary part of transmission time could be interpreted as a shift in the momentum of the pointer. Then the problem could be overcome by using the weak value formalism [STEINBERG 1995] [see section 18.8].[16] negative.

[16]See also [D. ROHRLICH *et al.* 1995] [AHARONOV *et al.* 1988] [AHARONOV *et al.* 1990] [AHARONOV *et al.* 1993b].

Part X

INFORMATION AND QUANTUM MECHANICS

Introduction to Part X The *tenth part* is devoted to the most recent QM domain — and surely one of the most promising ones: Quantum information.

- *Chapter 42* is centered on the relation between information and entropy. As we shall see, many themes and areas which have already been discussed (particularly problems connected with Measurement) can be treated in a new form in this context.
- *Chapter 43* is devoted to quantum cryptography and teleportation, another very interesting and recent area.
- *Chapter 44* deals with an entirely new method of computation: quantum computation, based on the quantum-mechanical principles and aiming to overcome the limitations of classical computers.

Chapter 42

INFORMATION AND ENTROPY IN QUANTUM MECHANICS

Introduction and contents Entropy, a concept born in the context of thermodynamics, and successively generalized in terms of information theory, is very useful, for example, when we have to handle problems of: intrinsic uncertainty of systems; Measurement (which is an acquiring of information); mutual information or relationships between observables (pertaining to different subsystems).

Section 42.1 is devoted to a formal and general analysis of quantum entropy. Then the relationship between uncertainty and entropy is discussed [section 42.2]. The problem of Measurement and of open systems is discussed again by using the concept of entropy [section 42.3]. Section 42.4 is devoted to the determination of different bounds for quantum information and entropy. In section 42.5 the problem of the reproducibility of information is discussed. Then special topics are again discussed in terms of entropy. Section 42.6 is centered on the relationship between which-path and information. Finally Bell inequalities [section 42.7] and the determination of a measure of entanglement [section 42.8] are examined in this context.

42.1 General Features of Quantum Entropy

42.1.1 Von Neumann Entropy

The entropy of a state describing a physical system $\mathcal{S}$ is a quantity expressing the randomness of $\mathcal{S}$. It was introduced in 1877 by Boltzmann and can be defined as follows [see Eq. (2.4)]:

$$H_{\mathrm{B}} := k_{\mathrm{B}} \ln w_E. \tag{42.1}$$

Shannon [SHANNON 1948] regarded this randomness as the amount of information carried by the state of the system. In fact if $\mathcal{S}$ has a large uncertainty and one receives information on $\mathcal{S}$, then the so-obtained information is more valuable than that received about a system having less uncertainty [*OHYA/PETZ* 1993, 6, 15]. Therefore the Shannon entropy can be defined as follows:

$$H_{\mathrm{S}} = -\sum \wp_k \ln \wp_k, \tag{42.2}$$

which is a general formula of wide applicability.

However, the concept of entropy and that of information must be kept carefully distinct: in fact, a gain in information would signify a net decrease of entropy (or of the uncertainty of the system). Hence the information I corresponds rather to a negative form of entropy, the *negentropy* — a measure of the quality of the energy (of its availability for work) [*BRILLOUIN* 1962, 116–17, 152–53] [*ZEH* 1989, 58]. Hence in the context of entropy it is better to speak of *potential information.*

Now we deduce Eq. (14.3) from the definition of the Shannon entropy [VON NEUMANN 1927c] [*VON NEUMANN* 1932, 202] [*VON NEUMANN* 1955, 3789–81].[1] First we define an orthonormal basis $\{|\psi_k\rangle\}$ of eigenvectors of the density operator $\hat{\rho}$ for a system $\mathcal{S}$ such that

$$\hat{\rho}|\psi_n\rangle = w_n|\psi_n\rangle, \tag{42.3}$$

where the weights w_n are the eigenvalues of $\hat{\rho}$. In the case of a premeasurement, the optimal test which minimizes the entropy is exactly the one which diagonalizes the density matrix $\hat{\rho}$ that expresses the state of $\mathcal{S}$. Hence Eq. (42.2) can be written for a QM state $\hat{\rho}$ as follows [*OHYA/PETZ* 1993, 4–5] :

$$H_{\mathrm{VN}}(\hat{\rho}) = -\sum_j w_j \ln w_j = -\mathrm{Tr}(\hat{\rho}\ln\hat{\rho}), \tag{42.4}$$

which is Eq. (14.3). In fact the density matrix can be seen as the operator which carries maximal information about the state of the system [BLANKENBECLER/PARTOVI 1985, 373–74] — though, as we know [section 28.5] and shall see again below [subsection 42.4.1], this maximal information cannot be extracted by a measurement on a single system. H_{VN} is referred to as the *von Neumann entropy.* As it is clear from the mathematical notation, the entropy is a function of the state [WEHRL 1978, 222]. But note, as it is already clear from the deduction of the formula, that the last expression is formally analogous to the Shannon entropy only if we can perform a decomposition $\hat{\rho} = \sum_j w_j \hat{P}_j$, where $\hat{P}_j = |j\rangle\langle j|$, i.e. only if the state vectors $|j\rangle$'s are orthogonal to each other and hence eigenstates of $\hat{\rho}$ [B. SCHUMACHER 1995, 2738]: normally we have $H_{\mathrm{VN}} < H_{\mathrm{S}}$.

42.1.2 Properties of the von Neumann Entropy

The properties of the von Neumann entropy are the following [*LINDBLAD* 1983, 20–21] [WEHRL 1978, 236–37, 242]:

- *Non-negativity*:
$$H_{\mathrm{VN}}(\hat{\rho}) \geq 0, \tag{42.5}$$
where the equality holds iff $\hat{\rho}$ is a pure state. H_{VN} expresses the degree of mixing [see p. 42]; hence pure states are states of evanescent entropy, i.e. highly ordered states [*OHYA/PETZ* 1993, 5] . Araki and Lieb [ARAKI/LIEB 1970] proved that the entropy of a compound system $\hat{\rho}^{1+2}$, with subsystems $\hat{\rho}^1, \hat{\rho}^2$, is subjected to following inequality: $|H^1_{\mathrm{VN}} - H^2_{\mathrm{VN}}| \leq H^{1+2}_{\mathrm{VN}} \leq H^1_{\mathrm{VN}} + H^2_{\mathrm{VN}}$, which shows that, if the system $\hat{\rho}^{1+2}$ is a pure state, then the two subsystems have equal entropies.

- *Unitary invariance.* Since unitary transformations do not change the spectrum of any operator, we have [see Eq. (3.72) and Eq. (14.2)]:
$$H_{\mathrm{VN}}(\hat{\rho}) = H_{\mathrm{VN}}(\hat{U}_t\hat{\rho}\hat{U}_t^\dagger). \tag{42.6}$$

[1] See also [*PERES* 1993, 263–64].

- *Concavity*: for all $\hat{\rho}_j \in \mathcal{H}^{\mathrm{Tr}=1;\geq 0}, w_j \geq 0; \sum_j w_j = 1$:

$$\sum_j w_j H_{\mathrm{VN}}(\hat{\rho}_j) \leq H_{\mathrm{VN}}(\sum_j w_j \hat{\rho}_j). \tag{42.7}$$

The concavity means that a mixture has always more entropy than the sum of the corresponding pure states [see also theorem 14.2: p. 222].

- *Subadditivity*: if we have $\hat{\rho} \in \mathcal{H}^{\mathrm{Tr}=1;\geq 0}$, where $\mathcal{H} = \mathcal{H}_1 \otimes \mathcal{H}_2$ and reduced states: $\hat{\rho}_1 = \mathrm{Tr}_2\hat{\rho} \in \mathcal{H}_1^{\mathrm{Tr}=1;\geq 0}, \hat{\rho}_2 = \mathrm{Tr}_1\hat{\rho} \in \mathcal{H}_2^{\mathrm{Tr}=1;\geq 0}$, then:

$$H_{\mathrm{VN}}(\hat{\rho}) \leq H_{\mathrm{VN}}(\hat{\rho}_1 \otimes \hat{\rho}_2) = H_{\mathrm{VN}}(\hat{\rho}_1) + H_{\mathrm{VN}}(\hat{\rho}_2), \tag{42.8}$$

where equality holds iff $\hat{\rho} = \hat{\rho}_1 \otimes \hat{\rho}_2$ [LIEB/RUSKAI 1973a] [LIEB/RUSKAI 1973b].[2]

Conditions (42.6) and (42.8), together with a very weak continuity condition, suffice to define completely the function $H_{\mathrm{VN}}(\hat{\rho})$ [OCHS 1975].

42.2 Uncertainty and Entropy

It is possible to express UP in terms of the von Neumann entropy. In the following we consider several formulations which partly give answers to different problems.

42.2.1 Deutsch Entropic Uncertainty

Following considerations similar to those developed earlier [see subsection 7.5.2] about the 'variance' formulations of UP[3], Deutsch [DEUTSCH 1983] [4] developed an entropic formulation of UP for arbitrary (discrete) observables $\hat{O}_1, \hat{O}_2$ and arbitrary state $|\psi\rangle$:

$$H_{\hat{O}_1}(|\psi\rangle) + H_{\hat{O}_2}(|\psi\rangle) \geq 2 \ln \frac{2}{1 + \mathrm{Max}\{|\langle o_1|o_2\rangle|\}}, \tag{42.9}$$

where $H_{\hat{O}_j}(|\psi\rangle), j = 1, 2$ means the von Neumann entropy (the subscript has been dropped for the sake of simplicity) associated with the observable $\hat{O}_j$ in the state $|\psi\rangle$, and $\{\langle o_1|o_2\rangle\}$ means the set of inner products between eigenstates of the two observables.

Note that expression (42.9) has the same mathematical structure as the Statistical URs [see for example Eq. (7.28)].

42.2.2 Kraus Entropic Uncertainty

The Entropic UR can be improved [KRAUS 1987] to:

$$H_{\hat{O}_1}(|\psi\rangle) + H_{\hat{O}_2}(|\psi\rangle) \geq -2 \ln \mathrm{Max}\{|\langle o_1|o_2\rangle|\}. \tag{42.10}$$

[2] See also [*OHYA/PETZ* 1993, 23–24] .

[3] Remember that the problem is that the RHS of Eq. (7.28) is not a fixed lower bound but depends on the state $|\psi\rangle$.

[4] Antecedents can be found in [EVERETT 1973, 51–52] [MAMOJKA 1974] [BIALYNICKI-B./MYCIELSKI 1975]. See also [*BUSCH et al.* 1995a, 139–43].

In both formulations (42.9) and (42.10) the RHSs are independent from the state $|\psi\rangle$. It is important because it makes UP more similar to the CCRs. Hence both formulations yield nontrivial information on the probability distributions for $\hat{O}_1, \hat{O}_2$ as long as $\text{Max}\{|\langle o_1|o_2\rangle|\} < 1$, that is when $\hat{O}_1, \hat{O}_2$ do not share any common eigenvector (this means that they do not commute). However, both formulations present the problem that they are inappropriate for continuous observables.

42.2.3 Maassen/Uffink Entropic Uncertainty

A generalization of such inequalities is possible [MAASSEN/UFFINK 1988] so that we can partly solve the previous difficulty. If $\wp$ is an arbitrary probability distribution over a set of N possible outcomes, the expression

$$M_r(\wp) = \left[\sum_j (\wp)^{1+r}\right]^{\frac{1}{r}}, \tag{42.11}$$

where $-1 < r < 0$ or $0 < r$, has the following properties:

- it is invariant under a relabeling of the set of possible outcomes;
- it is convex in $\wp$;
- if we have two probability distributions $\wp^{(1)}(\wp_1^{(1)}, \wp_2^{(1)}, \ldots, \wp_N^{(1)})$ and $\wp^{(2)}(\wp_1^{(2)}, \wp_2^{(2)}, \ldots, \wp_M^{(2)})$, and $\wp_{jk}^{(12)} = \wp_j^{(1)} \wp_k^{(2)}$, then $M_r(\wp^{(12)}) = M_r(\wp^{(1)}) M_r(\wp^{(2)})$;
- $M_r(\wp)$ is a continuous non-decreasing function of r for $-1 \leq r \leq \infty$ provided one defines the following limiting values:

  i) $M_0(\wp) = e^{-H(\wp)}$,

  ii) $M_{-1}(\wp) = \frac{1}{N'}$, where N' denotes the number of possible outcomes with a probability > 0,

  iii) $M_\infty(\wp) = \text{Max}_j \wp_j$.

Hence $M_r(\wp)$ can be seen as a measure of the average peakedness of $\wp$. For arbitrary operators $\hat{O}_1, \hat{O}_2$ the following inequality can be derived:

$$\boxed{M_r(\wp(\hat{O}_1)) M_s(\wp(\hat{O}_2)) \leq (\text{Max}\{|\langle o_1|o_2\rangle|\})^2} \tag{42.12}$$

for $s \geq 0, r = -s/(2s+1)$. Interchange of M_r, M_s gives rise to no difference.

Inequality (42.12) is better than inequality (42.9) or (42.10) for four reasons:

- it gives better lower bounds of entropy;
- it can be used for mixtures;
- for a reasonable continuity requirement — the measure of uncertainty of $\wp$ should approach its minimal value whenever the distribution $\wp$ approaches a δ distribution, where the total probability is concentrated in a single point — it can be shown that the von Neumann entropy cannot fulfill it, whereas $-\ln M_r(\wp)$ does it for $r > 0$;
- in the case of unbounded discrete spectrum, on the contrary to preceding inequalities, Eq. (42.12) is valid if at least one observable satisfies the continuity requirement.

42.2.4 POVM Entropic Uncertainty

A different approach is centered on state-dependent formulations on the ground that non-commuting observables which have trivial maxima (= 0) can also have very useful state-dependent maxima [BUSCH/LAHTI 1987]. In particular a stochastic (based on POVM) entropic UR can be derived [BUSCH/LAHTI 1987, 904–905].

The most important contribution is the idea of a maximal joint knowledge or maximal joint information of two non-commuting effects [BUSCH/LAHTI 1987, 900–901] [BUSCH *et al.* 1987, 2868–69], which, for two arbitrary effects $\hat{E}, \hat{E}'$ in the state $|\psi\rangle$, is proved to satisfy the uncertainty relation:

$$\boxed{(\Delta\hat{E})^2_\psi \cdot (\Delta\hat{E}')^2_\psi \leq \left[\Delta(\hat{E},\hat{E}')\right]^2_\psi} \tag{42.13}$$

where we define the *covariance* of two observables $\hat{O}_1, \hat{O}_2$ as follows:

$$\left[\Delta(\hat{O}_1,\hat{O}_2)\right]^2 = \left\langle \left(\hat{O}_1 - \left\langle \hat{O}_1 \right\rangle\right)\left(\hat{O}_2 - \left\langle \hat{O}_2 \right\rangle\right)\right\rangle, \tag{42.14}$$

which is formally identical with Eq. (3.110) for $\hat{O}_1 = \hat{O}_2$.

Since by a maximal information the variances have to be finite [BUSCH *et al.* 1987, 2871] — and the POVMs are informationally complete [see theorem 18.9: p. 304] —, this shows that the concept of variance, which in general has been rejected for the interpretation of URs [see subsection 7.5.2], is useful here.

Following the above proposals, other inequalities have been deduced [STEANE 1996a] which concern the minimum distance between code 'words' in different basis of a QM system, very useful for error correction in QM information [see subsection 44.5.2].

42.2.5 Uncertainty and Measurement

An application of Deutsch's formula (42.9) to the problem of Measurement is due to Partovi [PARTOVI 1983].[5] The problem, as we have seen, is that Deutsch's proposal neither accounts for continuous spectrum nor for discrete bounded spectrum. Including a measuring device $\mathcal{A}$, the probability for obtaining the value o_j (in the subset) for observable $\hat{O}$ in state $|\psi\rangle$ is given by:

$$\wp_j^{\hat{O}}(\psi|\mathcal{A}^{\hat{O}}) = \frac{\langle\psi|\hat{P}_j^{\hat{O}}|\psi\rangle}{\langle\psi|\psi\rangle}. \tag{42.15}$$

The entropy associated to such a measurement is:

$$H(\psi|\mathcal{A}^{\hat{O}}) = -\sum_j \wp_j^{\hat{O}}(\psi|\mathcal{A}^{\hat{O}})\ln[\wp_j^{\hat{O}}(\psi|\mathcal{A}^{\hat{O}})]; \tag{42.16}$$

and the uncertainty for the measurement of two observables $\hat{O}, \hat{O}'$ is the sum of the corresponding entropies: $H(\psi|\mathcal{A}^{\hat{O}}) + H(\psi|\mathcal{A}^{\hat{O}'})$, which in turn is equal to $-\sum_{j,k}\wp_j^{\hat{O}}\wp_k^{\hat{O}'}\ln(\wp_j^{\hat{O}}\wp_k^{\hat{O}'})$. But, for any $|\psi\rangle$ such that $\langle\psi|\psi\rangle = 1$, we have:

$$\begin{aligned} \parallel \hat{P}_j^{\hat{O}} + \hat{P}_k^{\hat{O}'} \parallel^2 &\geq \langle\psi|(\hat{P}_j^{\hat{O}} + \hat{P}_k^{\hat{O}'})^2|\psi\rangle \\ &\geq \langle\psi|(\hat{P}_j^{\hat{O}} + \hat{P}_k^{\hat{O}'})|\psi\rangle^2 \\ &= \langle\psi|\hat{P}_j^{\hat{O}}|\psi\rangle^2 + \langle\psi|\hat{P}_k^{\hat{O}'}|\psi\rangle^2 + 2\langle\psi|\hat{P}_j^{\hat{O}}|\psi\rangle\langle\psi|\hat{P}_k^{\hat{O}'}|\psi\rangle; \end{aligned} \tag{42.17}$$

[5] But Partovi thought that the uncertainty in QM is dependent upon perturbation due to a measurement, whereas, as we know [section 7.3], the URs are basic relations of the quantum formalism itself.

which implies that

$$\wp_j^{\hat{O}} \wp_k^{\hat{O}'} \leq \frac{1}{4} \parallel \hat{P}_j^{\hat{O}} + \hat{P}_k^{\hat{O}'} \parallel^2 . \tag{42.18}$$

Hence the uncertainty $\Delta(\hat{O}, \hat{O}')_M^{\psi}$ in the measurement of both observables in the state $|\psi\rangle$ is:

$$\Delta(\hat{O}, \hat{O}')_M^{\psi} \geq -\sum_{j,k} \wp_j^{\hat{O}} \wp_k^{\hat{O}'} \ln \left(\frac{\parallel \hat{P}_j^{\hat{O}} + \hat{P}_k^{\hat{O}'} \parallel^2}{4} \right) \geq 2 \ln \left(\frac{2}{\mathrm{Max}_{j,k} \parallel \hat{P}_j^{\hat{O}} + \hat{P}_k^{\hat{O}'} \parallel} \right) . \tag{42.19}$$

Since $\hat{P}_j^{\hat{O}}, \hat{P}_k^{\hat{O}'}$ are projectors, it follows that $1 \leq \parallel \hat{P}_j^{\hat{O}} + \hat{P}_k^{\hat{O}'} \parallel \leq 2$, where the upper bound is attained iff the two projectors share at least one eigenvector — which means that they commute[6].

42.3 Quantum Entropy for Open Systems

As we know [see section 14.1], Eq. (42.6) makes it impossible to use the von Neumann entropy for problems of irreversibility, which characterizes open systems in general and measurements in particular [*LINDBLAD* 1983, 27]: the von Neumann entropy is mixing preserving while a measurement augments the mixing and hence the entropy of the object system [*SCHROECK* 1996, 63, 65–66] [ZUREK 1991, 39]. In this section we aim to prove theorem 14.2 [p. 222], which was previously only stated.

42.3.1 Relative Entropy

We now introduce the concept of *relative entropy*, i.e. the entropy of a state $|\varsigma\rangle$ relative to another state $|\psi\rangle$ [LINDBLAD 1973] [UHLMANN 1977]:[7]

$$H(\varsigma||\psi) = \mathrm{Tr}[\hat{\rho}_\varsigma(\ln \hat{\rho}_\varsigma - \ln \hat{\rho}_\psi)], \tag{42.20}$$

provided that $\hat{\rho}_\psi, \hat{\rho}_\varsigma$ have positive eigenvalues. The relative entropy has properties (42.5) and (42.6), and is characterized by a third property, given by the following theorem:

Theorem 42.1 (Lindblad) *For any completely positive active map* $\mathcal{T}$ [see subsection 18.1.1] *and for all* $\hat{\rho}, \hat{\rho}' \in \mathcal{H}_{\mathrm{Tr}=1;\geq 0}$, *we have:*

$$H(\mathcal{T}\hat{\rho}||\mathcal{T}\hat{\rho}') \leq H(\hat{\rho}||\hat{\rho}'). \tag{42.21}$$

If, after a measurement, we obtain probability distributions μ_ψ, μ_ς, we have:

$$H(\mu_\psi||\mu_\varsigma) \leq H(\psi||\varsigma), \tag{42.22}$$

[6] Further developments for the statistics of premeasurement can be found in [BLANKENBECLER/PARTOVI 1985] [PARTOVI/BLANKENBECLER 1986a] [PARTOVI/BLANKENBECLER 1986b]. The limits of such work are due to the fact that, despite its generality, it can only account for a statistical description of the Measurement.

[7] See also [*LINDBLAD* 1983, 21–22] [WEHRL 1978, 250–52] [BENOIST *et al.* 1979] [DONALD 1985] [HIAI/PETZ 1991] [*PERES* 1993, 254–65].

where in the case of equality we speak of *sufficient measurement*. Sufficient measurement do not exist for non-commuting states, which is another form of quantum uncertainty. Eq. (42.21) is a version of the strong subadditivity property [Eq. (42.8)]. The larger $H(\varsigma||\psi)$, the more information for discriminating between the two states can be obtained from an observation. Hence the theorem expresses also a general property of loss of information in the sense that an evolution obeying a certain form of Markovian equation makes the two states less distinguishable [*LINDBLAD* 1983, 22–23] [see below subsection 42.3.4]. Loss of relative entropy is a form of uncertainty during measurement [*OHYA/PETZ* 1993, 8–9, 15–16]. We can define the von Neumann entropy by means of the relative one:

$$H_{\mathrm{VN}}(\psi) = \sup\{\sum_j c_j H(b_j||\psi) : |\psi\rangle = \sum_j c_j|b_j\rangle\}. \tag{42.23}$$

The relative entropy represents only a part of the entropy change which we need in order to describe a measurement, because, on the one hand, it is convex and not concave (as is the von Neumann entropy) [*OHYA/PETZ* 1993, 20], but, on the other, it does not account for any entropy increase.

42.3.2 Increasing of Entropy

Formalism An increase of entropy can be obtained by discarding the irrelevant potential information [*LINDBLAD* 1983, 27–28, 44–46, 59]: it consists exactly in the correlations which exist between subsystems of the observed system or between the observed system and the apparatus. The useful potential information H_I, instead, is defined as the sum of the partial entropies of the subsystems (for example of an object system and an apparatus).

In order to obtain such a result, we presume there is an initial total state so that the two systems are uncorrelated (a product state). The initial entropy is then equal to the sum of the entropies of the systems, and this entropy will not change during measurement. But, during measurement, the two systems may become correlated and now we can discard the correlation and obtain a final useful piece of potential information (sum of the final entropies of the two systems) which is greater then the sum of the initial partial entropies.

Formally, for an initial state $\hat{\rho}_i = \hat{\rho}^1 \otimes \hat{\rho}^2$ and a final state $\hat{\rho}_f$, we have:

$$\boxed{H(\hat{\rho}_i) = H(\hat{\rho}_i^1) + H(\hat{\rho}_i^2) = H(\hat{\rho}_f) \leq H(\hat{\rho}_f^1) + H(\hat{\rho}_f^2) = H_I(\hat{\rho}_f),} \tag{42.24}$$

if we can discard the correlation measured by [*ZEH* 1989, 88–89]:

$$\mathrm{C}^{12}(\hat{\rho}_f) := H(\hat{\rho}_f^1 \otimes \hat{\rho}_f^2) - H(\hat{\rho}_f) \geq 0. \tag{42.25}$$

What is presupposed by the whole process is that the two subsystems are open systems so that the correlation can be discarded in a larger system. Hence, as we know, we have to introduce the concept of environment[8].

[8] An interesting and partly related question is the amount of average entropy of a subsystem $\mathcal{S}_1$ of dimension $m \leq n$ which is part of a system $\mathcal{S}_{12}$ of dimension mn. Page [PAGE 1993b] conjectured that it is:

$$\langle H_1\rangle = \left(\sum_{k=n+1}^{mn} \frac{1}{k}\right) - \frac{m-1}{2n}, \tag{42.26}$$

where $H_1 = -\mathrm{Tr}\hat{\rho}_1 \ln \hat{\rho}_1$ and $\hat{\rho}_1$ is the reduced density matrix obtained by tracing out the other subsystem. Page

A model We can study a very easy model to better understand the meaning of Eq. (42.24) [PARTOVI 1989b, 447–48]. Take the initial density matrix $\hat{\rho}^{\mathcal{SAE}}(t_0)$, with $\hat{\rho}^{\mathcal{SAE}}(t_0) = \hat{\rho}^{\mathcal{SA}}(t_0)\hat{\rho}^{\mathcal{E}}(t_0)$, evolving to $\hat{\rho}^{\mathcal{SAE}}(t)$. For the strong subadditivity[9] we have:

$$H_{\mathcal{SAE}}(t) + H_{\mathcal{A}}(t) \leq H_{\mathcal{SA}}(t) + H_{\mathcal{AE}}(t). \tag{42.28}$$

Since the evolution of the density matrix of the total system is unitary, it follows that $H_{\mathcal{SAE}}(t_0) = H_{\mathcal{SAE}}(t)$. Due to the initial state (uncoupling between $\mathcal{E}$ on the one hand and $\mathcal{S}$ and $\mathcal{A}$ on the other), we have: $H_{\mathcal{SAE}}(t_0) = H_{\mathcal{SA}}(t_0) + H_{\mathcal{E}}(t_0)$. We suppose now that $\mathcal{S}$ is spectator while $\mathcal{A}$ and $\mathcal{E}$ interact for the time t; hence we have: $H_{\mathcal{S}}(t) = H_{\mathcal{S}}(t_0)$ and also $H_{\mathcal{AE}}(t) = H_{\mathcal{AE}}(t_0)$. The lack of correlation between $\mathcal{A}$ and $\mathcal{E}$ at t_0 implies that $H_{\mathcal{AE}}(t_0) = H_{\mathcal{A}}(t_0) + H_{\mathcal{E}}(t_0)$. Combining all these relations we obtain

$$\mathrm{C}_{\mathcal{SA}}(t) \leq \mathrm{C}_{\mathcal{SA}}(t_0), \tag{42.29}$$

where $\mathrm{C}_{\mathcal{SA}} = H_{\mathcal{S}} + H_{\mathcal{A}} - H_{\mathcal{SA}}$.

Therefore we download in the environment the correlations which are due to the entanglement [ZUREK 1982, 1874–78]. However, the correlations are not useful potential information because we cannot obtain a determinate value, which is the aim of a measurement. Mathematically we can obtain H_I by tracing out a reservoir (or generally the environment). Thermodynamically we can say that the correlations are inconsequential for the work function [*LINDBLAD* 1983, 60].[10] In other words for both cases of Measurement and of spontaneous decoherence, quantum useful potential information is increased since prediction is easier if a part of the complicated global system is neglected [*OHYA/PETZ* 1993, 7].

42.3.3 Increasing of Information

However, the process so far as we have described it here is not the whole Measurement process: it represent only the passage from a pure state to a mixture, which, in the case of a measurement, must be of a Lüders form [see Eq. (14.16)]. Evidently this passage is determined by a coupling with the environment such that it produces an increase of entropy in the observed system $\mathcal{S}$. But this signifies anyway a determined result: as we know we need a result such as that of Eq. (14.17). The solution is provided by the fact that the information corresponds to a negative entropy [see p. 710]. In other words we suppose [*BRILLOUIN* 1962, 153–54] that the effective gain of information obtained by a measurement process equals the difference between the initial entropy and the final one:

$$I = H_i - H_f, \tag{42.30}$$

or

$$H_f = H_i - I, \tag{42.31}$$

from which we see that the information appears as a negative term in the global final entropy of the system. In other words, each gain in the information happens at the expense of the entropy of the system. This is exactly the case if we think that the final state is an eigenstate of the

shows that for $1 \ll m \leq n$ one has:

$$\langle H_1 \rangle = \ln m - \frac{m}{2n}. \tag{42.27}$$

Such a conjecture has been proved by Sen [SEN 1996]..

[9] As regards the subadditivity [Eq. (42.8)] for three systems see [WEHRL 1978, 248]. For a proof of inequality (42.28) see [ARAKI/LIEB 1970, 161–62].

[10] See also [LLOYD 1997b, 3375–76].

measured observable, which in turn is a pure state, and we know that a pure state has a lower entropy than a mixture. Naturally this can happen only if simultaneously the entropy of some other system (the apparatus or the environment or both) increases, which is a very reasonable assumption.

42.3.4 Is the Information Conserved?

Now we wish to examine the following problem: does the downloading of correlations in the environment represent a net loss of potential information? It is generally presupposed that the potential information is only dispersed and not eliminated [ZUREK 1982, 1873–74]. Hence the global potential information (or entropy) of the system plus the environment remains constant. That it is conserved in the environment, can be shown by the fact that, in a Poincaré cycle, it can in principle be reconstructed. But if this is true in general terms, it is also evident that one can never exactly reconstruct the same potential information of the initial state[11], because one should reconstruct exactly the same state of the entire universe (due to the correlations between the object system, apparatus and environment), which is not only far away from each calculability but surely impossible, in particular if we take into account the aspects shown by the delayed choice [see chapter 26]. We must distinguish here two aspects:

- the conservation, or the increasing, of the global entropy or potential information of the 'entire universe',
- the availability of the potential information.

The first is very reasonable. Particularly Zeh and Joos [JOOS/ZEH 1985] [ZEH 1993] [JOOS 1996, 40–47] have insisted that the off-diagonal terms never disappear, and that it is only a practical, or local inaccessibility to them which characterizes this 'loss' [see theorem 17.6: p. 271]. On the other hand, a full availability of potential information is not possible if, with 'environment', we understand the 'rest of the world' and not a controllable physical reservoir. In fact we can prove the following theorem:

Theorem 42.2 (Availability of downloaded information) *The potential information (correlations) which has been downloaded in the 'rest of the world', can never be made into useful or active information.*

Proof
It is evident that the possibility to acquire information from the correlations downloaded in the so-understood environment presupposes the use of an universal wave function so that we can define the system plus all the rest of the world [see chapters 15 and 25, and particularly theorem 26.1: p. 452]: without such a definition it is impossible to acquire the desired information. But we have shown that it is impossible to define such a universal wave function [see corollary 26.3: p. 455]. Hence this impossibility excludes the recovering of downloaded (potential) information[12]. Q. E. D.

[11] For a classical example see [CRUTCHFIELD *et al.* 1986].
[12] The model studied by Wiseman and Milburn [WISEMAN/MILBURN 1993] does not represent a piece of evidence against the theorem, because they suppose a large but measurable system, and not the 'rest of the world' [see also chapter 27].

Similar considerations are valid for the information already made active. Consider the detection of a photon in an arm of an interferometer. Surely we can reconstruct the fact that the photon is passed through that arm. But as such the event and the photon no longer exist, because the latter has been absorbed by the detector.

In other words information is a process in which the initial informational content has been transformed (absorption of the photon) and the receptor has also been transformed (for example, if the detector is a two-level atom, it is passed from ground state to the excited state). In this process, unavoidably affected by dispersion, there is always some dispersion of the active information as well, i.e. we are never able to extract all the information which we could extract: as we shall discuss below, theorem 42.3 [p. 719] provides an upper bound for the available information. On the other hand, we have already shown that the relative entropy between two subsystems is decreasing. This represents a degradation of the information [see theorem 42.1: p. 714, and comments]. In other words: an exchange of information always supposes some 'interpretation' in consequence of which a part of the original informational content has been lost [see also subsection 28.4.2].

This degradation of the active information has already been theorized on a classical level by Brillouin [*BRILLOUIN* 1962, 154–55]. In fact he assumed that, in any irreversible process, the variation of the sum of the initial negentropy and of the gained information is less than zero:

$$\Delta(-H_i + I) < 0. \tag{42.32}$$

where $-H_i$ is the initial negentropy. This equation is formally analogous to the quantum expression (42.39).

42.3.5 Expectation value of Final Entropy

In conclusion we are interested to know the expectation value of the obtained final entropy conditional upon a measurement. After a measurement is performed with result r, the *a posteriori* probability for preparation i is given by Bayes formula [Eq. (11.5)]:

$$\wp(i|r) = \wp(r|i)\frac{\wp(i)}{\wp(r)}, \tag{42.33}$$

where $\wp(r) = \sum_i \wp(r|i)\wp(i)$ is the *a priori* probability for result r (the $\wp(r|i)$ are known and the *a priori* $\wp(i)$ are assumed) [PERES/WOOTTERS 1991, 1119–20]. The expectation of the final entropy is then[13]:

$$\left\langle H_I^f \right\rangle = -\sum_r \wp(r) \left[\sum_i \wp(i|r) \ln \wp(i|r) \right]. \tag{42.34}$$

42.4 Bounds for Information

42.4.1 Upper Bound

As we have said, a problem of great interest for measurement problems is the existence of bounds for the information gain. Holevo [HOLEVO 1973] proved the existence of an upper bound for

[13]Another interesting aspect is the following: the Measurement process is a gain of information. Hence, as we have seen, there must be some connection between Measurement statistics and mutual entropy (change in the entropy of the observed subsystem). This connection has been explicitly established and quantities calculated in the case of photon counting, with the result that the entropy change reflects the photon statistics [BAN 1997b].

the information one can extract from a quantum system. The importance of the proof derives from the fact that, on the contrary to the classical case, in the quantum case an infinite amount of information transfer would also require an infinite entropy of the system measured — which is certainly a more realistic situation. Hence, if there is such a bound, it is excluded that one can gain an infinite amount of information. The proof was then refined by Yuen and Ozawa [YUEN/OZAWA 1993] — Holevo's proof is restricted to the finite-dimensional case[14].

The *mutual information* between an input η and an output $\mathbf{x}$ is (classically) given by:

$$I_M(\eta : \mathbf{x}) := H(\eta) - H(\eta|\mathbf{x}), \tag{42.35}$$

where $H(\eta|\mathbf{x}) := -\int d\mathbf{x}\wp(\mathbf{x})\int d\eta\wp(\eta|\mathbf{x})\ln\wp(\eta|\mathbf{x})$ is the conditional entropy. Suppose now an effect $\hat{E}(d\mathbf{x})$ so that, when it is measured in a state $\hat{\rho}$, the output probability distribution $\mu_p[\hat{\rho}]$ of $\mathbf{x}$ is given by

$$\mu_p[\hat{\rho}](d\mathbf{x}) = \mathrm{Tr}[\hat{E}(d\mathbf{x})\hat{\rho}]. \tag{42.36}$$

Suppose now that η has a probability distribution $\wp(d\eta)$ on the alphabet H that parameterizes the state $\hat{\rho}_\eta$ $(\eta \in H)$. Hence we write the output conditional distribution as: $\wp(d\mathbf{x}|\eta) := \mu_p[\hat{\rho}_\eta](d\mathbf{x})$. Now we define a mixture of parameterized states $\hat{\rho}_\eta$:

$$\tilde{\hat{\rho}} := \int \hat{\rho}_\eta \wp(d\eta). \tag{42.37}$$

Then we state:

Theorem 42.3 (Holevo) *The information which can be obtained from the considered system is subjected to the following bound:*

$$\boxed{I_M(\eta : \mathbf{x}) \leq H(\tilde{\hat{\rho}}) - \int H(\hat{\rho}_\eta)\wp(d\eta)} \tag{42.38}$$

where the equality holds if all the $\hat{\rho}_\eta$'s commute.

We omit here the proof of the theorem[15]. Since we have $H(\hat{\rho}_\eta) \geq 0$, inequality (42.38) implies in particular:

$$I_M(\eta : \mathbf{x}) \leq H(\tilde{\hat{\rho}}), \tag{42.39}$$

which means that, independently of the quantum measurement one can perform, the related information transfer is never larger than the entropy of the mixture one obtains. The importance of this point is that it closely connects the von Neumann entropy with the Shannon (potential) information formulation — see also theorem 44.4 [p. 770]. Note that in this respect the theorem is also a better and more quantum-mechanical formulation of the negentropy principle of Brillouin (42.32). Inequality (42.38) satisfies our intuitive notions.

For example a spin-$\frac{1}{2}$ system has an information capacity of one bit — which is also proved by quantum computation techniques [see chapter 44]. But, as we know, there are an infinite number of states of a spin-$\frac{1}{2}$ system, one for each point on the Poincaré sphere [see figure 3.7 and comments: p. 59].

---

[14] See also [DAVIES 1978] [CAVES/DRUMMOND 1994, 502–505].

[15] See the original article, which is in turn based on a result of Uhlmann [UHLMANN 1977]. See also [OHYA 1989].

Hence an infinite amount of information can be coded in a spin state — which is again proved by the quantum computation. Nevertheless, the quantum state of the spin is not an observable [see subsection 28.4.3]. Hence, as we have said at the beginning, the upper bound is a new concept, typically quantum-mechanical, and gives us the amount of *Measurement-accessible information*, which, for the example considered, is one bit [B. SCHUMACHER 1990, 32]. Hence Holevo's theorem is a clear confirmation of the impossibility to measure (by a single measurement) the density matrix of a system — which represents, as already stated, the maximal amount of information contained in a physical state — and of the impossibility to measure the universal wave function, since one could gain an infinite amount of information. Once more we face the duality between state and observables.

42.4.2 Lower Bound

As we know, we can express the von Neumann entropy as $H_{\rm VN}(\hat{\rho}) = -\sum_j w_j \ln w_j$ [Eq. (42.4)], where the w_j's are the eigenvalues of $\hat{\rho}$. Jozsa, Robb and Wootters introduced the concept of *subentropy*, understood as the lower limit for information transfer [JOZSA *et al.* 1994].

This limit is defined by the following Eq. (we omit the proof):

$$\boxed{\mathrm{Q}(\hat{\rho}) := -\sum_j \left(\prod_{k \neq j} \frac{w_j}{w_j - w_k} \right) w_j \ln w_j} \tag{42.40}$$

The importance of the limit is the following: as we know a given density matrix admits different decompositions [see corollary 3.3: p. 46, and figure 3.4]. **For example** in a two-dimensional space, the density matrix $\hat{\rho} = \hat{I}/2$ can be decomposed in an ensemble of two states $(1,0), (0,1)$ (with equal probability), or in an ensemble of three states $(1,0), (1/2, \sqrt{3}/2), (1/2, -\sqrt{3}/2)$ (with equal probability). But in the former case the Measurement-accessible information is 1 bit, in the second case only 0.585. Hence it is an important question to know what is the lower bound for such a state.

42.4.3 Systematic Comparison Between the Different Bounds

Fuchs and Caves [C. FUCHS/CAVES 1994] [C. FUCHS/CAVES 1995] further developed this research. The Measurement-accessible information I and the entropy of the mixed state $H(\tilde{\hat{\rho}})$ are both downwardly convex, which can be seen by their derivatives. We show it by calculating the second derivatives. Let a binary channel be specified by $\hat{\rho}_0, \hat{\rho}_1$ with weights $1-w, w$, respectively. We write as usual [see Eq. (3.48)] $\tilde{\hat{\rho}} = w\hat{\rho}_1 + (1-w)\hat{\rho}_0$. The second derivative of the accessible information is given by:

$$I''(w) = -\sum_j \frac{\left[\mathrm{Tr}((\hat{\rho}_1 - \hat{\rho}_0)\hat{E}_j)\right]^2}{\mathrm{Tr}(\tilde{\hat{\rho}}\hat{E}_j)}, \tag{42.41}$$

for some result $\hat{E}_j$. If we represent $H(\tilde{\hat{\rho}})$ by the contour integral

$$H(\tilde{\hat{\rho}}) = \left(-\frac{1}{2\pi\imath}\right) \oint_{\mathcal{C}} (z \ln z) \mathrm{Tr}[(z\hat{I} - \tilde{\hat{\rho}})^{-1}] dz, \tag{42.42}$$

where the curve $\mathcal{C}$ encloses all nonzero eigenvalues of $\tilde{\hat{\rho}}$, one finds:

$$H''(w) = -\sum_{\{i,k|\lambda_i + \lambda_k \neq 0\}} \frac{\ln \lambda_i - \ln \lambda_k}{\lambda_i - \lambda_k} |\langle i|(\hat{\rho}_1 - \hat{\rho}_0)|k\rangle|^2, \tag{42.43}$$

where $|k\rangle$ is the eigenvector corresponding to the λ_k eigenvalue. Now we have that for all POVM $\hat{E}_j$:

$$\boxed{H''(w) \leq I''(w) \leq 0} \tag{42.44}$$

In the derivation of his results Holevo used the following function:

$$L''(w) := -\sum_{\{i,k|\lambda_i+\lambda_k\neq 0\}} \left(\frac{2}{\lambda_i + \lambda_k}\right) |\langle i|(\hat{\rho}_1 - \hat{\rho}_0)|k\rangle|^2. \tag{42.45}$$

The function $L(w)$ represents the ensemble-dependent upper bound. Symmetrically Fuchs and Caves calculates an ensemble-dependent lower bound $M(w)$, which for our example takes the form:

$$M(w) = \mathrm{Tr}\left[(1-w)\hat{\rho}_0 \ln[L_{\hat{\rho}}(\hat{\rho}_0)] + w\hat{\rho}_1 \ln[L_{\hat{\rho}}(\hat{\rho}_1)]\right], \tag{42.46}$$

where:

$$L_{\hat{\rho}} := \sum_{\{i,k|\lambda_i+\lambda_k\neq 0\}} \left(\frac{2}{\lambda_i + \lambda_k}\right) \langle i|(\hat{\rho}_1 - \hat{\rho}_0)|k\rangle |i\rangle\langle k|. \tag{42.47}$$

The relations between I(w), H(w), Q(w), L(w), M(w) are shown in figure 42.1.

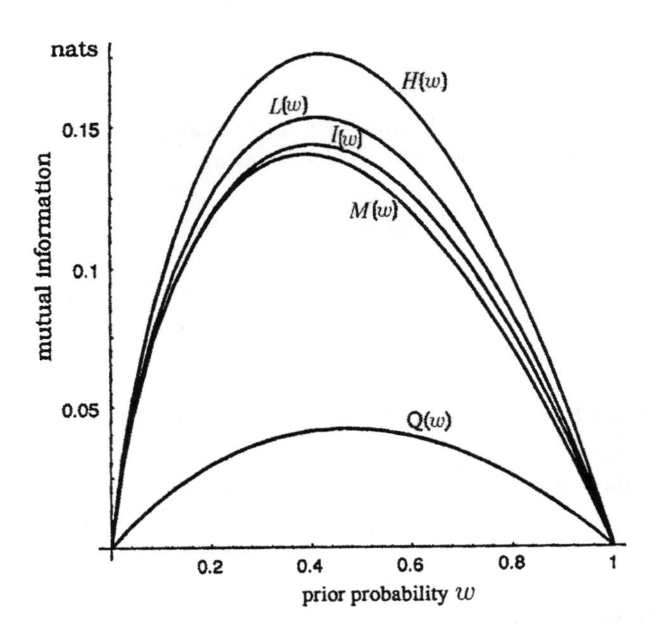

Figure 42.1: Behaviour of $I(w), H(w)$, Q(w), $L(w), M(w)$. The signal state is a pure Bloch vector **a** aligned along the z–axis and a mixed state **b** (with $b = 2/3$) aligned at an angle of $\pi/3$ away from the z–axis, with respective probabilities for the messages $1-w, w$ — from [C. FUCHS/CAVES 1994, 3050].

42.5 No-Cloning, Broadcasting and Copying

42.5.1 The Problem

By the examination of subsection 42.3.4, a question may arise: is it possible to amplify or arbitrarily reproduce the QM (potential or active) information ? The answer is no, because in QM one cannot clone initial states [WOOTTERS/ZUREK 1982]:

Theorem 42.4 (No-Cloning) *In QM no information can be cloned in the case of a number of states equal or greater than three.*

Proof

Suppose that such an amplification is possible. Then, in the case of a single base element, for an initial 'ready' apparatus state $|\mathcal{A}_0\rangle$, an incoming photon with vertical polarisation $|\updownarrow\rangle$ and a generic state of a third system, which could be the vacuum $|0\rangle$, we would have:

$$|\mathcal{A}_0\rangle|0\rangle|\updownarrow\rangle \rightsquigarrow |\mathcal{A}_\updownarrow\rangle|\updownarrow\updownarrow\rangle; \tag{42.48a}$$

and, for the same initial state of the apparatus and vacuum and incoming photon with horizontal polarisation, we would have:

$$|\mathcal{A}_0\rangle|0\rangle|\leftrightarrow\rangle \rightsquigarrow |\mathcal{A}_\leftrightarrow\rangle|\leftrightarrow\leftrightarrow\rangle. \tag{42.48b}$$

Now, giving the (42.48) assumptions, and supposing an incoming photon in a linear superposition $c'|\updownarrow\rangle + c''|\leftrightarrow\rangle$, the result of the interaction will be the following:

$$|\mathcal{A}_0\rangle|0\rangle\left(c'|\updownarrow\rangle + c''|\leftrightarrow\rangle\right) \rightsquigarrow c'|\mathcal{A}_\updownarrow\rangle|\updownarrow\updownarrow\rangle + c''|\mathcal{A}_\leftrightarrow\rangle|\leftrightarrow\leftrightarrow\rangle. \tag{42.49}$$

Now it is easy to recognize that this result is in no way a clone of the superposition represented by the state of the incoming photon in the LHS of Eq. (42.49). In fact such a cloning should, instead, be represented by:

$$|\mathcal{A}_0\rangle|0\rangle\left(c'|\updownarrow\rangle + c''|\leftrightarrow\rangle\right) \rightsquigarrow |\mathcal{A}_?\rangle\left(c'|\updownarrow\rangle + c''|\leftrightarrow\rangle\right)\left(c'|\updownarrow\rangle + c''|\leftrightarrow\rangle\right). \tag{42.50}$$

Q.E.D. Note that, if a quantum state could be cloned, then we would violate Einstein Locality [HERBERT 1982].

42.5.2 Violation of Unitarity

The previous proof shows a violation of the superposition principle only in the case of three arbitrary states. The proof has been generalized by Yuen and D'Ariano [D'ARIANO/YUEN 1996, 2832–33] to two non-orthogonal states, where a violation is shown of the unitarity of QM evolutions — while it is quite possible to duplicate any set of orthogonal states[16]:

[16] The article of D'Ariano and Yuen uses the results of [YUEN 1986].

Theorem 42.5 (D'Ariano/Yuen) *A cloning of two non-orthogonal states violates the unitarity of evolution.*

Proof

Suppose that to produce $n > 1$ copies of a generic state $|\psi\rangle$ from a given set of possible states, it is subjected to a unitary evolution of the form:

$$|\mathcal{A}\rangle \otimes |\psi\rangle \otimes |b_1\rangle \otimes \cdots \otimes |b_{n-1}\rangle \mapsto |\mathcal{A}'(\psi)\rangle \otimes |\psi\rangle \otimes \cdots \otimes |\psi\rangle, \tag{42.51}$$

where $|\mathcal{A}\rangle$ represents the state of the apparatus or the environment, the $|\psi\rangle$ states are present n times on the RHS, $|b_1\rangle \otimes \cdots \otimes |b_{n-1}\rangle$ are the state preparation of the modes which support the clones. $|\mathcal{A}\rangle$ is the initial state of sufficiently enough other modes (environment and others) so that the transformation is unitary. Now take two non-orthogonal states $|\psi\rangle, |j\rangle$, with $0 < |\langle\psi|j\rangle| < 1$, and suppose we know *a priori* that the system is in any one of them. We know that the transformation (42.51) must preserve the scalar product in order to be unitary [see theorem 3.7: p. 60]. Now multiply the LHS of Eq. (42.51) for $|\psi\rangle, |j\rangle$ and equate the result to the scalar product of the two corresponding states on the right: $\langle\psi|j\rangle = \langle\mathcal{A}'(\psi)|\mathcal{A}'(j)\rangle(\langle\psi|j\rangle)^n$. From this expression it immediately follows that: $\langle\mathcal{A}'(\psi)|\mathcal{A}'(j)\rangle(\langle\psi|j\rangle)^{n-1} = 1$, which would in turn require that $|\langle\mathcal{A}'(\psi)|\mathcal{A}'(j)\rangle| > 1$ for $n > 1$. Q. E. D.

In conclusion, also as a result of a preceding discussion [see p. 494], we state the following corollary of this theorem:

Corollary 42.1 (D'Ariano/Yuen) *The knowledge of the prior state of a single system violates the unitarity of the evolution.*

42.5.3 Broadcasting

Barnum *et al.* have recently shown [BARNUM *et al.* 1996] that generally there are no physical means for broadcasting an arbitrary mixed state $\tilde{\hat{\rho}}$ onto two separate quantum systems. By *broadcasting* it is meant that the marginal density operator of each of the separate systems is the same as the input state $\tilde{\hat{\rho}}$. The no-cloning theorem forbids broadcasting for pure states because the only way to broadcast a state $|\psi\rangle$ is to obtain a product state of the form $|\psi\rangle \otimes |\psi\rangle$, i.e. to clone it. But there are many ways to broadcast $\tilde{\hat{\rho}}$ without the joint state being in the product form $\tilde{\hat{\rho}} \otimes \tilde{\hat{\rho}}$. For instance, if the mixed state has spectral decomposition

$$\tilde{\hat{\rho}} = \sum_k c_k |k\rangle\langle k|, \tag{42.52}$$

a potential broadcasting state is the highly correlated joint state $\tilde{\hat{\rho}}_{\rm S} = \sum_k c_k |k\rangle|k\rangle\langle k|\langle k|$, which, though not a product state, reproduces the correct marginal density operators. From here we see that cloning is a stronger form of broadcasting. Now Barnum *et al.* have proved that, assuming a system $\mathcal{S}$ is in one of two possible but unknown states $\hat{\rho}_0, \hat{\rho}_1$,

- one can broadcast the state of $\mathcal{S}$ iff $\hat{\rho}_0, \hat{\rho}_1$ commute;
- one can clone the state of $\mathcal{S}$ iff $\hat{\rho}_0, \hat{\rho}_1$ are identical or orthogonal.

42.5.4 Universal Copying Machine

The question here is the following: given that one cannot perfectly clone a state, is it possible to copy it imperfectly? A study of Bužek and Hillery [BUŽEK/HILLERY 1996] is devoted to this problem.

State-Dependence of Wooters/Zurek's Copying Machine

Let us rewrite Eqs. (42.48) in the following form [BUŽEK/HILLERY 1996, 1844–49]:

$$|0\rangle_a|?\rangle|\mathcal{A}\rangle_x \rightsquigarrow |0\rangle_a|0\rangle_b|\mathcal{A}_0\rangle_x, \tag{42.53a}$$
$$|1\rangle_a|?\rangle|\mathcal{A}\rangle_x \rightsquigarrow |1\rangle_a|1\rangle_b|\mathcal{A}_1\rangle_x, \tag{42.53b}$$

where $|?\rangle$ is an unknown state of a third system and the subscript a marks the original (input) state and b the copy. As we now show, such a machine is input-state dependent. Let the input state be:

$$|i\rangle_a = c_0|0\rangle_a + c_1|1\rangle_a, \tag{42.54}$$

such that [similarly to Eq. (42.49)]:

$$|i\rangle_a|?\rangle|\mathcal{A}\rangle_x \rightsquigarrow c_0|0\rangle_a|0\rangle_b|\mathcal{A}_0\rangle_x + c_1|1\rangle_a|1\rangle_b|\mathcal{A}_1\rangle_x \equiv |\Psi\rangle^f_{abx}, \tag{42.55}$$

where orthonormality of the apparatus' basis is assumed and that, without loss of generality, c_0, c_1 are real: $c_0^2 + c_1^2 = 1$. Under the first assumption, the reduced density operator $\hat{\rho}^{ab}_f$ input + copy reads:

$$\hat{\rho}^{ab}_f = \mathrm{Tr}_x[\hat{\rho}^{abx}_f] = c_0^2|00\rangle\langle 00| + c_1^2|11\rangle\langle 11|, \tag{42.56}$$

where $\hat{\rho}^{abx}_f = |\Psi\rangle^f_{abx}{}^f_{abx}\langle\Psi|$ and $|jj\rangle = |jj\rangle_a|jj\rangle_b (j = 0, 1)$. Density operators describing quantum states of the original mode and of the copy mode are, respectively:

$$\hat{\rho}^a_f = \mathrm{Tr}_b[\hat{\rho}^{ab}_f] = c_0^2|0\rangle_{aa}\langle 0| + c_1^2|1\rangle_{aa}\langle 1|, \tag{42.57a}$$
$$\hat{\rho}^b_f = \mathrm{Tr}_b[\hat{\rho}^{ab}_f] = c_0^2|0\rangle_{bb}\langle 0| + c_1^2|1\rangle_{bb}\langle 1|. \tag{42.57b}$$

As we know the Hilbert/Schmidt norm of an operator $\hat{O}$ is $\parallel \hat{O} \parallel = [\mathrm{Tr}(\hat{O}^\dagger\hat{O})]^{\frac{1}{2}}$ which for operator $\hat{O}, \hat{O}'$ has the property $|\mathrm{Tr}(\hat{O}^\dagger\hat{O}')| \leq \parallel \hat{O} \parallel\parallel \hat{O}' \parallel$.

Now the distance between density operators $\hat{\rho}_1$ and $\hat{\rho}_2$ can be written as:

$$\delta_{12} := (\parallel \hat{\rho}_1 - \hat{\rho}_1 \parallel)^2, \tag{42.58}$$

and the distance between the input density operator $\hat{\rho}^a_i$ of the original mode (where the superscript id stands for 'ideal') and the output density operator $\hat{\rho}^a_f$ of the original mode as

$$\delta_a := \mathrm{Tr}[\hat{\rho}^a_i - \hat{\rho}^a_f]^2, \tag{42.59}$$

where $\hat{\rho}^a_i$ [corresponding to Eq. (42.54)] reads:

$$\hat{\rho}^a_i = c_0^2|0\rangle_{aa}\langle 0| + c_1^2|1\rangle_{aa}\langle 1| + c_0c_1|0\rangle_{aa}\langle 1| + c_1c_0|1\rangle_{aa}\langle 0|. \tag{42.60}$$

Now the distance between the density operator (42.57a) and the density operator (42.60) is given by:

$$\delta_a = 2c_0^2c_1^2 = 2c_0^2(1 - c_0^2), \tag{42.61}$$

which reflects the fact that the states $|0\rangle_a$ ($c_0 = 1$) and $|1\rangle_a$ ($c_0 = 0$) are copied perfectly, i.e. for these states $\delta_a = 0$, while the pure superposition states $|i\rangle_a = (|0\rangle_a \pm |1\rangle_a)/\sqrt{2}$ are copied badly, i.e. $\delta_a = \frac{1}{2}$. From Eq. (42.61) we see that the Zurek/Wootters' copying machine is input-state dependent (some states are copied well and others badly). We expect the mean distance in this case to be:

$$\overline{\delta}_a = \int_0^1 dc_0^2 \delta_a(c_0^2) = \frac{1}{3}. \tag{42.62}$$

Entanglement in the Copying Machine

Moreover, as follows from Eq. (42.56), the output states are highly entangled, while, from a perfect copying machine, we would expect a factorization:

$$\hat{\rho}_i^{ab} = \hat{\rho}_i^a \otimes \hat{\rho}_i^b, \tag{42.63}$$

where both $\hat{\rho}_i^a$ and $\hat{\rho}_i^b$ are expressed by Eq. (42.60) and, by introducing vectors

$$|+\rangle = \frac{1}{\sqrt{2}}(|10\rangle + |01\rangle), \tag{42.64a}$$

$$|-\rangle = \frac{1}{\sqrt{2}}(|10\rangle - |01\rangle), \tag{42.64b}$$

we have:

$$\begin{aligned}\hat{\rho}_i^{ab} &= c_0^4|00\rangle\langle 00| + c_1^4|11\rangle\langle 11| \\ &+c_0^2c_1^2|00\rangle\langle 11| + c_0^2c_1^2|11\rangle\langle 00| + 2c_0^2c_1^2|+\rangle\langle +| \\ &+\sqrt{2}c_0^3c_1|00\rangle\langle +| + \sqrt{2}c_0^3c_1|+\rangle\langle 00| + \sqrt{2}c_0c_1^3|11\rangle\langle +| + \sqrt{2}c_0c_1^3|+\rangle\langle 11|.\end{aligned} \tag{42.65}$$

Now we write the distance between the actual two-mode density operator $\hat{\rho}_f^{ab}$ and the direct product of $\hat{\rho}_f^a$ and $\hat{\rho}_f^b$:

$$\delta_{ab}^{(1)} = \mathrm{Tr}[\hat{\rho}_f^{ab} - \hat{\rho}_f^a \otimes \hat{\rho}_f^b]^2, \tag{42.66}$$

for which we find:

$$\delta_{ab}^{(1)} = \delta_{ab}, \tag{42.67}$$

where δ_a (δ_b) is given by Eq. (42.61). Analogously we evaluate the distance between $\hat{\rho}_f^{ab}$ and $\hat{\rho}_i^{ab}$:

$$\delta_{ab}^{(2)} = \mathrm{Tr}[\hat{\rho}_f^{ab} - \hat{\rho}_i^{ab}]^2 = \delta_a + \delta_b. \tag{42.68}$$

This result together with (42.67) implies that the output state is most entangled when the performance of the copying machine is worst. Finally we calculate the distance between $\hat{\rho}_i^{ab}$ and the direct product of the single-mode output density operators:

$$\delta_{ab}^{(3)} = \mathrm{Tr}[\hat{\rho}_i^{ab} - \hat{\rho}_f^a \otimes \hat{\rho}_f^b]^2 = \delta_a + \delta_b - \delta_{ab}^{(1)}. \tag{42.69}$$

We expect to have the following relationship:

$$\delta_{ab}^{(1)} \leq \delta_{ab}^{(3)} \leq \delta_{ab}^{(2)}. \tag{42.70}$$

In fact the input-state-averaged distances [see Eqs. (42.62) and (42.68)] are: $\overline{\delta}_{ab}^{(2)} = 2/3, \overline{\delta}_{ab}^{(1)} = 2/15, \overline{\delta}_{ab}^{(3)} = 8/15$. Note that a Wootters/Zurek's copying machine also produces entanglement in the case of mixtures and while it preserves the mean value of some operators it completely destroys the mean value of others. For instance, it preserves the mean value of $\hat{\sigma}_z = (|1\rangle\langle 1| - |0\rangle\langle 0|)/2$:

$$\langle\hat{\sigma}_z\rangle_a^{(\text{in})} = \langle\hat{\sigma}_z\rangle_a^{(\text{out})} = -\frac{1}{2}\cos(2\phi), \tag{42.71a}$$

where we have used the parameterization $c_0 = \cos\phi$; and it destroys the mean value of $\hat{\sigma}_x = (|1\rangle\langle 0| + |0\rangle\langle 1|)/2$:

$$\langle\hat{\sigma}_x\rangle_a^{(\text{out})} = 0 \neq \langle\hat{\sigma}_x\rangle_a^{(\text{in})} = \frac{1}{2}\sin(2\phi), \tag{42.71b}$$

where we have used $c_1 = \sin\phi$.

Universal Copying Machine

We now wish to find a copying machine which is not input-state dependent, i.e. a universal machine which allows to copy well superposition states for any value of c_0 in the sense that distances (42.59) and (42.68) do not depend on c_0. Instead of Eqs. (42.53), the most general quantum-copying transformation can be written as follows:

$$|0\rangle_a|?\rangle|\mathcal{A}\rangle_x \rightsquigarrow \sum_{k,l=0}^{1} |k\rangle_a|l\rangle_b|\mathcal{A}_{kl}\rangle_x, \tag{42.72a}$$

$$|1\rangle_a|?\rangle|\mathcal{A}\rangle_x \rightsquigarrow \sum_{m,n=0}^{1} |m\rangle_a|n\rangle_b|\mathcal{A}_{mn}\rangle_x, \tag{42.72b}$$

where the states $|\mathcal{A}_{mn}\rangle_x$ are not necessarily orthonormal. As a particular realization Bužek and Hillery propose the following:

$$|0\rangle_a|?\rangle|\mathcal{A}\rangle_x \rightsquigarrow |0\rangle_a|0\rangle_b|\mathcal{A}_0\rangle_x + [|0\rangle_a|1\rangle_b + |1\rangle_a|0\rangle_b]|Y_0\rangle_x, \tag{42.73a}$$

$$|1\rangle_a|?\rangle|\mathcal{A}\rangle_x \rightsquigarrow |1\rangle_a|1\rangle_b|\mathcal{A}_1\rangle_x + [|0\rangle_a|1\rangle_b + |1\rangle_a|0\rangle_b]|Y_1\rangle_x. \tag{42.73b}$$

Due to the unitarity of the transformation, the following relations hold:

$${}_x\langle\mathcal{A}_j|\mathcal{A}_j\rangle_x + 2_x\langle Y_j|Y_j\rangle_x = 1, \quad j = 0, 1, \tag{42.74a}$$

$${}_x\langle Y_0|Y_1\rangle_x = {}_x\langle Y_1|Y_0\rangle_x = 0. \tag{42.74b}$$

In order to reduce the number of parameters we assume that the copying machine state vectors $|\mathcal{A}_j\rangle_x, |Y_j\rangle_x$ are mutually orthogonal and that ${}_x\langle\mathcal{A}_0|\mathcal{A}_1\rangle_x = 0$. Under these assumptions we find that the output density operator of original + copy mode after copying the superposition (42.54) is:

$$\begin{aligned}\hat{\rho}_f^{ab} = {} & c_0^2|00\rangle\langle 00|_x\langle\mathcal{A}_0|\mathcal{A}_0\rangle_x + c_1^2|11\rangle\langle 11|_x\langle\mathcal{A}_1|\mathcal{A}_1\rangle_x + [c_{0x}^2\langle Y_0|Y_0\rangle_x + 2c_{1x}^2\langle Y_1|Y_1\rangle_x]|+\rangle\langle +| \\ & +\sqrt{2}c_0c_1|00\rangle\langle +|_x\langle Y_1|\mathcal{A}_0\rangle_x + \sqrt{2}c_0c_1|+\rangle\langle 00|_x\langle\mathcal{A}_0|Y_1\rangle_x + \sqrt{2}c_0c_1|+\rangle\langle 11|_x\langle\mathcal{A}_1|Y_0\rangle_x \\ & +\sqrt{2}c_0c_1|11\rangle\langle +|_x\langle Y_0|\mathcal{A}_1\rangle_x.\end{aligned} \tag{42.75}$$

The density operator describing the a mode is obtained by tracing out b:

$$\begin{aligned}\hat{\rho}_f^a = {} & |0\rangle_{aa}\langle 0|[c_0^2 + c_{1x}^2\langle Y_1|Y_1\rangle_x - c_{0x}^2\langle Y_0|Y_0\rangle_x)] + |1\rangle_{aa}\langle 1|[c_1^2 + c_{0x}^2\langle Y_0|Y_0\rangle_x - c_{1x}^2\langle Y_1|Y_1\rangle_x)] \\ & +|0\rangle_{aa}\langle 1|c_0c_1[{}_x\langle\mathcal{A}_1|Y_0\rangle_x + {}_x\langle Y_1|\mathcal{A}_0\rangle_x)] + |1\rangle_{aa}\langle 0|c_0c_1[{}_x\langle\mathcal{A}_0|Y_1\rangle_x + {}_x\langle Y_0|\mathcal{A}_1\rangle_x)],\end{aligned} \tag{42.76}$$

and analogously for $\hat{\rho}^b_f$. Note that the original mode is not the same before the copying [Eq. (42.60)] and after the copying [Eq. (42.76)], i.e. the original state is distorted by the copying. In order to quantify such a distortion we evaluate the distance between these density operators:

$$\delta_a = 2\xi^2(4c_0^4 - 4c_0^2 + 1) + 2c_0^2(1 - c_0^2)(\eta - 1)^2, \tag{42.77}$$

where $\xi := {}_x\langle Y_0|Y_0\rangle_x = {}_x\langle Y_1|Y_1\rangle_x$ and $\eta/2 := {}_x\langle Y_0|\mathcal{A}_1\rangle_x = {}_x\langle \mathcal{A}_0|Y_1\rangle_x = {}_x\langle \mathcal{A}_1|Y_0\rangle_x = {}_x\langle Y_1|\mathcal{A}_0\rangle_x$, with $0 \leq \xi \leq 1/2$, $0 \leq \eta \leq 2\xi^{\frac{1}{2}}(1-2\xi)^{\frac{1}{2}} \leq 1/\sqrt{2}$. Looking for a machine independent from the parameter c_0 (or c_0^2), we can determine either the parameter ξ or the parameter η by the condition:

$$\frac{\partial}{\partial c_0^2}\delta_a = 0, \tag{42.78}$$

in consequence of which we find:

$$\eta = 1 - 2\xi. \tag{42.79a}$$

Then the distance δ_a is state independent and takes the value

$$\delta_a = 2\xi^2. \tag{42.79b}$$

Taking into account the relations (42.79) and (42.74), we can rewrite the operators $\hat{\rho}^{ab}_f$ [Eq. (42.75)] and $\hat{\rho}^a_f$ [Eq. (42.76)], respectively, as:

$$\begin{aligned}\hat{\rho}^{ab}_f \;=\;& c_0^2(1-2\xi)|00\rangle\langle 00| + c_1^2(1-2\xi)|11\rangle\langle 11| + 2\xi|+\rangle\langle +| \\ &+\frac{c_0c_1}{\sqrt{2}}(1-2\xi)|00\rangle\langle +| + \frac{c_0c_1}{\sqrt{2}}(1-2\xi)|+\rangle\langle 00| \\ &+\frac{c_0c_1}{\sqrt{2}}(1-2\xi)|+\rangle\langle 11| + \frac{c_0c_1}{\sqrt{2}}(1-2\xi)|11\rangle\langle +|,\end{aligned} \tag{42.80}$$

and

$$\begin{aligned}\hat{\rho}^a_f \;=\;& |0\rangle_{aa}\langle 0|[c_0^2 + \xi(c_1^2 - c_0^2)] + |1\rangle_{aa}\langle 1|[c_1^2 + \xi(c_0^2 - c_1^2)] \\ &+|0\rangle_{aa}\langle 1|c_0c_1(1-2\xi) + |1\rangle_{aa}\langle 0|c_0c_1(1-2\xi).\end{aligned} \tag{42.81}$$

Final Considerations

We find the optimal value of ξ by assuming that the distance between $\hat{\rho}^{ab}_f$ and $\hat{\rho}^{ab}_i$ (factorised) is input-state independent, that is we solve the Eq.

$$\frac{\partial}{\partial c_0^2}\delta^{(2)}_{ab} = 0. \tag{42.82}$$

Then we obtain:

$$\delta^{(2)}_{ab} = (f_{11})^2 + 2(f_{12})^2 + 2(f_{13})^2 + (f_{22})^2 + (f_{23})^2 + (f_{33})^2, \tag{42.83}$$

where

$$\begin{aligned} f_{11} &= c_0^4 - c_0^2(1-2\xi), & f_{33} &= c_1^4 - c_1^2(1-2\xi), \\ f_{13} &= c_0^2c_1^2, & f_{22} &= 2c_0^2c_1^2 - 2\xi, \\ f_{12} &= \sqrt{2}c_0c_1[c_0^2 - \frac{1}{2}(1-2\xi)], & f_{23} &= \sqrt{2}c_0c_1[c_1^2 - \frac{1}{2}(1-2\xi)]. \end{aligned} \tag{42.84}$$

By solving Eq. (42.82) we find $\xi = 1/6$. For this value the distance $\delta^{(2)}_{ab}$ is c_0^2 independent and its value is 2/9. Now it can be proved [BUŽEK/HILLERY 1996, 1849–52] that such a copying machine is input-state independent — with a fidelity [see Eq. (43.24)] of $\sqrt{\frac{5}{6}}$ for all input-states —, with average distances of $\overline{\delta}_a = \frac{1}{3}, \overline{\delta}_{ab} = \frac{2}{3}$ respectively — much better than with the Wootters/Zurek copying machine. Moreover the universal copying machine preserves the mean value of $\hat{\sigma}_z, \hat{\sigma}_x$. On the other hand, in order to make such a machine useful, we must be allowed to perform a measurement on mode b without significantly disturbing mode a. The problem is that, also with the universal copying machine, output states of these subsystems are entangled. Now the authors show that any measurement of a projection in the b mode leaves the a mode near to the ideal output state, i.e. the input state, and can be used to find the expectation value of any a–mode operator in the ideal output state. In addition, the b–mode measurement provides information about the original input state. Hence, even if perfect cloning is impossible, a satisfactory copying can be found[17].

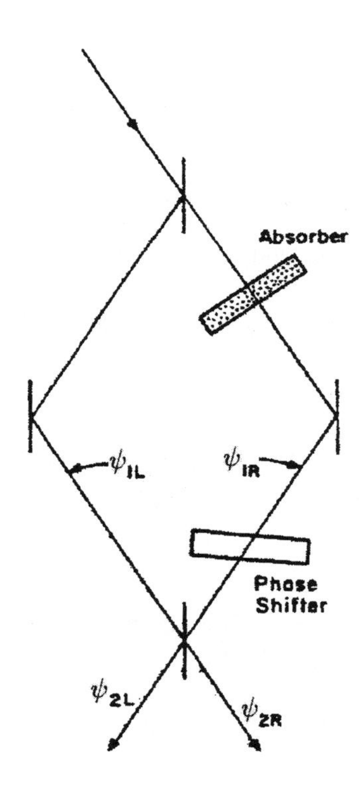

Figure 42.2: Principle sketch of the arrangement to obtain partial information about the particle's path through the interferometer — from [ZEILINGER 1986a, 238].

[17]Recently a quantum copying network has been proposed [BUŽEK *et al.* 1997c] and an upper bound for the fidelity of a universal copying machine has been calculated [BRUß *et al.* 1998].

42.6 Which Path and Information

42.6.1 Zeilinger's Analysis

The work of Wootters and Zurek [WOOTTERS/ZUREK 1979, 451–52] [see section 30.1] was further developed by Zeilinger [ZEILINGER 1986a, 238–41] [ZEILINGER 1986b], in connection with the problem of information. It is a very interesting aspect because it is exactly the informational content of the which-path prediction which gave the hint to conceive the complementarity as a not jump-like relation. Suppose there is an experimental arrangement (Interfer. Exp. 27) as shown in figure 42.2. The total lack of information about a particle's path through an interferometer (before the beams' recombination) can be expressed as:

$$H_1 = -\wp_{1L} \ln \wp_{1L} - \wp_{1R} \ln \wp_{1R}, \tag{42.85}$$

where H_1 is the entropy before the BM and $\wp_{1L}, \wp_{1R}$ are the probabilities that the particle takes the left and the right arm respectively. These probabilities are given by: $\wp_{1L} = |\psi_{1L}|^2, \wp_{1R} = |\psi_{1R}|^2$, respectively. We now calculate the maximal and minimal amplitudes after the merging of the beams:

$$|\psi_{2Max}| = \frac{1}{\sqrt{2}}(|\psi_{1L}| + |\psi_{1R}|), \tag{42.86a}$$

$$|\psi_{2min}| = \frac{1}{\sqrt{2}}(|\psi_{1L}| - |\psi_{1R}|). \tag{42.86b}$$

Therefore, if we pose $\wp_{2Max} = |\psi_{2Max}|^2, \wp_{2min} = |\psi_{2min}|^2$, we have, as a measure of the informational content of the interference pattern, the following quantity:

$$H_2 = -\wp_{2Max} \ln \wp_{2Max} - \wp_{2min} \ln \wp_{2min}. \tag{42.87}$$

The two types of information characterized by Eqs. (42.85) and (42.87) are interdependent since they are connected via Eqs. (42.86). The resulting functional dependence of H_2 on H_1 is shown in figure 42.3.

42.6.2 Relationship Between Zeilinger's Measure and that of Greenberger/Yasin

Greenberger and Yasin have shown that Zeilinger's measure of interference and their one [see subsection 30.2.1] is approximately the same [GREENBERGER/YASIN 1988a] [see figure 42.4].

42.6.3 Another Model

Very similar results are also obtained by the stochastic model proposed by Mittelstaedt, Prieur and Schieder [see subsection 30.3.1] where the lack of information (entropy) about the predictability of the path $H(\eta, \mathcal{P})$ and about the visibility of interference $H(\eta, \mathcal{V})$ are calculated — we remember that η is the transparency parameter of the II BS in the proposed experimental arrangement [see figure 30.7]. The results are shown in figure 42.5, where the sum $H(\eta, \mathcal{P}) + H(\eta, \mathcal{V})$ is also represented in the spirit of Deutsch entropic UR [see subsection 42.2.1]: $H(\eta, \mathcal{P}) + H(\eta, \mathcal{V}) \geq \ln 2$, with the lower limit given by $H_D = 2[\ln 2 - \ln(1 + \sup\{|\langle \eta, \delta\psi | \eta', \delta\psi' \rangle|\})]$. The maximal value is when $H(\eta, \mathcal{P}) = H(\eta, \mathcal{V})$.

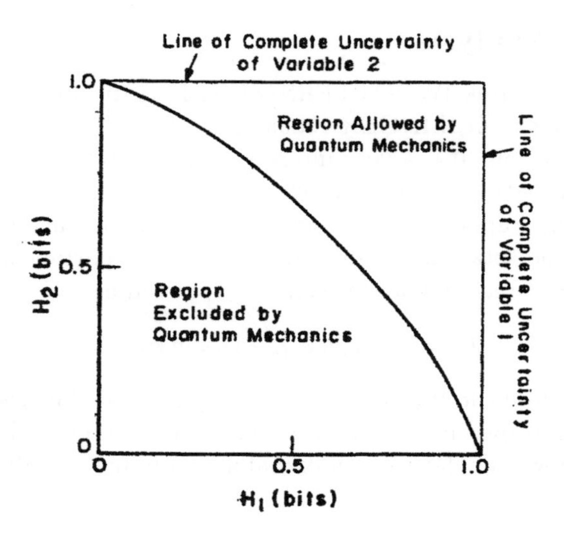

Figure 42.3: The continuous quantitative complementarity principle: the lack of information H_2 contained in the interference pattern in relationship to the lack of information H_1 about the particle's path through the interferometer — from [ZEILINGER 1986a, 240].

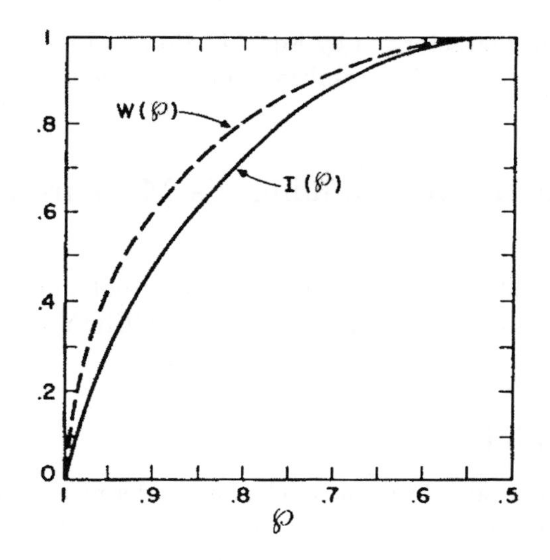

Figure 42.4: $W(\wp)$ represents the measure of the wave-like properties of the beam following Greenberger and Yasin's model (dotted curve). $I(\wp)$ (solid curve) represents an information theory measure of the same quantity — from [GREENBERGER/YASIN 1988a, 393].

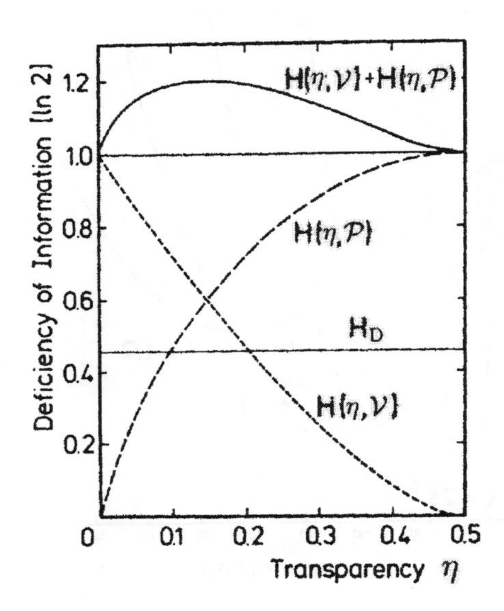

Figure 42.5: The deficiencies of information $H(\eta, \mathcal{P})$ and $H(\eta, \mathcal{V})$ and their sum in Interfer. Exp. 18 [p. 635] — from [MITTELSTAEDT *et al.* 1987, 894] .

42.7 Entropy and Bell Inequalities

42.7.1 Information-Theoretic Bell Inequality

We can also give an interpretation to Bell inequalities in terms of information and entropy. Starting with Bell inequalities for an arbitrary s–spin proposed by Mermin [MERMIN 1980] [see section 39.5], Braunstein and Caves [BRAUNSTEIN/CAVES 1988] [BRAUNSTEIN/CAVES 1990, 45–46] formulated an information-theoretic Bell inequality — corresponding to the III Bell inequality [the (35.27)]:

$$H(\hat{O}^a|\hat{O}^b) \leq H(\hat{O}^a|\hat{O}^{b'}) + H(\hat{O}^{b'}|\hat{O}^{a'}) + H(\hat{O}^{a'}|\hat{O}^b). \tag{42.88}$$

Inequality (42.88) is deduced from the expression

$$H(\hat{O}^a|\hat{O}^b) \leq H(\hat{O}^a) \leq H(\hat{O}^a, \hat{O}^b), \tag{42.89}$$

where the LHS inequality means that removing a condition never decreases the information carried by a quantity, and the RHS inequality that two quantities never carry less information than either separately; and from the following generalization of inequality (42.89):

$$H(\hat{O}^a, \hat{O}^b) \leq H(\hat{O}^a, \hat{O}^{a'}, \hat{O}^b, \hat{O}^{b'}) = H(\hat{O}^a|\hat{O}^{a'}, \hat{O}^b, \hat{O}^{b'}) + H(\hat{O}^{b'}|\hat{O}^{a'}, \hat{O}^b) + H(\hat{O}^{a'}|\hat{O}^b) + H(\hat{O}^b), \tag{42.90}$$

where the expansion of the RHS of inequality (42.90) is an application of Bayes's rule [Eq. (11.5)].

Proof

In order to deduce Eq. (42.88) from inequality (42.90) one needs to substitute the first term of the RHS of inequality (42.90) with the first term of the RHS of inequality (42.88) by means of the

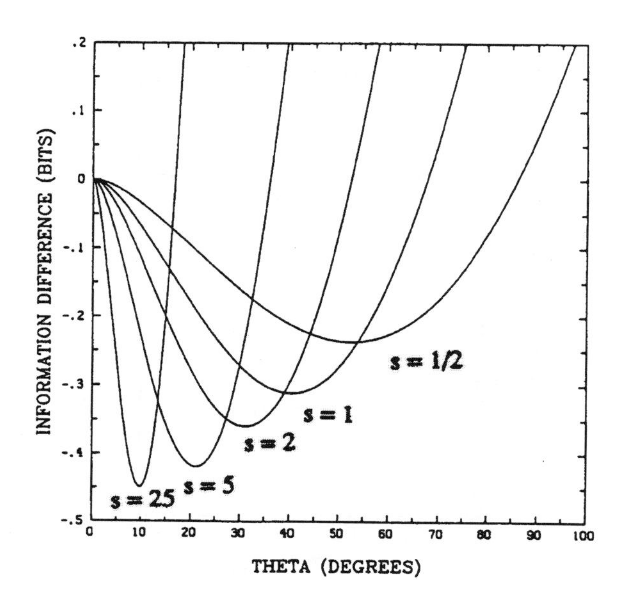

Figure 42.6: Information difference in bits versus angle θ in degrees for $s = \frac{1}{2}, 2, 5$, and 25. The maximum information deficit for $s = \frac{1}{2}$ is - 0.2369 bits at 52.31; for $s = 25$ - 0.4493 bits at 9.798 — from [BRAUNSTEIN/CAVES 1988, 664].

following application of inequality (42.89): $H(\hat{O}^a|\hat{O}^{a'}, \hat{O}^b, \hat{O}^{b'}) \leq H(\hat{O}^a|\hat{O}^{b'})$; the second term of the RHS of inequality (42.90) with the second term of the RHS of inequality (42.88) by means of another application of inequality (42.89): $H(\hat{O}^{b'}|\hat{O}^{a'}, \hat{O}^b) \leq H(\hat{O}^{b'}|\hat{O}^{a'})$; finally one leaves the third term of the RHS of inequality (42.90), and one substitutes the LHS term $H(\hat{O}^a, \hat{O}^b)$ with the term $H(\hat{O}^a|\hat{O}^b)$ by means of another application of Bayes's rule $H(\hat{O}^a, \hat{O}^b) = H(\hat{O}^a|\hat{O}^b) + H(\hat{O}^b)$, and finally one eliminates the fourth term, $H(\hat{O}^b)$ of the RHS of inequality (42.90). Q.E.D.

Inequality (42.88) means that the four quantities cannot carry less information than any two of them. The authors show that inequality (42.88) is violated by QM and that the information deficit increases with increasing spin-number, even if the angle between spin orientation becomes smaller [see figure 42.6].

42.7.2 Quadrilateral Information-Distance Bell Inequality

It is possible to arrive at a result formally similar to the preceding ones but by using different conceptual instruments [B. SCHUMACHER 1991].[18] As we shall see, there are also far reaching consequences. The instruments are those of algorithmic complexity[19]. We express the correlation or mutual information [see Eq. (42.35)] between two generic variables ξ, ξ' as follows:

$$\begin{aligned} H(\xi : \xi') &= H(\xi) - H(\xi|\xi') \\ &= H(\xi') - H(\xi'|\xi) \\ &= H(\xi) + H(\xi') - H(\xi, \xi'). \end{aligned} \tag{42.91}$$

[18] See also [BRAUNSTEIN/CAVES 1990, 46–48] .
[19] On this point see [ZUREK 1989a] [ZUREK 1989b].

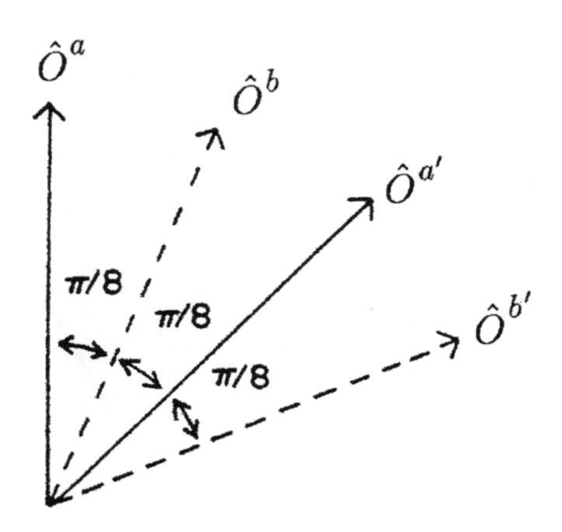

Figure 42.7: Spin measurement axes yielding a violation of the information distance quadrilateral inequality for a singlet state — from [B. SCHUMACHER 1991, 7050].

We remember that two generic variables are correlated in an information-theoretic sense if one provides information about the other. We now define the informational distance $\delta(\xi,\xi')$ between ξ and ξ' as:

$$\begin{aligned}\delta(\xi,\xi') &= H(\xi|\xi') + H(\xi'|\xi) \\ &= H(\xi,\xi') - H(\xi:\xi') \\ &= 2H(\xi,\xi') - H(\xi) - H(\xi').\end{aligned} \tag{42.92}$$

The informational distance is positive definite and symmetric. Note that the concept of informational distance is an informational expression of the already analysed concept of state-separation [see definition 39.1: p. 659].

It can be proved [B. SCHUMACHER 1990, 7049–50] that the informational distance satisfies the following triangular inequality for three observables:

$$\delta(\hat{O}_1,\hat{O}_2) + \delta(\hat{O}_2,\hat{O}_3) \geq \delta(\hat{O}_1,\hat{O}_3), \tag{42.93}$$

which means that the informational distance from $\hat{O}_1$ to $\hat{O}_3$ cannot be bigger than the informational distance from both when the path is taken over a $\hat{O}_2$.[20] Note that inequality (42.93) is exactly the same as the (39.2), as already anticipated.

Similarly we can state a quadrilateral inequality, the quadrilateral information-distance Bell inequality:

$$\delta(\hat{O}_1,\hat{O}_2) + \delta(\hat{O}_2,\hat{O}_3) + \delta(\hat{O}_3,\hat{O}_4) \geq \delta(\hat{O}_1,\hat{O}_4), \tag{42.94}$$

which is exactly the same as Eq. (39.3).

Now it suffices to take $\hat{O}_1 = \hat{O}^a, \hat{O}_2 = \hat{O}^b, \hat{O}_3 = \hat{O}^{a'}, \hat{O}^{b'}$ to return to our original spin model. If we fix the information distance between $\hat{O}^a$ and $\hat{O}^b$ (separated by angle θ) to be:

$$\delta(\hat{O}^a,\hat{O}^b) = 2f\left(\frac{\theta}{2}\right), \tag{42.95}$$

[20] The same conclusion, but with other methods, can be found in [TYAPKIN/VINDUSHKA 1991].

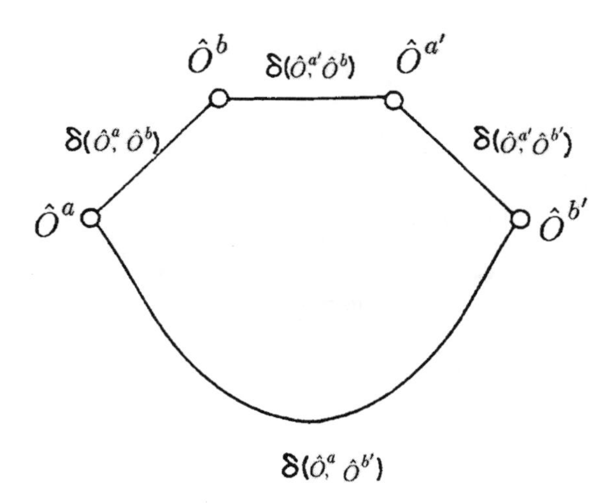

Figure 42.8: Schematic representation of quantum non-separability. The information distance $\delta(\hat{O}^a, \hat{O}^{b'})$ is greater than is allowed by the classical metric properties of informational distance — from [B. SCHUMACHER 1991, 7050].

where $f(\phi) = -\cos^2\phi \ln \cos^2\phi - \sin^2\phi \ln \sin^2\phi$, and take the angle between $\hat{O}^a$ and $\hat{O}^b$, between $\hat{O}^b$ and $\hat{O}^{a'}$, between $\hat{O}^{a'}$ and $\hat{O}^{b'}$ to be equal to $\pi/8$ [see figure 42.7], we arrive at the following values:

$$\delta(\hat{O}^a, \hat{O}^b) = \delta(\hat{O}^{a'}, \hat{O}^b) = \delta(\hat{O}^{a'}, \hat{O}^{b'})2f\left(\frac{\pi}{16}\right) = 0.323, \quad (42.96a)$$

$$\delta(\hat{O}^a, \hat{O}^{b'}) = 2f\left(\frac{3\pi}{16}\right) = 1.236. \quad (42.96b)$$

But since $0.323 + 0.323 + 0.323 \leq 1.236$ it is evident that the Bell inequality is violated, which in turn signifies that either $\hat{O}^a$ and $\hat{O}^{b'}$ are less correlated than the others or that the other pairs are more correlated than they should be in the framework of classical information theory [see figure 42.8].[21]

42.7.3 Covariance-Distance Bell Inequalities

A natural question is how much two subsystems must be correlated in order to violate Bell inequalities [B. SCHUMACHER 1990, 7050–7052] — in fact, as we know, the singlet state is a state of maximal violation [see section 38.3]. The following examination is a generalization on an informational level of the discussion of mixed states [see chapter 38].

Remembering the definition (42.14) of covariance, we now define the covariance distance as

$$\delta(\hat{O}_1, \hat{O}_2)_{\Delta^2} = 1 - [\Delta(\hat{O}_1, \hat{O}_2)]^2. \quad (42.97)$$

By this means it is also possible to define a quadrilateral covariance-distance Bell inequality equivalent to inequality (42.94):

$$\delta(\hat{O}_1, \hat{O}_2)_{\Delta^2} + \delta(\hat{O}_2, \hat{O}_3)_{\Delta^2} + \delta(\hat{O}_3, \hat{O}_4)_{\Delta^2} \geq \delta(\hat{O}_1, \hat{O}_4)_{\Delta^2}. \quad (42.98)$$

[21] See also [BRAUNSTEIN/CAVES 1995] for further developments: here two new URs are also introduced.

Now consider a state of the form [see Eq. (38.32)]:

$$|\psi(\theta)\rangle = \cos\theta|b_1^1, b_2^1\rangle - \sin\theta|b_1^2, b_2^2\rangle, \tag{42.99}$$

where $|b_1^1\rangle, |b_1^2\rangle$ form a set of basis states for subsystem $\mathcal{S}_1$, and $|b_2^1\rangle, |b_2^2\rangle$ for subsystem $\mathcal{S}_2$. As we know, for $\theta = 0$, $|\psi(\theta)\rangle$ is a product state, hence with no violation at all of the Bell inequalities. We have the singlet state for $\theta = \pi/4$ and identical eigenvalues for each pair $|b_1^1\rangle, |b_2^2\rangle$ and $|b_2^1\rangle, |b_1^2\rangle$, and opposed between both pairs. For a generic $\{|b^1\rangle, |b^2\rangle\}$ basis we have:

$$|+\rangle = \cos\phi|b^1\rangle + \sin\phi|b^2\rangle, \tag{42.100a}$$
$$|-\rangle = -\sin\phi|b^1\rangle + \cos\phi|b^2\rangle, \tag{42.100b}$$

from which we obtain (we fix ϕ_1, ϕ_2 for observables of $\mathcal{S}_1, \mathcal{S}_2$):

$$\langle +_1 +_2 |\psi(\theta)\rangle = \cos\theta\cos\phi_1\cos\phi_2 + \sin\theta\sin\phi_1\sin\phi_2, \tag{42.101a}$$
$$\langle +_1 -_2 |\psi(\theta)\rangle = -\cos\theta\cos\phi_1\sin\phi_2 + \sin\theta\sin\phi_1\cos\phi_2, \tag{42.101b}$$
$$\langle -_1 +_2 |\psi(\theta)\rangle = -\cos\theta\sin\phi_1\cos\phi_2 + \sin\theta\cos\phi_1\sin\phi_2, \tag{42.101c}$$
$$\langle -_1 -_2 |\psi(\theta)\rangle = \cos\theta\sin\phi_1\sin\phi_2 + \sin\theta\cos\phi_1\cos\phi_2. \tag{42.101d}$$

Let us now fix the following values for ϕ: $\phi_{\hat{O}^a} = -\phi_{\hat{O}^{b'}} \neq 0 = \phi$ and $\phi_{\hat{O}^{a'}} = \phi_{\hat{O}^b} = 0$. Then the covariance distances are:

$$\delta(\hat{O}^{a'}, \hat{O}^b)_{\Delta^2} = 0, \tag{42.102a}$$
$$\delta(\hat{O}^{a}, \hat{O}^b)_{\Delta^2} = \delta(\hat{O}^{a'}, \hat{O}^{b'})_{\Delta^2} = 2\sin^2\phi, \tag{42.102b}$$
$$\delta(\hat{O}^{a}, \hat{O}^{b'})_{\Delta^2} = 4(\cos\theta + \sin\theta)^2\cos^2\phi\sin^2\phi, \tag{42.102c}$$

from which the quadrilateral covariance distance Bell inequality can be derived:

$$4\sin^2\phi \geq 4(1 + 2\cos\theta\sin\theta)\cos^2\phi\sin^2\phi; \tag{42.103a}$$

which can be also expressed as:

$$1 \geq [1 + \sin(2\theta)]\cos^2\phi. \tag{42.103b}$$

It is easy to see that this quadrilateral covariance distance Bell inequality is violated for each value of θ in $(0, \pi/4]$ and some values of ϕ. Hence violations of Bell inequalities can be found for each non-product state.

42.7.4 Generalization of Information-Distance Bell Inequalities

It is possible to completely parallelize the probability structure developed by Maczyński and Beltrametti [see section 39.3] by entropic means [CERF/ADAMI 1997]. In analogy with Eq. (42.91) we may write for three observables $\hat{O}_1, \hat{O}_2, \hat{O}_3$:

$$H(\hat{O}_1 : \hat{O}_2) = H(\hat{O}_1 : \hat{O}_2|\hat{O}_3) + H(\hat{O}_1 : \hat{O}_2 : \hat{O}_3), \tag{42.104}$$

in consequence of which we may write as follows the piece of information which is shared by a third observable:

$$H(\hat{O}_1 : \hat{O}_2 : \hat{O}_3) = H(\hat{O}_1) + H(\hat{O}_2) + H(\hat{O}_3) - H(\hat{O}_1, \hat{O}_2) - H(\hat{O}_1, \hat{O}_3) - H(\hat{O}_2, \hat{O}_3) + H(\hat{O}_1, \hat{O}_2, \hat{O}_3). \tag{42.105}$$

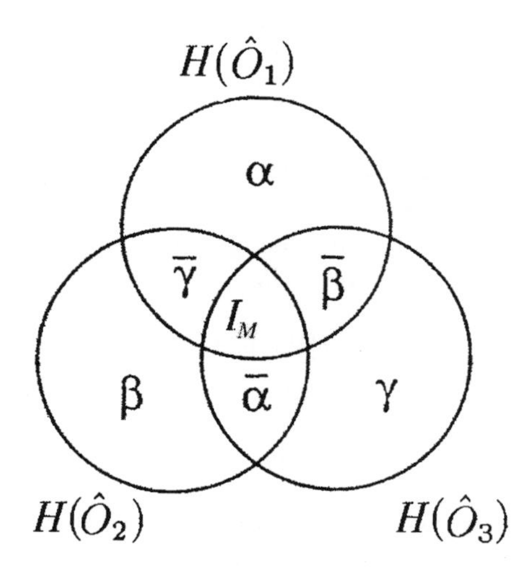

Figure 42.9: Ternary entropy diagram for three observables $\hat{O}_1, \hat{O}_2, \hat{O}_3$ — from [CERF/ADAMI 1997, 3373].

Let us use α, β, γ for writing the conditional entropies — i.e. $\alpha = H(\hat{O}_1|\hat{O}_2 + \hat{O}_3)$ — and $\overline{\alpha}, \overline{\beta}, \overline{\gamma}$ to write the conditional informations — i.e. $\overline{\alpha} = H(\hat{O}_2 : \hat{O}_3|\hat{O}_1)$ — and I_M to write the mutual information between the three observables — i.e. $I_M = H(\hat{O}_1 : \hat{O}_2 : \hat{O}_3)$ [see figure 42.9]. In a classical system all entries except I_M are nonnegative. Hence we can write:

$$0 \leq \alpha + \overline{\alpha} = H(\hat{O}_1) + H(\hat{O}_2 : \hat{O}_3) - H(\hat{O}_1 : \hat{O}_2) - (\hat{O}_1 : \hat{O}_3), \tag{42.106a}$$

$$0 \leq \beta + \overline{\beta} = H(\hat{O}_2) + H(\hat{O}_1 : \hat{O}_3) - H(\hat{O}_1 : \hat{O}_2) - H(\hat{O}_2 : \hat{O}_3), \tag{42.106b}$$

$$0 \leq \gamma + \overline{\gamma} = H(\hat{O}_3) + H(\hat{O}_1 : \hat{O}_2) - H(\hat{O}_1 : \hat{O}_3) - H(\hat{O}_2 : \hat{O}_3), \tag{42.106c}$$

which are exactly equivalent to inequalities (39.16b)–(39.16e).

A similar treatment can be developed for the CH or CHSH inequality, from which the entropic inequalities developed by Braunstein/Caves and Schumacher can also be obtained. Note that a violation of inequalities (42.106) implies that one of the sum of type $\alpha + \overline{\alpha}$ is less then zero. But since $\overline{\alpha}, \overline{\beta}, \overline{\gamma}$ are always ≥ 0, it follows that one of the conditional entropies α, β, γ must be negative. Therefore, the application of a HV local model to QM leads to negative entropies.

42.8 Measure of Entanglement

It is astonishing that only in the 1990's the problem of a measure of entanglement has arisen. Perhaps, in analogy with the yes/no interpretation of the complementarity [see section 8.1], one thought that entanglement was a property which systems could or could not have. However, the difficulty in defining a measure of the entanglement is due to the fact that there can be different physical conditions which characterize the amount of entanglement. Hence there may be no unique measure.

42.8.1 Barnett and Phoenix's Proposal

We have already discussed the purification procedure [see section 38.3]. From here it follows that there is the possibility to give a measure of the entanglement E. A first attempt to measure quantum optical correlations using von Neumann entropy was developed by Barnett and Phoenix [BARNETT/PHOENIX 1989] [BARNETT/PHOENIX 1991]. They proposed to consider as a measure of the correlation between two subsystems (generalization follows by mathematical induction) the following index of entanglement:

$$\mathrm{E}_I = I_{12} - (I_1 + I_2), \tag{42.107}$$

where I_{12} is the informational content of the full density operator, given by $I_{12} = H_{\mathrm{Max}} - H$ (where H_{Max} is the maximum possible entropy for the system and H its actual entropy), and I_1, I_2 are the information content of the reduced density operator of systems 1, 2, respectively. Since it can be difficult to calculate such a index, the authors proposed an observable correlation index. If

$$I_1(\hat{O}_1) = H_{\mathrm{Max}}(\hat{O}_1) + \sum_j \langle j|\hat{\rho}^1|j\rangle \ln\langle j|\hat{\rho}^1|j\rangle, \tag{42.108}$$

represents the informational content of the observable $\hat{O}_1$ on subsystem 1 (where $|j\rangle$ are its eigenvectors), then the joint Shannon entropy between two observables pertaining to the two subsystems, respectively, is:

$$H_{12}(\hat{O}_1, \hat{O}_2) = -\sum_j \sum_k \langle j|\langle k|\hat{\rho}|k\rangle|j\rangle \ln\langle j|\langle k|\hat{\rho}|k\rangle|j\rangle, \tag{42.109}$$

where the $|k\rangle$'s are eigenstates of $\hat{O}_2$. From this we obtain:

$$\mathrm{E}_I(\hat{O}_1, \hat{O}_2) = H_1(\hat{O}_1) + H_2(\hat{O}_2) - H_{12}(\hat{O}_1, \hat{O}_2), \tag{42.110}$$

where $H_1(\hat{O}_1), H_2(\hat{O}_2)$ are the entropies of observables $\hat{O}_1, \hat{O}_2$ on subsystems 1 and 2, respectively. The use of use of mutual information for measuring entanglement was also in [ZUREK 1983]. Barnett and Phoenix have then shown that this measure is less than or equal to E_I given by Eq. (42.107).

42.8.2 Bennett and Co-workers' Proposal

Bennett and co-workers [BENNETT *et al.* 1996c] made also use of the von Neumann entropy for two subsystems $\mathcal{S}_1, \mathcal{S}_2$:

$$\mathrm{E}(\hat{\rho}_{1+2}) := \min \sum_j \wp_j H_{\mathrm{VN}}(\hat{\rho}_1^j), \tag{42.111}$$

where the minimum is taken over all the possible realizations of the state $\hat{\rho}_{1+2} = \sum_k \wp_k |\psi_k\rangle\langle\psi_k|$ with $\hat{\rho}_1^j = \mathrm{Tr}_2(|\psi_k\rangle\langle\psi_k|)$. The problem with such a formulation is that it fails to distinguish between classical and QM correlations. This formulation is but good if is only for pure states (it is right? ask Bennett).

42.8.3 Vedral and Co-workers' Measure

Another, more stringent formulation, has been given by Vedral and co-workers [VEDRAL *et al.* 1997 2276–77]. They establish three conditions which such a measure has to fulfil for a state $\hat{\rho}$ (where we drop the subscript '1 + 2'):

- $\mathrm{E}(\hat{\rho}) = 0$ iff $\hat{\rho}$ is separable [see definition 3.2: p. 42].

- Local unitary operations leave $\mathrm{E}(\hat{\rho})$ invariant, i.e.: $\mathrm{E}(\hat{\rho}) = \mathrm{E}\left(\hat{U}_1 \otimes \hat{U}_2 \hat{\rho} \hat{U}_1^\dagger \otimes \hat{U}_2^\dagger\right)$.

- The measure of the entanglement $\mathrm{E}(\hat{\rho})$ cannot increase under local measurement and classical correlation [see Eq. (38.33)]: if we represent such an operation by $\mathcal{T}$, then $\mathrm{E}(\mathcal{T}\hat{\rho}) \leq \mathrm{E}(\hat{\rho})$. If we add the requirement of postselection [Eq. (38.34)], we can also write, in a more general form:

$$\mathrm{Tr}\left(\hat{O}_j \hat{\rho} \hat{O}_j^\dagger\right) \mathrm{E}\left[\frac{\hat{O}_j \hat{\rho} \hat{O}_j^\dagger}{\mathrm{Tr}\left(\hat{O}_j \hat{\rho} \hat{O}_j^\dagger\right)}\right] \leq \mathrm{E}(\hat{\rho}), \tag{42.112}$$

where $\mathcal{T}\hat{\rho} = \sum_j \hat{O}_j \hat{\rho} \hat{O}_j^\dagger$ [see Eq. (18.5a)].

The reason for the first requirement is evident (separable states contain no entanglement and hence cannot be purified); the reason for the second requirement is that local unitary transformations only represent local changes of basis [see theorem 17.6: p. 271] and leave quantum correlations unchanged; the reason for condition III is that each increase in correlations achieved by $\mathcal{T}$ is classical in nature, and hence entanglement is not increased: since each form of operation is local, correlations cannot be increased by this means. On the other hand, there cannot be a purification procedure without some loss of entanglement, i.e. such a procedure is irreversible [VEDRAL/PLENIO 1998, 1631].

Let us now consider the set of all density matrices of two quantum subsystems $\mathcal{S}_1, \mathcal{S}_2$: E + D, which has two subsets: D, which is the set of all disentangled states, and E, which is the set of all entangled ones [see figure 42.10]. Note that the sets E + D and D are convex [see theorem 3.2: p. 41] but not E.

Now we define the entanglement of a density operator $\hat{\rho}_\mathrm{E}$ as follows:

$$\boxed{\mathrm{E}(\hat{\rho}_\mathrm{E}) := \min H(\hat{\rho}_\mathrm{E} \parallel \hat{\rho}_\mathrm{D})} \tag{42.113}$$

where $\hat{\rho}_\mathrm{D}$ is an arbitrary density operator $\in$ D, and the distance between it and $\hat{\rho}_\mathrm{E}$ is expressed in terms of relative entropy [VEDRAL *et al.* 1997b] [see subsection 42.3.1].[22] The difficulty is that the relative entropy is not symmetric, as should be the case for a proper distance. However, it can be reasonable to define quantum entanglement as a distance from (relative to) a disentangled state. Therefore, we can formulate the following theorem [VEDRAL/PLENIO 1998, 1629–30]:

Theorem 42.6 (Vedral/Plenio) *The probability of not distinguishing two quantum states $\hat{\rho}$ and $\hat{\rho}'$ after N measurements is:*

$$\wp(\hat{\rho} \mapsto \hat{\rho}') = e^{-NH(\hat{\rho}' \| \hat{\rho})}. \tag{42.114}$$

Our I requirement is satisfied for $\hat{\rho}_\mathrm{E} = \hat{\rho}_\mathrm{D}$, the II one is automatically satisfied, because D is invariant under unitary transformations, and for the III one it is sufficient that

$$H\left(\mathcal{T}\hat{\rho}_\mathrm{E} \parallel \mathcal{T}\hat{\rho}_\mathrm{D}\right) \leq H\left(\hat{\rho}_\mathrm{E} \parallel \hat{\rho}_\mathrm{D}\right), \tag{42.115}$$

[22] For a similar definition see also [SHIMONY 1995, 676].

which can be satisfied, if, one suppose that $\hat{\rho}^*$ satisfies the minimum required by Eq. (42.113), because of $\mathcal{T}\mathrm{D} \subset \mathrm{D}$ — where the operation $\mathcal{T}$ is defined by theorem 18.1 [p. 292] —, we have:

$$\mathrm{E}(\hat{\rho}_\mathrm{E}) := H(\hat{\rho}_\mathrm{E} \parallel \hat{\rho}^*) \geq (\mathcal{T}\hat{\rho}_\mathrm{E} \parallel \mathcal{T}\hat{\rho}^*) \quad (42.116)$$

$$\geq \min_{\hat{\rho}_\mathrm{D} \in \mathrm{D}} H(\mathcal{T}\hat{\rho}_\mathrm{E} \parallel \hat{\rho}_\mathrm{D}) = \mathrm{E}(\mathcal{T}\hat{\rho}_\mathrm{E}). \quad (42.117)$$

Therefore, the amount of entanglement given by Eq. (42.113) can be interpreted as finding a state $\hat{\rho}^*$ in D that is the closest to $\hat{\rho}_\mathrm{E}$ under the 'measure' $H(\hat{\rho}_\mathrm{E} \parallel \hat{\rho}_\mathrm{D})$. $\hat{\rho}^*$ approximates the classical correlations of $\hat{\rho}_\mathrm{E}$ as closely as possible. In this way we are able to divide the correlations in the QM component $\mathrm{E}(\hat{\rho}_\mathrm{E})$ and the classical one $H(\hat{\rho}^* \parallel \hat{\rho}_1^* \otimes \hat{\rho}_2^*)$.

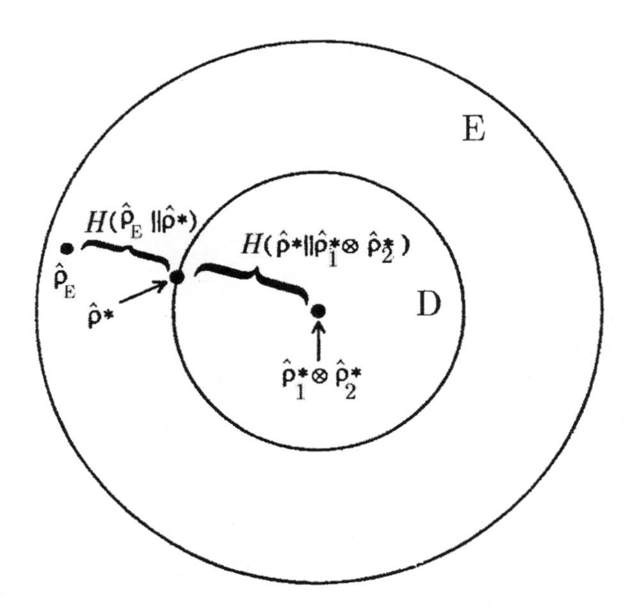

Figure 42.10: The set of all density matrices is represented by E + D. Its subset, a set of the disentangled states D is represented by the inner circle. A state $\hat{\rho}_\mathrm{E}$ belongs to the entangled states, and $\hat{\rho}^*$ is the disentangled state which minimizes the distance $\delta(\hat{\rho}_\mathrm{E} \parallel \hat{\rho}_\mathrm{D})$, thus representing the amount of QM correlations in $\hat{\rho}_\mathrm{E}$. A state $\hat{\rho}_1^* \otimes \hat{\rho}_2^*$ is obtained by tracing $\hat{\rho}^*$ over $\mathcal{S}_1, \mathcal{S}_2$. $\delta(\hat{\rho}^* \parallel \hat{\rho}_1^* \otimes \hat{\rho}_2^*)$ represents the classical part of correlations in state $\hat{\rho}_\mathrm{E}$ — from [VEDRAL *et al.* 1997a, 2277].

42.8.4 Schlienz and Mahler's Measure

Another interesting proposal is due to Schlienz and Mahler [SCHLIENZ/MAHLER 1995]. Though not so general as the preceding proposal, it is a powerful formalism. The conditions posed by the authors on E are the same as those posed by Vedral and co-workers. They follow a method of Hioe and Eberly [HIOE/EBERLY 1981]: since any Hermitian operator on a discrete N–dimensional Hilbert space can be expanded in the unit operator and the generators of the SU(N) algebra, one can specify any density operator by the coefficients of these generators. In order to obtain the generators, let us introduce the transition-projectors:

$$\hat{P}_{jk} = |j\rangle\langle k|, \quad (42.118)$$

and construct the $N^2 - 1$ operators:

$$\begin{aligned} \hat{w}_l &= -\sqrt{\frac{2}{l(l+1)}}\,(\hat{P}_{11} + \hat{P}_{22} + \ldots + \hat{P}_{ll} - l\hat{P}_{l+1,l+2}), & (42.119a) \\ \hat{w}^+_{jk} &= \hat{P}_{jk} + \hat{P}_{kj}, & (42.119b) \\ \hat{w}^-_{jk} &= \imath(\hat{P}_{jk} - \hat{P}_{kj}), & (42.119c) \end{aligned}$$

where $1 \leq l \leq N, 1 \leq j < k \leq N$. Then the following set:

$$\{\hat{\lambda}_j\} := \{\hat{P}^+_{12}, \hat{P}^+_{13}, \ldots, \hat{P}^-_{12}, \hat{P}^-_{13}, \ldots, \hat{P}_1, \hat{P}_2, \ldots, \hat{P}_{N-1}\}, \quad (42.120)$$

where $j = 1, \ldots, N^2 - 1$, fulfilling the relations

$$\begin{aligned} \mathrm{Tr}\{\hat{\lambda}_j\} &= 0, & (42.121a) \\ \mathrm{Tr}\{\hat{\lambda}_j\hat{\lambda}_k\} &= 2\delta_{jk}, & (42.121b) \end{aligned}$$

is the required generator of the SU(N) algebra. In the case $N = 2$ these operators can be represented by the Pauli matrices (3.125). Then a density operator can be represented as follows:

$$\hat{\rho} = \frac{1}{N}\hat{I} + \frac{1}{2}\sum_{j=1}^{N^2-1} \lambda_j\hat{\lambda}_j, \quad (42.122)$$

where $1/N$ is a normalization factor in order to obtain the required $\mathrm{Tr}(\hat{\rho}) = 1$ and $\lambda_j = \mathrm{Tr}(\hat{\lambda}_j\hat{\rho})$. λ is a generalized Bloch vector.

In the case of a two-particle system on Hilbert spaces $\mathcal{H}_1, \mathcal{H}_2$ with dimensions N_1, N_2 respectively, the density operator can be written:

$$\begin{aligned} \hat{\rho} &= \frac{1}{N_1N_2}(\hat{I} \otimes \hat{I}) \\ &+\frac{1}{2N_2}\sum_{j=1}^{N_1^2-1} \lambda_j(1)(\hat{\lambda}_j \otimes \hat{I}) + \frac{1}{2N_1}\sum_{j=1}^{N_2^2-1} \lambda_j(2)(\hat{I} \otimes \hat{\lambda}_j) \\ &+\frac{1}{4}\sum_{j=1}^{N_1^2-1}\sum_{j=1}^{N_2^2-1} \mathrm{C}_{jk}(1,2)(\hat{\lambda}_j \otimes \hat{\lambda}_k), & (42.123) \end{aligned}$$

where

$$\begin{aligned} \lambda_j(1) &= \mathrm{Tr}(\hat{\rho} \cdot \hat{\lambda}_j \otimes \hat{I}), & (42.124a) \\ \lambda_j(2) &= \mathrm{Tr}(\hat{\rho} \cdot \hat{I} \otimes \hat{\lambda}_j); & (42.124b) \end{aligned}$$

while the second-rank tensor

$$\mathrm{C}_{jk}(1,2) = \mathrm{Tr}(\hat{\rho} \cdot \hat{\lambda}_j \otimes \hat{\lambda}_k) \quad (42.125)$$

accounts for correlations — it is the discarded part by a measurement [see section 42.3]. Then one can write the entanglement in terms of the tensor $\mathrm{E}(1,2)$ such that:

$$\mathrm{E}_{jk}(1,2) := \mathrm{C}_{jk}(1,2) - \lambda_j(1)\lambda_k(2). \quad (42.126)$$

A calculation shows that, for the problem examined, a good measure of entanglement is

$$\mu_{\mathrm{E}} = \frac{N^2}{4(N^2-1)}\,\mathrm{Tr}(\mathrm{E}^T\mathrm{E}), \quad (42.127)$$

where $0 \leq \mu_{\mathrm{E}} \leq 1$, as required.

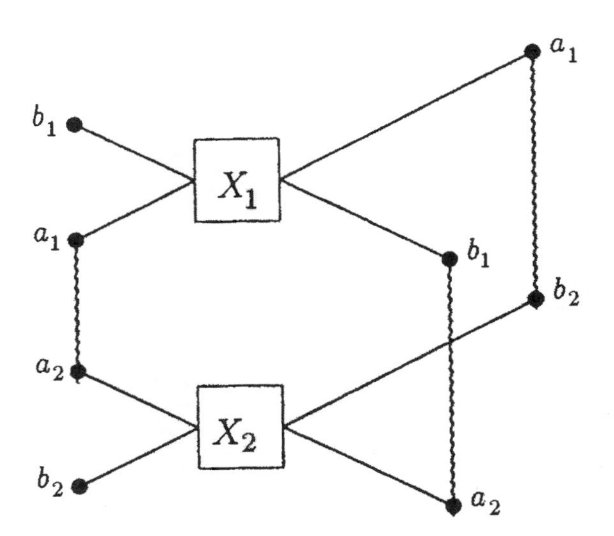

Figure 42.11: Scheme for decompression of information. The input is on the left (entanglement between a_1 and a_2) and the output is on the right (entanglement between a_1 and b_2 and between b_1 and a_2) — from [BUŽEK *et al.* 1997a, 3327].

42.8.5 Popescu and Rohrlich's Measure

Another measure has been proposed by Popescu and Rohrlich [POPESCU/D. ROHRLICH 1997]:

$$\mathrm{E}(|\Psi_{1+2}\rangle) = \lim_{n,k\longrightarrow\infty} \frac{n}{k}\mathrm{E}(|\Psi^0_{1+2}\rangle), \tag{42.128}$$

where k is the initial number of pairs of subsystems in an entangled state and n the number of pairs of subsystems in an entangled state after some operation (purification), and $\mathrm{E}(|\Psi^0_{1+2}\rangle)$ measures the entanglement of a singlet pair. It has the advantage of being unique (due to the fact that the authors allow collective local operations on the entangled pairs), but also the disadvantage that it is dependent on the measure of entanglement of a singlet — which is taken as a basic one.

42.8.6 Decompression

In order to control and to determine the amount of entanglement a technique of *decompression* of the information has been developed [BUŽEK *et al.* 1997a]. It is the inverse operation with respect to the purification [see section 38.3 and subsection 44.3.1]: by starting with a number of highly entangled pairs shared by two distant parties, we end up (by local operations) with a greater number of pairs with a lower entanglement.

Such an operation is a local copying [see section 42.5] of non-local QM correlations. By this means we can control the entanglement, because if we are able to optimally split the original entanglement of a single pair into two equally entangled pairs (having the same state), we have a means of defining half the entanglement of the original pair, which, for example, can be a maximally entangled state like the singlet state [see figure 42.11].

Chapter 43

QUANTUM CRYPTOGRAPHY AND TELEPORTATION

Contents The subjects of this chapter are similar treatments of information (both make use of non-local correlations): quantum cryptography [section 43.1] and teleportation [section 43.2]. The first one concerns the possibility of a perfect secret encoding. The second one exploits the possibility of transmitting quantum information by using Bell correlations. Finally [section 43.3] we discuss some strategies of eavesdropping.

43.1 Quantum Cryptography

43.1.1 Some Basic Concepts and a Brief History

The aim of cryptography is to study procedures for the encryption and decryption of messages so that an eavesdropper cannot decrypt a secret message. Such procedures traditionally required a prior agreement on a key — a string of secret random digits necessary for decryption. The greatest problem here is how to perfectly transmit and store a key. In order to solve the problem of storage, several methods were studied[1]. However, classically there cannot be a perfect safe transmission procedure: a macroscopical object can be measured without significant disturbance[2].

Wiesner was the first physicist to propose (in 1970) the use of QM in order to overcome classical methods. Unfortunately, his results remained unpublished until 1983. A first model, which makes use of them was proposed by Bennett and Brassard in 1984 [it will be discussed in the following subsection]. In 1991 Ekert proposed a model making use of EPR correlations [it will be discussed in subsection 43.1.3].

43.1.2 Bennett/Brassard Model

Bennett and Brassard [BENNETT/BRASSARD 1984] were the first proponents of a quantum cryptography model. They proposed a scheme in which a sender (conventionally called 'Alice') transmits information coded in photon states chosen at random among a set of four possible

[1] For example that of Vernam in 1918, by which a key cannot be reused to send another message. For a general account of quantum cryptography and a brief history of it see [BENNETT *et al.* 1992b].

[2] Other methods, for example that of public-key cryptography (proposed in the 1970's), are also not completely safe for this reason.

polarization states:

$$\{|\updownarrow\rangle, |\leftrightarrow\rangle, |\nearrow\swarrow\rangle, |\searrow\nwarrow\rangle\}, \tag{43.1}$$

where $|\updownarrow\rangle, |\leftrightarrow\rangle$ are states of 90° (vertical) and 0° (horizontal) polarizations, respectively, and $|\nearrow\swarrow\rangle, |\searrow\nwarrow\rangle$ are states of 45° and 135° polarizations, respectively. The receiver (conventionally called 'Bob'), for each photon, chooses at random a measurement either in the rectilinear, i.e. $\{|\updownarrow\rangle, |\leftrightarrow\rangle\}$ basis or in the diagonal, i.e. $\{|\nearrow\swarrow\rangle, |\searrow\nwarrow\rangle\}$ basis [see figure 43.1.(a)]. By using the first basis, rectilinearly polarised photons preserve their polarization direction, whereas diagonally polarised photons randomly change their polarization to one of the two rectilinear polarizations. By using the second basis, diagonally polarised photons preserve their polarization direction, whereas rectilinearly polarised photons randomly change their polarization to one of the two diagonal polarizations. Later on Bob publicly announces what type of measurement he performed, and Alice tells him which measurements were of the correct type. Thereafter, these cases are used as a key. Similar procedures can be used for detecting if some eavesdropping took place [see figure 43.1.(b)]. However, in this way we have not completely solved the problem of key storage. In order to guarantee the security of both key distribution and key storage, models based on EPR correlations were proposed.

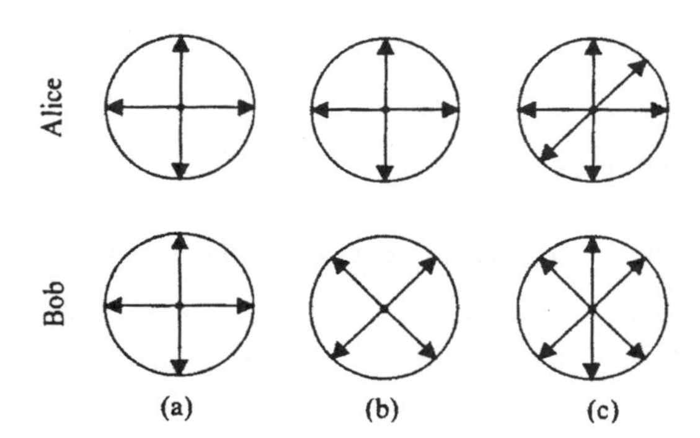

Figure 43.1: In the Bennett/Brassard model Alice and Bob use the same bases to prepare and measure their particles.
(a) Representation of their states on the Poincaré sphere.
(b) A similar setup, but with Bob's bases rotated by 45°, can be used to test the violation of a Bell inequality.
(c) In the Ekert protocol, Alice and Bob may use the violation of the Bell inequality to test for the eavesdropping — from [GISIN/HUTTNER 1997, 14].

43.1.3 Ekert's Model

As we have already said, it was Ekert who first proposed a quantum cryptographic procedure based on EPR correlations [EKERT 1991]. Consider a source which emits pairs of spin–$\frac{1}{2}$ particles in singlet state, and following orientation of $\mathbf{a}_i, \mathbf{b}_j, i, j = 1, 2, 3$ (where $\mathbf{a}$ is 'Alice' and $\mathbf{b}$ 'Bob'): $\phi_{a_1} = 0, \phi_{a_2} = \frac{1}{4}\pi, \phi_{a_3} = \frac{1}{2}\pi, \phi_{b_1} = \frac{1}{4}\pi, \phi_{b_2} = \frac{1}{2}\pi, \phi_{b_3} = \frac{3}{4}\pi$. The quantity composed by correlation coefficients — which appear in the III Bell inequality [the (35.27)] —

$$\mathrm{C}_{\mathbf{a}_i,\mathbf{b}_j} = \mathrm{C}(\mathbf{a}_1, \mathbf{b}_1) - \mathrm{C}(\mathbf{a}_1, \mathbf{b}_3) + \mathrm{C}(\mathbf{a}_3, \mathbf{b}_1) + \mathrm{C}(\mathbf{a}_3, \mathbf{b}_3) \tag{43.2}$$

has a value [see Eq. (35.29)]:

$$C_{\mathbf{a}_i,\mathbf{b}_j} = -2\sqrt{2}. \tag{43.3}$$

After transmission has taken place, Alice and Bob publicly announce the orientation of the analyzers they have chosen for each measurement — they discard all null results. Successively if they publicly reveal only the results for different analyzers' orientations, they can evaluate the quantity of Eq. (43.2), which, as said, should be $= -2\sqrt{2}$. Then they convert, in a secret string — the key —, the results obtained by the same polarisers' orientations (anticorrelation): $C(\mathbf{a}_2, \mathbf{b}_1) = C(\mathbf{a}_3, \mathbf{b}_2) = -1$ [see Eq. (35.6)]. An eavesdropper (c onventionally called 'Eve') can do nothing because, even if she intercepts particles in transit from the source to Alice or Bob, they contain no useful information. Even if Eve tries to disturb the communication, it is possible to know that this manipulation was made because the correlations change. In fact this situation can be described by the following equation, instead of Eq. (43.2):

$$\begin{aligned} C_{\mathbf{a}_i,\mathbf{b}_j} = & \int d\mathbf{n}_a d\mathbf{n}_b \wp(\mathbf{n}_a, \mathbf{n}_b) \\ & \times [(\mathbf{a}_1 \cdot \mathbf{n}_a)(\mathbf{b}_1 \cdot \mathbf{n}_b) - (\mathbf{a}_1 \cdot \mathbf{n}_a)(\mathbf{b}_3 \cdot \mathbf{n}_b) + (\mathbf{a}_3 \cdot \mathbf{n}_a)(\mathbf{b}_1 \cdot \mathbf{n}_b) + (\mathbf{a}_3 \cdot \mathbf{n}_a)(\mathbf{b}_3 \cdot \mathbf{n}_b)], \end{aligned}$$

where $\mathbf{n}_a, \mathbf{n}_b$ are oriented along the axes for which the eavesdropper acquired information about the spin component of the corresponding particle (a or b). The normalized probability measure $\wp(\mathbf{n}_a, \mathbf{n}_b)$ describes the eavesdropper strategy. Choosing as a special case $-\mathbf{n}_a = \mathbf{n}_b$, a simple calculation for a given orientation of $\mathbf{a}_1, \mathbf{a}_3$ or $\mathbf{b}_1, \mathbf{b}_3$ gives:

$$C_{\mathbf{a}_i,\mathbf{b}_j} = \int \wp(\mathbf{n}_a, \mathbf{n}_b) d\mathbf{n}_a d\mathbf{n}_b \left[\sqrt{2}\mathbf{n}_a \cdot \mathbf{n}_b\right], \tag{43.4}$$

which implies

$$-\sqrt{2} \leq C_{\mathbf{a}_i,\mathbf{b}_j} \leq \sqrt{2}, \tag{43.5}$$

which contradicts Eq. (43.3).

If several EPR–correlated particles are distributed to Alice and Bob (an element of a correlated pair to Alice and the other to Bob), just before a key is needed they can measure and compare some particles in order to test for eavesdropping. As already said, such a procedure can guarantee the security of both key distribution and key storage. In other words QM-correlations offer the possibility to apparently encode and transmit information without the possibility of interception [see figure 43.1].[3]

Problems Such a scheme presupposes that the quantum channel used to transmit cryptographed messages is noiseless — this is a general problem of quantum transmission [see subsection 44.3.4]. The problem is that it is difficult to distinguish the entanglement with an eavesdropper from the entanglement with the environment caused by 'innocent' noise, some of which is presumably always present [DEUTSCH *et al.* 1996].

[3] See also [BENNETT *et al.* 1992a] [EKERT *et al.* 1992]. Grunhaus, Popescu and Rohrlich [GRUNHAUS *et al.* 1996] have shown something like the complementary effect: by non-local jamming one can destroy at distance quantum correlations — but there is no violation of the Lorentz invariance. For practical proposals of communication using the quantum cryptography see [MÜLLER *et al.* 1995].

43.2 Teleportation

43.2.1 General Formalism

As we know, an instantaneous transfer of information is not possible [see theorem 36.3: 634]. But it is possible to 'teleport' some information by exploiting EPR correlations and without violating the Einstein locality (the locality *strictu sensu*) [BENNETT/WIESNER 1992] [BENNETT *et al.* 1993]. Suppose that 'Alice' wishes to give 'Bob' some information about a quantum system (a particle named '1') prepared in state $|j\rangle_1$ unknown to her (this is named the *ancilla*). For this purpose Alice lets the particle interact with a couple $\{2, 3\}$ of particles in EPR state, of which 2 is used by her and 3 is previously given to Bob. Now Alice performs a measurement on the compound system $\mathcal{S}_1 + \mathcal{S}_2$, so that, by telling Bob her result, the latter can reconstruct the same state of 1 on 3. To express the procedure formally, let particles 2, 3 be in the EPR state:

$$|\Psi_{23}^-\rangle = \frac{1}{\sqrt{2}}(|\uparrow\rangle_2|\downarrow\rangle_3 - |\downarrow\rangle_2|\uparrow\rangle_3). \tag{43.6}$$

Now Alice performs the measurement (on 1, 2) of the Bell operator [see Eq. (38.25)], introduced by Braunstein, Mann and Revzen, whose basis is given by:

$$|\Psi_{12}^-\rangle = \frac{1}{\sqrt{2}}(|\uparrow\rangle_1|\downarrow\rangle_2 - |\downarrow\rangle_1|\uparrow\rangle_2), \tag{43.7a}$$

$$|\Psi_{12}^+\rangle = \frac{1}{\sqrt{2}}(|\uparrow\rangle_1|\downarrow\rangle_2 + |\downarrow\rangle_1|\uparrow\rangle_2), \tag{43.7b}$$

$$|\Phi_{12}^-\rangle = \frac{1}{\sqrt{2}}(|\uparrow\rangle_1|\uparrow\rangle_2 - |\downarrow\rangle_1|\downarrow\rangle_2), \tag{43.7c}$$

$$|\Phi_{12}^+\rangle = \frac{1}{\sqrt{2}}(|\uparrow\rangle_1|\uparrow\rangle_2 + |\downarrow\rangle_1|\downarrow\rangle_2), \tag{43.7d}$$

which is a complete orthonormal basis of the Hilbert spaces of particles 1 and 2.

If we write the unknown state of the ancilla in the form

$$|j\rangle_1 = c|\uparrow\rangle_1 + c'|\downarrow\rangle_1, \tag{43.8}$$

with $|c|^2 + |c'|^2 = 1$, then the complete state of $\mathcal{S}_1 + \mathcal{S}_{23}$ before the measurement is:

$$|\Psi_{123}\rangle = \frac{c}{\sqrt{2}}(|\uparrow\rangle_1|\uparrow\rangle_2|\downarrow\rangle_3 - |\uparrow\rangle_1|\downarrow\rangle_2|\uparrow\rangle_3) + \frac{c'}{\sqrt{2}}(|\downarrow\rangle_1|\uparrow\rangle_2|\downarrow\rangle_3 - |\downarrow\rangle_1|\downarrow\rangle_2|\uparrow\rangle_3). \tag{43.9}$$

In this Eq. the direct product $|1\rangle|2\rangle$ (because 1, 2 are not entangled) can be written in terms of basis (43.7), which gives:

$$\begin{aligned}|\Psi_{123}\rangle = \frac{1}{2}\Big[&|\Psi_{12}^-\rangle(-c|\uparrow\rangle_3 - c'|\downarrow\rangle_3) + |\Psi_{12}^+\rangle(-c|\uparrow\rangle_3 - c'|\downarrow\rangle_3) \\ &+|\Phi_{12}^-\rangle(c|\downarrow\rangle_3 + c'|\uparrow\rangle_3) + |\Phi_{12}^+\rangle(c|\downarrow\rangle_3 - c'|\uparrow\rangle_3)\Big].\end{aligned} \tag{43.10}$$

Now after Alice's measurement the particle 3 is projected into one of the four pure states superposed in Eq. (43.10), and this depends on the measurement outcome. The possible output states for particle 3 (Bob's one) are very simply related to the original state $|j\rangle_1$ which Alice wishes to teleport:

$$|l\rangle_3 = \hat{U}_i|j\rangle_1, \quad i = 1, 2, 3, 4, \tag{43.11}$$

where:

$$\hat{U}_1 := \begin{bmatrix} -1 & 0 \\ 0 & -1 \end{bmatrix}; \ \hat{U}_2 := \begin{bmatrix} -1 & 0 \\ 0 & 1 \end{bmatrix}; \ \hat{U}_3 := \begin{bmatrix} 0 & 1 \\ 1 & 0 \end{bmatrix} |l\rangle_3; \ \hat{U}_4 \begin{bmatrix} 0 & -1 \\ 1 & 0 \end{bmatrix} |l\rangle_3. \tag{43.12}$$

In the case of the first outcome (the singlet one), the state of particle 3 is the same as of particle 1 except for an irrelevant phase factor, so that Bob need do nothing further to reproduce the state of 1. In the three other cases, Bob must apply one of the unitary operators $\hat{U}_2, \hat{U}_3, \hat{U}_4$ — corresponding, respectively, to 180° rotations around the z, x, y axes — in order to convert the state of 3 into the state of 1. What Bob must do, obviously depends on the (classical) communication of Alice's result.

In conclusion teleportation is based on two channels: a *classical channel*, by which Alice communicates the result of her measurement, and a *quantum channel*, by means of which, using the EPR entanglement, Bob instantaneously recovers — after having received the classical information, and hence without violating the locality [see theorem 36.3: p. 634] — the state of $|j\rangle_1$. However, the original $|j\rangle_1$ is destroyed in accordance with the no-cloning theorem [42.4: p. 722]. If we define an *ebit* as the amount of entanglement between a maximally entangled pair of two-state systems (for example two spin-$\frac{1}{2}$ particles in singlet state), then by teleportation we transmit a quantum bit (= qubit) [see section 44.2] by means of a shared ebit and a two-bit piece of classical information. Note that an ebit is a weaker resource than a qubit [BENNETT *et al.* 1996b, 2046]: in fact the transmission of a qubit can always be used to create one ebit, while the sharing of one ebit or many ebits does not suffice to transmit a qubit (we also need classical information).

43.2.2 Teleportation is the Reverse of an Operation

We have seen an interesting generalization about reversible measurements [subsection 27.4.4]. Nielsen and Caves show [NIELSEN/CAVES 1997, 2551–54] that teleportation is the reverse of a QM operation. Suppose that the initial state for the three particles of the preceding analysis is written in the form:

$$\hat{\rho}_1^{\mathrm{T}} \otimes \hat{\rho}^{23}, \tag{43.13}$$

where $\hat{\rho}_1^{\mathrm{T}}$ is the state to be teleported. In the three-particle case we need unitary reversible operations. With a fourth particle it is possible to abandon the requirement of unitarity. We denote the one-to-one correspondence between particles 1 and 3 by:

$$|l\rangle_3 \leftrightarrow |l^{\mathrm{T}}\rangle_1, \tag{43.14}$$

which means the transport of an unchanged system from location 1 to location 3. Hence we can also write (in the product basis for 2 and 3):

$$|\Psi_{23}\rangle = \sum_{k,l} c_{kl}|k\rangle_2|l\rangle_3 \leftrightarrow \sum_{k,l} c_{kl}|l^{\mathrm{T}}\rangle_1|k\rangle_2 = |\Psi_{12}^{\mathrm{T}}\rangle. \tag{43.15}$$

The map can be extended to the global system:

$$|\Psi_{123}\rangle \leftrightarrow |\Psi_{123}^{\mathrm{T}}\rangle = \hat{U}_{13}|\Psi_{123}\rangle, \tag{43.16}$$

where the unitary ('swap') operator acts on the product states as follows:

$$\hat{U}_{13}|j^{\mathrm{T}}\rangle_1|k\rangle_2|l\rangle_3 = |l^{\mathrm{T}}\rangle_1|k\rangle_2|j\rangle_3. \tag{43.17}$$

Now, supposing that Alice performs a measurement on systems 1 and 2, we can describe it by the operators $(\hat{O}_{12}^{rs})_{\rm T} \otimes \hat{I}_3$, where r is the result of the measurement. By supposing such a result, the unnormalized final state of the target system 3 is:

$$\hat{\rho}_3^r = {\rm Tr}_{12}\left\{\sum_s \left[\left(\hat{O}_{12}^{rs}\right)_{\rm T} \otimes \hat{I}_3\right]\left[\hat{\rho}_1^{\rm T} \otimes \hat{\rho}_{23}\right]\left[\left(\hat{O}_{12}^{rs}\right)_{\rm T}^{\dagger} \otimes \hat{I}_3\right]\right\}. \tag{43.18}$$

As a consequence of Eq. (43.17) we may write:

$$\hat{\rho}_3^r = {\rm Tr}_{12}\left\{\sum_s \left[\left(\hat{O}_{12}^{rs}\right)_{\rm T} \otimes \hat{I}_3\right]\left[\hat{U}_{13}\left(\hat{\rho}_{12}^{\rm T} \otimes \hat{\rho}_3\right)\hat{U}_{13}^{\dagger}\right]\left[\left(\hat{O}_{12}^{rs}\right)_{\rm T}^{\dagger} \otimes \hat{I}_3\right]\right\}. \tag{43.19}$$

This Eq. describes what follows: after the composite system evolves under the action of the unitary swap operator, a measurement is performed on 1 and 2, and then both systems are discarded. This having been done, now the problem of the teleportation for Bob is to recover the initial state $\hat{\rho}_3$ from the output state $\hat{\rho}_3^r = \mathcal{T}_r\hat{\rho}_3$, i.e. he must apply an inverse deterministic operation $\mathcal{T}_r^{-1}$ such that:

$$\mathcal{T}_r^{-1}\left(\frac{\mathcal{T}_r\hat{\rho}_3}{{\rm Tr}[\mathcal{T}_r\hat{\rho}_3]}\right) = \hat{\rho}_3. \tag{43.20}$$

As we have already said, if we have four particles at our disposal, the reverse operation needs not be unitary, though it is deterministic, while in the three particle case, it must have the form of Eq. (27.20):

$$\hat{U}_r\left(\frac{\mathcal{T}_r\hat{\rho}_3}{{\rm Tr}[\mathcal{T}_r\hat{\rho}_3]}\right)\hat{U}_r^{\dagger} = \hat{\rho}_3. \tag{43.21}$$

Sufficient conditions for such a reverse operation are:

- that the state of system 1 is unknown, while the states of 2 and 3 are known to be pure states;

- that we perform an ideal measurement on the joint system 1 and 2 that gives complete information about the posterior state of that system — state 1-2 is left in a pure state — but giving no information about the prior state $\hat{\rho}_1^{\rm T}$ of the system 1 [see also subsection 28.4.3].

43.2.3 Experimental Proposals and Realizations

Cavity Exp. 7 Haroche and co-workers [DAVIDOVICH *et al.* 1994] have developed an experimental proposal. They made use of two microcavities (Cavity Exp. 7) [see figure 43.2]. Here we synthetically describe an outline of the experiment. The cavities are initially empty, and the three atomic beams (A, B, C) are made of identical two-level atoms ($|g\rangle$ and $|e\rangle$: they are circular Rydberg levels with adjacent principal quantum number). The $|e\rangle \rightsquigarrow |g\rangle$ transition is close to resonance with the cavity mode frequency. After switching on beam C the first atom c of this beam crosses the two cavities establishing a nonlocal correlation between them. The atom a (pertaining to A), which is to be duplicated, only crosses C_1. Then its state is reconstructed on b (of beam B), which only crosses C_2. Using this proposal, it has been recently shown that it is possible to interchange two unknown states of two QM systems [MOUSSA 1997].

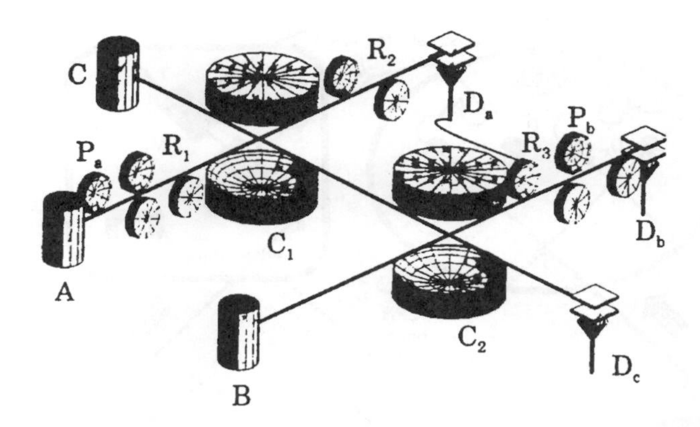

Figure 43.2: Scheme of Cavity Exp. 7 — from [DAVIDOVICH *et al.* 1994, R896].

Interferometry Braunstein and Mann proposed [BRAUNSTEIN/MANN 1995] an interferometry scheme for teleportation[4]. De Martini and co-workers [BOSCHI *et al.* 1998] carried out an experiment which presented the limit that no physical state enters the device from the outside — due to the fact that photons were used. A very important experimental realization was performed by Zeilinger and co-workers [BOUWMEESTER *et al.* 1997]. However, it presents the limit that the teleported state is destroyed at Bob's station.

In a recent experiment performed by Kimble and co-workers [FURUSAWA *et al.* 1998], and based on a preceding theoretical study [BRAUNSTEIN/KIMBLE 1998], a genuine quantum teleportation was realized which made use of optical fields (Interfer. Exp. 28) — which also had the advantage of being a generalization for infinite-dimension Hilbert spaces (due to the continuous character of the observables involved): in fact finite-dimensional systems (with discrete observables) can be considered as subsystems of infinite-dimensional systems. Kimble and co-workers obtained a fidelity [see subsection 43.3.1], i.e. the match between input and output states, of 0.58 ± 0.02, a value not attainable without using quantum correlations. The experiment can be summarized as follows. Two independent squeezed beams are combined at 50%/%50 BS [see figure 43.3]. Squeezed beams are produced by SPDC in an optical parametric oscillator (= OPO). Beam 2 propagates to Alice's sending station, where it is combined at a 50%/%50 BS with $|\psi_i\rangle$, which is in a coherent state of amplitude $\psi_i = \hat{q}_i + \imath \hat{p}_i$ — where $(\hat{q}, \hat{p})$ are quadrature phase amplitudes. Alice uses the set of balanced homodyne detectors $(\mathrm{D}_x, \mathrm{D}_p)$ to make a 'Bell-state' measurement of the amplitudes $\hat{q}_{12} = \frac{\hat{q}_i - \hat{q}_2}{\sqrt{2}}$, $\hat{p}_{12} = \frac{\hat{p}_i - \hat{p}_2}{\sqrt{2}}$ for the input states and the EPR field. Because of the entanglement between 2 and 3, Alice's Bell-state detection collapses Bob's field 3 into a state conditioned on the measurement outcome (i_x, i_p). Bob can then construct a teleported state $\hat{\rho}_f$ via a simple phase-space displacement of the EPR field 2. Finally, a third party, Victor, verifies the fidelity through measurement of the quadrature phase amplitudes.

43.2.4 Superdense Coding

Bennett and Wiesner [BENNETT/WIESNER 1992] studied the property of *superdense coding*: a two-bit classical message is encoded in a particle of an EPR pair by performing one of the

[4]See also [WEINFURTER 1994] for further analysis.

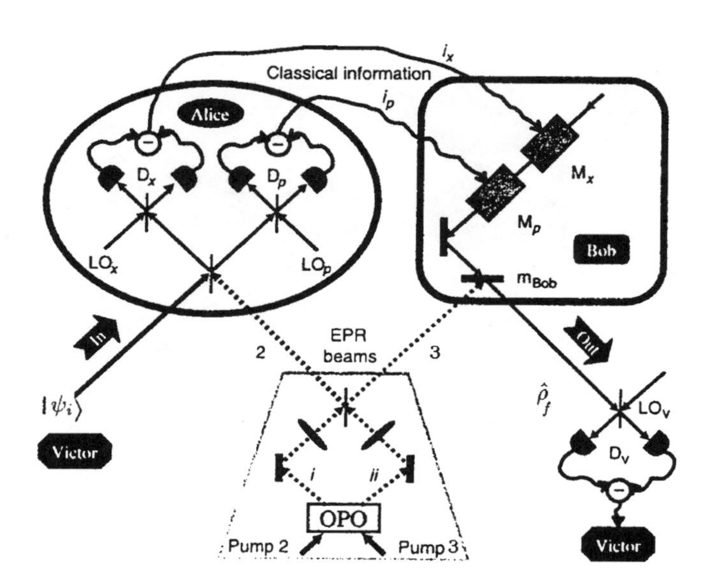

Figure 43.3: Schematic description of Interfer. Exp. 28. An unknown quantum state $|\psi_i\rangle$ (of particle 1) is teleported from Alice's sending station to Bob's receiving terminal by means of the classical information (i_x, i_p) sent from Alice to Bob and the shared entanglement between particles 2 and 3. A third party, Victor, verifies the fidelity through measurement of the quadrature phase amplitudes — from [FURUSAWA *et al.* 1998].

four unitary operations (43.12), so that the whole EPR system is placed in one of the four Bell states. Now the treated particle is given to the receiver — which already has the untreated one —, and, by measuring both, the receiver can recover both bits of the classical message. Hence, while by teleportation one transmits a quantum two-state system by means of two bits of classical information plus a pair of entangled (EPR) two-state particles, by superdense coding one transmits 2 bits of classical information using a QM two-state system plus a pair of entangled (EPR) two-state particles [BENNETT 1995, 26–27] [see figure 43.4].

Kwiat, Mattle and Zeilinger [MATTLE *et al.* 1996] have performed a dense-coding experiment: they have transmitted one of three messages (1 trit = 1.58 bit) by manipulating only one of two entangled particles.

43.3 Fidelity, Disturbance and Information

The following discussion is strictly related to the universal cloning machine [see subsection 42.5.4] and the measure of entanglement [see section 42.8].

43.3.1 Fidelity of Teleportation

Let [POPESCU 1994] [POPESCU 1995] [GISIN 1996b] $\mathbf{F}(\Psi_{23})$ indicate the fidelity of the teleportation, i.e. the mean distance between $|j\rangle_1$ and $|l\rangle_3$ (after Bob's operation) corresponding to the optimal strategy [see Eq. (18.55) and also subsection 44.3.1]:

$$\mathbf{F}(\Psi_{23}) = \langle C \rangle_{\hat{m}} \left[\overline{{}_1\langle j|l_3|j\rangle_1}\right]. \tag{43.22}$$

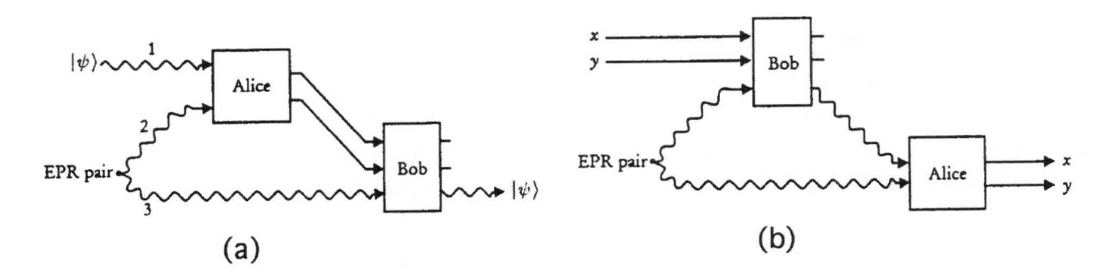

Figure 43.4: (a) Quantum teleportation uses a two-bit classical message (solid lines) as well as an entangled EPR pair of particles to disembody an unknown quantum state $|\psi\rangle$ from one particle and to 'reincarnate' it in another.
(b) Quantum superdense coding reliably transmits two classical bits (x, y) through an entangled pair of particles, even though only one member of the pair is handled by the sender — from [BENNETT 1995, 27].

In the case chosen by Bennett and co-workers [BENNETT *et al.* 1992a] (singlet state) the fidelity is maximal (= 1).

If $|\Psi_{23}\rangle$ is a product state, Alice can only measure $|j\rangle_1$ along an arbitrary but given direction $\mathbf{n}$ and tell Bob the result — up or down — and Bob prepares $|l\rangle_3$ in the up or down state. The corresponding fidelity is $= 2/3$ and can be regarded as the fidelity of a 'classical' teleportation scheme. For local HV-Theories we have a fidelity of $\simeq 0.87$. If we now take the Werner mixed state [see Eq. (38.12)]

$$\tilde{\hat{\rho}} = w\hat{P}_{\Psi_0} + (1-w)\hat{I}, \tag{43.23}$$

where $|\Psi_0\rangle$ is the singlet state, we have a fidelity $= 1/2(w+1)$. For $w < 1/\sqrt{2}$ this fidelity is larger than that of product states. In conclusion we have two bounds, one for local HV-Theories and the other for local quantum states (product states) and between both for non-local mixed QM states. Hence we have confirmed the results of theorem 38.5 [p. 658], that product states cannot be purified.

An almost equivalent form to treat the fidelity is by using density operators [JOZSA 1994]. Using a formula introduced by Uhlmann [UHLMANN 1976] for describing transition probabilities for mixed states, he defined the concept of fidelity $\mathbf{F}$ between two density operators $\hat{\rho}_0, \hat{\rho}_1$ as follows[5]:

$$\mathbf{F}(\hat{\rho}_0, \hat{\rho}_1) = \mathrm{Tr}\left(\sqrt{\hat{\rho}_0^{\frac{1}{2}}\hat{\rho}_1\hat{\rho}_0^{\frac{1}{2}}}\right). \tag{43.24}$$

Its range is between 0 ($\hat{\rho}_0, \hat{\rho}_1$ are orthogonal) and 1 ($\hat{\rho}_0 = \hat{\rho}_1$).

43.3.2 Relationship Between Information and Disturbance

As we have already said, quantum cryptography is nearly sure for communication through noiseless channels; but, for noisy channels (where a signal in initial pure state can arrive in a mixture state due to dispersion), eavesdropping can be possible. In the following subsections we study the problem on a theoretical and practical level.

[5]See also [BARNUM *et al.* 1996, 2818] .

Eve Fuchs and Peres [C. FUCHS/PERES 1996] worked on the problem of a trade-off between the information gain and the disturbance. Consider a teleportation scheme, where, as usual, Alice is the sender, Bob the receiver, and Eve the eavesdropper. We wish to know how much information Eve can gain with minimal disturbance. Alice sends Bob one of the orthonormal states $\{c_m|b_m\rangle\}$ $(m = 1, \ldots, N)$, while Eve prepares a probe in one of the states $\{|v_\theta\rangle\}$ and lets it interact unitarily with the particle sent by Alice to Bob [see figure 43.5]:

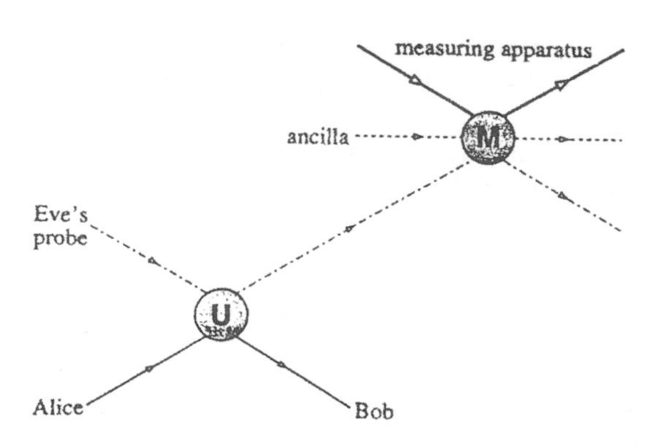

Figure 43.5: Eve's probe interacts unitarily (U) with the particle sent by Alice to Bob and is then subjected to a QND-Measurement (M) — from [C. FUCHS/PERES 1996, 2039].

$$c_m|b_m, v_\theta\rangle \mapsto \hat{U} c_m|b_m, v_\theta\rangle = \sum_{n,\phi} c_m c_{mn\theta\phi}|b_n, v_\phi\rangle, \tag{43.25}$$

where the coefficients $c_{mn\theta\phi}$ (assuming here to be real) are matrix elements of $\hat{U}$:

$$c_{mn\theta\phi} = \langle b_n, v_\phi|\hat{U}|b_m, v_\theta\rangle. \tag{43.26}$$

Since the final optimized results are independent of the choice of the initial state (any pure state can be unitarily transformed into any other pure state), then the index θ is unnecessary:

$$\sum_{n,\phi} c_{mn\phi} c_{m'n\phi} = \delta_{mm'}. \tag{43.27}$$

Note that there can be no more than N^2 linearly independent vectors (pure states of the probe) $\sum_\phi c_{mn\phi}|v_\phi\rangle$; hence we prefer to replace ϕ by the latin index rs and write c_{mnrs}.

Some parameters Now we restrict ourselves to the case where the sent system has two dimensions: 0, 1. Hence the coefficients run from c_{0000} to c_{1111}. This quadruple index can be considered as a single binary number, so that we introduce the notation:

$$c_{mnrs} \mapsto X_K, \quad K = 0, \ldots, 15. \tag{43.28}$$

In order to further simplify, assume that Alice prepares one of the pure states with equal probability [see figure 43.6]:

$$|0\rangle = \cos\theta|b_0\rangle + \sin\theta|b_1\rangle, \tag{43.29a}$$
$$|1\rangle = \cos\theta|b_1\rangle + \sin\theta|b_0\rangle, \tag{43.29b}$$

where their scalar product is denoted by:

$$c_{\langle 0|1\rangle} = \langle 0|1\rangle = \sin(2\theta). \tag{43.30}$$

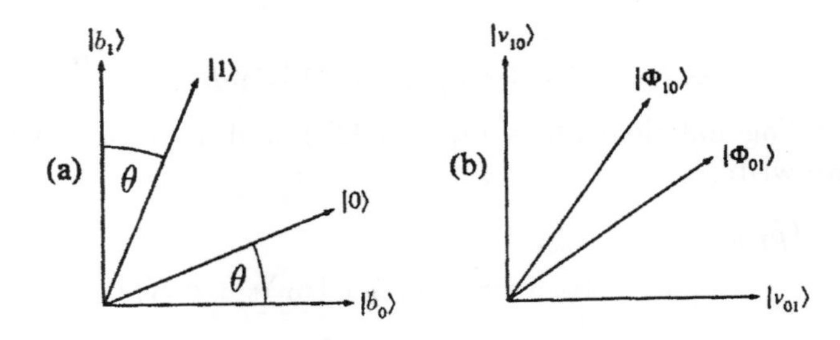

Figure 43.6: Choice of basis for (a) signal states and (b) probe states — from [C. FUCHS/PERES 1996, 2040].

By some algebra and other assumptions (without loss of generality) the parameters X_K become 6 and can be fixed as follows (for four independent parameters λ, μ, η, ξ):

$$X_0 = \sin\lambda\cos\mu, \qquad X_3 = \sin\lambda\sin\mu,$$
$$X_1 = \cos\lambda\cos\eta\cos\xi, \qquad X_2 = \cos\lambda\cos\eta\sin\xi,$$
$$X_5 = \cos\lambda\sin\eta\cos\xi, \qquad X_6 = -\cos\lambda\sin\eta\sin\xi. \tag{43.31}$$

Reduced states If we write $Y_{n\phi} := \sum_m c_m c_{mn\phi}$, then the reduced density matrix of the system received by Bob after Eve's disturbance (Bob knows nothing about Eve) $\hat{\rho}'_{\rm B} = {\rm Tr}_{\rm Eve}(\hat{\rho}')$ (where $\hat{\rho}'$ is the output combined state) is:

$$(\hat{\rho}'_{\rm B})_{mn} = \sum_{\phi} Y_{m\phi}Y_{n\phi}; \tag{43.32a}$$

and the reduced output state of the probe is:

$$(\hat{\rho}'_{\rm Eve})_{\theta\phi} = \sum_{m} Y_{m\theta}Y_{m\phi}. \tag{43.32b}$$

Let us now define the disturbance as the difference between the unity and the fidelity:

$$\mathbf{D} = 1 - \langle\psi_n|\hat{\rho}_f|\psi_n\rangle, \tag{43.33}$$

where $|\psi_n\rangle$ is the input state and $\hat{\rho}_f$ the output state.

Now the discrepancy observed by Bob is:

$$\mathbf{D} = 1 - \sum_{m,n} c_n Y_{n\phi}. \tag{43.34}$$

Information and Disturbance Under the assumption that Alice prepares the equiprobable pure states (43.29) and by the 01 symmetry, Eve's two possible density matrices can be written as:

$$\hat{\rho}'_0 = \begin{pmatrix} a & f \\ f & b \end{pmatrix}, \quad \hat{\rho}'_1 = \begin{pmatrix} b & f \\ f & a \end{pmatrix}, \tag{43.35}$$

with $a + b = 1$ and where explicit reference to Eve is dropped. The two matrices have the same determinant $d = ab - f^2 \geq 0$. Then the mutual information [see Eq. (42.35) and also subsection 42.4.3] that can be extracted is given by [C. FUCHS/CAVES 1994]:

$$I = \frac{1}{2}[(1+z)\ln(1+z) + (1-z)\ln(1-z)], \tag{43.36}$$

where:

$$z = [1 - 2d - \mathrm{Tr}(\hat{\rho}'_0\hat{\rho}'_1)]^{\frac{1}{2}} = (1 - 4ab)^{\frac{1}{2}}. \tag{43.37}$$

Hence we only need the diagonal elements in Eq. (43.32b), and, by virtue of Eqs. (43.31) and of the definition of $Y_{n\gamma}$, we write:

$$\begin{aligned} (\hat{\rho}'_E)_{01,01} &= Y^2_{n01} \\ &= (c_0X_1 + c_1X_6)^2 + (c_0X_5 + c_1X_2)^2 \\ &= \frac{1}{2}(1 + \cos 2\theta \cos 2\xi); \end{aligned} \tag{43.38a}$$

and likewise:

$$(\hat{\rho}'_E)_{10,10} = \frac{1}{2}(1 - \cos 2\theta \cos 2\xi); \tag{43.38b}$$

from which we obtain:

$$z = \cos 2\theta \cos 2\xi. \tag{43.39}$$

The discrepancy rate $\mathbf{D}$ can now be expressed as:

$$\mathbf{D} = \sin^2\eta - \left(\frac{c_{\langle 0|1\rangle}}{2}\right)\sin 2\eta \cos 2\xi + \left(\frac{c^2_{\langle 0|1\rangle}}{2}\right)\cos 2\eta(1 - \sin 2\xi). \tag{43.40}$$

For each ξ, the angle $\eta = \eta_0$ making $\mathbf{D}$ minimal is such that $\tan 2\eta_0 = c_{\langle 0|1\rangle}\cos 2\xi/[1 - c^2_{\langle 0|1\rangle}(1 - \sin 2\xi)]$, and the minimal discrepancy is

$$\mathbf{D}_0 = \frac{1}{2} - \frac{1}{2}\left\{c^2_{\langle 0|1\rangle}\cos^2 2\xi + [1 - c^2_{\langle 0|1\rangle}(1 - \sin 2\xi)]^2\right\}^{\frac{1}{2}}. \tag{43.41}$$

Now take the value $\xi = 0$; the maximal value of I is given by:

$$I_{\mathrm{Max}} = \ln 2 + \cos^2\theta \ln(\cos^2\theta) + \sin^2\theta \ln(\sin^2\theta), \tag{43.42}$$

and the minimal disturbance corresponding to this value is:

$$\mathbf{D}_1 = \frac{1}{2}\left[1 - (1 - c^2_{\langle 0|1\rangle} + c^4_{\langle 0|1\rangle})^{\frac{1}{2}}\right], \tag{43.43}$$

while the relationship between the maximal mutual information and the minimal disturbance is given by:

$$z = \cos 2\theta \left\{1 - \left[1 - \sqrt{\frac{\mathbf{D}_0(1-\mathbf{D}_0)}{\mathbf{D}_1(1-\mathbf{D}_1)}}\right]^2\right\}^{\frac{1}{2}}, \tag{43.44}$$

which completely determines the searched trade-off. Figure 43.7 shows such a trade-off for three values of θ.

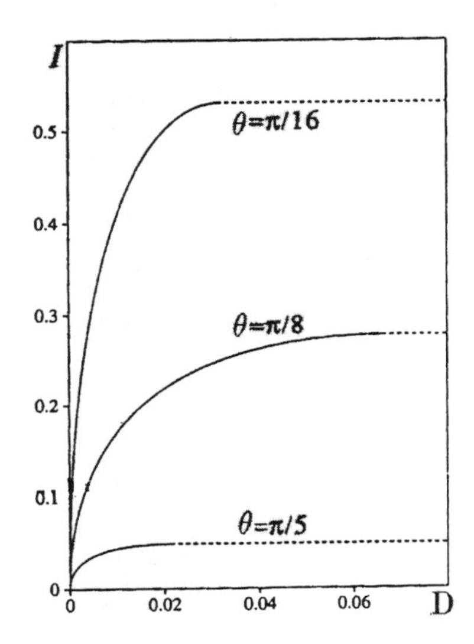

Figure 43.7: Maximal mutual information obtainable for a given disturbance **D**, for two equiprobable input signals. The dashed lines represent the maximal obtainable information I, which cannot be exceeded by accepting a further increase of **D** — from [C. FUCHS/PERES 1996, 2043].

43.3.3 Eavesdropping

Optimal Strategy As we have seen, some strategies have been developed in order to eavesdrop on the exchanged messages. In general eavesdropping consists of letting a third QM system (the probe) interact with the message sent by Alice. Pioneering works in this direction are [EKERT *et al.* 1994] [BUŽEK/HILLERY 1996] [see subsection 42.5.4], where a possibility to almost 'clone' a state is shown, and [C. FUCHS/PERES 1996] [see subsection 43.3.2], where the trade-off between the information gain and the disturbance is calculated.

Eavesdropping strategies can be divided into symmetric and non-symmetric ones. The *symmetric* strategies treat all possible messages sent by Alice on the same footing (for a more rigorous formulation, see below). It can be proved [CIRAC/GISIN 1997, 6–7] that, given a non-symmetric strategy, one can always find a symmetric one which reproduces the result of the non-symmetric strategy. Hence we limit ourselves to the simpler case of symmetric strategies.

Symmetric strategies can be further divided into incoherent and coherent ones: *incoherent* strategies are characterized by the fact that each of Eve's probes interacts independently with each qubit — a two state system [see subsection 44.2.1] — sent by Alice; while, in a *coherent* strategy, Eve lets each probe interact with more than one of Alice's qubits at the same time. It can be shown [CIRAC/GISIN 1997, 4–6] that a coherent strategy cannot improve Eve's information gain, but it can improve the probability that Eve correctly guesses the entire key, at the cost that Eve's guess is wrong on each qubit. Here we limit ourselves to the analysis of incoherent strategies [CIRAC/GISIN 1997, 1–3].

We supposes that Alice sends a random sequence of states of the form (43.1) — in the following we use $|0\rangle$ instead of $|\downarrow\rangle$ and $|1\rangle$ instead of $|\uparrow\rangle$. The initial state of Eve's probe is represented by $|E\rangle$. We suppose that her strategy consists of a unitary evolution (interaction between probe and Alice's message) and of a measurement (extraction of the information from the probe). The unitary evolution is of the form:

$$\hat{U} : |E\rangle \otimes |0\rangle_z \mapsto |E^z_{0,0}\rangle|0\rangle_z + |E^z_{0,1}\rangle|1\rangle_z, \tag{43.45a}$$

$$\hat{U} : |E\rangle \otimes |1\rangle_z \mapsto |E^z_{1,0}\rangle|0\rangle_z + |E^z_{1,1}\rangle|1\rangle_z, \tag{43.45b}$$

where $|E^z_{j,k}\rangle$ represents Eve's state in case Alice sends a j $(= 0, 1)$ bit, and Bob detects a k $(= 0, 1)$ bit. Similar expressions can be written for the x polarisation.

Now we analyse Eve's measurement. The quantities of interest are scalar products such as $\langle E^z_{i,j}|E^z_{k,l}\rangle$, which, for the above assumption of symmetry, need to be invariant

- with respect to the 0 and 1 interchange,
- and with respect to the z and x interchange.

The *first condition* means:

$$\mathbf{F} = \langle E_{0,0}|E_{0,0}\rangle = \langle E_{1,1}|E_{1,1}\rangle, \tag{43.46a}$$

$$\mathbf{D} = \langle E_{0,1}|E_{0,1}\rangle = \langle E_{1,0}|E_{1,0}\rangle, \tag{43.46b}$$

$$\mathbf{F}_O = \langle E_{0,0}|E_{1,1}\rangle = \langle E_{1,1}|E_{0,0}\rangle, \tag{43.46c}$$

$$\mathbf{D}_O = \langle E_{0,1}|E_{1,0}\rangle = \langle E_{1,0}|E_{0,1}\rangle, \tag{43.46d}$$

where $\mathbf{F}$ is the fidelity (Bob detects correctly Alice's bit) [see subsection 44.3.1] and $\mathbf{D}$ the disturbance (the probability of wrong detection) — and $\mathbf{F}_O, \mathbf{D}_O$ the relative overlaps.

The *second condition* (together with the normalization condition) gives:

$$\mathbf{F} + \mathbf{D} = 1, \quad \mathbf{F} - \mathbf{D} = \mathbf{F}_O + \mathbf{D}_O. \tag{43.47}$$

Hence the number of independent parameters is reduced to two (for example $\mathbf{D}, \mathbf{D}_O$). Fixing the disturbance $\mathbf{D}$, one can determine the eavesdropping strategy as a function of only one parameter. With this notation Eqs. (43.45) can be rewritten in the form:

$$\hat{U} : |E\rangle \otimes |0\rangle_z \mapsto \sqrt{\mathbf{F}} E^z_{0,0}\rangle|0\rangle_z + \sqrt{\mathbf{D}}|E^z_{0,1}\rangle|1\rangle_z. \tag{43.48}$$

By Eve's measurement (and limiting ourselves to the z case) we have the following reduced states:

$$\hat{\rho}_{\mathrm{E}}(0) = |E^z_{0,0}\rangle\langle E^z_{0,0}| + |E^z_{0,1}\rangle\langle E^z_{0,1}|, \tag{43.49a}$$

$$\hat{\rho}_{\mathrm{E}}(1) = |E^z_{1,1}\rangle\langle E^z_{1,1}| + |E^z_{1,0}\rangle\langle E^z_{1,0}|. \tag{43.49b}$$

The optimal measurement strategy is based on the fact that the four states (43.49) fall into two mutually orthogonal sets: $S_{\mathbf{F}} = \{|E^z_{0,0}\rangle, |E^z_{1,1}\rangle\}$ and $S_{\mathbf{D}} = \{|E^z_{0,1}\rangle, |E^z_{1,0}\rangle\}$, which therefore can be deterministically distinguished. The set $S_{\mathbf{F}}$ occurs with probability $\mathbf{F}$, in which case Eve has to extract information about two states with the same *a priori* probability and overlap $\cos\alpha := \mathbf{F}_O/\mathbf{F}$. The set $S_{\mathbf{D}}$ occurs with probability $\mathbf{D}$ and state overlap $\cos\beta := \mathbf{D}_{O/\mathbf{D}}$. Hence Eve's information is:

$$I_{\mathrm{E}} = 1 + \mathbf{F}H(\wp_\alpha) + \mathbf{D}H(\wp_\beta), \tag{43.50}$$

where $H(\wp) = \wp\ln(\wp) + (1-\wp)\ln(1-\wp)$ is the Shannon entropy and $\wp_\alpha = \frac{1+\sin\alpha}{2}$. The probability that Eve finds out the state sent by Alice is:

$$\wp_{\mathrm{E}} = \mathbf{F}\wp_\alpha + \mathbf{D}\wp_\beta. \tag{43.51}$$

For disturbance below 5%, as happens in experiments [see figure 43.8], the information gain is proportional to the disturbance $I_{\mathrm{E}} \simeq 2\ln(2)\mathbf{D}$.

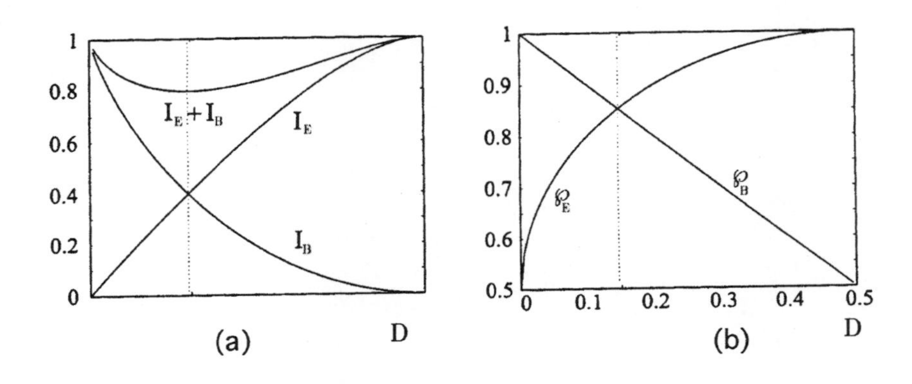

Figure 43.8: (a) Eve's I_{E} information, Bob's I_{B} information and $I_{\mathrm{E}} + I_{\mathrm{B}}$ as functions of the disturbance **D**.
(b) Eve's probability $\wp_{\mathrm{E}}$ and Bob's probability $\wp_{\mathrm{B}}$ of obtaining the message sent by Alice as a function of **D** — from [CIRAC/GISIN 1997, 3].

Counterstrategies As we have already said, a counterstrategy used by Alice and Bob in order to combat the presence of the eavesdropper makes use of the violation of Bell inequalities as a measure of eavesdropping [GISIN/HUTTNER 1997, 18–19]: in fact if the inequalities are violated, then Alice and Bob know that Eve has less mutual information than themselves, and hence they can in principle perform classical information processing to obtain a secret key. If the inequality is not violated, then Eve may have more information, and so a secret key cannot be distilled by classical means. Anyway purification procedures [see section 38.3] are also possible in order to obtain larger entanglements.

Chapter 44

QUANTUM INFORMATION AND COMPUTATION

Introduction and contents Quantum computation is a real theoretical development which is able to use superposition (like teleportation uses entanglement), i.e. an aspect of the theory which seemed puzzling or was sometimes considered as a limitation, in order to obtain results which are not possible within the classical theory framework (of information and communication).

In this chapter, after some definitions [section 44.1], four problems will be examined: how to build an elementary coding system [section 44.2]; how to find a fidelity criterion in order to transmit and encode/decode the signals [section 44.3]; how to build a computation procedure so that we can utilize the specific quantum features [section 44.4]; what problem the decoherence represents for quantum computation and how it could be possible to overcome it [section 44.5]. In section 44.6 we summarize a powerful generalization, and, finally, in section 44.7, concluding remarks follow.

44.1 Classical and Quantum Computation

Before we discuss quantum computation, and in order to understand its specific features, let us summarize, in a few words, some classical concepts [EKERT/JOZSA 1996, 734].[1] A classical *Turing machine* is a recursive device composed of a processing unit in the form of a write/read head and a memory with unlimited storage capacity in the form of an infinite tape divided into cells. Each cell can have a symbol from a finite alphabet. The tape is scanned, one cell at time, from the read/write head. The head can be in one of a finite set of states. The machine action is made up of discrete steps and each step is determined by two initial conditions: the current state of the head and the symbol that occupies the cell currently scanned. Given these two conditions, the machine receives a three-part instruction for its next step, which specifies: the next state of the head; the symbol to write into the scanned cell; and whether the head has to move (left or right) along the tape or to stop. Now we state a theorem proved by Bennett [BENNETT 1973], concerning the reversibility of classical computation:[2]

Theorem 44.1 (Bennett) *Given any Turing machine, there is another reversible one which is able to perform the same computation.*

[1]For a recent detailed study concerning the whole subject of quantum computation see [EKERT/MACCHIAVELLO 1998b]

[2]See also [EKERT/JOZSA 1996, 735].

The importance of the theorem derives (as we shall see) from the reversibility of quantum computation (unitary operations). However, it is not an integral part of the theory, because it is a general theorem about the computation (classical or quantum).

We define [DEUTSCH 1989] a *classical gate* (or Boolean gate) as a Boolean operation [see section 9.1] with two input bits and two output bits.

Quantum computation, on the other hand, positively makes use of the Superposition principle. The first idea of a quantum computation able to exploit the power of the Superposition principle is due to Deutsch [DEUTSCH 1985b], but first steps are due to Feynman [FEYNMAN 1982], who had already proposed a simulation of physical processes on a quantum computer, and to Benioff [BENIOFF 1982], who showed that a Turing machine can be simulated by the unitary evolution of a QM system. In general the MWI [see chapter 15] helped to see the superposition as a positive factor and not as a limit, as was traditionally believed. However, note that a Feynman computer can present the problem of unintended reversals in the computational process, while the more recent proposals (after Deutsch's work) are clocked externally [LANDAUER 1996, 190]. The most important recent developments in quantum computation are dependent on a theorem of Shor [SHOR 1994] [see subsection 44.4.1]. Quantum computation is perhaps the branch of QM which currently has the greatest scientific and technological potentialities.

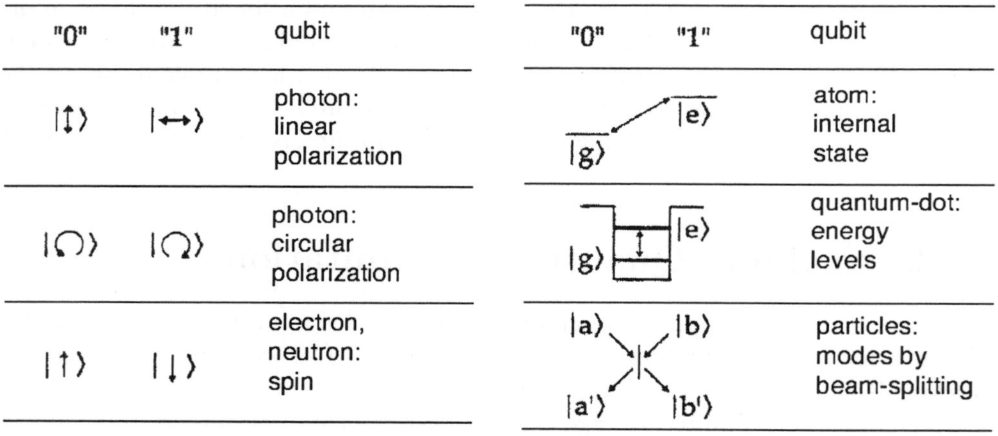

"0"	"1"	qubit
$\vert\updownarrow\rangle$	$\vert\leftrightarrow\rangle$	photon: linear polarization
$\vert\circlearrowleft\rangle$	$\vert\circlearrowright\rangle$	photon: circular polarization
$\vert\uparrow\rangle$	$\vert\downarrow\rangle$	electron, neutron: spin

"0"	"1"	qubit
$\vert g\rangle$	$\vert e\rangle$	atom: internal state
$\vert g\rangle$	$\vert e\rangle$	quantum-dot: energy levels
$\vert a\rangle$, $\vert a'\rangle$	$\vert b\rangle$, $\vert b'\rangle$	particles: modes by beam-splitting

Figure 44.1: Different possibilities to experimentally realize a qubit. With a photon one can apply either linear polarised or circular polarised light, with an electron or neutron one can take advantage of the spin, with an atom the quantum (dot) ground state and excited state, or with a radiation the two modes before and after a BS — from [WEINFURTER/ZEILINGER 1996, 219].

44.2 Quantum Information

44.2.1 Classical Operations

We discuss here the first problem: the coding (i.e. to find a support for 'writing' information in symbolic form). In general terms a *quantum computer* is a physical system where strings of bits (software) are realized by strings of dichotomous elements such as spins (hardware) — it can be shown that systems of n two-state subsystems can simulate the behaviour of any other QM system [DEUTSCH 1985b] . Hence we must establish some connection between a bit of information and a two-state system (*qubit*) [see figure 44.1; see also subsection 43.2.1].

Now we give some simple examples of the reproduction of classical logical connections by quantum states [see figure 44.2].

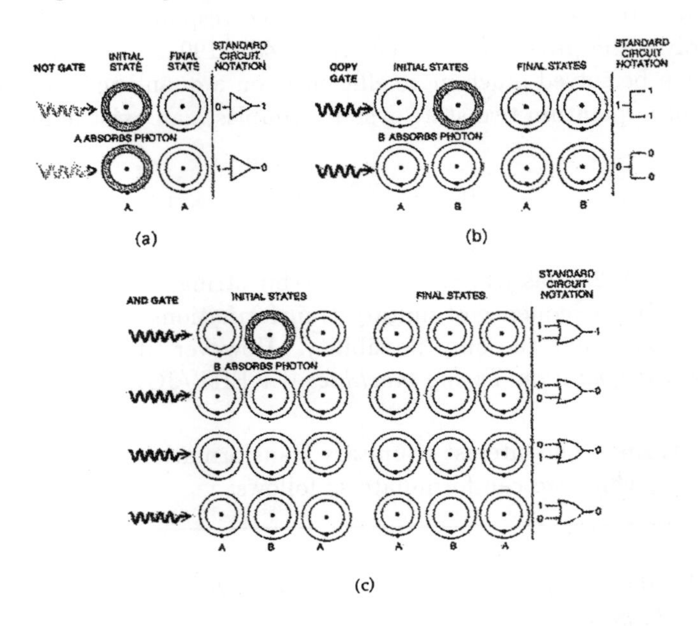

Figure 44.2: Quantum logic gates. We can use atoms or other quantum systems to perform the elementary Boolean operations NOT, COPY, and AND.
(a) NOT involves nothing more than bit flipping, as the notation on the right shows: if A is 0, makes it a 1, and vice versa. With atoms this can be obtained by applying a pulse whose energy equals the difference between A's ground state (shown as the inner ring) and its excited state (outer ring). Unlike classical NOT gates, quantum ones can also flip bits only halfway.
(b) COPY can be realized in the quantum world by the interaction between two atoms. Imagine A storing either 0 or 1, sitting next to B, in its ground state. The difference in energy between the states of B will be a certain value if A is 0, and a different one if A is 1. Now apply a pulse of light whose photons have an energy equal to the latter amount. If the pulse is of the right intensity and duration, and if A is 1, then B will absorb a photon and flip (top row); if A is 0, B cannot absorb a photon from the pulse and stays unchanged (bottom row). Then if A is 1, B becomes 1; if A is 0, B remains 0.
(c) AND can also rely on atomic interactions. Imagine three atoms, of type A, B, and A, respectively, sitting next to each other. The difference in energy between the ground and excited states of B is a function of the states of the two A's. Suppose B is 0. Now apply a pulse whose energy equals the difference between the two states of B only when the neighboring A atoms are both 1. If in fact both A's are 1, this pulse will flip B (top row); otherwise it will leave B unchanged (all other rows) — from [LLOYD 1995b, 46–47].

44.2.2 Qubits

Superposition The interest of a quantum coding and processing is due to the fact that, while classical computers can only be in one of two states (0 or 1), quantum computers can be in an arbitrary superposition of both states [DEUTSCH 1985b] [DEUTSCH 1989] and thus carrying much more information than is allowed in the classical case because, instead of taking only one path, the computation (the intermediate state between input and output) can follow distinct paths in a superposition [BENNETT 1995, 24–25]. Practically, while a classical computer, in order to solve a problem, needs to evaluate a function stepwise, this is not necessary on a quantum computer, due to the superposition of states corresponding to different answers, the collection of which solves the problem [DEUTSCH/JOZSA 1992, 554]. This is the real reason why some problems can be solved much more efficiently on a quantum computer.

Formally a string of n quantum two-state systems (qubits) can exist in any state of the form:

$$|\Psi\rangle = \sum_{\xi=00...0}^{11...1} c_\xi |\xi\rangle, \tag{44.1}$$

where the ξ ranges over all 2^n classical values of a n–bit string. Currently quantum data processing consists of applying a sequence of unitary transformations to the state vector $|\Psi\rangle$. The unitarity of the evolution is not the only possibility. However, it is the most studied and it produces what are named *Quantum Turing machines* [EKERT/JOZSA 1996, 736].

Unitary transformations The interest of unitary transformation is due to a theorem of Lloyd [LLOYD 1995a, 346–47], which we can formulate as follows:

Theorem 44.2 (I Lloyd) *It is possible to generate any unitary transformation by an appropriate choice of a pair of Hamiltonians.*

Proof

Let a set of quantum observables evolve according to a Hamiltonian $\hat{H}_I$ on a n–dimensional Hilbert space. Suppose that one can 'turn on' and 'off' another Hamiltonian $\hat{H}_{II}$ for desired time intervals. Then one lets the observables evolve according to $\hat{H}_I$ for a time t_1, then apply $\hat{H}_{II}$ for a time t_2, then another time $\hat{H}_I$ for a time t_3, and so on, a process corresponding to the unitary operator:

$$\hat{U} := \cdots e^{\imath \hat{H}_I t_3} e^{\imath \hat{H}_{II} t_2} e^{\imath \hat{H}_I t_1}. \tag{44.2}$$

Now, which $\hat{U}$ can be created in this manner? Any $\hat{U} = e^{\imath \hat{O} t}$ [see theorem 3.6: p. 47], where $\hat{O}$ is a member of the algebra $\mathcal{L}_{\hat{H}_I, \hat{H}_{II}}$ generated from $\hat{H}_I, \hat{H}_{II}$ through commutation. This algebra is a set of Hermitian matrices, regarded as vectors, spanned by $\hat{H}_I, \hat{H}_{II}, \imath[\hat{H}_I, \hat{H}_{II}], [\hat{H}_I, [\hat{H}_I, \hat{H}_{II}]], \ldots$. The result can be proved in different ways. Here we report the most simple one[3]. In fact we can build up infinitesimal transformations using the fact that at the limit one obtains:

$$\lim_{n \longrightarrow \infty} \left(e^{-\imath \hat{H}_{II} \sqrt{t/n}} e^{-\imath \hat{H}_I \sqrt{t/n}} e^{\imath \hat{H}_{II} \sqrt{t/n}} e^{\imath \hat{H}_I \sqrt{t/n}} \right)^n = e^{[\hat{H}_I, \hat{H}_{II}]t}. \tag{44.3}$$

Q. E. D.

Experimentally, it was shown by Zeilinger and co-workers [RECK *et al.* 1994] that it is possible to optically construct any discrete unitary operator.

[3] Even if it has the inconvenience that it can require arbitrarily large times. For other methods see the original article [LLOYD 1995a].

Universality of quantum gates Theorem 44.2 now allows the proof of the following theorem, again due to Lloyd [LLOYD 1995a, 348]:

Theorem 44.3 (II Lloyd) *Almost any quantum logic gate with two or more inputs (not necessarily binary) is universal.*

Following Lloyd's definition, we say that a quantum gate is universal if it can be wired (where a wire is a device that takes an output variable from one gate and moves it to the input of another gate) together to make circuits to evaluate any desired classical logic function and to enact any desired unitary transformation on a set of quantum observables.

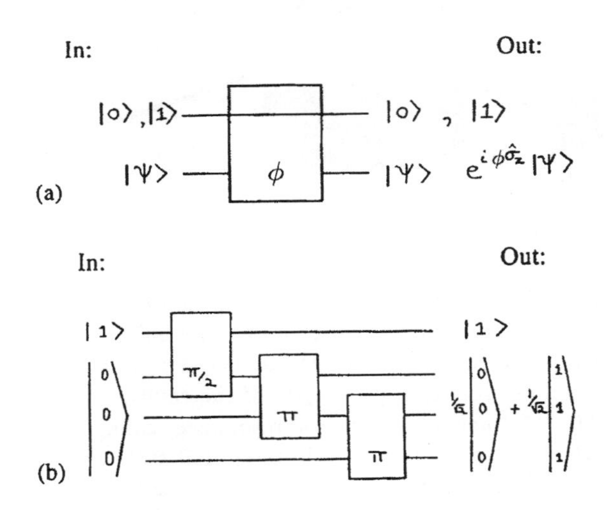

Figure 44.3: (a) Quantum controlled rotation gate. If the first input is 1, the second input is rotated by ϕ about the z axis and multiplied by a phase factor. When ϕ/π is irrational, this gate is a quantum universal logic gate.
(b) Construction of a GHZ state [see section 36.4] by means of a quantum controlled rotation gate. Controlled rotation gates suffice to build a quantum computer — from [LLOYD 1995a, 7]

Proof
Suppose that one can apply each $\hat{H}_I, \hat{H}_{II}$ for predetermined time intervals t_I, t_{II} respectively. Hence one can generate operators of the form:

$$\hat{U} = \cdots \hat{U}^{c_3}_{\hat{H}_I} \hat{U}^{c_2}_{\hat{H}_{II}} \hat{U}^{c_1}_{\hat{H}_I}, \tag{44.4}$$

where $\hat{U}_{\hat{H}_I} = e^{\imath \hat{H}_I t_I}$ and $\hat{U}_{\hat{H}_{II}} = e^{\imath \hat{H}_{II} t_{II}}$, and c_j are integers. Now which $\hat{U}$ can be realized? As we know from the preceding theorem, by applying sequences of Hamiltonians for predetermined time intervals, one can get arbitrarily close to any of the $\hat{U}$ one can create before using continuous times. The reason is that as long as the eigenvalues of $\hat{U}_{\hat{H}_I}$ have phases that are irrationally related to π [see figure 44.3], then for all t there exists some c such that $\| \hat{U}^c_{\hat{H}_I} - e^{\imath \hat{H}_I t} \| < \epsilon$. Here c is of the order of ϵ^{-n}. Similar reasons are also valid for $\hat{H}_{II}$. That is each term of Eq. (44.2) can be approximated with accuracy

ϵ by iterating one of the operations a number of times of the order of ϵ^{-n}. Since the iterations can be chosen so that errors tend to be cancelled out in the product of Eq. (44.4), the total number of iterations in Eq. (44.4) required to realize an arbitrary $\hat{U}$ to accuracy ϵ is of the order of $n^2\epsilon^{-n}$. Now let $\hat{U}_{\hat{H}_I}$ correspond to the action of some arbitrary gate of N inputs, and let $\hat{U}_S$ be the unitary operator corresponding to switching two of the gate's inputs. The set of such $\hat{U}_{\hat{H}_I}, \hat{U}_{\hat{H}_{II}} = \hat{U}_{\hat{H}_I}\hat{U}_S$ that do not obey the criteria established is of zero measure, and $\hat{U}_S, \hat{U}_{\hat{H}_I}$ can be iterated to realize a universal gate such as a controlled-rotation gate on two of the inputs. But a gate that can enact a universal gate is itself universal. Q. E. D.[4]

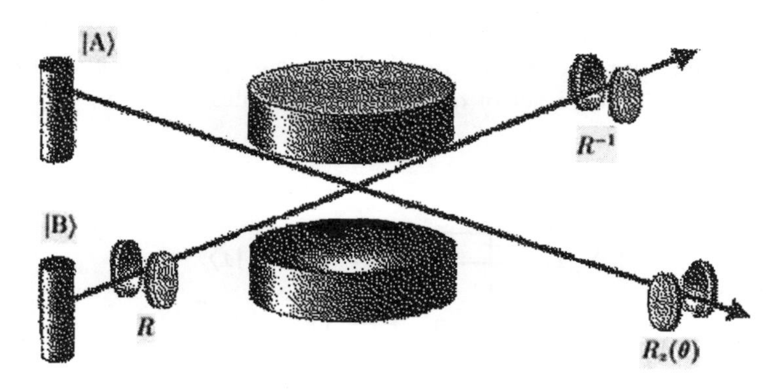

Figure 44.4: Example of realization of universal Quantum logic gate with high-finesse microwave cavities and two-level atoms. All possible quantum computations can be built up with a network of such gates. Passing atom A in the state $|A\rangle$ through the empty cavity on resonance transfers the atom's state, $|A\rangle = c_1|g\rangle + |e\rangle$, to the photon occupancy $c_1|0\rangle + c_2|1\rangle$ of the cavity. Passing atom B in the state $|B\rangle$ through the Ramsey zones (R, R^{-1}) and the cavity (off resonance) changes its state to the exclusive-OR (XOR) — defined by $aXORb := (a \vee b) \wedge \neg(a \wedge b)$ — of the initial state $|A\rangle \oplus |B\rangle$. The original state $|A\rangle$ can now be transferred back to an atom initially in the ground state $|g\rangle$ passing along the path taken by atom A. More general output states can be obtained by adjusting the atom-cavity interactions and adding another Ramsey zone $R_z(\theta)$, which introduces a phase shift — from [BENNETT 1995, 25].

An extension of Boolean operations In other words, we can represent the quantum logic operations as an extension of classical Boolean ones to superpositions of input states. Practically we reproduce on the level of computation the logico-algebraic structure of QM: an orthomodular lattice characterized by choice negation, choice disjunction and the Sasaki arrow [see sections 9.2 and 9.5]. **For example** the Boolean NOT is represented by the unitary matrix:

$$\hat{\sigma}_x = \begin{pmatrix} 0 & 1 \\ 1 & 0 \end{pmatrix}, \tag{44.5}$$

[4]See also [DEUTSCH *et al.* 1995] and [BARENCO *et al.* 1995b] , where it is shown that any unitary operation on n arbitrary bits can be expressed as a composition of one-bit quantum gates and two-bit XOR (exclusive disjunction) gates. On the practical realisability of universal quantum logic gates see [CIRAC/ZOLLER 1995] [SLEATOR/WEINFURTER 1995b] [CERF *et al.* 1998] [MONROE *et al.* 1995b] [MONROE *et al.* 1997] [DOMOKOS *et al.* 1995] . Note that in general quantum computation technologies are strictly dependent on experimental results in the generation of S-Cats [chapter 24]: in fact the two latter quoted articles are due to the same research teams which have studied the latter problem. For a recent study concerning the implementation of a universal set of one- and two-quantum-bit gates using the spin states of a coupled single electron quantum dots, see [LOSS/DIVINCENZO 1998].

which flips the Boolean state of a single bit. But transformations as the following:

$$\frac{1}{\sqrt{2}}\begin{pmatrix} 1 & -1 \\ 1 & 1 \end{pmatrix}, \tag{44.6}$$

which corresponds to a 45° rotation of the polarisation, are intrinsically non-classical, because they transform Boolean states in superpositions. While in classical processing the wires carry bits and the gates perform deterministic Boolean operations on them, in quantum information processing the wires carry qubits and the gates perform unitary operations [see figure 44.4].[5]

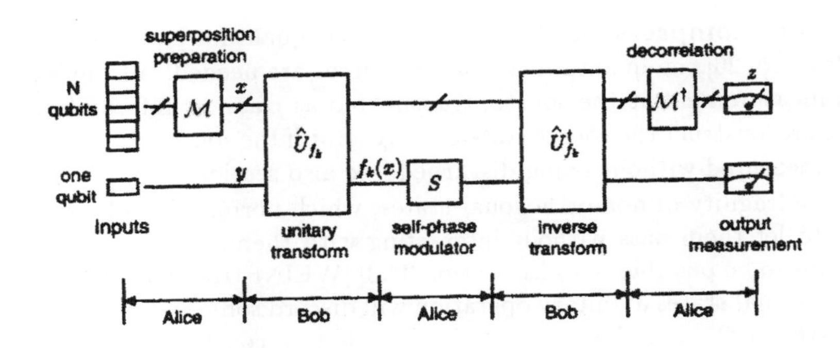

Figure 44.5: Scheme of the Algorithm proposed by Deutsch and Jozsa — from [CHUANG/YAMAMOTO 1995, 3490].

44.2.3 First Example of Quantum Information-Processing

A remarkable **example** of quantum processing is originally due to Deutsch [DEUTSCH 1985b] (therefore it is named *Deutsch's problem*) and was successively generalized by Deutsch and Jozsa [DEUTSCH/JOZSA 1992, 555–57].[6] Alice selects a number x from 0 to $2N-1$ and sends it to Bob. Bob calculates some function $f(x)$ and replies with the result, which is either 0 or 1. Now Bob agrees to use only functions which are either constant for all values of x, or which are equal to 1 for half of all the possible x (balanced). Alice's mission is to discover which type of function, constant or balanced, Bob has chosen. Classically this problem can only be solved with $N+1$ queries.

Now suppose we perform the operation by qubits [see figure 44.5]. Alice envoys $N+1$ two-level atoms, N of which, representing x, are in a superposition of the two states, and one (y) is put in its ground state. Bob calculates $f(x)$ by using a unitary transformation $\hat{U}_f$ and uses a sequence of electromagnetic pulses in order to put y in the state $f(x)$. Since x is in a superposition of all values $[0, 2^N-1]$, y is left in a superposition of all possible values of $f(x)$.

Now Alice receives the answer and gives y a π phase-shift relative to x; then she sends the qubits to Bob again. Bob now inverts its operation (performs $\hat{U}_f^\dagger$), leaving y in its ground state.

[5]See also [BARENCO *et al.* 1995a].

[6]But we follow the easier version of Chuang and Yamamoto [CHUANG/YAMAMOTO 1995, 3489–90].

Since y and x are entangled, the operation leaves the N qubits x with a special relative phase, so that those values of x for which $f(x)$ is even are 180° out of phase with respect to the others. Now, performing an interference experiment, Alice can determine with certainty the type of Bob's function, and this can be done by using only two queries, which, for N large, is a striking speed-up with respect to the classical function evaluation.

44.3 Quantum Transmission

44.3.1 Fidelity of a Quantum Transmission

Difficulty by data compression If the transmitted quantum states $\{|\psi_j\rangle\}$ are orthogonal [BENNETT 1995, 25–26], no quantum channel resources are needed: all the information can be extracted by a measurement at the source, transmitted as classical information to the receiver, and used there to reconstruct the source state exactly. But if the source states are non-orthogonal, they cannot be measured without being disturbed [see also section 18.9].

Because of the fragility of non-orthogonal states, which normally, if observed, are disturbed, it seems better to let them pass without interacting with them. In this case no quantum data compression seems to be possible [see also section 38.3] [WEINFURTER/ZEILINGER 1996, 223–24] : by non-orthogonal states a unitary operation will discard some states. However, Schumacher [B. SCHUMACHER 1995] has shown that it is possible to reconstruct discarded states[7].

Cloning and transposing We know that we cannot violate the no-cloning theorem [42.4: p. 722] However, we must distinguish here between cloning and transposing. By *cloning* we would have a duplication of the original state, and the quoted theorem shows it as impossible. But *transposing* is the transfer of the original state of a system to another system without leaving copies behind. It can happen with a normal unitary transformation — teleportation [see section 43.2] is a particular type of transposition which allows the transmission of a qubit by means of a shared ebit and two-bit classical information.

Encoding and decoding Suppose [B. SCHUMACHER 1995, 2741–42] that we wish to pass some information from a system $\mathcal{S}_1$ to another system $\mathcal{S}_2$ (coding), which we suppose is composed of two subsystems: a channel $\mathcal{S}_\mathcal{C}$ and an extra part $\mathcal{S}_\mathcal{E}$, which we wish or are obliged to discard. Hence we first perform the transformation:

$$\hat{U} : \mathcal{S}_1 \mapsto \mathcal{S}_\mathcal{C} + \mathcal{S}_\mathcal{E}; \tag{44.7}$$

and then

$$\mathcal{S}_\mathcal{C} + \mathcal{S}_\mathcal{E} \mapsto \mathcal{S}_\mathcal{C}. \tag{44.8}$$

Normally, in order to transmit in unitary form, the necessary and sufficient condition is that the Hilbert space of the receiving system has a dimension at least as large as the subspaces of the Hilbert space of the transmitting system spanned by the signal states. Here we have evidently assumed that in general the dimension of the Hilbert space of the channel is less than the dimension of $\mathcal{H}_{\mathcal{S}1}$: it is an approximate transposition. Now we wish to invert the transformation

[7]For preceding work see also [CAVES/DRUMMOND 1994].

and transmit the signal back to the source (decoding). In order to accomplish the task we add an auxiliary system $\mathcal{S}_{\mathcal{E}}'$ to $\mathcal{S}_C$:

$$\mathcal{S}_C \mapsto \mathcal{S}_C + \mathcal{S}_{\mathcal{E}}', \tag{44.9}$$

and we send everything back by another transformation, which we suppose here to be unitary, but it need not necessarily be:

$$\hat{U}^\dagger : \mathcal{S}_C + \mathcal{S}_{\mathcal{E}}' \mapsto \mathcal{S}_1'. \tag{44.10}$$

If the original state to be transmitted is some projection $\hat{P}_j = |j\rangle\langle j|$, then the passage (44.7) can be written as $\hat{P}_{C+\mathcal{E}} = \hat{U}\hat{P}_j\hat{U}^\dagger$. In order to exclude $\mathcal{S}_{\mathcal{E}}$ — the passage (44.8) — we need a partial trace: $\mathrm{Tr}_{\mathcal{S}_{\mathcal{E}}}\hat{P}_{C+\mathcal{E}}$. After $\mathcal{S}_{\mathcal{E}}'$ (which we suppose is in some state $|0_{\mathcal{S}_{\mathcal{E}}'}\rangle$) has been adjoined — passage (44.9) — we obtain: $\hat{\rho}_j = \mathrm{Tr}_{\mathcal{S}_{\mathcal{E}}}(\hat{P}_{C+\mathcal{E}}) \otimes |0_{\mathcal{S}_{\mathcal{E}}'}\rangle\langle 0_{\mathcal{S}_{\mathcal{E}}'}|$. Finally the unitary decoding (44.10) is of the form: $\hat{\rho}_f^j = \hat{U}^\dagger\hat{\rho}_j\hat{U}$.

Fidelity Schumacher [B. SCHUMACHER 1995, 2742] defines the *fidelity of quantum transmission* [see Eq. (43.24)] as the overall probability that a signal which is transmitted from $\mathcal{S}_1$ and then decoded to $\mathcal{S}_1'$ passes a validation test (a filter) — conducted by someone (conventionally called Victor) who knows what the input was — comparing it to its original. In our case $\mathrm{Tr}(\hat{P}_j\hat{\rho}_f^j)$ is the probability that the final state $\hat{\rho}_f^j$ passes the validation test, so that the fidelity is given by:

$$\begin{aligned}\mathbf{F} &= \sum_j \wp(j)\mathrm{Tr}(\hat{P}_j\hat{\rho}_f^j) \\ &= \sum_j \wp(j)\mathrm{Tr}[\hat{U}^\dagger\hat{P}_{C+\mathcal{E}}\hat{U}\hat{U}^\dagger\hat{\rho}_j\hat{U}] \\ &= \sum_j \wp(j)\mathrm{Tr}(\hat{P}_{C+\mathcal{E}}\hat{\rho}_j),\end{aligned} \tag{44.11}$$

so that one can calculate the fidelity of the whole transmission process by only considering the signal states of $\mathcal{S}_C + \mathcal{S}_{\mathcal{E}}$ and $\mathcal{S}_C + \mathcal{S}_{\mathcal{E}}'$.

44.3.2 Two Lemmas*

Now we prove two lemmas which are needed in order to prove the central Schumacher theorem [B. SCHUMACHER 1995, 2742–44].

I lemma The first lemma can be formulated as follows:

Lemma 44.1 (I Schumacher) *Suppose that the dimension of $\mathcal{H}_{\mathcal{S}_C}$ is d, and suppose that the ensemble of signals in $\mathcal{S}_1$ described by an initial density matrix $\hat{\rho}_i = \sum_j \wp(j)\hat{P}_j$ has the property that, for every projector $\hat{P}_d$ onto a d-dimensional subspace of $\mathcal{H}_{\mathcal{S}_1}$, we have*

$$\mathrm{Tr}\left(\hat{\rho}_i\hat{P}_d\right) < \eta, \tag{44.12}$$

for some fixed η. Then the fidelity $\mathbf{F} < \eta$.

Proof

Let $|\psi_k\rangle$ for $k = 1, \ldots, d$ be the orthogonal basis for this subspace that is composed of eigenstates of $\hat{\rho}_f^j$. We can write:

$$\hat{\rho}_f^j = \sum_k w_k^f |\psi_k\rangle\langle\psi_k|, \tag{44.13}$$

where the w_k^f are the eigenvalues of $\hat{\rho}_f^j$, including all of the nonzero ones. Clearly $w_k^f \leq 1$. Then:

$$\hat{P}_d = \sum_k |\psi_k\rangle\langle\psi_k|. \tag{44.14}$$

Now consider the term $\mathrm{Tr}(\hat{P}_j \hat{\rho}_f^j)$ appearing in the fidelity [Eq. (44.11)]:

$$\begin{aligned}
\mathrm{Tr}\left(\hat{P}_j \hat{\rho}_f^j\right) &= \mathrm{Tr}\left[\hat{P}_j\left(\sum_k w_k^f |\psi_k\rangle\langle\psi_k|\right)\right] = \sum_k w_k^f \mathrm{Tr}\left(\hat{P}_j |\psi_k\rangle\langle\psi_k|\right) \\
&\leq \sum_k \mathrm{Tr}\left(\hat{P}_j |\psi_k\rangle\langle\psi_k|\right) = \mathrm{Tr}\left[\hat{P}_j\left(\sum_k |\psi_k\rangle\langle\psi_k|\right)\right] \\
&= \mathrm{Tr}\left(\hat{P}_j \hat{P}_d\right).
\end{aligned} \tag{44.15}$$

Then the fidelity is:

$$\begin{aligned}
\mathbf{F} &= \sum_j \wp(j) \mathrm{Tr}\left(\hat{P}_j \hat{\rho}_f^j\right) \\
&\leq \sum_j \wp(j) \mathrm{Tr}\left(\hat{P}_j \hat{P}_d\right) \\
&= \mathrm{Tr}\left(\hat{\rho}_i \hat{P}_d\right),
\end{aligned} \tag{44.16}$$

and therefore $\mathbf{F} < \eta$. Q.E.D.

II lemma The second lemma is the following:

Lemma 44.2 (II Schumacher) *Suppose that the dimension of $\mathcal{H}_{\mathcal{S}_C}$ is d, and suppose that there exists a projector $\hat{P}_d$ onto a d–dimensional subspace of $\mathcal{H}_{\mathcal{S}1}$ such that*

$$\mathrm{Tr}\left(\hat{\rho}_i \hat{P}_d\right) > 1 - \eta. \tag{44.17}$$

Then there exists a transposition scheme with fidelity $\mathbf{F} > 1 - 2\eta$.

Proof

We suppose that the dimension of $\mathcal{H}_{\mathcal{S}1}$ is N, and we divide the eigenstates of $\hat{\rho}_i$ in such a way that the eigenstates $|1\rangle, \ldots, |d\rangle$ span $\mathcal{H}_\#$ and $|d+1\rangle, \ldots, |N\rangle$ span the orthogonal $\mathcal{H}_\#{}^\perp$.[8] Then we write:

$$\sum_{n=1}^{d} |n\rangle\langle n| = \hat{P}_d, \quad \sum_{n=1}^{d} w_n^i > 1 - \eta, \quad \sum_{n=d+1}^{N} w_n^i < \eta, \tag{44.18}$$

where w_n^i are the eigenvalues of $\hat{\rho}_i$.

Let us write the mapping (44.7) as follows:

$$|n\rangle \mapsto \begin{array}{lll} |n_{\mathcal{C}}, 0_{\mathcal{E}}\rangle, & n = & 1, \ldots, d \\ |0_{\mathcal{C}}, n_{\mathcal{E}}\rangle, & n = & d+1, \ldots, N \end{array}, \tag{44.19}$$

where $\{|n_{\mathcal{C}}\rangle\}, \{|n_{\mathcal{E}}\rangle\}$ are orthogonal sets of states of $\mathcal{S}_{\mathcal{C}}$ and $\mathcal{S}_{\mathcal{E}}$, respectively, and $|0_{\mathcal{C}}\rangle, |0_{\mathcal{E}}\rangle$ are fixed null states. We require that $|0_{\mathcal{E}}\rangle$ is orthogonal to each $|n_{\mathcal{E}}\rangle$ for $n = d+1, \ldots, N$. Correspondingly we write the two-step mapping (44.8)–(44.9) as (we require that $\mathcal{S}'_{\mathcal{E}}$ is initially in $|0_{\mathcal{E}'}\rangle$):

$$|n\rangle \mapsto \begin{array}{lll} |n_{\mathcal{C}}, 0_{\mathcal{E}'}\rangle, & n = & 1, \ldots, d \\ |0_{\mathcal{C}}, 0_{\mathcal{E}'}\rangle, & n = & d+1, \ldots, N \end{array}. \tag{44.20}$$

[8]See also [GRAHAM 1973].

Suppose that any state $|j\rangle$ of $\mathcal{S}_1$ is a superposition of states:

$$|j\rangle = c_j|k_j\rangle + c_j^\perp|l_j^\perp\rangle, \tag{44.21}$$

where $|k_j\rangle \in \mathcal{H}_\#$, $|l_j^\perp\rangle \in \mathcal{H}_\#{}^\perp$, and $|c_j|^2 + |c_j^\perp|^2 = 1$. The states $|k_j\rangle$ and $|l_j^\perp\rangle$ can be expanded in terms of the basis eigenstate of $\hat{\rho}_i$:

$$|j\rangle = c_j\left(\sum_{n=1}^{d}\langle n|k_j\rangle|n\rangle\right) + c_j^\perp\left(\sum_{n=d+1}^{N}\langle n|l_j^\perp\rangle|n\rangle\right). \tag{44.22}$$

Then we write the mapping (44.7) as:

$$|j\rangle \mapsto c_j|k_j^{\mathcal{C}}, 0_{\mathcal{E}}\rangle + c_j^\perp|0_{\mathcal{C}}, l_j^{\mathcal{E}}\rangle. \tag{44.23}$$

The projector $\hat{P}_{\mathcal{C}+\mathcal{E}}$ can be written as follows:

$$\begin{aligned}\hat{P}_{\mathcal{C}+\mathcal{E}} &= |c_j|^2|k_j^{\mathcal{C}}, 0_{\mathcal{E}}\rangle\langle k_j^{\mathcal{C}}, 0_{\mathcal{E}}| + |c_j^\perp|^2|0_{\mathcal{C}}, l_j^{\mathcal{E}}\rangle\langle 0_{\mathcal{C}}, l_j^{\mathcal{E}}| \\ &\quad + c_j^* c_j^\perp|0_{\mathcal{C}}, l_j^{\mathcal{E}}\rangle\langle k_j^{\mathcal{C}}, 0_{\mathcal{E}}| + c_j(c_j^\perp)^*|k_j^{\mathcal{C}}, 0_{\mathcal{E}}\rangle\langle 0^{\mathcal{C}}, l_j^{\mathcal{E}}|.\end{aligned} \tag{44.24}$$

Then the partial trace [corresponding to transition (44.8)] can be written as:

$$\mathrm{Tr}_{\mathcal{S}_\mathcal{E}}\hat{P}_{\mathcal{C}+\mathcal{E}} = |c_j|^2|k_j^{\mathcal{C}}\rangle\langle k_j^{\mathcal{C}}| + |c_j^\perp|^2|0_{\mathcal{C}}\rangle\langle 0^{\mathcal{C}}|. \tag{44.25}$$

Adjoining the system $\mathcal{S}_\mathcal{E}{}'$ in the state $|0_{\mathcal{E}'}\rangle$ yields $\hat{\rho}_j$:

$$\hat{\rho}_j = |c_j|^2|k_j^{\mathcal{C}}, 0_{\mathcal{E}'}\rangle\langle k_j^{\mathcal{C}}, 0_{\mathcal{E}'}| + |c_j^\perp|^2|0_{\mathcal{C}}, 0_{\mathcal{E}'}\rangle\langle 0^{\mathcal{C}}, 0_{\mathcal{E}'}|. \tag{44.26}$$

Since we decode this signal by using $\hat{U}^\dagger$, the overall fidelity is $\mathbf{F} = \sum_j \wp(j)\mathrm{Tr}\left(\hat{P}_{\mathcal{C}+\mathcal{E}}\hat{\rho}_j\right)$ [Eq. (44.11)]. For a given signal we have:

$$\begin{aligned}\mathrm{Tr}\left(\hat{P}_{\mathcal{C}+\mathcal{E}}\hat{\rho}_j\right) &= |c_j|^4 + |c_j|^2|c_j^\perp|^2|\langle k_j^{\mathcal{C}}|0_{\mathcal{C}}\rangle|^2 \\ &\geq |c_j|^4 = (1 - |c_j^\perp|^2)^2 \\ &\geq 1 - 2|c_j^\perp|^2.\end{aligned} \tag{44.27}$$

Hence the fidelity is:

$$\mathbf{F} \geq 1 - 2\sum_j \wp(j)|c_j^\perp|^2. \tag{44.28}$$

We required $\mathrm{Tr}\left(\hat{\rho}_i\hat{P}_d\right) > 1 - \eta$. On the other hand, we have:

$$\begin{aligned}\mathrm{Tr}\left(\hat{\rho}_i\hat{P}_d\right) &= \sum_j \wp(j)\mathrm{Tr}\left(\hat{P}_j\hat{P}_d\right) = \sum_j \wp(j)|c_j|^2 = \sum_j \wp(j)(1 - |c_j^\perp|^2) \\ &= 1 - \sum_j \wp(j)|c_j^\perp|^2.\end{aligned} \tag{44.29}$$

Hence our requirement on $\mathrm{Tr}\left(\hat{\rho}_i\hat{P}_d\right)$ amounts to requiring that $\sum_j \wp(j)|c_j^\perp|^2 < \eta$, which implies that $\mathbf{F} > 1 - 2\eta$. Q.E.D.

44.3.3 Transmission, Entropy and Information

Now we can prove the theorem [B. SCHUMACHER 1995, 2745] [JOZSA/B. SCHUMACHER 1994] about quantum noiseless coding — where for *noiseless* we intend a channel which is isolated from the environment —:

Theorem 44.4 (Schumacher) *Let $\mathcal{S}_1$ be a quantum signal source with a signal ensemble described by $\hat{\rho}_i$ and let $\delta, \epsilon > 0$.*

- *Suppose that $H(\hat{\rho}_i) + \delta$ qubits are available for the signal $\mathcal{S}_1$. Then for sufficiently large N, groups of N signals from the source can be transposed via the available qubits with fidelity $\mathbf{F} > 1 - \epsilon$.*

- *Suppose that $H(\hat{\rho}_i) - \delta$ qubits are available for the $\mathcal{S}_1$ signal. Then for sufficiently large N, if groups of N signals from the source are transposed via the available qubits, then the fidelity $\mathbf{F} < \epsilon$.*

Proof

The first part is proved as follows. If the quantum channel $\mathcal{S}_C$ has $N[H(\hat{\rho}_i) + \delta]$ qubits Q, the dimension of $\mathcal{H}_{\mathcal{S}_C}$ is $2^{N[H(\hat{\rho}_i)+\delta]}$. It is possible to show[9] that, for large N, the number of likely eigenstate sequences $f \leq 2^{N[H(\hat{\rho}_i)+\delta]}$ and the sum of the remaining eigenvalues of $(\hat{\rho}_i)^N$ can be made less than $\epsilon/2$. Adding some additional eigenstate sequences we can bring the total to exactly $f = \dim\mathcal{H}_{\mathcal{S}_C}$, and this will not increase the sum of the remaining eigenvalues. If $\hat{P}_f$ is a projector onto the f–dimensional subspace $\mathcal{H}_\#$, then $\mathrm{Tr}[(\hat{\rho}_i)^N \hat{P}_f] > 1 - \epsilon/2$. By lemma 44.2 there is a transposition scheme with fidelity $\mathbf{F} > 1 - \epsilon$.

For the second part, it is possible to prove that, for large N, none of the $2^{N[H(\hat{\rho}_i)+\delta]}$ eigenstate sequences for $\mathcal{S}_1^N$ has eigenvalues whose sum is as large as ϵ. Therefore, for each projection $\hat{P}_C$ onto a subspace of dimension $2^{N[H(\hat{\rho}_i)+\delta]} = \dim\mathcal{H}_{\mathcal{S}_C}$, we have $\mathrm{Tr}[(\hat{\rho}_i)^N \hat{P}_C] < \epsilon$. Then by lemma 44.1 every transposition scheme has fidelity $\mathbf{F} < \epsilon$. Q.E.D.

44.3.4 Discussion

The importance of the Schumacher theorem is the following: Holevo Theorem [42.3: p. 719] showed that the (classical) information which one can extract from a system is generally smaller than the von Neumann entropy (of the mixture of the final state obtained after a measurement) [see particularly Eq. (42.39)]. Hence the Holevo theorem does not guarantee an interpretation of the von Neumann entropy in the framework of classical information theory, while Schumacher has shown that the von Neumann entropy is just the mean number of qubits necessary to encode the states in the ensemble in an ideal coding scheme — exactly as, classically, the Shannon entropy [see subsection 42.1.1] is the limit for the number of bits per signal to which one should approach in order to transmit faithfully. On the other hand, by performing a measurement, it is Holevo's result which should be used because we cannot know, with a single measurement, the state of a system, in which, if a pure state, an infinite amount of information can be encoded.

However, Shannon's classical result is more powerful than Schumacher's result because it also establishes a connection between potential information and entropy for channels with noise, while

[9] For the proof see [B. SCHUMACHER 1995, 2739–40].

Schumacher's theorem is valid only for noiseless ones [on the problem see section 43.3]. In fact, due to the universality of the quantum logic gates [see theorem 44.3: p. 763], any channel which transmits quantum information can be used to transmit classical information as well. But there are channels which are able to transmit classical information without being able to transmit quantum information [LLOYD 1997a]. It depends on the amount of noise in the channel — the noise can in fact produce the result that an initial pure state arrives in mixed state. We shall return to the problem of the von Neuman entropy and capacity in subsection 44.5.3.

Recently methods have been developed [BENNETT *et al.* 1996b] [DEUTSCH *et al.* 1996] in order to overcome the problem of noise in quantum channels by the method of purification [see section 38.3].[10] Further developments will be analysed in subsection 44.5.3.

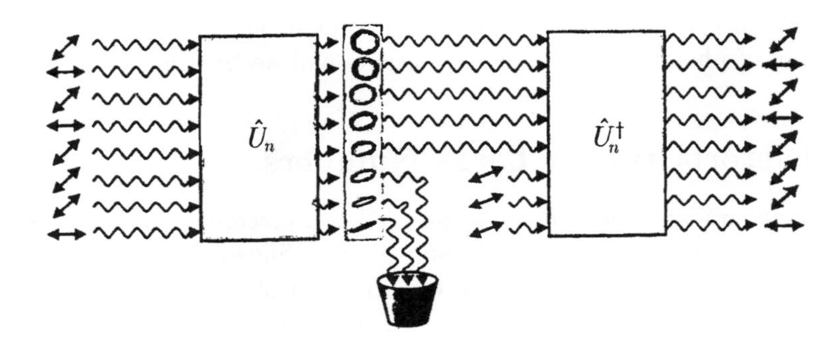

Figure 44.6: Quantum data compression. A unitary operation $\hat{U}_n$ on n unknown qubits from a known nonorthogonal source ensemble (here photons at random $| \leftrightarrow\rangle$ or $| \nearrow\!\swarrow\rangle$ polarisation) concentrates most of the quantum information into some of the qubits, allowing others to be discarded. At the receiving end of the channel, the discarded qubits are replaced by standard ones (here, 22.5° polarised photons corresponding to the major eigenvector of the source's density matrix) and the unitary operation is undone, resulting in a very good, but slightly entangled, approximation to the original input state — from [BENNETT 1995, 26].

44.3.5 An Example of Quantum Data Compression

Consider the **example** shown in figure 44.6 [BENNETT 1995, 25–26]. The source has a quantum kind of redundancy neither consisting in unequal probabilities of its signals, nor in correlations between signals emitted at different times, but rather in the fact that the two signal states are non-orthogonal and hence not wholly distinct as physical states [see also section 18.9]. The redundancy can be expressed by the density matrix

$$\tilde{\hat{\rho}} = \frac{1}{2}(| \leftrightarrow\rangle\langle\leftrightarrow | + | \nearrow\!\swarrow\rangle\langle\nearrow\!\swarrow |), \tag{44.30}$$

which has two unequal eigenvalues, $\lambda^n_{\text{Max}} = \cos^2(\pi/8) \simeq 0.854$, and $\lambda^n_{\text{min}} = \sin^2(\pi/8) \simeq 0.146$, with respective eigenvectors in the $22.5°, 112.5°$ polarisation directions: the mixture has a von Neumann entropy of $H_{\text{VN}}(\hat{\rho}) \simeq 0.601$ bits per photon, or about 0.399 bits per photon less than if the two signal states were orthogonal. To exploit the source's redundancy, a block of n photons is unitarily transformed

[10]Explicit algorithms for data compressing have been considered by Cleve and DiVincenzo [CLEVE/DIVINCENZO 1996].

into a basis of products of eigenvectors of $\tilde{\hat{\rho}}$, arranged in order of decreasing eigenvalues from λ^n_{Max} to λ^n_{min}, and approximately $[1 - H_{\mathrm{VN}}(\hat{\rho})]n$ of the highest order photons are discarded. These photons contain little information, being polarised in the 22.5° direction with high probability. By contrast, the low-order photons, which are retained, contain almost all the information of the original state, and would appear almost entirely depolarised if examined individually. The discarded photons are finally replaced by 22.5° pure photons and the unitary transformation is undone, resulting in a very good, but slightly entangled, approximation to the original input state.

44.4 More on Quantum Computation

The interest of quantum computation is not only of a technological type: it is also of a foundational and theoretical type to the extent that it can lead (and has actually led thus far) to a better understanding of phenomena like superpositions and decoherence processes.

44.4.1 The Factorization of Large Numbers

As we have said, quantum superpositions can be used to overcome problems for which classical computation based on Boolean codes and gates only gives a solution after a very long computation time. The most known problem is that of the factorization of large integer numbers, for which classically there is no known fast algorithm[11] — one needs 15 billion years ca. with actual computation standards for factorising a number of 60 figures. One says that an algorithm is *fast* if the time taken to execute it increases no faster than a polynomial function of the size of the input (polynomial-time solution) [BENNETT 1995, 27–28].[12] Now the problem of factorising a number can be transformed in that of finding the period of a periodic function.

For example suppose we wish to factorise the number $N = 15$ [BARENCO *et al.* 1996]. We take an arbitrary number $\xi < 15$ which has no common prime factors with N, say $\xi = 7$. Now we define a function

$$f(x) = \xi^x \,\mathrm{mod} 15, \tag{44.31}$$

where the x's are positive integers. $f(x)$ is the periodic function of which we wish to find the period, and mathematically it is the rest of the division of ξ^x by N. For $x = 0, 1, 2, 3, 4, 5, 6, \ldots$ we find that $f(x) = 1, 7, 4, 13, 1, 7, 4, \ldots$, i.e. the period of $f(x)$ is $r = 4$. Now we find the largest common divisors of N and $\xi^{\frac{r}{2}} \pm 1 = 50, 48$, which are respectively $5, 3$, which are the desired factors of N.

Classically a calculation like this is very long (one has to check every integer from 1 to $\sqrt{N}$), and for very large integers it is practically not performable — the number of computational steps increases exponentially as does the binary dimension of the input. But in QM the solution to this problem is possible thanks to the superposition: in fact a QM system can be in a superposition of all: $f(0), f(1), f(2), f(3), f(4), \ldots$ But it is evident that, if we perform a measurement, we cannot obtain all the values together [see postulates 3.3: p. 37, and 3.4: p. 38]. But if we are able to evaluate the wave function, we can perform a Fourier transform so that we obtain the period r: in fact the different values $f(0), f(1), f(2), f(3), f(4), \ldots$ are related to r as the fine structure of the interference pattern to the envelope of the same (Overall width) [see section 7.6].

This is the important result of the following theorem stated and proved by Shor [SHOR 1994]:

[11] Other possible applications are briefly discussed in [DEUTSCH/EKERT 1998].

[12] On this point see also [DIVINCENZO 1995a] [CHUANG *et al.* 1995] [EKERT/JOZSA 1996, 736–37].

Theorem 44.5 (Shor) *By evaluating a wave function on a superposition of exponentially many arguments, each one representing a value of the requested periodic function of the type* (44.31), *computing a parallel Fourier transform on the superposition, and finally sampling the Fourier power spectrum, one obtains the searched period* [see figure 44.7].

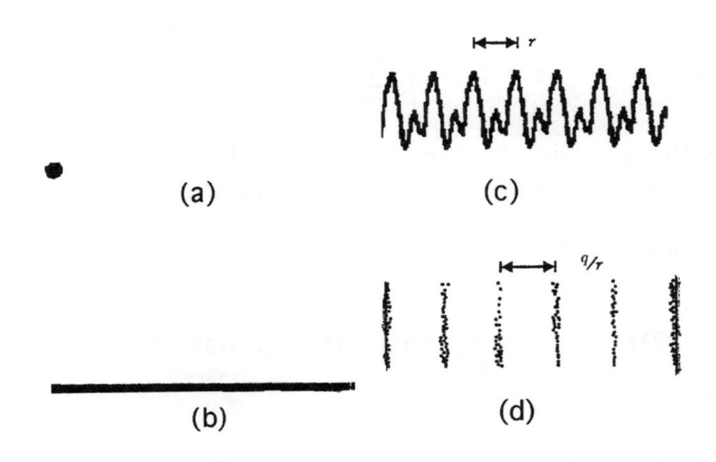

Figure 44.7: Shor's superfast quantum Fourier sampling uses quantum interference to measure the period r of a periodic function $f(x)$. The period may be exponentially larger than the number of qubits involved in the computation.
(a) The computer starts in the state $|x, f(x)\rangle = |0, 0\rangle$.
(b) the x–register is put in a superposition of all possible values [see Eq. (44.32)].
(c) the value $f(x)$ is computed in the y–register simultaneously for all x values [see Eq. (44.33)].
(d) A Fourier transform of the x register is performed [see Eq. (44.34)].
Measuring the Fourier transform of x yields a result $c = \lambda\frac{q}{r}$ from which the period r can be deduced [see Eq. (44.36)] — from [BENNETT 1995, 28].

We choose a number ξ at random and find a period r such that $\xi^r \equiv 1 \bmod N$, where N is the number to factorise. Now we choose a smooth number q (a number with small prime factors) such that $N^2 < q < 2N^2$ and build the initial or input state:

$$|\psi_i\rangle = \frac{1}{\sqrt{q}} \sum_{x=0}^{q-1} |x, 0\rangle, \tag{44.32}$$

from which, with a quantum computation step, one can obtain:

$$|\psi_f\rangle = \frac{1}{\sqrt{q}} \sum_{x=0}^{q-1} |x, \xi^x \bmod N\rangle. \tag{44.33}$$

Now we Fourier-transform this pure state to obtain:

$$|\tilde{\psi}_f\rangle = \frac{1}{q} \sum_{m=0}^{q-1} \sum_{x=0}^{q-1} e^{\imath 2\pi x m/q} |m, \xi^x \bmod N\rangle, \tag{44.34}$$

where use has been made of the discrete Fourier transform [EKERT/JOZSA 1996, 740]:

$$\hat{U}_{\mathrm{DF}} : |x\rangle \mapsto \frac{1}{\sqrt{q}} \sum_{m=0}^{q-1} e^{\imath 2\pi x m/q} |m\rangle. \tag{44.35}$$

We can now measure both arguments of the superposition (44.34), obtaining the value c for m in the first one, and some ξ^k as the answer to the second one (k being any number between 0 and r). The probability for such results will be:

$$\wp(c, \xi^k) = \left| \frac{1}{q} \sum_{x=0}^{q-1}{}' e^{\imath 2\pi x c/q} \right|^2, \tag{44.36}$$

where the prime indicates a restricted sum over values which satisfy $\xi^x \equiv \xi^k \bmod N$. $\wp(c, x^k)$ is periodic in c with period q/r. A measurement gives, with high probability, $c = \lambda \frac{q}{r}$ — where λ is an integer corresponding to one of the peaks shown in figure 44.7.(c)–(d). But since we know q, we can determine r with few trials.

44.5 Decoherence and Quantum Information

44.5.1 General Problems

There are several problems connected with quantum computation. One is that Hamiltonian computers of the type described provide no insights into their internal communication links [LANDAUER 1996, 191].

But the greatest problem one encounters in the realization of such a quantum computation is the decoherence time [given by the formula (21.102)], which is normally very small, generally much smaller than the relaxation time, so that the superposition can decohere before the end of the calculation [CHUANG *et al.* 1995] [EKERT/JOZSA 1996, 745–46].[13] As a consequence we must repeat the operation an exponentially large number of times in order to be sure of the result.

By introducing the environment (the world external to the computer) the input state (44.32) may be rewritten as:

$$|\Psi_i^{\mathcal{S}+\mathcal{E}}\rangle = \frac{1}{\sqrt{q}} \sum_{x=0}^{q-1} |x, 0\rangle \otimes |\mathcal{E}_i\rangle, \tag{44.37}$$

where $\mathcal{E}_i$ is the initial state of the environment. It is initially uncorrelated with the quantum computer; but the bits necessary for the calculation of ξ^x mod N will most likely interact with it, so that the the next state [see Eq. (44.33)] is given by:

$$|\Psi_f^{\mathcal{S}+\mathcal{E}}\rangle = \frac{1}{\sqrt{q}} \sum_{x=0}^{q-1} |x, \xi^x \bmod N\rangle \otimes |\mathcal{E}_x\rangle. \tag{44.38}$$

Now we calculate now the reduced density matrix

$$\mathrm{Tr}_{\mathcal{E}}(\hat{\rho}^{\mathcal{S}+\mathcal{E}}) = \frac{1}{\sqrt{q}} \sum_{x=0}^{q-1}{}' \sum_{x'=0}^{q-1}{}' (1 - \eta_{xx'}) |x\rangle\langle x'|, \tag{44.39}$$

[13] For very detailed studies see [PALMA *et al.* 1996] [EKERT/MACCHIAVELLO 1998a].

where $\hat{\rho}^{\mathcal{S}+\mathcal{E}} = |\Psi_f^{\mathcal{S}+\mathcal{E}}\rangle\langle\Psi_f^{\mathcal{S}+\mathcal{E}}|$, and $(1-\eta_{xx'}) = |\langle\mathcal{E}_x|\mathcal{E}_{x'}\rangle|^2$ is a measure of the extent to which the state of the environment has become correlated with the states of the quantum computer. If $|x\rangle, |x'\rangle$ are qubit register states diagonal in the pointer basis, then we may take:

$$1 - \eta_{xx'} \simeq \exp\left[-\zeta \cdot (x \text{ XOR } x')\right], \tag{44.40}$$

where the XOR (exclusive disjunction) function gives the distance between x and x', and ζ depends on the particular realization of a quantum computer. It is now clear that the states of the quantum computer — initially in superposition — become (in a very short time) uncorrelated. In fact, if the states $|\mathcal{E}_x\rangle, |\mathcal{E}_{x'}\rangle$ are almost orthogonal, then state (44.39) almost represents a mixture. This is the biggest obstacle for the practical realization of a quantum computer. However [as we shall see in the following subsection], solutions are also possible.

44.5.2 Quantum Error-Correcting Codes

Cleve and Cerf's model This problem can be handled in the case of errors which depend on decoherence occurring at known places [CERF/CLEVE 1997]. They are called *erasures*. The classical condition to correct erasures is that the entire information must be found in the unchanged bits. If we name the input $\mathcal{X}$ and divide the output $\mathcal{Y}$ in a part, $\mathcal{Y}_u$, which remains unchanged, and another part, $\mathcal{Y}_e$, which is erased, we may express such a condition — in terms of mutual [see Eq. (42.35)] and conditional entropy — as [see figure 44.8]:

$$H(\mathcal{X} : \mathcal{Y}_e|\mathcal{Y}_u) = 0; \tag{44.41}$$

or

$$H(\mathcal{X} : \mathcal{Y}_u) = H(\mathcal{X} : \mathcal{Y}) = I, \tag{44.42}$$

where I represents the entire information about $\mathcal{X}$. Quantum-mechanically we have a reference R which plays the role of $\mathcal{X}$, a quantum channel Q, which plays the role of $\mathcal{Y}$ and an environment $\mathcal{E}$. The quantum equivalent of Eq. (44.42) expressed in terms of the von Neumann entropy is:

$$H_{\text{VN}}(R : Q_u) = H_{\text{VN}}(R : Q) = I_{\text{QM}}; \tag{44.43}$$

or in analogy with Eq. (44.41):

$$H_{\text{VN}}(R : Q_e|Q_u) = 0. \tag{44.44}$$

Supposing now that the tripartite system RQ_eQ_u is in a pure state, then we have $H_{\text{VN}}(R : Q_e : Q_u) = 0$ (because the ternary mutual entropy of any entangled tripartite system vanishes), from which, together with Eq. (44.44), we obtain:

$$H_{\text{VN}}(R : Q_e) = H_{\text{VN}}(R : Q_e|Q_u) + H_{\text{VN}}(R : Q_e : Q_u) = 0. \tag{44.45}$$

Let us now write the tripartite system reference + QM channel + environment after the decoherence in the form $R'Q'\mathcal{E}'$ — where $H_{\text{VN}}(R) = H_{\text{VN}}(R')$. The loss L of the QM channel can be defined as [see figure 44.9] [B. SCHUMACHER/NIELSEN 1996]:

$$\begin{aligned} L &:= H_{\text{VN}}(R' : \mathcal{E}'|Q') \\ &= H_{\text{VN}}(RQ') + H_{\text{VN}}(\mathcal{E}'Q') - H_{\text{VN}}(Q') - H_{\text{VN}}(RQ'\mathcal{E}') \\ &= H_{\text{VN}}(\mathcal{E}') + H_{\text{VN}}(R) - H_{\text{VN}}(Q'). \end{aligned} \tag{44.46}$$

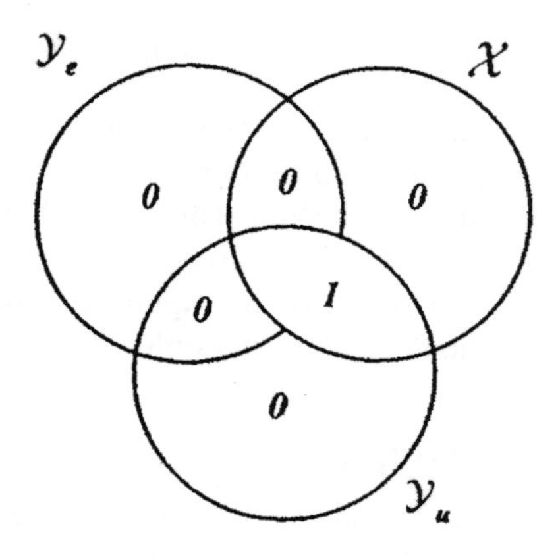

Figure 44.8: Entropy diagram for a classical erasure-correcting code — from [CERF/CLEVE 1997, 1724].

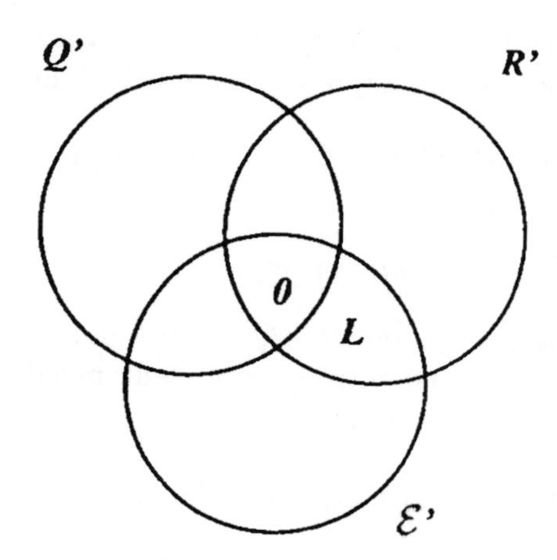

Figure 44.9: Entropy diagram summarizing relations between Q', $R' = R$ and $\mathcal{E}'$ after decoherence — from [CERF/CLEVE 1997, 1725].

Hence the necessary and sufficient condition for the existence of a perfect quantum decoding scheme is:

$$L = H_{\rm VN}(\mathcal{E}') + H_{\rm VN}(R) - H_{\rm VN}(Q') = 0. \tag{44.47}$$

In other words, when $L = 0$, the state of Q' becomes entangled separately with $\mathcal{E}$ (bad entanglement) and R (good entanglement), allowing this bad entanglement to be transferred to an ancilla while recovering only the good one [see figure 44.10].

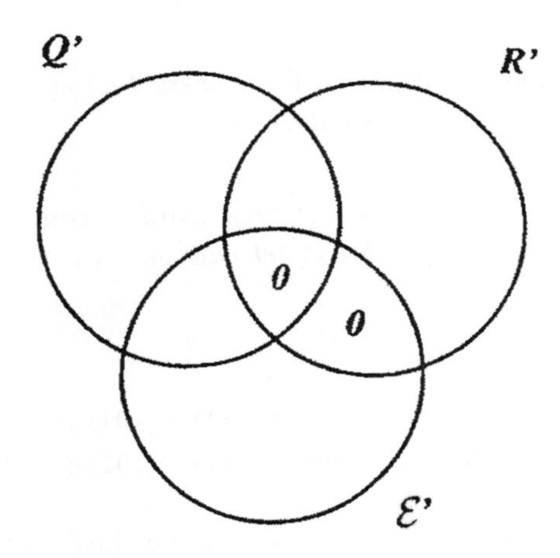

Figure 44.10: Entanglement between $Q', R = R'$ and $\mathcal{E}'$ in a lossless ($L = 0$) quantum channel. Q' is entangled 'separately' with R' and $\mathcal{E}'$ — from [CERF/CLEVE 1997, 1726].

Naturally the problem of overcoming the decoherence is increasingly difficult as the number of processed bits increases.

Shor's proposal Other methods have also been developed. Shor proposed [SHOR 1995] to encode each qubit into nine qubits as follows:

$$|0\rangle \mapsto \frac{1}{2\sqrt{2}}\left(|000\rangle + |111\rangle\right)\left(|000\rangle + |111\rangle\right)\left(|000\rangle + |111\rangle\right), \tag{44.48a}$$

$$|1\rangle \mapsto \frac{1}{2\sqrt{2}}\left(|000\rangle - |111\rangle\right)\left(|000\rangle - |111\rangle\right)\left(|000\rangle - |111\rangle\right), \tag{44.48b}$$

in consequence of which we pass from a probability $\wp$ to decohere for each 'normally encoded' qubit and a probability $(1 - \wp)^k$ that k qubits do not decohere, to a probability that our $9k$ qubits can be decoded to give the original quantum state of approximately $(1 - 36\wp^2)^k$.

At any rate, solutions of this type show a fundamental limit of quantum computation today: in order to overcome decoherence one needs an amount of information and complexity which grows exponentially with the number of the processed bits. However, employing a simplified version of Shor's proposal, Vaidman and co-workers [VAIDMAN *et al.* 1996] have shown that the error correction exhibits manifestation of the Zeno effect, which, under certain circumstances, allows a protection of unknown quantum state.

Steane's proposal If we write a product state $|0\rangle \otimes |0\rangle \otimes |1\rangle$ in the form $|001\rangle$, we can codify each system defined by the basis $\{|0\rangle, |1\rangle\}$ (basis 1) in a new basis (basis 2) as follows [STEANE 1996a]:[14]

$$|0\rangle = |0\rangle + |1\rangle, \quad |1\rangle = |0\rangle - |1\rangle, \quad |00\rangle = |00\rangle + |01\rangle + |10\rangle + |11\rangle, \tag{44.49}$$

and so on. The transformations are known as the *Hadamard transform*, a special case of the Fourier transform. By defining a *word* as a unique string of bits, we state the following theorems:

Theorem 44.6 (I Steane) *The word $|00\cdots 0\rangle$ in basis 1 is equal to a superposition of all 2^n possible words in basis 2, with equal coefficients.*

Theorem 44.7 (II Steane) *If the j–th bit of each word is complemented $(0 \leftrightarrow 1)$ in basis 1, then all words in basis 2 in which the jth bit is set change sign.*

For example:

$$\begin{aligned} |000\rangle + |111\rangle &= |000\rangle + |011\rangle + |101\rangle + |110\rangle, &(44.50a)\\ |001\rangle + |110\rangle &= |000\rangle - |011\rangle - |101\rangle + |110\rangle. &(44.50b)\end{aligned}$$

Though both valid, the theorems are not integrated in the theory because they are too technical and go beyond the scope of this book. A *code* is a set of words, all of the same length. Now we define the *Hamming distance* between two words (of the same length) as the number of places where they differ. The *minimum distance* of a code is the smallest Hamming distance between any two code words in the code. A code of minimum distance d allows $(d-1)/2$ errors to be corrected. The price of this error correction is that only code words (a subset of the 2^n possible n–bit words) may be transmitted.

Now the fundamental problem, which has no general solution, is to find codes having the maximum number $M(n, d)$ of code words for a given length n and a minimum distance d. From the entropic uncertainty relations [see section 42.2] the following inequality can be derived:

$$|m_1\rangle|m_2\rangle \geq 2^n, \tag{44.51}$$

where we have a superposition of product states $|m_1\rangle$ and $|m_2\rangle$ respectively in basis 1 and basis 2. Since by definition $|m_j\rangle \leq M(n, d_j)$ $(j = 1, 2)$, then we have:

$$M(n, d_1)M(n, d_2) \geq 2^n. \tag{44.52}$$

In other words, we have a kind of 'complementarity' between d_1 and d_2. Then the error correction might be performed by the use of a repetition code, on a line similar to that followed by Shor[15].

Another problem, recently examined [GOTTESMAN 1998], is the fault-tolerant implementations of a universal set of gates in order to avoid a catastrophic spread of existing errors.

[14] See also [STEANE 1996b] for further developments.

[15] See also [LAFLAMME *et al.* 1996] [CALDERBANK/SHOR 1996] [KNILL/LAFLAMME 1997] [BRAUNSTEIN 1998]Braunstein, Samuel L.. For a recent review see [DIVINCENZO/TERHAL 1998].

44.5.3 Von Neumann Capacity

The preceding results have been used by Cerf and Adami [ADAMI/CERF 1997]. As we have seen, a quantum channel can be used either in order to transmit classical information or in order to transmit quantum information. Recently Wootters and co-workers [HAUSLADEN *et al.* 1996] proved that the capacity for the transmission of classical information through quantum channels is almost equal to the von Neumann entropy [see also subsection 44.3.4] for the transmission of pure states; Holevo [HOLEVO 1998] proved the equality for the general case, which includes the transmission of mixed states.

Adami and Cerf show that by defining, as above [see Eq. (44.43)], the mutual quantum information (the mutual entanglement) between systems Q and R as $I_{\rm QM} = H_{\rm VN}(R : Q)$, and supposing that the entropies of Q and R before the entanglement are the same — i.e. $H_{\rm VN}(Q) = H_{\rm VN}(R)$ —, one can derive that the loss [see Eq. (44.46)] has bounds:

$$0 \leq L \leq 2H_{\rm VN}(R), \ \ 0 \leq L \leq 2H_{\rm VN}(\mathcal{E}), \tag{44.53}$$

where $\mathcal{E}$ is the environment, as is usual. From these inequalities one can deduce:

$$0 \leq L \leq 2{\rm min}[H_{\rm VN}(R) H_{\rm VN}(\mathcal{E})]. \tag{44.54}$$

In other words, the initial mutual entanglement $2H_{\rm VN}(R)$ is split, through the action of $\mathcal{E}$, into a piece shared with Q' — $H_{\rm VN}(Q' : R') = 2H_{\rm VN}(R) - L$ (where Q', R' are the final systems —, and a piece shared with the environment (the remaining loss L), according to the relation [see again Eq. (44.46)]:

$$H_{\rm VN}(R' : Q') + H_{\rm VN}(R' : \mathcal{E}'|Q') = H_{\rm VN}(R' : \mathcal{E}'Q') = H_{\rm VN}(R : Q), \tag{44.55}$$

where the first and second term on the LHS are the mutual entanglement and the loss respectively:

$$I_{\rm QM} + L = 2H_{\rm VN}(R). \tag{44.56}$$

Now the authors define the von Neumann capacity $C_{\rm VN}$ as the mutual entanglement transmitted by the quantum channel, maximized over the density operator of the input channel:

$$C_{\rm VN} = {\rm Max}_{\hat{\rho}_Q} I_{\rm QM}. \tag{44.57}$$

The authors prove that the von Neumann capacity reduces to Holevo's result [HOLEVO 1998] for the transmission of classical information if the quantum state is measured before the transmission, i.e. it is known (the specifical quantum entanglement vanishes). In other words: the von Neumann capacity encompasses both the capacity for processing classical and quantum information. But the fact that the von Neumann capacity can be achieved by quantum coding (which would establish a full equivalence between the Shannon noisy coding theorem and the quantum case) could not be proved.

44.6 A Generalization

Deutsch's problem for evaluating functions [see subsection 44.2.3] and Shor's factorization algorithm [see subsection 44.4.1] are not the only algorithms which can be developed. Another algorithm is Grover's database searching algorithm. He proved [GROVER 1996] [GROVER 1997] that, while with classical methods one needs to look at a minimum of $\frac{N}{2}$ entries, **for example** names in a phone directory, in order to find someone's phone number with a probability of $\frac{1}{2}$, using quantum methods, i.e. a transformation of type $\frac{1}{\sqrt{2}}\begin{bmatrix} 1 & 1 \\ 1 & -1 \end{bmatrix}$ on the single input qubits, followed by an application of the Hadamard transform and by a transformation of type $\begin{bmatrix} e^{\imath\phi_1} & 0 \\ 0 e^{\imath\phi_2} & \end{bmatrix}$, and by a final measurement, one can obtain the same result with a number of steps of the order of $\sqrt{N}$. Even though this does not represent an exponential speed-up ($\sqrt{N}$ instead of $\frac{N}{2}$ steps) as in Shor's factorization algorithm, it is a sensible improvement with respect to classical computation.

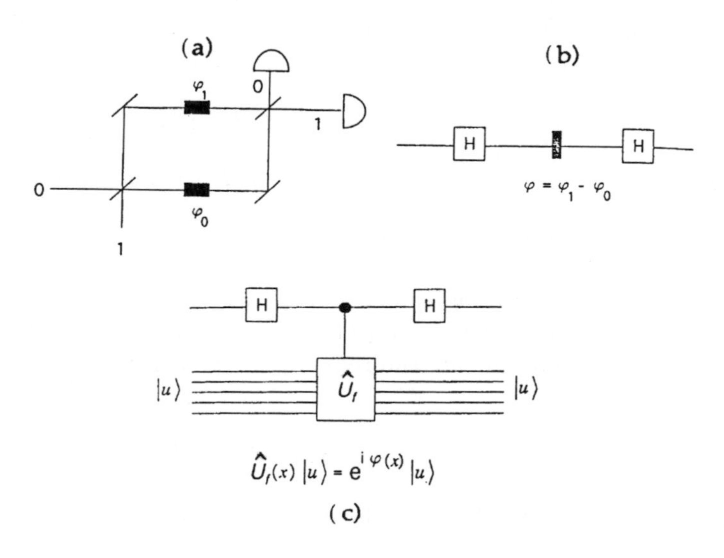

Figure 44.11: Interferometry as muster of any quantum algorithm. (a) Scheme of MZ interferometer with two phase shifters. The interference pattern depends on the difference between the phase shift in different arms of the interferometer.
(b) The corresponding quantum network representation.
(c) Network representation for the phase shift following the Hadamard transform (here x is the label for the state of the first qubit) — from [CLEVE *et al.* 1998].

However, recently Ekert and co-workers [CLEVE *et al.* 1998] have provided a powerful generalization. They showed that any quantum algorithm can be interpreted in terms of a multiparticle interferometry. Let us propose an interferometry scheme with two phase shifters [see figure 44.11.(a)] instead of only one. After passing the phase shifters the state of the photon is a superposition given by $\frac{1}{\sqrt{2}}(|0\rangle + e^{\imath(\varphi_1-\varphi_0)}|1\rangle$. Then the particle is detected in $|0\rangle$ and $|1\rangle$ with the following respective probabilities:

$$\frac{1}{2}(1+\cos\varphi), \quad \frac{1}{2}(1-\cos\varphi). \tag{44.58}$$

Let us now use the Hadamard transform (44.49). We can interpret the BS as a single qubit Hadamard transform and the quantum shifter can be viewed as a single-qubit gate [see figure 44.11.(b)]. Then the phase shift can be computed with the help of an auxiliary qubit described by $|u\rangle$, on which a unitary shift transformation of the form: $\hat{U}|u\rangle = e^{\imath\varphi}|u\rangle$ acts [see figure 44.11.(c)].

The controlled unitary operation must be followed by another Hadamard transformation which brings all the computational paths together. The whole process can then be described by:

$$\begin{aligned}|0\rangle|u\rangle &\mapsto \frac{1}{\sqrt{2}}(|0\rangle + |1\rangle)\,|u\rangle \\ &\mapsto \frac{1}{\sqrt{2}}(|0\rangle + e^{\imath\varphi}|1\rangle)\,|u\rangle \\ &\mapsto \left(\cos\frac{1}{2}\varphi|0\rangle - \imath\sin\frac{1}{2}\varphi|1\rangle\right)e^{\imath\frac{\varphi}{2}}|u\rangle, \end{aligned} \tag{44.59}$$

which is a perfect simulation of the MZ interferometer. The application of such a method allows us to solve Deutsch's problem [see subsection 44.2.3] with a single function evaluation instead of two (it suffices to previously subject an input state to a Hadamard transform). Ekert and co-workers also showed that the generation of arbitrary interference between computational paths is possible, a fundamental result in order to control such a process.

44.7 Concluding Remarks

The last word about quantum computation cannot be said. Much work has been done, but much more has still to be realized. It is most likely that the next ten years will tell us whether quantum computers can be built. In fact, this is the most rapidly changing field of quantum theory. However, it has already proved to be a very useful playground for a better understanding of the foundations of QM, especially for a better insight into the relationship between superposition and decoherence, and between superposition or entanglement and the interference between several possible 'paths'.

Part XI

CONCLUSIONS

Introduction to part XI As we have said [see p. 2], the two points which we need as preliminary ones for the accomplishment of the task of a global understanding of the theory are a better development of QM's *foundations* and a first global *interpretation.* We have already dealt with the historical developments concerning the interpretation [mainly in chapters 7 and 8] and the foundations [mainly in chapter 9]. After a long examination covering different areas and problems, we can now summarize our conclusions by trying to give a general orientation about both issues.

- In *chapter 45* we discuss the problem of foundations. We try to present a foundational synthesis of a theory whose constituent parts are a logico-probabilistic structure and a theory of open systems.
- In *chapter 46* we discuss the problem of the interpretation of QM. In the introduction [p. 3] we distinguished between the physical interpretation of the theory and the philosophical interpretation or philosophical aspects of the interpretational problem. Here we try to summarize a consistent physical interpretation and to see some general philosophical consequences. On the other hand, we are not essentially concerned here with specific metatheoretical problems — principles or other assumptions which have been made during the examination.

Chapter 45

A FOUNDATIONAL SYNTHESIS

45.1 Introduction and Contents

QM can be divided in the following way:

QM = {QM-Formalism, QM-Foundations, QM-Interpretation}.

In turn QM-Formalism consists of Basic-QM plus the already discussed extensions (Relativistic-QM, Q-Optics, and other areas discussed in this book):

QM-Formalism = {Basic-QM, QM-Extensions}.

Almost all of what has been discussed in the first part of the book — particularly definitions, corollaries, postulates, principles, and theorems which are *between horizontal lines*, and *squared* equations, constitutes the QM-Formalism. Here we are interested in the problem of their foundations.

As we have said [see section 9.4 and subsection 26.4.1], we still do not have, and it is certainly very difficult to obtain, an axiomatic system for QM of the type of classical mechanics. On the one hand, we have a basic duality in the theory between observables and states and, on the other, we have two constitutive pieces of it: the ('reversible') formalism and the theory of open systems. Therefore, it is suitable, in this present situation, not to 'close the circle' and to preserve a certain pluralism inside the theory. In other words[1], a lot of formalism which we have discarded as generalized solutions to fundamental problems, can be useful, on a formal level, in order to describe specific situations. It depends on the context and on the object we pursue. For example we think here about the GRW model [section 23.2], or the Bohm model of channels by measurement [subsection 32.6.2], or also some classical stochastic methods [section 23.1].

On the other hand, a certain 'pluralism' in the theory and specifically in the interpretation stems from the impossibility to eliminate theoretical terms, because what QM shows is that we always experience partial elements of a reality which in principle cannot be dominated as a whole [see figure 45.1].

But, as we have already said, the most important aspect is that quantum theory presents, at a foundational level, a 'duality' between the mathematical structure which depends on the superposition principle (and the entanglement), and the Measurement theory (or more generally the theory of irreversible interactions).

Therefore, we divide the problem of foundations into three subjects:

[1]On this point see [VAN FRAASSEN 1994, 8–9].

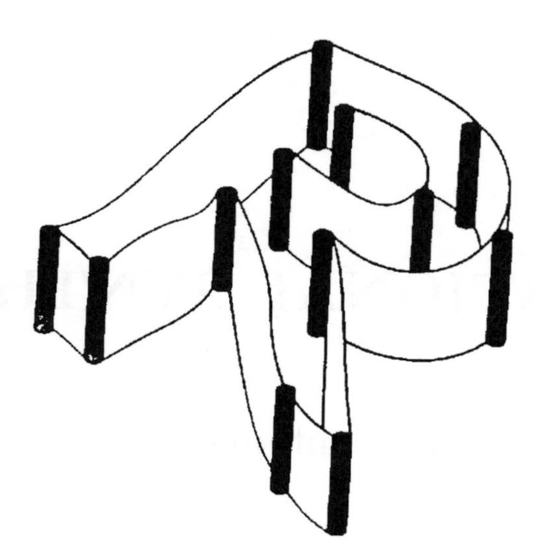

Figure 45.1: What we call reality, symbolized by the letter 'R' in the diagram, consists in an elaborate papier-maché construction of imagination and theory fitted in between a few iron posts of observation — from [WHEELER 1983, 195].

- The logico-algebraic structure [section 45.2], which mainly derives from the superposition principle, plus the POVM.
- The Measurement theory [section 45.3].
- The complementarity which-path/visibility — the bridge between these two, and itself part of the interpretation [section 45.4].

45.2 Foundations of Quantum Mechanics

45.2.1 Logico-Algebraic Structure

Comparison Between Different Basic Principles and Properties

There are several principles and properties which in QM are basic (in the sense that other aspects depend upon them):

- The CCRs [see Eqs. (3.14)]: they consist in an abstract mathematical relationship between arbitrary conjugate observables (represented here by operators) independently from any consideration of the state (but obviously not of the system of reference).
- UP [principle 7.1: p. 120] or CP1 [p. 137]: it is a mathematical relation between conjugate observables acting on the same state – its generalized mathematical formulation is the Hilgevoord/Uffink UR (7.44).
- The Superposition principle [3.3: p. 39]: it is a relationship between eigenstates of some reference observable(s).

- The Entanglement [definition 3.3: p. 43]: it is a property shared by several (sub-)systems and it is state-dependent.
- CP, which we assume here to be basically CP4 [principle 8.4: p. 139], and whose mathematical content is given by inequality (30.19): the Complementarity which-path/visibility (of the fringes of interference) is equally valid for observables and states and it is ultimately a piece of the interpretation because it is a relationship between a system considered as closed (subjected to superposition) and the same as an open system (due to the measurement in order to obtain the which-path information). Hence it will be discussed below [in section 45.4].

The CCRs and UP are not exactly the same i) regarding the mathematical content, ii) regarding the fact that the CCRs are between arbitrary conjugate observables represented by operators while UP is not necessarily so [see subsection 7.5.1], iii) and finally because, differently from UP, no consideration of the state enters the CCRs.

On the other hand, as we know [see section 7.3], UP is not a principle depending on an observation act but a basic feature of the theory as well like the CCRs are. For the second point note that we have already shown [see sections 10.2, 10.3, and 10.4] that it is possible to find more complex mathematical definitions of operators able to represent observables as time, position in Relativistic-QM and angle. It is true that up to now there is no general method able to produce an operator for a given observable, but nevertheless there is also no constraint so that we could not define some operator for a given observable. For the third point (the problem of the state), we have already seen [inequalities (42.9) and (42.10), or (42.12)] that entropic formulations of the URs are possible in which no reference to the state is made. Therefore, though UP and the CCRs are not exactly the same regarding the mathematical contents they can, to a certain extent, be considered as expressing the same basic aspect of the theory.

Regarding the relationship between UP and the Superposition principle, we know [see subsection 7.1.4] that they have different contents and ranges, so that they must be kept distinct.

A comparison between the Superposition principle and the property of entanglement [see subsections 3.5.2 and 9.4.1] shows that entanglement is not observable-dependent in the sense that it consists in a non-factorizability which is valid for every choice of basis. But, in another sense, it is observable-dependent, since, if we we wish to entangle two given systems, this is in general possible by allowing them to interact — or by allowing other particles to interact which are entangled with them [see SPDC Exp. 11: p. 692] —, which in turn signifies that they become entangled 'through' some observable. Therefore, to a certain extent, they can be considered as expressing the same fundamental aspect of the theory. But, by handling non-local features of the theory, it is better to discuss entanglement apart.

Superposition

Therefore, we assume that the two principles which the abstract structure of the theory depends on are the Superposition principle [3.3: p. 39] and UP or CP1 [p. 137].

How can we summarize their consequences for the general characteristics of the theory? On a logico-algebraic level it is an orthomodular lattice [theorem 9.2: p. 162], characterized by the atomicity and covering property [theorem 9.3: p. 163], whose logical calculus acknowledges a choice negation and a choice disjunction and hence presents value-gaps [theorem 9.10: p. 171]. Therefore, the implication which characterizes QM is the Sasaki arrow [see formula (9.20), and, for POVMs, see also (29.1)]. On the one hand, the Hilbert spaces are not a necessity for the theory [theorem 9.6: p. 166] and, on the other, it is possible to order the multiplicity of different possible representations of its algebraic structure [theorem 9.7: p. 167].

The Probability Structure of QM is made up of irreducible subsets of events pertaining to some Borel set [theorem 11.1: p. 197], parallelling its algebraic structure constituted of irreducible Boolean subalgebras [theorem 9.5: p. 164].

The I Gleason theorem [11.2: p. 200] establishes the validity of the use of projectors for calculating quantum-mechanical expectation values for Hilbert spaces with dimension $d \geq 3$. As a corollary we have that in QM there cannot be dispersion-free states or observables [32.4: p. 548], as proved positively by the I Bell theorem [32.3: p. 549] and the Kochen/Specker theorem [32.4: p. 551], which are, on the other hand, a confirmation of the existence of value-gaps in QM. This situation is also a good reason for introducing effects in the place of projectors.

As a consequence of the Superposition principle, there are phase relations (geometric phase) in QM which are unknown in classical mechanics: in particular, the taken path is of central relevance here and not only the points of departure and arrival [theorem 12.2: p. 205]: after a cycle — e.g. on the Poincaré sphere — a system may not return to the point of departure.

Entanglement

QM is essentially a non-local theory due to entanglement [definition 3.3: p. 43], a manifestation of the correlations in multiparticle systems. However, information can never be exchanged instantaneously [theorem 36.3: p. 634]: non-locality is a general term intended to unify different aspects of the theory — precisely those for which a local (or a classical) description fails.

The non-local structure of QM is first of all characterized by a violation of the separability [theorems 35.1: p. 591, 36.1: p. 622, 40.1: p. 685, 38.1: p. 651, 38.5: p. 658]. This non-separability is a confirmation of the importance of the context in which we not only factually perform but could also perform some experiments — as it is clear from a theorem of the Measurement theory [17.2: p. 267].

The specificity of these features is reflected in the non-classical features of quantum probability [theorem 39.1: p. 659; 39.3: p. 661; 39.4: p. 663; 39.5: p. 663; 39.6: p. 665; 39.7: p. 665].

We also showed that Bell inequalities can be dealt with in informational terms [see inequalities (42.88), (42.94) and (42.98)].

45.2.2 POVM Theory

The POVM theory originated, on the one hand, from the absence of dispersion-free observables in QM, and, on the other, from the localization problems which QM poses, particularly in a relativistic context, [theorem 10.1: p. 187]. Furthermore, some operators associated with observables which in Basic-QM or in its extensions are only variables — for example the relativistic stochastic position [Eq. (33.55)] — are only symmetric. Now only the POVM theory guarantees their measurability by the Naimark Theorem [18.5: p. 301] and the Holevo corollary [18.2: p. 301]. The latter theorem and its corollary establish the relationship between a PVM and a POVM as a coarse-graining one. In other words, the POVM theory is an extension of the traditional sharp QM theory. On an interpretational level we must admit that the stochastic theory (unsharp observables) is more correct [see section 33.1]. However, the Naimark Theorem and its corollary also guarantee the inverse passage, in the sense that one can accept a large part of the mathematical results of the traditional QM theory and integrate them in a stochastic generalization.

Here the most important definitions are those of operation [18.1: p. 292] and effect [18.2: p. 293] with related theorem [18.1: p. 292]. The operation is a state transformation which can also be entropy increasing. The effect is an equivalence class of operations, which is characterized by assuming all values between 0 and 1, even if it is not decomposable (it does not represent a mixture).

A POVM is mathematically defined by Eq. (18.10) and it characterizes a new class of (unsharp) observables.

The Informational Completeness [definition 18.6: p. 302] is the ability to distinguish between different states. Its necessary condition is some non-commutativity [theorem 18.6: p. 302]; but it is not a sufficient one, because non-commuting sharp observables are not informationally complete [theorem 18.7: p. 303]. On the contrary, pairs of partially non-commuting unsharp observables can be informationally complete [theorem 18.9: p. 304], which already proves the importance of the stochastic theory.

The POVM theory is also very important in order to determine that the relationship between non-commuting observables — UP [p. 137] — is a smooth Complementarity [Eqs. (7.44), (29.1) and (30.19)]. This is also of maximal importance for the Measurement theory because the diagonalization of the measured observable by a concrete measurement is never perfect.

45.3 Measurement Theory

One may wonder why we consider here the Measurement theory as a foundational aspect, given the fact that it presents enormous interpretational problems and implications. But the Measurement theory could be part of the interpretation of QM only to the extent that it would still be an open problem, and we think that this is largely no longer the case.

The centrality of the Measurement theory is due to: corollary 26.2 [p. 452], after which in QM there are never really isolated or closed systems; and to postulate 27.2 [p. 466], after which the reversible behaviour is only a limiting case of the irreversible one. Closely linked is the important postulate 27.1 (CP6) [p. 461] which establishes that the Complementarity Reversibility/Irreversibility is smooth.

45.3.1 General Features of the Measurement Process

After the basic definitions of the Measurement [13.1: p. 215] and the Premeasurement [13.2: p. 216] the specificity of the Measurement process in QM is pointed out by the II theorem of von Neumann [14.2: p. 222]: it does not conserve the quantum entropy, contrary to other forms of evolution (unitary ones). However, the Measurement process is subjected to constraints which come from the conservation laws [theorem 14.5: p. 234].

The specific problem of QM — which is absent in classical mechanics — is the objectification problem [definition 14.1: p. 235], with the consequence that even a hypothetical assignment of values to a generic observable is not always possible [theorem 14.6: p. 235].

On a technical level a measurement must satisfy, on the one hand, the probability reproducibility condition [theorem 14.7: p. 237], and, on the other, the objectification requirement, which has in turn two aspects: the pointer objectification [theorem 14.9: p. 239], and the value objectification [theorem 14.10: p. 241; theorem 14.11: p. 242; theorem 14.12: p. 242]. The conclusion is that there can be no unitary measurement which is able to satisfy these requirements [theorem 14.13: p. 243].

45.3.2 Theory of Measurement

Necessary and sufficient conditions for satisfying the probability reproducibility condition and the objectification requirement — as stated by theorem 17.7 [p. 273] — are the following ones: the relative-state theorem [15.1: p. 246], or better still (considering the constraints posed by conservation laws) the Davies theorem [18.3: p. 298], which establishes a correlation between the apparatus and the object system; the theorem of Machida/Namiki [16.2: p. 256; see also corollary 16.1: p. 256], following which the pointer can be understood as a statistical average on quantum states; the theorem of Watanabe/Machida/Namiki on the necessity of the apparatus [14.3: p. 230].

Here two aspects of the theory are of maximal interest:

- The theorem 17.6 [p. 271] and its corollary 17.2 [p. 272] guarantee that there cannot be reduction at zero of off-diagonal components. Hence in QM we cannot have measurements with perfect results. This explains the importance of the POVM because it allows a partial 'presence', in the measurement result, of other non-commuting observables. Hence the transition from microphysics to macrophysics is also a smooth one and the superposition is never completely lost — see the generation of S-Cats [chapter 24].

  Therefore, we do not assume quantum jumps as a basic form of reality, but as being derived from the existence of extended particles and unsharp observables [postulate 33.1: p. 567]; hence we do not admit, as classical mechanics does, the theoretical reducibility of the uncertainty in the measurement to zero [postulate 33.2: p. 568].

  It is the theorem of Busch [18.2: p. 296] and its corollary [18.1: p. 297] which specifically assure then the joint measurability of non-commuting unsharp observables.

- The theorem of Zeh/Zurek [17.4: p. 269; see also corollary 17.1: p. 269], on the importance of the environment, so as to obtain an eigenstate of the measured observable [see also theorem 17.3: p. 268], assures an irreversible operation because, once we have obtained an 'unsharp' diagonalization in a determinate observable basis by triorthogonal decomposition, we cannot decompose the state in any other way. We remember here that the triorthogonal decomposition is only necessary in order to obtain a determined value in the measurement process, but that there can be spontaneous decoherence with the sole coupling between a system and a (sufficiently large) reservoir. In this case there can be partial reversibility.

  Theorem 42.2 [p. 717] asserts the practical impossibility to make information which has been downloaded in our 'universe' available.

  The theorem of Shea/Scully/McCullen/Zurek [17.5: p. 271] is centered on the reduced density matrix as the mathematical operation which allows the required mixture.

Theorem 17.2 [p. 267] assures that there can also be a 'reduction' due to the sole presence of a detector without a performed measurement — a result which is confirmed by the interaction-free Measurement theory.

Theorem 42.3 [p. 719] gives the maximal amount of information accessible to a measurement and theorem 44.4 [p. 770] establishes an equivalence between potential information and entropy for the quantum case. Note that from this theorem and the proved impossibility of measuring a density matrix with a single measurement [theorem 28.2: p. 497] follows that we can never acquire the whole information contained in a state, which justifies the basic duality between observables and states.

45.3.3 Operational Definitions of Observable and State

How can we define elementary concepts as those of *observable* or of *state*? In the case of the observables we can first of all provide a probabilistic definition [11.5: p. 197], and then an operational definition [18.4 p. 293]: it is an equivalence class of measurements. For the state there is also an operational definition [18.3: p. 293]: it is an equivalence class of preparations. Both operational definitions presuppose the theory of Measurement and specifically an operational approach, which necessarily goes together with the assumption of such a centrality of the Measurement theory.

45.4 Complementarity

45.4.1 CP4 Again

Here we only discuss the Complementarity (part of the interpretation) to the extent to which it represents a bridge between the two parts of the foundational synthesis already dealt with — hence only CP4 [principle 8.4: p. 139].

In the which-path/visibility experiments [see chapter 30] we experience a Complementarity between the superposition of the system (the visibility of interference fringes) and a (partial) localization due to a measurement. In other words, CP4 has the following features:

- it joins the two fundamental aspects of the theory (formalism and Measurement theory) and acknowledges that both are fundamental for QM;
- simultaneously it points out that both aspects are exclusive;
- finally its mathematical content [inequality (30.19)] expresses the fact that this Complementarity is a smooth one, which allows us to consider closed and reversible systems as a special case of open and irreversible systems.

45.4.2 POVM Again

Is it possible to unify the two aspects of the theory further? Perhaps, given that the POVM theory is a generalization of the PVM formalism. It is not by chance that the Complementarity which-path/visibility can be very well expressed in terms of the POVM theory [see section 30.3]. It could be very useful to work out a generalized measure of entanglement [see section 42.8] which would allow us to quantify the relationship between measurements (or interactions) and entanglements in a general form.

Chapter 46

OUTLINE OF AN INTERPRETATION OF QUANTUM MECHANICS

Introduction and contents The enormous work which has been done in the last twenty years makes it possible perhaps to clearly distinguish between technical instruments and devices and a physical interpretation or an ontological reference of the theory for the first time. We have here a clear confirmation of the meta-principle 6.1 [p. 103]. Hence we may be able to overcome Feynman's dictum, which asserts that we can only know how QM works and not explain it [*FEYNMAN et al.* 1965, 1-1].

In section 46.1 we go back to the fundamental relationship between totality and locality. In section 46.2 we discuss the mode of interaction, while in section 46.3 we summarize the results of the preceding discussion. In section 46.4 we discuss the relationship between the two different parts of the foundations of the theory, and in section 46.5 the relationship between microworld and macroworld. In section 46.6 we briefly make some final philosophical considerations.

46.1 Locality and Totality

Non-locality consists of instantaneous correlations which, due to the relativity, are equivalent to a negation of spatial distance [FRANSON 1986, 128] [*SCHROECK* 1996, 221], in the sense that the distance between two entangled systems is irrelevant. On the other hand, the locality of interactions consists precisely of a space-time order. This locality is the same as 'individuality': in fact, when we force a system to 'choose' one state among an infinity of possible states in a superposition, we 'individualize' that system in the sense that we determine a univocal space-time relation to us and to the surroundings. From the point of view of the Quantum field theory the universal correlation can be thought of as the vacuum, and the individualisation processes as the continuous creation — or destruction — of particles and antiparticles with both directions of time[1]. The fact that the physical entities 'oscillate' between interaction and non-local correlations is, to a certain extent, another way of saying that they 'oscillate' between individuality and interconnection.

But if there would be no individualisation — i.e. no physical interaction —, our universe would be a whole without distinctions at all, where space, time, structures and distinctions would have

[1]See also [*JAMMER* 1974, 430].

would have no place. And it is exactly the individuality of physical interactions which forbids us to consider it as a whole — or, better still, the 'whole' is the other, intractable, inconceivable 'face' of our universe. *Per definitionem* the processes of localization are affected by a 'particularism' which is impossible to eliminate [see also corollary 26.3: p. 455]. The first to understand that our universe — from our point of view — cannot be considered as a whole, was Bell [BELL 1981a, 125]. This is also evident by the delayed choice experiments [chapter 26]: it is the experiment or the individual interaction which determines what the things are. Philosophically it is interesting to note that here we encounter the same essential tenet as Kant, when he said that we can never dominate all the possible experiences [*KANT* 1787, 346–47].

One can also see this from another point of view: pure states are states of evanescent entropy [see section 42.1]. If everything were interconnected, and hence in a superposition, we would have a zero-entropy universe, i.e. a universe where no information, active or potential (and hence, as already said, no structure and no organization) would be possible.

46.2 Dynamics

Once more we stress that this has nothing to do with some form of subjectivism or idealism [see section 26.4]. It only signifies that the basic form of reality in our universe is not the static being but the interaction (in our case the physical interaction).[2] Now it is not necessary that there are interacting humans: we know in fact that there are also spontaneous transitions to decoherence [chapter 21]. Hence QM provides us with a new form of thinking which is neither 'realistic' (in the classical sense of the term) nor 'subjectivistic'.

We can also state: *to be is to interact*, and: *individuals are operations* [see subsection 26.4.1 and section 30.4]. Remember here the concept of participation, which we developed in subsection 26.4.4. More than two thousand years of Platonism and Aristotelism are at an end. The first thinker to have searched for a new path in this direction was Whitehead [*WHITEHEAD* 1929], not surprisingly quoted by some quantum physicists, such as Stapp [STAPP 1975] [STAPP 1979] or Haag [HAAG 1996]. It is true that Aristotle spoke of a *dynamics*; but, as already said [see subsection 29.2.2], it was conceived either as a univocal evolution — for example the development of a tree from the seed — in the sense of a program, or as a pure possibility depending upon the subjective action of a human being — for example the marble block which can become a statue or a bank or some other thing. He did not conceive the interaction as the mode of being itself: for him the basic beings are the substances. Here, on the contrary, the ground entities are 'events', i.e. interaction modes [HAAG 1990] [*HAAG* 1992, 309–322] [STAPP 1979] [STAPP 1997].

Therefore, we can understand what Zurek says: the correlations are ontologically precedent to individuals [ZUREK 1982, 1876, 1878]: in fact individuals are 'extracted' from entanglement and superpositions. In this sense *to know is to separate*. This is the measurement (or subjective) form of *to be is to interact*. Hence the knowledge perfectly corresponds to the dynamics of being: it is an aspect of this dynamics [SHIMONY 1993b, 40]. Regarding the level of definition we already need a choice or a strategy [subsection 18.7.1]. Another way of saying the same thing is that we are, in principle, not completely separated from the rest of the world. But if we wish to know we must separate. In conclusion the knowledge is not only 'knowledge of interactions' but is itself an operation: To measure a system is to single out some degrees of freedom — in order to define and observe it [JOOS 1996, 36–37]. There is no longer place for a detached sensorial

[2] See also [*FINKELSTEIN* 1996, 13, 26].

experience of the world[3].

With Wheeler's words one can also say: *it from bit* [WHEELER 1990]. We understand this point not as a subjective proposition, but as the fundamental idea that the organization is ultimately information, and that without some form of information there would be no place for physical interaction.

As we have already said, here we have the ultimate justification of an operational point of view which we have supported throughout this book. At an epistemological level one may ask if an operational interpretation is less universal and less fundamental than an 'objective' one[4]. If so, then QM would imply a fundamental new limitation with respect to classical mechanics. As a matter of fact a part of the Copenhagen interpretation goes in this direction [see postulate 8.1: p. 137], and surely the three-valued logic approach as well [see chapter 29]. But the problem is that one usually conceives the operational approach as centered on the measurement[5]. On the contrary, as said, we interpret the measurement process as a particular realization of an 'ontology' based on interactions. For the interpretation here proposed, the operations are 'objective' processes in the sense that they do not necessarily depend on the human consciousness. It is another form of objectivity, which is not 'classic'.

46.3 Three 'Levels'

In summing up, we can say that we have three different aspects of the 'world': the correlations, the (quasi-)individuals, and, between these two extremes, the interactions or the dynamics. The interactions (measurements, spontaneous decoherence, and so on) are a junction between the extremes and are at the same time complementary to both extremes: both forms can be expressed by CP4 [principle 8.4: p. 139], since the localization can be understood in a dynamical way (like the measurement process itself) or as the final result (individualisation) of measurement. It is the dynamics which, on the one hand, limits the correlations and, on the other, makes the emergence of individuals possible. But it is also the dynamics which prevents a full individualisation — here we remember [see subsection 33.1.3] that also in classical mechanics kinematics is to a certain extent complementary to dynamics. It is interesting to observe that, as a consequence of the latter point, if the individuals were perfectly determined beings (hence uncorrelated), as was believed in classical mechanics, then the world would be 'frozen' and no dynamics would be possible.

From the perspective of a quantum system, the individualisation is a noise: in fact it is a decoherence process by means of which some information is lost — here we remember that some channels which transmit classical information cannot transmit quantum information [see subsection 44.3.4] and the decoherence process in quantum computation [subsection 44.5.1]. But, also from the perspective of an individual, quantum correlations appear as noise — see the discussion about the S-Cat [see subsection 24.5.3]. In the first case it is a 'classical' noise; in the second case a 'quantum' noise — here we remember the distinction between measurement error and quantum uncertainty (perturbation) [see sections 7.3 and 19.5].

It is also interesting to note that, historically speaking, the three 'levels' of the world have been studied and discovered in this sequence: individuals, dynamics and correlations. In fact

[3] On this point see [J. SCHUMACHER 1984, 104] .

[4] Bunge poses the problem and answers 'yes' [BUNGE 1965, 202].

[5] See again [BUNGE 1965, 209]. Consequently he thinks that, due to the problems of the Measurement theory, an operational interpretation is unsound.

the first physical science to be born was kinematics — with Galilei, Descartes, Newton. It has always remained as the basis of classical mechanics. It was kinematics that introduced the idea of perfect individuals and conceived physical states as the sum of perfectly determined properties. Then dynamics — with Newton, Lagrange, Hamilton— followed, but nobody thought that such a new development could radically change the basic assumptions of kinematics — the only one was Leibniz[6]. Finally Maxwell discovered a new reality: the fields, which are in open conflict with classical mechanics — as was acknowledged by Einstein [EINSTEIN 1936, 77] [EINSTEIN 1949b] [*EINSTEIN* MW, 159–62]. However, with only QM, and especially after the II Bell theorem [35.1: p. 591], the correlations have acquired a central theoretical status. Note that historically Leibniz was the first philosopher to discuss the problem of the correlations (*Harmonia praestabilita*), but he gave exclusive importance (though he was a nominalist in his youth) to the correlations by eliminating the level of individuals[7]. On the other hand, he understood, and very well, that the perfection of our world consists of a compromise between the unity of laws (correlations, QM basic laws) and the variety of beings (individuals) [*LEIBNIZ* 1686, 431].[8]

46.4 Reversibility and Irreversibility

We have seen that on a foundational level we have two parts of the theory which traditionally seemed in conflict since one is founded on reversible laws, while the other (the Measurement theory) on irreversibility. The interpretation must further clarify this point and provide a bridge between them [see section 45.4].

Now the space-time order that is created by interactions is not a universal connection, but local, as is evident by the existence of antiparticles which fly in the other direction of time [see subsection 4.2.3] or by the delayed choice — we never have to deal with a past defined once and for all [postulate 26.1 [p. 446] —, following which not only prediction, but also retrodiction is impossible [theorem 26.1: p. 452]. In other words: each order is a local topological structure. But, and this is the most important thing, though the order is absolutely local and relative, nobody can change it arbitrarily: in other words, nobody can go forward and backward 'in the time', because the time originates from interactions which *per definitionem* are irreversible — if not, there would be no individualisation and no order at all. Hence we must distinguish the theoretical possibility of going in one or the other direction of time, and the possibility to physically change the time direction. While the latter is impossible, because we cannot change 'our point of view' with that of antimatter, the former is exactly the content of the 'reversible' laws of QM and of physics in general. More exactly one should not speak of reversibility, but of 'time isotropic laws'. And the theory of delayed choice is exactly the physical content of this isotropy.

It is true that the fundamental laws of QM (for example the Schrödinger Eq.) are reversible. But this is only true if assuming isolated systems, which, as we have shown [see section 26.4], is never really the case. One can say that an individual system, in order to exist, must have interacted with something, and hence *per definitionem* is not isolated. Even the interesting experiments on QM reversibility can never completely eliminate some form of dispersion. Hence QM systems can go 'forward' or 'backward', but every time they do so they produce some form of irreversibility (dispersion).

[6]See several writings contained in the V volume of [*LEIBNIZ* MS]. See also [*GEROULT* 1967].
[7]On this point see examination in chapters III and IV of [*AULETTA* 1994b].
[8]For other references see [*AULETTA* 1994b, 142–43].

In conclusion QM implies a more radical form of ontological relativity than the one we are accustomed to admit by the relativity theory, though in the same spirit; and Bohr was really disappointed that Einstein did not see the analogy between the consequences of QM and that of the relativity theory [*MURDOCH* 1987, 146].

Note that the partial fragmentation of the theory can reflect, to a certain extent, the 'ontological datum' of the locality of spatio-temporal orders (there is no single order in nature).

46.5 Micro and Macro

The same is also valid when we speak of scale structures: there is nothing at all that forbids the 'matter' to be analysed in smaller parts than the ones we have reached [see section 33.1]. There are no last bricks, of which all is done. Surely the universe has, each time, a 'minimum' of subdivision — in the sense that, as we know from Aristotle, the matter cannot be infinitely subdivided —, but nothing forbids us or other beings from overcoming that subdivision. Naturally the same is valid for macrostructures, but generally they are outside of our technical possibilities. Therefore, an extreme form of reductionism to 'last elements' is surely false.

It is only the belief in static objects which creates the artificial distinction between 'natural' things and 'artificial' ones: if the things are what they are due to interactions, then all is artificial (and hence also natural, in the sense that this is how nature works). We are claiming that the naturalistic point of view [SHIMONY 1993b, 40–41] must be integrated by the opposite point of view: the human race is not only a natural system interacting with other natural systems in a world in which natural systems are already given, but it modifies what 'natural systems' can be, exactly as all other 'natural' systems do.

On the other hand, our normal macroworld — that of experience — is only approximately non-quantic, exactly as is the case for relativity. This is the only true correspondence principle [see principle 3.2: p. 29], in that it does not acknowledge our normal world as the rule or as the basic form of reality. On the contrary, it is a specific mode of being in a more general one. The problem is that we are accustomed to arbitrarily generalize this mode and not to see that also in our world there is some indefiniteness — see the problem of S-Cat [chapter 24]. Naturally the traces of the superposition are macroscopically so negligible that nobody considers them to be relevant. However, the level of practical purposes should be kept clearly distinct from the level of theoretical generalization. We can say that the notion of causality and the desire for certainty are imposed by us on Nature; and the maturity of science is in increasing knowledge of inaccuracy and randomness [REECE 1977, 9, 11].

Hence Bohr, though perhaps justified for his time in assuming the world of our experience as the context in which we must always deal with QM phenomena [postulate 8.2: p. 146], is, in this respect, largely superseded by recent developments. In fact, it is the opposite: with QM we have left, for the first time since prehistory, the garden in which we were born and have opened our eyes upon things that we had never imagined. But this is probably only the beginning, and the Copernican revolution will perhaps be small if compared with the new possibilities which are open to us.

46.6 Final Philosophical Considerations

46.6.1 Epistemology

The fact that the interpretation is not insignificant can be seen from an epistemological point of view. The great development of QM would have been impossible without Einstein (EPR), Schrödinger (S-Cat) or Bohm (HV theory) who tried a 'reasonable' interpretation of the theory. Surely their proposals have been shown to be generally incorrect, but science and philosophy learn much more from systematical errors (especially if done by great thinkers) than from accidentally right answers. Hence the search for philosophical and interpretative models is at least methodologically fundamental.

For this reason it is not a good pedagogical method when physics is taught with no reference at all to past discussions and problems. This has obviously to do with the formation of a critical attitude (and without it science becomes a technical device), but goes much further: the knowledge of the (critically examined) history of a science is to a certain extent an integral part of theory itself because it is the only way to show, in the right light, what the basis of the theory is (why basic assumptions were made); why a determined interpretation can be formulated; and what its limits are.

46.6.2 Philosophy and Quantum Mechanics

If so, the problem of physical interpretation cannot be disjointed from a philosophical work. Without philosophy, science would lose its critical spirit and would eventually become a technical device. Every theory can be overcome, and perhaps almost all of the basic assumptions of a theory. Without such a critical attitude, to only quote some examples, no modern physical science would have arisen. We remember the experiments in the Middle-Ages by the Franciscan Oxford school concerning Aristotle's 'dynamics' [*KOYRÉ* 1966] —; no Copernican revolution would have existed — because technical mending is always possible [*KUHN* 1957] —; non-Euclidian geometries would not have been formulated — for the first time a non-Euclidian geometry was developed by the Italian jesuit Girolamo Saccheri in XVIII century [*SACCHERI* 1733].

If philosophy is or can be important for science, then, *vice-versa*, all the lessons discussed in these chapters are enough to explain the importance of QM (and of science in general) for philosophy. In fact detailed attention to science is necessary, not only in order to solve epistemological problems [SHIMONY 1981, I, 4], but also for metaphysical, logical and general theoretical problems. Obviously, metaphysics cannot be developed in a systematic way as in the past. It should be closely linked to science and directed toward the generalization of what science assesses and toward the generation of useful 'paradigms' or general hypotheses which can allow science to 'search' in new directions.

If philosophy is the art of mediation between the universal domain and the empirical one, one can conclude by saying that there is no easy 'deductive' or 'inductive' passage between them.

BIBLIOGRAPHY

General Information Concerning the Use of the Bibliography

— In all quotations in the book the first number (which together with the author's or authors' name makes up the code) indicates the work's date and the following number(s) the page(s).

— If the authors of a determined work are more than two, normally the citation code (between square parenthesis) only contains the first name followed by "*et al.*"; in the bibliography order all co-works come at the end of the list under the first name.

- If one of two co-authors has a composed name, normally in the citation code only the first is quoted completely together with the first letter of the second.

— In all quotations in the book, code names in italics refer to a book. In plain text they refer to an article.

— The editions quoted in the text are always the last ones of the respective entry of the bibliography

— The reviews which are mostly quoted in the bibliography (more than 15 quotations) are abbreviated with the following codes:

AJP = American Journal of Physics
AP = Annals of Physics
CMP = Communications in Mathematical Physics
FP = Foundations of Physics
HPA = Helvetica Physica Acta
JMP = Journal of Mathematical Physics
JP = Journal of Physics
NA = Nature
NC = Nuovo Cimento
PL = Physics Letters
PR = Physical Review
PRep = Physics Reports
PRL = Physical Review Letters
PRSL = Proceedings of the Royal Society of London
PS = Philosophy of Science
PT = Physics Today
PTF = Progress of Theoretical Physics
RMP = Review of Modern Physics
ZP = Zeitschrift für Physik

— After a full stop and between { }, all titles are assigned areas with the following codes (codes 3-10 correspond to parts III-X of the book):

1 = Handbooks of Physics and Mathematics
2 = Handbooks of QM and of Special topics of QM
3 = Beginnings and Foundations of QM

4 = QM Measurement Problem
5 = Micro-/Macrophysics
6 = Delayed Choice
7 = Wave/Particle Dualism and Welcher Weg Problem
8 = Completeness of QM
9 = Non-Locality
10 = Information in QM
11 = Relativistic QM and Field Theory
12 = Interpretation of QM
13 = General or Other Physical Problems
14 = Epistemology, and General or Other Philosophical Problems
15 = Mathematical Problems

— In the case of a wide exposition of an argument or of a simplified report we use the letters 'W' and 'S', respectively.

Bibliography

[ÅBERG 1985] ÅBERG, Claes, "A Bell-Type Argument for the Double-Slit Experiment", in [*LAHTI/MITTELSTAEDT* 1985, 695–712]. {7}

[*ABRAHAM et al.* 1983] ABRAHAM, R./MARSDEN, J. E./RATIU, T., *Manifolds, Tensor Analysis, and Applications*, Reading Mass., Addison-Wesley, 1983. {1, 15}

[ACCARDI 1981] ACCARDI, Luigi "Topics in Quantum Probability", PRep **77** (1981): 169–192. {3}

[*ACCARDI* 1994a] — (ed.), *The Interpretation of Quantum Theory*, Roma, Istituto Enciclopedia Italiana, 1994. {12}

[ACCARDI 1994b] —, "Einstein-Bohr: One All", in [*ACCARDI* 1994a, 95–116]. {12}

[ACCARDI/FEDULLO 1982] ACCARDI, L./FEDULLO, A., "On the Statistical Meaning of Complex Numbers in Quantum Mechanics", *Lettere a Nuovo Cimento* **34** (1982): 161–72. {3, 8, 9}

[ADAMI/CERF 1997] ADAMI, C./CERF, N. J., "Von Neumann Capacity of Noisy Quantum Channels", PR **56** (1997): 3470–83. {10}

[AERTS 1982a] AERTS, Dirk, "Description of Many Separated Physical Entities Without the Paradoxes Encountered in Quantum Mechanics", FP **12** (1982): 1131–70. {9}

[AERTS 1985] —, "The Physical Origin of EPR Paradox and How to Violate Bell Inequalities by Macroscopical Systems", in [*LAHTI/MITTELSTAEDT* 1985, 305–320]. {9}

[AERTS 1986] —, "A Possible Explanation for the Probabilities of Quantum Mechanics", JMP **27** (1986): 202–210. {12}

[AERTS 1988] —, "The Description of Separated Systems and Quantum Mechanics and a Possible Explanation for the Probabilities of Quantum Mechanics", in [*VAN DER MERWE et al.* 1988a, 391–407]. {3}

[AGARWAL 1971] AGARWAL, G. S., "Brownian Motion of a Quantum Oscillator", PR **A4** (1971): 739–47. {5}

[AGARWAL 1995] —, "Interference in Complementary Spaces", FP **25** (1995): 219–28. {7}

[AGARWAL *et al.* 1994] AGARWAL, G. S./GRAF, M./ORSZAG, M./SCULLY, M. O./WALTHER, H., "State Preparation via Quantum Coherence and Continuous Measurement" PR **A49** (1994): 4077–84. {3, 4}

[AGARWAL *et al.* 1997] AGARWAL, G. S./PURI, R. R./SINGH, R. P., "Atomic Schrödinger Cat States", PR **A56** (1997): 2249–54. {5}

[AHARONOV/ALBERT 1981] AHARONOV, Y./ALBERT, D. Z., "Can We Make Sense Out of the Measurement Process in Relativistic Quantum Mechanics?", PR **D24** (1981): 359–70. {4, 11, 12}

[AHARONOV/ALBERT 1984] —, "Is the Usual Notion of Time Evolution Adequate for Quantum-Mechanical Systems?", PR **D29** (1984): 223–27, 228–34. {3, 4, 12}

[AHARONOV/BOHM 1959] AHARONOV, Y./BOHM, D., "Significance of Electromagnetic Potentials in the Quantum Theory", PR **115** (1959): 485–91; rep. in [*SHAPERE/WILCZEK* 1989, 104–110]. {3, 12}

[AHARONOV/BOHM 1961] —, "Time in the Quantum Theory and the Uncertainty Relation for Time and Energy", PR **122** (1961): 1649–58; rep. in [*WHEELER/ZUREK* 1983, 715–24]. {3}

[AHARONOV/BOHM 1964] —, "Answer to Fock Concerning the Time Energy Indeterminancy Relation", PR **134B** (1964): 1417–18. {3}

[AHARONOV/SUSSKIND 1967] AHARONOV, Y./SUSSKIND, L., "Charge Superselection Rules", PR **155** (1967): 1428–31. {3}

[AHARONOV/VAIDMAN 1989] AHARONOV, Y./VAIDMAN, L., "A New Characteristic of a Quantum System Between Two Measurements — A 'Weak Value"', in [*KAFATOS* 1989, 17–22]. {4}

[AHARONOV/VAIDMAN 1990] —, "Properties of a Quantum System During the Time Interval Between Two Measurements", PR **A41** (1990): 11–20. {4}

[AHARONOV/VAIDMAN 1993] —, "Measurement of the Schrödinger Wave of a Single Particle", PL **178A** (1993): 38–42. {4, 7}

[AHARONOV/VAIDMAN 1995] —, "Protective Measurements", in [*GREENBERGER/ZEILINGER* 1995a, 361–73]. {4}

[AHARONOV/VARDI 1980] AHARONOV, Y./VARDI, M., "Meaning of an Individual 'Feynman Path"', PR **D21** (1980): 2235–40. {4, 7, 12}

[AHARONOV *et al.* 1964] AHARONOV, Y./BERGMANN, P. G./LEBOWITZ, J. L., "Time Symmetry in the Quantum Process of Measurement", PR **B134** (1964): 1410–16. {4, 5, 12}

[AHARONOV *et al.* 1987] AHARONOV, Y./ALBERT, D. Z./CASHER, A./VAIDMAN, L., "Surprising Quantum Effects", PL **124A** (1987): 199–203. {4, 12}

[AHARONOV *et al.* 1988] AHARONOV, Y./ALBERT, D. Z./VAIDMAN, L., "How the Result of a Measurement of a Component of the Spin of a Spin-1/2 Particle Can Turn Out to be 100", PRL **60** (1988): 1351–54. {4}

[AHARONOV *et al.* 1990] AHARONOV, Y./ANANDAN, J./POPESCU, S./VAIDMAN, L., "Superposition of Time Evolutions of a Quantum System and a Quantum Time-Translation Machine", PRL **64** (1990): 2965–68. {3, 4, 9}

[AHARONOV *et al.* 1993a] AHARONOV, Y./ANANDAN, J./VAIDMAN, L., "Meaning of the Wave Function", PR **A47** (1993): 4616–26. {3, 4, 7, 12}

[AHARONOV *et al.* 1993b] AHARONOV, Y./POPESCU, S./ROHRLICH, D./VAIDMAN, L., "Measurements, Errors, and Negative Kinetic Energy", PR **A48** (1993): 4084–90. {9, 12}

[AHARONOV *et al.* 1995] AHARONOV, Y./POPESCU, S./VAIDMAN, L., "Causality, Memory Erasing, and Delayed-Choice Experiments", PR **A52** (1995): 4984–85. {4, 6, 10, 12}

[AHARONOV *et al.* 1996a] AHARONOV, Y./ANANDAN, J./VAIDMAN, L., "The Meaning of Protective Measurements", FP **26** (1996): 117–26. {3, 4, 7}

[AHARONOV *et al.* 1996b] AHARONOV, Y./MASSAR, S./POPESCU, S./TOLLAKSEN, J./VAIDMAN, L., "Adiabatic Measurements on Metastable Systems", PRL **77** (1996): 983–87. {4}

[AHARONOV *et al.* 1998] AHARONOV, Y./KAUFHERR, T./POPESCU, S./REZNIK, B., "Quantum Measurement Backreaction and Induced Topological Phases", PRL **80** (1998): 2023–26. {3, 4}

[AHMAD/WIGNER 1975] AHMAD, S. M. W./WIGNER, E. P., "Invariant Theoretic Derivation of the Connection Between Momentum and Velocity", NC **28A** (1975): 1–11. {3}

[ALBERGOTTI 1973] ALBERGOTTI, Clifton J., "Uncertainty Principle Limited Experiments", *The Physics Teacher* **11** (1973): 19–23. {3}

[ALBERT 1983] ALBERT, David, "On Quantum-Mechanical Automata", PL **98A** (1983): 249–52. {4, 10}

[ALBERT 1984] —, "Measurement Theory and a Quantum Mechanical Automaton", in [*ROTH/INOMATA* 1984, 111–15]. {4, 10}

[ALBERT 1990] —, " The Quantum Mechanics of Self-Measurement", in [*ZUREK* 1990a, 471–76]. {4}

[*ALBERT* 1992] —, *Quantum Mechanics and Experience*, Cambridge Mass., Harvard University Press, 1992. {4, 12; S}

[ALBERT/VAIDMAN 1989] ALBERT, D./VAIDMAN, L., "On a Theory of the Collapse of the Wave Function", in [*KAFATOS* 1989, 1–6]. {4}

[ALBRECHT 1992] ALBRECHT, Andreas, "Investigating Decoherence in a Simple System" PR **D46** (1992): 5504–5520. {4, 5, 6}

[ALBRECHT 1993] —, "Following a 'Collapsing' Wave Function", PR **D48** (1993): 3768–78. {4, 5, 6}

[ALEKSIEJUNAS/IVASKA 1997] ALEKSIEJUNAS, R./IVASKA, V., "Geometric Properties for Scalar Wave Superpositions", PL **235A** (1997): 1–6. {3}

[ALI/DOEBNER 1976] ALI, S. T./DOEBNER, H.-D., "On the Equivalence of Nonrelativistic Quantum Mechanics Based Upon Sharp and Fuzzy Measurements", JMP **17** (1976): 1105–111. {3, 4, 8}

[ALI/DOEBNER 1990] —, "Ordering Problem in Quantum Mechanics: Prime Qantization and a Physical Interpretation", PR **A41** (1990): 1199–210. {3}

[ALI/EMCH 1974] ALI, S. T./EMCH, G. G., "Fuzzy Observables in Quantum Mechanics", JMP **15** (1974): 176–82. {3, 8}

[ALI/PRUGOVEČKI 1977a] ALI, S. T./PRUGOVEČKI, E., "Systems of Imprimitivity and Representations of Quantum Mechanics on Fuzzy Phase Spaces", JMP **18** (1977): 219–28. {3, 8}

[ALI/PRUGOVEČKI 1977b] —, "Classical and Quantum Statistical Mechanics in a Common Liouville Space", *Physica* **89A** (1977): 501–521. {3, 8}

[ALI/PRUGOVEČKI 1981] —, "Self-Consistent Relativistic Model for Extended Spin-$\frac{1}{2}$ Particles in External Fields", NC **63A** (1981): 171–203. {8, 11}

[ALI/PRUGOVEČKI 1986a] —, "Mathematical Problems of Stochastic Quantum Mechanics: Harmonic Analysis and Systems of Covariance for Phase Space Representations of the Poincaré Group", *Acta Applied Mathematics* **6** (1986): 1–18. {3, 8}

[ALI/PRUGOVEČKI 1986b] —, "Extended Harmonic Analysis of Phase Space Representations of the Galilei Group", *Acta Applied Mathematics* **6** (1986): 19–45. {3, 8}

[ALI/PRUGOVEČKI 1986c] —, "Harmonic Analysis and Systems of Covariance for Phase Space Representations for the Galilei Group", *Acta Applied Mathematics* **6** (1986): 47–62. {3}

[ALI *et al.* 1988] ALI, S. T./BROOKE, J. A./BUSCH, P./GAGNON, R./SCHROECK, F. E. Jr., "Current Conservation as a Geometric Property of Space-Time", *Canadian Journal of Physics* **66** (1988): 238–44. {3}

[ALICKI 1984] ALICKI, Robert, "Dynamical Violation of the Superposition Principle for Open Quantum Systems", PL **103A** (1984): 247–49. {3, 4, 5}

[*ALICKI/LENDI* 1987] ALICKI, R./LENDI, K., *Quantum Dynamical Semigroups and Applications*, Berlin, Springer, 1987. {3, 4, 5}

[ALICKI/MESSER 1983] ALICKI, R./MESSER, J., "Nonlinear Quantum Dynamical Semigroups for Many-Body Open Systems", *Journal of Statistical Physics* **32** (1983): 299–312. {3, 4, 5}

[ALLCOCK 1969] ALLCOCK, G. R., "The Time of Arrival in Quantum Mechanics. I–III", AP **53** (1969): 253–348. {3}

[ALLMAN *et al.* 1992] ALLMAN, B. E./CIMMINO, A./KLEIN, A. G./OPAT G. I./KAISER, H./WERNER S. A., "Scalar Aharonov-Bohm Experiment with Neutrons", PRL **68** (1992): 2409–412. {9}

[ALTENMÜLLER/SCHENZLE 1993] ALTENMÜLLER, T. P./SCHENZLE, A., "Dynamics by Measurement: Aharonov's Inverse Quantum Zeno Effect", PR **A48** (1993): 70–79. {4, 7}

[ALTER/YAMAMOTO 1995a] ALTER, O./YAMAMOTO, Y., "Inhibition of the Measurement of the Wave Function of a Single Quantum System in Repeated Weak Quantum Nondemolition Measurements", PRL **74** (1995): 4106–109. {4}

[ALTER/YAMAMOTO 1995b] —, "Can We Measure the Wave Function of a Single Wave Packet of Light?", in [*GREENBERGER/ZEILINGER* 1995a, 103–109]. {4, 7, 12}

[ALTER/YAMAMOTO 1996a] —, "Protective Measurement of the Wavefunction of a Single Squeezed Harmonic Oscillator State", PR **A53** (1996): R2911–14. {3, 7, 12}

[ALTER/YAMAMOTO 1996b] —, "The Quantum Zeno Effect of a Single System is Equivalent to the Indetermination of the Quantum State of a Single System", in [*DE MARTINI et al.* 1996, 539–44]. {4, 7}

[ALTER/YAMAMOTO 1997] —, "Quantum Zeno Effect and the Impossibility of Determining the Quantum State of a Single System", PR **A55** (1997): R2499–R2502. {4, 7}

[AMMANN *et al.* 1998] AMMANN, H./GRAY, R./SHVARCHUCK, I./CHRISTENSEN, N., "Quantum Delta-Kicked Rotor: Experimental Observation of Decoherence", PRL **80** (1998): 4111-15. {4, 5}

[ANANDAN/BROWN 1995] ANANDAN, J./BROWN, H. R., "On the Reality of Space-Time Geometry and the Wave Function", FP **25** (1995): 349–60. {3}

[ANDÅS 1992] ANDÅS, H. E., "Bell's Inequalities for Quantum Mechanics", PL **167A** (1992): 6–10. {9}

[ANGELIDIS/POPPER 1985] ANGELIDIS, T. D./POPPER, K. R., "Towards a Local Explanatory Theory of EPR-Bohm Experiment", in [*LAHTI/MITTELSTAEDT* 1985, 37–49]. {9, 12}

[ANGLIN/ZUREK 1996] ANGLIN, J. R./ZUREK, W. H., "Decoherence of Quantum Fields: Pointer States and Predictability", PR **D53** (1996): 7327–35. {3, 5, 5, 11}

[ANSARI 1997] ANSARI, Nadeem A., "Violation of Bell's Inequality in a Driven Three-Level Cascade Atomic System", PR **A55** (1997): 1639–46. {9}

[ANSARI/ZUBAIRY 1988] ANSARI, N. A./ZUBAIRY, M. S., "Violation of Cauchy-Schwarz and Bell's Inequalitiesin Four-Wave Mixing", PR **A43** (1988): 2380–85. {9}

[ARAKI 1964] ARAKI, Huzihiro, "On the Algebra of All Local Observables", PTF **32** (1964): 844–54. {3}

[ARAKI 1980] —, "A Remark on Machida-Namichi Theory of Measurement", PTF **64** (1980): 719–30. {4}

[ARAKI/LIEB 1970] ARAKI, H./LIEB, E. H., "Entropy Inequalities", CMP **18** (1970): 160–70. {3, 5, 10}

[ARAKI/YANASE 1960] ARAKI, H. A./YANASE, M.M., "Measurement of Quantum Mechanical Operators", PR **120** (1960): 622–26; rep. in [*WHEELER/ZUREK* 1983, 707-711]. {3, 4}

[ARDEHALI 1998] ARDEHALI, M., "Clauser-Horne-Shimony-Holt Correlation and Clauser-Horne Correlation Do not Lead to the Largest Violations of Bell's Inequality", PR **A57** (1998): 114–19. {9}

[*ARISTOTLE De Gen. et corr.*] ARISTOTLE, *De Generatione et Corruptione*, Oxford, Clarendon, 1949, 1986. {14}

[*ARISTOTLE De Int.*] —, *De Interpretatione*, Oxford, Clarendon, 1949, 1986. {14}

[*ARISTOTLE Phys.*] —, *Physica*, Oxford, Clarendon, 1950, 1988. {14}

[ARTHURS/KELLY 1965] ARTHURS, E./KELLY, J. L., "On the Symultaneous Measurement of a Pair of Conjugate Observables", *Bell System Technical Journal* **44** (1965): 725–29. {4, 12}

[ASPECT 1976] ASPECT, Alain, "Proposed Experiment to Test the Nonseparability of Quantum Mechanics", PR **D14** (1976): 1944–51; rep. in [*WHEELER/ZUREK* 1983, 435–42]. {9}

[ASPECT/GRANGIER 1985] ASPECT, A./GRANGIER, P., "Test of Bells Inequalities with Pairs of Low Energy Correlated Photons: An Experimental Realization of EPR-Type Correlations", in [*LAHTI/MITTELSTAEDT* 1985, 51–71]. {9}

[ASPECT *et al.* 1981] ASPECT, A./GRANGIER, P./ROGER, G., "Experimental Tests of Realistic Local Theories via Bell's Theorem", PRL **47** (1981): 460–63. {9}

[ASPECT *et al.* 1982a] —, "Experimental Realization of Einstein-Podolsky-Rosen-Bohm Gedankenexperiment", PRL **48** (1982): 91–94. {9}

[ASPECT *et al.* 1982b] ASPECT, A./DALIBARD, J./ROGER, G., "Experimental Tests of Bell's Inequalities", PRL **49** (1982): 1804–7. {9}

[AULETTA 1992] AULETTA, Gennaro, "Sémiotique et problématique corps-esprit chez Leibniz et les Occasionnalistes", *Histoire Epistemologie Language* **14.2** (1992): 85–106. {14}

[AULETTA 1994a] —, "Il rapporto tra i concetti di possibile ed esistente nel quadro della teoria leibniziana dei mondi possibili" *Filosofia* **45.3** (sept.–dec. 1994): 275–89. {14}

[*AULETTA* 1994b] —, *Determinismo e contingenza. Saggio sulla filosofia leibniziana delle modalità*, Napoli, Morano, 1994. {14}

[*AUROUX* 1990] AUROUX, Sylvain, *Barbarie et philosophie*, Paris, Presse Universitaire de France, 1990; it. tr.: *Contro la filosofia ristretta*, Roma, Editori Riuniti, 1993. {14}

[BADUREK *et al.* 1983a] BADUREK, G./RAUCH, H./SUMMHAMMER, J., "Time-Dependent Superposition of Spinors", PRL **51** (1983): 1015–18. {3}

[BADUREK *et al.* 1983b] BADUREK, G./RAUCH, H./SUMMHAMMER, J./KISCHKO, U./ZEILINGER, A., "Direct Verification of the Quantum Spin-State Superposition Law", JP **A16** (1983): 1133–39. {3, 12}

[BADUREK *et al.* 1986] BADUREK, G./RAUCH, H./TUPPINGER, D., "Neutron Interferometric Double-Resonance Experiment", PR **A34** (1986): 2600–608. {7}

[BAEZ 1987] BAEZ, John, "Bell's Inequalitiy for C*–Algebras", *Letters on Mathematical Physics* **13** (1987): 135–36. {3, 9}

[BALAZS 1980] BALAZS, N. L., "Weyl's Association, Wigner's Function and Affine Geometry", *Physica* **102A** (1980): 236–54. {3}

[BALIAN *et al.* 1986] BALIAN, R./ALHASSID, Y./REINHARDT, H., "Dissipation in Many-Body Systems: A Geometric Approach Based on Information Theory", PRep **131** (1986): 1–146. {3, 4, 5, 10; W}

[BALLENTINE 1970] BALLENTINE, Leslie E., "The Statistical Interpretation of Quantum Mechancis", RMP **42** (1970): 358–81. {3, 4, 12}

[BALLENTINE 1973] —, FP **3** (1973): 229. {4, 12}

[BALLENTINE 1984] —, "What is the Point of the Quantum Theory of Measurement?", in [*ROTH/INOMATA* 1984, 65–75]. {4, 12}

[BALLENTINE 1986a] —, "Probability Theory in Quantum Mechanics", AJP **54** (1986): 883–89. {3}

[BALLENTINE 1986b] —, "Probability in Quantum Mechanics", in [*GREENBERGER* 1986, 382–92]. {3}

[BALLENTINE 1987] —, "Realism and Quantum Flux Tunneling", PRL **59** (1987): 1493–95. {8, 9, 12}

[BALLENTINE 1990a] —, "Limitations of the Projection Postulate", FP **20** (1990): 1329–43. {4, 12}

[*BALLENTINE* 1990b] —, *Quantum Mechanics*, Prentice Hall, Englewood Cliffs, 1990. {2}

[BALLENTINE 1991a] —, "Failure of Some Theories of State Reduction", PR **A43** (1991): 9–12. {3, 4, 5}

[BALLENTINE 1991b] —, "Comments on 'Quantum Zeno Effect"', PR **A43** (1991): 5165–67. {4}

[BALLENTINE 1994] —, "The Outlook from the Quantum Theory of Measurement", [*ACCARDI* 1994a, 87–94]. {4, 12}

[BALLENTINE/JARRETT 1987] BALLENTINE, L. E./JARRETT, J., "Bell's Theorem: Does Quantum Mechanics Contradict Relativity?", AJP **55** (1987): 696–701. {9, 12}

[BAN 1997a] BAN, Masashi, "Quasicontinuous Measurements of Photon Number", PR **A49** (1994): 5078–85. {4, 5}

[BAN 1997b] —, "Entropy Change and Information Gain in Photon Counting Measurement", PL **235A** (1997): 209–216. {5, 10}

[BANASZEK/WÓDKIEWICZ 1996] BANASZEK, K./WÓDKIEWICZ, K., "Direct Probing of Quantum Phase Space by Photon Counting", PRL **76** (1996): 4344–47. {3, 7}

[BAND/PARK 1979] BAND, W./PARK, J. L., "Quantum State Determination: Quorum for a Particle in One Dimension", AJP **47** (1979): 188–91. {3, 4, 7}

[BARACCA *et al.* 1975] BARACCA, A./BOHM, D./HILEY, B. J./STUART, A. E. G., "On Some New Notions Concerning Locality and Nonlocality in the Quantum Theory", NC **28B** (1975): 453–65. {8, 9}

[BARCHIELLI/BELAVKIN 1991] BARCHIELLI, A./BELAVKIN, V. P., "Measurement Continuous in Time and *a posteriori* States in Quantum Mechanics", JP **A24** (1991): 1495–1514. {4, 5}

[BARCHIELLI *et al.* 1982] BARCHIELLI, A./LANZ, L./PROSPERI, G. M., "A Model for the Macroscopic Description and Continual Observations in Q.M.", NC **72B** (1982): 79–121. {4, 5}

[BARCHIELLI *et al.* 1983] —, "Statistics of Continuous Trajectories in Quantum Mechanics: Operation-Valued Stochastic Processes", FP **13** (1983): 779–812. {4, 5, 8}

[BARDROFF *et al.* 1995] BARDROFF, P. J./MAYR, E./SCHLEICH, W. P., "Quantum State Endoscopy: Measurement of the Quantum State in a Cavity", PR **A51** (1995): 4963–66. {4, 7}

[BARDROFF *et al.* 1996] BARDROFF, P. J./LEICHTLE, C./SCHRADE, G./SCHLEICH, W. P., "Endoscopy in the Paul Trap: Measurement of the Vibratory Quantum State of a Single Ion", PRL **77** (1996): 2198–201. {5}

[BARDROFF *et al.* 1999] BARDROFF, P. J./FONTENELLE, M. T./STENHOLM, S., "Simple Scheme for State Measurement", PR **59** (1999): R950-53. {4, 7}

[BARENCO *et al.* 1995a] BARENCO, A./DEUTSCH, D./EKERT, A./JOZSA, R., "Conditional Quantum Physics and Logic Gates", PRL **74** (1995): 4083–86. {10}

[BARENCO *et al.* 1995b] BARENCO, A./BENNETT, C. H./CLEVE, R./DIVINCENZO, D. P./MARGOLUS, N./SHOR, P./SLEATOR, T./SMOLIN, J./WEINFURTER, H., "Elementary Gates for Quantum Computation", PR **A52** (1995): 3457–67. {10}

[BARENCO *et al.* 1996] BARENCO, A./EKERT, A./MACCHIAVELLO, C., "Un saut d'échelle pour le calculateurs", *La Recherche* **292** (1996): 52–58. {10; S}

[BARNETT/PHOENIX 1989] BARNETT, S. M./PHOENIX, S. J. D., "Entropy as a Measure of Quantum Optical Correlation", PR **A40** (1989): 2404–409. {3, 10}

[BARNETT/PHOENIX 1991] —, "Information Theory, Squeezing, and State Correlations", PR **A44** (1991): 535–45. {3, 10}

[BARNETT/PHOENIX 1992] —, "Bell's Inequality and the Schmidt Decomposition", PL **167A** (1992): 233–37. {3, 9}

[BARNUM *et al.* 1996] BARNUM, H./CAVES, C. M./FUCHS, C. A./JOZSA, R./SCHUMACHER, B., "Noncommuting Mixed States Cannot Be Broadcast", PRL **76** (1996): 2818–21. {3, 7, 9, 10}

[BARREAU 1985] BARREAU, Hervé, "Reality in Classical Physics and Reality in Quantum Theory", in [*LAHTI/MITTELSTAEDT* 1985, 73–83]. {3, 12}

[BARRETTO B. F./SELLERI 1995] BARRETTO BASTOS FILHO, J./SELLERI, F., "Propensity, Probability, and Quantum Physics", FP **25** (1995): 701–716. {7, 8, 9, 12}

[BARTELL 1980] BARTELL, Lawrence S., "Complementarity in the Double-Slit Experiment: On Simple Realizable Systems for Observing Intermediate Particle-Wave Behavior", PR **D21** (1980): 1698–99; rep. in [*WHEELER/ZUREK* 1983, 455–56]. {7}

[*BARUT* 1980] BARUT, Asim O. (ed.), *Foundations of Radiation Theory and Quantum Electrodynamics*, New York, Plenum, 1980. {1, 2}

[BARUT 1984] —, "Theoretical Experiments on the Foundations of Quantum Theory", in [*ROTH/INOMATA* 1984, 33–40]. {3}

[BARUT 1994] —, "The Deterministic Wave Mechanics: A Bridge between Classical Mechanics and Probabilistic Quantum Theory", [*ACCARDI* 1994a, 57–66]. {3, 8, 12}

[BASS 1992] BASS, Jean, "Probability, Pseudoprobability, Mean Values", in [*SELLERI* 1992a, 1–18]. {3}

[*BASTIN* 1971] BASTIN, Ted (ed.), *Quantum Theory and Beyond*, Cambridge, University Press, 1971. {3, 4, 12}

[BASTIN 1976] —, "Probability in a Discrete Model of Particles and Observations", in [*SUPPES* 1976, 195–219]. {3, 4}

[*BATES* 1962] BATES, D. R. (ed.), *Quantum Theory*, New York, Academic Press, 1962. {3, 4, 12}

[BELAVKIN 1994] BELAVKIN, Viacheslav P., "Nondemolition Principle of Quantum Measurement and Filtering Theory", [*ACCARDI* 1994a, 67–86]. {4}

[BELAVKIN/MELSHEIMER 1996] BELAVKIN, V. P./MELSHEIMER, O., "A Stochastic Hamiltonian Approach for Quantum Jumps, Spontaneous Localizations, and Continuous Trajectories", *Quantum and Semiclassical Optics* **8** (1996): 167–87. {4, 5, 8, 12}

[*BELAVKIN et al.* 1995] BELAVKIN, V. P./HIROTA, O./HUDSON, R. L. (ed.s), *Quantum Communications and Measurement*, New York, Plenum, 1995. {4}

[*BELINFANTE* 1973] BELINFANTE, F. J., *A Survey of Hidden-Variables Theory*, Oxford, Pergamon, 1973. {3, 8, 9, 12; W}

[*BELINFANTE* 1975] —, *Measurements and Time Reversal*, Oxford, Pergamon, 1975. {3, 4, 12}

[BELL 1964] BELL, John S., "On Einstein Podolsky Rosen Paradox", *Physics* **1** (1964): 195–200; rep. in [*BELL* 1987b, 14–21]. {8, 9}

[BELL 1966] —, "On the Problem of Hidden Variables in Quantum Mechanics", RMP **38** (1966): 447–52; rep. in [*BELL* 1987b, 1–13]. {3, 8, 9}

[BELL 1971] —, "Introduction to the Hidden-Variable Question", in [*D'ESPAGNAT* 1971, 171–81]; rep. in [*BELL* 1987b, 29–39]. {8, 9}

[BELL 1973] —, "Subject and Object", in [*MEHRA* 1973]; rep. in [*BELL* 1987b, 40–44]. {4}

[BELL 1975] —, "On Wave Packet Reduction in the Coleman-Hepp Model", *Helvetica Physica Acta* **48** (1975): 93–98; rep. in [*BELL* 1987b, 45–51]. {4, 5}

[BELL 1976a] —, "The Theory of Local Beables", *Epistemological Letters* (1976); rep. in [*BELL* 1987b, 52–62]. {3, 9, 12}

[BELL 1976b] —, "Einstein-Podolsky-Rosen Experiments", *Proceedings of the Symposium on Frontier Problems in High Energy*, Pisa, 1976; rep. in [*BELL* 1987b, 81–92]. {9}

[BELL 1976c] —., "The Measurement Theory of Everett and de Broglie's Pilot Wave", in [*FLATO et al.* 1976, 11–17]; rep. in [*BELL* 1987b, 93–99]. {3, 4, 7}

[BELL 1980a] —, "Atomic-Cascade Photons and Quantum-Mechanical Non-Locality", *Comments on Atomic and Molecular Physics* **9** (1980): 121–26; rep. in [*BELL* 1987b, 105–110]. {9}

[BELL 1980b] —, "de Broglie-Bohm Delayed-Choice Double-slit Experiment, and Density Matrix", *International Journal of Quantum Chemistry* (1980): 155–59; rep. in [*BELL* 1987b, 111–16]. {4, 5, 6, 7, 12}

[BELL 1981a] —, "Quantum Mechanics for Cosmologists", in ISHAM/ PENROSE/ SCIAMA (ed.), *Quantum Gravity 2*, Clarendon, Oxford, 1981; rep. in [*BELL* 1987b, 117–38]. {3, 4, 12}

[BELL 1981b] —, "Bertlmann's Socks and the Nature of Reality", *Journal de Physique Supplementa* **3** (1981): 41–61; rep. in [*BELL* 1987b, 139–58]. {3, 8, 9, 12}

[BELL 1982] —, "On the Impossible Pilot Wave", FP **12** (1982): 989–99; rep. in [*BELL* 1987b, 159–68]. {3, 7, 8}

[BELL 1984a] —, "Speakable and Unspeakable in Quantum Mechanics", in [*BELL* 1987b, 169–72]. {3, 12}

[BELL 1984b] —, "Beables for Quantum Field Theory" in [*BELL* 1987b, 173–80]. {3, 11}

[BELL 1986a] —, "Six Possible Worlds of Quantum Mechanics", *Proceedings of the Nobel Symposium 65: Possible Worlds in Art and Sciences*; rep. in [*BELL* 1987b, 181–95]. {3, 12}

[BELL 1986b] —, "EPR Correlations and EPW Distributions", in [*GREENBERGER* 1986, 263–66]; rep. in [*BELL* 1987b, 196–200]. {5, 9}

[BELL 1987a] —, "Are There Quantym Jumps?", in [*KILMISTER* 1987]; rep. in [*BELL* 1987b, 201–12]. {3, 4, 9, 12}

[*BELL* 1987b] —, *Speakable and Unspeakable in Quantum Mechanics*, Cambridge, University Press, 1987, 1994. {3, 4, 5, 6, 7, 8, 9, 11, 12, 13}

[BELL/HALLETT 1982] BELL, J. S./HALLETT, M., "Logic, Quantum Logic, and Empiricism", PS **49** (1982): 355–79. {3, 12, 13, 14}

[BELL/NAUENBERG 1966] BELL, J. S./NAUENBERG, M., "The Moral Aspect of Quantum Mechanics", in DE SHALIT/FESHBACH/VAN HOVE (ed.), *Preludes in Theoretical Physics*, Amsterdam, North Holland, 1966; rep. in [*BELL* 1987b, 22–28]. {3, 4, 9}

[BELTRAMETTI 1985] BELTRAMETTI, Enrico G., "The Non-unique Decomposition of Mixtures: Some Remarks", in [*LAHTI/MITTELSTAEDT* 1985, 85–95]. {3}

[*BELTRAMETTI/CASSINELLI* 1981] BELTRAMETTI, E./CASSINELLI, G., *The Logic of Quantum Mechanics*, Redwood City, Addison-Wesley, 1981. {3; W}

[BELTRAMETTI/CASSINELLI 1994] "Quantum Measurement and Probability', [*ACCARDI* 1994a, 49–56]. {3, 4}

[BELTRAMETTI/MACZYŃSKI 1991] BELTRAMETTI, E./MACZYŃSKI, M. J., "On a Characterization of Classical and Nonclassical Probabilities", JMP **32** (1991): 1280–86. {3, 9}

[*BELTRAMETTI/VAN FRAASSEN* 1981] BELTRAMETTI, E. G./VAN FRAASSEN, B. (ed.s), *Current Issues in Quantum Logic*, New York, Plenum, 1981. {3}

[BELTRAMETTI *et al.* 1990] BELTRAMETTI, E./CASSINELLI, G./LAHTI, P., "Unitary Measurements of Discrete Quantities in Quantum Mechanics", JMP **31** (1990): 91–98. {3, 4}

[BELTRAMETTI *et al.* 1993] BELTRAMETTI, E./DEL NOCE, C./MACZINSKI, M. J., "Characterization and Deduction of Bell-Type Inequalities", in [*GAROLA/ROSSI* 1993]. {9}

[BENIOFF 1972a] BENIOFF, Paul A., "Operator Valued Measures in Quantum Mechanics: Finite and Infinite Processes", JMP **13** (1972): 231–42.{3, 4}

[BENIOFF 1972b] —, "Decision Procedures in Quantum Mechanics", JMP **13** (1972): 908–915.{3, 4}

[BENIOFF 1972c] —, "Procedures in Quantum Mechanics without von Neumann's Projection Axiom", JMP **13** (1972): 1347–53.{3, 4}

[BENIOFF 1982] —, "Quantum Mechanical Models of Turing Machines That Dissipates No Energy", PRL **48** (1982): 1581–85. {10}

[BENNETT 1973] BENNETT, Charles H., *IBM Journal Res Dev* **17** (1973): 525. {10}

[BENNETT 1990] —, "How to Define Complexity in Physics and Why" in [*ZUREK* 1990a, 137–48]. {12, 13}

[BENNETT 1992] —, "Quantum Cryptography Using Any Two Nonorthogonal States", PRL **68** (1992): 3121–24. {9, 10}

[BENNETT 1995] —, "Quantum Information and Computation", PT **48** (Oct. 1995): 24–30. {10; W}

[BENNETT 1996] —, "Classical and Quantum Information Transmission and Interactions", in [*DE MARTINI et al.* 1996, 41–56]. {10}

[BENNETT/BRASSARD 1984] BENNETT, C. H./BRASSARD, G., *Proceedings IEEE International Conference on Computer Systems*, New York, IEEE, 1984: 175–79. {10}

[BENNETT/WIESNER 1992] BENNETT, C. H./WIESNER, S. J., "Communication via One- and Two-Particle Operators on EPR States", PRL **69** (1992): 2881–84. {9, 10}

[BENNETT *et al.* 1992a] BENNETT, C. H./BRASSARD, G./MERMIN, N. D., "Quantum Cryptography without Bell's Theorem", PRL **68** (1992): 557–59. {9, 10}

[BENNETT *et al.* 1992b] BENNETT, C. H./BRASSARD, G./EKERT, A., "Quantum Cryptography", *Scientific American* **267.4** (1992): 26–33. {9, 10; S}

[BENNETT *et al.* 1993] BENNETT, C. H./BRASSARD, G./CREPEAU, C./JOZSA, R./PERES, A./WOOTTERS, W. K., "Teleporting an Unknown Quantum State via Dual Classical and EPR Channels", PRL **70** (1993): 1895–99. {9, 10}

[BENNETT *et al.* 1996a] BENNETT, C. H./BRASSARD, G./POPESCU, S./SCHUMACHER, B./SMOLIN, J. A./WOOTTERS, W. K., "Purification of Noisy Entanglement and Faithful Teleportation via Noisy Channels", PRL **76** (1996): 722–25. {9, 10}

[BENNETT *et al.* 1996b] BENNETT, C. H./BERNSTEIN, H. J./POPESCU, S./SCHUMACHER, B., "Concentrating Partial Entanglement by Local Operations", PR **A53** (1996): 2046–52. {9, 10}

[BENNETT *et al.* 1996c] BENNETT, C. H./DIVINCENZO, D. P./SMOLIN, J. A./WOOTTERS, W. K., "Mixed-State Entanglement and Quantum Error Correction", PR **A54** (1996): 3824–51. {9, 10; W}

[BENOIST *et al.* 1979] BENOIST, R. W./MARCHAND, J.-P./WYSS, W., "A Note on Relative Entropy", *Letters in Mathematical Physics* **3** (1979): 169–73. {3, 10}

[BENSON *et al.* 1994] BENSON, O./RAITHEL, G./WALTHER, H., "Quantum Jumps of the Micromaser Field: Dynamic Behavior Close to Phase Transition Points" PRL **72** (1994): 3506–509. {5}

[BERGMANN/LEBOWITZ 1955] BERGMANN, P. G./LEBOWITZ, J. L., "New Approach to Nonequilibrium Processes", PR **99** (1955): 578–87. {4, 5, 13}

[BERGQUIST *et al.* 1986] BERGQUIST, J. C./R. G. HULET/ITANO, W. M./WINELAND, D. J., "Observation of Quantum Jumps in a Single Atom", PRL **57** (1986): 1699–1702. {5}

[BERGSTRÖM 1984] BERGSTRÖM, Lars "Underdetermination and Realism", *Erkenntnis* **21** (1984): 394. {14}

[*BERKELEY* 1710] BERKELEY, George, *A Treatise Concerning the Principles of Human Knowledge*, Dublin, 1710, London, 1734; Penguin, 1988. {14}

[BERNDL/S. GOLDSTEIN 1994] BERNDL, K./GOLDSTEIN, S., "Comment on 'Quantum Mecahnics, Local Realistic Theories, and Lorentz-Invariant Realistic Theories"' PRL **72** (1994) 780. {3, 8, 12}

[BERRY 1984] BERRY, Michael V., "Quantal Phase Factors Accompanying Adabiatic Changes", PRSL **A392** (1984): 45–57; rep. in [*SHAPERE/WILCZEK* 1989, 124–36]. {3}

[BERRY 1987] —, "The Adiabatic Phase and the Pancharatnams Phase for Polarized Light", *Journal of Modern Optics* **34** (1987): 1401–407; rep. in [*SHAPERE/WILCZEK* 1989, 67–73]. {3, 9}

[BERRY 1989] —, "The Quantum Phases, Five Years After", in [*SHAPERE/WILCZEK* 1989, 7–28]. {3}

[BERRY 1995] —, "Two-State Quantum Asymptotics", in [*GREENBERGER/ZEILINGER* 1995a, 303–317]. {3, 4, 5}

[BERTLMANN 1990] BERTLMANN, R. A., "Bells Theorem and the Nature of Reality", FP **20** (1990): 1191–1212. {9; W}

[BERTRAND/BERTRAND 1987] BERTRAND, J./BERTRAND, P., "A Tomographic Approach to Wigner's Function", FP **17** (1987): 397–405. {7}

[BHANDARI/SAMUEL 1988] BHANDARI, R./SAMUEL, J., "Observation of Topological Phase by Use of a Laser Interferometer", PRL **60** (1988): 1211–13. {3}

[*BHAT* 1981] BATH, B. R., *Modern Probability Theory*, New York, Wiley, 1981. {1}

[BIALYNICKI-B./MYCIELSKI 1975] BIALYNICKI-BIRULA, I./MYCIELSKI, J., "Uncertainty Relations for Information Entropy in Wave Mechanics", CMP **44** (1975): 129–32. {3, 10}

[BIALYNICKI-B./MYCIELSKI 1976] —, "Nonlinear Wave Mechanics", AP **100** (1976): 62–93. {3, 7, 12}

[*BIALYNICKI-B. et al.* 1992] BIALYNICKI-BIRULA, I./CIEPLAK, M./KAMINSKI, J., *Theory of Quanta*, Oxford, University Press, 1992. {2}

[BIHAM *et al.* 1996] BIHAM, E./HUTTNER, B./MOR, T., "Quantum Cryptographic Network Based on Quantum Memories", PR **A54** (1996): 2651–58. {9, 10}

[BIRKHOFF/VON NEUMANN 1936] BIRKHOFF, G./VON NEUMANN, J., "The Logic of Quantum Mechanics", *Annals of Mathematics* **37** (1936): 823–43; rep. in [*HOOKER* 1975, 1–26]. {3}

[*BJORKEN/DRELL* 1964] BJORKEN, J. D./DRELL, S. D., *Relativistic Quantum Mechanics*, New York, McGraw-Hill, 1964. {2}

[*BJORKEN/DRELL* 1965] —, *Relativistic Quantum Fields*, New York, McGraw-Hill, 1965. {2, 11}

[*BLACK et al.* 1992] BLACK, T./NIETO, M./PILLOFF, H./SCULLY, M. O./ SINCLAIR, R. (eds.), *Foundations of Quantum Mechanics*, Singapore, World Scientific, 1992. {3}

[BLANCHARD/JADCZYK 1993] BLANCHARD, P./JADCZYK, A., "On the Interaction Between Classical and Quantum Systems", PL **175A** (1993): 157–64. {4, 5}

[BLANCHARD/JADCZYK 1995] —, "Event Enhanced Quantum Theory and Piecewise Deterministic Dynamics", *Annalen der Physik* **4** (1995): 583–99. {3, 4, 5}

[BLANCHARD *et al.* 1998] BLANCHARD, P./PASQUINI, M./SERVA, M., "Localization Induced by Noise and Non Linearity", preprint. {4}

[BLANKENBECLER/PARTOVI 1985] BLANKENBECLER, R./PARTOVI, M. H., "Uncertainty, Entropy, and the Statistical Mechanics of Microscopic Systems", PRL **54** (1985): 373–76. {3, 4, 10}

[BLÄSI/HARDY 1995] BLÄSI, B./HARDY, L., "Realism and Time Symmetry in Quantum Mechanics", PL **207A** (1995): 119–25. {3, 12}

[J. BLATT 1959] BLATT, J. M., "An Alternative Approach to the Ergodic Problem", PTF **22** (1959): 745–56. {4, 5, 13}

[R. BLATT *et al.* 1995] BLATT, R./CIRAC, J. I./ZOLLER, P., "Trapping States of Motion with Cold Ions", PR **A52** (1995): 518–24. {3}

[BLOCH 1946] BLOCH, F., "Nuclear Induction", PR **70** (1946): 460–74. {3}

[BLOCK/BERMAN 1991] BLOCK, E./BERMAN, P. R., "Quantum Zeno Effect and Quantum Zeno Paradox in Atomic Physics", PR **A44** (1991): 1466–72. {4}

[*BLOKHINTSEV* 1964] BLOKHINTSEV, Dimitrii I., *Principles of Quantum Mechanics*, Dordrecht, Reidel, 1964. {2, 3}

[*BLOKHINTSEV* 1965] —, *The Philosophy of Quantum Mechanics*, 1965, engl. trans. Dordrecht, Reidel, 1968. {12}

[*BLOKHINTSEV* 1973] —, *Space and Time in the Microworld*, Dordrecht, Reidel, 1973. {3, 12}

[BLOKHINTSEV 1976] —, "Statistical Ensembles in Quantum Mechanics", in [*FLATO et al.* 1976, 147–58]. {3, 4, 12}

[BOCCHIERI/LOINGER 1959] BOCCHIERI, P./LOINGER, A., "Ergodic Foundation of Quantum Statistical Mechanics", PR **114** (1959): 948–51. {4, 5, 12}

[*BÖHM* 1979] BÖHM, A., *Quantum Mechanics*, New York, Springer, 1979. {2}

[*BOHM* 1951] BOHM, David, *Quantum Theory*, New York, Prentice-Hall, 1951. {2, 9}

[BOHM 1952] —, "A Suggested Interpretation of the Quantum Theory in Terms of 'Hidden Variables"', PR **85** (1952): 166–93; rep. in [*WHEELER/ZUREK* 1983, 369–96]. {3, 4, 8, 9, 12}

[BOHM 1953a] —, "Proof that Probability Density Approaches $\psi^*\psi$ in Causal Interpretation of the Quantum Theory", PR **89** (1953): 458–66. {3, 5}

[BOHM 1953b] —, "A Discussion of Certain Remarks by Einstein in Borns Interpretation of the ψ–function", in *Scientific Papers presented to Max Born*, Oliver and Boyd, London, 1953. {12}

[BOHM 1953c] —, "Comments on an Article of Takabayasi Concerning the Formulation of Quantum Mechanics with Classical Pictures", PTF **9** (1953): 273–87. {8, 12}

[*BOHM* 1957] —, *Causality and Chance in Modern Physics*, Princeton, Van Nostrand, 1957. {12}

[BOHM 1976] —, "On the Creation of a Deeper Insight into What May Underlie Quantum Physical Law", in [*FLATO et al.* 1976, 1–10]. {8, 12}

[*BOHM* 1980] —, *Wohleness and the Implicate Order*, London, Routledge, 1980, 1994. {8, 9, 12, 14}

[BOHM 1987] —, "Hidden Variables and Implicate Order", in [*HILEY/PEAT* 1987a, 33–45]. {12}

[BOHM/AHARONOV 1957] BOHM, D./AHARONOV, Y., "Discussion of Experimental Proof for the Paradox of Einstein, Rosen and Podolsky", PR **108** (1957): 1070–76. {9}

[BOHM/AHARONOV 1959] —, "Significance of Electromagnetic Potentials in the Quantum Theory", PR **115** (1959): 485–91. {3, 9, 12}

[BOHM/BUB 1966a] BOHM, D./BUB, J., "A Proposed Solution of the Measurement Problem in Quantum Mechanics by a Hidden Variable Theory", RMP **38** (1966): 453–69. {3, 4, 12}

[BOHM/BUB 1966b] —, "A Refutation of the Proof by Jauch and Piron that Hidden Variables Can Be Excluded in Quantum Mechanics", RMP **38** (1966): 470–75. {3, 9}

[BOHM/HILEY 1981] —, "Nonlocality in Quantum Theory Understood in Terms of Einstein's Nonlinear Field Approach", FP **11** (1981): 529–46. {3, 9, 11, 12}

[BOHM/HILEY 1984] —, "Measurement Understood Through the Quantum Potential Approach", FP **14** (1984): 255–74. {3, 4}

[BOHM/HILEY 1987] —, "An Ontological Basis for the Quantum Theory I: Non Relativistic Particle Systems", PRep **144** (1987): 323–48. {8, 12}

[BOHM/HILEY 1989] —, "Non-Locality and Locality in the Stochastic Interpretation of Quantum Mechanics", PRep **172** (1989): 93–122. {9}

[*BOHM/HILEY* 1993] —, *The Undivided Universe. An Ontological Interpretation of Quantum Theory*, London, Routledge, 1993. {3, 4, 5, 7, 8, 9, 10, 11, 12, 13}

[BOHM/SCHILLER 1955] BOHM, D./SCHILLER, R., "A Causal Interprerpretation of Pauli Equation. II", *Nuovo Cimento Supplementa* **1** (1955): 67–91. {3, 12}

[BOHM/VIGIER 1954] BOHM, D./VIGIER, J. P., "Model of the Causal Interpretation of Quantum Theory in Terms of a Fluid with Irregular Fluctuations", PR **96** (1954): 208–16. {3, 12}

[BOHM/VIGIER 1958] —, "Relativistic Hydrodynamics of Rotating Fluid Masses", PR **109** (1958): 1881–89. { 3, 12}

[BOHM *et al.* 1955] BOHM, D./SCHILLER, R./TIOMNO, J., "A Causal Interprerpretation of Pauli Equation. I", *Nuovo Cimento Supplementa* **1** (1955): 48–66. {3, 12}

[BOHM *et al.* 1987] BOHM, D./HILEY, B. J./KALOYEROU, P. N., "An Ontological Basis for the Quantum Theory II: A Causal Interpretation of Quantum Fields", PRep **144** (1987): 349–75. {3, 8, 11, 12}

[*BOHR* CW] BOHR, Niels, *Collected Works*, Amsterdam, North Holland, 1985–96. {3, 7, 8, 9, 11, 12, 13, 14}

[*BOHR* PhW] —, *Philosophical Writings*, Woodbridge (Connecticut), Ox Bow Press, 1958–63, 1987. {12, 13, 14}

[BOHR 1913] —, "On the Constitution of Atoms and Molecules", *Philosophical Magazine* 26 (1913): 1–25, 476–502, 857–75; rep. in [*BOHR* CW, II, 161–233]. {3}

[BOHR 1920] —, "Über die Linienspektren der Elemente", ZP **2** (1920): 423–69. {3}

[BOHR 1927a] —, "The Quantum Postulate and the Recent Development of Atomic Theory" in [*BOHR* CW, VI, 113–36]. {3, 12}

[BOHR 1927b] "Discussione sulla comunicazione di Bohr", *Congresso Internazionale dei Fisici, 1927*; in [BOHR 1927a, VI, 137–46]. {12}

[BOHR 1928] —, "The Quantum Postulate and the Recent Development of Atomic Theory", NA **121** (1928); rep. in [*BOHR* CW, VI, 148–58]. {3, 12, 13, 14}

[BOHR 1929] —, "Wirkungsquantum und Naturbeschreibung", *Die Naturwissenschaften* **17** (1929): 483–86; rep. in [*BOHR* CW, VI, 203–206]. {3, 12, 13}

[BOHR 1930] —, "Die Atomtheorie und die Naturbeschreibung", *Naturwissenschaften* **18** (1930): 73–78. {12, 13, 14}

[BOHR 1931] —, "Maxwell and the Modern Theoretical Physics", NA **28** (1931): 691–92; rep. in [*BOHR* CW, VI, 359–60]. {12, 13}

[BOHR 1935a] —, "Quantum Mechanics and Physical Reality", NA **136** (1935): 65; rep. in [*BOHR* CW, VII, 290]. {8, 9, 12}

[BOHR 1935b] —, "Can Quantum-Mechanical Description of Physical Reality Be Considered Complete?", PR **48** (1935): 696–702; rep. in [*BOHR* CW, VII, 292–96]. {8, 9, 12}

[BOHR 1936] —, "Kausalität und Komplementarität", *Erkenntnis* **6** (1936): 293. {3, 12, 14}

[BOHR 1937] —, "Causality and Complementarity" PS **4** (1937): 289–98; rep. in [*BOHR* CW, X, 37-48]. {12, 13, 14}

[BOHR 1939] "The Causality Problem in Atom Physics", *New Theories in Physics*, Paris, 1939; rep. in [*BOHR* CW, VII, 299–322]. {12, 13, 14}

[BOHR 1948] —, "On the Notions of Causality and Complementarity", *Dialectica* **1** (1948): 312–19; rep. in [*BOHR* CW, VII, 325–37]. {12, 13, 14}

[BOHR 1949] —, "Discussion With Einstein on Epistemological Problems in Atomic Physics", in [*SCHILPP* 1949, 201–241]. {3, 12; W}

[BOHR 1955a] —, "Unity of Knowledge", in *The Unity of Knowledge*, New York, Doubleday, 1955 rep. in [*BOHR* PhW, II, 67–82]. {12, 13, 14}

[BOHR 1955b] —, "Atoms and Human Knowledge", in [*BOHR* PhW, II, 83–93]. {12, 13, 14}

[BOHR 1958a] —, "Quantum Physics and Philosophy", in R. Kilbansky (ed.), *Philosophy in the Mid-Century*, Firenze, La Nuova Italia, 1958; rep. in [*BOHR* PhW, III, 1–7]. {12, 13, 14}

[*BOHR* 1958b] —, *Atomic Physics and Human Knowledge*, New York, Wiley, 1958. {12, 13, 14}

[BOHR 1961b] —, "The Genesis of Quantum Mechanics", in [*BOHR* PhW, III, 74–78]. {12}

[BOHR 1961c] —, "The Solvay Meetings and the Development of Quantum Physics", in [*BOHR* PhW, III, 79–100]. {12}

[*BOHR* 1961d] —, *Atomic Theory and the Description of Nature*, Cambridge, University Press, 1961. {12, 13, 14}

[BOHR *et al.* 1924] BOHR, N./KRAMERS, H. A./SLATER, J. C., "Über die Quantentheorie der Strahlung", ZP **24** (1924): 69–87. {3, 12}

[BOLLINGER *et al.* 1989] BOLLINGER, J. J./HEINZEN, D. J./ITANO, W. M./GILBERT, S. L./WINELAND, D. J., "Test of the Linearity of Quantum Mechanics by rf Spectroscopy of the 9Be+ Ground State", PRL **63** (1989): 1031–34. {7}

[*BOOLE* 1854] BOOLE, George, *An Investigation of the Laws of Thought, on which re Founded the Mathematical Theories of Logic and Probabilities*, London, 1854; New York, Dover, 1958. {15}

[BOPP 1947] BOPP, Fritz, "Quantenmechanische Statistik und Korrelationsrechung", *Zeitschrift für Naturforschung* **2a** (1947): 202–216. {3, 12}

[BOPP 1952] —, "Ein für di Quantenmechanik bemerkenswerter Satz der Korrelationsrechnung", *Zeitschrift für Naturforschung* **7a** (1952): 82–87. {3, 12}

[BOPP 1953] —, "Ein statistisches Modell für den Grundproze in der Quantentheorie der Teilchen", *Zeitschrift für Naturforschung* **8a** (1953): 228–33. {3, 12}

[BOPP 1954] —, "Korpuskularstatistische Begründung der Quantenmechanik", *Zeitschrift für Naturforschung* **9a** (1954): 579–600. {3, 4, 12}

[BOPP 1955a] —, "Quantenmechanische und stochastische Prozesse", *Zeitschrift für Naturforschung* **10a** (1955): 789–93. {3, 12}

[BOPP 1955b] —, "Einfaches Beispiel aus der stochastischen Quantenmechanik", ZP **143** (1955): 233–38. {3, 12}

[*BOPP* 1961a] — (ed.), *Werner Heisenberg und die Physik unserer Zeit*, Braunschweig, Vieweg, 1961. {3, 12}

[BOPP 1961b] —, "Statistische Mechanik bei Störung des Zustandes eines physikalischen Systems durch die Beobachtung", in [BOPP 1961b , 128–49]. {3, 4, 12}

[BOPP 1963] —, "Zur Quantenmechanik relativistischer Teilchen bei gegebenen Hilbert-Raum", ZP **171** (1963): 90–115. {3, 12}

[BOPP 1966] —, "Elementarvorgänge der Quantenmechanik in stochastischer Sicht", *Annalen der Physik* **17** (1966): 407–414. {4, 12}

[*BORN* AA] BORN, Max, *Ausgewhlte Abhandlungen*, Göttingen, Vandehoeck and Ruprecht, 1963. {3, 4, 7, 8, 12, 13, 14}

[BORN 1926a] —, "Zur Quantenmechanik der Stoßvorgänge", ZP **37** (1926): 863–67; rep. in [*BORN* AA, II, 228–32]. {3}

[BORN 1926b] —, "Quantenmechanik der Stoßvorgänge", ZP **38** (1926): 803–27; rep. in [*BORN* AA, II, 233–57]. {3, 12}

[BORN 1927] —, "Quantenmechanik und Statistik", *Naturwissenschaften* **15** (1927): 238–42; rep. in [*BORN* AA, II, 299–309]. {3}

[BORN 1938] —, "A Suggestion for Unifying Quantum Theory and Relativity", PRSL **A165** (1938): 291–303; rep. in [*BORN* AA, II, 560–78]. {3, 12, 13}

[*BORN* 1949b] —, *Natural Philosophy of Chance and Cause*, Oxford, Clarendon, 1949, New York, 1964. {14}

[BORN 1949c] —, "Reciprocity Theory of Elementary Particles", RMP **21** (1949): 463–73. {8, 12, 13}

[BORN 1949d] —, "Elementary Particles and the Principle of Reciprocity", NA **163** (1949): 207. {8, 12, 13}

[BORN 1953] —, "Physical Reality", *Philosophical Quarterly* **3** (1953): 139–49. {12, 13, 14}

[BORN 1954] —, "Die statistische Deutung der Quantenmechanik", Nobelvortrag; rep. in [*BORN* AA, II, 430–41]. {3, 12}

[BORN 1955a] —, "Continuity, Determinism and Reality", *Dan. Mat. Fys. Medd.* **30** (1955); rep. in [*BORN* AA, I, 196–219]. {3, 8, 12, 13, 14}

[BORN 1955b] —, "Zur Frage des Determinismus", *Physikalische Blätter* **11** (1955): 314–15. {8, 12, 13}

[BORN 1961] —, "Bemerkungen zur statistischen Deutung der Quantenmechanik", in [*BOPP* 1961a, 103–118]; rep. in [*BORN* AA, II, 454–69]. {12}

[*BORN* 1971] — (ed.), *The Born-Einstein-Letters*, London, Mcmillan, 1971. {12, 13, 14}

[BORN/BIEM 1968] BORN, M./BIEM, W., "Dualism in Quantum Theory", PT **21.8** (1968): 51–55. {12}

[BORN/JORDAN 1925] BORN, M./JORDAN, P., "Zur Quantenmechanik", ZP **34** (1925): 858–88; rep. in [*BORN* AA, II, 124–54]. {3}

[BORN/WIENER 1926] BORN, M./WIENER, N., "Eine neue Formulierung der Quantengesetzte fr periodische und nichtperiodische Vorgänge", ZP **36** (1926): 174–87; rep. in [*BORN* AA, II, 214–27]. {3}

[BORN *et al.* 1926] BORN, M./HEISENBERG, W./JORDAN, P., "Zur Quantenmechanik II", ZP 35 (1926): 557–615; rep. in [*BORN* AA, II, 155–213]. {3}

[BOSCHI *et al.* 1997a] —, "Test of the Violation of Local Realism in Quantum Mechanics Without Bell Inequalities", PL **228A** (1997): 208–214. {9}

[BOSCHI *et al.* 1997b] BOSCHI, D./BRANCA, S./DE MARTINI, F./HARDY, L., "A Ladder Proof of Nonlocality Without Inequalities: Theoretical and Experimental Results", preprint. {9}

[BOSCHI *et al.* 1998] BOSCHI, D./BRANCA, S./DE MARTINI, F./HARDY, L./POPESCU, S., "Experimental Realisation of Teleporting an Unknown Pure Quantum State via Dual Classical and EPR Channels", PRL **80** (1998): 1121–25. {9, 10}

[S. BOSE *et al.* 1997] BOSE, S./JACOBS, K./KNIGHT, P. L., "Preparation of Nonclassical States in Cavities with a Moving Mirror", PR **A56** (1997): 4175–86. {5}

[S. BOSE *et al.* 1998a] BOSE, S./VEDRAL, V./KNIGHT, P. L., "Multiparticle Generalization of Entanglement Swapping", PR **A57** (1998): 822–27. {9}

[S. BOSE *et al.* 1998b] —, "A Scheme to Probe the Decoherence of a Macroscopic Object", preprint; cod. 03.65.Bz,42.50.Vk,42.50.Dv. {4, 5}

[BOTHE/GEIGER 1924] BOTHE, W./GEIGER, H., "Ein Weg zur experimentellen Nachprüfung der Theorie von Bohr, Kramers and Slater", ZP **26** (1924): 44. {3}

[BOTHE/GEIGER 1925a] —, "Experimentelles zur Theorie von Bohr, Kramers and Slater", *Naturwissenschaften* **13** (1925): 440–441. {3}

[BOTHE/GEIGER 1925b] —, "Über das Wesen des Comptoneffekts; ein experimenteller Beitrag zur Theorie der Strahlung", ZP **32** (1925): 639–63. {3}

[BOTT/TU 1982] BOTT, R./TU, L., *Differential Forms in Algebraic Topology*, New York, Springer, 1982. {1}

[BOUWMEESTER *et al.* 1997] BOUWMEESTER, D., PAN, J.-W./MATTLE, K./EIBL, M./WEINFURTER, H./ZEILINGER, A., "Experimental Quantum Teleportation", NA **390** (1997): 575–79. {10}

[BOYD 1983] BOYD,, "On Current Status of Scientific Realism", *Erkenntnis* **19** (1983): 65. {14}

[BRAGINSKY 1968] BRAGINSKY, Vladimir B., "Classical and Quantum Restrictions on the Detection of Weak Disturbances of a Macroscopic Oscillator" (engl. transl.), *Soviet Physics JEPT* **26** (1968): 831–34. {4}

[BRAGINSKY 1988] —, *Usp. Fiz. Nauk.* **156** (1988): 93–115; "Resolution in Macroscopic Measurements: Progress and Prospects", engl. tr. *Soviet Physics-Uspekhi* **31** (1989): 836–49. {4}

[BRAGINSKY/KHALILI 1990] BRAGINSKY, V. B./KHALILI, F. Y., "Gravitational Wave Antenna with QND Speed Meter", PL **147A** (1990): 251–56. {4}

[BRAGINSKY/KHALILI 1991] —, "Friction and Fluctuations Produced by the Quantum Ground State", PL **161A** (1991): 197–201. {4}

[*BRAGINSKY/KHALILI* 1992] —, *Quantum Measurement*, Cambridge, University Press, 1992. {3, 4; W}

[BRAGINSKY/KHALILI 1996] —, "Quantum Nondemolition Measurements: the Route from Toys to Tools", RMP **68** (1996): 1–11. {4}

[BRAGINSKY/VORONTSOV 1974] BRAGINSKY, V.B./VORONTSOV, Y.I., *Usp. Fiz. Nauk.* **114** (1974): 41–53; "Quantum-Mechanical Limitations in Macroscopic Experiments and Modern Experimental Technique" (engl. tr.), *Soviet Physics Uspekhi* **17** (1975): 644–50. {4}

[BRAGINSKY/VYATCHANIN 1988] BRAGINSKY, V. B./VYATCHANIN, S. P., "Quadratic Scattering of an Electron for QND Measurement of Energy", PL **132A**: 206–208. {4}

[BRAGINSKY *et al.* 1977] BRAGINSKY, V.B./VORONTSOV, Y.I./KHALILI, F. Y., *Zh. Eksp. Teor. Fiz.* **73** (1977): 1340–43; "Quantum Singularities of a Ponderomotive Meter of Electromagnetic Energy" (engl. trans.): *Soviet Physics JETP* **46** (1977): 705–706. {4}

[BRAGINSKY *et al.* 1978] —, *Pis'ma Zh. Eksp. Teor. Fiz.* **27** (1978): 296–301; "Optimal Quantum Measurements in Detectors of Gravitation Radiation" (engl. tr.),: *Soviet Physics JETP Letters* **27** (1978): 276–80. {4}

[BRAGINSKY *et al.* 1980] BRAGINSKY, V.B./VORONTSOV, Y. I./THORNE, K.S., "Quantum Nondemolition Measurements", *Science* **209** (1980): 547–57; rep. in [*WHEELER/ZUREK* 1983, 749–68]. {4}

[BRAGINSKY *et al.* 1981] BRAGINSKY, V.B./VORONTSOV, Y.I./KHALILI, F. Y., "Small Vibration Detector for a Gravitational Antenna", *Soviet Physics-JETP Letters* **33** (1981): 405–407. {4}

[BRAUNSTEIN 1992] BRAUNSTEIN, Samuel L., "Quantum Limits on Precision Measurement of Phase", PRL **69** (1992): 3598–3601. {3, 4}

[BRAUNSTEIN 1994] —, "Some Limits to Precision Phase Measurement", PR **A49** (1994): 69–75. {3, 4}

[BRAUNSTEIN 1998] —, "Quantum Error Correction for Communication with Linear Optics", NA **394** (1998): 47–49. {10}

[BRAUNSTEIN/CAVES 1988] BRAUNSTEIN, S. L./CAVES, C. M., "Information-Theoretic Bell Inequalities", PRL **61** (1988): 662–65. {9, 10}

[BRAUNSTEIN/CAVES 1989] —, "Chained Bell Inequalities", in [*KAFATOS* 1989, 27–36]. {9}

[BRAUNSTEIN/CAVES 1990] —, "Wringing Out Better Bell Inequalities", AP **202** (1990): 22–56. {9}

[BRAUNSTEIN/CAVES 1994] —, "Statistical Distance and the Geometry of Quantum States", PRL **72** (1994): 3439–43. {3, 10}

[BRAUNSTEIN/CAVES 1995] —, "Geometry of Quantum States", in [*GREENBERGER/ZEILINGER* 1995a, 786–97]. {3, 10}

[BRAUNSTEIN/KIMBLE 1998] BRAUNSTEIN, S. L./KIMBLE, H. J., "Teleportation of Continuous Quantum Variables", PRL **80** (1998): 869–72. {10}

[BRAUNSTEIN/MANN 1993] BRAUNSTEIN, S. L./MANN, A., "Noise in Mermin's n–Particle Bell Inequality", PR **A47** (1993): R2427–30. {9}

[BRAUNSTEIN/MANN 1995] —, "Measurement of the Bell Operator and Quantum Teleportation", PR **A51** (1995): R1727–R1730. {9, 10}

[BRAUNSTEIN *et al.* 1992a] BRAUNSTEIN, S. L./MANN, A./REVZEN, M., "Maximal Violation of Bell Inequalities for Mixed States", PRL **68** (1992): 3259–61. {9, 10}

[BRAY/M. MOORE 1982] BRAY, A. J./MOORE, M. A., "Influence of Dissipation on Quantum Coherence", PRL **49** (1982): 1545–49. {4, 5}

[BREITENBERGER 1985] BREITENBERGER, Ernst, "Uncertainty Measures and Uncertainty Relations for Angle Observables", FP **15** (1985): 353–64. {3, 12}

[BRENDEL *et al.* 1992a] BRENDEL, J./MOHLER, E./MARTIENSSEN, W., "Time-Resolved Dual-Beam Two-Photon Interferences with High Visibility", PRL **66** (1991): 1142–45. {9}

[BRENDEL *et al.* 1992b] —, "Experimental Test of Bell's Inequality for Energy and Time", *Europhysics Letters* **20** (1992): 575–80. {9}

[H. BREUER/PETRUCCIONE 1995a] BREUER, H.-P./PETRUCCIONE, F., "Stochastic Dynamics of Quantum Jumps", PR **E52** (1995): 428–41. {3, 4, 5}

[H. BREUER/PETRUCCIONE 1995b] —, "Reduced System Dynamics as a Stochastic Process in Hilbert Space", PRL **74** (1995): 3788–91. {3, 4, 5}

[H. BREUER/PETRUCCIONE 1996a] —, "Quantum Measurement and the Transformation from Quantum to Classical Probabilities", PR **A54** (1996): 1146–53. {3, 4, 5}

[H. BREUER/PETRUCCIONE 1996b] —, "A Stochastic Wave Function Approach to Quantum Measurement", PL **220A** (1996): 315–19. {3, 4, 5}

[H. BREUER *et al.* 1997] BREUER, H.-P./KAPPLER, B./PETRUCCIONE, F., "Stochastic Wave-Function Approach to the Calculation of Multitime Correlation Functions of Open Quantum Systems", PR **A56** (1997): 2334–51. {4, 5}

[*T. BREUER* 1997] BREUER, Thomas, *Quantenmechanik — Ein Fall für Gödel?*, Heidelberg, Spektrum, 1997. {4, 6, 12, 13, 14}

[BRIDGMAN 1929] BRIDGMAN, P. W., "The New Vision of Science", *Harpers Magazine* **158** (1928): 443–51. {14}

[*BRILLOUIN* 1960] BRILLOUIN, Léon, *Wave Propagation and Group Velocity*, New York, Academic, 1960. {1}

[*BRILLOUIN* 1962] —, *Science and Information Theory*, New York, Academic, 1962. {1, 10}

[BRODSKY/DRELL 1980] BRODSKY, S. J./DRELL, S. D., "Anomalous Magnetic Moment and Limits on Fermion Substructure", PR **D22** (1980): 2236–43. {12, 13}

[BRODY/DE LA PEÑA-A. 1979] BRODY, T. A./DE LA PEÑA-AUERBACH, L., "Real and Imagined Nonlocalities in Quantum Mechanics", NC **54B** (1979): 455–62. {8, 9}

[H. BROWN 1986] BROWN, Harvey R., "The Insolubility Proof of the Quantum Measurement Problem", FP **16** (1986): 857–70 .{4}

[*H. BROWN/HARRÉ* 1988] BROWN, H. R./HARRÉ, R. (ed.s), *Philosophical Foundations of Quantum Field Theory*, Oxford, Clarendon, 1988. {3, 11, 12}

[H. BROWN/SVETLICHNY 1990] BROWN, H. R./SVETLICHNY, G., "Nonlocality and Gleason's Lemma. Part I: Deterministic Theories", FP **20** (1990): 1379– . {9}

[H. BROWN *et al.* 1995] BROWN, H. R./DEWDNEY, C./HORTON, G., "Bohm Particles and Their Detection in the Light of Neutron Interferometry", FP **25** (1995): 329–47. {7, 8}

[*L. BROWN* 1993a] BROWN, Laurie M. (ed.), *Renormalization*, Berlin, Springer, 1993. {3, 11, 12}

[L. BROWN 1993b] —, "Renormalization 1930–1950", in [*L. BROWN* 1993a, 1–27]. {3, 11}

[BRUN 1997] BRUN, Todd A., "Quantum Jumps as Decoherent Histories", PRL **78** (1997): 1833–37. {5, 6}

[BRUN 1998] —, "Continuous Measurements, Quantum Trajectories, and Decoherent Histories", e-print, code quant-ph/9710021. {4, 5, 6}

[BRUNE *et al.* 1990] BRUNE, M./HAROCHE, S./LEFEVRE, V./RAIMOND, J. M./ZAGURY, N., "Quantum Nondemolition Measurement of Small Photon Numbers by Rydberg-Atom Phase-Sensitive Detection", PRL **65** (1990): 976–79. {5}

[BRUNE *et al.* 1992] BRUNE, M./HAROCHE, S./RAIMOND, J. M./DAVIDOVICH, L./ZAGURY, N., "Manipulation of Photons in a Cavity by Dispersive Atom-Field Coupling: Quantum-Nondemolition Measurements and Generation of 'Schrödinger Cat' States", PR **A45** (1992): 5193–214. {5}

[BRUNE *et al.* 1994] BRUNE, M./NUSSENZWEIG, P./SCHMIDT-KALER, F./ BERDARDOT, F./MAALI, A./RAIMOND, J. M./HAROCHE, S., "From Lamb Shift to Light Shifts: Vacuum and Subphoton Cavity Fields Measured by Atomic Phase Sensitive Detection", PRL **72** (1994): 3339–42. {5, 10}

[BRUNE *et al.* 1996a] BRUNE, M./SCHMIDT - KALER, F./MAALI, A./DREYER, J./HAGLEY, E./RAIMOND, J. M./HAROCHE, S., "Quantum Rabi Oscillation: A Direct Test of Field Quantization in Cavity", PRL **76** (1996): 1800–803. {5}

[BRUNE *et al.* 1996b] BRUNE, M./HAGLEY, E./DREYER, J./MAÎTRE, X./MAALI, A./WUNDERLICH, C./RAIMOND, J. M./HAROCHE, S., "Observing the Progressive Decoherence of the 'Meter' in a Quantum Measurement", PRL **77** (1996): 4887–90. {4, 5}

[BRUß *et al.* 1998] BRUß, D./EKERT, A./MACCHIAVELLO, C., "Optimal Universal Quantum Cloning and State Estimation", e-print; code: 03.65.Bz, 03.67.-a. {7, 10}

[BUB 1968a] BUB, Jeffrey, "Hidden Variables and the Copenhagen Interpretation A Reconciliation", *British Journal for the Philosophy of Science* **19** (1968): 185–210. {8, 12}

[BUB 1968b] —, "The Daneri/Loinger/Prosperi Quantum Theory of Measurement", NC **57B** (1968): 503–20. {4, 5}

[BUB 1973] —, "On the Completeness of Quantum Mechanics", in [*HOOKER* 1973a, 1–65]. {8, 12}

[*BUB* 1974] —, *The Interpretation of Quantum Mechanics*, Dordrecht, Reidel, 1974. {12}

[BUB 1976a] —, "Hidden Variables and Locality", FP **6** (1976): 511–25. {8, 9, 12}

[BUB 1976b] —, "Randomness and Locality in Quantum Mechanics", in [*SUPPES* 1976, 397–420]. {8, 9, 12}

[BUB 1977] —, "Von Neumann's Projection Postulate as a Probability Conditionalization Rule in Quantum Mechanics", *Journal of Philosophical Logic* **6** (1977): 381–90. {3, 4, 12}

[BUB 1979] —, "Conditional Probabilities in Non-Boolean Possibility Structures", in [*HOOKER* 1979, 209–26]. {3}

[BUB 1982] —, "Quantum Logic, Conditional Probability, and Interference", PS **49** (1982): 402–421. {3, 12}

[BUB 1988] —, "How to Solve the Measurement Problem of Quantum Mechanics?", FP **18** (1988): 701–722. {4}

[BUB 1989a] —, "On Bohrs Response to EPR: A Quantum Logical Analysis", FP **19** (1989): 793–805. {3, 8, 9, 12}

[BUB 1989b] —, "On the Measurement Problem of Quantum Mechanics", in [*KAFATOS* 1989, 7–16]. {4}

[BUB 1992] —, "Quantum Mechanics Without the Projection Postulate", FP **22** (1992): 737. {4, 12}

[BUB 1994a] —, "On the Structure of Quantal Proposition Systems", FP **24** (1994): 1261–79. {3}

[BUB 1994b] —, "The Measurement Problem", [*ACCARDI* 1994a, 15–24]. {3, 4, 5, 12}

[BUB 1996] —, "Schütte's Tautology and the Kochen-Specker Theorem", FP **26** (1996): 787–806. {8}

[*BUB* 1997] — *Interpreting the Quantum World*, Cambridge, University Press, 1997. {4, 8, 9, 12}

[BUB/CLIFTON 1996] BUB, J./CLIFTON, R. A., "A Uniqueness Theorem for No Collapse Interpretations of Quantum Mechanics", *Studies in History and Philosophy of Modern Physics* (1996). {4, 12}

[BUNGE 1956] BUNGE, Mario, "Survey of the Interpretations of Quantum Mechanics", AJP **24** (1956): 272–86. {12}

[BUNGE 1965] —, "Physics and Reality", *Dialectica* **19** (1965): 195–222. {12, 14}

[*BUNGE* 1967a] — (ed.), *Quantum Theory and Reality*, New York, Springer, 1967. {12}

[BUNGE 1967b] —, "A Ghost-Free Axiomatization of Quantum Mechancis", in [*BUNGE* 1967a, 105–117]. {3}

[BUNGE 1970] —, "The So-Called Fourth Indeterminancy Relation", *Canadian Journal of Physics* **48** (1970): 1410–11. {3}

[*BUNGE* 1971] — (ed.), *Problems in the Foundations of Physics*, New York, Springer, 1971. {3}

[*BUNGE]* –, *Treatise on Basic Philosophy*, VII, Dordrecht, Kluwer, 1985. {12, 13, 14}

[BURNHAM/D. WEINBERG 1970] BURNHAM, D. C./WEINBERG, D. L., "Observation of Simultaneity in Parametric Production of Optical Photon Pairs", PRL **25** (1970): 84–87. {3}

[*BURT* 1981] BURT, Philip B., *Quantum Mechanics and Nonlinear Waves*, Chur, Harwood Academic, 1981. {7}

[BUSCH 1984] BUSCH, Paul, "On Joint Lower Bounds of Position and Momentum Observables in Quantum Mechanics", JMP **25** (1984): 1794–97. {3}

[BUSCH 1985a] —, "Elements of Unsharp Rality in the EPR Experiment", in [*LAHTI/MITTELSTAEDT* 1985, 343–58]. {8, 9}

[BUSCH 1985b] —, "Can Quantum Theoretical Reality Be Considered Sharp?", in [*MITTELSTAEDT/STACHOW* 1985, 81–101]. {8, 9}

[BUSCH 1985c] —, "Momentum Conservation Forbids Sharp Localisation", JP **A18** (1985): 3351–54. {3, 4, 8}

[BUSCH 1985d] —, "Indeterminancy Relations and Simultaneous Measurements in Quantum Theory", *International Journal of Theoretical Physics* **24** (1985): 63–92. {3, 4, 12}

[BUSCH 1986] —, "Unsharp Reality and Joint Measurements for Spin Observables", PR **D33** (1986): 2253–61. {3, 4}

[BUSCH 1987a] —, "Some Realizable Joint Measurements of Complementary Observables", FP **17** (1987): 905–937. {4}

[BUSCH 1987b] —, "Quantum Observables: Compatibility versus Commutativity and Maximal Information", JMP **28** (1987): 2866–72. {3, 10, 12}

[BUSCH 1988] —, "Surprising Features Of Unsharp Quantum Measurements", PL **130A** (1988): 323–29. {4}

[BUSCH 1990] "On the Energy-Time Uncertainty Relation I-II", FP **20** (1990): 1–32; 33–43. {3, 12}

[BUSCH 1994] —, "The Status of Quantum Mechanics in the Light of the Objectification Problem", [*ACCARDI* 1994a, 205–218]. {4, 12}

[BUSCH/LAHTI 1984] BUSCH, P./LAHTI, P. J., "On Various Joint Measurement of Position and Momentum Observables in Quantum Theory", PR **D29** (1984): 1634–46. {4}

[BUSCH/LAHTI 1987] —, "Minimal Uncertainty and Maximal Information for Quantum Position and Momentum", JP **A20** (1987): 899–906. {3, 10}

[BUSCH/LAHTI 1989] —, "The Determination of the Past and the Future of a Physical System in Quantum Mechanics", FP **19** (1989): 633–78. {3, 4, 8}

[BUSCH/LAHTI 1990a] —, "Some Remarks on Unsharp Quantum Measurements, Quantum Non-Demolition, and All that", *Annalen der Physik* **47** (1990): 369–82. {4}

[BUSCH/LAHTI 1990b] —, "Completely Positive Mappings in Quantum Dynamics and Measurement Theory", FP **20** (1990): 1429–39. {3, 4, 5}

[BUSCH/LAHTI 1996a] —, "Individual Aspects of Quantum Measurements", JP **A29** (1996): 5899–907. {4}

[BUSCH/LAHTI 1996b] —, "Correlation Properties of Quantum Measurements", JMP **37** (1996): 2585–601. {4}

[BUSCH/LAHTI 1996c] —, "The Standard Model of Quantum Measurement Theory: History and Applications", FP **26** (1996): 875–93. {4}

[BUSCH/MITTELSTAEDT 1991] BUSCH, P./MITTELSTAEDT, P., "The Problem of Objectification in Quantum Mechanics", FP **21** (1991): 889–904. {3, 4}

[BUSCH/SCHROECK 1989] BUSCH, P./SCHROECK, F. E., "On the Reality of Spin and Helicity", FP **19** (1989): 807–872. {3, 4; W}

[BUSCH *et al.* 1987] BUSCH, P./SCHONBEK, T. P./SCHROECK, F. E. Jr., "Quantum Observables: Compatibility versus Commutativity and Maximal Information", JMP **28** (1987): 2866–72. {3, 10}

[BUSCH *et al.* 1990] BUSCH, P./CASSINELLI, G./LAHTI, P., "On the Quantum Theory of Sequential Measurements", FP **20** (1990): 757–75. {4}

[*BUSCH* et al. 1991] BUSCH, P./LAHTI, P./MITTELSTAEDT, P., *The Quantum Theory of Measurement*, Berlin, Springer, 1991, II ed. 1996. {3, 4, 8, 12; W}

[BUSCH *et al.* 1992] —, "Weak Objectification, Joint Probabilities, and Bell Inequalities in Quantum Mechanics", FP **22** (1992): 949–62. {3, 4, 9}

[BUSCH *et al.* 1993] BUSCH, P./KIENZLER, P./LAHTI, P./MITTELSTAEDT, P., "Testing Quantum Mechanics Against a Full Set of Bell Inequalities", PR **A47** (1993): 4627–31. {8, 9}

[*BUSCH* et al. 1995a] BUSCH, P./GRABOWSKI, M./LAHTI, P. J., *Operational Quantum Physics*, Springer, Berlin, 1995. {3, 4, 8, 12; W}

[BUSCH *et al.* 1995b] —, "Repeatable Measurements in Quantum Theory: Their Role and Feasibility", FP **25** (1995): 1239–66. {4}

[BUSCH *et al.* 1995c] —, "Who Is Afraid of POV Measures? Unified Approach to Quantum Phase Observables", AP **237** (1995): 1–11. {3, 4}

[BUSSEY 1982] BUSSEY, P. J., "'Superluminal Communication' in EPR Experiments", PL **90A** (1982): 9–12. {9}

[BÜTTIKER 1983] BÜTTIKER, M., "Larmor Porcession and the Traversal Time for Tunneling", PR **B27** (1983): 6178–88. {3, 9}

[BÜTTIKER/LANDAUER 1982] BÜTTIKER, M./LANDAUER, R., "Traversal Time for Tunneling", PRL **49** (1982): 1739–42. {3, 9}

[BUŽEK/HILLERY 1996] BUŽEK, V./HILLERY, M., "Quantum Copying: Beyond the No-Cloning Theorem", PR **A54** (1996): 1844–52. {3, 7, 10}

[BUŽEK/KNIGHT 1995] BUŽEK, V./KNIGHT, P. L., "Quantum Interference, Superposition States of Light, and Nonclassical Effects", *Progress in Optics* (1995): 1–158. {5; W}

[BUŽEK *et al.* 1992a] BUŽEK, V./VIDIELLA-BARRANCO, A./KNIGHT, P. L. "Superposition of Coherent States: Squeezing and Dissipation", PR **A45** (1992): 6570–85. {5}

[BUŽEK *et al.* 1992b] BUŽEK, V./MOYA-CESSA, H./KNIGHT, P. L./PHOENIX, S. J. D., "Schrödinger-Cat States in the Resonant Jaynes-Cummings Model: Collapse and Revival of Oscillations of the Photon-Number Distribution" PR **A45** (1992): 8190–8203. {5, 6}

[BUŽEK *et al.* 1995] BUŽEK, V./KEITEL, C. H./KNIGHT, P. L., "Sampling Entropies and Operational Phase-Space Measurement. I–II", PR **A51** (1995): 2575–93, 2594–601. {3, 7, 10}

[BUŽEK *et al.* 1996a] BUŽEK, V./ADAM, G./DROBNÝ, G., "Quantum State Reconstruction and Detection of Quantum Coherences on Different Observation Levels", PR **A54** (1996): 804–820. {4, 7; W}

[BUŽEK *et al.* 1997a] BUŽEK, V./VEDRAL, V./PLENIO, M. B./KNIGHT, P. L./HILLERY, M., "Broadcasting of Entanglement via Local Copying", PR **A55** (1997): 3327–32. {3, 9, 10}

[BUŽEK *et al.* 1997b] BUŽEK, V./DROBNÝ, G./KIM, M. S./ADAM, G./KNIGHT, P. L., "Cavity QED with Cold Trapped Ions", PR **A56** (1997): 2352–60. {5, 9}

[BUŽEK *et al.* 1997c] BUŽEK, V./BRAUNSTEIN, S. L./HILLERY, M./BRUß, D., "Quantum Copying: A Network", PR **A56** (1997): 3446–52. {10}

[CABELLO/GARCÍA-A. 1998] CABELLO, A./GARCÍA-ALCAINE, G., "Proposed Experimental Tests of the Bell-Kochen-Specker Theorem", PRL **80** (1998): 1797–99. {8, 9}

[CABELLO *et al.* 1996a] CABELLO, A./ESTEBARANZ, J. M./GARCÍA-ALCAINE, G., "Bell-Kochen-Specker Theorem: A Proof with 18 Vectors", PL **212A** (1996): 183–87. {8}

[CABELLO *et al.* 1996b] —, "New Variants of the Bell-Kochen-Specker Theorem", PL **218A** (1996): 115–18. {8}

[CAHILL/GLAUBER 1969] CAHILL, K. E./GLAUBER, R. J., "Density Operators and Quasiprobability Distributions", PR **177** (1969): 1882–1902. {3; W}

[CALDEIRA/LEGGETT 1981] CALDEIRA, A./LEGGETT, A., "Influence of Dissipation on Quantum Tunneling in Macroscopic Systems", PRL **46** (1981): 211–14. {5, 9}

[CALDEIRA/LEGGETT 1983a] —, "Quantum Tunnelling in a Dissipative System", AP **149** (1983): 374–456. {5, 9}

[CALDEIRA/LEGGETT 1983b] —, "Path Integral Approach to Quantum Brownian Motion", *Physica* **121A** (1983): 587–616. {3, 5}

[CALDEIRA/LEGGETT 1985] —, "Influence of Damping On Quantum Interference: An Exactly Soluble Model", PR **A31** (1985): 1059–66. {5}

[CALDERBANK/SHOR 1996] CALDERBANK, A. R./SHOR, P. W., "Good Quantum Error-Correcting Codes Exist", PR **A54** (1996): 1098–1105. {10}

[CALZETTA *et al.* 1992] CALZETTA, E./CASTAGNINO, M./SCOCCIMARRO, R., "Coarse-Graining Approach to Quantum Cosmology", PR **D45** (1992): 2806-813. {6}

[*CAMPBELL* 1928] CAMPBELL, Norman, *An Account of the Principles of Measurement and Calculation*, London, Longmans and Green, 1928. {13}

[CAMPOS *et al.* 1989] CAMPOS, R. A./SALEH, B. E. A./TEICH, M. C., "Quantum-Mechanical Lossless Beam Splitter: SU(2) Simmetry and Photon Statistics", PR **A40** (1989): 1371–84. {3; W}

[CANTONI 1975] CANTONI, V., "Generalized 'Transition Probability"', CMP **44** (1975): 125–28. {3}

[CANTRELL/SCULLY 1978] CANTRELL, C. D./SCULLY, M. O., "The EPR Paradox Revisited", PL **43C** (1978): 499–508. {8, 9}

[CAPASSO *et al.* 1970] CAPASSO, V./FORTUNATO, D./SELLERI, F., "Von Neumann's Theorem and Hidden-Variable Models", NC **2** (1970): 149–99. {8}

[CAPASSO *et al.* 1973] –, "Sensitive Observables of Quantum Mechanics", *International Journal of Modern Physics* **7** (1973): 319–26. {4}

[CARELLI *et al.* 1995] CARELLI, P./CASTELLANO, M. G./CHIARELLO, F./CHIATTI, L./CIRILLO, M./COSMELLI, C./DIAMBRINI PALAZZI, G./FARGION, D./LEONI, R./ROTOLI, G./SCARAMUZZI, F./TORRIOLI, G., "MQC: An Experiment for Detecting Macroscopic Quantum Coherence", in [*GREENBERGER/ZEILINGER* 1995a, 845–47]. {5, 9}

[*CARMICHAEL* 1993] CARMICHAEL, Howard J., *An Open Systems Approach to Quantum Optics*, Heidelberg, Springer, 1993. {3, 4, 5; W}

[CARMICHAEL 1997] —, "Coherence and Decoherence in the Interaction of Light with Atoms", PR **A56** (1997): 5065–99. {3, 4, 5, 12; W}

[CARMICHAEL/WALLS 1975] CARMICHAEL, H. J./WALLS, D. F., "A Comment on the Quantum Treatment of Spontaneous Emission from Strongly Driven Two-Level Atom", JP **B8** (1975): L77–81. {3, 5}

[CARMICHAEL/WALLS 1976a] CARMICHAEL, H. J./WALLS, D. F., "A Proposal for the Measurement of the Resonant Stark Effect by Photon Correlation Techniques", JP **B9** (1975): L43–46. {3, 5}

[CARMICHAEL/WALLS 1976b] CARMICHAEL, H. J./WALLS, D. F., "A Quantum-Mechanical Master Equation Treatment of the Dynamical Stark Effect", JP **B9** (1975): 1199–1219. {3, 5}

[CARMICHAEL *et al.* 1989] CARMICHAEL, H. J./SINGH, S./VYAS, R./RICE, P. R., "Photoelectron Waiting Times and Atomic State Reduction in Resonance Fluorescence", PR **A39** (1989): 1200–218. {4, 5}

[CARRUTHERS/NIETO 1968] CARRUTHERS, P./NIETO, M. M., "Phase and Angle Variables in Quantum Mechanics", RMP **40** (1968): 411–40. {3}

[CARTWRIGHT 1976] CARTWRIGHT, Nancy D. "Superposition and Macroscopic Observation", in [*SUPPES* 1976, 221–34]. {12, 14}

[CASER 1992] CASER, Serge "Local Vacua", in [*SELLERI* 1992a, 19–36]. {8, 9}

[CASSINELLI/BELTRAMETTI 1977] CASSINELLI, G./BELTRAMETTI, E. G., "Quantum Logics and Ideal Measurements of the First Kind", in [*LOPES/PATY* 1977, 63–67]. {3, 4}

[CASSINELLI/LAHTI 1989] CASSINELLI, G./LAHTI, P. J., "The Measurement Statistics Interpretation of Quantum Mechanics: Possible Values and Possible Measurement Results of Physical Quantities", FP **19** (1989): 873–90. {4, 12}

[CASSINELLI/LAHTI 1993a] —, "Conditional Probability in the Quantum Theory of Measurement", NC **108B** (1993): 45–56. {3, 4}

[CASSINELLI/LAHTI 1993b] —, "Spectral Properties of Observables and Convex Mappings in Quantum Mechanics", JMP **34** (1993): 5468–75. {3}

[CASSINELLI *et al.* 1994] CASSINELLI, G./DE VITO, E./LATHI, P. J./LEVRERO, A., "Geometric Phase and Sequential Measurements in Quantum Mechanics", PR **A49** (1994): 3229–33. {3}

[CATLIN 1968] CATLIN, Donald E., "Spectral Theory in Quantum Logic", *International Journal of Theoretical Physics* **1** (1968): 285–97; rep. in [*HOOKER* 1979, 3–16]. {3}

[CAVES 1982] CAVES, Carlton M., "Quantum Limits on Noise in Linear Amplifiers", PR **D26** (1982): 1817–39. {3, 4}

[CAVES 1986] —, "Quantum Mechanics of Measurements Distributed in Time. A Path-Integral Formulation", PR **D33** (1986): 1643–65. {3, 4}

[CAVES 1987a] —, "Quantum Mechanics of Measurements Distributed in Time. II. Connections Among Formulations", PR **D35** (1987): 1815–30. {3, 4}

[CAVES 1987b] —, "Measurements Distributed in Time", in [*PIKE/SARKAR* 1987, 195–207]. {4}

[CAVES 1990] —, "Entropy and Information: How Much Information is Needed to Assign a Probability?" in [*ZUREK* 1990a, 91–115]. {3, 10}

[CAVES/DRUMMOND 1994] CAVES, C. M./DRUMMOND, P. D., "Quantum Limits on Bosonic Communication Rates", RMP **66** (1994): 481–537. {10; W}

[CAVES/MILBURN 1987] CAVES, C. M./MILBURN, G. J., "Quantum-Mechanical Model for Continuous Position Measurements", PR **A36** (1987): 5543–55. {4, 5}

[CAVES/SCHUMAKER 1985] CAVES, C. M./SCHUMAKER, B. L., "New Formalism for Two-Photon Quantum Optics", PR **A31** (1985): 3068–92. {3}

[CAVES *et al.* 1980] CAVES, C. M./THORNE, K. S./DREVER, R. W. P./SANDBERG, V. D./ZIMMERMAN, M., "On the Measurement of a Weak Classical Force Coupled to a Quantum-Mechanical Oscillator. I", RMP **52** (1980): 341–92 .{4; W}

[CELLUCCI 1990] CELLUCCI, Carlo, "La logica e la rappresentazione delle conoscenze", *Rivista di Filosofia* **81** (1990): 19–55. {14}

[CELLUCCI 1992] —, "Dalla logica teoretica alla logica pratica", *Rivista di Filosofia* **83** (1992): 169–207. {14}

[CELLUCCI 1993] —, "From Closed to Open Systems", in J. CZERMAK (ed.), *Philosophy of Mathematics: Proceedings of the 15th Wittgenstein-Symposium*, Wien, Hölder Pchler-Tensky, 1993: 206–220. {14}

[CERECEDA 1995] CERECEDA, José L., "The Kochen-Specker Theorem and Bell's Theorem: An Algebraic Approach", FP **25** (1995): 925–49. {8, 9}

[CERECEDA 1997] —, "Two-Particle Entanglement as a Property of Three-Particle Entangled States", PR **A56** (1997): 1733–38. {3, 9}

[CERECEDA 1998] —, "Generalized Probability for Hardy's Nonlocality Contradiction", PR **A57** (1998): 659–62. {9}

[CERF/ADAMI 1997] CERF, N./ADAMI, C., "Entropic Bell Inequalities", PR **A55** (1997): 3371–74. {9, 10}

[CERF/CLEVE 1997] CERF, N. J./CLEVE, R., "Information-Theoretic Interpretation of Quantum Error-Correcting Codes", PR **A56** (1997): 1721–32. {10}

[CERF *et al.* 1998] CERF, N. J./ADAMI, C./KWIAT, P. G., "Optical Simulation of Quantum Logic", PR **A57** (1998): R1477–80. {10}

[CHAKRAVARTY 1986] CHAKRAVARTY, Supid, "Quantum Mechanics on a Macroscopic Scale", in [*GREENBERGER* 1986, 25–35]. {5}

[CHAKRAVARTY/LEGGETT 1984] CHAKRAVARTY, S./LEGGETT, A. J., "Dynamics of the Two-State System with Ohmic Dissipation", PRL **52** (1984): 5–8. {5}

[CHANDLER *et al.* 1992] CHANDLER, C./COHEN, L./LU, C./SCULLY, M. O./WÓDKIEWICZ, K., "Quasi-Probability Distribution for Spin 1/2 Particles", FP **22** (1992): 867–78. {3}

[CHAPMAN *et al.* 1995] CHAPMAN, M. S./EKSTROM, C. R./HAMMOND, T. D./RUBENSTEIN, R. A./SCHMIEDMAYER, J./WEHINGER, S./PRITCHARD, D., "Optics and Interferometry with Na2 Molecules", PRL **74** (1994): 4783–86. {4, 5}

[CHIAO/WU 1986] CHIAO, R. Y./WU, Y.-S., "Manifestation of Berry's Topological Phase for the Photon", PRL **57** (1986): 933–36. {3}

[CHIAO/STEINBERG 1997] CHIAO, R. Y./STEINBERG, A. M., "Tunneling Times and Superluminality", *Progress in Optics* **37** (1997): 345–405. {9; W}

[CHIAO *et al.* 1988] CHIAO, R. Y./ANTARAMIAN, A./GANGA, K. M./JIAO, H./WILKINSON, S. R./NATHEL, H., "Observation of a Topological Phase by Means of a Nonplanar Mach-Zehnder Interferometer", PRL **60** (1988): 1214–17. {3}

[CHIAO *et al.* 1994] CHIAO, R. Y./KWIAT, P. G./STEINBERG, A. M. "The Photonic Tunneling Time and The Superluminal Propagation of Wave Packets", in [*DE MARTINI et al.* 1994, 258–73]. {9}

[CHIAO *et al.* 1995a] CHIAO, R. Y./BOYCE, J./GARRISON, J. C., "Superluminal (but Causal) Effects in Quantum Physics", in [*GREENBERGER/ZEILINGER* 1995a, 400–416]. {9}

[CHIAO *et al.* 1995b] CHIAO, R. Y./KWIAT, P. G./STEINBERG, A. M., "Quantum Nonlocality in Two-photon Experiments at Berkeley", *Quantum and Semiclassical Optics* **7** (1995): 259–78. {9; W}

[CHIU *et al.* 1977] CHIU, C. B./MISRA, B./SUDARSHAN, E. C. G., "Time Evolution of Unstable Quantum States and a Resolution of Zeno's Paradox", PR **D16** (1977): 520–529. {4}

[CHO *et al.* 1993] CHO, G.-C./KASARI, H./YAMAGUCHI, Y., "The Time Evolution of Unstable Particles", PTF **90** (1993): 803–816. {3}

[*CHOMSKY* 1966] CHOMSKY, Noam, *Cartesian Linguistic. A Chapter in the History of Rationalist Thought*, Lanham, Univesity Press of America, 1966. {14}

[CHUANG/YAMAMOTO 1995] CHUANG, I. L./YAMAMOTO, Y., "Simple Quantum Computer", PR **A52** (1995): 3489–96. {10}

[CHUANG *et al.* 1995] CHUANG, I. L./LAFLAMME, R./SHOR, P. W./ZUREK, W. H., "Quantum Computers, Factoring and Decoherence", *Science* **270** (1995): 1633–35. {5, 10}

[CHYLINSKI 1965] CHYLINSKI, Zygmunt, "Uncertainty Relation Between Time and Energy", *Acta Physica Polonica* **28** (1965): 631–38. {3, 12}

[CINI 1983] CINI, Marcello, "Quantum Theory of Measurement Without Wave Packet Collapse", NC **73B** (1983): 27–54. {4, 5, 12}

[CINI 1999] —, "Quantum Theory Without Waves: A Way of Eliminating Quantum Mechanical Paradoxes", preprint. {3, 7, 12}

[*CINI/LÉVY-L.* 1990] CINI, M./LÉVY-LEBLOND, J.-M. (ed.s), *Quantum Theory Without Reduction*, Bristol, Hilger, 1990. {4, 5, 12}

[CINI/SERVA 1990] CINI, M./SERVA, M., "State Vector Collpase as a Classical Statistical Effect of Measurement", in [*CINI/LÉVY-L.* 1990, 103–121]. {4, 5, 12}

[CINI/SERVA 1992] —, "Measurement in Quantum Mechanics and Classical Statistical Mechanics", PL **167A** (1992): 319–25. {4, 12}

[CINI *et al.* 1979] CINI, M./DE MARIA, M./MATTIOLI, G./NICOLÓ, F., "Wave Packet Reduction in Quantum Mechanics: A Model of a Measuring Apparatus", FP **9** (1979): 479–500. {4, 12}

[CIRAC/GISIN 1997] CIRAC, J. I./GISIN, N., "Coherent Eavesdropping Strategies for the Four State Quantum Cryptography Protocol", PL **229A** (1997): 1–7. {10}

[CIRAC/PARKINS 1994] CIRAC, J. I./PARKINS, A. S., "Schemes for Atomic-State Teleportation", PR **A50** (1994): R4441–44. {10}

[CIRAC/ZOLLER 1994] CIRAC, J. I./ZOLLER, P., "Preparation of Macroscopic Superpositions in Many-atoms Systems", PR **A50** (1994): R2799–802 .{4, 9}

[CIRAC/ZOLLER 1995] —, "Quantum Computation with Cold Trapped Ions", PRL **74** (1995): 4091–94. {10}

[CIRAC *et al.* 1993a] CIRAC, J. I./PARKINS, A. S./BLATT, R./ZOLLER, P., "'Dark' Squeezed States of the Motion of a Trapped Ion", PRL **70** (1993): 556–59. {3}

[CIRAC *et al.* 1993b] CIRAC, J. I./BLATT, R./PARKINS, A. S./ZOLLER, P., "Preparation of Fock States by Observations of Quantum Jumps in an Ion Trap", PRL **70** (1993): 762–65. {5}

[CLARK 1987] CLARK, T. D., "Macroscopic Quantum Objects", in [*HILEY/PEAT* 1987a, 121–50]. {5, 12}

[CLAUSER 1976] CLAUSER, John F., "Measurement of the Circular-Polarization Correlation in Photons from an Atomic Cascade", NC **33B** (1976): 740–46. {9}

[CLAUSER/DOWLING 1996] CLAUSER, J. F./DOWLING, J. P., "Factoring Integers with Youngs N–slit Interferometer", PR **A53** (1996): 4587–90. {3, 10}

[CLAUSER/HORNE 1974] CLAUSER, J. F./HORNE, M. A., "Experimental Consequences of Objective Local Theories", PR **D10** (1974): 526–35. {8, 9}

[CLAUSER/SHIMONY 1978] CLAUSER, J. F./SHIMONY, A., "Bell's Theorem: Experimental Tests and Implications", *Reporting Progress Physics* **41** (1978): 1881–927. {9; W}

[CLAUSER *et al.* 1969] CLAUSER, J. F./HORNE, M. A./SHIMONY, A./HOLT, R. A., "Proposed Experiment to Test Local Hidden-Variable Theories", PRL **23** (1969): 880–84; rep. in [*WHEELER/ZUREK* 1983, 409–413]. {8, 9}

[CLEVE/DIVINCENZO 1996] CLEVE, R./DIVINCENZO, D. P., "Schumacher's Quantum Data Compression as a Quantum Computation", PR **A54** (1996): 2636–50. {10}

[CLEVE *et al.* 1998] CLEVE, R./EKERT, A. K./MACCHIAVELLO, C./MOSCA, M., "Quantum Algorithms Revisited", PRSL **A454** (1998): 339-54. {3, 10}

[CLIFTON 1995a] CLIFTON, Rob, "Independently Motivating the Kochen-Dieks Modal Interpretation of Quantum Mechanics", *British Journal for the Philosophy of Science* **46** (1995): 33–57. {8}

[CLIFTON 1995b] —, "Making Sense of the Kochen-Dieks 'No-Collapse' Interpretation of Quantum Mechanics Independent of the Measurement Problem", in [*GREENBERGER/ZEILINGER* 1995a, 570–78]. {4, 8}

[*CLIFTON* 1995c] — (ed.), *Perspectives on Quantum Reality. Non-Relativistic, Relativistic, and Field-Theoretic*, Dordrecht, Kluwer, 1995. {11, 12}

[CLIFTON 1996] —, "The Properties of Modal Interpretation of Quantum Mechanics", *British Journal for the Philosophy of Science* **47** (1996): 371–98. {8}

[CLIFTON/NIEMANN 1992] CLIFTON, R./NIEMANN, P., "Locality, Lorentz Invariance, and Linear Algebra; Hardy's Theorem for Two Entangled Spin-*s* Particles", PL **166A** (1992): 177–84. {8, 9}

[*CLIFTON et al.* 1992] CLIFTON, R./PAGONIS, C./PITOWSKY, I., *Relativity, Quantum Mechanics and EPR*, Philosophy of Science Association, 1992. {9, 11, 12}

[CLIFTON *et al.* 1991] CLIFTON, R./REDHEAD, M./BUTTERFIELD, J. N., "Generalization of the Greenberger-Horne-Zeilinger Algebraic Proof of Nonlocality", FP **21** (1991): 149–84. {9}

[L. COHEN 1966a] COHEN, Leon, "Can Quantum Mechanics be Formulated as a Classical Probability Theory?", PS **33** (1966): 317–22. {3}

[L. COHEN 1966b] —, "Generalized Phase-Space Distribution Functions", JMP **7** (1966): 781–86. {3}

[L. COHEN 1973] —, "Joint Probability Distributions in Quantum Mechanics", in [*HOOKER* 1973a, 66–69]. {3}

[L. COHEN 1986] —, "Joint Quantum Probabilities and the Uncertainty Principle", in [*GREENBERGER* 1986, 283–91]. {3}

[L. COHEN/SCULLY 1986] COHEN, L./SCULLY, M. O., "Joint Wigner Distribution for Spin-1/2 Particles", FP **16** (1986): 295–310. {3}

[O. COHEN 1997] COHEN, Oliver, "Nonlocality of the Original Einstein-Podolsky-Rosen State", PR **56** (1997): 3484–92. {8, 9}

[O. COHEN 1998] —, "Unlocking Hidden Entanglement with Classical Information", PRL **80** (1998): 2493–96. {9, 10}

[O. COHEN/HILEY 1995] COHEN, O./HILEY, B. J., "Retrodiction in Quantum Mechanics, Preferred Lorentz Frames, and Nonlocal Mesurements", FP **25** (1995): 1669–98. {4, 8, 9, 11}

[O. COHEN/HILEY 1996] —, "Elements of Reality, Lorentz Invariance, and the Product Rule", FP **26** (1996): 1–15. {8, 9, 11}

[COHEN-T./DALIBARD 1986] COHEN-TANNOUDJI, C. N./DALIBARD, J., "Single-Atom Laser Spectroscopy. Looking for Dark Periods in Fluorescence Light", *Europhysics Letters* **1** (1986): 441–48. {5}

[COHEN-T./PHILLIPS 1990] COHEN-TANNOUDJI, C. N./PHILLIPS, W. D., "New Mechanisms for Laser Cooling", PT **43.10** (1990): 33–40. {3, 5; S}

[*COHEN-T. et al.* 1977] COHEN-TANNOUDJI, C. N./DIU, B./LALO, F., *Quantum Mechanics*, New York, Wiley, 1977. {2}

[COLLETT 1988] COLLETT, M. J., "Exact Density-Matrix Calculations for Simple Open Systems", PR **A38** (1988): 2233–47. {4, 5}

[*COLODNY* 1972] COLODNY, R. G. (ed.), *Paradigms and Paradoxes*, Pittsburgh, University Press, 1972. {8, 9, 12, 13, 14}

[COMBOURIEU/RAUCH 1992] COMBOURIEU, M.-C./RAUCH, H., "The Wave-Particle Dualism in 1992: A Summary", FP **22** (1992): 1403–433. {3, 7, 12}

[COMISAR 1965] COMISAR, G. G., "Brownian-Motion Model of Nonrelativistic Quantum Mechanics", PR **B138** (1965): 1332–37. {5}

[COMPTON 1923] COMPTON, Arthur H., "A Quantum Theory of the Scattering of X-rays by Light Elements", PR **21** (1923): 483–502. {3}

[COOK 1988] COOK, R. J., "What are Quantum Jumps?", *Physica Scripta* **T21** (1988): 49–51. {5, 12}

[COOK/KIMBLE 1985] COOK, R. J./KIMBLE, H. J., "Possibility of Direct Observation of Quantum Jumps", PRL **54** (1985): 1023–26. {10}

[J. COOPER 1950] COOPER, Jacob L. B., "The Paradox of Separated Systems in Quantum Theory", *Proceedings of the Cambridge Philosophical Society* **46** (1950): 620–25. {9}

[L. COOPER/VAN VECHTEN 1969] COOPER, L.N./VAN VECHTEN, D., "On the Interpretation of Measurement within the Quantum Theory", AJP **37** (1969): 1212–220. {4}

[CORMIER-DELANOUE 1995] CORMIER-DELANOUE, Christian, "Wave-Corpuscle Duality of Light Reconsidered", FP **25** (1995): 465–79. {7, 12}

[CORMIER-DELANOUE 1996] —, "Strangeness of Matter Waves", FP **26** (1996): 95–103. {7, 12}

[COSTANTINI/GARIBALDI 1989] COSTANTINI, D./GARIBALDI, U., "Classical and Quantum Statistics as Finite Random Processes", FP **19** (1989): 743–55. {3}

[CRAIG 1953] CRAIG, W., "On Axiomatizability Within a System", *Journal of Symbolic Logic* **18** (1953): 30–32. {14}

[CROCA 1987] CROCA, J. R., "Neutron Interferometry Can Prove (or Refute) the Existence of de Broglie's Waves", FP **17** (1987): 971–80. {7}

[CRUTCHFIELD *et al.* 1986] CRUTCHFIELD, J. P./FARMER, J. D./PACKARD, N. H./SHAW, R. S., "Chaos", *Scientific American* (dec. 1986): 46–57. {13}

[CUFARO-P./GUERRA 1995] CUFARO-PETRONI, N./GUERRA, F., "Quantum Mechanical States as Attractors for Nelson Processes", FP **25** (1995): 297–315. {5, 7, 8}

[CUFARO-P./VIGIER 1979] CUFARO-PETRONI, N./VIGIER, J.-P., "Causal Superluminal Interpretation of the EPR Paradox", *Lettere a Nuovo Cimento* **26** (1979): 149–54. {8, 9}

[CUFARO-P./VIGIER 1982] —, "Stochastic Model for the Motion of Correlated Photon Pairs", PL **88A** (1982): 272–74. {8}

[CUFARO-P. *et al.* 1984a] CUFARO-PETRONI, N./GUÉRET, P./VIGIER, J.-P., "Form of a Spin-Dependent Quantum Potential", PR **D30** (1984): 495–97. {8}

[CUFARO-P. *et al.* 1984b] —, "A Causal Stochastic Theory of Spin–1/2 Fields", NC **B81** (1984): 243–59. {8}

[CUFARO-P. *et al.* 1984c] CUFARO-PETRONI, N./KYPRIANIDIS, A./MARICZ, A./SARDELIS, D./VIGIER, J.-P., "Causal Stochastic Interpretation of Fermi-Dirac Statistics in Terms of Distinguishable Non-Locally Correlated Particles", PL **101A** (1984): 4–6. {8}

[CUMMINGS 1965] CUMMINGS, F. W., "Stimulated Emission of Radiation in a Single Mode", PR **A140** (1965): 1051–56. {3}

[CUSHING 1991] CUSHING, James T., "A Bohmian Response to Bohrs Complementarity", in [*FAYE/FOLSE* 1994, 57–76]. {8, 12}

[CUSHING 1992] —, "Causal Quantum Theory: Why a Nonstarter?", in [*SELLERI* 1992a, 37–68]. {8, 12}

[CUSHING 1995] —, "Quantum Tunneling Times: A Crucial Test for the Causal Program?", FP **25** (1995): 269–280. {8, 9}

[*CUSHING/MCMULLIN* 1989] CUSHING, J./MCMULLIN, E. (ed.s), *Philosophical Consequences in Quantum Theory: Reflections on Bells Theorem*, Notre Dame, University Press, 1989. {8, 9, 12, 14}

[*CUSHING et al.* 1996] CUSHING, J.T./FINE, A./GOLDSTEIN, S. (ed.s), *Bohmian Mechanics and Quantum Theory: An Appraisal*, Dordrecht, Kluwer, 1996. {7, 8, 12}

[DAKNA *et al.* 1997] DAKNA, M./ANHUT, T./OPATRNÝ, T./KNÖLL, L./WELSCH, D.-G., "Generating Schrödinger-Cat-Like States by Means of Conditional Measurements on a Beam Splitter", PR **A55** (1997): 3184–94. {5}

[DALIBARD *et al.* 1992] DALIBARD, J./CASTIN, Y./MØLMER, K., "Wave-Function Approach to Dissipative Processes in Quantum Optics", PRL **68** (1992): 580–583. {5}

[DALLA CHIARA 1977] DALLA CHIARA, Maria-Luisa, "Lolgical Self-reference, Set-Theoretical Paradoxes, and the Measurement Problem in Quantum Mechanics", *Journal of Philosophical Logic* **6** (1977): 331–47. {4, 12, 13}

[DALLA CHIARA/GIUNTINI 1989] DALLA CHIARA, M.-L./GIUNTINI, R., "Paraconsistent Quantum Logics", FP **19** (1989): 891–904. {3, 14}

[DALLA CHIARA/TORALDO DI F. 1985] DALLA CHIARA, M.-L./TORALDO DI FRANCIA, G., "'Individuals', 'Properties' and 'Truths' in the EPR-Paradox", in [*LAHTI/MITTELSTAEDT* 1985, 379–402]. {9, 14}

[DANERI *et al.* 1962] DANERI, A./LOINGER, A./PROSPERI, G. M., "Quantum Theory of Measurement", *Nuclear Physics* **33** (1962): 297–319; rep. in [*WHEELER/ZUREK* 1983, 657–79]. {4, 5}

[DANERI *et al.* 1966] —, "Further Remarks on the Relations Between Statistical Mechanics and Quantum Theory of Measurement", NC **44B** (1966): 119–28. {4, 5}

[DANIEL/MILBURN 1989] DANIEL, D. J./MILBURN, G. J., "Destruction of Quantum Coherence in a Nonlinear Oscillator via Attenuation and Amplification", PR **39** (1989): 4628–40. {5}

[D'ARIANO/SACCHI 1997] D'ARIANO, G. M./SACCHI, M. F., "Optical von Neumann Measurement", PL **231A** (1997): 325–30. {4, 5}

[D'ARIANO/YUEN 1996] D'ARIANO, G. M./YUEN, H. P., "Impossibility of Measuring the Wave Function of a Single Quantum System", PRL **76** (1996): 2832–35. {3, 4, 7}

[D'ARIANO *et al.* 1994] D'ARIANO, G. M./MACCHIAVELLO, C./PARIS, M. G. A., "Detection of the Density Matrix Through Optical Homodyne Tomography Without Filtered Back Projection", PR **A50** (1994): 4298–302. {7}

[D'ARIANO *et al.* 1995a] D'ARIANO, G. M./FORTUNATO, M./TOMBESI, P., "Isotropic Phase Number Squeezing and Macroscopic Quantum Coherence", NC **110B** (1995): 1127–36. {5}

[D'ARIANO *et al.* 1995b] D'ARIANO, G. M./LEONHARDT, U./PAUL, H., "Homodyne Detection of the Density Matrix of the Radiation Field", PR **A52** (1995): R1801–804. {7}

[DATZEFF 1958] DATZEFF, Asséne B., "Sur linterpretation de la mécanique quantique", *Comptes Rendus l'Academie des Sciences* **246** (1958): 1502–505. {12}

[DATZEFF 1988] —, "How The Nonlinear Phenomena Impose the Reinterpretation of Quantum Mechanics", in [*VAN DER MERWE et al.* 1988a, 195–205]. {7, 12}

[DAVIDOVICH *et al.* 1994] DAVIDOVICH, L./ZAGURY, N./BRUNE, M./RAIMOND, J. M./HAROCHE, S., "Teleportation of an Atomic State Between two Cavities Using Nonlocal Microwave Fields", PR **A50** (1994): R895–R898. {9, 10}

[DAVIDOVICH *et al.* 1996] DAVIDOVICH, L./BRUNE, M./RAIMOND, J. M./HAROCHE, S., "Mesoscopic Quantum Coherences in Cavity QED: Preparation and Decoherence Monitoring Schemes", PR **A53** (1996): 1295–309. {5}

[DAVISSON/GERMER 1927] DAVISSON, C. J./GERMER, L. H., "Diffraction of Electrons By a Crystal of Nickel", PR **30** (1927): 705–740. {3}

[DAVIES 1969] DAVIES, E. B., "Quantum Stochastic Processes", CMP **15** (1969): 277–304. {3, 4}

[DAVIES 1970a] DAVIES, E. B., "Quantum Stochastic Processes II", CMP **19** (1970): 83–105. {3, 4}

[DAVIES 1970b] DAVIES, E. B., "Quantum Stochastic Processes III", CMP **22** (1970): 51–70. {3, 4}

[DAVIES 1970c] —, "On Repeated Measurements of Continuous Observables in Quantum Mechanics", *Journal of Functional Analysis* **6** (1970): 318–46. {4}

[DAVIES 1974] —, "Markovian Master Equations", CMP **39** (1974): 91–110. {3, 5}

[*DAVIES* 1976] —, *Quantum Theory of Open Systems*, London, Academic Press, 1976. {3, 4, 5}

[DAVIES 1978] —, "Information and Quantum Measurement", *IEEE Transactions on Information Theory* **IT–24** (1978): 596–99. {3, 4, 5, 10}

[DAVIES/J. LEWIS 1970] DAVIES, E. B./LEWIS, J. T., "An Operational Approach to Quantum Probability", CMP **17** (1970): 239–60. {3, 4, 5}

[DE BROGLIE 1926] DE BROGLIE, Louis, "Sur la possibilit é de relier les phénomènes dinterference et de diffraction à la théorie des quanta de lumière", *Comptes Rendus à l'Academie des Sciences* **183** (1926): 447–48; rep. in [*DE BROGLIE* 1953, 25–27]. {3, 12}

[DE BROGLIE 1927a] —, "La structure de la matière et du rayonnement et la mécanique ondulatoire", *Comptes Rendus à l'Academie des Sciences* **184** (1927): 273–74; rep. in [*DE BROGLIE* 1953, : 27–29]. {3, 12}

[DE BROGLIE 1927b] —, *Comptes Rendus à l'Academie des Sciences* **185** (1927): 380–82. {7, 8, 12}

[DE BROGLIE 1927c] —, *Journal de Physique* **8** (1927): 225–41. {7, 8, 12}

[DE BROGLIE 1928] —, "La mécanique ondulatoire et la structure atomique de la matière et du rayonnement", *Journal de Physique et du Radium* **8** (1928): 225–41; rep. in [*DE BROGLIE* 1953, 29–54]. {3, 12}

[*DE BROGLIE* 1932] —, *Sur une forme plus restrictive de Relations d'Incertitude*, Paris, Hermann, 1932. {3, 12}

[DE BROGLIE 1951] —, "Remarques sur la note precedente de M. Vigier", *Comptes Rendus à l'Academie des Sciences* **233** (1951): 1012–13. {7, 8}

[*DE BROGLIE* 1953] —, *La Physique Quantique: Restera-t-elle Indeterministe?*, Paris, Gauthier Villars, 1953. {7, 8, 12}

[DE BROGLIE 1955] —, "Une interprétation nouvelle de la mécanique ondulatoire est-elle possible?", NC **1** (1955): 37–50. {7, 8, 12}

[*DE BROGLIE* 1956] —, *Une tentative d'Interpretation Causale et Non-linéaire de la Mécanique ondulatoire*, Paris, Gauthier-Villars, 1956. {3, 7, 12}

[*DE BROGLIE* 1957] —, *La Théorie de la Mesure en Mécanique Ondulatoire*, Paris, Gauthiers-Villars, 1957. {3, 4, 12}

[DE BROGLIE 1969] —, "Sur l'interpretation de relations dincertitude", *Comptes Rendus à l'Academie des Sciences* **268** (1969): 277–80. {3, 12}

[DE CARO/GARUCCIO 1996] DE CARO, L./GARUCCIO, A., "Bells Inequality, Trichotomic Observables, and Supplementary Assumptions", PR **A54** (1996): 174–81. {9}

[DEHMELT 1975] DEHMELT, Hans, *Bullettin of American Physical Society* **20** (1975): 60. {3, 12}

[DEHMELT 1989] —, "Triton, ... Electron, ... Cosmon, ...: An Infinite Regression?", *Proceedings of National Academy of Sciences USA* **86** (1989): 8618–19. {12}

[DEHMELT 1990] —, "Less is More: Experiments with an Individual Atomic Particle at Rest in Free Space", AJP **58** (1990): 17–27. {3, 4, 12; W}

[DEKKER 1980] DEKKER, H., "Classical and Quantum Mechanics of the Damped Harmonic Oscillator", PRep **80** (1980): 1–112. {3, 5; W}

[DE LA PEÑA-A. 1967] DE LA PEÑA-AUERBACH, Luis, "A Simple Derivation of the Schrödinger Equation from the Theory of Markoff Processes", PL **24A** (1967): 603–604. {3, 5}

[DE LA PEÑA-A. 1968] —, "A New Formulation of Stochastic Theory and Quantum Mechanics", PL **27A** (1968): 594–95. {3, 5}

[DE LA PEÑA-A. 1969] —, "New Formulation of Stochastic Theory and Quantum Mechanics", JMP **10** (1969): 1620–30. {3, 5}

[DE LA PEÑA-A. 1970] —, "Stochastic Quantum Mechanics for Particles with Spin", PL **31A** (1970): 403–404. {3, 5}

[DE LA PEÑA-A./CETTO 1969] DE LA PEÑA-AUERBACH, L./CETTO, A. M., "Lagrangian Form of Stochastic Equations and Quantum Theory", PL **29A** (1969): 562–63. {3, 5}

[DE LA PEÑA-A./CETTO 1971] —, "Self-Interaction Corrections in a Nonrelativistic Stochastic Theory of Quantum Mechanics", PR **D3** (1971): 795–800. {3, 5}

[*DE LA PEÑA-A./CETTO* 1995] —, *The Quantum Dice. An Introduction to Stochastic Electrodynamics*, Dordrecht, Kluwer. {2, 3, 5}

[DE LA PEÑA-A./GARCIA-C. 1968a] DE LA PEÑA-AUERBACH, L./GARCIA-COLIN, L. S., "Possible Interpretation of Quantum Mechanics", JMP **9** (1968): 916–21. {5, 12}

[DE LA PEÑA-A./GARCIA-C. 1968b] —, "Simple Generalization of Schrödingers Equation", JMP **9** (1968): 922–27. {3, 5}

[DE LA PEÑA-A. *et al.* 1968] DE LA PEÑA-AUERBACH, L./BRAUN, E./GARCIA-COLIN, L. S., "Quantum Mechanical Description of a Brownian Particle", JMP **9** (1968): 668–74. {3, 5}

[DELGADO/MUGA 1996] DELGADO, V./MUGA, J. G., "Are Anomalously Short Tunnelling Times Measurable?", AP **248** (1996): 122–33. {3, 9}

[DELGADO/MUGA 1997] —, "Arrival Time in Quantum Mechanics", PR **A56** (1997): 3425–35. {3, 9}

[*DE MARTINI et al.* 1994] DE MARTINI, F./DENARDO, G./ZEILINGER, A. (ed.s), *Quantum Interferometry. Proceedings of the Adriatico Workshop, Trieste 1993*, Singapore, World Scientific, 1994. {3, 4, 5, 9, 12}

[*DE MARTINI et al.* 1996] DE MARTINI, F./DENARDO, G./SHIH, Y. (ed.s), *Quantum Interferometry. Proceedings of the Adriatico Workshop, Trieste 1996*, Weinheim, VCH, 1996. {3, 4, 5, 9, 12}

[*DE MARTINO et al.* 1996] DE MARTINO, S./DE NICOLA, S./DE SIENA, S./FEDELE, R./MIELE, G. (ed.s), *New Perspectives in the Physics of Mesoscopic Systems*, Singapore, World Scientific, 1996. {5}

[DE MATOS F./W. VOGEL 1996] DE MATOS FILHO, R. L./VOGEL, W., "Even and Odd Coherent States of the Motion of a Trapped Ion", PRL **76** (1996): 608–11. {5}

[DEMOPOULOS 1980] DEMOPOULOS, W., "Locality and the Algebraic Structure of Quantum Mechanics", in [*SUPPES* 1980, 119–44].{3, 9}

[DE MUYNCK 1999] DE MUYNCK, Willem M., "Preparation and Measurement: Two Independent Sources of Uncertainty in Quantum Mechanics", e-print quant-ph/9901010. {3, 4, 12}

[DE MUYNCK *et al.* 1991] DE MUYNCK, W. M./STOFFELS, W. W./MARTENS, H., "Joint Measurement of Interference and Path Observables in Optics and Neutron Interferometry", *Physica* **B175** (1991): 127–32. {3, 4, 7}

[DERKA *et al.* 1998] DERKA, R./BUŽEK, V./EKERT, A. K., "Universal Algorithm for Optimal Exstimation of Quantum States from Finite Ensembles via Realizable Generalized Measurement", PRL **80** (1998): 1571–75. {4}

[*D'ESPAGNAT* 1971] D'ESPAGNAT, Bernard (ed.), *Foundations of Quantum Mechanics*, New York, Academic Press, 1971. {3, 4, 12}

[*D'ESPAGNAT* 1976] —, *Conceptual Foundations of Quantum Mechanics*, 1976, II ed., Reading, Mass., Addison-Wesley, 1989. {3, 4, 8, 12}

[D'ESPAGNAT 1979] —, "The Quantum Theory and Reality", *Scientific American* (nov. 1979): 128–40. {8, 12, 13}

[*D'ESPAGNAT* 1983] —, *In Search of Reality* (engl. tr.), New York, Springer, 1983. {12, 13, 14}

[D'ESPAGNAT 1984] —, "Nonseparability and the Tentative Description of Reality": PRep **110** (1984): 201–264. {3, 9, 12}

[D'ESPAGNAT 1987] —, "Consistent Histories and the Measurement Problem", PL **124A** (1987): 204–206. {4, 6, 12}

[D'ESPAGNAT 1989] —, "Are There Realistically Interpretable Local Theories?", *Journal of Statistical Physics* **56** (1989): 747–66. {8, 9, 12}

[*D'ESPAGNAT* 1995] —, *Veiled Reality*, Reading, Mass., Addison/Wesley, 1995. {5, 7, 8, 9, 12}

[DESTOUCHES-F. 1937a] DESTOUCHES-FEVRIER, Paulette, "Les relations d'incertitude de Heisenberg et la logique", *Comptes Rendue* **204** (1937): 481–83. {3, 7, 12}

[DESTOUCHES-F. 1937b] —, "Sur une forme générale de la définition d'une logique", *Comptes Rendue* **204** (1937): 958–59. {3, 7}

[*DESTOUCHES-F.* 1951] —, *La Structure des Théories Physiques*, Paris, Presse Universitaire de France, 1951. {14}

[*DESTOUCHES-F.* 1956] —, *Linterprétation physique de la mécanique ondulatoire et des théories quantiques*, Paris, Gauthier-Villars, 1956. {12}

[DEUTSCH 1983] DEUTSCH, David, "Uncertainty in Quantum Mechanics", PRL **50** (1983): 631–33. {3, 10}

[DEUTSCH 1985a] —, "Quantum Theory as a Universal Physical Theory", *International Journal of Theoretical Physics* **24** (1985): 1–41. {6, 12}

[DEUTSCH 1985b] —, "Quantum Theory, the Church-Turing Principle and the Universal Quantum Computer", PRSL **A400** (1985): 97–117. {10}

[DEUTSCH 1989] —, "Quantum Computational Networks", PRSL **A425** (1989): 73–90. {10}

[DEUTSCH/EKERT 1998] DEUTSCH, D./EKERT, A., "Quantum Computation", *Phyiscs World* (March 1998): 47–51. {10; S}

[DEUTSCH/JOZSA 1992] DEUTSCH, D./JOZSA, R., "Rapid Solution of Problems by Quantum Computation", PRSL **A439** (1992): 554–58. {10}

[DEUTSCH *et al.* 1995] DEUTSCH, D./BARENCO, A./EKERT, A., "Universality in Quantum Computation", PRSL **A449** (1995): 669–77. {10}

[DEUTSCH *et al.* 1996] DEUTSCH, D./EKERT, A. K./JOZSA, R./MACCHIAVELLO, C./POPESCU, S./SANPERA, A., "Quantum Privacy Amplification and the Security of Quantum Cryptography over Noisy Channels", PRL **77** (1996): 2818–21. {9, 10}

[DEWDNEY 1985] DEWDNEY, C., "Particle Trajectories and Interference in a Time-Dependent Model of Neutron Single Crystal Interferometry", PL **109A** (1985): 377–84. {7, 8}

[DEWDNEY/HILEY 1982] DEWDNEY, C./HILEY, B. J., "A Quantum Potential Description of One-Dimensional Time-Dependent Scattering From Square Barriers and Square Wells", FP **12** (1982): 27–48. {7, 8}

[DEWDNEY *et al.* 1984] DEWDNEY, C./GARUCCIO, A./KYPRIANIDIS, VIGIER, J.-P., "Energy Consevation and Complementarity in Neutron Single Crystal Interferometry", PL **104A** (1984): 325–28. {7, 8}

[DEWDNEY *et al.* 1985] DEWDNEY, C./HOLLAND, P. H./KYPRIANIDIS, A./VIGIER, J.-P., "Causal Action at a Distance in a Relativistic System of two Bound Charged Spinless Particles: Hydrogenlike Models", PR **D31** (1985): 2533–38. {7, 8, 9}

[DEWDNEY *et al.* 1993] DEWEDNEY, C./HARDY, L./SQUIRES, E., "How Late Measurements of Quantum Trajectories Can Fool a Detector", PL **184A** (1993): 6–11. {4, 8}

[*DEWEY* 1929] DEWEY, John, *Experience and Nature*, 1929, New York, Dover, 1958. {14}

[DEWITT 1970] DEWITT, Bryce S., "Quantum Mechanics and Reality", PT *23* (1970); rep. in [*DEWITT/GRAHAM* 1973, 155–65]. {4, 6, 12}

[DEWITT 1971] —, "The Many-Universes Interpretation of Quantum Mechanics", in [*D'ESPAGNAT* 1971]; rep. in [*DEWITT/GRAHAM* 1973, 167–218]. {4, 6, 12}

[*DEWITT/GRAHAM* 1973] DEWITT, Bryce S./GRAHAM, Neill, (ed.s) *The Many World Interpretation of Quantum Mechanics*, Princeton, University Press, 1973. {4, 6, 12}

[DICKE 1981] DICKE, Robert H., "Coherence in Spontaneous Radiation Processes", PR **93** (1954): 99–110. {3, 4, 5}

[DICKE 1981] —, "Interaction-Free Quantum Measurements: A Paradox?", AJP **49** (1981): 925–30. {4, 10, 12}

[DICKE 1989] —, "Quantum Measurements, Sequential and Latent", FP **19** (1989): 385–95. {4, 12}

[DIEKS 1982] DIEKS, Dennis, "Communication by EPR Devices", PL **92A** (1982): 271–72. {3, 9}

[DIEKS 1985] —, "Quantum Theory as the First Probabilistic Theory Giving Really Probabilistic Correlations", in [*LAHTI/MITTELSTAEDT* 1985, 403–415]. {3}

[DIEKS 1988] —, "Overlap and Distinguishability of Quantum States", PL **126A** (1988): 303–306. {3, 4, 5}

[DIEKS 1989] —, "Quantum Mechanics Without the Projection Postulate and Its Realistic Interpretation", FP **19** (1989): 1397–23. {4, 8, 12}

[DIEKS 1994] —, "Modal Interpretation of Quantum Mechanics, Measurements, and Macroscopic Behavior", PR **A49** (1994): 2290–299. {4, 5, 8, 12}

[DI GIUSEPPE *et al.* 1997a] DI GIUSEPPE, G./DE MARTINI, F./BOSCHI, D., "Test of the Violation of Local Realism in Quantum Mechanics with no Use of Bell's Inequalities", *Erkenntnis* **45** (1997): 367–77. {9}

[DI GIUSEPPE *et al.* 1997b] —, "Experimental Test of the Violation of Local Realism in Quantum Mechanics without Bell Inequalities", PR **A56** (1997): 1–6. {8, 9}

[DI GIUSEPPE *et al.* 1997c] DI GIUSEPPE, G./HEIBERGER, L./DE MARTINI, F./SERGIENKO, A. V., "Quantum Interference and Indistinguishability With Femtosecond Pulses", preprint. {9}

[DILWORTH 1984] DILWORTH, Craig, "On Theoretical Terms", *Erkenntnis* **21** (1984): 405–21. {14}

[DINER 1984] DINER, S., "The Wave-Particle Duality as an Interplay Between Order and Chaos", in [*DINER et al.* 1984, 215–30]. {7, 12}

[*DINER et al.* 1984] DINER, S./FARGUE, D./LOCHAK, G./SELLERI, F. (ed.s), *The Wave-Particle Dualism*, Dordrecht, Reidel, 1984. {4, 7, 8, 12}

[DIÓSI 1987] DIÓSI, Lajos, "Exact Solution for Particle Trajectories in Modified Quantum Mechanics", PL **122A** (1987): 221–25. {4, 5}

[DIÓSI 1988] —, "Continuous Quantum Measurement and Itô Formalims", PL **129A** (1988): 419–23. {4, 5}

[DIÓSI 1989] —, "Models for Universal Reduction of Macroscopic Quantum Fluctuations", PR **A40** (1989): 1165–74. {5, 7}

[DIÓSI 1990a] —, "Landau's Density Matrix in Quantum Electrodynamics", FP **20** (1990): 63–70. {5, 6}

[DIÓSI 1990b] —, "Relativistic Theory for Continuous Measurement of Quantum Fields", PR **A42** (1990): 5086–92. {5, 6, 11}

[DIÓSI 1993a] —, "On High-Temperature Markovian Equation for Quantum Brownian Motion", *Europhysics Letters* **22** (1992): 1–3. {3, 5}

[DIÓSI 1993b] —, "Caldeira-Leggett Master Equation and Medium Temperatures", *Physica* **199A** (1993): 517–26. {3, 5}

[*DIÓSI/LUKÁCZ* 1994] DIÓSI, L./LUKÁCZ, B., *Stochastic Evolution of Quantum States in Open Systems and in Measurement Processes*, Singapore, World Scientific, 1994. {4, 5}

[DIÓSI/STRUNZ 1997] DIÓSI, L./STRUNZ, W. T., "The Non-Markovian Stochastic Schrödinger Equation for Open Systems", PL **235A** (1997): 569–73. {3, 4, 5}

[DIÓSI *et al.* 1994] DIÓSI, L./GISIN, N./HALLIWELL, J. J./PERCIVAL, I. C., "Decoherent Histories and Quantum State Diffusion", PRL **74** (1994): 203–207. {3, 4, 5, 6}

[DIRAC 1925] DIRAC, Paul, "The Fundamental Equations of Quantum Mechanics", PRSL **A109** (1925): 642–53. {3}

[DIRAC 1926a] —, "Quantum Mechanics and a Preliminary Investigation of the Hydrogen Atom", PRSL **A110** (1926): 561–79. {3}

[DIRAC 1926b] —, "Relativity, Quantum Mechanics with an Application to Compton Scattering", PRSL **A111** (1926): 405–423. {3, 11}

[DIRAC 1926c] —, "The Physical Interpretation of Quantum mechanics", PRSL **A113** (1926): 621–41. {3, 12}

[DIRAC 1926d] —, "On Quantum Algebra", *Proceedings of the Cambridge Philosophical Society* **23** (1926): 412–18. {3}

[DIRAC 1928a] —, "The Quantum Theory of the Electron I", PRSL **A117** (1928): 610–24. {3}

[DIRAC 1928b] —, "The Quantum Theory of the Electron II", PRSL **A118** (1928): 351–61. {3}

[DIRAC 1930a] —, "The Proton", NA **126** (1930): 605–606. {3}

[DIRAC 1930b] —, "A Theory of Electrons and Protons", PRSL **126** (1930): 360–65. {3, 12}

[*DIRAC* 1930c] —, *The Principles of Quantum Mechanics*, IV ed., Oxford, Clarendon, 1958, 1993. {2, 3}

[DIRAC 1937] —, "Physical Science and Philosophy", NA **139** (1937): 1001–2. {13, 14}

[DIRAC 1951a] —, "Is There an AEther?", NA **168** (1951): 906–7. {11, 12}

[DIRAC 1951b] —, "A New Classical Theory of Electrons, I", PRSL **A209** (1951): 291–96. {3}

[DIRAC 1952a] —, "Is There an AEther?", NA **169** (1952): 146–47; 702. {11, 12}

[DIRAC 1952b] —, "A New Classical Theory of Electrons, II", PRSL **A212** (1952): 330–39. {3}

[DIRAC 1954] —, "A New Classical Theory of Electrons, III", PRSL **A223** (1954): 438–45. {3}

[DIRAC 1957] —, "The Vacuum in Quantum Electrodynamics", *Nuovo Cimento Supplementa* **6** (1957): 322–39. {3, 11}

[DIRAC 1978] —, "The Mathematical Foundations of Quantum Theory", in [*MARLOW* 1978a, 1–8]. {3, 11}

[DITCHBURN 1930] DITCHBURN, Robert W., "The Uncertainty Principle in Quantum Mechanics", *Proceedings of the Royal Irish Academy* **39** (1930): 73–80. {3}

[DIVINCENZO 1995a] DIVINCENZO, David P., "Quantum Computation", *Science* **270** (1995): 255–61. {10; W}

[DIVINCENZO 1995b] —, "Two-Bit Gates Are Universal for Quantum Computation", PR **A51** (1995): 1015–22. {10}

[DIVINCENZO/TERHAL 1998] DIVINCENZO, D./TERHAL, B., "Decoherence: The Obstacle to Quantum Computation", *Physics World* (March 1998): 53–57. {10; S}

[DOMOKOS *et al.* 1995] DOMOKOS, P./RAIMOND, J. M./BRUNE, M./HAROCHE, S., "Simple Cavity-QED Two-Bit Universal Quantum Logic Gate: The Principle and Expected Perfprmances", PR **A52** (1995): 3554–59. {5, 10}

[DOMOTOR 1976] DOMOTOR, Zoltan, "The Probability Structure of Quantum-Mechanical Systems", in [*SUPPES* 1976, 147–77]. {3}

[DONALD 1985] DONALD, Matthew J., "On the Relative Entropy", CMP **105** (1985): 13–34. {10, 13}

[DONALD 1992] —, "A *Priori* Probability and Localized States", FP **22** (1992): 1111–72. {3, 4, 11, 12, 13}

[DOWLING *et al.* 1991] DOWLING, J. P./SCHLEICH, W. P./WHEELER, J. A., "Interference in Phase Space", *Annalen der Physik* **48** (1991): 423–502. {3; W}

[DRAGOMAN 1998] DRAGOMAN, D., "The Wigner Distribution Function in Optics and Optoelectronics", *Progress in Optics* **37** (1997): 1–56. {3; W}

[DRESDEN 1993] DRESDEN, Max, "Renormalization in Historical Perspective", in [*L. BROWN* 1993a, 29–55]. {11}

[*DRIESCHNER* 1979] DRIESCHNER, Michael, *Voraussage Wahrscheinlichkeit Objekt. Über die begrifflichen Grundlagen der Quantenmechanik*, Berlin, Springer, 1979. {3, 12, 14}

[DRUMMOND 1983] DRUMMOND, P. D., "Violation of Bell's Inequality in Cooperative States", PRL **50** (1983): 1407–410. {9}

[DUANE 1923] DUANE, William, "The Transfer in Quanta of Radiation Momentum to Matter", *Proceedings of the National Academy of Sciences* **9** (1923): 158–64. {3}

[DUMMETT 1969] DUMMETT, Michael, "The Reality of the Past", *Proceedings of the Aristotelian Society* **69** (1968–69): 223–38. {14}

[*DUMMETT* 1978] —, *Truth and Other Enigmas*, London, Duckworth, 1978. {14}

[*DUMMETT* 1991] —, *Logical Basis of Metaphysics*, London, Duckworth, 1991, 1995. {14}

[DUMONT/MARCHIORO 1993] DUMONT, R. S./MARCHIORO II, T. L., "Tunneling-Time Probability Distribution", PR **A47** (1993): 85–97. {9}

[*DURAND* 1958] DURAND III, L., *On the Theory and Interpretation of Measurement in Quantum Mechanical Systems*, 1958. {4, 12}

[DÜRR *et al.* 1992a] DÜRR, D./GOLDSTEIN, S./ZANGHI, N., "Quantum Equilibrium and the Origin of Absolute Uncertainty", *Journal of Statistical Physics* **67** (1992): 843–97. {3, 5, 8}

[DÜRR *et al.* 1992b] —, "Quantum Chaos, Classical Randomness, and Bohmian Mechanics", *Journal of Statistical Physics* **68** (1992): 259–70. {3, 5, 8}

[DURT 1997] DURT, Thomas, "Three Interpretations of the Violation of Bell's Inequalities", FP **27** (1997): 415–34. {9}

[*EARMAN* 1986] EARMAN, John, *A Primer on Determinism*, Dordrecht, Reidel, 1986. {14}

[EBERHARD 1977] EBERHARD, Philippe H., "Bell's Theorem Without Hidden Variables", NC **38B** (1977): 75–80.{9}

[EBERHARD 1978] —, "Bell's Theorem and the Different Concepts of Locality", NC **46B** (1978): 392–419. {9}

[EBERHARD 1989a] —, "The EPR Paradox. Roots and Ramification", in [*SCHOMMERS* 1989a, 49–88]. {9}

[EBERHARD 1989b] —, "A Realistic Model for Quantum Theory With a Locality Property", in [*SCHOMMERS* 1989a, 169–215]. {9}

[EBERHARD 1993] —, "Background-Level and Counter Efficiencies Required for a Loop-Free EPR Experiment" PR **A47** (1993): R747–50. {9}

[EBERHARD/ROSSELET 1995] EBERHARD, P. H./ROSSELET, P., "Bell's Theorem based on a Generalized EPR Criterion of Reality", FP **25** (1995): 91–111. {9}

[EBERLY/L. SINGH 1973] EBERLY, J. H./SINGH, L. P. S., "Time Operators, Partial Stationarity, and the Energy-Time Uncertainty Relation", PR **D7** (1973): 359–62. {3}

[EBERLY *et al.* 1980] EBERLY, J. H./NAROZHNY, N. B./SANCHEZ-MONDRAGON, J. J., "Periodic Spontaneous Collapse and Revival in a Simple Quantum Model", PRL **44** (1980): 1323–26. {6}

[EDER/ZEILINGER 1976] EDER, G./ZEILINGER, A., "Interference Phenomena and Spin Rotation of Neutrons by Magnetic Materials", NC **34B** (1976): 76–90. {3}

[EDWARDS 1970] EDWARDS, C. M., "The Operational Approach to Algebraic Quantum Theory I", CMP **16** (1970): 207–230. {3}

[EHRENFEST 1932] EHRENFEST, Paul, "Einige die Quantenmechanik betreffende Erkundigungsfragen", ZP **78** (1932): 555–59. {12}

[EICHMANN *et al.* 1993] EICHMANN, U./BERGQUIST, J. C./BOLLINGER, J. J./GILLIGAN, J. M./ITANO, W. M./WINELAND, D. J./RAIZEN, M. G., "Young's Interference Experiment With Light Scattered from Two Atoms", PRL **70** (1993): 2359–62. {4}

[*EINSTEIN* CP] EINSTEIN, Albert, *The Collected Papers of Albert Einstein*, Princeton, University Press, 1993 e ss.. {3, 4, 8, 9, 12, 13}

[*EINSTEIN* AV] —, *Akademie Vorträge*, Berlin, Akademie Verlag. {13}

[EINSTEIN 1905] —, "Über einen die Erzeugung und Verwandlung des Lichtes betreffenden heuristischen Gesichtspunkt", *Annalen der Physik* **17** (1905): 132–48. {3, 12}

[EINSTEIN 1906] —, "Zur Theorie der Lichterzeugung und Lichtabsorption", *Annalen der Physik* **20** (1906): 199–206. {3, 12}

[EINSTEIN 1909] —, "Über die Entwicklung unserer Anschauungen über das Wesen und die Konstitution der Strahlung", *Physikalische Zeitschrift* **10** (1909): 817–25. {3, 12}

[EINSTEIN 1917] —, "Zur Quantentheorie der Strahlung", *Physikalische Zeitschrift* **18** (1917): 121–28. {3, 12}

[EINSTEIN 1924] —, "Quantentheorie des einatomigen idealen Gases. I", *Sitzungsberichte der preußischen Akademie der Wissenschaften* (1924): 261–67; rep. in [*EINSTEIN* AV]. {3}

[EINSTEIN 1925] —, "Quantentheorie des einatomigen idealen Gases. II", *Sitzungsberichte der preußischen Akademie der Wissenschaften* (1924): 3–14; rep. in [*EINSTEIN* AV]. {3, 12}

[EINSTEIN 1928] —, in *Rapports et discussions du V Conseil*, Paris, Institut Internaltional de Physique Solvay, 1928: 253. {12}

[EINSTEIN 1936] —, "Physik und Realität", *Journal of Franklin Institute* **221** (1936): 313–47; eng. tr. in [*EINSTEIN* 1956, 59–97]. {12, 13, 14}

[EINSTEIN 1940] —, "The Fundaments of Theoretical Physics",*Science* (1940); rep. in [*EINSTEIN* 1956, 98–110]. {12, 13, 14}

[EINSTEIN 1948] —, "Quantenmechanik und Wirklichkeit", *Dialectica* **2** (1948): 320–24. {12, 13, 14}

[EINSTEIN 1949a] —, "Autobiographisches", in [*SCHILPP* 1949, 2–94]. {3, 12, 13, 14}

[EINSTEIN 1949b] —, "Remarks Concerning the Essays Brought Together in This Cooperative Volume", in [*SCHILPP* 1949, 665–88]. {3, 12, 13}

[EINSTEIN 1953a] —, "Einleitende Bemerkungen über Grundbegriffe", in [*GEORGE* 1953, 4–14]. {12}

[EINSTEIN 1953b] —, "Elementare Überlegungen zur Interpretation der Grundlagen der Quantenmechanik", in Scientific Papers Presented to Max Born, London, Oliver and Boyd, 1953: 33–40. {12}

[*EINSTEIN* 1956] —, *Out of My Later Years, Estate of A. Einstein*, 1956, 1993: New York, Wings Books. {3, 12, 13, 14}

[*EINSTEIN* MW] —, *Mein Weltbild*, Frankfurt, Ullstein, 1993. {3, 12, 13, 14}

[EINSTEIN *et al.* 1931] EINSTEIN, A./TOLMAN, R. C./PODOLSKY, B., "Knowledge of Past and Future in Quantum Mechanics", PR **37** (1931): 780–81. {6, 8, 12}

[EINSTEIN *et al.* 1935] EINSTEIN, A./PODOLSKY, B./ROSEN N., "Can Quantum-Mechanical Description of Physical Reality be Considered Complete?": PR **47** (1935): 777-80; rep. in [*WHEELER/ZUREK* 1983, 138–41]. {3, 4, 8, 9, 12, 13, 14}

[EKERT 1991] EKERT, Artur K., "Quantum Cryptography Based on Bell's Theorem", PRL **67** (1991): 661–63. {9, 10}

[EKERT/JOZSA 1996] EKERT, A. K./JOZSA, R., "Quantum Computation and Shor's Factorizing Algorithm", RMP **68** (1996): 733–753. {10; W}

[EKERT/KNIGHT 1995] EKERT, A. K./KNIGHT, P. L., "Entangled Quantum Systems and the Schmidt Decomposition", AJP **63** (1995): 415–23. {3, 4; W}

[EKERT/MACCHIAVELLO 1996] EKERT, A. K./MACCHIAVELLO, C., "Quantum Error Correction for Communication", PRL **77** (1996): 2585–88. {10}

[EKERT/MACCHIAVELLO 1998a] —, "Against Quantum Noise", *Acta Physica Polonica* **A93** (1998): 63–76. {5, 10; W}

[EKERT/MACCHIAVELLO 1998b] —, "An Overview of Quantum Computing", CALUDE/CASTI/DINNEEN (ed.s), *Unconventional Models of Computation*, Singapore, Springer, 1998: 19–44. {10; W}

[EKERT/PALMA 1994] EKERT, A. K./PALMA, G. M., "Quantum Cryptography with Quantum Interferometry" in [*DE MARTINI et al.* 1994, 274–86]. {10}

[EKERT *et al.* 1992] EKERT, A. K./RARITY, J. G./TAPSTER, P. R./PALMA, G. M., "Practical Quantum Cryptography Based on Two-Photon Interferometry", PRL *69* (1992): 1293–95. {10}

[EKERT *et al.* 1994] EKERT, A. K./HUTTNER, B./PALMA, G. M./PERES, A., "Eavesdropping on Quantum-Cryptographical Systems", PR **A50** (1994): 1047–56. {10}

[ELBY 1990] ELBY, Andrew, "Nonlocality and Gleason's Lemma. Part II: Stochastic Theories", FP **20** (1990): 1389–97. {9}

[ELBY/BUB 1994] ELBY, A./BUB, J., "Triorthogonal Uniqueness Theorem and its Relevance to the Interpretation of Quantum Mechanics", PR **A49** (1994): 4213–16. {3, 4}

[ELBY *et al.* 1993] ELBY, A./BROWN, H. R./FOSTER, S., "What Makes a Theory Physically 'Complete'?", FP **23** (1993): 971–85. {13, 14}

[ELITZUR/VAIDMAN 1993] ELITZUR, A. C./VAIDMAN, L., "Quantum Mechanical Interaction -Free Measurements", FP **23** (1993): 987–97. {4}

[ELITZUR *et al.* 1992] ELITZUR, A. C./POPESCU, S./ROHRLICH, D., "Quantum Nonlocality for Each Pair in an Ensemble", PL *162A* (1992): 25–28. {9}

[*ELLIS* 1966] ELLIS, B., *Basic Concepts of Measurement*, Cambridge, University Press, 1966. {4, 13}

[EMCH 1963] EMCH, Gérard, "Mécanique quantique quaternionienne et relativité restreinte", *Helvetica Physica Acta* **36** (1963): 739–88. {3}

[*EMCH* 1972] —, *Algebraic Methods in Statistical Mechanics and Quantum Field Theory*, New York, Wiley, 1972. {3, 11}

[EMCH 1982] —, "Quantum and Classical Mechanics on Homogeneous Riemann Manifolds", JMP **23** (1982): 1785–91. {3}

[EMCH 1983] —, *International Journal of Theoretical Physics* **22** (1983): 397–420. {3}

[EMCH/PIRON 1963] EMCH, G./PIRON, C., "Symmetry in Quantum Theory", JMP **4** (1963): 469–73. {3}

[ENDERS/NIMTZ 1993] ENDERS, A./NIMTZ, G., "Zero-Time Tunneling of Evanescent Mode Packets", *Journal de Physique* **I.3** (1993): 1089-92. {9}

[ENGELKE/ENGELKE 1995] ENGELKE, C. E./ENGELKE, C. W., "'Which Path' Experiments", in [*GREENBERGER/ZEILINGER* 1995a, 850–54]. {7}

[ENGELMANN/FICK 1959] ENGELMANN, F./FICK, E., "Die Zeit in der Quantenmechanik", *Nuovo Cimento Supplementa* **12** (1959): 63–72. {3}

[ENGLERT 1996a] ENGLERT, Berthold-Georg, "Fringe Visibility and Which-Way Information: An Inequality", PRL **77** (1996): 2154–57. {3, 7, 12}

[ENGLERT 1996b] —, "Duality in the Ramsey Interferometer", *Acta Physica Slovaca* **46** (1996): 249–58. {3, 7, 12}

[ENGLERT *et al.* 1988] ENGLERT, B.-G./SCHWINGER, J./SCULLY, M. O., "Is Spin Coherence Like Humpty-Dumpty?", FP **18** (1988): 1045–56. {6}

[ENGLERT *et al.* 1993] ENGLERT, B.-G./SCULLY, M. O./SÜSSMAN, G./WALTHER, H., "Surrealistic Bohm Trajectories", *Zeitschrift für Naturforschung* **47a** (1993): 1175–86. {7, 8}

[ENGLERT *et al.* 1994a] ENGLERT, B.-G/SCULLY, M. O./WALTHER, H., "The Duality in Matter and Light", *Scientific American* (dec. 1994): 56–61. {3, 4, 7, 10, 12; S}

[ENGLERT *et al.* 1994b] ENGLERT, B.-G/FEARN, H./SCULLY, M. O./WALTHER, H., "The Micromaser Welcher-Weg Detector Revisited", in [*DE MARTINI et al.* 1994, 103–119]. {3, 7}

[ENGLERT *et al.* 1995] ENGLERT, B.-G/SCULLY, M. O./WALTHER, H., "Complementarity and Uncertainty", NA **375** (1995): 367–68. {3, 4, 7, 10, 12}

[EPSTEIN 1945] EPSTEIN, Paul S., "The Reality Problem in Quantum Mechanics", AJP **13** (1945): 127–36. {8, 12}

[EPSTEIN/EHRENFEST 1924] EPSTEIN, P. S./EHRENFEST, P., "The Quantum Theory of the Fraunhofer Diffraction", *Proceedings of the National Academy of Sciences* **10** (1924): 133–39. {3, 7, 12}

[ERBER 1995] ERBER, T., "Testing the Randomness of Quantum Mechanics: Nature's Ultimate Cryptogram?", in [*GREENBERGER/ZEILINGER* 1995a, 748–56]. {12}

[EVERETT 1957] EVERETT, Hugh III, "'Relative State' Formulation of Quantum Mechanics", RMP **29** (1957): 454–62; rep. in [*DEWITT/GRAHAM* 1973, 141–49]. {3, 4, 6, 12}

[EVERETT 1973] —, "The Theory of the Universal Wave Function", in: [*DEWITT/GRAHAM* 1973, 3–140]. {3, 4, 6, 10, 12; W}

[FANO 1957] FANO, U., "Description of States in Quantum Mechanics by Density Matrix and Operator Techniques", RMP **29** (1957): 74–93. {3; W}

[FARACI *et al.* 1974] FARACI, G./GUTKOWSKI, D./NOTARIGO, S./PENNISI, A. R., "An Experimental test of the EPR Paradox", *Lettere a Nuovo Cimento* **9** (1974): 607–611. {9}

[FAVELLA 1967] FAVELLA, L. F., "Brownian Motions and Quantum Mechanics", *Annales de l'Institut Henri Poincaré* **7** (1967): 77–94. {5}

[*FAYE* 1991] FAYE, Jan, *Niels Bohr: His Heritage and Legacy. An Anti-realist view of Quantum Mechanics*, Dordercht, Kluwer, 1991. {12, 13, 14}

[FAYE 1994] "Non-Locality or Non-Separability?", in FAYE/FOLSE (ed.s) 1994: 97–118. {8, 9}

[*FAYE/FOLSE* 1994] FAYE, J./FOLSE, H. J. (ed.s), *Niels Bohr and Contemporary Philosophy*, Dordrecht, Kluwer, 1994. {3, 12}

[FELLER 1957] FELLER, W., *An Introduction to Probability Theory and its Applications*, New York, Wiley, 1957. {1}

[FÉNYES 1952a] FÉNYES, Imre, "Eine wahrscheinlichkeitstheoretische Begründung und Interpretation der Quantenmechanik, ZP **132** (1952): 81–106. {3, 5}

[FÉNYES 1952b] —, "Stochastischer Abhängigkeitscharakter der Heisenbergschen Ungenauigkeitsrelationen, *Naturwissenschaften* **39** (1952): 568. {3, 5}

[FERMI 1926] FERMI, Enrico, "Zur Wellenmechanik des Stossvorganges", ZP **40** (1926): 399–402. {3}

[FERRERO/SANTOS 1997] FERRERO, M./SANTOS, E., "Empirical Consequences of the Scientific Construction: The Program of Local Hidden-Variables Theories in Quantum Mechanics", FP **27** (1997): 765–800. {3, 8, 9, 12, 13, 14}

[FERRERO *et al.* 1990] FERRERO, M./MARSHALL, T. W./SANTOS, E., "Bell's Theorem: Local Realism versus Quantum Mechanics", AJP **58** (1990): 683–88. {8, 9, 12, 13}

[FERTIG 1990] FERTIG, H. A., "Traversal-Time Distribution and the Uncertainty Principle in Quantum Tunneling", PRL **65** (1990): 2321–24. {3, 9}

[FEYERABEND 1957a] FEYERABEND, Paul, "Die Quantentheorie der Messung", ZP **148** (1957): 551. {4, 12}

[FEYERABEND 1957b] —, "On the Quantum Theory of Measurement", in [*KÖRNER/PRYCE* 1957, 121–30]. {4, 12}

[FEYERABEND 1958a] —, "Complementarity", *Supplementary Volume 32 of Proceedings of the Aristotelian Society* (1958): 75–104. {12}

[FEYERABEND 1958b] —, "Reichenbach's Interpretation of Quantum Mechanics", *Philosophical Studies* **9** (1958): 49–59; rep. in [*HOOKER* 1975, 109–21]. {7, 12}

[FEYERABEND 1962] —, "Problems of Microphysics", in COLODNY (ed.), *Frontiers of Science and Philosophy*, London, Allen and Unwin, 1962. {12}

[FEYERABEND 1967] —, "Review of Reichenbach's *Philosophic Foundations of Quantum Mechanics*", *British Journal for the Philosophy of Science* **17** (1966): 326–28. {7, 12, 14}

[FEYERABEND 1968] —, "On a Recent Critique of Complementarity. I", PS **35** (1968): 309–31. {12}

[FEYERABEND 1969] —, "On a Recent Critique of Complementarity. II", PS **36** (1969): 82–105. {12}

[FEYERABEND 1981] —, *Realism, Rationalism and Scientific Method: Philosophical Papers*, I, Cambridge, university Press, 1981. {12, 14}

[FEYNMAN 1948] FEYNMAN, Richard P., "Space-Time Approach to Non-relativistic Quantum Mechanics", RMP **20** (1948): 367–87. {3}

[FEYNMAN 1949] —, "The Theory of Positrons", PR **76** (1949): 749–59. {3}

[FEYNMAN 1951] —, "The Concept of Probability in Quantum Mechanics", *Proceeding II Berkeley Symposium in Mathematical Statistics and Probability,* Berkeley, 1951: 533–41. {3}

[FEYNMAN 1982] —, "Simulating Physics with Computers", *International Journal of Theoretical Physics* **21** (1982): 467–88. {10}

[FEYNMAN 1986] —, "Quantum Mechanical Computers", FP **16** (1986): 507–531. {10}

[FEYNMAN 1987] —, "Negative Probability", in [*HILEY/PEAT* 1987a, 235–48]. {3}

[FEYNMAN/HIBBS 1965] FEYNMAN, R./HIBBS, A. R., *Quantum Mechanics and Path Integrals*, New York, McGraw-Hill, 1965. {2, 3; W}

[FEYNMAN/VERNON 1963] FEYNMAN, R. P./VERNON, F. L., "The Theory of a General Quantum System Interacting with a Linear Dissipative System", AP **24** (1963): 118–73. {3, 5}

[*FEYNMAN et al.* 1965] FEYNMAN, R./LEIGHTON, R. B./SANDS, M., *Lectures on Physics*, III v., Reading (Massachussetts), Addison-Wesley, 1965, 1966. {2}

[*FICK* 1969] FICK, Eugene, *Einführung in die Grundlagen der Quantentheorie*, Leipzig, Akademische Verlagsgesellschaft, 1969. {2, 3}

[FICK/ENGELMANN 1963a] FICK, E./ENGELMANN, F., "Quantentheorie der Zeitmessung. I", ZP **175** (1963): 271–82. {3, 4}

[FICK/ENGELMANN 1963b] —, "Quantentheorie der Zeitmessung. II", ZP **178** (1964): 551–62. {3, 4}

[A. FINE 1970] FINE, Arthur, "Insolubility of the Quantum Measurement Problem", PR **D2** (1970): 2783–87. {4, 12}

[A. FINE 1972] —, "Some Conceptual Problems of Quantum Theory", in [*COLODNY* 1972, 3–31]. {12}

[A. FINE 1974] —, "On the Completeness of Quantum Theory", *Synthese* **29** (1974): 257–89. {8}

[A. FINE 1976] —, "On the Completness of Quantum Theory", in [*SUPPES* 1976, 249–81]. {8}

[A. FINE 1979] —, "Counting Frequencies: A Primer for Quantum Realists", *Synthese* **42** (1979): 145–54. {4, 8}

[A. FINE 1982a] —, "Hidden Variables, Joint Probability, and the Bell Inequalities", PRL **48** (1982): 291–95. {9}

[A. FINE 1982b] —, "Joint Distributions, Quantum Correlations, and Commuting Observables", JMP **23** (1982): 1306–10. {3}

[A. FINE 1982c] —, "Reply to Garg and Mermin", PRL **49** (1982): 243. {9}

[*A. FINE* 1986] —, *The Shaky Game: Einstein, Realism and the Quantum Theory,* Chicago, University Press, 1986. {12, 14}

[A. FINE 1989] —, "Do Correlations Need to be Explained?", in [*CUSHING/MCMULLIN* 1989, 175–94]. {9, 12}

[A. FINE 1994] —, "Pragmatism and Quantum Theory", [*ACCARDI* 1994a, 159–70]. {12, 14}

[A. FINE/TELLER 1978] FINE, A./TELLER, P., "Algebraic Costraints on Hidden Variables", FP **8** (1978): 629–37. {3, 8}

[T. FINE 1976] FINE, Terence L., "Towards a Revised Probabilistic Basis for Quantum Mechanics", in [*SUPPES* 1976, 179–93]. {3}

[FINKELSTEIN 1962] FINKELSTEIN, David R., "The Logic of Quantum Physics", *Transactions of the New York Academy of Sciences* **25** (1962): 621–37. {3}

[FINKELSTEIN 1969] —, "Matter, Space and Logic", *Boston Studies in the Philosophy of Science* **5** (1969): 199–215; rep. in [*HOOKER* 1979, 123–40]. {3, 12, 14}

[FINKELSTEIN 1972] —, "The Physics of Logic", in [*COLODNY* 1972, 141–60]; rep. in [*HOOKER* 1979, 141–60]. {12, 14}

[FINKELSTEIN 1977] —, "The Leibniz Project", *Journal of Philosophical Logic* **6** (1977): 425–39; rep. in [*HOOKER* 1979, 423–37]. {12, 14}

[FINKELSTEIN 1991] —, "Theory of Vacuum", in [*SAUNDERS/H. BROWN* 1991, 251–74]. {12}

[*FINKELSTEIN* 1996] —, *Quantum Relativity*, Berlin, Springer, 1996. {3, 11, 12}

[FINKELSTEIN *et al.* 1959] FINKELSTEIN, D./JAUCH, J. M./ SPEISER, D, "Notes on Quaternion Quantum Mechanics",*CERN Report* **59.7** (1959); rep. in [*HOOKER* 1979, 367–422]. {3}

[FINKELSTEIN *et al.* 1962] FINKELSTEIN, D./JAUCH, J. M./SCHIMINOVICH, S./SPEISER, D., "Foundations of Quaternion Quantum Mechanics", JMP **3** (1962): 207–220. {3}

[FINKELSTEIN *et al.* 1963a] —, "Quaternionic Representations of Compact Groups", JMP **4** (1963): 136–40. {3}

[FINKELSTEIN *et al.* 1963b] —, "Principle of Generale Q Covariance", JMP **4** (1963): 788–96. {3}

[FITCHARD 1979] FITCHARD, E. E., "Proposed Experimental Test of Wave Packet Reduction and the Uncertainty Principle", FP **9** (1979): 525–35. {3, 4, 12}

[FIVEL 1995a] FIVEL, Daniel I., "The Lattice Dynamics of Completely Entangled States and Its Application to Communication Schemes", in [*GREENBERGER/ZEILINGER* 1995a, 687–97]. {3}

[FIVEL 1995b] —, "Remarkable Phase Oscillations Appearing in the Lattice Dynamics of EPR States", PRL **74** (1995): 835–38. {3, 8, 9}

[FIVEL 1997] —, "Dynamical Reduction Theory of EPR Correlations and a Possible Origin of *CP* Violations", PR **A56** (1997): 146–56. {5, 9}

[FLATO 1976] FLATO, M., "Quantum Mechanics and Determinism", in [*FLATO et al.* 1976, 19–31]. {8, 12}

[*FLATO et al.* 1976] FLATO, M./MARIC, Z./MILOJEVIC, A./STERNHEIMER, D./VIGIER, J.-P. (ed.s), *Quantum Mechanics, Determinism, Causality, and Particles*, Dordrecht, Reidel, 1976. {4, 7, 8, 9, 12, 14}

[FLEMING 1973] FLEMING, G. N., "A Unitarity Bound on the Evolution of Nonstationary States", NC **A16** (1973): 232–40. {3}

[FOCK 1932] FOCK, V. A., "Konfigurationsraum und zweite Quantelung", ZP **75** (1932): 622–47. {3}

[FOCK 1962] —, "Criticism of an Attempt to Disprove the Uncertainty Relation Between Time and Energy", *Soviet Physics JETP* **15** (1962): 784–86. {3}

[FOCK 1966] —, "More About the Energy-Time Uncertainty Relation", *Soviet Physics Uspekhi* **8** (1966): 628–29. {3}

[FOCK/KRYLOV 1947] FOCK, V. A./KRYLOV, N., *Journal of Physics (USSR)* **11** (1947): 112. {3, 12}

[FOLDY/WOUTHUYSEN 1950] FOLDY, L. L./WOUTHUYSEN, S. A., "On the Dirac Theory of Spin–$\frac{1}{2}$ Particles and its Non-Relativistic Limit", PR **78** (1950): 29–36. {3, 12}

[*FOLSE* 1985] FOLSE, Henry J., *The Philosophy of Niels Bohr. The Framework of Complementarity*, Amsterdam, North-Holland, 1985. {4, 7, 12, 13, 14}

[FOLSE 1989] —, "Complementarity and Space-Time Description", in [*KAFATOS* 1989, 251–60]. {12}

[FONDA *et al.* 1983] FONDA, L./GHIRARDI, G.-C./WEBER, T., "On Proton Decay", PL **131B** (1983): 309–312. {4}

[FONSECA R./NEMES 1997] FONSECA ROMERO, K. M./NEMES, M. C., "Quantum Decoherence without Damping", PL **235A** (1997): 432–37.{3, 5}

[FORMAN 1971] FORMAN, P., "Weimar Culture, Causality and Quantum Theory 1918–1927", in R. MCCORMACH (ed.), *Historical Studies in Physical Sciences 3*, Philadelphia, University of Pennsylvania Press, 1971. {14}

[*FORTUNATO* 1993] FORTUNATO, Mauro, *Studio teorico della produzione di stati quantistici macroscopicamente distinguibili*, Ph. D., 1993 (unpublished). {4, 5}

[FORTUNATO 1996] —, "Comments on 'Creating Metastable Schrödinger Cat States"', PRL **77** (1996). {5}

[FORTUNATO *et al.* 1996] FORTUNATO, M./SCHLEICH, W. P./KURIZKI, G., "Quantum Controll of Chaos Inside a Cavity", *Acta Physica Slovaca* **46** (1996): 381–86. {5}

[FORTUNATO *et al.* 1998a] FORTUNATO, M./HAREL, G./KURIZKI, G., "Recovering Coherence via Conditional Measurements", *Optics Communications* **147** (1998): 71–77. {5, 6}

[FORTUNATO *et al.* 1998b] FORTUNATO, M./TOMBESI, P./SCHLEICH, W. P., "Endoscopic Tomography and Quantum Non-Demolition", preprint (mauro). {4}

[FORTUNATO *et al.* 1999] FORTUNATO, M./RAIMOND, J.M./TOMBESI, P./VITALI, D., "Autofeedback Scheme for Schrödinger cat state preservation in microwave cavities", e-print quant-ph/9902071 (PR **A** in press). {5}

[FOULIS 1989] FOULIS, David J., "Coupled Physical Systems", FP **19** (1989): 905–22. {3, 9}

[FOULIS/M. BENNETT 1994] FOULIS, D. J./BENNETT, M. K., "Effect Algebras and Unsharp Quantum Logic", FP **24** (1994): 1331–52. {3}

[FOULIS/RANDALL 1972] FOULIS, D. J./RANDALL, C. H., "Operational Statistics. I. Basic Concepts", JMP **13** (1972): 1667–75. {3}

[FOULIS *et al.* 1983] FOULIS, D. J./PIRON, C./RANDALL, C. H., "Realism, Operationalism, and Quantum Mechanics", FP *13* (1983): 813–41. {8, 12}

[FRAGET/FERT 1957] FRAGET, J./FERT, C., "Diffraction et interférence en optique électronique", *Cahiers de Physique* **11** (1957): 285–96.

[FRANK 1936] FRANK, Philipp, "Philosophische Deutungen und Missdeutungen der Quantentheorie", *Erkenntnis* **6** (1936): 303–17. {12, 14}

[FRANSON 1982] FRANSON, J. D., "Exstension of the EPR Paradox and Bell's Theorem", PR **D26** (1982): 787–800. {9}

[FRANSON 1984] —, "Optical Interferometer Data in Support of Local Theories", in [*ROTH/INOMATA* 1984, 159–68]. {8, 9}

[FRANSON 1985] —, "Bell's Theorem and Delayed Determinism", PR **D31** (1985): 2529–32. {6, 8, 9}

[FRANSON 1986] —, "An Experimental Test of Locality Using a Single-Photon Interferometer", in [*GREENBERGER* 1986, 127–32]. {3, 9}

[FRANSON 1989] —, "Bell Inequality for Position and Time", PRL **62** (1989): 2205–208. {9}

[FRANSON 1991] —, "Violations of a Simple Inequality for Classical Fields", PRL **67** (1991): 290–93. {9}

[FRANSON 1993] —, "Nonlocal Interferometry with High-Intensity Fields", PR **A48** (1993): 4610–616. {9}

[FRANSON 1994a] —, "Nonlocal Reduction of the Wave Function by Quantum Phase Measurements", PR **A49** (1994): 3221–28. {4, 9}

[FRANSON 1994b] —, "Realization of the EPR-Paradox Using High-Intensity Fields", in [*DE MARTINI et al.* 1994, 150–58]. {9}

[FRANSON 1995] —, "The Entanglement of Virtual Photons", in [*GREENBERGER/ZEILINGER* 1995a, 654–63]. {3, 9}

[FRANSON 1996] —, "Coherent Splitting of Single Photons by an Ideal Beam Splitter", PR **A53** (1996): 3756–60. {3}

[FRANSON 1997] —, "Experimental Observation of the Splitting of Single Photons by a Beam Splitter", PR **A56** (1997): 1800–805. {3}

[FRANSON/POTOCKI 1988] FRANSON, J. D./POTOCKI, K. A., "Single-Photon Interference Over Large Distances", PR **A37** (1988): 2511–15. {3, 9, 12}

[FREEDMAN/CLAUSER 1972] FREEDMAN, S.J./CLAUSER, J.F., "Experimental Test of Local Hidden-Variable Theories", PRL **28** (1972): 938–41; rep. in [*WHEELER/ZUREK* 1983, 414–17]. {9}

[FREEDMAN *et al.* 1976] FREEDMAN, S. J./HOLT, R. A./PAPALIOLIOS, C., "Experimental Status of Hidden Variable Theories", in [*FLATO et al.* 1976, 43–59]. {8, 9}

[FRENKEL 1990] FRENKEL, Andor, "Spontaneous Localizations of the Wave Function and Classical Behavior", FP **20** (1990): 159–88. {5, 7}

[FRERICHS/SCHENZLE 1991] FRERICHS, V./SCHENZLE, A., "Quantum Zeno Effect without Collapse of the Wave Packet", PR **A44** (1991): 1962–68. {3, 4}

[FREYBERGER/HERKOMMER 1994] FREYBERGER, M./HERKOMMER, A. M., "Probing a Quantum State via Atomic Deflection", PRL **72** (1994): 1952–55. {7}

[FRIEDMAN/GLYMOUR 1972] FRIEDMAN, M./GLYMOUR, C., *Journal of Philosophical Logic* **1** (1972): 16–28. {3, 7}

[FRIEDMAN/PUTNAM 1978] FRIEDMAN, M./PUTNAM, H., "Quantum Logic, Conditional Probability and Inference", *Dialectica* **32** (1978): 305–15. {12, 13, 14}

[FROISSART 1981] FROISSART, M., "Constructive Generalization of Bell's Inequalities", NC **B64** (1981): 241–51. {9}

[FRY 1995] FRY, Edward S., "Bell Inequalities and two Experimental Tests with Mercury", *Quantum Optics* **7** (1995): 229–58. {9; W}

[FRY/THOMPSON 1976] FRY, E. S./THOMPSON, R. C., "Experimental Test of Local Hidden-Variable Theories", PRL **37** (1976): 465–68; rep. in [*WHEELER/ZUREK* 1983, 418–21]. {9}

[FRY *et al.* 1995] FRY, E. S./WALTHER, T./LI, S., "Proposal for a Loophole-Free Test of the Bell Inequalities", PR **A52** (1995): 4381–95. {9}

[C. FUCHS/CAVES 1994] FUCHS, C. A./CAVES, C. M., "Ensemble-Dependent Bounds for Accessible Information in Quantum Mechanics", PRL **73** (1994): 3047–50. {3, 10}

[C. FUCHS/CAVES 1995] —, "Bounds for Accessible Information in Quantum Mechanics", in [*GREENBERGER/ZEILINGER* 1995a, 706–714]. {3, 10}

[C. FUCHS/PERES 1996] FUCHS, C. A./PERES, A., "Quantum-State Disturbance versus Information Gain: Uncertainty Relations for Quantum Information", PR **A53** (1996): 2038–45. {3, 4, 7, 10}

[*J. FUCHS/SCHWEIGERT* 1997] FUCHS, J./SCHWEIGERT, C., *Symmetries, Lie Algebras and Representations*, Cambridge, University Press, 1997. {1, 3, 13}

[FURRY 1936] FURRY, Wendell H., "Note on Quantum-Mechanical Theory of Measurement", PR **49** (1936): 393–99. {4, 8, 9}

[FÜRTH 1933] FÜRTH, Reinhold, "Über Einige Beziehung zwischen klassischer Statistik und Quantenmechanik", ZP **81** (1933): 143–62. {5, 12}

[FURUSAWA *et al.* 1998] FURUSAWA, A./SØRENSEN, J. L./BRAUNSTEIN, S. L./FUCHS, C. A./KIMBLE, H. J./POLZIK, E. S., "Unconditional Quantum Teleportation", *Science* **282** (1998): 706–709. {9, 10}

[GABRIELSE *et al.* 1986] GABRIELSE, G./FEI, X./HELMERSON, K./ROLSTON, S. L./TJOELKER, R./TRAINOR, T. A./KALINOWSKY, H./HAAS, J./KELLS, W., "First Capture of Antiprotons in a Penning Trap: A Kiloelectronvolt Source", PRL **57** (1986): 2504–507. {3}

[GÄHLER *et al.* 1981] GÄHLER, R./KLEIN, A. G./ZEILINGER, A., "Neutron Optical Tests of Nonlinear Wave Mechanics", PR **A23** (1981): 1611–17. {7}

[GALE *et al.* 1968] GALE, W./GUTH, E./TRAMMEL, G. T., "Determination of the Quantum State by Measurements", PR **165** (1968): 1434–36. {3, 4, 7}

[*GALILEI* S] GALILEI, Galileo, *Il Saggiatore*, Opere v. IV. {13, 14}

[GALINDO 1984] GALINDO, "Phase and Number", *Letters in Mathematical Physics* **8** (1984): 495–500. {3}

[GALLAVOTTI 1986a] GALLAVOTTI, Giovanni, "Meccanica statistica classica", in *Dizionario delle scienze fisiche*, Roma, Istituto dell'Enciclopedia Italiana, 1986. {8, 12, 13}

[GALLAVOTTI 1986b] GALLAVOTTI, Giovanni, "Equiripartizione e critica dellas Meccanica statistica classica", in *Dizionario delle scienze fisiche*, Roma, Istituto dell'Enciclopedia Italiana, 1986. {8, 12, 13}

[GALLIS 1993] GALLIS, Michael R., "Models for Local Ohmic Quantum Dissipation", PR **A48** (1993): 1028–34. {5}

[GALLIS 1996] —, "Emergence of Classicality via Decoherence Described by Lindblad Operators", PR **A53** (1996): 655–60. {5}

[GALLIS/FLEMING 1991] GALLIS, M. R./FLEMING, G. N., "Comparison of Quantum Open-Systems Models with Localization", PR **A43** (1991): 5778–86. {4, 5}

[GAO 1997] GAO, Shiwu, "Dissipative Quantum Dynamics with a Lindblad Functional" PRL **79** (1997): 3101–104. {5}

[GARCÍA DE P. 1996] GARCÍA DE POLAVIEJA, Gonzalo, "A Causal Quntum Theory in Phase Space", PL **220A** (1996): 303–314. {8}

[*GARDINER* 1991] GARDINER, Crispin W., *Quantum Noise*, Berlin, Springer, 1991. {2, 3, 5}

[GARG 1996] GARG, Anupam, "Decoherence in Ion Trap Quantum Computers", PRL **77** (1996): 964–67. {5, 10}

[GARG/MERMIN 1982a] GARG, A./MERMIN, N. D., "Comment on 'Hidden Variables, Joint Probability and Bell Inequalities", PRL **49** (1982): 242. {8, 9}

[GARG/MERMIN 1982b] —, "Bell Inequalities With Range of Violation that Does not Diminuish as the Spin Becomes Arbitrarly Large", PRL **49** (1982): 901–904. {3, 5, 9, 12}

[GARG/MERMIN 1982c] —, "Correlation Inequalities and Hidden Variables", PRL **49** (1982): 1220–23. {8, 9}

[GARG/MERMIN 1983] —, "Local Realism and Measured Correlations in the Spin-s EPR Experiment", PR **D27** (1983): 339–48. {8, 9}

[GARG/MERMIN 1984] —, "Farka's Lemma and the Nature of Reality: Statistical Implications of Quantum Correlations", FP **14** (1984): 1–39. {3, 5, 9, 12}

[GAROLA 1989] GAROLA, Claudio, "Classical Foundations of Quantum Logic", *International Journal of Theoretical Physics* **30** (1991): 1–52. {3}

[GAROLA 1992] —, "Truth versus Testability in Quantum Logic", *Erkenntnis* **37** (1992): 197–222. {3}

[GAROLA 1993] —, "Questioning Nonlocality: An Operation Critique to Bells Theorem", in [*GAROLA/ROSSI* 1993]. {9}

[GAROLA 1998a] —, "Against 'Paradoxes': A New Quantum Philosophy for Quantum Mechanics", in D. AERTS/J. PYKACZ (ed.s), *Quantum Physics and the Nature of Reality*, 1998. {3, 12}

[GAROLA 1998b] —, "Is Quantum Physics Contextual?", preprint. {3, 9}

[*GAROLA/ROSSI* 1993] GAROLA, C./ROSSI, A. (ed.s), *The Foundations of Quantum Mechanics. Historical Analysis and Open Questions*, Dordrecht, Kluwer, 1993. {3, 4, 5, 8, 9, 12}

[GAROLA/SOLOMBRINO 1996] GAROLA, C./SOLOMBRINO, L., "The Theoretical Apparatus of Semantic Realism: A New Language for Classical and Quantum Physics", FP **26** (1996): 1121–64. {3}

[GARRAWAY/KNIGHT 1994] GARRAWAY, B. M./KNIGHT, P. L., "Comparison of Quantum-State Diffusion and Quantum-Jump Simulations of Two-Photon Processes in a Dissipative Environment", PR **A49** (1994): 1266–74. {5, 12}

[GARUCCIO 1995] GARUCCIO, A., "Parametric Down-conversion Photon Sources, Beam-Splitters, and Bell's Inequality", in [*GREENBERGER/ZEILINGER* 1995a, 632–40]. {9}

[GARUCCIO/DE CARO 1993] GARUCCIO, A./DE CARO, L., "Bells Inequality for Trichotomic Observables", in [*GAROLA/ROSSI* 1993]. {9}

[GARUCCIO/RAPISARDA 1981] GARUCCIO, A./RAPISARDA, V., "Bells Inequalities and the Four-Coincidence Experiment", NC **A65** (1981): 269–97. {9}

[GARUCCIO/SELLERI 1984] GARUCCIO, A./SELLERI, F., "Enhanced Photon Detection in EPR Type Experiments", PL **103A** (1984): 99–103. {8, 9}

[GARUCCIO *et al.* 1982] GARUCCIO, A./RAPISARDA, V./VIGIER, J.-P., "Next Experimental Set-Up for the Detection of de Broglie Waves", PL **90A** (1982): 17–19. {7, 8}

[GARUCCIO *et al.* 1984a] GARUCCIO, A./KYPRIANIDIS, A./SARDELIS, D./VIGIER, J.-P., "Possible Experimental Test of Wave Packet Collapse", *Lettere al Nuovo Cimento* **39** (1984): 225–33. {4, 8}

[GARUCCIO *et al.* 1984b] GARUCCIO, A./KYPRIANIDIS, A./VIGIER, J.-P., "Relativistic Quantum Potential: the N-Body Case", NC **B83** (1984): 135–44. {8, 11}

[GARUCCIO *et al.* 1990] GARUCCIO, A./LEPORE, V. L./SELLERI, F., "Probabilities for Correlated Systems", FP **20** (1990): 1173–89. {8, 9}

[GARUCCIO *et al.* 1996] GARUCCIO, A./TORGESON, J. R./MONKEN, C./BRANNING, D./MANDEL, L., "Experimental Test of Selleri's Variable Photodetection-Probability Model", PR **A53** (1996): 2944–47. {9}

[GEA-BANACLOCHE 1990] GEA-BANACLOCHE, Julio, "Collapse and Revival of the State Vector in the Jaynes-Cummings Model: An Example of State Preparation by a Quantum Apparatus", PRL **65** (1990): 3385–88. {5, 6}

[GEA-BANACLOCHE 1998] —, "Qubit-Qubit Interaction in Quantum Computers", PR **A57** (1998): R1–4. {10}

[GELL-MANN/HARTLE 1989] GELL-MANN, M./HARTLE, J. B., "Quantum Mechanics in the Light of Quantum Cosmology", in [*KOBYASHI* 1989]. {6}

[GELL-MANN/HARTLE 1990] —, "Quantum Mechanics in the Light of Quantum Cosmology" in [*ZUREK* 1990a, 425–69]. {5, 6, 12}

[GELL-MANN/HARTLE 1993] —, "Classical Equations For Quantum Systems", PR **D47** (1993): 3345–82. {5, 6, 12}

[GELL-MANN/HARTLE 1994] —, "Time Symmetry and Asymmetry in Quantum Mechanics and Quantum Cosmology", in [*HALLIWELL et al.* 1994, 311–45]. {6, 12}

[*GEORGE* 1953] GEORGE, André (ed.), *Louis de Broglie. Physicien et Penseur*, Paris, Albin, 1953. {3, 4, 7, 12}

[GERHARDT *et al.* 1974] GERHARDT, H./BÜHLER, U./LIFTINE, G., "Phase Measurement of a Microscopic Radiation Field", PL **49A** (1974): 119–20. {4}

[GERLACH/O. STERN 1922a] GERLACH, W./STERN, O., "Der experimentelle Nachweis des magnetischen Moments des Silberatoms", ZP **8** (1922): 110–111. {3}

[GERLACH/O. STERN 1922b] —, "Der experimentelle Nachweis der Richtungsquantelung im Magnetfeld", ZP **9** (1922): 349–52. {3}

[GERLACH/O. STERN 1922c] —, "Das magnetische Moment des Silberatoms", ZP **9** (1922): 353–55. {3}

[GERJOUY 1973] GERJOUY, E., "Is the Principle of Superposition Really Necessary?", in [*HOOKER* 1973a, 114–42]. {3}

[*GEROULT* 1967] GEROULT, Martial, *Leibniz. Dynamique et Metaphysique*, Paris, Aubier, 1967. {14}

[GERRY 1996] GERRY, Christopher C., "Nonlocality of a Single Photon in Cavity QED", PR **A53** (1996): 4583–86. {9}

[GERRY 1997] —, "Generation of Schrödinger Cats and Entangled Coherent States in the Motion of a Trapped Ion by a Dispersive Interaction", PR **A55** (1997): 2478–81. {5}

[GERRY/KNIGHT 1997] GERRY, C. C./KNIGHT, P. L., "Quantum Superpositions and Schrödinger Cat States in Quantum Optics", AJP **65** (1997): 964–74. {5}

[GHABOUSSI 1988] GHABOUSSI, Farhad, "On the Equivalence Between the Superposition Principle and a Global Gauge Invariance", in [*VAN DER MERWE et al.* 1988a, 207–211]. {3, 12}

[GHIRARDI 1994] GHIRARDI, Gian Carlo, "An Attempt at a Macroscopic Quantum Worldview", [*ACCARDI* 1994a, 25–48]. {12}

[GHIRARDI 1995] —, "Spontaneous Wave Packet Reduction", in [*GREENBERGER/ZEILINGER* 1995a, 506–524]. {4, 5, 12}

[GHIRARDI *et al.* 1979] GHIRARDI, G. C./OMERO, C./WEBER, T./RIMINI, A., "Small-Time Behaviour of Quantum Nondecay Probability and Zeno's Paradox in Quantum Mechankics", NC **52A** (1979): 421–442. {3, 4}

[GHIRARDI *et al.* 1980] GHIRARDI, G. C./RIMINI, A./WEBER, T., "A General Argument Against Superluminal Transmission Through the Quantum Mechanical Measurement Process", *Lettere a Nuovo Cimento* **27** (1980): 293–98. {4, 9, 12}

[GHIRARDI *et al.* 1981] GHIRARDI, G. C./MIGLIETTA, F./RIMINI, A./WEBER, T., "Limitations on Quantum Measurements. I–II", PR **D24** (1981): 347–58. {4, 5}

[GHIRADI *et al.* 1982] GHIRARDI, G. C./RIMINI, A./WEBER, T., "Quantum Evolution in the Presence of Additive Conservations Laws and the Quantum Theory of Measurement", JMP **23** (1982): 1792–96. {4, 5}

[GHIRARDI *et al.* 1986] —, "Unified Dynamics for Microscopic and Macroscopic Systems", PR **D34** (1986): 470–91. {4, 5, 7, 12}

[GHIRADI *et al.* 1995] GHIRARDI, G. C./GRASSI, R./BENATTI, F., "Describing the Macroscopic World: Closing the Circle within the Dynamical Reduction Program", FP **25** (1995): 5–38. {5, 12}

[GHOSE/HOME 1992] GHOSE, P./HOME, D., "Wave-Particle Duality of Single-Photon States", FP **22** (1992): 1435–47. {3, 7, 8, 9, 12}

[GHOSE/HOME 1995] —, "An Analysis of the Aharonov-Anandan-Vaidman Model", FP **25** (1995): 1105–109. {7}

[GHOSE/HOME 1996] —, "The Two-Prism Experiment and the Wave-Particle Duality of Light", FP **26** (1996): 943–53. {4, 7}

[GHOSE/ROY 1991] GHOSE, P./ROY, S. M. N., "Confronting the Complementarity Principle in an Interference Experiment", PL **161A** (1991): 5–8. {7, 12}

[GHOSE *et al.* 1991] GHOSE, P./HOME, D./AGARWAL, G. S., "An Experiment to Throw More Light on Light", PL **153A** (1991): 403–406. {4, 7}

[GHOSE *et al.* 1992] —, "An 'Experiment to Throw More Light on Light': Implications", PL **168A** (1992): 95–99. {4, 7}

[GHOSH/MANDEL 1987] GOSH, R./MANDEL, L., "Observation of Nonclassical Effects in the Interference of two Photons", PRL **59** (1987): 1903–905. {3, 7, 12}

[GHOSH *et al.* 1986] GHOSH, R./HONG, C. K./OU, Z. Y./MANDEL, L., "Interference of Two Photons in Parametric Down Conversion", PR **A34** (1986): 3962–68. {3}

[*GIBBINS* 1987] GIBBINS, Peter, *Particles and Paradoxes: The Limits of Quantum Logic*, Cambridge, University Press, 1987, 1994. {3}

[GILES 1970] GILES, Robin, "Foundations for Quantum Mechanics", JMP **11** (1970): 2139–60; rep. in [*HOOKER* 1979, 277–322]. {3}

[GILSON 1968] GILSON, James G., "On Stochastic Theories of Quantum Mechanics", *Proceedings of the Cambridge Philosophical Society* **64** (1968): 1061–70. {5}

[GISIN 1984] GISIN, Nicolas, "Quantum Measurement and Stochastic Processes", PRL **52** (1984): 1657–60. {5, 7, 12}

[GISIN 1986] —, "Generalisation of Wigner's Theorem for Dissipative Quantum Systems", JP **A19** (1986): 205–210. {5, 7, 12}

[GISIN 1989] —, "Stochastic Quantum Dynamics and Relativity", HPA **62** (1989): 363–71. {5}

[GISIN 1990] —, "Weinberg's Non-linear Quantum Mechanics and Supraluminal Communications", PL **143A** (1990): 1–2. {7, 9}

[GISIN 1991a] —, "Bell's Inequality Holds for all Non-product States", PL **154A** (1991): 201–202. {9}

[GISIN 1991b] —, "Can Quantum Entangled States Collapse Spontaneously?", PL **155A** (1991): 445–49. {4, 5}

[GISIN 1996a] —, "Hidden Quantum Nonlocality Revealed by Local Filters", PL **210A** (1996): 151–56. {8, 9}

[GISIN 1996b] —, "Nonlocality Criteria for Quantum Teleportation", PL **210A** (1996): 157–59. {9, 10}

[GISIN/HUTTNER 1997] GISIN, N./HUTTNER, B., "Quantum Cloning, Eavesdropping and Bell's Inequality", PL **228A** (1997): 13–21. {9, 10}

[GISIN/PERCIVAL 1992a] GISIN, N./PERCIVAL, I. C., "Wave-Function Approach to Dissipative Processes: Are There Quantum Jumps?", PL **167A** (1992): 315–18. {5, 12}

[GISIN/PERCIVAL 1992b] —, "The Quantum-State Diffusion Model Applied to Open Systems", JP **A25** (1992): 5677–91. {5, 12}

[GISIN/PERCIVAL 1993a] —, "The Quantum State Diffusion Picture of Physical Processes" JP **A26** (1993): 2245–60. {5, 12}

[GISIN/PERCIVAL 1993b] —, "Stochastic Wave Equations Versus Parallel World Components", PL **175A** (1993): 144–45. {4, 5, 12}

[GISIN/PERES 1992] GISIN, N./PERES, A., "Maximal Violation of Bells Inequality for Arbitrarily Large Spin", PL **162A** (1992): 15–17. {9}

[GISIN/PIRON 1981] GISIN, N./PIRON, C., "Collapse of the Wave Packet Without Mixture", *Letters in Mathematical Physics* **5** (1981): 379–85. {4, 5, 12}

[GIULINI/KUPSCH 1996] GIULINI, D./KUPSCH, J., "Superselection Rules and Symmetries", in [*GIULINI et al.* 1996, 187–222]. {3}

[GIULINI *et al.* 1995] GIULINI, D./KIEFER, C./ZEH, H. D., "Symmetries, Superselection Rules, and Decoherence", PL **199A** (1995): 291–98. {3, 4}

[*GIULINI et al.* 1996] GIULINI, D./JOOS, E./KIEFER, C./KUPSCH, J./STAMATESCU, I.-O./ZEH, H. D. (ed.s), *Decoherence and the Appearance of a Classical World in Quantum Theory*, Berlin, Springer, 1996. {4, 5, 6, 12}

[*GIUNTINI* 1991] GIUNTINI, Roberto, *Quantum Logic and Hidden Variables*, Mannheim, Bibliographisches Institut, 1991. {3, 8}

[GIUNTINI/GREULING 1989] GIUNTINI, R./GREULING, H., "Toward a Formal Language for Unsharp Properties", FP **19** (1989): 931–45. {3, 4}

[GLAUBER 1963a] GLAUBER, Roy J., "Photon Correlations", PRL **10** (1963): 84–86. {3}

[GLAUBER 1963b] —, "The Quantum Theory of Optical Coherence", PR **130** (1963): 2529–39. {3}

[GLAUBER 1963c] —, "Coherent and Incoherent States of the Radiation Field", PR **131** (1963): 2766–88. {3}

[GLAUBER 1966] —, "Classical Behavior of Systems of Quantum Oscillators", PL **21** (1966): 650–52. {3}

[GLEASON 1957] GLEASON, Andrew M., "Measures on the Closed Subspaces of a Hilbert Space", *Journal of Mathematics and Mechanics* **6** (1957): 885–93; rep. in [*HOOKER* 1975, 123–33]. {3, 8, 15}

[*GLIMM/JAFFE* 1981] GLIMM, J./JAFFE, A. *Quantum Physics. A Functional Integral Point of View*, New York, Springer, 1981, II ed. 1987. {2, 11}

[*GNEDENKO* 1969] GNEDENKO, B. V., *The Theory of Probability*, Moscow, Mir, 1969, 1988. {1}

[GÖDEL 1931] GÖDEL, Kurt, "Über formal unentscheidbar Sätze der *Principia Mathematica* und verwandter Systeme. I", *Monatshefte für Mathematik und Physik* **38** (1931): 173–98. {15}

[GÖDEL 1949] —, "A Remark About the Relationship Between Relativity Theory and Idealistic Philosphy", in [*SCHILPP* 1949, 557–62]. {13, 14}

[GOETSCH *et al.* 1996] GOETSCHE, P./GRAHAM, R./HAAKE, F., "Microscopic Foundation of a Finite-Temperature Stochastic Schrödinger Equation", *Quantum and Semiclassical Optics* **8** (1996): 157–65. {5}

[*GOLD'MAN et al.* 1960] GOLD'MAN, I. I./KRIVCHENKOV, V. D./KOGAN, V. I./GALITSKIJ, V.M., *Problems in Quantum Mechanics*, London, 1960, Mass., Addison-Wesley, 1961. {3, 12}

[*H. GOLDSTEIN* 1950] GOLDSTEIN, Herbert, *Classical Mechanics*, Massachussets, Addison-Wesley, 1950, 1965. {1}

[S. GOLDSTEIN 1994] GOLDSTEIN, Sheldon, "Nonlocality without Inequalities for Almost All Entangled States for Two Particles", PRL **72** (1994): 1951. {9}

[GORDON 1927] GORDON, W., "Der Comptoneffekt nach der Schrödingerschen Theorie", ZP **40** (1926): 117–33. {3}

[GORINI *et al.* 1976] GORINI, V./KOSSAKOWSKI, A./SUDARSHAN, E. C. G., "Completely Positive Dynamical Semigroups of N–Level", JMP **17** (1976): 821–25. {3, 4, 5}

[GOTTESMAN 1998] GOTTESMAN, Daniel, "Theory of Fault-Tolerant Quantum Computation", PR **A57** (1998): 127–37. {10}

[GOU *et al.* 1997] GOU, S.-C./STEINBACH, J./KNIGHT, P. L., "Generation of Mesoscopic Superposition of Two Squeezed States of Motion for a Trapped Ion" PR **A55** (1997): 3719–23. {5}

[GRABOWSKI 1987] GRABOWSKI, Marian, "Entropic Uncertainty Relations for 'Phase-Number of Quanta' and 'Time-Energy"', PL **124A** (1987): 19–21. {3, 10}

[GRABOWSKI 1989] —, "What is an Observable?", FP **19** (1989): 923–30. {3, 4}

[GRACIA-B./VÁRILLY 1988] GRACIA-BONDIÁ, José M./VÁRILLY, Joseph C., "Phase-Space Representation for Galilean Quantum Particles of Arbitrary Spin", JP **A21** (1988): L879–93. {3}

[GRAHAM 1973] GRAHAM, Neill, "The Measurement of Relative Frequency", in [*DEWITT/GRAHAM* 1973, 229–53]. {4, 12}

[GRANGIER *et al.* 1986a] GRANGIER, P./ROGER, G./ASPECT, A., "Experimental Evidence for a Photon Anticorrelation Effect on a Beam Splitter: A New Light on Single-Photon Interferences", *Europhysucs Letters* **1** (1986): 173–79. {3, 7, 9}

[GRANGIER *et al.* 1986b] —, "A New Light on Single Photon Interferences", in [*GREENBERGER* 1986, 98–107]. {3, 7, 9}

[GRANGIER *et al.* 1986c] GRANGIER, P./ROGER, G./ASPECT, A./HEIDMANN, A./REYNAUD, S., "Observation of Photon Antibunching in Phase-Matched Multiatom Resonance Fluorescence", PRL **57** (1986): 687–90. {5}

[GRANGIER *et al.* 1988] GRANGIER, P./POTASEK, M. J./YURKE, B., "Probing the Phase Coherence of Parametrically Generated Photon Pairs: A New Test of Bell's Inequalities", PR **A38** (1988): 3132–35. {9}

[GRANGIER *et al.* 1991] GRANGIER, P./ROCH, J.-F./ROGER, G., "Observation of Backaction-Evading Measurement of an Optical Intensity in a Three-Level Atomic Nonlinear System", PRL **66** (1991): 1418–21. {4}

[GRANGIER *et al.* 1992] GRANGIER, P./COURTY, J.-M./REYNAUD, S., "Characterization of Nonideal Quantum Non-Demolition Measurements", *Optics Communications* **89** (1992): 99–106. {4}

[GRANIK 1997] GRANIK, A., "Why Quantum Mechanics Indeed?", FP **27** (1997): 511–32. {3, 8, 12, 13}

[GREECHIE 1976] GREECHIE, Richard J., "Some Results from the Combinatorial Approach to Quantum Logic", in [*SUPPES* 1976, 105–119]. {3}

[GREECHIE 1978] —, "Another Nonstandard Quantum Logic (and How I Found It)", in [*MARLOW* 1978a, 71–85]. {3}

[GREECHIE/GUDDER 1973] GREECHIE, R. J./GUDDER, S. P., "Quantum Logics", in [*HOOKER* 1973a, 143–73]. {3}

[H. GREEN 1958] GREEN, H. S., "Observation in Quantum Mechanics", NC **9** (1958): 880–89. {4, 12}

[T. GREEN/LANFORD 1960] GREEN, T. A./LANFORD, O. E., "Rigorous Derivation of the Phase Shift Formula for the Hilbert Space Scattering Operator of a Single Particle", JMP **1** (1960): 139–48. {3}

[GREENBERGER 1983] GREENBERGER, Daniel M., "The Neutron Interferometer as a Device for Illustrating the Strange Behavior of Quantum Systems", RMP **55** (1983): 875–905. {3, 7; W}

[GREENBERGER 1984] —, "The Neutron Interferometer and the Quantum Mechanical Superposition Principle", in [*ROTH/INOMATA* 1984, 117–30]. {3, 7}

[*GREENBERGER* 1986] — (ed.), *New Techniques and Ideas in Quantum Measurement Theory*, New York, Academy of Sciences, 1986. {4, 5, 6, 7, 8, 9}

[GREENBERGER 1994] —, "Evidence for the 'Law of the Excluded Muddle"' in [*DE MARTINI et al.* 1994, 245–53]. {4, 7, 12}

[GREENBERGER 1995] —, "Two-Particle versus Three Particle EPR Experiments", in [*GREENBERGER/ZEILINGER* 1995a, 585–99]. {9}

[GREENBERGER/YASIN 1986] GREENBERGER, D. M./YASIN, A., "The Haunted Measurement in Quantum Theory", in [*GREENBERGER* 1986, 449–57]. {3, 4, 12}

[GREENBERGER/YASIN 1988a] —, "Simultaneous Wave and Particle Knowledge in a Neutron Interferometer", PL **128A** (1988): 391–94. {4, 7, 12}

[GREENBERGER/YASIN 1988b] —, "The Haunted Measurement in Quantum Theory", in [*VAN DER MERWE et al.* 1988a, 15–32]. {3, 4, 6, 12}

[GREENBERGER/YASIN 1989] —, "'Haunted' Measurements in Quantum Theory", FP **19** (1989): 679–704. {3, 4, 6, 12}

[*GREENBERGER/ZEILINGER* 1995a] GREENBERGER, D. M./ZEILINGER, A. (ed.s), *Fundamental Problems in Quantum Theory*, New York, Academy of Sciences, 1995. {3–13}

[GREENBERGER/ZEILINGER 1995b] —, "Quantum Theory: Still Crazy After All These Years", *Physics World* (sept. 1995): 33–38. {4, 7, 8, 9, 12; S}

[GREENBERGER *et al.* 1989] GREENBERGER, D. M./HORNE, M. A./ZEILINGER, A., "Going Beyond Bell's Theorem", in [*KAFATOS* 1989, 69–72]. {9}

[GREENBERGER *et al.* 1990] GREENBERGER, D. M./HORNE, M. A./SHIMONY, A./ZEILINGER, A., "Bell's Theorem Without Inequalities", AJP **58** (1990): 1131–43. {9, 10, 12}

[GREENBERGER *et al.* 1993] —, "Multiparticle Interferometry and the Superposition Principle", PT (aug. 1993): 22–29. {3; W}

[GREENBERGER *et al.* 1995] GREENBERGER, D. M./HORNE, M. A./ZEILINGER, A., "Nonlocality of a Single Photon?", PRL **75** (1995): 2064. {7, 9}

[GREENBERGER *et al.* 1996] GREENBERGER, D. M./HORNE, M. A./ZEILINGER, A., "Tangled Concepts about Entangled States", in [*DE MARTINI et al.* 1996, 119–34]. {3}

[GREENWOOD/PRUGOVEČKI 1984] GREENWOOD, D. P./PRUGOVEČKI, E., "Stochastic Microcausality in Relativistic Quantum Mechanics", FP **14** (1984): 883–906. {3, 11}

[GRIFFITHS 1984] GRIFFITHS, Robert B., "Consistent Histories and the Interpretation of Quantum Mechanics", *Journal of Statistical Physics* **36** (1984): 219–72. {4, 6, 12}

[GRIFFITHS 1986] —, "Making Consistent Inferences from Quantum Measurements", in [*GREENBERGER* 1986, 512–17]. {4, 6, 12}

[GRIFFITHS 1987] —, "Correlation in Separated Quantum Systems: A Consistent History Analysis of the EPR Problem", AJP **55** (1987): 11–17. {4, 9, 12}

[GRIFFITHS 1993] —, "Consistent Interpretation of Quantum Mechanics Using Quantum Trajectories", PRL **70** (1993): 2201–204. {4, 6, 12}

[GRIFFITHS 1998] —, "Choice of Consistent Family, and Quantum Incompatibility", PR **A57** (1998): 1604–18. {4, 9, 12}

[*GRIGOLINI*] GRIGOLINI, P., *Quantum Mechanical Irreversibility and Measurement*, Singapore, World Scientific, 1993. {4, 5}

[GROENEWOLD 1946] GROENEWOLD, Hilbrand J., "On the Principles of Elementary Quantum Mechanics", *Physica* **12** (1946): 405–460. {3, 5, 12}

[E. GROSS 1987] GROSS, Eugene P., "Collective Variables in Elementary Quantum Mechanics", in [*HILEY/PEAT* 1987a, 46–65]. {3}

[H. GROSS 1990] GROSS, Herbert, "Hilbert Lattices: New Results and Unsolved Problems", FP **20** (1990): 529–59. {3}

[GRÖSSING 1995] GRÖSSING, Gerhard, "An Experiment to Decide Betwen the Causal and the Copenhagen Interpretation", in [*GREENBERGER/ZEILINGER* 1995a, 438–44]. {8, 12}

[GROT *et al.* 1996] GROT, N./ROVELLI, C./TATE, R. S., "Time-of-Arrival in Quantum Mechanics", PR **A54** (1996): 4676–90. {3}

[GROVER 1996] GROVER, Lov. K., "A Fast Quantum Mechanical Algorithm for Database Searching", in *Proceedings STOC 1996*, Philadelphia: 212–19. {10}

[GROVER 1997] —, "Quantum Mechanics Helps in Searching for a Needle in a Haystack", PRL (1997). {10}

[GRÜNBAUM 1957] GRÜNBAUM, Arthur, "Complementarity in Quantum Physics and Its Philosophical Generalizations", *Journal of Philosophy* **54** (1957): 713–27. {12, 14}

[GRUNHAUS *et al.* 1996] GRUNHAUS, J./POPESCU, S./ROHRLICH, D., "Jamming Nonlocal Quantum Correlations", PR **A53** (1996): 3781–84. {9}

[GUDDER 1968a] GUDDER, Stanley P., "Hidden Variables in Quantum Mechanics Reconsidered", RMP **40** (1968): 229–34. {8}

[GUDDER 1968b] —, "Dispersion-free States and the Exclusion of Hidden Variables", *Proceedings of the American Mathematical Society* **19** (1968): 319–24. {8}

[GUDDER 1970] —, "On Hidden-Variable Theories", JMP **11** (1970): 431–36. {8}

[GUDDER 1971] —, "Representations of Groups as Automorphisms on Orthomodular Lattices and Posets", *Canadian Journal of Mathematics* **23** (1971): 659–73; rep. in [*HOOKER* 1979, 31–48]. {3}

[GUDDER 1972] —, "Hidden-Variable Models for Quantum Mechanics", NC **10B** (1972): 518–22. {8}

[GUDDER 1973] —, "Quantum Logics, Physical Space, Position Observables and Symmetry", *Reports Mathematical Phisics* **4** (1973): 193–202. {3}

[GUDDER 1977] —, "Four Approaches to Axiomatic Quantum Mechanics", in [*W. PRICE/CHISSICK* 1977, 247–76]. {3}

[GUDDER 1978] —, "Some Unsolved Problems in Quantum Logics", in [*MARLOW* 1978a, 87–103]. {3}

[GUDDER 1979] —, "A Survey of Axiomatic Quantum Mechanics", in [*HOOKER* 1979, 323–63]. {3; W}

[GUDDER 1981] —, "Comparison of the Quantum Logic, Convexity, and Algebraic Approaches to Quantum Mechanics", in [*NEUMANN* 1981, 125–31]. {3}

[GUDDER 1988a] —, "A Theory of Amplitudes", JMP **29** (1988): 2020–2035. {3}

[*GUDDER* 1988b] —, *Quantum Probability*, Boston, Academic Press, 1988. {3}

[GUDDER 1989a] —, "Fuzzy Amplitude Densities and Stochastic Quantum Mechanics", FP **19** (1989): 293–317. {3}

[GUDDER 1989b] —, "Realism in Quantum Mechanics", FP **19** (1989): 949–70. {8, 12}

[GUDDER 1990] —, "Quantum Probability and Operational Statistics", FP **20** (1990): 499–526. {3}

[GUÉRET 1992] GUÉRET, Philippe, "Dualism Within Dualism: Open Questions", in [*SELLERI* 1992a, 97–108]. {7, 12}

[GUÉRET/VIGIER 1982] GUÉRET, P./VIGIER, J.-P., "De Broglie's Wave Particle Duality in the Stochastic Interpretation of Quantum Mechanics: A Testable Physical Assumption", FP **12** (1982): 1057–83. {7}

[GUERRA 1981] GUERRA, Francesco, "Structural Aspects of Stochastic Mechanics and Stochastic Field Theory", PRep **77** (1981): 263–312. {4, 5, 11}

[GUERRA 1993] "Introduction to Nelson Stochastic Mechanics as a Model for Quantum Mechanics", in [*GAROLA/ROSSI* 1993]. {5}

[GUERRA, F./MORATO, L.] GUERRA, F./MORATO, L., "Quantization of Dynamical Systems and Stochastic Control Theory", PR **D27** (1983): 1774–86. {5}

[*GUILLEMIN/STERNBERG* 1991] GUILLEMIN, V./STERNBERG, S., *Symplectic Techniques in Physics*, Cambridge, University Press, 1991. {1}

[GUZ 1984] GUZ, W., "Stochastic Phase Spaces, Fuzzy Sets, and Statistical Metric Spaces", FP **14** (1984): 821–48. {3}

[HAAG 1990] HAAG, Rudolf, "Fundamental Irreversibility and the Concept of Events", CMP **132** (1990): 245–51. {5, 6, 12}

[*HAAG* 1992] —, *Local Quantum Physics. Fields, Particles, Algebras*, Berlin, Springer, 1992, II ed. 1996. {2, 3, 11}

[HAAG 1996] —, "An Evolutionary Picture for Quantum Physics", CMP **180** (1996): 733–43. {12, 13, 14}

[HAAG/KASTLER 1964] HAAG, R./KASTLER, D., "An Algebraic Approach to Quantum Field Theory", JMP **5** (1964): 848–61. {3, 4, 5, 11}

[HAAG/SCHROER 1962] HAAG, R./SCHROER, B., "Postulates of Quantum Field Theory", JMP **3** (1962): 248–56. {3}

[*HAAK* 1974] HAAK, Susan, *Deviant Logic*, Cambridge, University Press, 1974, 1977, 1988. {14}

[*HAAKE* 1973] HAAKE, Fritz, "Statistical Treatment of Open Systems by Generalized Master Equations", in G. HÖHLER (ed.), *Quantum Statistics in Optics and Solid State Physics*, Berlin, 1973: 98–168. {5}

[*HAAKE* 1991] —, *Quantum Signatures of Chaos*, Berlin, Springer, 1991. {3, 13}

[HAAKE/WALLS 1987] HAAKE, F./WALLS, D. F., "Overdamped and Amplifying Meters in the Quantum Theory of Measurement", PR **A36** (1987): 730–39. {4, 5}

[HAAKE/ŻUKOWSKI 1993] HAAKE, F./ŻUKOWSKI, M., "Classical Motion of Meter Variables in the Quantum Theory of Measurement", PR **A47** (1993): 2506–517. {4, 5}

[HABIB *et al.* 1998] HABIB, S., SHIZUME, K./ZUREK, W. H., "Decoherence, Chaos, and the Correspondence Principle", PRL **80** (1998): 4361–4365. {3, 4, 5, 12}

[HACYAN 1997] HACYAN, S., "Test of Bell's Inequalities with Harmonic Oscillator", PR **A55** (1997): R2492–94. {9}

[HADJISAVVAS 1981] HADJISAVVAS, N., "Properties of Mixtures on Non-Orthogonal States", *Letters in Mathematical Physics* **5** (1981): 327–32. {3, 4}

[HAFNER/SUMMHAMMER 1997] HAFNER, M./SUMMHAMMER, J., "Experiment on Interaction-Free Measurement in Neutron Interferometery" PL **235A** (1997): 563–68. {4}

[HAGLEY *et al.* 1997] HAGLEY, E./MAÎTRE, X./NOGUES, G./WUNDERLICH, C./BRUNE, M./RAIMOND, J. M./HAROCHE, S., "Generation of EPR Pairs of Atoms", PRL **79** (1997): 1–5. {9}

[HAJI-HASSAN *et al.* 1987] HAJI-HASSAN, T./DUNCAN, A. J./PERRIE, W./BEYER, H. J./KLEINPOPPEN, H., "Experimental Investigation of the Possibility of Enhanced Photon Detection in EPR Type Experiments", PL **123A** (1987): 110–114. {9}

[A. R. HALL 1963] HALL, A. Rupert, *From Galilei to Newton*, New York, Dover, 1963, 1981. {13, 14}

[M. HALL/O'ROURKE 1993] HALL, M. J. W./O'ROURKE, M. J., "Realistic Performance of the Maximum Information Channel", *Quantum Optics* **5** (1993): 161–80. {10}

[R. L. HALL 1983] HALL, Richard L., "A Geometrical Theory of Energy Trajectories in Quantum Mechanics", JMP **24** (1983): 324–35. {3}

[R. L. HALL 1989] —, "Spectral Geometry of Power-Law Potentials in Quantum Mechanics", PR **A39** (1989): 5500–507. {3}

[HALLIWELL 1993] HALLIWELL, Jonathan J., "Quantum-Mechanical Histories and the Uncertainty Principle: Information-Theoretic Inequalities", PR **D48** (1993): 2739–52. {3, 6, 12}

[HALLIWELL 1995a] HALLIWELL, J. J., "An Operator Derivation of the Path Decomposition Expansion", PL **207A** (1995): 237–42. {3, 6}

[HALLIWELL 1995b] —, "A Review of the Decoherent Histories Approach to Quantum Mechanics", in [*GREENBERGER/ZEILINGER* 1995a, 726–40]. {6; W}

[*HALLIWELL* *et al.* 1994] HALLIWELL, J.J./PEREZ-MERCADER, J./ZUREK, W.H. (ed.s), *Physical Origins of Time Asymmetry*, Cambridge, University Press, 1994. {3, 5, 6, 12, 13}

[*HALMOS* 1950] HALMOS, Paul R., *Measure Theory*, New York, Van Nostrand, 1950. {1, 13}

[*HALMOS* 1951] —, *Introduction to Hilbert Space*, New York, Chelsea, 1951, II ed. 1957. {1, 15}

[HALPERN 1952] HALPERN, Otto, "A Proposed Re-Interpretation of Quantum Mechanics", PR **87** (1952): 389. {12}

[HALPERN 1966] "On the Einstein-Bohr Ideal Experiment", *Acta Physica Austriaca* **24** (1966): 274–79. {12}

[HÄNSCH/SCHAWLOW 1975] HÄNSCH, T./SCHAWLOW, A., "Cooling of Gases by Laser Radiation", *Optics Communication* **13** (1975): 68–69. {5}

[HANSON 1959] HANSON, Norwood R., "Copenhagen Interpretation of Quantum Theory", AJP **27** (1959): 1–15. {12}

[HARDEGREE 1979] HARDEGREE, Gary M., "The Conditional in Abstract and Concrete Quantum Logic", in [*HOOKER* 1979, 49–108]. {3}

[HARDY 1991a] HARDY, Lucien, "N–Measurement Bell Inequalities, N–Atom entangled States, and the Nonlocality of one Photon", PL **160A** (1991): 1–8. {9}

[HARDY 1991b] —, "Can Classical Wave Theory Explain the Photon Anticorrelation Effect on a Beam Splitter?", *Europhysics Letter* **15** (1991): 591–95. {3, 9}

[HARDY 1991c] —, "A New way to Obtain Bell Inequalities" PL **161A** (1991): 21. {9}

[HARDY 1992a] —, "Quantum Mechanics, Local Realistic Theories, and Lorentz-Invariant Realistic Theories", PRL **68** (1992): 2981–84. {8, 9, 11, 12, 13}

[HARDY 1992b] —, "On the Existence of Empty Waves in Quantum Theory", PL **167A** (1992): 11–16. {7, 12}

[HARDY 1992c] —, "A Quantum Optical Experiment to test Local Realism", PL **167A** (1992): 17–23. {8, 9}

[HARDY 1992d] —, "Reply to 'Empty Waves: Not Necessarily Effective"', PL **169A** (1992): 222–23. {7}

[HARDY 1993a] —, "Nonlocality for Two Particles Without Inequalities for Almost All Entangled States", PRL **71** (1993): 1665–68. {9}

[HARDY 1993b] —, "Reply to Żukowski's Comment", PL **175A** (1993): 259–60. {7}

[HARDY 1994a] —, "Nonlocality of a Single Photon Revisited", PRL **73** (1994): 2279–83. {7, 9}

[HARDY 1994b] —, "Replies to Berndl/Goldstein", PRL **72** (1994): 781. {9}

[HARDY 1994c] —, "Replies to Schauer", PRL **72** (1994): 783. {9}

[HARDY 1995a] —, "Replies to Vaidman and Greenberger/Horne/Zeilinger", PRL **75** (1995): 2065–66. {9}

[HARDY 1995b] —, "The EPR Argument and Nonlocality without Inequalities for a Single Photon", in [*GREENBERGER/ZEILINGER* 1995a, 600–615]. {7, 9}

[HARDY 1996] —, "Properties of Particles" in [*DE MARTINI et al.* 1996, 439–49]. {7, 8, 9, 12}

[HARDY/SQUIRES 1992] HARDY, L./SQUIRES, E. J., "On the Violation of Lorentz-Invariance in Deterministic Hidden-Variable Interpretations of Quantum Theory", PL **168A** (1992): 169–73. {8, 9}

[HAROCHE/KLEPPNER 1989] HAROCHE, S./KLEPPNER, D., "Cavity Quantum Electrodynamics", PT **42** (jan. 1989): 24–30. {2; W}

[HAROCHE/RAIMOND 1996] HAROCHE, S./RAIMOND, J.-M., "Quantum Computing: Dream or Nightmare?", PT (1996): 51–52. {10}

[HARRIS *et al.* 1967] HARRIS, S. E./OSHMAN, M. K./BYER, R. L., "Observation of Tunable Optical Parametric Fluorescence", PRL **18** (1967): 732–34. {3}

[HARTLE 1968] HARTLE, James B., "Quantum Mechanics of Individual Systems", AJP **36** (1968): 704–12. {3, 12}

[HARTLE 1988a] —, "Quantum Kinematics of Spacetime. I. Nonrelativistic Theory", PR **D37** (1988): 2818–32; *Erratum*: PR **D43** (1991): 1434E–35E. {3}

[HARTLE 1988b] —, "Quantum Kinematics of Spacetime. II. A Model Quantum Cosmology with Real Clocks", PR **D38** (1988): 2985–99. {3, 6, 11}

[HARTLE 1991] —, "Spacetime Coarse Grainings in Nonrelativistic Quantum Mechanics", PR **D44** (1991): 3173–96. {3}

[HARTLE/HAWKING 1983] HARTLE, J. B./HAWKING, S. W., "Wave Function of the Universe", PR **D28** (1983): 2960–75. {3, 6, 13}

[HASEGAWA/EZAWA 1980] HASEGAWA, H./EZAWA, H., *Progress of Theoretical Physics Supplementa* **69** (1980): 41. {5}

[HASS/BUSCH 1994] HASS, K./BUSCH, P., "Causality of Superluminal Barrier Traversal", PL **185A** (1994): 9–13. {9}

[HASSELBACH 1992] HASSELBACH, Franz, "Recent Contributions of Electron Interferometry to Wave-Particle Duality", in [*SELLERI* 1992a, 109–126]. {7}

[HASSELBACH *et al.* 1995] HASSELBACH, F./SCHÄFER, A./WACHENDORFER, H., "Particle Spectroscopy by Application of Michelson's 'Light-Wave Analysis' Technique to de Broglie Waves", in [*GREENBERGER/ZEILINGER* 1995a, 374–82]. {7}

[HAUGE/STØVNENG 1989] HAUGE, E. H./STØVNENG, J. A., "Tunneling Times: A Critical Review", RMP **61** (1989): 917–936. {9; W}

[HAUS/KÄRTNER 1996] HAUS, H. A./KÄRTNER, F. X., "Optical Quantum Nondemolition Measurements and the Copenhagen Interpretation", PR **A53** (1996): 3785–91. {4, 12}

[HAUSLADEN *et al.* 1995] HAUSLADEN, P./SCHUMACHER, B./WOOTTERS, W. K./WESTMORELAND, M., "Sending Classical Bits via Quantum Its", in [*GREENBERGER/ZEILINGER* 1995a, 698–705]. {10}

[HAUSLADEN *et al.* 1996] HAUSLADEN, P./JOZSA, R./SCHUMACHER, B./WESTMORELAND, M./WOOTTERS, W. K., "Classical Information Capacity of a Quantum Channel", PR **A54** (1996): 1869–76. {10}

[HAWKING 1984] HAWKING, Stephen W., "The Quantum State of the Universe", *Nuclear Physics* **B239** (1984): 257–76. {6, 13}

[D. HEALEY/SCHROECK 1995] HEALEY, D. M., Jr./SCHROECK, F. E. Jr., "On Informational Completeness of Covariant Localization Observables and Wigner Coefficients", JMP **36** (1995): 453–507. {3, 4, 10}

[R. HEALEY 1979] HEALEY, R., "Quantum Realism: Naiveté is No Excuse", *Synthese* **42** (1979): 121–44. {8, 12}

[R. *HEALEY* 1989] —, *The Philosophy of Quantum Mechanics*, Cambridge, University Press, 1989. {12, 14}

[R. HEALEY 1995] —, "Dissipating the Quantum Measurement Problem", *Topoi* **14** (1995): 55–65. {4, 8}

[HEELAN 1970a] HEELAN, Patrick A., "Complementarity, Context Dependence, and Quantum Logic", FP **1** (1970): 95–100; rep. in [*HOOKER* 1979, 161–80]. {3, 12}

[HEELAN 1970b] —, "Quantum and Classical Logic: Their Respective Roles", *Synthese* **21** (1970): 1–33

[HEGERFELDT 1974] HEGERFELDT, Gerhard C., "Remarks on Causality and Particle Localization", PR **D10** (1974): 3320–21. {3, 8}

[HEGERFELDT 1985] —, "Violation of Causality in Relativistic Quantum Theory?", PRL **54** (1985): 2395–98. {3}

[HEGERFELDT/RUIJSENAARS 1980] HEGERFELDT, G. C./RUIJSENAARS, S. N. M., "Remarks on Causality, Localization, and Spreading of Wave Packets", PR **D22** (1980): 377–84. {3, 8, 12}

[HEGERFELDT *et al.* 1968] HEGERFELDT, G. C./KRAUS, K./WIGNER, E. P., "Proof of the Fermion Superselection Rules Without the Assumption of Time-Reversal Invariance", JMP **9** (1968): 2029–31. {3}

[*HEISENBERG* GW] HEISENBERG, Werner, *Gesammelte Werke*, Mnchen/Berlin, Piper/Springer, 1984. {3, 4, 7, 12, 13, 14}

[HEISENBERG 1925] —, "Über quantentheoretische Umdeutung kinematischer und mechanischer Beziehungen", ZP **33** (1925): 879–93; rep. in [*HEISENBERG* GW, A.I, 382–96]. {3}

[HEISENBERG 1927] —, "Über den anschaulichen Inhalt der quantentheoretischen Kinematik und Mechanik", ZP **43** (1927): 172–98; rep. in [*HEISENBERG* GW, A.I, 478–504] {3, 4, 12}

[HEISENBERG 1929] —, "Die Entwicklung der Quantentheorie 1918–1928", *Naturwissenschaften* **17** (1929): 490–96; rep. in [*HEISENBERG* GW, B, 109–15]. {12}

[*HEISENBERG* 1930a] —, *Physikalische Prinzipien der Quantentheorie*, Leipzig, 1930; Stuttgart, Hirzel V., 1958; Mannheim, Wissenschaftsverlag, 1991. {3, 4, 12}

[*HEISENBERG* 1930b] —, *The Physical Principles of Quantum Theory*, engl. tr. of [*HEISENBERG* 1930a], Chicago, University Press, 1930; rep. in [*HEISENBERG* GW, B, 117–66]. {3, 4, 12}

[HEISENBERG 1931a] —, "Die Rolle der Unbestimmtheitsrelationen in der modernen Physik", *Monatshefte fr Mathematik und Physik* **38** (1931): 365–72. {3, 12}

[HEISENBERG 1931b] —, "Kausalgesetz und Quantenmechanik", *Erkenntnis* **2** (1931): 172–82. {12, 13}

[HEISENBERG 1942] —, "Die 'beobachtbaren Größen' in der Theorie der Elementarteilchen. I–II", ZP **120** (1943): 513–38, 673–702. {3, 4, 11}

[*HEISENBERG* 1952] —, *Philosophical Problems of Quantum Physics*, Woodbridge (Connecticut), Ox Bow Press, 1952, 1979. {12, 13, 14}

[*HEISENBERG* 1958] —, *Physics and Philosophy*, New York, Harper. {7, 12, 13, 14}

[*HEISENBERG* 1959] —, *Physik und Philosophie* (ger. transl. of [*HEISENBERG* 1958]); rep. in [*HEISENBERG* GW, C.II, 3–201]. {7, 12, 13, 14}

[*HEISENBERG* 1961] —, *On Modern Phyiscs*, New York, Clarkson N. Potter, 1961.

[HEISENBERG 1972] —, "Indefinite Metric in State Space", in [*SALAM/WIGNER*, 129–36]. {3}

[HEISENBERG 1976] —, "The Nature of Elementary Particles", PT **29.3** (1976): 32–39. {7, 12}

[HEISENBERG 1977] —, "Remarks on the Origin of the Relations of Uncertainty", in [*W. PRICE/CHISSICK* 1977, 3–6]. {12, 13}

[HELLMANN 1982] HELLMANN, G., "Stochastic Einstein-Locality and the Bell Theorems", *Synthese* **53** (1982): 461–504. {8, 9}

[HELLMUTH *et al.* 1985] HELLMUTH, T./ZAJONC, A. G./WALTHER, H., "Realization of a 'Delayed Choice', Mach-Zehnder Interferometer", in [*LAHTI/MITTELSTAEDT* 1985, 417–22]. {6}

[HELLMUTH *et al.* 1986] —, "Realizations of 'Delayed Choice' Experiments", in [*GREENBERGER* 1986, 108–114]. {6}

[HELLMUTH *et al.* 1987] HELLMUTH, T./WALTHER, H./ZAJONC, A. G./SCHLEICH, W. P., "Delayed-Choice Experiments in Quantum Interference", PR **A35** (1987): 2532–41. {6}

[HELLWIG 1981] HELLWIG, K.-E., "Conditional Expectations and Duals of Instruments", in [*NEUMANN* 1981, 113–24]. {3}

[HELLWIG/KRAUS 1970] HELLWIG, K.-E./KRAUS, K., "Formal Description of Measurements in Local Quantum Field Theory", PR **D1** (1970): 566–71. {4, 11}

[*HELSTROM* 1976] HELSTROM, Carl W., *Quantum Detection and Estimation Theory*, New York, Academic, 1976. {3, 4, 10}

[HEMPEL 1945] HEMPEL, Carl G., "Review of Reichenbach's *Philosophic Foundations of Quantum Mechanics*", *Journal of Symbolic Logic* **10** (1945): 97–100. {7, 12, 14}

[*HEMPEL* 1953] —, *Methods of Concept Formation Science*, Chicago, University of Chicago Press, 1953. {14}

[*HENDRY* 1984] HENDRY, John, *The Creation of Quantum Mechanics and the Bohr-Pauli Dialogue*, Dordrecht, Reidel, 1984. {12, 14}

[HEPP 1972] HEPP, Klaus, "Quantum Theory of Measurement and Macroscopic Observables", *Helvetica Physica Acta* **45** (1972): 237–48. {4, 5}

[HEPP/LIEB 1973] HEPP, K./LIEB, E. H., "Phase Transitions in Reservoir-Driven Open Systems with Applications to Lasers and Superconductors", *Helvetica Physica Acta* **46** (1973): 573–603. {5}

[HERBERT 1982] HERBERT, Nick, "FLASH A Superluminal Communicator Based Upon a New Kind of Quantum Measurement", FP **12** (1982): 1171–79. {4, 9}

[HERBURT/VUJICIC 1985] HERBURT, F./VUJICIC, M., "Distant Correlations Between Identical Particles and Inoperativeness of Pauli Principle", in [*LAHTI/MITTELSTAEDT* 1985, 423–33]. {3, 9}

[HERKOMMER *et al.* 1996] HERKOMMER, A. M./CARMICHAEL, H. J./SCHLEICH, W. P., "Localization of an Atom by Homodyne Measurement", *Quantum and Semiclassical Optics* **8** (1996): 189–203. {5, 7}

[HERZBERG/LONGUET-H. 1963] HERZBERG, G./LONGUET-HIGGINS, H. C., "Interaction of Potential Energy Surfaces in Polyatomic Molecules", *Disc. Faraday Society* **35** (1963): 77–82; rep. in [*SHAPERE/WILCZEK* 1989, 74–79]. {3}

[HERZOG *et al.* 1994] HERZOG, T. J./RARITY, J. G./WEINFURTER, H./ZEILINGER, A., "Frustrated Two-Photon Creation via Interference", PRL **72** (1994): 629–32. {3}

[HERZOG *et al.* 1995] HERZOG, T. J./KWIAT, P. G./WEINFURTER, H./ZEILINGER, A., "Complementarity and the Quantum Eraser", PRL **75** (1995): 3034–37. {3, 6, 7, 12}

[HEYWOOD/REDHEAD 1983] HEYWOOD, P./REDHEAD, M. L. G., "Non-Locality and the Kochen-Specker Paradox", FP **13** (1983): 481–99. {8, 9}

[HIAI/PETZ 1991] HIAI, F./PETZ, D., "The Proper Formula for Relative Entropy and its Asymptotics in Quantum Probability", CMP **143** (1991): 99–114. {10}

[HILBERT *et al.* 1927] HILBERT, D./VON NEUMANN, J./NORDHEIM, , "Über die Grundlagen der Quantenmechanik", *Mathematische Annalen* **98** (1927). {3}

[HILEY 1989] HILEY, B. J., "Cosmology, EPR Correlations and Separability", in [*KAFATOS* 1989, 181–90]. {9, 13}

[HILEY 1991] "Vacuum or Holomovement", in [*SAUNDERS/H. BROWN* 1991, 217–50]. {7, 8, 9}

[*HILEY/PEAT* 1986] HILEY, B. J./PEAT, F. D. (ed.s), *Quantum Theory and Beyond*, London, Routledge, 1986. {3, 12}

[*HILEY/PEAT* 1987a] — (ed.s), *Quantum Implications: Essays in Honour of David Bohm*, Routledge, London, 1987, 1991, 1994. {4, 8, 9, 12, 13}

[HILEY/PEAT 1987b] —, "The Development of David Bohms Ideas From the Plasma to the Implicate Order", in [*HILEY/PEAT* 1987a, 1–32]. {8, 9, 12, 14}

[*HILEY/PEAT* 1987c] — (ed.s), *Negative Probability in Quantum Mechanics*, London, Routledge, 1987. {3, 4, 5, 7, 8, 9, 12}

[HILGEVOORD/UFFINK 1983] HILGEVOORD, J./UFFINK, J. B. M., "Overall Width, Mean Peak Width and the Uncertainty Principle", PL **95A** (1983): 474–76. {3, 12}

[HILGEVOORD/UFFINK 1988] —, "The Mathematical Expression of the Uncertainty Principle", in [*VAN DER MERWE et al.* 1988a, 91–114]. {3}

[HILLERY/YURKE 1995] HILLERY, M./YURKE, B., "Bell's Theorem and Beyond", *Quantum Optics* **7** (1995): 215–27. {3, 9; W}

[HILLERY *et al.* 1984] HILLERY, M. A./O'CONNELL, R. F./SCULLY, M. O./WIGNER, E., "Distribution Functions in Physics: Fundamentals", PRep **106** (1984): 121–67. {3; W}

[HIOE/EBERLY 1981] HIOE, F. T./EBERLY, J. H., "N-Level Coherence Vector and Higher Conservation Laws in Quantum Optics and Quantum Mechanics", PRL **47** (1981): 838–41. {3}

[*HIRSCH/SMALE* 1974] HIRSCH, M. W./SMALE, S., *Differential Equations, Dynamical Systems, and Linear Algebra*, San Diego, Academic Press, 1974. {1}

[HOLEVO 1973] HOLEVO, A. S., *Problems Information Transmission* **9** (1973): 177. {10}

[*HOLEVO* 1982] —, *Probabilistic and Statistical Aspects of Quantum Theory* (engl. tr.), Amsterdam, North Holland, 1982. {3, 4, 10}

[HOLEVO 1998] —, "The Capacity of Quantum Channel with General Signal States", e-print, code: quant-ph/9611023. {10}

[*HOLLAND* 1993] HOLLAND, Peter R., *The Quantum Theory of Motion*, Cambridge, University Press, 1993, 1995. {4, 7, 8, 12}

[HOME 1992] HOME, Dipankar, "Optical Tunneling of Single Photon Stats: Wave-Particle Complementarity Revisited", *4th International Symposium on the Foudations of QM*, Tokyo, 1992. {9}

[HOME/BOSE 1996] HOME, D./BOSE, S., "Standard Quantum Mechanics with Environment Induced Decoherence and Wavefunction Collapse: Possibility of an Empirical Discrimination Using Neutron Interferometry", PL **217A** (1996): 209–14. {4, 5, 12}

[HOME/CHATTOPADHYAYA 1996] HOME, D./CHATTOPADHYAYA, R., "DNA Molecular Cousin of Schrödinger's Cat: A Curious Example of Quantum Measurement", PRL **76** (1996): 2836–39. {4, 5, 12}

[HOME/SELLERI 1992] HOME, D./SELLERI, F., "The Aharonov-Bohm Effect from the Point of View of Local Realism", in [*SELLERI* 1992a, 127–37]. {8, 9}

[HOME/WHITAKER 1986a] HOME, D./WHITAKER, M. A. B., "Reflections on the Quantum Zeno Paradox", JP **A19** (1986): 1847–54. {4}

[HOME/WHITAKER 1986b] —, "The Ensemble Interpretation and Context-Dependence in Quantum Systems", PL **115A** (1986): 81–83. {4, 5, 12}

[HOME/WHITAKER 1992] —, "Ensemble Interpretations of Quantum Mechanics. A Modern Perspective", PRep **210** (1992): 223–317. {4, 5, 12; W}

[HONG/MANDEL 1986] HONG, C. K./MANDEL, L., "Experimental Realization of a Localized One-Photon State", PRL **56** (1986): 58–60. {3}

[HONG *et al.* 1987] HONG, C. K./OU, Z. Y./MANDEL, L., "Measurements of Subpicosecond Time Intervals Between Two Photons by Interference", PRL **59** (1987): 2044–46. {4}

[HONG *et al.* 1988] —, "Interference Between a Fluorescent Photon and a Classical Field: An Example of Nonclassical Interference", PR **A37** (1988): 3006–3009. {3, 4, 7, 12}

[*HONNER* 1987] HONNER, J., *The description of Naure: Niels Bohr and the Philosophy of Quantum Mechanics*, Oxford, Clarendon, 1987. {12, 14}

[HOOKER 1970] HOOKER, C. A., "Concerning EPR's Objection to Quantum Theory", AJP **38** (1970): 851–57. {8, 9}

[HOOKER 1971] —, "Sharp and the Refutation of the EPR Paradox", PS **38** (1971): 224–33. {8, 9}

[HOOKER 1972] —, "The Nature of Quantum Mechanical Reality", in [*COLODNY* 1972, 67–302]. {8, 9, 12}

[*HOOKER* 1973a] — (ed.), *Contemporary Research in the Foundations and Philosophy of Quantum Theory*, Dordrecht, Reidel, 1973. {3, 12, 14}

[HOOKER 1973b] —, "Metaphysics and Modern Physics", in [*HOOKER* 1973a, 174–304]. {14}

[*HOOKER* 1975] — (ed.), *The Logico-Algebraic Approach to Quantum Mechanics. Historical Evolution*, Dordrecht, Reidel, 1975. {3}

[*HOOKER* 1979] — (ed.), *The Logico-Algebraic Approach to Quantum Mechanics. Contemporary Consolidation*, Dordrecht, Reidel, 1979. {3, 8}

[HORGAN 1992] HORGAN, John, "Quantum Philosophy", *Scientific American* (july 1992): 72-80. {12, 13, 14; S}

[HORGAN 1994] —, "Particle Metaphysics", *Scientific American* (february 1994): 70–78. {12, 13, 14; S}

[*HÖRMANDER]* HÖRMANDER, L., *The Analysis of Linear Partial Differential Operators*, 1985, Berlin, Springer, 1985. {1}

[HORNE/SHIMONY 1995] HORNE, M. A./SHIMONY, A., "Multipath Interferometry of the Biphoton", in [*GREENBERGER/ZEILINGER* 1995a, 664–74]. {3}

[HORNE/ZEILINGER 1985] HORNE, M. A./ZEILINGER, A., "A Bell-Type EPR Experiment Using Linear Momenta", in [*LAHTI/MITTELSTAEDT* 1985, 435–39]. {9}

[HORNE *et al.* 1989] HORNE, M. A./SHIMONY, A./ZEILINGER, A., "Two-Particle Interferometry", PRL **62** (1989): 2209–12. {9}

[R. HORODECKI 1994] HORODECKI, Ryszard, "Informationally Coherent Quantum Systems", PL **187A** (1994): 145–50. {3, 9, 10}

[R. HORODECKI/M.HORODECKI 1996] —, "Information-Theoretic Aspects of Inseparability of Mixed States", PR **A54** (1996): 1838–43. {3, 9, 10}

[R. HORODECKI/P.HORODECKI 1994] HORODECKI, R./HORODECKI, P., "Quantum Redundancies and Local Realism", PL **194A** (1994): 147–52. {3, 8, 9, 10}

[R. HORODECKI *et al.* 1995] HORODECKI, R./HORODECKI, P./HORODECKI, M., "Violating Bell Inequality by Mixed Spin-$\frac{1}{2}$ States: Necessary and Sufficient Condition", PL **200A** (1995): 340–344. {9}

[R. HORODECKI *et al.* 1996a] —, "Quantum a-Entropy Inequalities: Independent Condition for Local Realism?", PL **210A** (1996): 377–81. {8, 9}

[R. HORODECKI *et al.* 1996b] —, "Separability of Mixed States: Necessary and Sufficient Conditions", PL **223A** (1996): 1–8. {3, 9, 10}

[R. HORODECKI *et al.* 1997] —, "Inseparable Two Spin-$\frac{1}{2}$ Density Matrices Can Be Distilled to a Singlet Form", PRL **78** (1997): 574–77. {3, 9, 10}

[R. HORODECKI *et al.* 1998] —, "Distillability of Inseparable Quantum Systems", e-print, code: quant-ph/9607009. {3, 9, 10}

[*HORWICH* 1987] HORWICH, Paul, *Asymmetries in Time. Problems in the Philosophy of Science*, Cambridge Mass., MIT Press, 1987, 1988. {13, 14}

[HORWITZ/KATZNELSON 1983] HORWITZ, L. P./KATZNELSON, E., "Is Proton Decay Measurable?", PRL **50** (1983): 1184–86. {4}

[HÖRZ 1960] HÖRZ, Herbert, "Die philosophische Bedeutung der Heisenbergschen Unbestimmtheitsrelationen", *Deutsche Zeitschrift fr Philosophie* **8** (1960): 702–709. {12}

[HOWARD 1985] HOWARD, Don, "Einstein on Locality and Separability", *Studies in the History and Philosophie of Science* **16** (1985): 171–201. {8, 9, 12, 13, 14}

[HOYNINGEN-HUENE 1991] HOYNINGEN-HUENE, P., "Theorie antireduktionistischer Argumente: Fallstudie Bohr", *Deutsche Zeitschrift fr Philosophie* **39** (1991): 194–204. {12, 14}

[HRADIL 1997] HRADIL, Z., "Quantum-State Estimation", PR **A55** (1997): R1561–64. {4, 7, 10}

[HU *et al.* 1992] HU, B. L./PAZ, J. P./ZHANG, Y., "Quantum Brownian Motion in a General Environment: Exact Master Equation with Nonlocal Dissipation and Colored Noise", PR **D45** (1992): 2843–61. {5}

[H. HUANG/AGARWAL 1994] HUANG, H./AGARWAL, G. S., "General Linear Transformations and Entangled States", PR **A49** (1994): 52–60. {3, 9}

[*K. HUANG* 1963] HUANG, Kerson, *Statistical Mechanics*, New York, Wiley, 1963, 1987. {1}

[HÜBNER 1992] HÜBNER, Matthias, "Explicit Computation of the Bures Distance for Density Matrices", PL **163A** (1992): 239–42. {3, 10}

[HÜBNER 1993] —, "Computation of Uhlmann's Parallel Transport for Density Matrices and the Bures Metric on the Three-Dimensional Hilbert Space", PL **179A** (1993): 226–30. {3, 10}

[HUELGA *et al.* 1995a] HUELGA, S. F./FERRERO, M./SANTOS, E., "Loophole-Free Test of the Bell Inequality", PR **A51** (1995): 5008–5011. {8, 9}

[HUELGA *et al.* 1995b] —, "Proposed New Polarization Correlation Test of Local Realism", in [*GREENBERGER/ZEILINGER* 1995a, 429–37]. {8, 9}

[HUELGA *et al.* 1995c] HUELGA, S. F./MARSHALL, T. W./SANTOS, E., "Proposed Test for Realist Theories Using Rydberg Atoms Coupled to a High-Q Resonator", PR **A52** (1995): R2497–500. {8, 9}

[*HUGHES* 1989] HUGHES, R., *The Structure and Interpretation of Quantum Mechanics*, Cambridge, Harvard University Press, 1989. {2}

[HULET/KLEPPNER 1983] HULET, R. G./KLEPPNER, D., "Rydberg Atoms in 'Circular' States", PRL **51** (1983): 1430–33. {3}

[HUME *E*] HUME, David, *Enquiries Concerning Human Understanding and Conserning the Principles of Morals*, Oxford, Clarendon, 1777, 1902, III ed. 1975, 1992. {14}

[HUSEMOLLER 1975] HUSEMOLLER, D., *Fiber Bundles*, New York, Springer, 1975. {1}

[HUSIMI 1937] HUSIMI, Kodi, "Studies in the Foundations of Quantum Mechanics", *Proceedings of the Physico-Mathematical Society of Japan* **19** (1937): 766–89. {3}

[*HUSSERL* 1936] HUSSERL, Edmund, *Die Krisis der europäischen Wissenschaften und die transzendentale Phänomenologie*, *Philosophia* **1** (1936): 77–176, rep. Hamburg, Meiner, 1977, 1982. {14}

[*IAGOLNITZER* 1978] IAGOLNITZER, D., *The S-matrix*, Amsterdam, North-Holland, 1978. {2; W}

[IL KIM *et al.* 1996] IL KIM, J./NEMES, M. C./DE TOLEDO PIZA, A. F. R./BORGES, H. E., "Perturbative Expansion for Coherence Loss", PRL **77** (1996): 207–210. {5}

[IMAMOḠLU 1993] IMAMOḠLU, A., "Logical Reversibility in Quantum-Nondemolition Measurements", PR **A47** (1993): R4577–80. {4, 5, 6, 7}

[IMAMOḠLU *et al.* 1997] IMAMOḠLU, A./LEWENSTEIN, M./YOU, L., "Inhibition of Coherence in Trapped Bose-Einstein Condensates", PRL **78** (1997): 2511–14. {5}

[IMOTO/SAITO 1989] IMOTO, N./SAITO, S., "Quantum Nondemolition Measurement of Photon Number in a Lossy Optical Kerry Medium", PR **A39** (1989): 675–82. {4}

[IMOTO *et al.* 1990] IMOTO, N./UEADA, M./OGAWA, T., "Microscopic Theory of the Continuous Measurement of Photon Number", PR **A41** (1990): 4127–30. {4, 5}

[IRANI 1984] IRANI, K. D., "Philosophical Problems in the Interpretation of Quantum Mechanics", in [*ROTH/INOMATA* 1984, 309–317]. {12}

[ISAKSON 1927] ISAKSON, A., "Zum Aufbau der Schrödingerschen Gleichung", ZP **44** (1927): 893–99. {3}

[ISAR *et al.* 1991] ISAR, A./SANDALESCU, A/SCHEID, W., "Quasiprobability Distributions for Open Systems within the Lindblad Theory", JMP **32** (1991): 2128–34. {3, 5}

[ISAR *et al.* 1993] —, "Density Matrix for the Damped Harmonic Oscillator within the Lindblad Theory", JMP **34** (1993): 3887–3900. {5}

[ITANO *et al.* 1987] ITANO, W.H./BERGQUIST, J. C./HULET, R. G./WINELAND, D. J., "Radiative Decay Rates in Hg^+ from Observations of Quantum Jumps in a Single Ion", PRL **59** (1987): 2732–35. {5}

[ITANO *et al.* 1990] ITANO, W. H./HEINZEN, D. J./BOLLINGER, J. J./WINELAND, D. J., "Quantum Zeno Effect", PR **A41** (1990): 2295–300. {4}

[ITANO *et al.* 1991] ITANO, W. H./HEINZEN, D. J./BOLLINGER, J. J./WINELAND, D. J., "Reply to 'Comment on 'Quantum Zeno Effect''", PR **A43** (1991): 5168–69. {4}

[ITÔ 1951] ITÔ, K., *Memoirs American Mathematical Society* **4** (1951). {15}

[IVANOVIĆ 1981] IVANOVIĆ, I. D., "Geometrical Description of Quantal State Determination", JP **A14** (1981): 3241–45. {3, 4, 7}

[IVANOVIĆ 1983] —, "Formal State Determination", JMP **24** (1983): 1199–1205. {3, 4, 7}

[IVANOVIĆ 1987] —, "How to Differentiate between Non-Orthogonal States", PL **123A** (1987): 257–59. {3, 4}

[JACK 1995] JACK, Colin, "Sherlock Holmes Investigates the EPR Paradox", *Physics World* (apr. 1995): 39–42. {9; S}

[JACKIW 1988] JACKIW, R., "Three Elaborations on Berrys Connection, Curvature and Phase", *International Journal of Modern Physics* **A3** (1988): 285–97; rep. in [*SHAPERE/WILCZEK* 1989, 29–41]. {3}

[JADCZYK 1995a] JADCZYK, A., "Particle Tracks, Events and Quantum Theory", PTF **93** (1995): 631–46. {5, 6, 12}

[JADCZYK 1995b] —, "On Quantum Jumps, Events, and Spontaneous Localization Models", FP **25** (1995): 743–62. {3, 4, 5, 12}

[JAEGER/SHIMONY 1995] JAEGER, G./SHIMONY, A., "Optimal Distinction Between Two Non-Orthogonal Quantum States", PL **197A** (1995): 83–87. {3, 4}

[*JAMMER* 1966] JAMMER, Max, *The Conceptual Development of Quantum Mechanics*, New York, McGraw-Hill, 1966; II ed.: Thomas Pub., 1989. {3, 12, 13, 14}

[*JAMMER* 1974] —, *The Philosophy of Quantum Mechanics. The Interpretation of Quantum Mechanics in Historical Perspective*, New York, Wiley, 1974. {3, 4, 7, 8, 12, 13, 14}

[JAMMER 1985] —, "The EPR Problem in its Historical Development", in [*LAHTI/MITTELSTAEDT* 1985, 129–49]. {8, 9}

[JÁNOSSY 1962] JÁNOSSY, L., "Zum hydrodynamischen Modell der Quantenmechanik", ZP **169** (1962): 79–89. {5, 12}

[JÁNOSSY/NAGY 1956] JÁNOSSY, L./NAGY, K., "Über eine Form des Einsteinschen Paradoxes der Quantentheorie", *Annalen der Physik* **17** (1956): 115–21. {8}

[JÁNOSSY/NARAY 1958] JÁNOSSY, L./NARAY, *Nuovo Cimento Supplementa* **9** (1958): 588. {3, 9}

[JAPHA/KURIZKI 1996] JAPHA, Y./KURIZKI, G., "Superluminal Delays of Coherent Pulses in Nondissipative Media: A Universal Mechanism", PR **A53** (1996): 586–90. {3, 9}

[JARRETT 1984a] JARRETT, Jon P., "On the Physical Significance of the Locality", *Noûs* **18** (1984): 573. {8, 9}

[JARRETT 1984b] —, "An Analysis of the Locality Assumption in the Bell Argument", in [*ROTH/INOMATA* 1984, 21–28]. {8, 9}

[JAUCH 1960] JAUCH, Joseph M., "Systems of Observables in Quantum Mechanics", *Helvetica Physica Acta* **33** (1960): 711–726. {3}

[JAUCH 1964] —, "The Problem of Measurement in Quantum Mechanics", *Helvetica Physica Acta* **37** (1964): 293–316. {4}

[*JAUCH* 1968] —, *Foundations of Quantum Mechanics*, Massachussets, Addison-Wesley, 1968. {3}

[JAUCH 1972] —, "On Bras and Kets", in [*SALAM/WIGNER*, 137–67]. {3}

[JAUCH 1974] —, *Synthese* **29** (1974): 131. {3, 4}

[JAUCH 1976] —, "The Quantum Probability Calculus', in [*SUPPES* 1976, 123–46]. {3}

[JAUCH 1977] —, "The Quantum Probability Calculus", in [*LOPES/PATY* 1977, 39–62]. {3}

[JAUCH/PIRON 1963] JAUCH, J. M./PIRON, C., "Can Hidden Variables be Excluded in Quantum Mechanics?", *Helvetica Physica Acta* **36** (1963): 827–37. {3, 8, 12}

[JAUCH/PIRON 1968] —, "Hidden Variables Revisited", RMP **40** (1968): 228–29. {3, 8, 9}

[JAUCH/PIRON 1969] —, "On the Structure of Quantal Proposition Systems", *Helvetica Physica Acta* **42** (1969): 842–48; rep. in [*HOOKER* 1975, 427–36]. {3}

[JAUCH *et al.* 1967] JAUCH, J. M./WIGNER, E. P./YANASE, M. M., "Some Comments Concerning Measurements in Quantum Mechanics", NC **48B** (1967): 144–51. {4, 12}

[JAVANAINEN 1986] JAVANAINEN, J., "Possibility of Quantum Jumps in a Three-Level Atom", PR **A33** (1986): 2121–23. {5}

[JAYNES 1972] JAYNES, Edwin T., "Survey of the Present Status of Neoclassical Radiation Theory", *III Rochester Conference on Coherence and Quantum Optics*, 1972. {3}

[JAYNES 1990] —, "Probability in Quantum Theory" in [*ZUREK* 1990a, 381–403]. {3, 10, 12}

[JAYNES/CUMMINGS 1963] JAYNES, E. T./CUMMINGS, F. W., "Comparison of Quantum and Semiclassical Radiation Theories with Application to the Beam Maser", *Proceedings IEEE* **51** (1963): 89–109. {3}

[JEFFERTS *et al.* 1995] JEFFERTS, S. R./MONROE, C./BELL, E. W./WINELAND, D. J., "Coaxial-Resonator-Driven rf (Pauli) Trap for Strong Confinement", PR **A51** (1995): 3112–16. {5}

[JENSEN 1934] JENSEN, Paul, "Kausalität, Biologie und Psychologie", *Erkenntnis* **4** (1934): 165–214. {14}

[JONA-LASINIO *et al.* 1981] JONA-LASINIO, G./MARTINELLI, F./SCOPPOLA, E., "The Semiclassical Limit of Quantum Mechanics: A Qualitative Theory via Stochastic Mechanics", PRep **77** (1981): 313–28. {5}

[K. JONES 1994] JONES, K. R. W., "Fundamental Limits Upon the Measurement of State Vector", PR **A50** (1994): 3682–99. {3, 4, 7, 10, 12}

[R. JONES/ADELBERGER 1994] JONES, R. T./ADELBERGER, E. G., "Quantum Mechanics and Bell's Inequalities", PRL **72** (1994): 2675–77. {9}

[JOOS 1984] JOOS, E., "Continuous Measurement: Watchdog Effect Versus Golden Rule", PR **D29** (1984): 1626–33. {4}

[JOOS 1986a] —, "Why we do Observe a Classical Space-Time?", PL **116A** (1986): 6–8. {5, 12, 13}

[JOOS 1986b] —, "Quantum Theory and the Appearance of a Classical World", in [*GREENBERGER* 1986, 6–13]. {5, 12, 13}

[JOOS 1987] —, "Comment on 'Unified Dynamics for Microscopic and Macroscopic Systems"', PR **D36** (1987): 3285–86. {4, 5}

[JOOS 1996] —, "Decoherence Through Interaction with the Environment", in [*GIULINI et al.* 1996, 35–136]. {4, 5, 12; W}

[JOOS/ZEH 1985] JOOS, E./ZEH, H. D., "The Emergence of Classical Properties Through Interaction with the Environment", ZP **B59** (1985): 223–43. {4, 5, 12; W}

[P. JORDAN 1926] JORDAN, Pascual, "Über kanonische Transformationen in der Quantenmechanik", ZP **37** (1926): 383–86. {3}

[P. JORDAN 1927] —, "Über eine neue Begrndung der Quantenmechanik", ZP **40** (1927): 809–38. {3, 12}

[P. JORDAN 1932] —, "Über eine Klasse nichtassoziativer hyperkomplexer Algebren", *Göttinger Nachrichten* (1932): 569–75. {3, 15}

[P. JORDAN 1934] —, "Quantenphysikalische Bemerkungen zur Biologie and Psychologie", *Erkenntnis* **4** (1934): 215–52. {12, 14}

[P. JORDAN 1935a] —, "Versuch, den Vitalismus quanten mechanisch zu retten", *Erkenntnis* **5** (1935): 56–64. {4, 12, 13, 14}

[P. JORDAN 1935b] —, "Ergänzende Bemerkungen über Biologie und Quantenmechanik", *Erkenntnis* **5** (1935): 348–52. {4, 12, 13, 14}

[P. JORDAN 1949a] —, "On the Process of Measurement in Quantum Mechanics", PS **16** (1949): 269–78. {4}

[P. JORDAN 1949b] —, "Zur Quantum Logik", *Archiv der Mathematik* **2** (1949): 166–71. {3}

[P. JORDAN 1952] —, "Algebraische Betrachtungen zur Theorie des Wirkungsquantum und der Elementarlänge", *Abhandlungen aus dem Mathematischen Seminar der Universitt Hamburg* **18** (1952): 99–119. {3}

[P. JORDAN 1959] —, "Quantenlogik und das kommutative Gesetz", in HENKIN/SUPPES/TARSKI (ed.s), *The Axiomatic Method*, Amsterdam, North-Holland, 1959. {3}

[P. JORDAN/KLEIN 1927] JORDAN, P./KLEIN, O., "Zum Mehrkörperproblem der Quantentheorie", ZP **45** (1927): 751–56. {3}

[P. JORDAN/WIGNER 1928] JORDAN, P./WIGNER, E. P., "Über das Paulische Äquivalenzverbot", ZP **47** (1928): 631–51. {3}

[P. JORDAN *et al.* 1934] JORDAN, P./VON NEUMANN, J./WIGNER, E. P., "On a Algebraic Generalization of the Quantum Mechanical Formalism", *Annals of Mathematics* **35** (1934): 29–64. {3}

[T. JORDAN 1983] JORDAN, Thomas F., "Quantum Correlations Do Not Transmit Signals", PL **94A** (1983): 264. {9, 10, 12}

[T. JORDAN 1993] —, "Disappearance and Reapparance of Macroscopic Quantum Interference", PR **A48** (1993): 2449–50. {5}

[T. JORDAN 1994] —, "Testing EPR Assumptions without Inequalities with Two Photons or Particles with Spin $\frac{2}{2}$", PR **A50** (1994): 62–66. {9}

[JOSEPHSON 1962] JOSEPHSON, Brian D., "Possible New Effects in Superconductive Tunnelling", PL **1** (1962): 251–53. {3}

[JOSEPHSON 1988] —, "Limits to the Universality of Quantum Mechanics", FP **18** (1988): 1195–204. {12, 13, 14}

[JOSEPHSON 1991] JOSEPHSON, B. D./PALLIKARI-VIRAS, F., "Biological Utilization of Quantum Nonlocality", FP **21** (1991): 197–207. {9, 12, 13, 14}

[JOST 1972] JOST, Res, "Foundation of Quantum Field Theory", in [*SALAM/WIGNER*, 61–77]. {3, 11}

[JOZSA 1994] JOZSA, Richard, "Fidelity for Mixed Quantum Systems", *Journal of Modern Optics* **41** (1994): 2315–23. {3, 10}

[JOZSA/B. SCHUMACHER 1994] JOZSA, R./SCHUMACHER, B., "A New Proof of the Quantum Noiseless Coding Theorem", *Journal of Modern Optics* **41** (1994): 2343–49. {10}

[JOZSA *et al.* 1994] JOZSA, R./ROBB, D./WOOTTERS, W. K., "Lower Bound for Accessible Information in Quantum Mechanics", PR **A49** (1994): 668–77. {10}

[JUDGE 1963] JUDGE, D., "On the Uncertainty Relation for L_z and ϕ", PL **5** (1963): 189. {3}

[JUDGE 1970] —, "On the Uncertainty Relation for Angle Variables", NC **31** (1970): 332–40. {3}

[JUDGE/J. LEWIS 1963] JUDGE, D./LEWIS, J. T., "On the Commutator $[L_z, f]$", PL **5** (1963): 190. {3}

[*KAFATOS* 1989] KAFATOS, Menas (ed.), *Bell's Theorem, Quantum Theory, and Conceptions of the Universe*, Dordrecht, Kluwer, 1989. {9}

[KAILA 1950] KAILA, Eino, "Zur Metatheorie der Quantenmechanik", *Acta Philosophica Fennica* **5** (1950): 1–98. {12, 14}

[KAK 1996] KAK, Subhash C., "Information, Physics, and Computation", FP **26** (1996): 127–37. {10, 13}

[KAKAZU 1991] KAKAZU, Kiyotaka, "Equivalence Classes and Generalized Coherent States in Quantum Measurement", NC **B106** (1991): 1173–85. {3}

[KAKAZU/MATSUMOTO 1990] KAKAZU, K./MATSUMOTO, S., "Stability of Particle Trajectories and Generalized Coherent States", PR **A42** (1990): 5093–102. {3}

[*KANT* 1787] KANT, Immanuel, *Kritik der reinen Vernunft*, II ed. 1787, Akademieausgabe, Berlin, W. de Gruyter, 1968. {14}

[KAR/ROY 1995] KAR, G./ROY, S., "Unsharp Spin–$\frac{1}{2}$ Observable and CHSH Inequalities", PL **199A** (1995): 12–14. {4, 9}

[KASDAY *et al.* 1970] KASDAY, L. R./ULLMAN, J./WU, C.S., "The EPR Argument: Positron Annihilation Experiment", *Bulletin American Physical Society* **15** (1970): 586. {9}

[KELLEY/KLEINER 1964] KELLEY, P. L./KLEINER, W. H., "Theory of Electromagnetic Field Measurement and Photoelectron Counting", PR **A136** (1964): 316–34. {3}

[KELLER/MAHLER 1996] KELLER, M./MAHLER, G., "Stochastic Dynamics and Quantum Measurement", *Quantum and Semiclassical Optics* **8** (1996): 223–35. {4, 5}

[KEMBLE 1935] KEMBLE, Edwin C., "The Correlation of Wave Functions with the States of Physical Systems", PR **47** (1935): 973–74. {3}

[KEMBLE 1951] —, "Reality, Measurements, and the State of the System in Quantum Mechanics", PS **18** (1951): 273–99. {4, 7}

[KENNARD 1927] KENNARD, E. H., "Zur Quantenmechanik einfacher Bewegungstypen", ZP **44** (1927): 326–52. {3}

[KENNEDY/WALLS 1988] KENNEDY, T. A. B./WALLS, D. F., "Squeezed Quantum Fluctuations and Macroscopic Quantum Coherence", PR **A37** (1988): 152–57. {5}

[KERNAGHAN 1994] KERNAGHAN, Michael, "Bell-Kochen-Specker Theorem for 20 Vectors", JP **A27** (1994): L829–30. {8, 9}

[KERSHAW 1964] KERSHAW, David, "Theory of Hidden Variables", PR **136B** (1964): 1850–56. {8}

[KETZMERICK *et al.* 1997] KETZMERICK, R./KRUSE, K., KRAUT, S./GEISEL, T., "What Determines the Spreading of a Wave Packet?", PRL **79** (1997): 1959–63. {3}

[KHALFIN 1982] KHALFIN, L. A., "Proton Nonstability and the Nonexponentiality of the Decay Law", PL **112B** (1982): 223–25. {3}

[KHALFIN 1990a] —, "The Quantum-Classical Correspondence in Light of Classical Bell's and Quantum Tsirelson's Inequalities", in [*ZUREK* 1990a, 477–93]. {3, 8, 9, 12}

[KHALFIN 1990b] —, *Usp. Fiz. Nauk.* **160** (1990): 185–88; "Zeno's Quantum Effect" (engl. tr.), *Soviet Physics-Uspekhi* **33** (1991): 868–69. {4}

[KHALFIN/TSIRELSON 1985] KHALFIN, L. A./TSIRELSON, B. S., "Quantum and Quasi-Classical Analogs of Bell Inequalities", in [*LAHTI/MITTELSTAEDT* 1985, 441–60]. {5, 9, 12}

[KHALFIN/TSIRELSON 1987] —, "A Quantitative Criterion of Applicability of the Classical Description Within the Quantum Theory", in [*LAHTI/MITTELSTAEDT* 1987, 369–401]. {5, 9, 12}

[KHALFIN/TSIRELSON 1992] —, "Quantum/Classical Correspondence in the Light of Bell's Inequalities", FP **22** (1992): 879–948. {3, 5, 8, 9, 12}

[*KHINCHIN* 1957] KHINCHIN, A. I., *Mathematical Foundations of Information Theory* (engl. tr.), New York, Dover, 1957. {1}

[KIEFER 1992] KIEFER, Claus, "Decoherence in Quantum Electrodynamics and Quantum Gravity", PR **D46** (1992): 1658–70. {5, 6}

[KIEFER 1996a] —, "Decoherence in Quantum Field Theory", in [*GIULINI et al.* 1996, 137–56]. {5, 6, 11}

[KIEFER 1996b] —, "Consistent Histories and Decoherence", in [*GIULINI et al.* 1996, 157–86]. {5, 6, 12}

[KIEFER 1998] —, "Wigner Function and Decoherence", e-print; code: quant-ph/9711012. {3, 4, 5}

[KIEFER/ZEH 1995] KIEFER, C./ZEH, H. D., "Arrow of Time in a Recollapsing Quantum Universe", PR **D51** (1995): 4145–53. {5, 6, 12, 13}

[KIESS *et al.* 1993] KIESS, T. E./SHIH, Y. H./SERGIENKO, A. V./ALLEY, C. O., "Einstein-Podolsky-Rosen-Bohm Experiment Using Pairs of Light Quanta Produced by Type-II Parametric Down-Conversion", PRL **71** (1993): 3893–97. {9}

[KIJOWSKI 1974] KIJOWSKI, J., *Reports in Mathematical Physics* **6** (1974): 361. {3}

[*KILMISTER* 1987] KILMISTER, C. W. (ed.), *Schrödinger. Centenary Celebration of a Polymath*, Cambridge, University Press, 1987. {3, 4, 8, 9, 12}

[KIM/WIGNER 1987] KIM, Y. S./WIGNER, E. P., "Covariant Phase-Space Representation for Localized Light Waves", PR **A36** (1987): 1293–97. {3}

[KIMBLE/MANDEL 1976] KIMBLE, H. J./MANDEL, L., "Theory of Resonance Fluorescence", PR **A13** (1976): 2123–44. {5}

[KIMBLE *et al.* 1977] KIMBLE, H. J./DAGENAIS, M./MANDEL, L., "Photon Antibunching in Resonance Fluorescence", PRL **39** (1977): 691–95. {5}

[KIMBLE *et al.* 1986] KIMBLE, H. J./COOK, R. J./WELLS, A. L., "Intermittent Atomic Fluorescence", PR **A34** (1986): 3190–195. {5}

[A. KLEIN 1981] KLEIN, Abel G., "Stochastic Processes Associated With Quantum Systems", PRep **77** (1981): 329–37. {3, 5}

[A. KLEIN 1995] —, "Experimental Tests of the Foundations of Quantum Mechanics Using Neutrons: The Scalar A-B Effect", in [*GREENBERGER/ZEILINGER* 1995a, 288–92]. {9}

[A. KLEIN/WERNER 1983] KLEIN, A. G./WERNER, S. A., *Rep Prog Phys* **46** (1983): 259–335. {7}

[O. KLEIN 1926] KLEIN, Oscar, "Quantentheorie und fünfdimensionale Relativitätstheorie", ZP **37** (1926): 895–906. {3}

[O. KLEIN 1927] —, "Elektrodynamik und Wellenmechanik vom Standpunkt des Korrespondenzprinzips", ZP **41** (1927): 407–442. {3, 12}

[KLYSHKO 1995] KLYSHKO, D. N., "Quantum Optics: Quantum, Classical, and Metaphysical Aspects", in [*GREENBERGER/ZEILINGER* 1995a, 13–27]. {3, 12, 14}

[KLYSHKO 1996] —, "The Bell Theorem and the Problem of Moments", PL **218A** (1996): 119–27. {9}

[KNIGHT 1992] KNIGHT, Peter L., "Pratical Schrödinger's Cats", NA **357** (1992): 438–39. {5}

[KNILL/LAFLAMME 1997] KNILL, E./LAFLAMME, R., "Theory of Quantum Error-Correcting Codes", PR **A55** (1997): 900–911. {10}

[*KOBYASHI* 1989] KOBYASHI, (ed.), *Proceedings 3rd International Symposium Foundations of Quantum Mechanics*, Tokyo, Physical Society of Japan, 1989. {3}

[KOCHEN 1985] KOCHEN, Simon, "A New Interpretation of Quantum Mechanics", in [*LAHTI/MITTELSTAEDT* 1985, 151–69]. {12}

[KOCHEN/SPECKER 1965] KOCHEN, S./SPECKER, E., "Logical Structures Arising in Quantum Theory", in ADDISON/HENKIN/TARSKY (ed.s), *Theory of Models*, Amsterdam, North-Holland, 1965; rep. in [*HOOKER* 1975, 263–76]. {3}

[KOCHEN/SPECKER 1967] —, "The Problem of Hidden Variables in Quantum Mechanics", *Journal of Mathematics and Mechanics* **17** (1967): 59–87; rep. in [*HOOKER* 1975, 293–328]. {3, 8, 9, 14}

[KOH 1992] KOH, Yujiro, "Are Two-Beam Self-Interferences Mass-Independent?", in [*SELLERI* 1992a, 139–56]. {7}

[*KOLMOGOROV* 1956] KOLMOGOROV, A. N., *Foundations of the Theory of Probability*, New York, Chelsea, 1956. {1}

[*KOLMOGOROV/FOMIN* 1980] KOLMOGOROV, A. N./FOMIN, S. V., *Elementi di teoria delle funzioni e di analisi funzionale*, (it. tr.), Moscow, Mir, 1980. {1}

[KOMAR 1964] KOMAR, A., "Undecidability of Macroscopically Distinguishable States in Quantum Field Theory", PR **B133** (1964): 542–44. {11}

[KOMAR 1973] —, "The General Relativistic Quantization Program", in [*HOOKER* 1973a, 305–27]. {3, 11}

[KONO *et al.* 1996] KONO, N./MACHIDA, K./NAMIKI, M./PASCAZIO, S., "Decoherence and Dephasing in a Quantum Measurement Process", PR **A54** (1996): 1064–86. {4, 5}

[KORN 1927] KORN, Arthur, "Schrödingers Wellenmechanik und meine mechanische Theorien", ZP **44** (1927): 745–53. {12}

[*KÖRNER/PRYCE* 1957] KÖRNER, S./PRYCE, M. H. L. (ed.s), *Observation and Interpretation in the Philosophy of Physics – With Special Reference to Quantum Mechanics*, London, Constable and Company, 1957. {4, 12, 13}

[KOSSAKOVSKI 1972] KOSSAKOVSKI, Andrzej, *Reports on Mathematical Physics* **3** (1972): 247. {5}

[KOUTZETSOV 1965] KOUTZETSOV, Boris, "Einstein and Bohr",*Organon* **2** (1965): 105–121. {12, 14}

[*KOYRÉ* 1966] KOYRÉ, Alexandre, *Etudes galiléennes*, Paris, Hermann, 1966, 1980. {13, 14}

[KRÄMER *et al.* 1995] KRÄMER, D. S./VOGEL, K./AKULIN, V. M./SCHLEICH, W. P., "Quantum Interference, State Engineering, and Quantum Eraser", in [*GREENBERGER/ZEILINGER* 1995a, 545–59]. {7}

[KRAUS 1977a] KRAUS, Karl, "General State Changes in Quantum Theory", AP **64** (1971): 311–35. {3, 4, 5, 12}

[KRAUS 1977b] —, "Position Observables of the Photon", in [*W. PRICE/CHISSICK* 1977, 293–320]. {3, 8}

[KRAUS 1981] —, "Measuring Processes in Quantum Mechanics. I. Continuous Observation and the Wtchdog Effect", FP **11** (1981): 547–76. {3, 4}

[*KRAUS* 1983] —, *States, Effects and Operations*, Berlin, Springer, 1983. {3, 4, 5, 10, 12; W}

[KRAUS 1985] —, "Quantum Theory, Causality, and EPR Experiments", in [*LAHTI/MITTELSTAEDT* 1985, 461–80]. {8, 9}

[KRAUS 1987] —, "Complementary Observables and Uncertainty Relations", PR **D35** (1987): 3070–3075. {3, 10, 12}

[KRENN/ZEILINGER 1995] KRENN, G./ZEILINGER, A, "Entangled Entanglement", in [*GREENBERGER/ZEILINGER* 1995a, 873–76]. {3, 9}

[KRENN/ZEILINGER 1996] —, "Entangled Entanglement", PR **A54** (1996): 1793–97. {3, 9}

[KRENN/ZEILINGER 1997] —, "Reply to 'Comment on 'Entangled Entanglement' ", PR **A56** (1997): 4336. {3, 9}

[KRENN *et al.* 1996] KRENN, G./SUMMHAMMER, J./SVOZIL, K., "Interaction-Free Preparation", PR **A53** (1996): 1228–31. {4}

[KRIPS 1969] KRIPS, H., "Two Paradoxes in Quantum Mechanics", PS **36** (1969): 145–52. {4, 8, 9, 12}

[KRIPS 1974] —, "Foundations of Quantum Theory", FP **4** (1974): 181–93. {3}

[KRIPS 1977] —, "Quantum Theory and Measures on Hilbert Space", JMP **18** (1977): 1015–21. {3, 8}

[*KRIPS* 1987] —, *The Metaphysics of Quantum Theory*, Oxford, Clarendon, 1987. {12}

[KRIPS 1989] —, "Conditionals, Probability and Bell's Theorem", in [*KAFATOS* 1989, 105–117]. {3, 9}

[KRIPS 1994] —, "A Critique of Bohr's Local Realism", in [*FAYE/FOLSE* 1994, 269–78]. {8, 9, 12}

[KRONFLI 1971] KRONFLI, N. S., "Atomicity and Determinism in Boolean Systems", *International Journal of Theoretical Physics* **4** (1971): 141–43; rep. in [*HOOKER* 1975, 509–12]. {3}

[KRUSZYŃSKI/DE MUYNCK 1987] KRUSZYŃSKI, P./DE MUYNCK, W. M., "Compatibility of Observables Represented by Positive Operator-Valued Measures", JMP **28** (1987): 1761–63. {4}

[KRYLOV/FOCK 1947] KRYLOV, N. S./FOCK, V. A., "On the Uncertainty Relation Between Time and Energy", *Journal of Physics (USSR)* **11** (1947): 112–20. {3}

[KÜBLER/ZEH 1973] KÜBLER, O./ZEH, H. D., "Dynamics of Quantum Correlations", AP **76** (1973): 405–418. {4, 5}

[KÜHN *et al.* 1994] KÜHN, H./WELSCH, D.-G./VOGEL, W., "Determination of Density Matrices from Field Distributions and Quasiprobabilities", *Journal of Modern Optics* **41** (1994): 1607–613. {3, 7}

[KUDAKA/KAKAZU 1992] KUDAKA, S./KAKAZU, K., "The Wigner-Araki-Yanase Theorem and its Extension in Quantum Measurement with Generalized Coherent States", PTF **87** (1992): 61–76. {4}

[*KUHN* 1957] KUHN, Thomas S., *The Copernican Revolution. Planetary Astronomy in the Development of Western Thought*, Cambridge Mass., Harvard U. P., 1957. {14}

[*KUHN* 1962] —, *The Structure of Scientific Revolutions*, Chicago, University Press, 1962, 1970. {14}

[*KUHN* 1978] —, *Black-Body Theory and the Quantum Discontinuity. 1894–1912*, Oxford, Clarendon, 1978. {3, 12, 13}

[KÜMMEL 1955a] KÜMMEL, Hermann, "Zur quantentheoretischen Begründung der klassischen Physik. I — Dynamik der Gase und Flüssigkeiten", NC **1** (1955): 1057–1077. {5}

[KÜMMEL 1955b] "Zur quantentheoretischen Begründung der klassischen Physik. II — Statistische Mechanik und Thermodynamik", NC **2** (1955): 877–97

[KUPSCH 1996] KUPSCH, J., "Open Quantum Systems", in [*GIULINI et al.* 1996, 223–47]. {4, 5}

[KURYSHKIN 1977] KURYSHKIN, Vassili V., "Uncertainty Principle and the Problems of Joint Coordinate-Momentum Probability Density in Quantum Mechanics", in [*W. PRICE/CHISSICK* 1977, 61–83]. {3}

[KWIAT/CHIAO 1991] KWIAT, P. G./CHIAO, R. Y., "Observation of a Nonclassical Berry's Phase for the Photon", PRL **66** (1991): 588–91. {3}

[KWIAT *et al.* 1992] KWIAT, P. G./STEINBERG, A. M./CHIAO, R. Y., "Observation of a 'Quantum Eraser': A Revival of Coherence in a Two-Photon Interference Experiment" PR **A45** (1992): 7729–39. {4, 6, 12}

[KWIAT *et al.* 1993] —, "High-Visibility Interference in a Bell-Inequality Experiment for Energy and Time", PR **A47** (1993): R2472–75. {9}

[KWIAT *et al.* 1994a] KWIAT, P. G./EBERHARD, P. H./STEINBERG, A. M./CHIAO, R. Y., "Proposal for a Loophole-Free Bell Inequality Experiment", PR **A49** (1994): 3209–20. {9}

[KWIAT *et al.* 1994b] KWIAT, P. G./STEINBERG, A. M./CHIAO, R. Y., "Three Proposed 'Quantum Erasers"', PR **A49** (1994): 61–68. {4, 6}

[KWIAT *et al.* 1995a] KWIAT, P./WEINFURTER, H./HERZOG, T./ZEILINGER, A./KASEVICH, M., "Interaction-Free Measurement", PRL **74** (1995): 4763–66. {4, 10}

[KWIAT *et al.* 1995b] —, "Experimental Realization of Interaction-Free Measurements", in [*GREENBERGER/ZEILINGER* 1995a, 383–93]. {4, 10}

[KWIAT *et al.* 1995c] KWIAT, P./MATTLE, K./WEINFURTER, H./ZEILINGER, A./SERGIENKO, A. V./SHIH, Y., "New High-Intensity Source of Polarization-Entangled Photon Pairs", PRL **75** (1995): 4337–41. {9}

[KWIAT *et al.* 1996] KWIAT, P./WEINFURTER, H./ZEILINGER, A., *Scientific American* (nov. 1996). {4}

[KYPRIANIDIS *et al.* 1984] KYPRIANIDIS, A./SARDELIS, D./VIGIER, J.-P., "Causal Non-Local Character of Quantum Statistics", PL **100A** (1984): 228–30. {8, 9, 12}

[LAFLAMME *et al.* 1996] LAFLAMME, R./MIQUEL, C./PAZ, J. P./ZUREK, W. H., "Perfect Quantum Error Correcting Code", PRL **77** (1996): 198–201. {10}

[LAHTI 1994] LAHTI, Pekka J., "Limitations of Measurability in Quantum Mechanics", [*ACCARDI* 1994a, 233–42]. {3, 4}

[LAHTI/MACZYŃSKI 1987] LAHTI, P. J./MACZYŃSKI, M. J., "Heisenberg Inequality and the Complex Field in Quantum Mechanics", JMP **28** (1987): 1764–69. {3}

[*LAHTI/MITTELSTAEDT* 1985] LAHTI, P./MITTELSTAEDT, P. (ed.s), *Symposium on the Foundations of Modern Physics 1985*, Singapore, World Scientific, 1985. {3–14}

[*LAHTI/MITTELSTAEDT* 1987] — (ed.s), *Symposium on the Foundations of Modern Physics 1987*, Singapore, Worl Scientific, 1987. {3–14}

[*LAHTI/MITTELSTAEDT* 1989] — (ed.s), Symposium on the Foundations of Modern Physics 1989, Singapore, Worl Scientific, 1989. {3–14}

[LAHTI/YLINEN 1987] LAHTI, P./YLINEN, K., "On Total Noncommutativity in Quantum Mechanics", JMP **28** (1987): 2614–17. {3, 4}

[LALOVIĆ *et al.* 1992] LALOVIĆ, D./DAVIDOVIĆ, D. M./BIJEDIĆ, N., "Quantum Mechanics in Terms of Non-Negative Smoothed Wigner Functions", PR **A46** (1992): 1206–211. {3}

[LAMEHI-R./MITTIG 1976] LAMEHI-RACHTI, M./MITTIG, W., "QM and Hidden Variables: A Test of Bell's Inequality", PR **D14** (1976): 2543–55; rep. in [*WHEELER/ZUREK* 1983, 422–34]. {9}

[LAMB 1969] LAMB, Willis E. jr, "Nonrelativistic Quantum Mechanics", PT **22** (apr. 1969): 23–28. {4}

[LAMB 1987] —, "Sequential Measurements in Quantum Mechanics", in [*PIKE/SARKAR* 1987, 183–93]. {4}

[H. LANDAU/POLLAK 1961] LANDAU, H. J./POLLAK, H. O., *Bell System Technical Journal* (1961): 65–84. {3, 12}

[Lw. LANDAU 1987a] LANDAU, Lawrence J., "On the Violation of Bell's Inequality in Quantum Theory", PL **120A** (1987): 54–56. {3, 9}

[Lw. LANDAU 1987b] —, "Experimental Tests of General Quantum Theories", *Letters on Mathematical Physics* **14** (1987): 33–40. {3}

[Lw. LANDAU 1987c] —, "On the Non-Classical Structure of the Vacuum", PL **123A** (1987): 115–18. {9, 11}

[Lw. LANDAU 1988] —, "Empirical Two-Point Correlation Functions", FP **18** (1988): 449–60. {3, 9}

[Lev LANDAU 1927] LANDAU, Lev D, "Das Dämpfungsproblem in der Wellenmechanik", ZP **45** (1927): 430–41. {3}

[*Lev LANDAU/LIFSTITS* 1976a] LANDAU, Lev D./LIFSTITS, E., *Meccanica* (it. tr.), Roma, Editori Riuniti, 1976, 1991. {1}

[*Lev LANDAU/LIFSTITS* 1976b] —, *Meccanica Quantistica. Teoria non relativistica* (it. tr.), Roma, Editori Riuniti, 1976, 1994. {2}

[*Lev LANDAU/LIFSTITS* 1978] —, *Teoria quantistica relativistica* (it. tr.), Roma, Editori Riuniti, 1978. {2}

[*Lev LANDAU/LIFSTITS* 1994] —, *Teoria dei campi* (it. tr.), Roma, Editori Riuniti, 1994. {1, 11}

[Lev LANDAU/PEIERLS 1931] LANDAU, L. D./PEIERLS, R., "Erweiterung des Unbestimmtheitsprinzips für die relativistische Quantentheorie", ZP **69** (1931): 56–69; engl. tr. in [*WHEELER/ZUREK* 1983, 465–76]. {3}

[LANDAUER 1993] LANDAUER, Rolf, "Statistical Physics of Manichery: Forgotten Middle-Ground", *Physica* **A194** (1993): 551–62. {10, 13}

[LANDAUER 1996] —, "The Physical Nature of Information", PL **217A** (1996): 188–93. {10}

[LANDAUER/MARTIN 1994] LANDAUER, R./MARTIN, T., "Barrier Interaction Time in Tunneling", RMP **66** (1994): 217–28. {3, 9; W}

[*LANDÉ* 1951] LANDÉ, Alfred, *Quantum Mechanics*, London, Pitman, 1951. {2}

[LANDÉ 1952] —, "Quantum Mechanics and Thermodynamic Continuity", AJP **20** (1952): 353–58. {3, 8, 12}

[LANDÉ 1953a] —, "Quantum Mechanics – A Thermodynamic Approach", *American Scientist* **41** (1953): 439–48. {3, 8, 12}

[LANDÉ 1953b] —, "Continuity, a Key to Quantum Mechanics", PS **20** (1953): 101–109. {3, 8, 12}

[LANDÉ 1954] —, "Quantum Mechanics and Thermodynamic Continuity II", AJP **22** (1954): 82–87. {3, 8, 12}

[*LANDÉ* 1955] —, *Foundations of Quantum Theory. A Study in Continuity and Simmetry*, New Haven, Yale U. P., 1955. {3, 8, 12}

[LANDÉ 1957] —, "Non-Quantal Foundations of Quantum Theory", PS **24** (1957): 309–320. {3, 8, 12}

[*LANDÉ* 1960] —, *From Dualism to Unity in Quantum Theory*, London, Cambridge University Press, 1960. {3, 8, 12}

[LANDÉ 1961a] —, "Unitary Interpretation of Quantum Theory", AJP **29** (1961): 503–507. {3, 8, 12}

[LANDÉ 1961b] —, "Ableitung der Quantenregeln auf nicht-quantenmässiger Grundlage", ZP **162** (1961): 410–412. {3, 8, 12}

[LANDÉ 1961c] —, "Warum interferieren die Wahrscheinlichkeiten?", ZP **164** (1961): 558–62. {3, 12}

[LANDÉ 1961d] —, "Dualism, Wissenschaft und Hypothese", in [*BOPP* 1961a, 119–27]. {12}

[LANDÉ 1962] —, "Answer to Mr. Stopes-Roe", NA **193** (1962): 1277. {12}

[*LANDÉ* 1965a] —, *New Foundations of Quantum Mechanics*, Cambridge, University Press, 1965. {3, 8, 12}

[LANDÉ 1965b] —, "Quantum Fact and Fiction. I", AJP **33** (1965): 123–27. {3, 12}

[LANDÉ 1966] —, "Quantum Fact and Fiction. II", AJP **34** (1966): 1160–63. {3, 12}

[LANDÉ 1969] —, "Quantum Fact and Fiction. III", AJP **37** (1969): 541–48. {3, 12}

[*LANDÉ* 1973] —, *Quantum Mechanics in a New Key*, New York, Exposition Press, 1973. {3, 8, 12}

[LANDÉ 1975] —, "Quantum Fact and Fiction. IV", AJP **43** (1975): 701–704. {3, 12}

[LANDSBERG 1988] LANDSBERG, P. T., "Why Quantum Mechancis?", FP **18** (1988): 969–82. {3}

[LANDSBERG 1990] —, "Uncertainty and Measurement", in [*CINI/LÉVY-L.* 1990, 161–67]. {3, 4, 10}

[LANDSBERG/HOME 1987] LANDSBERG, P. T./HOME, D., "An Analysis of Wavefunction Collapse Using the Ensemble Interpretation", AJP **55** (1987): 226–30. {4, 12}

[LANFORD/RUELLE 1969] LANFORD, O. E./RUELLE, D., "Observables at Infinity and States with Short Range Correlations in Statistical Mechanics", CMP **13** (1969): 194–215. {3, 4, 5}

[*LANGEVIN* 1934] LANGEVIN, Paul, *La Notion de Corpuscule et d'Atome*, Paris, Hermann, 1934. {12}

[LANZ 1977] LANZ, Ludovico, "The Problem of Measurement in Quantum Mechanics", in [*W. PRICE/CHISSICK* 1977, 87–108]. {4, 5}

[LA PORTA *et al.* 1989] LA PORTA, A./SLUSHER, R. E./YURKE, B., "Back-Action Evading Measurements of an Optical Field Using Parametric Down Conversion", PRL **62** (1989): 28–31. {4, 5}

[LARCHUK *et al.* 1993] LARCHUK, T. S./CAMPOS, R. A./RARITY, J. G./TAPSTER, P. R./JAKEMAN, E./SALEH, B. E. A./TEICH, M. C., "Interfering Entangled Photons of Different Colors", PRL **70** (1993): 1603–606. {9}

[LARSEN 1990] LARSEN, Ulf, "Superspace Geometry: The Exact Uncertainty Relationship Between Complementary Aspects", JP **A23** (1990): 1041–1061. {3, 12}

[*LAUDAN* 1977] LAUDAN, Larry, *Progress and Its Problems*, Berkeley, University of California Press, 1977. {14}

[LAURIKAINEN 1985] LAURIKAINEN, K. V., "Wolfgang Pauli and the Copenhagen Philosophy", in [*LAHTI/MITTELSTAEDT* 1985, 273–87]. {12, 13, 14}

[*LAURIKAINEN* 1988] —, *Beyond the Atom. The Philosophical Thought of Wolfgang Pauli*, Heidelberg, Springer, 1988. {12, 13, 14}

[LEAVENS 1993] LEAVENS, C. R., "Arrival Time Distributions", PL **178A** (1993): 27–32. {3, 9}

[LEAVENS 1995] —, "Bohm Trajectory and Feynman Path Approaches to the 'Tunneling Time Problem"', FP **25** (1995): 229–68. {3, 8, 9, 12; W}

[LEGGETT 1980] LEGGETT, A. J., "Macroscopic Quantum Systems and the Quantum Theory of Measurement", *Progress of Theoretical Physics Supplementa* **69** (1980): 80–100. {4, 5}

[LEGGETT 1986] —, "Quantum Mechanics and Realism at the Macroscopic Level: Is an Experimental Discrimination Feasible?", in [*GREENBERGER* 1986, 21–24]. {5, 12}

[LEGGETT 1987] —, "Reflections on Quantum Measurement Paradox", in [*HILEY/PEAT* 1987a, 85–104]. {4, 12}

[LEGGETT 1995] —, "Is 'Relative Quantum Phase' Transitive?", FP **25** (1995): 113–22. {3}

[LEGGETT/GARG 1985] LEGGETT, A. J./GARG, A, "Quantum Mechanics Versus Macroscopic Realism: Is the Flux There when Nobody Looks?, PRL **54** (1985): 857–60. {5}

[LEGGETT *et al.* 1987] LEGGETT, A. J./CHAKRAVARTY, S./DORSEY, A. T./FISHER, M. P. A./GARG, A./ZWERGER, W., "Dynamics of the Dissipative Two-State System", RMP **59** (1987): 1–85. {4, 5; W}

[LEIBFRIED *et al.* 1996] LEIBFRIED, D./MEEKHOF, D. M./KING, B. E./MONROE, C./ITANO, W. M./WINELAND, D. J., "Experimental Determination of the Motional Quantum State of a Trapped Ion", PRL **77** (1996): 4281–85. {5, 7}

[*LEIBNIZ* MS] LEIBNIZ, Gottfried W., *Mathematische Scriften* (ed. Gerhardt), Halle, 1860; rep. Hildesheim, Olms, 1971. {13, 15}

[*LEIBNIZ* PS] —, *Philosophische Scriften* (ed. Gerhardt), Halle, 1875; rep. Hildesheim, Olms, 1978. {14}

[*LEIBNIZ* 1686] —, *Discours de Metaphysique*, in [*LEIBNIZ* PS, IV, 427–63]. {14}

[LEICHTLE *et al.* 1996] LEICHTLE, C./AVERBUKH, I. S./SCHLEICH, W. P., "Generic Structure of Multilevel Quantum Beats", PRL **77** (1996): 3999–4002. {5, 6}

[LEITER 1969] LEITER, D., "Classical Elementary Measurement Electrodynamics", AP **51** (1969): 561–75. {4}

[LENZEN 1945] LENZEN, Victor F., "The Concept of Reality in Physical Theory", *Proceedings and Adresses of the American Philosophical Association* **18** (1945): 321–44. {8, 12, 14}

[LEONHARDT 1995] LEONHARDT, Ulf, "Quantum-State Tomography and Discrete Wigner Function", PRL **74** (1995): 4101–105. {3, 4, 7}

[LEONHARDT/RAYMER 1996] LEONHARDT, U./RAYMER, M. G., "Observation of Moving Wave Packets Reveals their Quantum State", PRL **76** (1996): 1985–89. {7}

[LEONHARDT/SCHNEIDER 1997] LEONHARDT, U./SCHNEIDER, S., "State Reconstruction in One-Dimensional Quantum Mechanics: The Continuous Spectrum", PR **A56** (1997): 2549–56. {7}

[LEONHARDT *et al.* 1996] LEONHARDT, U./MUNROE, M./KISS, T./RICHTER, T./RAYMER, M. G., "Sampling of Photon Statistics and Density Matrix Using Homodyne Detection", *Optics Communications* **127** (1996): 144–60. {3, 4, 7}

[LEUBNER/KIENER 1985] LEUBNER, C./KIENER, C., "Improvement of the Eberly-Singh Time-Energy Inequality by Combination with the Mandelstamm-Tamm Approach", PR **A31** (1985): 483–85. {3}

[LEVENSON *et al.* 1986] LEVENSON, M. D./SHELBY, R. M./REID, M./WALLS, D. F., "Quantum Nondemolition Detection of Optical Quadrature Amplitudes", PRL **57** (1986): 2473–76. {4}

[LÉVY-LEBLOND 1963] LÉVY-LEBLOND, Jean-Marc, "Galilei Group and Nonrelativistic Quantum Mechanics", JMP **4** (1963): 776–88. {3}

[LÉVY-LEBLOND 1974] —, NC **4** (1974): 99. {7}

[LÉVY-LEBLOND 1976] —, "Who is Afraid of Nonhermitian Operators? A Quantum Description of Angle and Phase", AP **101** (1976): 319–41. {3, 4}

[LÉVY-LEBLOND 1986] —, "Correlation of Quantum Properties and the Generalized Heisenberg Inequality" AJP **54** (1986): 135–36. {3}

[LEWENSTEIN/SANPERA 1998] LEWENSTEIN, M./SANPERA, A., "Separability and Entanglement of Composite Quantum Systems", PRL **80** (1998): 2261–64. {8, 9, 10}

[*D. LEWIS* 1973] LEWIS, David, *Counterfactuals*, Oxford, Blackwell, 1973. {14}

[D. LEWIS 1976] —, "The Paradoxes of Time Travel", *American Philosophical Quarterly* **13** (1976): 145–52. {6, 14}

[*D. LEWIS* 1986] —, *On the Plurality of Worlds*, Oxford, Blackwell, 1986. {4, 14}

[J. LEWIS/FRIGERIO 1981] LEWIS, J. T./FRIGERIO, A., "Quantum Stochastic Processes", PRep **77** (1981): 339–58. {3, 5}

[P. LEWIS 1997] LEWIS, Peter J., "Quantum Mechanics, Orthogonality, and Counting", *British Journal for Philosophy of Science* **48** (1997): 313–28. {3, 12}

[LICHTNER/GRIFFIN 1976] LICHTNER, P. C./GRIFFIN, J. J., "Evolution of a Quantum System: Lifetime of a Determinant", PRL **37** (1976): 1521–24. {5}

[LIEB/RUSKAI 1973a] LIEB, E. H./RUSKAI, M. B., "Proof of the Strong Subadditivity of Quantum-Mechanical Entropy", JMP **14** (1973):1938–41. {3, 10}

[LIEB/RUSKAI 1973b] —, "A Fundamental Property of Quantum-Mechanical Entropy", PRL **30** (1973): 434–36. {3, 10}

[LINDBLAD 1973] LINDBLAD, Göran, "Entropy, Information and Quantum Measurement", CMP **33** (1973): 305–322. {4, 5, 10}

[LINDBLAD 1974] —, "Expectations and Entropy Inequalities for Finite Quantum Systems", CMP **39** (1974): 111–19. {3, 10}

[LINDBLAD 1975] —, "Completely Positive Maps and Entropy Inequalities", CMP **40** (1975): 147–51. {3, 10}

[LINDBLAD 1976] —, "On the Generators of Quantum Dynamical Semigroups", CMP **48** (1976): 119–30. {3, 5}

[LINDBLAD 1979] —, "Non-Markovian Quantum Stochastic Processes and Their Entropy", CMP **65** (1979): 281–94. {3, 5}

[*LINDBLAD* 1983] —, *Non-Equilibrium Entropy and Irreversibility*, Dordrecht, Reidel, 1983. {5, 10; W}

[*LINDEMANN* 1932] LINDEMANN, Frederick A., *The Physical Significance of the Quantum Theory*, Oxford, Clarendon, 1932. {12}

[LIU/SUN 1995] LIU, X.-J./SUN, C.-P., "Generalization of Cini's Model for Quantum Measurement and Dynamical Realization of Wavefunction Collapse", PL **198A** (1995): 371–77. {4}

[LLOYD 1993] LLOYD, Seth, "A Potentially Realizable Quantum Computer", *Science* **261** (1993): 1569–71. {10; S}

[LLOYD 1994] —, "Envisioning a Quantum Computer", *Science* **263** (1994): 695. {10}

[LLOYD 1995a] —, "Almost Any Quantum Logic Gate is Universal", PRL **75** (1995): 346–49. {10}

[LLOYD 1995b] —, "Quantum-Mechanical Computers", *Scientific American* **273** (oct. 1995): 44–50. {10; S}

[LLOYD 1997a] —, "Capacity of a Noisy Quantum Channel", PR **A55** (1997): 1613–22. {10}

[LLOYD 1997b] —, "Quantum-Mechanical Maxwell's Demon", PR **A55** (1997): 3374–82. {4, 5, 6, 10}

[LLOYD 1998] —, "Microscopic Analogs of the Greenberger-Horne-Zeilinger Experiment", PR **A57** (1998): R1473–76. {9}

[LO/SHIMONY 1981] LO, T. K./SHIMONY, A., "Proposed Molecular Test of Local Hidden-Variables Theories", PR **A23** (1981): 3003–12. {9}

[LOCHAK 1984] LOCHAK, Georges, "De Broglies Initial Conception of de Broglie Waves", in [*DINER et al.* 1984, 1–26]. {7}

[LOCHAK/DUTHEIL 1992] LOCHAK, G./DUTHEIL, R., "Wave Mechanics and Relativity", in SELLERI (ed.) 1992b: 157–67. {12, 13}

[LOINGER 1968] LOINGER, A., "Comments on a Recent Paper Concerning the Quantum Theory of Measurement", *Nuclear Physics* **108** (1968): 245–49. {4}

[LONDON 1926] LONDON, F., "Winkelvariable und kanonische Transformationen in der Undulationsmechanik", ZP **40** (1926): 193–210. {3}

[*LONDON/BAUER* 1939] LONDON, F./BAUER, E., *La théorie de l'observation en mécanique quantique*, Paris, Hermann, 1939. {4, 12}

[*LOPES/PATY* 1977] LOPES, J. L./PATY, M. (ed.s), *Quantum Mechanics, a Half-century Later*, Dordrecht, Reidel, 1977. {3, 4}

[*LORENZEN* 1955] LORENZEN, Paul, *Einführung in die operative Logik und Mathematik*, Berlin, Springer, 1955. {1, 14, 15}

[LORENTZ 1910] LORENTZ, Hendrik A., "Die Hypothese der Lichtquanten", *Physikalische Zeitschrift* **11** (1910): 349–54. {3}

[LOSS/DIVINCENZO 1998] LOSS, D./DIVINCENZO, D. P., "Quantum Computation with Quantum Dots", PR **A57** (1998): 120–26. {10}

[*LOUDON* 1973] LOUDON, Rodney, *The Quantum Theory of Light*, London, Clarendon, 1973, II ed. 1983. {2}

[LOUISELL 1963] LOUISELL, W. H, "Amplitude and Phase Uncertainty Relations", PL **7** (1963): 60–61. {3}

[LOUISELL *et al.* 1961] LOUISELL, W. H./YARIV, A./SIEGMAN, A. E., "Quantum Fluctuations and Noise in Parametric Processes. I", PR **124** (1961): 1646–54. {3, 4}

[LÜDERS 1951] LÜDERS, Gerhart, "Über die Zustandsänderung durch Meßprozeß", *Annalen der Physik* **8** (1951): 322–28. {3, 4}

[LUDWIG 1953] LUDWIG, Günther, "Der Messprozess", ZP **135** (1953): 483–511. {3, 4}

[*LUDWIG* 1954] —, *Die Grundlagen der Quantenmechanik, Berlin*, Springer, 1954. {2, 3}

[LUDWIG 1958a] —, "Zum Ergodensatz und zum Begriff dder makroskopischen Observablen. I", ZP **150** (1958): 346–74. {3, 4, 5}

[LUDWIG 1958b] —, "Zum Ergodensatz und zum Begriff dder makroskopischen Observablen. II", ZP **152** (1958): 98–115. {3, 4, 5}

[LUDWIG 1961] —, "Gelöste und ungelöste Probleme des Meßprozesses in der Quantenmechanik", in [*BOPP* 1961a, 150–81]. {4}

[LUDWIG 1964] —, "Versuch einer axiomatischen Grundlegung der Quantenmechanik und allgemeinerer physikalischen Theorien", ZP **181** (1964): 233–60. {3}

[LUDWIG 1967] —, "Attempt of an Axiomatic Foundation of Quantum Mechanics and More General Theories. II", *Communication in Mathematical Physics* **4** (1967): 331–48. {3}

[LUDWIG 1968] —, "Attempt of an Axiomatic Foundation of Quantum Mechanics and More General Theories III", *Communication in Mathematical Physics* **9** (1968): 1–12. {3}

[*LUDWIG* 1970] —, *Deutung des Begriffs 'physikalische Theorie' und axiomatische Grundlegung der Hilbertraumstruktur der Quantenmechanik durch Hauptsätze des Messens*, Berlin, Springer, 1970. {3, 12}

[LUDWIG 1977] —, "A Theoretical Description of Single Microsystems", in [*W. PRICE/CHISSICK* 1977, 189–226]. {3, 4}

[LUDWIG 1981] —, "An Axiomatic Basis of Quantum Mechanics", in [*NEUMANN* 1981, 49–70]. {3}

[LUDWIG 1985] —, *An Axiomatic Basis for Quantum Mechanics*, Berlin, Springer, 1985. {3}

[LUDWIG 1989] —, "Atoms: Are They Real or Are They Objects?", FP **19** (1989): 971–83. {12}

[LUDWIG/NEUMANN 1981] LUDWIG, G./NEUMANN, H., "Connections Between Different Approaches to the Foundations of Quantum Mechanics", in [*NEUMANN* 1981, 133–43]. {3}

[LUKASIEWICZ 1973] LUKASIEWICZ, Jan, "Über den Determinisums", *Studia Leibnitiana* **5** (1973): 5–25. {14}

[LYONS 1983] LYONS, Louis, "An Introduction to the Possible Substructure of Quarks and Leptons", *Progress in Particle and Nuclear Physics* **10** (1983): 227–304. {5, 12}

[MAASSEN/UFFINK 1988] MAASSEN, H./UFFINK, J. B. M., "Generalized Entropic Uncertainty Relations", PRL **60** (1988): 1103–106. {3, 10}

[MABUCHI/ZOLLER 1996] MABUCHI, H./ZOLLER, P., "Inversion of Quantum Jumps in Quantum Optical Systems under Continuous Observation", PRL **76** (1996): 3108–111. {3, 5, 6}

[MACHIDA/NAMIKI 1980] MACHIDA, S./NAMIKI, M., "Theory of Measurement in Quantum Mechanics. Mechanism of Reduction of Wave Packet I–II", PTF **63** (1980): 1457–73, 1833–47. {4}

[MACHIDA/NAMIKI 1984] —, "Many-Hilbert-Space Description of Measuring Apparatus and Reduction of the Wave Packet", in [*ROTH/INOMATA* 1984, 77–92]. {4}

[*MACKEY* 1960] MACKEY, George W., *Lecture Notes on the Mathematical Foundations of Quantum Mechanics*, Cambridge Mass., Harvard University Press, 1960. {3}

[*MACKEY* 1963] —, *The Mathematical Foundations of Quantum Mechanics*, New York, Benjamin, 1963. {3}

[*MACKEY* 1968] —, *Induced Representations and Quantum Mechanics*, New York, Benjamin, 1968. {3}

[*MACKINNON* 1982] MACKINNON, E. M., *Scientific Explanation and Atomic Physics*, Chicago, University Press, 1982. {11, 12, 13}

[MACKMANN/SQUIRES 1995] MACKMANN, S./SQUIRES, E., "Lorentz Invariance and the Retarded Bohm Model", FP **25** (1995): 391–97. {8, 12}

[MACLAREN 1965] MACLAREN, M. D., "Notes on Axioms for Quantum Mechanics", *AEC Research and Development Report* **ANL-7065** (1965): 11. {3}

[MACZYŃSKI 1971] MACZYŃSKI, Maciej J., "Boolean Properties of Observables in Axiomatic Quantum Mechanics", *Reports on Mathematical Physics* **2** (1971): 135–50. {3}

[MACZYŃSKI 1972] —, "Hilbert Space Formalism of Quantum Mechanics Without the Hilbert Space Axiom", *Reports on Mathematical Physics* **3** (1972): 209–19. {3}

[MACZYŃSKI 1973] —, "The Field of Real Numbers in Axiomatic Quantum Mechanics", JMP **14** (1973): 1469–71. {3}

[MACZYŃSKI 1974] —, "When the Topology of an Infinite-Dimensional Banach Space Coincides with a Hilbert Space Topology", *Studia Math* **49** (1974): 149–52. {3}

[MADELUNG 1926] MADELUNG, Erwin, "Quantentheorie in hydrodynamischer Form", ZP **40** (1926): 322–26. {3, 5, 12}

[MADELUNG 1930] —, "Geschehen, Beobachten und Messen im Formalismus der Wellenmechanik", ZP **62** (1930): 721–25. {3, 4, 5}

[MAMOJKA 1974] MAMOJKA, B., *International Journal of Theoretical Physics* **11** (1974): 73. {10}

[MANCINI/TOMBESI 1997] MANCINI, S./TOMBESI, P., "Macroscopic Coherence for a Trapped Electron", PR **A56** (1997): R1679–81. {4, 5}

[MANCINI *et al.* 1997a] MANCINI, S./MAN'KO, V. I./TOMBESI, P., "Ponderomotive Control of Quantum Macroscopic Coherence", PR **A55** (1997): 3042–50. {5}

[MANCINI *et al.* 1997b] —, "Classical-Like Description of Quantum Dynamics by Means of Symplectic Tomography", FP **27** (1997): 801–24. {3, 5, 7}

[MANDEL 1958a] MANDEL, Leonard, "Fluctuations of Photon Beams and their Correlation", *Proceedings of the Physical Society, London* **72** (1958): 1037–48. {3, 5}

[MANDEL 1959] —, "Fluctuations of Photon Beams: The Distribution of the Photo-Electrons", *Proceedings of the Physical Society, London* **74** (1959): 233–43. {3, 5}

[MANDEL 1983] —, "Photon Interference and Correlation Effects Produced by Independent Quantum Sources", PR **A28** (1983): 929–43. {3, 9}

[MANDEL 1991] —, "Coherence and Indistinguishability", *Optics Letters* **16** (1991): 1882–83. {3, 7}

[MANDEL 1995a] —, "Indistinguishability in One-Photon and Two-Photon Interference" FP **25** (1995): 211–18. {4, 7}

[MANDEL 1995b] —, "Two-Photon Downconversion Experiments", in [*GREENBERGER/ZEILINGER* 1995a, 1–12]. {7, 9}

[MANDELSTAM/TAMM 1945] MANDELSTAM, L./TAMM, I., "The Uncertainty Relation Between Energy and Time in Non-Relativistic Quantum Mechanics", *Journal of Physics (USSR)* **9** (1945): 249–54. {3}

[MANN *et al.* 1992] MANN, A./NAKAMURA, K./REVZEN, M., "Bell's Inequality for Mixed States", JP **A25** (1992): L851–L854. {9}

[MARGENAU 1934] MARGENAU, Henry, "On the Application of Many-Valued Systems of Logic to Physics", PS **1** (1934): 118–21. {7, 12, 13, 14}

[MARGENAU 1936] —, "Quantum Mechanical Description", PR **49** (1936): 240–42. {12}

[MARGENAU 1937] —, "Critical Points in Modern Physical Theory", PS **4** (1937): 337–70. {12, 13}

[MARGENAU 1949] —, "Einstein's Conception of Reality", in [*SCHILPP* 1949, 245–68]. {12, 13, 14}

[*MARGENAU* 1950] —, *The Nature of Physical Reality*, New York, McGraw-Hill, 1950. {12, 13, 14}

[MARGENAU 1954] —, "Advantages and Disadvantages of Various Interpretations of the Quantum Theory", PT **7.10** (1954): 6–13. {12}

[MARGENAU 1958] —, "Philosophical Problems Concerning the Meaning of Measurement in Physics", PS **25** (1958): 23–33. {4, 12, 13, 14}

[MARGENAU 1963] —, "Measurements in Quantum Mechanics", AP **23** (1963): 469–85. {4}

[MARGENAU/HILL 1961] MARGENAU, H./HILL, R. W., "Correlation Between Measurements in Quantum Theory", PTF **26** (1961): 722–38. {3, 4}

[MARGENAU/WIGNER 1962] MARGENAU, H./WIGNER, E. P., "Comments on Professor Putnams Comments", PS **29** (1962): 292–93. {12, 13}

[MARGOLUS 1990] MARGOLUS, Norman, "Parallel Quantum Computation", in [*ZUREK* 1990a, 273–87]. {10}

[MARINOV/SEGEV 1997] MARINOV, M. S./SEGEV, B., "Barrier Penetration by Wave Packets and the Tunneling Times", PR **A55** (1997): 3580–85. {9}

[*MARLOW* 1978a] MARLOW, A. R. (ed.), *Mathematical Foundations of Quantum Theory*, New York, Academic, 1978. {3}

[MARLOW 1978b] —, "Orthomodular Structures and Physical Theory", in [*MARLOW* 1978a, 59–69]. {3, 13}

[MARSHALL 1991] MARSHALL, Trevor W., "What Does Noise Do to the Bell Inequalities?", FP **21** (1991): 209–219. {8, 9, 12}

[MARSHALL/SANTOS 1985] MARSHALL, T. W./SANTOS, E., "Local Realist Model for the Coincidence Rates in Atomic-Cascade Experiments", PL **107A** (1985): 164–68. {8, 9, 12}

[MARSHALL/SANTOS 1987] —, "Comment on 'Experimental Evidence for a Photon Anticorrelation Effect on a Beam Splitter: a New Light on Single-Photon Interferences"', *Europhysics Letters* **3** (1987): 293–96. {3, 7}

[MARSHALL *et al.* 1983] MARSHALL, T. W./SANTOS, E./SELLERI, F., "Local Realism Has not Been Refuted by Atomic Cascade Experiments", PL **98A** (1983): 5–9. {9, 12}

[MARTENS/DE MUYNCK 1990a] MARTENS, H./DE MUYNCK, W. M., "Nonideal Quantum Measurements", FP **20** (1990): 255–81. {4}

[MARTENS/DE MUYNCK 1990b] —, "The Inaccuracy Principle", FP **20** (1990): 357–80. {3, 4}

[MARTENS/DE MUYNCK 1992] —, "Disturbance, Conservation Laws and Uncertainty Principle", JP **A25** (1992): 4887–901. {3, 4}

[MARTINIS *et al.* 1987] MARTINIS, J. M./DEVORET, M. H./CLARKE, J., "Experimental Tests for the Quantum Behavior of a Macroscopic Degree of Freedom: The Phase Difference Across a Josephson Junction", PR **B35** (1987): 4682–98. {2, 5}

[MASSAR/POPESCU 1995] MASSAR, S./POPESCU, S., "Optimal Extraction of Information from Finite Quantum Ensembles", PRL **74** (1995): 1259–63. {3, 10}

[MATTLE *et al.* 1996] MATTLE, K./WEINFURTER, H./KWIAT, P./ZEILINGER, A., "Dense Coding in Experimental Quantum Communication", PRL **76** (1996): 4656–59. {9, 10}

[*MAUDLIN* 1994] MAUDLIN, T., *Quantum Non-Locality and Relativity*, Cambridge, Blackwell, 1994. {8, 9, 11, 12}

[MCELWAINE 1997] MCELWAINE, Jim, "Maximum Information and Quantum Prediction Algorithms", PR **A56** (1997): 1756–66. {10}

[MCKENNA/WAN 1984] MCKENNA, I. H./WAN, K. K., "The Role of the Connection in Geometric Quantization", JMP **25** (1984): 1798–1803. {3}

[MCKINSEY/SUPPES 1954] MCKINSEY, J. C. C./SUPPES, P., "Review of Destouches-Frèvier's *La Structure des Théories Physiques*", *Journal of Symbolic Logic* **19** (1954): 52–55. {7, 12}

[*MCWEENY* 1972] MCWEENY, R., *Quantum Mechanics: Principles and Formalism*, Oxford, Pergamon, 1972. {2}

[MEAD/TRUHLAR 1979] MEAD, C. A./TRUHLAR, D. G., "On the Determination of Born-Oppenheimer Nuclear Motion Wave Functions Including Complications due to Conical Intersections and Identical Nuclei", *Journal of Chemical Physics* **70** (1979): 2284–96; rep. in [*SHAPERE/WILCZEK* 1989, 80–103]. {3}

[MECOZZI/TOMBESI 1987] MECOZZI, A./TOMBESI, P., "Distinguishable Quantum States Generated via Nonlinear Birefringence", PRL **58** (1987): 1055–58. {4}

[MEEKHOF *et al.* 1996] MEEKHOF, D. M./MONROE, C./KING, B. E./ITANO, W. M./WINELAND, D. J. "Generation of Nonclassical Motional States of a Trapped Atom", PRL **76** (1996): 1796–99. {5}

[MEHRA 1972] MEHRA, "'The Golden Age of the Theoretical Physics': Diracs Scientific Work from 1924 to 1933", in [*SALAM/WIGNER*, 17–59]. {12}

[*MEHRA* 1973] — (ed.), *The Physicist's Conception of Nature*, Dordrecht, Reidel, 1973. {4, 5, 12, 13}

[MEIS 1997] MEIS, C., "Photon Wave Particle Duality and Virtual Electromagnetic Waves", FP **27** (1997): 865–73. {7}

[MENSKY 1994] MENSKY, Michael B., "Continuous Quantum Measurements and the Action Uncertainty Principle", [*ACCARDI* 1994a, 171–86]. {3, 4}

[MENSKY 1996] —, "A Note on Reversibility of Quantum Jumps", PL **222A** (1996): 137–40. {3, 5, 6}

[MERMIN 1980] MERMIN, N. David, "Quantum Mechanics vs Local Realism near Classical Limits: A Bell Inequality for Spin *s*", PR **D22** (1980): 356–61. {5, 8, 9}

[MERMIN 1981] —, "Bringing Home the Atomic World. Quantum Mysteries for Anybody", AJP **49** (1981): 940–43. {8, 9, 12}

[MERMIN 1984] —, "Generalizations of Bells Theorem to Higher Spins and Higher Correlations", in [*ROTH/INOMATA* 1984, 7–20]. {3, 9}

[MERMIN 1986] —, "The EPR Experiment — Thoughts About the 'Loophole"', in [*GREENBERGER* 1986, 422–27]. {9}

[MERMIN 1990a] —, "Extreme Quantum Entanglement in a Superposition of Macroscopically Distinct States", PRL **65** (1990): 1838–40. {5, 9}

[MERMIN 1990b] —, "Simple Unified Form for the Major No-Hidden-Variables Theorems", PRL **65** (1990): 3373–76. {9}

[MERMIN 1990c] —, "Quantum Mysteries Revisited", AJP **58** (1990): 731–34. {8, 9, 12}

[MERMIN 1990d] —, "What's Wrong With These Elements of Reality?", *Phisics Today* **43** (1990): 9–11. {8, 9; S}

[MERMIN 1993] —, "Hidden Variables and the Two Theorems of John Bell", RMP **65** (1993): 803–15. {8, 9}

[MERMIN 1994] —, "Quantum Mysteries Refined", AJP **62** (1994): 880–887. {8, 9, 12}

[MERMIN 1995a] —, "The Best Version of Bells Theorem", in [*GREENBERGER/ZEILINGER* 1995a, 616–23]. {9}

[MERMIN 1997] —, "Nonlocality and Bohr's Reply to EPR", e-print; code PHY9722065. {8, 9, 12}

[MERMIN/SCHWARZ 1982] MERMIN, N. D./SCHWARZ, G. M., "Joint Distributions and Local Realism in the Higher Spin EPR Experiment", FP **12** (1982): 101–135. {3, 9}

[*MESSIAH* 1961] MESSIAH, A., *Quantum Mechanics*, Amsterdam, North-Holland, 1961. {2}

[*MEYSTRE/SCULLY* 1983] MEYSTRE, P./SCULLY, M. O. (ed.s), *Quantum Optics, Experimental Gravity, and Measurement Theory*, New York, Plenum, 1983. {2, 3, 4}

[MIELNIK 1968] MIELNIK, Bogdan, "Geometry of Quantum States", CMP **9** (1968): 55–80. {3}

[MIELNIK 1969] —, "Theory of Filters", *Communication in Mathematical Physics* **15** (1969): 1–46. {3}

[MIELNIK 1974] —, "Generalized Quantum Mechanics", *Communication in Mathematical Physics* **37** (1974): 221–56. {3}

[MILBURN 1986] MILBURN, G. J., "Quantum and Classical Liouville Dynamics of the Anharmonic Oscillator", PR **A33** (1986): 674–85. {3, 13}

[MILBURN/HOLMES 1986] MILBURN, G. J./HOLMES, C. A., "Dissipative Quantum and Classical Liouville Mechanics of the Anharmonic Oscillator", PRL **56** (1986): 2237–40. {5}

[MILBURN/WALLS 1983] MILBURN, G. J./WALLS, D. F., "Quantum Nondemolition Measurements via Quantum Counting", PR **A28** (1983): 2646–48. {4, 5}

[MILBURN/WALLS 1988] —, "Effect of Dissipation on Interference in Phase Space", PR **A38** (1988): 1087–90. {5}

[MILLS 1993] MILLS, Robert, "Tutorial on Infinities in QED", in [*L. BROWN* 1993a, 57–85]. {11}

[MISRA 1967] MISRA, B., "When Can Hidden Variables Be Excluded in Quantum Mechanics?", NC **A47** (1967): 841–59. {8}

[MISRA/SUDARSHAN 1977] MISRA, B./SUDARSHAN, E. C. G., "The Zeno's Paradox in Quantum Theory", JMP **18** (1977): 756–63. {4}

[MISRA *et al.* 1979] MISRA, B./PRIGOGINE, I/COURBAGE, M., "Lyapounov variable: Entropy and Measurement in Quantum Mechanics", *Proceedings of the National Academy of Sciences of U.S.A.* **76** (1979): 4768–72; rep. in [*WHEELER/ZUREK* 1983, 687–91]. {5}

[MITTELSTAEDT 1959] MITTELSTAEDT, Peter, "Untersuchungen zur Quantenlogik", *Sitzungsberichte der Bayerischen Akademie der Wissenschaften* (1959): 321–86. {3}

[MITTELSTAEDT 1960] —, "Über die Gültigkeit der Logik in der Natur", *Naturwissenschaften* **47** (1960): 385–91. {3, 14}

[MITTELSTAEDT 1961] —, "Quantenlogik", *Fortschritte der Physik* 9 (1961): 106–47. {3}

[*MITTELSTAEDT* 1963] —, *Philosophische Probleme der modernen Physik*, Mannheim, Bibliographische Institut, 1963; engl. trans.: *Philosophical Problems of Modern Physics*, Dordrecht, Reidel, 1976. {3, 13, 14}

[MITTELSTAEDT 1968] —, "Verborgene Parameter und beobachtbare Grössen in physikalischen Theorien", *Philosophia naturalis* **10** (1968): 468–82. {3, 13}

[*MITTELSTAEDT* 1978] —, *Quantum Logic*, Dordrecht, Reidel, 1978. {3}

[MITTELSTAEDT 1981] —, "The Concepts of Truth, Possibility and Probability in the Language of Quantum Physics", in [*NEUMANN* 1981, 71–94]. {3, 14}

[MITTELSTAEDT 1985] —, "EPR-Paradox, Quantum Logic and Relativity", in [*LAHTI/MITTELSTAEDT* 1985, 171–86]. {9}

[MITTELSTAEDT 1992] —, "Unsharp Particle-Wave Duality in Double-Slit Experiments", in [*SELLERI* 1992a, 169–86]. {7, 12}

[*MITTELSTAEDT* 1998] —, *The Interpretation of Quantum Mechanics and the Measurement Process*, Cambridge, University Press, 1998. {4, 12}

[*MITTELSTAEDT/STACHOW* 1985] — (ed.s), *Recent Developments in Quantum Logic*, Mannheim, Bibliographisches Institut, 1985. {3}

[MITTELSTAEDT *et al.* 1987] MITTELSTAEDT, P./PRIEUR, A./SCHIEDER, R., "Unsharp Particle-Wave Duality in a Photon Split-Beam Experiment", FP **17** (1987): 891–903. {4, 7, 12}

[MIZOBUCHI/OHTAKÉ 1992] MIZOBUCHI, Y./OHTAKÉ, Y., "An 'Experiment to Throw More Light on Light"', PL **168A** (1992): 1–5. {7}

[MØLMER 1997] MØLMER, Klaus, "Optical Coherence: A Convenient Fiction", PR **A55** (1997): 3195–203. {3, 12}

[MONROE *et al.* 1995a] MONROE, C./MEEKHOF, D. M./KING, B. E./JEFFERTS, S. R./ITANO, W. M./WINELAND, D. J./GOULD, P., "Resolved-Sideband Raman Cooling of a Bound Atom to 3D Zero-Point-Energy", PRL **75** (1995): 4011–14. {5}

[MONROE *et al.* 1995b] MONROE, C./MEEKHOF, D. M./KING, B. E./ITANO, W. M./WINELAND, D. J., "Demonstration of a Fundamental Quantum Logic Gate", PRL **75** (1995): 4714–17. {5, 10}

[MONROE *et al.* 1996] MONROE, C./MEEKHOF, D. M./KING, B. E./WINELAND, D. J., "A 'Schrödinger Cat' Superposition State of an Atom", *Science* **272** (1996): 1131–36.{5}

[MONROE *et al.* 1997] MONROE, C./LEIBFRIED, D./KING, B. E./MEEKHOF, D. M./ITANO, W. M./WINELAND, D. J., "Simplified Quantum Logic with Trapped Ions", PR **A 55** (1997): R2489–91. {10}

[MOUSSA 1997] MOUSSA, M. H. Y., "Teleportation with Identity Interchange of Quantum States", PR **A55** (1997): R3287–90. {9, 10}

[MOZYRSKY/PRIVMAN 1998] MOZYRSKY, D./PRIVMAN, V., "Adiabatic Decoherence" e-print; code: quant-ph/9709020. {5}

[MOYAL 1949] MOYAL, Jose E., "Quantum Mechanics as a Statistical Theory", *Proceedings of the Cambridge Philosophical Society* **45** (1949): 99–124. {3, 12}

[MÜCKENHEIM 1992] MÜCKENHEIM, Wolfgang, "Some Arguments Against the Existence of De Broglie Waves", in [*SELLERI* 1992a, 187–91]. {7}

[MUGUR-S. 1992] MUGUR-SCHÄCHTER, M., "Toward a Factually Induced Space-Time Quantum Logic", FP **22** (1992): 963–94. {3}

[MÜLLER *et al.* 1995] MÜLLER, A./ZBINDEN, H./GISIN, N., "Underwater Quantum Coding", NA **378** (1995): 449. {10}

[MUNRO/REID 1993] MUNRO, W. J./REID, M. D., "Violation of Bell's Inequality by Microscopic States Generated Via Parametric Down-Conversion", PR **A47** (1993): 4412–21. {9}

[*MURDOCH* 1987] MURDOCH, Dugald, *Niels Bohr's Philosophy of Physics*, Cambridge, University Press, 1987. {12, 13, 14}

[NAGEL 1945] NAGEL, Ernest, "Review of Reichenbach's *Philosophic Foundations of Quantum Mechanics*", *Journal of Philosophy* **42** (1945): 437–44. {7, 12}

[NAGEL 1946] —, "Professor Reichenbach on QM: A Rejoinder", *Journal of Philosophy* **43** (1946): 247–50. {7, 12, 14}

[NAGELS *et al.* 1997] NAGELS, B./HERMANS, L. J. F./CHAPOVSKY, P. L., "Quantum Zeno Effect Induced by Collision", PRL **79** (1997): 3097–3100. {4}

[NAGOURNEY *et al.* 1986] NAGOURNEY, W./SANDBERG, J./DEHMELT, H., "Shelved Optical Electron Amplifier: Observation of Quantum Jumps", PRL **56** (1986): 2797–99. {5}

[NAKATSUJI/YASUDA 1996] NAKATSUJI, H./YASUDA, K., "Direct Determination of the Quantum-Mechanical Density Matrix Using the Density Equation", PRL **76** (1996): 1039–42. {7}

[NAKAZATO/PASCAZIO 1993] NAKAZATO, H./PASCAZIO, S., "Solvable Dynamical Model for a Quantum Measurement Process", PRL **70** (1993): 1–4. {4}

[NAKAZATO/PASCAZIO 1994] —, "A Coherent Understanding of Solvable Models for Quantum Measurement Processes", PL **192A** (1994): 169–74. {4, 5}

[NAKAZATO *et al.* 1994] NAKAZATO, H./NAMIKI, M./PASCAZIO, S., "Exponential Behavior of a Quantum System in Macroscopic Medium", PRL **73** (1994): 1063–66. {4, 5}

[NAKAZATO *et al.* 1995] NAKAZATO, H./NAMIKI, M./PASCAZIO, S./RAUCH, H., "On the Quantum Zeno Effect", PL **199A** (1994): 27–32. {4}

[NAMIKI 1986] NAMIKI, Mikio, "Quantum Mechanics of Macroscopic Systems and the Measurement Process", in [*GREENBERGER* 1986, 78–88]. {4, 5}

[NAMIKI 1988a] —, "Many-Hilbert-Spaces Theory of Quantum Measurements", FP **18** (1988): 29–55. {4; W}

[NAMIKI 1988b] —, "Quantum Mechanics of Macroscopic Systems and Measurement Processes", in [*VAN DER MERWE et al.* 1988a, 3–14]. {4, 5}

[NAMIKI/MUGIBAYASHI 1953] NAMIKI, M./MUGIBAYASHI, N., "On the Radiation Damping and the Decay of an Excitate State", PTF **10** (1953): 474–76. {3}

[NAMIKI/PASCAZIO 1990] NAMIKI, M./PASCAZIO, S., "On a Posssible Reduction of the Interference Term Due to Statistical Fluctuations", PL **147A** (1990): 430–34. {4, 5}

[NAMIKI/PASCAZIO 1991] —, "Wave-Function Collapse by Measurement and its Simulation", PR **A44** (1991): 39–53. {4}

[NAMIKI/PASCAZIO 1993] —, "Quantum Theory of Measurement Based On the Many-Hilbert-Space Approach", PRep **232** (1993): 301–411. {4; W}

[NAMIKI/PASCAZIO 1994] —, "The Many-Hilbert-Space Approach to Quantum Measurements", [*ACCARDI* 1994a, 339–67]. {4, 5}

[*NAMIKI et al.* 1987] NAMIKI e al (ed.s), *Foundations of Quantum Mechanics*, Tokyo, Physical Society of Japan, 1987. {3–9}

[NAUENBERG *et al.* 1994] NAUENBERG, M./STROUD, C./YEAZELL, J., "The Classical Limit of an Atom", *Scientific American* **270**.6 (1994): 24–29. {5; S}

[NELSON 1966] NELSON, Edward, "Derivation of the Schrödinger Equation from Newtonian Mechanics", PR **150** (1966): 1079–85. {3, 5, 12}

[*NELSON* 1967] —, *Dynamical Theories of the Brownian Motion*, Princeton, University Press, 1967. {13}

[*NELSON* 1985] —, *Quantum Fluctuations*, Princeton, University Press, 1985. {3, 5, 12}

[NELSON 1986] —, "The Locality Problem in Stochastic Mechanics", in [*GREENBERGER* 1986, 533–38]. {5, 8, 9}

[NEMES/DE TOLEDO P. 1986] NEMES, M. C./DE TOLEDO PIZA, A. F. R., "Effective Dynamics of Quantum Subsystems", *Physica* **A137** (1986): 367–88. {3, 5}

[*NEUMANN* 1981] NEUMANN, H. (ed.), *Interpretation and Foundations of Quantum Theory*, Mannheim/Wien/Zrich, Bibliographisches Institut, 1981. {3, 12}

[R. NEWTON 1980] NEWTON, R. G., “Quantum Action-Variables for Harmonic Oscillators”, AP **124** (1980): 327–46. {3}

[T. NEWTON/WIGNER 1949] NEWTON, T. D./WIGNER, E. P., “Localized States for Elementary Particles”, RMP **21** (1949): 400–406. {3, 8, 11}

[NG/VAN DAM 1995] NG, Y. J./VAN DAM, H., “Limitation to Quantum Measurements of Space-Time Distances”, in [*GREENBERGER/ZEILINGER* 1995a, 579–84]. {3}

[NICHOLSON 1954] NICHOLSON, A. F., “On a Theory due to I. Fényes”, *Australian Journal of Physics* **7** (1954): 14–21. {8, 12}

[NIELSEN/CAVES 1997] NIELSEN, M. A./CAVES, C. M., “Reversible Quantum Operations and their Application to Teleportation”, PR **A55** (1997): 2547–56. {6, 7, 9, 10}

[NIMTZ *et al.* 1994] NIMTZ, G./ENDERS, A./SPIEKER, , “Photonic Tunneling Times”, *Journal de Physique* **I.4** (1994): 565–70. {9}

[NOEL/STROUD 1996] NOEL, M. W./STROUD, C. R., Jr., “Excitation of an Atomic Electron to a Coherent Superposition of Macroscopically Distinct States”, PRL **77** (1996): 1913–16. {5}

[OCHS 1972] OCHS, Wilhelm, “On Gudder’s Hidden-Variable Theorem”, NC **10B** (1972): 172–84. {3, 8}

[OCHS 1975] —, *Reports on Mathematical Physics* **8** (1975): 109. {10}

[OCHS 1977] —, “On the Strong Law of Large Numbers in Quantum Probability Theory”, *Journal of Philosophical Logic* **6** (1977): 473–80. {3}

[*OCKHAM* OP] OCKHAM, Guillelmi de, *Opera Philosophica*, New York, St. Bonaventure, 1974. {14}

[*OCKHAM* SL] —, *Summa Logicæ*, in [*OCKHAM* OP, I]. {14}

[*OCKHAM* ELP] —, *Expositio Super VIII Libros Physicorum*, in [*OCKHAM* OP]. {14}

[O’CONNELL 1983] O’CONNELL, R. F., “The Wigner Distribution Function — 50th Birthday”, FP**13** (1983): 83–92. {3}

[O’CONNELL *et al.* 1986] O’CONNELL, R. F./SAVAGE, C. M./WALLS, D. F., “Decay of Quantum Coherence Due to the Presence of a Heath Bath: Markovian Master-Equation Approach”, in [*GREENBERGER* 1986, 267–74]. {5}

[OGAWA *et al.* 1991] OGAWA, T./UEDA, M./IMOTO, N., “Measurement-Induced Oscillations of a Highly Squeezed State between Super and Sub-Poissonian Photon Statistics”, PRL **66** (1991): 1046–49. {4, 5}

[OHIRA/PEARLE 1988] OHIRA, T./PEARLE, P., “Perfect Disturbing Measurements”, AJP **56** (1988): 692–95. {4, 12}

[OHYA 1989] OHYA, M., *Reports on Mathematical Physics* **27** (1989): 18. {10}

[*OHYA/PETZ* 1993] OHYA, M./PETZ, D., *Quantum Entropy and its Use*, Berlin, Springer, 1993. {2, 3, 10}

[OLKHOVSKY/RECAMI 1992] OLKHOVSKY, V. S./RECAMI, E., “Recent Developments in the Time Analysis of Tunnelling Processes”, PRep **214** (1992): 339–56. {3; W}

[OLKHOVSKY *et al.* 1974] OLKHOVSKY, V. S./RECAMI, E./GERASIMCHUK, A. J., “Time Operator in Quantum Mechanics”, NC **22A** (1974): 263–78. {3}

[*OMELJANOVSKII* 1962] OMELJANOVSKII, Moisey A., *Philosophische Probleme der Quantenmechanik*, Berlin, Deutscher Verlag der Wissenschaften, 1962. {12, 14}

[OMNÈS 1988] OMNES, Roland, "Logical Reformulation of Quantum Mechanics. I–III", *Journal of Statistical Physics* **53** (1988): 893–975. {3, 6, 12}

[OMNÈS 1989] —, "Logical Reformulation of Quantum Mechanics. IV", *Journal of Statistical Physics* **57** (1989): 357–382. {3, 6, 12}

[OMNÈS 1990] —, "From Hilbert Space to Common Sense: A Synthesis of Recent Progress in the Interpretation of Quantum Mechanics", AP **201** (1990): 354–447. {12}

[OMNÈS 1991] —, "About the Notion of Truth in Quantum Mechanics", *Journal of Statistical Physics* **62** (1991): 841–861. {3, 12, 14}

[OMNÈS 1992] —, "Consistent Interpretation of Quantum Mechanics", RMP **64** (1992): 339–82. {12}

[*OMNÈS* 1994] —, *Interpretation of Quantum Mechanics*, Princeton, University Press, 1994. {4, 5, 6, 12, 14}

[OMNÈS 1995] —, "A New Interpretation of Quantum Mechanics and Its Consequences in Epistemology", FP **25** (1995): 605–29. {4, 5, 6, 12, 14}

[OPATRNÝ *et al.* 1997] OPATRNÝ, T./WELSCH, D.-G./VOGEL, W., "Least-Square Inversion for Density-Matrix Reconstruction" PR **A56** (1997): 1788–99. {3, 7}

[*OTT* 1993] OTT, Edward, *Dynamical Systems*, Cambridge, University Press, 1993. {1}

[OU 1988] OU, Z. Y., "Quantum Theory of Fourth-Order Interference", PR **A37** (1988): 1607–19. {2}

[OU 1996] —, "Complementarity and Fundamental Limit in Precision Phase Measurement", PRL **77** (1996): 2352–55. {3}

[OU 1997] —, "Nonlocal Correlation in the Realization of a Quantum Eraser", PL **226A** (1997): 323–26. {6, 12}

[OU/MANDEL 1988a] OU, Z. Y./MANDEL, L., "Violation of Bell's Inequality and Classical Probability in a Two-Photon Corrrelation Experiment", PRL **61** (1988): 50–53. {9}

[OU/MANDEL 1988b] —, "Observation of Spatial Quantum Beating with Separated Photodetectors", PRL **61** (1988): 54–57. {9}

[OU *et al.* 1990a] OU, Z. Y./WANG, L. J./ZOU, X. Y./MANDEL, L., "Evidence for Phase Memory in Two-Photon Down Conversion Through Entanglement with the Vacuum", PR **A41** (1990): 566–68. {3, 9}

[OU *et al.* 1990b] OU, Z. Y./ZOU, X. Y./WANG, L. J./MANDEL, L., "Observation of Nonlocal Interference in Separated Photon Channels", PRL **65** (1990): 321–24. {9}

[OZAWA 1984] OZAWA, Masanao, "Quantum Measuring Processes of Continuous Observables", JMP **25** (1984): 79–87. {3, 4}

[OZAWA 1986] —, "Concepts of Conditional Expectations In Quantum Theory", JMP **26** (1985): 1948–55. {3}

[OZAWA 1986] —, "On Information Gain by Quantum Measurements of Continuous Observables", JMP **27** (1986): 759–63. {3, 4, 10}

[OZAWA 1991] —, "Does a Conservation Law Limit Position Measurements?", PRL **67** (1991): 1956–59. {3, 4}

[OZAWA 1992] —, "Cat Paradox for C*-Dynamical Systems", PTF **88** (1992): 1051–64. {3, 5}

[OZAWA 1993a] —, "Canonical Approximate Quantum Measurements", JMP **34** (1993): 5596–624. {3, 4}

[OZAWA 1993b] —, "Wigner-Araki-Yanase Theorem for Continuous Observables", in DOEBNER, H. D./SCHERER, W./SCHROECK, F. Jr. (eds.), *Classical and Quantum Systems*, Singapore, World Scientific, 1993: 224–27. {3, 4}

[OZAWA 1998] —, "Quantum State Reduction: An Operational Approach", preprint. {3, 4}

[PAGE 1984] PAGE, Don N., "Information Basis of States for Quantum Measurements", in [*ROTH/INOMATA* 1984, 53–64]. {4, 12}

[PAGE 1993a] —, "No Time Asymmetry from Quantum Mechanics", PRL **70** (1993): 4034–37. {4, 5, 6}

[PAGE 1993b] —, "Average Entropy of a Subsystem", PRL **71** (1993): 1291–94. {4, 10}

[PAGONIS 1992] PAGONIS, Constantine, "Empty Waves: Not Necessarily Effective", PL **169A** (1992): 219–21. {7}

[PAGONIS/CLIFTON 1992] PAGONIS, C./CLIFTON, R., "Hardy's Nonlocality Theorem for n spin-$\frac{1}{2}$ Particles", PL **168A** (1992): 100–102. {9}

[PAGONIS/CLIFTON 1995] —, "Unremarkable Contextualism: Dispositions in the Bohm Theory", FP **25** (1995): 281–96. {8}

[PAGONIS *et al.* 1991] PAGONIS, C./REDHEAD, M./CLIFTON, R., "The Breakdown of Quantum Non-Locality in the Classical Limit", PL **155A** (1991): 441–44. {9}

[PAIS 1972] PAIS, Abraham, "Einstein and the Quantum Theory", RMP **51** (1979): 863–914. {12}

[PAIS 1972] —, "The Early History of the Theory of the Electron: 1897–47", in [*SALAM/WIGNER*, 79–93]. {12}

[PALMA *et al.* 1996] PALMA, G. M./SUOMINEN, K.-A./EKERT, A., "Quantum Computers and Dissipation", PRSL **A452** (1996): 567–84. {10}

[PANCHARATNAM 1956] PANCHARATNAM, S., "Generalized Theory of Interference, and Its Applications", *Proceedings of Indian Academy of Sciences* **A44** (1956): 247–62; rep. in [*SHAPERE/WILCZEK* 1989, 51–66]. {3}

[PAPP 1977] PAPP, Erhardt W. R., "Quantum Theory of the Natural Space-Time Units", in [*W. PRICE/CHISSICK* 1977, 29–50]. {12, 13}

[PAPP 1986] —, "Quasiclassical Approach to the Virial Theorem and to the Evaluation of the Ground-State Energy", PRep **136** (1986): 103–151. {3, 5}

[PARAMANANDA/BUTT 1987] PARAMANANDA, V./BUTT, D. K., "Quantum Correlations, Nonlocality and the EPR Paradox", JP **G13** (1987): 449–52. {9}

[*PARISI* 1988] PARISI, Giorgio, *Statistical Field Theory*, Redwood C. (California), Addison-Wesley, 1988. {1, 11}

[D. PARK 1984] PARK, David, "Time in Quantum Mechanics", in [*ROTH/INOMATA* 1984, 263–78]. {3, 12}

[J. PARK 1968] PARK, James L., "Nature of Quantum States", AJP **36** (1968): 211–26. {3, 12}

[J. PARK/MARGENAU 1968] J. PARK, J. L./MARGENAU, H., "Simultaneous Measurability in Quantum Theory", *International Journal of Theoretical Physics* **1** (1968): 211–83. {4}

[PARTHASARATHY 1992] PARTHASARATHI, K. R., *An Introduction to Quantum Stochastic Calculus*, Basel, Birkhäuser Verlag, 1992. {2}

[PARTOVI 1983] PARTOVI, Hossein M., "Entropic Formulation of Uncertainty for Quantum Measurement", PRL **50** (1983): 1883–885. {3, 4, 10}

[PARTOVI 1989a] —, "Quantum Thermodynamics", PL **137A** (1989): 440–44. {5}

[PARTOVI 1989b] —, "Irreversibility, Reduction and Entropy Increase in Quantum Measurements", PL **137A** (1989): 445–50. {4, 5, 10}

[PARTOVI 1990] —, "Entropy and Quantum Mechanics", in [*ZUREK* 1990a, 357–66]. {10}

[PARTOVI/BLANKENBECLER 1986a] PARTOVI, M. H./BLANKENBECLER, R., "Time in Quantum Measurement", PRL **57** (1986): 2887–90. {4, 10}

[PARTOVI/BLANKENBECLER 1986b] —, "Quantum Limit for Successive Position Measurements", PRL **57** (1986): 2891–93. {4, 10}

[PASCAZIO 1996] PASCAZIO, Saverio "Quantum Zeno Effect and Inverse Zeno Effect", in [*DE MARTINI et al.* 1996, 525–38]. {3, 4}

[PASCAZIO/NAMIKI 1994] PASCAZIO, S./NAMIKI, M., "Dynamical Quantum Zeno Effect", PR **A50** (1994): 4582–92. {3, 4}

[PASCAZIO/NAMIKI 1995] —, "Quantum Zeno Effect as a Purely Dynamical Process", in [*GREENBERGER/ZEILINGER* 1995a, 335–52]. {3, 4}

[PASCAZIO *et al.* 1993] PASCAZIO, S./NAMIKI, M./BADUREK, G./RAUCH, H., "Quantum Zeno Effect with Neutron Spin", PL **179A** (1993): 155–60. {4}

[PATY 1995] PATY, Michel, "The Nature of Einstein's Objections to the Copenhagen Interpretation of Quantum Mechanics", FP **25** (1995): 183–204. {12, 13}

[PAUL 1962] PAUL, Harry, "Über quantenmechanische Zeitoperatoren", *Annalen der Physik* **9** (1962): 252–61. {3}

[PAUL *et al.* 1996] PAUL, H./TÖRMÄ, P./KISS, T./JEX, I., "Photon Chopping: New Way to Measure the Quantum State of Light", PRL **76** (1996): 2464–67. {7}

[PAULI 1927] PAULI, Wolfgang, "Zur Quantenmechanik des magnetischen Elektrons", ZP **43** (1927): 601–23. {3}

[PAULI 1933a] —, "Einige die Quantenmechanik betreffende Erkundigungsfragen", ZP **80** (1933): 573–86. {12}

[PAULI 1933b] —, "Die allgemeinen Prinzipien der Wellenmechanik", in GEIGER/SCHEEL (ed.s), *Handbuch der Physik*, XXIV, Berlin, Springer, 1933: 83–272. {2, 3}

[PAULI 1949] —, "Einsten's Contribution to Quantum Theory", in [*SCHILPP* 1949, 147–60]. {12, 13}

[PAULI 1953] —, "Remarques sur le problème des paramètres cachés dans la mécanique quantique et sur la théorie de l'onde pilote", in [*GEORGE* 1953, 32–42]. {7, 12}

[*PAULI* 1980] —, *General Principles of Quantum Mechanics*, Berlin, Springer, 1980; engl. trans. of [PAULI 1933b]. {2, 3}

[PAVIČIĆ 1989] PAVIČIĆ, Mladen, "Unified Quantum Logic", FP **19** (1989): 999–1016. {3}

[PAVIČIĆ 1996] —, "Resonance Energy-Exchange-Free Detection and 'welcher Weg' Experiment", PL **A223** (1996): 241–45. {7}

[PAZ/ZUREK 1993] PAZ, J. P./ZUREK, W. H., "Environment-Induced Decoherence, Classicality, and Consistency of Quantum Histories", PR**D48** (1993): 2728–38. {4, 5, 6, 12}

[PAZ *et al.* 1993] PAZ, J. P./HABIB, S./ZUREK, W. H., "Reduction of the Wave Packet: Preferred Observable and Decoherence Time Scale", PR **D47** (1993): 488–501. {4, 5}

[PEARLE 1967] PEARLE, Philip, "Alternative to the Ortodox Interpretation of Quantum Theory", AJP **35** (1967): 742–53. {4, 12}

[PEARLE 1976] —, "Reduction of the State Vector by a Nonlinear Schrödinger Equation", PR **D13** (1976): 857–68. {4, 7, 12}

[PEARLE 1984] —, "Statevector Reduction as a Dynamical Process", in [*ROTH/INOMATA* 1984, 41–52]. {4, 12}

[PEARLE 1986] —, "Stochastic Dynamical Reduction Theories and Superluminal Communication", PR **D33** (1986): 2240–52. {3, 4, 5, 7, 9}

[PEARLE 1993] —, "Ways to Describe Dynamical State-Vector Reduction", PR **A48** (1993): 913–23. {4, 12}

[PEARLE/SQUIRES 1994] PEARLE, P./SQUIRES, E., "Bound State Excitation, Nucleon Decay Experiments, and Models of Wave Function Collapse", PRL **73** (1994): 1–5. {3, 4, 5, 12}

[PEAT 1973] PEAT, David "Quantum Physics and General Relativity: the Search for a Deeper Theory", in [*HOOKER* 1973a, 328–45]. {3}

[PEGG 1991] PEGG, D. T., "Wavefunction Collapse Time", PL **153A** (1991): 263–64. {4, 5}

[PEGG/KNIGHT 1988a] PEGG, D. T./KNIGHT, P. L., "Interrupted Fluorescence, Quantum Jumps, and Wave-Function Collapse", PR **A37** (1988): 4303–308. {5}

[PEGG/KNIGHT 1988b] —, "Interrupted Fluorescence, Quantum Jumps, and Wave-Function Collapse. II", JP **D21** (1988): S128–30. {5}

[PEGG *et al.* 1986] PEGG, D. T./LOUDON, R./KNIGHT, P. L., "Correlations in Light Emitted by Three-Level Atoms", PR **A33** (1986): 4085–91. {5}

[PEIERLS 1985] PEIERLS, Rudolf, "Observations in Quantum Mechanics and the 'Collapse of the Wave Function"', in [*LAHTI/MITTELSTAEDT* 1985, 187–96]. {3, 4}

[*PEIRCE* W] PEIRCE, Charles S., *Writings*, Bloomington, Indiana Univeristy Press, 1986. {13, 14, 15}

[*PEIRCE* 1872-73] —, *Toward a Logic Book*, in [*PEIRCE* W, III, 13–108]. {14}

[PERCIVAL 1994] PERCIVAL, Ian C., "Primary State Diffusion", PRSL **A447** (1994): 189–209. {5, 12}

[PERCIVAL/STRUNZ 1997] PERCIVAL, I. C./STRUNZ, W., "Detection of Spacetime Fluctuations by a Model Matter Interferometer", PRSL **A453** (1997): 431–46. {3}

[PERES 1978a] PERES, Asher, "Unperformed Experiments Have no Results", AJP **46** (1978): 745–47. {4}

[PERES 1978b] —, "Pure States, Mixtures and Compounds", in [*MARLOW* 1978a, 357–63]. {3}

[PERES 1980a] —, "Can We Undo Quantum Measurements?", PR **D22** (1980): 879–83; rep. in [*WHEELER/ZUREK* 1983, 692–96]. {4, 12}

[PERES 1980b] —, "Measurement of Time by Quantum Clocks", AJP **48** (1980): 552–57. {3}

[PERES 1980c] —, "Zeno Paradox in Quantum Theory", AJP **48** (1980): 931–32. {4}

[PERES 1984a] —, "On Quantum-Mechanical Automata", PL **101A** (1984): 249–50. {4, 10}

[PERES 1984b] —, "The Classical Paradoxes of Quantum Theory", FP **14** (1984): 1131–45.

[PERES 1985] —, "Reversible Logic and Quantum Computers", PR **A32** (1985): 3266–76. {10}

[PERES 1986a] —, "When is a Quantum Measurement?", in [*GREENBERGER* 1986, 438–48]. {4}

[PERES 1986b] —, "Existence of 'Free Will' as a Problem of Physics", FP **16** (1986): 573–84. {4, 9, 12, 13, 14}

[PERES 1986c] —, "When is a Quantum Measurement?", AJP **54** (1986): 688–92. {4}

[PERES 1988a] —, "How to Differentiate Between Non-Orthogonal States", PL **128A** (1988): 19. {3}

[PERES 1988b] —, "Quantum Limitations on Measurement of Magnetic Flux", PRL **61** (1988): 2019–21. {3, 4}

[PERES 1989] —, "The Logic of Quantum Nonseparability", in [*KAFATOS* 1989, 51–60]. {9}

[PERES 1990a] —, "Incompatible Results of Quantum Measurements", PL **151A** (1990): 107–108. {4, 9}

[PERES 1990b] —, "Thermodynamic Constraints on Quantum Axioms", in [*ZUREK* 1990a, 345–56]. {3, 4, 5, 7, 10}

[PERES 1990c] —, "Consecutive Quantum Measurements", in [*CINI/LÉVY-L.* 1990, 122–39]. {4}

[PERES 1991] —, "Two Simple Proofs of the Kochen-Specker Theorem", *Jornal of Physics* **A24** (1991): L175–78. {8}

[*PERES* 1993] —, *Quantum Theory. Concepts and Methods*, Dordrecht, Kluwer, 1993. {2, 8, 9, 10}

[PERES 1994] —, "Classification of Quantum Paradoxes: Nonlocality vs. Contextuality", [*ACCARDI* 1994a, 117–36]. {8, 9}

[PERES 1995a] —, "Nonlocal Effects in Fock Space", PRL **74** (1995): 4571. {9}

[PERES 1995b] —, "Relativistic Quantum Measurements", in [*GREENBERGER/ZEILINGER* 1995a, 445–50]. {4}

[PERES 1996a] —, "Separability Criterion for Density Matrices", PRL **77** (1996): 1413–15. {3, 9, 10}

[PERES 1996b] —, "Generalized Kochen-Specker Theorem", FP **26** (1996): 807–812. {8}

[PERES 1998] —, "Interpreting the Quantum World", preprint. {12}

[PERES/RON 1988] PERES, A./RON, A., "Cryptodeterminism and Quantum Theory", in [*VAN DER MERWE et al.* 1988b, 115–23]. {8}

[PERES/RON 1990] —, "Incomplete 'Collapse' and Partial Quantum Effect Zeno", PR **A42** (1990): 5720–22. {4}

[PERES/WOOTTERS 1991] PERES, A./WOOTTERS, W. K., "Optimal Detection of Quantum Information", PRL **66** (1991): 1119–22. {9, 10}

[PERES/ZUREK 1982] PERES, A./ZUREK, W. H., "Is Quantum Theory Universally Valid?", AJP **50** (1982): 807–10. {4, 5}

[PERIWAL 1998] PERIWAL, Vipul, "Quantum Hamilton-Jacobi Equation", PRL **80** (1998): 4366–69. {3}

[PERRIE *et al.* 1985] PERRIE, W./DUNCAN, A. J./BEYER, H. J./KLEINPOPPEN, H., "Polarization Correlation of the Two Photons Emitted by Metastable Atomic Deuterium: A Test of Bell's Inequality", PRL **54** (1985): 1790–793. {9}

[PESHKIN 1995] PESHKIN, Murray, "On the Aharonov-Bohm Effect with Neutrons", in [*GREENBERGER/ZEILINGER* 1995a, 330–34]. {9}

[*PESHKIN/TONOMURA* 1989] PESHKIN, M./TONOMURA, A., *The Aharonov-Bohm Effect*, Berlin, Springer, 1989. {2, 3, 9; W}

[PESHKIN *et al.* 1961] PESHKIN, M./TALMI, I./TASSIE, L. J., "The Quantum-Mechanical Effects of Magnetic Fields Confined to Inaccesible Regions", AP **12** (1961): 426–35. {9}

[PESLAK 1979] PESLAK, John Jr., "Comparison of Classical and Quantum Mecahnical Uncertanties", AJP **47** (1979): 39–45. {3, 8, 12, 13}

[*PETERSEN* 1968] PETERSEN, A., *Quantum Physics and Philosophical Tradition*, Cambridge Mass., MIT Press, 1968. {12, 14}

[PETROSKY *et al.* 1990] PETROSKY, T./TASAKI, S./PRIGOGINE, I., "Quantum Zeno Effect", PL **151A** (1990): 109–13. {4}

[PFAU *et al.* 1994] PFAU, T./SPÄLTER, S./KURTSIEFER, C./EKSTROM, C. R./MLYNEK, J., "Loss of Spatial Coherence by a Single Spontaneous Emission", PRL **73** (1994): 1223–26. {5, 7}

[PFLEEGOR/MANDEL 1967a] PFLEEGOR, R. L./MANDEL, L., "Interference Effects at the Single Photon Level", PL **24A** (1967): 766–67. {3, 7, 9}

[PFLEEGOR/MANDEL 1967b] —, "Interference of Independent Photon Beams", PR **159** (1967): 1084–88. {7, 9}

[PHILIPP 1949] PHILIPP, Frank G., "Einstein, Mach and Logical Positivism", in [*SCHILPP* 1949, 271–86]. {13, 14}

[PHILIPPIDIS *et al.* 1979] PHILIPPIDIS, C./DEWDNEY, C./HILEY, B. J., "Quantum Interference and the Quantum Potential", NC **52B** (1979): 15–28. {4, 7, 8, 12}

[PHILIPPIDIS *et al.* 1982] PHILIPPIDIS, C./BOHM, D./KAYE, R. D., "The Aharonov-Bohm Effect and the Quantum Potential", NC **71B** (1982): 75–87. {4, 7, 8, 9, 12}

[PHIPPS 1992] PHIPPS, Thomas E. jr., "On the Completeness of Quantum Mechanics", in [*SELLERI* 1992a, 193–205]. {8}

[PHOENIX/KNIGHT 1991] PHOENIX, S. J. D./KNIGHT, P. L., "Comments on 'Collapse and Revival of the State Vector in the Jaynes-Cummings Model: An Example of State Preparation by a Quantum Apparatus"', PRL **66** (1991): 2833. {5, 6}

[PIERCE 1978] PIERCE, John R., "Optical Channels: Practical Limits With Photon Counting", *IEEE Transactions on Communications* **COM-26** (1978): 1819–21; rep. in [*WHEELER/ZUREK* 1983, 743–48]. {10}

[*PIKE/SARKAR* 1987] PIKE, E. R./SARKAR, S. (ed.s), *Quantum Measurement and Chaos*, New York, Plenum, 1987. {4, 13}

[PIKE *et al.* 1987] PIKE, E. R./SARKAR, S./SATCHELL, J. S., "The Dynamic of the Pure State Density Matrix", in [*PIKE/SARKAR* 1987, 119–46]. {3}

[PIPKIN 1978] PIPKIN, F. M., "Atomic Physics Tests of the Basic Concepts in Quantum Mechanics", *Advances in Atomic and Molecular Physics* **14** (1978): 281–340. {3, 12}

[PIRON 1964] PIRON, Constantin, "Axiomatique quantique", *Helvetica Physica Acta* **37** (1964): 439–68. {3}

[PIRON 1969] —, "Les règles de superselection continues", *Helvetica Physica Acta* **42** (1969): 330–38. {3, 4}

[PIRON 1972] —, "Survey of General Quantum Physics", FP **2** (1972): 287–314; rep. in [*HOOKER* 1975, 513–43]. {3}

[PIRON 1976] —, *Foundations of Quantum Physics*, Reading, Massachussets, Benjamin, 1976. {3}

[PIRON 1978] —, "The Lorentz Particles: A New Model for the $\frac{1}{2}$-Spin Particle and the Hydrogen Atom", in [*MARLOW* 1978a, 49–58]. {3}

[PIRON 1981] —, "A Unified Concept of Evolution in Quantum Mechanics", in [*NEUMANN* 1981, 109–112]. {3}

[PIRON 1985] —, "Quantum Mechanics: Fifty Years Later", in [*LAHTI/MITTELSTAEDT* 1985, 207–11]. {3, 12}

[PITOWSKY 1982] PITOWSKY, Itamar, "Resolution of the EPR and Bell Paradoxes", PRL **48** (1982): 1299–302. {8, 9}

[PITOWSKY 1983] —, "Deterministic Model of Spin and Statistics", PR **D27** (1983): 2316–35. {8}

[PITOWSKY 1986] —, "The Range of Quantum Probability", JMP **27** (1986): 1556–65. {3}

[PITOWSKY 1989a] —, "From George Boole to John Bell — The Origins of Bell's Inequality", in [*KAFATOS* 1989, 37–50]. {9, 14}

[*PITOWSKY* 1989b] —, *Quantum Probability Quantum Logic*, Berlin, Springer, 1989. {3, 8, 9}

[PITOWSKY 1991] —, "The Relativity of Quantum Predictions", PL **156A** (1991): 137. {3, 9}

[PITTMAN *et al.* 1996] PITTMAN, T. B./STREKALOV, D. V./MIGDALL, A./RUBIN, M. H./SERGIENKO, A. V./SHIH, Y. H., "Can Two-Photon Interference be Considered the Interference of Two Photons?", PRL **77** (1996): 1917–20. {3, 7}

[*PIZZI* 1978a] PIZZI, Claudio, *Leggi di natura, modalità, ipotesi*, Milano, Feltrinelli, 1978. {14}

[PIZZI 1978b] —, "Introduzione", in [*PIZZI* 1978a, 11–92]. {14}

[PLAGA 1997] PLAGA, R., "On a Possibility to Find Experimental Evidence for the Many-Worlds Interpretation of Quantum Mechanics", FP **27** (1997): 559–77. {4, 12}

[PLANCK 1900a] PLANCK, Max, "Über die Verbesserung der Wien'schen Spektralgleichung", *Verhandlungen der Deutchen Physikalischen Gesellschaft* **2** (1900): 202–204. {3, 13}

[PLANCK 1900b] —, "Zur Theorie des Gesetzes der Energieverteilung im Normalspektrum", *Verhandlungen der Deutchen Physikalischen Gesellschaft* **2** (1900): 237–45. {3, 13}

[PLANCK 1906] —, *Vorlesungen über die Theory der Wärmestrahlung*, Leipzig, Barth, 1906, II ed. 1913. {3, 13}

[PLANCK 1912] —, "La loi du rayonnement noir et l'hypothèse des quantités élémentaires d'action", in LANGEVIN, P./DE BROGLIE, M. (ed.s), *La Théorie du Rayonnement et les Quanta – Rapports et Discussions de la Réunion Tenue à Bruxelles*, Pairs, Gauthier-Villars, 1912: 93–114. {3, 13}

[PLENIO *et al.* 1996] PLENIO, M. B./VEDRAL, V./KNIGHT, P. L., "Computers and Communication in th Quantum World", *Physics World* **9** (Oct. 1996): 19–20. {10; S}

[POOL 1966] POOL, James C. T., "Mathematical Aspects of the Weyl Correspondence", JMP **7** (1966): 66–76. {3, 7}

[POOL 1968] —, "Baer*-Semigroups and the Logic of Quantum Mechanics", CMP **9** (1968): 118–41; rep. in [*HOOKER* 1975, 365–94]. {3}

[POPESCU 1994] POPESCU, Sandu, "Bell's Inequality versus Teleportation: What is Nonlocality?", PRL **72** (1994): 797–99. {9, 10}

[POPESCU 1995] —, "Bell's Inequalities and Density Matrices: Revealing 'Hidden' Nonlocality", PRL **74** (1995): 2619–22. {9}

[POPESCU/D. ROHRLICH 1992a] POPESCU, S./ROHRLICH, D., "Generic Quantum Nonlocality", PL **166A** (1992): 293–97. {9}

[POPESCU/D. ROHRLICH 1992b] —, "Quantum Nonlocality as an Axiom", FP **24** (1994): 379–85. {3, 9}

[POPESCU/D. ROHRLICH 1997] —, "Thermodynamics and the Measure of Entanglement", PR **56** (1997): R3319–21. {3, 10}

[POPESCU *et al.* 1995] POPESCU, S./AHARONOV, Y./COLEMAN, S./GOLDHABER, A. S./NUSSINOV, S./REZNIK, B./ROHRLICH, D./VAIDMAN, L. "Interplay of Aharonov-Bohm and Berry Phases for a Quantum Cloud of Charge", in [*GREENBERGER/ZEILINGER* 1995a, 882–87]. {3, 9}

[*POPPER* 1934a] POPPER, Karl R., *Logik der Forschung*, Wien, Springer, 1934, VIII ed. Tübingen, Mohr, 1984 {14}

[POPPER 1934b] —, "Zur Kritik der Ungenuigkeitsrelationen", *Naturwissenschaften* **22** (1934): 807–808. {12}

[POPPER 1957] —, "The Propensity Interpretation of the Calculus of Probability, and Quantum Mechanics", in KÖRNER, S./PRYCE, M. H. L. (ed.s), *Observation and Interpretation in the Philosophy of Physics — With Special Refernce to Quantum Mechanics*, London, Constable and Company, 1957, New York, Dover, 1962: 65–70. {7, 12}

[POPPER 1959] —, "The Propensity Interpretation of Probability", *British Journal for Philosophy of Science* **10** (1959): 25–42. {14}

[POPPER 1967] —, "Quantum Mechanics Without 'the Observer"', in [*BUNGE* 1967a, 7–44]. {4, 12}

[POPPER 1968] —, "Birkhoff and von Neumanns Interpretation of Quantum Mechanics", NA **219** (1968): 682–85. {3, 14}

[*POPPER* 1982] —, *Quantum Theory and the Schism in Physics*, London, Unwin Hyman Ltd, 1982; Routledge, 1992, 1995. {12, 13, 14}

[*POPPER* 1990] —, *A World of Propensities*, Bristol, Thoemmes, 1990, 1995. {13, 14}

[POWER/KNIGHT 1996] POWER, W. L./KNIGHT, P. L., "Stochastic Simulations of the Quantum Zeno Effect", PR **A53** (1996): 1052–59. {5, 7}

[PRESILLA *et al.* 1997] PRESILLA, C./ONOFRIO, R./PATRIARCA, M., "Classical and Quantum Measurements of Position", JP **A30** (1997)/ 7385–411. {4, 5}

[J. PRICE 1983] PRICE, John F., "Inequalities and Local Uncertainty Principles", JMP **24** (1983): 1711–14. {3}

[*W. PRICE/CHISSICK* 1977] PRICE, W. C./CHISSICK, S. S. (ed.s), *The Uncertainty Principle and Foundations of Quantum Mechanics*, New York, Wiley, 1977. {3}

[PRIGOGINE/ELSKENS 1987] PRIGOGINE, I./ELSKENS, Y., "Irreversibility, Stochasticity, and Non-Locality in Classical Dynamics", in [*HILEY/PEAT* 1987a, 205–223]. {5, 9}

[*PRIMAS* 1983] PRIMAS, Hans, *Chemistry, Quantum Mechanics and Reductionism*, Berlin, Springer, 1983. {12, 13, 14}

[PRIMAS 1990] —, "The Measurement Process in the Individual Interpretation of Quantum Mechanics", in [*CINI/LÉVY-L.* 1990, 49–68]. {4, 12}

[PRIOR 1953] PRIOR, A. N., "Three-Valued Logic and Future Contingents", *Philosophical Quarterly* **3** (1953): 317–26. {14}

[PROSPERI/SCOTTI 1959] PROSPERI, G. M./SCOTTI, A., "Ergodic Theorem in Quantum Mechanics", NC **13** (1959): 1007–1012. {4, 5}

[PROSSER 1976a] PROSSER, R. D., "The Interpretation of Diffraction and Interference in Terms of Energy Flow", *International Journal of Theoretical Physics* **15** (1976): 169–180. {3, 12}

[PROSSER 1976b] —, "Quantum Theory and the Nature of Interference", *International Journal of Theoretical Physics* **15** (1976): 181–93. {4}

[PRUGOVEČKI 1967] PRUGOVEČKI, Eduard, "A Theory of Measurement of Incompatible Observables in Quantum Mechanics", *Canadian Journal of Physics* **45** (1967): 2173–219. {3, 4}

[*PRUGOVEČKI* 1971] —, *Quantum Mechanics in Hilbert Space*, 1971, II ed., New York, Academic Press, 1981. {2, 3}

[*PRUGOVEČKI* 1976b] —, "Probability Measures on Fuzzy Events in Phase Space", JMP **17** (1976): 517–23. {3, 8}

[PRUGOVEČKI 1977a] —, "Information-Theoretical Aspects of Quantum Measurement, *International Journal of Theoretical Physics* **16** (1977): 321–33. {4, 10}

[PRUGOVEČKI 1977b] —, "On Fuzzy Spin Spaces", JP **A10** (1977): 543–49. {3, 8}

[PRUGOVEČKI 1978] —, "Consistent Formulation of Relativistic Dynamics for Massive Spin-Zero Particles in External Fields", PR **D18** (1978): 3655–75. {3, 11}

[PRUGOVEČKI 1981a] —, "A Self-Consistent Approach to Quantum Field Theory for Extended Particles" FP **11** (1981): 355–82; 501. {3, 11}

[PRUGOVEČKI 1981b] —, "Quantum Action Principle and Functional Integration over Paths in Stochastic Phase Space", NC **61A** (1981): 85–118. {3, 8}

[PRUGOVEČKI 1981c] —, "Born's Reciprocity Principle in Stochastic Phase Space", *Lettere a Nuovo Cimento* **32** (1981): 272–76. {3, 8}

[PRUGOVEČKI 1981d] —, "Quantum Space-Time Excitons as Eigenstates of Relativistic Harmonic Oscillator", *Lettere a Nuovo Cimento* **32** (1981): 481; **33** (1981): 481–85. {3, 11}

[PRUGOVEČKI 1984a] —, "The Stochastic Quantum Mechanics Approach to the Unification of Relativity and Quantum Theory", FP **14** (1984): 1147–62. {3, 8, 11, 12}

[*PRUGOVEČKI* 1984b] —, *Stochastic Quantum Mechanics and Quantum Spacetime*, Dordrecht, Reidel, 1984. {3, 4, 8, 12}

[PRUGOVEČKI 1985] —, "Quantum Geometry and the EPR Gedankenexperiment", in [*LAHTI/MITTELSTAEDT* 1985, 525–39]. {9}

[*PRUGOVEČKI* 1992] —, *Quantum Geometry. A Framework for Quantum General Relativity*, Dordrecht, Kluwer, 1992. {3, 8}

[PRUGOVEČKI 1994] —, "On Foundational and Geometric Critical Aspects of Quantum Electrodynamics", FP **24** (1994): 335–62 . {3, 8}

[*PRUGOVEČKI* 1996] —, *Principles for Quantum General Relativity*, Singapore, World Scientific, 1996. {3, 8}

[*PUTNAM* PP] PUTNAM, Hilary, *Philosophical Papers*, Cambridge, University Press, 1975–94. {7, 12, 14}

[PUTNAM 1957] —, "Three-Valued Logic", *Philosophical Studies* **8** (1957): 73–80; rep. in [*PUTNAM* PP, I, 166–73]. {7, 12, 14}

[PUTNAM 1961] —, "Comments on the Paper of David Sharp", PS **28** (1961): 234–37. {7, 12}

[PUTNAM 1964] —, "Discussion: Comments on Comments on Comments: A reply to Margenau and Wigner", PS **31** (1964): 1–6; rep. in [*PUTNAM* PP, I, 159–65]. {7, 12, 14}

[PUTNAM 1965] —, "A Philosophers Looks at Quantum Mechanics", in R. COLODNY (ed.), *Beyond the Edge of Certainty*, New Jersey, Prentice-Hall, 1965; rep. in [*PUTNAM* PP, I, 130–58]. {12}

[PUTNAM 1968] —, "Is Logic Empirical?", in COHEN/WARTOFSKI (ed.s), *Boston Studies in the Philosophy of Science*,Dordrecht, Reidel, 1968; rep. as "The Logic of Quantum Mechanics" in [*PUTNAM* PP, I, 174–97]. {7, 12, 14}

[PUTNAM 1974] —, "How to Think Quantum-Logically", *Synthese* **29** (1974): 55–61. {7, 12, 14}

[PUTNAM 1981] —, "Quantum Mechanics and the Observer", *Erkenntnis* **16** (1981): 193–219; rep. in [*PUTNAM* PP, III, 248–70]. {4, 12}

[PYKACZ/SANTOS 1991] PYKACZ, J./SANTOS, E., "Hidden Variables in Quantum Logic Approach reexamined", JMP **32** (1991): 1287–92. {3, 8, 9}

[QUADT 1989] QUADT, Ralf, "The Nonobjectivity of Past Events in Quantum Mechanics", FP **19** (1989): 1027–35. {6}

[QUADT/BUSCH 1994] QUADT, R./BUSCH, P., "Coarse Graining and the Quantum-Classical Connection", *Open Systems and Information Dynamics* **2** (1994): 129–55. {4, 5}

[QUINE 1948] QUINE, Willard van Orman, "On Waht There Is", *Review of Metaphysics* (1948) rep. in [*QUINE* 1953, 1–19]. {14}

[QUINE 1951] —, "Two Dogmas of Empiricism", *Philosophical Review* (1951), rep. in [*QUINE* 1953, 20–46]. {14}

[*QUINE* 1953] —, *From a Logical Point of View*, Cambridge, Mass., Harvard University Press, 1953, II ed. 1961, 1980. {14}

[*QUINE* 1969] —, *Ontological Relativity and Other Essays*, New York, Columbia University Press, 1969. {14}

[RAIMOND *et al.* 1997] RAIMOND, J. M./BRUNE, M./HAROCHE, S., "Reversible Decoherence of a Mesoscopic Superposition of Field States", PRL **79** (1997): 1964–67. {4, 5, 6}

[RANDALL/FOULIS 1973] RANDALL, C. H./FOULIS, D. J., "Operational Statistics. II. Manuals of Operations and their Logics", JMP **14** (1973): 1472–80. {3, 4}

[RANDALL/FOULIS 1981] —, "Operational Statistics and Tensor Products", in [*NEUMANN* 1981, 21–28]. {3}

[RANDALL/FOULIS 1983] —, "Properties and Operational Propositions in Quantum Mechanics", FP **13** (1983): 843–57. {3}

[RANFAGNI *et al.* 1993] RANFAGNI, A./FABENI, P./PAZZI, G. P./MUGNAI, D., "Anomalous Pulse Delay in Microwave Propagation: A Palusible Connection to the Tunneling Time", PR **E48** (1993): 1453–60. {9}

[RANKIN 1965] RANKIN, Bayard, "Quantum Mechanical Time", JMP **6** (1965): 1057–71. {3}

[RARITY/TAPSTER 1990] RARITY, J. G./TAPSTER, P. R., "Experimental Violation of Bell's Inequality Based on Phase and Momentum", PRL **64**: 2495–98. {9}

[RARITY/TAPSTER 1994] —, "Interference and indistinguishability of Photons", in [*DE MARTINI et al.* 1994, 140–49]. {9}

[RARITY *et al.* 1990] RARITY, J. G./TAPSTER, P. R./JAKEMAN, E./LARCHUK, T./CAMPOS, R. A./TEICH, M. C./SALEH, B. E. A., "Two-Photon Interference in a Mach-Zehnder Interferometer", PRL **65** (1990): 1348–51. {9}

[RASTALL 1985] RASTALL, Peter, "Locality, Bell's Theorem, and Quantum Mechanics", FP 15 (1985): 963–72. {9}

[RAUCH 1987] RAUCH, Helmut, "Quantum Measurement in Neutron Interferometry", in [*NAMIKI et al.* 1987, 3–17]. {4}

[RAUCH 1992] —, "Neutron Interferometric Tests of Quantum Mechanics", in [*SELLERI* 1992a, 207–34]. {4, 7}

[RAUCH 1993] —, "Phase Space Coupling in Interference and EPR Experiments", PL **173A** (1993): 240–42 . {9}

[RAUCH 1995] —, "More Quantum Information due to Postselection in Neutron Interferometry", [*GREENBERGER/ZEILINGER* 1995a, 263–87]. {7, 12}

[RAUCH/VIGIER 1990] RAUCH, H./VIGIER, J.-P., "Proposed Neutron Interferometry Test of Einstein's 'Einweg' Assumption in the Bohr-Einstein Controversy", PL **151A** (1990): 269–75. {7, 8, 12}

[RAUCH *et al.* 1975] RAUCH, H./ZEILINGER, A./BADUREK, G./WILFING, A., "Verification of Coherent Spinor Rotation of Fermions", PL **54A** (1975): 425–27. {4}

[RAUCH *et al.* 1990] RAUCH, H./SUMMHAMMER, J./ZAWISKY, M./JERICHA, E., "Low-Contrast and Low-Counting-Rate Measurements in Neutron Interferometry", PR **A42** (1990): 3726–32. {4, 5, 7}

[RAYSKI/RAYSKI 1977] RAYSKI, J./RAYSKI, J. M. Jr., "On the Meaning of the Time-Energy Uncertainty Relation", in [*W. PRICE/CHISSICK* 1977, 13–20]. {3}

[RECAMI 1977] RECAMI, Erasmo, "A Time Operator and the Time-Energy Uncertainty Relation", in [*W. PRICE/CHISSICK* 1977, 21–28]. {3}

[RECK *et al.* 1994] RECK, M./ZEILINGER, A./BERTANI, P./BERNSTEIN, H., "Experimental Realization of Any Discrete Unitary Operator", PRL **73** (1994): 58–61. {3}

[REDEI 1989] REDEI, Miklos, "Quantum Field Theory, Bell's Inequalities and the Problem of Hidden Variables", in [*KAFATOS* 1989, 73–76]. {9, 11}

[REDHEAD 1986] REDHEAD, Michael, "Relativity and Quantum Mechanics Conflict or Peaceful Coexistence", in [*GREENBERGER* 1986, 14–20]. {12, 13}

[REDHEAD 1987a] —, "Whither Complementarity?", in N. RESCHER (ed.), *Scientific Inquiry in Philosophical Perspective*, Lanham, University Press of America, 1987. {12}

[*REDHEAD* 1987b] —, *Incompleteness, Nonlocality and Realism*, Oxford, Clarendon, 1987, 1992. {8, 9, 12, 14}

[*REDHEAD* 1995] —, *From Physics to Metaphysics*, Cambridge, University Press, 1995, 1996. {12, 14; S}

[REECE 1977] REECE, Gordon, "In Praise of Uncertainty", in [*W. PRICE/CHISSICK* 1977, 7–12]. {12, 14}

[REICHENBACH 1929] REICHENBACH, Hans, "Ziele und Wege der physikalischen Erkenntnis", in GEIGER/SCHEEL (ed.s), Handbuch der Physik, IV, Berlin, Springer, 1929: 78. {14}

[REICHENBACH 1932] —, "Wahrscheinlichkeitslogik", *Berliner Berichte* (1932): 476–88. {14}

[*REICHENBACH* 1938] —, *Experience and Prediction*, Chicago, University Press, 1938. {14}

[*REICHENBACH* 1944] —, *Philosophic Foundations of Quantum Mechanics*, Berkeley, University of California Press, 1944, 1982. {4, 7, 8, 12, 14}

[REICHENBACH 1946] —, "Reply to Ernest Nagels Criticism of My View on Quantum Mechanics", *Journal of Philosophy* **43** (1946): 239–47. {7, 12, 14}

[REICHENBACH 1953] —, "La signification philosophique du dualisme ondes-corpuscules", in [*GEORGE* 1953, 117–34]. {4, 7, 12}

[REID/DRUMMOND 1988] REID, M. D./DRUMMOND, P. D., "Quantum Correlations of Phase in Nondegenerate Parametreic Oscilation", PRL **60** (1988): 2731–33. {9}

[REID/KRIPPNER 1993] REID, M. D./KRIPPNER, L., "Macroscopic Quantum Superposition States in Nondegenerate Parametric Oscillation", PR **A47** (1993): 552–55. {5}

[REID/MUNRO 1992] REID, M. D./MUNRO, W. J., "Macroscopic Boson States Exhibiting the Greenberger-Horne-Zeilinger Contradiction with Local Realism", PRL **69** (1992): 997–1001. {5, 9}

[REID/WALLS 1984] REID, M. D./WALLS, D. F., "Violation of Bell's Inequalities in Quantum Optics", PRL **53** (1984): 955–57. {9}

[REID/WALLS 1986] —, "Violations of Classical Inequalities in Quantum Optics", PR **A34** (1986): 1260–276. {9; W}

[REMPE *et al.* 1987] REMPE, G./WALTHER, H./KLEIN, N., "Observation of Quantum Collpase and Revival in a One-Atom Maser", PRL **58** (1987): 353–56. {5, 6}

[RENNINGER 1960] RENNINGER, Mauritius K., "Messungen ohne Störung des Meßobjekts", ZP **158** (1960): 417–21. {4}

[*RENYI* 1970] RENYI, A., *Foundations of Probability*, San Francisco, Holden-Day, 1970. {1}

[RESTA 1998] RESTA, Raffaele, "Quantum-Mechanical Position Operator in Extended Systems", PRL **80** (1998): 1800–803. {3}

[REVZEN/MANN 1996] REVZEN, M./MANN, A., "Bell's Inequality for a Single Particle", FP **26** (1996): 847–50. {9}

[RICHTER/WÜNSCHE 1996] RICHTER, T./WÜNSCHE, A., "Determination of Occupation Probabilities from Time-Averaged Position Distributions", PR **A53** (1996): R1974–77. {7}

[ROBERTSON 1929] ROBERTSON, H. P., "The Uncertainity Principle", PR **34** (1929): 163–64; rep. in [*WHEELER/ZUREK* 1983, 127–28]. {3}

[ROBINSON 1969] ROBINSON, M. C., "A Thought Experiment Violating Heisenbergs Uncertainty Principle", *Canadian Journal of Physics* **47** (1969): 963–67. {3, 12}

[D. ROHRLICH *et al.* 1995] ROHRLICH, D./AHARONOV, Y./POPESCU, S./VAIDMAN, L. , "Negative Kinetic Energy Between Past and Future State Vectors", in [*GREENBERGER/ZEILINGER* 1995a, 394–99]. {4, 9}

[F. ROHRLICH 1985] ROHRLICH, Fritz, "Schrödingers Criticism of Quantum Mechanics. Fifty Years Later", in [*LAHTI/MITTELSTAEDT* 1985, 555–72]. {12}

[F. ROHRLICH 1986] "Reality and Quantum Mechanics", in [*GREENBERGER* 1986, 373–81]. {8, 12}

[RONCADELLI 1996] RONCADELLI, Marco, "Random Path Approach to Quantum Mechanics", in [*DE MARTINO et al.* 1996, 252–63]. {12}

[RONCADELLI/DEFENDI 1992] RONCADELLI, Marco/DEFENDI, Antonio, *I cammini di Feynman*, Pavia, Università degli studi, 1992. {2}

[ROSEN 1945] ROSEN, Nathan, "On Waves and Particles", *Journal of the Elisha Mitchell Scientific Society* 61 (1945): 67–73. {7, 12}

[ROSEN 1947] —, "Statistical Geometry and Fundamental Particles", PR **72** (1947): 298–303. {3, 13}

[ROSEN 1962] —, "Quantum Geometry", AP **19** (1962): 165–72. {3}

[ROSEN 1985] —, "Quantum Mechanics and Reality", in [*LAHTI/ MITTELSTAEDT* 1985, 17–33]. {8, 12}

[ROSENBAUM 1969] ROSENBAUM, D. M., "Super Hilbert Space and the Quantum Mechanical Time-Operator", JMP **10** (1969): 1127–44. {3}

[ROSENFELD 1953] ROSENFELD, Léon, "L'évidence de la complémentarité", in [*GEORGE* 1953, 43–65]. {12}

[ROSENFELD 1961] —, "Foundations of Quantum Theory and Complementarity", NA **190** (1961): 384–88. {12}

[ROSS 1976] ROSS, David J., "Operator-Observables Correspondence", in [*SUPPES* 1976, 365–96]. {3}

[*ROTH/INOMATA* 1984] ROTH, L. M./INOMATA, A. (ed.s), *Fundamental Questions in Quantum Mechanics*, New York, Gordon and Breach, 1984. {3–5, 8-12}

[ROVELLI 1998a] ROVELLI, Carlo, "What Is a Gauge Transformation in Quantum Mechanics?", PRL **80** (1998): 4613–16. {3}

[ROVELLI 1998b] —, "'Incerto tempore, incertisque loci': Can We Compute the Exact Time at Which a Quantum Mesurement Happens?", quant-ph/9802020 v3. {3, 4}

[ROY/SINGH 1989] ROY, S. M./SINGH, V., "Hidden Variable Theories Without Non-Local Signaling and Their Experimental Tests", PL **139A** (1989): 437–41. {8, 9}

[ROY/SINGH 1990] —, "Generalized Beable Quantum Field Theory", PL **234B** (1990): 117–20. {8, 11, 12}

[ROYER 1985] ROYER, Antoine, "Measurement of the Wigner Function", PRL **55** (1985): 2745–48. {3, 4, 7, 12}

[ROYER 1989] —, "Mesurement of Quantum States and the Wigner Function", FP **19** (1989): 3–30. {3, 4, 7, 12}

[ROYER 1994] —, "Reversible Quantum Measurements on a Spin 1/2 and Measuring the State of a Single System", PRL **73** (1994): 913–17. {4, 7}

[ROYER 1995] —, "Erratum to 'Reversible Quantum Measurements on a Spin 1/2 and Measuring the State of a Single System"', PRL **74** (1995): 1040. {4, 7}

[ROYER 1996] —, "Reduced Dynamics with Initial Correlations, and Time-Dependent Environment and Hamiltonians", PRL **77** (1996): 3272–75. {5}

[RÜTTIMANN 1970] RÜTTIMANN, G. T., "On the Logical Structure of Quantum Mechanics", FP **1** (1970): 173–82; rep. in [*HOOKER* 1979, 109–19]. {3}

[RÜTTIMANN 1981] —, "Detectable Properties and Spectral Quantum Logics", in [*NEUMANN* 1981, 35–47]. {3}

[RYFF 1992] RYFF, Luiz C., "Gedanken Experiments on Duality", in [*SELLERI* 1992a, 235–52]. {7, 12}

[RYLOV 1971] RYLOV, Yuri A., "Quantum Mechanics as a Theory of Relativistic Brownian Motion", *Annalen der Physik* **27** (1971): 1–11. {5, 11}

[RYLOV 1977] "The Correspondence Principle and Measurability of Physical Quantities in Quantum Mechanics", in [*W. PRICE/CHISSICK* 1977, 109–146]. {4, 12}

[*SACCHERI* 1733] SACCHERI, Girolamo, *Euclides ab omni naevo vindicatus*, 1733. {14, 15}

[SACHS 1973] SACHS, Mendel, "On the Nature of Light and the Problem of Matter", in [*HOOKER* 1973a, 346–68]. {3}

[*SALAM/WIGNER]* SALAM, A./WIGNER, E. P. (ed.s), *Aspects of Quantum Theory*, Cambridge, University Press, 1972. {3, 11, 12}

[SALECKER/WIGNER 1958] SALECKER, H./WIGNER, E. P., "Quantum Limitations of the Measurement of Space-Time Distances", PR **109** (1958): 571–77. {3, 4}

[SAMUEL/BHANDARI 1988] SAMUEL, J./BHANDARI, R., "General Setting for Berry's Phase", PRL **60** (1988): 2339–42; rep. in [*SHAPERE/WILCZEK* 1989, 149–52]. {3}

[SANDERS 1992] SANDERS, Barry C., "Entangled Coherent States", PR **A45** (1992): 6811–15; *Erratum*: PR **A46** (1992): 2966. {9}

[SANTHANAM 1977] SANTHANAM, Thalanayar S., "Quantum Mechanics of Bounded Operator", in [*W. PRICE/CHISSICK* 1977, 227–43]. {3}

[SANTOS 1986] SANTOS, Emilio, "The Bell Inequalities as Tests of Classical Logic", PL **115A** (1986): 363–65. {3, 9}

[SANTOS 1991a] —, "Does Quantum Mechanics Violate the Bell Inequalities?", PRL **66** (1991): 1388–90; Errata: 3227. {9}

[SANTOS 1991b] —, "Interpretation of Quantum Formalism and Bell's Theorem", FP **21** (1991): 221-41. {8, 9, 12}

[SANTOS 1992a] —, "Comment on 'Nonlocality of a Single Photon"', PRL **68** (1992): 894. {9}

[SANTOS 1992b] —, "Reply to Ben-Ayreh, Postan and Rae", PRL **68** (1992): 2702–703. {8, 9}

[SANTOS 1992c] —, "Critical Analysis of the Empirical Tests of Local Hidden-Variable Theories", PR **A46** (1992): 3646–56. {8, 9}

[SANTOS 1995] —, "Constraints for the Violation of the Bell Inequality in EPRB Experiments", PL **200A** (1995): 1–6. {8, 9}

[SANTOS 1996] —, "Unreliability of Performed Tests of Bell's Inequality Using Parametric Down-Converted Photons", PL **212A** (1996): 10–14. {9}

[*SARGENT et al.* 1973] SARGENT, M./SCULLY, M. O./LAMB, W., *Laser Physics*, Massachussets, Addison-Wesley, 1973.{1, 2, 4}

[*SAUNDERS/H. BROWN* 1991] SAUNDERS, S./BROWN, H. R. (ed.s), *The Philosophy of Vacuum*, Oxford, Clarendon, 1991. {8, 9, 12, 13, 14}

[SAUTER *et al.* 1986] SAUTER, T./NEUHAUSER, D./BLATT, R./TOSCHEK, P. E., "Observation of Quantum Jumps", PRL **57** (1986): 1696–98. {5}

[SAVAGE/WALLS 1985a] SAVAGE, C. M./WALLS, D. F., "Damping of Quantum Coherence: The Master-Equation Approach", PR **A32** (1985): 2316–23. {5}

[SAVAGE/WALLS 1985b] —, "Quantum Coherence and Interference of Damped Free Particles", PR **A32** (1985): 3487–92. {5}

[SAVAGE *et al.* 1990] SAVAGE, C. M./BRAUNSTEIN, S. L./WALLS, D. F., "Macroscopic Quantum Superpositions by Means of Single-Atom Dispersion", *Optics Letters* **15** (1990): 628–30. {5}

[SCHACK 1998] SCHACK, Rüdiger, "Using a Quantum Computer to Investigate Quantum Chaos", PR **A57** (1998): 1634–35. {10, 13}

[SCHAFIR 1997] SCHAFIR, R. L., "Comment on 'Entangled Entanglement"', PR **A56** (1997): 4335. {3, 9}

[SCHAUER 1994] SCHAUER, Daniel L., "Comment on 'Quantum Mecahnics, Local Realistic Theories, and Lorentz-Invariant Realistic Theories"' PRL **72** (1994): 782. {8, 9}

[*SCHEIBE* 1964] SCHEIBE, E., *Die kontingenten Aussagen in der Physik*, Frankfurt/M., Athenäum, 1964. {12, 13, 14}

[*SCHEIBE* 1973] —, *The Logical Analysis of Quantum Mechanics*, Oxford, Pergamon, 1973. {3}

[SCHENZLE/BREWER 1986] SCHENZLE, A./BREWER, R. G., "Macroscopic Quantum Jumps in a Single Atom", PR **A34** (1986): 3127–42. {5}

[SCHENZLE *et al.* 1986] SCHENZLE, A./DEVOE, R. G./BREWER, R. G., "Possibility of Quantum Jumps", PR **A33** (1986): 2127–30. {4, 5}

[*SCHIFF* 1955] SCHIFF, Leonard, *Quantum Mechanics*, 1955, III ed., New York, McGraw-Hill, 1968. {2}

[SCHILLER *et al.* 1996] SCHILLER, S./BREITENBACH, G./PEREIRA, S. F./MÜLLER, T./MLYNEK, J., "Quantum Statistics of the Squeezed Vacuum by Measurement of the Density Matrix in the Number Representation", PRL **77** (1996): 2933–36. {5, 7}

[*SCHILPP* 1949] SCHILPP, Arthur (ed.), *Albert Einstein. Philosopher-Scientist*, La Salle (Illinois), Open Court, 1949, III ed. 1988. {3, 12, 13, 14}

[*SCHLEICH/BARNETT* 1993] SCHLEICH, W. P./BARNETT, S. M. (ed.s), *Quantum Phase and Phase Dependent Measurements*, Stockholm, Royal Swedish Society, 1993. {3, 4}

[SCHLEICH *et al.* 1988] SCHLEICH, W./WALLS, D. F./WHEELER, J. A., "Area of Overlap and Interference in Phase Space Versus Wigner Pseudoprobabilities", PR **A38** (1988): 1177–86. {3}

[SCHLESINGER 1996] SCHLESINGER, George N., "The Power of Thought Experiments", FP **26** (1996): 467–82. {14}

[SCHLICK 1931] SCHLICK, Moritz, "Die Kausalität in der gegenwärtigen Physik", *Naturwissenschaften* **19** (1931): 145–62. {13, 14}

[SCHLIEDER 1968] SCHLIEDER, S., "Einige Bemerkungen zur Zustandsänderung von relativistischen quantenmechanischen Systemen durch Messungen und zur Lokalitätsforderung", CMP **7** (1968): 305–331. {4, 11}

[SCHLIEDER 1969] —, "Einige Bemerkungen über Projektionsoperatoren", CMP **13** (1969): 216–25. {3, 4}

[SCHLIENZ/MAHLER 1995] SCHLIENZ, J./MAHLER, G., "Description of Entanglement", PR **52** (1995): 4396–4404. {9, 10}

[E. SCHMIDT 1907] SCHMIDT, E., "Zur Theorie des linearen und nichtlinearen Integralgleichungen. I Teil: Entwicklung willkürlicher Funktionen nach Systemen vorgeschriebener", *Mathematische Annalen* **63** (1907): 433–76. {15}

[M. SCHMIDT 1992] SCHMIDT, Michael, "Wind Effect of Empty Waves in a Pfleegor-Mandel-Type Experiment for Electrons", in [*SELLERI* 1992a, 253–76]. {7}

[SCHNEIDER 1994] SCHNEIDER, C., "Two Interpretations of Objective Probabilities", *Philosophia Naturalis* **31** (1994): 107–31. {3, 12}

[*SCHOMMERS* 1989a] SCHOMMERS, Wolfram (ed.), *Quantum Theory and Pictures of Reality*, Berlin, Springer, 1989. {3, 8, 12}

[SCHOMMERS 1989b] —, "Evolution of Quantum Theory", in [*SCHOMMERS* 1989a, 1–48]. {3, 12}

[SCHOMMERS 1989c] —, "Space-Time and Quantum Phenomena", in [*SCHOMMERS* 1989a, 217–77]. {12, 13}

[*SCHRÖDINGER* GA] SCHRÖDINGER, Erwin, *Gesammelte Abhandlungen*, Wien, Verlag der österreichischen Akademie der Wissenschaften, 1984. {3, 4, 5, 7, 8, 9, 12, 13, 14}

[SCHRÖDINGER 1926a] —, "Quantisierung als Eigenwertproblem. I–II", *Annalen der Physik* **79** (1926): 361–76 and 489–527; rep. in [*SCHRÖDINGER* GA, III, 82–136]. {3}

[SCHRÖDINGER 1926b] —, "Quantisierung als Eigenwertproblem. III", *Annalen der Physik* **80** (1926): 437–90; rep. in : [*SCHRÖDINGER* GA, III, 166–219]. {3}

[SCHRÖDINGER 1926c] —, "Quantisierung als Eigenwertproblem. IV", *Annalen der Physik* **81** (1926): 109–39; rep. in [*SCHRÖDINGER* GA, III, 220–50]. {3}

[SCHRÖDINGER 1926d] —, "Der stetige Übergang von der Mikro- zur Makromechanik", *Naturwissenschaften* **14** (1926): 664–66; rep. in [*SCHRÖDINGER* GA, III, 137–42]. {5, 12}

[SCHRÖDINGER 1926e] —, "Über das Verhältnis der Heisenberg-Born-Jordanschen Quantenmechanik zu der meinen", *Annalen der Physik* **79** (1926): 734–56; rep. in [*SCHRÖDINGER* GA, III, 143–65]. {3}

[SCHRÖDINGER 1926f] —, "An Undulatory Theory of the Mechanics of Atoms and Molecules", PR **28** (1926): 1049–70; rep in [*SCHRÖDINGER* GA, III, 280–301]. {12}

[SCHRÖDINGER 1927a] —, "Über den Comptoneffekt", *Annalen der Physik* **82** (1927): 257–64; rep. in [*SCHRÖDINGER* GA, III, 251–58]. {12}

[SCHRÖDINGER 1927b] —, "Der Energieimpulssatz der Materiewellen", *Annalen der Physik* **82** (1927): 265–73; rep. in [*SCHRÖDINGER* GA, III, 259–66]. {3}

[SCHRÖDINGER 1927c] —, "Energieaustausch nach der Wellenmechanik", *Annalen der Physik* **83** (1927): 956–68; rep. in [*SCHRÖDINGER* GA, III, 267–79]. {3}

[SCHRÖDINGER 1929] —, "Die Erfassung der Quantengesetze durch kontinuierliche Funktionen, *Naturwissenschaften* **17** (1929): 486–89; rep. in [*SCHRÖDINGER* GA, III, 326–29]. {12}

[SCHRÖDINGER 1930a] —, "Zum Heisenbergschen Unschärfeprinzip", *Sitzungsberichte der Preussischen Akademie der Wissenschaften* (1930): 296–303; rep. in [*SCHRÖDINGER* GA, III, 348–56]. {3, 12}

[SCHRÖDINGER 1930b] —, "Über die kräftefreie Bewegung in der relativistischen Quantenmechanik", *Sitzungsberichte der Preussischen Akademie der Wissenschaften* (1930): 418–28; rep. in [*SCHRÖDINGER* GA, III, 357–68]. {3, 11}

[SCHRÖDINGER 1931a] —, "Über die Umkehrung der Naturgesetzte", *Sitzungsberichte der Preussischen Akademie der Wissenschaften* (1931): 144–53; rep. in [*SCHRÖDINGER* GA, I, 412–22]. {12}

[SCHRÖDINGER 1931b] —, "Spezielle Relativitätstheorie und Quantenmechanik", *Sitzungs-Berichte der Preussischen Akademie der Wissenschaften* (1931): 238–48; rep. in [*SCHRÖDINGER* GA, III, 380–90]. {12, 13}

[SCHRÖDINGER 1932] —, "Sur la théorie relativiste de l'électron et l'interpretation de la mécanique quantique", *Annales de l'Institut Henri Poincaré* **2** (1932): 269–310; rep. in [*SCHRÖDINGER* GA, III, 394–435]. {12}

[SCHRÖDINGER 1935a] —, "Die gegenwärtige Situation in der Quantenmechanick. I–III", *Naturwissenschaften* **23** (1935): 807–12, 823–28, 844–49; rep. in [*SCHRÖDINGER* GA, III, 484–501]. {5, 12}

[SCHRÖDINGER 1935b] —, "Discussion of Probability Relations Between Separated Systems", *Proceedings of the Cambridge Philosophical Society* **32** (1936): 446–52; rep. in [*SCHRÖDINGER* GA, I, 433–39]. {3, 8, 9}

[SCHRÖDINGER 1936] —, "Probability Relations Between Separated Systems", *Proceedings of the Cambridge Philosophical Society* **31** (1935): 555–63; rep. in [*SCHRÖDINGER* GA, I, 424–32]. {3, 8, 9}

[SCHRÖDINGER 1952] —, "Are There Quantum Jumps? I–II", *The British Journal for the Philosophy of Science* **3** (1952): 109–123, 233–42; rep. in [*SCHRÖDINGER* GA, IV, 478–502]. {12}

[SCHRÖDINGER 1953] —, "The Meaning of Wave Mechanics", in [*GEORGE* 1953, 16–30]; rep. in [*SCHRÖDINGER* GA, III, 694–708] {12}

[SCHRÖDINGER 1954] —, "Measurement of Length and Angle in Quantum Mechanics", NA **173** (1954): 442; rep. in [*SCHRÖDINGER* GA, III, 723]. {4}

[SCHRÖDINGER 1955] —, "The Philosophy of Experiment", NC **1** (1955): 5; rep. in [*SCHRÖDINGER* GA, IV, 558–68]. {14}

[SCHRÖDINGER 1958] —, "Might Perhaps Energy Be a Merely Statistical Concepts?", NC **9** (1958): 162–70; rep. in [*SCHRÖDINGER* GA, I, 502–510]. {12}

[SCHROECK 1978] SCHROECK, Franklin E., Jr., "Measures with Minimum Uncertainty on Non-Commutative Algebras with Application to Measurement Theory in Quantum Mechanics", in [*MARLOW* 1978a, 299–327]. {3}

[SCHROECK 1981] —, "A Model of a Quantum Mechanical Treatment of Measurement With a Physical Interpretation", JMP **22** (1981): 2562–72. {4, 12}

[SCHROECK 1982a] —, "On the Stochastic Measurement of Incompatible Spin Components", FP **12** (1982): 479–97. {3, 4}

[SCHROECK 1982b] —, "The Transitions Among Classical Mechanics, Quantum Mechanics, and Stochastic Quantum Mechanics", FP **12** (1982): 825–41. {4, 5, 8, 12}

[SCHROECK 1985a] —, "On the Nonoccurence of two Paradoxes in the Measurement Scheme of Stochastic Quantum Mechanics", FP **15** (1985): 279–302. {4, 8}

[SCHROECK 1985b] —, "The Dequantization Programme for Stochastic Quantum Mechanics", JMP**26** (1985): 306–310. {5, 12}

[SCHROECK 1985c] —, "Entropy and Measurement in Stochastic Quantum Mechanics", in [*LAHTI/MITTELSTAEDT* 1985, 573–90]. {4, 10, 12}

[SCHROECK 1985d] —, "Compatible Observables That Do not Commute", FP **15** (1985): 677–81. {3, 4}

[SCHROECK 1989] —, "On the Entropic Formulation of Uncertainty for Quantum Measurements", JMP **30** (1989): 2078–2082. {3, 10}

[SCHROECK 1994] —, "Quantum Mechanics on Phase Space: an Overview", [*ACCARDI* 1994a, 327–38]. {8}

[*SCHROECK* 1996] —, *Quantum Mechanics on Phase Space*, Dordrecht, Kluwer, 1996. {3, 4, 10, 12}

[J. SCHUMACHER 1984] SCHUMACHER, John A., "The Quantum Mecahnics of Vision", in [*ROTH/INOMATA* 1984, 93–109]. {12, 14}

[B. SCHUMACHER 1990] SCHUMACHER, Benjamin W., "Information from Quantum Measurements", in [*ZUREK* 1990a, 29–37]. {10}

[B. SCHUMACHER 1991] —, "Information and Quantum Nonseparability", PR **A44** (1991): 7047–52. {9, 10}

[B. SCHUMACHER 1995] —, "Quantum Coding", PR **A51** (1995): 2738–47. {10}

[B. SCHUMACHER 1996] —, "Sending Quantum Entanglement Through Noisy Quantum Channels", PR **A54** (1996): 2614–28. {10}

[B. SCHUMACHER/NIELSEN 1996] SCHUMACHER, B. W./NIELSEN, M. A., "Quantum Data Processing and Error Correction", PR **A54** (1996): 2629–35. {10}

[SCHUMAKER/CAVES 1985] SCHUMAKER, B. L./CAVES, C. M., "New Formalism for Two-Photon Quantum Optics. II", PR **A31** (1985): 3093–3111. {3}

[*SCHWEBER* 1961] SCHWEBER, Silvan S., *An Introduction to Relativistic Quantum Field Theory*, New York, Harper and Row, 1961. {2}

[SCHWEBER 1993] —, "Chancing Conceptualization of Renormalization Theory", in [*L. BROWN* 1993a, 135–66]. {11, 12}

[SCHWINBERG *et al.* 1981] SCHWINBERG, P. B./VAN DYCK, R. S. Jr./DEHMELT, H., "Trapping and Thermalization of Positrons for Geonium Spectroscopy", PL **81A** (1981): 119–20. {3, 12}

[SCHWINGER 1960] SCHWINGER, Julian, *Proceedings of the National Academy of Sciences U.S.A.* **46** (1960): 570–79. {3}

[SCULLY 1983] SCULLY, Marlan O., "How to Make Quantum Mechanics Look Like a Hidden-Variable Theory and vice versa", PR **D28** (1983): 2477–84. {8, 9}

[SCULLY/COHEN 1986] SCULLY, M. O./COHEN, L., "EPRB and the Wigner Distribution for Spin-$\frac{1}{2}$ Particles", in [*GREENBERGER* 1986, 115–17]. {9}

[SCULLY/DRÜHL 1982] SCULLY, M. O./DRÜHL, K., "Quantum Eraser: A Proposed Photon Correlation Experiment Concerning Observation and 'Delayed Choice' in Quantum Mechanics", PR **A25** (1982): 2208–13. {4, 6}

[SCULLY/WALTHER 1989] SCULLY, M. O./WALTHER, H., "Quantum Optical Test of Observation and Complementarity in QM", PR **A39** (1989): 5229–36. {12}

[SCULLY *et al.* 1978a] SCULLY, M. O./SHEA, R./MCCULLEN, J. D., "State Reduction in Quantum Mechanics: A Calculation Example", PRep **43C** (1978): 485–98. {4, 5, 12}

[SCULLY *et al.* 1989] SCULLY, M. O./ENGLERT, B.-G./SCHWINGER, J., "Spin Coherence and Humpty-Dumpty; III. The Effects of Observation", PR **A40** (1989): 1775–84. {4, 6}

[SCULLY *et al.* 1991] SCULLY, M. O./ENGLERT, B.-G./WALTHER, H., "Quantum Optical Tests of Complementarity", NA **351** (1991): 111–16. {4, 6, 7, 10, 12}

[SCULLY *et al.* 1994a] SCULLY, M. O./WALTHER, H./SCHLEICH, W., "Feynman's Approach to Negative Probability in Quantum Mechanics", PR **A49** (1994): 1562–66. {3}

[SEGAL 1947] SEGAL, Irving E., "Postulates for General Quantum Mechanics", *Annals of Mathematics* **48** (1947): 930–48. {3}

[SELLERI 1969] SELLERI, Franco, "On the Wave Function of Quantum Mechanics", *Lettere a Nuovo Cimento* **1** (1969): 908–910. {7, 8}

[SELLERI 1971] —, "Realism and Wave-Function of Quantum Mechanics", in [*D'ESPAGNAT* 1971, 398–406]. {7, 8, 12}

[SELLERI 1982] —, "On the Direct Observability of Quantum Waves", FP **12** (1982): 1087–112. {7, 12}

[SELLERI 1985a] —, "Local Realistic Photon Models and EPR-Type Experiments", PL **108A** (1985): 197–202. {8, 9}

[SELLERI 1985b] —, "New Variable Detection Probability Model for EPR-Type Experiments", in [*LAHTI/MITTELSTAEDT* 1985, 591–607]. {8, 9}

[*SELLERI* 1988] — (ed.), *Quantum Mechanics versus Local Realism*, New York, Plenum, 1988. {8, 9}

[SELLERI 1989] —, "Wave-Particle Duality: Recent Proposals for the Detection of Empty Waves", in [*SCHOMMERS* 1989a, 279–332]. {7}

[*SELLERI* 1992a] — (ed.), *Wave-Particle Duality*, New York, Plenum, 1992. {3, 7, 12}

[SELLERI 1992b] —, "Two-Photon Interference and the Question of Empty Waves", in [*SELLERI* 1992a, 277–90]. {7}

[SELLERI 1997] —, "Incompatibility Between Local Realism and Quantum Mechanics for Pairs of Neutral Kaons", PR **56** (1997): 3493–506. {8, 9}

[SEN 1996] SEN, Siddhartha, "Average Entropy of a Quantum Subsystem", PRL **77** (1996): 1–3. {3, 10}

[SENITZKY 1960] SENITZKY, I. R., "Dissipation in Quantum Mechanics. The Harmonic Oscillator", PR **119** (1960): 670–679. {3, 5}

[SENITZKY 1961] —, "Dissipation in Quantum Mechanics. The Harmonic Oscillator. II", PR **124** (1961): 642–48. {3, 5}

[SENITZKY 1981] —, "Classical Statistics Inherent in Pure Quantum States", PRL **47** (1981): 1503–1506. {3, 5}

[SHANNON 1948] SHANNON, C. E., "A Mathematical Theory of Communication", *Bell System Technical Journal* **27** (1948): 379–423; 623–56. {15}

[*SHAPERE/WILCZEK* 1989] SHAPERE, A./WILCZEK, F. (ed.s), *Geometric Phases in Physics*, Singapore, World Scientific, 1989. {3}

[SHAPIRO/SHEPARD 1991] SHAPIRO, J. H./SHEPARD, S. R., "Quantum Phase Measurement: A System-Theory Perspective", PR **A43** (1991): 3795–818. {3, 4}

[SHAPIRO *et al.* 1989] SHAPIRO, J. H./SHEPARD, S. R./WONG, N. W., "Ultimate Quantum Limits on Phase Measurement", PRL **62** (1989): 2377–80. {3, 4}

[SHARP 1961] SHARP, David H., "The EPR Paradox Re-Examined", PS **28** (1961): 225–33. {8, 9, 12}

[SHE/HEFFNER 1966] SHE, C. Y./HEFFNER, H., "Simultaneous Measurement of Non-commuting Observables", PR **152** (1966): 1103–10. {4}

[SHERMAN 1956] SHERMAN, Seymour, "On Segal's Postulates for General Quantum Mechanics", *Annals of Mathematics* **64** (1956): 593–601. {3}

[SHERRY/SUDARSHAN 1978] SHERRY, T. N./SUDARSHAN, E. C. G., "Interaction Between Classical and Quantum Systems: A New Approach to Quantum Measurement. I", PR **D18** (1978): 4580–89. {3, 4}

[SHERRY/SUDARSHAN 1979] —, "Interaction Between Classical and Quantum Systems: A New Approach to Quantum Measurement. II", PR **D20** (1979): 857–68. {3, 4}

[SHIH/ALLEY 1988] SHIH, Y. H./ALLEY, C. O., "New Type of EPR-Bohm Experiment Using Pairs of Light Quanta Produced by Optical Parametric Down Conversion", PRL **61** (1988): 2921–24. {9}

[SHIH/RUBIN 1993] SHIH, Y. H./RUBIN, M. H., "Four Photon Interference Experiment for the Testing of the Greenberger-Horne-Zeilinger Theorem", PL **182A** (1993): 16–22. {9}

[SHIH/SERGIENKO 1994a] SHIH, Y. H./SERGIENKO, A. V., "Observation of Quantum Beating in a Simple Beam-Splitting Experiment: Two Particle Entanglement in Spin and Space-Time", PR **A50** (1994): 2564–68. {9}

[SHIH *et al.* 1993] SHIH, Y. H./SERGIENKO, A. V./RUBIN, M. H., "EPR State for Space-Time Variables in a Two-Photon Interference Experiment", PR **A47** (1993): 1288–93. {9}

[SHIH *et al.* 1995a] SHIH, Y. H./SERGIENKO, A. V./PITTMAN, T. B./RUBIN, M. H., "EPR and Two-Photon Interference Experiments Using Type-II Parametriuc Downconversion", in [*GREENBERGER/ZEILINGER* 1995a, 40–60]. {9}

[SHIH *et al.* 1995b] SHIH, Y. H./SERGIENKO, A. V./PITTMAN, T. B./STREKALOV, D. V./KLYSHKO, D. N., "Two-Photon 'Ghost' Image and Interference-Diffraction", in [*GREENBERGER/ZEILINGER* 1995a, 121–32]. {3, 9}

[SHIMONY 1963] SHIMONY, Abner, "Role of the Observer in Quantum Theory", AJP **31** (1963): 755–73; rep. in [*SHIMONY* 1993a, II, 3–33]. {4, 12}

[SHIMONY 1966] —, "Review of Landé's New Foundations of Quantum Mechanics", PT **19** (sept. 1966): 85–86. {8, 12}

[SHIMONY 1971a] —, "Filters with Infinitely Many Components", FP **1** (1971): 325–28; rep. in [*SHIMONY* 1993a, II, 68–71]. {3}

[SHIMONY 1971b] —, "Experimental Test of Local Hidden-Variable Theories", in [*D'ESPAGNAT* 1971, 182–94]; rep. in [*SHIMONY* 1993a, II, 77–89]. {3, 8, 9, 12}

[SHIMONY 1978] —, "Metaphysical Problems in the Foundations of Quantum Mechanics", *International Philosophical Quarterly* **18** (1978): 3–17. {12}

[SHIMONY 1979] —, "Proposed Neutron Interferometer Test of Some Non-linear Variants of Wave Mechanics", PR **A20** (1979): 394–96; rep. in [*SHIMONY* 1993a, II, 48–54]. {3, 7, 12}

[SHIMONY 1981] —, "Integral Epistemology", in M. Brewer/B. Collins (eds.), *Scientific Inquiry and Social Sciences*, San Francisco, Jossey-Bass Inc., 1981; rep. in [*SHIMONY* 1993a, I, 3–20]. {14}

[SHIMONY 1984a] —, "Contextual Hidden Variables Theories and Bell's Inequalities", *British Journal for the Philosophy of Science* **35** (1984): 25–45; rep. in [*SHIMONY* 1993a, II, 104–129]. {3, 7, 12}

[SHIMONY 1984b] —, "Response to "Comment on 'Proposed molecular Test of Local Hidden-Variables Theories"" ', PR **A30** (1984): 2130–31. {9}

[SHIMONY 1984c] —, "The Significance of Jarret's Completeness Condition", in [*ROTH/INOMATA* 1984, 29–31]. {8, 9}

[SHIMONY 1988] —, "Issues in the Bohr-Einstein Debate", in FESHBACH/MATSUI/OLESON (ed.s), *Niels Bohr: Physics and the World*, Chur, Harwood, 1988, rep. in [*SHIMONY* 1993a, II, 171–87]. {9, 12, 13}

[SHIMONY 1989] —, "Search for a Worldview which Can Accommodate our Knowledge of Microphysics", in [*CUSHING/MCMULLIN* 1989]; rep. in [*SHIMONY* 1993a, I, 62–76]. {12, 14}

[*SHIMONY* 1990] —, *An Exposition of Bell's Theorem*, New York, Plenum, 1990. {8, 9}

[SHIMONY 1991] —, "Desiderata for a Modified Quantum Dynamics", in A. Fine/M. Forbes/L. Wessels, *Phylosophy of Science Association 1990*, Michingan, 1991: II, 49–59; rep. in [*SHIMONY* 1993a, II, 55–67]. {3, 4, 5, 7, 12}

[*SHIMONY* 1993a] —, *Search for a Naturalistic Point of View*, Cambridge, University Press, 1993. {4, 5, 6, 7, 8, 9, 12, 13, 14}

[SHIMONY 1993b] —, "Reality, Causality, and Closing the Circle", in [*SHIMONY* 1993a, I, 21–61]. {14}

[SHIMONY 1995] —, "Degree of Entanglement", in [*GREENBERGER/ZEILINGER* 1995a, 675–79]. {3, 9, 10}

[SHIMONY *et al.* 1985] SHIMONY, A./HORNE, M. A./CLAUSER, J. F., "Comment on Bell's Theory", *Dialectica* **39** (1985): 97–102; rep. in [*SHIMONY* 1993a, II, 163–68]. {8, 9}

[SHIRKOV 1993] SHIRKOV, Dmitri V., "Historical Remarks on the Renormalization Group", in [*L. BROWN* 1993a, 167–86]. {11}

[SHIZUME 1995] SHIZUME, Kousuke, "Heat Generation Required by Information Erasure", PR **E52** (1995): 3495–99. {3, 10, 13}

[SHOR 1994] SHOR, Peter W., "Algorithms for Quantum Computation: Discrete Log and Factoring", in *Proceedings of the 35th Annual Symposium on the Foundations of Computer Science*, Los Alamos, IEEE Computer Society Press, 1994: 124. {10}

[SHOR 1995] —, "Scheme for Reducing Decoherence in Quantum Computer Memory", PR **A52** (1995): R2493–96. {5, 10}

[SHULL *et al.* 1980] SHULL, C. G./ATWOOD, D. K./ARTHUR, J./HORNE, M. A., "Search For a Nonlinear Variant of the Schrödinger Equation by Neutron Interferometry", PRL **44** (1980): 765–68. {7}

[SIEGEL/WIENER 1956] SIEGEL, A./WIENER, N., "'Theory of Measurement' in Differential-Space Quantum Theory", PR **101** (1956): 429–32. {4}

[SIMON 1983] SIMON, Barry, "Holonomy,the Quantum Adiabatic Theorem, and Berry's Phase", PRL **51** (1983): 2167–70; rep. in [*SHAPERE/WILCZEK* 1989, 137–40]. {3}

[SIMONIUS 1978] SIMONIUS, Markus, "Spontaneous Symmetry Breaking and Blocking of Metastable States", PRL **40** (1978): 980–83. {4, 5}

[SINHA 1997] SINHA, Supurna, "Decoherence at Absolute Zero", PL **228A** (1997): 1–6. {5}

[SJÖQVIST 1995] SJÖQVIST, Erik, "Geometric Phase Shifts in Pilot-Wave Theory", in [*GREENBERGER/ZEILINGER* 1995a, 898–99]. {3, 7, 8}

[SKIRMS 1985] SKIRMS, Brian, "EPR and the Metaphysics of Causation", in [*LAHTI/MITTELSTAEDT* 1985, 609–23]. {9, 14}

[SKLAR 1970] SKLAR, Lawrence, "Is Probability a Dispositional Property?", *Journal of Philosophy* **67** (1970): 355–66. {14}

[SLAWIANOWSKI 1977] SLAWIANOWSKI, Jan S., "Uncertainty, Correspondence and Quasiclassical Compatibility", in [*W. PRICE/CHISSICK* 1977, 147–88]. {3, 5, 12}

[SLEATOR/WEINFURTER 1995a] SLEATOR, T./WEINFURTER, H., "Quantum Teleportation and Quantum Computation", in [*GREENBERGER/ZEILINGER* 1995a, 715–25]. {10}

[SLEATOR/WEINFURTER 1995b] —, "Realizable Universal Quantum Logic Gates", PRL **74** (1995): 4087–90. {10}

[SLOSSER *et al.* 1989] SLOSSER, J. J./MEYSTRE, P./BRAUNSTEIN, S. L., "Harmonic Oscillator Driven by a Quantum Current", PRL **63** (1989): 934–37. {5}

[SLOSSER *et al.* 1990] SLOSSER, J. J./MEYSTRE, P./WRIGHT, E. M., "Generation of Macroscopic Superpositions in a Micromaser", *Optics Letters* **15** (1990): 233–35. {5}

[SLUSHER *et al.* 1985] SLUSHER, R. E./HOLLBERG, L. W./YURKE, B./MERTZ, J. C./VALLEY, J. F., "Observation of Squeezed States Generated by Four-Wave Mixing in an Optical Cavity", PRL **55** (1985): 2409–412. {3}

[F. SMITH 1960] SMITH, Felix T., "Lifetime Matrix in Collision Theory", PR **118** (1960): 349–56. {3}

[*P. SMITH* 1981] SMITH, Peter, *Realism and the Progress of Science*, Cambridge, University Press, 1981. {14}

[SMITHEY *et al.* 1993] SMITHEY, D. T./BECK, M./RAYMER, M. G./FARIDANI, A., "Measurement of the Wigner Distribution and the Density Matrix of a Light Mode Using Optical Homodyne Tomography: Application to Squeezed States and the Vacuum", PRL **70** (1993): 1244–47. {7}

[SNEED 1966] SNEED, Joseph D., "Von Neumanns Argument for the Projection Postulate", PS **33** (1966): 22–39. {7, 12}

[*SNIATYCKI* 1980] SNIATYCKI, J., *Geometric Quantization and Quantum Mechanics*, Berlin, Springer, 1980. {3}

[SOKOLOVSKI 1998] SOKOLOVSKI, D., "From Feynman Histories to Observables", PR **A57** (1998): R1469–72. {3}

[SOKOLOVSKI/BASKIN 1987] SOKOLOVSKI, D./BASKIN, L. M., "Traversal Time in Quantum Scattering", PR **A36** (1987): 4604–611. {3, 9}

[SOKOLOVSKI/CONNOR 1990] SOKOLOVSKI, D./CONNOR, J. N. L., "Path-Integral Analysis of the Time Delay for Wave-Packet Scattering and the Status of Complex Tunneling Times", PR **A42** (1990): 6512–25. {3, 9}

[SOKOLOVSKI/CONNOR 1991] —, "Negative Probability and the Distributions of Dwell, Transmission, and Reflection Times for Quantum Tunneling", PR **A44** (1991): 1500–504. {3, 9}

[SOKOLOVSKI/CONNOR 1993] —, "Quantum Interference and Determination of the Traversal Time", PR **A47** (1993): 4677–80. {3}

[SOMMERFELD 1912] SOMMERFELD, Arnold, "Rapport sur l'application de la théorie de l'élément d'action aux phénomènes moléculaires non périodiques", in LANGEVIN, P./DE BROGLIE, M. (ed.s), *La Théorie du Rayonnement et les Quanta – Rapports et Discussions de la Réunion Tenue à Bruxelles*, Pairs, Gauthier-Villars, 1912: 313–72. {3}

[SONG *et al.* 1990] SONG, S./CAVES, C. M./YURKE, B., "Generation of Superpositions of Classical Distinguishable Quantum States from Optical Back-Action Evasion", PR **A41** (1990): 5261–64. {5}

[SPILLER 1994] SPILLER, T. P., "The Zeno Effect: Measurement versus Decoherence", PL **192A** (1994): 163–68 . {4}

[SPOHN 1980] SPOHN, Herbert, "Kinetic Equations from Hamiltonian Dynamics: Markovian Limits", RMP **53** (1980): 569–615. {3, 5, 13; W}

[SQUIRES 1987] SQUIRES, Euan J., "Many views of One World an Interpretation of Quantum Theory", *European Journal of Physics* **8** (1987): 171–73. {4, 12}

[SQUIRES 1990] —, "An Attempt to Understand the Many-Worlds Interpretation of Quantum Theory", in [*CINI/LÉVY-L.* 1990, 151–60]. {4}

[SQUIRES 1995] —, "Quantum Theory, Relativity, and the Bohm Model", in [*GREENBERGER/ZEILINGER* 1995a, 451–63]. {8, 11, 12}

[SRINIVAS 1976] SRINIVAS, M. D., "Foundations of a Quantum Probability Theory", JMP **16** (1976): 1672–85; rep. in [*HOOKER* 1979, 227–60]. {3}

[SRINIVAS/DAVIES 1981] SRINIVAS, M. D./DAVIES, E. B., "Photon Counting Probabilities in Quantum Optics", *Optica Acta* **28** (1981): 981–96. {4, 5}

[SRINIVAS/DAVIES 1982] —, "What Are the Photon Counting Probabilities for Open Systems — A Reply to Mandel's Comments", *Optica Acta* **29** (1982): 235–38. {4, 5}

[STACHOW 1981] STACHOW, Ernst-Walter, "The Propositional Language of Quantum Physics", in [*NEUMANN* 1981, 95–107]. {3}

[STACHOW 1985] —, "Structures of Quantum Language for Compound Systems", in [*LAHTI/MITTELSTAEDT* 1985, 625–36]. {3}

[STAIRS 1983] STAIRS, Allen, "Quantum Logic, Realism, and Value Definiteness", PS **50** (1983): 578–602. {3, 12, 14}

[STAMATESCU 1996] STAMATESCU, I.-O., "Stochastic Collapse Models", in [*GIULINI et al.* 1996, 249–67]. {4, 5}

[*STAPP* 1968] STAPP, Henry P., *Correlation Experiments and the Nonvalidity of the Ordinary Ideas about the Physical World*, Berkeley, 1968. {9, 12, 13, 14}

[STAPP 1971] —, "S-Matrix Interpretation of Quantum Theory", PR **D3** (1971): 1303–20. {9, 12}

[STAPP 1972] —, "The Copenhagen Interpretation", AJP **40** (1972): 1098–116. {12}

[STAPP 1975] —, "Bell's Theorem and World Process', NC **B29** (1975): 270–76. {9, 12, 13, 14}

[STAPP 1977a] —, "Are Superluminal Connections Necessary?", NC **40B** (1977): 191–205. {9, 12}

[STAPP 1979] —, "Whiteheadian Approach to Quantum Theory and the Generalized Bell's Theorem", FP **9** (1979): 1–25. {9, 12, 13}

[STAPP 1980] —, "Locality and Reality", FP **10** (1980): 767–95. {9}

[STAPP 1985a] —, "On the Unification of Quantum Theory and Classical Physics", in [*LAHTI/MITTELSTAEDT* 1985, 213–21]. {12}

[STAPP 1985b] —, "EPR: What Has it Taught Us?", in [*LAHTI/MITTELSTAEDT* 1985, 637–52]. {9}

[STAPP 1989a] —, "Quantum Nonlocality and the Description of Nature", in [*CUSHING/MCMULLIN* 1989]. {9, 12}

[STAPP 1989b] —, "Quantum Ontologies", in [*KAFATOS* 1989, 269–78]. {12}

[*STAPP* 1993] —, *Mind, Matter, and Quantum Mechanics*, Berlin, Springer, 1993. {4, 5, 9, 12}

[STAPP 1997] —, "Nonlocal Character of Quantum Theory", AJP **65** (1997): 300–304. {9}

[STEANE 1996a] STEANE, A. M., "Error Correcting Codes in Quantum Theory", PRL **77** (1996): 793–97. {10}

[STEANE 1996b] —, "Simple Quantum Error-Correcting Codes", PR **A54** (1996): 4741–51. {10}

[STEGMÜLLER 1969] STEGMÜLLER, W., *Probleme und Resultate der Wissenschaftstheorie und analytischen Philosophie*, Berlin, Springer, 1969. {14}

[STEIN 1984] STEIN, Howard, "The Everett Interpretation of Quantum Mechanics: Many Worlds ofr None?", *Noûs* **18** (1984): 635–52. {4, 12}

[STEINBERG 1995] STEINBERG, Aephraim M., "Conditional Probabilities in Quantum Theory, and the Tunnelling Time Controversy", PR **A52** (1995): 32–42. {3, 4, 9}

[STEINBERG/CHIAO 1994] STEINBERG, A. M./CHIAO, R. Y., "Tunneling Delay Times in One and two Dimensions", PR **A49** (1994): 3283–95. {9}

[STEINBERG *et al.* 1992] STEINBERG, A. M./KWIAT, P. G./CHIAO, R. Y., "Dispersion Cancellation in a Measurement of the Single-Photon Propagation Velocity in Glass", PRL **68** (1992): 2421–24. {9}

[STEINBERG *et al.* 1993] —, "Measurement of the Single-Photon Tunneling Time", PRL **71** (1993): 708–711. {9}

[STEINHAUSER *et al.* 1980] STEINHAUSER, K.-A./STEYERL, A./SCHECKENHOFER, H./MALIK, S. S., "Observation of Quasibound States of the Neutron in Matter", PRL **44** (1980): 1306–309. {3}

[STENGER 1977] STENGER, William, "Intermediate Problems for Eigenvalues in Quantum Theory", in [*W. PRICE/CHISSICK* 1977, 277–92]. {3}

[A. STERN *et al.* 1990] STERN, A./AHARONOV, Y./IMRY, Y., "Phase Uncertainty and Loss of Interference: a General Picture", PR **A41** (1990): 3436–48. {5}

[STEUERNAGEL/VACCARO 1995] STEUERNAGEL, O./VACCARO, J. A., "Reconstructing the Density Operator via Simple Projectors", PRL **75** (1995): 3201–205. {7}

[STONE 1949] STONE, M. H., "Postulates for Barycentric Calculus", *Annals of Mathematics Pura Applied* **(4) 29** (1949): 25–30. {15}

[STOPES-ROE 1962] STOPES-ROE, Harry V., "Interpretation of Quantum Physics", NA **193** (1962): 1276–77. {12}

[STOREY *et al.* 1994a] STOREY, E. P./TAN, S. M./COLLETT, M. J./WALLS, D. F., "Path Detection and the Uncertainty Principle", NA **367** (1994): 626–28. {3, 7, 12}

[STOREY *et al.* 1994b] —, "Physical Origins of Loss of Interference in Welcher-Weg Detection", in [*DE MARTINI et al.* 1994, 120–29]. {3, 7, 12}

[STOREY *et al.* 1995] —, "Reply to Complementarity and Uncertainty of Englert/Scully/Walther", NA **375** (1995): 368. {3, 7, 12}

[STRAUSS 1936] STRAUSS, Martin, "Zur Begründung der statistischen Transformationstheorie der Quantenphysik", *Berliner Berichte* (1936): 382–98; engl. trans. in [*HOOKER* 1975, 27–40]. {3}

[STRAUSS 1937] —, "Mathematics as Logical Syntax A Method to Formalize the Language of a Physical Theory", *Erkenntnis* **7** (1937–38): 147–53; rep. in [*HOOKER* 1975, 45–52]. {3}

[STRAUSS 1967] —, "Grundlagen der modernen Physik", in LEY/LOTHER (ed.s), *Mikrokosmos-Makrokosmos*, Berlin, 1967; engl. trans. in [*HOOKER* 1975, 351–63]. {3}

[STRAUSS 1971] —, "Postscript to The Logic of Complementarity", in [*HOOKER* 1975, 41–44]. {3}

[STRAUSS 1973] —, "Two Concepts of Probability in Physics, in SUPPES et al. (ed.s), *Logic, Methodology and Philosophy of Science*, IV, Amsterdam, North-Holland, 1973; rep. in [*HOOKER* 1979, 261–74]. {3}

[STREKALOV/SHIH 1997] STREKALOV, D. V./SHIH, Y. H., "Two-Photon Geometrical Phase", PR **A56** (1997): 3129–33. {3}

[STREKALOV *et al.* 1998] STREKALOV, D. V./PITTMAN, T. B./SHIH, Y. H., "What we Can Learn About Single Photons in a Two-Photon Interference Experiment", PR **A57** (1998): 567–70. {3}

[STREKALOV *et al.* 1996] STREKALOV, D. V./PITTMAN, T. B./SERGIENKO, A. V./SHIH, Y. H., "Postselection-Free Energy-Time Entanglement", PR **A54** (1996): R1–R4. {9}

[STUECKELBERG 1960] STUECKELBERG, Ernst C. G., "Quantum Theory in Real Hibert Space", *Helvetica Physica Acta* **33** (1960): 727–52. {3}

[STUECKELBERG/GUENIN 1961] STUECKELBERG, E. C. G./GUENIN, M., "Quantum Theory in Real Hibert Space II (Addenda and Errats)", *Helvetica Physica Acta* **34** (1961): 621–28. {3}

[STUECKELBERG/GUENIN 1961] STUECKELBERG, E. C. G./GUENIN, M., "Quantum Theory in Real Hibert Space III: Fields of the I Kind (Linear Field Operators)", *Helvetica Physica Acta* **34** (1961): 675–98. {3}

[STUECKELBERG 1962] —, "Thèorie des quanta dans l'espace de Hilbert rèel IV: Champs de II espéce", *Helvetica Physica Acta* **35** (1962): 673–95. {3}

[SUDARSHAN 1963] SUDARSHAN, E. C. G., "Equivalence Between Semiclassical and Quantum Mechanical Descriptions of Statistical Light Beams", PRL **10** (1963): 277–79. {3, 5, 12}

[SUDARSHAN *et al.* 1995] SUDARSHAN, E. C. G./CHIU, C. B./BHAMATHI, G., "Generalized Uncertainty Relations and Characteristic Invariants for the Multimode States", PR **A52** (1995): 43–54. {3, 10, 12}

[SUMMERS/R. WERNER 1985] SUMMERS, S./WERNER, R., "The Vacuum Violates Bell's Inequalities", PL **110A** (1985): 257–59. {9, 11}

[SUMMERS/R. WERNER 1987a] —, "Bell's Inequalities and Quantum Field Theory. I–II", JMP **28** (1987): 2440–56. {9, 11}

[SUMMERS/R. WERNER 1987b] —, "Maximal Violation of Bell's Inequalities is Generic in Quantum Field Theory", CMP **110** (1987): 247–59. {9, 11}

[SUMMHAMMER *et al.* 1982] SUMMHAMMER, J./BADUREK, G./RAUCH, H./KISCHKO, U., "Explicit Experimental Verification of Quantum Spin-State Superposition", PL **90A** (1982): 110–112. {7}

[SUMMHAMMER *et al.* 1983] SUMMHAMMER, J./BADUREK, G./RAUCH, H./KISCHKO, U./ZEILINGER A., "Direct Observation of Fermion Spin Superposition by Neutron Interferometry", PR **A27** (1983): 2523–32. {3}

[SUN 1993] SUN, Chang-Pu, "Quantum Dynamical Model for Wave-Function Reduction in Classical and Macroscopic Limit", PR **A48** (1993): 898–906. {4, 5}

[SUPPES 1965] SUPPES, Patrick, "Logics Appropriate to Empirical Theories", in ADDISON/HENKIN/TARSY (ed.s), Theory of Models, Amsterdam, North-Holland, 1965; rep. in [*HOOKER* 1975, 329–40]. {3, 14}

[SUPPES 1966] —, "The Probabilistic Argument for a Non-Classical Logic of Quantum Mechanics", PS **33** (1966): 14–21; rep. in [*HOOKER* 1975, 341–50]. {3}

[*SUPPES* 1976] — (ed.), *Logic and Probability in Quantum Mechanics*, Dordrecht, Reidel, 1976. {3}

[*SUPPES* 1980] — (ed.), *Studies in the Foundations of Quantum Mechanics*, East Lansing, Michigan, PSA, 1980. {3}

[SUPPES/ZANOTTI 1976a] SUPPES, P./ZANOTTI, M., "Stochastic Incompleteness of Quantum Mechanics", in [*SUPPES* 1976, 303–322]. {3, 8}

[SUPPES/ZANOTTI 1976b] —, "On the Determinism of Hidden Variable Theories with Strict Correlation and Conditional Statistical Indipendence of Observables", in [*SUPPES* 1976, 445–55]. {3, 8}

[SUPPES/ZANOTTI 1984] —, "Causality and Simmetry", in [*DINER et al.* 1984, 331–40]. {12, 13}

[SUPPES/ZANOTTI 1991] —, *Foundations of Physics Letters* **4** (1991): 101. {9}

[SUTHERLAND 1997] SUTHERLAND, Roderick I., "Phase Space Generalization of the de Broglie-Bohm Model", FP **27** (1997): 845–63. {7, 8}

[SVETLICHNY 1987] SVETLICHNY, G., "Distinguishing Three-Body from Two-Body Nonseparability by a Bell-Type Inequality", PR **D35** (1987): 3066–69. {9}

[SVETLICHNY *et al.* 1988] SVETLICHNY, G./REDHEAD, M. L. G./BROWN, H. R./BUTTERFIELD, J., "Do the Bell Inequalities Require the Existence of Joint Probabilities?", *Phylosophy of Science* **55** (1988): 387–401. {8, 9}

[SZILARD 1929] SZILARD, Leo, "Über die Entropieverminderung in einem thermodynamischen System bei Eingriffen intelligenter Wesen", ZP **53** (1929): 840–56; engl. tr. in [*WHEELER/ZUREK* 1983, 539–48]. {13, 14}

[TAKUMA *et al.* 1995] TAKUMA, H./SHIMIZU, K./SHIMIZU, F., "Observations of the Wave Nature of an Ultracold Atom", in [*GREENBERGER/ZEILINGER* 1995a, 217–26]. {3}

[TAMESHTIT/SIPE 1996] TAMESHTIT, A./SIPE, J. E., "Positive Quantum Brownian Evolution", PRL **77** (1996): 2600–603. {3, 5}

[TAN *et al.* 1991] TAN, S. M./WALLS, D. F./COLLETT, M. J., "Nonlocality of a Single Photon", PRL **66** (1991): 252–55. {7, 9}

[TAN *et al.* 1992] —, "Reply to Santos' Comment', PRL **68** (1992): 895. {7, 9}

[TAROZZI 1985] TAROZZI, Gino, "Experimental Tests of the Properties of the Quantum-Mechanical Wave Function", *Lettere a Nuovo Cimento* **42** (1985): 438–42. {12}

[TAROZZI 1993] —, "On the Different Forms of Quantum Acausality", in [*GAROLA/ROSSI* 1993]. {12}

[TAROZZI 1996] —, "Quantum Measurements and Macroscopical Reality: Epistemological Implications of a Proposed Paradox", FP **26** (1996): 907–917. {4, 5, 12}

[TASSIE/PESHKIN 1961] TASSIE, L. J./PESHKIN, M., "Symmetry Theory of the AB-Effect: Quantum Mechanics in a Multiply Connected Region", AP **16** (1961): 177–84. {3, 9}

[TAUBES 1996] TAUBES, Gary, "Schizofrenic Atom Doubles as Schrödinger Cat or Kitten", *Science* **272** (1996): 1101. {5; S}

[*A. TAYLOR/LAY* 1958] TAYLOR, A. E./LAY, D. C., *Introduction to Functional Analysis*, New York, Wiley, 1958, II ed. 1980, rep. Malabar, Krieger, 1986. {1}

[M. TEICH/SALEH 1990] TEICH, M. C./SALEH, B. E. A., "Squeezed and Antibunched Light", PT (june 1990): 26–34. {3, 12; S}

[W. TEICH/MAHLER 1992] TEICH, W. G./MAHLER, G., "Stochastic Dynamics of Individual Qauntum Systems: Stationary Rate Equations", PR **A45** (1992): 3300–318. {5}

[TELLER 1961] TELLER, E., "Der quantenmechanischer Meßprozeßund die Entropie", in [*BOPP* 1961a, 90–92]. {4, 10}

[TERSOFF/BAYER 1983] TERSOFF, J./BAYER, D., "Quantum Statistics for Distinguishable Particles", PRL **50** (1983): 553–54. {3}

[TESCHE 1986] TESCHE, C. D., "Schrödingers Cat: A Realization in Superconducting Devices", in [*GREENBERGER* 1986, 36–50]. {5}

[TESCHE 1990] —, "Can a Noninvasive Measurement of Magnetic Flux be Performed with Superconducting Circuits?", PRL **64** (1990): 2358–61. {5, 12}

[THOMASON 1970] THOMASON, Richmond H., "Indeterminist Time and Truth-Value Gaps", *Theoria* **36** (1960): 264–81. {3}

[THORNE *et al.* 1978] THORNE, K. S./DREVER, R. W. P./CAVES, C.M./ZIMMERMANN, M./SANBERG, V.D., "Quantum Nondemolition Measurements of Harmonic Oscillators", PRL **40** (1978): 667–71. {4}

[TITTEL *et al.* 1998] TITTEL, W./RIBORDY, G./GISIN, N., "Quantum Cryptography", *Phyiscs World* (March 1998): 41–45. {10; S}

[TOFFOLI 1990] TOFFOLI, Tommaso, "How Cheap Can Mechanics' First Principles Be?" in [*ZUREK* 1990a, 301–318]. {3, 12}

[TOMBESI/VITALI 1996] TOMBESI, P./VITALI, D., "All-Optical Model for the Generation and the Detection of Macroscopic Quantum Coherence", PRL **77** (1996): 411–15. {5}

[TONOMURA 1984] TONOMURA, Akira, "Experimental Confirmation of the Aharonov-Bohm Effect by Electron Holography", in [*ROTH/INOMATA* 1984, 169–76]. {9}

[TONOMURA 1992] —, "Experiments on Aharonv-Bohm Effect", in [*SELLERI* 1992a, 291–99]. {9}

[TONOMURA 1995] —, "Recent Advances in Electron Interferometry", in [*GREENBERGER/ZEILINGER* 1995a, 227–40]. {3}

[TONOMURA *et al.* 1986] TONOMURA, A./OSAKABE, N./MATSUDA, T./KAWASAKI, T./ENDO, J./YANO, S./YAMADA, H., "Evidence for Aharonov-Bohm Effect with Magnetic Field Completely Shielded form Electron Wave", PRL **56** (1986): 792–95. {9}

[TONOMURA *et al.* 1989] TONOMURA, A./ENDO, J./MATSUDA, T./KAWASAKI, T./EZAWA, H., "Demonstration of Single-Electron Buildup of an Interference Pattern", AJP **57** (1989): 117–20. {3}

[*TORALDO DI F.* 1976] TORALDO DI FRANCIA, Giuliano, *L'indagine del mondo fisico*, Torino, Einaudi, 1976. {13, 14}

[TORGESON *et al.* 1995] TORGESON, J. R./BRANNING, D./MONKEN, C. H./MANDEL, L., "Experimental Demonstration of the Violation of Local Realism Without Bell Inequalities", PL **204A** (1995): 323–28. {9}

[TREDER 1970] TREDER, Hans-Jürgen, "Das Einstein-Bohrsche Kasten-Experiment", *Monatsberichte der Deutschen Akademie der Wissenschaften zu Berlin* **12** (1970): 180–84. {12}

[TREDER/SCHRÖDER 1997] TREDER, H.-J./SCHRÖDER, W., "Magnetohydrodynamics Corresponding with Wave Mechanics", FP **27** (1997): 875–79. {3, 7, 11}

[TSIRELSON 1980] TSIRELSON, B. S., "Quantum Generalizations of Bell's Inequality", *Letters in Mathematical Physics* **4** (1980): 93–100. {3, 9}

[TURNER 1988] TURNER, Edwin L., "Gravitational Lens", *Scientific American* (1988). {13}

[TURCHETTE *et al.* 1995] TURCHETTE, Q. A./HODD, C. J./LANGE, W./MABUCHI, J./KIMBLE, H. J., "Measurement of Conditional Phase Shifts for Quantum Logic", PRL **75** (1995): 4710–13. {3}

[TYAPKIN/VINDUSHKA 1991] TYAPKIN, A. A./VINDUSHKA, M., "The Geometrical Aspects of the Bell Inequalities", FP **21** (1991): 185–95. {3, 9}

[TZARA 1987] TZARA, C., "Fuzzy Measurements in Quantum Mechanics and the Representation of Macroscopic Motion", NC **B98** (1987): 131–43. {4, 8}

[TZARA 1988] —, "Emergence of a Classical Motion from a Quantum State: A Further Test of a Theory of Fuzzy Measurements", *Physics Letetrs* **A127** (1988): 247–250. {4, 8}

[UEDA/IMOTO 1994] UEDA, M./IMOTO, N., "Anomalous Commutation Relation and Modified Spontaneous Emission Inside a Microcavity", PR **A50** (1994): 89–92. {4, 5}

[UEDA/KITAGAWA 1992] UEDA, M./KITAGAWA, M., "Reversibility in Quantum Measurement Processes", PRL **68** (1992): 3424–27. {4, 6, 7}

[UEDA *et al.* 1992] UEDA, M./IMOTO, N./NAGAOKA, H./OGAWA, T., "Continuous Quantum-Nondemolition Measurement of Photon Number", PR **A46** (1992): 2859–69. {3, 4}

[UFFINK/HILGEVOORD 1985] UFFINK, J. B. M./HILGEVOORD, J., "Uncertainty Principle and Uncertainty Relations", FP **15** (1985): 925–44. {3, 12}

[UHLMANN 1970] UHLMANN, A., *Reports on Mathematical Physics* **1** (1970): 147–59. {3, 10}

[UHLMANN 1976] —, *Reports on Mathematical Physics* **9** (1976): 273. {3, 10}

[UHLMANN 1977] —, "Relative Entropy and the Wigner-Yanase-Dyson-Lieb Concavity in an Interpolation Theory", CMP **54** (1977): 21–32. {10}

[ULLMO 1949] ULLMO, Philipp F., "La mécanique quantique e la causalité", *Revue Philosophique* **139** (1949): 257–87, 441–73. {12, 14}

[ULLMO 1951] —, "Le théoreme de von Neumann et la causalité", *Revue de Métaphysique et de Morale* **56** (1951): 143–70. {12, 14}

[ULLMO 1963] —, "La philosophie d'Heisenberg", *La Nouvelle Revue Franaise* **22** (1963): 296–308. {14}

[UNRUH 1978] UNRUH, W. G., "Analysis of Quantum-Nondemolition Measurement, PR **D18** (1978): 1764–72. {4}

[UNRUH 1979] —, "Quantum Nondemolition and Gravity-wave Detection", PR *D19* (1979): 2888–96. {4}

[UNRUH 1986] —, "Quantum Measurement", in [*GREENBERGER* 1986, 242–49]. {4}

[UNRUH 1994] —, "Comment: Reality and Measurement of the Wave Function", PR **A50** (1994): 882–87. {4, 7}

[UNRUH 1995a] —, "Varieties of Quantum Measurement", in [*GREENBERGER/ZEILINGER* 1995a, 560–569]. {4}

[UNRUH 1995b] —, "Maintaining Coherence in Quantum Computers", PR **A51** (1995): 992–97. {5, 10}

[UNRUH 1998] —, "Is Quantum Mechanics Non-Local?", e-print: code: quant-ph/9710032. {9, 12}

[UNRUH/ZUREK 1989] UNRUH, W. G./ZUREK, W. H., "Reduction of a Wave Packet in Quantum Brownian Motion", PR **D40** (1989): 1071–94. {4, 5}

[VAIDMAN 1988] VAIDMAN, Lev, "Meaning and Measurability of Nonlocal Quantum States", in [*VAN DER MERWE et al.* 1988a, 81–88]. {4, 8, 9}

[VAIDMAN 1995] —, "Nonlocality of a Single Photon Revisited Again", PRL **75** (1995): 2063. {7, 9}

[VAIDMAN 1996] —, "Weak-Measurement Elements of Reality", FP **26** (1996): 895–906. {4, 12}

[VAIDMAN *et al.* 1996] VAIDMAN, L./GOLDENBERG, L./WIESNER, S., "Error Prevention Scheme with Four Particles", PR **A54** (1996): R1745–48. {10}

[*VAN DER MERWE et al.* 1988a] VAN DER MERWE, A./SELLERI, F./TAROZZI, G. (ed.s), *Microphysical Reality and Quantum Formalism*, I, Dordrecht, Kluwer, 1988. {3-9}

[*VAN DER MERWE et al.* 1988b] — (ed.s) *Microphysical Reality and Quantum Formalism*, II, Dordrecht, Kluwer, 1988. {3, 9}

[VAN DYCK *et al.* 1986] VAN DYCK, R./MOORE, F./FARNHAM, D./SCHWINBERG, P., "Single Proton Isolated and Resolved in a Penning Trap", *Bullettin of American Physical Society* **31** (1986): 974. {3}

[VAN FRAASSEN 1966] VAN FRAASSEN, Bas C., "Singular Terms, Truth-Value Gaps, and Free Logic", *Journal of Philosophy* **63** (1966): 481–95. {14}

[VAN FRAASSEN 1969] —, "Presuppositions, Supervaluations, and Free Logic", in K. LAMBERT (ed.), *The Logical Way of Doing Things*, New Haven, Yale University Press, 1969. {14}

[VAN FRAASSEN 1973] —, "Semantic Analysis of Quantum Logic", in [*HOOKER* 1973a, 80–113]. {3}

[VAN FRAASSEN 1974] —, "The Labyrinth of Quantum Logic", *Boston Studies in the Philosophy of Science* **13** (1974): 224–54; rep. in [*HOOKER* 1975, 545–75]. {3}

[VAN FRAASSEN 1976] —, "The EPR Paradox", in [*SUPPES* 1976, 283–301]. {9}

[VAN FRAASSEN 1979a] —, "Hidden Variables and the Modal Interpretation of Quantum Mechanics", *Synthese* **42** (1979): 155–65. {8}

[VAN FRAASSEN 1979b] —, "Foundations of Probability: a Modal Frequency Interpretation", in TORALDO DI FRANCIA (ed.), *Problems in the Foundations of Physics*, Amsterdam, North-Holland, 1979: 344–94. {3, 8}

[VAN FRAASSEN 1982] —, "The Charidibis of Realism: Epistemological Implications of Bell's Inequality", *Synthese* **52** (1982): 885. {8, 9, 12}

[VAN FRAASSEN 1985] —, "EPR: When is a Correlation not a Mistery?", in [*LAHTI/MITTELSTAEDT* 1985, 113–28]. {8, 9, 12}

[*VAN FRAASSEN* 1991] —, *Quantum Mechanics. An Empiricist View*, Oxford, Clarendon, 1991, 1995. {4, 8, 12, 14}

[VAN FRAASSEN 1994] —, "Interpretation of Quantum Mechanics: Parallels and Choices", in [*ACCARDI* 1994a, 7–14]. {12}

[VAN KAMPEN 1954] VAN KAMPEN, N. G., "Quantumstatistics of Irreversible Processes", *Physica* **20** (1954): 603–622. {5}

[VAN HOVE 1955] VAN HOVE, Léon, "Quantum-Mechanical Perturbations Giving Rise to a Statistical Transport Equation", *Physica* **21** (1955): 517–40. {4, 5}

[VAN HOVE 1957] —, "The Approach to Equilibrium in Quantum Statistics", *Physica* **23** (1957): 441–80. {4, 5}

[VAN HOVE 1959] —, "The Ergodic Behaviour of Quantum Many-Body Systems", *Physica* **25** (1959): 268–76; rep. in [*WHEELER/ZUREK* 1983, 648–56]. {4, 5}

[VAN VLECK 1941] VAN VLECK, John H., "Note on Liouville's Theorem and the Heisenberg Uncertainty Principle", PS **8** (1941): 275–79. {12, 13}

[VARADARAJAN 1962] VARADARAJAN, V. S., "Probability in Physics and a Theorem on Simultaneous Observability", *Communications in Pure and Applied Mathematics* **15** (1962): 189–217; rep. in [*HOOKER* 1975, 171–203]. {3}

[*VARADARAJAN* 1968] —, *Geometry of Quantum Theory*, I, Princeton, Van Nostrand, 1968. {3}

[*VARADARAJAN* 1970] —, *Geometry of Quantum Theory*, II, Princeton, Van Nostrand, 1970. {3}

[*VARADARAJAN* 1984] —, *Lie Groups, Lie Algebras, and Their Representations*, Berlin, Springer, 1984. {3}

[VEDRAL/PLENIO 1998] VEDRAL, V./PLENIO, M. B., "Entanglement Measures and Purification Procedures", PR **A57** (1998): 1619–33. {3, 9, 10}

[VEDRAL *et al.* 1997a] VEDRAL, V./PLENIO, M. B./RIPPIN, M. A./KNIGHT, P. L., "Quantifying Entanglement", PRL **78** (1997): 2275–79. {3, 9, 10}

[VEDRAL *et al.* 1997b] VEDRAL, V./PLENIO, M. B./JACOBS, K./KNIGHT, P. L., "Statistical Inference, Distinguishability of Quantum States, and Quantum Entanglement", PR **A56** (1997): 4452–55. {9, 10}

[VERMAAS 1997] VERMAAS, Pieter E., "A No-Go Theorem for Joint Property Ascription in Modal Interpretations of Quantum Mechanics", PRL **78** (1997): 2033–37. {3, 8, 12}

[VERMAAS/DIEKS 1995] VERMAAS, P. E./DIEKS, D., "The Modal Interpretation of Quantum Mechanics and Its Generalization to Density Operators", FP **25** (1995): 145–58. {8, 12}

[VIGIER 1951] VIGIER, Jean-Pierre, "Introduction géometrique de londe pilote en théorie unitaire affine", *Comptes Rendus l'Academie des Sciences* **233** (1951): 1010–12. {7}

[VIGIER 1952] —, *Comptes Rendus l'Academie des Sciences* **235** (1952): 1107. {7}

[*VIGIER* 1956] —, *Structures des Micro-objets dans l'interpretation causale de la Thorie des quanta*, Paris, Gauthier-Villars, 1956. {7, 12}

[VIGIER 1976] —, "Possible Implications of de Broglie's Wave-Mechanical Theory of Photon Behaviour", in [*FLATO et al.* 1976, 237–49]. {7}

[VIGIER 1979] —, "Superluminal Propagation of the Quantum Potential in the Causal Interpretation of Quantum Mechanics", *Lettere a Nuovo Cimento* **24** (1979): 258–64. {7, 8, 12}

[VIGIER 1982] —, *Astron Nachr* **303** (1982): 55. {7, 12}

[VIGIER 1985] —, "Causal, Non-Local Interpretation of Neutron Interferometry Experiments, EPR Correlations and Quantum Statistics", in [*LAHTI/MITTELSTAEDT* 1985, 653–75]. {7}

[VIGIER 1986] —, "Trajectories, Spin, and Energy Conservation in Time-Dependent Neutron Interferometry", in [*GREENBERGER* 1986, 503–511]. {7, 12}

[VIGIER 1988] —, "EPR Version pf Wheeler's Delayed Choice Experiment", in [*VAN DER MERWE et al.* 1988b, 207–223]. {6, 7, 8, 9}

[VIGIER 1991] —, "Explicit Mathematical Construction of Relativistic Nonlinear de Broglie Waves Described by Three-Dimensional (Wave and Electromagnetic) Solitons 'Piloted' (Controlled) by Corresponding Solutions of Associated Linear Klein-Gordon and Schrödinger Equations", FP **21** (1991): 125–47. {7, 8, 12; W}

[VIGIER 1997] —, "Possible Consequences of an Extended Charged Particle Model in Electromagnetic Theory" PL **235A** (1997): 419–31. {7, 12}

[VIGIER *et al.* 1987] VIGIER, J.-P./DEWDNEY, C./HOLLAND, P. R./KYPRIANIDIS, A., "Causal Particle Trajectories and the Interpretation of Quantum Mechanics", in [*HILEY/PEAT* 1987a, 169–204]. {7, 8}

[VIOLA/ONOFRIO 1997] VIOLA, L./ONOFRIO, R., "Measured Quantum Dynamics of a Trapped Ion", PR **A55** (1997): R3291–94. {5}

[VITALI *et al.* 1997] VITALI, D./TOMBESI, P./MILBURN, G. J., "Controlling the Decoherence of a 'Meter' via Stroboscopic Feedback", PRL **79** (1997): 2442–45. {5}

[K. VOGEL/RISKEN 1989] VOGEL, K./RISKEN, H., "Determination of Quasiprobability Distributions in Terms of Probability Distributions for the Rotated Quadrature Phase", PR **A40** (1989): 2847–49. {3, 7}

[*K. VOGEL/SCHLEICH* 1992] VOGEL, K./SCHLEICH, W. P., *More on Interference in Phase Space*, Les Houches, Elsevier Science Publishers, 1992. {3}

[VOGT 1989] VOGT, Andrew, "Bell's Theorem and Mermin's Gedankenexperiment", in [*KAFATOS* 1989, 61–64]. {9}

[*VON NEUMANN* CW] VON NEUMANN, John, *Collected Works*, Oxford, Pergamon, 1963. {3, 4, 8, 10, 12, 13, 14, 15}

[VON NEUMANN 1927a] —, "Mathematische Begründung der Quantenmechanik", *Göttinger Nachrichten* (1927); rep. in[*VON NEUMANN* CW, I, 151–207]. {3}

[VON NEUMANN 1927b] —, "Wahrscheinlichkeitstheoretischer Aufbau der Quantemechanik", *Göttinger Nachrichten* (1927); rep. in [*VON NEUMANN* CW, I, 208–35]. {3}

[VON NEUMANN 1927c] —, "Thermodinamik quantmechanischer Gesamtheiten", *Göttinger Nachrichten* (1927); rep. in [*VON NEUMANN* CW, I, 236–54]. {3}

[VON NEUMANN 1929] —, "Beweis des Ergodensatzes und des H-Theorems in der neuen Mechanik", ZP **57** (1929): 30-70; rep. in [*VON NEUMANN* CW, I, 558–98]. {3}

[VON NEUMANN 1931] —, "Die Eindeuditigkeit der Schrödingerschen Operatoren", *Mathematische Annalen* **104** (1931): 570–78; rep. in [*VON NEUMANN* CW, II, 221–29]. {3}

[*VON NEUMANN* 1932] —, *Mathematische Grundlagen der Quantenmechanik*, Berlin, Springer, 1932, 1968, 1996. {2, 3, 4, 10, 12, 13, 14, 15}

[VON NEUMANN 1936] —, "Continuous Geometry", *Proceedings of the National Academy of Science (USA)* **22** (1936): 92–108; rep. in [*VON NEUMANN* CW, IV, 126–34]. {3}

[VON NEUMANN 1937] —, "Quantum Logics (Strict and ProbàbilityLogics)", in [*VON NEUMANN* CW, 195–97]. {3} ????

[*VON NEUMANN* 1955] —, *Mathematical Foundations of Quantum Mechanics*, Princeton, University Press, 1955; engl. trans. of [*VON NEUMANN* 1932]. {2, 3, 4, 10, 12, 13, 14, 15}

[VON WEIZSÄCKER 1931] VON WEIZSÄCKER, CARL F., "Ortsbestimmung eines Elektrons durch ein Mikroskop", ZP **70** (1931): 114–30. {3, 12}

[VON WEIZSÄCKER 1941] —, "Zur Deutung der Quantenmechanik", ZP **118** (1941): 489–509. {12}

[VON WEIZSÄCKER 1955] —, "Komplementarität und Logik", *Naturwissenschaften* **42** (1955): 521–29. {3, 7}

[VON WEIZSÄCKER 1958] —, "Die Quantentheorie der einfachen Alternative", *Zeitschrift für Naturforschung* **13a** (1958): 245–53. {7, 12}

[*VON WEIZSÄCKER* 1971a] —, *Die Einheit der Natur*, Mnchen, Hanser, 1971; DTV, 1974. {13, 14}

[VON WEIZSÄCKER 1971b] —, "Niels Bohr and Complementarity: The Place of the Classical Language", in [*BASTIN* 1971, 23–31]. {12}

[*VON WEIZSÄCKER* 1971c] —, *Voraussetzungen des naturwissenschaftlichen Denkens*, Mnchen Hanser, 1971. {14}

[VON WEIZSÄCKER 1985] —, "Quantum Theory and Space-Time", in [*LAHTI/MITTELSTAEDT* 1985, 223–37]. {12}

[VON WEIZSÄCKER 1987] —, "Heisenbergs Philosphy", in [*LAHTI/MITTELSTAEDT* 1987, 277–96]. {7, 12, 14}

[VON WEIZSÄCKER *et al.* 1958] VON WEIZSÄCKER, C. F./SCHEIBE, E./SÜSSMANN, G., "Komplementarität und Logik", *Zeitschrift fr Naturforschung* **13a** (1958): 705–21. {3, 7}

[VORONTSOV/KHALILI 1982] VORONTSOV, Y. I/KHALILI, F. Y., *Zh. Eksp. Teor. Fiz.* **82** (1982): 72–76; "Detection of a Force Acting on an Oscillator by Measuring the Integral of its Coordinate" (engl. tr.), *Soviet Physics-JETP* **55** (1982): 43–45. {4}

[VUJICIC/HERBURT 1985] VUJICIC, M./HERBURT, F., "Distant Correlations in Quantum Mechanics", in [*LAHTI/MITTELSTAEDT* 1985, 677–89]. {9}

[WAGH/RAKHECHA 1990] WAGH, A. G./RAKHECHA, V. C., "Geometric Phase in Neutron Interferometry", PL **148A** (1990): 17–19. {3}

[WAGH *et al.* 1997] WAGH, A. G./RAKHECHA, V. C./SUMMHAMMER, J./BADUREK, G./WEINFURTER, H./ALLMAN, B. E./KAISER, H./HAMACHER, K./JACOBSON, D. L./WERNER, S. A., "Experimental Separation of Geometric and Dynamical Phases Using Neutron Interferometry", PRL **78** (1997): 755–59. {3}

[WAKITA 1960] WAKITA, H., "Measurement in Quantum Mechanics. I", PTF **23** (1960): 32–40. {4}

[WAKITA 1962a] —, "Measurement in Quantum Mechanics. II", PTF **27** (1962): 139–44. {4}

[WAKITA 1962b] —, "Measurement in Quantum Mechanics. III", PTF **27** (1962): 1156–64. {4}

[WALLENTOWITZ/W. VOGEL 1995] WALLENTOWITZ, S./VOGEL, W., "Reconstruction of the Quantum Mechanical State of a Trapped Ion", PRL **75** (1995): 2932–25. {4, 7}

[WALLENTOWITZ/W. VOGEL 1996] —, "Unbalanced Homodyning for Quantum State Measurement", PR **A53** (1996): 4528–33. {4, 7}

[WALLNER 1970] WALLNER, L. G., "Hydrodynamical Analogies to Quantum Mechanics", *Symposium Report, International Atomic Energy Agency*, Wien, 1970. {5, 12}

[WALLS/MILBURN 1985] WALLS, D. F./MILBURN, G. J., "Effect of Dissipation on Quantum Coherence", PR **A31** (1985): 2403–408. {5}

[*WALLS/MILBURN* 1994] —, *Quantum Optics*, Berlin, Springer, 1994. {2}

[WALLS *et al.* 1985] WALLS, D. F./COLLET, M. J./MILBURN, G. J., "Analysis of a Quantum Measurement", PR **D32** (1985): 3208–15. {4, 5}

[WANG *et al.* 1991] WANG, L. J./ZOU, X. Y./MANDEL, L., "Induced Coherence Without Induced Emission", PR **A44** (1991): 4614–22. {4, 7, 9, 10}

[*WATANABE* 1969] WATANABE, Michael Satosi, *Knowing and Guessing*, New York, Wiley, 1969. {14}

[WATANABE 1976] —, "Conditional Probability in Wave Mechanics", in [*FLATO et al.* 1976, 159–65]. {3}

[WEHRL 1978] WEHRL, Alfred, "General Properties of Entropy", RMP **50** (1978): 221–60. {10, 13; W}

[WEIHS *et al.* 1996] WEIHS, G./RECK, M./WEINFURTER, H./ZEILINGER, A., "Two-Photon Interference in Optical Fiber Multiports", PR **A54** (1996): 893–97. {7}

[S. WEINBERG 1989a] WEINBERG, Steven, "Precision Tests of Quantum Mechanics", PRL **62** (1989): 485–88. {7}

[S. WEINBERG 1989b] —, "Testing Quantum Mechanics", AP **194** (1989): 336–86. {7; W}

[*S. WEINBERG* 1992] —, *Dreams of a Final Theory*, New York, Pantheon, 1992. {12, 13, 14}

[WEINFURTER 1994] WEINFURTER, Harald, "Experimental Bell-State Analysis", *Europhysics Letters* (1994): 559–64. {9, 10}

[WEINFURTER/ZEILINGER 1996] WEINFURTER, H./ZEILINGER, A., "Informationsübertragung und Informationsverarbeitung in der Quantenwelt", *Physikalische Blätter* **52** (1996): 219–24. {10; S}

[WEINFURTER *et al.* 1995] WEINFURTER, H./HERZOG, T./KWIAT, P. G./RARITY, J. G./ZEILINGER, A./ŻUKOWSKI, M., "Frustrated Downconversion: Virtual or Real Photons?", in [*GREENBERGER/ZEILINGER* 1995a, 61–72]. {3, 12}

[WEINFURTER *et al.* 1999] WEINFURTER, H./BOUWMEESTER, /ZEILINGER, PRL (1999). {9}

[WEISSKOPF/WIGNER 1930a] WEISSKOPF, V./WIGNER, E. P., "Berechnung der natürlichen Linienbreite auf Grund der Diracschen Lichttheorie", ZP **63** (1930): 54–73. {3}

[WEISSKOPF/WIGNER 1930b] —, "Über die natürliche Linienbreite der Strahlung des harmonischen Oszillators", ZP **65** (1930): 18–29. {3}

[WEIZEL 1953a] WEIZEL, W., "Ableitung der Quantentheorie aus einem klassischen, kausal determinierten Modell", ZP **134** (1953): 264–85. {5, 12}

[WEIZEL 1953b] —, "Ableitung der Quantentheorie aus einem klassischen, kausal determinierten Modell. II", ZP **135** (1953): 270–73. {5, 12}

[WENTZEL 1949] WENTZEL, G., *Quantum Theory of Fields*, New York, Interscience, 1949.{2, 11}

[R. WERNER 1983] WERNER, Reinhard, "Physical Uniformities on the State Space of Nonrelativistic Quantum Mechanics", FP **13** (1983): 859–81. {3}

[R. WERNER 1986] —, "Screen Observables in Relativistic and Nonrelativistic Quantum Mechanics", JMP **27** (1986): 793–803. {4}

[R. WERNER 1989] —, "Quantum States with Einstein-Podolsky-Rosen Correlations Admitting a Hidden-Variable Model", PR **A40** (1989): 4277–81. {3, 8, 9}

[S. WERNER 1995] WERNER, Samuel A., "Neutron Interferometry Tests of Quantum Theory", [*GREENBERGER/ZEILINGER* 1995a, 241–62]. {3}

[*WEYL* 1931] WEYL, Hermann, *Gruppentheorie und Quantenmechanik*, 1931, engl. tr.: *The Theory of Groups and Quantum Mechanics*, Dover Publ., 1950. {3, 13, 15}

[WHEELER 1937] WHEELER, John A., "On the Mathematical Description of Light Nuclei by the Method of Resonating Group Structure", PR **52** (1937): 1107–1122. {3}

[WHEELER 1957] —, "Assessment of Everett's 'Relative State' Formulation of Quantum Theory": RMP **29** (1957): 463–65; rep. in [*DEWITT/GRAHAM* 1973, 151–53]. {4, 12}

[WHEELER 1968] —, "Superspace and Quantum Geometrodynamics", in DEWITT/WHEELER (ed.s), *Battelle Rencontres*, New York, Benjamin, 1968. {3, 6, 7}

[WHEELER 1978b] —, "The 'Past' and the 'Delayed-Choice' Double-Slit Experiment", in [*MARLOW* 1978a, 9–48]. {6, 7, 9}

[WHEELER 1983] —, "Law Without Law", in [*WHEELER/ZUREK* 1983, 182–213]. {4, 6, 7, 12}

[WHEELER 1985] —, "Franck-Condon Effect and Squeezed-State Physics as Double-Source Interference Phenomena", *Letters on Mathematical Physics* **10** (1985): 201. {3}

[WHEELER 1986] —, "How Come the Quantum?", in [*GREENBERGER* 1986, 304–316]. {3, 12, 13, 14}

[WHEELER 1988] —, "The 2–Photon Elementary Quantum Phenomena and the Spacetime Structure", in [*VAN DER MERWE et al.* 1988a, 215–31]. {3, 11, 12, 13}

[WHEELER 1990] —, "Information, Physics, Quantum: The search for Links", in: [*ZUREK* 1990a, 3–28]. {10, 12, 13, 14}

[*WHEELER/ZUREK* 1983] WHEELER, J. A./ZUREK, W. (ed.s), *Quantum Theory and Measurement*, Princeton, University Press, 1983. {4, 5, 6, 7, 8, 9, 10, 11, 12}

[WHITE *et al.* 1998] WHITE, A. G./MITCHELL, J. R./NAIRZ, O./KWIAT, P. G., "'Interaction-Free' Imaging", PR **A58** (1998): 605–613. {4}

[*WHITEHEAD* 1929] WHITEHEAD, Alfred N., *Process and Reality*, London, Macmillan, 1929, 1957, corrected ed. 1978. {14}

[WICK *et al.* 1952] WICK, G. C./WIGHTMAN, A. S./WIGNER, E.P., "The Intrinsic Parity of Elementary Particles", PR **88** (1952): 101–105. {3, 4}

[WICKES *et al.* 1983] WICKES, W.C./ALLEY, C. O./JAKUBOWICZ, O., "A 'Delayed-Choice' QM Experiment", in [*WHEELER/ZUREK* 1983, 457–61]. {6}

[WIENER/SIEGEL 1955] WIENER, N./SIEGEL, A., *Nuovo Cimento Supplementa* **2** (1955): 304. {8}

[WIGHTMAN 1962] WIGHTMAN, A. S., "On the Localizability of Quantum Mechanical Systems", RMP **34** (1962): 845–72. {3}

[WIGHTMAN 1972] —, "The Dirac Equation", in [*SALAM/WIGNER*, 95–115]. {3, 11}

[WIGNER 1932] WIGNER, Eugene P., "On the Quantum Correction for Thermodynamic Equilibrium", PR **40** (1932): 749–59. {3}

[WIGNER 1939] —, "On Unitary Representations of the Inhomogeneous Lorentz Group", *Annals of Mathematics* **40** (1939): 149–204; rep. in [*WIGNER* CP, I, 334–89]. {3, 11}

[WIGNER 1952] —, "Die Messung quantenmechanischer Operatoren", ZP **133** (1952): 101–108; rep. in [*WIGNER* CP, III, 415–22]. {3}

[WIGNER 1955a] —, "Lower Limit for the Energy Derivative of the Scattering Phase Shift", PR **98** (1955): 145–47. {3}

[WIGNER 1955b] —, *Helvetica Physica Acta Supplementa* **4** (1955): 210. {3, 8}

[*WIGNER* 1959] —, *Group Theory*, New York, Academic, 1959. {3, 15}

[WIGNER 1961] —, "Remarks on the Mind-Body Question", in GOOD (ed.), *The Scientist Speculates*, London , Heinemann, 1961; rep. in [*WHEELER/ZUREK* 1983, 168–81]. {4, 12}

[WIGNER 1963] —, "The Problem of Measurement", AJP **31** (1963): 6–15; rep. in [*WHEELER/ZUREK* 1983, 324–41]. {4, 12}

[WIGNER 1964] —, "Two Kinds of Reality", *Monist* **48** (1964): 248–64; rep. in [*WIGNER* CP, VI, 33–47]. {4, 12, 13, 14}

[*WIGNER* 1967] —, *Symmetries and Reflections*, Indiana, University Press, 1967. {4, 12, 13, 14}

[WIGNER 1970] —, "On Hidden Variables and Quantum Mechanical Probabilities", AJP **38** (1970): 1005–9. {8, 9}

[WIGNER 1971] "The Subject of our Discussion. Foundations of Quantum Mechanics", in D'ESPAGNAT (ed.), Internal School of Physics 'Enrico Fermi 1970, New York, Academic Press, 1971: 122-24, rep. in [*WIGNER* CP, VI, 199–217]. {3, 12}

[WIGNER 1972] —, "On the Time-Energy Uncertainty Relation", in [*SALAM/WIGNER*, 237–47]. {3}

[WIGNER 1973] —, "Epistemological Perspective on Quantum Theory", in [*HOOKER* 1973a, 369–85]. {12, 14}

[WIGNER 1983a] —, "Interpretation of Quantum Mechanics", in [*WHEELER/ZUREK* 1983, 260–314]. {12}

[WIGNER 1983b] —, "Review of the Quantum-Mechanical Measurement Problem", in [*MEYSTRE/SCULLY* 1983, 43–63]; rep. in [*WIGNER* CP, VI, 225–44]. {4, 12}

[WIGNER 1986] —, "The Non-Relativistic Nature of the Present Quantum Mechanical Measurement Theory", in [*GREENBERGER* 1986, 1–5]. {4}

[*WIGNER* CP] —, *Collected Papers*, Berlin, Springer, 1993. {3, 4, 8, 9, 11, 12, 13, 14, 15}

[WILSON/KOGUT 1974] WILSON, K. G./KOGUT, J., "The Renormalization Group and the e Expansion", PRep **12** (1974): 75–200. {11, 13}

[WINELAND/DEHMELT 1975] WINELAND, D. J./DEHMELT, H., "Proposed $10^{14}\Delta\nu < \nu$ Laser Fluorescence Spectroscopy on $T1^{+}$ Mono-Ion Oscillator III", *Bulletin of the American Physical Society* **20** (1975): 637. {3}

[WINELAND/ITANO 1987] WINELAND, D. J./ITANO, W. M., "Laser Cooling", PT (june 1987): 34–40. {5}

[WINELAND *et al.* 1973] WINELAND, D. J./EKSTROM, P./DEHMELT, H., "Monoelectron Oscillator", PRL **31** (1973): 1279–82. {12}

[WINELAND *et al.* 1994] WINELAND, D. J./BOLLINGER, J. J./ITANO, W. M./HEINZEN, D. J., "Squeezed Atomic States and Projection Noise in Spectroscopy", PR **A50** (1994): 67–88. {3}

[WINTER 1961] WINTER, Rolf G., ''Evolution of a Quasi-Stationary State", PR **123** (1961): 1503–1507. {3}

[WISEMAN 1996] WISEMAN, H. M., "Quantum Trajectories and Quantum Measurement Theory", *Quantum and Semiclassical Optics* **8** (1996): 205–222. {4, 5}

[WISEMAN/MILBURN 1993] WISEMAN, H. M./MILBURN, G. J., "Interpretation of Quantum Jump and Diffusion Processes Illustrated on the Bloch Sphere", PR **A47** (1993): 1652–66. {3, 5}

[WÓDKIEWICZ 1986a] WÓDKIEWICZ, K., "Operational Approach to Phase-Space Measurements in Quantum Mechanics", PRL **52** (1986): 1064–67. {3}

[WÓDKIEWICZ 1986b] —, "Propensity and Probability in Quantum Mechanics", PL **115** (1986): 304–306. {3, 12}

[WÓDKIEWICZ 1988a] —, "On the Equivalence of Nonlocality and Nonpositivity of Quasi-Distributions in EPR Correlations", PL **129** (1988): 1–3. {3, 8, 9, 12}

[WÓDKIEWICZ 1988b] —, "Quantum Jumps and the EPR Correlations", PL **129** (1988): 415–18. {5, 9}

[WOLFE 1936] WOLFE, Hugh C., "Quantum Mechanics and Physical Reality", PR **49** (1936): 274. {8, 9}

[WOLINSKY/CARMICHAEL 1988] WOLINSKY, M./CARMICHAEL, H. J., "Quantum Noise in the Parametric Oscillator: From Squeezed States to Coherent-State Superpositions", PRL **60** (1988): 1836–39. {5}

[WOO 1981] WOO, C. H., "Consciousness and Quantum Interference — An Experimental Approach", FP **11** (1981): 933–44. {12}

[WOO 1986] —, "Why the Classical-Quantal Dualism is Still With Us", AJP **54** (1986): 923–28. {4, 5, 12, 13}

[WOOTTERS 1981] WOOTTERS, William K., "Statistical Distance and Hilbert Space", PR **D23** (1981): 357–62. {3, 10}

[WOOTTERS 1984] —, "Is Spacetime a Bookkeeping Device for Quantum Correlations?", in [*ROTH/INOMATA* 1984, 279–9O]. {3, 9, 10}

[WOOTTERS 1986a] —, "Quantum Mechanics without Probability Amplitudes", FP **16** (1986): 391–405. {3, 4, 7, 12}

[WOOTTERS 1986b] —, "The Discrete Wigner Function", in [*GREENBERGER* 1986, 275–82]. {3}

[WOOTTERS 1987] —, "A Wigner-Function Formulation of Finite-State Quantum Mechanics", AP **176** (1987): 1–21. {3, 4, 7}

[WOOTTERS 1990] —, "Local Accessibility of Quantum States" in [*ZUREK* 1990a, 39–46]. {3}

[WOOTTERS/FIELDS 1989] WOOTTERS, W. K./FIELDS, B. D., "Optimal State-Determination by Mutually Unbiased Measurements", AP **191** (1989): 363–81. {3, 4, 7}

[WOOTTERS/ZUREK 1979] WOOTTERS, W./ZUREK, W. H., "Complementarity in the Double-Slit Experiment: Quantum Nonseparability and a Quantitative Statement of Bohr's Principle", PR **D19** (1979): 473–84; rep. in [*WHEELER/ZUREK* 1983, 443–54]. {3, 4, 7, 12}

[WOOTTERS/ZUREK 1982] —, "A Single Quantum Cannot Be Cloned", NA **299** (1982): 802–803. {3, 4, 10}

[WRIGHT *et al.* 1996] WRIGHT, E. M./WALLS, D. F./GARRISON, J. C., "Collapses and Revivals of Bose-Einstein Condensates Formed in Small Atomic Samples", PRL **77** (1996): 2158–61. {5, 6}

[WU/YANG 1975] WU, T. T./YANG, C. N., "Concept of Nonintegrable Phase Phactors and Global Formulation of Gauge Fields", PR **D12** (1975): 3845–57. {3}

[YAFFE 1982] YAFFE, Laurence G., "Large N Limits as Classical Mechanics", RMP **54** (1982): 407–435. {5}

[YAMADA/TAGAKI 1991a] YAMADA, N./TAGAKI, S., "Quantum Mechanical Probabilities on a General Space-Time Surface", PTF **85** (1991): 985–1012. {3, 6}

[YAMADA/TAGAKI 1991b] —, "Quantum Mechanical Probabilities on a General Space-Time Surface. II", PTF **86** (1991): 599–615. {3, 6}

[YAMADA/TAGAKI 1992] —, "Spacetime Probabilities in Nonrelativistic Quantum Mechanics", PTF **87** (1992): 77–91. {3}

[YAMAMOTO *et al.* 1986] YAMAMOTO, Y./IMOTO, N./MACHIDA, S., "Amplitude Squeezing in a Semiconductor Laser Using Quantum Nondemolition Measurement and Negative Feedback", PR **A33** (1986): 3243–61. {4}

[YANASE 1961] YANASE, Michael M., "Optimal Measuring Apparatus", PR **123** (1961): 666–68; rep. in [*WHEELER/ZUREK* 1983, 712–14]. {4}

[YA'SIN/GREENBERGER 1986] YA'SIN, A./GREENBERGER, D. M., "Simultaneous Particle and Wave Knowledge in Quantum Theory", in [*GREENBERGER* 1986, 622–24]. {3, 4, 7, 12}

[YEAZELL/STROUD 1991] YEAZELL, J. A./STROUD, C. R., Jr., "Observation of Fractional Revivals in the Evolution of a Rydberg Atomic Wave Packet", PR **A43** (1991): 5153–56. {5, 6}

[YLINEN 1985] YLINEN, Karl, "On a Theorem of Gudder on Joint Distributions of Observables", in [*LAHTI/MITTELSTAEDT* 1985, 691–94]. {3}

[YOUSSEF 1995] YOUSSEF, Saul, "Is Quantum Mechanics an Exotic Probability Theory?", in [*GREENBERGER/ZEILINGER* 1995a, 904–905]. {3}

[YU CAO 1993] YU CAO, Tian, "New Philosophy of Renormalization", in [*L. BROWN* 1993a, 87–133]. {3, 11}

[YUEN 1976] YUEN, Horace P., "Two-Photon Coherent States of the Radiation Field", PR **A13** (1976): 2226–43. {3}

[YUEN 1986] —, "Amplifiction of Quantum States and Noiseless Photon Amplifiers", PL **113A** (1986): 405–407. {5}

[YUEN/CHAN 1983] YUEN, H. P./CHAN, V. W. S., "Noise in Homodyne and Heterodyne Detection", *Optics Letters* **9** (1983): 177–79. {4, 5, 7}

[YUEN/OZAWA 1993] YUEN, H. P./OZAWA, M., "Ultimate Information Carrying Limit of Quantum Systems", PRL **70** (1993): 363–66. {10}

[YUEN/SHAPIRO 1980] YUEN, H. P./SHAPIRO, J. H., "Optical Communication with Two-Photon Coherent States — Part III", *IEEE Transactions on Information Theory* **IT-26** (1980): 78–92. {3, 7}

[YURKE 1985a] YURKE, Bernard, "Optical Back-Action-Evading Amplifiers", *Journal of Optical Society of America* **B2** (1985): 732–38. {4}

[YURKE 1985b] —, "Squeezed-Coherent-State Generation via Four-Wave Mixers and Detection via Homodyne Detectors", PR **A32** (1985): 300–310. {3}

[YURKE 1985b] —, "Wideband Photon Counting Homodyne Detection", PR **A32** (1985): 311–23. {4, 7}

[YURKE/STOLER 1986] YURKE, B./STOLER, D., "Generating Quantum Mechanical Superpositions of Macroscopically Distinguishable States via Amplitude Dispersion", PRL **57** (1986): 13–16. {5}

[YURKE/STOLER 1992a] —, "EPR Effects from Independent Particle Sources", PRL **68** (1992): 1251–54. {9}

[YURKE/STOLER 1992b] —, "Bell's Inequality Experiments Using Independent-Particle Sources", PR **A46** (1992): 2229–34. {9}

[YURKE/STOLER 1995] —, "Bell's-Inequality Experiment Employing Four Harmonic Oscillators", PR **A51** (1995): 3437–43. {9}

[YURKE *et al.* 1990] YURKE, B./SCHLEICH, W. P./WALLS, D. F., "Quantum Superpositions Generated by Quantum Nondemolition Measurements", PR **A42** (1990): 1703–711. {4, 5}

[ZAHEER/ZUBAIRY 1989] ZAHEER, K./ZUBAIRY, M. S., "Phase Sensitivity in Atom-Field Interaction via Coherent Superposition", PR **A39** (1989): 2000–2004. {5}

[ZAJONC *et al.* 1991] ZAJONC, A. G./WANG, L. J./ZOU, X. Y./MANDEL, L., "Quantum Eraser", NA **353** (1991): 507–508. {4, 12}

[ZAWIRSKI 1932] ZAWIRSKI, Zygmunt, "Les logiques nouvelles et le champ de leur application", *Revue de Mtaphysique et de Morale* **39** (1932): 503–19. {14}

[ZEH 1970] ZEH, H. D., "On the Interpretation of Measurement in Quantum Theory", FP **1** (1970): 69–76; rep. in [*WHEELER/ZUREK* 1983, 342–49]. {4, 5, 12}

[ZEH 1979] —, "Quantum Theory and Time Asymmetry", FP **9** (1979): 803–18.{12, 13}

[ZEH 1986] —, "Emergence of Classical Time from a Universal Wave Function", PL **116A** (1986): 9–12. {3, 6, 13}

[*ZEH* 1989] —, *The Physical Basis of the Direction of Time*, Berlin, Springer, 1989, II ed., 1992. {6, 12, 13}

[ZEH 1990] —, "Quantum Measurements and Entropy", in [*ZUREK* 1990a, 405–422]. {4, 10}

[ZEH 1993] —, "There Are No Quantum Jumps, Nor Are There Particles!", PL **172A** (1993): 189–92. {4, 6}

[ZEH 1996] —, "The Program of Decoherence: Ideas and Concepts", in [*GIULINI et al.* 1996, 5–34]. {4, 6}

[ZEH/JOOS 1996] ZEH, H. D./JOOS, E., "Related Ideas and Concepts", in [*GIULINI et al.* 1996, 269–82]. {4, 6}

[ZEILINGER 1986a] ZEILINGER, Anton, "Complementarity in Neutron Interferometry", *Physica* **137B** (1986): 235–44. {4, 7}

[ZEILINGER 1986b] —, "Three Gedankenexperiments on Complementarity in Double-Slit Diffraction", in [*GREENBERGER* 1986, 164–74]. {4, 12}

[ZEILINGER 1990] —, "Experiment and Quantum Measurement Theory", in [*CINI/LÉVY-L.* 1990, 9–26]. {4}

[ZEILINGER 1998] —, "Fundamentals of Quantum Information", *Phyiscs World* (March 1998): 35–40. {10; S}

[ZEILINGER *et al.* 1994] ZEILINGER, A./ŻUKOWSKI, M./HORNE, M. A./BERNSTEIN, H. J./GREENBERGER, D. M., "EPR Correlations in Higher Dimensions", [*DE MARTINI et al.* 1994, 159–69]. {9}

[ZEILINGER *et al.* 1997] ZEILINGER, A./HORNE, M. A./WEINFURTER, H./ŻUKOWSKI, M., "Three-Particle Entanglements from Two Entangled Pairs", PRL **78** (1997): 3031–34. {9, 10}

[ZIERLER 1961] ZIERLER, Neal, "Axioms for Non-Relativistic Quantum Mechanics", *Pacific Journal of Mathematics* **11** (1961): 1151–69; rep. in [*HOOKER* 1975, 149–70]. {3}

[ZIERLER/SCHLESSINGER 1965] ZIERLER, N./SCHLESSINGER, M., "Boolean Embeddings of Orthomodular Sets and Quantum Logic", *Duke Mathematical Journal* **32** (1965): 251–62; rep. in [*HOOKER* 1975, 247–62]. {3, 14}

[ZILSEL 1935] ZILSEL, E., "P. Jordans Versuch, den Vitalismus quantenmechanisch zu retten", *Erkenntnis* **5** (1935): 56–64. {4, 12, 13, 14}

[ZOLLER *et al.* 1987] ZOLLER, P./MARTE, M./WALLS, D. F., "Quantum Jumps in Atomic Systems", PR **A35** (1987): 198–207. {5}

[ZOU *et al.* 1991] ZOU, X. Y./WANG, L. J./MANDEL, L., "Induced Coherence and Indistinguishability in Optical Interference", PRL **67** (1991): 318–21

[ZUCCHETTI *et al.* 1996] ZUCCHETTI, A./VOGEL, W./TASCHE, M./WELSCH, D.-G., "Direct Sampling of Density Matrices in Field-Strength Bases", PR **A54** (1996): 1678–81. {7}

[ŻUKOWSKI 1993] ŻUKOWSKI, Marek, "On the Existence of Empty Waves in Quantum Theory: A Comment", PL **175A** (1993): 257–58. {7}

[ŻUKOWSKI 1994] —, "Exstensions of Bell Theorem", [*DE MARTINI et al.* 1994, 178–225]. {9}

[ŻUKOWSKI/KASZLIKOWSKI 1997] ŻUKOWSKI, M./KASZLIKOWSKI, D., "Critical Visibility for N–Particle Gree,berger-Horne-Zeilinger Correlations to Violate Local Realism", PR **A56** (1997): R1682–85. {9}

[ŻUKOWSKI *et al.* 1993] ŻUKOWSKI, M./ZEILINGER, A./HORNE, M. A./EKERT, A. K., "'Event-Ready-Detectors' Bell Experiment via Entanglement Swapping", PRL **71** (1993): 4287–90. {9}

[ŻUKOWSKI *et al.* 1995] ŻUKOWSKI, M./ZEILINGER, A./WEINFURTER, H., "Entangling Photons Radiated by Independent Sources", [*GREENBERGER/ZEILINGER* 1995a, 91–102]. {9}

[ŻUKOWSKI *et al.* 1997] ŻUKOWSKI, M./ZEILINGER, A./HORNE, M. A., "Realizable Higher-Dimensional Two-Particle Entanglements via Multiport Beam Splitters", PR **A55** (1997): 2564–79. {8, 9}

[ZUREK 1981] ZUREK, Wojciech H., "Pointer Basis of Quantum Apparatus: Into What Mixture Does the Wave Packet Collapse?", PR **D24** (1981): 1516–25. {4, 5, 12}

[ZUREK 1982] —, "Environment-induced Superselection Rules", PR **D26** (1982): 1862–80. {3, 4, 5, 10, 12}

[ZUREK 1983] —, in [*MEYSTRE/SCULLY* 1983, 87–116]. {4, 10, 12}

[ZUREK 1984a] —, "Reversibility and Stability of Information Processing Systems", PRL **53** (1984): 391–94. {10}

[ZUREK 1984b] —, "Destruction of Coherence in Nondemolition Monitoring: Quantum 'Watchdog Effect' in Gravity Wave Detectors", in [*DINER et al.* 1984, 515–28]. {4, 5}

[ZUREK 1986a] —, "Reduction of Wave Packet and Environment-Induced Superselection", in [*GREENBERGER* 1986, 89–97]. {4, 5}

[ZUREK 1986b] —, "Reduction of the Wavepacket: How Long Does It Take?", in MOORE, G. T./SCULLY, M. O. (ed.s), *Frontiers of Nonequilibrium Statistical Physics*, New York, Plenum Press, 1986: 145–51. {4, 5, 12}

[ZUREK 1989a] —, "Thermodynamic Cost of Computation, Algorithmic Complexity and the Information Metric", NA **341** (1989): 119–24. {10}

[ZUREK 1989b] —, "Algorithmic Randomness and Physical Entropy", PR **A40** (1989): 4731–51. {10}

[*ZUREK* 1990a] — (ed.), *Complexity, Entropy and the Physics of Information*, Redwood City, Addison-Wesley, 1990. {4, 10, 12, 13}

[ZUREK 1990b] —, "Algorithmic Information Content, Church-Turing Thesis, Physical Entropy, and Maxwell's Demon", in [*ZUREK* 1990a, 73–89]. {10, 13}

[ZUREK 1991] —, "Decoherence and the Transition from Quantum to Classical", PT **44**.10 (1991): 36–44. {4, 5, 10, 12; S, W}

[ZUREK 1993a] —, "Preferred Stats, Predictability, Classicality, and the Environment-Induced Decoherence" PTF **89** (1993): 281–312. {4, 5, 12; W}

[ZUREK 1993b] —, "Negotiating the Tricky Border Between Quantum and Classical", PT (april 1993): 13–15, 81–90. {4, 5, 12}

[ZUREK/PAZ 1994] ZUREK, W. H./PAZ, J. P., "Decoherence, Chaos, and the Second Law", PRL **72** (1994): 2508–11. {4, 13}

[ZUREK *et al.* 1993] ZUREK, W. H./HABIB, S./PAZ, J. P., "Coherent States via Decoherence", PRL **70** (1993): 1187–90. {5}

[ZWICKY 1933] ZWICKY, Fritz, "On a New Type of Reasoning and Some of its Possible Consequences", PR **43** (1933): 1031–33. {7, 14}

LIST OF SYMBOLS

In the following we only list the main symbols which are employed in at least two different contexts or which are of general interest. It may occur that a symbol appears in a context with a different meaning from that reported here: in this case its 'local' meaning is always specified.

The pages indicated in the following are the places where the introduction of the corresponding symbol can be found.

Letters

First we list small letters with all their modifications, then capital ones with all their modifications.

Latin Letters

a Arbitrary element of a set or of a group, proposition: p. 10
a' Complement of a: p. 61
a_j, $(j = 1, 2, \ldots)$ Arbitrary propostions (elements of a logic): p. 158
$\hat{a}$ Annihilation operator: p. 72
$\hat{a}^\dagger$ Creation operator: p. 72
$\{|a_k\rangle\}$, $(k = 1, 2, \ldots)$ Basis of apparatus' state or, generic state vector: p. 223
$\mathbf{a}, \mathbf{a}'$ polarizer's orientations: p. 589
A Result of the measurement on polarizer with direction $\mathbf{a}$: 589
A Arbitrary point: p. 121
$|\mathrm{A}\rangle$ State representing an atom,a box or some other device in a fixed point: p. 355
$\mathbf{A}$ Vector Potential
$\mathcal{A}$ Measuring apparatus: p. 23
$|\mathcal{A}\rangle$ State vector representing the apparatus: p. 223
$|\mathcal{A}(s_k)\rangle$ Relative state of $\mathcal{A}$ for $|s_k\rangle$: p. 246

b Arbitrary element of a set or of a group, proposition: p. 10
$\hat{b}, \hat{b}^\dagger$ Two-photon operators or generic annihilation and creation operators: p. 81
$\{|b_j\rangle\}$, $(j = 1, 2, \ldots)$ Generic basis for a state: p. 34
$\mathbf{b}, \mathbf{b}'$ polarizer's orientations: p. 589
B Result of the measurement on polarizer with direction $\mathbf{b}$: p. 589
B Arbitrary point: p. 121
$|\mathrm{B}\rangle$ State representing an atom, a box or some other device in a fixed point: p. 355
$\mathbf{B}$ Magnetic Field: p. 55
$\mathcal{B}, \mathcal{B}'$ Borel sets: p. 88

c Light speed: p. 18
c Arbitrary element of a set or of a group: p. 61
$c_j, j = 1, \ldots$ Arbitrary complex coefficients: p. 34
$c_{\mathrm{T}}, c_{\mathrm{R}}$ Coefficients of transmission and reflection: p. 388
$\hat{c}, \hat{c}^\dagger$ Generic annihilation and creation operators: p. 348
C, C' Arbitrary Constants: p. 10

C_{jk} Cost when Hypothesis $\mathrm{H_j}$ is chosen being true hypothesis $\mathrm{H_k}$: p. 305
$\hat{C}$ Commutation operator: p. 171
$\mathrm{C}, \mathrm{C}^{(j)}$ $(j = 1, 2, \ldots)$ Correlation function, j-th order correlation fucntion: p. 82
$\hat{\mathrm{C}}$ Correlation operator: p. 644
$\mathcal{C}$ Cycle, closed curve: p. 206

d Number of dimensions (of Hilbert space, of problems, etc.): p. 200
$\hat{d}, \hat{d}^\dagger$ Generic annihilation and creation operators: p. 348
D Diffusion matrix: p. 404
$\hat{D}$ Displacement operator for electromagnetic field: p. 73
D Disentanglement: p. 738
D, D_j $(j = 1, 2, \ldots)$ Detector(s): p. 93
D Disturbance (discrepancy): p. 753
$\mathcal{D}$ Decohering factors (decoherence functional): p. 350

$|e\rangle$ Excited state: p. 83
$\{|e_j\rangle\}$ Basis for the Environment or generic state vector: p. 277
$\hat{e}, \hat{e}^\dagger$ Generic annihilation and creation operators: p. 348
e Electric charge: p. 694
$\mathbf{e}_j, j = 1, 2, 3$ Unitary vectors: p. 59
E Energy: p. 18
E_i Initial energy: p. 25
$\overline{E}$ Average energy: p. 19
E Measure of entanglement: p. 224
$\mathcal{E}$ Environment: p. 269
$|\mathcal{E}\rangle$ State representing the environment: p. 270

f Arbitrary function: p. 10
$\hat{f}, \hat{f}^\dagger$ Generic annihilation and creation operators: p. 348
$|f\rangle$ Final (output) state of a system or generic state vector: p. 35
F Field of (complex) numbers: p. 3
F Fidelity: p. 750
$\mathcal{F}$ Wigner friend: p. 231
$\mathcal{F}_{<x}$ Distribution function: p. 51
$\mathcal{F}_\alpha$ Stochastic probability distribution when α is obtained: p. 569

g Gravitational constant: p. 124; arbitrary function: p. 10
$g_{\mathcal{SA}}$ Coupling Constant between $\mathcal{A}$ and $\mathcal{S}$: 224
g_{AF} Coupling atom-field: p. 415
$\hat{g}_c$ Coupling operator: p. 519
$|g\rangle$ Ground state: p. 83
$\hat{g}, \hat{g}^\dagger$ Generic annihilation and creation operators: p. 348
G Green function or propagator: p. 63
$\mathcal{G}(b, \mathbf{a}, \mathbf{v}, R)$ Galilei group: p. 61

h Planck constant: p. 17
$\hat{h}, \hat{h}^\dagger$ Generic annihilation and creation operators: p. 348
$\hbar = h/2\pi$: p. 31
H Classical Hamiltonian: p. 9
H_{B} Boltzmann entropy: p. 18
H_{S} Shannon entropy : p. 709
H_{VN} von Neumann entropy (sometimes simply H): p. 710
H_n n-th Hermitian polinominial: p. 35
$\hat{H}$ Hamiltonian operator: p. 34
$\hat{H}_0$ Interaction-free Hamiltonian: p. 49

$\hat{H}_i$ Generic interaction Hamiltonian: p. 49
$\hat{H}_{\mathcal{A}}$ Measuring Apparatus' Hamiltonian: 254
$\hat{H}_{\mathcal{E}}$ Hamiltonian of the environment: p. 278
$\hat{H}_{\mathcal{S}}$ Object System's Hamiltonian: p. 178
$\hat{H}_{\mathcal{A}\mathcal{E}}$ Interaction Hamiltonian between $\mathcal{A}, \mathcal{E}$: p. 269
$\hat{H}_{\mathcal{S}\mathcal{A}}$ Interaction Hamiltonian between $\mathcal{A}, \mathcal{S}$: p. 224
$\hat{H}_{\mathcal{S}+\mathcal{A}}$ Hamiltonian of object system plus apparatus: p. 217
$\hat{H}_{\mathcal{S}\mathcal{E}}$ Interaction Hamiltonian between $\mathcal{S}, \mathcal{E}$: p. 272
H_j Hypothesis j: p. 304
$\mathcal{H}$ Hilbert space: p. 36
$\mathcal{H}_{\mathcal{A}}, \mathcal{H}_{\mathcal{S}}, \mathcal{H}_{\mathcal{S}\oplus\mathcal{A}}$ Apparatus' Hilbert space, object system's Hilbert space, Hilbert space for object system plus apparatus: p. 223
$\mathcal{H}_{\mathcal{S}\oplus\mathcal{E}}$ Hilbert space for object system plus environment: p. 277
$\mathcal{H}_j$ j-th Hilbert subspace: p. 37
$\mathcal{H}^{\mathrm{Tr}=1;\geq 0}$ Hilbert space of density operators (non negative operators of trace class one): p. 40

$|i\rangle$ Initial (input) state of a system: p. 35
I Intensity: p. 82; information: p. 716
I_c Intesity of the emitted radiation (classically): p. 18
$\hat{I}$ Identity operator: p. 31
$\mathcal{I}, \mathcal{I}_{\mathcal{M}}$ Generic state transformer: p. 237

$|j\rangle$ Arbitrary state: p. 35
$\mathbf{J}$ Angular momentum: p. 9
$\hat{\mathbf{J}}$ Total angular momentum operator: p. 56
$\hat{\mathcal{J}}$ Jump superoperator: p. 398

k Propagation vector for one-dimensional problems:
k_{B} Boltzmann's constant: p. 18
$\hat{k}, \hat{k}^{\dagger}$ Generic annihilation and creation operators: p. 348
$|k\rangle$ Momentum eigenstate or a rbitrary state vector: p. 35
$\mathbf{k}$ Propagation vector: p. 50
K Path Integral's Kernel: p. 63

l Arbitrary lenght: p. 122
$|l\rangle$ Arbitrary state vector: p. 726
L Lagrangian: p. 10
$\hat{L}_j$ j-component of the Orbital angular momentum: p. 56
$\hat{L}_{\pm}$ Raising and lowering operators: p. 56
L_{φ} Field Lagrangian: p. 71
$|\mathrm{L}\rangle$ Left arm (of an interferometer or other measuring apparatus): p. 382
$\hat{\mathbf{L}}$ Orbital angular momentum: p. 56
$\mathcal{L}$ Arbitrary set; Set of all proposition (yes/no experiments) pertaining to a physical system
$\mathcal{L}_{QM}$ Logical calculus of QM: p. 171
$\mathcal{L}_{\sigma}$ σ-algebra: p. 294
$\mathcal{L}_{\preceq}$ Partially ordered set (of all propositions pertaining to a physical system): p. 156
$\mathcal{L}_{\preceq,\vee\wedge}$ Lattice: p. 156
$\mathcal{L}_{\succeq,\vee\wedge,\perp}$ Orthocomplemented lattice: p. 157
$\mathcal{L}_{\succeq,\perp,a_n}$ Logic: p. 158
$\mathcal{L}(\mathcal{H})$ Set of Bounded Linear Operators on $\mathcal{H}$: p. 197
$\hat{\mathcal{L}}^{\dagger}, \hat{\mathcal{L}}$ Lindblad superoperators (transformations induced by a Reservoir on an observable and on a state respectively): p. 366

m Mass: p. 9
$m_{\hat{\sigma}}$ Number of components of spin: p. 57

$|m\rangle$ "Middle state" in a transition (metastable level) or arbitrary state vector: p. 43
$\hat{m}_{\mathbf{R}}$ Lagrange operator: p. 306
M Mirror: p. 92
$\mathcal{M}$ (Pre-)Measurement: p. 32

n Number of physical components (systems, freedom degrees, elements, etc.): p. 9
$|n\rangle$ Arbitrary state: p. 43
$|n_k\rangle$ Number (Fock) state or arbitrary state: p. 79
$\mathbf{n}$ Arbitrary direction (unitary vector): p. 52
N Number of physical components (systems, freedom degrees, elements, etc.); normalization constant: p. 64
$\hat{N}_k, \hat{N}$ Number operator: p. 72
$\hat{\mathcal{N}}_1$ First noise operator: p. 298
$\hat{\mathcal{N}}_2$ Second noise operator: p. 298

$o, o_j,\ j = 1, 2, \ldots$ Eigenvalues of an observable $\hat{O}$: p. 34
$|o\rangle$ Eigenstate of an observable $\hat{O}$: p. 249
$\hat{O}, \hat{O}'$ Arbitrary (generally conjugate) observables: p. 32
$\hat{O}_j,\ (j = 1, 2, \ldots)$ Arbitrary ensemble of observable (generally on different subsystems): p. 222
$\hat{O}^{\mathcal{A}}$ Pointer observable: p. 225
$\hat{O}^{\mathcal{S}}$ Observable of the object system, measured observable: p. 224
$\hat{O}^a, \hat{O}^{a'}, \hat{O}^b, \hat{O}^{b'}$ Bell observables (where generally a, a' and b, b' represent alternative settings of apparata, generally polarizers, a, b, respectively): p. 615
$\hat{O}_{\mathrm{ND}}$ QND Observable: p. 331
$\mathcal{O}$ Observer: p. 226

$\hat{p}_j, (j = 1, \ldots, n)$ Momentum component for a system with n degrees of freedom: p. 9
p_{M} Generic result of a measurement of the momentum: p. 506
$\dot{p_k}$ Time derivative of the k-th momentum component: p. 9
$\mathbf{p}$ Momentum: p. 9
$\hat{\mathbf{p}}$ Momentum operator: p. 34
$\hat{\mathbf{p}}_{\mathcal{A}}$ Momentum of the apparatus: p. 224
$\wp$ Probability function: p. 52
$\wp_{fi}$ Transition probability from initial state i to final f: p. 82
$\wp_n$ Probability of obtaining n detections: p. 84
P QM conjugate to classical observable: p. 254
$\hat{P}$ Projection operator: p. 44
$\hat{P}_{o_j}$ Projection operator on the j-th eigenvector of an observable $\hat{O}$: p. 45
P P-function: p. 86
$\hat{\mathbf{P}}$ Momentum operator after some (unitary) transformation: p. 47
$\mathcal{P}$ Predictability of the Path: p. 517
$\mathcal{P}(\mathcal{H})$ Space of Projectors: p. 164

$\hat{q}_j, (j = 1, \ldots, n)$ Position component for a system with n degrees of freedom: p. 9
q_{M} Generic result of a measurement of the position: p. 324
$\dot{q_k}$ Time derivative of the k-th position component: 9
$\mathbf{q}$ Position: p. 9
$\hat{\mathbf{q}}$ Position operator: p. 37
$\hat{\mathbf{q}}_{\mathcal{A}}$ Position of the pointer: p. 224
Q Classical observable: p. 254
$\hat{Q}$ Center of mass position operator: p. 408
Q Q-function: p. 85
$\hat{\mathbf{Q}}$ Position operator after some (unitary) transformation: p. 47
$\mathcal{Q}$ Poincaré group: 67

$\mathbf{r}$ Position ray-vector: p. 33

$\mathbf{r}_0$ Fixed position on a screen: p. 131
R_j Rate at which particles of beam j are detected: p. 110
$R^{(n)}$ n-fold coincidence rate: p. 82
$R(\theta)$ Count rate corresponding to probability $\wp_{12}(\theta)$: p. 602
R_0 Count rate when both analyzers are absent: p. 602
$|\mathrm{R}\rangle$ Right arm (of an interferometer or other measuring apparatus): p. 382
$\mathbf{R}$ Rotation group: p. 61 $\mathcal{R}_{\mathcal{S}}, \mathcal{R}_{\mathcal{A}}, \mathcal{R}_{\mathcal{E}}$ Reduced density matrices respectively for objects system, apparatus, environment: p. 237
$\Re$ Set of Real Numbers: p. 9
Re Real part of a complex quantity: p. 27

s spin number: p. 57
s_{D} Sensitivity of the detector: p. 82
$\{|s_j\rangle\}$ $(j = 1n, \ldots)$ Basis of an object system: p. 223
$S_{\mathbf{q}}$ Spectral density of the noise added to position: p. 337
$S_{\mathbf{F}}$ Spectral density of the noise added to the force: p. 337
$\hat{S}$ Scattering matrix: p. 65
$\hat{S}_{fi}$ Scattering matrix from initial state to final state: p. 65
S Action: p. 10
$\mathbf{S}$ Arbitrary surface, area: p. 17
$\hat{\mathrm{S}}$ S-matrix: p. 65
$\mathcal{S}$ Physical system, observed system: p. 9
$\mathcal{S}_k$ k-th Subsystem of a compound one: p. 37
$\mathcal{S}_{jk}$ System compounded of $\mathcal{S}_j$ and $\mathcal{S}_k$: p. 37
$\mathcal{S}_{\mathrm{QP}}$ Quantum probe: p. 326
$|\varsigma\rangle$ Generic state vector; especially state representing the object system: p. 223
$|\varsigma_1\rangle, |\varsigma_2\rangle$ (Separated) components of an (object or of a generic) system: p. 482

t Time variable (parameter): p. 9
$\hat{t}$ Time operator: p. 177
T Temperature: p. 17
T_E Kinetic energy: p. 10
T Transmission : p. 388; transposed matrix: p. 32
$\mathcal{T}$ Operation (not necessarily unitary): p. 292

$|u\rangle$ State on an arm of an interferometer or generic state vector: p. 484
$\mathbf{u}$ Arbitrary vector: p. 44
$\hat{U}, \hat{U}^{\dagger}$ Arbitrary unitary operators: p. 47
$\hat{U}^{-1}$ Inverse of $\hat{U}$: p. 47
$\hat{U}_F$ Fourier transform (operating on coordinate representation): p. 50
$\hat{U}_F^{-1}$ Inverse of Fourier transform (operating on momentum representation): p. 50
$\hat{U}_t$ Time-evolution unitary operator: 48
$\hat{U}(t_2, t_1)$ Time evolution unitary operator from t_1 to t_2: p. 65
$\hat{U}_{\mathbf{a}}$ Space translation unitary operator: 61
$\hat{U}_{\mathbf{v}}$ Boost unitary Operator: p. 61
$\hat{U}_{\mathbf{R}}$ Rotation unitary operator: p. 61

v Velocity for one-dimensional problems: p.
v_g Group velocity: p. 461
$|v\rangle$ State on an arm of an interferometer or generic state vector: p. 484
v($\cdot$) Value of obs $\cdot$: p. 312
$\mathbf{v}$ arbitrary vector: p. 44; velocity, momentum translations (in Galilei group): p. 61
V Potential energy: p. 9
V Volume, generic region of space: p. 56
$\mathcal{V}$ Fringes Visibility: p. 82

w, w_j $(j = 1, 2, \ldots)$ Weight functions: p. 41
w_E Number of different distributions of the energy between the resonators: p. 18
w_Δ Measure of the interference: p. 132
W Wigner function: p. 86
W_Δ Measure of the bulk width: p. 131

$\hat{x}$ Linear combination of quadratures: p. 328
X Fixed coordinate: p. 129
X_j Macroscopic coordinate: p. 280
$\hat{X}_j$, $j = 1, 2$ Quadratures: p. 81
$\mathcal{X}$ Arbitrary (sub-)set: p. 156

$\mathcal{Y}$ Arbitrary (sub-)set

$\mathcal{Z}$ Minkowsky space: p. 68

Greek Letters

α_k Dirac matrices: p. 70
$|\alpha\rangle$ Coherent state: p. 84

β Dirac matrix: p. 70
$|\beta\rangle$ Coherent state: p. 85

γ Damping factor: p. 370
γ_ν, γ^μ Dirac matrices: p. 70
$|\gamma\rangle$ Generic photon state: p. 93
$|\gamma_i\rangle, |\gamma_s\rangle$ Idler and signal photons: p. 94
$|\gamma_0\rangle$ Input photon: p. 93
Γ Cut-off: p. 382; gamma-function: p. 382
$\mathit{\Gamma}$ Phase space: p. 9

δ Dirac function, small quantity: p. 31
δ Measure of the distance between observables or states: p. 724
Δ Uncertainty or spread of a quantity: p. 118
$(\Delta\cdot)_{\mathrm{A}}$ Additional Error measurement of observable $\cdot$: p. 334
$(\Delta\cdot)_{\mathrm{M}}$ Uncertainty of Measurement of observable $\cdot$: p. 126
$(\Delta\cdot)_{\mathrm{P}}$ Uncertainty due to the perturbation of observable $\cdot$: 126
$(\Delta\cdot)_{\mathrm{ND}}$ QND minimal measurement error of observable $\cdot$: p. 334
$(\Delta\cdot)_{\mathrm{S}}$ Standard Measurement error of observable $\cdot$: p. 334
$(\Delta\cdot)_{\mathrm{SQL}}$ SQL in the measurement of observable $\cdot$: p. 328
$(\Delta\cdot)_{\mathrm{T}}$ Total measurement error of observable $\cdot$: p. 334

ϵ Discrete element of a partition; small quantity: p. 18

ζ Different types of parameters, variables, numbers: p. 30

η Different types of parameters, variables, numbers: p. 53

θ Angle: p. 44
ϑ Amplitude: p. 27
$\vartheta_{\mathcal{S}\mathcal{A}}$ Correlation amplitude between $\mathcal{S}$ and $\mathcal{A}$: p. 275
$\hat{\vartheta}$ Amplitude operator: p. 325

$|\iota\rangle$ Internal state of an atom: p. 140

λ Wavelenght: p. 26
Λ Lagrangian density (for fields): p. 71
$\lambda_{\rm HV}$ Hidden Parameter: p. 543
$\mathit{\Lambda}$ Lorentz proper transformation: p. 68
$\mathit{\Lambda}_{\nu}$ Lorentz boost: p. 68

μ_p Probability measure: 164

ν Frequency: p. 17

ξ Arbitrary variable, random variable: p. 30
$|\xi\rangle$ State by quantum computation: p. 762
Ξ Degree of mixtureness or of separation measure between subsystems: p. 42

$\hat{\pi}$ Momentum conjugate to the field: p. 72

ρ Classical state: p. 11
$\rho(\lambda_{\rm HV})$ Probability distribution of hidden parameters $\lambda_{\rm HV}$: 590
$\hat{\rho}$ Density Operator: p. 40
$\hat{\rho}_i, \hat{\rho}_f$ Initial and final state: p. 222
$\hat{\rho}_k, \hat{\rho}^k, (k = 1, 2, \ldots)$ States of subsistems $\mathcal{S}_1, \mathcal{S}_2, \ldots$: p. 41
$\hat{\rho}_0$ Density matrix at initial time t_0: p. 48
$\hat{\rho}_t$ Density matrix after an unitary evolution at time t: p. 48
$\hat{\rho}^{\mathcal{A}}$ Density matrix of the apparatus: p. 223
$\hat{\rho}^{\mathcal{S}}$ Density matrix of the object system: p. 223
$\hat{\rho}^{\mathcal{S}+\mathcal{A}}$ Density matrix of the compound system $\mathcal{S} + \mathcal{A}$: p. 280
$\hat{\rho}^{\mathcal{S}+\mathcal{E}}$ Density matrix of the compound system $\mathcal{S} + \mathcal{E}$: p. 278
$\hat{\rho}^{\mathcal{S}+\mathcal{A}+\mathcal{E}}$ Density matrix of the compound system $\mathcal{S} + \mathcal{A} + \mathcal{E}$: p. 270
$\hat{\rho}^{1+2}_{\rm Tr2}$ Tracing out of the subsystem $\mathcal{S}_2$: p. 52
$\tilde{\hat{\rho}}$ Mixture: p. 41
$\hat{\rho}_{\rm W}$ Werner state: p. 653
ϱ Classical flux density: p. 404

$\hat{\sigma}$ Spin Operator: p. 56
$\hat{\sigma}_k,\ k = x, y, z$ Spin components: p. 56
$\hat{\sigma}^+, \hat{\sigma}^-$ Raising and lowering operators: p. 83

τ Arbitrary time interval: p. 20

Υ: Spectrum (eigenvalues of an operator): p. 38

ϕ angle: p. 59
ϕ_e Magnetic flux: p. 694
$\mathbf{\Phi}$ Wave function of a compound system: p. 656
φ Phase: p. 34
φ_d Dynamical Phase: p. 205
$\varphi(\mathbf{C})$ Geometric phase: p. 207
$\delta\varphi$ Phase difference: p. 205
$\hat{\varphi}$ Field operator '(second quantization): p. 71

χ Characteristic Function: p. 53

$\psi, |\psi\rangle$ Generic wave function, vector: p. 33
$|\psi^i\rangle, |\psi^f\rangle$ (or $|\psi^i\rangle, |\psi^f\rangle$) Initial and final wave vector: p. 65
$|\psi_j\rangle$, $(j = 1, 2, \ldots)$ Subsistem of a compounded system described by $|\Psi\rangle$: p. 100
$\psi(\mathbf{r})$ Wave function in coordinate representation:p. 33
$\tilde{\psi}(\mathbf{k})$ = Wave function in momentum representation (fourier transform of $\psi(\mathbf{r})$):p.50
$\psi_S(t)$ Wave function in Schrödinger picture: p. 49
$\psi_H(t)$ Wave function in Heisenberg picture: p. 49
$\overline{\psi}$ Adjoint spinor: p. 74
$\psi^{(-)}$ Negative energy electron: p. 75
$\psi^{(+)}$ Positive energy electron: p. 74
$|\Psi\rangle$ Wave vector of a compound system: p. 43
$|\Psi_0\rangle$ Wave vector of a multi-particle system in singlet state: p. 60
$|\Psi^{\mathcal{A}+\mathcal{S}}\rangle$ Wave vector of $\mathcal{A} + \mathcal{S}$:
$|\Psi_{pw}\rangle$ De Broglie function of wave + particle: p. 474 ω Angular frequency:p. 35
Ω Angular velocity: p. 206
$\hat{\Omega}$ Integrated intensity: p. 396
$\mathit{\Omega}$ Set of events: p. 194

Other Symbols

Logic

$\rightarrow$ Logical implication ("if ..., then ...") = sufficient condition: p. 156
$\leftrightarrow$ Logical equivalence: p. 198
$a \vdash b$ 'b' is deducible from 'a' (metalanguaged implication): p. 156
$\Rightarrow$ QM implication: p. 172
$\triangleright$ Quasi-implication (three-valued-logic implication): p. 506
$\Uparrow$ Infinite-valued implication: p. 508
$\forall$ For all: p. 39
$\exists$ There is almost one: p. 39
$\vee, \wedge$ Disjunction, conjunction (of set elements or propositions): p. 155

Mathematics

$\longrightarrow$ Tends to: p. 22
$\longrightarrow_w$ Weak convergence: p. 264
$\mapsto$ A map (unitary transformation) from to: p. 48
$:=$ ... is defined as ... :p. 31
$\equiv$ Mathematical equivalence: p. 51
$\langle\cdot\rangle$ Expectation or mean value of an observable $\cdot$: p. 51
$\Delta\cdot$ Standard deviation of $\cdot$: p. 118
$\langle\cdot|\cdot\rangle$ Scalar product in Dirac algebra formalism: p. 35
$\otimes$ Tensorial product: p. 37
$\oplus$ Tensioral sum: p. 52
$\circ$ Group binary operation: p. 61
$\perp$ Orthogonal: p. 157
$\Box$ Dalemebertian: p. 69
$\triangle$ Probability triangle: p. 197
$[,], [,]_+$ Commutator and anticommutator: p. 31

Set theory

$\cup, \cap$: Addition and intersection between sets or spaces: p. 171
$\preceq$ Order relation between set elements: p. 155
$\subset, \subseteq$ Order relation between sets: p. 155
$\in$ An element is in the set ...: p. 155
$\mathbf{0}, \mathbf{1}$ Null and unit elements: p. 156

Physics

$\leadsto$ A (not necessarily unitary) trasformation from ... to ...: p. 220
$\uparrow, \downarrow$ Spin up, down: 56
$|\uparrow\rangle, |\downarrow\rangle$ States of spin up and spin down fo a chosen direction: p. 281
$\updownarrow, \leftrightarrow$ Vertical and horizontal polarization: p. 613
$|+\rangle, |-\rangle$ Alternative two-state basis: p. 273
$\odot$ ...is compatible (= contemporary measurable) with ...: p. 296
$\sqcap$ Successive measurements: 32
$\{,\}$ Poisson brackets: 9
$|0\rangle, |1\rangle$ Vacuum, presence of one photon; or, by quantum computation, ground and excited state: p. 84
$|\leftrightarrow\rangle, |\updownarrow\rangle$ Photon horizontal and vertical polarization states: p. 613
$|\nearrow\swarrow\rangle, |\searrow\nwarrow\rangle$ Photon states of 45° and 135° polarization: p. 744

LIST OF ABBREVIATIONS

AB-Effect = Aharonov-Bohm Effect
BBO = β–$Ba_zB_2O_4$: beta-barium-borate
BS = Beam Splitter(Half-silvered mirror)
BS-Exp. = Beam-Splitting Experiment
CCR = Canonical Commutation Relation
CH = Clauser/Horne
CHSH = Clauser/Horne/Shimony/Holt
CP = Complementarity Principle
CR = Correspondence Rules (of a Theory)
CROC = Canonically Relatively Orthocomplemented (Lattice)
ebit = Amount of entanglement for EPR state
EPR = Einstein/Podoslky/Rosen
EPRB = Einstein/Podoslky/Rosen/Bohm
Eqs. = Equation(s)
F = Formalism (of a Theory)
GHSZ = Greenberger/Horne/Shimony/Zeilinger
GHZ = Greenberger/Horne/Zeilinger
GRW = Ghirardi/Rimini/Weber
HV = Hidden Variable
iff = if and only if
Interfer. Exp. = Interferometry Experiment
i-photon = Idler Photon
JCM = Jaynes/Cummings Model
KS = Kochen/Specker
LHS = Left-Hand Side (of an equation)
I-Measurement: Measurement of the first kind
II-Measurement: Measurement of the second kind
Meta-QM = Metatheory on Quantum Mechanics
MWI = Many Worlds Interpretation
MZ = Mach-Zehnder interferometer
NL = Non-Linear
PDC = Pumped Downconversion
POSet = Partially Ordered Set
POVM = Positive Operator Valued Measure
PVM = Projector Valued Measure
Q.E.D. = Quod erat demonstrandum
QM = Quantum Mechanics (or 'Quantum-Mechanical', depending on the context)
QM-Extension: Extension of Quantum Mechanics
QM-Interpretation = Interpretation of Quantum Mechanics
QM-Optics = Quantum Optics
Q-Optics = Quantum optics
QP = Quantum Postulate
qubit: Quantum bit
Relativistic-QM = Relativistic Quantum Mechanics
rf = radio-frequency
RHS = Right-Hand Side (of an equation)
S-Cat = Schrödinger cat
SGM = Stern/Gerlach Magnet
s-photon = Signal Photon
SQL = Standard Quantum Limit
SSR = Superselection Rule
Statistical-UR = Statistical Uncertainty Relation
Two-Slit-Exp. = Two-Slit-Experiment
UP = Uncertainty Principle
UR = Uncertainty Relation
WAY = Wigner/Araki/Yanase
W-Function = Wigner Function
WKB = Wentzel/Kramers/Brillouin
WP-Dualism = Wave/particle dualism

LIST OF COROLLARIES, DEFINITIONS, LEMMAS, POSTULATES, PRINCIPLES, PROPOSITIONS, AND THEOREMS

In the following list all definitions, etc., which are assumed (in the context of QM: hence not those from classical mechanics or general mathematics) by the author, are starred (in the text they are between two horizontal lines). The theorems which are included by more general ones are never starred. No Lemma is starred.

Corollaries

Definitions

Lemmas

Postulates

Principles

Propositions

Theorems

LIST OF EXPERIMENTS

Of the following experiments we give the name of the proponents and, if realized by others, also that of the performers. If the realizations are many (as for EPRB), we give the original proposed experiment separately. Many of the reported experiments are only very general *Gedankenexperimenten*.

Theoretical studies and models are generally not considered as experiments.

Only those experiments which are described in more detail are numbered.

Atomic-Level Experiments

Coincidence Cascade Experiments

Interferometry

Microcavity Experiments

One-Hole-Experiments

Parametric Down Conversion

SGM-Experiments

Tunnelling

Two-Slit-Experiments

Other Experiments

Index

Index

List of Figures

List of Tables